HIGH T$_c$ SUPERCONDUCTOR THIN FILMS

INTERNATIONAL CONFERENCE ON ADVANCED MATERIALS
-ICAM 91-

Organised by the European Materials Research Society
Under the auspices of the
International Union of Materials Research Societies

Part of the
EMRS 1991 Spring Meeting

HIGH T$_c$ SUPERCONDUCTOR THIN FILMS

Proceedings of Symposium A1 on
High Temperature Superconductor Thin Films
of the International Conference on Advanced Materials – ICAM 91
Strasbourg, France, 27–31 May, 1991

Edited by:

L. CORRERA
Istituto LAMEL - CNR
Bologna, Italy

1992

NORTH-HOLLAND
AMSTERDAM • LONDON • NEW YORK • TOKYO

North-Holland
ELSEVIER SCIENCE PUBLISHERS B.V.
Sara Burgerhartstraat 25
P.O. Box 211
1000 AE Amsterdam
The Netherlands

Distributors for the United States and Canada:

ELSEVIER SCIENCE PUBLISHING COMPANY INC.
655 Avenue of the Americas
New York, N.Y. 10010
U.S.A.

Library of Congress Cataloging-in-Publication Data

Symposium A1 on High Tc Superconductor Thin Films (1991 : Strasbourg,
  France)
    High Tc superconductor thin films : proceedings of Symposium A1 on
  High Tc Superconductor Thin Films of the International Conference on
  Advanced Materials--ICAM 91, Strasbourg, France, 27-31 May, 1991 /
  edited by L. Correra.
       p.   cm.
    Includes bibliographical references and index.
    ISBN 0-444-89353-9
    1. Thin films--Congresses. 2. High temperature superconductors-
  -Congresses. 3. Epitaxy--Congresses.   I. Correra, L.
  II. International Conference on Advanced Materials--ICAM 91 (1991 :
  Strasbourg, France) III. Title.
  QC176.82.S88 1991
  621.381'52--dc20   91-44392
                                                                      CIP

ISBN 0 444 89353 9

Printed in The Netherlands

PREFACE

This Proceedings collects the papers presented at the symposium on High T_c Superconductor Thin Films, one of the four symposia of the International Conference on Advanced Materials – ICAM 1991 (27-31 May 1991, Strasbourg).

The symposium focused on interdisciplinary research on super-conducting oxides, dealing with several aspects of the thin film field, from fundamental properties to applications. Over 40 contributed papers, 110 poster contributions and 21 invited talks were presented during the meeting, averaging 130 attendees. Most of the papers dealt with the 1-2-3 system, indicating that the Y-Ba-Cu-O and related compounds are still the most intensively studied materials by far. However, interesting results for the Bi system were also presented.

To develop a better understanding of the relationships between microstructure and properties, four sessions were devoted to physical properties and two to structural characteristics. Superconductivity and two-dimensional transport behavior in Y-Ba-Cu-O and Bi-Sr-Ca-Cu-O thin films were widely discussed, as were optical, magnetic and microwave properties. Film growth mechanisms were studied by a variety of techniques, including RBS-channeling, x-ray diffraction and high-resolution TEM, this latter being a powerful technique to study layer-by-layer epitaxy.

The main deposition techniques covered were laser ablation, evaporation, sputtering and chemical vapor deposition. Complete control over deposition conditions is required for growing complex structures and for compatability with device fabrication technology. Considerable progress was reported on high-quality thin film preparation by CVD and organometallic CVD.

The papers devoted to processing emphasized strategical subjects such as deposition of epitaxial and smooth films on large area substrates and control of the film orientation. The importance of methods for film deposition on silicon was highlighted. Surface and interface studies, including growth and properties of buffer layers, were discussed in a session devoted to substrates and multilayers.

Two sessions on application illustrated significant progress in the fabrication of microwave devices, IR detectors, SQUID magnetometers and optoelectronic devices. However, many challenging problems are still to be solved for junction-based technology.

I would like to thank the Program Committee for the help given in finding excellent speakers and in undertaking the selection of papers. I would also like to thank all those who contributed to the success of the symposium and, in particular, the invited speakers, the session chairpersons and the referees of the papers. I must mention that some – not many – papers were published without the language improvements suggested by the adviser. This was necessary as it was impossible to contact the authors and have the revised version in time for publication, and I did not consider it correct to exclude scientifically valid work from the proceedings. Finally, I am indebted to the E-MRS staff for their support in the organization and smooth running of the symposium. Many thanks are also due to Elsevier Science Publishers for the assistance and the timely reproduction of these proceedings.

SYMPOSIUM INFORMATION

This Conference was held under the auspices of:

- The Council of Europe
- The Commission of the European Communities

It is our pleasure to acknowledge with gratitude the financial assistance provided by:

- Banque Populaire (France)
- Centre de Recherches Nucléaires (France)
- Centre National de la Recherche Scientifique (France)
- Service de Documentation Touristique du Palais des Congrès de Strasbourg (France)
- The Commission of the European Communities
- The Council of Europe
- The European Parliament
- The Brewery Kronenbourg (France)

and in particularly with respect to Symposium A1:

- The Consiglio Nazionale delle Ricerche (Italy)
- The Balzers S.p.A. (Italy)

Chairpersons:

L. Correra (Italy)
T. Kawai (Japan)
T. Venkatesan (USA)

Symposium Programme Committee:

A. Barone (Italy), L. Correra (Italy), J.E. Evetts (UK),
H.-U. Habermeier (Germany), J. Klein (France), L. Shultz (Germany),
A.I. Usoskin (USSR), R. Vaglio (Italy), T. Venkatesan (USA)

ICAM 91 International Organizing Committee:

Chairman: M. Balkansi (France)

K.J. Bachmann (USA), L. Correra (Italy), A.T. di Benedetto (USA),
H.L. Hwang (Taiwan, China), T. Kawai (Japan), L. Nicolais (Italy),
C. Schwab (France), T. Takahashi (Japan), H.L. Tuller (USA),
T. Venkatesan (USA), E. Yasuda (Japan)

TABLE OF CONTENTS

Part II. Film Growth & Processing

Part III. Substrates & Multilayers

Part IV. Structural Characterization

Part V. Applications

Part I
Properties

High T_c Superconductor Thin Films
L. Correra (Editor)

Recent progress in cuprate superconductors

W. Y. Liang

IRC in Superconductivity, Madingley Road, Cambridge CB3 0HE, U.K.

Abstract
    The parent compounds of all copper oxide based superconductors are antiferromagnetic insulators despite the presence of a single electron in the non-magnetic unit cell, and charge carriers are introduced by means of partial substitution of the cations or by changing the oxygen content. Doping produces carriers in either the charge reservoir layers or in the CuO_2 layers. Increasing the concentration of mobile carriers in the CuO_2 layers initially induces superconductivity until a maximum T_c is reached which corresponds to approximately 0.2 hole carriers per copper atom. Thereafter T_c decreases and the non-superconducting metallic state is reached at a carrier concentration which is somewhat below ~0.5 hole carriers per copper atom. We have carried out a study of the relationship between the insulating, the superconducting and the non-superconducting metallic in detail on a number of systems. A particular example will be given of the family of $(Ca_{1-x}Y_x)Sr_2(Tl_{1-z}Pb_zCu_2)O_7$ compounds where both x and z can be varied; the highest T_c is 108K while the non-superconducting metallic member is given by the formula $CaSr_2(TlCu_2)O_7$ and the insulating state is given by the composition with z=0.5 and y>0.6. Other combinations are also possible, giving this family of materials their inherently very interesting properties for investigative studies as well as for practical applications. In addition to the variation in T_c as a function of carrier concentration and determined from the temperature dependent resistivity, other measurements on the structure, Hall coefficients, thermo-electric power, magnetisation and specific heat have revealed further interests relating to superconducting order parameters, critical currents, dimensionality and other superconductivity properties in these materials. We have found that variations in the fluctuation superconductivity, thermo-power and magneto-resistance strongly suggest that two qualitatively different regimes exist within the superconducting range of the composition. In the low carrier density regime the Cooper pairs may be compared with the real space spin bi-polarons. In the high carrier density regime strongly interacting BCS behaviour dominate. The implications on the practical uses of the systematic variation of superconducting properties with composition will be discussed.

1. INTRODUCTION

    High temperature superconductors have made very significant advances in a number of directions since their discovery nearly four years ago. These include the further discoveries of many more oxide superconducting systems in addition to the La_2CuO_4 based family, a better control of the fabrication conditions for bulk materials to produce single phase, a better control for the fabrication of thin epitaxial films including those with buffer layers and heterojunctions, a better understanding of the chemical and physical processes, and many others. This paper will give a brief overview of the materials, and will then use mainly one of the oxide systems to illustrate some of the remarkable electronic properties. We hope to bring out those physical and chemical properties, if properly understood, should apply generally to most if not all the oxide superconductors. Finally the measurements of magnetoresistance have led us recently to pose an interesting (old) question: Can high temperature superconductors in their ceramic form be useful in high current applications?

2. MATERIALS

We are concern only with the p-type oxide superconductors whose charge carriers are hole-like. These are presented generally either as families of La-Cu-O, Y-Ba-Cu-O, Tl-Ba-Ca-Cu-O and Bi-Sr-Ca-Cu-O, or as compounds containing one, two or more of CuO_2 active layers in the formula unit. There are also further divisions according to the number of atomic layers which separate the CuO_2 active layers. The latter forms the charge reservoir which can provide a means of moderating the number of carriers in the active layers without introducing scattering centres for carriers but it can also affect the degree of electrical and magnetic anisotropy which may bring the unwelcome effect of increased superconducting fluctuations. Figure 1 shows our view of the structural division between the active layer block and the charge reservoir block for a few typical compounds.

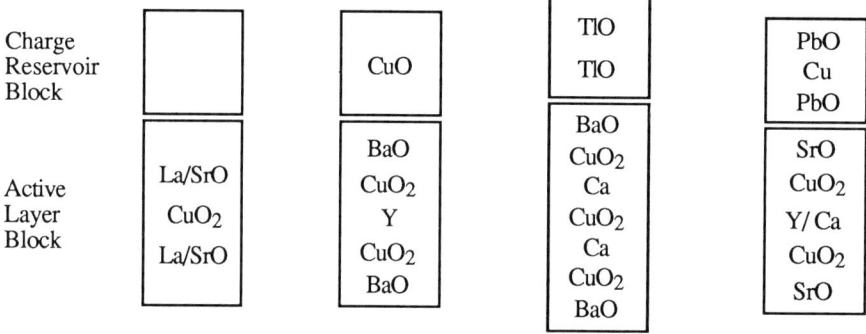

(a) $(La,Sr)_2CuO_4$ (b) $YBa_2Cu_3O_7$ (c) $Tl_2Ba_2Ca_2Cu_3O_{10}$ (d) $Pb_2(Y,Ca)Sr_2Cu_3O_8$

Figure 1. Block view of oxide superconductors in terms of the active layer block which contains the superconducting CuO_2 layers, and the charge reservoir block which may be conducting, insulating or highly disordered. $(La,Sr)_2CuO_4$ does not possess charge reservoir.

The distribution of charge and the effects on the doping efficiency will be discussed in the next section. In Table 1 we list typical oxide systems together with the range of T_c reported for each of the system. This T_c range includes all the values obtainable by means of cation doping, by varying oxygen content, by hydrogen annealing [1] as well as by the application of high pressure [2]. One of the more interesting developments concerning materials preparation in recent years has been the discovery that certain phases could be considerably stabilized by the addtion of a small amount of tracer element. This way not only the high T_c phase can be preserved but also it iseasier to fabricate single phase materials, for examples In stabilizes the 60K phase of $(Tl,V,In)Sr_2(Y,Ca)Cu_2O_7$ [3] and Bi stabilizes the 120K phase of $(Tl,Pb,Bi)Sr_2Ca_2Cu_3O_{10}$ [4].

Table 1
Families of p-type oxide superconductors and the range of T_c observed.

No. of CuO_2 layers in active block	No. of atomic layers in reservoir block	Examples	High-est T_c, K
1	0	$(La,Sr)_2CuO_4$	45
	1	$Tl(Ba,La)_2CuO_{5-\partial}$, $Bi(Sr,La)_2CuO_{5-\partial}$	42
	2	$Tl_2Ba_2CuO_6$, $Bi_2Sr_2CuO_6$	85
2	0	$(La,Sr)_2CaCu_2O_6$	60
	1	$YBa_2Cu_3O_7$, $TlBa_2CaCu_2O_8$, $BiSr_2CaCu_2O_8$	110
	2	$YBa_2Cu_4O_8$, $Tl_2Ba_2CaCu_2O_8$, $Bi_2Sr_2CaCu_2O_8$	115
	3	$Pb_2(Y,Ca)Sr_2Cu_3O_8$	80
3	1	$TlBa_2Ca_2Cu_3O_9$, $BiSr_2Ca_2Cu_3O_9$	120
	2	$Tl_2Ba_2Ca_2Cu_3O_{10}$, $(Bi,Pb)_2Sr_2Ca_2Cu_3O_{10}$	125

3. DISTRIBUTION OF CHARGE CARRIERS

Superconductivity in the layered cuprates requires the presence of carriers in the active CuO_2 layers but prefers not to have impurity ions into these layers. The carriers can be introduced via the charge reservoir block or the partial substitution between Ca and Y in the active block as shown in Figure 1. This provides the freedom to select specific sites in which to place the doping atoms in order to optimise the effect, similar to the process of modulation doping in semiconductor hetero-junction multilayers. However unlike doping in semiconductors, the amount of impurity ion substitution is usually large leading to changes in the lattice potential. This will have the effect of shifting the band density of states of the layer with respect to the Fermi level and to that of the other layers. It is well known that doping does not always lead to a direct increase in the number of carriers in the CuO_2 layers. We note here that adding oxygen ions or a partial substitution of a less positive ion such as Ca^{2+} for a more positive ion such as Y^{3+} can provide hole carriers but also raises the energy of the receiving layer or block as illustrated in Figure 2 for $YBa_2Cu_3O_6$. This example shows that although $YBa_2Cu_3O_{6.22}$ and $(Y_{.75}Ca_{.25})Ba_2Cu_3O_{6.1}$ both have about 0.45 holes per formula unit, the former is a long way from becoming superconducting while the latter has a T_c of 15K [5]. Another more extreme example is provided by the compound $Pb_2(Y,Ca)Sr_2Cu_3O_8$ which does not superconduct without Ca substitution despite increasing the oxygen concentration to 9.8, yet a modest 0.25 Ca substitution for Y will produce a T_c of ~80K [6]. The explanation in both examples is that when oxygen enters the charge reservoir block, it tends to first oxidize the local atoms such as Cu^{1+} in the case of $YBa_2Cu_3O_6$ and Pb^{2+} in the case of $Pb_2(Y,Ca)Sr_2Cu_3O_8$ before releasing the 'holes' to the active block.

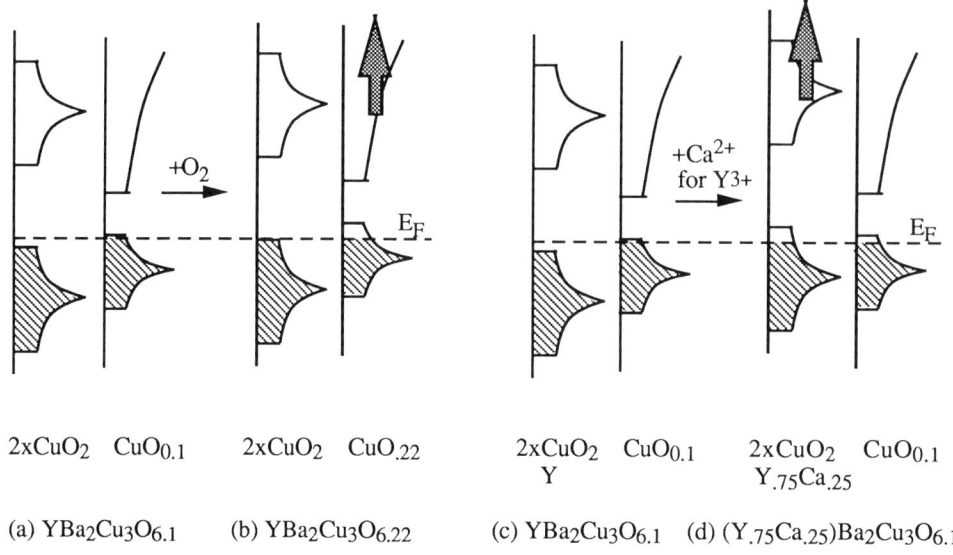

2xCuO$_2$ CuO$_{0.1}$ 2xCuO$_2$ CuO$_{.22}$ 2xCuO$_2$ CuO$_{0.1}$ 2xCuO$_2$ CuO$_{0.1}$
                                                          Y   Y$_{.75}$Ca$_{.25}$

(a) YBa$_2$Cu$_3$O$_{6.1}$ (b) YBa$_2$Cu$_3$O$_{6.22}$ (c) YBa$_2$Cu$_3$O$_{6.1}$ (d) (Y$_{.75}$Ca$_{.25}$)Ba$_2$Cu$_3$O$_{6.1}$

Figure 2. Schematic density of states diagrams shown separately for CuO$_2$ and CuO layers in YBCO. In (a) and (c) Cu atoms in the chain layer are slightly oxidized with O$_{6.1}$; in (b) the Fermi level just touches the top of the CuO$_2$ band with O$_{6.22}$ as shown by NMR data [7], the energy of the Cu-O layer is raised as it takes on oxygen; in (d) the energy of the CuO$_2$ layer is raised as Y is partially substituted by Ca expediting the creation of holes in the layers.

4. THE SEPTENARY SYSTEM (Y$_x$Ca$_{1-x}$)Sr$_2$(Pb$_z$Tl$_{1-z}$)Cu$_2$O$_{7-\partial}$

The septenary (Y$_x$Ca$_{1-x}$)Sr$_2$(Pb$_z$Tl$_{1-z}$)Cu$_2$O$_{7-\partial}$ system can be prepared in an uncomplicated way to produce single phase samples nearly right through the compositional range [8]. This uses the method of first forming the precursor without Tl and Pb by calcining stoichiometric proportions of cations of high purity CaCO$_3$, Y$_2$O$_3$, SrCO$_3$ and CuO powder at 970 °C for 12 hours in air. The precursor was then mixed with Tl$_2$O$_3$ and PbO, ground, and pressed into a pellet form which was wrapped in a gold foil to prevent loss of Tl during heating. The sintering was carried out at 950°C in flowing oxygen for 3 hours, followed by a fairly rapid cooling (5°C/min.) to room temperature. The T$_c$ (R=0) over a whole range has been determined and this is plotted in Figure 3 as functions of x and z. This demonstrates metal (CaSr$_2$TlCu$_2$O$_{7-\partial}$) -superconductor - insulator (e.g. YSr$_2$TlCu$_2$O$_{7-\partial}$ and CaSr$_2$PbCu$_2$O$_{7-\partial}$) transitions for a wide choice of values of x and z [9], thus making it an extremely useful model system for detailed investigations. The subsequent results have been obtained from the series (Y$_x$Ca$_{1-x}$)Sr$_2$(Pb$_{.5}$Tl$_{.5}$)Cu$_2$O$_{7-\partial}$.

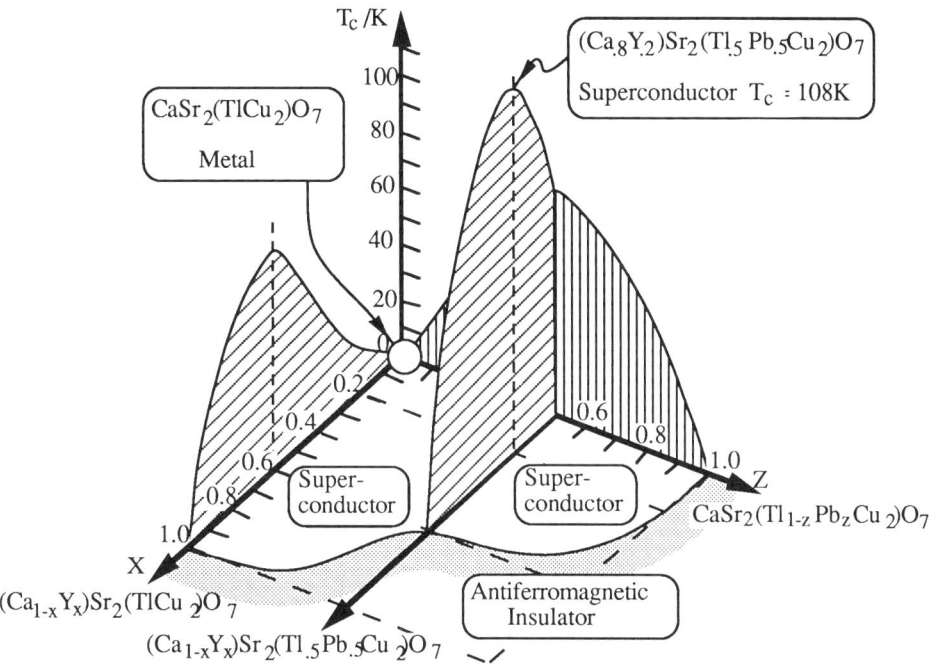

Figure 3. The phase diagram for metal-superconductor-insulator transitions in the septenary $(Y_xCa_{1-x})Sr_2(Pb_zTl_{1-z})Cu_2O_{7-\partial}$ system as functions of composition x and z.

5. RESISTIVITY

It was found that as x is increased with Ca^{2+} being substituted by Y^{3+} in the inter-copper planar sites, the net effect is a decrease in the overall hole concentration. At 0% yttrium the transition temperature is roughly 80K rising smoothly to a maximum at around x = 0.2 when T_c = 108K. Further increase of Y^{3+} results in the lowering of T_c until at x = 0.6 superconductivity completely disappears. Identifying x = 0.2 as the 'optimum' doping case then $0.6 \geq x \geq 0.2$ corresponds to the regime of under-doping (of carriers), and $x \leq 0.2$ to the regime of over-doping (of carriers). Figure 4 shows the resistivity curves of the full series of the compounds. It can be seen that as the normal resistivity increases with increasing Y concentration indicating decreasing carrier concentrations, T_c goes through a maximum value (of 108K) when x = 0.2.

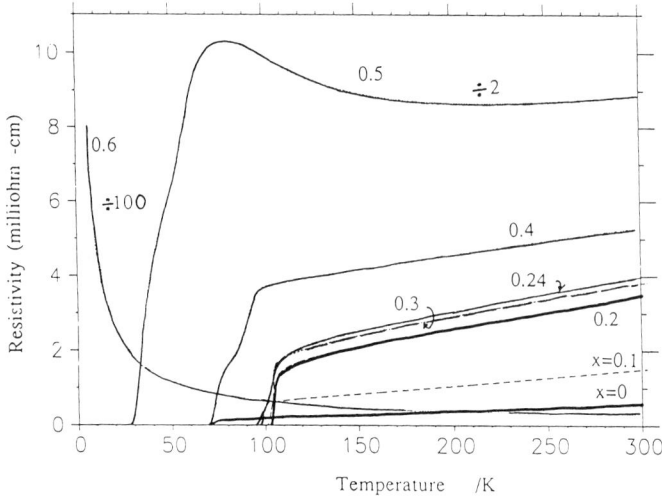

Figure 4. The series of resistivity curves of $(Y_xCa_{1-x})(Tl_{.5}Pb_{.5})Sr_2Cu_2O_{7-\delta}$ where $\partial \sim 0$ as deter-mined from TGA. The samples were prepared by R. S. Liu and the measurements were taken by J. R. Cooper.

6. THERMOELECTRIC POWER

The thermoelectric coefficient has negative values for all overdoped samples and has positive values for all underdoped samples, despite the fact that the carriers are hole-like throughout the range. This is shown in Figure 5. It turns out that the maximum T_c corresponds to the first set of curves when the thermoelectric power has just become positive. This result appears to be quite general in that a similar sign change in the thermopower corresponding to maximum T_c composition or hole carrier density has also been observed with high precision in YBCO and BSCCO irrespective of the means by which the carrier density is changed (oxygen variation or cation substitution). This suggests that the measurements can be used to accurately pinpoint the composition which gives the highest T_c. The behaviour is adaquately described by the formula [10]

$$S = (\kappa / 2e) \ln\{(1-z)/z\} \tag{1}$$

where z is the ratio of carriers to sites. This depends on the interpretation that when all the carriers provided by doping form bosons whose dimensions are such that the space in the solid is just 'filled' by these bosons without their wavefunctions overlapping, z would have a value close to 0.5, making S vanishingly small. For a smaller density of carriers, i.e. when x > 0.2, z < 0.5 and S turns positive. On the other hand the opposite is true when the density of carriers is so large that there is a strong overlap between such bosons. This occurs when x < 0.2 and z > 0.5. Figure 6 shows schematically the carrier concentration in the three regimes.

7. SPECIFIC HEAT MEASUREMENTS

The electronic specific heat of the same series of compounds has been reported by Loram and Mirza [12] and their results are shown in Figure 7. The most striking feature is that the shape and evolution of the specific heat anomaly are very different in the underdoped and in

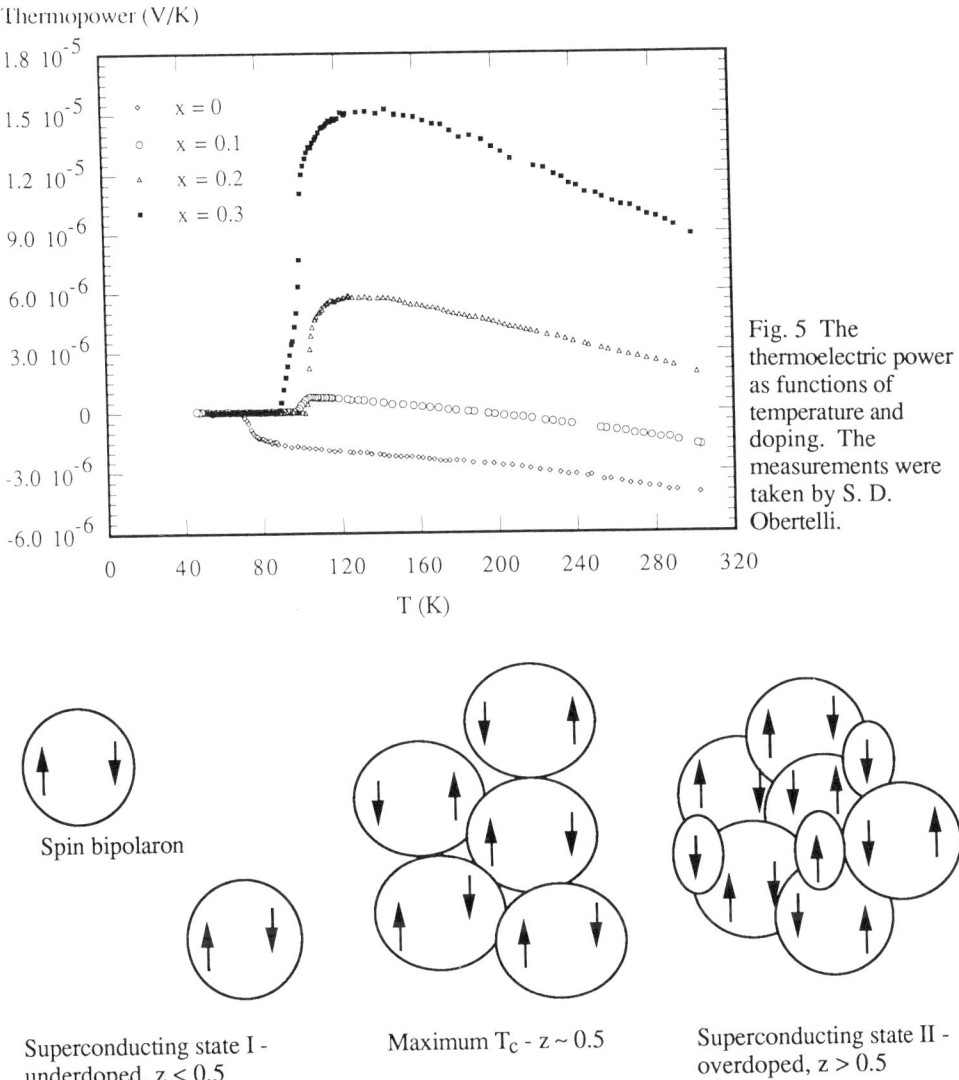

Fig. 5 The thermoelectric power as functions of temperature and doping. The measurements were taken by S. D. Obertelli.

Spin bipolaron

Superconducting state I - underdoped, z < 0.5

Maximum T_c - z ~ 0.5

Superconducting state II - overdoped, z > 0.5

Figure 6. Real space representation of spin-bipolarons [11] with concentrations corresponding to the three regimes of under-, optimum and over-doping.

the overdoped regimes. In the low carrier density regime (state I), the anomaly is broad and extending to temperatures well beyond T_c. At the same time, the anomaly grows rapidly in area and sharpness with increasing carrier density suggesting an increased range of spatial coherence, while the onset temperature remains almost unchanged at around 120K. The largest anomaly is found to correspond to maximum T_c. On increasing the carrier density beyond that which gives maximum T_c and into the state II, the anomaly retains the same sharpness but its position falls steadily to low temperature. This confirms that superconducting fluctuations are very strong in the underdoped regime and can be related with the decreasing coherence length and a low density of carriers. The behaviour in the overdoped regime resembles that of a BCS type superconductors. This is perhaps not surprising since the coherence length has increased sufficiently that each coherence volume now contains a significant number of Cooper pairs.

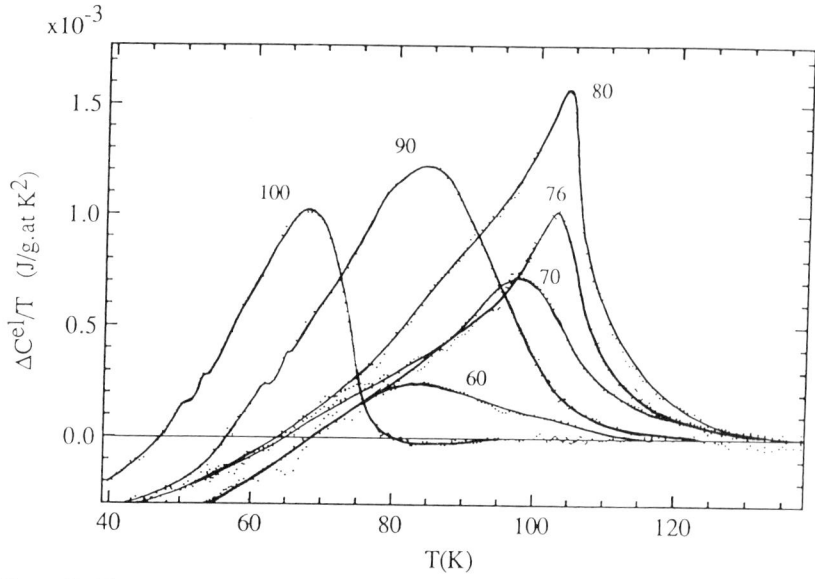

Figure 7. Electronic specific heat anomalies near the superconducting transition for the series $(Y_xCa_{1-x})(Tl_{.5}Pb_{.5})Sr_2Cu_2O_{7-\delta}$, the number next to each curve corresponds to the concentration of Y as x = 1- number/100.

8. MAGNETORESISTANCE

Another striking result is the effect of a moderately high field (7 T) on the resistance of ceramic samples. We examined the resistivity of three samples: one with x = 0.2 which has the highest T_c at zero field and two others with $T_c \sim 70$ K but with over- and under-doping, Figure 8. The results clearly distinghish between the overdoped (state II) sample which is least sensitive to magnetic fields and the underdoped (state I) sample whose T_c is strongly lowered under the same magnetic field. Since the samples used were ceramic with density of the order of 55 to 65% of the theoretical density, the measured resistance should correspond to that for intergrain currents. One possible explanation for this could be an increase in the superconducting coherence length associated with a corresponding increase in carrier density. This is also likely to be the explanation for the shift from 3-D to 2-D type fluctuations; as the carrier concentration decreases the coherence length shrinks to a point where superconductivity is unable to span the distances between CuO_2 planes. Understanding this behaviour is important in bulk applications of oxide superconductors.

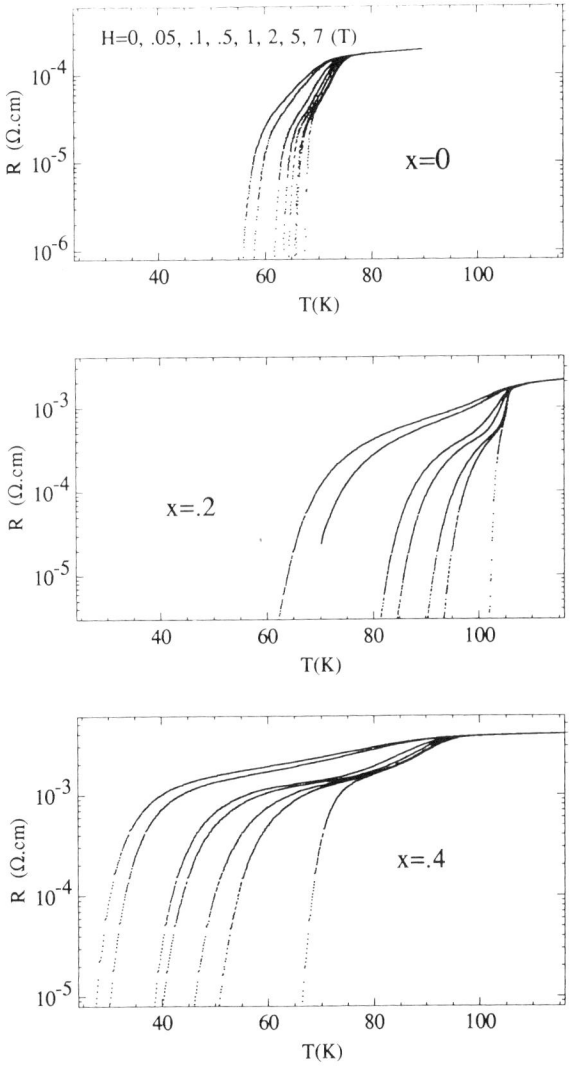

Figure 8.
Magnetoresistance of three samples representing the maximum T_c, over- and under-doping showing BCS and Bose condensate behaviour of oxide superconductors determined by doping. The measurements were taken by J M Wade.

9. CONCLUSION

Since the first observation of superconductivity in $SrTiO_{3-\partial}$ ($T_c \sim 0.7$ K) in 1964, we have been aware of the fact that the characteristics of an oxide superconductor depend sensitively on the density of carriers of the material. We now found that the coherence length is also a sensitive function of the carrier concentration. This appears to dictate the behaviour of the superconductors from one of BCS like in the overdoped regime to one of Bose condensate in the low density regime.

10. REFERENCES

[1]   A. Maignan, C. Martin, M. Huve, J. Provost, M. Hervieu, C. Michel and B. Raveau, Physica C, 170 (1990) 351

[2]   N. Mori, H. Takahashi and C. Murayama, Supercond. Sci. and Techn., 4 (1991) S439

[3]   R. S. Liu, P. P. Edwards and P. T. Wu, Adv. Mat., 2 (1990) 369

[4]   T. Kaneko, T. Wada, H. Yamauchi and S. Tanaka, Appl. Phys. Lett. 56 (1990) 1281

[5]   R. S. Liu, J. R. cooper, J. W. Loram, W. Zhou, W. Lo, P. P. Edwards, W. Y. Liang and L. S. Chen, Sol. Stat. Comm., 76 (1990) 679

[6]   R. J. Cava, Science, 247 (1990) 656

[7]   P. Mendels, H. Alloul, J.F. Marucco, J. Arabski and G. Collin, Physica C (1991)

[8]   J. M. Liang, R. S. Liu, Y. T. Huang, S. F. Wu, P. T. Wu and L. J. Chen, Physica C, 165 (1990), 347

[9]   R. S. Liu, P. P. Edwards, Y. T. Huang, S. F. Wu and P. T. Wu, Sol. Stat. Chem., 86 (1990) 334

[10] R. R. Heikes and R. W. Ure, "Thermoelectricity" (1961) 81

[11] N. F. Mott, Adv. in Phys. 39 (1990) 55

[12] J. W. Loram and K. A. Mirza, Supercond. Sci. and Techn., 4 (1991) S286

High T_c Superconductor Thin Films
L. Correra (Editor)

Superconducting Properties and Two Dimensional Transport Behavior in YBa$_2$Cu$_3$O$_7$-based Superlattices

Qi Li, X. X. Xi, T. Venkatesan[a]

Center for Superconductivity Research, Department of Physics, University of Maryland, College Park, MD 20742, USA

[a]Also Department of Electrical Engineering, University of Maryland, College Park, MD 20742, USA

Abstract

The superconductivity of YBa$_2$Cu$_3$O$_7$-based superlattices, especially YBa$_2$Cu$_3$O$_7$/PrBa$_2$Cu$_3$O$_7$ (YBCO/PrBCO) superlattices, have been studied extensively recently [1-5]. It has been shown that a YBCO layer with nominal thickness of that of a single unit cell is superconducting. However the critical temperature of a thin YBCO layer decreased when the thickness was reduced to less than that of ~ 4 unit cells, accompanied by a broadened resistive transition and a lower zero resistance temperature. We found that the mean field T_c starts to decrease when the sheet resistance of YBCO layer reaches about a few KΩ/□ as a result of reducing the YBCO layer thickness. This number is of the same order as the "critical sheet resistance" value of $\hbar/e^2$ at which a similar T_c reduction was observed in a variety of ordinary two dimensional superconducting thin films. This suggests that the reduction of the mean field T_c for thin YBCO layer is of the same origin as in other two dimensional systems. Detailed studies of R-T and I-V characteristics of thin decoupled YBCO samples showed behaviors consistent with the prediction of Kosterlitz-Thouless transition. We thus conclude that the zero resistance temperature and the resistance well below mean field T_c in decoupled thin YBCO layer are associated with the dissipation of the vortex-antivortex pair excitations. These results suggest that the origin of the transition characteristics in YBCO layer for thicknesses less than a few unit cells is primarily related to the reduced dimensionality of these thin layers. Enhanced critical current anisotropy in the presence of a magnetic field was also observed.

1. INTRODUCTION

Recently, epitaxial growth of high temperature superconducting YBa$_2$Cu$_3$O$_7$-based superlattices, especially YBa$_2$Cu$_3$O$_7$/PrBa$_2$Cu$_3$O$_7$

(YBCO/PrBCO) superlattices, have been obtained by using both pulsed laser deposition and sputtering [1-5]. Superconductivity of these superlattices has been studied systematically to address some of the intrinsic properties of YBCO system, such as the interlayer coupling effect and superconducting properties in a single unit cell or thin layers. In this paper we present an overview of the recent results on superconducting properties of YBCO/PrBCO superlattices, especially the properties of the decoupled thin YBCO layers and discuss the possible reasons for their superconducting transition behavior.

One of the characteristic features of high T_c materials is that they are all layered structures with Cu-O planes separated by other planes. In YBCO compound, the two Cu-O planes within a unit cell are separated by ~3 Å which is comparable to the coherence length of 1~5 Å in the same direction and therefore is mostly considered to be closely coupled. The distance between the two Cu-O planes in the adjacent unit cells is ~8 Å and it is not clear at the present time as to whether there is any interaction between the adjacent unit cells and how it affects the superconducting properties in this system. In another words, whether the superconductivity of a unit cell thick layer would be more like two-dimensional or similar to that of a bulk sample, is still unresolved.

Superlattices of YBCO/PrBCO have been used in an attempt to answer these questions. The advantages of using PrBCO layers are that PrBCO, contrary to other rare earth doped "123" structures, is not superconducting but is similar to YBCO in terms of crystal structure and oxygen composition. The orthorhombic PrBCO has lattice parameters very close to those of YBCO (with lattice mismatch of ~1%) thus providing a much better lattice match to YBCO than most of the substrates.

It has been observed that the YBCO layer with nominal thickness of that of single unit cell is superconducting [2, 5]. This indicates that a Cu-O bilayer is sufficient for achieving superconductivity. However, the T_c, including the onset transition temperature, was found to be lower than the bulk value when the film thickness was reduced to less than that of ~4 unit cells and the transition was broadened.

The reduction of T_c in a few cell thick layers of YBCO has raised different explanations. The first possibility would be material imperfection. As we have discussed in Ref.3 and also Ref.4, it is not possible to explain the superconducting properties of the superlattices by using purely material problems, such as interdiffusion between the two layers or defects induced by the interfaces. Based on the evidences in transport properties, we had speculated that the reduction of T_c is related to the reduced dimensionality in the unit cell thick layers due to the decoupling of the YBCO layers [2]. Recently, an alternative explanation for the T_c reduction using a model of "hole filling" by the PrBCO in the superlattices has been proposed by R. F. Wood [6] which fits to the T_c data of YBCO/PrBCO superlattices very well. However, it is inconsistent with our result that the ultrathin YBCO films grown on $SrTiO_3$ (100) crystals show similar T_c changes to that of YBCO films sandwiched between PrBCO layers [2,7]. It is also not consistent with the fact that the sheet resistance does not change at all with the thickness of the

PrBCO layer [2]. In this paper we will summarize and present some of the results on studies of resistive transition, I-V characteristics, and anisotropy effect, mostly in superlattices with decoupled thin YBCO layer.

2. SAMPLE PREPARATION AND CHARACTERIZATION

Superlattice samples of YBCO/PrBCO with different thicknesses for each layers were made by pulsed-laser deposition using an excimer laser with 248 nm wavelength, 30 ns pulses and typical energy density of 1.7 J/cm^2 in a multitarget deposition system. The deposition rate was controlled at ~0.2 Å/pulse and the repetition rate of 3 Hz. The substrate was held at 760°C, measured at the heater surface during deposition in 100 mTorr oxygen, and after deposition the sample was allowed to cool to room temperature directly in 200 Torr oxygen ambient. The details of the preparation and characterization of epitaxially grown YBCO and YBCO/PrBCO structures on SrTiO$_3$ (100), LaAlO$_3$ (100), and MgO (100) have been published in Ref.2-3, Ref.8-10. Briefly, the structure and the interface properties of YBCO/PrBCO structures have been characterized by X-ray diffraction, ion channeling and Rutherford backscattering, secondary ion mass spectroscopy (SIMS), Auger analysis, and cross sectional TEM. These results indicate that the layers grow epitaxially on one another with good crystallinity and abrupt interfaces and without additional disorder induced by the superlattice preparation process. The result has also been proved by measuring the superconducting transition of YBa$_2$Cu$_3$O$_7$/DyBa$_2$Cu$_3$O$_7$ [1] and YBa$_2$Cu$_3$O$_7$/Y$_y$Pr$_{1-y}$Ba$_2$Cu$_3$O$_7$ (y=0.8) [3] with each layer of only one unit cell thick, where the other species besides YBCO is also superconducting and no degradation of the T$_c$ due to the interfaces was found.

Fig. 1 shows a X-ray diffraction spectrum in the region of the (001) and (002) peaks from the sample with the period of 3 cell thick YBCO and 9 cell thick PrBCO. The superlattice has c axis normal to the substrate surface, and

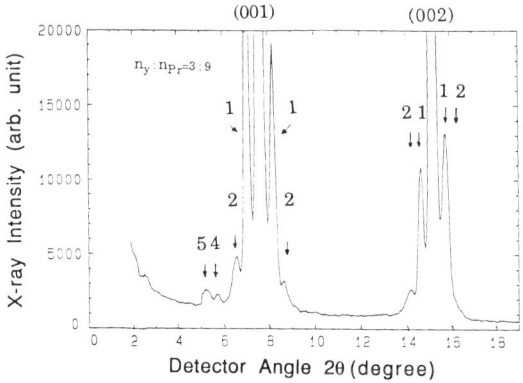

Fig. 1. X-ray diffraction spectra of a superlattice sample.

the satellite peaks due to the superlattice modulation were observed up to the fifth order peak as expected from the number of unit cells in a period, indicating a high degree of structure ordering in the superlattice. The satellite peaks have been observed even when each layer was as thin as 12 Å. Our recent studies on the interdiffusion effect between Y and Pr reveals a remarkably low diffusion coefficient in Y site [11]. We thus believe that the remaining imperfection is mainly from the surface (interface) atomic steps observed by cross sectional TEM images [12].

3. SUPERCONDUCTING TRANSITION

Superconducting critical temperatures of the YBCO/PrBCO superlattices have been measured by resistance as well as ac susceptibility measurement as both YBCO and PrBCO layer thicknesses were varied. The resistive transition curves for different thicknesses of YBCO with 100 Å PrBCO layers are shown in Fig.2. As shown in the next figure, in this case Fig. 2 corresponds to the superconducting transitions of the isolated YBCO layers with the same layer thicknesses. The resistive transition midpoint temperature T_c^{mid} for three YBCO layer thicknesses, 12 Å, 24 Å, and 48 Å, are shown in Fig.4 as a function of the PrBCO layer thickness. The transition onset and zero resistance temperature behave in a similar way to that of the T_c^{mid}. The zero resistance temperatures will be discussed in the next section.

Ideally we should use the mean field transition temperature T_c^{mf} for our discussion, but as we know, by using standard fluctuation conductivity analysis [13] in YBCO thin films the T_c^{mf} varied according to the chosen fitting parameters. In particular, the normal state resistance vs. temperature curves of ultrathin YBCO films showed convex behavior in a wide temperature range (see also Ref.7) which makes this type of analysis more inaccurate. Nevertheless, according to our estimation based on the standard fluctuation conductivity the T_c^{mf} had a value, in most of the cases, in the temperature range of 10 % to 70 % of the normal state resistance. Therefore T_c

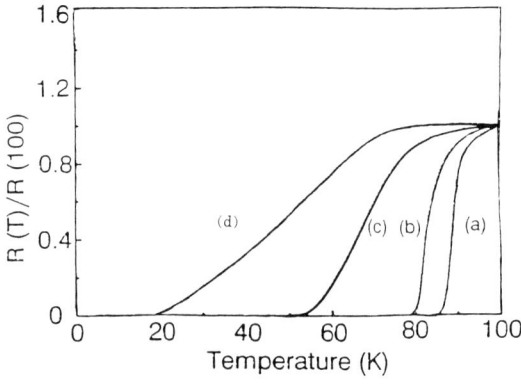

Fig.2. Superconducting transition curves for various decoupled YBCO layer: (a) 100 Å, (b) 48 Å, (c) 24 Å, (d) 12 Å

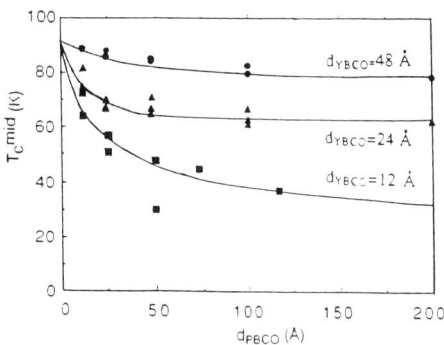

Fig. 3. The midpoint T_c vs PrBCO layer thickness for superlattices with YBCO layer thickness of 12 Å, 24 Å, 48 Å.

midpoint was used in this paper to reflect the changes of T_c^{mf}. This is in contrast to the result implied in Ref. 14, where the T_c^{mf} is identified as the zero resistance temperature. We will address this problem again in next section.

As shown in Fig. 3, for each YBCO layer thickness, T_c remains almost constant when PrBCO layer is thicker than about 4 unit cells. In this case, YBCO layers are believed to be isolated from each other by intercalated PrBCO layers and the superconductivity can be considered as that of a single thin YBCO layer sandwiched between the PrBCO layers. This conclusion has also been supported by the similarity of the T_c's of ultrathin single YBCO films on SrTiO₃ single crystal substrates and on PrBCO buffer layers (except for 12 Å thick layer). As seen in Fig.2 and Fig.3, a decoupled YBCO layer with nominal thickness of 12 Å (one unit cell thick) is indeed superconducting.

The T_c for isolated YBCO layers is plotted as a function of YBCO layer thickness in Fig.4, which indicates that T_c decreases when YBCO layer thickness is less than about 4 unit cells. By comparing the T_c's obtained by different groups using different techniques and also using different nonsuperconducting layers rather than the PrBCO, reasonably good

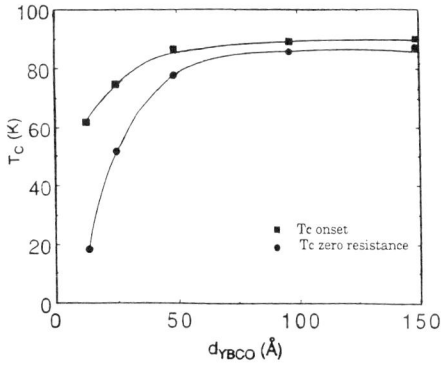

Fig. 4. T_c as a function of the YBCO layer thickness.

agreement on T_c for YBCO layers with a few unit cell thickness is found except that $T_c(R=0)$ points for 12 Å thick layers exhibit a large scatter, ranging from zero to about 30 K, although the T_c onset temperatures are still close to each other at around 60 K. We believe that at the present time for a one unit cell layer of YBCO, $T_c(R=0)$ is still affected by some nonideal conditions, especially the atomic step structures observed in one unit cell thick layer [12].

Regardless of the deposition techniques and nonsuperconducting layers used, the general features of the superconducting transition for YBCO layers with thickness less than that of a few cells are the same: i.e., reduced midpoint T_c (or T_c^{mf}) compared with bulk value and broadened resistive transition with a lower zero resistance temperature.

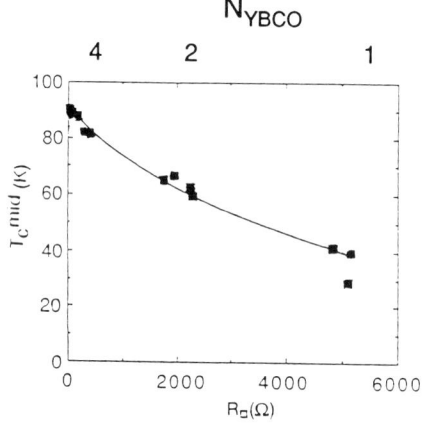

Fig. 5. The midpoint T_c of the decoupled YBCO layer vs. the sheet resistance per YBCO layer.

To further explore the nature of the T_c decrease with film thickness, the midpoint T_c of the decoupled YBCO layers are plotted as a function of sheet resistance per YBCO/PrBCO layer at 100 K as shown in Fig. 5. The resistivity of PrBCO at such a temperature is orders of magnitude larger than that of YBCO. Thus the sheet resistance quoted in the figure corresponds to the sheet resistance per single YBCO layer. The behavior of T_c as a function of sheet resistance in Fig. 5 is very similar to that observed in conventional two-dimensional superconducting thin films like Bi, Pb [15], and amorphous Mo-Ge [16]. As can be seen from Fig.5 T_c is reduced when the sheet resistance reaches a few KΩ/☐ as a result of reducing the film thickness, which is of the same order as the "critical sheet resistance" value $\hbar/e^2$. The sheet resistance in the order of $\hbar/e^2$ has been identified as a "universal" critical sheet resistance, at which the system will experience a superconducting to insulating transition. The exact value of the critical sheet resistance depends upon the system studied, but this effect has been widely observed [15-19]. This suggests that the reduction of T_c for the YBCO layer in approaching the single cell thickness might have the same origin as in the ordinary two-dimensional superconducting films, i.e. the effect of approaching the critical sheet resistance for the appearance of superconductivity. The existence of a "universal" critical sheet resistance and its correspondence with superconductivity have been discussed theoretically by many authors but have not been completely understood yet [19].

The normal state resistivity of standard YBCO films is about 50 $\mu\Omega$ cm - 100 $\mu\Omega$ cm. The sheet resistance of one or two cell thick layers measured is a few times larger than that estimated using the above resistivity number. We did not find additional microscopic defects in the superlattices compared to regular YBCO film which could contribute to increased impurity scattering and the Hall effect measured by O. Fisher et al [14] showed no changes of the carrier number with the changes of YBCO layer thickness. Taking into account that the electron mean free path is about tens of angstroms which is larger than the lattice constant of c direction, and the anisotropy factor in YBCO is only on the order of tens, we believe the larger sheet resistances in one or two cell thick layers are mainly due to the interface boundary scattering.

It should also be noted that there are substantial differences between the conventional two-dimensional superconductors and the YBCO films. The ordinary superconductors studied are all in the "dirty limit", i.e., the electron mean free path is much less than the intrinsic coherence length, and the film thicknesses are also smaller than that, while YBCO film is a relatively "clean" system and the intrinsic coherence length is shorter than the layer thickness. Nevertheless, the similarity of the dependence of T_c on sheet resistance in significantly different types of systems illustrates again the universal nature of the relation between these two quantities in two dimensional systems.

4. CURRENT VOLTAGE CHARACTERISTICS

One of the characteristic features of two-dimensional superconductivity is the Kosterlitz-Thouless transition which has been observed in a variety of uniform two dimensional films [20]. According to the K-T phase transition theory thermally excited vortex-antivortex pairs which are bounded at low temperature dissociate into free vortices at a temperature T_{KT}. The thermally dissociated free vortices give rise to a finite resistance above T_{KT}.

The K-T transitions have been observed in single crystals of $Bi_2Sr_2CaCu_2O_8$ [21], which confirms the nearly two-dimensional nature of $Bi_2Sr_2CaCu_2O_8$. There are reports on observations of K-T transition in YBCO system but the results are still controversial [22,23]. On the other hand, K-T transition has been observed in the superlattices with the thin decoupled YBCO layers (for details see S.Vadlamannati et al Ref.24 and C.T.Rogers Ref.25).

Several characteristic features of K-T transition have been predicted by the theory. Resulting from thermally activated unbinding of vortex-antivortex pairs the resistance in zero magnetic field near T_{KT} well below $T_c{}^{mf}$ should follow the temperature dependence described by $\rho/\rho_n \sim a\exp[-2(b\tau_c/\tau)^{1/2}]$, where $\tau_c=(T_c^{mf}/T_{KT}-1)$ and $\tau=(T/T_{KT}-1)$, ρ_n is the normal state resistance, and a and b are nonuniversal parameters. T_{KT} should correspond to the truely zero resistance temperature $T_c(R=0)$. Magnetic fields and currents will induce breaking of vortex pairs. In the case of current induced vortex pair breaking, the I-V characteristics below T_{KT} will follow a power law relation, $V \sim I^{a(T)}$, where $a(T)=1+\pi K$ and πK is proportional to the 2-D superfluid density $\pi K \sim n_s^{2D}$. At T_{KT} a drop of $a(T)$ from 3 to 1 corresponding to the superfluid density jump will be observed [26].

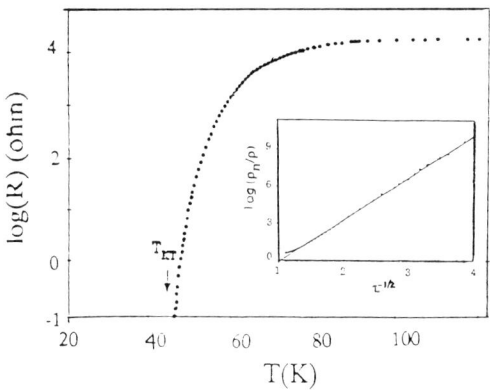

Fig.6. Resistive transition curve for a sample of 24 Å/ 100 Å in semilog scale. Insert: $\ln(\rho/\rho n)$ vs. $\tau^{-1/2}=(T/T_{KT}-1)$

Fig. 6 shows the resistance transition of a 24 Å/100 Å superlattice sample. The resistance as a function of temperature near the zero resistance temperature region is plotted in accordance with the expression described above. A good agreement with the expected behavior is found, shown in the insert. The transition temperature of the 24 Å thick YBCO layer from ac susceptibility measurement is consistent with the T_c of a 24 Å thick single YBCO layer grown on $SrTiO_3$ substrate with the PrBCO buffer layers. This proves that the 24 Å thick YBCO layers in our specimen are completely decoupled from each other by the PrBCO layers.

Fig.7 shows the I-V curves of a 24 Å/100 Å superlattices sample. The linear curves at the low voltage level indicate the power law behavior of V and I $(V \sim I^{a(T)})$. The power of the exponential $a(T)$ has been plotted in Fig.8 as a function of temperature. As seen from the figure, $a(T)$ shows a clear jump as predicted from the value of around 3 to 1. The T_{KT} value for this sample is identified as the temperature at which $a(T)=3$ as marked in the figure. The T_{KT} value obtained from the I-V characteristics and from the resistance measurement are in a good agreement with each other within several tenths of degrees. Fig.8 has been fitted to the Ginzburg-Landau expression for the superfluid density $n_s^{2D}(T)=n_s^{2D}(0)(1-(T/T_c^{mf})^4)$ below T_{KT} and plotted as the solid line in the figure. The T_c^{mf} has also been obtained by extrapolating the solid line to the temperature axis as shown in the figure. Using the values obtained for T_{KT} and T_c^{mf} (see ref.24), the parameter $\tau_{KT}=(T_c^{mf}-T_{KT})/T_c^{mf}$

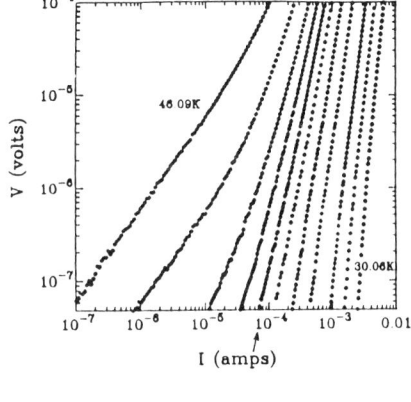

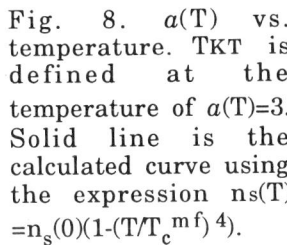

Fig. 8. $a(T)$ vs. temperature. TKT is defined at the temperature of $a(T)=3$. Solid line is the calculated curve using the expression $n_s(T)$ $=n_s(0)(1-(T/T_c^{mf})^4)$.

Fig. 7. I-V characteristics for a 24 Å/100 Å sample

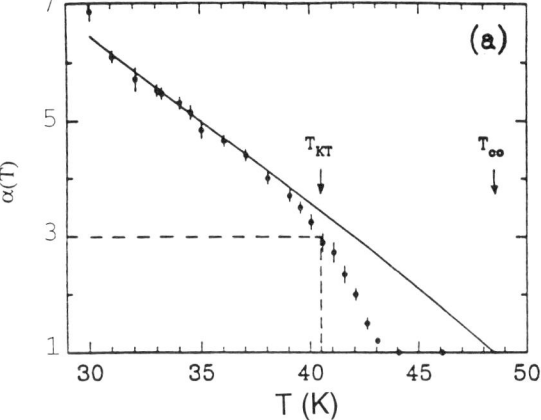

obtained for 24 Å and 48 Å YBCO layer is ~0.17 and ~0.05, which is an order of magnitude larger than in single crystals. If we use the approximation form for dirty superconductors, $T_{KT}/T_c^{mf}= (1+0.17R\Box/Rc)^{-1}$ where $Rc=\hbar/e^2=4100\,\Omega$, the calculated τ_{KT} is 0.1 and 0.05 for the 24 Å and 48 Å YBCO layers, which is in a reasonably good agreement with the experimental data. It should be noted that the YBCO films are not in the "dirty limit", therefore the above formula may not be applied quantitively to YBCO system.

The resistive transition for the 12 Å YBCO layer is still much broader than expected from the K-T theory. We believe it is related to the structure features mentioned early in this paper.

We have also measured our standard high quality films for comparison and no $a(T)$ jump near the predicted value was observed at the same measurement current limit. The $a(T)$ increased with decreasing temperature very rapidly in the thick film. There may be two reasons for not observing K-T transition in thick film. First, because the coupling in YBCO system is too strong the K-T transition does not exist in bulk samples as discussed by S. Martin in Ref.20. Secondly, it is not possible to observe the K-T transition in thick YBCO films within our voltage measurement limit.

From the comparison between the properties of standard high quality film and those of very thin YBCO layers and the fact of existence of K-T transition and large τ_{KT} in the latter case, it is clear that the T_c (R=0) in such films and the finite resistance above T_{KT} are associated with the dissipation of vortex-antivortex pair excitations.

5. ANISOTROPY IN SUPERLATTICES

Among the high T_c cuprates, $Bi_2Sr_2CaCu_2O_8$ showed much stronger anisotropy in the electrical resistance and critical current than YBCO when a magnetic field was applied parallel and perpendicular to ab-planes. It is believed at present time that this is an intrinsic effect and related to the fact that the distance between the CuO_2 bilayer is large which makes $Bi_2Sr_2CaCu_2O_8$ a weakly coupled and a nearly two dimensional system. If we believe that certain amount of coupling exists in the YBCO system, as we decouple the YBCO thin layers by intercalating insulating material, an enhanced anisotropy effect should be expected as a result of reducing the dimensionality of the system.

Anisotropy of resistive transition in thin layers of YBCO has been studied earlier (see X. X. Xi, Ref. 27, S. Schauwer, Ref. 28) and a similar effect has also been observed in YBCO/PrBCO superlattices [4,9]. It has been found that the broadening of the resistive transition diminished with reduced thickness of YBCO when the magnetic field was applied parallel to the ab-plane up to 13 T and disappeared when the layer thickness is very thin. The critical currents of the superlattices has been measured as a function of the angle between the field and the c direction for comparison with some earlier results on single thick layer of YBCO.

The result of a sample of 48 Å /100 Å superlattice is shown in Fig. 9 (b). For comparison, the angular dependence of the critical current of a thick YBCO film and a $Bi_2Sr_2CaCu_2O_8$ film are also shown in Fig.9 (a) and (c) measured by B. Roas et al [29,30]. The magnetic fields applied as well as the reduced temperature are slightly different in the three cases as shown in the figures.

As we can see from the figure, the anisotropy of the critical current in the superlattice is much stronger than that in the YBCO film at such a temperature (note the log scale of J_c for (b) and (c) and linear scale for (a)). It is very interesting to see that the angular dependence of the critical current of the 48 Å/100 Å superlattice sample is very similar to BSCCO rather than that of the YBCO film. It indicates that as we decouple the YBCO layers the system becomes more like BSCCO (nearly two dimensional) than YBCO in terms of critical current anisotropy. This confirms that as we reduce the dimensionality of the YBCO layers the anisotropy factor is largely enhanced. This result also demonstrates that the anisotropy of the high T$_c$ cuprate can be controlled by changing the distance between the superconducting planes.

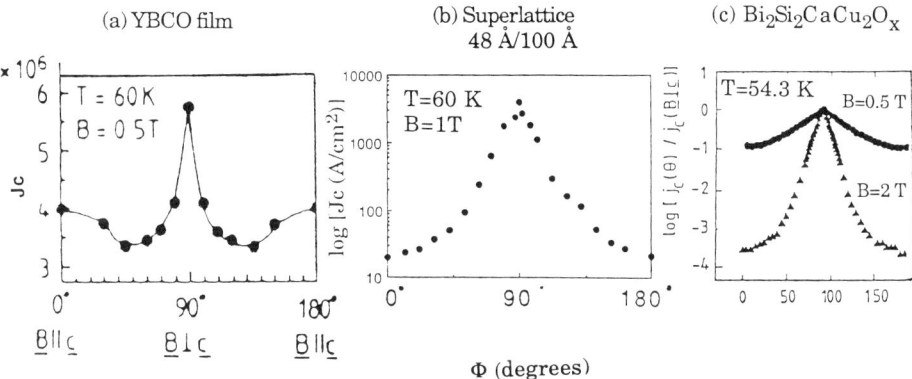

Fig. 9. Comparison of the angular dependence of the critical current between (a). YBCO film , (b). 48 Å/ 100 Å superlattice, and (c). Bi$_2$Sr$_2$CaCu$_2$O$_8$ film.

6. CONCLUSION

In summary, it has been demonstrated that single unit cell thick YBCO layer is superconducting, but the T$_c$ of the thin YBCO layer is lower than the bulk value with a broadened resistive transition. The reduced zero resistance T$_c$ and the broadening of the transition are shown to arise mainly from the reduced dimensionality of the thin films as indicated by the enhanced Kosterlitz-Thouless transition observed in the system, except for the 12 Å YBCO layer where the transition is still much broader than expected from the K-T theory. In addition, the correspondence between the observation of a reduction of mean field T$_c$ at nearly the same sheet resistance of the YBCO layers as for the different ordinary superconductors suggests the universal nature of these two quantities. Finally, the YBCO/PrBCO superlattices is a model system to study the nearly two dimensional and strong anisotropic nature of the cuprate superconductors.

Acknowledgement
The authors would like to acknowledge the collaborations with S. Vadlamannati, W. L. McLean, S. Schwarz, R. Ramesh, T. Paastza, P. Lindenfeld, H. Zhang, and J. Lynn in conducting the experiments and A. T. Fiory, A. F. Hebard, C.T. Rogers, and C. Lobb for helpful comments and discussions.

REFERENCES

1. J.-M. Triscone, M. G. Karkut, L. Antognazza, O. Brunner, and O. Fisher, Phys. Rev. Lett. 63, (1990) 1016.
2. Qi Li, et al, Phys. Rev. Lett. 64, (1990) 3086
3. X. D. Wu et al, Appl. Phys. Lett. 56, (1990) 400.
4. J.-M. Triscone et al Phys. Rev. Lett. 64, (1990) 804.
5. D. H. Lowndes, D. P. Norton, and J. D. Budai, Phys. Rev. Lett. 65, (1990) 1160.
6. R. F. Wood, Phys. Rev. Lett. 66, (1991) 829.
7. X. X. Xi, J. Geerk, G. Linker, Qi Li, O. Meyer, Appl. Phys. Lett. 54 (1989) 2367.
8. A. Inam et al., Appl. Phys. Lett. 53, (1988) 908.
9. Qi Li, et al. IEEE Tran Magn. 27, (1991) 2472.
10. T. Venkatesan, Appl. Phys. Lett. 56, (1990) 391.
11. S. Schwarz, Qi Li, X. X. Xi, and T. Venkatesan, unpublished
12. J.Pennycook et al, Preprint. R. Ramash, private conversation.
13. L. G. Aslamazov and A. I. Larkin, Phys. Lett., 26, (1968).
14. O. Fisher, J. -M. Triscone, O. Brunner, L. Antognazza, M. Affronte, and L. Mieville, Proc. of E-MRS, Strasbourg,(1990) and O. Fisher, Preprint.
15. D. B. Haviland, Y. Liu, and A. M. Goldman, Phys. Rev. Lett. 62, (1989) 2180.
16. J. M. Graybeal and M. R. Beasley, Phys. Rev. Lett. 29, (1984) 4167.
17. A. F. Hebard et al, Proc. of MRS Meeting, Boston, 1990.
18. T. Pang, Phys. Rev. Lett., 62, (1989) 2176.
19. "Percolation, localization, and superconductivity", edited by A. M. Goldman and S. A. Wolf (Plenum, New York,1983)
20. J. M. Kosterlitz and D. J. Thouless, J. Phys. C 6, (1973) 1181.
21. S. Martin, A. T. Fiory, R. M. Fleming, G. P. Espinosa, and A. S. Cooper, Phys. Rev. Lett. 62 (1989) 677.
22. N. C. Yeh and C. C. Tsuei, Phys. Rev. B 39, (1989) 9708.
23. Q. Y. Ying and H. S. Kwok, Phys. Rev. B 42, (1990) 2242.
24. S. Vadlamannati, Qi Li, T. Venkatesan, W. L. Mclean, and P. Lindenfeld, submitted to Phys. Rev. B.
25. C. T. Rogers et al, preprint.
26 A. M. Kadin, K. Epstein, and A. M. Goldman, Phys. Rev. B, 27, (1983) 6691.
27. X. X. Xi, S. Schauer, V. Windte, G. Linker, Q. Li, G. Geerk, and O. Meyer, Proc. of MRS Meeting, Boston, 1989.
28. S. Schauer, X. X. Xi, V. Windte, O. Meyer, G. Linker, Q. Li, and G. Geerk, Cryogenics 30, (1990) 586.
29. B. Roas, L. Schultz and G. Saemann-Ischenko, Phys. Rev. Lett. 64, (1990) 479.
30. L. Shultz et al, IEEE Tran. Mag. 27, (1991) 990

High T$_c$ Superconductor Thin Films
L. Correra (Editor)

25

2D CONDUCTIVITY IN THE NORMAL STATE OF THE 123 SUPERCONDUCTING COPPER OXIDES

J.C. Ousset [a], S. Askénazy [a], H. Rakoto [a], J.M. Broto [a], J.F Bobo [b], M.S. Osofsky [c]
I R.J. Soulen [c], S.A. Wolf [c].

(a)   *Service des Champs Intenses, INSA, Avenue de Rangueil, 31077 Toulouse-Cedex (France)*

(b)   *Laboratoire CNRS-St-Gobain, CRPAM BP 109, 54704 Pont à Mousson-Cedex (France)*

(c)   *Naval Research Laboratory, Washington, DC 20375 (USA)*

ABSTRACT

It is now well known that the physical properties of the high T_c 123 superconducting oxides are strongly anisotropic.

We performed high field (pulsed field up to 40 Tesla) magnetoresistance measurements on $YBa_2CO_3O_7$ films which were laser ablated, in situ, from a stoichiometric target on MgO and on $GdBa_2Cu_3O_7$ films deposited on ZrO_2 using three resistively heated boats (containing Gd, BaF_2 and Cu) and post-annealed. X-ray diffraction showed all the films to be oriented with the c-axis normal to the substrate. By applying a field which orientation varies respect to the c-axis, we show that the classical magnetoresistance only depends on the normal component of the field. This is the direct proof that the electronic band has a real 2D character in these materials. That property can easily be extended to the whole 123 family.

1. INTRODUCTION

The high-T_c superconducting oxides have a layered structure [1], with either one or a small number of parallel CuO_2 planes in each unit cell. In the case of the 123 family $(RE)Ba_2Cu_3O_7$ (RE = Y, Sm, Gd,...) these planes go by pairs. It is now usually accepted that the electronic band has a 2D shape. Such assumption, also based on electronic structure arguments, is supported by indirect experimental data. The carriers are more mobile in the CuO_2 planes than in the perpendicular direction : the resistivity and the plasmon frequency in the normal state are very anisotropic [2,3]. The anisotropy ratio for the 123 family is about equal to or higher than 25. A strong anisotropy is also expected in the superconducting state for the penetration depth λ and coherence length ξ.

So it should be reasonable to consider the electrons as being itinerant within each one of the CuO_2 planes.

All the 123 compounds synthesized after the discovery of $YBa_2Cu_3O_7$ are isostructural and the band structure is the same for this class of materials.

The analysis of our measurements of the anisotropy in the normal state of $GdBa_2Cu_3O_7$ and $YBa_2Cu_3O_7$ will lead to the conclusion that the conductivity is two-dimensional. This essential property is a characteristic of the whole family $(RE)Ba_2Cu_3O_7$.

2. EXPERIMENTAL

2.1. Characterization of the samples

The studies were performed on $YBa_2Cu_3O_{7-y}$ films which were laser ablated, in situ, from a stoichiometric target onto MgO and $GdBa_2Cu_3O_{7-y}$ films which were deposited onto ZrO_2 using three resistively heated boats (containing Gd, BaF_2 and Cu) and post-annealed. Analysis by x-ray diffraction showed all of the films to be oriented with the c-axis normal to the substrate (see fig. 1).

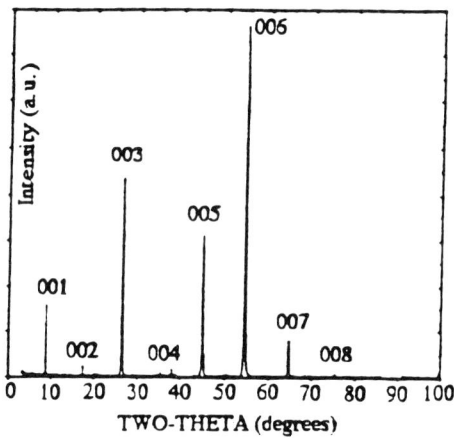

FIGURE 1. X-Ray diffraction on a GdBaCuO film.

Rocking curves allowed us to evaluate a desorientation of about one degree. Zero resistance was reached over 85 K for all of our samples. We choose the best giving R = 0 at 90 K for the magnetoresistance measurements.

2.2.1. Magnetoresistance measurements and discussion.

The magnetoresistance measurements were performed at the Service des Champs Magnétiques Intenses in Toulouse. A standard four probe method was used to measure the resistance with a current density of 10 mA/cm². In order to get the magnetoresistance of the sample, we used an ac (100 kHz) technique with a selective amplifier and a digital storage triggered by the magnetic field. The increasing and decreasing times of the 40 T pulse are 70 ms and 800 ms, respectively. Data were recorded during both periods, thus avoiding any spurious effect like a temperature drift or transient effects.

In this metallic conductor, whose anisotropy was underlined in the introduction, we can assume that $\mu B << 1$ even at maximum field. Indeed the resistivity is about 20 $\mu.\Omega$.cm at 110 K and for an electronic band half-filled that is coherent with Hall effect measurements [5] one finds $\mu = 2.5.10^3$ m²/v.s and $\mu B \approx 10^{-1}$ at 40 T. This implies that the Lorentz forces contribute to the transverse magnetoresistance through a positive term increasing like B² as we are in the low magnetic-field limit. Let us call "classical" magnetoresistance this contribution and this term will be generally written in the form

$$\left[\frac{\Delta R}{R_0} (B,\theta) \right]_c = aB^2 G(\theta) , \quad \theta = (c,B)$$

$G(\theta)$ is an anisotropy factor . In a real 2D conductor its expression is

$$G(\theta) = \cos^2 \theta.$$

The other contributions to the magnetoresistance are involved in a term that we call transverse "polarization" magnetoresistance :

$$\left[\frac{\Delta R}{R_0} (B,\theta) \right]_P$$

We present on fig. 2 the magnetoresistance data for a GdBaCuO film. They clearly indicate the contribution of two terms. The first one is isotropic and negative, we attribute it to the magnetic ordering of Cu and Gd ions which are in an isotropic paramagnetic state at 110 K [6]. The second term is illustrated by the positive upturn at higher fields. This component increases as the angle between B and C-axis gets lower.

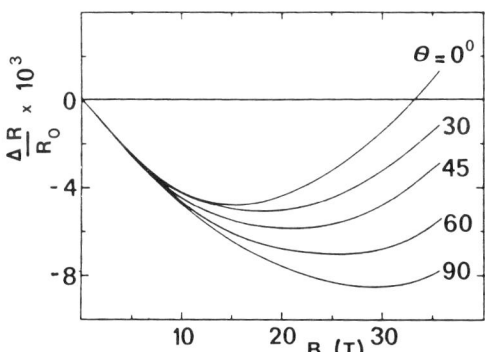

FIGURE 2. Magnetoresistance curves obtained at 110K for various orientations of a GdBaCuO film.

Assuming a 2D behaviour for the carriers the $\theta = 90°$ curve reprensents the polarization term. It exhibits a slight upturn over B = 32 T which could be, the signature of a weak coupling between consecutive CuO_2 planes. Such interaction is necessary to explain the high Tc transition in a weak coupling electron-phonon model [1]. By substracting the $\theta = 90°$ curve minus its upturn at high fields to all the other ones we get the "classical" 2D magnetoresistance which dependence on $B^2\cos^2\theta$ is clearly observed on figure 3.

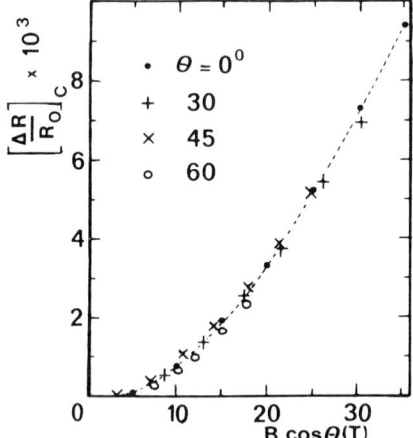

FIGURE 3. Plot of the classical magnetoresistance versus $B.\cos\theta$ for a GdBaCuO film. Evidence for a 2D-behaviour.

Same experiments were performed on YBaCuO films at 110 K up to 36 T. Qualitatively we observe similar results. The only difference is a lower negative magnetoresistance for $\theta = 90°$ (polarization contribution) as shown on figure 4.

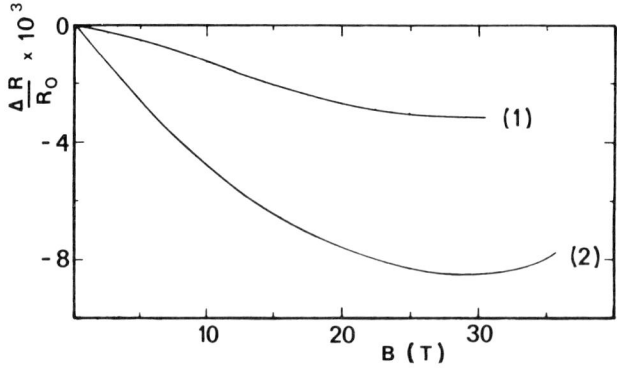

FIGURE 4. Polarization contribution to magnetoresistance ($\theta = 90°$) for YBaCuO(1) and GdBaCuO(2) films at T = 110K.

The use of magnetic (Gd) or non magnetic (Y) rare earth does'nt affect the transition temperature. However the orientation by the field of the gadolinium paramagnetic moments induces a polarization magnetoresistance. This explains the difference of the data for GdBaCuO and YBaCuO films.

In figure 5 we present the data of classical magnetoresistance versus $B \cos \theta$ after substracting this polarization contribution for YBaCuO. They clearly exhibit a 2D behaviour of the carriers in the CuO_2 planes, exactly as for Gd BaCuO films.

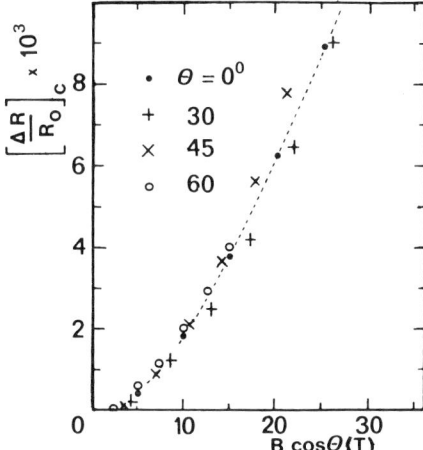

FIGURE 5. Plot of the classical magnetoresistance versus $B \cos \theta$ for a YBaCuO film.

3. CONCLUSION

Using a long pulsed magnetic field, we could measure the magnetoresistance up to 36 T on textured films of $GdBa_2Cu_3O_7$ and $YBa_2Cu_3O_7$ for various orientations of the field with respect to the c- axis. Two contributions are observed. The first one isotropic and negative can unambiguously be attributed to magnetic ordering arising from copper ions in the case of YBaCuO and from both copper and gadolinium ions in the case of GdBaCuO. The other is the "classical" magnetoresistance arising from Lorentz forces. We show in this paper that this last one only depends on the component of the magnetic field parallel to the c-axis giving a direct proof of the two-dimensional character in the CuO_2 plane of the conductivity in the normal state. Having in mind the crystal structure of the "123" superconducting oxides, one can generalize this property to the whole family.

J.C. Ousset et al.

REFERENCES

1 WU M.K.,ASHBURM J.R., TORNG C.J., HOR P.H., MENG R.L., GAO L., HUANG Z.J., WANG Y.Q. and CHU C.W., *Phys. Rev. Lett., 58 (1987) 908.*

2 TOZU S.M., KLEINSASSER A.W., PENNEY T., KAISER D. and HOLZBERG F., *Phys. Rev. Lett., 58 (1987) 1768.*

3 MARTIN C., MICHEL C., MAIGNAN A., HERVIEU M. and RAVEAU B., *C.R. Acad. Sci. Paris II, 307 (1988) 27.*

4 OUSSET J.C., RAKOTO H., BROTO J.M., MATZEN G., BOBO J.F., LABBE J., ASKENAZY S., *Europhys. Lett. 14 (1991) 581.*

5 NICHOLS T.R., MURATA K., FORTUNE N.A., KOMAZAKI T., NISHIHARA Y. and ITOZAKI H., *Preprint Conf. L.T. 19, Brighton (1990).*

6 GUILLOT M., THOLENCE J.L., POTEL M., GOUJON P., NOEL H. and LEVET J.C., *private communication.*

7 LABBE J., *Phys. Scr., 29 (1989) 82.*

High T$_c$ Superconductor Thin Films
L. Correra (Editor)
© 1992 Elsevier Science Publishers B.V. All rights reserved.

Anisotropic critical current density and two-dimensionality of Bi$_2$Sr$_2$CaCu$_2$O$_{8+x}$ epitaxial thin films

P. Schmitt[*,+], L. Schultz[+] and G. Saemann-Ischenko[*]

[+]Siemens Research Laboratories, Erlangen, Germany

[*]Physikalisches Institut Universität Erlangen, Germany

Abstract

The critical current anisotropy $j_c(\theta)$ of an epitaxial Bi$_2$Sr$_2$CaCu$_2$O$_{8+x}$ film with T$_c \approx$ 80K grown by pulsed laser deposition has been measured at T = 60K as a function of the angle θ between magnetic field B and the c-axis of the film at different magnetic fields. In the whole field range 49mT < B < 2T $j_c(\theta)$ is determined by the field component B$_\parallel$ parallel to the c-axis of the film. An influence of a 3D-to-2D crossover expected at a magnetic field B$_{2D} \approx$ 0.3T below a critical temperature T$_{cr}$ on $j_c(\theta)$ has not been detected.

1. Introduction

The extremely short coherence length ξ_c of Bi$_2$Sr$_2$CaCu$_2$O$_{8+x}$ results in a very weak Josephson-coupling [1-3] of the superconducting CuO$_2$ layers and therefore 2D behaviour of this material. For this reason, according to Kes et al. [3], the transport properties of Bi$_2$Sr$_2$CaCu$_2$O$_{8+x}$ are independent of a magnetic field parallel to the CuO$_2$ layers, whereas for a field inclined by some angle with respect to the layers only the field component B$_\parallel$ parallel to the c-axis influences the transport properties. In fact, in a previous letter [4] we have shown that the anisotropic $j_c(\theta)$ behaviour can fully be accounted for by this assumption well below the 3D-to-2D crossing temperature T$_o$ = [1-(2ξ_{ab}(0)/d$_{CuO}$)^2]/Γ [1] where ξ_{ab} is the coherence length in the ab-plane, d$_{CuO}$ the separation of the layers and Γ the effective mass ratio Γ = m$_c$/m with m$_c$ and m denoting the effective masses perpendicular and parallel to the planes, respectively. At this temperature, the coherence length ξ_c perpendicular to the CuO$_2$ planes becomes shorter than half the distance d_{CuO} of these planes resulting in a decoupling of the superconducting layers as discussed by Lawrence and Doniach [1]. The crossing temperature determined in our experiment is, however, about 10 K below T_c, whereas the formula yields $T_0 \approx T_c - 0.5K$ which means that decoupling of the layers is observed only for $\xi_c \ll d_{CuO}$.

The Josephson-coupling of the superconducting CuO_2-layers is not the only mechanism controlling the dimensionality of a layered system. In addition, below a characteristic magnetic field parallel to the c-axis $B_{2D} = \Phi_0/\Gamma \cdot d_{CuO}^2$ with $\Phi_0 = h/2e$ denoting the flux quantum, magnetic coupling between vortices in adjacent CuO_2-layers becomes important, resulting in a more 3D character of $Bi_2Sr_2CaCu_2O_{8+x}$ [5,6] below this field B_{2D}, which amounts to $B_{2D} \approx 0.3T$ [5] for this compound. A dimensional crossover from 3D to 2D with increasing field B should therefore be observeable below some critical temperature $T_{cr}(B) \propto B^{1/2}/\ln B$ [5].

In our previous letter [4], we have shown that the critical current anisotropy $j_c(\Theta,B,T)$ is a useful tool to investigate the dimensional crossover of $Bi_2Sr_2CaCu_2O_{8+x}$ as a function of temperature. In this letter, we present $j_c(\Theta,B)$ measurements in external magnetic fields $0.049T < B < 2T$ at $T = 60K$ made in order to substantiate the predicted field-induced dimensional crossover. This temperature has been selected for the following reasons: i) at this temperature, the variation of j_c with Θ is appreciable even at very small external magnetic fields, ii) this temperature is well below the 3D-2D crossover temperature T_0.

2. Experimental

Thin films of the high temperature superconductor $Bi_2Sr_2CaCu_2O_{8+x}$ have reproducibly been prepared by pulsed laser deposition [7] from a stoichiometric, rotating sinter target onto (100) $SrTiO_3$ substrates using a Siemens XP2020 excimer laser (XeCl, $\lambda = 308nm$, 60ns pulse duration) with $T_c \approx 80K$ and $j_c(4.2K,B=0) > 2 \cdot 10^6 A/cm^2$. During deposition, the films were kept at a temperature about 5K below the decomposition temperature of the $Bi_2Sr_2CaCu_2O_{8+x}$ phase (about 750-800°C) in an oxygen pressure of $p_{O2} = 0.35mbar$. The energy density of the laser beam has been adjusted to $2.7J/cm^2$ in order to minimize surface roughness of the films. After depositon, the films were kept at the deposition temperature in $p_{O2} = 1mbar$ for 1 hour, resulting in an improvement of both T_c and the inductively measured transition width ΔT_c. The films were cooled to room temperature

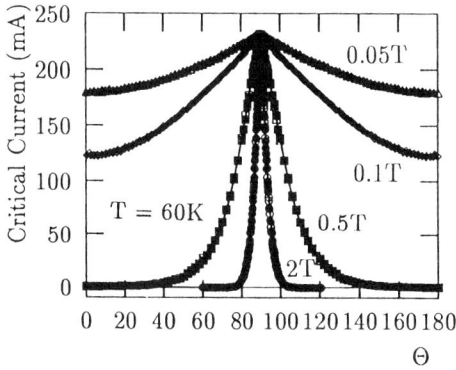
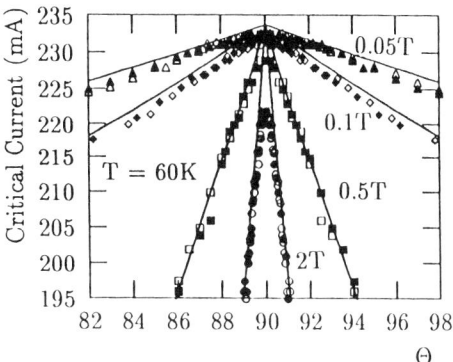

Figure 1a. Critical Current as a function of the angle Θ between the magnetic field direction and the c-axis of the film.

Figure 1b. Same data as in fig. 1a but on an elarged Θ-scale.

within 1 h by switching off the heater without changing the oxygen pressure. The epitaxial growth of the films has been confirmed by TEM investigations [8]. There is, however, some scatter of the c-axis in the films, which follows from the width of the (00$\underline{10}$) rocking curve ranging from 0.35° to 0.43° as well as from RBS channeling experiments yielding a X_{min} of 0.34.

The films were patterned mechanically by scratching typically 1mm long and 0.1mm wide striplines along the a- or b-axis of the film using a plotter equipped with a stainless steel needle. The critical current density $j_c(\Theta,B,T)$ has been measured as a function of temperature T, magnetic field B and the angle Θ between the c-axis of the film and the magnetic field direction, keeping magnetic field and current perpendicular to each other. An electrical field criterion of $5\mu V/cm$ has been adopted for the j_c measurements. The angular resolution of the setup used in these experiments is about 0.01°, while the absolut angle was accurate to 1° or 2°. Calibration of the absolute angle was done by making use of the symmetry of $j_c(\Theta)$ with respect to Θ = 90°, which is evident from the data shown in figures 1a and 1b. In these figures, solid symbols represent data measured at the angle Θ, whereas open symbols represent data measured at an angle $\Theta^* = |90°-\Theta|$.

3. Results and discussion

In our previous letter [4], we have shown that the critical current anisotropy $j_c(\Theta,B)$ is determined by the field component $B_\parallel$ parallel to the c-axis of the film and can therefore be written

$$j_c(\Theta,B,T) = j_{c\parallel}(B \cdot \cos\Theta, T) \tag{1}$$

with $j_{c\parallel}(B,T)$ denoting the field dependence of j_c in the $\underline{B \parallel c}$ direction at temperature T. The solid lines drawn in figures 1a and 1b have been calculated from a polynominal fit to the $j_{c\parallel}(B)$ data shown in Fig. 2 using equation 1. In Fig. 2, measured data are represented by solid symbols, the polynomial fit is drawn as solid line.

Close to the $\underline{B \perp c}$ direction, as evident from Fig. 1b, the measured $j_c(\Theta)$ values are appreciably smaller than those calculated from equation 1, as even for perfect alignment of the sample plane parallel to the magnetic field there exists always a magnetic field component parallel to the c-axis because of the mosaic spread [3,9] of about 0.2^0 which is evident from the width of the rocking curve of the (00$\underline{10}$) peak. In fact, as has been shown in our previous letter, at temperatures well below the 3D-2D crossing temperature T_0, $j_c(B)$ scales with $j_{c\parallel}(B)$ as

$$j_{c\perp}(B) = j_{c\parallel}(B \cdot \sin\Phi) \tag{2}$$

with $j_{c\perp}(B)$ defined analogous to $j_{c\parallel}(B)$ as the field dependence of j_c for the $\underline{B \perp c}$ direction and Φ denoting some effective texture angle, which is close to the half width of the rocking curve of the (001) peaks.

At temperatures close to T_c, due to the increase of ξ_c, a 2D-3D crossover occurs and the scaling behaviour (equation 2) is no more observed [4]. This means, that the critical current anisotropy $j_c(\Theta)$ can be used as an indicator of the dimensionality of the system. In fact, the typical 2D behaviour described in equations (1) and (2) has been observed both in $Bi_2Sr_2CaCu_2O_{8+x}$ [4,9,10] and $Tl_2Sr_2Ca_2Cu_3O_{10+x}$ [11], which are regarded as 2D materials because of their large effective mass ratios ($\Gamma \approx 2500$ in case of $Bi_2Sr_2CaCu_2O_{8+x}$) [12]. In case of $YBa_2Cu_3O_7$, which shows a smaller effective mass ratio $\Gamma \approx 25$ [12] and may therefore be described as a more 3D superconductor, equations (1) and (2) fail in describing $j_c(\Theta)$ as measured by Roas et al. [13].

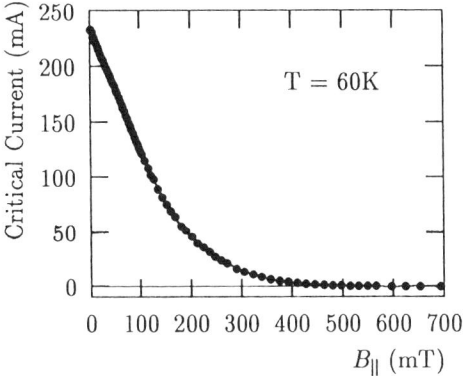

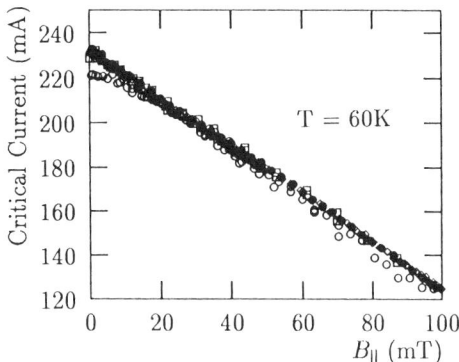

Figure 2. Critical current as a function of a magnetic field $\underline{B}\|\underline{c}$.

Figure 3. Critical current as a function of a magnetic field $\underline{B}\|\underline{c}$. Solid symbols represent measured data, data shown as open symbols have been calculated from $j_C(\Theta,B)$ (triangles: B=0.05T, rhombs: B=0.1T, squares: B=0.5T, circles: B=2T.)

In Figure 1b, the calculated $j_C(\Theta,B)$ curves (solid lines) seem to fit the measured $j_C(\Theta,B)$ data much better for B = 2T and B = 0.5T than for B = 0.1T and B = 0.049T. It is tempting to identify this behaviour with the 2D-3D crossover predicted at B≈0.3T (5). However, when calculating $j_{C\|}(B)$ from the four $j_C(\Theta,B)$ curves by making use of equation (1), the deviation from the measured $j_{C\|}(B)$ data is even largest for the $j_C(\Theta,B)$ values measured at 2T as evident from Fig. 3. The apparent crossover deduced from Fig. 1b is just an optical delusion caused by the much steeper decrease of j_C with Θ at 2T than at 0.1T or 0.049T. The small (<1%) deviation of the measured $j_C(\Theta)$ from the calculation is presumably caused by a small detoriation of the sample during measurement. In fact, measurements have been done in the sequence $j_{C\|}(B)\equiv j_C(0,B)$ / $j_C(\Theta,0.5T)$ / $j_C(\Theta,0.1T)$ / $j_C(\Theta,0.049T)$/ $j_C(\Theta,2T)$ and the deviation of the calculated $j_{C\|}(B)$ curves from the measured one increases in the same sequence. For this reason, we conclude that equation (1) describes $j_C(\Theta)$ accurately in the whole field range 0.049mT < B < 2T at T = 60K. This means, that $T_{Cr}(B)$ should be smaller than 60K in this field range.

Acknowledgements

The authors gratefully acknowledge helpful discussions with R. Busch and G. Ries as well as technical assistance by M. Kühnl. This work was supported by the Bundesminister für Forschung und Technologie.

References

1 W.E. Lawrence and S. Doniach, in Proceedings of the Twelfth International Conference on Low Temperature Physics, Kyoto, 1970, edited by E. Kanda (Keigaku, Tokyo, 1971), p. 361

2 M. Tachiki and S. Takahashi, Solid State Commun. $\underline{70}$ (1989) 291

3 P.H. Kes, J. Aarts, V.M. Vinokur and C.J. van der Beek, Phys. Rev. Lett. $\underline{64}$ (1990) 1063

4 P. Schmitt, P. Kummeth, L. Schultz and G. Saemann-Ischenko, submitted to Phys. Rev. Lett.

5 V.M. Vinokur, P.H. Kes and A.E. Koshelev, Physica C $\underline{168}$ (1990), 29

6 M.V. Feigel'man, V.B. Geshkenbein and A.I. Larkin, Physica C $\underline{167}$ (1990), 177

7 P. Schmitt, L. Schultz and G. Saemann-Ischenko, Physica C $\underline{168}$ (1990) 475

8 X.F. Zhang, B. Kabius, K. Urban, P. Schmitt, L. Schultz and G. Saemann-Ischenko, to be published in Physica C

9 H. Raffy, S. Labdi, O. Laborde and P. Monceau, Supercond. Sci. Technol. $\underline{4}$ (1991) 100

10 H. Raffy, S. Labdi, O. Laborde and P. Monceau, Physica B $\underline{165/166}$ (1990) 1423

11 T. Nabatame, Y. Saito, K. Aihara, T. Kamo and S.-P. Matsuda, preprint

12 T.T.M. Palstra, B. Batlogg, L.F. Schneemeyer and J.V. Waszczak, Phys. Rev. B $\underline{43}$ (1991)

13 B. Roas, L. Schultz and G. Saemann-Ischeko, Phys. Rev. Lett. $\underline{64}$ (1990) 479

High T$_c$ Superconductor Thin Films
L. Correra (Editor)

Superconducting transition and fluctuation conductivity of Bi(Pb)-Sr-Ca-Cu-O thin films made by sputtering.

S. Labdi, S. Megtert and H. Raffy

Laboratoire de Physique des Solides, Bât 510, Université Paris-Sud, 91405 Orsay (France).

Abstract
    Thin films of BiSrCaCuO (2212 and 2223 phases) have been prepared and their excess conductivity $\Delta\sigma(T)$ above T_c have been analyzed. In all cases $\Delta\sigma(T)$ displays a two- dimensional behaviour in a wide range of temperature. The results are consistent with the 2D limit of the Lawrence- Doniach model.

1 INTRODUCTION

    In this paper, we report on a study of the fluctuation conductivity of c-axis oriented BiSrCaCuO thin films, either composed of the 2212 phase or predominantly composed of the 2223 phase and for various sample thicknesses (1000 Å to 8000Å)
    Firstly, we describe the preparation and the characterization of the samples and particularly the case of the high T_c samples (T_c>100 K).
    Secondly, we develope an analysis of the paraconductivity, with a comparison to the Lawrence-Doniach model for superconducting layers coupled by Josephson effect [1]. BiSrCaCuO high Tc superconductors have a very anisotropic crystalline structure which can be viewed as a layered system with superconducting layers $(CuO_2)_n$ (n=2 or 3). The anisotropy of the superconducting properties and the 2D behaviour of Bi oxide (2212) films for T<T$_c$, were studied by transport measurements under magnetic field at various orientations of the magnetic field and reported in [2-3]. In the present study, the dimensionality of the system is probed by looking at the excess conductivity extracted from the R(T) curves of our films .

2. FILM PREPARATION AND CHARACTERIZATION

    In all cases the films were deposited on MgO(100) substrates at room temperature by dc triode sputtering with a sintered single target with a pressure of 1mTorr of Argon and post annealed. R(T) measurements were done by a classical four probe dc technique.
    a) For *2212 single phase films* with T_{co} (R=0) in the range 85 K to 89 K, the target nominal cationic (Bi,Pb):Sr:Ca:Cu composition was (1.4, 0.6):1.5:1.5:2.0 with 30% lead. The post-annealing treatment was done at 810°-820°C in an oxygen depleted atmosphere (7%O$_2$). By this method, the lead is completely lost during the annealing. More details concerning this preparation were reported elsewhere [4,5]. The films are *highly c-axis oriented* and their composition is homogeneous.
    b) For the films *predominantly* composed of the *2223 phase* with T_{c0} >100K, the target cationic compositions used were (1.57,0.53):2:2:2.9 and (2.0,0.0):2.0:2.0:3.0 (target without lead). The post annealing treatment was done at 860°C for several hours [6]. The loss of lead in this range of temperature is very important. So the

annealing must be performed in a lead doped atmosphere in order to promote the high T_C phase crystallisation. This was obtained by putting the samples in the immediate neighborhood of a ceramic pellet containing lead. To favour the formation of the 2223 phase, the lead concentration in the films, $x=Pb/(Bi + Pb)$, had to be in the range : $0.15<x<0.2$. This result does not depend either on the film thickness or on the composition of the deposits. It is interesting to note that the 2223 phase films were as well obtained from samples deposited by using a target without lead. In this case the lead was only incorporated during the annealing process.

Structural characterization and *phase identification* were done by X-ray diffraction studies. As it was found that the high T_C phase and the low T_C phase were equally well oriented in the deposits, their relative amount was estimated by using the peak intensities of the low T_C phase (L) and of the high T_C phase (H), from the ratio : $r= H(0014)/(H(0014)+L(0012))$. It was found that the maximum value of r was decreasing with decreasing sample thickness (Fig.1). In Fig.1-c we have plotted the maximum value of r as the function of the film thickness t : it appeared that it was not possible to get more than 50% of the high T_C phase for film thicknesses less than 2000Å. Moreover it was shown that more than 50% of 2223 phase are needed to achieve a complete superconducting transition above 100K [6]. Consequently $T_{C0}(R=0)$ values larger than 100K were essentially observed for films with a thickness equal or larger than 2000Å.

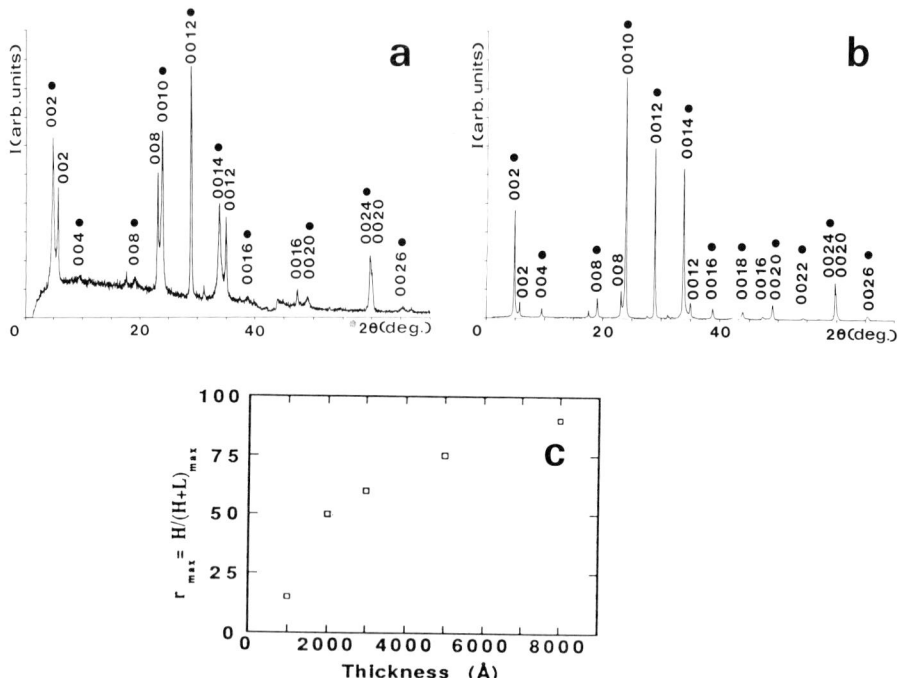

Figure 1. (a) X- ray diffraction patterns for a 2000Å thick film, (b) for a 8000Å film
(c) Maximum amount r of the 2223 phase as a function of film thickness.
● indicates the 2223 phase.

3. EXCESS CONDUCTIVITY ANALYSIS

This study was done for samples with thicknesses ranging from 1000Å to 8000Å and of both phases: 2223 and 2212

3.1. Theoretical model

For the data analysis, we used the Lawrence-Doniach model [3] for superconducting multilayers coupled by Josephson tunneling. In such a model the excess conductivity along the plane of the layers is given by :

$$\Delta\sigma = \sigma - \sigma_n = \frac{e^2}{16\hbar d} \varepsilon^{-1/2} \, (\varepsilon + 4 \, (\frac{\xi_c(0)}{d})^2)^{-1/2} \qquad (1)$$

σ_n is the normal state conductivity, $\varepsilon=(T-T_c)/T_c$ the reduced temperature, d the layer interspacing and ξ_c is the superconducting coherence length along the c-axis.

It is important to note that the L.D. model includes only the direct Aslamazov-Larkin contribution [7] which arises from the superconducting pair acceleration. In our analysis we have not taken into account another indirect contribution, calculated by Maki [8] and Thompson [9], since the inelastic scattering time is presumably very short in this compound. This approximation is also supported by the fact that the experimental data are in good agreement with the L.D. model, as shown later. Hikami and Larkin have calculated the importance of each term : they found the indirect M.T. term to be smaller than the A.L. one for $T\tau_\phi < 10^{-12}$ s.K.

For a weak coupling between the superconducting layers (2D behaviour), expression (1) becomes :

$$\Delta\sigma_{2D} = \frac{e^2}{16\hbar d} \varepsilon^{-1} \qquad (2)$$

For a strong coupling or 3D behaviour, (1) becomes :

$$\Delta\sigma_{3D} = \frac{e^2}{32\hbar\xi_c(0)} \varepsilon^{-1/2} \qquad (3)$$

The 2D-3D crossover temperature is given by: $T_0=T_c(1+4(\xi_c(0)/d)^2)$ (4)

3. 2. Normal state resistivity

All the films exhibit a linear temperature dependence of their resistance at high temperature. The normal state resistance R_n was determined by fitting the experimental data to a linear law : $R_n(T)=AT+B$, from $2T_c$ to 300K. The extrapolation of this linear dependence at low temperature, $T_c<T<2T_c$, gave us the normal state resistances in the fluctuation regime. For the 2223 phase films, $R_n(T)$ intercepts the resistance axis very close to the origin.

3.3. Determination of the critical temperature T_c

The determination of the thermodynamical critical temperature Tc is crucial for this analysis. We have no experimental access to it. Many more or less arbitrary criteria have been used in the literature : defining T_c by: $R(T_c)= R_n(T_c)/2$ [12] or taking the temperature corresponding to the inflexion point, $d^2R(Tc)/dT^2=0$, in the R(T) curves [13]. Oh et al.in their study on YBaCuO films [14] chose T_c to be the temperature where the linear extrapolation of $\Delta\sigma^{-2}$ (T) (3D regime) crosses the T axis. In the case of BiSrCaCuO compounds, there is no evidence of this 3D regime. So we used the linear extrapolation of $\Delta\sigma^{-1}(T)$ (2D regime), as in the inset of Fig.2d.This procedure gave us the best and most consistent results.

3.4. Results

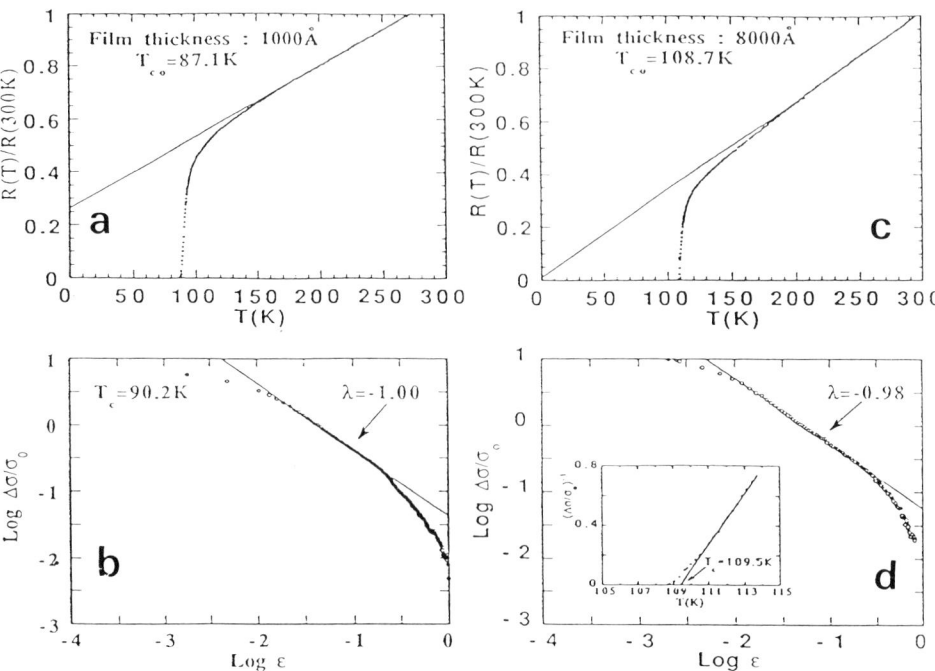

Figure 2. (a) : R(T) curve and (b) : fluctuation conductivity of sample 2
  (c): R(T) curve and (d) : fluctuation conductivity of sample 3
The inset in Fig. 2 (d) shows $\Delta\sigma^{-1}$ vs. T to determine T_C. The solid line in Fig.(b) and (d) is the best linear fit with a slope λ.

We have plotted $\Delta\sigma/\sigma_0$ as a function of the reduced temperature ε in a Log-Log plot where σ_0 is the conductivity at T=300K. The slopes λ of these curves give us directly the dimensionality of the system . By comparison with the L.D. model, we can estimate a value of the superconducting layer spacing d. Fig.2b and 2d are two typical examples of such plots : for a 1000Å thick, 2212 single phase film (sample 2) and for a 8000Å thick film with 90% of the 2223 phase, 10% of the 2212 phase (sample 3). The first thing which clearly appears is that we have a two dimensional behaviour (λ very close to -1) for temperatures ranging from 92K to 115K for the 2212 phase and 112K to 150 K for the 2223 phase. Table I summarizes the parameters obtained for six samples. H% and L% denote the 2223 and 2212 phases respectively.

Table 1

n°	L%	H%	t(Å)	T_{c0}(K)	T_c(K)	T 2D range	λ	ρ_0/d(Ω/□)	d(Å)
1	100	-	3000	88.5	90.2	93-120	-1.02	3550	18
2	100	-	1000	87.1	90.2	92-113	-1.00	2850	19
3	10	90	8000	108.7	109.5	111-144	-0.98	3700	15
4	25	75	5000	107.2	109.3	111-140	-1.03	2100	16
5	40	60	3000	106.9	109.8	112-145	-0.96	2400	15
6	50	50	2000	105.0	109.8	112-150	-0.95	2550	12

From this table we have observed clearly in all cases a 2D behaviour (λ~ -1) in a wideT range. Near T_c, our curves show a departure from the 2D behaviour. However it does not appear to correspond to a 3D regime as there is no clear evidence of a linear region in the curves $\Delta\sigma^{-2}$(T) and it is known that the contribution of inhomogeneities can be important near T_c. This is in contrast with YBaCuO compounds where a 3D region has been clearly shown [14].

The values of d estimated from (2) are almost independent of the film thickness. d is found to lie between 12 and 19Å. It is in agreement with the $(CuO_2)_n$ block (n=2 or 3) spacing deduced from the cristallographic structure of these materials (c/2~15.3Å for n=2 and 18.5Å for n=3). It is important to note that the accuracy of the determination of d is limited by the accuracy in the determination of T_c and ρ_0 the normal state resistivity. For the almost single phase samples, the typical precision we had, was about 15%, which implies that Δd=3Å. The incertitude on d is more important when the percentage of the 2212 phase increases in the high T_c films and it would be necessary to know the repartition of both phases.

Fig. 3 (b) shows the excess conductivity of a 2000Å film containing 50% of the

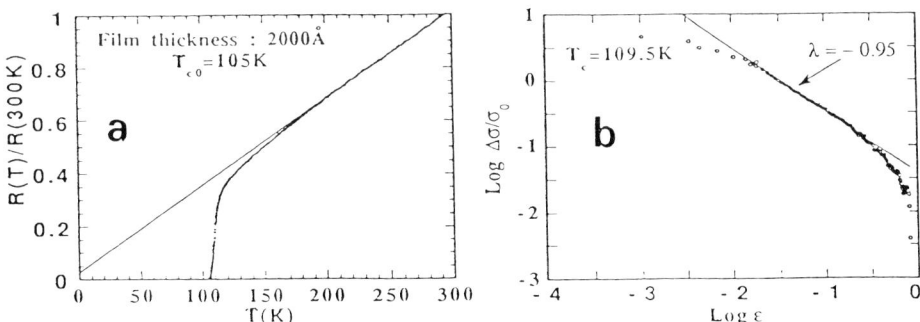

Figure 3. (a) : R(T) curve and (b) : fluctuation conductivity of sample 6. The solid line is the best linear fit.

2223 phase and 50% of the 2212 (sample 6) .We can see that even with only 50% of the high T_c phase, we still essentially observe the fluctuations of the high T_c phase, because the T range of the fluctuation regime of each phase practically does not overlap.

4. CONCLUSION

We have studied the temperature dependence of the excess conductivity of BSCCO films. We observed a *2D behaviour* in a wide range of temperature from $1.05T_c$ to $1.30T_c$ for the 2212 phase and from $1.02T_c$ to $1.40T_c$ for the 2223 one. By comparison with the L.D. model, the superconducting layer interspacing d deduced from this analysis is consistent with the value of the cristallographic parameter c of this compound. These results are consistent with our results on the dependence of the critical currents and of the magnetoresistance under magnetic field for various orientations of H . These results had shown for $T<T_c$ that the BSCCO compound is very anisotropic and has a 2D behaviour.

It would be interesting to study the MR in the fluctuation regime. Since one has to take into account the M.T. contribution, it would be then possible to have an estimation of $\xi_c(0)$ and of τ_ϕ the inelastic scattering time.

5. REFERENCES

1   W.E.Lawrence and S.Doniach, Proc.of the 12th Int. Conf. on Low Temp. Phys. Kyoto (1970), ed. E.Kanda (Academic Press of Japan (1971)) 361
2   H.Raffy, S.Labdi, O.Laborde and P.Monceau, Physica B, 165&166 (1990) 1423; Supercond. Sci. Technol., 4 (1991) S100.
3   H.Raffy, S.Labdi, O.Laborde, P.Monceau, Phys.Rev.Lett., 66 (1991) 2515.
4   H.Raffy, A.Vaurès, J.Arabski, S.Megtert, F.Rochet, J.Perrière, Solid State Comm., 68 (1988) 235
5   H.Raffy, A.Vaurès, J.Arabski, S.Megtert, J.Perrière, Physica C, 162-164 (1989) 613
6   S.Labdi, H.Raffy, A.Vaurès, J.Arabski, S.Megtert, P.Tremblay, J.Less Comm. Met., 164&165 (1990) 687
7   L.G. Aslamazov and A.I. Larkin, Phys.Lett. A26 (1968) 238
8   K.Maki, Progr.Theor.Phys., 39 (1968) 897
9   R.S. Thompson, Phys. Rev.B1, (1970) 327
10 S.Hikami and A.I.Larkin, Mod.Phys.Lett.B, 2 (1988) 693
11 G.Kumm and K.Winzer, Physica B 165&166 (1990) 1361
12 G. Balestrino, A.Nigro , Phys.Rev B, 39 (1989) 122
13 F.Vidal et al., J.Less Comm. Met.,151(1989)165
14 B.Oh et al., Phys.Rev.B, 37 (1988)7861

High T$_c$ Superconductor Thin Films
L. Correra (Editor)
© 1992 Elsevier Science Publishers B.V. All rights reserved.

DIRECT DETERMINATION OF THE SUPERCONDUCTING GAP IN YBa$_2$Cu$_3$O$_{7-\delta}$

M.Balkanski and A.Sacuto

Laboratoire de Physique des Solides, associé au CNRS, Université Pierre et Marie Curie

Tour 13, 4 Place Jussieu, 75252 Paris Cedex 05, France

O. Gorochov and R.Suryanarayanan

Laboratoire de Physique des Solides de Bellevue, CNRS,1 Place Aristide Briand

92195 Meudon Principal Cedex, France

L.Correra

CNR, Instituto Lamel,Via Castagnoli,1- 40126 Bologna, Italy

ABSTRACT :

Raman scattering experiments have been performed on two different YBa$_2$Cu$_3$O$_{7-\delta}$ single crystals (δ=0; δ=0.3) and on YBa$_2$Cu$_3$O$_7$ thin films in order to investigate superconducting gap excitations. Studies carried out in Z(x,x+y)Z configurations allow one to detect a drastic change in the electronic Raman scattering below the B$_{1g}$ mode at the transition temperature, whereas the electronic Raman scattering above the B$_{1g}$ mode is independant of the transition temperature T$_c$. Subtracting the electronic Raman spectra at temperatures just above and well into the superconducting phase, the energy gap can be estimated to be 308+20 cm^{-1} for δ = 0 and to 210+20 cm^{-1} for δ = 0.3, giving a coupling constant 2Δ / K$_b$T$_c$ which lies beetween 4 and 5. On the other hand the broad electronic band centered at 530 cm^{-1} can be attributed to electronic interband transitions near the S symmetry point of the Brillouin zone.

1. INTRODUCTION

In the past few years, considerable effort has been devoted to an understanding of the superconducting state of the new high T$_c$ superconductors. Many experimental studies on Raman spectroscopy[1-7], infrared reflectivity[8-9] and tunneling[10] have been performed to evaluate the superconducting gap. The values estimated of the energy gap given by different authors are very diverse, ranging from $2\Delta = 0$ to $2\Delta = 10\ K_b T_c$. The first Raman spectroscopy studies[1-4] have shown evidence of strong electron-phonon coupling characterised by Fano line shapes of the 117 cm^{-1} A_g normal mode assigned to the Ba atom motion and the 336 cm^{-1} B_{1g} normal mode assigned to the O2-O3 atoms in out of phase motion. Moreover, two strong contributions to the electronic Raman spectrum in the superconducting state have been detected in B_{1g} and A_g symmetries for light polarised within the Cu-O plane, and attributed to a gap anisotropy with respectively two quasi partical pair breaking peaks at $\omega = 530$ cm^{-1} and $\omega = 340$ cm^{-1}. However Recent observations[6], do not find evidence on the transition temperature dependance of the feature centered at 530 cm^{-1}, this peak persisting well above T_c. In this paper we confirm the the transition temperature independance of the electronic continuum peak centered at 530 cm^{-1} suggesting that this electronic feature cannot be related to a superconducting energy gap, but can be attributed to electronic interband transitions between two narrow bands near the S symmetry point in the Brillouin zone as observed in band structure calculations[11-12]. Subtracting the electronic Raman spectra just above T_c from that well into the superconducting phase we show that the superconducting gap can be evaluated taking into account the theory of Raman scattering in superconductors[13], to $\omega_1 = 308 + 20$ cm^{-1} for $\delta = 0$ and $\omega_2 = 210 + 20$ cm^{-1} for $\delta = 0.3$, corresponding to a coupling constant $2\Delta / k_b T_c$ included between 4 and 5.

2. EXPERIMENTS

Raman spectroscopy investigations have been performed on $YBa_2Cu_3O_{7-\delta}$ single crystals with two different oxygen contents $\delta = 0$ and $\delta = 0.3$ (giving respectively Tc = 92K and T_c=65 K) and $YBa_2Cu_3O_7$ thin films with 1800A of thickness and epitaxially grown on a $SrTiO_3$ substrate. Measurements were carried out on an U.1000 I.S.A Jobin Yvon double monochromator, using a 514.52 nm line of an argon laser. An I.T.T F.W 130, photomultiplicator was connected to a micro computer for data collection. A liquid Helium cryostat was used to cool the samples and a heating resistance connected to a regulation RGP 3000, was used to control the temperature. A near backscattering geometry with the incident light polarized perpendicular to the C axis allowed studies in Z (x, x+y)Z configurations coupling Ag and B_{1g} symmetry excitations. The intensity of the incident laser beam was maintained below roughly 10 mw/cm^2 in order to minimize possible thermal damage.

3. RESULTS AND DISCUSSION

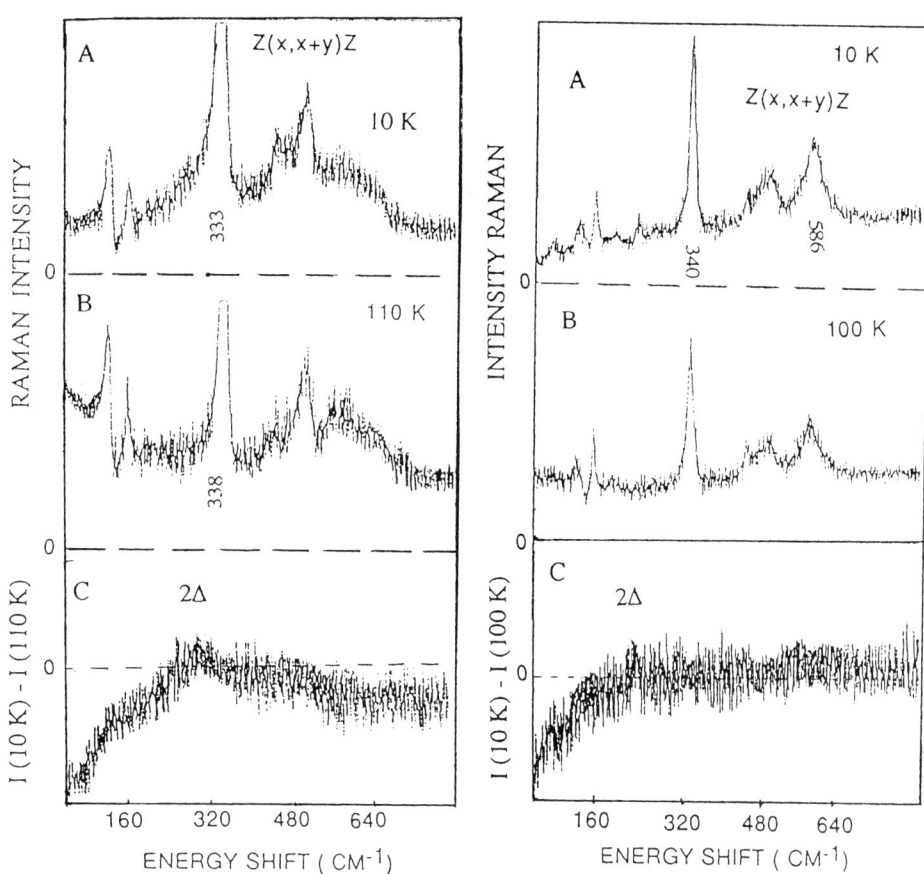

Figure 1: Raman spectra of YBa$_2$Cu$_3$O$_7$ single crystal A : 10K, B: 110K, C: subtraction of B from A spectrum.

Figure 2: Raman spectra of YBa$_2$Cu$_3$O$_{6.7}$ single crystal A: 10K, B: 100K, C: substraction of B from A spectrum.

Figures 1A and 1B illustrate respectively the Raman spectra in the Z (x, x + y) Z configuration of a $YBa_2Cu_3O_7$ single crystal with T_c = 92 K at T =10 K, and T=110 K. We see clearly that the electronic background strongly interferes with the optical phonons at 117 cm^{-1} and 336 cm^{-1}. These interference effects are reflected by the asymmetric line shapes as well as in the strong antiresonance on the high-energy side of the 117 cm^{-1} normal mode and the 336 cm^{-1} normal mode. This Fano line shape is more evident for the 117 cm^{-1} normal mode above T_c and, on the contrary more intense for the 336 cm^{-1} B_{1g} normal mode below T_c. We can also observe that the normal B_{1g} mode at 338 cm^{-1} in the normal state shifts to lower frequency well below T_c and the 437 cm^{-1} normal mode assigned to the O2- O3 oxygen atoms in phase motion stiffens to 443 cm^{-1} in the superconducting phase[15]. The 117, 151 and 500 cm^{-1} normal modes don't change in frequency above and below the transition temperature. Moreover, we notice that the broad electronic continuum centered around 530 cm^{-1} is unchanged below T_c . On the other hand, the electronic continuum below the B_{1g} mode changes drastically at the transition temperature and a strong decrease of the electronic Raman scattering is then detected below T_c. In the same way a drasic fall of the electronic continuum below T_c is also observed in $YBa_2Cu_3O_7$ thin films.

In figures 2A and B the Raman spectra of the δ = 0.3 compound for T=10 K and T=100 K are respectively represented.The intense peak detected at 586 cm^{-1} in this compound can be explained as a reorganisation of oxygen in the Cu-O chains for low oxygen content allowing some of the k=0 infrared active modes to become Raman active[16].

As the broad electronic continuum centered at 530 cm^{-1} is independant of the transition temperature it cannot be assigned to an energy gap but it can be attributed to interband transitions. Indeed near the S symmetry point of the Brillouin zone two narrow bands above and below the Fermi level exist [11-12] which allow electronic transitions in a range of 400 and 700 cm^{-1} in agreement with the width of the electronic feature centered at 530 cm^{-1}.

According to the electronic Raman scattering theory in superconductors based on the Bardeen, Cooper, Schrieffer model, there is no scattering on free electrons unless the energy transfered ω_i - ω_d exceeds 2Λ. Moreover, when the superconducting coherence is small compared to the optical penetration depth $\xi << \delta$ as it is in the case of the $YBa_2Cu_3O_7$ with ξ = 10 A [17] and δ = 1000 A [18] for 514.52 nm wave length, the limit of small momentum transfer q holds[13] the internal cross section per unit volume is egual to 0 for $0 < \omega < 2\Lambda$ and varies as $\Lambda^2 / (\omega^2 (\omega^2-4\Lambda^2))^{1/2}$ for $\omega > 2\Lambda$, thus predicting no electronic Raman scattering below 2Λ, an electronic continuum peak for ω = 2Λ and a decrease of the electronic continuum above 2Λ in the superconducting phase.

By subtracting the electronic continuum spectra of the YBa$_2$Cu$_3$O$_7$ at 110 K, just above the transition temperature, from the one at 10 K well into the superconduting phase, and truncating the 336 asymmetric peak, a drastic electronic continuum decrease appears in the superconducting state compared to the normal state below $\omega_1 = 308$ cm^{-1} as illustrated in Fig.1C. An electronic scattering lower in the superconducting phase than in a normal one allow one to attribute the ω_1 frequency to a superconducting gap energy. The electronic coupling constant can be evaluated to be 2Λ / K$_b$T$_c$ = 4.6 + 0.3. In the same way we have performed the substraction of the electronic continuum spectra of the δ = 0.3 compound below and above the transition temperature T$_c$= 65 K as represented in Fig. 2C. In this case the superconducting energy gap has been evaluated to 2Λ = 210+20cm^{-1} giving a coupling constant :

2Λ / K$_b$T$_c$ = 4.6 +0.4.

These results show the proportionality of the energy gap with T$_c$ and the coupling constant is found to be in the range of : 4<2Λ/K$_b$T$_c$<5. This would suggest a strong coupling B.C.S limit in the YBa$_2$Cu$_3$O$_{7-\delta}$ superconductor compound in agreement with the recent determination of the superconducting gap from the imaginary part of the self energy, estimated to be 316 cm^{-1} giving 2Λ/K$_b$T$_c$ = 4.9 + 0.1. However, B.C.S model implies a sharp drop of the electronic scattering in the superconducting state below 2Λ, whereas we observe a continuous decrease of the light scattering from the superconducting energy gap. This clearly contradicts the prediction from a classical B.C.S theory and needs an adequate explanation.

ACKNOWELDGEMENT

It's a pleasure to thank M.A.Kanehisa , R.F.Wallis and J.Deppe for their suggestions and advice.

REFERENCES

1. K.B Lyons, S.H Liou, M.Hong, H.S Cheu, J.Kwo and T.J Negran, Phys. Rev. B **36**, 5592 (1987).

2. S. L Cooper, M.V Klein, B.G Pazol, J.P Rice and D.M Ginsberg, Phys. Rev. B **37**, 5920 (1988).

3. R. Hackl and W. Glaser, Phys. Rev. B **38**, 7133 (1988).

4. S. L Cooper, F. Slakey, M. V Klein, J.P Rice, E D Bukowshi and D.M Ginsberg, Phys. Rev. B **38**, 11934 (1988).

5. F. Slakey, S. L Cooper, M.V Klein, J. P Rice and D. M Ginsberg Phys. Rev. B **39**, 2781 (1989).

6. M. V Klein, F. Sacley, D. Reznik, J. P Rice and D. M Ginsberg Phys. Rev. B **42**, 2643 (1990).

7. K. F. Mc Carty, J. F Liu, R. N.Shelton and H.B.Radousky, Phys. Rev. B **42**, 9973 (1990).

8. G.A.Thomas, D.H.Rapkine, M.Capizzi, A.J.Millis, R.N.Bhatt, L.F.Schneemeyer and J.V. Waszczak. Phys.Rev.Lett.,**61**,1313 (1988).

9. R.T.Collins, Z.Schlesinger, F.Holtzbergand, C.Feild, Phys.Rev.letter **63**, 422 (1989).

10. M.Gurvitch, J.M.Valles, Jr, A.M.Cucolo, R.C.Dynes, J.P.Garno, L.F.Schneemeyer and J.V.Waszczak, Phys.Rev.Lett.**63**, 1008 (1989).

11. A.J.Freeman, J.Yu, S.Massida and D.D.Koelling, Phisica **148 B**, 212 (1987).

12. W.E.Pickett, Rev.Mod.Phys.**61**, 456 (1989).

13. M.V.Klein and S.B.Dierker, Phys.Rev.B **29**, 4976 (1984) and references cited therein.

14. B.Friedl, C.Thomsen and M.cardona, Phys.Rev.Lett.**65**,915 (1990) and references cited therein.

15. C.Thomsen, M.Cardona, BFriedl, C.O Rodriguez, I.I Mazinand, O.K.Andersen, Sol.State.Com.**75**, 219 (1990).

16. G.Burns, F.M.Dacol, C.Feild and F.Holtzberg, Sol.stat.Com.**75**, 893 (1990).

17. G.Deutscher,"*earlier and recent aspects of the sperconductivity*" Springer-Verlag, Berlin, Series in Sol-State Sciences, **90**,170 (1990).

18. J.Humlicek, M.Garriga, M.Cardona, B.Gegenheimer, E.Schonherr, P.Berberich and J.Tate, Solid.State Com.**66**, 1071 (1988).

High T$_c$ Superconductor Thin Films
L. Correra (Editor)
© 1992 Elsevier Science Publishers B.V. All rights reserved.   49

Ellipsometric studies on c-axis oriented high-T$_c$ superconducting YBa$_2$Cu$_3$O$_{7-x}$ thin films; thickness measurements and oxygen out-diffusion experiments

W.A.M. Aarnink, R.P.J. IJsselsteijn, J. Gao, A. van Silfhout and H. Rogalla,

University of Twente, P.O. Box 217, 7500 AE Enschede, the Netherlands

Abstract

Spectroscopic ellipsometry was used to study c-axis oriented high-T$_c$ superconducting YBa$_2$Cu$_3$O$_{7-x}$ thin films. The thickness and complex dielectric function $\tilde{\varepsilon} = \varepsilon_1 + i\varepsilon_2$ were determined on a set of layers in the photon energy range of 2.2 - 4.4 eV. Reliable values for YBa$_2$Cu$_3$O$_{7-x}$ film thicknesses up to 70 nm were obtained. No strong interface reactions between the Yttria Stabilized ZrO$_2$ substrates and YBa$_2$Cu$_3$O$_{7-x}$ layers could be observed. Also oxygen out-diffusion experiments were performed, using a rotating analyser ellipsometer (RAE) as a non-destructive optical in situ monitor. Effects of oxygen deficiency x on the complex dielectric function $\tilde{\varepsilon}$ of c-axis oriented YBa$_2$Cu$_3$O$_{7-x}$ thin films are investigated. For the description of the oxygen out-diffusion experiments a two step process is proposed. Initially an oxygen deficient top layer (x $\approx$ 0.8, thickness $\approx$ 3 nm) is formed. On further heat treatment, also the oxygen deficiency in the YBa$_2$Cu$_3$O$_{7-x}$ below this surface layer increases.

1. INTRODUCTION

Ellipsometry is an optical non-destructive technique, that can be used for the analysis of solid materials. In the case of Cu-O based ceramic high-T$_c$ superconducting materials, spectroscopic ellipsometry was used to determine the complex dielectric function $\tilde{\varepsilon}$ of sintered polycrystalline superconducting YBa$_2$Cu$_3$O$_{7-x}$ samples and YBa$_2$Cu$_3$O$_{7-x}$ single crystals [1,2]. The effect of oxygen deficiency x on the complex dielectric function $\tilde{\varepsilon}$ was investigated [2,3] and measurements were compared with results on related Cu-O based materials [4]. The effects of metallic overlayers [5] and ionic substitutions for Y in YBa$_2$Cu$_3$O$_{7-x}$ [6] were studied. The effect on $\tilde{\varepsilon}$ of cooling down the high-T$_c$ superconducting YBa$_2$Cu$_3$O$_{7-x}$ samples below the critical temperature T$_c$ was determined [7]. Also the effects of the high anisotropy of YBa$_2$Cu$_3$O$_{7-x}$ on $\tilde{\varepsilon}$ were studied on YBa$_2$Cu$_3$O$_{7-x}$ single crystals and on a- and c-axis oriented thin films [8,9]. Synchrotron radiation was used for spectroscopic ellipsometry on sintered polycrystalline superconducting YBa$_2$Cu$_3$O$_{7-x}$ samples [10]. Results of spectroscopic ellipsometry on YBa$_2$Cu$_3$O$_{7-x}$ are thoroughly discussed in ref. 11.

Here we present layer thickness measurements on c-axis oriented high-T$_c$ superconducting YBa$_2$Cu$_3$O$_{7-x}$ thin films, using spectroscopic ellipsometry. The YBa$_2$Cu$_3$O$_{7-x}$ films had a thickness in the range of 20-70 nm. These thicknesses enable the interface between the substrate and the YBa$_2$Cu$_3$O$_{7-x}$ thin film to be studied. Additionally to the layer thickness measurements, oxygen outdiffusion experiments were performed. A Rotating Analyser Ellipsometer (RAE) was used as a real time in situ monitor. The influence of oxygen deficiency on the complex dielectric function $\tilde{\varepsilon}$ of the $YBa_2Cu_3O_{7-x}$ thin films has been investigated. A two step process describing the oxygen out-diffusion in $YBa_2Cu_3O_{7-x}$ thin films is proposed.

2. LAYER THICKNESS MEASUREMENTS

When Δ and Ψ are measured on a substrate covered with a thin film of unknown complex dielectric function $\tilde{\varepsilon}_f$, they are a function of three independent quantities, the real and imaginary part of the complex dielectric function $\tilde{\varepsilon}_f$ of the film and the film thickness d [12]. To determine both $\tilde{\varepsilon}_f$ and d with ellipsometry, Δ and Ψ can be measured on a set of thin films with different film thicknesses, in our case c-axis oriented high-T_c superconducting $YBa_2Cu_3O_{7-x}$ thin films. We assume that the $YBa_2Cu_3O_{7-x}$ films in the set under investigation all have the same complex dielectric function $\tilde{\varepsilon}_f$. If N layers are measured, one obtaines Δ_i^{exp}, i = 1..N and Ψ_i^{exp}, i = 1..N, that is, 2N measured values.

Using a matrix representation for the single layer model [12], we obtain theoretical values for Δ and Ψ. This set of theoretical values depends on N+2 parameters, the thicknesses d_i, i = 1..N and the real and imaginary part of the complex dielectric function $\tilde{\varepsilon}_f$. With help of a modified Levenberg-Marquardt method [13], the parameters $\tilde{\varepsilon}_f$, d_i, i = 1..N can be optimized by minimizing the error function F. This error function is defined by:

$$F = \left[\sum_i (\Delta_i^{exp} - \Delta_i^{theo})^2 + \sum_i (\Psi_i^{exp} - \Psi_i^{theo})^2 \right] / (n - 1 - p), \qquad (1)$$

where n is the number of measurements and p the number of fit parameters.

3. EXPERIMENTAL

High-T_c superconducting $YBa_2Cu_3O_{7-x}$ thin films were deposited on Yttria Stabilized ZrO_2 (YSZ) (100) single crystals using a modified off-axis RF-magnetron sputtering technique [14]. X-Ray diffraction analysis (XRD) was used to determine the structure and orientation of the $YBa_2Cu_3O_{7-x}$ layers. The length of the c-axis of the different $YBa_2Cu_3O_{7-x}$ films was determined. Superconducting properties were derived from critical temperature T_c and critical current density j_c measurements.

A Debye-Scherrer diffractometer with a Cu K-α X-ray source was used for XRD analysis. Ellipsometry was performed using a Rotating Analyser Ellipsometer (RAE), similar to those described elsewhere [15,16]. The angle between photon path and normal to the sample surface typically equals 68.5°. In the two-zone measurements the polarizer angle P equaled $\pm$ 45°. Inside the UHV system, the YSZ substrates with on top a $YBa_2Cu_3O_{7-x}$ layer were mounted on a Si(111) substrate, p type, 2000 Ωcm. By resistively heating this Si substrate, the $YBa_2Cu_3O_{7-x}$ thin films can be given a controllable heat treatment.

Layer thickness measurements were performed on a set of 6 $YBa_2Cu_3O_{7-x}$ thin films, with thicknesses of 20 - 70 nm. With help of these thicknesses the complex dielectric function $\tilde{\varepsilon}$ of c-axis oriented high-T_c superconducting $YBa_2Cu_3O_{7-x}$ thin films was calculated. The thicknesses of the $YBa_2Cu_3O_{7-x}$ films are of the same order of magnitude as the optical penetration depth σ, which typically equals 60 nm for $YBa_2Cu_3O_{7-x}$ (x $\approx$ 0.0). Therefore, also the interface between substrate and $YBa_2Cu_3O_{7-x}$ thin film can be studied.

Additionally, on two other samples oxygen out-diffusion experiments were carried out. First the layer thicknesses were determined with ellipsometry using the results obtained on samples 1-6. The effect of oxygen deficiency x on the complex dielectric function $\tilde{\varepsilon}$ of c-axis oriented high-T_c superconducting $YBa_2Cu_3O_{7-x}$ thin films was investigated by carefully heating two samples in the UHV system. Because of this heat treatment, oxygen diffuses out of the YBa$_2$Cu$_3$O$_{7-x}$ thin film across the interface between the YBa$_2$Cu$_3$O$_{7-x}$ layer and vacuum, increasing x. The heat treatment was monitored in situ. Δ and Ψ were measured at incident photon energies of 4.0, 4.1 and 4.2 eV during the warming up of the samples, the oxygen out-diffusion and the cooling down of the YBa$_2$Cu$_3$O$_{7-x}$ thin films. Simultaneously to the RAE measurements, the temperature of a thermocouple clamped on the substrate was measured. After cooling down, the complex dielectric function $\tilde{\epsilon}$ of the YBa$_2$Cu$_3$O$_{7-x}$ films with different x was measured at room temperature. The length of the c-axis in YBa$_2$Cu$_3$O$_{7-x}$ is related to the value of x [17]. With XRD, the length of the c-axis in YBa$_2$Cu$_3$O$_{7-x}$ thin films can be determined and, therefore, an estimate of x can be obtained. In this way the complex dielectric function $\tilde{\epsilon}$ of YBa$_2$Cu$_3$O$_{7-x}$ as a function of oxygen deficiency x can be measured.

4. LAYER THICKNESS MEASUREMENTS AND OXYGEN OUT-DIFFUSION EXPERIMENTS

By means of the modified RF-magnetron sputtering technique, as mentioned in section 3, high-T$_c$ superconducting YBa$_2$Cu$_3$O$_{7-x}$ thin films with a transition temperature T$_{c,zero}$ of about 90 K were obtained routinely for film thicknesses of 8-300 nm. The critical current density j$_c$ at 77 K of these films is found to be higher than 1 x 10^6 A/cm^2. With X-ray diffraction analysis, besides the substrate reflections only the (00l) reflections could be observed [18].

The complex dielectric function of an YSZ (100) single crystal was determined on a clean, polished YSZ (100) substrate [12]. Almost no structure could be observed in the real and imaginary part of the dielectric function. A set of 6 high-T$_c$ superconducting YBa$_2$Cu$_3$O$_{7-x}$ thin films deposited on YSZ (100) single crystals was investigated. With a RAE, Δ and Ψ were measured at photon energies of 2.2, 2.3, ...,4.4 eV. Each set of 6 Δ's and 6 Ψ's, obtained at photon energy hv$_i$, was fitted simultaneously using the single layer model as outlined in section 2. At each photon energy hv$_i$ a set of 6 optimal thicknesses and 1 optimal complex dielectric function $\tilde{\epsilon} = \epsilon_1 + i\epsilon_2$ were found. The optimal thicknesses d$_i$, i = 1..N are given in fig. 1. One observes that for photon energies of 2.2-3.9 eV, the thicknesses do not strongly depend on the photon energy, although deviations of $\pm$ 10% may occur. Above 3.9 eV, deviations strongly increase, especially for the relatively thicker films with a thickness of 50-70 nm. The thicknesses found at photon energies of 2.2-3.9 eV have been averaged and the results are given in table 1. Also the critical temperatures of the different layers are given. Using the thicknesses listed in table 1, each set of 6 Δ's and 6 Ψ's measured at photon energy hv$_i$ was simultaneously fitted using the single layer model. In this case the only parameter was the complex dielectric function $\tilde{\epsilon}$ at the different photon energies. The result of this fit is given in fig. 2, where the real and imaginary part of the dielectric function of c-axis oriented high-T$_c$ superconducting YBa$_2$Cu$_3$O$_{7-x}$ thin films (x = 0.0) are given as a function of photon energy. Weak features can be seen near 2.7 and 3.9 eV. The optical penetration depth σ (which equals $\lambda/4\pi k$ and is defined as the depth at which the intensity of the light is reduced to 1/e of it's incidenting value), corresponding to this dielectric function typically equals 60 nm in the photon energy range used.

Structural properties of YBa$_2$Cu$_3$O$_{7-x}$ are known to depend very strong on the oxygen concentration in the material, the a-, b- and c-axis lengths change with oxygen deficiency x [17]. To investigate the effect of oxygen deficiency x on the complex dielectric function $\tilde{\epsilon}$ of YBa$_2$Cu$_3$O$_{7-x}$ thin films, two layers were carefully heated in a UHV system (background pressure: 10^{-9} mbar). First a quick warming up from roomtemperature to about 300 °C within 5 minutes is performed. After another 80 minutes

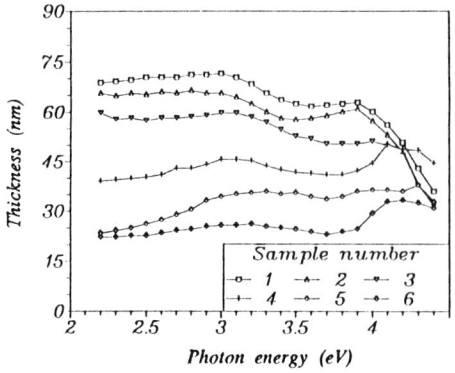

Fig.1: Layer thickness measurements. The deviations above 3.9 eV are ascribed to small differences in the oxygen concentrations in the different samples.

a quick cooling down to temperatures below the temperature were the oxygen out-diffusion starts is performed. The oxygen out-diffusion is followed using the ellipsometer as an in situ monitor.

The critical temperature of the layers after the oxygen out-diffusion was measured to investigate the effect of the heat treatment. XRD analysis was used to determine the structure of the $YBa_2Cu_3O_{7-x}$ layers before and after the heat treatment. From XRD, also the length of the c-axis c and an estimate for oxygen deficiency x were determined. Using the results obtained on samples 1-6, the thickness and complex dielectric function of layer 7 and 8 were determined. The results are shown in table 2 and fig. 2.

Table 1: Layer thickness measurements

Sample	Thickness (nm)	$T_{c,onset}$ (K)	$T_{c,zero}$ (K)
1	68 ± 5	91.0	88.5
2	63 ± 4	90.0	85.6
3	56 ± 5	92.0	86.0
4	42 ± 3	90.0	85.0
5	32 ± 6	91.0	85.5
6	24 ± 3	91.5	87.2

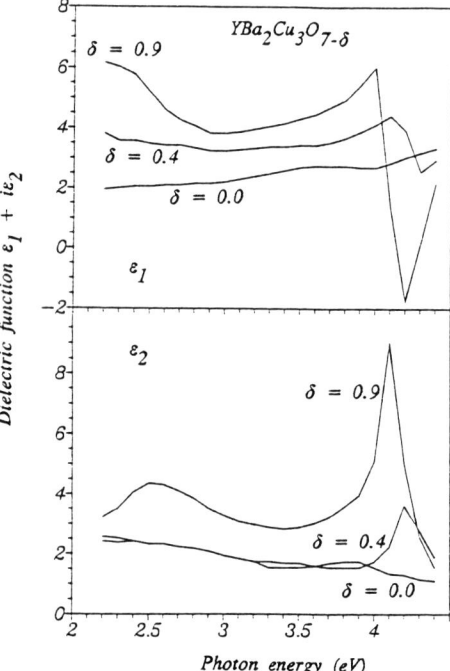

Fig. 2a, 2b: The real and imaginary part of the complex dielectric function $\tilde{\varepsilon} = \varepsilon_1 + i\varepsilon_2$ of c-axis oriented $YBa_2Cu_3O_{7-x}$ thin films for different values of x.

Table 2: Anneal of c-axis oriented high-T_c superconducting $YBa_2Cu_3O_{7-x}$ thin films. The indices i and f indicate before and after the heat treatment, respectively. The length of the c-axis is given by c, the oxygen deficiency by x.

Sample	Thickness (nm)	$T_{c,zero,i}$ (K)	$T_{c,zero,f}$ (K)	c_i (nm)	c_f (nm)	x_f
7	67 ± 7	85.7	-	1.167 ± 0.001	1.182 ± 0.002	0.9
8	40 ± 4	87.2	54	1.167 ± 0.001	1.173 ± 0.002	0.4

5 DISCUSSION AND CONCLUSIONS

Here we showed that for thicknesses in the range 20 - 70 nm, ellipsometry enables us to determine the thickness and complex dielectric function $\tilde{\varepsilon} = \varepsilon_1 + i\varepsilon_2$ of c-axis oriented high T$_c$ superconducting YBa$_2$Cu$_3$O$_{7-x}$ thin films non-destructively and accurately. With an α-step profiler the YBa$_2$Cu$_3$O$_{7-x}$ film thickness of sample 2 was determined to be (65 $\pm$ 5) nm, averaged over the analysis area. With Auger depth profiling we found d = (61 $\pm$ 2) nm for sample 2. These results well agree with the layer thickness measured using ellipsometry, where we found d = (63 $\pm$ 4) nm.

For the determination of the complex dielectric function $\tilde{\varepsilon}$ of c-axis oriented high-T$_c$ superconducting YBa$_2$Cu$_3$O$_{7-x}$ thin films (see fig. 2) we used a single layer model. We tried to use a two layer model to fit the measurements. This model describes a YSZ substrate with an intermediate layer of constant thickness for all the samples, representing a thin layer where the YSZ and the YBa$_2$Cu$_3$O$_{7-x}$ interacted, and on top a YBa$_2$Cu$_3$O$_{7-x}$ layer. In fitting the two layer model we found that the complex dielectric functions of the two layers equaled the same value on minimizing the error function F, defined in eq. (1). The thicknesses correlated very strongly, they were allowed to change over 100% without increasing the error function F. This means that no intermediate layer is needed to describe the ellipsometric measurements. Therefore, we conclude that the interface between the YSZ substrate and YBa$_2$Cu$_3$O$_{7-x}$ layer is very well defined and that no strong interface reactions occured. Interface studies confirm these results [19].

Oxygen diffusion experiments showed that the complex dielectric function of c-axis oriented high T$_c$ superconducting YBa$_2$Cu$_3$O$_{7-x}$ thin films very strong depends on oxygen deficiency x. Upon increasing x, a strong feature appears near 4.1 eV. We used this to monitor in situ the oxygen out-diffusion during the gentle heat treatment of a YBa$_2$Cu$_3$O$_{7-x}$ thin film. In this way we obtain Δ, Ψ and $\partial\Delta/\partial\Psi$ which are all independent quantities [20] and the oxygen out-diffusion for each sample can be fitted using different models for this out-diffusion. Results show that a two step process may be used to describe the in situ ellipsometric measurements. Initially an oxygen deficient surface layer is formed with a thickness of 3-4 nm and x $\approx$ 0.8. While the sample is heated further, also the oxygen deficiency in the YBa$_2$Cu$_3$O$_{7-x}$ layer below the surface layer increases [21].

For the YBa$_2$Cu$_3$O$_{7-x}$ material assignment of the features in its complex dielectric function in the incident photon energy range of 1-5 eV converge in the literature. Aspnes and Kelly summarized results in ref. 11. The broad adsorption feature near 2.6 eV is common to all Cu-O based high-T$_c$ superconducting materials and is assigned to transitions in the CuO$_2$ planes. A weak feature can be seen near 3.9 eV in fig. 2. It resembles those found on as grown c-axis oriented YBa$_2$Cu$_3$O$_{7-x}$ single crystals and c-axis oriented YBa$_2$Cu$_3$O$_{7-x}$ thin films [2,9]. Studies with a very surface sensitive technique, X-ray photoemission spectroscopy (XPS) show a top layer of about 1 nm thickness, consisting of Ba, Cu and O [22]. The feature in the dielectric function may well be explained by a toplayer consisting of BaCuO$_2$, which has a feature in the dielectric function near 3.9 eV [4]. Cation substitutions revealed that no unoccupied atomic levels in Y and Ba are involved in the relatively strong feature that appears near 4.1 eV on increasing oxygen deficiency x [2]. Anisotropy studies of the complex dielectric function showed that the 4.1 eV feature is predominantly present in the complex dielectric tensor component $\tilde{\varepsilon}_{a,b}$, perpendicular to the c-axis of the YBa$_2$Cu$_3$O$_{7-x}$ [8]. With the instrumental settings we used for ellipsometry, the complex dielectric function that was obtained from the single layer model yields $\tilde{\varepsilon}_{a,b}$ within 20% in the used photon energy range [9,23]. These results indicated that the 4.1 eV feature may be ascribed to transitions in O-Cu^{1+}-O complexes, resulting from a removal of oxygen from sites in the Cu-O chains [2,8,11], the O4 sites [17].

ACKNOWLEDGEMENTS

This work is part of the research program of the "Stichting voor Fundamenteel Onderzoek der Materie (FOM)", which is financially supported by the "Nederlandse organisatie voor Wetenschappelijk Onderzoek (NWO)". The authors want to acknowledge J. Zoller for performing introductory measurements.

REFERENCES

[1]	J. Humlíček, M. Carriga, M. Cardona, B. Gegenheimer, E. Schönherr, P. Berberich and J. Tate, Solid State Commun. 66 (1988) 1071.
[2]	M.K. Kelly, P. Barboux, J.-M. Tarascon, D.E. Aspnes, W.A. Bonner and P.A. Morris, Phys. Rev. B 38 (1988) 870.
[3]	M. Carriga, J. Humlíček, M. Cardona and E. Schönherr, Solid State Commun. 66 (1988) 1231.
[4]	M.K. Kelly, P. Barboux, J.-M. Tarascon and D.E. Aspnes, Phys. Rev B 40 (1989) 6797.
[5]	M.K. Kelly, S.-W. Chan, K. Jenkin II, D.E. Aspnes, P. Barboux and J.-M Tarascon, Appl. Phys. Lett. 53 (1988) 2333.
[6]	M. Carriga, J. Humlíček, J. Barth, R.L. Johnson and M. Cardona, J. Opt. Soc. Am. B6 (1989) 470.
[7]	A. Bjørneklett, A. Borg and O. Hunderi, Physica A 157 (1989) 164.
[8]	J. Kircher, M. Alouani, M. Carriga, P. Murugaraj, J. Maier, C. Thomsen, M. Cardona, O.K. Andersen and O. Jepsen, Phys. Rev. B 40 (1989) 7368.
[9]	I. Bozovic, K. Char, S.J.B. Yoo, A. Kapitulnik, M.R. Beasly, T.H. Geballe, Z.Z. Wang, S. Hagen, N.P. Ong, D.E. Aspnes and M.K. Kelly, Phys. Rev. B 38 (1988) 5077.
[10]	R.L. Johnson, J. Barth, M. Cardona, D. Fuchs and A.M. Bradshaw, Rev. Sci. Instrum. 60 (1989) 2209.
[11]	D.E. Aspnes and M.K. Kelly, IEEE J. of Quantum Elec. 25 (1989) 2378.
[12]	R.M.A. Azzam, N.M. Bashara, Ellipsometry and polarized light, North-Holland (1979) Amsterdam.
[13]	D.W. Marquardt, J. Soc. Indust. Appl. Math. 11 (1963) 431.
[14]	J. Gao, B. Häuser and H. Rogalla, J. Appl. Phys. 67 (1990) 2512.
[15]	D.E. Aspnes and A.A. Studna, J. Appl. Opt. 14 (1975) 220.
[16]	A.H.M. Holtslag, PhD Thesis, University of Twente (1986) Enschede.
[17]	J.D. Jorgensen, B.W. Veal, A.P. Paulikas, L.J. Nowicki, G.W. Crabtree, H. Claus and W.K. Kwok, Phys. Rev. B 41 (1990) 1863.
[18]	J. Gao, W.A.M. Aarnink, G.J. Gerritsma and H. Rogalla, Appl. Surf. Sci. 46 (1990) 74.
[19]	J. Gao, W.A.M. Aarnink, G.J. Gerritsma, A.J.H.M. Rijnders, H. Rogalla, F. Hakkens, W. Coene and M.A.M. Gijs, submitted to Physica C.
[20]	J.M.M. de Nijs, PhD Thesis, University of Twente (1989) Enschede.
[21]	W.A.M. Aarnink, R.P.J. IJsselsteijn, J. Gao, A. van Silfhout and H. Rogalla, to be published.
[22]	W.A.M. Aarnink, J. Gao, H. Rogalla and A. van Silfhout, submitted for publication.
[23]	D.E. Aspnes, J. Opt. Soc. Am. 70 (1980) 1275.

High T$_c$ Superconductor Thin Films
L. Correra (Editor)

Temperature dependence of activation energy for flux lines in $(Nd_{1-x}Ce_x)_2CuO_{4-y}$

T. Fukami[a], K. Hayashi[b], T. Nishizaki[b], Y. Horie[b], V. Soares,
T. Aomine[b] and L. Rinderer

Institut de Physique Expérimentale, Université de Lausanne, CH-1015
Lausanne-Dorigny, Switzerland

[a]On leave from Department of Physics, Kyushu University, Fukuoka 812,
Japan

[b]Department of Physics, Kyushu University, Fukuoka 812, Japan

Abstract

We prepared superconducting $(Nd_{1-x}Ce_x)_2CuO_{4-y}$ films by laser ablation
technique. After two stages of thermal process, these films showed a metallic
conduction and superconductivity with decreasing temperature. We
measured the temperature dependence of resistivity in magnetic fields and the
data are analyzed using an activation type formula. The temperature
dependence of the activation energy for **H** // the basal plane can be explained
well in terms of the intrinsic pinning model given by Tachiki and Takahashi.

1. INTRODUCTION

Since the electron-doped superconductor $(Nd_{1-x}Ce_x)_2CuO_{4-y}$ (NCC) was
found by Tokura et al. [1], its electronic and magnetic properties have been
investigated [1-4]. The electron carriers in this superconductor are donated by
Ce^{4+} which replaces Nd^{3+}. Since in the process of preparing this oxide the
concentration of Ce^{4+} is easy to have a gradient in bulk samples, thin films
have been prepared by sputtering [4-6] and other methods [7,8] to decrease the
gradient of concentration of Ce^{4+}. The electric resistivity $\rho(T,H)$ vs.
temperature, T, for the high T_c oxides in magnetic fields, H, have been
analyzed using the Arrhenius law,

$$\rho(T,H) = \rho(0) \exp[-U(T,H)/k_BT], \qquad (1)$$

Where $\rho(0)$ is a constant and $U(T,H)$ an activation energy for one magnetic
flux line. Until now many measurements of $\rho(T,H)$ vs. T for high T_c oxide
superconductors in magnetic fields have been carried out and most of them
have been analyzed by using eq. (1) and assumed

$$U(T,H) = U(0,H)(1-t)^n, \tag{2}$$

where $t = T/T_c(H)$ with a critical temperature in magnetic fields, $T_c(H)$, and n is a constant exponent. Palstra et al. obtained $n=0$ from $\rho(T,H)$ vs. T for single crystalline samples of $Bi_2Sr_2CaCu_2O_{8+x}$ (BSCC) in the state of thermally assisted flux flow [9]. On the other hand, we showed for BSCC thin films prepared by the laser ablation technique [10] that the relation of log $[\rho(T,H)]$ versus $(1-t)/T$ reproduced well the experimental results for $\boldsymbol{H}$ // the basal plane [11]. For $YBa_2Cu_3O_{7-y}$ (YBC), the values around 1.5 have been obtained for n with $\boldsymbol{H}$ // the c-axis [12]. Recently, Suzuki and Hikita obtained $n=2$ for $\boldsymbol{H}$ // the basal plane and $n=3$ for $\boldsymbol{H}$ // the c-axis for NCC films prepared by RF sputtering technique [4]. However, at present we cannot understand the reasons why the values of the exponent are different for different oxide superconductors.

In the present paper we report that the temperature dependence of the activation energy has values of $n=3$ both for $\boldsymbol{H}$ // the c-axis and for $\boldsymbol{H}$ // the basal plane for NCC films and we show that the different values of n for NCC, BSCC and YBC for $\boldsymbol{H}$ // the basal plane can be obtained systematically in terms of the intrinsic pinning model for flux lines given by Tachiki and Takahashi [13,14].

2. EXPERIMENTAL RESULTS

Films were prepared by laser ablation technique on single crystalline $SrTiO_3$ (100) substrates kept at room temperature. The target had a nominal composition of $(Nd_{0.925}Ce_{0.075})_2CuO_4$, and the Ar-F gas excimer pulse laser with an initial energy of about 70 mJ/pulse was focused on the surface of the target. As-deposited films were processed at 1045 °C for 1 hour in air and successively processed at 890 °C for 2 hours in N_2 gas. The c-axis of NCC is oriented preferably along the vertical direction of the surface of the substrate.

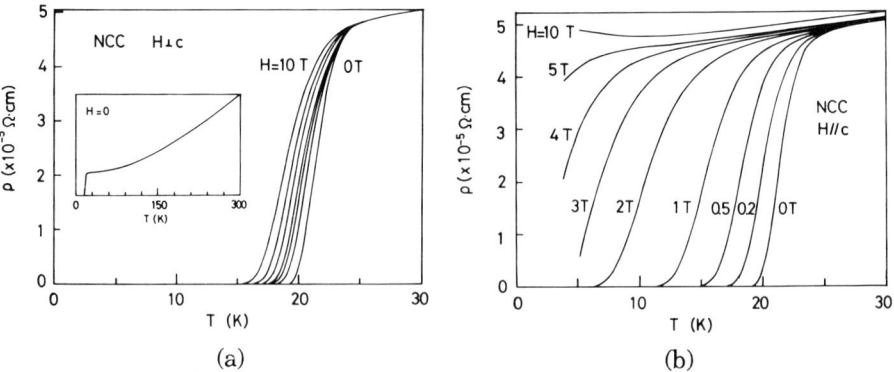

(a)   (b)

Fig. 1 $\rho(T,H)$ vs. T for NCC; (a) for $\boldsymbol{H}$ // the basal plane and (b) for $\boldsymbol{H}$ // the c-axis. The inserted figure shows the result in $H=0$ from room temperature.

Figure 1 shows the temperature dependence of $\rho(T,H)$ for NCC in magnetic fields up to 10 T for two configurations of H // the c-axis and H // the basal plane keeping H perpendicularly to the current density J which was flowed in the basal plane. For H // the basal plane, the behaviors of $\rho(T,H)$ vs. T are very similar to those ones of other oxide superconductors.

Figure 2 shows the normalized critical current density $J_c(\theta)/J_c(0)$ vs. the angle θ between H and the basal plane keeping H perpendicularly to J. The dots, circles and triangles are experimental values and solid lines are ones calculated using the intrinsic pinning model which is described below.

3. ANALYSIS AND DISCUSSION

The experimental results shown in Fig. 1 were analyzed by the same method as that one used by Suzuki and Hikita using the eqs. (1) and (2). Taking ρ_0, $U(0,H)$, n and $T_c(H)$ as adjustable parameters, we determined these parameters so as to obtain the best fit of eq. (1) to the data. In order to confirm the reasonableness of the parameters, we plotted $k_BT \log [\rho(T,H)/\rho_0]/U(0,H)$ vs. $(1-t)$ on a double logarithmic scale and the results are shown in Fig. 3. From the slope of the data points, the best fits are obtained for $n=3$ both for H // the c-axis and

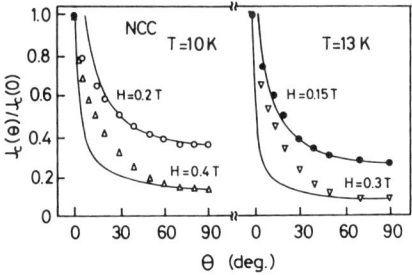

Fig. 2 Critical current density J_c vs. θ for $T = 10$ K and 13 K. The solid lines are calculated ones.

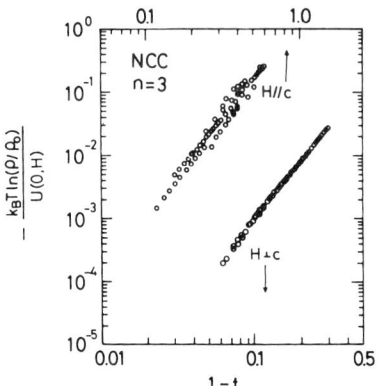

Fig.3 Double-logarithmic $k_BT \log [\rho/\rho_0] /U(0,H)$ vs. 1-t.

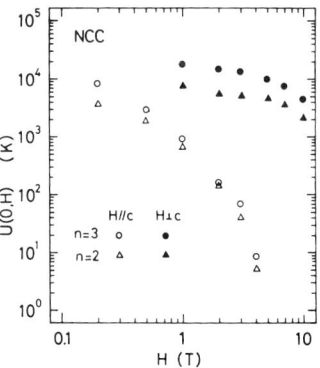

Fig. 4 Plots of $U(0,H)$ vs. H for $n=3$ and $n=2$ for NCC.

H // the basal plane. In our case, we could not obtain the best fit with $n=2$ which had been obtained for H // the basal plane by Suzuki and Hikita [4]. From the field dependence of $T_C(H)$ obtained by fitting using $n=3$ just below $T_C(0)$, we could calculate the coherence lengths as $\xi_{ab}(0) = 80$ Å and $\xi_c(0) = 2.3$ Å. Furthermore, in Fig. 4 we show the field dependence of the activation energy, $U(0,H)$, for two cases of $n=3$ and $n=2$. $U(0,H)$ for H // the basal plane seems to be proportional to $H^{-1/2}$, but for H // the c-axis it changes in proportion to around $H^{-5/2}$ in the low field region. A rapid decrease of $U(0,H)$ in the high field region suggests the disappearance of the activation energy. Until now several groups have explained the origin of different values for n in terms of Tinkham model [15] or Josephson junction [16]. But as long as these models are used it is not so easy to explain the difference of n for different oxides.

By the way, according to the intrinsic pinning model given by Tachiki-Takahashi [14], J_C (T,H) vs. θ under the condition of $J \perp H$ is represented by

$$J_C(T,H)/J_{c1}(T,H) = [J_{c2}(T,H)/J_{c1}(T,H)]/|\sin\theta|^{1/2}, \qquad (3)$$

where $J_{c1}(T,H)$ and $J_{c2}(T,H)$ are critical current densities when magnetic fields were applied to the basal plane and the c-axis, respectively. $J_C(T,H)/J_{c1}(T,H)$ vs. θ was calculated for several magnetic fields and temperatures. The results are shown by solid lines in Fig. 2. These data show that the angular dependence of $J_C(T,H)$ seems to obey qualitatively the intrinsic pinning model.

Thus it would be reasonable that we use the cohesive energy difference deduced in [13] as an activation energy for one magnetic flux line. Here, we summarize their results briefly to use for the explanation. Since oxide superconductors have a layered structure consisting of alternate superconducting CuO_2 and weak superconducting planes, it may be assumed that the superconducting order parameter $\Delta(r)$ is modulated with a period given by the inter-distance of CuO_2 planes. It was assumed there that $\Delta(r)$ changes as a cosine function of space coordinate r. The x and y axes are taken parallel to the a and b axes in a CuO_2 plane, respectively, and the z-axis parallel to the c-axis. Under the situation that one magnetic flux line with coherence lengths ξ_{ab} and ξ_c is inserted along the b-axis at $(0,0,z_0)$, the difference of cohesive energy, $U(z_0)$, between states without and with one flux line in the layer is written as

$$U(z_0) = (H_c^2/4\pi)\int dxdz[1 + \delta\cos(2\pi z/a_c)]^2 \mathrm{sech}^2[(x/\xi_{ab})^2 + \{(z-z_0)/\xi_c\}^2]^{1/2}. \qquad (4)$$

Here $H_c(T)$ is the thermodynamic critical field, δ is a positive parameter which measures the degree of the modulation and is smaller than 1, and a_c is the inter-distance between successive CuO_2 superconducting planes. Using the temperature dependences of $(1-t^2)$ for $H_c(T)$ and of $(1-t)^{-1/2}$ for the coherence lengths, we calculated the temperature dependence of the potential difference $\Delta U(T)$ between the maximum and the minimum of $U(z_0)$, namely

$$\Delta U(T) = U(z_0=0) - U(z_0=a_c/2). \tag{5}$$

Here $z=0$ is taken on a CuO_2 plane. The half of the lattice constant, $c=12.108$ Å, along the c-axis was used as a_c [3], because NCC has two sheets of CuO_2 plane in the crystallographic unit cell. Figure 5 shows the curves of log $[4\pi\Delta U(T)/H_c^2(0)]$ vs. log $[1-t]$ for $\xi_c(0) = 2.1, 2.3$ and 2.7 Å with $\xi_{ab}(0) = 80$ Å and $\delta=0.6$. These curves are not linear but we can see that the dependence seems to have a value near 3 around $1-t = 0.3$.

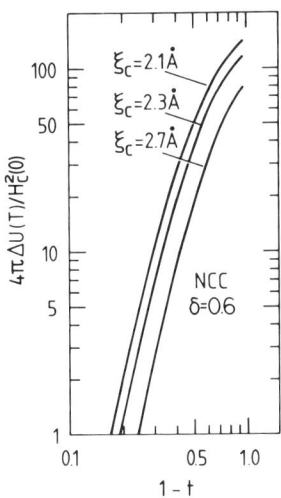

Fig. 5 $4\pi\Delta U(T)/H_c^2(0)$ vs. $1-t$ calculated for NCC.

Similarly, we calculated $\Delta U(T)$ vs. T for H // the basal plane for BSCC ($\xi_{ab}(0) = 15$ Å, $\xi_c(0)=1.0$ Å) and for YBC ($\xi_{ab}(0) = 15$ Å, $\xi_c(0)=2.5$ Å). Since BSCC and YBC have two pairs and one pair of $CuO2$ planes in the crystallographic unit cell, respectively, we modified eq. (4) a little and used $a_c = c/2= 15.4$ Å for BSCC [17] and $a_c = c = 11.7$ Å for YBC [18]. The temperature dependence of $\Delta U(T)$ were also not linear but we could see that the dependence seemed to have values around $n=1$ near $1-t=0.3$ for BSCC; $n= 2$ near $1-t=0.3$, and $n=3$ near $1-t=0.1$ for YBC [19]. For YBC the experimental value of n have not been estimated for H // the basal plane, probably because the resistive transition curves are too complicated to analyze with eqs. (1) and (2) and the width is very narrow. It is, however, important that clearly different values of n can be obtained depending on the values of $\xi_c(0)/a_c$.

Let us consider the temperature dependence of $\Delta U(T)$ semi-qualitatively. According to the model, the elementary pinning force, f_{pM}, is represented by,

$$f_{pM}=(H_c^2/8\pi)2\pi a_c(\xi_{ab}/\xi_c)\eta_M, \tag{6}$$

where η_M is the maximum value of the next $\eta(z_0)$ with respect to z_0,

$$\eta(z_0)=(4\delta/a_c^2)\int dxdz\{1+\delta\cos[2\pi(z+z_0)/a_c]\}\sin[2\pi(z+z_0)/a_c]\mathrm{sech}^2[(x^2+z^2)^{1/2}/\xi_c]. \tag{7}$$

Since the temperature dependence of $\Delta U(T)$ may be given approximately by $\Delta U(T) \approx (a_c/2)f_{pM} \propto H_c^2(T)\eta_M(T)$, it is enough only to consider the temperature dependences of $H_c(T)$ and $\eta_M(T)$. Let us refer Fig. 2 of ref. 13. Since for NCC with $\xi_c(0)/a_c=0.38$ its $\eta_M(T)$ decreases monotonically with the increase of the $\xi_c(T)/a_c$, it increases with increasing $(1-t)$. Therefore, the temperature dependence of $\Delta U(T)$ becomes stronger than that of $H_c^2(T)$, $(1-t^2)^2$. On the other hand, for BSCC with $\xi_c(0)/a_c = 0.08$, $\eta_M(T)$ increases with decreasing $1-t$. Thus the temperature dependence of $\Delta U(T)$ becomes weaker than $(1-t^2)^2$. For YBC with $\xi_c(0)/a_c = 0.21$, its $\eta_M(T)$ changes near its maximum peak with the increase of the temperature. Therefore, the dependence of $\Delta U(T)$ on $(1-t)$ has an intermediate character between ones for NCC and BSCC.

4. REFERENCES

1   Y. Tokura, H. Takagi and S. Uchida, Nature, 337 (1989) 345.
2   H. Takagi, S. Uchida and Y. Tokura, Phys. Rev. Lett., 62 (1989) 1197.
3   Y. Hidaka and M. Suzuki, Nature, 338 (1989) 635.
4   M. Suzuki and M. Hikita, Phys. Rev., B 41 (1990) 9566.
5   H. Adachi, S. Hayashi, K. Setsune, S. Hatta, T. Mitsuya and K. Wasa, Appl. Phys. Lett., 54 (1989) 2713.
6   S. Saitho, M. Hiratani and K. Miyauchi, Jpn. J. Appl. Phys., 28 (1989) L975.
7   K. Kamigaki, H. Terauchi, T. Terashima, K. Iijima, K. Hirata, K. Yamamoto, K. Hayashi and Y. Bando, Jpn J. Appl. Phys., 28 (1989) L2207.
8   A. Gupta, G. Koren, C. C. Tsuei, A. Segmüller and T. R. McGuire, Appl. Phys. Lett., 55 (1989) 1795.
9   T. T. M. Palstra, B. Batlogg, L. F. Schneemeyer and J. V. Waszczak, Phys. Rev. Lett., 61(1988) 1662.
10 T. Fukami, T. Kamura, A.A.A. Youssef, Y. Horie and S. Mase, Physica, C 159 (1989) 427.
11 T. Fukami, T. Kamura, A.A.A. Youssef, Y. Horie and S. Mase, Physica, C 160 (1989) 391.
12 E. Zeldov, N. M. Amer, G. Koren, A. Gupta, M. W. McElfresh and R. J. Gambino, Appl. Phys. Lett., 56 (1990) 680.
13 M. Tachiki and S. Takahashi, Solid State Commun., 70 (1989) 291.
14 M. Tachiki and S. Takahashi, Solid State Commun., 72 (1989) 1083.
15 M. Tinkham, Phys. Rev. Lett., 61 (1988) 1658.
16 D. H. Kim, K. E. Gray, R. T. Kampwirth and D. M. McKay, Phys. Rev., B 42 (1990) 6249.
17 T. Kajitani, K. Kusaba, M. Kikuchi, N. Kobayashi, Y. Shono, T. B. Williams and M. Hirabayashi, Jpn. J. Appl. Phys., 27 (1988) L587.
18 F. Izumi, H. Asano, T. Ishigaki, A. Ono and F. P. Okamura, Jpn. J. Appl. Phys., 26 (1987) L611.
19 T. Fukami, V. Soares and L. Rinderer, to be published in Helv. Phys. Acta., (1991)

High T$_c$ Superconductor Thin Films
L. Correra (Editor)

Hall effect in 2212- and 2223-BiSrCaCuO thin films

I. Khassanov[a,b], R. Hopfengärtner[a], H. Nakano[c], and G. Saemann-Ischenko[a]

a) Physikalisches Institut, Universität Erlangen-Nürnberg, Erwin-Rommel-Str. 1, D-8520 Erlangen, BRD

b) On leave from the Solid State Physics Institute of the USSR Academy of Sciences, Chernogolovka, USSR

c) Department of Physics, College of General Education, Nagoya University, Nagoya, Japan

Abstract

The Hall effect and the electrical resistivity in magnetic fields up to 5T in the normal state as well as in the vicinity of the superconducting phase transition have been measured in Bi-Sr-Ca-Cu-O thin films deposited on MgO substrates by a rf-magnetron sputtering method. The samples which have been investigated have $T_c \approx 107$ K (2223-phase) and $T_c \approx 80$ K (2212-phase), respectively. The Hall coefficient R_H (the magnetic field B perpendicular and the transport current parallel to the surface plane of the films) above T_c is positive and strongly temperature dependent ($R_H \sim 1/T$). Near T_c we observe a sign reversal of R_H varying with the magnetic field and a pronounced broadening of the resistivity curves.

1.INTRODUCTION

Hall effect measurements in various cuprate oxide samples – single crystals, films and polycrystalline bulk material – belonging to different HTSC systems demonstrate rather common features, which seem to be intrinsic, characteristic of this new generation of superconductors [1]. A strong temperature dependance of the Hall coefficient R_H in the normal state far above T_c ($R_H \sim 1/T$) and its sign reversal in the vicinity of T_c are very likely to be the manifestations of the peculiarities of superconductivity in the highly anisotropic layered compounds, as are ReBaCuO-, Bi-, and Tl-systems.

The Bi-Sr-Ca-Cu-O system, which has three different stoichiometric compositions 2201, 2212, 2223 and correspondingly one, two and three CuO_2 planes (mainly responsible for superconductivity), separated by adjacent Bi-O planes, offers a good opportunity for modelling the electronic transport properties of the new HTSC materials with different degrees of anisotropy. Hall effect measurements have been reported for 2212-single crystals

[2], thin films [3] and polycrystalline samples [4, 5]. Certain information is available also for 2223- polycrystalline material [4] and films [6]. In this report we present data of the Hall effect and resistivity measurements for both 2212- and 2223-films.

2. SAMPLE PREPARATION

The films have been deposited on (100)-MgO substrates by rf-magnetron sputtering and subsequently annealed according to the procedure described elsewhere [7]. The samples (thickness $\approx 0.5\mu$m) for the resistivity and the Hall effect measurements have been patterned mechanically by scratching in bridges 4 mm long and 0.25 mm wide. Silver contacts have been evaporated and copper leads have been attached by silver paste.

The films used in this work are characterized as polycrystalline with a high degree of granularity [7]. X-ray diffraction patterns (Fig.1) show the presence of a small part of the 2212-phase in the 2223-phase samples and vice versa.

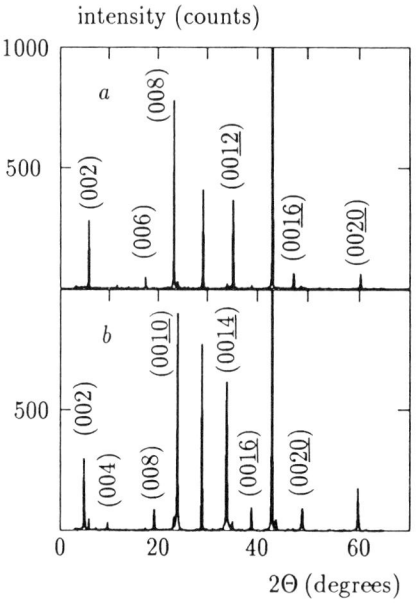

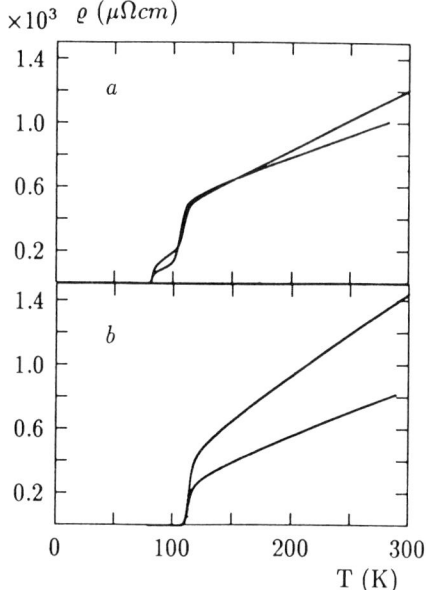

Fig.1: X-ray diffraction patterns of 2212- (a) and 2223-samples (b).

Fig.2: $\varrho(T)$ vs T of 2212- (a) and 2223-samples (b).

3. EXPERIMENTAL RESULTS AND DISCUSSION

Resistivity measurements of the 2212- and 2223-samples reveal a T_c of 80K and 107K, respectively (Fig.2). Depending on the fraction of the high-T_c phase (Fig.1) a smaller or a larger shoulder can be seen on the resistivity curves of the 80K samples. At higher temperatures the resistivity $\varrho(T)$ changes linearly and follows the dependance $\varrho(T) = \varrho_0 + \alpha T$ with $\varrho_0 \approx 0$ for 2223-samples, $\varrho_0 \approx 100 - 250 \, \mu\Omega$cm for 2212-samples and $\alpha \approx 2.6 - 4.8 \, \mu\Omega$cm/K for both types of samples.

The Hall voltage has been measured in constant magnetic fields sweeping the temperature as well as at constant temperatures (in the transition region) with a variation of the magnetic field. In the normal state, above 130 K, $R_H(T)$ nicely follows a $1/T$-dependance for both 2212- and 2223-samples (Fig.3) in good agreement with the results of other authors [1]. This temperature dependance is clearly seen on the plot of the Hall number $n_H(T)$ ($n_H = 1/R_He$) (Fig 4). It is worth to note that due to the strong temperature dependance of R_H a simple interpretation of n_H as the carrier concentration is questionable.

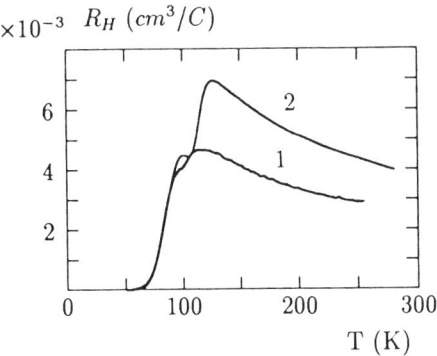

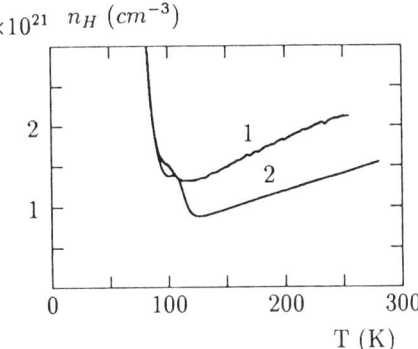

Fig.3: Temperature dependance of the Hall coeffiecient $R_H(T)$ of 2212-(1) and 2223-samples(2).

Fig.4: Temperature dependance of the Hall concentration $n_H(T)$ of 2212-(1) and 2223-samples(2).

In Fig.5(a) we present the temperature dependence of the Hall coefficient in the transition region for different magnetic fields. In the vicinity of T_c the Hall coefficient drops and passes through a minimum, which in lower fields ($B = 1$T) lies below zero, and increases once again to a local maximum at lower temperatures and finally becomes zero.

To get a hint on the mechanism of the current transport in the mixed state we have measured also the resistivity for the same magnetic fields (Fig.5(b)). Corresponding curves demonstrate a broad widening of the transition region. It is well established that the Bi-compounds, having much higher degree of anisotropy [8] in comparison to e.g. the 123-material, exhibit flux flow in a very wide temperature range below T_c [9], which is due to the small activation energy in comparison to k_BT. This temperature range extends down to the temperature $T \approx 50$ K, so that the low temperature slope of the R_H-curves might be accounted for by the flux flow mechanism. One more argument in favour of flux flow mechanism is provided by the magnetoresistivity measurements at constant temperatures (Fig.6(a)). The resistivity of the samples at the lowest temperature $T = 72.5\,K$ changes linearly with the magnetic field.

Recently other promising attempts have been published that explain the sign reversal of the Hall coefficient in terms of superconducting fluctuations caused by magnetic skew-scattering [10].

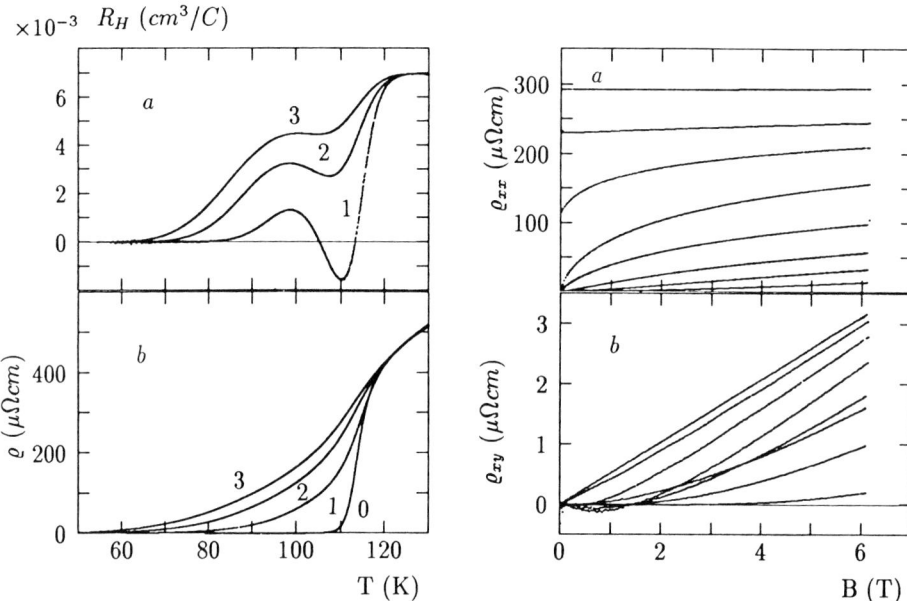

Fig.5:Temperature dependance of the Hall coefficient (a) and the resistivity (b) of the 2223-sample in the transition region for different magnetic fields: 1T (1), 3T (2), and 5T (3). The resistivity in zero magnetic field on plot (b) is labeled by (0).

Fig.6:Magnetic field dependance of the longitudinal (a) and the Hall resistivity (b) of the 2223 sample at different temperatures: T= 72.5, 84.8, 94.8, 104.8, 106.9, 112.9, 117.2, and 125.3 K, starting from lower curves.

4. CONCLUSION

We have measured the Hall coefficient $R_H(T)$ and the electrical resistivity of 2212- and 2223 films of the Bi-Sr-Ca-Cu-O system. Far above T_c the Hall coefficient is positive and strongly temperature dependent ($R_H(T) \sim 1/T$). In the vicinity of the phase transition we observe a sign reversal of the Hall coefficient depending on the magnetic field. The low temperature slope of $R_H(T)$ might be accounted for by flux flow.

One of us (I. Kh.) greatly acknowledges the financial support of the Deutscher Akademischer Austauschdienst (DAAD) and the helpful assistance from the staff of the Low Temperature Physics group of the Physikalisches Institut der Universität Erlangen.

5. REFERENCES

1 N. P. Ong, in Physical Properties of High Temperature Superconductors II, edited by D. M. Ginsberg, World Scientific Publishing Co. Pte. Ltd. (1990) 459.

2 S. N. Artemenko, I. G. Gorlova, and Yu. I. Latyshev, Physics Letters A **138** (1989) 428.

3 Y. Iye, S. Nakamura, and T. Tamegai, Physica C **159** (1989) 616.

4 J. Clayhold, S. J. Hagen, N. P. Ong, J. M. Tarascon, and P. Barboux, Phys. Rev. B **39** (1989) 7320.

5 M. Galffy, Solid State Commun.**72** (1989) 589.

6 M. Mukaida, K. Kuroda, S. Miyazawa, Appl. Phys. Lett. **55** (1989) 1129.

7 H. Nakano, S. Nicoud, M. Suzuki, G. Burri, and L. Rinderer, in High T_c Superconductor Materials, North Holland, Amsterdam, Proceedings of E-MRS 1990 Spring Meeting, May 29-June 1, 1990, Strasbourg, France (1990) 679.

8 P. Schmitt, P. Kummeth, L. Schultz, and G. Saemann- Ischenko, submitted to Phys.Rev.Lett..

9 T. T. M.Palstra, B. Batlogg, L. F. Schneemeyer, and J. V. Waszczak, Phys. Rev. B **43** (1991) 3756.

10 A. G. Aronov and S. Hikami, Phys. Rev. B **41** (1990) 9548.

High T$_c$ Superconductor Thin Films
L. Correra (Editor)
1992 Elsevier Science Publishers B.V.   67

Hydrogen-Effects in YBaCuO Thin Films

J. Erxmeyer, J. Steiger and A. Weidinger

Hahn–Meitner–Institut, Bereich Schwerionenphysik, Glienickerstr. 100,
D–1000 Berlin 39, FRG

Abstract
$Y_1Ba_2Cu_3O_7$ thin films were charged with hydrogen from the gas phase. The depth profiles of the H–concentration is measured by a resonant nuclear reaction. The effect of H–doping is investigated by Hall–effect and AC–susceptibility measurements.

1. INTRODUCTION

Hydrogen absorption changes the properties of high–T$_c$ materials drastically. Extensive investigations[1-6] were performed on polycrystalline samples or powders of $Y_1Ba_2Cu_3O_7$. However the role of hydrogen in $Y_1Ba_2Cu_3O_7$ is not clear yet. For example, the possible formation of a hydride phase and the structure of it are still not known with certainity. Similarily the change of the electronic structure with increasing hydrogen concentration is not understood. An interesting aspect is that hydrogen may donate its electron into the conduction band and thereby compensate the (hole) charge carriers in $Y_1Ba_2Cu_3O_7$.

In the present experiments we used thin films instead of bulk materials since their properties are better defined. The $Y_1Ba_2Cu_3O_7$ films were prepared by laser ablation. We measured the hydrogen depth profile of H–charged films, the change of the Hall–number and of the AC–susceptibility.

2. EXPERIMENTAL DETAILS AND RESULTS

The films were prepared by laser ablation on single crystalline $SrTiO_3$–substrates at the KFA Jülich[7]. They were c–axis oriented and had a T$_c$(onset) of 89 K with a transition width of about 1 K.

The hydrogen charging was performed in a small vacuum–chamber from the gas phase at a pressure of 100 mbar. The sample temperature during charging was 460 K. The hydrogen uptake was monitored by in situ resistivity measurements in the van der Pauw geometry. For this purpose gold contacts were evaporated at the four corners of the films.

Fig. 1 shows the resistivity of the film as a function of time. After the

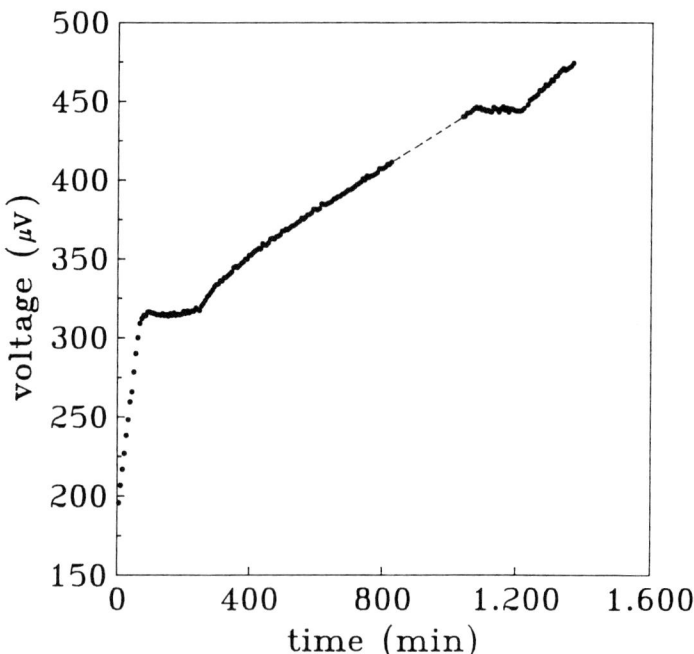

Figure 1. Resistivity of a $Y_1Ba_2Cu_3O_7$ thin film during hydrogen charging. It is plotted the voltage proportional to the resistivity. For details see text.

chamber has been evacuated the film was heated up accompanied by an increase of its resistivity. After reaching the charging temperature the resistivity remained constant. 300 minutes later the chamber was vented with hydrogen and the increasing resistivity indicated the hydrogen uptake into the film. Finally after 1000 minutes hydrogen was pumped out. The fact that the resistivity remained constant, as shown in fig. 1, shows that no hydrogen left the sample even at this temperature. After 1200 minutes the H–charging was continued. Again the increase of the resistivity indicates the uptake of hydrogen.

The hydrogen depth profiles were measured by the resonant $^1H(^{15}N,\alpha\gamma)^{12}C$ nuclear reaction method[8]. The γ–yield of this reaction measured by a 6"×6" NaI(Tl) detector is recorded as a function of the incident beam energy. Knowing the stopping power (dE/dx) of the ^{15}N–ions the energy scale can be converted into a depth scale. dE/dx was calculated to 2.7 keV/nm using the computer code TRIM[9].

Charging the samples for different times lead to different hydrogen concentrations (fig. 1). A remarkable feature of the profiles shown in fig. 2 is the increase of the H–concentration with the increasing depth (open circles and squares). The average concentration in both films is [H]/cell = 2.1 and [H]/cell = 1.5 respectively. In earlier experiments we always found a decreasing concentration from the surface to the substrate.

The film with the higher hydrogen concentration (open circles) was annealed for 2 hours at 460 K in vacuum. Then the hydrogen depth profile was measured

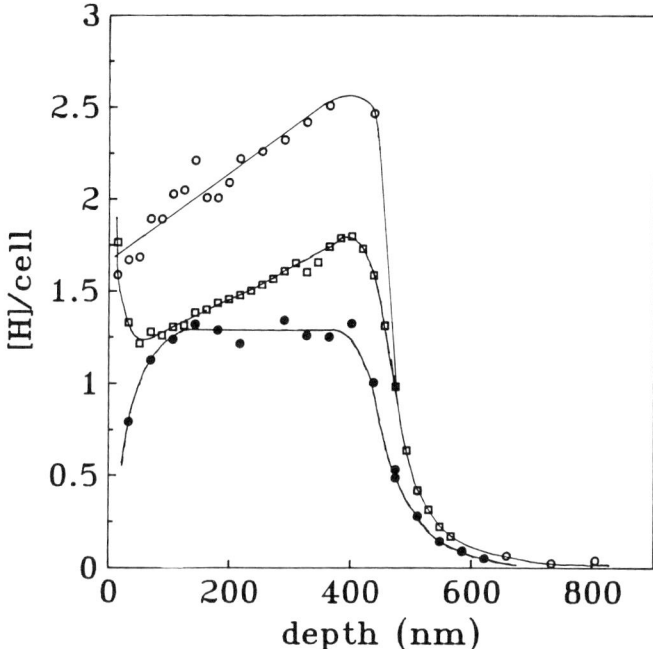

Figure 2. Depth profile of the hydrogen concentration. Sample 1 (open circles) and sample 2 (rectangles) were charged for different times. The closed circles show sample 1 after vacuum annealing at 460 K.

again. The result is shown in fig. 2 marked by closed circles. Now the hydrogen concentration is found to be constant nearly over the whole film and is lowered to [H]/cell = 1.3. This is a very surprising behavior. In earlier experiments we found for lower hydrogen contents of the films no decrease of the concentration at the charging temperature. This measurements show that at higher H–concentrations (in this case [H]/cell > 1.3) an outdiffusion of hydrogen at the charging temperature is possible. For technical reasons we were unable to monitor the resistivity of this film during annealing. Therefore we do not know wether the resistivity remains constant or not. Fig. 1 only shows the change of the resistivity for low hydrogen concentrations.

After the ^{15}N–measurement the irradiated sites of the film look brighter and in some cases even transparent, i. e. the crystal–structure of the films has been changed by irradiation. But this has no significant influence on the measured depth profile as can be shown by profile measurements in reverse order. After annealing the film with the higher hydrogen concentration as described above, the depth profile was measured on a non irradiated site of the film.

For hydrogen concentrations up to [H]/cell = 0.3 Hall–effect measurements were performed in the van der Pauw geometry. The Hall–number n_h/cell can be calculated from the measured Hallcoefficient R using the relation n_h/cell = V_0/eR, where V_0 = 0.173 nm^3 is the volume of the unit cell of $Y_1Ba_2Cu_3O_7$.

Fig. 3 shows the Hall–number n_h/cell as a function of temperature for an

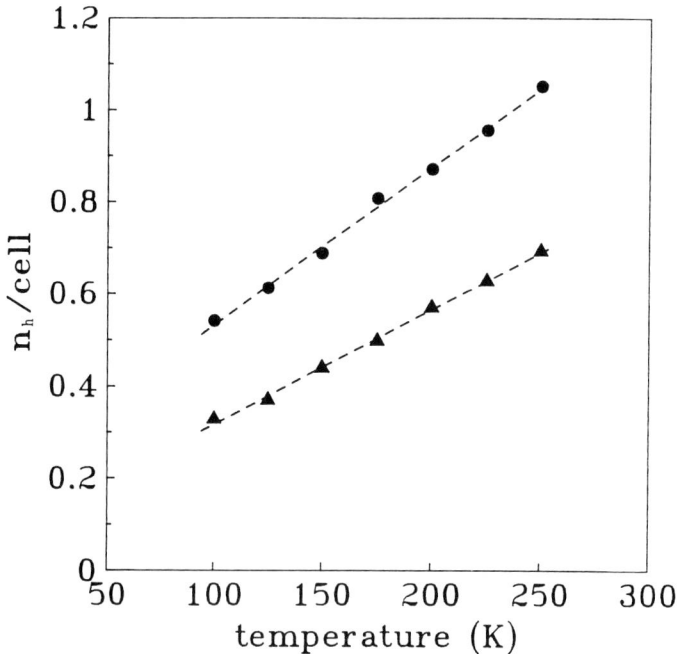

Figure 3. Dependence of the Hall–number from the temperature before (circles) and after (triangles) charging. The hydrogen concentration for the charged sample is [H]/cell=0.3. The slope dn_h/dT decrease with increasing hydrogen concentration.

uncharged film (circles) and after charging with a hydrogen concentration of [H]/cell = 0.3 (triangles). Both films, the charged and uncharged one, show a linear decrease of dn_h/dT with temperature. The relative reduction of the Hall–number $(1-n_h(T,c_h)/n_h(T,c_h=0)$, where c_h is the H–concentration) is nearly constant for all measured temperatures. A possible explanation for this behavior is a reduction of the carrier concentration by H–doping. We found that 0.6 H–atoms per unit cell were necessary to get a value of 1 for the relative reduction of the Hall–number, i. e. $n_h(T,c_h=0.6)=0$.

The change of dn_h/dT together with a decrease of the transition temperature seems to be a feature of all high–T_c superconductors[10]. As shown in fig. 3 this behavior can be verified for doping with hydrogen too. For this film a decrease of 3 K of the transition temperature due to the hydrogen charging was found.

To study the change of the superconducting behavior with increasing hydrogen concentration AC–susceptibility measurements were performed. In fig. 4 the voltage which is proportional to the real part of the susceptibility is plotted as a function of temperature for an uncharged film (a) and after charging with hydrogen (b). The ^{15}N–measurement yields a H–concentration of [H]/cell = 0.3. As can be seen in fig. 4 the transition temperature decreases with increasing hydrogen concentration. For this film there is a change of nearly 10 K. In the case of sintered $Y_1Ba_2Cu_3O_7$ samples a decrease of T_c was not found[3]. Further

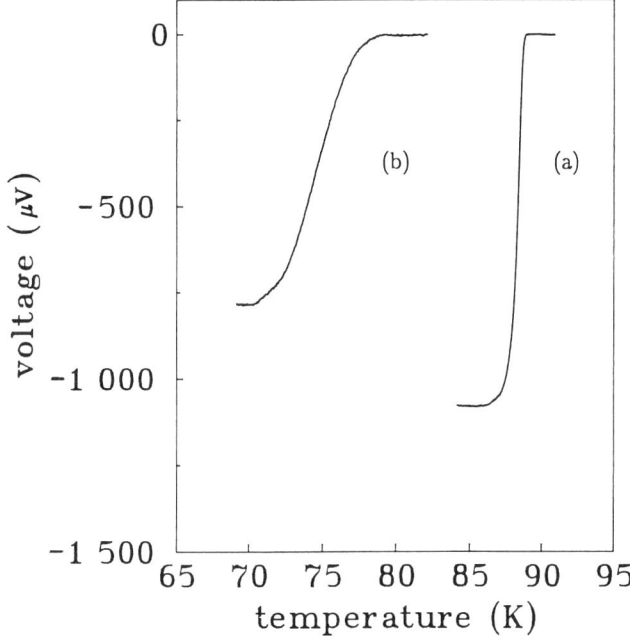

Figure 4. Real part of the AC–susceptibility before (a) and after (b) charging. It is plotted the measured voltage which is proportional to the real part of the susceptibility. The hydrogen concentration is [H]/cell=0.3 for the charged sample.

the width of the transition from the normal to the superconducting state increases with increasing H–concentration. This could indicate that, due to the hydrogen charging, regions with different transition temperatures are formed. Finally the real part of the susceptibility changes for $T < T_c$ due to hydrogen doping. This could be interpreted as a decrease of the superconducting volume with increasing hydrogen concentration.

3. CONCLUSION

Hydrogen depth profiles were measured for H–charged $Y_1Ba_2Cu_3O_7$ films. A remarkable feature is that the hydrogen concentration increases with the depth of the film. It is possible that for higher concentrations the hydrogen uptake is reversible at the charging temperature. The decrease of the hydrogen content of the film after a prolonged annealing period in vacuum indicates this. After annealing the depth profile of the film was found to be constant nearly over the whole film with a hydrogen concentration of [H]/cell = 1.3. At this concentration no more hydrogen left the film at the charging temperature. Depth profile measurements of films with lower hydrogen concentrations indicate an irreversible hydrogen uptake.

AC–susceptibility measurements show a decreasing transition temperature with increasing hydrogen concentration. Additionally we found a broadening of the transition from the normal to the superconducting state. In earlier experiments we found the same behavior by measuring the resistivity[11]. This indicates that, due to the hydrogen charging, regions with different transition temperatures are formed, i. e. the hydrogen uptake is inhomogeneous. Finally we found as in the case of sintered samples[3] a change of the real part of the susceptibility with increasing hydrogen concentration, which could be interpreted as a decrease of the superconducting volume.

In the investigated temperature range Hall–effect measurements show a nearly constant relative reduction of the Hall–number for a hydrogen concentration up to $[H]/cell = 0.3$. The hydrogen charging does not change the linear temperature dependence of the Hall–number. A reduction of the carrier concentration with increasing hydrogen concentration is a possible explanation of the experimental results.

ACKNOWLEDGEMENT

We would like to thank B. Stritzker and W. Zander for the preparation of the $Y_1Ba_2Cu_3O_7$–films. Our work was kindly supported by the Bundesminister für Forschung und Technologie.

REFERENCES

1   J.J. Reilly, M. Suenaga, J.R. Johnson, P. Thompson, A.R. Moodenbaugh, Phys. Rev. B36 (1987) 5694.
2   H. Fujii, H. Kawanaka, W. Ye, S. Orimo, H. Fukuba, Jpn. J. Appl. Phys. 27 (1988) L525
3   M. Nicolas, J.N. Daou, I. Vajda, J. Burger, L. Lesueur, L. Dumoulin, Solid State Commun. 66 (1988) 1157
4   D. Fruchart, J.L. Soubeyroux, D. Tran Qui, C. Pique, C. Rillo, F. Lera, V. Orera, J. Flokstra, D.H.A. Blank, J. Less–Common Met. 157 (1990) 233
5   Ch. Niedermayer, H. Glückler, R. Simon, A. Golnik, M. Rauer, E. Recknagel, A. Weidinger, J.I. Budnick, W. Paulus, R. Schöllhorn, Phys. Rev. B40 (1989) 11386
6   T. Takabatake, W. Ye, S. Orimo, T. Tamagai, H. Fujii, Physica C 162–164(1989) 65
7   J. Fröhlingsdorf, W. Zander, B. Stritzker, Solid State Commun. 67 (1988) 965
8   W.A. Lanford, Nucl. Instr. and Meth. 149 (1988) 1
9   J.P. Biersack, L.G. Haggmark, Nucl. Instr. and Meth. 174 (1980) 257
10 J. Clayhold, N.P. Ong, Z.Z. Wang, J.M. Tarascon, P. Barboux, Phys. Rev. B39 (1989) 7324
11 H. Glückler, Ch. Niedermayer, G. Nowitzke, E. Recknagel, J. Erxmeyer, A. Weidinger, J.I. Budnick, Europhys. Lett., in press

High T$_c$ Superconductor Thin Films
L. Correra (Editor)

Interpretation of I-V characteristics measured on epitaxially grown thin films of YBa$_2$Cu$_3$O$_7$

R. Wördenweber, and M.O. Abd-El-Hamed

Forschungszentrum Jülich, Institut für Schicht- und Ionentechnik, Postfach 1913, W5170 Jülich, FRG

Abstract
    The determination of the critical current is usually based upon a voltage criterion, which is used as an indication that the critical state has been reached in the superconducting regime. Resistive measurements of the I-V characteristics are carried out on patterned thin films of epitaxially grown and c-axis oriented YBa$_2$Cu$_3$O$_7$. Bridges with widths ranging between 1 μm and 100 μm and lengths between a few μm and 7 mm are characterized. The volume pinning force is evaluated for different voltage criteria. Although the absolute values of the volume pinning force change with the value of the criteria, the field and temperature scalings are not affected. They agree with a flux-flow mechanism dominated by flux-line shear. Model calculations on the basis of a modified flux-line shear mechanism are carried out, which are in good agreement with the experimentally measured I-V characteristics and the resulting volume pinning force.

1. INTRODUCTION

    In this paper a comparison between I-V curves calculated using a model for flux motion based on a modified flux-line shear mechanism (FLS) and experimental results obtained from resistive dc-measurements of epitaxially grown YBa$_2$Cu$_3$O$_7$ (YBCO) films is given. YBCO films are known to possess very strong pinning potentials. In contrast to for instance Bi- or Tl-compounds thermally activated flux motion seems to play a minor role except for temperatures and fields close to the transition to the normal state. This might be the reason why this model, which does not account for thermal activation, can be applied to thin films of YBCO. A detailed despription of the sample preparation and characteristics is given in Ref.1.

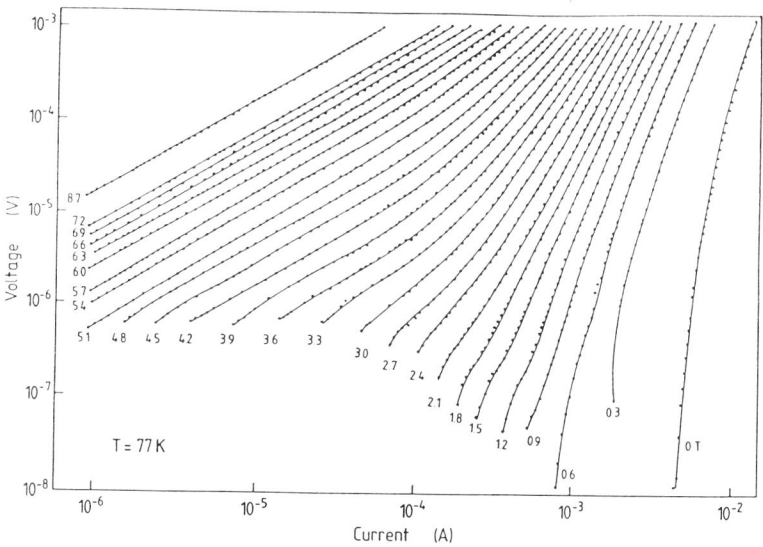

Fig.1: Double-logarithmic plot of a set of I-V curves of an epitaxially grown YBa$_2$Cu$_3$O$_7$ film on MgO. The resistive measurements are executed on a bridge of a width of 5μm.

2. EXPERIMENTAL RESULTS

Fig. 1 represents a typical set of I-V curves obtained for our epitaxial thin YBa$_2$Cu$_3$O$_7$-films with the crystallographic c-axis parallel to the direction of the applied magnetic induction, B. The double logarithmic plot shows the power law dependence of the voltage on the applied current. Similar features have been reported in the literature /2/. At large currents flux flow should become dominant, resulting in a linear dependence of V on I. This regime is not measured due to heating effects at such large dissipations, which might destroy the bridge. At low voltages thermal activation might become dominant. An exact determination of small voltages signal renders more difficulties. Small offsets in the dc-voltage are difficult to eliminate in the dc-measuring technique.

An I-V curve can intercept a voltage criterion, which is commonly used to define the depinning current, I_c. Fig. 2 represents the resulting values for the pinning force, F_p, and the temperature dependence of C(T) (see Eq.(3)) obtained for different voltage criteria, V_c, ranging from 10^{-5} to 10^{-8} V. The high field behaviour of F_p is usually a good indication for the pinning mechanism (e.g. summation of the elementary pinning interactions) in type II superconductors. In contrast to the linear decrease of $F_p \alpha$ (1-b) close to B_{c2}, for instance observed for NbTi, a number of conventional superconductors, e.g. NbN, exhibit a decrease $F_p \alpha$ (1-b)^2, where b = B/B$_{c2}$.

This behaviour is generally ascribed to the flux-line shear /3/. In the flux-line sheer (FLS) model the onset of dissipation is defined by flux motion along weakly pinning regions that form channels across the sample. Once the Lorentz force exceeds the flow stress of the flux-line lattice (FLL) at the channel edge the flux lines (FL) in the channels start to move

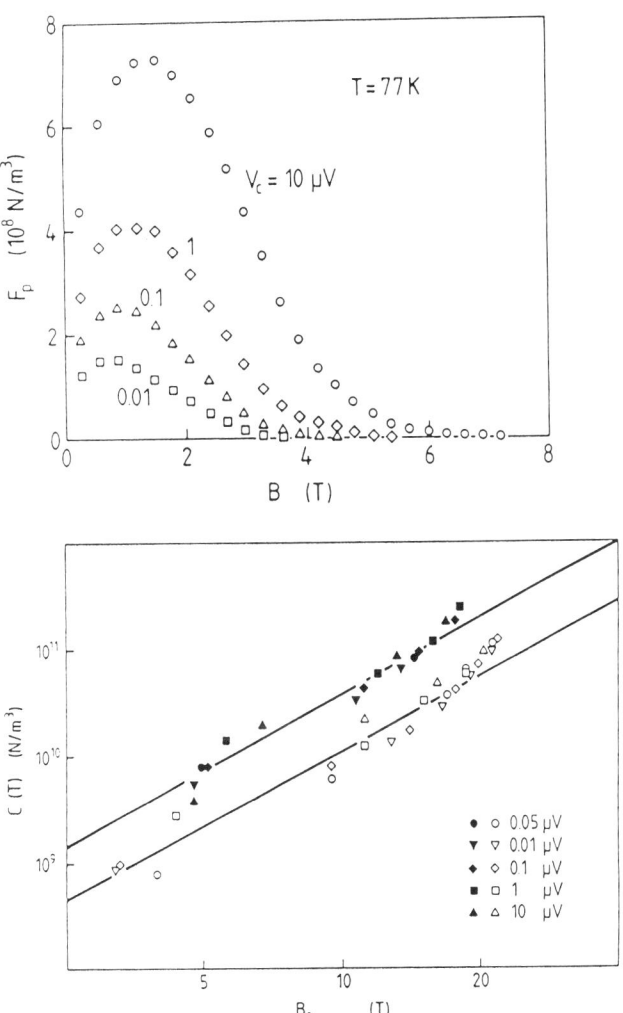

Fig. 2: In a: Plot of the volume pinning force vs. applied magnetic induction obtained of various voltage criteria, V_c. In b: Double-logarithmic plot of $C(T)$ as a function of irreversibility field, B_0, for a 7mm long bridge of $YBa_2Cu_3O_7$ on MgO (open symbols) and a 2mm long bridge of $YBa_2Cu_3O_7$ on $SrTiO_3$ (solid symbols).

and cause dissipation. Therefore, the driving force, which is necessary to originate this motion, is determined by the elastic properties of the FLL. According to the FLS mechanism, the volume pinning force for high-κ material is given by /4/

$$F_p = [\, 0.094 \ G \ / \ 4 \ \mu_o \,] \ (B_c{}^2/w) \ b \ (1-b)^2 \ (1-0.29b) \ . \qquad (1)$$

The prefactor G describes the enhancement factor due to the orientation of the channels with respect to the Lorentz force, w is the effective width of the channels and B_c denotes the thermodynamical critical field.

One of the most crucial parameters is the effective channel width, w. On one hand it is defined by the morphology of the sample. E.g. weak links or areas with reduced critical properties, which form a percolative path across the sample, can form such channels. The geometrical extensions of these channels specify the width, w. On the other hand, the FLL has to be able to adopt the geometry and size of the channel. Therefore, w also depends on the FL-spacing /4,5/ and the elastic properties of the FLL. Finally, the measurements reveal a voltage V, which represents the sum of the contributions of all channels between the voltage contacts. Therefore, a field and temperature scaling according to

$$F_p \ \ \alpha \ \ b^p(1-b)^2 \ (1-0.29b) \qquad\qquad (2)$$

and

$$C(T) = [0.094 \ G \ / \ 4 \ \mu_o] \ (B_c{}^2 \ / \ w) \ \alpha \ B_{c2}{}^q \ \alpha \ B_o{}^q \qquad (3)$$

is expected, where $1 \leq p \leq 1.5$ and $2 \leq q \leq 2.5$.

The above scaling laws are able to explain the experimental data obtained for epitaxially grown thin films of YBCO up to some extend /6/. However, deviations are observed (see Fig. 2) and discussed in the literature. E.g. B_{c2} has to be replaced by B_o, which represents a definition of the irreveribility line. Furthermore, it is shown, that the deviations in the field dependence increase with increasing inhomogeniety of the samples /7/. Therefore, it seems necessary to reconsider the model in order to get a better agreement between experiment and theory.

3. MODIFIED FLS MODEL

One of the most critical issues of the FLS model is the assumption, that all channels are identical. In artificial systems /4/ this might be the case. For real systems differences in the channel properties have to be expected. It has been discussed above, that different geometrical arrangements of the channels cause different scaling properties. Moreover, bearing in mind the small coherence length of YBCO, large variations of the critical field have to be expected in the channels due to different oxygen

concentrations. This effect probably dominates the variations
in the channel properties. Thus, the resulting critical
properties of the thin films can be evaluated.

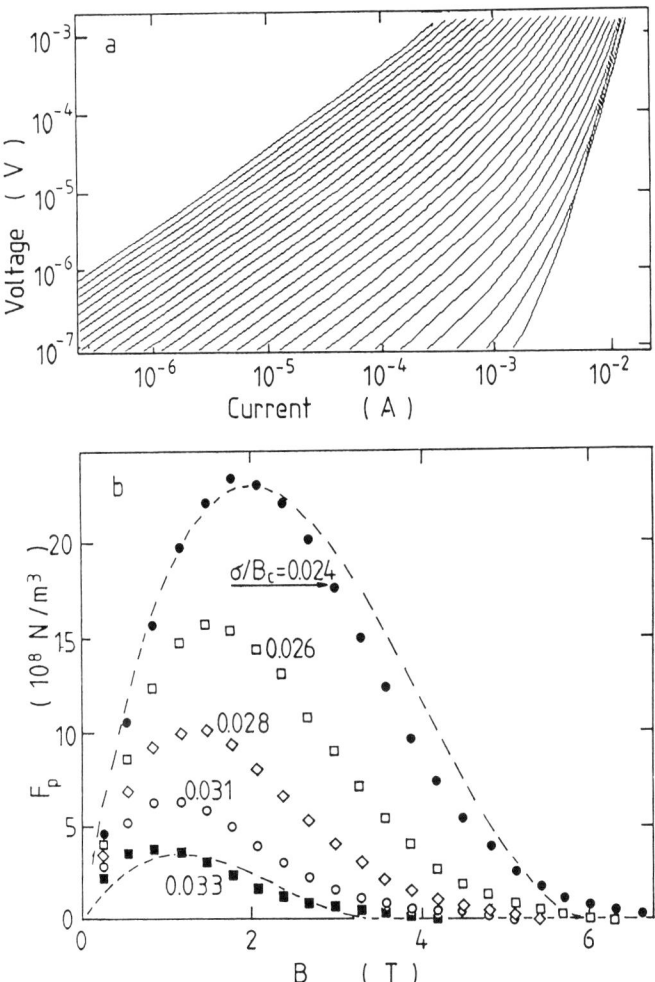

Fig. 3: FLS-model calculations. In a) set I-V curves at 77 K
with $\sigma = 0.033$ B_c. The experimental parameters of the sample
shown in Figs. 1 and 2 are adopted. The applied fields are
increased in steps of 0.3T starting from 0.3T at the right to
8.7T. In b) resulting F_p-values vs. B for the 1-μV criterion
and different σ values. The dashed lines represent typical
field dependence of the shear module, c_{66}.

Assuming a statistical distribution of the critical field in the channels around $B_{c,mid}$ (as obtained from the measured transition curves), the expected contributions of the channels to the electric field can be calculated according to

$$E_{channel} = (J - J_{c,channel}) \, \rho_n \, [B/B_{c2}(0K)] \qquad \text{for } J > J_{c,channel}$$

and

$$E_{channel} = 0 \qquad \text{for } J < J_{c,channel}.$$

ρ_n is the normal sate resistance. Using the empirical relation of Kim et al. /8/ for the flux-flow resistanceis, $J_{c,channel}$ can be obtained from Eq.(1) inserting a value for the upper critical field in the channel. Assuming that only single rows of flux lines are moving, and that the channel orientation is statistical, i.e. G=2, all parameters can be obtained from the experiment, except for the width of the field distribution, σ. Fig. 3 shows the evaluation for different σ values. In a) the I-V curves are given. The model calculations of the I-V curves show the same features as the experimental values in Fig.1. The deviations in the field dependence can also be explained by the model. Fig. 3b represents the resulting volume pinning force together with the field dependence expected from the FLS model. The same deviations are observed. In conclusion, the modified FLS model can explain the experimentally observed I-V characteristics in the strong pinning system YBCO.

We like to acknowlege the helpful discussions with C. Heiden and A. Braginski, and the technical support of R. Kutzner, U. Krüger, and J. Schneider.

4. REFERENCES

1. A. Höhler, H. Neeb, and C. Heiden, J. of the Less-Comm. Met., 151, 341 (1989); A. Höhler, D. Guggi, H. Neeb, and C. Heiden, Appl. Phys. Lett., 54 (1989) 1066, R. Wördenweber, J. Schneider, T. Göddenhenrich, and U. Krüger, this issue.
2. see e.g.: R.H. Koch V. Foglietti, W.J. Gallagher, G. Koren, A. Gupta, and M.P.A. Fisher, Phys. Rev. Lett., 63 (1989) 1511.
3. E.J. Kramer, J. Appl. Phys., 44 (1976) 1360.
4. A. Pruymboom, P.H. Kes, E. van der Drift, and S. Radelaar, Appl. Phys. Lett., 52 (1988) 662; Phys.Rev.Lett., 60 (1988) 662
5. R. Wördenweber, J. Schneider, M.O. Abd-El-Hamed, R. Lehmann, and D. Guggi, in Proceedings of the ICMC '90, Garmisch-Partenkirchen (W.-Germany), 1990.
6. see e.g.: C. Schlenker, C.J. Liu, R. Buder, J. Schubert, and B. Stritzker, in Proceedings of "Int. Workshop om HTSC Thin Films, Properties and Applications", 4.91 in Rome.
7. Publication in preparation.
8. see e.g. A.R. Strnad, C.F. Hempstead, and Y.B. Kim, Phys. Rev. Lett., 13 (1964) 794;ibidem, Phys. Rev., 139 (1965) A1163.

High T$_c$ Superconductor Thin Films
L. Correra (Editor)
1992 Elsevier Science Publishers B.V.

Magnetic properties of superconducting YBa2Cu3O7-x CVD thin films

J. Fick[a], E. Mossang[a], O. Thomas[a], F. Weiss[a], D. Boursier[a], R. Madar[a],
J.P. Senateur[a], S.K. Agarwal[b], C. Schlenker[b]

[a]Laboratoire des Matériaux et du Génie Physique, URA CNRS 1109, ENSPG,
BP 46, 38402 St Martin d'Hères Cedex, France

[b]Laboratoire d'Etudes des Propriétés Electroniques des Solides, CNRS, BP 166X,
38042 Grenoble Cedex, France

Abstract

Thin films of YBa2Cu3O7-x have been grown by chemical vapor deposition at
825 °C (5 Torr). The starting materials are tetramethylheptanedionates of Y, Ba and
Cu. The superconducting properties of the films obtained on monocrystalline MgO
(100) and SrTiO3 (100) have been studied by AC resistivity, AC susceptibility and
magnetization measurements. Superconducting layers with zero resistance at 85 K
and a transition width lower than 2 K are routinely obtained on MgO. On SrTiO3 the
critical temperature is 90 K and the transition width is less than 1 K.The morphology
of the layers has been investigated by scanning electron microscopy; their structure
has been determined via x-ray diffraction. The films are highly orientated with their
c-axis perpendicular to the substrate plane, and the pole figures show a very good
in-plane orientation. The zero field critical currents as deduced from the remanent
magnetization or from the screening properties of the films are a strong function of
the nature of the substrate: J_c (10 K) = 1.2 10^7 A cm^{-2} on SrTiO3 and J_c (10 K) =
1.5 10^6 A cm^{-2} on MgO. Pinning potentials U_0 as calculated from the thermal
dependance of remanent magnetization come out to be in the range of 60 to 70
meV.

1. Introduction

Since the discovery of superconducting oxides, an enormous amount of
research has been conducted on the preparation of high-T_c superconducting films.
The great majority of these films are made by physical vapor deposition techniques.
Chemical Vapor Deposition (CVD) is also a very attractive deposition method. The
possibility of conformal coverage and the ability of deposition under a high oxygen
partial pressure without post deposition annealing are only some of the advantages
of CVD.
The most commonly quoted parameters indicating the quality of a superconducting
material are the transition temperature and the critical current density. These
parameters have been determined by ac and dc measurement techniques. In this

paper we report and compare the results of these experiments performed on thin films prepared by CVD.

2. Experimental Procedures

Since the details of the experimental set up are given elsewhere [1], we will give only a short description. The precursors, tetramethylheptanedionates (tmhd) of yttrium, barium and copper are evaporated in separated furnaces, mixed and led into the reactor with argon. Oxygen is injected into the reactor with a separate gas line. The substrate temperature is 825° C and the total gas pressure is fixed at 5 torr. For a deposition time of 30 min the thickness of the films is in the range of 2000-3000 Å. After the deposition the films are immediately cooled down to room temperature under 1 atm. of pure oxygen. (100) orientated MgO, $SrTiO_3$ and Y-stabilized ZrO_2 single crystals are used as substrates.

The resistivity of the samples is measured by a standard four-point technique with an ac-current of 0.1 mA at a frequency of 8 Hz.

The screening properties as a function of temperature down to 20 K are determined with a home-built apparatus. The sample is placed between a driving and a pick-up coil. Magnetic fields between 20 mOe and 16 Oe at a frequency of 800 Hz are used [2].

The magnetization is measured with a vibrating sample magnetometer between 10 K and 100 K in fields up to 6 T. The sensitivity of the apparatus is 10^{-6} emu. This comparatively high sensitivity is obtained with pick-up coils immerged into liquid helium and located close to the sample [3]. The results reported here were obtained with the applied field perpendicular to the film.

X-ray diffraction spectra were obtained in the Bragg-Brentano geometry. For some films on MgO and $SrTiO_3$ precession photographs were taken, using a Buerger precession camera. The crystalline orientation was also investigated with a pole figure goniometer in the Schultz geometry.

3. Results and Discussion

The x-ray spectra show a nearly perfect c-axis orientation. The pole figure [4] of a film on MgO and the precession photographs of films on MgO and $SrTiO_3$ show a very strong in-plane orientation. The <100> or <010> directions of these films are parallel to <100> of the substrates, as evidenced by the precession photographs.

The resistivity versus temperature curve (figure 1) shows a sharp transition at T_C for all the samples and a metallic behavior in the normal state. The film resistivity at 100 K is in the range of 300 $\mu\Omega$ cm on $SrTiO_3$ and 600 $\mu\Omega$ cm on MgO. The values of T_{cr} (offset temperature) and ΔT_{cr} are reported in table 1 with a 10 - 90 % criterion.

The screening behavior of the films is shown in figure 1. The in-phase (s') and the 90° out-of-phase signal (s") are recorded. At T_{cs} the in-phase signal drops down to zero. At the same temperature the out-of-phase signal shows a very sharp peak (where ΔT_{cs} is the f.w.h.m. of the peak).

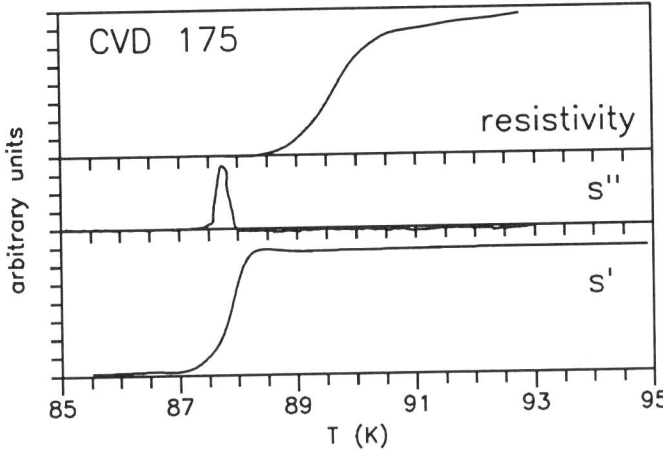

Figure 1. Resistivity, in-phase (s') and 90° out-of-phase (s'') signal from the screening measurement for the film CVD 175 on SrTiO₃.

The peak doesn't correspond directly to the critical temperature T_{cr}, determined by resistivity measurements. T_{cs} lies typically 2 - 5 K lower than T_{cr} (table 1). One can say, that this peak corresponds to the transition between almost reversible flux flow above T_{cs} and strong irreversible screening caused by flux pinning below T_{cs}. So we can determine three regions in figure 1: the normal state above T_{cr}, the thermally assisted flux flow (TAFF) [5] state between T_{cr} and T_{cs}, and the flux creep state below T_{cs}. The sharp transition and the absence of step-like anomalies in s' indicate a homogenous superconducting film. Similar results were obtained by Guilloux-Viry et. al. on sputtered films [6].

Table 1. Critical temperatures and transition withs determined using resistance (T_C), screening (T_{cs}) and magnetization (T_{cm}) measurements.

	T_{cr}	ΔT_{cr}	T_{cs}	ΔT_{cs}	T_{cm}
CVD167 / MgO	83.0 K	2.0 K	79.2 K	1.1 K	82 K
CVD187 / MgO	81.9 K	1.9 K	76.8 K	1.3 K	80 K
CVD175 / SrTiO₃	89.0 K	0.9 K	87.9 K	0.4 K	90 K
CVD181 / SrTiO₃	89.4 K	1.0 K	86.8 K	0.9 K	90 K

The transition temperature T_{cm} is determined by the measurement of the magnetization M as a function of temperature. After a zero field cooling (ZFC) the samples are heated under a field of 10 Oe. T_{cm} corresponds to the onset of diamagnetism. The values of T_{cm} and T_{cr} are comparable, so we can say that they correspond both to the transition between the normal and the superconducting state.

The critical current density J_{CS} is derived from screening measurements. The experimental geometry is comparable to a normal transformer where the sample corresponds to a third winding. In this model, we assume in a first approximation that the screening current density J is homogeneous over the whole sample. This leads to the relation:

$$J_{CS} = \frac{f_{corr} I_{coil}}{r\ d}$$

(f_{corr}: correction factor; I_{coil}: current in the driving coil; r, d: radius and thickness of the sample). With our apparatus J_{CS} measurements are only possible in a narrow temperature range near T_{CS}. These measurements, however, are very sensitive and non destructive; the shape of the curves allows to compare the superconducting properties of the films.

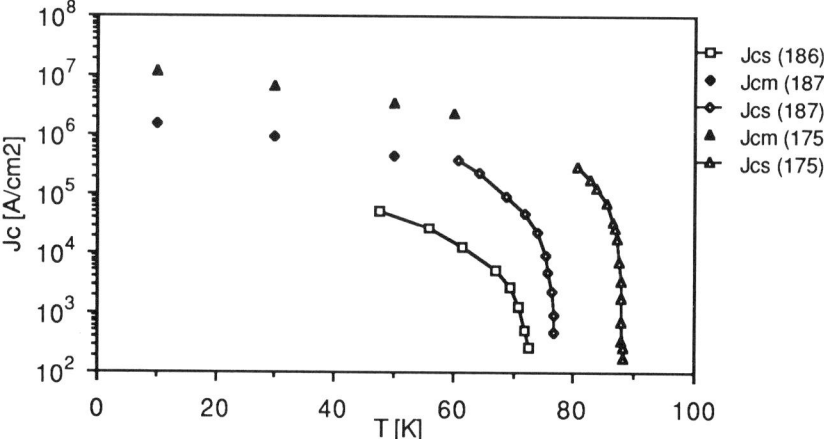

Figure 2. Critical current density for films on ZrO_2 (CVD186), MgO (CVD187) and $SrTiO_3$ (CVD175).

The critical current density J_{cm} at zero field was also calculated for several temperatures from magnetization measurements using the Bean formula [9]:

$$J_{cm} = 30\ \frac{M_r}{r}$$

(M_r: remanent magnetization after a field sweep with $H_{max} \gg H_{c1}$; r: radius of the sample). In figure 2 the results for these two different measurements are plotted. The correction factor f_{corr} has been chosen in order for the J_{CS} values to scale with the J$_{cm}$ values. The absolute values derived from the two experimental techniques may, however, differ. This is due to the two different experimental procedures, the different critera to determine Jc and the different field and current distribution in the two experiments.

The critical current is a strong function of the nature of the substrate. The zero field values determined from the magnetization measurement are:

J$_{cm}$ (10 K) = 1.2 10^7 A cm^{-2} on SrTiO$_3$ and J$_{cm}$ (10 K) = 1.5 10^6 A cm^{-2} on MgO. An extrapolation of J$_{cm}$ gives values comparable to results obtained by other groups using transport measurement techniques for thin films deposited with as different techniques as CVD or sputtering [7,10].

The temperature dependence of the remanent magnetization M$_r$ was also measured. After a field sweep with H$_{max}$ = 1 T at 10 K the samples were heated up to 100 K under zero field. The results are shown in figure 3 as normalized data for films on MgO and SrTiO$_3$.

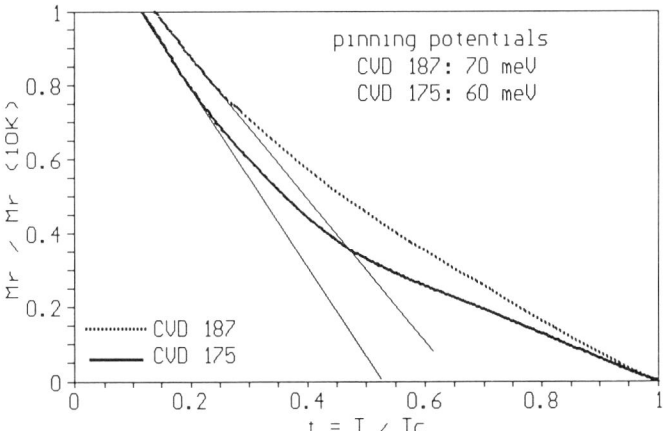

Figure 3. Remanent magnetization versus temperature for a film on SrTiO$_3$ (CVD175) and a film on MgO (CVD187).

The curves obtained for H$_{max}$ = 1 T can be described by two regimes, a low temperature one with a large slope and a high temperature one with a smaller slope. To calculate the vortex pinning potential U$_0$, we use the conventional flux-creep relation [9]:

$$M_r(T) = M_r(0)\left[1 - \frac{kT}{U_0} \ln\left(\frac{\omega_0}{\omega}\right)\right]$$

where ω_0 is a frequency characteristic of vortex vibration in the potential wells due to pinning centers and ω a frequency related to the time scale of the measurement.

ω_0 is usually assumed to be in the range 10^5 - 10^{11} sec^{-1} and $\omega \approx 1$ sec^{-1} in our case. Taking $\ln(\omega_0/\omega) = 20$, we obtain values of vortex pinning energies for our films of $U_0 = 70$ meV (on MgO) and $U_0 = 60$ meV (on SrTiO$_3$).
Similar values of U_0 for thin films are found by other groups [3,10-13]. Measurements of U_0 for YBa$_2$Cu$_3$O$_7$ single crystals give values in the range of 20 - 30 meV. If we consider the linear relation $J_cB = U_0/aV$ (a: size of the potential well; V: activation volume) between the critical current density and the pinning potential, the difference in the pinning potential for single crystals and thin films could be one explanation for the higher critical current density in the films.

4. Conclusion

We have performed magnetic and resistivity measurements on YBa$_2$Cu$_3$O$_7$ superconducting thin films prepared by CVD. The high J_c values of 1.2 10^7 A cm^{-2} on SrTiO$_3$ and 1.5 10^6 A cm^{-2} on MgO at 10 K and the sharp transition in the resistivity and screening behavior ($<$ 1K) show that these films are of a good quality and comparable to films produced by other deposition techniques.

5. References

1   O. Thomas, A. Pisch, E. Mossang, F. Weiss, R. Madar, J.P. Senateur, J. Less Comm. Met. 164 & 165 (1990) 444.
2   F. Weiss, E. Senet, M. Langlet, O. Thomas, A. Pisch, R. Madar, J.C. Joubert, J.P. Senateur, J. Less Comm. Met. 164 & 165 (1990) 1393.
3   Ch.J. Liu, R. Buder, C. Schlenker, J. Schubert, W. Zander, B. Stritzer, J. Less Comm. Met. 164-165 (1990) 1285.
4   O. Thomas, E. Mossang, J. Fick, F. Weiss, R. Madar, J.P. Senateur, M. Ingold, P. Germi, M. Pernet, F. Labrize, L. Hubert, presented at Int. Workshop on HTCS Thin Films Rome 1991, to be published in Physica C.
5   P.H. Kes, J. Aarts, J. van der Berg, C.J. van der Beek, J.A. Mydosh, Sup. Sci. Tech. 1 (1989) 242.
6   M. Guilloux-Viry, M.G. Karkut, A. Perrin, O. Peña, J. Padiou, M. Sergent, Physica C 166 (1990) 105.
7   H. Yamane, H. Kurosuwa, T. Hirai, K. Watanabe, H. Iwasaki, N. Kobayashi, Y. Muto, Sup. Sci. Tech. 2 (1989) 115.
8   C.P. Bean, Phys. Rev. Lett.,8 (1962) 250; Rev. Mod. Phys., 36 (1964) 31.
9   P.W. Anderson and Y.B. Kim, Rev. Mod. Phys., 36 (1964) 39.
10 N. Savvides ,Physica C 165 (1990) 371.
11 J. Mannhart, P. Chaudhari, D. Dimos, C.C. Tsuei, T.R. McGuire, Phys Rev. Lett. 61 (1988) 2476.
12 K. Enpuku, T. Kisu, R. Sako, K. Yoshida, M. Takeo, K. Yamafuji, Jpn. J. Appl. Phys. 28 (1989) L991.
13 J.W.C. de Vries, G.M. Stollman, M.A.M. Gijs, Physica C 157 (1989) 406.

High T$_c$ Superconductor Thin Films
L. Correra (Editor)
1992 Elsevier Science Publishers B.V.

CORRELATION BETWEEN MICROSTRUCTURE AND PROPERTIES OF HIGH
Tc FILMS: APPLICATION TO MICROWAVE PROPERTIES

V.Z. Kresin[a] S.A. Wolf[b], G. Deutscher[c]

[a]Lawrence Berkeley Laboratory, University of California, Berkeley CA
94720

[b]Naval Research Laboratory, Washington, DC   20375-5000

[c]Department of Physics and Astronomy, Tel Aviv University, Ramat
Aviv, Israel

ABSTRACT

The Y-Ba-Cu-O compound contains two conductive subsystems and,
as a result, displays a two gap structure. Oxygen ordering affects the
value of the smaller gap. Charge transfer leads to the induced
superconducting state in the chains. The short coherence length,
anisotropy and the presence of different structural units lead to
peculiar microwave properties of high Tc films. In this paper we are
going to describe our approach in theory of high Tc and then focus on
their microwave properties.

In our previous papers [1,2] we described the Y-Ba-Cu-O compound
as containing two conductive subsystems planes (α) and chains (β) and
correspondingly, having two energy gaps. Recent data provides
convincing experimental support for this concept. NMR data [3], the
observation of the plane part of the Fermi surface by positron-
annihilation [4], thermopower data [5], and measurements of the

penetration depth [6] show that the chains, besides providing carrier doping, represent a conductive metallic subsystem.

The charge transfer between these subsystems can be provided by 1) transitions accompanied by radiation (absorption) of phonons; this channel is similar to that in the two-band model [1,2] and 2) intrinsic proximity effect; this channel can be studied by analogy with the well known McMillan tunneling model [7]. The total Hamiltonian can be written in the form:

$$
H = H_o^\alpha + H_o^\beta + \sum_{\alpha\beta\gamma} g_{\alpha\beta\gamma} \, a_\alpha^+ \, a_\beta \, b_\gamma + \sum_{\alpha\beta} T_{\alpha\beta} \, a_\alpha^+ \, a_\beta
$$

(1)

The first two terms describe the isolated α and β subsystems, including the intrinsic pairing interaction; for example H_o^α contains the term

$$
\sum_{\alpha\alpha'\gamma} \lambda_\alpha \, a_\alpha^+ \, a_{\alpha'} \, b_\gamma
$$

where λ is an electron-phonon coupling parameter. The last two terms describe the charge transfer.

Let us study the superconducting state of the system. In our model the chains form an intrinsically normal metallic subsystem; in other words $\lambda_\beta = 0$. This property has been discussed by us in [2]. Of course it does not mean that the chains are not in the superconducting state, but this state is induced by the phonon mediated channel and by the intrinsic proximity effect.

The equations for the pairing order parameters Δ^α and Δ^β have the form at $T = T_C$:

$$
\Delta_n^\alpha \, Z_\alpha = \lambda_\alpha \sum_{n'} (D_{nn'} - \tilde{\mu}) \frac{\Delta_{n'}^\alpha}{|2n' + 1|} + \lambda_{\alpha\beta} \sum_{n'} D_{nn'} \frac{\Delta_{n'}^\beta}{|2n' + 1|} + \gamma_{\alpha\beta} \frac{\Delta_n^\beta}{|2n + 1|}
$$

(3)

$$
\Delta_n^\beta \, Z_\beta = \lambda_{\beta\alpha} \sum_{n'} D_{nn'} \frac{\Delta_{n'}^\alpha}{|2n' + 1|} + \gamma_{\beta\alpha} \frac{\Delta_n^\alpha}{|2n + 1|}
$$

(4)

We are using the method of the thermodynamic Green's function (cf [8]). In Eqs. (3), (4) $D_{nn'} = \Omega^2 [\Omega^2 + (\omega_n - \omega_{n'})^2]^{-1}$ is the phonon Green's function, Ω is the average phonon frequency, $\omega_n = (2n + 1)\pi T$; λ_α, $\lambda_{\alpha\beta}$ and $\lambda_{\beta\alpha}$ are electron-phonon coupling constants (see e.g. [2]) $\gamma_{\alpha\beta}$

$= \Gamma_{\alpha\beta}/\pi T_C$, where $\Gamma_{\alpha\beta}$ is the McMillan proximity parameter. Note also that $\lambda_{\alpha\beta}(\gamma_{\alpha\beta}) = \lambda_{\beta\alpha}(\gamma_{\beta\alpha})\nu_{\alpha}/\nu_{\beta}$; $\nu_{\alpha}, \nu_{\beta}$ are the densities of states in the two bands respectively, $\tilde{\mu}$ describes the Coulomb screening and contribution of high energy modes.

The first term in Eq. (3) corresponds to the usual Eliashberg equation and to the intrinsic pairing in the Cu-O plane. The second term describes the phonon mediated $\beta \to \alpha$ charge transfer and the last term represent the intrinsic proximity effect. The terms in Eq. (4) have a similar meaning. We think that a major contribution to the pairing comes from the soft optical modes (for Y-Ba-Cu-O, $\Omega_{opt.} \cong 300K$); this view is supported by recent experimental data on tunneling [9].

In order to calculate T_C one should exclude the function Δ_n^{β} from the system (3) - (4). It is convenient to use the matrix method developed in [10]. As a result, we arrive at the following expression for T_C(cf. [11])

$$T_c \cong (2\pi)^{-1}(\Lambda_{eff})^{1/2}\Omega \tag{5}$$

where

$$\Lambda_{eff} = \frac{\lambda_{\alpha} + 2\lambda_{\alpha\beta}\gamma_{\beta\alpha}}{1 + \gamma_{\alpha\beta}(1+\gamma_{\beta\alpha})^{-1} + 2\tilde{\mu}} \tag{6}$$

Let us study Λ_{eff}. If we neglect the phonon mediated channel that is we put $\lambda_{\alpha\beta}=0$ then T_C will be depressed by the proximity effect (factor proportional to $\gamma_{\alpha\beta}$ in the denominator). On the other hand the interference channel proportional to $(\lambda_{\alpha\beta}\,\gamma_{\beta\alpha})$ is favorable for superconductivity. This means that the intrinsic proximity effect combined with inelastic charge transfer can be a positive factor which moves T_C up.

Therefore, there are two competitive channels. This peculiar property is an important factor which determines behavior of T_C as a function of the oxygen ordering (see below)

It is important to note that Eq. (5) has a form similar to that obtained by Allen and Dynes in Ref. [12] (see also [11]); this is not surprising because of the large value of the ratio $(2\pi T_C/\Omega)^2$ (as was noted a more detailed evaluation, which includes higher order correction will be described elsewhere). However, Eq. (5), represents the equation which allows us to determine T_C because the right hand side depends on T_C (remember that $\gamma_{\alpha\beta}=\Gamma_{\alpha\beta}/\pi T_C$).

Let us turn to the evaluation of the induced energy gap $\varepsilon_\beta=\varepsilon_\beta(0)$. It can be evaluated from an equation, similar to Eq. (3), but at T=0K, the energy gap can be determined as a root of the equation $\omega=\Delta_r^\beta(-i\omega)$ where Δ_r is an analytical continuation of the thermodynamic order parameter. In this case the energy gap ε_β can be induced by both, proximity and phonon-mediated, channels. Unlike the behavior of T_C, both channels induce the energy gap ε_β and there is no cancellation effect. One can prove based on the strong coupling theory [13], that the major contribution to ε_β (it $\varepsilon_\beta<<\varepsilon_\alpha$) comes from the proximity term, and we obtain ([7]).

$$\varepsilon_\beta \cong \Gamma_{\beta\alpha} = \pi \, \gamma_{\beta\alpha} T_C \qquad\qquad (7)$$

The material has a two gap structure. It means that the electronic density of states has two peaks. It is important that, because of the short coherence length ($\xi_0 < \ell$, ℓ is a mean free path), averaging [14] does not occur and unlike the case of conventional superconductors, one can observe the two-gap structure (in the presence of two conductive subsystems). The two gap structure in Y-Ba-Cu-O has been observed experimentally in [3, 6, 15]. In particular, interesting data has been described in [6], where the analysis of the penetration depth allows one to determine the value of ε_β.

The oxygen ordering affects the value of the coupling constants and the proximity parameter $\gamma_{\beta\alpha}$. The decrease in the oxygen content leads to a decrease in the values of these parameters for a number of reasons (e.g., a decrease in the delocalization of the chain electronic wavefunction, an increase in the c-lattice parameter and therefore a decrease in the overlap of the wave functions). However, one can see directly from Eqs. (5) and (6) that the decrease in $\gamma_{\beta\alpha}$ and $\gamma_{\alpha\beta}$ affects the value of the induced gap more drastically than it does the value of T_C. T_C changes slowly because of the interplay of the different channels.

Let us take an example where we have chosen reasonable parameters for $YBa_2Cu_3O_7$. The smallest energy gap $\varepsilon_\beta \cong 110K$ [6]. It means (see Eq. (6)) that $\Gamma_{\beta\alpha} \cong 110K$. According to band structure calculations [16], $v_\beta/v_\alpha \cong 2$; therefore $\Gamma_{\alpha\beta} \cong 220$ K. We also assume that

$\lambda_{\beta\alpha} \cong 0.5$ (this corresponds to an estimation obtained in [17]) and $\lambda_\alpha \cong 3$. The presence of a layered structure leads to the presence of an acoustic plasmon mode which provides an additional attraction. The estimation of this interaction leads to the value $\tilde{\mu} \cong -0.2$ [18]. As a result, we obtain from Eqs. (5), (6) the value $T_C = 86K$. Of course, there is some uncertainty in the above values. We would like to emphasize that we are trying to illustrate our general concept of charge transfer with two channels and its effect on the superconducting properties.

Note that if we neglect the inelastic processes ($\lambda_{\beta\alpha} = 0$) then T_C will be depressed by the pure proximity effect and would be equal to 74 K. Therefore, the interference channel plays an important role.

A decrease in the oxygen content leads to both a decrease in the metalization of the chain and an increase in the c-lattice parameter. As a result the overlap of the plane's and chain's wavefunctions gets smaller, and this leads to a decrease of the parameters $\Gamma_{\alpha\beta}$, $\lambda_{\beta\alpha}$, etc. If for example, $\Gamma_{\beta\alpha} \cong 80K$ ($\varepsilon_\beta \cong 80$ K), $\lambda_{\beta\alpha} = 0.3$, we find $T_C = 85K$. If $\Gamma_{\beta\alpha} \cong 60K$ ($\varepsilon_\beta \cong 60$ K), $\lambda_{\beta\alpha} = 0.1$, then $T_C \cong 85$ K and $2\varepsilon_\beta/T_C \cong 1$. Therefore, indeed, we are dealing with a drastic change in the value of the smaller energy gap, whereas T_C changes slowly. Of course, the change in the oxygen content eventually effects the doping and leads to a decrease of λ_α [19] and T_C. For the last example if $\lambda_\alpha \cong 2$ instead of 3 then $T_C \cong 70$ K, but here we mainly focus on the strong effect of charge transfer on the induced energy gap.

There have been several experiments that have demonstrated that there is anisotropic transport in the a-b plane of Y-Ba-Cu-O. Resistivity measurements [20] on untwinned single crystals have shown that the conductivity of the chains exceeds that of the planes. Optical reflectivity data [21] are also consistent with this fact and also clearly show that a large gap is associated with the planes. Finally, the strongly negative thermopower data [5] in samples with the nearly ideal oxygen stoichiometry of 7 clearly show that the transport of entropy in these samples is dominated by the chains. These results are consistent with our model which has assigned a vanishingly small value of the electron-phonon coupling constant to the chains themselves (λ_β) and a small value (approximately 0.5) for the electron-phonon coupling parameter $\lambda_{\beta\alpha}$. In fact the data [17] which has been used as evidence for weak coupling, by virtue of the fact that the slope of the high temperature resistivity is consistent with a small value of λ, is easily explained in the context of our model since the resistivity is dominated by the more highly conducting chains which in effect short out the planes which have lower conductivity and stronger electron-phonon scattering.

The concept of charge transfer we described also allows us to explain a number of experiments. Recent measurements of the penetration depth [6], carried out by one of us (G. Deutscher) and his colleagues show that the oxygen disorder leads to a noticeable decrease of the smaller gap, whereas T_c is almost constant. This behavior is totally consistent with the theory presented here. The increase in the residual microwave losses with decrease of the oxygen content, as well as in the existence of a zero-bias anomaly, is connected with the behavior of the induced gap.

The intrinsic proximity effect has been also studied in [22]; however, the authors introduce the pure proximity effect channel only. Then the conductivity of the chains would only depress T_c. The rigorous introduction of the inelastic channel (see above) leads to an entirely different picture.

We have focused here on the Y-Ba-Cu-O compound. We think that the concept described above is also applicable to other high T_c cuprates (Bi and Tl based compounds; e.g, charge transfer between Cu-O and Bi-O or Tl-O planes).

MICROWAVE PROPERTIES

Let us discuss the microwave properties of high Tc films. This problem is of definite current interest because it is related both to the basic properties of the materials and to various applications. At present there exist a number of experimental data on the temperature and frequency dependencies of surface, resistance. The usual Mattis Bardeen approach [23], which is successful in the analysis of conventional superconductors is strictly not applicable to the cuprates. The cuprates are characterized by short coherence lengths much smaller than the penetration depth δ. Whereas the opposite is true in conventional superconductors. Moreover the cuprates are highly anisotropic systems, and in addition some of them (e.g., Y-Ba-Cu-O) have a two gap structure. These crucial factors need to be taken into account in any theory. The response of high T_c films to an AC field for the one gap case La-Sr-Cu-O system, has been studied by one of the authors in [24].

Consider a layered superconductor. We shall study the situation when the layers (ab-plane) are perpendicular to the boundary; a c axis oriented film has already been studied by Chaub and Scalopino[25]..

Assume that the magnetic component of the external field is perpendicular to the layers; the vector potential $\vec{A}(\vec{r}, t)$ and, consequently, the electric field is parallel to the ab-plane. In the first approximation, we neglect interlayer hopping.

We consider frequencies much smaller than 2Δ; in this case the incident radiation does not create quasiparticle excitations. The current is due only to the existing thermal excitations. As a result we obtain the following expression for the conductivity.

$$\sigma = i\,(c/\omega)\,Q;\; Q = Q_1 + i\,Q_2$$

$$Q_1 = n_s e^2/mc$$

$$Q_2 = (e^2 P_F{}^2/\pi cm)\,L^{-1}\,(\frac{\Delta}{T})\,\frac{\omega}{\kappa V_F}\,\ell n\,\frac{\omega}{\kappa V_F} \tag{8}$$

where n_S is the carrier concentration, k is the in-plane momentum and L is the interlayer distance.

The impedance defined by the relation

$$Z = -\,(4\pi i\omega/c^2)\,A(0)/H(0)$$

can be evaluated from the expression:

$$Z = -\,i(8\omega/c^2)\,\int d\kappa\,[\kappa^2 + (\frac{4\pi}{C})Q]^{-1}$$

As a result the term Z_1 which describes the losses has the form:

$$Z_1 \propto \gamma\,(\frac{\omega}{\Delta})^2\,(\frac{\Delta}{T})^2\,\cosh^{-2}\,(\frac{\Delta}{T}); \gamma = (\frac{\delta}{\xi_o})^2 \tag{9}$$

The dependence of the losses on different parameters is described by Eq. (9). First we would like to stress the presence of the large term γ. The large value of this term in the cuprates is due to the short coherence length in these materials.   The presence of this term explains the high surface resistance observed experimentally   Indeed the surface resistance of the oxides is about five to six orders of magnitude larger than in conventional superconductors, such as Nb_3Sn, with similar values of T_c/T (or Δ/T, see Eq. (2)).

The frequency dependence of the losses appears to be similar to that in the usual superconductors ($\alpha\ \omega^2$). In the region $T << \Delta$ the impedance   decreases   exponentially   with   temperature.

Eq. (9) describes losses in pure layered cuprates where the mean free path is larger than the penetration depth.   The presence of impurities does not affect the temperature or frequency dependences of the impedance, but the parameter γ should be replaced by a function of δ / ℓ ; so that $\gamma \to \tilde{\gamma} = \gamma\, f\, (\delta / \ell)$; $f(x) = x[(1+x^{-2})^{-1/2} -1]$. If $\delta << \ell$ then $f = 1$ and we arrive at Eq. (9). In the opposite case of $\delta >> \ell$ we have $\gamma = \delta\ 1/2$.

It is interesting   to note that a superconductor is capable of absorbing energy even in the absence of impurities.   Landau damping is the mechanism of such losses.

In the presence of a two gap structure the impendence represents a sum of two terms (in the low temperature region we have a sum of two exponents).   As a result, the surface resistance decreases more slowly with temperature than in the one gap case.   Such is the experimental situation for Y-Ba-Cu-O [26].

Oxygen disorder leads to decrease of the smaller gap.   As a result situation when $\varepsilon_\alpha > h\omega > \varepsilon_\beta$ is perfectly realsitic.   Then one should observe residual losses due to absorption by rains.   Therefore, there is a direct correlation between structure and the effect of residual losses. The ordering leads to decreases in the losses.   This effect has been observed experimentally in Ref. [27].

In summary, we have developed a new concept for charge transfer in the high Tc Cuprates.   This transfer leads to peculiar two-gap superconductivity.   Surface resistance and the, residual of losses are greatly affected by the two-gap structure.

This research of VZK is supported by the Office of Naval Research under Contract No. N00014-89-F0006 and carried out at the Lawrence Berkeley Laboratory under Contract No. DE-AC03-76S-F00098.

REFERENCES

1. V. Kresin and S. Wolf, Solid State Comm. 63, 1141 (1987); J. of Superconductivity 1, 143 (1988); V. Kresin, Solid State Comm. 63, 725 (1987)
2. V. Kresin and S. Wolf, Phys Rev. B 41, 4278 (1990); Physica C 169, 476 (1990)
3. S. Barrett et al. Phys. Rev. B 41, 6283 (1990)
4. H. Hajhighi et al. Bull Amer. Phys. Soc., 36, 376 (1991); Preprint
5. J. Cohn et al., Phys. Rev. Lett., 66, 1098 (1991)
6. S. M. Anlage, et al., Phys. Rev. Lett. (To be published)
7. W. McMillan, Phys. Rev. 175, 537 (1968)
8. V. Kresin, Phys. Rev. B 25, 157 (1982)
9. R. Dynes, Private Communication
10. C. Owen and D. Scalapino, Physica 55, 691 (1971)
11. V. Kresin, H. Gutfreund, W. A. Little, Solid State Comm. 51, 339 (1984); V. Kresin, Phys. Lett, 122, 434 (1987)
12. P. Allen and R. Dynes, Phys. Rev. B 12, 905 (1975).
13. P. Morel and P. Anderson, Phys. Rev. 125, 1203 (1962); B. Geilikman, V. Kresin, and M. Masharov, J. of Low Temp. Phys. 18, 241 (1975)
14. P. Anderson, J. Phys. Chem. Sol. 11, 26 (1959)
15. M. Gurvitch et al., Phys. Rev. Lett. 63, 1008 (1989)
16. W. Pickett, Rev. Mod. Phys. 61, 433 (1989)
17. M. Gurvitch and A.T. Fiory, Phys. Rev. Lett. 59, 1337 (1987)
18. V. Kresin and H. Morawitz, Phys. Rev. B 37, 7854 (1988); J. of Superconductivity 1, 108 (1988); Preprint
19. V. Kresin and H. Morawitz, Solid State Comm. 74, 1203 (1990)
20. T. A. Friedman, et. al., Phys. Rev. B39, 4258 (1989)
21. Z. Schlesinger et. al., Phys. Rev. Lett. 65, 801 (1990)
22. S. Takahashi and M. Tachiki, Physica B, 165-166 1067 (1990); Physica C, 170, 505 (1990)
23. D. Mattis and J. Bardeen, Phys. Rev. 111, 412 (1958); A. Abrikowev, L.Gorkov and I. Khalatniker, Sov. Phys. JETP 8, 182 (1959).
24. V. Kresin, J. of Superconductivity 3, 177 (1990)
25. J. Chang and D. Scalapino, Phys. Rev. B 40, 4299 (1989)
26. H. Piel et al. Physica C 153-155, 1604 (1988)
27. N. Klein et al. (to be published)

High T$_c$ Superconductor Thin Films
L. Correra (Editor)
95

LOW MICROWAVE LOSS AT HIGH SURFACE MAGNETIC FIELDS IN LARGE EPITAXIAL YBa$_2$Cu$_3$O$_{7-\delta}$ THIN FILMS

M. HEIN, S. HENSEN, G. MÜLLER, S. ORBACH, H. PIEL, M. STRUPP
Fachbereich Physik, Bergische Universität Wuppertal, D-5600 Wuppertal 1, Germany

N. G. CHEW, J. A. EDWARDS, S. W. GOODYEAR, J. S. SATCHELL, R. G. HUMPHREYS
DRA Electronics Division (RSRE), Malvern, Worcs. WR14 3PS, United Kingdom

Abstract

YBa$_2$Cu$_3$O$_{7-\delta}$ thin films have been grown by electron beam coevaporation in the presence of atomic oxygen. The films were up to 700 nm thick, deposited on (001) MgO substrates 10 mm square or 25 mm diameter. The structure of these films has been extensively characterised by X-ray diffractometry, and current densities up to $6 \cdot 10^6$ A/cm^2 at 77 K have been measured nondestructively by dc magnetisation. The surface impedance of five small and two large samples was measured in cylindrical host cavities of copper and niobium at 87 GHz and 21.5 GHz, respectively. At 87 GHz and low field levels, surface resistance values down to 1 mΩ have been achieved even for a 700 nm thick film. Surprisingly, the normally observed plateau-like residual resistance decreases, in some cases significantly, at temperatures below 40 K. This result gives first evidence for a complete pair condensation of charge carriers in superconducting YBa$_2$Cu$_3$O$_{7-\delta}$. At 21.5 GHz and 4.2 K, a residual resistance of about 0.1 mΩ has been obtained for both a 300 nm and a 700 nm thick film. The most remarkable result, however, is its power independence up to surface fields of 120 Oe which seems to be limited only by imperfections of the niobium host cavity.

1. INTRODUCTION

The application of high-T$_C$ superconductors for planar linear microwave devices requires homogeneous films of 300 nm to 1 μm in thickness on low loss dielectric substrates of large size [1]. Their effective surface impedance at high power levels should be smaller than that of copper at temperatures above 20 K and at frequencies in the GHz range. At present, YBa$_2$Cu$_3$O$_{7-\delta}$ (YBCO) thin films epitaxially grown on singlecrystalline MgO or LaAlO$_3$ substrates with c-axis orientation perpendicular to the surface provide the lowest microwave loss, which is most often nearly constant for temperatures below 60 K [2]. However, most of the very good results for the surface resistance have been achieved only with YBCO samples of a maximum size of 1 cm^2 and only at low magnetic surface fields.

Coevaporation of the metals with atomic oxygen incident on the heated substrate is one of the most promising in situ film preparation techniques. It enables uniform coating even of large substrates with precise composition control [3]. Therefore, we have started to optimize this technique for the production of large epitaxial YBCO thin films and to investigate their structural and transport properties as described in the next chapter. As a first step, we have studied the influence of the film thickness and the deposition parameters on the low field surface impedance at 87 GHz. Moreover, we will give in the main chapter first results on the high field microwave performance at 21.5 GHz of YBCO films of 1 inch in diameter and up to 700 nm in thickness.

2. PREPARATION AND CHARACTERISATION OF THE THIN FILMS

Details of the film growth process have been described elsewhere [3, 4], and in another paper at this conference [5]. Briefly, the films were grown by coevaporation of the metals from electron beam heated hearths in the presence of atomic oxygen. The substrates were polished (001)-oriented MgO single crystals. The growth rate was 0.12 nm/s. The oxygen pressure during growth was in the region of 10^{-4} mbar, of which approximately 10 % was atomic. The pressure was increased by a factor of about 50 after growth, and the sample cooled to 200°C in 1 hour. The large source to substrate spacing (420 mm) combined with substrate rotation (0.5 Hz) ensures excellent uniformity of composition even on 1 inch diameter samples. As discussed in [5], it is believed that YBCO is thermodynamically unstable under these growth conditions, and only forms because it is kinetically stabilised. This is therefore a very different regime to other growth techniques. The samples have been grown at intervals spread over nearly a year, and a number of detailed changes in the growth process have been made in this time, mainly small but significant differences in the oxygen pressure and the cooldown rate. All the samples were superconducting as removed from the evaporator, but all were annealed at 500°C in 1 bar oxygen for the sake of consistency.

Considerable care has been taken to achieve stable and reproducible control of the film composition. This has a marked effect [3] on both the morphology and critical current density J_C. Small changes of composition have less effect on J_C in films with an excess of copper than in those with a deficiency. This was explained in [3] as being due to efficient segregation of the excess Cu into precipitates, thus leading to less disruption of the growth of the superconductor. Surface impedance measurements on thermally coevaporated YBCO films [6] have shown a broad transition curve and a high residual loss for a copper deficiency of 6 % but no additional microwave loss for Cu-O passivation layers. All the films used in the present work were therefore grown slightly Cu rich (<1%), and a low density of submicron copper oxide lumps were visible. Surface features on a finer scale are observed in the scanning electron microscope, the most obvious of which can be identified as a-oriented grains. The residual roughness fell as the film thickness increased.

One object of the present work is to identify which other material properties correlate with the microwave loss. Therefore, the YBCO films have been characterised extensively with nondestructive techniques. Details of the films are listed in Tab. 1. The samples for the 21.5 GHz measurements were trepanned into 25 mm diameter circles, while those for the 87 GHz measurements remained 10 mm square. The growth temperatures quoted as T_{gr} are estimates of the actual substrate temperature, based on calibration experiments using a substrate with a thermocouple inside it [4]. In the early (≈ < 400) experiments, T_{gr} fell by up to 20°C during growth, due to changes in emissivity. This is compensated for in the more recent samples. The film thickness d was varied between 172 nm and 690 nm because of contradictory reports about enhanced [7] or constant [8] microwave loss for increasing d. The structural data given in Tab. 1 were obtained using a Siemens X-ray diffractometer. The films were predominantly c-oriented, with lattice parameters in the region of 1.174 nm. In the Bragg-Brentano geometry, weak diffraction peaks were detectable due to the substrate and small amounts of a-oriented material. The "% a" column has been determined from the relative intensities of the 200 and 006 reflections, making the approximation that these two lines have equal structure factors. Surface impedance measurements on predominantly a-oriented films have shown two orders of magnitude higher residual losses than for c-oriented films [9]. The "Δc_{007}" column is the line width of the 007 diffraction peak expressed as an uncertainty in lattice parameter. Analysis of the dependence of Δc on diffraction order suggests that it is mainly due to inhomogeneous broadening. The rocking curve width "$\Delta \omega_{006}$" gives a measure of the mosaic spread of the c-axis orientation over the sample. These values indicate reasonable quality,

Tab. 1: Growth conditions, structural data and transport properties of the YBCO films fabricated for microwave measurements. All symbols are explained in the text.

#	size [mm]	T_{gr} [°C]	d [nm]	% a	Δc_{007} [nm]	$\Delta\omega_{006}$ [°]	% 45°	T_c [K]	$J_c(77)$ [A/cm^2]	S(77)
289	25	670	345	–	–	–	–	88.5	2.2·10^6	–
363	10	690	690	<.02	.0030	.78	.2	82.6	1.5·10^6	.027
364	10	690	345	1.0	.0030	.60	<.05	87.1	4.4·10^6	.020
365	10	690	172	.3	.0024	.52	.1	86.0	3.8·10^6	.021
424	25	670	690	<.02	.0040	.50	<.03	89.8	–	–
476	10	670	345	1.3	.0020	.45	<.05	90.7	6.0·10^6	.022
477	10	740	345	2.9	.0024	.51	<.05	89.3	4.0·10^6	.016

but are larger than those quoted by groups who have taken care to optimise this property [10]. The rocking curve was often asymmetric, reflecting a slight accidental misorientation of the substrate [11, 12] which also enlarged $\Omega\omega_{006}$. The "% 45°" column gives an estimate of the percentage of the film the a-b axes of which are rotationally misoriented with respect to the corresponding substrate directions. The presence of such grains has been associated with poor microwave properties [13]. The levels found in the present films are at the low end of the range studied in ref. 13, but samples 363 and 365 would be expected to have significant loss due to the misoriented grains.

The remaining columns of Tab. 1 refer to transport properties of the films measured by a dc magnetisation technique as described in ref. 5. Quoted as T_c is the temperature at which our magnetometer can resolve an unequivocal signal ($J_c = 2 \cdot 10^3$ A/cm^2). The values of T_c for some of these films are rather low. Several groups have reported films which are otherwise of high quality to have reduced T_c. Although the details are not well understood, it appears, at least for our films, to be associated with insufficient oxygen being present either during growth or on subsequent cooling. The critical current density J_c is deduced from the magnetisation assuming that the film carries a uniform current over its whole area. This has been shown to yield $J_c(T)$ values very comparable to the results of transport measurements. The flux creep rate $S = -d(\log M)/d(\log t)$ has been determined from the decay of the persistent current between 12 ms and 500 ms following the removal of an applied field. For most samples, S does not vary too much with temperature except for a divergence within a few degrees of T_c. At 77 K, the J_c values scatter much more than the S values mainly due to the different T_c values. The detailed mechanism of flux creep is still a controversial subject, and we include this data simply as a phenomenological measure of an aspect of the pinning which we hope to correlate with the microwave loss. It is important to recognise that strong pinning often implies many regions of reduced order parameter in the superconductor, and that poor pinning, i.e. low J_c and high S, may be expected to be characteristic of material with low microwave loss.

3. SURFACE IMPEDANCE RESULTS AND DISCUSSION

The effective surface impedance of the YBCO films was measured by mounting them as an endplate of cylindrical cavities. The cavity used for the 10 mm square samples resonated near 87 GHz in the TE$_{013}$ and TE$_{021}$ modes, and that for the 25 mm diameter samples at 21.5 GHz in the TE$_{011}$ mode. All of these modes provide high sensitivity for residual losses of the superconducting samples due to the vanishing surface field at the joint to the cavity. Assuming a homogenous surface impedance, the contribution of one endplate to the total microwave losses of these cavities amounts 40 % in the TE$_{013}$, 5 % in the TE$_{021}$ and 23 % in the TE$_{011}$ mode. Besides the size-dependent choice of operating frequency, different aspects of the film quality are investigated with our test systems. At 87 GHz we use a copper cavity, which enables accurate measurements of the temperature dependence of the surface resistance R_S and of the penetration depth λ as long as the sensitivity limits of ± 3 mΩ and ± 10 nm set by the losses and thermal expansion of the host cavity are not reached [14]. Moreover, the homogeneity of the samples can be checked by switching from the TE_{013} to the TE_{021} field distribution. This is usually performed near T_C because of the limited power (10 mW) and dynamic range of the synthesizer-based millimeter wave system. The peak magnetic surface field H_S at the sample never exceeds 20 A/m ($\triangleq .25$ Oe) due to the losses in the copper cavity. In contrast, $R_S(T)$ and $\lambda(T)$ of high quality YBCO films cannot be resolved at 21.5 GHz with a copper cavity, since the superconductor becomes much better than the normal conductor at this frequency [2]. Therefore, we use a niobium host cavity at temperatures between 1.5 K and 4.2 K, the sensitivity of which is limited by the quality of the indium-sealed joint or by the BCS surface resistance of niobium (~ 75 $\mu\Omega$ at 4.2 K) to about $\pm (50-80$ $\mu\Omega$). A further advantage of our lower frequency system is that much higher power (25 W) is available. For $R_S < 10$ $\mu\Omega$, this is sufficient for H_S values of some hundred Oe to be achieved provided neither the YBCO film nor the niobium cavity reaches its field limitation.

The results of the $R_S(T)$ measurements of the five small films are shown in Fig. 1. It should be recognised that plotted are not the originally measured effective values which depend on the relative film thickness d/λ, but the analysed intrinsic values which result from impedance transformations [15]. For the discussion of the film quality, this correction is necessary at least for $d < \lambda(T)$, i.e. especially in the regime close to or above T_C. All of our coevaporated YBCO films demonstrate by their steep microwave transition curves and low residual values of R_S a quality, which has been first observed for epitaxially grown films made by the laser ablation technique [14]. The values for $\lambda(0)$ resulting from a fit of $\lambda(T)$ to the weak-coupling BCS-theory in the clean limit vary slightly between 140 nm (# 364 and # 476) and 160 nm (# 363)

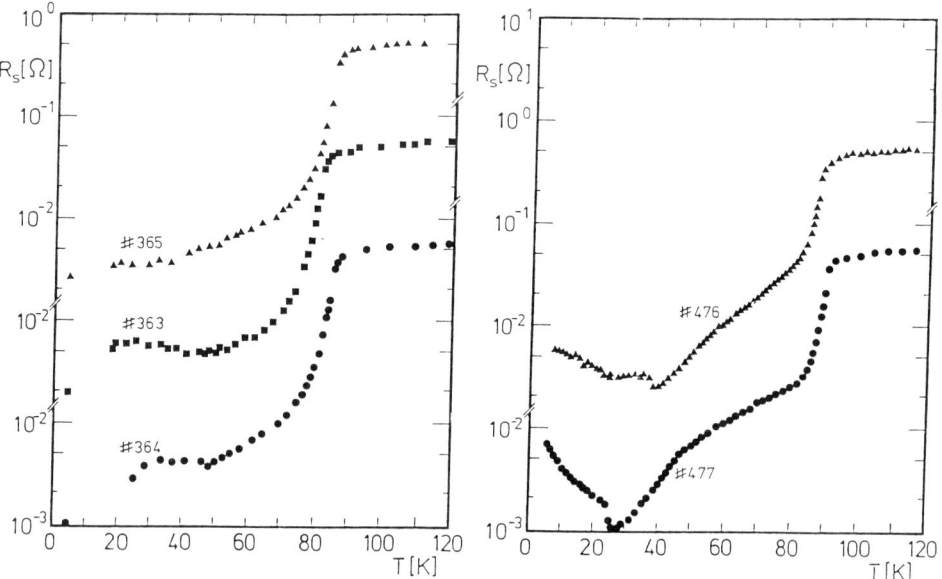

Fig. 1: Temperature dependence of the intrinsic surface resistence of the small YBCO films in Tab. 1 at 87 GHz. Please note the broken scales.

and are similar to those for the laser ablated films [15]. For the resistivity ρ in the normal state, average values between $72\,\mu\Omega\text{cm}$ (**477**) and $82\,\mu\Omega\text{cm}$ (**364**) at 100 K and between $177\,\mu\Omega\text{cm}$ (**365**) and $243\,\mu\Omega\text{cm}$ (**476**) at 290 K with a maximum slope $\rho(290)/\rho(100)$ of 3.3 (**477**) can be derived from the R_S data using the well-known skin effect formula. These data reveal a trend to lower $\rho(100)$ values and larger slopes for the more recent (**> 400**) films, in accordance with the improvements in the growth process and the significantly higher T_C values. The latter, as listed in Tab. 1, are clearly confirmed by the onset of the microwave transitions in Fig. 1. The continuity of the R_S data near T_C measured in different modes demonstrate excellent homogeneity of the films. Obviously, the main difference between the $R_S(T)$ curves of these films occurs at temperatures below 50 K. While **365** yields nearly temperature independent residual losses as generally observed for epitaxially grown films [2], **363** and **364** have shown for the first time, at least to our knowledge, a significant drop of R_S at very low temperatures. This feature has since been confirmed in a much more pronounced form with YBCO films sputtered on LaAlO₃ [16]. Motivated by this success, we have started to vary the deposition parameters again and to investigate the structural and transport properties of the films extensively, as listed in Tab. 1. Surprisingly, both the more recent films show a combination of old and new features. The more gradual decrease of $R_S(T)$ below $0.9\,T_C$, which might be correlated with an increasing amount of granularity (**% a**), has been often observed during optimisation of epitaxial YBCO film growth. New is the occurence of a minimum R_S around 30 K and a significant increase of R_S below 20 K, the origin of which is still unclear. Nevertheless, we would like to point out that a better control of the film growth and composition seem to be the key issues. Since these results argue against an intrinsic presence of nonpairing charge carriers [17], they are very encouraging for the further improvement of epitaxial YBCO films for microwave applications below $T_C/2$ [1].

In Fig. 2, the results of the $R_S(H_S)$ measurements on the two large YBCO films at 21.5 GHz are shown. Both films yield an extraordinary quality up to peak magnetic surface fields of 120 Oe. The absolute R_S values of about $100\,\mu\Omega$ at 4.2 K are close to the sensitivity limit of the niobium cavity, and special care has been taken to avoid

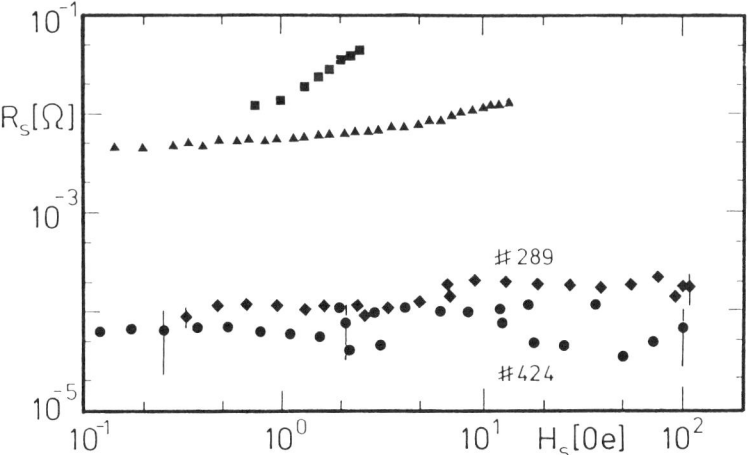

Fig. 2: Magnetic surface field dependence of R_S at 4.2 K and 21.5 GHz for the two large coevaporated YBCO films in Tab. 1 compared to results for electrophoretically deposited [18] untextured (squares) and c-axis textured YBCO layers (triangles).

additional losses due to the joint. Therefore, the measurements were repeated several times to minimize the joint losses. In the best case, a further reduction of $R_S(T)$ between 4.2 K and 1.5 K of about a factor of three could be resolved which confirms the trend of the $R_S(T)$ data of the small films #363 and #364 in Fig. 1. Therefore, we have to improve our measurement techniques for more precise determination of the absolute values of R_S and for further optimization of YBCO films. Nevertheless, it is remarkable that the measured values of $R_S(4.2 K)$ at 21.5 GHz and 87 GHz scale about quadratically with frequency as expected below 100 GHz from a survey of the best R_S data [2], and that the quality of the coevaporated YBCO films does not degrade with the film thickness up to 700 nm. The most important result, however, is the indepence of R_S on the microwave power level up to H_S values of 120 Oe ($\approx 10^4$ A/m), which is in clear contrast to the increase of losses in polycrystalline YBCO layers (Fig. 2) caused by weak link effects [19]. At the maximum field, a limitation occurs even if the sample is replaced with a niobium disc. Therefore, we don't believe to have reached already the intrinsic microwave field limitation of this superconductor.

This work has been funded in part by the German Federal Minister for Research and Technology (BMFT) under the contract number 13N5502.

4. REFERENCES

[1] H. Chaloupka and G. Müller, subm. to Physica C (1991).

[2] H. Piel and G. Müller, IEEE Trans. Magn. **MAG-27**, 854 (1991).

[3] N.G. Chew, S.W. Goodyear, J.A. Edwards, J.S. Satchell and R.G. Humphreys, Appl. Phys. Lett. **57**, 2016 (1990).

[4] R.G. Humphreys, J.S. Satchell, N.G. Chew, J.A. Edwards, S.W. Goodyear, S.E. Blekinsop, O.D. Dosser and A.G. Cullis, Supercond. Sci. Technol. **3**, 38 (1990).

[5] R.G. Humphreys, N.G. Chew, J.S. Satchell, J.A. Edwards, S.W. Goodyear, this conf.

[6] S. Orbach, N. Klein, G. Müller, H. Piel, P. Berberich and H. Kinder, J. Less-Common Metals **164&165**, 1261 (1990).

[7] N. Klein, G. Müller, S. Orbach, H. Piel, H. Chaloupka, B. Roas, L. Schultz, U. Klein and M. Peiniger, Physica **C 162-164**, 1549 (1989).

[8] T.H. Kuhlemann and J. Hinken, IEEE Trans. Magn. **MAG-27**, 872 (1991).

[9] T.L. Hylton, M.R. Beasley, A. Kapitulnik, J.P. Carini, L. Drabeck and G. Grüner, IEEE Trans. Magn. **MAG-25**, 810 (1989).

[10] Q. Li, O. Meyer, X.X. Xi, J. Geerk and G. Linker, Appl. Phys. Lett. **55**, 310 (1989).

[11] S.S. Laderman, R. Taber, R.D. Jacowitz, J.L. Moll, C.B. Eom, L.T. Hylton, A.F. Marshal, T.H. Geballe and M.R. Beasley, Phys. Rev. **B 43**, 2922 (1991).

[12] S.K. Streiffer, B.M. Lairson and J.C. Bravman, Appl. Phys. Lett. **57**, 2501 (1990).

[13] S.E. Russek, B. Jeanneret, D.A. Rudman and J.W. Ekin, IEEE Trans. Magn. **MAG-27**, 931 (1991).

[14] N. Klein, G. Müller, H. Piel, B. Roas, L. Schultz, U. Klein and M. Peiniger, Appl. Phys. Lett. **54**, 757 (1989).

[15] N. Klein, H. Chaloupka, G. Müller, S. Orbach, H. Piel, B. Roas, L. Schultz, U. Klein and M. Peiniger, J. Appl. Phys. **67**, 6940 (1990).

[16] N. Klein et al., to be publ. in Physica C.

[17] G. Müller, N. Klein, A. Brust, H. Chaloupka, M. Hein, S. Orbach, H. Piel and D. Reschke, J. Superconduct. **3**, 235 (1990).

[18] M. Hein, S. Kraut, E. Mahner, G. Müller, D. Opie, H. Piel, L. Ponto, D. Wehler, M. Becks, U. Klein and M. Peiniger, J. Superconduct. **3**, 323 (1990).

[19] M. Hein et al., to be publ. in J. Magn. Magn. Mat..

High T$_c$ Superconductor Thin Films
L. Correra (Editor)
© 1992 Elsevier Science Publishers B.V. All rights reserved.

Harmonic generation by field modulation of the microwave absorption of YBa$_2$Cu$_3$O$_7$ and granular lead thin films

S. Revenaz[a], J. Dumas[a], A. Gerber[b], J. Schubert[c]

[a]Laboratoire d'Etudes des Propriétés Electroniques des Solides, CNRS, BP 166, 38042 Grenoble Cedex France

[b]Laboratoire Louis Néel, CNRS, BP 166, 38042 Grenoble Cedex France

[c]ISI - KFA, Julich, 5170 Julich, Germany

Abstract

Low frequency harmonic generation and real time oscillations in the reflected microwave power by application of an ac magnetic field has been investigated in granular, textured, epitaxial and He$^+$ irradiated YBa$_2$Cu$_3$O$_7$ thin films as well as in granular lead thin films. All samples show even harmonics below T$_c$. In granular films, the fundamental is observed just below T$_c$. The amplitude of the harmonics as a function of a superimposed dc field as well as their anisotropy is reported. The results are discussed in relation with the microstructure and in terms of viscous flux motion.

1. INTRODUCTION

The study of electrodynamic properties of high -T$_c$ copper oxide superconductors provides useful information not only for a basic understanding of their superconducting state but also about their possible performances for passive microwave devices [1]. These properties include microwave absorption in low magnetic field [3-4] and non linear magnetic response to small applied ac fields [5-6] There is now considerable evidence that the granularity of the sample plays a crucial role in these phenomena.The interplay between intergranular and intragranular properties remains controversial. Harmonic generation in the low frequency spectrum of the microwave power transmitted through high T$_c$ textured films in the presence of an ac field has been investigated recently [7]. In this paper, we report harmonic generation phenomena in the microwave power reflected from a conventional EPR cavity loaded with the sample under study driven by a weak ac magnetic field.

Non linear response is not only observed in high -T$_c$ materials, it is also present in low -T$_c$ type II superconductors. We compare the harmonic contents in granular, textured unirradiated and He$^+$ irradiated and in epitaxial YBa$_2$Cu$_3$O$_7$ thin films with their c-axis normal to the substrate to those obtained on a conventional low -T$_c$ superconductor, namely granular lead thin films. The results are discussed in terms of low field flux penetration and viscous flux motion.

2. EXPERIMENTAL TECHNIQUES

The YBa$_2$Cu$_3$O$_7$ films used in this study were prepared by rf sputtering [8] and laser ablation [9-10]. Sputtered films (Z) were textured and laser ablated film (C) was epitaxial with c-axis normal to the substrate, as ascertained by x-ray diffraction studies. Laser ablated film (B) showed a high degree of a,b orientation. He$^+$ irradiation procedure of B-like films is given in Ref.[11]; irradiation leads to a decrease of T_c and a broadening of the transition. Granular Pb films were obtained by electron beam evaporation of Pb on a preevaporated layer of Ge at room temperature on a glass substrate [12]. In order to prevent oxidation in air, they were covered by an additional 400 Å thick layer of Ge. The cluster size of the Pb films could be adjusted in a controller manner by varying the thickness of the film. Three films with thickness of 200, 350 and 10^4 Å respectively were investigated. The sheet resistance $R_\square$ of these films decreases as the thickness increases. While 200 and 350 Å films were granular, with an average grain size of 200 Å, the 10^4 Å film was continuous. All the films used in this study had nearly the same rectangular dimensions 3 mm x 1.5 mm. Table I summarizes some characteristics of the films.

Table 1
Film characteristics.

Film	Z	B	C	Pb-I	Pb-II	Pb-III
Preparation method	rf sputtering	laser ablation	laser ablation		electron beam evaporation	
Substrate	ZrO_2	ZrO_2	$LaAlO_3$		Ge	
Thickness	1 μm	0.8 μm	0.3 μm	200Å	350Å	10^4Å
T_c onset	90 K	90 K	94 K		7.2 K	
ΔT_c	~ 3 K	3 K	< 1 K		< 0.2 K	
Ref.	8	9	10		12	

The microwave absorption measurements were carried out using a 9.4 GHz (X-band) Bruker spectrometer equipped with a rectangular cavity and a continuous helium gas flow cryostat. The microwave magnetic field H_1 is parallel to the surface of the film placed at the center of a rectangular TE_{102} cavity where H_1 has a maximum. The field H_1 is normal to the externally applied field $H_a = H + h_m \sin \omega t$. Zero field cooling was obtained by appropriate shielding with a μ-metal box. Below T_c, when the system is driven by the modulation field h = $h_m \sin\omega t$, the Fourier analysis of the direct signal P(H + h), detected at the output of the microwave bridge, has been performed using a HP 70000 spectrum analyzer. The reflected power P(H + h) can be expanded in the following form :

$$P = \sum_n \left(h^n/n!\right) \sin^n \omega t \cdot \left(d^n P/dH^n\right) = \sum_n c_{2n} \cos 2n\, \omega t + c'_{2n+1} \sin(2n + 1)\, \omega t \quad \text{with}$$

$c_{2n} = \sum_p a_{2p}\left(d^{2p}P/dH^{2p}\right)$ and similar form for c'_{2n+1}. P is proportional to the surface resistance of the film. Our measured quantity S defined below, is proportional to hdP/dH. For some samples, the signal P(H + h) has been analyzed in the time domain using a HP 3561 A spectrum analyzer.

3. EXPERIMENTAL RESULTS

3.1. Temperature dependence of harmonic generation
In Figure 1a, we show real time oscillations in the reflected power P(h) when the sample (C) is driven by an ac field. In Figure 1b, we show the corresponding Fourier analysis recorded simultaneously. Only even harmonics are observed.

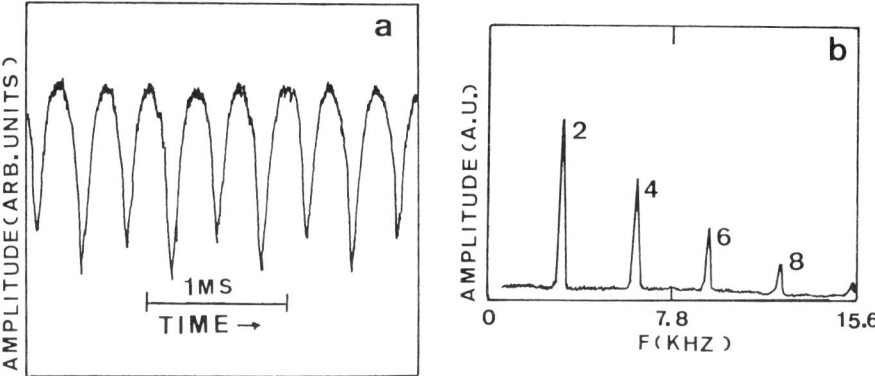

Figure 1: (a) Real time oscillations in the reflected power P(h); (b): Corresponding Fourier spectrum recorded simultaneously. Film (C); h_m=5G; ω=1.56 kHz; T/ T_c = 0.98.

A common feature to samples B, C and Pb-I is the generation, below T_c, of even harmonics of the ac magnetic field fundamental frequency. In films (C) and (B) the amplitude of the harmonics vanishes a few degrees below T_c. The average dissipated power defined as

$S = \sqrt{\sum_{n} |c_{2n}|^2}$, where c_{2n} are the individual amplitudes, is shown in Figure 2 as a function

of temperature for samples (C) and Pb-I.

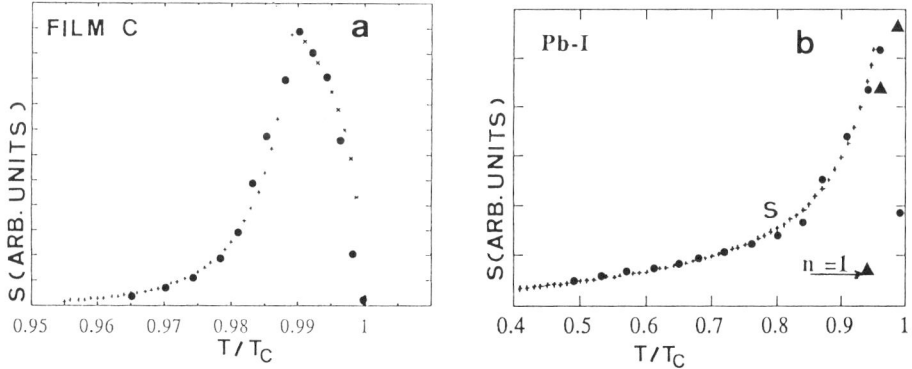

Figure 2: (a)Temperature dependence of S (See text); (+) fit according to Ref. 16; (b) same as in (a) for film Pb-I. The amplitude of the n=1 component is also shown (▲); (+) fit according to Ref. 16.

The temperature range of existence of harmonic seems comparatively large in Pb-I film since they are detected down to T ~ 0.4 T_c. In He⁺ irradiated film (B) the even harmonics are observed in a more extended temperature range than in the non irradiated film, as shown in Figure 3. While in unirradiated films B,C harmonics appear just below T_c, they are generated with an onset temperature $T_c' < T_c^{irr}$, typically 3K below T_c^{irr}.

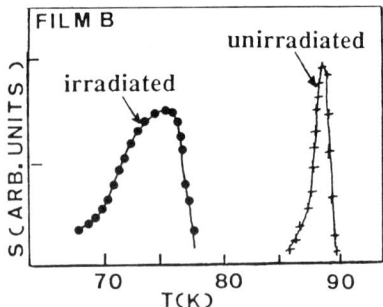

Figure 3:Dissipated power as a function of reduced temperature T/T$_c$ for: (+) pure film (B);
(•): He$^+$ irradiated film (B) with an irradiation dose 5.10^{15} He $^+$/ cm^2 (Ref.11). T$_c$ = 89K in
pure film and T$_c^{irr}$ = 79K.

In films (B,C) the fundamental n = 1 is barely measurable; it is detected just below T$_c$ in
Pb films, prior to the onset of even harmonics. In a granular film (Z), the fundamental first
appears, with an onset temperature T$_c^{'}$ < T$_c$, then, at a temperature T$_c^{"}$ < T$_c^{'}$, additional odd
and even harmonics are found at lower temperatures, down to the lowest temperatures explored
(25 K) [13]. While Pb-I and Pb-II films give similar results, in continuous film Pb-III and in
bulk Pb, only the fundamental is observed in a more narrow range of temperature.

3.2. Magnetic field dependence

At fixed temperature, the amplitude of the harmonics in high and low-T$_c$ films depends
strongly on the magnitude of superimposed dc field H on the ac field h.We have previously
shown [14] that S decreases rapidly at low dc fields For films Pb-I, Pb-II, S decreases and
vanishes for a value which depends on temperature and film orientation. For this critical field
value, the surface resistance reaches its normal state value. In Pb-III, the maximum in the S vs
H curve occurs for H = H$_c$. Therefore, Pb-III sample has to be viewed as a type-I
superconductor.
For our highest available fields (H= 6kG), the surface resistance of a film C, with a much
larger critical field H$_{c2}$, is still much lower than its normal state value and a long tail in the S vs
H curve is found at large fields.Figure 4 shows the field dependence of S for films C and Pb-I.

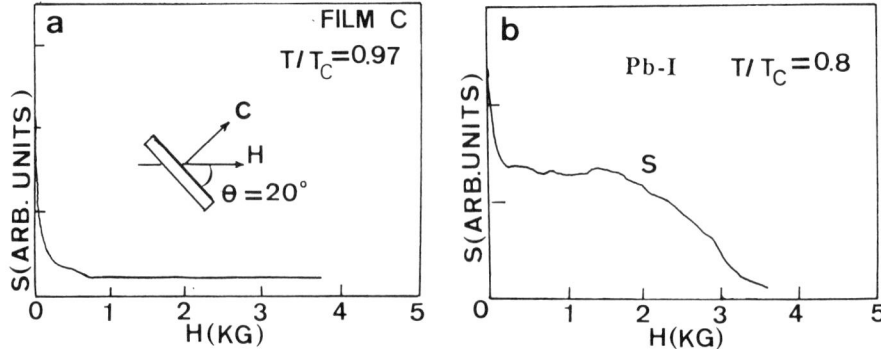

Figure 4: S as a function of a dc field H for a film C (a) and for a film Pb-I (b). The orientation
of H with respect to the plane of the film is indicated.

3.3. Angular dependence

At fixed temperature, when a film   is rotated about a vertical axis, we observe an angular variation of the amplitude of the harmonics. For all samples, no harmonics are found when H is parallel to the film. For a film C, with the c-axis in a horizontal plane, S increases rapidly, then structures are observed ,as illustrated in Figure 5a. In a granular film Pb-II, a strong anisotropy is also observed, as shown in Figure 5b, when a dc field is applied. For H_0 = 1800G, S drops rapidly above a maximum at θ_0 ; the component $H_0 \sin \theta_0$ normal to the film corresponds to the critical field $H_c^\perp \approx 1400$ G found by Gerber at al.[12]. No anisotropy in Pb-III film is found.

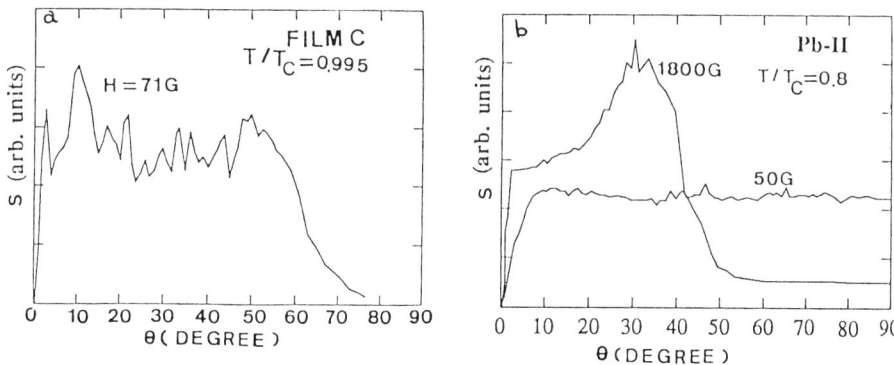

Figure 5. (a) Angular variation of S with a dc field H_0=71G for film C; (b) same as in (a) for film Pb-II for dc fields 1800G and 50G.

4. DISCUSSION

We have observed coherent harmonic generation effects in low ac fields both in high-T_c granular, textured (pure and irradiated) or epitaxial thin films and in low-T_c granular lead thin films. The observed strong low frequency non linearities indicate that a modulation of the magnetic field does not bring necessarily a simple field derivative dP/dH in variance with the model proposed in Ref. 3. The similarity in the field dependence of harmonics amplitude in films (C) and Pb-I is rather surprising [14]. For film Pb-I, ξ/d, where ξ is the coherence length and d the film thickness, is larger than 1.8, value above which the two dimensional behavior should be observed [15]. In our case, ξ is larger than the average grain size. In YBa₂Cu3O₇, ξ_c (T = 0) $\approx$ 3Å ; a ratio ξ_c (T = 0.95 T_c)/d $\approx$ 1.8 would imply d $\approx$ 10Å. This effective thickness has to be compared to the separation between Cu-O planes. These results, in conjunction with the fact that textured film (B) , epitaxial film (C) and granular lead show similar field dependence, raise the question of the role of the low dimensionality and granularity in harmonic generation close to T_c. In epitaxial film (C), planar defects could play a major role in harmonic generation. In this context, the similarities in the S vs H curves at low fields for Pb-I and Pb-II films suggest that the nature of the intergranular couplings is relevant. The absence of harmonics for h parallel to the film may indicate the absence of flux penetration for this orientation.

In films C and Pb-I, a good fit to the S vs T curves is obtained using the phenomenological flux flow model at low field for microwave surface resistance described in Ref. [16]. In this model, the only adjustable parameter is the ratio U_0/T_c where U_0 is a pinning energy. From the fit to the data obtained on film C, in the temperature range where S increases with temperature, an average pinning energy $U_0 \sim 60$ meV is obtained, in good agreement with the value found by remanent magnetization measurements [17]. Above the maximum, one obtains $U'_0 \sim 6$ meV.
At low dc field, the drop in the S vs H curve may reflect the magnetic decoupling of weak links in the material [18].

In summary, we have reported that the low frequency coherent harmonic generation effects are strongly correlated to the film granularity. The temperature interval where harmonics are observed is much larger in granular films than in epitaxial films. This contactless and very sensitive microwave technique may be very useful in understanding the origin of microwave losses in high-T_c superconductors.

5. ACKNOWLEDGEMENTS

The authors wish to thank B. Roas and L. Schultz for laser ablated films, J. Marcus for sputtered films, B. Pannetier for lead films and M. Trunin, M. Golosovsky for useful discussions. This work was supported in parts by a ECC- Science Contract SC1-0038C.

6. REFERENCES

1　G. Müller, N. Klein, A. Brust, H. Chaloupka, M. Hein, S. Orbach, H. Piel, D. Reschke, J. of Supercond. 3 (1990) 235.
2　K.W. Blazey, Physica Scripta T 29 (1989) 92.
3　K. Moorjani, J. Bohandy, B.F. Kim, F.J. Adrian, Solid State Commun. 74 (1990) 497.
4　B. Ravkin, M. Pozek, M. Paljevic, N. Brnicevic, Solid State Commun. 70 (1989) 729.
5　R.B. Goldfarb, A.F. Clark, A.I. Braginski, A.J. Panson, Cryogenics 27 (1987) 475.
6　C. Jeffries, Q.H. Lam, Y. Kim, L.C. Bourne, A. Zettl, Phys. Rev. B39 (1989) 11526.
7　M. Golosvsky, D. Davidov, E. Farber, T. Tsach, M.Schieber, Physica A 168 (1990 353.
8　J. C. Bruyère, P.L. Reydet, C. Filippini, C. Schlenker, Mat. Res. Bull. 23 (1988) 429.
9　J. Frölingsdorf, W. Zander, B. Stritzker, Solid State Commun. 67 (1988) 965.
10　B. Roas, L. Schultz, G. Endres, Appl. Phys. Lett. 53 (1988) 1557.
11　S.K. Agarwal, O. Muller, J. Schubert, C. Schlenker, Physica C, to be published.
12　A. Gerber, G. Deutscher, Phys. Rev. Lett. 63 (1989) 1184.
13　S. Revenaz, J. Dumas, C.J. Liu, C. Schlenker, S. Orbach, C. Müller, N. Klein, H. Piel, J. Less Common Metals 164-165 (1990) 1252.
14　S. Revenaz, J. Dumas, A. Gerber, Physica C, to be published.
15　M. Tinkham, Introduction to superconductivity (Mc Graw Hill, New York, 1975).
16　R. Marcon, R. Fastampa, M. Giura, E. Silva, Phys. Rev. B 43 (1991) 2940.
17　C. Schlenker, C. J. Liu, R. Buder, J. Schubert, Physica C, to be published.
18　J.R. Clem, Physica C 153-155 (1988) 50.

High T$_c$ Superconductor Thin Films
L. Correra (Editor)
© 1992 Elsevier Science Publishers B.V. All rights reserved.

107

EFFECT OF ION AND NEUTRON IRRADIATION ON MICROWAVE LOSSES OF EPITAXIALLY GROWN YBa$_2$Cu$_3$O$_{7-\delta}$ FILMS

S. ORBACH, S. HENSEN, G. MÜLLER, H. PIEL
Fachbereich Physik, Bergische Universität Wuppertal, D-5600 Wuppertal, Germany

M. LIPPERT, W. SCHINDLER, G. SAEMANN-ISCHENKO
Physikalisches Institut, Universität Erlangen, D-8520 Erlangen, Germany

B. ROAS
Siemens AG, Forschungszentrum Erlangen, D-8520 Erlangen, Germany

We have investigated the effect of point-like defects on the microwave properties of high quality YBa$_2$Cu$_3$O$_{7-\delta}$ thin films which were epitaxially grown on $\langle 100 \rangle$ LaAlO$_3$ substrates by the laser ablation technique. For this purpose four films of about 250 nm in thickness have been irradiated stepwise with 25 MeV ^{16}O-ions at 77 K or with neutrons of at least 0.1 MeV at 300 K. After each irradiation dose, the surface impedance of the films was measured at 87 GHz between 4 and 120 K and at 300 K. For integral fluences between 10^{14} and 2×10^{15} ions/cm^2 we have observed the expected decrease of T_c and both an increase of the resistivity in the normal conducting state and of the penetration depth λ in the superconducting state. However, the residual surface resistance R_{res} decreases for small fluences and shows even for large ion fluences only a small increase, which is further enhanced by annealing at 500° C for 64 h. In comparison, a fluence of 10^{17} neutrons/cm^2 has no effect on T_c and λ, but leads to a much stronger degradation of R_{res}. The different influence of ion and neutron irradiation on the microwave losses will be discussed on the basis of the two-fluid model.

1. INTRODUCTION

Crystallographic defects [1] are one of the most probable causes for the anomalously high residual microwave losses of high-T_c superconductors [2]. The negative influence of grain boundaries on the microwave absorption has been shown by comparative studies of YBa$_2$Cu$_3$O$_{7-\delta}$ (YBCO) thick and thin films prepared by different techniques [3]. Irradiation with light ions or fast neutrons has been established as a useful method to create in a controlled way point-like defects in high-T_c thin films [4,5,6]. In the test series reported here, we have started to investigate the effect of ion and neutron irradiation on the microwave losses of epitaxially grown YBCO films prepared by laser ablation [7]. For the applied ^{16}O-ion fluences Φ between 10^{14} and 10^{15}/cm^2, it is known that irradiation induced defects lead to a decrease of T_c and to an increase of the dc-resistivity ρ, while the transport critical current density j_c at magnetic fields of about 2 T is enhanced for the smaller Φ-values due to additional pinning centers, but significantly reduced for Φ above 10^{15}/cm^2 [4,5]. Neutron fluences of 10^{17}/cm^2 lead to an enhancement of j_c even at vanishing magnetic fields in epitaxial thin films [6,8]. In contrast, for any kind of inhomogeneities increased microwave losses are expected like for classical superconductors. In order to clarify the role of point-like defects for the residual microwave losses of YBCO, we have measured the surface impedance of $1 \times 1\,cm^2$ samples between 4.2 K and 300 K with high sensitivity at 87 GHz [9]. In the first part the details of the fabrication and irradiation of the samples are given and their structure before and after irradiation is shown by X-ray diffraction data. In the main part the results of the surface impedance measurements will be presented and discussed with respect to the structural changes.

2. FABRICATION, CHARACTERISATION AND IRRADIATION

Five epitaxial, c-axis oriented YBCO thin films have been deposited on $\langle 100 \rangle$ LaAlO$_3$ substrates by an excimer laser ablation process at Siemens AG, Erlangen [7]. The ion irradiations were performed with 25 MeV ^{16}O-ions at the low temperature irradiation facility of the Erlangen Tandem-Van de Graaff accelerator at 77 K. The stopping range of the ^{16}O-ions in YBCO is estimated to about 10 μm, so oxygen implantation in the films is negligible. Two of the films have been irradiated stepwise with fluences between 1×10^{14} and 2×10^{15}/cm^2. Two further samples were irradiated with fast neutrons ($E > 0.1$ MeV) at the Forschungsreaktor München at 300 K with fluences of 1×10^{17} and 3×10^{17}/cm^2. As a last step the ion irradiated films have been annealed in oxygen at 500° C for 64 h. The film thickness d was determined with an optical interference microscope at a pattern, which was wet etched three millimeters from one edge of the samples. The thickness varies along this edge about ± 10 %. The quality of the about 250 nm thick films has been checked before and after irradiation and annealing by X-ray diffraction and by measurements of the surface impedance, as described in chapter 3. Detailed informations about the samples as well as structural and microwave data are summarized in Tab.1.

Tab.1 : Overview of the structural and microwave data of the YBCO films before and after irradiation with ^{16}O and neutrons (n). All symbols are explained in the text.

#	d [nm]	integral Φ [cm^{-2}]	c [Å]	T_c [K]	$\rho_1(100\,K)$ [$\mu\Omega$ cm]	$\rho_1(300\,K)$ [$\mu\Omega$ cm]	$\lambda(0)$ [nm]	R_{res} [mΩ]	ρ_{res} [$\mu\Omega$ cm]
1	250	—	11.678	90	128	284	180	9	15
2	210	—	11.685	91	72	220	150	9	9
		1×10^{14} ^{16}O	—	89	79	201	150	3	26
		2×10^{14} ^{16}O	—	87.5	86	246	165	13	11
		4×10^{14} ^{16}O	—	86	100	197	200	97	2
		& annealing	11.72	—	—	—	—	—	—
3	270	—	11.676	91	80	210	160	5	10
		1×10^{15} ^{16}O	—	80	140	252	190	3.5	35
		2×10^{15} ^{16}O	—	69	235	323	360	7	130
		& annealing	11.707	76	160	309	250	78	10
4	270	—	11.681	91	54	158	130	20	3
		1×10^{17} n	11.684	90	148	404	—	~ 1000	0.05
5	270	—	11.677	91	148	375	180	10	13
		3×10^{17} n	11.684	90	310	482	180	~ 1000	0.1

The low values of the resistivity $\rho_1(100\,K)$, the penetration depth $\lambda(0)$ and the residual surface resistance R_{res} at 4.2 K demonstrate the high quality of the unirradiated films, but they show also that the reproducibility is not optimal. Sample 1 was not irradiated and its quality has been checked again after two months at the end of all experiments.

Within the measurement accuracy the results remained unchanged, so we are sure to measure only the effect of irradiation for the other samples. Before and after irradiation the X-ray diffraction patterns revealed c-axis oriented single-phase YBCO with a full-width of half-maximum for the rocking curve of the <005> peak of about 0.3°. The length of the c-axis was determined from the evaluation of seven <00ℓ> reflections. Before irradiation the films exhibit lattice parameters in the region of $c = 11.68$ Å, which demonstrates the correct oxygen content [10]. The lengthening of the c-axis after ion irradiation [4] shows even after annealing a significant oxygen deficiency, while the c-axis after neutron irradiation remains nearly unchanged. The amount of 45° misoriented grains in the a-b-plane is increased from an initially not measurable value to about 2 ⁰/₀₀ after ion irradiation and annealing. The effect of the irradiation on the microwave properties is described in the next chapter.

3. SURFACE IMPEDANCE MEASUREMENTS

The effective surface impedance of the films was measured between 4.2 K and 120 K and at 300 K by mounting them as an endplate of a cylindrical cavity, which is excited near 87 GHz in TE$_{013}$-mode for $R_{eff} \leq 0.5\,\Omega$ and in the TE$_{021}$-mode for $R_{eff} \geq 0.5\,\Omega$ [9]. The sensitivity and accuracy for the determination of the effective surface resistance R_{eff} and penetration depth λ_{eff} of the films is limited by the losses and thermal expansion of the copper host cavity to about $\pm 3\,\mathrm{m}\Omega$ and $\pm 10\,\mathrm{nm}$, respectively. According to the available microwave power, the peak magnetic surface field at the sample never exceeded 20 A/m. The intrinsic values for the surface resistance R_s and the penetration depth λ of the films, which describe the material properties, are somewhat smaller than the effective ones depending on the relative film thickness d/λ. They were determined by an analysis of the film-substrate sandwich based on impedance transformations [11]. The accuracy for the absolute values of R_s and λ is limited apart from the measurement accuracy by the variation of the film thickness to about $\pm 10\,\%$.

For YBCO a local relationship between the electromagnetic fields and the current density exists. Therefore, the complex surface impedance $Z_s = R_s + i\omega\mu_0\lambda$ can be described in terms of the complex conductivity $\sigma = \sigma_1 - i\sigma_2$ as $Z_s = (i\omega\mu_0/\sigma)^{1/2}$, where $\sigma_2 = 1/\omega\mu_0\lambda^2$ describes the supercurrent density. In the limit $\sigma_1 \ll \sigma_2$, which is fulfilled for our films for $T < 0.8\,T_c$, $R_s = 0.5\omega^2\mu_0^2\lambda^3\sigma_1$ results [12]. For normal conductors σ_2 vanishes, and one obtains the well-known skin effect formula for $Z_s = (\omega\mu_0/2\sigma_1)^{1/2}(1+i)$.

The resistivity $\rho_1 = 1/\sigma_1$ determined from microwave measurements provides information about the quality of the whole sample. In Fig.1, the $\rho_1(T)$ data in the normal conducting regime are compared for the ion irradiated films. In good agreement to earlier results of the dc-resistivity [4], we observe a reduction of T_c and an increase of ρ_1 proportional to the irradiation dose, while the slope of $\rho_1(T)$ below 120 K is unchanged and the metallic behaviour is still observed. However, the ratio $\rho_1(300)/\rho_1(100)$ decreases with increasing irradiation dose (Tab.1). Annealing at 500° C recovers partially the above described effects. A fluence of 10^{17} neutrons/cm^2 lead to a much stronger degradation of the resistivity, while T_c is nearly unchanged (Tab.1).

From the measurement of the resonant frequency shift we obtain only the change of the penetration depth with temperature. The absolute $\lambda(0)$ values are determined from fits to the weak-coupling BCS-theory in the clean limit as shown in Fig.2. The penetration depth also increases with increasing ^{16}O ion fluence and decreases after annealing. Surprisingly, the neutron irradiated samples show no change in $\lambda(T)$ for sample 5 (Tab.1) and a completely different, not BCS-like $\lambda(T)$ shape for sample 4.

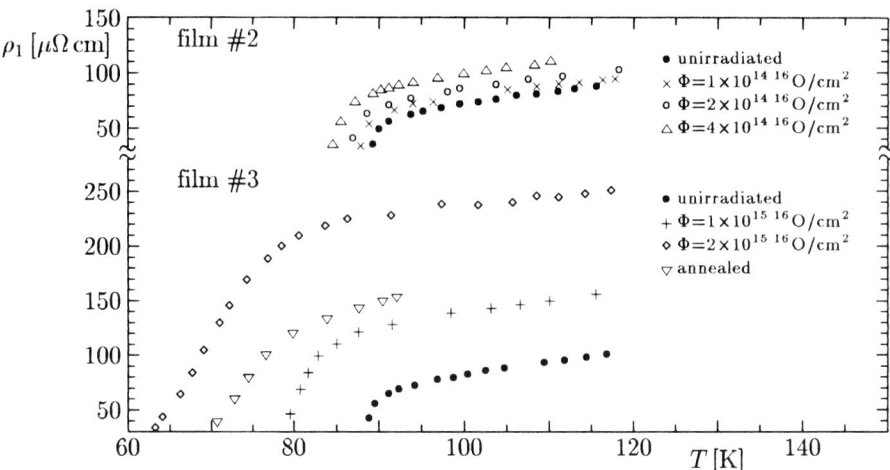

Fig.1 : $\rho_1(T)$ in the normal conducting state of the samples 2 and 3 as a function of
the irradiation fluence with 25 MeV ^{16}O and after annealing.

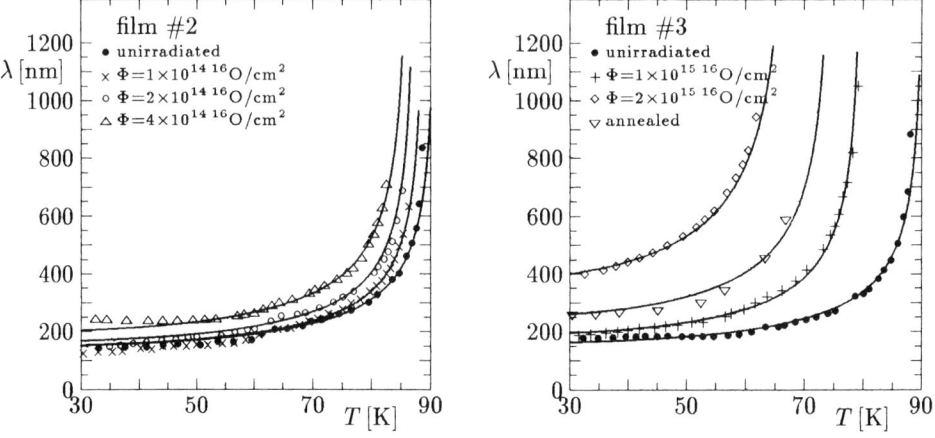

Fig.2 : Temperature dependence of the penetration depth for the same films and fluences as in Fig.1. The lines result from fits to the $\lambda_{BCS}(T)$ dependence.

In contrast to T_c, ρ_1 and $\lambda(0)$, the surface resistance below 70 K of the ion irradiated samples shows a much more complicate behaviour, as shown in Fig.3. For both
samples the initially nearly constant R_s decreased most surprisingly after the first irradiation, while further irradiations caused an increase of R_{res}. It is remarkable that
sample 2 shows increased R_{res} already for $\Phi \leq 4 \times 10^{14}/\text{cm}^2$, while sample 3 provides an
improved R_{res} for $\Phi = 1 \times 10^{15}/\text{cm}^2$. However, there is a clear difference between the
effect of ion doses below (#2) and above $10^{15}/\text{cm}^2$ (#3) in the shape of the $R_s(T)$ curve.
While sample 2 shows steep transition curves with residual $R_s(T)$ curves of different
slope, sample 3 shows a gradual, monotonous decrease of $R_s(T)$. An also remarkable
result is the strong increase of R_{res} after annealing, although the other properties (ρ_1,

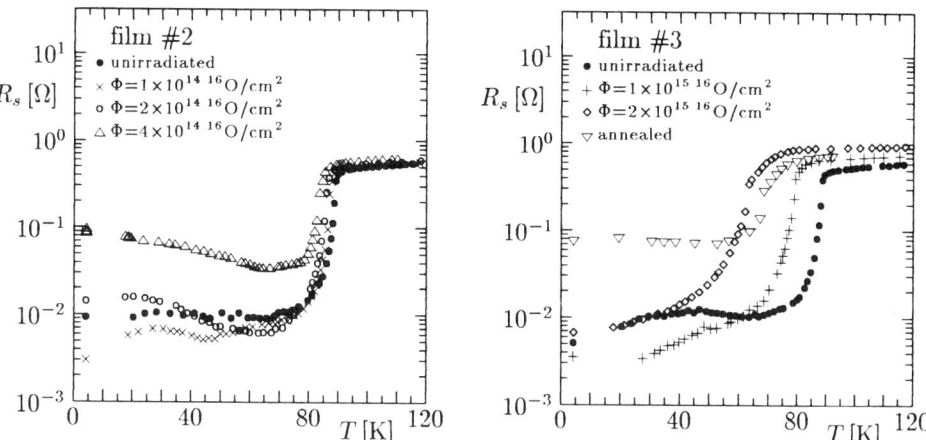

Fig.3 : $R_s(T)$ at 87 GHz for the same films and fluences as in Fig.1.

T_c and $\lambda(0)$) have been improved. In comparison to the ion irradiation, a fluence of 10^{17} neutrons/cm² lead to a much stronger degradation of R_{eff} at all temperatures as shown in Fig.4, as expected from similar results on neutron irradiated YBCO ceramics [13]. For the further interpretation of the data it is important to note that the irradiation has produced uniformly distributed defects in the films. This is proven by the continuity of the surface resistance data near T_c, where the microwave field distribution is switched from the TE_{013} to the TE_{021}-mode.

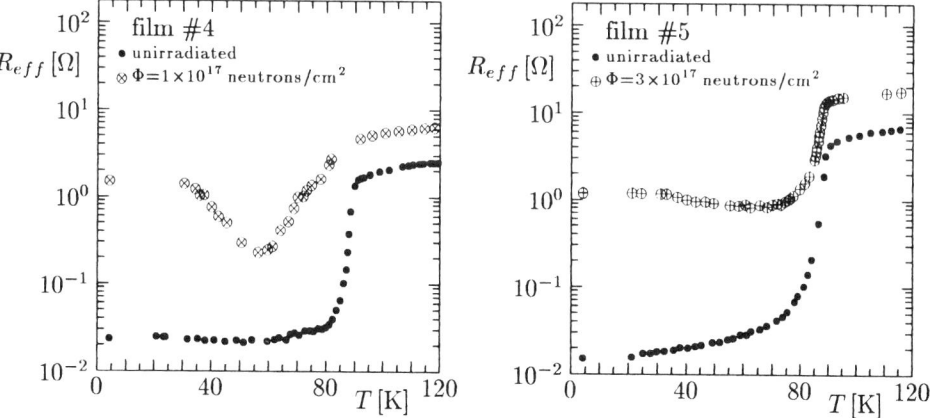

Fig.4 : $R_{eff}(T)$ at 87 GHz for the neutron irradiated films.

4. DISCUSSION

For all irradiated samples the point-like nature of the produced defects has been demonstrated by nearly unchanged X-ray diffraction patterns. The uniform distribution of the defects is proven by measurements with different microwave field distributions, which lead to the same R_s. In most cases steep transitions have been obtained, suggesting that granularity is not the dominant loss mechanism after irradiation.

In case of the ion irradiation these defects are at least partially due to oxygen deficiency, which is shown by the lengthening of the c-axis and a corresponding T_c reduction even after annealing. These point-like defects act as scattering centers, which lead to the observed increase of ρ_1 above T_c proportional to Φ. All kinds of normal conducting, semiconducting or isolating defects lead to a locally enlarged penetration depth. Therefore, one expects that the measured average λ increases with Φ, as observed. The residual surface resistance R_{res} shows a different behaviour, which seems not to be correlated with Φ, but the two-fluid model $(R_{res} \sim \lambda^3/\rho_{res})$ suggests a monotonous decrease of the density of non pairing charge carriers [14] or a decrease of their mean free path with increasing Φ at least for the strongly irradiated sample (Tab.1). Annealing recovers partially the irradiation induced effects on T_c, λ and ρ_1, but increases the surface resistance below T_c, so there seems to be an additional loss mechanism, which affects mainly the superconducting state. Possibly these losses are due to the increased amount of $45°$ misoriented grains, which have been associated with poor microwave properties [15]. Moreover we cannot exclude pollution by interdiffusion during the long annealing time.

The neutron irradiation seems to produce other defects than the ion irradiation. Above T_c the expected increase of ρ_1 is obtained, but we observe completely different effects in the superconducting state. This was not expected from the j_c measurements, which have shown j_c enhancement even in zero field. The penetration depth and T_c remain constant, while R_s shows a strong increase, what in terms of the two-fluid model suggests a strong increase of the density of non pairing charge carriers or an increase of their mean free path.

We have shown that ion and neutron irradiation are good methods to influence the microwave losses in different ways. Especially it is possible to reduce the losses, what proves that there exists a non intrinsic loss mechanism besides the irradiation induced defects. Further systematic microwave measurements on irradiated samples with initially lower losses are planned to confirm and to understand these remarkable results. In addition, measurements in superposed dc-magnetic fields can provide informations about the nature of the loss mechanism.

The work has been funded in parts by the German Federal Minister for Research and Technology (BMFT) under the contract numbers 13N5502 and TK03360.

REFERENCES

1 O. Eibl and B. Roas, J. Mat. Rev. **5**, 2620 (1990).

2 H. Piel and G. Müller, IEEE Trans. Magn. **MAG-27**, 854 (1991).

3 J. Dumas et al., J. Less-Common Metals, 164 & 165, 1252 (1990).

4 B. Roas et al., Appl. Phys. Lett. **54**, 1051 (1989).

5 W. Schindler et al., Physica C 169, 117 (1990).

6 B. Hensel et al., Phys. Rev. **B42**, 4318 (1990).

7 B. Roas et al., Appl. Phys. Lett. **53**, 1557 (1988).

8 W. Schindler et al., subm. to J. Appl. Phys.

9 N. Klein et al., Appl. Phys. Lett. **54**, 757 (1989).

10 B. Bucher et al., J. Less-Common Metals, 164 & 165, 20 (1990).

11 N. Klein et al., J. Appl. Phys. **67**, 6940 (1990).

12 G. Müller et al., J. Superconductivity **3**, 235 (1990).

13 D. W. Cooke et al., Appl. Phys. Lett. **26**, 2462 (1990).

14 K. F. Renk et al., subm. to Europhys. Lett.

15 S. S. Laderman et al., Phys. Rev. **B43**, 2922 (1991).

High T$_c$ Superconductor Thin Films
L. Correra (Editor)
113

Isothermal sections in the system Y – Ba – Cu – O in the range 800 to 1100°C

W. Bieger[a], G. Krabbes[a], U. Wiesner[a], M. Ritschel[a], J. Hauck[b] and H. Altenburg[c]

[a]Zentralinstitut für Festkörperphysik und Werkstofforschung, Helmholtzstr. 20, O–8027 Dresden, FRG

[b]Institut für Festkörperforschung der KFA, W–5170 Jülich, FRG

[c]Fachhochschule Münster, W–4430 Steinfurt, FRG

Abstract

On the base of phase equilibrium investigations along the sections Y_2BaCuO_5 – CuO_x and $YBa_2Cu_3O_{7-\delta}$ – CuO_x and literature data a set of isothermal sections of the quasiternary system $YO_{1.5}$ – BaO – CuO_x in air atmosphere has been derived. Preliminary results concerning the influence of oxygen partial pressure will be discussed.

1. INTRODUCTION

The knowledge of phase relations in the system Y – Ba – Cu – O is an important precondition to optimize the preparation conditions of ceramic materials as well as thin films of the superconducting phases. Much work has been focussed to clarify the phase relations along several sections of the quasiternary $YO_{1.5}$ – BaO – CuO_x system (see e.g. [1-5]) using DTA/TG investigations. Since the melting behaviour seems to be very complicated there remained many uncertainties and contradicting results. Attempts to systematisize the behaviour have been published by Aselage and Keefer [6] as well as Byeong-Joo Lee and Dong Nyung Lee [7]. The consideration in ref. [7] is based on a computer calculation using the concept of free–energy minimization supposing the occurrence of only one type of melt.

The aim of this paper is a consistent reflection of the melting process at 20 kPa O_2 in both, a set of isothermal sections as well as phase relations along the binary sections Y_2BaCuO_5 – CuO_x and $YBa_2Cu_3O_{7-\delta}$ – CuO_x. DTA/TG measurements have been supported by soaking experiments [8] which are very useful for melting behaviour investigations.

It was shown e.g. by Ahn et al. [9] that the phase relations depend on the oxygen partial pressure. A first attempt to clarify the phase equilibrium composition along the section Y_2BaCuO_5 – CuO_x in argon atmosphere on the base of soaking experiments will be briefly touched.

2. EXPERIMENTS

DTA/TG investigations have been carried out with a heating rate of 10°C/min from room temperature to 1100°C using a MOM derivatograph. The same heating rate was used at the DSC study in a Netzsch STA 429 equipment for the thermal analysis up to 1420°C.

The soaking method which has been proposed by Nevriva et al. [8] is based on a separation of the melt from the residual solid. A pellet of a mixture the melting behaviour of which has to be investigated is put on a porous substrate tablet (usually $YO_{1.5}$). Both tablets are heated two days at a defined temperature, and the arising melt is soaked by the substrate tablet. The residual solid is characterized by XRD.

3. THE MELTING BEHAVIOUR IN AIR

To clarify the melting behaviour of the quasiternary $YO_{1.5}$ – BaO – CuO_x system several soaking and DTA/TG investigations have been carried out. Especially the quasibinary sections $YBa_2Cu_3O_{7-\delta}$ – CuO_x and Y_2BaCuO_5 – CuO_x have been studied in detail. Further information has been taken from literature. As shown in Figure 1, $YBa_2Cu_3O_{7-\delta}$ and CuO_x are coexisting at 920°C. But already at 940°C this coexistence is displaced by two three phase equilibria, 211+ 123+ L and 211+ CuO+ L, respectively, separated by the two phase region 211+ L. This result corresponds to DTA/TG investigations which indicate an endothermic peak at 940°C. In addition to CuO, $Y_2Cu_2O_5$ was found by XRD analysis of the residual solids from CuO rich initial mixtures at 960°C. This means that $Y_2Cu_2O_5$ forms a further equilibrium with the liquid phase at about 950°C. Probably the phase relations will be changed at about 1030°C. Except that CuO is reduced to Cu_2O at 1025°C, the Y_2O_3+ L equilibrium joins the considered section. XRD of the residual solid of soaking experiments at 1060°C indicates Y_2O_3. The number of performed experiments is not sufficient yet to get a quantitatively proved diagram. The broken lines indicate probable phase equilibria.

The phase relations according to soaking experiments along the section Y_2BaCuO_5 – CuO_x are represented in Figure 2. The substitution of the Y_2BaCuO_5 – CuO_x coexistence by several multiphase equilibria at about 950°C is caused by the formation of the $Y_2Cu_2O_5$+ L equilibrium. The value observed by DTA/TG (968°C) is somewhat higher. Up to about 1030°C soaking experiments indicate only three multiphase equilibria, namely $Y_2Cu_2O_5$+ 211+ L, $Y_2Cu_2O_5$+ L and $Y_2Cu_2O_5$+ CuO+ L. At higher temperatures $YO_{1.5}$ joins the considered section and considerable changes of phase equilibria have been observed. Nearly at the same temperature copper oxide will be reduced to Cu_2O. Soaking experiments up to 1080°C show that the $YO_{1.5}$+ L two phase equilibrium becomes broader and the $Y_2Cu_2O_5$+ L equilibrium is shifting towards the CuO_x phase. Possible phase equilibria at higher temperatures which result from stability limits of pure phases and binary mixtures discussed in literature (see e.g. [10]) are represented as broaken lines in Figure 2.

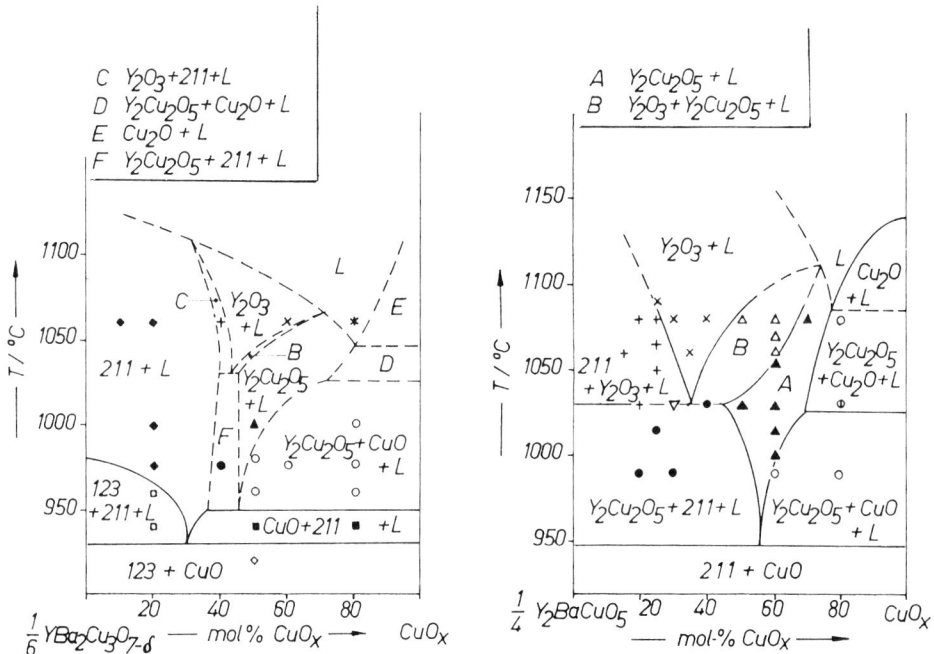

Figure 1. Soaking experiments along the section $YBa_2Cu_3O_{7-\delta}$ – CuO_x and deduced possible phase equilibria
($\Diamond$: 123+ CuO, $\square$: 123+ 211+ L, $\blacksquare$: CuO+ 211+ L, $\blacklozenge$: 211+ L, $\bullet$: $Y_2Cu_2O_5$+ 211+ L, $\blacktriangle$: $Y_2Cu_2O_5$+ L, $\circ$: $Y_2Cu_2O_5$+ CuO_x+ L, $+$: Y_2O_3+ 211+ L, $\times$: Y_2O_3+ L, $*$: L)

Figure 2. Soaking experiments along the section Y_2BaCuO_5 – CuO_x and deduced possible phase equilibria
(symbols see Figure 1, $\triangle$: Y_2O_3+ $Y_2Cu_2O_5$+ L, $\triangledown$: Y_2O_3+ $Y_2Cu_2O_5$+ 211+ L, Φ : $Y_2Cu_2O_5$+ CuO+ Cu_2O+ L)

Starting from these investigations and taking into account the DTA/TG study of the BaO – CuO_x partial system [11] it becomes possible to construct a consistent picture of isothermal sections in the CuO_x rich part of the $YO_{1.5}$ – BaO – CuO_x system. Some characteristic changes of phase relations at defined temperatures are represented in Figure 3. At 930°C and 950°C, respectively, the section $YBa_2Cu_3O_{7-\delta}$ – CuO_x is intersected by the formation of the 211+ L and $Y_2Cu_2O_5$+ L equilibria. $YO_{1.5}$ forms an equilibrium with the melt at about 1030°C. A more detailed discussion will be given in [10].

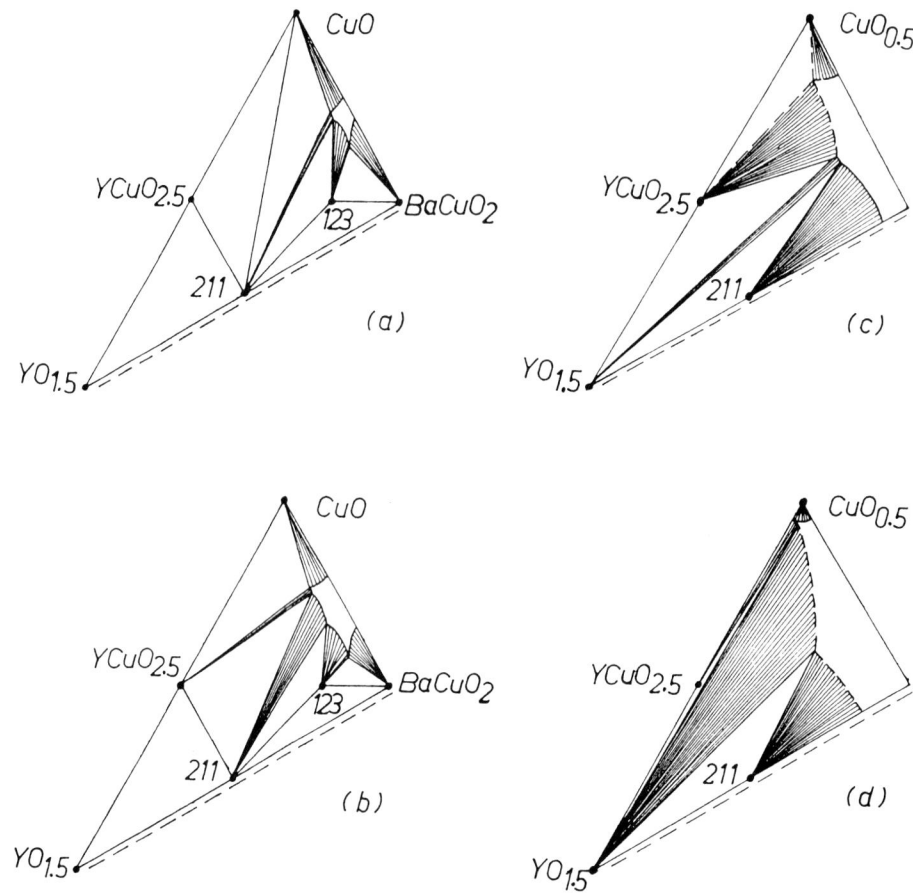

Figure 3. Three dimensional projections $YO_{1.5}$ – BaO – CuO_x of isothermal sections at (a) 930°C, (b) 950°C, (c) 1030°C, and (d) ≈ 1110°C

However, it is worth noting that the melting process is always combined with an oxygen release. Therefore, the figures have to be considered as three dimensional projections of a complicated quaternary system. The question of a possible existence of $BaCu_2O_2$ at higher temperatures in air could not be decided.

At about 800°C the existence of further ternary compounds has to be discussed. $YBa_2Cu_4O_8$, first detected in thin film preparation is also observed in bulk materials.

The phase which has been synthesized by Kaldis et al. [12] at high oxygen pressures can also be obtained in air. A precondition of its formation is the usage of fine pulverized $YBa_2Cu_3O_{7-\delta}$ and CuO. Owing to a kinetic hindrance the reaction is not successful if annealed initial compounds are used. The most convenient synthesis temperature is 830°C, the phase is existent up to about 840 to 850°C. The thermodynamic stability of $YBa_2Cu_4O_8$ and $Y_2Ba_4Cu_7O_{15}$ is not uncontradictedly evident until now.

4. PHASE EQUILIBRIA AT DIMINISHED OXYGEN PRESSURES

Whereas the thermodynamic stability of the $YBa_2Cu_3O_{7-\delta}$ phase is strongly dependent on oxygen partial pressure, the influence of $p(O_2)$ on the Y_2BaCuO_5 phase is only weak. The phase relations are changing significantly at diminished oxygen pressures as it was principially shown by Ahn et al. [9] at 850°C. Preliminary investigations along the section Y_2BaCuO_5 – Cu_2O in argon ($\approx$ 1 Pa O_2) indicate a partial decomposition of mixtures at about 925°C without the occurrence of a melt. Already at 935°C considerable quantities of the solid are melting. The remaining condensed phases are Y_2O_3 and Y_2BaCuO_5 from an initial mixture with less than 40 mol-% Cu_2O, $YCuO_2$ and Y_2O_3 for 40 to 60 mol-% Cu_2O and $YCuO_2+$ Cu_2O for more than 60 mol-% Cu_2O. Up to 975°C the phase equilibrium fields are shifting but no additional coexistence regions could be observed. The melting process is also accompanied by a mass loss.

5. CONCLUSIONS

The ternary eutectic temperature of the Y – Ba – Cu – O system is about 890°C in air which is much lower than the peritectic decomposition temperature of $YBa_2Cu_3O_{7-\delta}$ (980°C). Therefore, inhomogeneities in the samples cause the local melting already at lower temperatures.

Melt and coexisting condensed phases do not belong to the same section in view of the oxygen content. Because the melting process is accompanied by an oxygen loss a quasiternary $YO_{1.5}$ – BaO – CuO_x representation can only be considered as an approximation.

The phase stabilities and phase relations are drastically changing by diminishing the oxygen partial pressure. Already at temperatures where no melt appears the phase equilibria in argon are completely different from that in air.

The authors wish to express their thanks to Mr. F. Hansch for careful experimental work, Mr. G. Kunze for DTA/TG measurements, Dr. G. Leitner for DSC investigations as well as Mrs. A. Teresiak and Mrs. C. Heße for XRD analysis. This work is a part of a project granted by the Federal Minister of Science and Technology (BMFT) under N° 13 N 5897.

REFERENCES

1 F. Licci, P. Tissot and H. J. Scheel, J. Less-Common Met. 150 (1989) 201
2 K. Oka, K. Nakane, M. Ito, M. Saito and H. Unoki, Jap. J. Appl. Phys. 27 (1988) L 1065
3 N. Nevriva, E. Pollert, J. Sestak, L. Matejkova and A. Triska, Thermochim. Acta 136 (1988) 263
4 G. A. Mikirticheva, V. I. Shitova, O. G. Zhigareva, S. K. Kuchaeva, L. Yu. Grabovenko and R. G. Grebenshchikov, Zh. Neorg. Khim. 35 (1990) 223
5 Wei Zhang, Kozo Osamura and Shojiro Ochiai, J. Am. Ceram. Soc. 73 (1990) 1958
6 T. Aselage and K. Keefer, J. Mater. Res. 3 (1988) 1279
7 Byeong-Joo Lee and Dong Nyung Lee, J. Am. Ceram. Soc. 72 (1989) 314
8 N. Nevriva, P. Holba, S. Durcok, D. Zemanova, E. Pollert and A. Triska, Physica C 157 (1989) 334
9 B. T. Ahn, V. Y. Lee, R. Beyers, T. M. Gür and R. A. Huggins, Physica C 167 (1990) 529
10 G. Krabbes, U. Wiesner, W. Bieger, M. Ritschel and A. Teresiak, (in preparation)
11 G. Krabbes, U. Wiesner, W. Bieger, M. Ritschel and G. Kunze, in MASHTEC'90, Intern. Symposium Dresden, Apr. 24-27, 1990, Materials Science Forum 62-64 (1990) 75
12 J. Karpinski, S. Rusiecki, E. Kaldis, B. Bucher and E. Jilek, Physica C 160 (1989) 449

High T$_c$ Superconductor Thin Films
L. Correra (Editor)

Superconductivity and Electron-Phonon Interactions in Thin Films

R.P.Djajić
Faculty of Technical Sciences, University of Novi Sad, V.Vlahovića 3,
21000 Novi Sad, Yugoslavia

Abstract

In previous paper it was shown that electrons in thin films have a gap in harmonic approximation. Analyzing phonon characteristics in the film it was observed that the displacements in the film are much greater than in the bulk structure, causing the constant of the effective electron-electron interaction arising from the virtual phonon interchange, to be much bigger than the same constant connected with the bulk structure. Applying the BCS procedure on the T$_c$ calculations it was shown that due to the gap in harmonic spectra of electrons and greater interaction constant, T$_c$ in a film can be bigger for a whole order of magnitude than the T$_c$ in the bulk. Since perovskites act like layered structures with practically independent behavior of layer, the obtained result can be used to explain the high T$_c$ of the structures.

Here the effort will be made to explain the high T$_c$ in the superconducting ceramics using the fact that they practical behave as a system of independent thin layers of small grained structures [1-4]. The entire analysis will be related to one thin layer which will further be refered to thin film.

In [5] we investigated on electronic spectra in sputtered thin films and showed that electrons in harmonic approximation have a gap of such a size that only slightly changes the chemical potential with respect to the ideal structure.

In [6-9] we analyzed phonon spectra in thin films. We used the model of parabolic deformation due to sputtering and the equations of motion were solved in continual approximation. There, we did not calculate the ion displacements and therefore this problem will be analyzed here in more detail.

Since perovskites have the lattice constant along the c-axis (here it will be assumed that this direction corresponds to the z-axis in cubic structure) few times bigger then the planer lattice constants it is realistic to assume that the sputtered atoms would locate themselves along the z-axis. Consequently, along the z-axis we have a change in the lattice itself and the Hook's constant which in turn depends upon the same lattice constant. The distribution of masses changes as well along the z-axis. Assuming that the film is symmetrically deformed on both boundaries, in continual approximation the following laws can be given for the change in lattice constant and change in mass along the z-axis:

$$a_z(z)=a_z[1 - \frac{4}{L^2} \frac{n_0}{n_0+1}(z - \frac{L}{2})^2], \quad M(z)=\mu_m[1 + \frac{4}{L^2}n_0\frac{\mu_s}{\mu_m}(z - \frac{L}{2})^2] \tag{1}$$

In these expressions L is the thickness of the film, n_0 is the number of sputtered atoms between the two boundary layers of the film, μ_m are the masses of perovskites matrix and μ_s are the masses of sputtered atoms. Since the Hook's constant is of the form $C \simeq a_z^{-(p+2)}$,where p is the molecular interaction, it is evident that the Hook's constant along the z-axis depends on z.

Based on the given modeled assumptions using equations of motion the vibrational characteristics of the system are analyzed. On the details of the calculations as well as on the analysis of phonon spectra we will not take time here, since it was given in detail in [6-9]. Here, it should be emphasized only the expression for molecular displacements used:

$$u_{\alpha;n_xn_y}(\xi,\theta) = \Sigma_{k_x,k_y,\nu_z} F(k,\theta)\psi_{\alpha,\nu_z}(\xi)e^{ia(n_xk_x+n_yk_y)}(b_{\alpha;k_x,k_y,\nu_z} + b^+_{\alpha;-k_x,-k_y,\nu_z})$$

$$\tag{2}$$

where

$$z - \frac{L}{2} = \Lambda_{\nu_z}\xi, \quad \Lambda_{\nu_z} = (\frac{a_z L\Omega_{\alpha\alpha}}{2\omega_{\alpha;k_x,k_y,n_z}})^{1/2}g^{-1/4}.$$

$$g = n_0(\frac{p+2}{n_0+1} - \frac{\mu_s}{\mu_m}) , \quad \Omega_{\alpha\alpha}= \sqrt{C_{\alpha\alpha}/\mu_m}$$

and

$$\psi_{\nu_z}(\xi) = \frac{e^{-t^2/2}}{\sqrt{2^{\nu_z}\nu_z!\pi^{1/2}}} ; \quad H_{\nu_z}(\xi) = (-1)^{\nu_z}e^{\xi^2}\frac{d\nu_z}{d\xi^{\nu_z}}(e^{-\xi^2}) \tag{3}$$

Operators b^+ and b create and annihilate the phonons and $F(k,\theta)$ is the displacement amplitude which in mentioned references was not determined. This amplitude was determined here by the usual method. Using (2) the Hamiltonian of the system was diagonialized and the following expression obtained:

$$F_\alpha(k,\theta) = [\ \frac{\bar{h}}{2N_x N_y N_x \mu_m \Omega_{\alpha\alpha} \chi(k,\theta)}\ ]^{1/2} \tag{4}$$

$$\chi(k,\theta) = \frac{3(n_0+1)}{2n_0+3}\ (\ \frac{dak}{g^{1/2}}\ )^{1/2}\ [1 + \frac{n_0}{2g}\ \frac{\mu_s}{\mu_m}\ (1 - \tan^2\frac{\theta}{2})]\ \cos\frac{\theta}{2}$$

where θ is the azimuthal angle, $d=(2n_0+3)[2(n_0+1)N_z]^{-1}$, and N_x, N_y, and N_z are numbers of molecules along x, y, and z direction, respectively. On the basis of this and well known expression for the displacement amplitudes in the ideal structure there is a conclusion that the displacement amplitudes in film $F(k,\theta)\equiv F_f$ and amplitudes in ideal structure F_{id} are related in a following manner:

$$F_f = N_z^{1/4} F_{id} \tag{5}$$

which means that the displacement amplitude in film are much greater than in ideal structure. This conclusion is crucial for explanation of the fact that the critical temperatures in film (de facto critical temperatures in layered perovskites) are of the whole order of magnitude bigger than the critical temperatures in ordinary superconductors which are of the bulk structure.

It is well known that the effective electron-electron interaction constant is proportional to the square of the displacement amplitudes (by the way, on the basis of this, the isotope effect in the BCS theory was explained). Using this fact and (5) the following relations can be given:

$$G_f/G_{id} \simeq N_z^{1/2}, \quad G_{id} << G_f \tag{6}$$

Assuming that for coper pairing and explanation of the superconducting effects in film the BCS approach can be applied completely, the superconduction gaps in film and in corresponding ideal structure, can be written as:

$$\Delta_{f/id} = 2\bar{h}\omega_D \exp(-const/G_{f/id}). \tag{7}$$

On the basis of result in [10] it is easy to conclude that the Debye's energies and the densities of states in film and in ideal structure are slightly different therefore taken here to be approximately the same.

From the well known results for the standard superconductors, it can be taken that $\Delta_{id} \simeq 40k_B$ and $\hbar\omega_D \simeq 200k_B$, then from (7) it follows that $\exp(\text{const}/G_{id}) \geq 10$. On the other hand on the basis of (7) the following expression can be written:

$$\Delta_f = \Delta_{id}\exp\left(\text{const}\frac{G_f - G_{id}}{G_f G_{id}}\right) \tag{8}$$

which on the basis of the fact that $G_f \ll G_{id}$ deduces to

$$\Delta_f = \Delta_{id}\exp(\text{const}/G_{id}) \geq 10\Delta_{id} \tag{9}$$

Since critical temperatures T_f and T_{id} are proportional to the corresponding gaps we come to a conclusion that the critical temperature in layered perovskites are at least for the order of magnitude greater than in the ordinary ideal metallic structure. (Factor 10 comes from the estimate

$$\frac{\mu_{id}}{\mu_m} H_z^{1/2} \approx \frac{1}{3} 30 = 10$$

where $H_z \simeq 1000$ and μ_{id} is the mass of some usual superconductor).

It should be noticed that the following question can arise: may BCS formula (7) be applied when $F_f \gg F_{id}$? Within the framework of the BCS approach, the order of magnitude of F is irrelevant for deriving the result (7). Since BCS theory has a starting point in Frohlich's calculation [11], the problem of magnitude of F is connected with Frohlich's expansion and with the convergence of the infinite series for equivalent Hamiltonian. The problem of convergence is solved by the introduction of the "cut-off" into the Hamiltonian of electron electron interaction [11]. Frohlich [11] refutes Wentrel's claim [12] that F has to be small in order to stabilize electron-phonon system, and shows that system remains stable even for comparatively high F.

On the basis of above discussion, it is our opinion that the estimate (9) which follows from (7) is quite suitable for $F_f \simeq 10 F_{id}$.

To sum up the exposed analysis the following can be said. Relevant characteristic for explanation of high T_c in perovskites is its layering which in turn is responsible for the greater amplitudes of oscillation and therefore the greater effective electron-electron interaction constants which on the other hand defines the value of the critical temperature T_c.

References

[1] C.W. Chu, P.H. Hou, R.L. Meng, L. Gao, Z.J. Huang, and Y.Q. Wang, Phys. Rev. Lett. 58(1987) 405.

[2] C. Politis, et al., Z.Phys. B66 (1987) 141,275.

[3] C.U. Segre, et al., Nature 329 (1987) 227.

[4] M.R. Dietrich et al., Z.Phys. B66 (1987) 283.

[5] J.P. Šetrajčić, D.Lj. Mirjanić, B.S. Tošić, and G.Knežević, Electron Spectra in Symmetrically Sputtered Structures, 11th General Conference of the Condensed Matter Division-Exeter, April 1991 (Programme and Abstracts PC108, p.322).

[6] B.S. Tošić, J.P. Šetrajčić, R.P. Djajić and D.Lj. Mirjanić, Phys.Rev. B36(1987) 9094.

[7] B.S. Tošić, J.P. Šetrajčić, R.P. Djajić and D.Lj. Mirjanić, Int.J.Mod Phys. B1 (1088) 919.

[8] D.Lj. Mirjanić, R.P. Djajić, B.S. Tošić, J.P. Šetrajčić, FZKAAA 21 (1989) 303.

[9] J.P. Šetrajčić, R.P. Djajić, D.Lj. Mirjanić, B.S. Tošić, Physica Scripta 42 (1990) 732.

[10] B.S. Tošić, J.P. Šetrajčić, D.Lj. Mirjanić, and Z.V. Bundalo, Low-Temperature Properties of Thin Films, submitted to J.Low Temp. Phys.

[11] H.Frolich, Proc.Roy.Soc. A215 (1952) 291.

[12] Wentzel G. Phys. Rev. 83 (1951), 168.

High T$_c$ Superconductor Thin Films
L. Correra (Editor)

STRUCTURAL CONVERSION IN HIGH T$_c$ SUPERCONDUCTORS

Pham V. Huong^{a*}, J.P.Chaminade^b, J.C.Frison^b, Y.K.Park^c, J.C.Park^c, K.H.Kim^d and J.S.Park^d.

^{a}Laboratoire de Spectroscopie Moléculaire et Cristalline, (URA 124 C.N.R.S.)
^{b} Laboratoire de Chimie du Solide (L.P.8661 C.N.R.S.)
Université de Bordeaux I, 351, Cours de la Libération 33405 TALENCE, France
*Fax (33) 56 84 66 45

^{c}Korea Standards Research Institute, P.O.Box 3, Tae Dock, TAEJON
305-606 - Korea

^{d}Dept of Chemistry, Yonsei University, SEOUL 120-749, Korea

Abstract
   The structural conversion of Y$_{124}$ above 840° C into YBa$_2$Cu$_3$O$_7$ is found to be similar to that observed for Tl$_2$Ba$_2$CaCu$_2$O$_8$ into TlBa$_2$CaCu$_2$O$_7$.
   All these superconducting materials were studied by several physical methods, in particular, by X-ray diffraction and micro-Raman spectroscopy.
   The preferential departure of one Cu-O chain in Y$_{124}$ is analogous to the preferential departure of one Tl-O layer in Tl$_{2212}$. This mechanism is interpreted by weaker interaction forces between the involved layer and the adjacent ones, in comparison to the cohesion between the remaining layers. This result permits to localize an eventual intercalation in Y$_{123}$ as well as in Tl$_{1212}$.
   The structural conversion can be extrapolated to Bi$_2$Sr$_2$CaCu$_2$O$_8$ under appropriate thermal treatment. The success of such a degradation could constitute a new preparative route for obtaining BiSr$_2$CaCu$_2$O$_7$.

1. INTRODUCTION

   The stability of YBa$_2$Cu$_4$O$_8$, Y$_{124}$ in a large scale of temperature [1] and the increase in the T$_c$ by doping, in particular with Ca [2,3] induce a growing interest in this series of superconductors. Its composition is quite similar to that of Tl$_2$Ba$_2$CaCu$_2$O$_8$, Tl$_{2212}$ [4-6] :

$$Y \quad Ba_2 \quad Cu_2 \quad Cu_2 \quad O_8$$
$$\overline{\hspace{4cm}}$$
$$Ca \quad Ba_2 \quad Tl_2 \quad Cu_2 \quad O_8$$

   Effectively, in YBa$_2$Cu$_4$O$_8$, the two BaO^1 layers are intercalated by two Cu^{1}O^4 chain layers while in Tl$_{2212}$, the two BaO^1 layers are separated by two TlO^4 chain layers (Figs.1 & 2). The same parallelism is also obvious when one considers the structures of YBa$_2$Cu$_3$O$_7$ and TlBa$_2$CaCu$_2$O$_7$ (Figs.1 & 2).
   Remember that the critical temperatures T$_c$ of these superconductors are : Tl$_{2212}$ (124 K), Tl$_{1212}$ (80 K), Y$_{123}$ (92 K), and Y$_{124}$ (80 K) respectively. When doped with Ca, T$_c$ of latter increases and reaches 90 K.

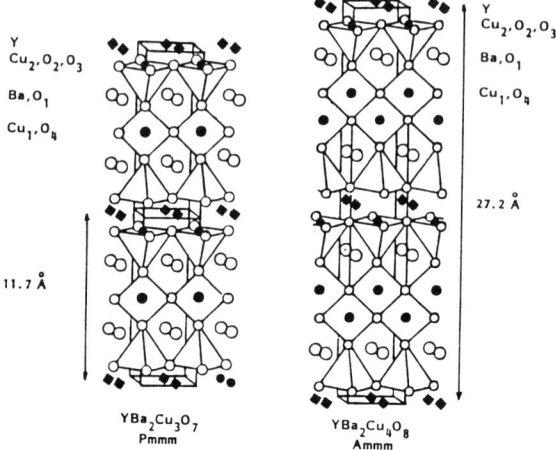

Figure 1 . Spatial structures of $YBa_2Cu_3O_7$ and $YBa_2Cu_4O_8$.

The stabilities of Y_{124} and Tl_{2212} will be described in this paper with respect to appropriate thermal treatment. The structural conversion involved will help to understand the difference of interaction forces between the layers of the materials and to deduce the intercalation mechanism in Y_{123} and Tl_{1212}.

2. EXPERIMENTAL

2.1. Sample preparation
YBa_2Cu_4O_8 was prepared by calcination of solid mixtures [Y_2O_3 + Ba(NO_3)_2 + CuO] with the proportion Y : Ba : Cu = 1 : 2 : 4 at 810° C in flowing oxygen and then sintering at 810° C for seven days in flowing oxygen with intermediate grindings.

The sample obtained presents a X-ray diffraction pattern (Fig.3) corresponding to $YBa_2Cu_4O_8$. Its resistivity *versus* temperature (Fig.4) gives R = 0 at 80 K.

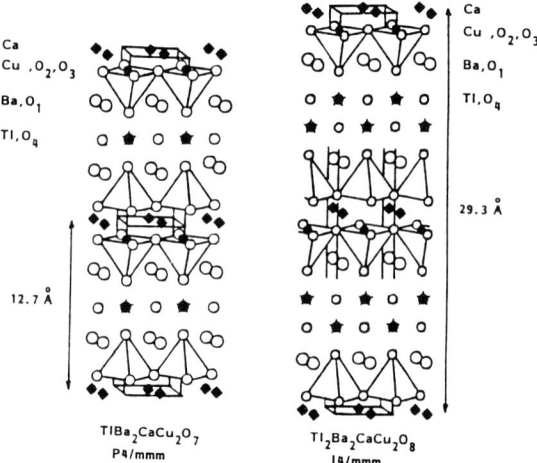

Figure 2 . Spatial structures of $TlCa_2CaCuO_7$ and $Tl_2Ba_2CaCu_2O_8$.

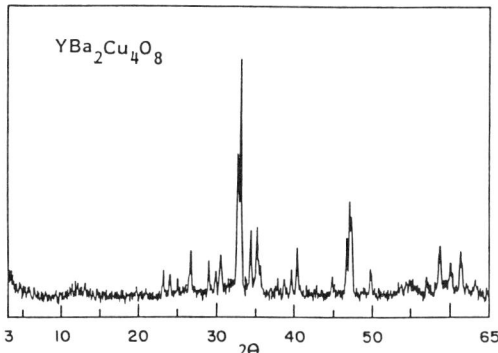

Figure 3 . X-ray diffraction pattern of $YBa_2Cu_4O_8$.

Single crystals of $Tl_2Ba_2CaCu_2O_8$ were prepared by a flux method using Tl_2O_3, BaO_2, CaO and CuO with an excess of BaO_2 and CuO. The best result was obtained with proportions $Tl : Ba : Ca : Cu = 2 : 2.6 : 2 : 4$. The mixture was put in a gold tube, itself sealed in a silica tube containing a low oxygen pressure, fired at 730° C for 24 hours, then at 955° C for 1 hour, and then cooled slowly (3° C per hour) to 850° C before cooling rapidly to room temperature. The single crystals obtained present a X-ray diffraction pattern (Fig.5) without impurities except Al and the plastic tape which came from the sample holder.

The T_c was determined, for the best crystals annealed under oxygen, at 107 K. This temperature is lower than those we observed with $Tl_2Ba_2Ca_2Cu_3O_{10}$ crystals at $T_c = 120$ K and powder pellet at $T_c = 125$ K.

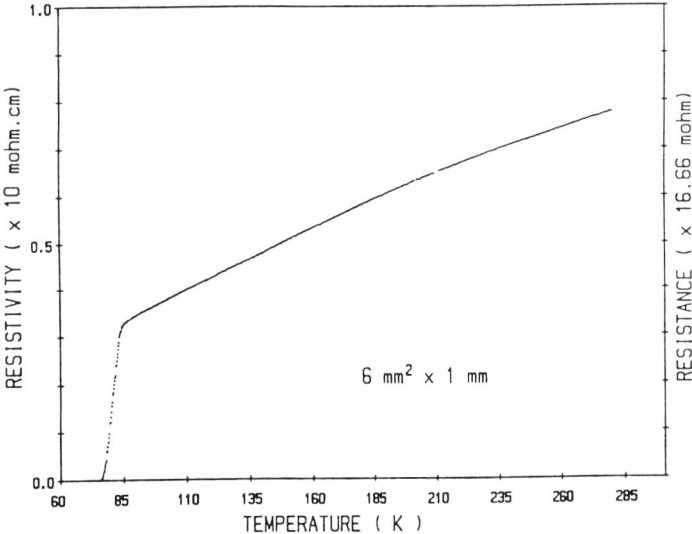

Figure 4 . Temperature dependence of resistivity of $YBa_2Cu_4O_8$ pellet.

2.2. The thermal treatment of both materials Y_{124} and Tl_{2212} was realized progressively and the characterization was made mainly by **X- ray diffraction** and by **micro-Raman spectroscopy**. The latter technique has been known to be very suitable for structural characterization of superconductors [7-12].

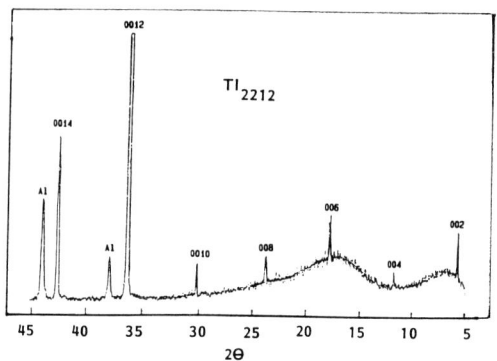

Figure 5 . X-ray diffraction pattern of $Tl_2Ba_2Ca_1Cu_2O_8$.

3 - RESULTS AND DISCUSSION

3.1. Raman spectra and structure

The Raman spectrum of $YBa_2Cu_4O_8$ is displayed on Figure 6. This spectrum is quite different from that of $YBa_2Cu_3O_{7-\delta}$. In particular, two Cu-O stretching vibrations are observed at 592 and 490 cm^{-1}, while $YBa_2Cu_3O_7$ has only one Cu^2-O^1 stretching observed along the c axis around 500 cm^{-1}. The Ba vibration appears at 98 cm^{-1}, while it is located at 118 cm^{-1} in $YBa_2Cu_3O_7$. The decreasing of this A_g vibrational frequency reflects an increasing of Ba-O^4 bond length and contributes to explain the expansion of the c parameter [13,14], already increased by the doubling of the CuO chain-layers.

This feature is similar to that of $Tl_2Ba_2CaCu_2O_8$ where two CuO stretching vibrations [10] were seen in ZZ polarization. Again, along the c axis,

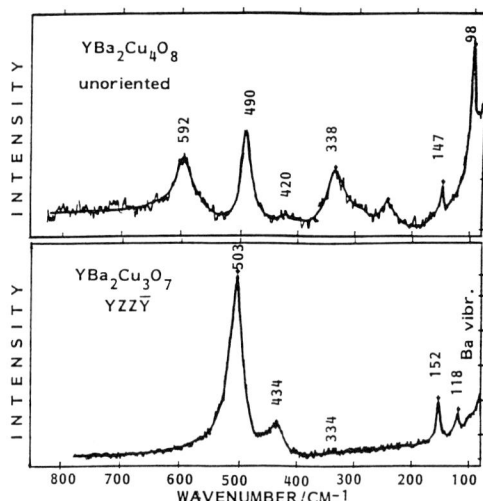

Figure 6 . Raman spectra of polycrystalline $YBa_2Cu_4O_8$ and $YBa_2Cu_3O_7$ single crystal.

$TlBa_2CaCu_2O_7$ presents only one Cu-O Raman band. One can notice that for the identification of Tl_2 as well as of Y_{124} it is necessary to look at ZZ polarization that is only specific, because along b or a axis, no doubling of Cu-O band is observed, analogous to Tl_1 and Y_{123} series.

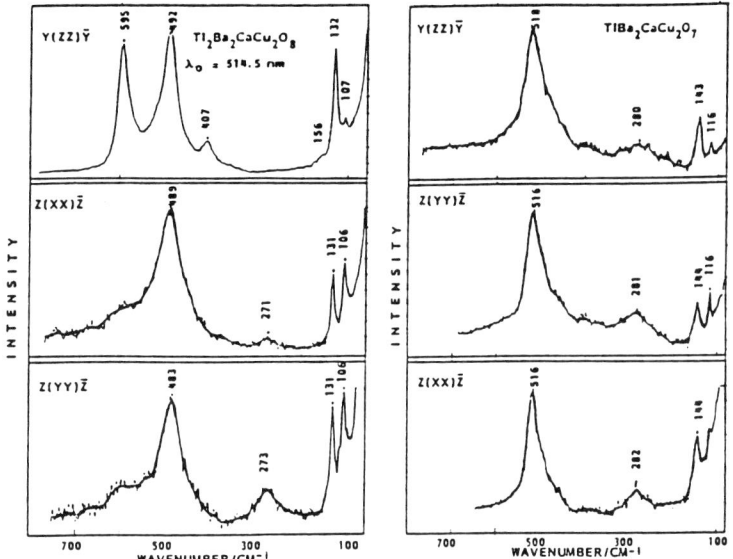

Figure 7. Polarized Raman spectra of single crystals $Tl_2Ba_2CaCu_2O_8$ and $TlBa_2CaCu_2O_7$.

3.2. Thermal treatments and structural conversion

Exposed at 940° C for 1.5 minutes, Y_{124} sample undergoes to a transformation. The resistivity curve presents two waves, the second appearing at higher temperature. The Raman spectrum as well as the X-ray diffraction patterns reveal the appearance of Y_{123} and CuO. This confirms an earlier observation that after an exposure at 940° C for 10 minutes, the sample gives only characterics of a single phase : the resistivity curve shows R = 0 at T_c = 90 K and the Raman spectrum corresponds to Y_{123} and CuO [15]. No trace of Y_{124} persists. Thus total conversion was recorded :

$$YBa_2Cu_4O_8 \rightarrow YBa_2Cu_3O_7 + "CuO"$$

The micro-Raman spectra recorded on different $YBa_2Cu_3O_{7-\delta}$ obtained after thermal treatment give a Cu^2-O^1 stretching frequency v_H varying between 493 and 499 cm^{-1}. When applying the relation x = 0.025 × v_H - 5.57 given by Huong [12,16] : , one can deduce that after thermal conversion, the resulting phase can oscillate between $YBa_2Cu_3O_{6.7}$ and $YBa_2Cu_3O_{6.9}$.

The additional departure of oxygen may not mask the main effect of the preferential departure of a whole Cu-O chain layer in the conversion.

A parallelism is observed in the thermal treatment of $Tl_2Ba_2CaCu_2O_8$. After a sintering at 670° C under argon for 24 hours and then under oxygen for 24 hours followed by an annealing at the same temperature under argon for 24 hours, the single crystal is destroyed and leads to an assembly of lamellas having their c axis slightly disoriented between them. The micro-Raman of both part of the resulting sample reveals the characteristics of Tl_{1212} phase, with only one Cu-O band at 520 cm^{-1} and its low frequency band at 116 cm^{-1}.

The X-ray pattern confirms the appearance of the Tl_{1212} phase on the basis of a c parameter equaling to 12.798 Å, while the initial Tl_{2212} gives a c parameter to 29.413 Å. The preferential and total departure of a TlO plane from the double layer of Tl_{2212} implies a glide of part of the unit cell by a vector (a + b)/2. This mechanical displacement can explain how the sample initially mono-crystalline transforms itself into an assembly of stacked lamellas.

This structural conversion

$$Tl_2Ba_2CaCu_2O_8 \rightarrow TlBa_2CaCu_2O_7 + "TlO"$$

is absolutely similar to that of Y_{124} into Y_{123}, "TlO" disappearing probably under the form of gaseous $(Tl_2O + \frac{1}{2}O_2)$ at the temperature of the sample treatment.

3.3. Interaction forces between layers and intercalations

The preferential departures of one TlO chain-layer in Tl_{2212} and of one Cu-O chain-layer in Y_{124} by thermal treatment allow to deduce that **the interaction forces between these layers and the adjacent layers is weaker than the forces linking other layers.** A calculation is underway. This result also allows to localize an eventual intercalation in Y_{123} or Tl_{1212} : it could be between the Cu^1-O^4 plane and the adjacent Ba-O^1 plane in Y_{123} and between the Tl-O^4 plane and the adjacent Ba-O^1 plane in Tl_{1212}.

3.4. Preparation route for obtaining $BiSr_2CaCu_2O_7$

Although big chemical differences exist between the above Y_{124}, Tl_{2212}, and Bi_{2212}, we can imagine some similarity between them.

By appropriate treatments, the following structural conversion could happen :

$$Bi_2Sr_2CaCu_2O_8 \rightarrow BiSr_2CaCu_2O_7 + "BiO"$$

This could be a preparation route for obtaining Bi_{1212} and other Bi_1 series.

4. REFERENCES

1   H.A.Ludwig, W.H.Fietz, M.C.Dietrich, H.Wuhl, J.Karpinski, E.Kaldis, S.Rusiecki,
     Physica C, **167** (1990) 335.
2   T.Miyatake, M.Kosoge, N.Koshizuka, H.Takahashi, N.Mori, S.Tanaka,
     Physica C, **167** (1990) 299.
3   R.G.Buckeley, J.L.Tallon, D.M.Pooke, M.R.Preslaud, Physica C, **165** (1990) 391.
4   Z.Z.Sheng, A.M.Hermani, Nature, **332** (1988) 138.
5   C.Martin, A.Maignan, J.Provost, C.Michel, M.Hervieu, R.Tournier,
     B.Raveau, Physica C, **168** (1990) 8.
6   M.Morosin, D.G.Ginley, E.L.Venturini, R.J.Baughman, C.P.Tigges,
     Physica C, **172** (1991) 213.
7   P.V.Huong, J.C.Bruyère, E.Bustarret, P.Granchamp , Solid State Comm., **72** (1989) 191.
8   P.V.Huong, E.Oh-Kim, K.H. Kim, D.Kim, J.S.Choi ,
     Materials Science and Engineering, **A109** (1989) 337.
9   P.V.Huong, J.Less.Common Metals, **165** (1990) 1193.
10   P.V.Huong, J.C.Frison, J.P.Chaminade,
     in "Modern Aspects of Superconductivity", Suryanayanan Ed., IITT Paris, 1989, p.211.
11   P.V.Huong, A.L.Verma, J.P.Chaminade, L.Nganga, J.C.Frison,
     Mat. Sciences and Engineering, **B5** (1990) 255.
12   E.T.Heyen, R.Liu, C.Thomsen, R.Kremer, M.Cardona, J.Karpinski, E.Kaldis, S.Rubiecki,
     Phys.Rev.B., **41**, (1990) 11058.
13   A.W.Hewat, P.Fisher, E.Kaldis, E.A.Hewat, E.Jlek, J.Karpinski,
     S.Rusiecki, J.Less Common Metals, **164** (1990) 39.
14   D.E.Morris, A.G.Markelz, B.Fayn, J.KH.Nickel, Physica C, **168** (1990) 153.
15   K.H. Kim,P.V.Huong, E.Oh-Kim,M.Lahaye, S.K.Cho, B.C.Kwak,
     J.Less.Common Metals, **164** (1990) 1201.
16   L.Nganga., P.V. Huong, J.P.Chaminade, P.Dordor, K.Frölich , M.Jergel ,
     J.Less.Common Metals, **164** (1990) 208.

High T$_c$ Superconductor Thin Films
L. Correra (Editor)
© 1992 Elsevier Science Publishers B.V. All rights reserved.   131

Phenomenological model for the resistance versus temperature behaviour of imperfect YBaCuO films

A. Kreisler, F. Hosseini Teherani, J.M. Depond and J. Baixeras

Laboratoire de Génie Electrique des Universités Paris 6 et Paris 11
Unité de Recherche Associée 127 du CNRS - Ecole Supérieure d'Electricité
Plateau de Moulon, 91192 GIF SUR YVETTE CEDEX, FRANCE.

Abstract
The resistive transition of YBaCuO granular thin films is interpreted in terms of a resistance network representing various superconducting and nonsuperconducting phases. The model capabilities are illustrated by examples pertaining to thin films, rf sputtered on substrates of industrial interest, and processed *ex situ* by the rapid thermal annealing technique.

1. INTRODUCTION

The resistance versus temperature variation of high-Tc superconducting films sometimes exhibit a slow decrease or varying slopes in the transition region. Such a behaviour may be encountered for instance when elements from the substrate have contaminated the film during a high temperature elaboration process, e.g. silicon in YBaCuO. It can also indicate the presence of several superconducting phases belonging to the same family, e.g. $YBa_2Cu_3O_{7-\delta}$ and $YBa_2Cu_4O_8$.

We have made use of a simple phenomenological model to describe such *R-T* variations, where each phase is figured by an electrical parallel circuit [1]. Each parallel cell is the image of two sub-phases, respectively superconducting and non-superconducting.This model has been exploited to interpret results on YBaCuO rf sputtered films processed by rapid thermal annealing (RTA)[2,3].

After giving some details on the sample preparation, we shall first present the model and show its ability to simulate various situations. We shall then proceed to use the model to fit typical experimental results, and finally present a short discussion, in relation with the identification of the various phases and sub-phases.

2. SAMPLE PREPARATION AND CHARACTERISTICS

Thin YBaCuO films were rf sputtered on two kinds of substrates, chosen for their potential use in various applications. Yttria-doped polycrystalline zirconia (YDPZ) is in fact interesting for developing cheap current limiters or switches [4], whereas silicon wafers (with SiO_2/Si_3N_4 buffer layers) are obvious candidates for microelectronics or optoelectronics devices [5]. *Ex situ* rapid thermal annealing was subsequently performed on the sputtered films, as previously described in detail [2-5].

A typical RTA cycle included a heating up step (for 10 s to 4 1/2 minutes) and high temperature dwell (for 5 s to 60s, at 830 to 950 °C) under flowing argon, and cooling down under flowing oxygen (for about 5 minutes). Moreover, cumulated RTA cycles were shown to improve the superconducting characteristics, as illustrated by figure 1 which refers to a silicon-based substrate [5].

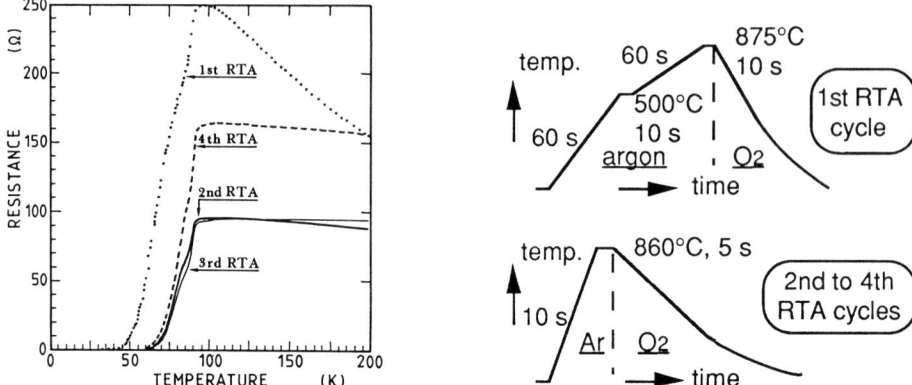

Figure 1. *Left*: R vs T plots for a 1.1 μm thick YBaCuO film sputtered on a $SiO_2(250Å)/Si_3N_4$ (900Å)/Si (100) substrate. For this sample, the cumulated RTA process allowed optimal characteristics to be reached after the third cycle [5]. *Right*: RTA cycles applied to the sample.

This procedure clearly shows how the semiconducting behaviour in the normal state, followed by a slow transition to the superconducting state, is improved by subsequent RTA cycles, to a nearly metallic behaviour in the normal state and a sharper superconducting transition. However, the growth of unwanted parasitic phases is detrimental to the superconducting phase, so that a specific number of RTA cycles is required to obtain an optimal material [5]. On the other hand, the use of cumulated RTA cycles has also been shown to improve the crystal orientation of the YBaCuO film, to obtain nearly textured layers on YDPZ substrates [3].

3. RESISTIVE TRANSITION MODELLING

3.1. Model equations

The main features which characterize such resistive transitions as those shown in Fig. 1 are i) the slope in the normal region, which can range from negative (semiconducting behaviour) to positive (metallic behaviour), with the "ideal" limit $R \propto T$; ii) the steepness of the superconducting transition to zero resistance; iii) the shape of this transition, where a "knee" can be observed (or even several knees in some cases).

The first feature may be related to intergrain insulating weak links, but also interpreted as indicating the presence of a parasitic nonsuperconducting oxide phase [1]. Its origin can be attributed to imperfect processing (e.g. Y_2BaCuO_5, $BaCuO_2$ or CuO) or to interaction with the substrate (e.g. Ba_2SiO_4)[5]. The resistance of this phase will be represented by an insulating type behaviour of the form

$$R_i = r_i \exp(T_i /T) , \tag{1}$$

where r_i is assumed to be constant. T is the sample temperature, $T_i = E_i/k_B$ (E_i is an activation energy and k_B is Boltzmann's constant).

The superconducting phase resistive behaviour can be ideally represented by a piecewise function. Above the superconducting onset temperature T_C, the behaviour is purely metallic,

$$R_s = r_s T \qquad (T \geq T_C) , \tag{2}$$

where r_s is also assumed constant. The decay to zero below the onset can be assumed to be approximately exponential, to represent the critical transport current behaviour observed in YBaCuO granular thin films [6] and ceramics [7], which is attributed to the S-N-S type tunneling at grain boundaries (or at twin boundaries within the grains). Thus

$$R_s = r_s T_C \exp[a(T-T_C)] \qquad (T \leq T_C), \qquad (3)$$

where the constant $a \approx 3/\Delta T_C$ (ΔT_C is the transition width, for $0.05 r_s T_C < R < 0.95 r_s T_C$).

3.2. Model capabilities

A simple way to illustrate the ability of the model to represent actual situations is to assume at first that only a single superconducting subphase and a single nonsuperconducting sub-phase are present, so that the current flowing along the film results from the contribution of both sub-phases. The film equivalent electrical circuit is therefore the association of the resistors R_s and R_i in parallel. The influence of the key parameters T_i and $\eta = r_s/r_i$ are shown in figures 2(a) and 2(b), respectively.

Secondly, when several superconducting and non-superconducting sub-phases are present, the corresponding parallel associations should be put into series. Figure 3 shows how two such groups can be fitted to interpret a very broad transition. It should be noted that the number of maxima of dR/dT corresponds to the number of sub-phase groups.

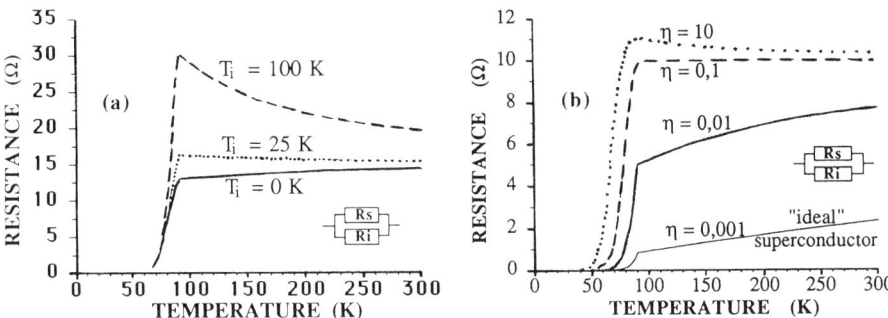

Figure 2. Resistive transitions simulated by a parallel association of resistors (Eqns. 1-3, with $T_C = 90K$, $a = 0.2$). The effect of the activation energy $E_i = k_B T_i$ in the parasitic subphase is shown in (a); large values of T_i give rise to a semiconducting behaviour ($r_s = 1$, $\eta = 1/15$). The effect of the ratio $\eta = r_s/r_i$ is shown in (b); large values of η enhance the effects of the parasitic subphase, leading to lower onset temperatures and broader transitions ($r_i = 10$, $T_i = 10K$).

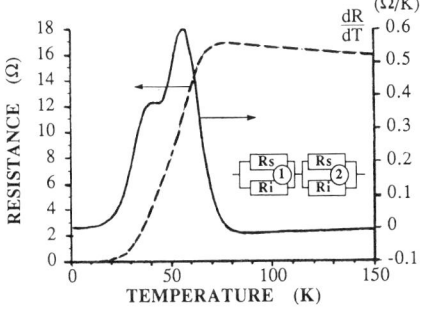

Figure 3. A broad transition is simulated by a resistor network representing two subphase groups. The maxima of the derivative are related to the onset temperatures of the superconducting subphases. ($T_{i1} = T_{i2} = 10K$, $r_{i1} = 10$, $r_{i2} = 5$, $T_{C1} = 90K$, $T_{C2} = 65K$, $r_{s1} = 100$, $r_{s2} = 50$, $a = 0.2$, $\eta_1 = \eta_2 = 10$).

4. APPLICATIONS OF THE MODEL

The model has been tested on YBaCuO films sputtered on both silicon based and YDPZ substrates. The fitting to the experimental data was performed by using the maximum likelihood estimator technique (i.e. chi-squared fitting), for the five variables r_i, T_i, r_s, T_C and a. Moreover, the number of phases (i.e. the number of parallel circuits in the equivalent network) was obtained by counting out the number of maxima of dR/dT after an adequate smoothing [8]. However, examples are given here for experimental data that could be well fitted with only one circuit.

Figure 4 illustrates fitting results for two silicon based substrates sputtered during the same run, belonging to the same family as in figure 1, but differing in the number of RTA cycles applied to the film (4 for sample 5103_6 and 5 for sample 5103_8). The parameter values are rather similar, showing the reproducibility of the process. The rather large values of $T_i > 100$ K ($E_i > 8$ meV) explain the semiconductor-like behaviour, and the low values of $\eta < 0.03$ characterize the detrimental influence of a presumably small content of impurity phase.

Figure 5 is relative to films sputtered on polycrystalline zirconia substrates. The low value of T_i for sample 5108_4 ($E_i \approx 3.5$ meV) is close to the value given by Dionne (4 meV[9]), and reflects the relative purity of the film (T_i is significantly smaller than T_C). The negative value of T_i for sample 5108_2 has no physical meaning, but is simply due to a numerical instability. It shows in fact that for this good quality film, the value of T_i of is close to zero, so that R_i is temperature independent and the impurity phase does not affect the sharpness of the transition.

Besides, it should be noted that the use of a piecewise definition of R_s generally induces an underestimation of T_C.

5. DISCUSSION ON MODEL IMPROVEMENT

This simple model is able to give some indications on the parasitic phase content of an imperfect sample. However, due to the various origins of the parasitic phases that can be envisaged, a detailed quantitative exploiting of the fitting results seems irrealistic within the framework of the present study. In fact, some of these phases can be identified as superconducting, but with a lower T_C (e.g. $YBa_2Cu_3O_{7-\delta}$ with a low oxygen content, or $YBa_2Cu_4O_8$), whereas insulating phases are characterized by their activation energy. Moreover, the influence of the weak link regions will be negligible for impure samples such as our films sputtered on silicon based substrates, but will be dominant for purer samples such as our films sputtered on YDPZ substrates. Finally, if the subphases could be identified and isolated, accurate measurements of the specific resistances r_i and r_s could be then performed separately, to obtain quantitatively the composition of the film.

To improve the model, more refined expressions for R_i and R_s should be used, with r_i as well as r_s being temperature dependent. In fact, for a hopping conductivity model, one would take $r \propto T$ [9], whereas $r \propto T^\alpha$ ($0 < \alpha < 1$) should be chosen to take into account the electron-phonon interaction [10].

Moreover, to avoid the numerical problems arising from the piecewise definition of R_s, a two fluid model could be used to represent the "ideal" superconducting subphase by two resistors in parallel, i.e. $R_N = r_{N0}T\exp(E_N/k_BT)$ and $R_S = R_{S0}\exp[(T-T_C)/\Delta T]$.

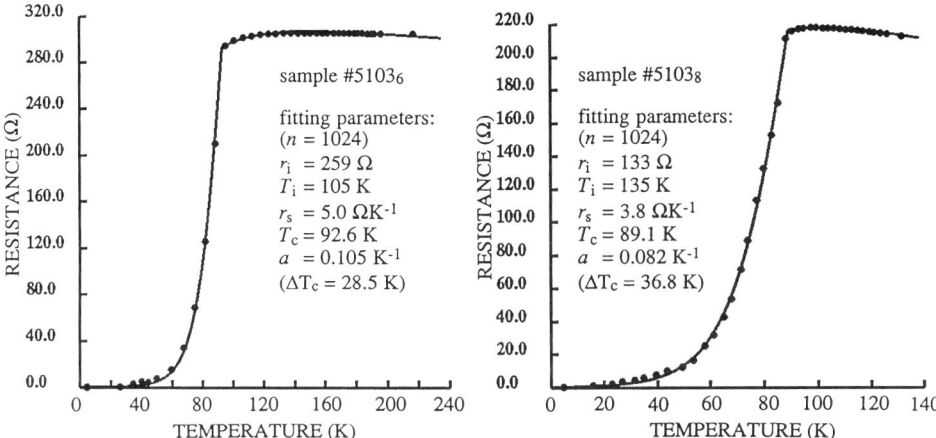

Figure 4. Examples of resistive transition experimental points, fitted with the model for two YBaCuO 1.1 μm thick films sputtered during the same run on $SiO_2(250Å)/Si_3N_4$ (900Å)/Si (100) substrates (for clarity, only ≈ 1/40 of the points are shown). Sample 5103_6 underwent four RTA cycles (such as those shown on figure 1), whereas sample 5103_8 underwent five RTA cycles (fifth cycle: same as in figure 1, bottom right). The fitting procedure was acheived with only one group of resistors in parallel, and the best fit values are indicated for each sample. The value of the decay constant $a ≈ 0.1$ K^{-1} is consistent with S-N-S type tunnelling through weak links [6]

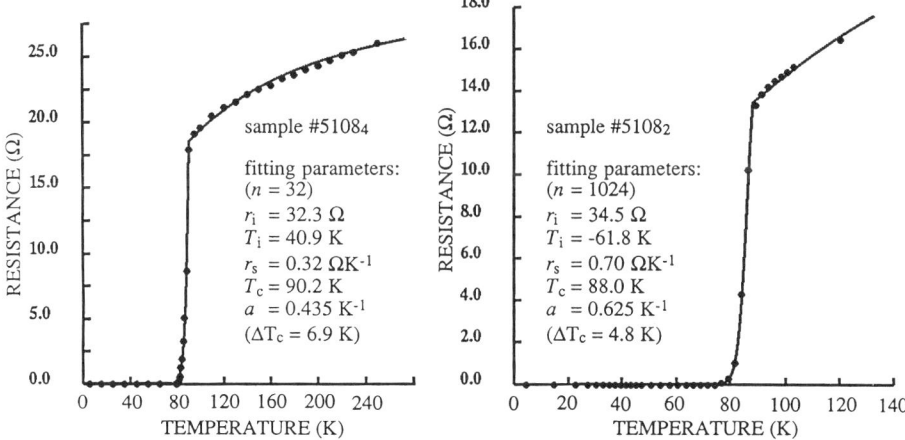

Figure 5. Examples of resistive transition experimental points, fitted with the model for two YBaCuO 0.7 μm thick films sputtered during the same run on YDPZ substrates. Both samples underwent three similar RTA cycles (heating up for 10 s, dwell at 920°C for 60s -both under argon-, cooling down under oxygen). The low value of T_i for sample 5108_4 reflects the relative purity of the film. For the good quality film of sample 5108_2 the negative value of T_i is of numerical origin; in fact $T_i ≈ 0$, and the value of R_i is temperature independent.

6. CONCLUSION

The resistive transitions of various YBaCuO granular films have been simulated by a resistor network representing both the superconducting and nonsuperconducting phases present in the sample. The examples given show that an accurate fit with experimental results can be obtained. The most significant parameters extracted from the fit are i) the activation energy for the conductivity in the nonsuperconducting phase (the value of which determines the metallic or semiconducting behaviour in the normal state) and ii) the resistance ratio of the two sub-phases at the superconducting onset temperature, the value of which can be related to the impurity phase content. Further improvements are sought to describe both resistance versus temperature variations more accurately, and to avoid numerical instabilities arising for sharp transitions.

ACKNOWLEDGEMENTS

The authors are grateful to E. Caristan and F. Carrié for their help in sample preparation and electrical characterization, and to Dr. T. Pech for stimulating discussions. The Ministère de la Recherche et de la Technologie is acknowledged for supporting this work in part under contract 87 A 0688.

REFERENCES

1   T. Itoh and H. Uchikawa, J. Appl. Phys., 66 (1989) 4900.
2   J. Baixeras, F. Carrié, F. Hosseini Teherani and A. Kreisler, J. Less-Common Metals, 164 & 165 (1990) 359.
3   J. Baixeras, F. Carrié, F. Hosseini Teherani and A. Kreisler, J. Less-Common Metals, 164 & 165 (1990) 366.
4   J. Baixeras, F. Carrié, J.P. Chabrerie, F. Hosseini Teherani, A. Kreisler, T. Pech, G. Poullain, J.F. Hamet, J. Muniesa, M. Rapeaux and J. Aymami, Proc. SPIE Conf. High Tc superconductivity: Thin films and applications, San Diego (March 1990), Vol. 1287, p. 104-115.
5   J. Baixeras, F. Hosseini Teherani, A. Kreisler, A. Straboni and K. Barla, Proc. SPIE Conf. Physical Concepts of Materials for Novel Optoelectronic Device Applications, Aachen (Oct. 1990), Vol. 1362, p. 117-126.
6   S.S. Yom, T.S. Hahn, Y.H. Kim, H. Chu and S.S. Choi, Appl. Phys. Lett., 54 (1989) 2370.
7   Ch. Laurent, M. Ausloos and S.K. Patapis, Revue Phys. Appl., 24 (1989) 501.
8   Adapted from W.H. Press, B.P. Flannery, S.A. Teukolsky and W.T. Vetterling, Numerical recipes in Pascal, Chapters 13 & 14, Cambridge University Press, New York, 1989.
9   G.F. Dionne, Proc. ASC 1990 Conf., paper MKP-2.05.
10 J.L. Cohn, S.A. Wolf, V. Selvamanickam and K. Salama, Phys. Rev. Lett., 66 (1991) 1098.

High T_c Superconductor Thin Films
L. Correra (Editor)
© 1992 Elsevier Science Publishers B.V. All rights reserved.

Temperature Properties of Films

J.P.Šetrajčić[a], D.Lj.Mirjanić[b], B.S.Tošić[a] and Z.V.Bundalo[c]

[a] Institute of Physics. Faculty of Sciences, University of Novi Sad, Trg D.Obradovića 4, 21000 Novi Sad, Yugoslavia
[b] Faculty of Technology, University of Banja Luka, D.Mitrova 63b, 78000 Banja Luka, Yugoslavia
[c] Electrotechnical Faculty, University of Banja Luka, Dr. V.Butozana 3, 78000 Banja Luka, Yugoslavia

Abstract

It is shown that phonons in thin films, "cut off" from simple lattice, possess the gap whose magnitude depends upon the thickness of the film. Due to the presence of the gap specific heat of the film for $T \approx 0$ changes approximately as $T^{-1} \exp(-const/T)$. This behaviour of specific heat can be used for experimental verification of the existence of phonon gap.

1. Phonon Gap

It was shown in a number of papers [1-4] that in thin films acoustical phonons connot arise even in the case when the film is "cut off" from the bulk structure of simple lattice. This, at the first sight, strong conclusion can be explained on the basic principle of quantum mechanics which is uncertainty principle.

Phonon energy is given by $E(p_x, p_y, p_z) = v(p_x^2 + p_y^2 + p_z^2)^{1/2}$ wherefrom it follows that energy uncertainty is given by $\Delta E = E(p_x + \Delta p_x, p_y + \Delta p_y, p_z + \Delta p_z) - E(p_x, p_y, p_z) = v(p_x \Delta p_x + p_y \Delta p_y + p_z \Delta p_z)/p$ where $\Delta p_\alpha (\alpha = x, y, z)$ are uncertainties of momentum components. Assuming that film is infinite in XY planes and finite in z - direction we can take $\Delta p_x \sim \hbar/2L_x \approx 0$, $\Delta p_y \sim \hbar/2L_y \approx 0$ and $\Delta p_z \sim \hbar/2L_z \neq 0$, since $L_z \ll L_x, L_y$. Consequently, the energy uncertainty becomes $\Delta E = \hbar v p_z/2pL_z = (\hbar v/2L_z) \cos \theta$. Taking into account that in film azimuthal angle changes from 0 to $\pi/2$, the mean values of $\cos \theta$ in this interval is $2/\pi$. So we can estimate finally:

$$\Delta E = \hbar v/\pi L_z \tag{1}$$

This value can be considered as a phonon gap in thin film. It is see from (1) that this gap decreases with the increase of film thickness $L_z = a_z N_z$ and tending to zero if L_z becomes of the order L_x or L_y. For $L_z = 10^{-6} cm$ and taking the velocity of sound v to be $4.10^3 ms^{-1}$ we find: $\Delta E = 1.34.10^{-23} J$, which means that excitation of phonons requires activation temperature of about $1K$.

On the basis of above considerations the phonon energies in thin films can be written as:

$$E(k) = (E_0^2 a^2 k^2 + \Delta^2)^{1/2}, \ E_0 \approx \hbar v/a, \ k^2 = k_x^2 + k_y^2, \ \Delta = \hbar v/\pi N_z a_z. \tag{2}$$

The direct measuring of phonon energies at extremely small k is practically impossible due to low resolution of instruments [5]. It should be noticed that recently [6] has been concluded that in perovskite layers only optical phonons appear although these structures are simple. It was concluded indirectly by the measuring of the ratio of $\hbar \omega_D / k_B T_C$ in these superconductors. Here we propose another indirect way of confirmation of existence of phonon gap in film. We believe that the measuring of low temperature specific heats in films and in corresponding bulk structures can answer the question: does phonon gap exist in films?

2. Specific Heat

So we shall evaluate the internal energy of thin film at low temperature and specific heat as its temperature derivate. The internal energy will be calculated by the formula (from [7],p.224]:

$$U = 3(N_z + 1) \sum_{k_x, k_y} E(k) \{\exp[-1 + E(k)/\theta]\}^{-1}, \tag{3}$$

where $\theta = k_B T$.

Going over from sum to integral and using (2) we reduce (3) to the form:

$$U = 3 N_x N_y (N_z + 1) \theta^3 (2\pi E_0^2)^{-1} \sum_{j=1}^{\infty} \int_{x_1}^{x_2} dx \, x^2 \exp(-jx), \tag{4}$$

$$x_1 = \Delta/\theta, \ x_2 = \sqrt{E_0^2 x_m^2 + \Delta^2}/\theta, \ x_m = ak_m = 2\sqrt{\pi}$$

Since $x_2 \gg x_1$ the contributions proportional to $\exp(-jx_2)$ will be neglected. In this approximation we finally obtained:

$$U = 3 N_f (2\pi)^{-1} + \theta^3 E_0^{-2} [x_1^2 Z_1(x_1) + 2x_1 Z_2(x_1) + 2Z_3(x_1)], \tag{5}$$

where $N_f = N_x N_y (N_z + 1)$ and $Z_r(x_1) = \sum_{m=1}^{\infty} m^{-r} \exp(-mx_1)$.

The specific heatsper elementary cell are given by the formulas:

a) for film

$$C_f = 3\Delta^2 (2\pi E_0^2)^{-1} [x_1 (e^{x_1} - 1)^{-1} + 3Z_1(x_1) + 6x_1^{-1} Z_2(x_1) + 6x_1^{-2} Z_3(x_1)], \tag{6}$$

b) for bulk structure

$$C_b = 36\pi^{-2} \theta^3 E_0^{-3} \zeta(4), \tag{7}$$

where $N_b = N_x N_y N_z^b$, $N_z^b \sim 10^8$ and ζ is the Rieman's function.

The numerical caclulations of C_f and C_b were caried out for the following set of parameters: $v = 3.10^3 ms^{-1}$. $a = 4.10^{-10} m$ and $L = 2\mu m$. The results of this analysis can be summarized as follows. The curves $C_f(T)$ and $C_b(T)$ have two intersection points at $T_1 = 2.2mK$ and $T_2 = 51.1K$. In the intervals $T < T_1$ and $T > T_2 : C_f < C_b$, while $C_f > C_b$ in the interval $T_1 < T < T_2$. This result can be experimentally examined since the measurements of C_b in the vicinity of the temperature T_2 can be easily carried out. The experiment has to be carried out on dielectric material since in the metallic structures the contributions to specific heat of electronic gas can appear.

3. Conclusion

Finally we can summarize the obtained results. The phonon spectra in thin film with the simple cells has the gap which decreases with the increase of the film thickness. The specific heat of thin film decreases exponentially to zero when temperature tends to zero, while the specific heat of the bulk structure goes to zero as T^3.

The experimental testing concerning existence of phonon gap can be carried out by the parallel measurement of specific heats of bulk structure and of the film which is "cut off" from the same bulk structure.

The presence of phonon gap and the corresponding activation temperature for exciting of phonons represents possible explanation for the fact that thin films have higher superconductive T_C, then the same bulk structure

The experimental data where transition temperatures in bulk and in the corresponding film are compared to is given in [8], table E-75, and here is seen that T_C of films is higher slightly as it is predicted in our approach.

References

[1] B.S.Tošić, J.P.Šetrajčić, R.P.Djajić and D.Lj.Mirjanić, Int. J. Mod. Phys. B1 (1988) 919.

[2] D.Lj.Mirjanić, R.P.Djajić, B.S.Tošić and J.P.Šetrajčić, FZKAAA 21 (1989) 303.

[3] J.P.Šetrajčić, R.P.Djajić, D.Lj.Mirjanić and B.S.Tošić, Physica Scripta 42 (1990) 732.

[4] B.S.Tošić, Lj.M.Ristovski, G.S.Davidović and M.M.Marinković, The Phonon Spectra of the Layered System with Grained Structure, submitted to J.Phys. C.

[5] Boyd Veal, Argonn National Labs. (private communication).

[6] A.P.Litvinchuk, C.Thomsen, P.Murugaraj and M.Cardona, Far-Infrared Spectroscopy of $YBa_2Cu_4O_8$: Optical Phonons and Superconducting Energy Gaps, 11th General Conference of the Condensed Matter Division - Exeter, April 1991 (Programme and Abstracts PA66, p. 147).

[7] C.Kittel, Introduction to Solid State Physics, Nauka, Moscow, 1978 (in Russian).

[8] C.R.C.Handbook of Chemisty and Physics (53 rd ed.), The Chemical Rubber Co., Cleveland-Ohio, 1973.

High T$_c$ Superconductor Thin Films
L. Correra (Editor)
1992 Elsevier Science Publishers B.V.

A.c. susceptibility harmonic analysis of the irreversibility line in YBa$_2$Cu$_3$O$_{7-\partial}$ thin film

L.A. Angurel, F. Lera, A. Badía, C. Rillo, R. Navarro, J. Bartolomé, J. Melero, J. Flokstra[#], R. P. J. IJsselsteijn[#]

I.C.M.A., C.S.I.C.-Universidad de Zaragoza. 50009 Zaragoza, Spain.

[#]University of Twente, 7500 AE Enschede, The Netherlands

Abstract

We propose as a model independent method to derive the irreversibility temperature on high T$_c$ superconductors the measurement of the onset of the a.c. susceptibility third harmonic. The method is applied to YBa$_2$Cu$_3$O$_{7-\partial}$ thin film and ceramic samples and compared with the results of other a.c. methods.

1. INTRODUCTION

One of the most controversial features of the new high T$_c$ superconductors (HTS) is the appearance of a wide region below H$_{c2}$(T) where the magnetization is reversible in field-cooled (FC) and zero-field cooled (ZFC) cycles. This was first observed in La-Ba-Cu-O ceramics by Müller et al [1] and since then has been also detected in most HTS materials as YBa$_2$Cu$_3$O$_{7-\partial}$ [2] and Bi-Sr-Ca-Cu-O [3] crystals, YBa$_2$Cu$_3$O$_{7-\partial}$ films [4-6] and Tl based ceramics [7,8].

For a given field, the onset of irreversibility occurs at a temperature T$_{irr}$(H) which scales with the magnetic field as:

$$H = A\left(1 - \frac{T_{irr}(H)}{T_c(0)}\right)^n \qquad (1)$$

In many cases the exponent n is equal to 3/2 but values ranging from 4/3 and 3 may be found in the literature.

Basis for the controversy are the different theories [9, 10 and references therein] that claim a proper understanding of the flux dynamics and thereafter of the experimental behavior. Here, without going into the discussion if there is a real phase transition at T$_{irr}$(H), we will propose a method to determine the so called irreversibility line.

The experimental difficulties in the assessment of the irreversibility line, that changes from one experimental technique to another, on the same group of samples [2], might be the origin of the discrepancies. Some of   the measurements commonly used are: FC and ZFC magnetization [1], a.c. susceptibility, maximum of the out-of-phase component $\chi''(H,T)$ [6], onset of χ'' [4] and harmonics [11], differential paramagnetic effect (DPE) [7] and electrical transport measurements [5].

Here we present a measurement technique that overcomes the existing difficulties in the a.c. methods and enables to derive values of $T_{irr}(H)$ independent of the a.c. exciting field amplitude, h_0. For an applied field $H = H_{dc} + h_0 \sin \omega t$, we associate $T_{irr}(H)$ to the temperature where the onset of the third harmonic modulus of the a.c. susceptibility, χ_3, appears. The method is applied to $YBa_2Cu_3O_{7-\partial}$ thin film and ceramic samples and compared with the results of other a.c. methods.

2. EXPERIMENTAL

The $YBa_2Cu_3O_{7-\partial}$ thin film was deposited on a (100) oriented Yttria stabilized ZrO_2 substrate using the pulsed laser deposition technique [12], and had a thickness of approximately 350 nm. The sample presents a polycrystalline structure, with the c-axis oriented perpendicular to the film plane. The laser beam was produced by an excimer laser ($\lambda = 308$ nm, $\tau = 20$ ns) and is focused (spot size = 7.3 mm^2) onto a rotating $YBa_2Cu_3O_x$ target, prepared by the citrate pyrolysis method [13]. The substrate temperature during deposition was 770 ºC, the laser frequency 5 Hz, the oxygen pressure during deposition 25 Pa and the target-substrate distance 45 mm. After deposition the layer was cooled down to 400 ºC in 10 minutes whereas the oxygen pressure was raised to 1 bar. After another 5 minutes the temperature of the layer was further reduced to room temperature in 30 minutes.

The ceramic sample was prepared by dissolving stoichiometric quantities of Y_2O_3, CuO and $Ba(NO_3)_2$ in 2N nitric acid. The blue solution was heated and the excess HNO_3 was removed by gentle boiling. The dried powder was decomposed at 400ºC during 2 hours and then calcined in oxygen at 900ºC

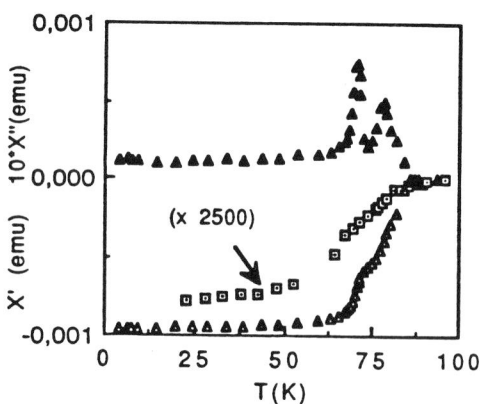

Figure 1: a.c. susceptibility of the thin film, with the applied field parallel (Δ,h_0=5.5 mOe) and perpendicular to the c-axis ($\Box$, h_0=11 Oe)

for 12 hours, furnace cooled, dry ground and pressed at 5 tons into bars. The bars were sintered in oxygen for 20 hours, followed by slow cooling (2 °C/min) and oxygen annealing at 450 °C for 10 hours. Afterwards, the bars were machined to become a cylinder of 2.5 mm diameter and 6 mm high.

The harmonic response of the susceptibility was measured in a computer controlled susceptometer. The reference of the lock-in was generated with a synchronous frequency multiplier.

The thin film was characterized by susceptibility measurements with h_0 applied parallel and perpendicular to the c-axis (figure 1). For h_0 perpendicular to the c-axis, an amplitude of 11 Oe was used. In the other geometry measurements with $h_0 = 5.5$ mOe are presented. According to the data $T_c(\chi') = 87.8K$, and χ'' shows the intra- and intergranular peaks.

In order to determine the irreversibility line, samples were cooled to 4.2K. Then an exciting field, with a frequency of 38 Hz and amplitude of 11 Oe, and the corresponding d.c. bias field were applied, both of them parallel to the c-axis. To minimize possible relaxation effects we waited for fifteen minutes until the temperature began to increase, measuring the third harmonic of the susceptibility. After finishing the measurement the fields were removed and the sample heated to room temperature before starting a new cycle.

Some points have been repeated with $h_0 = 2.75$ Oe and we cannot observe any difference on $T_{irr}(H)$. (figure 2). Even if the $\chi_3(T)$ curve is then different, the onset occurs at the same temperature within experimental resolution. So it seems that this definition of $T_{irr}(H)$ leads to a line which is independent of the a.c. field amplitude.

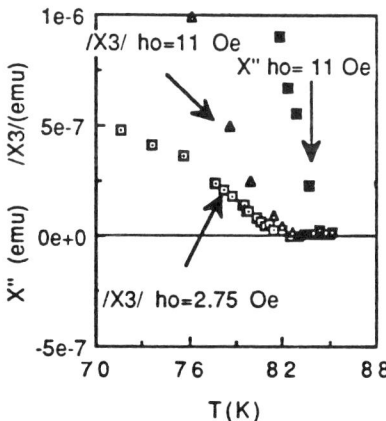

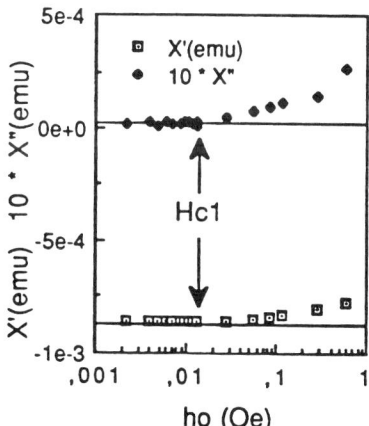

Figure 2: Onset of the third harmonic and the out-of-phase component of susceptibility in the thin film. The dc bias field is 0.3 T. Fields are applied parallel to the c-axis.

Figure 3: a.c field dependence of the susceptibility of the thin film at 4.2 K. The field is parallel to the c-axis and no demagnetization corrections have been done. Intergranular $H_{c1} = 15$ mOe.

It should be noted that there is a threshold a.c. field (see figure 3) for this method. At a fixed temperature, H_{c1} may be inferred from the dependence of the low frequency a.c. susceptibility data with h_0, in absence of d.c. field. It is the field for which χ' deviates from perfect shielding, and where χ'' and higher harmonics appear. In presence of high d.c. fields (up to 5 T) χ' shows perfect a.c. shielding, and no harmonics are present for h_0 lower than a given temperature dependent threshold value. This can be explained as reversible motion of the fluxoids or an insignificant number of unpinned vortices by the action of the a.c. field [14] and is the only limiting factor of the selection of the a.c. amplitude. Then, h_0 should be greater than this threshold to determine T_{irr}.

Figure 4 and figure 5 show the irreversibility line measured in the thin film and in the ceramic cylinder. The data can be fitted to the dependence predicted by equation 1 with n=1.9 in the thin film and n=1.4 in the ceramic sample. In table 1 we present the obtained values. Similar deviations from the 1.5 law can be found in the literature.

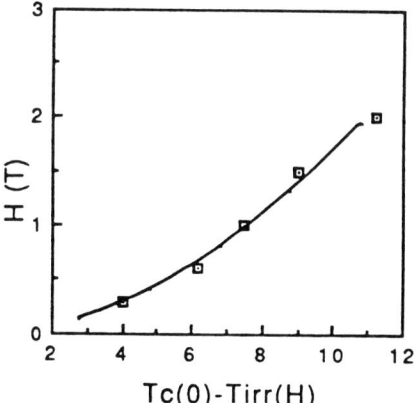
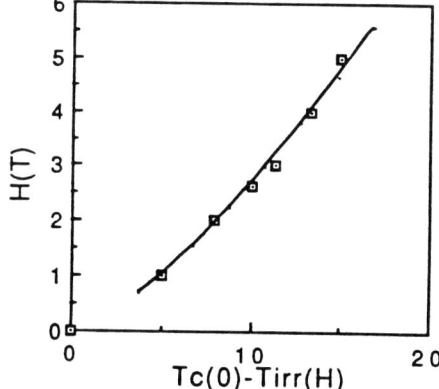

Figure 4: The irreversibility line of the thin film. Solid line is the best fit to equation 1.

FIGURE 5: The irreversibility line of the ceramic sample. Solid line is the best fit to equation 1.

Table 1
Parameters of the irreversibility line

	THIN FILM	CERAMIC
T_c (K)	87.8	91.8
n exponent	1.9	1.4
A constant (T)	110.5	61.75
Fields measured (T)	0,0.3,0.6,1,1.5,2	0,1,2,3,4,5

3. DISCUSSION

The irreversibility line was first detected in d.c. magnetization measurements and, for a given field, was defined as the temperature for which FC and ZFC magnetizations deviate one upon the other. The smooth joining requires precise measurements usually performed with SQUID magnetometers.

Other techniques are based on the response to a.c. magnetic fields. Under superimposed d.c. (H) and a.c. fields (h_0), $h_0 \ll H$, at a given temperature χ' shows step-like anomalies whereas χ'' has a maximum, usually identified with $T_{irr}(H)$ [2]. However this determination yields to different values, depending on the a.c. frequency and field amplitude.

The origin of the χ' and χ'' anomalies has been ascribed to different effects being most likely the hopping of the vortex over the pinning barriers. Thermally assisted flux flow may explain such behavior [15,16]. The maximum in the absorption may be related to skin effects and occurs when the a.c. penetration depth is of the order of the sample grain size. Moreover, phenomenological Critical State Models predict this behavior without taking into account the existence of a reversible region.

The presence of several effects superimposed in the first harmonic (a.c. susceptibility) may shadow the onset or end of irreversibility. Even more, the definition of $T_{irr}(H)$ from χ'' onset presents additional experimental difficulties: minute errors in phase setting can originate imprecisions coming from contamination of χ'' from the non-zero χ'. This difference between the onsets of the two components of the first harmonic of the a.c. susceptibility has been measured by authors investigating the DPE [7].

The above mentioned difficulties do not enable a proper determination of $T_{irr}(H)$ from a.c. susceptibility. However, in the measurements of critical fields, in addition to the a.c. susceptibility, higher harmonics of the a.c. response have been used in both classical [17] and HTS [11] materials.

For high enough a.c. amplitudes (higher than a threshold value) below $T_{irr}(H)$, the existence of an effective pinning gives rise to a.c. hysteresis loops whose high non-linearity produces the appearance of higher harmonics, that are zero in the reversible regime. Thus we should choose the onset of the third harmonic, because this is the more intense and will mark properly the pass through the irreversibility line. Also, the use of harmonic components has additional advantages: the phase setting problem disappears if higher harmonics are used because the irreversibility line can be determined with the onset of the modulus, making the measurement easier.

We have also measured χ'' (H,T) in the same conditions, in order to test the differences between the definitions of the line (figure 2). As it was pointed out by Shaulov and Dorman [11], the onset of the harmonics does not take place at the same temperature than the χ'' maximum. They also observed that losses may persist further up, beyond T_{irr} ($\chi_3 = 0$), due to flux flow resistance. In our measurements this behavior is reproduced, but with smaller differences between χ_3 and χ'' onsets. Further studies on the dependence of the third harmonic modulus at $T_{irr}(H)$ with the basic exciting frequency are in progress.

In summary, we propose that the onset of χ_3 is a good ac method to determine the irreversibility line in a h_0 and model independent way, with less experimental difficulties than other ac or dc techniques. Furthermore the results here presented prove the accuracy of the procedure.

4. ACKNOWLEDGEMENTS

The authors greatly appreciate J. Blasco for providing the ceramic sample. One of us (R. N.) would thank M.P. Maley (Los Alamos National Laboratory, USA) for helpful discussions. This research has been partially supported by the spanish projects CICYT (num. MAT90-0362) and MIDAS (num 89/3797) and by ECC projects (SCI-0036-F and SCI-0389-C).

5. REFERENCES

1   K.A. Müller, M. Takashige, J.G. Bednorz. Phys. Rev. Lett. 58, 408 (1987)
2   A.P. Malozemoff, T.K. Worthington, Y. Yeshurun, F. Holzberg, P.H. Kes. Phys. Rev. B 38, 7203 (1988)
3   Y. Yeshurun, A.P. Malozemoff, T.K. Worthington, R.M. Yandrofski, L. Krusin-Elbaum, F.H. Holtzberg, T.R. Dinger, G.V. Chandreshekhar. Cryogenics 29, 258 (1989)
4   L. Civale, T.K. Worthington, A. Gupta. Phys. Rev. B 43, 5425 (1991)
5   P.H. Koch, V. Foglietti, W.J. Gallagher, G. Koren, A. Gupta, M.P.A. Fisher. Phys. Rev. Lett. 63, 1511 (1989)
6   J.H.P.M. Emmen, G.M. Stollman, W.J.M. De Jonge. Physica C 169, 418 (1990)
7   M. Couach, A.F. Khoder, F. Monnier. Presented at the International Symposium on High Temperature Superconductors; Satellite Symposium of the 7th CIMTEC-World Ceramic Congress, Trieste, July 1990.
8   Y. Wolfus, Y. Yeshurun, I. Felner. Phys. Rev. B 39, 11690 (1989)
9   Y. Xu, M. Suenaga. Phys. Rev. B 43, 5516 (1991)
10 D.S. Fisher, M.P.A. Fisher, D.A. Huse. Phys. Rev. B 43, 130 (1991)
11 A. Shaulov, D. Dorman. Appl. Phys. Lett. 53, 2680 (1988)
12 D.H.A. Blank, D.J. Adelerhof, J. Flokstra, H. Rogalla, Physica C 167, 423 (1990)
13 D.H.A. Blank, H. Kruidhof, J. Flokstra, J. Phys. D 21, 226 (1988)
14 A.M. Campbell, J. Phys. C 2, 1492 (1969)
15 P.H. Kes, J. Aarts, J. van der Berg, C.J. van der Beek and J. A. Mydosh, Supercond. Sci. Technol. 1, 242 (1989)
16 V.B. Geshkenhein, V.M. Vinokur, R. Fehrenbacher, Phys. Rev. B 43, 3448 (1991)
17 S.A. Campbell, J.B. Ketterson, G.W. Crabtree. Rev. Sci. Instrum. 54, 1191 (1983)

High T$_c$ Superconductor Thin Films
L. Correra (Editor)

147

Search for deviations of time-reversal symmetry in YBa$_2$Cu$_3$O$_7$ using transport measurements

H.P. Assink[a], N.Y. Chen[a], H.U.Habermeier[b], and D. van der Marel[a,b]

[a]Delft University of Technology, Faculty of Applied Physics, Lorentzweg 1, 2628 CJ Delft, The Netherlands

[b]Max-Planck-Institut für Festkörperforschung, Heisenbergstrasse 1, D-7000 Stuttgart 80, Federal Republik of Germany

Abstract

We study the transport properties of thin films of YBa$_2$Cu$_3$O$_7$ using lithographically defined narrow crossings. One of the observable consequences of the possible breaking of the discrete space-time symmetries P and T in high-T$_c$ superconductors, as is predicted from anyon models, are anomalous transport properties similar to *e.g.* a Hall conductance. For example an off-diagonal component in the resistivity tensor would exist in the absence of an external magnetic field. We try to reduce possible cancellation effects due to domain formation, by using thin films with a lithografically defined sample area of a few square microns. The breaking of time reversal symmetry corresponds to an 'effective' magnetic field, which can be determined by calibrating the zero field off-diagonal resistivity with respect to the Hall voltage in an externally applied magnetic field. Our first experiments indicate that the 'effective' internal field due to time symmetry breaking is either absent or smaller than 75 Gauss.

1. INTRODUCTION

If the anyon model[1] is valid for High-T$_c$ Cu-O superconductors, it is expected that the Cu-O planes will have an intrinsic orbital magnetic moment perpendicular to each layer[2]. If the coupling between the layers is ferromagnetic, there should be a number of observable bulk effects, including anomalous transport properties analogous to a Hall-conductance, which would occur even in the absence of a magnetic field, as indicated by Halperin *et al.* [2]. In the same paper Halperin argues that the apparently unusual properties of the copper-oxide materials above T$_c$ suggests that the time reversal symmetry, T, is still present above T$_c$. Indeed effects of broken T persist above the superconducting transition in the simplest model of weakly interacting anyons.

As from a phenomenological point of view the anomalous properties of a two-dimensional system with spontaneously broken T symmetry are similar to the properties of a normal material in an externally applied magnetic field, it seems worthwhile to investigate experimentally the off-diagonal resistivity in the absense of such a field. Usually the Hall voltage is determined by measuring the transverse resistance of two contacts connected to opposite sides of a current carrying strip. As it is expermentally difficult to have a perfect alignment of the Hall contacts, one usually finds a non-vanishing voltage even for H=0 due to the presence of a finite longitudinal resistance. It is therefore common experimental practice to repeat the same measurement at a number of different magnetic field values, *e.g.* by repeating the measurement at a field H^{ext} and -H^{ext} so that the Hall voltage can be separated from the (assumedly field independent) longitudinal resistance. Obviously in our present investigation, the internal 'effective' field can not be controlled in this way. However, as has been shown by van der Pauw[3], the Hall resistance of a singly connected two dimensional sheet can also be determined by measuring the four terminal resistances R(1,2;3,4) and R(3,4;1,2) without changing the applied field. Here R(1,2;3,4)=(V(3)-V(4))/I(1,2) refers to a configuration in which the current I flows from contact 1 to contact 2, and the voltage V is measured between contacts 3 and 4. The latter two contacts have to be on opposite sides of the curve connecting contacts 1 and 2. For a sheet of homogeneous thickness, with sufficiently small contacts connected to the perifery, the off-diagonal sheet resistance R_{xy} then follows from

$$R_{xy} = (R(1,2;3,4) - R(3,4;1,2))/2$$

Gijs *et al.*[4] used this method on thin films with a thickness of 100 nm, and with a sample area of about 30 (mm^2) and found no significant deviation from the Onsager-Casimir symmetry relations[5], within their accuracy of 0.02%. As the effect of T-breaking may be masked by the presence of many small domains with random orientations, a more definite experimental statement about the presence or absence of T-breaking can be given when thinner samples are used with a smaller area.

2. EXPERIMENTAL SETUP

The measurements were done on a YBa$_2$Cu$_3$O$_{7-y}$ sample, with a thickness of about 254 nm, which was prepared with the pulsed laser deposition technique. The substrate temperature during growth was 775 °C with an O$_2$-pressure of 1 mbar. Then the sample was cooled down in 40 minutes in 1 atm. of pure oxygen. A microsopic pattern was lithographically defined by means of wet-chemical etching using hydrochloric acid in an aquous solution. The sample geometry is sketched in Fig. 1 and consists of a crossing of two 6 μm wide lines, connected to contact pads. Gold was evaporated onto these contact pads and contacts were made by means of bonding with gold wires. The sheet resistance (R_{xx}) was measured as a function of temperature using the van der Pauw method [3], *i.e.* by determining R(1,3;2,4) and R(3,2;4,1) and inverting numerically the equation

$$\exp\left(-\frac{\pi R(1,3;2,4)}{R_{xx}}\right) + \exp\left(-\frac{\pi R(3,2;4,1)}{R_{xx}}\right) = 1$$

The result is presented in Fig. 2, which shows a 1 K wide transition with the midpoint at 90.4 K, and zero resitivity at 89.5 K.

High precision measurements of the off-diagonal component of the 4-terminal resistivity (R_{xy}) were performed using an AC lockin-technique, with a frequency of about 100 Hz. A highly stable current source was used to provide an excitation current of 100 μA, which is not affected in our range of accuracy by the switching between the two current-voltage

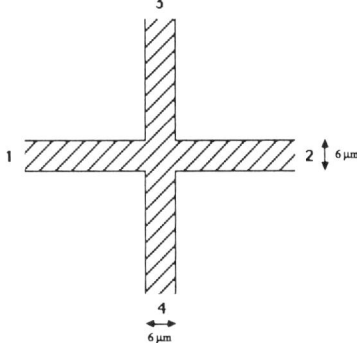

Figure 1. Sample layout.

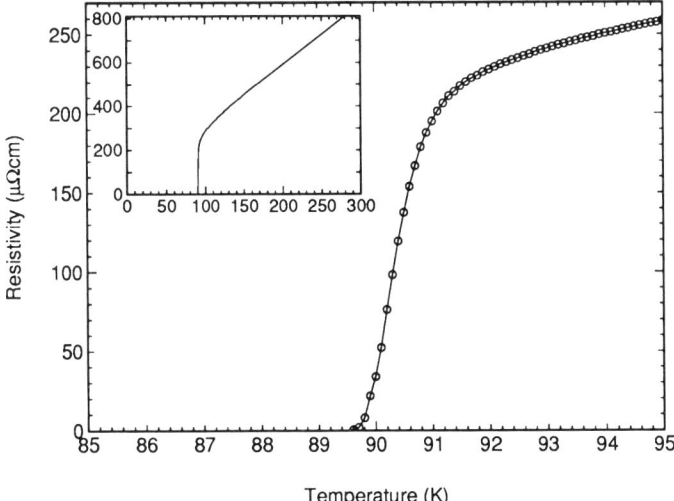

Figure 2. Resistivity as a function of temperature. Measured with the 'van der Pauw'-method. The inset shows the total curve, up to room-temperature.

geometries. The outputs of the two-phase lockin amplifier were fed into a personal computer, and the effect of stray capacitance of the measuring cables was eliminated by means
of a numerical calculation. This removes possible effects due to differences in the complex
impedance of contacts, sample and stray capacitance for the two current-voltage configu-
rarions R(1,2;3,4) and R(3,4;1,2). Effects due to a non-perfect common-mode rejection of
the signal amplifier were removed by interchanging the two voltage leads, which inverts
the differential voltage but leaves the common-mode voltage unaffected. After measuring
approximately 90 minutes, for R(1,2;3,4) and R(3,4;1,2) an accuracy was obtained of 42
$\mu\Omega$. This corresponds to 0.0004% of the sheet resistance R_{xx} of 10 Ω, which is the natural
resistance scale of the sample.

3. RESULTS

The sample temperature was stabilized at 91.8 K during the whole measurement. The
results are shown in Fig.3. R(1,2;3,4) and R(3,4;1,2) are plotted versus time. The first 22
points are without a magnetic field, then a magnetic field of 0.6 T was switched on during
the next 12 points, followed by 40 measurements without a magnetic field. In the inset
of Fig.3 an expanded view is given, clearly showing a drift as function of time. This drift
is of instrumental origine, and is due to small changes in temperature of the equipment.

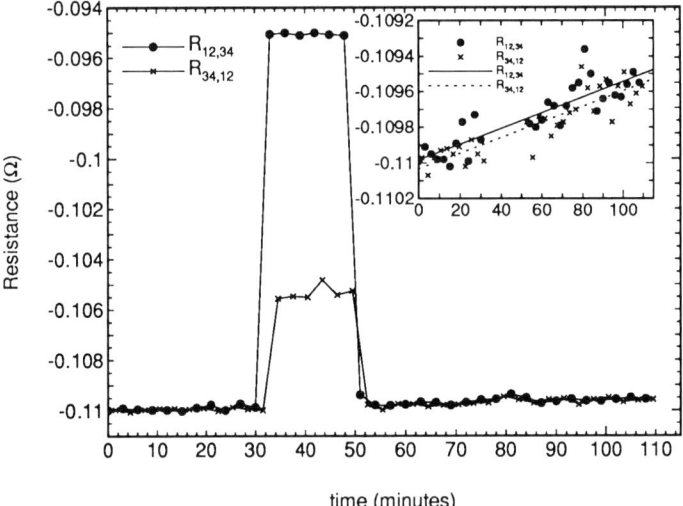

Figure 3. Four terminal Hall-resistances $R_{12,34}$ and $R_{34,12}$ as function of time. During six
points a magnetic field of 0.6 Tesla, perpendicular to the sample, was switched on. The inset
shows the drift of the measured points without a magnetic field. The two least-square fits
through these points were calculated to have the same slope.

When we assume that the drift is linear, and calculate the two most probable lines with the same slope from the zero-field points of Fig.3, we find for the distance between the lines:

$$R(1,2;3,4) - R(3,4;1,2) = 52 \pm 42\mu\Omega$$

In an externally applied magnetic field of 0.6 T perpendicular to the plane we measure

$$R(1,2;3,4) - R(3,4;1,2) = 10m\Omega$$

From the above results we calculate directly that the measured effect corresponds to an effective magnetic field of 31 ± 25 Gauss, the error being the 95% confidence range. This is of the order of the remanent magnetic field of our magnet (less than 80 Gauss). We can therefore indicate an upper limit of about 75 Gauss on the 'effective' field due to absence of time-reversal symmetry as predicted by anyon models. One must keep in mind here, that still a large number of differently oriented domains could contribute. If one for example assumes an in-plane coherence length for the domain structure of 100 nm and 1 nm in the c-direction, one would have rougly a million domains contributing to the sample which we used. If all domains are statistically frozen in random positions the 'effective' internal field is reduced by a factor of 1000. For this particular choice of values for the typical domain size our experimental results could still be compatible with an 'internal' effective field of (at most) 7.5 Tesla. To make further more definite statements it is therefor of importance that on the one hand reliable theoretical estimates are given of both the estimated internal effective field and the domain size, and on the other hand smaller and thinner samples are used for the experiments.

REFERENCES

1   V. Kalmeyer and R. B. Laughlin, Phys. Rev. Lett. 59 (1987), 2095.
2   B. I. Halperin, J. March-Russel, and F. Wilczek, Phys. Rev. B 40 (1989), 8726.
3   L. J. van der Pauw, Philips Res. Repts 13 (1958), 1.
4   M. A. M. Gijs, A. M. Gerrits, and C. W. J. Beenakker, Phys. Rev. B 42 (1990),10789.
5   S. R. de Groot, and P. Mazur, Non-Equilibrium Thermodynamics (Dover, New York, 1984).

High T$_c$ Superconductor Thin Films
L. Correra (Editor)
© 1992 Elsevier Science Publishers B.V. All rights reserved.

MICROWAVE CHARACTERIZATION OF YBaCuO THIN FILMS DEPOSITED ON (100)MgO

J.C.Carru, F.Mehri, D.Chauvel, Y.Crosnier and J.M.Wacrenier

Centre Hyperfréquences et Semiconducteurs, UA CNRS 287, Bâtiment P3, étage 3, Université de Lille 1, 59655, Villeneuve d'Ascq Cedex, France

Abstract
     This paper shows the feasability of the determination of the microwave conductivity in the temperature range 78-300K on YBaCuO films deposited on MgO. The method based on the measurement of the power transmission in the K-band offers the advantage of being non destructive. We also realized microstrip resonators operating near 7GHz. The surface resistance inferred from the resonance measurements at 78K is in agreement with the one deduced from the conductivity measurements.

1. INTRODUCTION

     Microwave characterizations of high Tc superconducting (HTSC) films are required for two reasons: firstly they are now used at 77K in applications such as passive components [1] and secondly data is needed to develop electromagnetic wave based models [2]. In this regard, the determination of the complex conductivity σ^* is of great interest. We show in this paper that with a waveguide measurement set up, it is possible to determine σ^* on YBaCuO thin films at least between 78K and 300K. In connection with these investigations, applications to microstrip resonators are also presented.

2. FILM PREPARATION AND CHARACTERIZATION

2.1. Preparation
     The films used in this study were realized at the LETI (Grenoble, France) with an inverted cylindrical magnetron DC sputtering [3]. The deposition is made under a high argon and oxygen pressure and the substrate temperature is kept at about 750°C. In usual conditions, the deposition rate is ≈600nm/h. The slow cooling to room temperature under a 1 atm. oxygen pressure allows the "in situ" deposition of YBaCuO superconducting films. At present the maximum size of the films is about 2x2cm which is enough for the microwave measurements.

2.2. Characterization
     The characterizations of the films have been done at the LETI and also at the Laboratoire de chimie minérale B (Université de Rennes 1, France). A X-ray diffraction spectra is given figure 1 for the K109 sample. It shows essentially the (00ℓ) planes corresponding to a c-axis orientation of the film and some small (h00) peaks due to a-axis orientation. A rocking curve done on the (005) peak has a FWHM of 1.2° which indicates the degree of misorientation of the c-axis.

A.C. magnetic susceptibility measurements [4] were also performed on K109
in order to determine the critical temperature. In figure 2, we give the χ'
and χ'' curves showing a T_c(onset)$\approx$79.5K and a $\Delta T_c \approx$2K. From the shape of the
curves it can be inferred that the quality and the homogeneity of the part
of the sample tested (7x7mm) is quite good.

3. MICROWAVE COMPLEX CONDUCTIVITY

A measurement method of the microwave complex conductivity was
initially proposed by M. Tinkham [5] on low Tc superconductors. It is based
on the determination of the power transmission through a superconducting
film located across a waveguide. To our knowledge, only a few studies have
been published [6-10] on high Tc superconductors. So, we have applied the
transmission technique to see if it is operational on HTSC films.

3.1. Theoretical bases

We consider the case of a guided wave propagating the TE_{10} mode
normally incident on the film. The film is characterized by its complex
conductivity $\sigma^* = \sigma_1 + j\sigma_2$ and the substrate by its relative dielectric
constant ε_r. The wave impedance of the TE_{10} mode is $Z_{10} = Z_0 \left[1 - (F_c/F)^2 \right]^{-\frac{1}{2}}$
where $Z_0 = 120\pi$ is the free space impedance, F and F_c are respectively the
operational and cut-off frequencies.

$$
\begin{array}{c|c|c|c|c}
 & 1 & 2 & 3 & \\
\hline
E_1\Big| & \begin{matrix} \rho_1 \\ \text{air} \end{matrix} & \begin{matrix} t_2 \quad \rho_2 \\ \text{film} \\ \sigma_1 + j\sigma_2 \end{matrix} & \begin{matrix} t_3 \quad \rho_3 \\ \text{substrate} \\ \varepsilon_r \end{matrix} & \text{air} \Big| E_4 \\
\end{array}
$$

$$<\!\!-\!\!-\!\!-a\!\!-\!\!-\!\!-\!\!>\!<\!\!-\!\!-\!\!-d\!\!-\!\!-\!\!-\!\!>$$

The ratio of the transmitted field E4 to the incident one E1 is the
coefficient S_{21} of the scattering matrix:

$$
S_{21} = \frac{(1+\rho_1)(1+\rho_2)(1+\rho_3)t_2 t_3}{1+\rho_1\rho_2 t_2^2 + \rho_3 t_3^2(\rho_2+\rho_1 t_2^2)} = |S_{21}| e^{j\arg(S_{21})} \quad \text{and} \quad S_{21}(dB)=20log|S_{21}| \quad (1)
$$

where ρ_1, ρ_2, ρ_3 are the reflexion coefficients at the interfaces and t_2, t_3
the transmission coefficients in the film and the substrate . An exact
determination of σ_1 and σ_2 seems difficult to obtain as $|S_{21}|$ and arg(S21)
are very complicated expressions. Therefore, we now consider a particular
case which corresponds to our experimental situation: *we neglect the
reflexions in the substrate* as its thickness is very inferior to the
wavelength. For instance, in our frequency range, the influence of a substrate
of MgO, 250μm thick is inferior to 5% on $|S_{21}|$.

In this condition, as $\rho_2 = -\rho_1, \rho_3 = 0, t_3 = 1$:

$$S_{21} = \frac{(1-\rho_1^2)t_2}{1-\rho_1^2 t_2^2} \quad \text{with} \quad \rho_1 = \frac{\alpha - \beta + j\gamma}{\alpha + \beta - j\gamma} \quad \text{and} \quad t_2 = e^{-a\sqrt{\pi\mu_0 F}(\gamma + j\beta)} \quad (2)$$

as $\quad \alpha = 2\left[\pi\varepsilon_0 F\left[1-(F_c/F)^2\right]\right]^{\frac{1}{2}}; \quad \beta = \left[\sqrt{\sigma_1^2 + \sigma_2^2} + \sigma_2\right]^{\frac{1}{2}}; \quad \gamma = \left[\sqrt{\sigma_1^2 + \sigma_2^2} - \sigma_2\right]^{\frac{1}{2}}$ (3)

An important simplification results when the thickness "a" of the film is small enough to verify the following condition:

$$a \ll \frac{1}{2}\left[\pi\mu_0 F\left[\sqrt{\sigma_1^2 + \sigma_2^2} - \sigma_2\right]\right]^{\frac{1}{2}} \quad (4)$$

Thus: $\quad |S_{21}| \approx \dfrac{1}{\left[\left[1 + \dfrac{\sigma_1 aZ_{10}}{2}\right]^2 + \left(\dfrac{\sigma_2 aZ_{10}}{2}\right)^2\right]^{\frac{1}{2}}} \quad$ and $\quad \arg(S_{21}) = -\tan^{-1}\left(\dfrac{\sigma_2 aZ_{10}}{2 + \sigma_1 aZ_{10}}\right)$ (5)

It should be noted that with equations 5, the determination of σ_1 and σ_2 is direct if the measurements of both the module and the phase are available.

3.2. Experimental results

The experimental set up consists of waveguide devices and operates in the K-band (18-26GHz) between 78K and 300K owing to the use of a home made cryostat [11]. We give, figure 3, the evolution of the transmission at room temperature as a function of frequency for the K109 sample (a=250nm, d=250μm). The conductivity σ_n in the normal state at 300K is evaluated with the formulae n°5 letting $\sigma_2 = 0$. A mean value of $\sigma_n = 3.2 \times 10^5$ S/m is then obtained.

We present, figure 4, the evolution of the transmission from 78K to 200K at different frequencies. The appearance of superconductivity is clearly seen at about 90K and there is a drastic decrease of the transmission from 80K to 78K in accordance with the χ' curve (see figure 2). In the normal state, the ratio $\sigma_n(100K)/\sigma_n(300K) \approx 1.9$ reflects some deviation from the metallic state. Moreover, it should be noted that this ratio is the same in microwaves frequencies and in DC (four probe technique).

3.3. Estimation of σ_1, σ_2, R_s at 78K and 22GHz

At the moment, we are able to determine the amplitude but not the phase of S_{21}. So, with only a single measurement, it is not possible to calculate both σ_1 and σ_2. Therefore, we assumed by simplicity that σ_1 follows the two fluid model :

$$\sigma_1 = \sigma_c \left[\frac{T}{T_c}\right]^4 \quad (6)$$

where σ_c is the normal conductivity at T_c. Taking arbitrarily $T_c = 90K$ (onset in microwave), this gives $\sigma_1 \approx 3.4 \times 10^5$ S/m and allows $|\sigma_2|$ to be determined

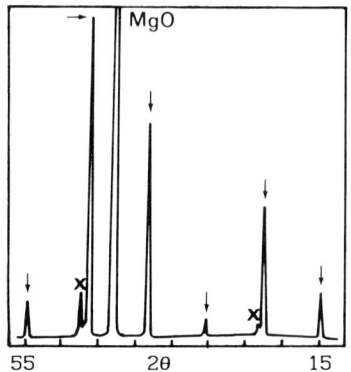

Figure 1. X-ray diffractogram of K109
(00ℓ) planes: →   (ℎ00) planes: **x**

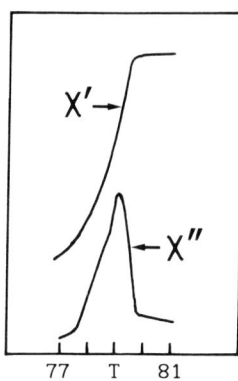

Figure 2. A.C.susceptibility at
119Hz vs temperature for K109

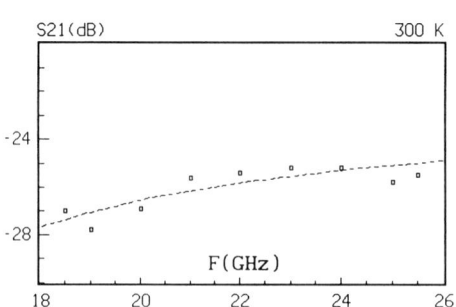

Figure 3. Transmission for K109 at 300K
(□) exp.; (---) th. curve for $\sigma_n = 3.2 \times 10^5$ S/m

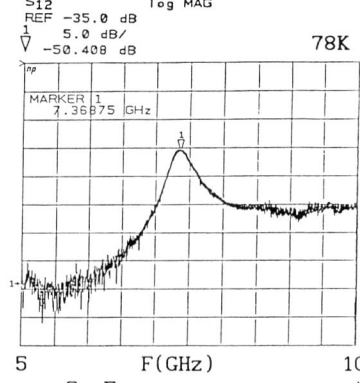

Figure 5. Frequency response at
78K of a YBaCuO resonator

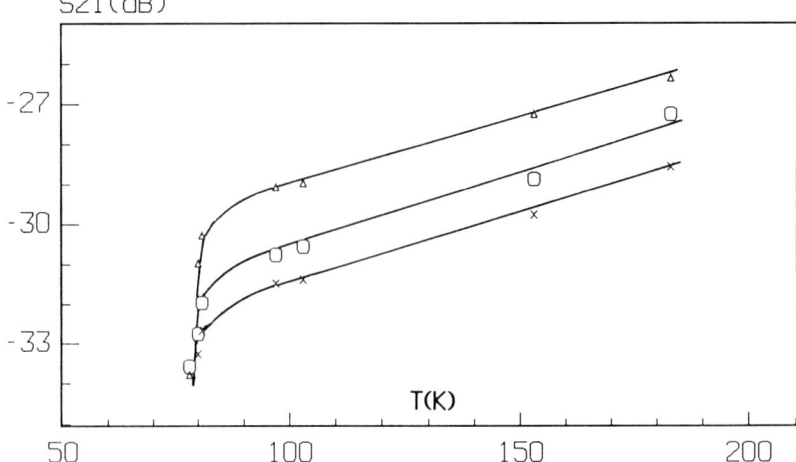

Figure 4. Transmission for K109: (-Δ-)18.5GHz; (-0-)22GHz; (-x-)25.5GHz

from $|S_{21}|$ (equation 5) leading to $|\sigma_2| \approx 7 \times 10^5 S/m$. These figures are only a rough estimate as the condition 4 which permits to make use of equation 5 is verified in a first approximation. However, our values of σ_1 and $|\sigma_2|$ are in close agreement with those of F.A.Miranda and coll.[12] at 28.5 GHz for an YBaCuO thin film (200nm) on MgO.

From σ_1, $|\sigma_2|$ and a, it is possible to infer the value of the square surface resistance $R_{s\square}$ with the formula of Kautz [13]:

$$Z_s = \sqrt{\frac{j2\pi\mu_0 F}{\sigma^*}} \ \coth\left[a\sqrt{j2\pi\mu_0 F\sigma^*}\right] \tag{7}$$

The calculation gives $R_{s\square} \approx 2.3\Omega$ at 78K and 22GHz.

The bulk surface resistance R_s can also be estimated with the previous formula letting $\coth\left[a\sqrt{j2\pi\mu_0 F\sigma^*}\right] = 1$. Thus we obtain $R_s = 0.1\Omega$ which is higher than the copper one by a factor 5.

4. RESONATOR APPLICATIONS

4.1. Microstrip resonators

Microstrip resonators are made by conventional photolithography. They are in the form of a linear stub and are weakly capacitively coupled to the external circuit by gaps. Resonance occurs when the circuit length ℓ verifies: $\ell = \frac{1}{2}n\lambda_g$ where n is an integer and λ is the guided wavelength.

The resonator is mounted in a measurement cell designed at our laboratory [11]. It comprises three parts: two coaxial to coplanar transitions with K connectors and an insert. The microstrip resonator is epoxy sealed on the insert which is thus used as a ground plane. The cell is connected to a HP8510A network analyzer using semi-rigid coaxial cables and immersed in a liquid nitrogen cryostat which allows the sample characterization between 78K and 300K in the frequency range 0.1-26GHz.

4.2. Results

Up to now, the cooling set up was limited to 78K so that measurements were carried out just where the superconducting transition started. As a consequence, all the quality factors Q remained modest. The Q and the resonance frequency were measured using the transmission parameter S_{21}. Figures 5 shows a typical example corresponding to a 0.8cm long resonator and a film thickness of 0.3μm. The conductor losses of the resonator can be extracted from Q_c, its unloaded quality factor. In a first approximation, the losses due to the coupling, to the dielectric and to the radiation are supposed negligible. Then, the square surface resistance $R_{s\square}$ of the film can be roughly estimated from the relations:

$$Q_c = \frac{L\omega}{R} \quad \text{with } L = \frac{Z_c\sqrt{\varepsilon_e}}{C_0} \quad \text{and } R_{s\square} = RW \tag{8}$$

L= microstrip inductance per meter , R= microstrip resistance per meter
$C_0 = 3 \times 10^8$ m/s , $Z_c = 50\Omega$, ε_e =effective permittivity , W= microstrip width

For instance, applying this procedure to the resonator of Figure 5 gives: $R_{s\square}$=0.33Ω at 7.4GHz .The film used to fabricate this resonator is similar to K109. Assuming a F^2 dependence of the surface resistance, this one would be equal to 2.9Ω at 22GHz, value close to the 2.3Ω found from the direct conductivity measurement. This result is satisfying given a possible degradation of the film during the photolithographic process (wet etching with H_3PO_4 solution).

5. CONCLUSION

We have shown that the conductivity of HTSC thin films can be measured between 78K and 300K in microwaves using a waveguide transmission technique. The surface resistance, a key parameter in view of future industrial applications, can be inferred and is in agreement with the one deduced from measurements on YBaCuO microstrip resonators. At present, work is in progress in order to realize measurements under 78K and also to determine the phase of S_{21}.

6. ACKNOWLEDGMENTS

One of us (D.C.) wishes to thank J.C.Villegier for the fabrication of the films with the facilities of the LETI (Grenoble).The authors are indebted to M.Guilloux-Viry and O.Peña (Université Rennes 1) for the X-ray diffraction and the a.c.susceptibility measurements.

7. REFERENCES

1   W.G.Lyons and R.S.Withers,Microwave Journal, 33 (1990) 85.
2   D.Kinowski,F.Huret,P.Pribetich and P.Kennis,Ann. Télécommun., 45 (1990) 334.
3   J.C.Villegier, H.Moriceau, H.Boucher, R.Chicault, L.Di Cioccio, A.Jäger, M.Schwerdtfeger,M.Vabre and C.Villard,IEEE Trans on Magnetics, 27 (1991) 1552.
4   A.Perrin,M.Guilloux-Viry,M.G.Karkut,J.Padiou and M.Sergent,J.Phys.III 1 (1991) 295.
5   R.E.Glover III and M.Tinkham,Phys.Rev., 108 (1957) 243.
6   W.Ho,P.J.Hood,W.F.Hall,P.Kobrin,A.B.Harker and R.E.De Wames, Phys.Rev.B, 38 (1988) 7029.
7   C.S.Nichols, N.S.Shiren, R.B.Laibowitz and T.G.Kazyaka, Phys.Rev.B, 38 (1988) 11970.
8   M.Golosovsky,D.Davidov,C.Rettori and A.Stern,Phys.Rev.B, 40 (1989) 9299.
9   P.H.Kobrin,W.Ho,W.F.Hall,P.J.Hood,I.S.Gergis and A.B.Harker, Phys.Rev.B, 42 (1990) 6259.
10 K.B.Bhasin,J.D.Warner,F.A.Miranda,W.L.Gordon and H.S.Newman, IEEE Trans. on magnetics, 27 (1991) 1284.
11 J.C.Carru,F.Mehri,D.Chauvel and Y.Crosnier, Infrared and Optoelectronic Materials and Devices,Proc. SPIE 1512 (to be published in 1991).
12 F.A.Miranda,W.L.Gordon,K.B.Bhasin,V.O.Heinen,J.D.Warner and G.J.Valco, Nasa Tech.Memo n°102345 (1989).
13 R.L.Kautz,J.Appl.Phys., 49 (1978) 308.

High T$_c$ Superconductor Thin Films
L. Correra (Editor)
© 1992 Elsevier Science Publishers B.V. All rights reserved.

Investigation of Magnetic Loss Mechanisms in High-T$_c$ Films by ac-Susceptibility Measurements

Ch. Heinzel, Ch. Neumann, Th. Ritzi, and P. Ziemann

Fakultät für Physik, Universität Konstanz, P.O. Box 5560, W-7750 Konstanz, FRG

Abstract

The lower critical field H$_{c1}$(T) has been determined for YBaCuO and EuBaCuO films on different substrates and of different c-axis orientations by an ac-susceptibility technique. It turns out that H$_{c1}$(0) can be taken as an excellent quality criterion. By studying geometrically structured films as well as the anisotropy of the magnetic response (external field parallel or perpendicular to the c-axis) it is concluded that the observed loss peak is mainly due to flux flow. For Tl$_2$Ba$_2$CaCu$_2$O$_8$ films, this interpretation of the experimental data suggests the possibility of flux-line entanglement.

1. INTRODUCTION

Besides their high transition temperature, the new high-T$_c$ superconductors exhibit a number of unusual phenomena like their diamagnetic response to external magnetic fields, which is of considerable interest also from an applications point of view. For fields above the lower critical field H$_{c1}$, these type-II superconductors show two different regimes within the Shubnikov state separated by an irreversibility line H*(T). In a dc-magnetization experiment, this line separates the hysteretic from the non-hysteretic M(H)-behavior [1]. If an ac-technique is used, within both regimes magnetic losses are observed, which can be classified as hysteretic and viscous, respectively. There is still a controversial discussion on the physical origin of the irreversibility line and quite different models have been suggested like the superconducting-glass model [1], flux-lattice melting [2, 3], flux-line glass-liquid transition [4] or flux-line entanglement transition [5]. The latter three models are applicable to perfect bulk superconductors and appear most attractive in the context of the large magnetic anisotropies of especially the Bi or Tl systems, while a glass model seems to be more appropriate for sintered samples. In this case, the magnetic behavior is dominated by the weak links providing the intergrain coupling. Due to this effect, the lower critical field of this effective "Josephson medium" is dramatically reduced as compared to the intragrain value found for e.g. single crystals. On the other hand, by comparing the experimentally determined H$_{c1}$-values to the corresponding single-crystalline data, a quality criterion for e.g. thin films can be provided as will be shown in the following. Furthermore, by measuring the magnetic ac-losses of high-T$_c$ films of different geometrical shapes and studying the influence of the field orientation relative to the crystal axes, information is obtained on the loss mechanism.

2. EXPERIMENTAL

The magnetic ac-losses of high-T_c films are studied by using a specially designed vector lock-in ac-susceptometer. It allows to measure the real (χ') and the imaginary (χ'') part of the susceptibility as a function of the temperature T, the field amplitude H_{ac} and the frequency f. The corresponding ranges are $5\,K \leq T \leq 300\,K$, $0.1\,Oe \leq H_{ac} \leq 50\,Oe$ and $31\,Hz \leq f \leq 3125\,Hz$. By moving the samples to different positions relative to the counterwise wound pick-up coils, one is able to preferentially determine the components of the magnetization perpendicular or parallel to the film surface. For a quantitative analysis, demagnetization effects must be taken into account. For this purpose, Pb films of the same size and thickness as the high-T_c films were used to calibrate each sample position. Details of this procedure as well as of the apparatus are described in ref. [6].

The results of the present work mostly refer to YBaCuO and EuBaCuO films on different substrates ($<100>$-SrTiO$_3$, $<110>$-SrTiO$_3$, $<100>$-MgO, sapphire) and of different orientations (c-axis perpendicular or parallel to the film surface). Additionally, results on $Tl_2Ba_2CaCu_2O_8$ films are reported. Details on the preparation of these films can be found in ref. [7, 8].

The analysis of the experimental data can be exemplified refering to fig. 1, where the temperature dependence of the imaginary part χ'' is shown for a c-axis oriented YBaCuO film on a $<100>$-SrTiO$_3$ substrate. A field amplitude of $H_{ac} = 30\,Oe$ oriented

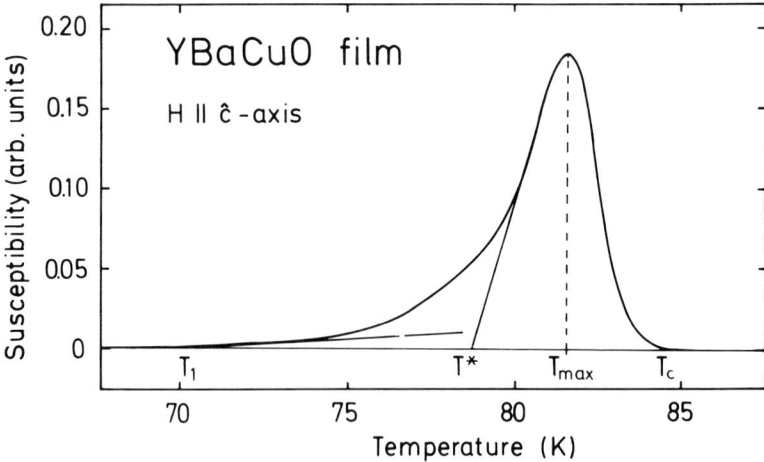

Figure 1. Temperature dependence of the imaginary part of the susceptibility obtained for a YBaCuO film on $<100>$-SrTiO$_3$ applying a field amplitude $H_{ac} = 30\,Oe$ (f = 1049 Hz) parallel to the c-axis.

parallel to the c-axis was applied at a frequency f = 1049 Hz. Clearly, different temperature regimes can be distinguished. At $T < T_1$ no losses can be observed within the experimental resolution. This is the regime of perfect shielding. At T_1 magnetic flux starts to penetrate the film leading to observable losses. This allows to determine the lower critical field H_{c1}, by setting $H_{c1}(T_1) = H_{ac}$ and measuring $\chi''(T)$ with different amplitudes H_{ac}. At T^* an additional loss mechanism sets in leading to a maximum of χ'' at T_{max}. As will be discussed in the next section, this mechanism is most probably due to flux flow. Eventually, at the transition temperature T_c, χ'' approaches zero.

3. RESULTS AND DISCUSSION

In fig. 2 the results of the H_{c1}-measurements are presented for YBaCuO and EuBaCuO films prepared on different substrates and with different c-axis orientations as given in the figure caption. To demonstrate the magnetic anisotropy, the H_{c1}-values obtained for the field orientations parallel or perpendicular to the c-axis are shown separately. For both orientations, demagnetization corrections were taken into account. The solid and dashed lines in fig. 2a, 2b represent calculations based on the relation $H_{c1}(t) = H_{c1}(0) \cdot (1 - t^2)$ with extrapolated values $H_{c1}(0)$ as indicated by the arrows (note the interrupted $t = T/T_c$-scale). These values have been chosen to indicate typical data reported in the literature for YBaCuO single crystals. Thus, these lines allow a comparison of films to single crystals. For H‖c-axis, the results of our best film (open squares in fig. 2a; c-axis-oriented on <100>-SrTiO$_3$, j_c(77K) = 5·10^6 A/cm^2 as deter-

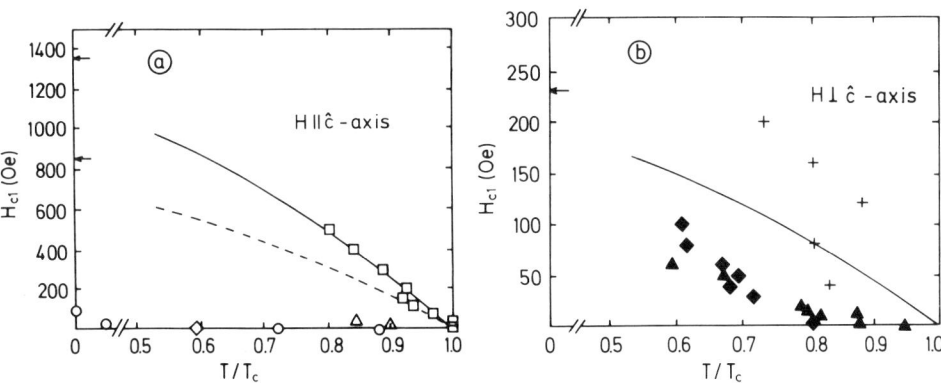

Figure 2. Lower critical fields H_{c1} for different YBaCuO/EuBaCuO films obtained by ac-susceptibility measurements. (a) Magnetic field parallel to the c-axis; □ YBaCuO on <100>-SrTiO$_3$, c-axis oriented; △ YBaCuO on <110>-SrTiO$_3$, c-axis oriented; ○ YBaCuO on <100>-SrTiO$_3$, c-axis oriented, non-optimized preparation conditions; ◇ YBaCuO on sapphire, polycrystalline. (b) Magnetic field perpendicular to the c-axis; ◆ EuBaCuO on <100>-MgO, a-axis oriented; ▲ YBaCuO on <110>-SrTiO$_3$, c-axis oriented; + YBaCuO on <100>-SrTiO$_3$, a-axis oriented, non-optimized conditions. Solid and dashed lines are typical of single crystals.

mined resistively) lie within the range observed for single crystals, while all other c-axis oriented or polycristalline films exhibit drastically reduced H_{c1}-values. This is a clear indication, that non-optimized preparation conditions including the choice of the substrate lead to weak links, which let the magnetic flux penetrate into the sample well below the intragrain H_{c1}-value. The results of the a-axis oriented films shown in fig. 2b all deviate from the single-crystalline behavior indicating the difficulty to optimize this growth mode. It is worth noting that a c-axis oriented film (closed triangle in fig. 2b) analyzed with $H \perp c$ shows the same behavior as an optimized a-axis oriented film (closed diamonds).

While the onset of magnetic losses at $T_1 \ll T_c$ can be used to determine H_{c1}, the mechanisms leading to the peak of χ'' at T_{max} are not clear. If interpreted in terms of hysteretic losses by applying e.g. Bean's model, the peak position should depend on the size of a sample [9]. For example, in case of cylindrical samples with the field parallel to the axis, a decreasing radius should shift the χ''-peak towards higher temperatures. To test this prediction, a YBaCuO film was geometrically structured as shown in the inset of fig. 3, i.e. two significantly different film sizes were prepared. At the indicated position, the perpendicular component of the magnetization is preferentially determined. According to Bean's model two χ''-peaks are expected for the above films in contrast to the experimental result shown in fig. 3, where only one peak is observed. This observation suggests flux creep/flow as the dominating loss mechanism. In this case, the peak position should exhibit a frequency dependence. For YBaCuO this dependence is too small (0.3 K per frequency decade [10]) to be observable within our resolution. The effect is much larger for the more anisotropic $Tl_2Ba_2CaCu_2O_8$ films, which will be discussed in the following.

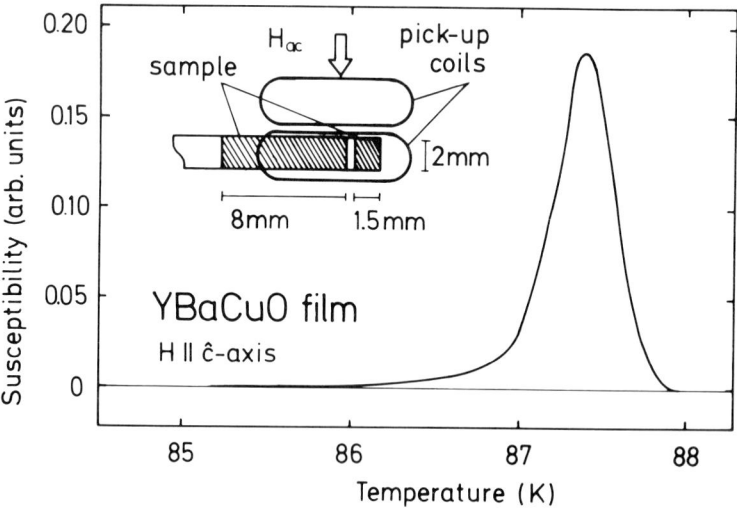

Figure 3. Temperature dependence of the imaginary part of the susceptibility observed for an YBaCuO film geometrically structured as shown in the inset. The applied field (30 Oe, f = 1049 Hz) is parallel to the c-axis.

The results of the susceptibility measurements are presented in fig. 4, where the temperature dependence of the real and imaginary part of χ are shown. By measuring at different film positions relative to the pick-up coils the different components of the magnetization are separated. Curves 1 (χ') and 4 (χ'') are dominated by the component perpendicular to the film surface. Here, for the frequency dependence of the loss peak a value of 1.5 K per decade is observed confirming the idea of flux flow as the most relevant loss mechanism. The behavior of the response of the magnetization parallel to the film surface is given by the curves 2 (χ') and 5 (χ''). Only a small loss peak is observed at a significantly lower temperature as compared to curve 4. Within our interpretation, this leads to the striking conclusion that the component of magnetic flux parallel to the CuO_2 planes starts to become mobile at much lower temperatures than the component perpendicular to these planes. Such a behavior could be favorable to flux-line entanglement. On the other hand, the parallel component of χ' (curve 2) is dominated by the loss-free penetration of the magnetic field within the penetration length $\lambda(T)$. This is demonstrated by the dashed curve 3 in fig. 4, which shows the theoretical behavior of χ', if only the $\lambda(T)$-effect is taken into account (here a slab geometry has been assumed with the experimental film thickness of 12 μm and a temperature dependence of λ according to the two-fluid model). In this way, a value of $\lambda(0) = 3640$ nm is obtained for fields parallel to the CuO_2 planes.

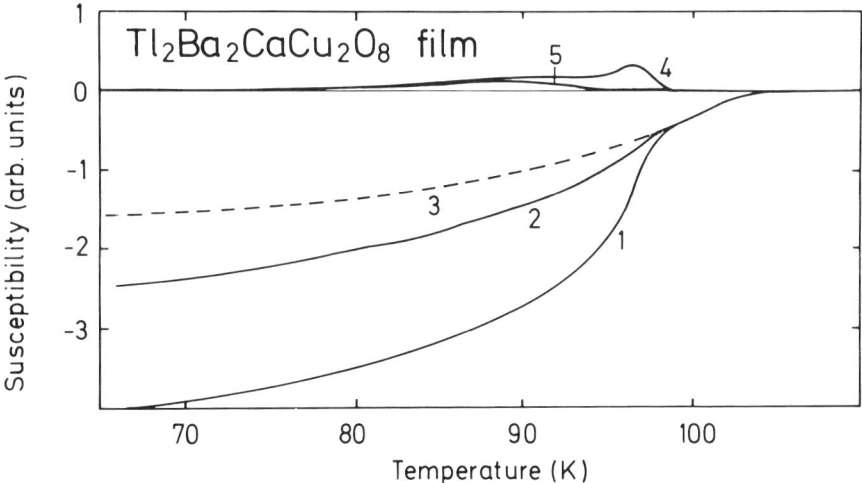

Figure 4. Temperature dependence of the real (χ') and imaginary (χ'') parts of the susceptibility found for a $Tl_2Ba_2CaCu_2O_8$ film (12 μm thick) with different field orientations relative to the c-axis. Curves 1 (χ') and 4 (χ''): field parallel to the c-axis; curves 2 (χ') and 5 (χ''): field perpendicular to the c-axis (i.e. field parallel to the film plane). The dashed curve 3 describes the theoretical behavior if only the temperature dependence of the penetration depth is taken into account for a field parallel to the film plane.

Acknowledgements

We thank our colleagues J. Geerk, G. Linker (GFK Karlsruhe) and W.Y. Lee (IBM, San Jose) for supplying us with HTSC films. This work was supported by Bundesministerium für Forschung und Technologie (BMFT) and Land Baden-Württemberg (MWK).

References

1   U.A. Müller, M. Takashige, and J.G. Bednorz, Phys. Rev. Lett. 58 (1987) 1143.
2   D.R. Nelson, Phys. Rev. Lett. 60 (1988) 1973.
3   E.H. Brandt, Phys. Rev. Lett. 63 (1989) 1106.
4   M.P.A. Fisher, Phys. Rev. Lett. 65 (1990) 923.
5   S.P. Obukhof and M. Rubinstein, Phys. Rev. Lett. 65 (1990) 1279.
6   Ch. Neumann, P. Ziemann, J. Geerk, and H.C. Li, J. Less-Comm. Met. 151 (1989) 363.
7   H.C. Li, G. Linker, F. Ratzel, R. Smithey, and J. Geerk, Appl. Phys. Lett. 52 (1988) 1098.
     R.L. Wang, J. Reiner, J. Remmel, E. Brecht, B. Rauschenbach, J. Geerk, O. Meyer, and G. Linker, this volume.
8   W.Y. Lee, V.Y. Lee, J. Salem, T.C. Huang, R. Savoy, D.C. Bullock, and S.S.P. Parkin, Appl. Phys. Lett. 53 (1988) 329.
9   D.-X. Chen and R.B. Goldfarb, J. Appl. Phys. 66 (1989) 2489.
10   L.T. Sagdahl, S. Gjolmesli, T. Laegreid, K. Fossheim, and W. Assmus, Phys. Rev. B 42 (1990) 6797.

High T$_c$ Superconductor Thin Films
L. Correra (Editor)
165

STRUCTURAL CHARACTERIZATION OF Bi-Sr-Ca-Cu-O and Y-Ba-Cu-O THIN FILMS BY MICRO-RAMAN SPECTROSCOPY.

Pham V.Huong[a], P.Bernstein[b], M.Viret[b], P.Paroli[c], M.Marinelli[c]

[a]) Laboratoire de Spectroscopie Moléculaire et Cristalline, (URA 124 C.N.R.S)
Université de Bordeaux I, Fax (33) 56 84 66 45
351, Cours de la Libération, 33405 Talence, France.

[b]) Ecole Supérieure de Physique Chimie Industrielle,
10, Rue Vauquelin, 75231 Paris, France

[c]) Dept.Ingegneria Meccanica, Universita degli Studi di Roma,
8 Via O.Raimondi, 00173 Roma, Italy.

Abstract

Thin films of Y-Ba-Cu-O and Bi-Sr-Ca-Cu-O on various substrates MgO, BeO, ZrO$_2$, SrTiO$_3$, etc... were examined by micro-Raman spectroscopy.

Based on correlations established between Raman spectra and structures of well oriented single crystals, structural information on superconductor thin films prepared by various methods has been deduced. In particular, the oxygen amount in YBa$_2$Cu$_3$O$_x$ can be evaluated through the relation $x = 0.025 \nu_H - 5.57$ where ν_H is the Raman frequency of the Cu^2-O^1 vibration along the c axis. Raman spectroscopy is also a method for characterizing Bi$_{2212}$ and Bi$_{2223}$.

The high Raman anisotropy efficiently helps to determine the orientation of Y-Ba-Cu-O and Ba-Sr-Ca-Cu-O surfaces, basing simply on the polarization of the Raman lines.

This method also allows to determine the micro-structure and orientation of the epitaxial deposits in function of their thickness, with step-etched samples.

Finally, perturbations of the Raman spectrum at the interface, with bevelled samples, reveal structural modifications at this level and can give the evidence of new chemicals resulting from the interaction between the superconducting layer and the substrate.

1. INTRODUCTION

The discovery of YBa$_2$Cu$_4$O$_8$ and doped superconductors enhances the interest in Y-Ba-Cu-O series. The potential use of these materials as well as of Bi$_2$Sr$_2$CaCu$_2$O$_8$, Bi$_{2212}$, Bi$_2$Sr$_2$Ca$_2$Cu$_3$O$_7$, Bi$_{2223}$ and their Pb doped materials imply realization of their thin films on various insulators or semiconductors. The knowledge of the micro-structure of such layers becomes necessary.

As the epitaxial quality of thin films strongly depends on the deposition parameters and on the nature of the substrate, it is necessary to control the micro-structure and the orientation of the superconducting films.

Micro-Raman spectroscopy is known to be helpful in such a characterization [1-3]. Its spatial resolution, less than 1 µm^2 also permits to verify the homogeneity of the surfaces as well as the epitaxial quality in function of the layer thickness. With bevelled or step-etched bi-layers, this technique can help to detect eventual structural modifications of the deposit and the substrate due to interactions at the interface and shows the appearance of eventual new chemical bonds [4].

In this paper, a method will be proposed to determine the microstructure and orientation of high T_c superconducting thin films, for both films having or not having grains greater than the wavelength of the laser used in the investigation ($\sim$ 0.5 μm).

2. EXPERIMENTAL

2.1. Thin films of $Bi_2Sr_2CaCu_2O_8$ and $Bi_{1.7}Pb_{0.3}Sr_2CaCu_2O_8$

Various B_{2212} and its Pb doped thin films on MgO and $LaGaO_3$ substrates were examined. The first one was prepared by laser ablation and *in situ* crystallization without further treatment.

$YBa_2Cu_3O_x$ thin layers were deposited on substrates of different nature : $SrTiO_3$, MgO, BeO and cubic ZrO_2, under various preparation conditions.

For these studies, it was also necessary to refer to Raman characteristics of single crystals. The orientation determination of unknown surfaces was based on a calibration using well-oriented surface of single crystals.

The Raman spectra were recorded on a Dilor instrument model OMARS, equipped with a multichannel photo-diode-array detector and a Spectra-Physics Model 165 ion argon laser source.

The other physical characterizations were performed by conventional methods.

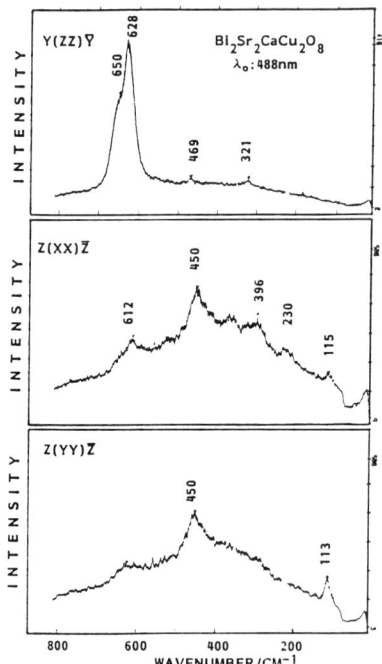

Figure 1. Polarized Raman spectra of $Bi_2Sr_2CaCu_2O_8$ single crystal.

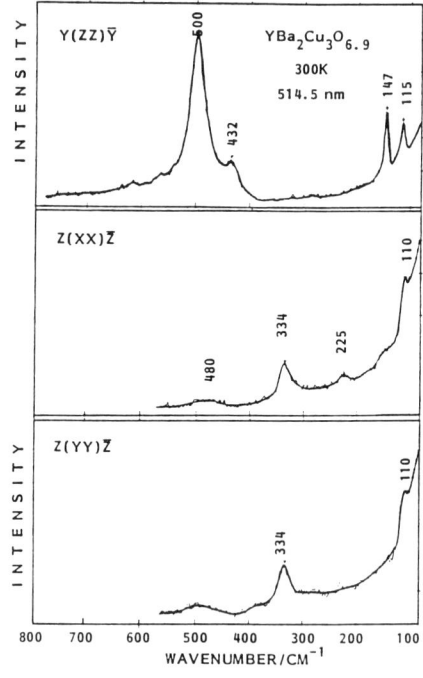

Figure 2. Polarized Raman spectra of $YBa_2Sr_2Cu_3O_{6.9}$ single crystal.

3. RESULTS AND DISCUSSION

3.1. Anisotropy in polarized Raman spectra of superconducting single crystals

The reference to single crystals is necessary. Bismuth cuprates as well as yttrium materials give strong anisotropy in Raman scattering [5].

For $Bi_2Sr_2CaCu_2O_8$ single crystal the ZZ polarized spectrum differs completely to those of other polarizations. On the spectrum (Fig.1), a strong feature corresponding to Cu-O stretching is observed : the band is sometimes double, one component is located at 650 cm^{-1} and the other at 628 cm^{-1}. Some samples present only one band at 654 cm^{-1}. Some others give a single band at 619 cm^{-1}. These Bi_{2212} samples generally contain an amount of Bi_{2223} : as the two different vibrational frequencies could generally correspond to two different distributions of Cu-O bonds, the observed double band could mean that there is some disorder in the structure [6]. The Pb doping tends to decrease the high frequency component and gives only a single band at 621 cm^{-1}. This fact can be connected with a bond ordering effect. Nevertheless, the Pb doping tends also to increase the amount of Bi_{2223} phase. Thus, the latter can also be responsible for the band at 620-630 cm^{-1}.

Anyhow, in the ZZ Raman spectrum, this Cu-O feature is very strong in comparison to the remaining part of the spectrum, in particular nearly nothing is observed at 450 cm^{-1}. In the contrary, for XX and YY spectra, a noticeable band appears at 450 cm^{-1}.

The same anisotropy has been observed with $YBa_2Cu_3O_x$, for that strong Cu^2-O^1 stretching vibration is located around 500 cm^{-1} in the ZZ Raman spectrum, while

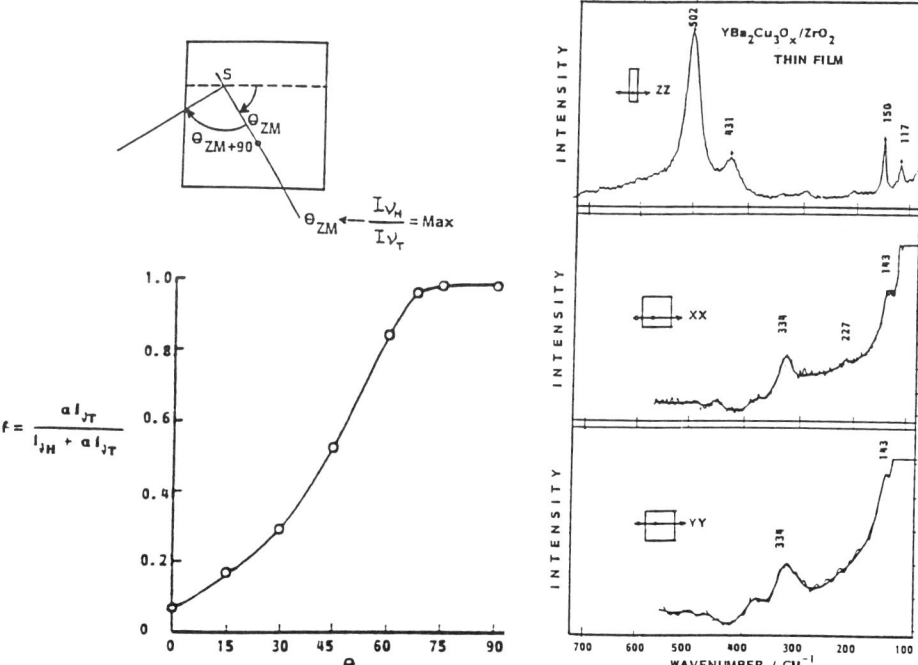

Figure 3. Anisotropy calibration and rotational screening of a surface spot.

Figure 4. Raman spectra of grains on $YBa_2Cu_3O_7$ thin film on cubic ZnO_2.

nearly no band is present around 334 cm^{-1}. On the XX and YY polarized spectra, a noticeable band appears around 330 cm^{-1} and is the most intense (Fig.2).

3.2. The method

This Raman anisotropy will help to determine the orientation of any surface or thin film [1,2]. Of course, a previous calibration has to be done. For instance, with known oriented surface of a $YBa_2Cu_3O_x$ single crystal, a calibration curve can be obtain for the relative intensity of the 330 band (v_T) and 500 band (v_H) (Fig.3).

For an unknown surface, a rotational screening around a spot can be applied to reach the maximum Raman anisotropy before recording the spectrum. Another record at right angle from this position avoids the ambiguity in the orientation [3].

For surface presenting grains larger than 1 μm^2, each grain can be considered as a single crystal under the microscope and with selected grains, results can be obtained similar to results on bigger crystals (Fig.4).

3.3. Bi$_{2212}$ and (Bi, Pb)$_{2212}$ thin films

Even without further treatment, a Bi$_{2212}$ layer deposited on MgO substrate by laser ablation and in situ crystalization gives a good epitaxy, with its c axis perpendicular to the surface, if we base on the Raman spectra recorded at two perpendicular directions of the film (Fig.5) which are quasi identical to the XX and YY spectra of a known oriented surface, as shown on fig.1.

This also holds the case of a thin film of $Bi_{1.7}Pb_{0.3}Sr_2CaCu_2O_8$ deposited on LaGaO$_3$ substrate (Fig.6). Notice that the bulk (BiPb)$_{2212}$ crystal gives a sharp band at 621 cm^{-1} indicating an ordering in the Cu-O of the material.

Of course, the experiment is repeated at several spots of the surface to average over the distribution of the orientation axis of the thin film.

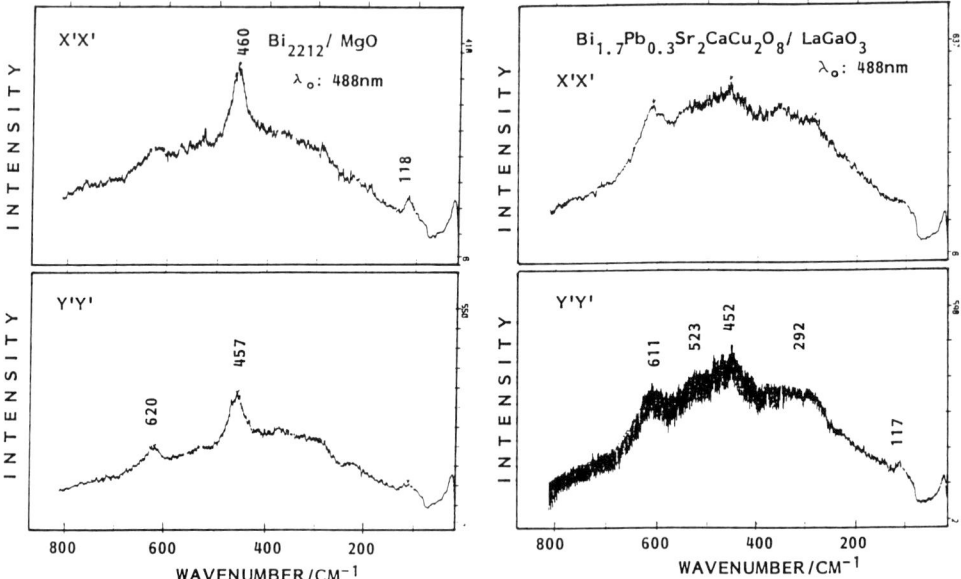

Figure 5. Raman spectrum of Bi$_{2212}$
thin film on MgO at two ⊥ orientations.

Figure 6. Raman spectrum of Bi$_{1.7}$Pb$_{0.3}$Sr$_2$CaCu$_2$O$_8$
on LaGaO$_3$ at two ⊥ orientations.

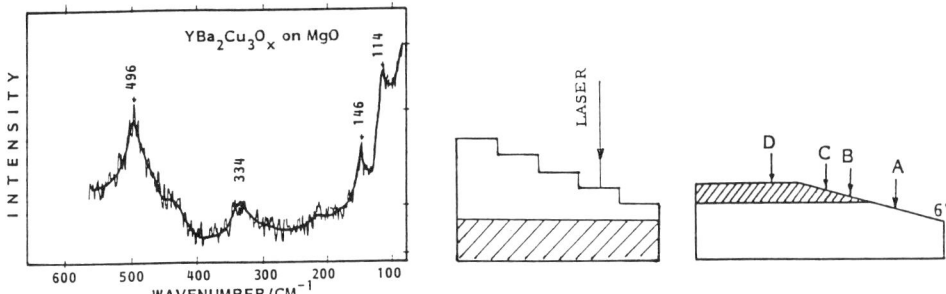

Figure 7. Raman spectrum of a $YBa_2Cu_3O_{6.8}$ thin film on MgO.

Figure 8. Step-etched and bevelled bilayer samples.

3.4. Thin films of $YBa_2Cu_3O_x$

Several correlations have been established between Raman spectra and structure of high T_c superconductors. Some vibrations are very sensitive to the oxygen content in the material. A relation given by Huong [7,8] permits a rapid evaluation of x in $YBa_2Cu_3O_x$: $x = 0.025$ v_H - 5.57 where v_H represents the Cu^2-O^1 vibration along the c axis. The accuracy is $\Delta x = \pm 0.05$. The variation of Cu^2-O^1 bond length in function of oxygen content gives a physical meaning for this empirical relation. This vibration v_H varies from 463 cm^{-1} in tetragonal $YBa_2Cu_3O_6$ to 503 cm^{-1} in orthorhombic $YBa_2Cu_3O_7$.

By applying this relation, we can easily determine the oxygen amount at any spot of a surface and deduce its homogeneity.

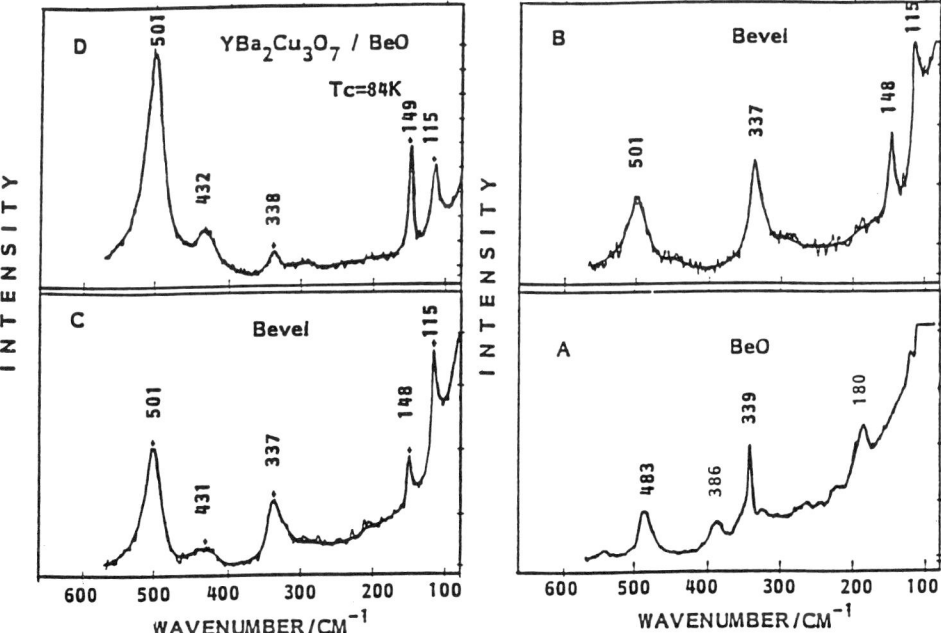

Figure 9. Raman spectrum of a $YBa_2Cu_3O_7$ thin film at various thicknesses on BeO substrate.

As a example, for a thin film of $YBa_2Cu_3O_x$ on MgO (Fig.7), the v_H band appears at 496 cm^{-1} allowing determination of its oxygen amount : x = 6.8. At this spot, the sample is not well oriented, as the relative intensity of the two bands at 334 and 496 cm^{-1} indicates an angle $\theta \simeq 30$ ° between the c axis and the substrate.

3.5. Micro structure in function of the thickness of the film

With step etched or bevelled bi-layers (Fig.8), micro-Raman technique allows to follow the epitaxial quality of the deposited layer at various thicknesses.

With a film of $YBa_2Cu_3O_x$ deposited on BeO, it is encouraging to see that the v_H frequency is always located at 501 cm^{-1} indicating that the oxygen content is everywhere nearly x = 7. On the contrary, the orientation is not good at any thickness (Fig.9). This result can help to modify the deposition conditions for obtaining better epitaxy.

3.6. Films containing grains

With superconducting films containing grains larger than 1 µm^2, one can operate micro-Raman spectroscopy on each grain like on any bigger single crystal [1,3]. For instance, an $YBa_2Cu_3O_x$ film deposited on a cubic zirconia substrate presents grains with two different shapes (Fig.4). The oxygen content was evaluated to be x = 7 and the determination of the orientation of the grains is trivial. Two orientations are seen : one having the c axis nearly parallel to the surface while the other series of grains have their c axis perpendicular to the substrate.

3.7. Interface

By illuminating a bevelled bi-layer, micro-Raman spectra can be recorded easily at different spots from the substrate to the outer surface, and also at the interface. For this superconducting $YBa_2Cu_3O_x$ material, the penetration depth is about 100 Å.

For the examined thin films, it was possible to detect structural modifications at the interface, due to interactions between film and substrate. Sometimes chemical reactions can occur. Such is the case of a sample of $YBa_2Cu_3O_7$ on cubic zirconia : a green phase has been detected [1]. This phase probably results from a reaction with Y_2O_3 used to stabilize the ZrO_2 substrate.

Acknowledgment : A grant from Indo-French Centre for the Promotion of Advanced Research is highly appreciated.

REFERENCES

1   P.V.Huong, J.C.Bruyère, E.Bustarret, P.Granchamp,.
     Solid State Comm., **72** (1989) 191.
2   P.V.Huong, J.C.Bruyère, in "Modern Aspects of Superconductivity",
     R.Suryanaryanan (Ed.), I.I.T.T., Paris, 1989, p.149.
3   Pham V.Huong, J.Less.Common Metals, **164** (1990) 1193.
4   P.V.Huong, A.L.Verma, J.P.Chaminade, L.Nganga, J.C.Frison,
     Materials Science and Engineering, **B5** (1990) 255.
5   P.V.Huong, E.Oh-Kim, K.H.Kim, D.Kim, J.S.Choi,
     Materials Research and Engineering, **A109** (1989) 337.
6   K.H.Kim, P.V.Huong, E.Oh-Kim, M.Lahaye, S.K.Cho, B.C.Kwak, J.Less.Common Metals,
     164 (1990) 1201.
7   Pham V.Huong, in Raman Spectroscopy, S.Jha et S.Banerjee Eds.,
     World Scientific Publ.Calcutta/Singapore, 1989, p.256.
8   L.Nganga., P.V. Huong, J.P.Chaminade, P.Dordor, K.Frölich , M.Jergel,
     J.Less.Common Metals, **164** (1990) 208.

High T$_c$ Superconductor Thin Films
L. Correra (Editor)
© 1992 Elsevier Science Publishers B.V. All rights reserved.

ON THE MAGNETIC MOMENT RELAXATION IN YBaCuO SINGLE CRYSTAL : SUB–SECOND RELAXATION TIMES

M. Jirsa, L. Půst and J. Pačes

Institute of Physics, CS Acad. Sci.,
Na Slovance 2, CS-18040 Praha 8, Czechoslovakia

Abstract

Magnetic moment relaxation at constant sweep rate $\dot{B}=dB_{ext}/dt$ (so called hysteresis loop relaxation) was measured in YBaCuO single crystal platelet by means of vibrating sample magneto-meter. In this way it is possible to analyze the range of very short effective relaxation times not allowed be conventional relaxation. In the range of high sweeping rates $\dot{B}$ corresponding to sub-second effective relaxation times deviation from $\ln(t)$ dependence was observed. This difference may be due to some other effect, probably viscosity of flux line lattice.

1. INTRODUCTION

Microscopic theories have not yet succeeded to explain sa-tisfactorily processes connected with large critical current relaxation in HTS. We are still dependent on phenomenological models well established in conventional superconductors and adopted to the field of HTS. One of them, Bean´s model, relates magnetic moment m induced in a superconductor by varying exter-nal magnetic induction B_{ext} to critical current density j_c |1|. Relaxation phenomena in superconductors are explained by other phenomenological model in terms of thermally activated flux creep |2,3|. According to this model magnetic moment m induced in constant external magnetic induction B_{ext} logarithmically decreases with time, in good agreement with most experiments performed in a wide range of times (from seconds to hours).

It can be expressed as

$$m(t) = m_1 - S \ln(t) \tag{1a}$$

where m_1 and S are independent of time,

$$m_1 = m_o (1+\frac{kT}{U_o} \ln(t_o)); \qquad S = m_o \frac{kT}{U_o} \tag{1b}$$

Here t_o is the mean time between two hopping attempts from one pinning center to another, m_o is the magnetic moment in absence of relaxation.

It has been shown in |4,5| that thermally activated flux motion can also explain magnetic moment dependence on sweep rate in the $\dot{B}$=const. regime (hysteresis loop relaxation, HLR). It is assumed that relaxation effect is here opposed by the rise due to Lorentz forces so that equilibrium is established for

$$m(\dot{B}) = D_m + S \ln|\dot{B}| \tag{2}$$

where D_m and S are independent of $\dot{B}$ and the relaxation rate S should be same in (1a) and (2) |4,5|.

In this model each hysteresis loop can be labeled with an effective time

$$t_{eff} = \frac{\mu_o S}{\chi \dot{B}} \tag{3}$$

where χ is differential susceptibility, $\chi = \mu_o \Delta m/\Delta B_{ext}$, and μ_o permeability of vacuum. Magnetic moment of a superconducting sample placed in the magnetic field sweeping with a constant $\dot{B}$ has the same value as after relaxation for time t_{eff} at constant B_{ext}. Eq. (3) can be used for comparison of HLR and conventional relaxation at B_{ext}=const. on the same time scale.

2. EXPERIMENTAL

Magnetic moment of the YBaCuO single crystal platelet of 2.12 mm^2 area (a-b plane), thickness 30 μm and mass 0.401 mg was measured by vibrating sample magnetometer at temperatures from 7K to 44K, fields +/-2T and sweep rates $\dot{B}$ from 89 to 0.08 mT/s. The overshooting of B_{ext} was eliminated even at the

highest sweep rate. Magnetic moment was measured with magnetic
field along c-axis. Differential susceptibility χ was measured
at each particular field by demagnetizing sample by means of ac
signal of logarithmically decreasing amplitude and then applying the ac signal of appropriately small constant amplitude.

Special attention was paid to well defined measurement conditions. Magnetic field was recorded simultaneously with magnetic moment for its later check. Field stability at the B_{ext}=const.
regime was typically +/-6.10^{-5}, field sweep linearity at
the $\dot{B}$=const. regime better than 2.10^{-4}. Temperature was kept
constant within +/- 0.1K.

3. RESULTS AND DISCUSSION

HLR results measured at T=21K and B_{ext}=+/-0.6T and +1T are
on the left-hand side of Fig. 1 (points connected by dashed
lines as guides for eye). The right-hand side of the figure
(full lines) shows logarithmic fits of measured conventional
relaxations. The significant upwards declination of measured
magnetic moment from logarithmic time dependence with increasing sweep rate (decreasing t_{eff}) is apparently due to increasing role of viscous forces opposing to relaxation. It qualitatively corresponds to theoretical predictions for the limiting
case of viscous flux flow without relaxation |6|. In this case
magnetic moment linearly depends on $\dot{B}$ so that using eq. (3)

$$m(\dot{B}) = A + \frac{C}{t_{eff}} \qquad (4)$$

where A and C are independent of $\dot{B}$.

The temperature dependence of the relaxation processes
is plotted in Fig. 2. Extensive measurements between 7K and 44K
showed that in this temperature range relaxation rate S is in
both types of experiments nearly the same function of temperature, approximately $S \sim T^{-1.5}$.

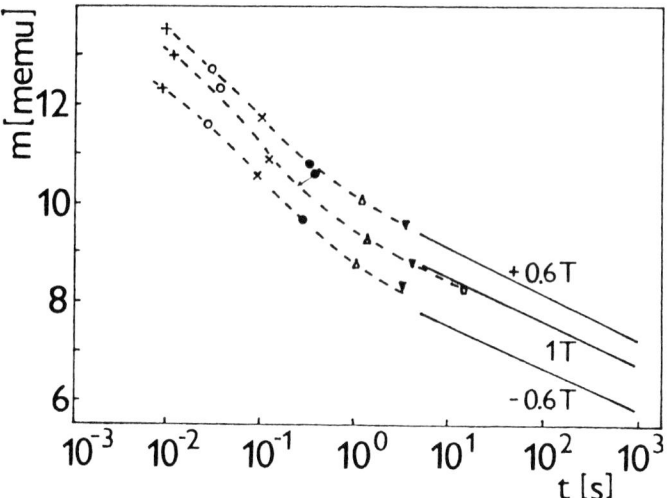

Figure 1. Hysteresis loop and conventional relaxations measured at T=21K and B_{ext}= +0.6T and +1T. The signs +, o, ×, •, ∆, ▼ and □ correspond to $\dot{B}$=89, 29, 8.9, 2.9, 0.88, 0.29 and 0.08mT/s, respectively.

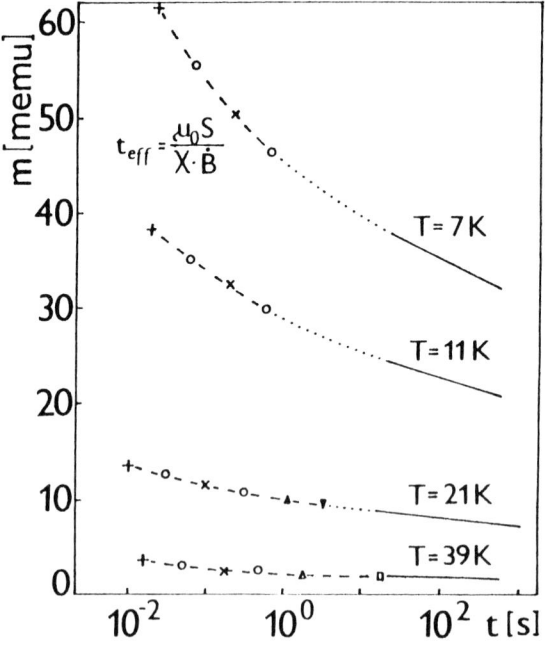

Figure 2. Hysteresis loop and conventional relaxations measured at B_{ext}=+0.6T for different temperatures from 7K to 39K. The signs +, o, ×, •, ∆, ▼ and □ correspond to $\dot{B}$=89, 29, 8.9, 2.9, 0.88, 0.29 and 0.08mT/s, respectively.

4. CONCLUSIONS

Magnetic moment relaxation in the vortex lattice exposed to the sweeping magnetic field is influenced by some additional mechanism dependent on sweep rate which retards the relaxation rate. The temperature dependence of the relaxation rate is nearly the same for both conventional and loop relaxation.

Acknowledgements

The authors would like express their thanks to dr. J. Kadlecová for valuable discussions and to J. Smolíková for technical help.

5. REFERENCES

1   C.P. Bean, Phys.Rev.Lett. 8 (1962) 250;
    Rev.Mod.Phys. 36 (1964) 31.
2   P.W. Anderson, Phys.Rev.Lett. 9 (1962) 309;
    Rev.Mod.Phys. 36 (1964) 39.
3   A.P. Malozemoff, in: Physical Properties of HTS,
    ed. D.M. Ginsberg (World Sci.Publ., Singapore 1989).
4   L. Půst, J. Kadlecová, M. Jirsa and S. Durčok: International
    Conference on Nature and Properties of the HTS, Wroclaw,
    June 1989; J.Low Temp.Phys. 78 (1990) 179;
    Cryogenics 30 (1990) 886.
5   L. Půst, Superc.Sci.Technol. 3 (1991) 598.
6   L. Půst and M. Jirsa, Proc.Int.Conf. on Magnetism,
    Edinburg, 2-6 September 1991.

High T$_c$ Superconductor Thin Films
L. Correra (Editor)

Effect of thickness on the magnetic properties of YBa$_2$Cu$_3$O$_7$ thin films

Ch.J. Liu[a], S.K. Agarwal[a*], R. Buder[a], J. Schubert[b] and C. Schlenker[a]

[a]CNRS-LEPES - BP 166 - 38042 Grenoble Cedex, France

[b]ISI-KFA Jülich - 5170 Jülich, Germany.

Abstract

The dc magnetic properties of YBa$_2$Cu$_3$O$_7$ thin films, prepared by laser ablation, have been measured with magnetic field perpendicular to the film, for thickness D ranging from 50 to 400 nm. The remanent magnetization and the critical current estimated from a Bean formula, measured at 10 K, decrease with D above 100 nm. The characteristic fields of vortex penetration in the films are increasing with D at 10 K. Close to T$_c$, the irreversible line is found to be displaced towards higher fields with increasing D. These results are discussed in relation with the existing models. The special properties of vortices in thin samples, for field perpendicular to the plane, are pointed out.

1. INTRODUCTION

YBa$_2$Cu$_3$O$_7$ thin films have now been extensively studied. Most of the available data concern their transport properties, with and without magnetic fields, in relation with determination of the critical current and of the so-called irreversible line, separating in a HT plane the regime of pinned vortex from that of reversible vortex motion [1,2]. Few results have been obtained by dc magnetic measurements on thin films, probably because of the small sample volume. They however provide convenient methods to evaluate the critical current through the Bean model [3]. We have previously reported detailed studies of the dc magnetic properties of textured and of epitaxial YBa$_2$Cu$_3$O$_7$ thin films, including data of the remanent magnetization as a function of the maximum applied field [4-5]. This last information gives some insight into the mechanism of vortex penetration in a thin film. The pinning force density and the irreversible line can also be determined through dc magnetization measurements [6]. We have recently shown that the variation of pinning force density versus temperature does not follow simple scaling laws [7] just below T$_c$ and that such studies could clarify the origin of the irreversible line [6].

However, the thin film geometry leads to specific problems for the properties of type II superconductors in the presence of a magnetic field. The first problem is related to the demagnetizing field effects, since most studies are performed with a field perpendicular to the film and therefore with a huge demagnetizing field in some cases. The second one is due to the fact that common thicknesses are in the range of 100 nm, which is precisely of the order of the London penetration depth. In the case of YBa$_2$Cu$_3$O$_7$ thin films oriented with the c-axis perpendicular to the substrate, vortex perpendicular to the films and therefore to the a b plane are expected to have a radius of the order of λ_{ab} which is at low temperatures ~1300 Å [8]. The vortex length would therefore be comparable to its diameter. Surface effects should then be dominant.

In this context, it is obvious that one should study the properties of thin films as a function of their thickness. Few systematic studies have been reported up to now. Recent results obtained by ac susceptibility measurements seem to show that the irreversible line does not depend on thickness above 100 nm [9]. We now report detailed studies of the magnetic properties including evaluation of critical current, field and temperature dependence of the remanent magnetization and the irreversible line for various thicknesses between 50 and 400 nm.

2. EXPERIMENT

The $YBa_2Cu_3O_7$ thin films have been prepared on $SrTiO_3(100)$ substrates by laser ablation, in situ, without post annealing, as described elsewhere [10]. They have been characterized by electrical resistivity, Rutherford back scattering (RBS) and channeling studies. The composition is found to be 1-2-3 within the accuracy of the RBS technique. The c-axis is perpendicular to the substrate. Zero resistivity is obtained at temperatures between 88 K and 90 K. Transport critical current is found to be of the order of 10^6 A cm^{-2} at 77 K. The films thickness ranges from 50 nm to 400 nm. It is measured by RBS with an accuracy of 5%.

Magnetic measurements are performed with a "home-built" vibrating sample magnetometer, between 10 K and 100 K in fields ranging from 0.1 mT to 6 T. High field is provided by a superconducting coil. For low field studies, zero field calibration is obtained by using a lead reference sample in its superconducting state. The high sensitivity of the apparatus, of 10^{-6} emu, is obtained with pick-up coils immersed in liquid helium and located close to the sample. The results reported here have been obtained with the magnetic field perpendicular to the sample. The in-plane film size was 4 x 4 mm^2 in most cases. For hysteresis loop and remanent magnetization measurements, the film was warmed up to 100 K between successive measurements and cooled again in zero field in order to suppress previously trapped flux [5].

The hysteresis loops have been studied at 10 K for different maximum applied fields. Table I summarizes some of the experimental data. The critical temperatures have been obtained from resistivity and from magnetic measurements. Small deviations are found between both sets of data. From the remanent magnetization obtained after large applied fields (M_r^{sat}), one can evaluate the critical current through the Bean formula ($J_c = 30 M_r/R$, see Table caption).

Table I
Characteristic data obtained on $YBa_2Cu_3O_7$ thin films of various thicknesses D.
The samples labeled S01, S07 to S09 and S15 to S18 correspond to 3 different sets of elaboration. T_c ($\rho = 0$) is the critical temperature obtained from the zero value of the resistivity. T_{cm} is obtained from the onset of the low field magnetization measured by decreasing the temperature. M_r^{sat} is the remanent magnetization obtained at 10 K after high applied fields ($\perp$ film). J_c is the critical current obtained from the Bean formula $J_c = 30 M_r/R$ with the average film radius in cm, M_r in emu cm^{-3}. H_B is the Bean full penetration field obtained from the maximum of the curve of the initial magnetization vs field. H_1 and H_2 are the fields corresponding to the onset and to the saturation of the magnetization vs maximum applied fields H_0^m.

Sample	D (nm)	$T_c(\rho=0)$ (K)	T_{cm} (K)	J_c (10^7 A cm^{-2})	M_r^{sat} (10^5 emu cm^{-3})	H_B (Oe)	H_1 (Oe)	H_2 (Oe)
S01	400	88.6	89.0	1.49	1.4	675	28	700
S07	100	89.0	88.0	1.28	1.2	200	8.0	200
S09	200	89.0	89.1	1.91	1.8	310	25	320
S15	200	89.4	91.8	2.13	2.0	645	32	780
S16	150	89.2	91.5	2.66	2.5	490	26	600
S17	100	89.8	91.1	2.81	2.1	400	16	480
S18	50	88.3	88.8	1.49	1.7	170	5.0	70

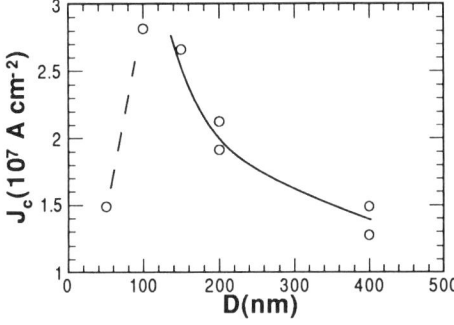

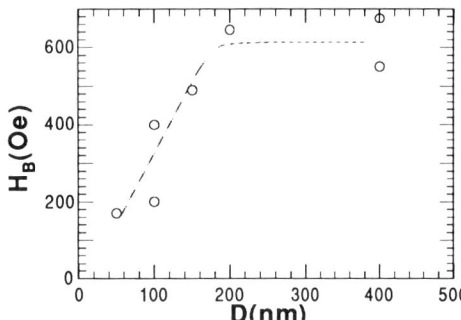

Fig. 1 Critical current evaluated from the remanent magnetization M_r through the Bean formula $J_c = 30 \, M_r/R$, (R is the average radius in cm, M_r in emu cm^{-3}), as a function of YBa2Cu3O7 film thickness.

Fig. 2 Bean full penetration field obtained from the maximum of the initial magnetization curve $M(H_0)$, as a function of film thickness.

This formula obtained for long cylindrical samples in principle does not apply for thin films in the perpendicular geometry. However, numerical calculations show that it is approximately valid for thicknesses larger than the penetration depth [11].

Fig. 1 shows the curve of the critical current J_c obtained from the Bean formula as a function of thickness. It seems to show a maximum for thickness D ~100 nm.

The curves of the initial magnetization obtained after zero field cooling, show a maximum at a field H_B which corresponds to the full Bean penetration field. Fig. 2 shows that H_B (measured at 10 K) is an increasing function of thickness below ~ 200 nm and seems to saturate above.

The remanent magnetization has been studied as a function of maximum applied field H_0^m for various thicknesses. Fig. 3a shows that the curves M_r/M_r^{sat} vs H_0^m are displaced towards higher fields for increasing thickness. One can define a threshold field H_1 for the onset of M_r, using a criterion related to the sensitivity of the magnetometer, chosen here as $M_r = 10^2$ emu cm^{-3} for $H_0^m = H_2$. The H_0^m value corresponding to the saturation of M_r is H_2. Fig. 3b and c show that H_1 and H_2 increase as a function of thickness and may saturate above 200 nm.

The irreversible line can be obtained from the hysteresis loops. H_{irr} is the field above which the magnetization is reversible [6]. The criterion used in this case corresponds to $\Delta M < 10$ emu cm^{-3}, ΔM being the difference of the M values obtained for increasing and decreasing fields. Using this definition of H_{irr}, we obtain a set of curves $H_{irr}(T)$ for different thicknesses, shown on Fig. 4 as a function of reduced temperature T/T_c. One notes that the curves tend to displace towards higher fields with increasing thickness.

3. DISCUSSION

The remanent magnetization measured at 10 K and therefore the critical current evaluated with the conventional Bean formula show a clear decrease for thicknesses larger than ~100 nm. The low thickness behaviour (D < 100 nm) is not clear at the moment since few samples have been studied. Also, in this case, the interaction with the substrate may be a dominant mechanism, leading to a possible poorer film quality.

If one first assumes that the conventional Bean formula applies for a thin film with D ~100 nm in the perpendicular geometry, the large thickness results would support a model in which the pinning decreases with increasing thickness. Two related mechanisms can be involved. First, the vortex may be pinned by the surface, either because of symmetry breaking or due to the presence on the surface of a thin layer of slightly different chemical composition, such as

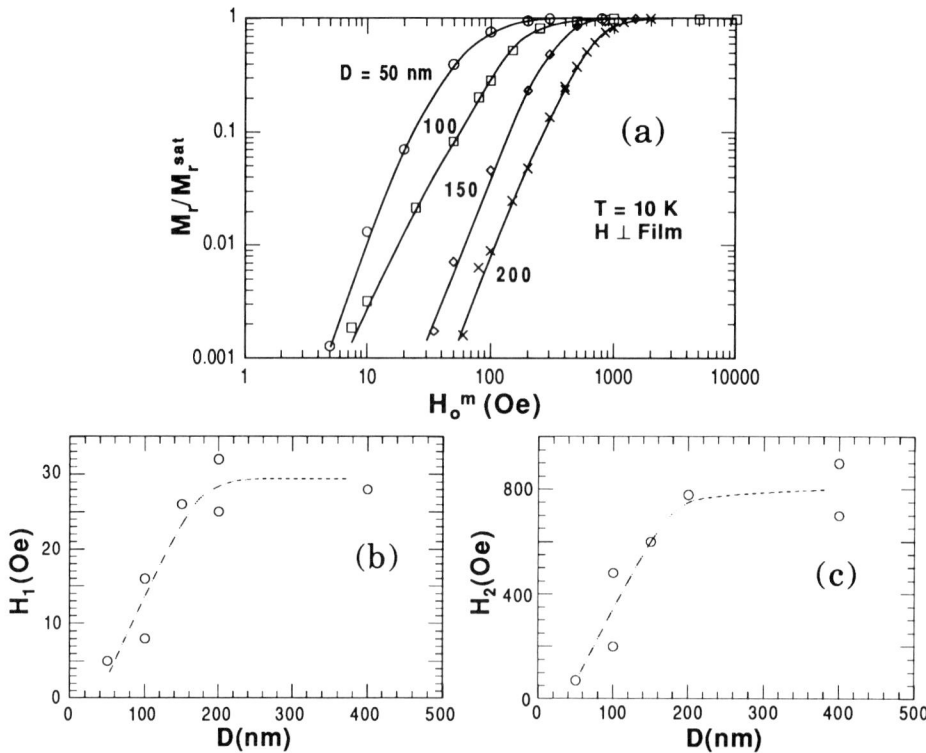

Fig. 3 (a) Remanent magnetization plotted in reduced units M_r/M_r^{sat} vs maximum applied field for several thicknesses. (b) Onset field for remanent magnetization vs thickness. (c) Saturation field corresponding to $M_r (H_2) = M_r^{sat}$ vs thickness.

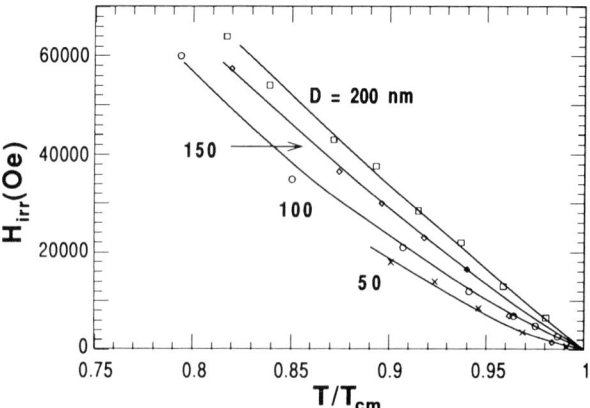

Fig.4 Irreversible line, H_{irr} vs T/T_{cm} for several film thicknesses. T_{cm} is the transition temperature obtained from the onset of the low field magnetization by decreasing temperature.

oxygen stoichiometry. Second, for D ~100 nm at 10 K, the vortex length is comparable to its radius. The pinning and other properties of such vortices may be quite different from those of vortices in bulk material. At higher temperature where the London penetration depth is larger, vortices must be still more spread out in the film plane and this effect should be more important. But the thickness dependence should be weaker since then $\lambda \gg$ D for commonly explored thicknesses. In this context, remanent magnetization and critical current measurements as a function of temperature and thickness should be performed.

In fact, the Bean formula established for infinite cylindrical samples, probably does not apply in this geometry to very thin samples. The short length of the vortices, their deformation close to the surface and their bending towards the sides of the film, make the validity of this formula very doubtful. Therefore, both the critical current J_c and the relation $M_r (J_c)$ may depend on the thickness. Measurements of magnetization and of J_c by transport should therefore be performed on the same samples to get more insight into these phenomena.

The thickness dependence of the full Bean penetration field H_B at 10 K, clearly shows indeed that the Bean formula does not apply for a thin film either for intermediate values of fields. The increase of H_B with thickness may be due to the geometry of the vortices and to the special properties of the critical state in this case. The available calculations apply to thicker samples and consider only the corrections involving demagnetizing field effects [11,12]. A theory adapted to a real thin film geometry would be welcome.

The thickness dependence of the onset field H_1, and of the "saturation" field H_2 for the remanent magnetization must involve the same mechanisms related to the vortex geometry. However, the onset field H_1 should be close to the first penetration field. In a first approximation in a thin film this is expected to be ~ε H_{c1} with ε ~D/L, where L is the in-plane dimension of the film. H_1 should then be roughly proportional to D. The experimental results are not too far from this estimation for D < 200 nm.

The "saturation" field H_2 should now be closely related to H_B. The difference between H_B and H_2 is probably due to the field dependence of the critical current. One notes indeed that H_B and H_2 follow a very similar thickness dependence.

Fig. 4 indicates that for a thickness range 50 to 400 nm, and for fields perpendicular to the film and to the ab plane, the irreversible line may be displaced towards higher fields. This result is surprising, since the critical current at 10 K is on the contrary decreasing with increasing thickness. Similar results have been found by ac susceptibility measurements for thickness range 20 - 1000 nm with some saturation above 100 nm. In our case, H_{irr} seems to increase with D at a given value of T/T_c even above 100 nm.

There is no clear indication at the moment that the film structural properties change in a consistent way as a function of thickness. Table I also indicates that there is no systematic variation of T_c with thickness. Therefore, it seems that for our films, in the perpendicular geometry for thicknesses larger than 100 nm, the critical current decreases and the irreversible field increases with increasing thickness. This would suggest that the irreversible line is not directly related to the pinning centers responsible for the critical current.

This result therefore supports a model in which the irreversible line is due to some intrinsic mechanism, such as a vortex lattice melting, independent or weakly dependent on the nature of the pinning centers [13]. In this context, the irreversibility line, similar to the melting line of a crystal, would be governed by the dynamic properties of the vortex lattice, including the elastic coefficients, such as the shear modulus c_{66}, in relation with the role of thermal fluctuations [14]. The melted phase could be stabilized at lower temperature in thinner samples, closer to a two-dimensional phase.

Other models proposed for the irreversibility line, such as the model of a heavily entangled flux liquid for the vortices [15], would probably also account for a decrease of the irreversible field with decreasing thickness.

However, one should stress that the properties of the vortex lattice for thin samples may be completely different from those of bulk samples. At $T \leq T_c$ vortices are expected to be spread out in the film plane, since the London penetration depth is much larger than the film thickness. Such vortices are "objects" totally distinct from conventional cylindrical vortices. At the moment, one would need a theory adapted to the geometry of thin films, with thickness in the range of 100 nm.

As a conclusion, we have reported that the role of thickness on the dc magnetic properties of $YBa_2Cu_3O_7$ thin films seems to be important for the thickness range 50 - 400 nm. While the critical current decreases above 100 nm, the fields characteristic of the vortex penetration in the sample are increasing with thickness. At the same time, the irreversible line is displaced towards higher fields. These results should be corroborated for a broader set of thicknesses. Models taking into account the special geometry of a thin film and the particular properties of vortices in this case would be welcome.

4. ACKNOWLEDGMENTS

The authors wish to thank K. Schroer for his participation in some of the measurements and J. Dumas for helpful discussions.

This work was partly supported by contracts from the Commission of the European Communities SCIENCE SC1-0038-CD and CI1-0340-F (Cooperation EC-India).

5. REFERENCES

* Permanent address : National Physical Laboratory, New Delhi 110012, India.
1   K.A. Müller, M. Takashige and J.G. Bednorz, Phys. Rev. Lett. 58 (1987) 1143.
2   R.H. Koch, V. Foglietti, W.J. Gallagher, G. Koren, A. Gupta and M.P.A. Fisher, Phys. Rev. Lett. 62 (1989) 1511.
3   C.P. Bean, Rev. Mod. Phys. 36 (1964) 31.
4   R. Buder, J. Dumas, C. Escribe-Filippini, H Guyot, Ch. J. Liu, J. Marcus, S. Revenaz, P.L. Reydet and C. Schlenker in "Studies of High Temperature Superconductors", Ed. A.V. Narlikar (Nova Science Publ.), Vol. 7 (1991) p. 223.
5   C.J. Liu, R. Buder, C. Schlenker, J. Schubert, W. Zander and B. Stritzker, J. Less Common Metals 164-165 (1990) 1285.
6   C. Schlenker, C.J. Liu, R. Buder, J. Schubert and B. Stritzker, Int. Workshop on HTCS Thin Films, Rome (April 1991), Physica C (to be published).
7   E.J. Kramer, J. Appl. Phys. 44 (2973) 1360.
8   See for example, A.P. Malozemoff in "Physical Properties of High Temperature Superconductors", Ed. D.M. Ginsberg (World Scientific Publ. Co., Singapore) (1989)p71.
9   L. Civale, T.K. Worthington and A. Gupta, Phys. Rev. B 43 (1991) 5425.
10   J. Fröhlingsdorf, W. Zander and B. Stritzker, Solid State Comm. 67 (1988) 1965.
11   L.W. Conner and A.P. Malozemoff, Phys. Rev. B 43 (1991) 402.
12   V.M. Krasnov, V.A. Larkin and V.V. Ryazanov, Physica C 174 (1991) 440.
13   P.L. Gammel, L.F. Schneemeyer, J.V. Waszczak and D.J. Bishop, Phys. Rev. Lett 61 (1988) 1666.
14   E.H. Brandt, Physica C 162-164 (1989) 1167.
15   D.R. Nelson, Physica C 162-164 (1989) 1156.

High T$_c$ Superconductor Thin Films
L. Correra (Editor)

Frequency dependent conductivity in $YBa_2Cu_3O_{7-\delta}$ ($\delta \approx 0$) thin films

P. Lunkenheimer[a], A. Loidl[a], C. Tomé-Rosa[b], and H. Adrian[b]

[a]Institut für Physik, Johannes Gutenberg Universität, D-6500 Mainz, Germany

[b]Institut für Festkörperphysik, Technische Hochschule, D-6100 Darmstadt, Germany

Abstract

The complex a.c. conductivity of $YBa_2Cu_3O_7$ thin films has been investigated in a frequency range $10^1 Hz \leq f \leq 10^9 Hz$ and for temperatures $10K \leq T \leq 300K$. The results were interpreted by network analysis using equivalent circuits taking into account contact resistances and a frequency dependent conductivity of the sample. For $T > T_c$ the frequency dependence of the conductivity depends strongly on the sample quality. Low quality samples can well be described by the universal law of dielectric relaxation, which indicates conduction by hopping mechanisms. High quality samples behaved purely metallic, with no traces of hopping conduction.

1. INTRODUCTION

The discovery of high-temperature superconductivity in the cuprates $La_{2-x}Sr_xCuO_4$ and $YBa_2Cu_3O_{7-\delta}$ (2:1:4 and 1:2:3 compounds, respectively) has generated great interest in the electrical transport properties not only in the superconducting but also in the normal state of these materials. In the semiconducting "parent" compounds ($x \approx 0$; $\delta \approx 1$) of these electronically highly correlated materials experimental evidence for conduction by hopping mechanisms has been found[1,2,3]. Here we report the frequency dependence of the complex conductivity in 1:2:3 compo-

unds with $\delta \approx 0$. Measurements of $\sigma(\omega)$ $(\omega=2\pi f)$ are an ideal tool to identify hopping conduction. At the same time these measurements provide
a direct and powerful test of the quality of films and contacts.

2. EXPERIMENTAL DETAILS

Thin films of $YBa_2Cu_3O_7$ were prepared by dc-sputtering on $SrTiO_3$
substrates.[4] As electrical contacts evaporated silver-pads were used.
Here we report measurements of the a.c.-conductivity for frequencies
$10^6Hz \leq f \leq 10^9Hz$ and for temperatures above the superconducting phase
transition temperature. The data were recorded using an HP 4191A impedance analyzer connected to a refrigerator system (CTI Cryogenics)
via an air line.[5] As this is a reflectometric method which uses two
point configuration the contact resistances contributed to the obtained
conductivity values. The contact geometry used in this experiment made
it difficult to exactly evaluate the cross-section of the samples. Therefore no absolute values for the specific conductivity are given. In addition the samples have been measured with a conventional four-point
technique in the frequency range $10Hz \leq f \leq 10^6Hz$ using a lock-in amplifier
and an automated low-frequency bridge (HP 4192 impedance analyzer).

3. RESULTS AND DISCUSSION

From a variety of samples investigated we present here two representative results: the low quality sample s123 was characterized by a
superconducting transition temperature of $T_c=80K$ and a large residual
resistivity $(\rho(300K)/\rho(100K) \approx 1.9)$. $\rho(T)$ in the high quality sample s194
$(T_c=87K)$ was almost "ideal", with a linear temperature dependence extrapolating to zero for $T \to 0K$ $(\rho(300K)/\rho(100K) \approx 3)$.

Fig. 1 shows the real and imaginary part of the conductivity (G' and
G'' respectively) vs. the logarithm of the measuring frequency f for the
low quality sample (s123) at different temperatures above T_c. As $G(\omega)$
was almost constant for f<1MHz, only the high frequency data are
shown in Fig. 1. Using network analysis the frequency dependence of
the conductivity (real *and* imaginary part) for $T>T_c$ can well be descri-

bed by a leaky capacitor ($R_K \| C$) in series with an inductance (L) and a parallel circuit consisting of a frequency dependent complex resistor R_1 and another (frequency independent) resistor $R_{D.C.}$. The leaky capacitor can be assigned to grain boundaries and/or complex contacts which behave like metal-to-semiconductor junctions in the normal-conducting state. The inductance L is caused by the leads (silver paint) used to connect the sample to the sample holder. For the description of G(ω) at temperatures below T_C (not shown in this work) the same equivalent circuit can be used with R_1 set to zero. Thus, in the superconducting state it is no longer necessary to include a frequency dependent sample resistance in the circuit. (The ω^2-dependence of G for T<T_C, as obtained from many microwave experiments[6], can hardly be determined with this two point measuring technique.) For the description of the conductivity of the sample in the normal-conducting state the universal law of

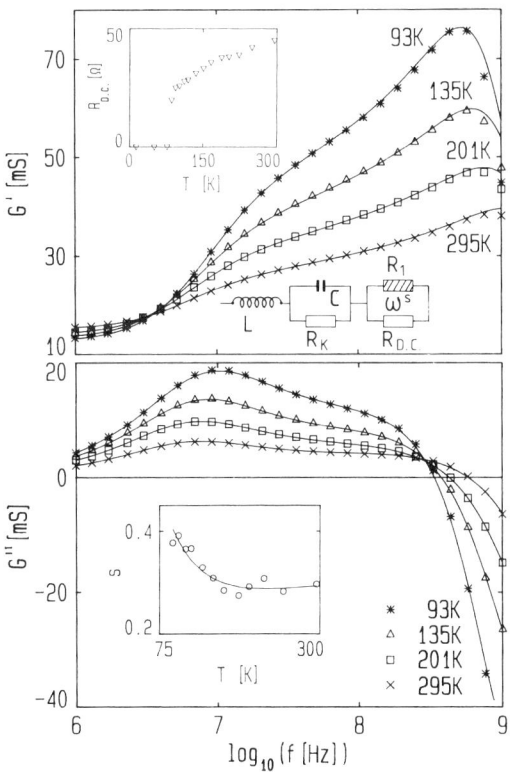

Figure 1. Real and imaginary part of the complex conductivity G=G'+i G'' of sample s123 vs. the logarithm of the measuring frequency f at different temperatures. The solid lines are calculated using the equivalent circuit indicated. The inset in the upper frame shows the temperature dependence of the d.c. resistance $R_{D.C.}$. The lower inset gives the temperature dependence of the frequency exponent s. The solid line is calculated using a model of overlapping large polarons.

dielectric relaxation was used, which has been introduced by Jonscher[7] and has been found in a wide variety of materials including La_2CuO_4[2]. Here the real part G' of the conductivity G = G'+ i G'' is proportional to ω^s, with s<1 and frequency independent. This implies that $G''\sim\omega^s$ as well and that the ratio of the imaginary to the real part of G is equal to tan($s\pi/2$) as follows from the Kramers-Kronig relation[7]. This universal power law is the most elementary way of describing the influence of hopping conduction on the frequency dependence of G, not taking into account any frequency dependence of the exponent s and any frequency dependent ratio G''/G' as obtained from most hopping theories[8]. Including these additional features would require much more complicated fitting procedures. However, from our experience in most cases a frequency independent s is sufficient to describe G(ω) caused by hopping conduction. The solid lines in Fig. 1 are the results of fits using the indicated equivalent circuit including the universal response behaviour for R_1. The first increase of G'(ω) at approximately 10MHz is due to the barrier, the increase at 100MHz due to hopping conduction. We want to emphasize that real and imaginary part are fitted simultaneously and both can consistently be explained within this model. The inset in the upper frame shows the d.c. resistance $R_{D.C.}$ as obtained from the fits. For T<T_c $R_{D.C.}$ is very small and dominated by the resistance of the leads. The temperature dependence of $R_{D.C.}$ for T>T_c agrees well with the results of a low frequency measurement at 26Hz which has been performed in four-point configuration therefore measuring the intrinsic resistivity of the sample without contacts. The temperature dependence of the frequency exponent s is shown as inset in the lower frame of Fig. 1. A possible explanation of s(T) can be found using a model of carrier transport via large overlapping polarons[9]. The solid line represents results of a fit using this model where tunneling of polarons is the dominant transport mechanism. The spatial extent of the polarons r_p is assumed to exceed the spacing between neighbouring sites thereby reducing the polaron hopping energy W_{HO}. Our fit of s(T) yields $W_{HO}\approx0.06eV$, $\alpha r_p\approx1.2$ and $\tau\approx3\cdot10^{-13}$ where α^{-1} is the spatial extent of the localized state wavefunction and τ is a characteristic relaxation time often taken to be of the order of an inverse phonon frequency.

As can be seen from Fig. 2, the high quality sample (s194) behaves

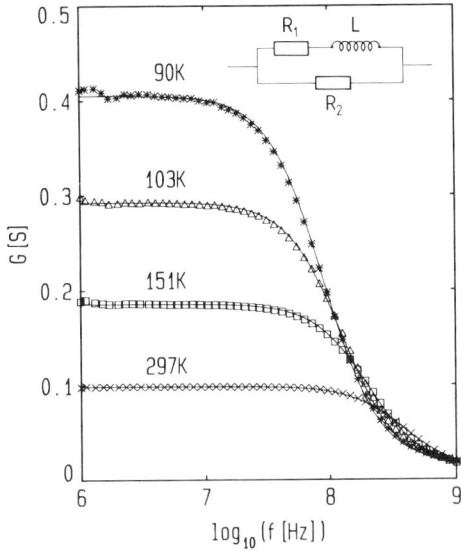

Figure 2. G vs. $\log_{10}(f)$ for sample s194 at different temperatures. The solid lines are calculated using the equivalent circuit indicated.

totally different. The results can be described using an ohmic resistor R_1 and an inductance in parallel with a second ohmic resistor R_2. At low frequencies the conductivity is determined by $G = 1/R_1 + 1/R_2$. At high frequencies $G = 1/R_2$, where R_2 is almost one order of magnitude larger than R_1. This frequency dependence of G indicates pure metallic behaviour, with no traces of hopping conduction or barriers. The fitted circuit diagram seams to provide evidence that two paths lead independently through the thin film, both of which percolate. However, we cannot exclude influences of the contact geometry that lead to this behaviour.

4. CONCLUSIONS

The interpretation of these results is straightforward: Sample s123 reveals traces of polaron hopping at high frequencies. At frequencies $f < 1$MHz the conductivity is dominated by the leakage of the barrier (R_K in Fig.1), which originates from grain boundaries and metal-to-semiconductor contacts. The high quality sample (s194) behaves metallic for $f < 1$GHz. For the semiconducting 1:2:3 compounds with $δ ≈ 1$, pure hopping conduction has been reported.[3] Thus, the polaron transport in s123 is the remainder of semiconducting behavior, which has totally vanished in s194. Here $G(T,\omega)$ can be described by pure metallic conductivity. From these results we conclude that polaron transport is not an important feature in high-quality samples of the doped cuprates.

Acknowledgement

This research was supported by the "Bundesministerium für Forschung und Technologie" under contract number 13N5705 and by SFB252 (Darmstadt/Frankfurt/Mainz/Stuttgart). One of us (C. T.-R.) was financially supported by the Conselho Nacional de Desenvolvimento Cientifico e Technologico do Brasil - CNPq.

References:
[1]   K. K. Som and B. K. Chaudhuri, Phys. Rev., B41 (1990) 1581.
      A. Gosh and D. Chakravorty, J. Phys Cond. Matt., 2 (1990) 649.
[2]   C. Y. Chen, N. W. Preyer, P. J. Picone, M. A. Kastner,
      H. P. Jenssen, D. R. Gabbe, A. Cassanho, and R. J. Birgeneau
      Phys. Rev. Lett., 63 (1990) 2307.
      C. Y. Chen, R. J. Birgeneau, M. A. Kastner, N. W. Preyer, and Tineke
      Thio, Phys. Rev., B43 (1991) 392.
[3]   G. A. Samara, W. F. Hammetter, and E. L. Venturini
      Phys. Rev., B41 (1990) 8974.
[4]   C. Tomé-Rosa, A. Walkenhorst, M. Maul, G. Jakob, H. Adrian,
      K. Haberle, P. Przyslupski, and G. Adrian,
      Physica B, 165&166 (1990) 1477.
[5]   R. Böhmer, M. Maglione, P. Lunkenheimer, and A. Loidl
      J. Appl. Phys., 65 (1989) 901.
[6]   see e.g. A. Inam, X. D. Wu et. al., Appl. Phys Lett., 56 (1990) 1178.
[7]   A. K. Jonscher, Dielectric Relaxations in Solids, Chelsea Dielectrics
      Press, London, 1983.
[8]   for a review see
      S. R. Elliott, Adv. Phys., 36 (1987) 135.
[9]   A. R. Long, Adv. Phys., 31 (1982) 553.

High T$_c$ Superconductor Thin Films
L. Correra (Editor)
© 1992 Elsevier Science Publishers B.V. All rights reserved.

Magnetoplasma-reflection of thin films of YBa$_2$Cu$_3$O$_7$

W. Markowitsch[a], W. Lang[a], H. Jodlbauer[a], P. Schwab[b], X.Z. Wang[b] and D. Bäuerle[b]

[a]Institut für Festkörperphysik der Universität Wien und Ludwig Boltzmann Institut für Festkörperphysik, Kopernikusgasse 15, A-1060 Wien, Austria

[b]Institut für Angewandte Physik, Johannes-Kepler-Universität Linz, A-4040 Linz, Austria

Abstract

Reflection and magnetoreflection measurements on thin films of Y-Ba-Cu-O are reported. The reflection spectra show an edge in the near infrared. This feature is interpreted as a plasma edge. Magnetoreflection measurements were carried out in Faraday-configuration using a He-Ne-laser with a wavelength of 1.15 μm (1.07 eV), i.e. in the region of the plasma edge. A change of the reflectance of circular polarized light was observed as a function of the external magnetic field. We interpret this effect as a shift of the plasma edge induced by the magnetic field (magnetoplasma effect). From this shift we determine the value of the effective carrier mass m$_{eff}$ = 4m$_0$. This result indicates a strong coupling of the carriers to other excitations.

1. INTRODUCTION

Intensive work was spended on the optical and infrared properties of high- temperature superconductors. Especially the reflection spectra of yttrium-barium- copper-oxide (YBCO) were discussed in a huge number of papers (a few examples are References 1 - 3). The main topics in this field are the superconducting energy gap and the properties of the free carriers in both the superconducting and normal states. Obviously these topics are closely connected. Due to the high standard in sample preparation that has been reached the experimental results are quite similar now. The reflection spectra of single crystals as well as of oriented films show a reflectivity edge rising from an onset energy between 1 eV and 2 eV to lower photon energies. There is general agreement that this feature cannot be interpreted as a simple Drude plasma edge, but the discussion about an alternative model is very controversial. The most simple description is to fit the spectra with a combination of a free carrier contribution and a Lorentzian oscillator [1]. The problem then is to explain the nature of the oscillator. Another possibility is the assumption of energy dependent scattering rates and effective masses [2]. This leads to the question about the nature of the excitations that interact with the free carriers. In this discussion we agree with Bozovic [3] that an explanation of the unusual features in the spectra should be based on the known properties of the cuprates, in particular the layered structure of these compounds.

In this work we report a new approach to determine the properties of the free carriers in the normal state of YBCO. Applying the method of "magnetoplasma-reflection" (see section 4) we yield the value of the effective carrier mass. This measurement offers the possibility to discriminate between the different models, because at the detection wavelength employed an magnetoplasma effect is expected only if the reflection edge in the near infrared arises from free carrier absorption. According to our results the edge is indeed a plasma edge induced by carriers with masses of about 4 times the free electron mass.

2. SAMPLE PREPARATION

The YBCO-films were grown by pulsed laser deposition in oxygen atmosphere. The oxygen pressure was 15 mbar. The substrate material employed was 100 MgO. The substrate temperature was 794 °C. For the reflection measurements samples of sufficient thickness had to be prepared, so that the spectra were not affected by interference effects. From the sputtering rate typical thicknesses of about 300 nm were determined. X-ray analysis showed that the films were single phase and c-oriented. In resistivity measurements the samples exhibited values of the critical temperature of 90 K.

3. REFLECTION MEASUREMENTS

3.1. Experimental techniques

Reflection spectra were taken in the energy region of 0.05 eV to 4 eV at nearly normal incidence using an Al mirror as a reference. For the measurements in the near IR and visible region we used a Hitachi spectrophotometer. In the mid-IR a Bruker Fourier transform spectrometer was utilized. Usually the two parts of the spectra showed coincidence within 10 percent (relative). To achieve continuous spectra we corrected the visible part by a appropriate factor, assuming that the IR part is less affected by surface roughness and should therefore be closer to the real values. Since measurements of the diffuse reflectance were not available the spectra may be uncertain at least at energies above 2 eV.

3.2. Results

Figure 1 shows the typical spectrum of an YBCO-film, that is in good agreement with recent results [5]. The reflection edge has an onset at about 1.3 eV. No phonon structures are visible at low energies indicating a good orientation of the film. The maximum at about 2 eV which does not appear in most of the spectra in literature is probably a consequence of the surface roughness which reduces the reflectance at higher energies.

The spectra of several other films studied showed similar reflection edges with a lower onset energies (~ 1.2 eV). Attempts to fit the spectra by a simple free carrier model give no satisfying results. A plasma edge of Drude type has a somewhat different shape as the experimental curves (see the dashed line in Figure 1). We agree with Bozovic [3] that the Drude model can only serve for a first approximation to the real behavior of the high-T_c superconductors.

We estimate the plasma energy of the sample in Figure 1 by means of a Kramers-Kronig analysis. Since the reflectance at the low energy limit of the measurement is about 0.8, a simple Hagen-Rubens extrapolation for $\omega \Rightarrow 0$ is not appropriate. Instead we took a weighted average of a linear extrapolation and a Hagen-Rubens behavior that served well with theoretical test spectra. For the high energy extrapolation we assumed a constant reflectance above 4 eV. This is justified by the investigations of Bozovic [3]. Being aware of extrapolation problems we will focus our attention on the features that are not influenced by the extrapolation procedures.

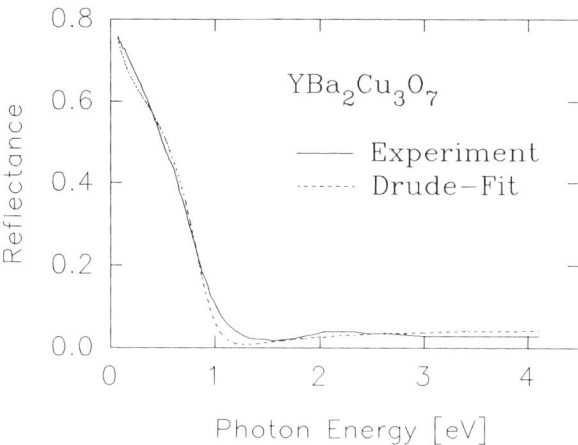

Figure 1. Reflection spectrum of a c-oriented YBCO film
Dashed line: best result of Drude fit

In Figure 2 the real and the imaginary part of the dielectric function are plotted. The real part decreases monotonically for $\omega \Rightarrow 0$ with a zero crossing at $\cong 0.9$ eV, whereas the imaginary part increases at the same time. This behavior is characteristic for free carriers. From the zero crossing of the real part we derive a plasma energy of about 0.9 eV.

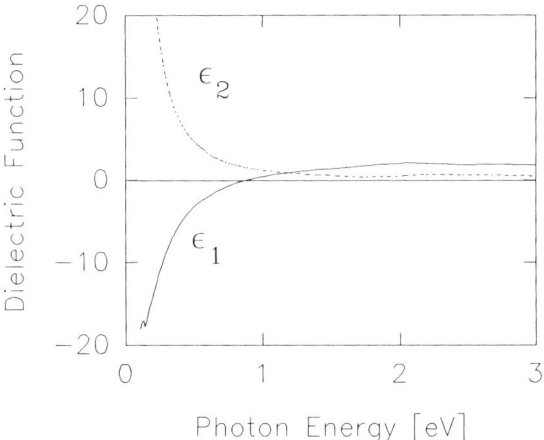

Figure 2. Real (ε_1) and imaginary (ε_2) part of the dielectric function from a Kramers-Kronig analysis

4. MAGNETOPLASMA-REFLECTION

4.1. Method and experimental technique

The method of magnetoplasma-reflection (MPR) has been used to determine the effective carrier mass in semiconductors [6] as well as in conducting polymers [7]. The basic effect is the influence of an external magnetic field B on the position of the plasma edge. In Faraday configuration the edge is shifted in opposite directions for left and right circular polarized light [6]. The displacement $\Delta\omega$ is approximately equal $\omega_c/2$ in weak magnetic fields ($\omega_c = q \cdot B/m^*$ is the cyclotron frequency). Hence, the measurement of the difference in reflectance of left and right circularly polarized light in the region of the plasma edge provides a possibility to determine the carrier mass m^*.

For the experimental setup it is easier to measure at a fixed wavelength and with variable magnetic field. In this case the method requires samples with a plasma edge near the measuring wavelength. In our experiments we used a Helium-Neon laser of wavelength 1.15 μm (1.07 eV). The samples were placed in a conventional electromagnet, mounted on a nonmagnetic plate, so that the distance of the pole pieces could by limited to 1.5 cm. In this way magnetic fields of up to 16 kG were achieved. The linear polarized laser beam was converted into circular polarized radiation of changing orientation by a KD*P pockels cell. The beam reflected from the sample was detected by a Ge diode detector. The modulated part of the reflected intensity ΔI was registrated by a PAR 5206 lock in amplifier. In addition the "bare" (dc-) intensity I of the reflected beam was monitored to correct for variations of the laser intensity.

For the evaluation of the magnetic spectra we assume a parallel shift of the plasma edge. This approximation is valid for weak magnetic fields and in the vicinity of the plasma energy. We find for the relative change of the reflectance $\Delta R/R$:

$$\Delta R/R = \Delta I/I = -\omega_c \cdot R^{-1} \cdot dR/d\omega \qquad (1)$$

In the weak field approximation $\Delta R/R$ depends linearly on B. To find m^* from the measured $\Delta R/R$, R^{-1} and $dR/d\omega$ (taken at the measuring wavelength) have to be determined from the reflection spectrum.

4.2. Results

For the sample of Figure 1 the measuring wavelength is positioned within the plasma edge. The magnetoreflection spectrum is shown in Figure 3. The relative change in reflectance $\Delta R/R$ is only about $2.5 \cdot 10^{-4}$ at the largest magnetic field available. Since R at this energy is about 10 %, the change in reflectance is of the order of 10^{-5}, which approaches the detection limit of our experimental arrangement. The main limiting factor here is the stability of our laser. Despite the scattering of the data points a linear least square fit to the data is well defined. From the slope of the theoretical curve a value of the effective carrier mass $m^* = 4.2\,m_0$ is calculated. The main source of error in this evaluation arises from the uncertainties of the reflection measurements. If we estimate these to be 10 %, then the error in m^* is 20 %. The calculated error of the least sqare fit is only about 3 %, so that the final result of the evaluation can be given as $m^* = (4.2 \pm 0.9)\,m_0$.

We have studied several other films with lower plasma energy, too. In this case the measuring wavelength was close to the minima in the reflection spectra. No MPR effect was observed within the limit of detection. This is expected, because from equation (1) follows $\Delta R/R \sim 0$ when $dR/d\omega \sim 0$.

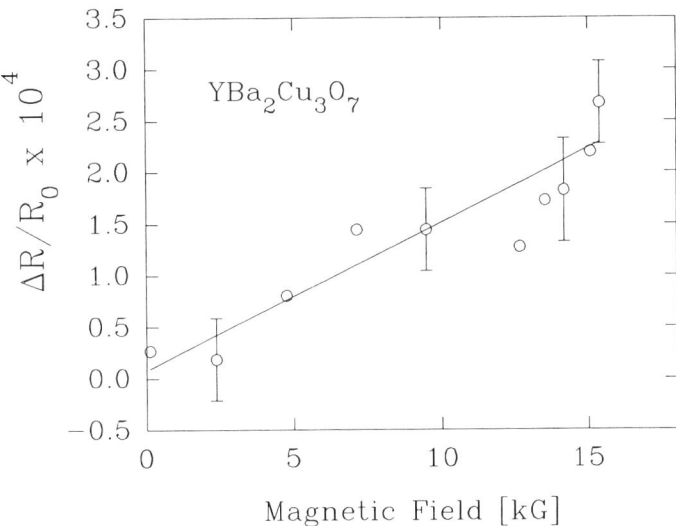

Figure 3. MPR spectrum of an YBCO film

5. DISCUSSION

Our evaluation avoids the usage of a special model for the reflectivity of YBCO. The only postulate is that the reflection edge in the near IR is caused by free carriers, which is plausible in a metallic system. Accepting this the interpretation of the magnetoreflection measurements is straightforward. Otherwise a new mechanism would have to be found to interprete the magnetic dependence of the reflectance. In the oscillator model, for example, the reflectance at 1.07 eV is dominated by interband or excitonic transitions [1] the definite nature of which is still unknown. For these transitions selection rules must be postulated so that left and right circularly polarized radiation lead to different transition rates. Even if this sort of explanation is possible, we find it preferable to base the interpretation on the known fact that YBCO contains "free" carriers. It is expected that these carriers differ from nearly free electrons for two reasons: a)YBCO is highly anisotropic, and b)a strong interaction with other excitations must exist that is responsible for superconductivity in this system. As a consequence the optical properties of YBCO are only approximated by the Drude model. Perhaps the "layered electron-gas" model proposed by Bozovic [5] will lead to a more quantitative description.

In the following we will discuss the results assuming that both the reflectivity edge and the magnetoreflectance spectrum are caused by charge carriers. The Kramers-Kronig analysis yields an approximate value of the plasma energy of 0.9 eV. Taking $m^* = 4 \cdot m_0$ from the MPR, a carrier concentration of $n \sim 6 \cdot 10^{21} \, cm^{-3}$ is calculated. This value is in good agreement with Hall effect data [7]. Since our extrapolation methods in the Kramers-Kronig analysis tend to underestimate the value of the plasma energy [5], the real value of n is probably larger by 10 - 20 %. The value of m^* is supported by the results of Fiory et al. [8]. They find $m^* = 5m_0$ by an electrostatic charge modulation technique, i.e. by a method based on a completely different effect.

The large effective mass of the carriers gives evidence for a strong coupling mechanism in YBCO. In the case of phonon coupling the relation $m^* = m \cdot (1 + \lambda)$ holds (m and λ mean the band mass and the electron-phonon coupling constant, respectively). Even if a different mechanism exists in YBCO, one can expect that a similar relation is valid with an alternative meaning of λ. $m^*/m_0 = 4$ implies $\lambda = 3$, which is a very large value. In strong coupling theories of superconductivity like McMillans [9], the critical temperature T_c is essentially a function of λ and of an average phonon frequency $<\omega>$. Large values of T_c require large λ and/or $<\omega>$. Taking the high energy phonons found in YBCO [10] into account phonon mediated pairing can no longer be excluded as a possible mechanism in high-T_c superconductors.

6. ACKNOWLEDGEMENTS

We wish to thank T.Pichler, J.Geißelbrecht and H.Kuzmany for experimental help and Prof.K.Seeger for his interest in our work. This work was supported by the "Fonds zur Förderung der Wissenschaft in Österreich", grant # P 8180.

7. REFERENCES

1 K.Kamaras, S.L.Herr, C.D.Porter, N.Tache, D.B.Tanner, S.Etemad, T.Venkatesan, E.Chase, A.Inam, X.D.Wu, M.S.Hedge and B.Dutta, Phys.Rev.Lett. 64, 84 (1990)
2 R.T.Collins, Z.Schlesinger, F.Holtzberg, P.Chaudhari and C.Feild,   Phys.Rev.B39, 6571 (1989)
4 I.Bozovic, Phys.Rev.B42,1969 (1990)
5 E.D.Palik, S.Teitler, B.W.Henvis, and R.F.Wallis, Proc.Int.Conf.Phys.Semicond. Exeter 1962 (A.G.Stickland,ed.), p.288. London: The Institute of Physics and the Physical Society. 1962
6 K.Seeger, W.Markowitsch, and F.Kuchar, Synthetic Metals 17, 527 (1987)
7 T.Penney, S.von Molnar, D.Kaiser, F.Holtzberg, and A.W.Kleinsasser, Phys.Rev.B38, 2918 (1988)
8 A.T.Fiory, A.F.Hebard, R.H.Eick, P.M.Mankiewich, R.E.Howard, and M.L.O'Malley, Phys.Rev.Lett.65, 3441 (1990)
9 W.L.McMillan, Phys.Rev.167, 331 (1968)
10 L.Genzel, A.Wittlin, M.Bauer, M.Cordona, E.Schönherr, and A.Simon, Phys.Rev.B40, 2170 (1989)

High T$_c$ Superconductor Thin Films
L. Correra (Editor)
© 1992 Elsevier Science Publishers B.V. All rights reserved.

195

Magnetooptic measurements of magnetic flux profiles in high-T$_C$ YBaCuO-films.

A.A. Polyanskii[a], L.A. Dorosinskii[a], M.V. Indenbom[a], V.I. Nikitenko[a],
Ju.A. Ossip'yan[a], V.K. Vlasko-Vlasov[a] and H.-U. Habermeier[b]

[a]Institute of Solid State Physics, the USSR Academy of Sciences, 142432 Chernogolovka, Moscow distr., USSR

[b]Max-Planck-Institut für Festkörperforschung, Heisenbergstrasse 1, D-7000 Stuttgart 80, Germany

Abstract

Magnetization of YBa$_2$Cu$_3$O$_{7-\delta}$ (YBaCuO) films is studied by direct magnetooptic observation using iron garnet indicator films. Magnetic flux profiles are measured, from which the critical current values are determined in a wide temperature range. Obtained results indicate existence of a complicated network of weak links in the HTSC films.

One of the most important characteristics of a superconductor is the maximum density of current that can flow through it without dissipation, i.e. the critical current density. Direct transport measurements of this value are often complicated due to difficulties in preparing good low ohm contacts and providing elimination of an overheating in the contact region. Therefore the critical current is most frequently determined indirectly from magnetization curves of superconductors. However the data obtained from macroscopic measurements are not easy to interpret in a simple way because of the inhomogeneity of studied samples. That is especially actual in the case of high-T$_c$ superconductors possessing extremely short coherence length, where structure imperfections can change drastically superconducting properties.

Recently a new method for direct control of the magnetic flux distribution in superconductors using magnetooptic iron garnet films was developed. It provides measurements of local values of the transition temperature, T$_c$, the lower critical field, H$_{c1}$, and the critical current density, J$_c$[1]. This method in contrast to magnetooptic techniques known earlier, that were working near the liquid helium point, has no temperature limitations and enables studies of high-T$_c$ superconductors up to T$_c$.

In present work the method was further developed. Together with

previously applied bubble films[1] new iron-garnet films with the in-plane
easy magnetization axis are used. They have no coercitivity for remagnetizing them normal to the film plane and provide working up to higher
magnetic fields with better spatial resolution. Due to dispersion of
magnetooptic effects color pattern of the flux distribution can be
observed. In addition using these films made it possible to measure exact
values of the local magnetic flux density on surfaces of superconducting
samples determined from the Faraday angle data and thus to build flux
profiles and to determine from them J_c values [2,3].

Details of the experimental technique are described in [1]. A sample,
placed in a miniature optic cryostat (on the finger cooled with the flowing
liquid helium) established on the stage of a polarizing microscope, is
covered by the indicator film[2]. Changes of the film magnetization revealed
due to the Faraday effect in the polarized reflected light double
transmitted through the garnet layer give a map of the normal component of
magnetic field above the sample. Bubble films are magnetizing by this field
due to motion of domain walls (domains magnetized oppositely to the
external field are shrinking and their neighbours magnetized along the
field are expanding). A spatial resolution in this case is limited by the
domain width which was down to 3 mkm in our films. Indicator films with the
in-plane anisotropy are magnetized in the normal field, H, by rotation of
the spontaneous magnetization declining from the film plane at an angle
$\Phi = arctg(H/K)$ (here K is the magnetic anisotropy constant). The minimum
scale of magnetic inhomogeneity in such a film has a scale of the order of
$2\pi\sqrt{A/K} \sim 0.1mkm$ (A is the exchange constant) and a spatial resolution for
visualizing magnetic fields is restricted for these films by their
thickness and the light wave length. For obtaining the Faraday rotation in
small regions cut in the image plane by the measuring diaphragm the setup
is supplied by an additional Faraday sell modulating the reflected light
polarization at the frequency of 70 Hz. Scanning the measuring diaphragm
across a sample the magnetic flux profiles are detected. The magnetooptic
signal values are calibrated by the indicator film response to external
fields in the absence of superconducting sample. Then exact values of
magnetic flux are obtained from the above measurements.

In present work the magnetooptic indicator garnet films are applied for
analyzing the magnetic flux behaviour in epitaxial $YBa_2Cu_3O_{7-\delta}$ films. These
superconducting films are manufactured by the laser sputtering on the
single crystal $SrTiO_3$ substrates. The films inherited the block structure
of substrates where blocks of approximately millimeter size with a
disorientation from several minutes up to one degree were observed by the
X-ray technique. Some smearing of the (104) reflex connected only with the
film indicated to substantial strains or, perhaps, to a small size twin
structure in the films.

Fig.1 illustrates the remagnetization process of the zero field cooled
YBaCuO film by the normal field observed using the in-plane indicator film.

----

[1] Bubble film is an iron-garnet film with normal magnetic anisotropy and
the labyrinth domain structure, where cylindrical domains ("bubble
domains") can be generated.

[2] The iron garnet films with a high magnetooptic quality were grown in the
scientific production firm "GAMMA".

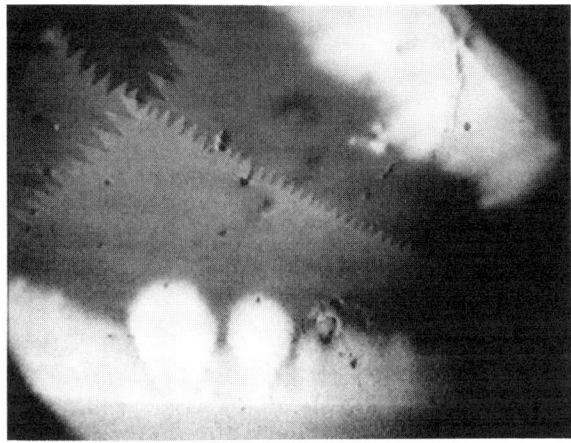

Figure 1. Magnetization patterns of an YBaCuO superconducting film, observed using the indicator garnet film with in-plane anisotropy.
Microscope polarizers are slightly uncrossed.

(a)Magnetic flux penetration (bright area) in the normal external field of 590 Oe (after zero field cooling down to 11 K).

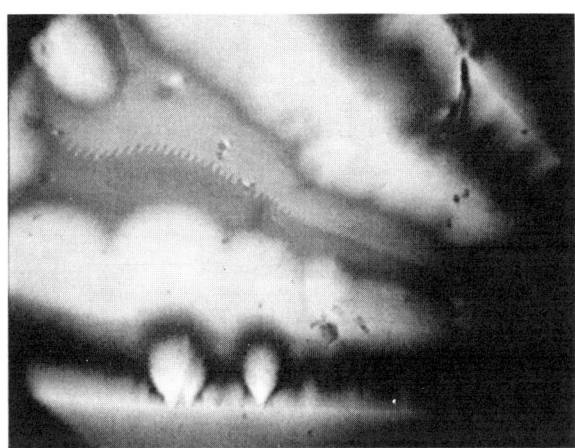

(b)Magnetic flux pattern after switching off the field 890 Oe. The flux is captured at the sample periphery and escaped at the edges were film is oppositely magnetized by the stray fields of the trapped flux.

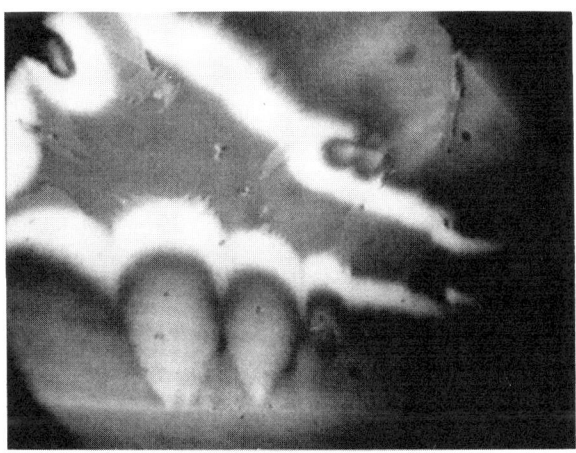

| 500 μkm |

(c) Remagnetization of the film by opposite external field of -490 Oe (See text for details).

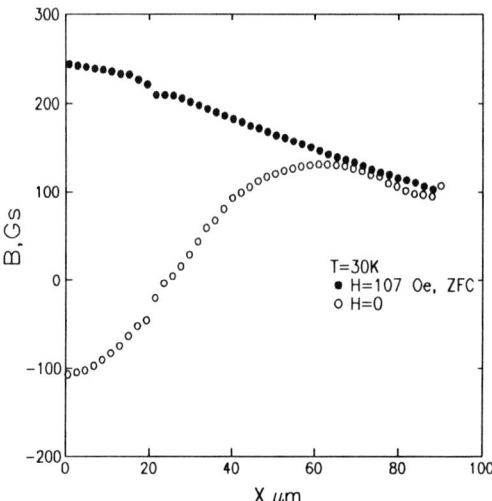

Figure 2. Magnetic flux profiles measured along the normal to the superconducting film edge at 30K. Black circles - H= 107 Oe after ZFC; open circles
- successive switching off the field.

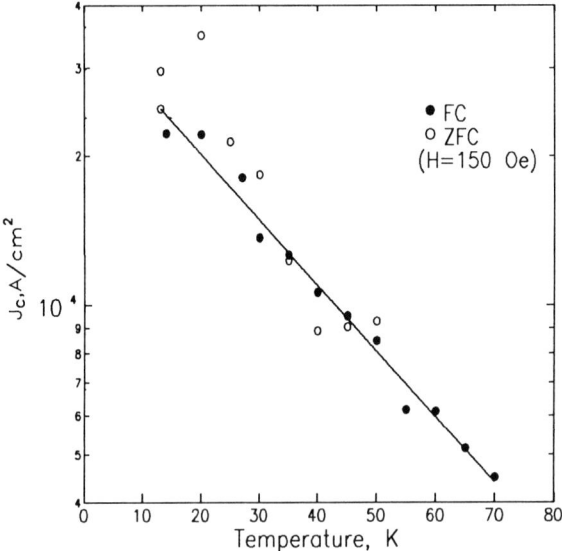

Figure 3. Temperature changes of the critical current density. Open points
obtained from profiles were measured in the field after ZFC; black — after
FC.

Bright coloring corresponds to presence of a field declining magnetization
in the indicator from its plane.

The flux penetrates in the superconducting film inhomogeneously in the
form of tongues moving from the sample edges (Fig.1a) and squeezing the
Meissner region in the central part. In the center a saw tooth like
boundary between in-plane domains in the indicator film is observed. That
in-plane domain structure is stipulated by horizontal fields enveloping the
superconductor. After switching off the external field the captured flux is
revealed (Fig.1b) remaining the boundary between the Shubnikov and Meissner
regions unchanged (Fig.1b). But at the edges the flux escapes and the
sample is remagnetized here in opposite direction by the stray fields of
the trapped vortex phase. That opposite remagnetization at the film edges
reveals the same peculiarities as in the case of direct magnetization
process (compare patterns at the lower edge in Fig.1a and 1b). When the
negative external field is applied (Fig.1c) the front of the flux penetration moves similar to that observed in the positive field. In Fig.1
opposite flux directions are revealed in different colors due to uncrossing
polarizers, which changes observation conditions for different signs of the
Faraday rotation [3]. The direct captured flux was seen yellow (light in
Fig.1c), and the negative remagnetizing flux was green (grey color with
changing intensity, reflecting the flux density variations in Fig.1c). The
Meissner region in the center was brown (homogeneously grey in Fig.1c). As
a whole the observed flux behaviour follows the classic Bean model.

To receive quantitative data the magnetic flux profiles were measured
along lines intersecting sample edges. Fig.2 shows an example of such
profiles, measured at the flux penetration in zero field cooled sample
(black circles) and after successive switching off the field (empty
circles). It is seen that at given temperature (30K) the flux is trapped at
the depth of $\geq$ 60 mkm, but it exits at the sample edge. Its density
decreases practically linearly towards the edge. At approximately 20 mkm
the field in the sample change the sign (c.f. also Fig.1c). Outside the
sample (x<0) the net stray field of the trapped flux, causing the opposite
remagnetization near the edge, is determined by lower curve. Values
$J_c[A/cm^2] = 0.8 \cdot \Delta B[G]/\Delta x[cm]$ derived from the slopes of measured B(x)
profiles for cases of ZFC and FC samples were coinciding within a good
accuracy in studied temperature range (Fig.3) in contrast to the single
crystals where they were different [3]. The critical current density
temperature dependence was well adapted by the exponential low: $J_c(T) =$
$J_{c0} \cdot \exp(-T/T_0)$, with $J_{c0} = 3.7 \cdot 10^4 A/cm^2$ and T_0 = 32K. Similar dependences
are usual for HTSC and were also obtained in our experiments with single
crystals of YBaCuO [4] where however T_0 was ~16K. It should be noted that
above values of J_c are lowered estimates of the critical current densities
due to curving vortices under the action of fields, enveloping the sample
in the case of the external field normal to the basal plane [4]. The
exponential decay of J_c at growing temperature can be connected with the
presence of weak links, as shown in [5] for traditional superconductors.

Following effect shows that weak links really exist in studied films.
Observation of motion of the magnetic flux using bubble films revealed that
before the flux penetration front circle shaped areas of the Shubnikov
phase arise (Fig.4), which are then merging with the main magnetization
wave moving from the edge. That vortex phase "lakes" appearance is

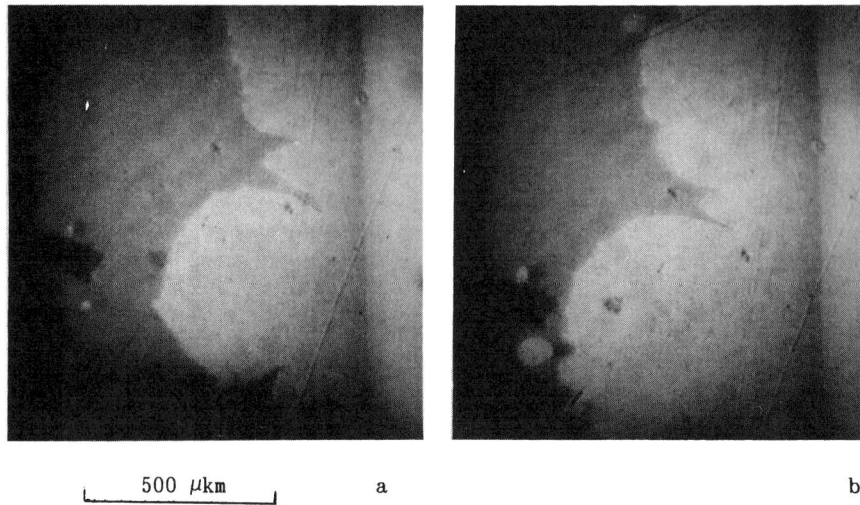

|_____ 500 μkm _____| a   b

Figure 4. Successive patterns of the magnetic flux moving from the sample edge under increasing field at 10K. Superconducting film is to the left hand, its edge is seen through the indicator bubble film as a vertical line. The Meissner region is revealed darker. (a) H = 280 Oe,(b) H = 350 Oe.

obviously caused by weak channels along which Josephson vortices penetrate into regions with small pinning inside the Meissner area. In that regions the flux spread in the form of Abrikosov vortices.

Thus both the penetration pattern and measurements of the critical current from the flux profiles manifest that even high quality YBaCuO films possess a complicated net of weak links. In addition the volume pinning forces are revealed to distribute inhomogeneously across the film area.

References

1 M.V.Indenbom, N.N.Kolesnikov, M.P.Kulakov, I.G.Naumenko, V.I.Nikitenko,
   A.A.Polyanskii, N.F.Vershinin, V.K.Vlasko-Vlasov,Physica C, 166 (1990)
   486.

2 V.K.Vlasko-Vlasov, L.A.Dorosinskii, M.V.Indenbom, V.I.Nikitenko,
   Yu.A.Ossip'yan, A.A.Polyanskii, to be published in Fizika Nizkikh
   Temperatur (1991).

3 L.A.Dorosinskii, M.V.Indenbom, V.I.Nikitenko, Ju.A.Ossip'yan,
   A.A.Polyanskii, V.K.Vlasko-Vlasov, to be published in Physica C (1991).

4 A. Polyanskii, L. Dorosinskii, M. Indenbom, V. Nikitenko, Yu. Ossipyan,
   V. Vlasko-Vlasov, J. Less-Common Metals, 164 & 165 (1990) 1300.

5 T.Y.Hsiang, D.K.Finnemore, Phys. Rev. B, 22 (1980) 154.

High T$_c$ Superconductor Thin Films
L. Correra (Editor)
© 1992 Elsevier Science Publishers B.V. All rights reserved.

Measurement of HTCS surface impedance in microwaves with the ring resonator method

M. Pyee[a], P. Meisse[b], M. Chaubet[c], D. Chambonnet[d] and D. Sinobad[d]

[a]L.D.I.M, Université P. et M. Curie, 4 Place Jussieu, 75005 Paris, France

[b]E.N.S.A.E, Laboratoire d'Electronique, 10 Av. E. Belin, 31400 Toulouse, France

[c]C.N.E.S, Département Hyperfréquences, 18 Av. E. Belin, 31055 Toulouse cédex, France

[d]Alcatel Alsthom Recherche, Route de Nozay, 91400 Marcoussis, France

ABSTRACT

We report on the measurement of low microwave losses of a high critical temperature superconductor (HTCS) by using a microstrip ring resonator deposited by laser ablation on MgO substrates. The unloaded quality factor Q_0 is determined and the surface resistance Rs is calculated. At 30 K and 5 GHz, Rs values of 1 mΩ have been determined.

1. INTRODUCTION

High temperature superconductors offer great possibilities in the area of microwave applications. Indeed, a recent review on the needs to realize HTCS components shows that superconducting losses, which are directly related to the surface resistance Rs at microwave frequencies, are less than those in a normal metal [1-2].

We report on the results of surface resistance measurements of a superconducting ring resonator deposited on the MgO substrate by laser ablation. By using a superconducting ring deposited on a substrate, we obtain a resonator which can function on the fundamental frequency and its harmonic frequencies. Thus, as it is possible to characterize the superconducting material between f_0 and 3 or 4 f_0, we can obtain variations of Rs versus frequency. Measurement of the kinetic inductance allows λ (London penetration depth), σ_n (conductivity corresponding to "normal" electrons) and σ_{sc} (conductivity corresponding to superconducting electrons) to be calculated. Results are compared to those previously reported.

2. DEVICE DESIGN

The resonant system is a microstrip ring resonator enclosed in a fixture (fig. 1a and 1b).

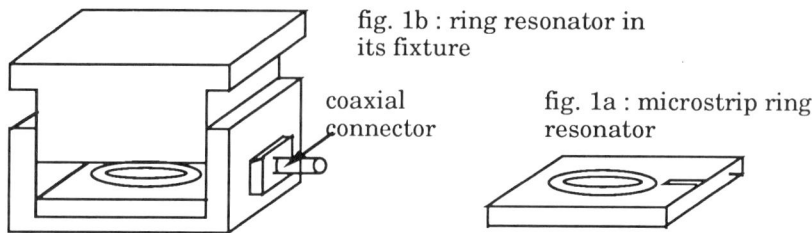

fig. 1b : ring resonator in its fixture

fig. 1a : microstrip ring resonator

coaxial connector

The ring resonator is used in reflection, and the Rs measurement is made by an indirect method. Indeed, from the measurement of the reflection coefficient S_{11}, we calculate Q_0 and the propagating constant γ. Then Rs is determined. The main advantages of this method are the following :
- measurement technique without direct electrical contact with the HTCS
- characterization is made in the volume of HTCS
- the method is relatively insensible to percolations
- current lines are parallel to the surface of superconductor, e.g. quite compatible with microwave components in thin film.

Dimensions of the circuit are given below :
- MgO substrate : 10 x 10 mm^2 x 0.5 mm
- superconducting film thickness : 150 nm
- silver ground plane thickness : 1.5 µm
- gap spacing : 80 - 200 µm
- ring radius : 4 mm

3. FILM PREPARATION

Thin films are deposited at 750 °C on (001) MgO substrates using the Pulsed Laser Deposition Technique (PLD).

In this study, a SOPRA type 510 excimer laser (λ = 380 nm, τ = 30 ns) is employed to evaporate a high density (90 to 95 %) $YBa_2Cu_3O_{7-x}$ target ; this laser delivers a pulse energy of typically 80 mJ at a repetition rate of 1 to 2 Hz. A laser fluence of 2-3 J/cm^2 is used to transfer the target material onto the substrates placed at the vertical of the target, at the distance of 5 to 7 cm. The laser beam, focused on the rotating target surface by means of an optical lens at a 45 ° impingement angle, can be scanned along a segment across the target surface.

Film deposition is made in a stainless steel growth chamber equipped with a high-vacuum pumping system. The deposition procedure is started as the residual pressure in the growth chamber reaches 10^{-7} mbar. During the heat-up and deposition steps, the ambient gas is pure oxygen at a 10^{-1} mbar pressure, which is injected through a nozzle directed towards the substrates.

Current runs are made at a pulse frequency of 2 Hz, which yields an average deposition rate of about 0.1 nm/s. As deposition is interrupted, the cooling sequence is started at a rate of 10 °C/min and the oxygen pressure is set at 300 mbar. Films thus made are not submitted to any ex-situ post-annealing.

The superconducting transition temperature Tc (R = 0), and critical current density Jc values are obtained using the classical four-probe technique.

The same laser technique is used for the patterning of bridges (for Jc measurements) and ring resonators. This technique allows the preparation of superconducting patterns which are a one-to-one replica of the metallic mirrors inserted on the path of a high-power laser beam ; the laser beam expels the irradiated part of the film whereas its complement , in the shadow of the mirror, is untouched and left with a virgin surface.

The laser used in this work is the excimer laser described as the above.

In these conditions, typical values of Tc (R = 0) in the 90-92 K range are obtained, leading to Jc values at 77 K in excess of 10^6 A/cm^2.

4. EXPERIMENTAL SYSTEM

The experimental set up is a conventional gas He cryostat. The temperature range is extended from 4 to 300 K. The cooled chamber has been especially designed to receive microwave coaxial connectors and the fixture which contains the superconducting sample [3]. Measurement device of reflection coefficient S_{11} is a network analyzer HP 8510. The complete system during a measurement is given in fig. 2 :

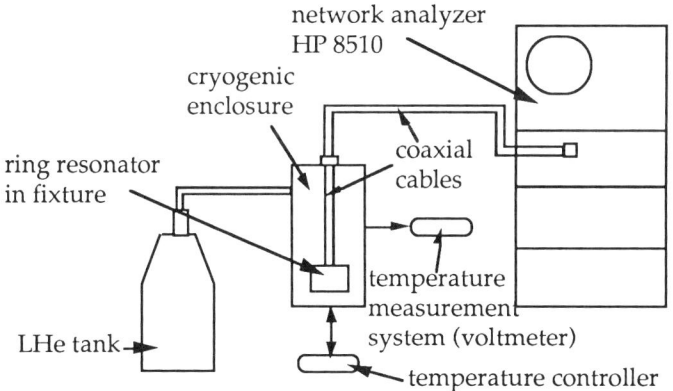

fig. 2 : measurement system

Calibration of the network analyzer is realized with a "home made calibration kit", consisting of coaxial open circuit, short circuit and 50 Ω load RIM-SMA which are used down to 4 K.

5. MODELLING

In order to take into account the complex conductivity of the superconductor, and the possible presence of buffer layers which is required in order to avoid reciprocal diffusion of atoms between substrate and HTCS, we have modelled the structure whose equivalent transmission line is shown in Fig. 3 :

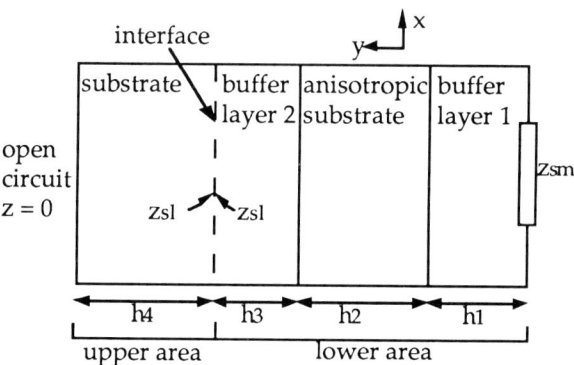

fig. 3 : equivalent transmission line for the structure (case t > 2 λ)

This analysis is based on the transverse resonance method [4]. This rigourous method considers a "fictitious" electromagnetic propagation in the transverse direction (along the y-axis). From this scheme, the equivalent circuit is determined in Fig. 4, given external conditions and continuity conditions at the interface of the microstrip line, with following notations :

j_{s1} (j_{s2}) : upper (lower) superficial current density on the strip

j_S : total current density on the strip

J_1 (J_2) : current density in the upper (lower) area

J : fictitious current density "source term"

E_1 (E_2) : electric field in the upper (lower) area

E : electric field on the strip

E' : fictitious electric field "source term"

Z_1 (Z_2) : impedance operator related to the upper (lower) area, expressed on the interface

z_{sl} : surface impedance of strip conductor

z_{sm} : surface impedance of ground plane

t : thickness of the strip metallization

For a superconductor :

$$z_{sl} = \frac{1}{\sigma_s \lambda} = Rs + j\,Xs \quad (\text{case } t > 2\,\lambda) \tag{1}$$

where σ_s is the superconducting conductivity (two-fluid model), Rs the surface resistance and Xs is related to the kinetic inductance.
we put :

$$J_1 = -j_{s1} \qquad (2)$$
$$J_2 = -j_{s2} \qquad (3)$$

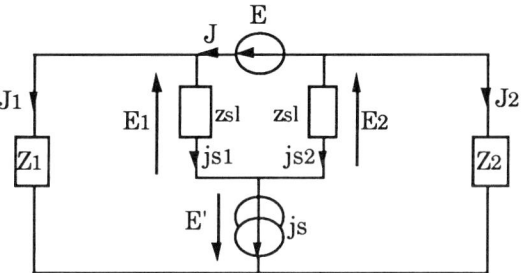

fig. 4 : equivalent electrical circuit (case t > 2 λ)

From this equivalent circuit, we write :

$$\begin{bmatrix} J \\ E' \end{bmatrix} = \widehat{H} \begin{bmatrix} E \\ j_s \end{bmatrix} = \begin{bmatrix} 0 \\ 0 \end{bmatrix} \qquad (4)$$

where H is an operator which depends on the circuit. J and E' are null because they are fictitious "source terms".

$\left\langle [\, E \ j_s \,] \widehat{H} \begin{bmatrix} E \\ j_s \end{bmatrix} \right\rangle$ being variational ($\langle \ \rangle$ denotes the scalar product of two functions), we must confirm that :

$$\left\langle [\, E \ j_s \,] \widehat{H} \begin{bmatrix} E \\ j_s \end{bmatrix} \right\rangle = 0 \qquad (5)$$

This relationship is a key point to calculate the resonance condition of the equivalent circuit and the propagating constant γ versus the conductivity σ of the material (or the surface resistance), or vice-versa. This method requires the use of a complete orthogonal base for expressing TE and TM modes, (cosine and sine functions), and a good test function on the strip conductor (which takes into account effects of "edges").

6. EXPERIMENTAL RESULTS

In fig. 5, the curve presents the variations of the resistance surface Rs versus frequency, at the temperature T = 30 K, for a YBaCuO thin film deposited on a MgO substrate, as described in paragraph 3. The frequency dependence of Rs is approximatively as f^2, which is in good agreement with the London's theory. Previously reported results give Rs values from 4 μΩ to 20 mΩ at 5 GHz and 4.2 K and from 2 μΩ to 70 mΩ at 10 GHz and 4.2 K [1-3]. From our measurements, we have also determined, according to the two-fluid model, λ and σ_n values which are of 210 nm and 54 10^6 Ω^{-1} m^{-1}, respectively.

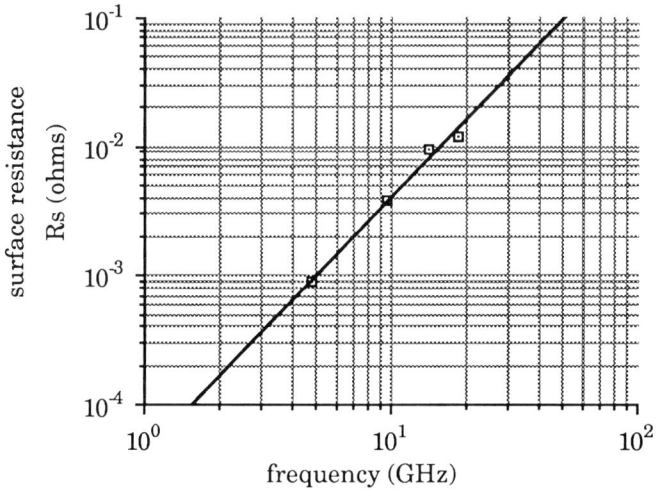

fig. 5 : variations of the surface resistance Rs at T = 30 K

7. CONCLUSION

We have described our method to characterize a superconducting film and we have explained the modelling technique. From this theorical study, we have made measurements of a YBaCuO thin ring resonator, elaborated by laser ablation as described and patterned also laser ablation, and the results, which are the first obtained, are in agreement with those previously reported.

8.ACKNOWLEDGMENTS

We would like to thank especially H. Lauvray, Rhône-Poulenc Company, and D. Parise, Alcatel Espace Company, for their help in this study

9. REFERENCES

1 H. Piel and G. Müller, IEEE Trans. on MAG-27, 854 (1991)
2 J. Talvacchio and G. R. Wagner, Proc. of the SPIE, 1292 (1990)
3 M. Pyee, P. Meisse, H. Baudrand, M. Chaubet, Proc. of the SPIE (1991), to be published
4 H. Baudrand, Ed. ENSEEIHT, Toulouse, 1985

High T$_c$ Superconductor Thin Films
L. Correra (Editor)

207

Critical current density and E-j characteristics of YBaCuO films deduced from inductive measurements

W. Schauer, V. Windte, M.Polak[a], J. Reiner, W. Maurer, A. Gurevich[b], and H. Wühl

Kernforschungszentrum Karlsruhe, Institut für Technische Physik,
7500 Karlsruhe, Germany

[a]Slovak Academy of Science, Institute of Electrical Engineering,
84239 Bratislava, Czech and Slovak Federal Republik

[b]Institute for High Temperatures, 127412 Moscow, USSR

Abstract

A contactless inductive method is used to determine the j$_c$(E,B) dependence of thin film ring shaped superconductors. For sputter deposited YBaCuO films the E-j characteristic could be determined over 6 decades using supplementary resistive measurements. In the region from 10^{-4} to 1 µV/cm an exponential E-j relation is observed.

1. INTRODUCTION

One of the most valuable tools in studying the performance of high-T$_c$ superconducting films is the determination of their critical current density j$_c$. The commonly used resistive j$_c$ measurement, however, has several severe disadvantages, essentially: (i) the lack in sensitivity which rarely falls below the 0.1 µV/cm limit, (ii) problems associated with the patterning of well defined narrow conducting traces to reduce the absolute current level, (iii) the difficulty to attach low ohmic contacts to the pads.

To avoid these complications we have developed a contactless inductive method to measure the critical current density and E-j characteristic in thin film samples [1]. For disk shaped samples several approaches to determine j$_c$ by magnetic, non-transport measurements have been reported [2]. Our method is based on measuring the magnetic field B$_s$ generated by shielding currents in the film which are induced by an external field sweep dB$_e$/dt. From the sample field B$_s$ the shielding currents and their dependence on the electric field E and the magnetic field B can be deduced. The j-E correlation is of special importance as it reflects the resistive dissipation mechanism due to the movement of magnetic flux. It is therefore a source of meaningful information in weighing different models describing the resistive state like flux creep or vortex glass behaviour.

The evaluation of j (E,B) is straightforward for a ring shaped sample with the ring width small compared to its diameter. In addition, film samples with a disk (or rectangular) geometry can also be used to determine j_c in a similar way [3]. - At present, our measurements are performed at 77 K in liquid nitrogen and in external fields up to 0.2 T.

It is important to show that the results of the inductive measuring technique described before overlap with those obtained from transport measurements. This is the subject of the present paper, in which we present measurements of the $E(j)$ characteristic over 6 orders of magnitude.

2. EXPERIMENT AND PROCEDURE OF MEASUREMENT

Thin YBaCuO films were prepared by hollow cathode magnetron sputtering from a cylindrical stochiometric sinter target onto (100) $SrTiO_3$ substrates [4]. Films were single crystalline with their c axes oriented perpendicular to the films surface as revealed from X-ray spectra showing (00ℓ) lines only and - more quantitatively - from extremly narrow rocking curves of $\sim 0.13°$ FWHM [5]. For magnetic measurements films were patterned to a ring or disk geometry using Ar ion milling and a photolithographically structured resist as sputter mask [6].

The experimental setup for magnetic measurements is shown in Figure 1. A detailed description of the experiment and the principle of the measurement is given in Ref. [1]. - The ring or disk sample is placed in the center of a copper solenoid which produces a magnetic field $B_e \leq 0.2$ T perpendicular to the film plane. The field can be ramped with a sweep rate between approximately 10^{-2} and 10^{-5} T/s. Circumferential screening currents are induced in the film by the field sweep dB_e/dt; they flow with the critical current density j_c and produce the sample field B_s. The axial (z-) component B_{sz} is measured by a Hall sensor usually positioned a few tenths of a millimeter above the center of the sample. To study flux profiles $B_{sz}(r,z)$ a mechanical stage is used to drive the Hall probe along the radial and axial direction. The lateral resolution of profiles is determined by the small active area of the Hall probe of 50 μm x 50 μm. Measurements are performed

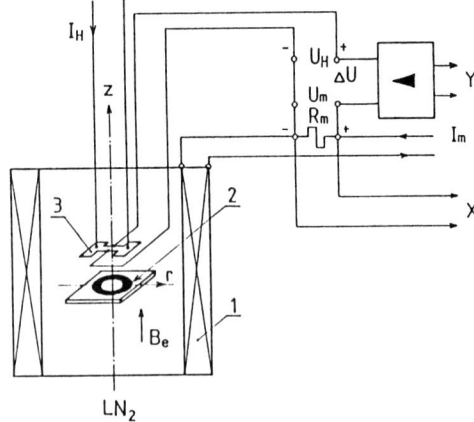

Figure 1. Experimental setup: (1) copper solenoid (2) thin film ring sample on the substrate (3) Hall sensor. The sketch shows the compensation circuit to suppress the Hall signal due to the external field B_e (see text).

in liquid nitrogen. - From the sample field on the ring axis, $B_{sz}(r=0,z)$, the magnetization and the critical current density of the sample can easily be calculated. The analysis is applicable if the external field B_e fully penetrates the sample. This condition is fulfilled beyond the saturation field for $B_e > B_s$, i.e. for $B_e > 1$ mT, as the sample field is of the order of 1 mT.

The measurement of B_s by the Hall sensor requires the external field to be compensated. This is achieved by subtracting a voltage $U_m = R_m \cdot I_m$ proportional to B_e from the Hall voltage U_H (Figure 1). The compensation $U_H-U_m=0$ is performed if no sample is present by adjusting the Hall control current. It works only if both the magnet voltage U_m and the Hall voltage are proportional to the external field. Therefore the use of an iron core solenoid or a superconducting magnet to achieve higher fields requires another compensation technique (e.g. by using a second Hall sensor) as hysteretic effetcs occur. The small non-linearity of $U_H(B_e)$ especially in the low field region is usually negligible and does not affect the j_c analysis.

Once the Hall signal of the external field is compensated the usual experimental procedure is to record a two quadrant hysteresis loop $B_s(B_e)$ of the sample (Figure 2). Doing this, the Hall sensor is positioned to measure the central axial field $B_{sz}(r=0, z)$ close to the film surface. Provided the magnetic field has completely penetrated the ring cross section ($B_e > B_s$) the critical current density j_c will be proportional to $B_{sz}(r=0, z)$ and can thus be determined from the height of the hysteresis loop. The proportionality constant is determined by the ring geometry and the Hall sensor position only. If the ring width w is small compared to the mean ring diameter $2 R_0$ the approximation of a current loop is justified leading to the following result [7] :

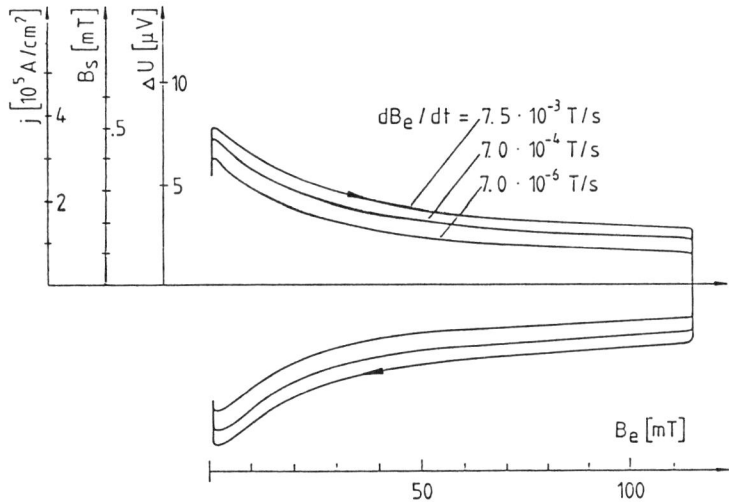

Figure 2. Hysteresis loops of an YBaCuO ring sample (ring diameter: 3.5 mm, ring width: 0.5 mm, film thickness: 1 μm). On the ordinate the Hall voltage ΔU, the sample field B_s ($r=0$, $z=0.8$ mm) and the critical current density j_c is given. The splitting of the loops due to the influence of dB_e/dt is demonstrated.

$$j = 2 \cdot B_s(r=0, z) \cdot (R_0{}^2 + z^2)^{3/2} / (\mu_0 R_0{}^2 \, w \, d) \qquad (1)$$

Here we have replaced the loop current I by the mean current density $j = I/(w \cdot d)$, where d is the film thickness. The $j_c(B)$ dependence can be deduced from the $B_s(B_e)$ loop height at different external fields B_e. To demonstrate the relative magnitude of the different quantities we have plotted the Hall voltage ΔU, the sample field $B_s(r=0, z=0.8\text{mm}) = \Delta U/(\sigma_H \cdot I_{H,\text{compens.}})$ and the critical current density j_c according to equation (1) on the ordinate of Figure 2.

The inductive method described gives information not only on the $j_c(B)$ dependence but also on the $j(E)$ relation. The electric field E driving the induced shielding current is determined by the external field sweep rate dB_e/dt. As the rate of change of the external field is much larger than for the sample field: $dB_e/dt \gg dB_s/dt$, the electric field is given by $E = (R_0/2) \cdot dB_e/dt$. Thus, as the sweep rate dB_e/dt is increased, the current density is determined for a larger E-field criterion and increases, too. This is clearly demonstrated in Figure 2, where a family of $B_s(B_e)$ curves is shown differing in the E criterion adjusted by the slope of the dB_e/dt ramp. In this way the E-j characteristic of the sample can be determined over a wide range of electric field values far below the E interval accessible in resistive measurements.

3.   RESULTS AND DISCUSSION

The aim of this investigation is to perform inductive and resistive measurements on the same sample and to look at the overlap of the E(j) characteristic using the method described in Chapter 2. To do this we used a thin film sample structured as shown in Figure 3. After the inductive $j_c(E,B)$ measurement had been done for the closed ring a cut was made in the smale ring segment at point A. Then resistive measurements could be performed using the current (I) and voltage (V) pads shown in Figure 3. The advantage of this pattern is that both measuring techniques use the same current path.

The results of the investigation are summarized in Figures 4 and 5. The E(j) characteristics at an external field of 30 mT, 80 mT and 130 mT deduced from the inductive and resistive measurements are shown in Figure 4. Both techniques

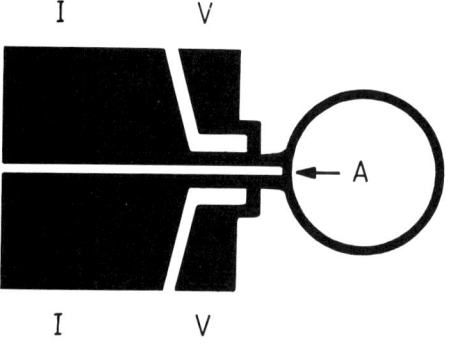

Figure 3. Film pattern for inductive and resistive determination of $j_c(E,B)$ in the ring. For the resistive measurements the ring is cut at point A and contacts are fixed to the current (I) and voltage (V) pads.

encompass an E interval of 6 decades. The E(j) relation is linear in the ^{10}log E vs j plot over 4 decades which correlates with the conventional flux creep model [8]:

$$E = E_0 \cdot \exp\left[-(U_0/kT)(1-j/j_{co})\right] \qquad (2)$$

Here j_{co} is an unrelaxed critical current density and U_0 is an activation energy. Notice that our preliminary results do not give evidence for both E-j curves of the form $E = E_0 \cdot \exp(-(j_0/j)^\mu)$ predicted by the vortex glass model [9], and a power dependence of E on j observed previously in thin high-T_c films [10].

The overlap of the inductive and the resistive E(j) determination at $\sim 10^{-1}$ $\mu V/cm$ is indicated by a clear translational displacement: the resistive j values are about 50 % smaller than those derived from the inductive method. Though this deviation seems to be unsatisfactory at a first glance, the overlap nevertheless shows that the physical relations merge continuously. The ratio j(30 mT):j(80 mT): j(130 mT) at 10^{-1} $\mu V/cm$ is the same for both curve triplets, i.e. a constant factor of 1.54 ± 0.03 scales from the resistive to the inductive set of curves. Also, the slope $d \ln E/dj = U_0/(kT j_{co})$ at $E = 10^{-1}$ $\mu V/cm$ is the same for both methods at the same magnetic field. These observations are consistent with our assumption that the

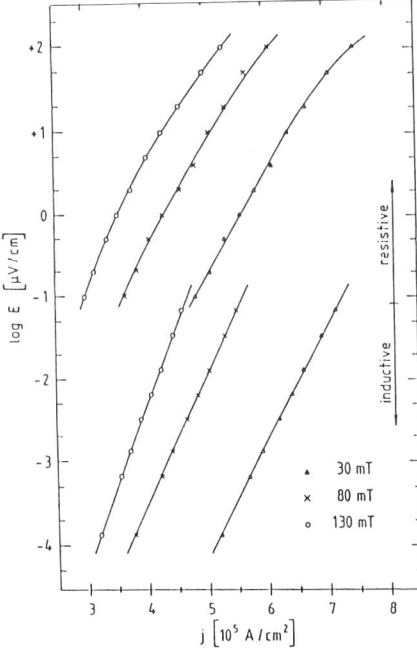

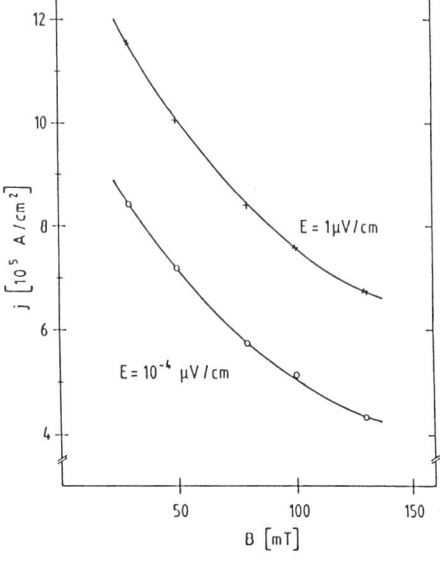

Figure 4. Inductive and resistive E(j) characteristics at 30 mT, 80 mT and 130 mT for an YBaCuO film 620 nm thick with $T_c(R=0) = 89.8$ K. The ring width is w = 0.25 mm, its mean diameter $2R_0 = 2.75$ mm.

Figure 5. Inductively determined critical current density dependence on B and E of an YBaCuO film ring (w = 0.25 mm, $R_0 = 2.75$ mm) 350 nm thick ($T_c(R=0) = 90.5$ K). Values at $E = 1$ $\mu V/cm$ are extrapolated.

active current carrying cross section of the ring has been accidentally reduced by
the manipulations for preparing the resistive measurements. The contact pads
were covered with a gold layer and cleaned before fixing the contact wires using a
methyl bromide etchant. Though the ring area was covered with a foil some of the
etchant could have damaged the small straight strip line connection from the pads
to the ring. Another critical procedure is to position the cut scalpel exactly at
point A and thereby to prevent the neighbouring film section to be hurt. - We
believe that some chemical or mechanical degradation reducing the active cross
section is the most probable explanation for the j_c offset. The field dependence of j_c
is shown in Figure 5. Values for $E = 1\mu V/cm$ have been extrapolated linearly from
the $E(j)$ characteristics. The steep slope of $^{10}\log E$ vs j results in a decrease of j_c of $\sim$
30 % only at a change in the E-criterion of 4 orders of magnitude.

The contactless inductive method presented enables us to determine the $j(E,B)$
dependence of ring shaped superconducting thin film samples. The method is
simple and versatile; it allows one in addition to measure flux profiles $B_{sz}(r)$ and
the time decay of the critical current due to flux creep. Together with
supplementary resistive measurements the $E(j)$ characteristic could be determined
over 6 decades in the electric field criterion confirming the exponential E-j flux
creep behaviour (Equation (2)) for $E < 1\ \mu V/cm$. The inductive $\rightarrow$ resistive
overlap seems to be capable of being improved.

4. REFERENCES

1   M. Polák, V. Windte, W. Schauer, J. Reiner, A. Gurevich, and H. Wühl:
     Physica C 174 (1991) 14.
2   D.J. Frankel: J. Appl. Phys. 50 (1979) 5402; M. Däumling and D.C.
     Larbalestier: Phys. Rev. B40 (1989) 9350; M.N. Kunchur and S.J. Poon: Phys.
     Rev. B43 (1991) 2916; M.J. Scharen, A.H. Cardona, J.Z. Sun, L.C. Bourne, and
     J.R. Schrieffer: Jap. J. Appl. Phys. 30 (1991) L15; T.L. Hylton, M.R. Beasley,
     and R.C. Taber (preprint), and references given in paper [1].
3   M. Polák, V. Windte, W. Schauer, J. Reiner, W. Maurer, and A. Gurevich: 6th
     Internat. Workshop on Critical Currents, Cambridge, England, July 1991.
4   J. Geerk, G. Linker, and O. Meyer: Mat. Sci. Rep. 4 (1989) 193.
5   W. Schauer, X.X. Xi, V. Windte, O. Meyer, G. Linker, Q. Li, and J. Geerk:
     Cryogenics 30 (1990) 586.
6   W.K. Schomburg, M. Heidinger, G. Nöther, J. Reiner, V. Windte, W. Schauer,
     and K. Kadel: Cryogenics, 31 (1991) 366.
7   J.D. Jackson: Classical Elektrodynamics (J. Wiley, New York, 1962), 143.
8   P.W. Anderson and Y.B. Kim: Rev. Mod. Phys. 36 (1964) 39; M. Tinkham:
     Introduction to Superconductivity (Mc Graw-Hill, New York, 1975).
9   M.P.A. Fisher: Phys. Rev. Lett. 62 (1989) 1415; R.H. Koch, V. Foglietti, W.J.
     Gallagher, G. Koren, A. Gupta, M.P.A. Fisher: Phys. Rev. Lett. 63 (1989) 1511
10   E. Zeldov, N.M. Amer, G. Koren, A. Gupta, R.J. Gambino, M.W. McElferesh:
     Phys. Rev. Lett. 62 (1989) 3093; E. Zeldow, N.M. Amer, G. Koren, A. Gupta:
     Appl. Phys. Lett. 56 (1990) 1700.

High T$_c$ Superconductor Thin Films
L. Correra (Editor)

Resistivity, Critical Current and Modulated Microwave Absorption of Ag$_2$O Doped YBa$_2$Cu$_3$O$_{7-\delta}$ Ceramics

R.R. Schulz and K.W. Blazey

IBM Research Division, Zurich Research Laboratory, 8803 Rüschlikon, Switzerland

Abstract

The critical current of ceramic YBa$_2$Cu$_3$O$_{7-\delta}$ is shown to be improved both by inclusion of silver in the intergranular spaces and by prolonged annealing without silver. Both processes improve the contact between the grains. The relation between the normal state resistivity, the magnetic critical current and the modulated microwave absorption show these composites behave like homogeneous dirty superconductors.

1. INTRODUCTION

The critical currents of ceramic cuprate superconductors are small because of the many intergranular Josephson junctions throughout the sample. Many attempts have been made to reduce the detrimental effect of these junctions and doping with Ag$_2$O has had some success.[1-4] Here we report the results on two series of YBa$_2$Cu$_3$O$_{7-\delta}$:Ag$_2$O composites that have been annealed for different times. The samples were characterized by resistivity, magnetization and microwave absorption measurements. In both series the normal state resistivity is reduced by the inclusion of silver. Furthermore the bulk transport critical current increases with silver content but the sample without silver annealed for the longer time showed the highest critical current of all samples. Thus long annealing can improve the critical current of the ceramic superconductors just as the inclusion of Ag does. But the magnetic critical current derived from the magnetization hysteresis decreases with increasing Ag content and is more closely related to the intragrain properties.

2. SAMPLE PREPARATION

The YBa$_2$Cu$_3$O$_{7-\delta}$ starting compound was prepared by the usual solid-state reaction method from Y$_2$O$_3$, CuO and BaCO$_3$. The powder was mixed, ground and reacted at 920 °C for 45 hours in air. After reaction the product was reground and sintered in flowing O$_2$ for 6 hours at 920 °C and 15 hours at 480 °C. From this superconducting YBa$_2$Cu$_3$O$_{7-\delta}$-powder two series were prepared, each doped with 0%, 2.5% and 5% Ag$_2$O powder. The mixed powders were pressed to form pellets of 8 mm diameter and 1 mm thickness. In both series all samples were heated in flowing O$_2$ to 280 °C, the dissociation temperature of Ag$_2$O, and held for 6 hours. After this they were heated up to 920 °C in 6 hours where the first series was annealed for 20 hours and the second series for 50 hours. Both series were then held for 20 hours at 480 °C before slowly cooling down to room temperature. X-ray diffraction analysis on the samples showed that all samples are of orthorhombic 1-2-3 phase and only additional diffraction peaks of pure Ag could be observed.[4-6]

3. EXPERIMENTAL TECHNIQUES

A conventional DC four-probe method was used to measure the temperature dependence of resistivity. A computer-controlled system with a digital nanovoltmeter was used to record the I-V characteristics. The critical current density was determined according to the 1 μV criterion. The samples were glued on a MgO single crystal substrate and ground to a thickness of $\approx$0.1 mm. The width was reduced to $\approx$0.25 mm. The typical length of the sample for the voltage measurement was about 0.3 mm. The connections were made with gold wires coated with silver paste bound to sputtered gold electrodes.

The DC magnetization measurements were made in a SQUID magnetometer as a function of temperature and applied magnetic field (0-45 kG). For the transition-curves the samples were field-cooled in 6 G from 95 K down to 5 K. The "magnetic" critical current densities were determined from the high-field (45 kG) hysteresis loops. The typical dimension for the samples were 1×1×6 mm. The samples were oriented with the long-axis along the magnetic field.

Modulated microwave absorption measurements[7] were made on both series of samples, all of approximately the same dimensions. Each sample was mounted in an Oxford Instrument ESR 900 continuous flow He gas cryostat attached to a Bruker ER 200D-SRC X band spectrometer.

4. RESULTS AND DISCUSSION

4.1 Resistivity and Critical Current Density

The temperature dependence of the resistivity showed all samples were superconducting below 92 K. The variation of the normal state resistivity at 100 K with Ag_2O content is shown in Fig. 1 for both series.[8,9] Inclusion of 5% Ag_2O reduces the normal state resistivity by an order of magnitude in the series, K, annealed for 20 hours but only marginally in the series, L, annealed for 50 hours. Thus while the addition of Ag produces a steep reduction of the resistivity, longer annealing also produces lower resistivity material with or without Ag. Accompanying this decrease in resistivity is an increase of the bulk transport critical current density as shown in Fig. 2 for both series. However, there is one notable exception, namely, the sample annealed for 50 hours without any Ag_2O which had the highest critical current of all the samples. The presence of silver improves the contact between the grains thus reducing the resistivity and forming intergranular SNS junctions instead of SIS junctions.[4,10-13] These SNS junctions increase in area with increasing Ag_2O content as seen by the increasing critical current. But their properties are apparently inferior to those formed by long annealing material without Ag_2O as evidenced by its large critical current.

A further effect of the addition of Ag_2O is to increase the superconducting transition temperature slightly as shown in Fig. 3.[4,9] A small but easily measurable shift of about 1 K was seen for the addition of 2.5% Ag_2O in both series not only in the resistivity measurements but also in the susceptibility measurements reported in the next section.

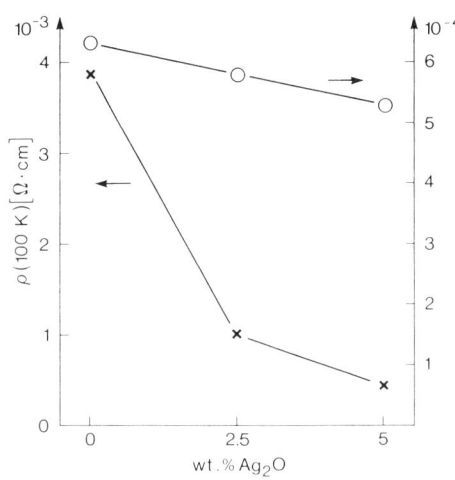

Fig. 1. Normal state resistivity at 100 K for series *K* (X) and series *L* (O) vs. wt.% Ag$_2$O doping.

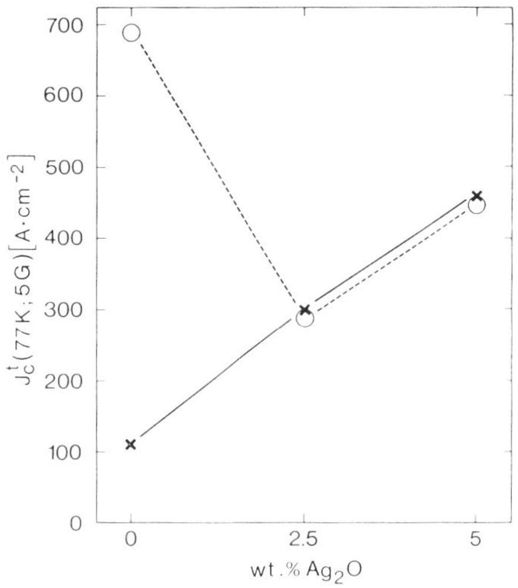

Fig. 2. Transport critical current density for series *K* (X) and series *L* (O) at 77 K and 5 G vs. wt.% Ag$_2$O doping.

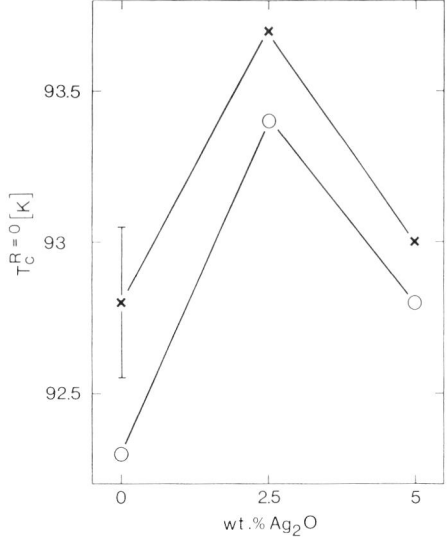

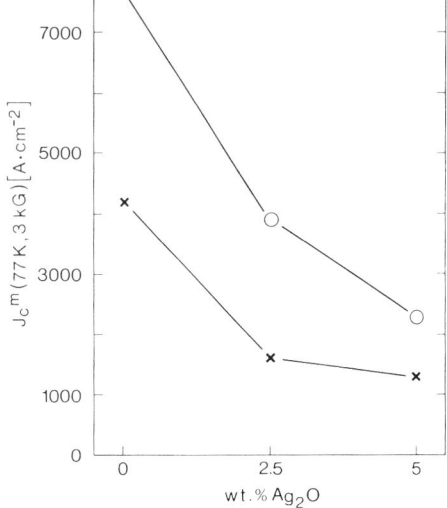

Fig. 3. Zero resistance temperature for series *K* (X) and series *L* (O) vs. wt.% Ag$_2$O doping.

Fig. 4. "Magnetic" critical current density for series *K* (X) and series *L* (O) at 77 K and 3 kG vs. wt.% Ag$_2$O doping.

4.2 Magnetic Critical Current

Using Bean's critical state model[14] a magnetic critical current was derived from the magnetization at 3 kG in the hysteresis loops.[15] Since similar values are obtained for ceramics and loose packed powders obtained by crushing the ceramics, these critical currents are taken to be intragranular values. An average grain size of ≈ 10 μm, used in the analysis, was determined by light scattering for all samples. The variation of these magnetic critical currents with silver content is presented in Fig. 4 which shows a clear improvement after long annealing.

The temperature dependences of the susceptibility are shown in Fig. 5 where the same shift of T_c with Ag content observed in the resistivity data is also clearly seen.

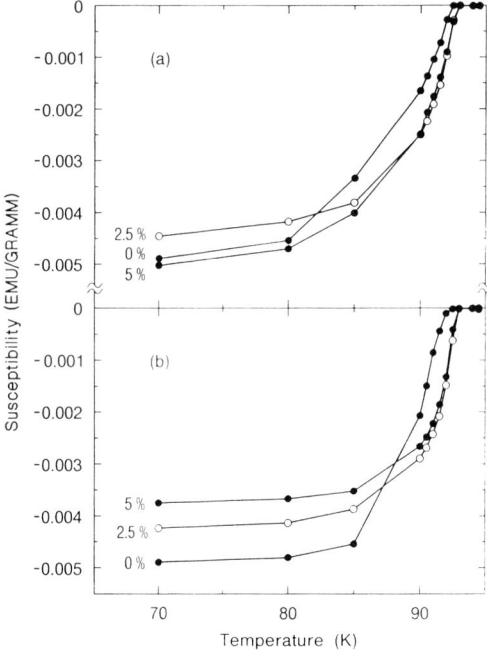

Fig. 5. Temperature dependence of dc susceptibility for series K (a) and L (b).

4.3 Microwave Absorption

The modulated microwave absorption of a granular superconductor shows open hysteresis loops as a function on an external magnetic field that gradually close with increasing modulation field.[16] The signals can be considered as the sum of two components, one due to the boundary current which dominates at small modulation fields before saturating and another due to viscous fluxon motion which increases linearly with modulation field.[17] Thus the variation of the modulated microwave absorption with modulation field shows a maximum at low fields and then a minimum at some critical field $2H_{c1}^*$, followed by a linear increase.[16] The critical field $2H_{c1}^*$ represents that field required to reverse the critical state at the surface and is given by $(4\pi/c) J_c \lambda$ where λ is an effective relaxation or penetration depth.[18] The temperature variation of $2H_{c1}^*$ was measured for both series and found to be linear.[16] A relation between the values of 2H$_{c1}$* at 30 K with the normal state resistivities and both kinds of critical currents was sought that satisfied the results on all samples.

For the transport critical current the best agreement was found when 2H$_{c1}$* was assumed to be proportional to the product $J_c^t \rho_n$, an unexpected result. On the other hand the magnetic critical current was related to 2H$_{c1}$* via the relation 2H$_{c1}$* $\propto J_c^m (\rho_n)^{1/2}$ (Fig. 6) which means $\lambda \propto (\rho_n)^{1/2}$ as expected for a homogeneous dirty superconductor.[19] This is a fair approximation of these ceramic samples where the grains also have many junctions due to the presence of twin domains.

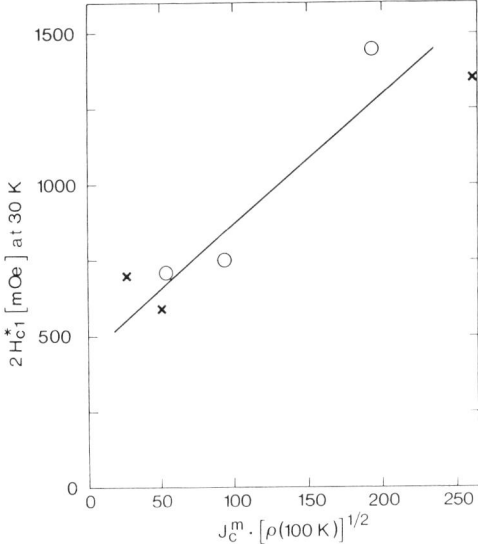

Fig. 6. Variation of 2H$_{c1}$* with the product $J_c^m \cdot [\rho\,(100\ K)]^{1/2}$.

5. CONCLUSIONS

By silver doping it is possible to make samples with lower normal state resistivity and higher critical transport current as shown in the first series. This may be due to the fact that the Josephson tunnel junctions (SIS) between the grains were replaced with a large number of normal-state proximity junctions (SNS). In the case of long sintering as in the second series we suggest that the (SIS) junctions in the undoped sample are intrinsically optimized. There the silver doping makes these (SNS) junctions with a lower critical current. A relation was found between the normal state resistivity, the magnetic critical current and the modulated microwave absorption indicating these composites behave as homogeneous dirty superconductors.

Stimulating discussions with A. Baratoff, J.G. Bednorz, G. Deutscher and K.A. Müller are gratefully acknowledged.

REFERENCES

[1] B. Dwir, M. Affronte and D. Pavuna, Appl. Phys. Lett. 55 (1989) 399.

[2] N. Imanaka, F. Saito, H. Imai and Gin-ya Adachi, Jpn. J. Appl. Phys. Pt. 2, 28 (1989) L580.

[3] M.K. Malik, V.D. Nair, A.R. Biswas, R.V. Raghavan, P. Chaddah, P.K. Mishra, G. Ravi Kumar and B.A. Dasannacharya, Appl. Phys. Lett. 52 (1988) 1525.

[4] J. Jung, M.A.-K. Mohamed, S.C. Cheng and J.P. Franck, Phys. Rev. B 42 (1990) 6181.

[5] Su Zhen-peng, Zhao Yong, Sun Shi-fang, Chen Zu-yao, Chen Xian-hui and Zhang Qi-rui, Solid State Commun. 69 (1989) 1067.

[6] B.R. Weinberger, L. Lynds, D.M. Potrepka, D.B. Snow, T. Burila, M.E. Eaton Jr., R. Cipolli, Z. Tan and J.I. Buduick, Physica C 161 (1989) 91.

[7] K.W. Blazey, in Earlier and Recent Aspects of Superconductivity, Eds. J.G. Bednorz and K.A. Müller (Springer, Heidelberg, 1990) 262; and references therein.

[8] D. Pavuna, M. Berger, M. Affronte, J. Van der Maas, J.J. Capponi, M. Guillot, P. Lejay and J.L. Tholence, Solid State Commun. 68 (1988) 535.

[9] S.M. Miller, S.L. Holder, J.D. Hunn and G.N. Holder, Appl. Phys. Lett. 54 (1989) 2256.

[10] F. Deslandes, B. Raveau, P. Dubots and D. Legat, Solid State Commun. 71 (1989) 407.

[11] C. Laubschat, M. Domke, M. Prietsch, T. Mandel, M. Bodenbach, G. Kaindl, M.J. Eickenbusch, R. Schoellhorn, R. Miranda, E. Moran, F. Garcia and M.A. Alario, Europhys. Lett. 6 (1988) 555.

[12] J. Moreland, R.M. Ono, J.A. Beall, M. Madden and A.J. Nelson, Appl. Phys. Lett. 54 (1989) 1477.

[13] M.A.M. Gijs, D. Scholten, Th. van Rooy and R. IJsselsteijn, Physica C 162-164 (1989) 1615.

[14] C.P. Bean, Rev. Mod. Phys. 36 (1964) 31, ibid, Phys. Rev. Lett. 8 (1962) 250.

[15] S. Senoussi, M. Oussence and S. Madjoudi, J. Appl. Phys. 63 (1988) 4176.

[16] K.W. Blazey, Physica Scripta T29 (1989) 92.

[17] A. Dulcic, B. Rakvin and M. Pozek, Europhys. Lett. 10 (1989) 593.

[18] K.W. Blazey, A.M. Portis and J.G. Bednorz, Solid State Commun. 65 (1988) 1153; M. Stalder, G. Stefanicki, M. Warden, A.M. Portis and F. Waldner, Physica C 153-155 (1988) 659.

[19] G. Deutscher, in Earlier and Recent Aspects of Superconductivity, Eds. J.G. Bednorz and K.A. Müller (Springer, Heidelberg, 1990) 183.

High T$_c$ Superconductor Thin Films
L. Correra (Editor)
© 1992 Elsevier Science Publishers B.V. All rights reserved.

Microstructure and microwave loss studies on epitaxial YBa$_2$Cu$_3$O$_x$ thin films

T.I. Selinder[a], Z. Han[a], U. Helmersson[a], S. Rudner[b], L.-D. Wernlund[b], and L.R. Wallenberg[c]

[a] Experimental Thin Film Physics, Linköping University,
  S-581 83 Linköping, Sweden

[b] National Defense Research Establishment,
  Box 1165, S-581 11 Linköping, Sweden

[c] National Centre for HREM, Inorganic Chemistry,
  Chemical Centre 2, Box 124, S-221 00 Lund, Sweden

Abstract

Thin YBa$_2$Cu$_3$O$_x$ films were grown on SrTiO$_3$ or LaAlO$_3$ single crystals by dc-magnetron reactive sputtering in argon/oxygen gas mixtures. Deposition temperatures ranged from 540 °C to 780 °C. After deposition and subsequent cool down, films grown above 680 °C are superconducting below 86-88 K. All films are oriented with the [001] direction parallel to the [001] substrate normal. The best films have critical current densities well above 10^6 A/cm^2, and effective 6 GHz surface resistance values below 300 μΩ, at 77 K. Despite good electrical properties, the films are littered with copper rich particles on a smooth and epitaxial single crystalline surface. Cross-section Transmission Electron Microscopy (X-TEM) was used to study the crystalline quality on a microscopic level, and the nature of particles occurring on/in the *as grown* film. Small misaligned YBa$_2$Cu$_3$O$_x$ grains often occur in the films. The large copper rich particles on the film surface seem to have nucleated at such grains on the film surface. Their occurrence can, at least partly, be explained by a nonstoichiometric flow to the substrate. The number density of the particles decreased with increasing growth temperature but the volume density seemed to be constant in the investigated temperature interval.

1. INTRODUCTION

The optimization of high temperature superconductive thin films for use in single layer, or more complex multilayer structures for high frequency applications requires a profound knowledge of the nucleation and growth process of these films. Especially multilayers and superlattices need a controlled epitaxy and atomically flat surface of each layer lest the crystalline quality and transport properties should deteriorate with increasing thickness. High quality films with low high frequency loss have been grown by laser ablation technique [1], which sometimes result in a granular film due to the island growth mode [2]. In other cases the film grew to produce a single crystal, but was still littered with particles, boulders and protruding outgrowths [3]. For sputtering, the growth conditions are quite different from laser ablation, but in this paper we show that these films have similar particles. None of the described defect structures are acceptable for multilayers, and more effort must be put in to explain and circumvent their occurrence. We have studied films grown by dc-magnetron sputtering at varying temperatures, on different substrates and to different thicknesses. Results from studies of the film surfaces by SEM and film microstructure by X-TEM, as well as measurements of the dc critical current density and the high frequency loss in the films are presented in this paper.

2. EXPERIMENTAL DETAILS

The films were prepared by dc-magnetron sputtering from a stoichiometric $YBa_2Cu_3O_x$ compound target in an argon/oxygen gas discharge. The substrates were mounted on a stainless steel block which was heated indirectly by a platinum strip heater. Temperatures quoted were measured by a thermocouple mounted at the stainless steel block. Possible effects of electron bombardment of the films during growth were minimized by keeping the substrate assembly electrically insulated (floating). The substrate surface was perpendicular to the target surface and located just outside the intensely glowing plasma region close to the target. The deposition parameters are summarized in Table 1.

Table 1

Sputtering parameters

Constant Current	200 mA
Cathode Voltage	100-120 V
Argon Pressure	19.0 Pa (142 mTorr)
Oxygen Pressure	3.0 Pa (22 mTorr)
Substrate Bias	Floating
Heater Temperature	540-780 °C
Substrates	1×1 cm $SrTiO_3$ or $LaAlO_3$

3. FILM CRYSTALLINITY AND SURFACE ROUGHNESS

At substrate temperatures which are far below the melting or formation temperature of a deposited compound, the *as grown* films will be amorphous. At higher temperatures the film will crystallize during growth and the structure will be more ordered. This is due to higher mobility of adsorbed atoms, and to annealing of defects in the grown layers. If the temperature is too high, the substrate/film reactions and interdiffusion will affect final film quality. The structure may also deteriorate due to melting and decomposition. For the growth of $YBa_2Cu_3O_x$, it is also necessary to keep a partial pressure of oxygen during deposition and cool-down, according to the requirements of thermodynamical stability of this phase [4]. In order to investigate how the substrate temperature affects the film microstructure, a series of depositions were made on $SrTiO_3$ single crystal (100) substrates at temperatures from 540 °C to 780 °C. X-ray diffraction show that all films are crystalline and oriented with the c-axis along the substrate normal. The films have considerable surface roughness, as can be seen in the

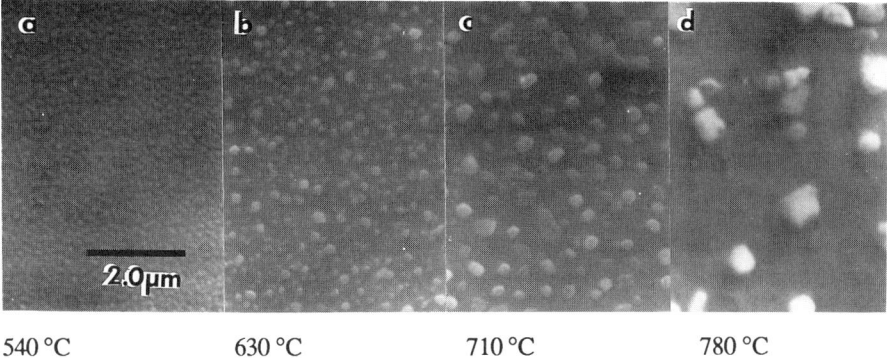

540 °C   630 °C   710 °C   780 °C

Figure 1. A sequence of secondary electron micrographs, showing the surface of 20 nm thick YBa₂Cu₃Oₓ films deposited on SrTiO₃ at temperatures from 540 °C to 780 °C.

secondary electron micrographs in Figure 1. A flat film surface is covered by large particles that grow larger at higher temperature. Their volume per unit area is constant in the temperature interval investigated. Figure 2 shows an example of a typical selected area channelling pattern from one of the films, taken from the flat region between the particles. The first order lines show that the films are well oriented but the absence of higher order lines indicate imperfect crystalline structure. By probing the whole surface of all films it was found that the films

Figure 2. A selected area channelling pattern obtained from the flat area between the particles. It has the [001] symmetry of YBa₂Cu₃Oₓ.

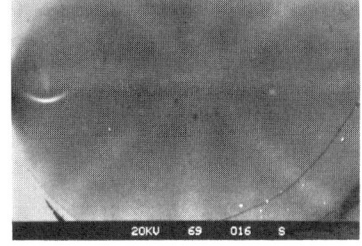

generally are single crystalline on both SrTiO₃ and LaAlO₃ substrates. A possible explanation of the film morphology is that (at least)two phases existed in thermodynamical equilibrium during film growth, and that their intermixing was negligible. Figure 3 a-b are cross section micrographs showing typical large particles on the flat epitaxial films. It is clear from the high resolution micrograph in Figure 3 c, that this particle has nucleated on the film and then continued to grow from this point. The contact point of the particles to the film surface is a misaligned small YBa₂Cu₃Oₓ grain. Thus it is weakly bonded to the film, and the largest particles could be removed by scrubbing. The nature of these outgrowths seem to be rather independent of growth temperature. Their size increases with growth temperature but the volume of the particles is approximately constant. Their appearance agrees with that previously observed for copper oxide, grown due to excess copper during growth [5].

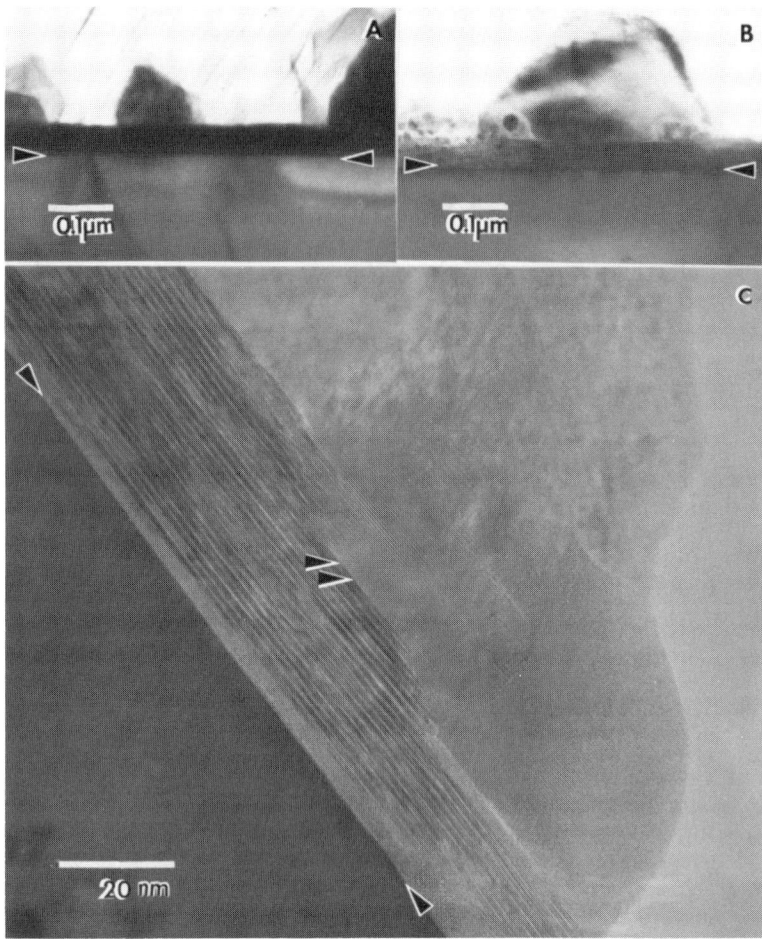

Figure 3. Cross section micrographs showing particles on the film surfaces. a) film on SrTiO$_3$ substrate, b) film on LaAlO$_3$ substrate, c) high resolution micrograph showing the nucleus of a particle. The nucleus is an a-axis oriented 1-2-3-grain (marked by double arrows) in the film . The film was grown on LaAlO$_3$ at 710 °C. The substrate/film interfaces are indicated by arrows.

The films in the temperature series are only 22-26 nm thick, so the particles on the epitaxial film surface has very large volume compared to the volume of the film. From the area density and size of the boulders it was estimated that their volume is more than 50 % of the total film volume. Energy dispersive X-ray analysis show that they are mainly constituted of copper. The natural assumption is that the flux of atoms to the substrate during growth was strongly overstoichiometric in copper. From analyzing a thicker, amorphous film deposited on a cold substrate it is, however, clear that this can not account for more than a small fraction of the copper rich particles. The copper metal fraction in the "cold" film was measured to 56 %, whereas the copper content of the films grown at elevated temperature is roughly 80 %, assuming particles are CuO.

4. ELECTRICAL PROPERTIES

The superconducting properties of the films show only weak dependence of deposition temperature, with exception for the samples grown at the two lowest temperatures, which are insulating and semiconducting, respectively. In the temperature range 680-780 °C, the films all have sharp resistive transitions above 86 K.

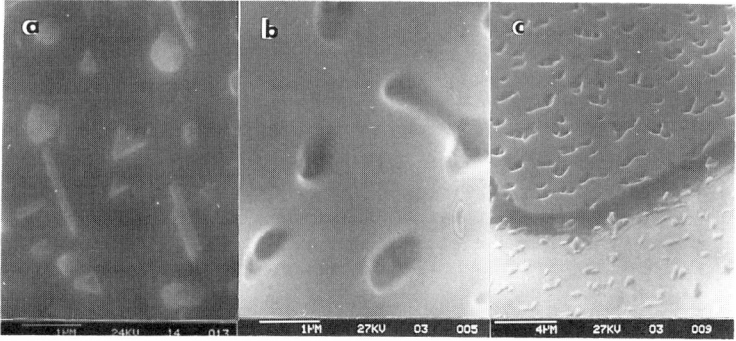

Figure 4. Micrographs showing the surface of one 20 nm and one 150 nm thick film on LaAlO₃ (a and b respectively). The surface of the thicker film is shown after patterning (c). The darker region is the unetched film.

To determine the high frequency loss of the films a coplanar waveguide resonator circuit was designed [6]. Films on LaAlO₃ were patterned and fitted in a closed cycle cryostate. Two of the films are shown in Figure 4 a-b, one 20 nm and one 150 nm in thickness. Both films were grown under comparable conditions. The flat surface of the thicker film is single crystalline, c-axis oriented film. The reason for the different morphology of thicker films is under investigation. The surface of the patterned 150 nm film is shown in Figure 4 c. By using the measured Q-value of the resonators the surface resistance of the films could be calculated. The effective surface resistance of the film in Figure 4 b, is 240 $\mu\Omega$ at 6 GHz, and 77 K. The resistance levels out at around 60 $\mu\Omega$ below 60 K. In agreement with recent results by Young et al. [7], we found a correlation between dc critical current density and the microwave surface resistance at 77 K. This is illustrated in Fig 5, for films grown on LaAlO₃ substrates.

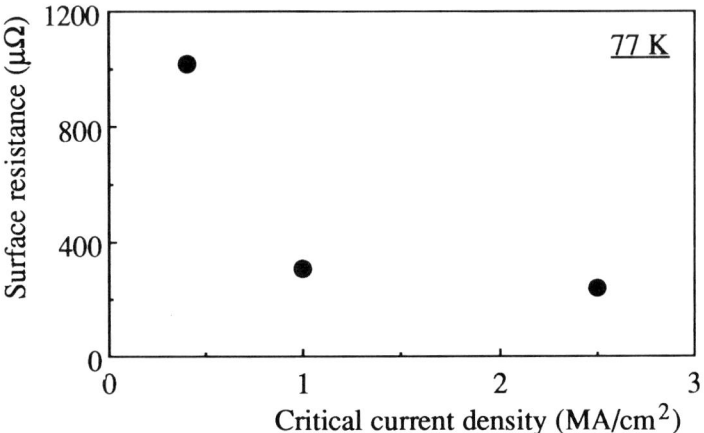

Figure 5. Effective 6 GHz surface resistance and critical current density for three films on LaAlO$_3$ substrates. The films have thicknesses 150-300 nm.

5. SUMMARY

In summary we have used dc magnetron sputtering to grow thin superconductive YBa$_2$Cu$_3$O$_x$ films that are well oriented single crystals. The resistive transition and the surface morphology of the films is insensitive to variations in deposition temperature. The surface of all films is littered with large copper rich particles, possibly CuO, nucleated at misoriented grains in the film. This cannot be explained simply by a non-stoichiometric flow of atoms to the substrate. In order to get the whole picture models accounting for, e.g., desorbtion phenomena has to be included. The film microstructure is far from perfect, still the dc critical current density is high and the microwave loss is low. With the optimization of crystalline quality even better high frequency properties should follow.

6. REFERENCES

[1] See, for example; S.S. Laderman, R.C. Taber, R.D. Jacowitz, J.L. Moll, C.B. Eom, T.L. Hylton, A.F. Marshall, T.H. Geballe, and M.R. Beasley, Phys. Rev. B **43**, 2922 (1991)

[2] M.G. Norton, L.A. Tietz, S.R. Summerfelt, and C.B. Carter, Accepted for publication in Appl. Phys. Lett.

[3] R. Ramesh, A. Inam, D.M. Hwang, T.D. Sands, C.C. Chang, and D.L. Hart, Appl. Phys. Lett. **58**, 14 (1991)

[4] R. Bormann and J. Nölting, Appl. Phys. Lett. **54**, 2148 (1989)

[5] J.A. Edwards, N.G. Chew, S.W. Goodyear, J.S. Satchell, S.E. Blenkinsop, and R.G. Humphreys, Proceedings from the E-MRS spring meeting May 29- June 1, 1990

[6] A.A. Valenzuela, B. Daalmans, and B. Roas, Electronics Lett. **25**, 1436 (1989)

[7] K.H. Young, G.V. Negrete, R.B. Hammond, A. Inam, R. Ramesh, D.L. Hart, and Y. Yonezawa, Appl. Phys. Lett. **58**, 1789 (1991)

High T_c Superconductor Thin Films
L. Correra (Editor)
© 1992 Elsevier Science Publishers B.V. All rights reserved.

Ellipsometrically measured infrared properties of sintered $YBa_2Cu_4O_8$-samples

E. Wold[a], J. Bremer[a], R. G. Buckley[b], O. Hunderi[a], F. Kong[c] and M.P. Staines[b]

[a]Department of Physics and Mathematics, Norwegian Institute of Technology, N-7030 Trondheim, Norway

[b]DSIR Physical Sciences, PO Box 31313, Lower Hutt, New Zealand

[c]Shanghai Jiao Tong University, Shanghai 200030, China.

Abstract

We report on the determination of infrared-optical properties for La-substituted bulk YBCO samples ($YBa_2Cu_4O_8$ and $YBa_{2-x}La_xCu_4O_8$). The real and imaginary part of the permittivity have been measured by a high-resolution ellipsometric method developed at our laboratory. A computer-controlled system of step-wise rotating polarizers and analyzers enables the polarization state of the reflected beam to be determined with high precision. Unphysical higher-harmonics are removed and physical meaningful phases and amplitudes are extracted. Reflectivity-coefficients at normal-incidence and 45° have been measured by two independent methods and corrected for diffuse scattering. Comparison between reflectivity and ellipsometry results have been made and show good agreement.

1. INTRODUCTION

Far infrared properties of high-T_c materials have been studied by numerous groups (see for example [1]) since their first discovery in 1987. Typically, the sintered surface is illuminated by unpolarized IR radiation with subsequent normalization against a gold sample. In order to correct for diffuse scattering, however, it has turned out that it is better to deposit a thin gold layer directly on the original sample surface. As a last step the dynamical conductivity can be obtained by Kramers Kronig-transforming the reflectivity data. This spectroscopic method has furnished valuable data on such topics as energy gaps, carrier concentration and phonon mode softening.

The above experimental procedure is unsatisfactory for a variety of reasons. First of all, it is unclear whether the method fully corrects for effects due to diffuse scattering. Secondly, in the case of $YBa_2C_3O_7$, for example, there are broad absorption bands at 8 eV and 15 ev [1]. The high oscillator-strengths of these lines are important for the absolute levels of the permittivity in the infrared region. Similarly, the Hagen-Rubens extrapolations towards low energies is crucial for the end result, but its validity has not been tested.

Very recently, the chain and plane contribution to the infrared conductivity of single crystalline $YBa_2Cu_3O_7$ was separatedly measured through the use of polarized reflectivity measurements [2]. The technique of ellipsometry, however, is always the natural choice when measuring optical constants. The reason is that both the amplitude ratio between the two polarization components as well as their phase difference are uniquely obtained. Until recently the method has been limited to the visible and near infrared range. Recenly, improvements have been made in the technique of Fourier transform IR ellipsometry [3,4]. Apart from the study of superlattices, thin films, surfaces and interfaces this non-destructive method holds great promise for the characterization of bulk matter like superconducting materials.

2. EXPERIMENTAL

The samples were prepared from a stoichiometric mix of oxides and nitrates as described elsewhere [5]. The well ground powder mix is held at approximately 700°C for 30 minutes in air to melt and decompose the nitrates. This is followed by further grinding and then pressed pellets are reacted at 940°C in 8 MPa of oxygen over night. This later step is repeated three times at which point XRD pure material results. After initial-measurements the samples, 10 mm in diameter, were surface treated with 4000 grinding paper.

The optical layout of the infrared ellipsometer [3] is briefly as follows: A commercial interferometer (Perkin Elmer 1760) is equipped with a baseplate that supports holders for IR polarizers, sample, step-motors and detectors. This instrument can be used in two modes. During the ellipsometric measurements of the YBCO samples the angle of incidence was kept at 70°. However, the instrument can also do conventional polarization-sensitive reflection measurements. Reflectivity coefficients R_s and R_p were measured at 45° and normalized against gold film data.

Normal-incidence reflectivity spectra of ceramic samples between 10 and 3000 cm^{-1} were collected on a Bomem DA3 FTIR-instrument. The resolution is 1 cm^{-1}. A number of detector and beam splitter combinations have been employed over this spectral range so that the signal-to-noise varies. To minimize the effect of light scattering from these samples the sample spectra have been normalized by spectra measured after the samples were gold coated.

3. RESULTS

3.1 Data analysis

In the absence of depolarization and diffuse scattering the polarized reflected intensity is given by $I=1+\alpha\cos(2\phi)+\beta\sin(2\phi)$. During the measurements, we have kept the analyzer at 45.0° when rotating the polarizer to the angle ϕ, and vice versa. The constants α and β contain sufficient information for the evaluation of amplitude and phase for the reflected wave, and the dielectric constant $\varepsilon=\varepsilon_1+i\varepsilon_2$ can therefore be obtained directly without Kramers Kronig-transforms. Experimentally, however, the intensity is affected also by such factors as interferometric polarization, dichroism in the detectors, and the IR-optical properties of the polarizers. As shown elsewhere [3], these factors are crucially important in IR ellipsometry. However, they can be effectively dealt with by writing I'=ID and Fourier-expanding I'(ϕ) and D(ϕ). Here, D(ϕ) is the azimuthal Fourier-signal involving the three optical effects mentioned above. It can be shown that $\alpha'=(\alpha+a)/(1+(\alpha a+\beta b)/2)$ and $\beta'=(\beta+b)/(1+(\alpha a+\beta b)/2)$. We eliminate the relative strong second order harmonics a and b of D(ϕ) in these equations through a numerical procedure. The basis for the method [3] is the experimental observation that coefficients higher than fourth order are typically three magnitudes lower than the zero'th order term.

3.2 Discussion

In general, our initial experiments have demonstrated that it is difficult to obtain full agreement with traditional normal-incidence methods. As an example of a case where the discrepancy is very small we present in Fig. 1 data on the optical properties of $YBa_{1.95}La_{0.05}Cu_4O_8$. The curves show the permittivity as function of energy. They are based on the ellipsometric measurements and the data analysis presented above. The upper, noiseless curve of Fig. 2 contains the recorded near-normal incidence spectra from an identical sample. Note that the phonon modes, clearly visible in Fig. 2, are much weaker in Fig. 1. This is mainly due to the noise level of the ellipsometric measurements and the limited number of scans for these measurements.

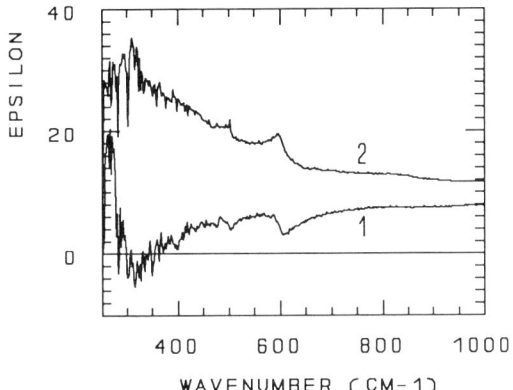

Fig. 1. Ellipsometry measurements of the permittivity for $YBa_{1.95}La_{0.05}Cu_4O_8$ as function of energy (1:real, 2:imaginary part). The noise in the spectra is due to low energy in the far infrared for these recordings.

One interesting question is whether the data in Fig. 1 are capable of reproducing the experimental reflection curves. The agreement should exist at all angles of incidence. Fig 2. contains also our calculated curve. It can be seen that the two sets of data very nearly overlap. For a variety of reasons this is not to be expected in the general case. The crystalline grain sizes and the surface-corrugation are for example roughly of the same order as the wavelength in the far IR region. A straight-forward analysis shows that the effects should manifest themselves differently in the reflectivity and ellipsometer measurements.

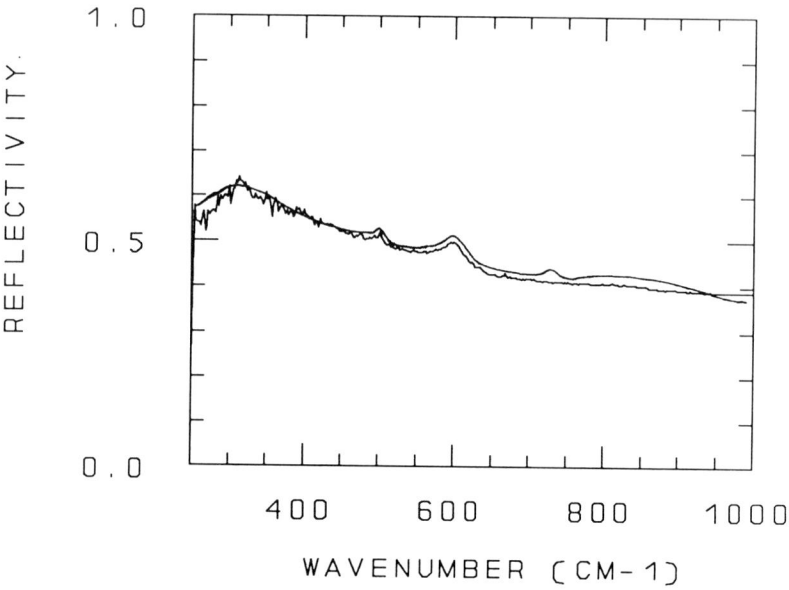

Fig. 2. Normal reflectivity for $YBa_{1.95}La_{0.05}Cu_4O_8$ obtained by direct measurement (upper curve) and by calculation from the ellipsometrically measured dielectric function (lower curve).

For example, depolarization effects can in principle perturb ellipsometric data while experimental reflections coefficients are left unaffected. In order to clarify this matter we have measured the reflected intensities with and without analyzer. Fig. 3 shows the experimentally measured reflection coefficients R_s, $R_{s,a}$ (upper curves) and R_p, $R_{p,a}$ (overlapping lower curves) for the sample. (The subscript 'a' denotes the presence of an analyzer during the measurements). All curves are normalized against similarly recorded gold data. Within the experimental error there is no significant change when removing the analyzer. Hence, it can be concluded that depolarization effects are weak for this particular high-T_c surface at 45°. Another important issue is the problem of diffuse scattering. In contrast to the depolarization effects discussed above these are probably important only when performing conventional reflectivity measurements. Thus, in the normal-incidence experiments the data were normalized against the gold-coated sample. If the fraction d of scattered photons is sufficiently large the most conspicous effect will be to reduce the intensity in the specularly reflected beam with a factor roughly equal to 1-d. It is important to verify that this effect can be neglected in the ellipsometry measurements. One way of looking for angular and energy dependent scattering effects is to set the angle of incidence to 45.0° and check whether the expression $R_s^2-R_p$ is non-zero. At this angle of incidence a reflected beam should give $R_s^2=R_p$ if there are no scattering effects. Our raw-data for this configuration is shown in Fig. 4. The curve is negative at high energies which shows that the degree of diffuse scattering become more serious with reduced wavelength/grain-size ratio.

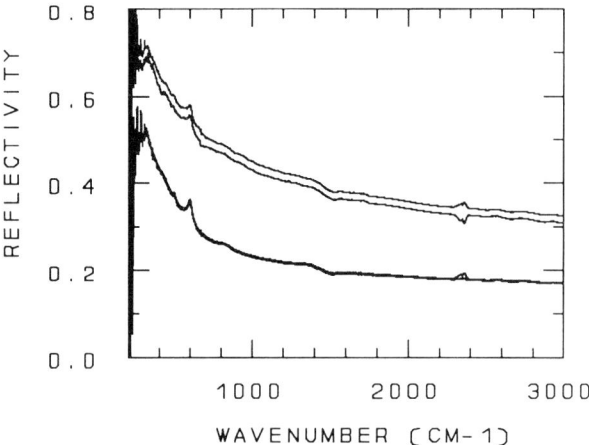

Fig. 3. Effect of depolarization on the measured reflectivities at 45°. The two sets of curves for R_s (upper curves) and R_p (overlapping lower curves) are measured with and without analyzer.

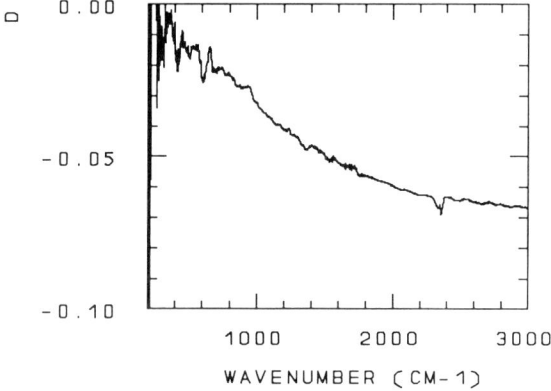

Fig. 4. The measured value of D (see text) at 45° angle of incidence against energy. The deviation from zero at longer wavelengths is mainly due to diffuse scattering.

A simplified analysis for the angular behaviour can be done as follows: Assume that the deviation from zero is D at a certain wave-number. We can then put $R_s^2(1-d)^2-R_p(1-d)=D$. Solving for d the amount of diffuse scattering can be followed as function of energy at 45°. The two lower curves of the two data sets in Fig. 5 shows R_s and R_p as measured without the assistance of any ellipsometric techniques. The two upper curves of the same sets are based on the ellipsometry data alone and are believed to be reliable. Knowing the values of d allows for correction of the lower curves. The curves in the middle are very close to the ellipsometry data and shows that they are consistent also at this angle.

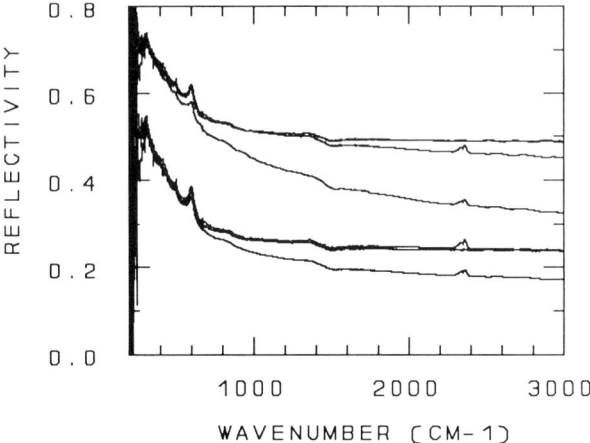

Fig. 5. Reflection coefficients R_s (upper triplet) and R_p (lower triplet) at 45°
angle of incidence. See text.

4.CONCLUSIONS

We have studied the optical properties of sintered $YBa_{1.95}La_{0.05}Cu_4O_8$ samples by means of IR
ellipsometry. It is demonstrated that the method enables both the real and imaginary component
of the permittivity to be obtained directly. Kramers-Kronig transformations and correction for
depolarization and diffuse scattering are not necessary. It is concluded that IR-ellipsometry
enables the important optical constants of ceramic high-T_c surfaces to be determined with high
precision.

5. REFERENCES

1   Z. Schlesinger, R.T. Collins, F. Holtzberg, C. Feild, G. Koren and A. Gupta,
      Phys. Rev. B, 41 (1990) 11237.

2   Z. Schlesinger, R.T. Collins, F. Holtzberg, C. Feild, S.H. Blanton, U. Welp,
      G.W. Crabtree and Y. Fang, Phys. Rev. Lett., 65 (1990) 801.

3   J. Bremer, F. Kong, O. Hunderi, T. Skauli and E. Wold, J. Appl. Opt.(submitted).

4   J. Bremer, O. Hunderi and Kong Fanping, Mat. Science and Eng., B5 (1990) 285.

5   R.G. Buckley, D.M. Pooke, J.L. Tallon, M.R. Presland, N.E. Flower, M.P. Staines,
      H.L. Johnson, M. Meylan, G. Williams and M. Bowden, Physica, C174 (1991) 383.

High T_c Superconductor Thin Films
L. Correra (Editor)
© 1992 Elsevier Science Publishers B.V. All rights reserved.

Non-destructive measurement of the critical current and the current-carrying length scale in superconducting films

M.A. Angadi[a], R. M. Bowman[b], A.D. Caplin[a], J.R. Laverty[a], A.L. de Oliveira[a] and C.M. Pegrum[b]

[a]   Centre for High Temperature Superconductivity, Blackett Laboratory, Imperial College, London SW7 2BZ, U.K.

[b]   Dept of Physics and Applied Physics, University of Strathclyde, Glasgow G4 0NG, U.K.

Abstract

Measurements of the irreversible magnetisation of superconducting thin films are often used to obtain the critical current density J_c, but the analysis depends on the assumption that the screening current flows around the entire sample. We show that, by examining a portion of the magnetic hysteresis loop in detail, *independent* measurements can be made of J_c and the current-carrying length scale Λ, and the presence of 'weak-links' within the film can be detected.

We report results on a laser-ablated highly-textured $YBa_2Cu_3O_7$ thin film at fields up to 8 T; these show that at temperatures below about 40 K and in not too high fields, Λ is indeed the sample dimension. As the temperature or field increases, the current pattern fragments, indicating the presence of field- and temperature-sensitive weak-links in the sample. However, these weak-links are much more robust than those associated with fabricated moderate angle grain boundaries in thin films, or those that occur inevitably in polycrystalline material.

1. INTRODUCTION

The general thrust of work on high temperature superconducting (HTS) films, particularly as far as applications are concerned, must be toward the reliable fabrication of films with high critical current densities. There has been considerable success along these lines with thin films of $YBa_2Cu_3O_7$. Although these films are usually highly textured, almost always with the c-axis perpendicular to the substrate, the in-plane alignment is much more difficult to control. Grain boundaries between crystallites, even between those with a common c-axis, degrade J_c severely when the misorientation angle exceeds a few degrees [1]. Weak magnetic fields, on the order of mT, depress J_c even further.

Consequently, characterisation of J_c in superconducting films, as a function of field and temperature, plays a central role, both in understanding the physics of flux pinning and granularity, and also as the most significant measure of film quality.

Transport measurements of J_c require the film to be patterned with a fairly narrow bridge, so that current levels can be kept to a reasonable level. Consequently, they are slow to perform, and not only may the patterning of the film render it unusable for other purposes, but there is always the concern that some aspect of the patterning process may have damaged the film. Also, the minimum detectable voltage criterion that is readily attainable corresponds to an electric field of order 10^{-4} V m^{-1}. For a film with J_c of 10^{11} A m^{-2}, this represents a resistivity of 10^{-15} ohm m, 10^7 times less than that of copper at room temperature, but still large enough that currents circulating in a ring of this material a few mm across would decay with a time constant measured in minutes. Such a level of dissipation is too high to be tolerable in, for example, a flux transformer.

The other common approach to measurement of J_c is to use the irreversible magnetisation ΔM of the film, as measured with a magnetometer. If the current can be assumed to be flowing uniformly through the sample, say a disc of radius R, then $J_c = 3 \Delta M / R$ (the Bean relationship, here expressed in SI units). Magnetic measurements require no patterning, and, because they can monitor the decay of circulating currents over long periods, are sensitive to extremely low levels of dissipation. However, this translation from ΔM to J_c requires confidence that the screening current is actually circulating on the scale R. If, because of grain boundary (or other) weak-links within the film, the current is confined within islands of some smaller scale Λ, naive use of the Bean relation will give an underestimate of J_c.

Recently, we have shown [2] that magnetic measurements of samples with large aspect ratio (i.e., those having a demagnetising factor close to unity) can be used to obtain *independent* estimates of Λ and J_c: A standard magnetic moment versus field hysteresis loop is performed. Then, under quite general conditions, the slope of the magnetization curve as the applied field is first reduced from its maximum value H_m is a direct measure of Λ:

$$\mathbf{dm/dH = f \; A \; \pi \; \Lambda \; / \; \Theta} \tag{1}$$

where m is the magnetic moment of the film. A is the film area and Θ a dimensionless parameter, typically 10, that depends logarithmically on the ratio of Λ to film thickness t; f is the fraction of sample area occupied by superconducting islands, and in a reasonably good film can be put equal to unity. Note that equation 1 defines Λ in terms of quantities that are measured directly.

As the magnetic field is further decreased, m approaches exponentially an asymptotic value m_∞:

$$\mathbf{m = m_\infty \; (1 - exp[(H - H_m)/\Gamma])} \tag{2}$$

where Γ is a scale field decrement equal to $2J_c t \, \Theta/3\pi$. This steep 'reverse leg' section of the magnetization loop corresponds to the macroscopic screening currents that were circulating at the field H_m having been reversed completely after a field decrement a few times Γ. These results hold provided that $H \gg \Gamma$.

m_∞ is related to J_c by:

$$\mathbf{J_c t = 3\pi \; m_\infty \; /2\Theta \, (dm/dH)} \tag{3}$$

In this paper we report detailed magnetic measurements on a laser-ablated $YBa_2Cu_3O_7$ film, illustrating how this analysis can be used to follow the evolution of granularity and weak-link behaviour as the field and temperature increase.

2. EXPERIMENTAL

The 200 nm thick $YBa_2Cu_3O_7$ film was laser-ablated from a stoichiometric target on to a MgO substrate, which had been covered with a 100 nm buffer layer of $SrTiO_3$. The film was oxygenated by cooling from 750 C in an oxygen pressure of 400 Torr. It is strongly textured with its c-axis normal to the substrate, and similar films have been shown to have a high degree of in-plane epitaxy. Details of film preparation and structural characterisation are given elsewhere [3].

For the magnetic measurements, a piece of the film approximately 6.1 mm x 6.2 mm was mounted with the field normal to the plane in a commercial magnetometer. A preliminary flux exclusion measurement, made by cooling in zero field, followed by warming toward the transition in a field of 5 mT, established T_c as 90±2 K, where the uncertainty is primarily that of the temperature difference between thermometer and sample.

3. RESULTS AND DISCUSSION

Figure 1 provides an overview of the magnetic behaviour at 20 K in a hysteresis loop extending out to 8 T. Even at a critical current density of 10^{11} A m^{-2}, the scale field Γ for a film of 200 nm thickness is about $4 \cdot 10^4$ A m^{-1} ($\equiv 50$ mT), so that the behaviour described in Section 1 should be observable for only rather small field decrements from the maximum field on the loop.

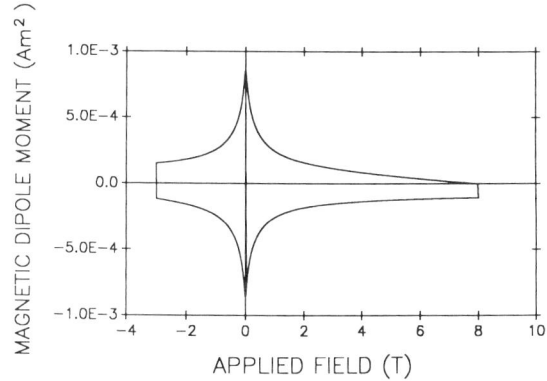

Figure 1. Overall magnetisation loop up to 8 T of $YBa_2Cu_3O_7$ thin film at 20 K.

Details of a series of hysteresis loops at 20 K are shown in Figure 2; in order for the 'reverse legs' to be traced properly, the field steps are 1 mT or less. For fields up to 1 T, the magnetisation follows closely the form indicated by equation 2, allowing the length scale Λ and J_c to be obtained directly. The data for the lowest field should be treated with some caution, as the applied field is not well into the limit H $\gg$ Γ. At fields above 1 T, a software constraint on the magnetometer imposes a minimum field step of 10 mT, which, because of the reduction of J_c with field, makes the steps comparable with Γ. Consequently, there is insufficient field resolution to check whether the magnetisation follows equation 2. An approximate estimate of Λ can still be obtained, but not a meaningful value of J_c.

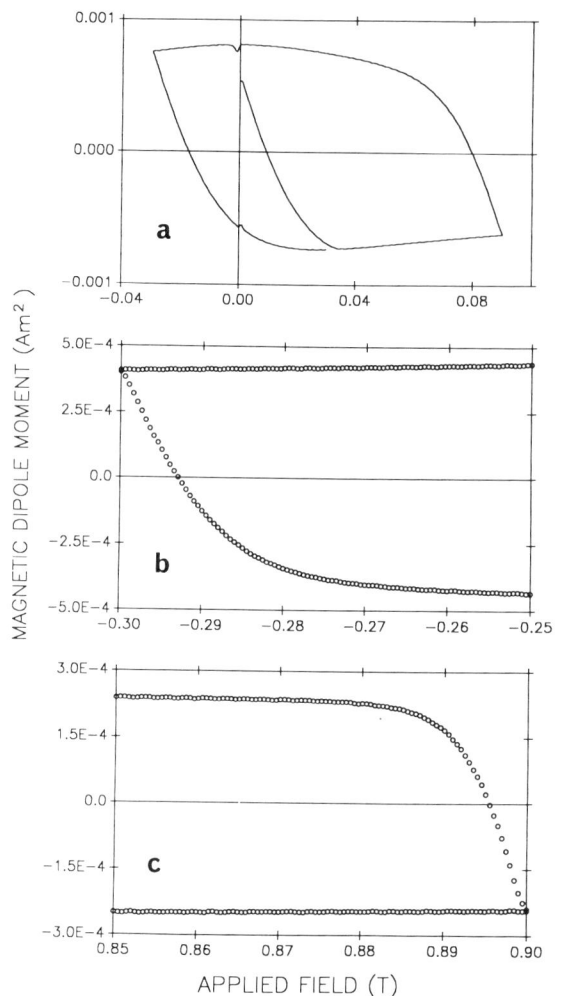

The values of Λ at low fields inferred from equation 1 are all close to 7 mm, about 30% *larger* than the semi-diagonal dimension of the sample, but this discrepancy is an instrumental artefact, caused by the large size of the sample compared with the magnetometer sense coils (the instrument is calibrated correctly for small samples, but the magnetic coupling between off-axis regions of the sample and the sense coils increases quite rapidly with radius. Analysis of the reverse leg magnetisation of small samples gives values of Λ consistent with sample dimensions). We can conclude that under these conditions, the screening currents do indeed circulate around the macroscopic periphery of the sample.

At higher temperatures, a deviation from equation 2 appears: the reverse leg magnetisation acquires a sigmoid shape (Figure 3), with an initial slope that is a factor of 3 or 4 times less than the maximum value. We interpret this behaviour as arising from the first flux penetration being into small regions of the film, perhaps those in the corners. With further increase in field or temperature, the length scale continues to diminish (Figure 4). We interpret the reduced length scale as indicating the presence of weak-links within the film, although they are weak links that are orders of magnitude stronger than those that occur at moderate angle grain boundaries in thin films [1], or in polycrystalline materials. These rather robust weak-links survive to fields of order 1 T. Possibly they arise from the small angle in-plane grain boundaries that must be present in this film.

Figure 2. Details of the 'reverse legs' of the hysteresis loops at 20 K for field amplitudes of (a) 30 and 90 mT; (b) 0.3 T; (c) 0.9 T.

As long as the islands are not much smaller than the film dimension, equation 3 can be used to estimate J_c; however, with many small islands, spanning a range of scales, it is not clear that equation 3 is applicable. The critical current density J_c, obtained in the regime of field and temperature where the measured magnetisation follows the behaviour described in Section 1, is shown in Figure 5. The gradual decrease with increasing temperature or increasing field is qualitatively in accord with the accepted behaviour of J_c in these materials.

4. CONCLUSIONS

These results demonstrate that much detailed information about the critical current behaviour of thin films can be derived from analysis of the reverse leg magnetisation. They show also that measurement of the irreversible magnetisation, with translation into a critical current density using the *sample* dimension in the Bean formula, can under some circumstances give a misleadingly low estimate of J_c.

We have shown that in a high quality thin film at low temperatures and at not too high fields, the screening current distribution is well-described by a simple Bean model, in which these currents circulate around the macroscopic sample.

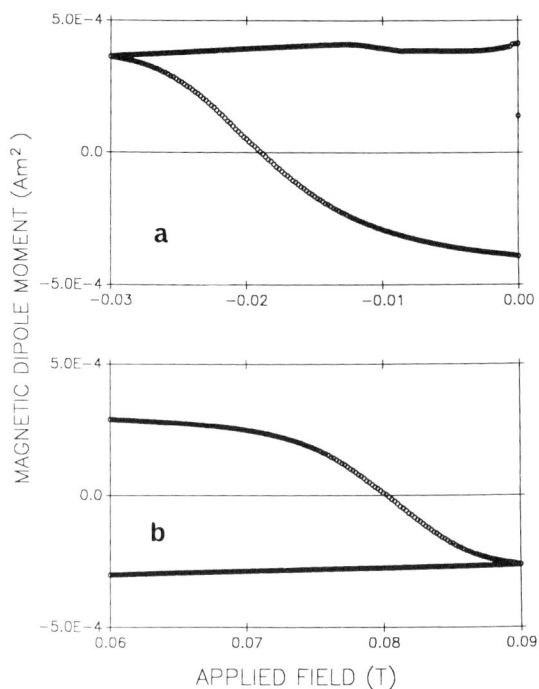

Figure 3. Reverse legs of magnetisation loops at 40 K, (a) 30 mT; (b) 90 mT.

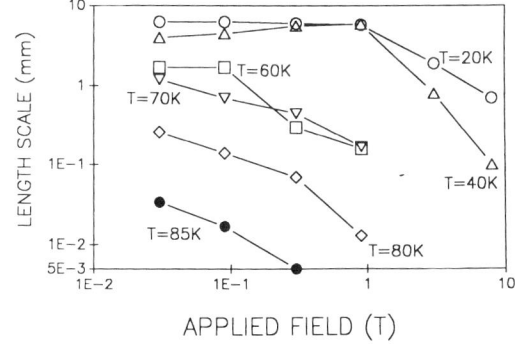

Figure 4. Length scale Λ of current flow, as derived from equation 1.

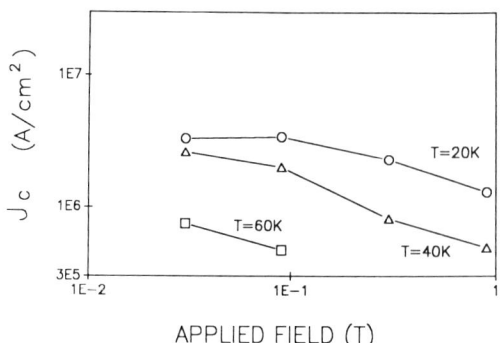

Figure 5. Critical current density as a function of field and temperature, as obtained from equation 3.

At temperatures of order half T_c, the pattern of current flow begins to break up, with some smaller islands, probably at the edge of the film, becoming effectively decoupled from the main body of the sample; this fragmentation increases steadily as the temperature or the applied field is raised further. Apparently, this film contains some 'weak links', but they are robust by comparison with the usual weak-links in HTS materials.

The reduction in irreversible magnetisation with increasing field that is always seen, as in figure 1, should not be attributed solely to a decrease in J_c at high fields, but in fact reflects also some fragmentation of the screening current pattern by weak links. Consequently, previous estimates of the suppression of J_c in thin films by large fields have almost certainly been unduly pessimistic.

5. REFERENCES

1   D. Dimos, P. Chaudhari and J. Mannhart, Phys. Rev. B 41 (1990) 4038.

2   M.A. Angadi, A.D. Caplin, J.R. Laverty and Z.X. Shen, Physica C (in press).

3   R M Bowman and C M Pegrum (to be published).

Part II
Film Growth & Processing

High T$_c$ Superconductor Thin Films
L. Correra (Editor)
239

LASER DEPOSITION OF HTSC FILMS ON DIFFERENT SUBSTRATES

A.I.Usoskin and I.N.Chukanova

Institute for Single Crystals, Kharkov, 310141, USSR

Abstract
    Different aspects of the problem of laser pulse deposition
of HTSC films on crystalline substrates, i. e. dependence of
structural properties of the film on geometrical conditions of
deposition process, thermal expansion effects, influence of
substrate imperfections, have been investigated.

1. INTRODUCTION

    Nowadays laser vacuum deposition presents the most
widespread experimental technique allowing to obtain high
temperature superconducting (HTSC) films with extremely high
transition parameters [1-3]. The obvious advantage of this
method seems to be the simplicity of reaching the identical
stoichiometric composition of the film with the composition of
the target. However there are considerable obstacles when films
with good surface uniformity, stable parameters and high
critical current density, i.e. films for technical applications
have to be obtained. Certain additional problems arise while it
is necessary to prepare best quality films on different
crystalline substrates such as sapphire, silicon etc. [4,5]. It
is apparent that the solution of these problems determines the
answer to the basic question whether there is any future for
the method of laser pulse deposition or it will be ousted by
other methods.
    In the present paper we have tried to study different
aspects of the problem of laser pulse deposition of HTSC films
on various crystalline substrates and to analyze those growing
mechanisms which lead to the limitation of film quality and
therefore could be used for reaching the superior level of
technology.

2. EXPERIMENTAL TECHNIQUE

    Experimental installation including a laser, target motion
device, substrate heater as well as devices for substrate

temperature control and for maintenance of constant oxygen pressure has been used for film obtaining. 1.06 μm IAG laser with pulse frequency repetition of 25 Hz, pulse duration of 10 ns and pulse energy of 0.4 J has been applied for ablation procedure. Film deposition on the ~ 800°C - heated substrate has been conducted under ~ 0.1 Torr oxygen pressure. More precise choice of substrate temperature depends on the type of substrate. For SrTiO$_3$ substrates with thickness of 0.5 mm this temperature at the initial stage of deposition was about 840°C (temperature of contact surface of heater). One of the essential factors leading to the reliable reproducibility of specimens was the application of pyrometer temperature control of substrate, which provided achievement of proper accuracy ($\pm$5°C) in oxygen environment. The complicated target motion due to which a laser beam makes imitation of threading over the cylindrical target surface has been used to provide the ablation uniformity.

Other important factor determining the possibility of manufacturing high quality HTSC films is the correlative position of substrate and target. In conventional geometry (Fig. 1 a) HTSC films deposited on the substrate contain numerous macroinclusions (Fig. 1 b) which are formed from particles ejected from the target surface, and also seemingly from microdrops formed when the adiabatically expended laser torch is overcooled. The discrepancy of widths of angular distributions of flux density for particles and for ion-molecular flow which is shown in Fig. 1 c can be used for producing films with a reduced content of microinclusions (see Fig. 1 d). However this method is not very efficient because film thickness becomes irregular on the surface. Another way of locating the substrate (see Fig. 1 e) providing the minimal concentration of inclusions is shown in Fig. 1 f. In this case the substrate is shaded with a screen from the side of the laser spot on the target surface, though it is affected by a laser torch re-reflected within the system of screens. The availability of the residual concentration of macroinclusions (Fig. 1 f) in this geometry is direct experimental evidence of the possibility of drop condensation in the process of laser torch expansion.

For measuring HTSC film parameters the 4-probe method of controlling the temperature dependence of resistance and the inductive method of measuring the real part χ' of magnetic susceptibility in magnetic fields up to 10 Oe have been used. In studying the structure and composition of films the scanning electron microscopy, optical microscopy, electron-probe and X-ray analysis analysis have been also applied.

3. EFFECT OF THERMAL EXPANSION

As in any other epitaxial growing process the substrates for HTSC film preparation must satisfy a great number of

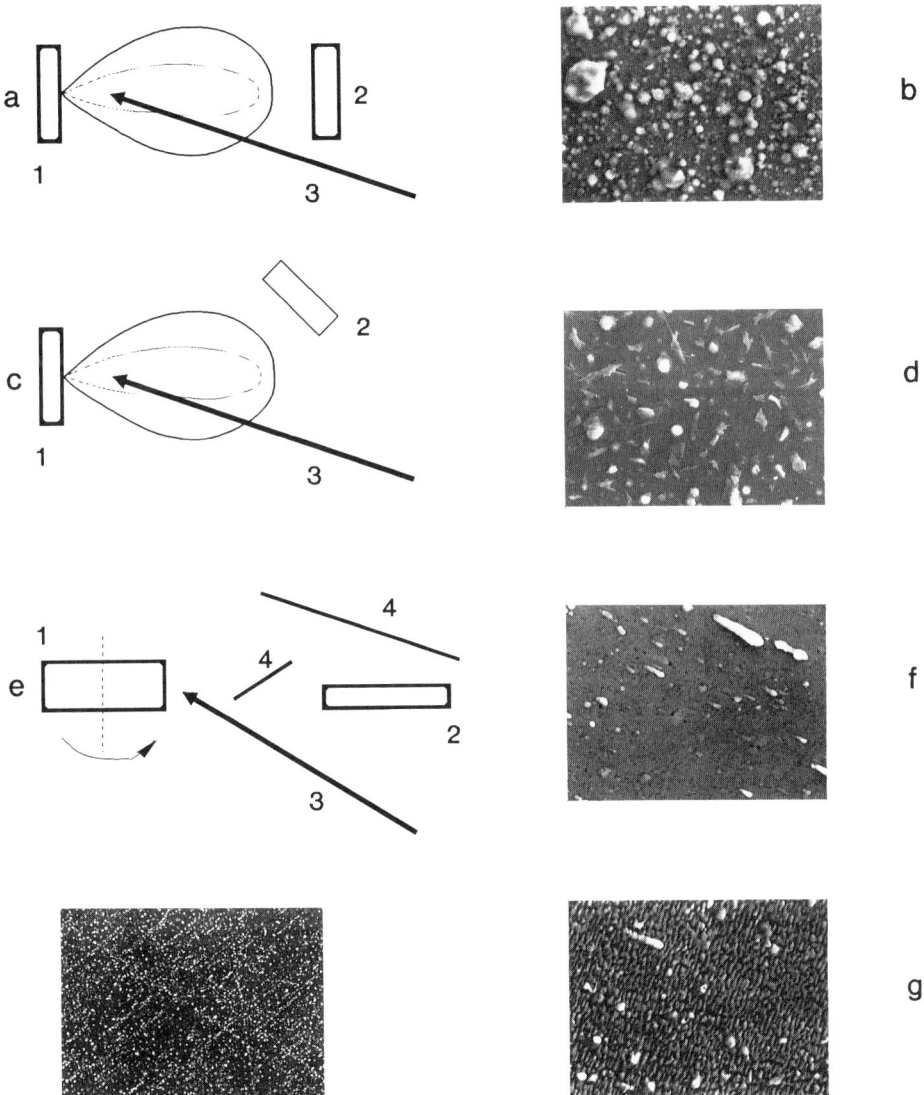

Figure 1. Schemes of target-substrate positions (a, c, e) and scanning-electron microphotographs (b, d, f, g, h) of $YBa_2Cu_3O_x$ films obtained with the help of these schemes (3000 X).
f, g - correspond to the substrate containing drop-type inclusion, f - film on inclusion-free surface, g - film on inclusion surface. 1 - target, 2 - substrate, 3 - laser beam, 4 - screens. Curves - angular distributions of flux density for ion-molecular flow (————) and for particles (— — —).

requirements, i.e. correspondence of lattice periods to those of films, high crystalline perfection, chemical inertness, etc. However, one of these numerous conditions seems to be particularly important. This is the mutual correspondence of thermal expansion coefficients of the film α_f and of the substrate α_s, deviation from which leads not only to the appearance of significant stresses within the film, but may also lead to the deterioration of contacts between HTSC crystallites when the film is cooled or it undergoes the thermal-cycling. If the crystallites of 1 m dimensions relatively weakly bound with the substrate then one can estimate the extreme deviation value $\Delta\alpha = \alpha_s - \alpha_f$: $|\Delta\alpha| < 10^{-6}$ deg^{-1}, obtained under the assumption that the characteristic dimensions of crystallites are about 1 m and distances between them must not be more than the coherent length of ~ 10 Å. In the extreme case of mosaic oriented structure, when the crystallites are in epitaxial contact with a substrate the value of $\Delta\alpha$ may be slightly higher. Temperature dependencies of $\alpha(T)$ for $YBa_2Cu_3O_x$ film [11] as well as for various crystalline substrates [12,13] are shown in Fig. 2. As follows from this figure the crystals considered above can not satisfy the condition formulated for $\Delta\alpha$. Moreover the estimations of relative deformations

$$\overline{\Delta l/l} = \int_{100\,K}^{800\,K} [\alpha_s(T) - \alpha_f(T)]\, dT \quad,$$

which the film undergoes on the substrate when the temperature changes from 800 K to 100 K, show that even the integral value of expansion coefficient difference can not satisfy the described condition which in this case acquires the form $\overline{\Delta l/l} < 10^{-3}$ (see Table 1). The exceptions present $SrTiO_3$ and sapphire,

Table 1
Average values of relative expansion

CRYSTALS	$SrTiO_3$	Sapphire	MgO	$LiNbO_3$	Si
$\overline{\Delta l/l}$, 10^{-4}	+3	-2, -11	+16	+34, -45	-44

though the first of these crystals leads to appearance of perceptible compression in the film, and the second one leads to obtaining of a stretched film. Perhaps these estimations allow to explain at least one of the reasons causing the better

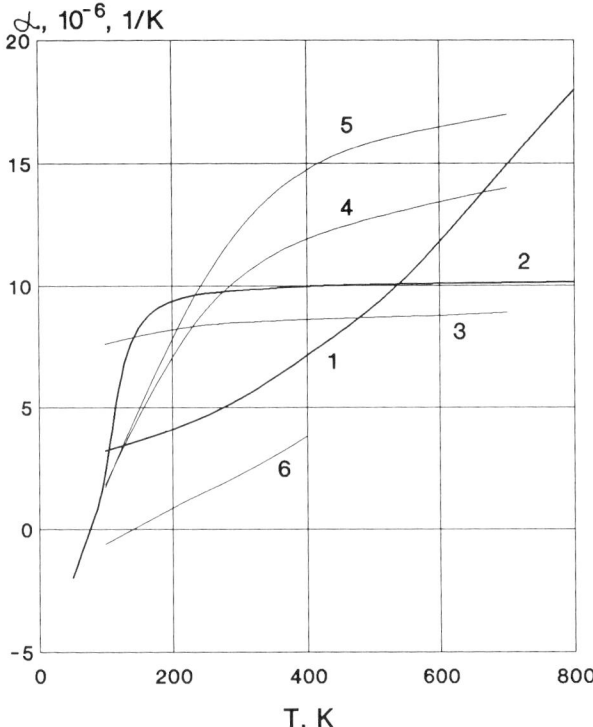

Figure 2. Temperature dependence of thermal expansion coefficient for $YBa_2Cu_3O_x$ film (1) and for crystalline substrates (2-6). 2- $SrTiO_3$, 3 - sapphire, 4 - MgO, 5 - $LiNbO_3$, 6 - Si.

quality of films deposited on $SrTiO_3$, since a certain degree of compression of the film when it is cooled should be the preferable choice for reaching the high current density.

It is important that as the limitation determined by basic property of crystalline substrates is under consideration, the possibility of obtaining of high quality HTSC films on the major of widely used substrates, which can not satisfy the $\Delta\alpha$-requirement, seems to be doubtful. It is also obvious, that the usage of thin buffer interlayers, which are practically unable to reduce such significant deformations, can not considerably simplify the solution of this problem.

One of the easily observed HTSC film defects whose appearance can be connected with mechanical stresses in the film, is macroopenings. Fig. 3 a shows the characteristic view of one of them (with ~ 40 μm diameter). Such openings are formed as a result of local film separating right after its obtaining or after the following temperature cycling. Defects of this type are always observed in $YBa_2Cu_3O_x$ films possessing high crystalline perfection. The typical distribution of dimensions of openings in epitaxial films with the surface

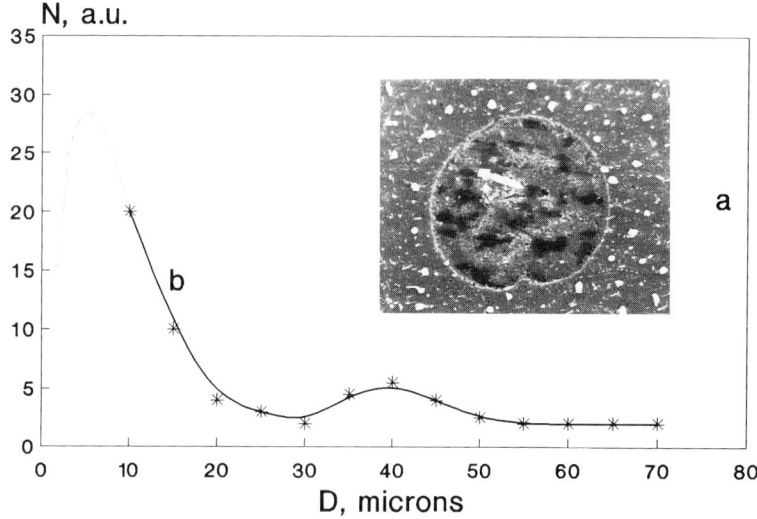

Figure 3. Microphotograph (a) and size distribution of openings (b) in $YBa_2Cu_3O_x$ film on (100) $SrTiO_3$ substrate (1500 X).

10×10 mm^2 is shown in Fig. 3 b. Such defects obviously present a significant obstacle for HTSC film technical applications. It has been found that surface concentration of openings considerably reduces if the process of film deposition is periodically interrupted with simultaneous cooling of the film. Complete elimination of the openings can be obtained when the amplitude of such thermal "modulation" exceeds ~ 300°C. This deposition procedure seems to be favourable for stepwise saturation of the film with the oxygen during the deposition pauses which lead to successive increase of film volume. The latter prevents the appearance of significant film stresses observed for conventional one-stage film growing. Moreover, thermal-cycling of this kind acts to a certain extent as a repeated annealing which also reduces the stresses within the film.

4. SUBSTRATE IMPERFECTION INFLUENCE

The influence of crystalline imperfection of the substrate on HTSC film morphology is shown in Fig. 4 f,g. In this example

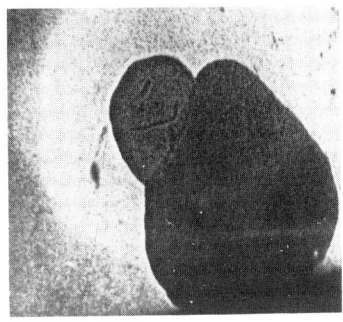

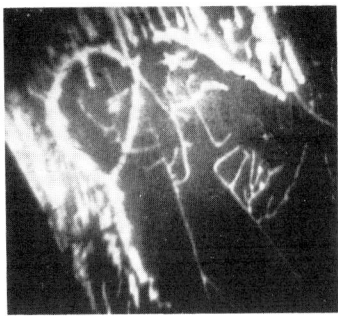

a   b

Figure 4. Optical image of $YBa_2Cu_3O_x$ film deposited on $SrTiO_3$ substrate with a drop-type inclusion (10 X). 1 - optical microscopy, 2 - magneto-optical image in reflected beam.

the film has been deposited on the substrate containing drop-type crystalline inclusions (see Fig. 4 a) which are typical for $SrTiO_3$. It can be seen (Fig. 1 g) that the film with fine granular structure forms over the crystallite having the deviation of $\delta = 10^o$ from (100) plane. At the same time the film is quite uniform over the main part of substrate with $\delta < 2^o$ (Fig. 1 f). Owing to this, the transition temperatures in these surface regions are considerably different: $T_c < 77$ K over the inclusion, and $T_c = 85$ K outside the inclusion. Similar experiments with different substrate inclusions or blocks have demonstrated a sharp structural and transition deterioration of the film when the angular misorientation of blocks increases up to $\delta > 7^o$. The film image obtained in the normal magnetic field (H = 200 Oe) with the help of transducer using magneto-optical Faraday effect is shown in Fig. 4 b. HTSC film's "weak" regions corresponding to the bright field in the photograph are very intensive along the boundary of the inclusion. This result seems to be quite natural since the structural disorder of the film as well as its maximal stresses must appear just over intercrystallite boundary.

The influence of surface scratches which are typical defects of substrates is shown in Fig. 1 h. Fine HTSC crystallites "decorating" the surface scratches lead to significant film imperfection and to the reduction of the critical current density.

5. FILMS ON THE PERFECT SUBSTRATES

HTSC characteristics of the $YBa_2Cu_3O_x$ films obtained under the conditions providing their high surface uniformity on the substrates possessing crystalline perfection are shown in Fig. 5. The best parameters was reached for the films deposited on (100) $SrTiO_3$: T_c = 90.8 K, T_c = 0.3 K (χ' measurements under the normal field H = 1 Oe). Under the field of H = 10 Oe the transition temperature reduces by ~ 1 K. At the same time in the case of resistance measurements the similar reduction of T_c is observed when H = $2\cdot10^3$ Oe while T_c decreases to 77 K only for the field of H = $6.6\cdot10^4$ Oe (6.6 Tl). It is important that the observed discrepancy between resistive and magnetic measurements points out the higher sensitivity of the latter. Critical current density of these films determined by resistive method using a 3-μm-wide film bridge corresponds to j_c= $8\cdot10^6-10^7$ A/cm^2.

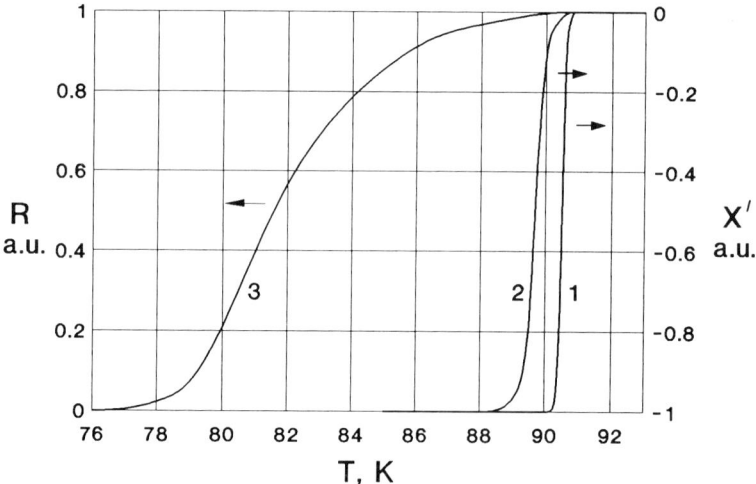

Figure 5. Magnetic (1, 2) and resistive (3) characteristics of $YBa_2Cu_3O_x$ film on perfect $SrTiO_3$ substrate.
1 - H = 1 Oe, 2 - H = 10 Oe, 3 - H = $6.6\cdot10^4$ Oe.

The films with such high parameters possess an epitaxial structure with a weak mosaic disorientation. The level of the latter does not exceed 0.5° not only for $\vec{c}^{\,>}$ axis but also for $\vec{a}^{\,>}$ and $\vec{b}^{\,>}$ axes. The longitudinal dimensions of HTSC crystallites are about 10 μm or more.

YBa$_2$Cu$_3$O$_x$ films deposited on sapphire substrates possess a lower level of crystalline ordering corresponding to angular deviation of 2° for $\vec{c}^{\,>}$ axis and of 5° for $\vec{a}^{\,>}$ and $\vec{b}^{\,>}$ axes. Transition width considerably increases (Table 2) and j$_c$ do not

Table 2
Parameters of superconducting transition for YBa$_2$Cu$_3$O$_x$

Substrate	SrTiO$_3$	sapphire	YAlO$_3$
T$_c$, K	90.8	89.6	88.9
T$_c$, K	0.3	2.8	2.0

exceed 10^5 A/cm^2 for these substrates. The films obtained on YAlO$_3$ substrates have a lower transition temperature (Table 2); they are characterized by only a texture property with a preferable orientation of $\vec{c}^{\,>}$ axis along the normal direction.

6. INFLUENCE OF SUPPLEMENTARY CONDITIONS

For obtaining of films with complicated surface configuration the shading masks are usually used in vacuum technique. However their application for HTSC film manufacturing leads, as a rule, to the deterioration of transition temperature and critical current density. Moreover, HTSC parameters become irregular over the film surface. The reason determining this deterioration follows from Fig. 6 a where the image of the "masked" film affected by the normal magnetic field H (similar to Fig. 4 b) is shown. However in this case white and black fields correspond accordingly to "strong" and "weak" parts of HTSC film (in contrast with Fig. 4 b). One can see that near the boundary of the non-masked surface, which is a 5-mm-diameter circle, a plenty of wedge-like sections possessing a weak HTSC properties is placed. Their longer axes are oriented preferably along radial direction. This type of film degradation near the boundary of the mask is seemingly determined by considerable temperature gradient which appears due to thermal-screening effect over the masked part of substrate surface. The radial dimension of degraded part of the film obtained on the substrate with a thickness of 0.5 mm corresponds to 1 mm. Therefore we can conclude that there are certain important limitations of masks usage for HTSC film deposition.

a

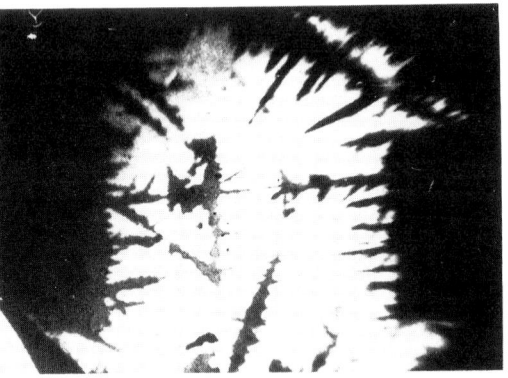

b

Figure 6. $YBa_2Cu_3O_x$ film with silver square layers right after the deposition trough the mask with a 5 mm opening (a) and after the keeping during 3 months (b). Magneto-optical image.

Next phenomenon which also leads to the film deterioration after prolonged sample keeping under normal conditions is an interaction of $YBa_2Cu_3O_x$ film with a silver layers. Fig. 6 b shows the result of change of $YBa_2Cu_3O_x$ film with square silver contact layers after 3 months keeping. This image has been obtained by the technique used for Fig. 6 a obtaining. As it follows from Fig. 6 b the contact squares after interaction acquire a weak HTSC properties. Moreover outside these squares supplementary wedge-like sections possessing weak superconducting parameters also exist. Direction of their main axes is approximately normal to the square boundary. The mechanism of the degradation of this kind are investigated now.

It is our pleasure to acknowledge Dr. A.Belyaeva and Dr. V.Yur'ev for magneto-optical Faraday measurements and for discussion of their results.

REFERENCES

1 J.Geerk, G.Linker, O.Meyer, Materials Science Reports, 4 (1989) 193.
2 L.Lynds, B.R.Weinberger, et al., Physica C, 159, No. 1-2 (1989) 61.
3 B.Roas, L.Schultz, G.Endres, J. Less - Common Metals, 151 (1989) 413.
4 S.Witanachchi, S.Patel, D.T.Shaw, H.S.Kwok, Materials Letters, 8, No. 1 (1989) 53.
5 S.I.Krasnosvobodtzev, E.V.Pechen, Proc. of 3-rd All-Union Conf. on High T_c Superconductivity, USSR, Kharkov, (1991) 63.
6 A.I.Golovashkin, E.V.Pechen, Obtaining and physical properties of HTSC films. Prepr. USSR Acad. Sci., Physical Institute, Moscow, No. 98 (1989) 1.
7 A.T.Matveev, V.F.Gremenyuk, V.P.Novikov et al., Sov. Tech. Phys. Lett. (USSR) , 16, No. 14 (1990) 89.
8 V.A.Antonov, P.A.Arsenev, I.G.Linda, V.L.Farshtendiker, Phys. Stat. Sol., A 15, No 1 (1973) K63.
9 G.Koren, A.Gupta, R.J.Baseman et al., Appl. Phys. Lett., 55 (1989) 2450.
10 A.Richter and G.Kessler, Solid State Communications, 70, No. 2 (1989) 1147.
11 I.K.Schuller, D.G.Hinks, M.A.Beno et al., Solid State Communications, 63, No. 5 (1987) 385.
12 M.P.Shaskol'skaya, Acoustic crystals, Science, Moscow, 1982.
13 A.A.Askochenskii, Optical materials for IR technique, Science, Moscow, 1965.

High T$_c$ Superconductor Thin Films
L. Correra (Editor)
251

DROPLETS AND OUTGROWTHS ON HIGH-Tc LASER ABLATED THIN FILMS

R.P.J. IJsselsteijn, D.H.A Blank, P.G. Out, F.J.G. Roesthuis, J. Flokstra, and H. Rogalla

University of Twente, P.O. Box 217, 7500 AE Enschede, The Netherlands

Abstract

YBa$_2$Cu$_3$O$_x$ thin films have been grown on silicon, SrTiO$_3$ and ZrO$_2$ substrates using the pulsed laser deposition technique. Special attention has been paid to droplets and outgrowths which appear on the thin films during the growth process. The droplet density was studied as a function of the laser spot size, the laser energy and the target density. The number of droplets could be reduced to 1 per 100 µm^2 for a 100 nm thick film, by taking a large laser spot size and low energy density. The droplet density does not depend on the target density in the range from 80 to 94 %. The number of outgrowths could be reduced to 1 per 100 µm^2 for a 100 nm thick film by reducing the deposition temperature or increasing the laser frequency. However, the critical temperature of these layers was reduced by 5 to 10 K. Using SAM no differences in composition between an outgrowth and the rest of the film could be detected.

1. INTRODUCTION

We report on a parameter study concerning the number of droplets (ball shaped particles coming from the target with a typical diameter of 1-2 µm) and outgrowths (0.3 µm centres peaking out of the film) per unit area. Droplet formation is a typical feature of the laser deposition technique. The outgrowths are a more common problem in the different deposition methods for c-axis oriented YBa$_2$Cu$_3$O$_{7-d}$ layers. In order to produce well defined multilayers these particles have to be eliminated.

In literature some attention has been paid to the reduction of the number of droplets by using a second laser beam parallel to the substrate [1], a high density target [2] or a large laser spot size [3]. However, no systematic study about the occurrence of droplets and outgrowths in relation to deposition parameters has been performed to our knowledge. Wu et al. [4] report that the outgrowths are crystalline centres with the c-axis parallel to the substrate. The outgrowths nucleate on second phases, like YBa$_3$Cu$_2$O$_7$ [5].

We report on the influence of the laser spot size, the laser pulse energy and the target density on the number of droplets coming from the target. The number of outgrowths is studied varying the substrate temperature, the laser repetition frequency and the number of pulses.

The composition of an outgrowth is compared with that of the rest of the YBa$_2$Cu$_3$O$_7$ thin film using Auger Measurements.

2. EXPERIMENTAL

Thin YBa$_2$Cu$_3$O$_x$ layers were deposited using the pulsed laser deposition technique. In the experiments an excimer laser (λ=308 nm, τ=20 ns) is focused via a quartz lens onto a rotating $YBa_2Cu_3O_x$ target which is placed inside a vacuum chamber [6]. The spot size of the laser beam can be adjusted by moving the lens on an optical rail and is varied in these experiments between 0.8 and 7.3 mm². We used laser pulse energies between 58 and 98 mJ. The $YBa_2Cu_3O_x$ targets were prepared by the citrate pyrolysis method [7]. The density of the targets varied between 80 and 94 % of the theoretical density and was measured by immersing the target in a mercurial bath, placed on a balance. Different target densities were obtained by changing the pressure at which the targets are pressed from the calcinated powder and by changing the sinter temperature.

The droplet density was studied using 2 inch silicon wafers at ambient temperature as a substrate. Because of the low deposition temperature no crystalline growth centres will appear in the film and all irregularities in the amorphous $YBa_2Cu_3O_x$ layer will be caused by clusters of material (droplets) coming from the target or possibly clusters of material created by collisions of the evaporated material in the plasma. The number of droplets is counted using optical microscope pictures. The deposition conditions were comparable with the ones used for the preparation of c-axis oriented thin $YBa_2Cu_3O_7$ layers. We used an oxygen pressure of 25 Pa, a laser frequency of 2 Hz and a target substrate distance of 35 mm.

Because the number of droplets is linear dependent on the amount of ablated material, the thickness of the deposited layer (not necessarily uniform) also has to be known. To determine the thickness profiles, an 18 x 18 grid consisting of 0.5 x 0.5 mm² holes, 2 mm apart, is etched in the amorphous layer using standard photoresist and diluted H_3PO_4 as etchant. The thickness is measured by an Alpha Stepper.

To study the outgrowths we used (100) oriented $SrTiO_3$ and ZrO_2 at elevated temperatures as substrates. The droplet density was minimized by using the optimized deposition conditions found in the experiments with the silicon substrates. The outgrowth density and critical temperature of the thin $YBa_2Cu_3O_{7-d}$ layers were determined as a function of deposition temperature and laser frequency. The deposition temperature varied between 700 and 770 °C, the laser frequency between 0.2 and 100 Hz. The number of pulses was varied between 120 and 3000, each pulse depositing 1 Å $YBa_2Cu_3O_{7-d}$. The number of outgrowths was counted using an optical microscope, equiped with a differential interference contrast filter. The critical temperature was measured using a standard 4 point DC measurement.

The composition of an outgrowth was compared with a smooth part of the $YBa_2Cu_3O_7$ thin film using Auger Measurements, on a PHI 600 multiprobe.

3. RESULTS

3.1 Droplets
Two series of layers were made at two different spot sizes: 1.45 x 0.55 mm² (I) and 2.8 x 0.9 mm² (II) at different laser energies. A third series of experiments was made at a constant laser energy of 90 mJ, but variable spot size (III).

In order to account for the linear proportionality between the number of droplets per unit area and the amount of deposited material the number of droplets should be corrected for the local film thickness. In table 1 the laser energy E, the laser spot size S, the maximum thickness d_{max}, and the normalized number of droplets Ñ of series I, II, and III are given. Ñ is obtained by dividing the number of droplets counted in an area of 100 μm² by the thickness, at the position of the maximum thickness of the layer.

It is found from table 1 that Ñ is in good approximation linear proportional to the laser energy E for the series I and II. Series III shows that Ñ varies with 1/S at constant E, so that in fact Ñ $\propto$ E/S or Ñ is linear dependent on the energy density. In figure 1 we present Ñ as a function of the energy density. A threshold value of approximately 1 J/cm² for the appearance of droplets is found.

Table 1
Droplet density for series I, II and III

series	E (mJ)	S (mm^2)	d$_{max}$ (nm)	Ñ (per 10 μm^3)
I	98	0.8	120	29.0
I	88	0.8	112	23.0
I	78	0.8	99	21.5
I	68	0.8	84	16.4
I	58	0.8	73	19.0
II	98	2.5	437	3.5
II	88	2.5	396	1.9
II	78	2.5	361	2.1
II	68	2.5	301	2.3
II	58	2.5	211	3.0
III	90	1.4	188	12.0
III	90	2.4	300	4.8
III	90	4.2	480	2.7
III	90	5.3	550	1.1
III	90	7.3	750	0.4

Table 2 shows the normalized number of droplets for different target densities. By increasing the target density no reduction in the normalized number of droplets is observed.

Table 2
Droplet density for different target densities

Target density (%)	Ñ (per 10 μm^3)
80	3.5
86	2.4
94	3.2

In figure 2 the distribution of the ablated material as well as the distribution of the droplets is given. No differences in distribution between the deposited material and the droplets can be seen.

3.2 Outgrowths

To study the outgrowths 100 nm thick YBa$_2$Cu$_3$O$_{7-d}$ layers have been deposited on (100) oriented SrTiO$_3$ and (100) oriented Yttria Stabilized ZrO$_2$ (YSZ) substrates. A series of layers were grown at different substrate temperatures ranging from 700 to 770 °C. The laser frequency was 2 Hz, the laser fluence 1.2 J/cm^2, the target substrate distance 45 mm, the spot size 7.3 mm^2 and the oxygen pressure during ablation 25 Pa. In figure 3 the number of outgrowths and the critical temperature of the layers on SrTiO$_3$ are given as a function of the deposition temperature. By lowering the deposition temperature both the number of outgrowths and the critical temperature reduce. The results for layers on YSZ show similar results, although the reduction of the critical temperature at low deposition temperature is higher.

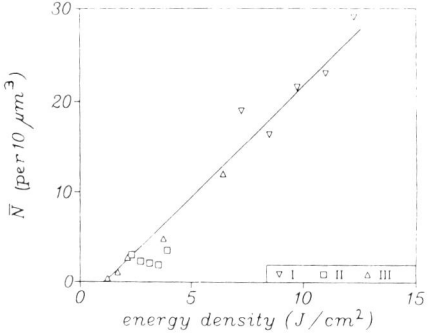

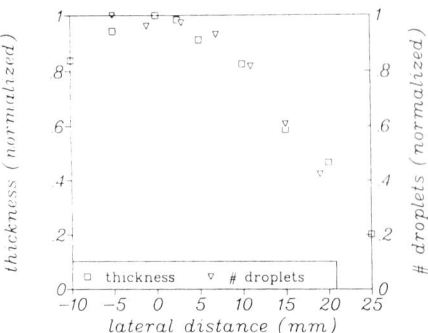

Figure 1. Normalized number of droplets as
a function of the laser intensity.

Figure 2. Distribution of the atomic
material and droplets.

A second series of layers is made by changing the pulse frequency between 0.2 and 100
Hz using a deposition temperature of 770 °C, a laser fluence of 1.2 J/cm², a target
substrate distance of 45 mm and an oxygen pressure of 25 Pa. The number of pulses during
each deposition was 960 what results in layers of about 100 nm thickness. In figure 4
the density of outgrowths and the critical temperature of the $YBa_2Cu_3O_{7-d}$ layers are
given as a function of the laser frequency. By increasing the frequency, the number of
outgrowths reduces to 0.4 per 100 μm^2 for 100 Hz. The critical temperature has a value
of 90 K for a laser frequency of 2 Hz and decreases to about 85 K for 100 Hz. The layer
prepared at 0.2 Hz was semiconducting.

Many small particles are found for very thin layers. By increasing the number of
pulses a substantial part of these particles disappear and outgrowths of sizes of 0.3
μm are formed. The number of outgrowths per unit area remains more or less constant when
film thicknesses of 100 nm are reached.

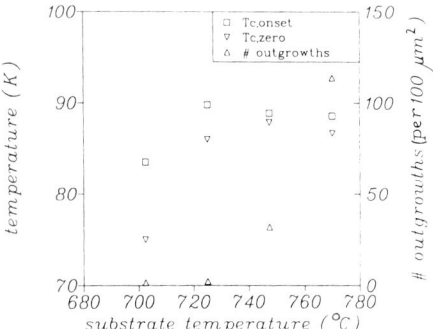

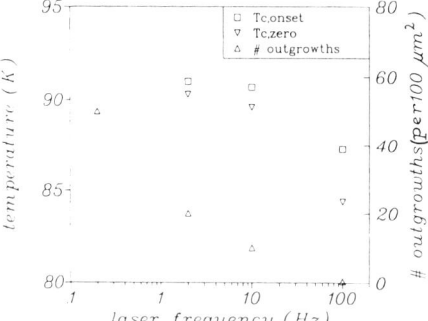

Figure 3. Outgrowth density and critical
temperature as a function of deposition
temperature.

Figure 4. Outgrowth density and critical
temperature as a function of laser repetition frequency.

The composition of an outgrowth of 300 nm is compared with that of the rest of the film using Auger measurements. The electron beam is focused to 100 nm. The results are given in table 3 for a film with a critical temperature of 89 K. The numbers in table 3 are not corrected for the sensitivity factors for the different elements. No relevant composition differences between an outgrowth and the rest of the film have been found.

Table 3
composition of an outgrowth

element	composition	
	outgrowth	rest of the film
Y	5	7
Ba	31	30
Cu	35	45
O	100	100

4. DISCUSSION AND CONCLUSIONS

As is demonstrated in figure 1 there is a linear relationship between the normalized droplet density and the laser fluence. The laser beam with an energy density above the threshold value of 1 J/cm^2 creates a very hot and dense gas just in front of the target that expands in all directions splashing out droplets from the molten surface layer of the target [8]. Apparently, mainly droplets are formed at these high energies and to a much lesser extent atomic material for layer growth. The appearance of the threshold value, reachable by decreasing the laser energy and/or the spot size, is a clear key to the elimination of the droplet problem. The number of droplets seems to be independent of the target density for densities ranging from 80 to 94 %. This is to be expected because the density at the surface of the target will be increased due to the solidification of the molten YBa$_2$Cu$_3$O$_x$ at the surface of the target. In this way a high density layer will be present at the outside of each target, independent of the target density. The fact that the normalized droplet density exhibits the same lateral behaviour on the substrate as the film thickness indicates that the spatial distribution of atomic material and droplets in the plasma is equivalent.

The number of outgrowths is reduced by lowering the deposition temperature at a constant oxygen pressure. This can be explained by the different growth rates of YBa$_2$Cu$_3$O$_{7-d}$ in the ab-plane and perpendicular to this plane at different temperatures. At high temperatures the growth rate in the ab-plane is higher than the growth rate perpendicular to this plane, resulting in a c-axis oriented film with a-axis oriented outgrowths. By lowering the deposition temperature the growth rate in the ab-plane decreases compared to the growth rate along the c-axis and a-axis nuclei will be covered with c-axis material.

By increasing the frequency both the number of outgrowths and the critical temperature decrease. The decrease of the critical temperature for high frequencies may be explained by a not sufficiently large oxygen take up from the plasma. Improvements can be obtained by increasing the oxygen pressure. The reduction of the outgrowth density at higher frequencies indicates that the a-axis growth in plane is favourable with respect to a-axis growth perpendicular to the plane of the substrate. Obviously the formation of large outgrowths needs more time.

In conclusion, by increasing the laser spot size and reducing the energy density to about 1.2 J/cm^2, the number of droplets coming of the target can be reduced to about 0.1 per μm^3. No differences can be found between the lateral distribution of the ablated material and the droplets. The number of outgrowths can be reduced by lowering the deposition temperature or by increasing the laser repetition frequency. This will also reduce the critical temperature of the thin films. No compositional differences between an outgrowth and the rest of the film were found.

5. ACKNOWLEDGEMENTS

This work is part of the research program of the "Stichting voor Fundamenteel Onderzoek der Materie (FOM)", which is financially supported by the "Nederlandse organisatie voor Wetenschappelijk Onderzoek (NWO)".

6. REFERENCES

1 G. Koren, R.J. Baseman, A. Gupta, M.I. Lutwyche, and R.B. Laibowitz, Appl. Phys. Lett. **56** (1990) 2144.
2 G. Koren, A. Gupta, R.J. Baseman, M.I. Lutwyche, and R.B. Laibowitz, Appl. Phys. Lett. **55** (1989) 2450.
3 J.J. Kingston, F.C. Wellstood, P. Lerch, A.H. Miklich, and J. Clarke. Appl. Phys. Lett. **56** (1990) 189.
4 X.D. Wu, R.E. Muenchausen, S. Foltyn, R.C. Estler, R.C. Dye, A.R. Garcia. N.S. Nogar, P. England, R. Ramesh, D.M. Hwang, T.S. Ravi, C.C. Chang, T. Venkatesan, X.X. Li, and A. Inam, Appl. Phys. Lett. **57** (1990) 523.
5 R. Ramesh, A. Inam, D.M. Hwang, T.D. Sands, C.C. Chang, and D.L. Hart, Appl. Phys. Lett. **58** (1991) 1557.
6 D.H.A. Blank, D.J. Adelerhof, J. Flokstra, and H. Rogalla, Physica C **167** (1990) 423.
7 D.H.A. Blank, H. Kruidhof, and J. Flokstra, J. Phys. D **21** (1988) 226.
8 D.H.A. Blank, R.P.J. IJsselsteijn, P.G. Out, H.J.H. Kuiper, J. Flokstra, and H. Rogalla, *High Tc-thin films prepared by laser ablation: material distribution and droplet problem,* accepted for publication in Mater. Sci. Eng. B.

High T$_c$ Superconductor Thin Films
L. Correra (Editor)
257

Cluster emission and the production of Bi–Pb–Sr–Ca–Cu–O thin films by laser ablation

F.M. Saba, J.A. Kilner, P. Sagoo+, A. Sajjadi*, F. Beech*, I.W. Boyd*

Department of Materials, Imperial College, London SW7 2BP, U.K.

+Cascade Scientific Ltd., ETC Building, Brunel University, Middlesex, U.K.

*Department of Electronic and Electrical Engineering, University College
   London, Torrington Place, London WC1E 7JE, U.K.

Abstract

   Thin films of Pb–doped Bi–Sr–Ca–Cu–O were deposited on MgO substrates by the laser ablation of bulk targets. The targets were characterised by energy–dispersive X–ray (EDX) analyses and were found to have nominal compositions close to the n=3 phase of the $(Pb,Bi)_2Sr_2Ca_{n-1}Cu_nO_x$ system of high temperature superconductors.

   The superconducting transition temperature, T_c, of the films was found to depend on the relative contents of the n = 2 and n = 3 phases in the targets. This suggested that laser ablation of the targets caused transmission of materials to the substrate which retains a "memory" of the phase composition of the target. This could be caused by the emission of clusters of the order of an unit cell (~ 1000 atomic mass units (amu)) of the target material which are then deposited on the substrates. These clusters would be large enough to form nucleation sites which could influence the subsequent growth of different phases in the films.

   In order to determine whether cluster emission and deposition occurred, tests were carried out using short deposition times in the laser ablation system. These were followed by scanning electron microscope (SEM) observations which showed the deposition of micron–sized particles. Laser ionisation mass analysis (LIMA) studies performed with a Pb–doped Bi–Sr–Ca–Cu–O target showed the presence of large molecular clusters in the ablation plume when the LIMA system was adjusted to simulate the conditions found in the ablation system.

1. INTRODUCTION

   The $Bi_2Sr_2Ca_{n-1}Cu_nO_x$ system of high T_c superconductors [1] has n = 2 and n = 3 phases [2] with transition temperatures of 80 K and 110 K respectively, which has prompted attempts to produce thin films of the 110 K phase. The 80 K phase forms more readily, but Pb–doping has been found to stabilise the 110 K phase in bulk fabrication [3].

   Laser ablation is an attractive technique for producing thin films because it promotes good stoichiometric transfer from the target to the substrate [4 – 7]. The process is carried out by focussing laser pulses on to a bulk target in a vacuum chamber. The high energy densities at the focus causes local evaporation with the ejection of molecular species with high velocities (10^5 ~ 10^6 cm/s [8]) in a highly forward–directed plume normal to the target surface [9, 10]. Films are deposited by placing a substrate in the ablation plume.

Excellent films of $YBa_2Cu_3O_{7-\delta}$ have been produced by this method [11], but it has so far proved difficult to produce repeatably good, single–phase Pb–doped $Bi_2Sr_2Ca_{n-1}Cu_nO_x$ films with $T_{c,zero}$ greater than 100 K [12, 13].

In our continuing programme to fabricate a single–phase 110 K film [14], we produced a number of films by laser ablation and found that the T_c of the films seemed to be related to the content of the 110 K phase in the target, as determined by X–ray diffractometry (XRD). This may possibly be explained if high molecular mass clusters of the order of a few unit cells of the super–conducting phases were being transferred from the target to the substrate early in the ablation process. These clusters could have acted as nucleation sites and affected the subsequent film growth [15].

Cluster emission is supported by the presence of macroscopic, micron–sized particles which are regularly seen on the surfaces of films produced by laser ablation [10, 16]. Perrière et al. [17] found that the density of such particles on thin films followed the angle–dependent variation of the film thickness, indicating that this phenomenon was closely related to the deposition process.

In order to determine whether macroscopic particles were present during the early stages of the deposition we decided to examine substrates with a scanning electron microscope (SEM) after very short deposition times. Molecular clusters, however, would only have been a few nanometres across, and would have been difficult to resolve with an SEM. We therefore planned to study the ablation plume with a mass spectrometer, but we realised that the mass of such clusters ($\sim$ 1000 amu) would be beyond the range of most quadrupole–based mass spectrometers. This suggested the use of a laser ionisation mass analysis (LIMA) system, because it is equipped with a time–of–flight mass spectrometer which enables it to analyse a wider range of masses. The LIMA technique uses a focussed high energy laser pulse to vaporise a micro–volume of sample surface material followed by a complete parallel mass analysis on the resulting secondary ions.

2. EXPERIMENTAL

The targets were prepared from Bi_2O_3 (99.999% purity), PbO (99.9995%), $SrCO_3$ (99.9965%), $CaCO_3$ (99.95%) and CuO (99.99%) powders which were ground together and cold–pressed (7000 kg) to form 13 mm diameter pellets. The initial nominal compositions were $Bi_{1.75}Pb_{0.25}Sr_2Ca_2Cu_3O_x$ for target A and $Bi_{1.68}Pb_{0.32}Sr_{1.75}Ca_{1.82}Cu_{2.75}O_x$ for targets B and C.

The pellets were calcined in a tube furnace, with the sample temperature held at 800±0.5°C for about 20 hours, and were quenched to room temperature. The pellets were then reground, reformed, and sintered in air according to the schedules in Table 1. Between each sintering, the pellets were reground and reformed, and after the final sintering, they were quenched to room temperature.

Table 1. Target sintering schedules

	First sinter	Second sinter	Third sinter
Target A	850°C , 25.5 hours	842°C , 48 hours	–
Target B	860°C , 45.5 hours	860°C , 48 hours	856°C , 24 hours
Target C	854°C , 46.1 hours	854°C , 47 hours	–

The targets were characterised by EDX, d.c. resistivity and XRD with the Cu

K_α emission. The resistivity results (Fig. 1) showed that targets A and B were superconducting with $T_{c,zero}$ greater than 100 K.

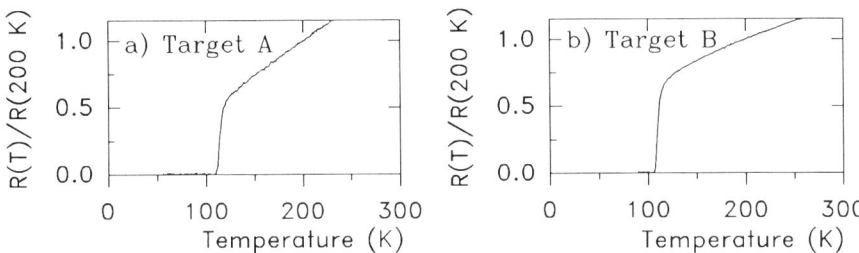

Figure 1. Temperature dependence of resistivity for targets A and B.

The laser ablation system consisted of a Nd:YAG laser, producing 532nm wavelength light pulses 4 ns wide at a rate of 10 Hz, and a vacuum chamber with a pressure of 10^{-4} mbar during deposition [14]. The laser was focussed down to a 1 mm spot on the target which was attached to a rotating holder, and the film was deposited on a substrate placed in the ablation plume.

Films were deposited on (100) MgO substrates at room temperature. They were then post–annealed in air for 20 minutes at 850˚C, quenched to room temperature, and characterised by four–point d.c. resistivity measurements carried out in a closed–cycle helium cryostat.

For the study of the deposition of macroscopic particles, films were produced on Si wafers at room temperature by firing 5 and 600 laser pulses at the target. They were then examined with a JEOL T220A SEM. Si wafers were used because of their flatness and because they cost less than MgO substrates.

The LIMA studies were carried out with target C in a Kratos LIMA which used a Nd:YAG laser producing single pulses of 266 nm wavelength light. The energy density at the target surface was adjusted by defocussing the laser.

3. RESULTS AND DISCUSSION

Films A1 and B1 were produced respectively from targets A and B. Film A1 had a $T_{c,onset}$ of about 90 K and $T_{c,zero}$ of 67.5 K, whereas film B1 had a $T_{c,onset}$ of about 115 K with a second transition at about 90 K and $T_{c,zero}$ of 65.3 K (Fig. 2). Table 2 shows that the EDX analyses for the targets were similar, although the XRD profiles (Fig. 3) show differences in the relative peak heights for the (002) lines for the n = 2 and n = 3 phases. Comparing the XRD results with those obtained by Briggs et al. [18], targets A and B contained 7 % and 87 % respectively of the n = 3 phase. It seems, given that the post–annealing conditions were the same, that the transition temperature of the film is related to the amount of n = 3 phase present in the target.

This may be explained if large molecular clusters were emitted from the target and were deposited on the growing surface during the ablation process. Pb-doped Bi–Sr–Ca–Cu–O films deposited at room temperature are generally amorphous and non–superconducting, but during the post–annealing stage, the superconducting film could have developed by nucleation and ordering around the embedded clusters. Iwai et al. [15] have shown that seeding can promote the growth of the n = 3 phase in bulk material. If elevated substrate temperatures had been used, nucleation and ordered film growth would probably have taken

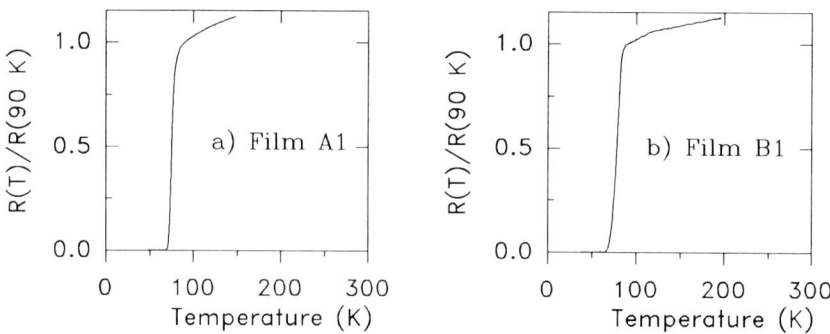

Figure 2. Temperature dependence of resistivity for films A1 and B1.

Table 2. EDX results – atomic percentages

	Bi	Pb	Sr	Ca	Cu
Target A	29.63	2.76	21.47	17.39	28.75
Target B	23.64	2.74	21.54	19.74	32.35
Target C	25.40	2.55	22.62	17.56	31.87

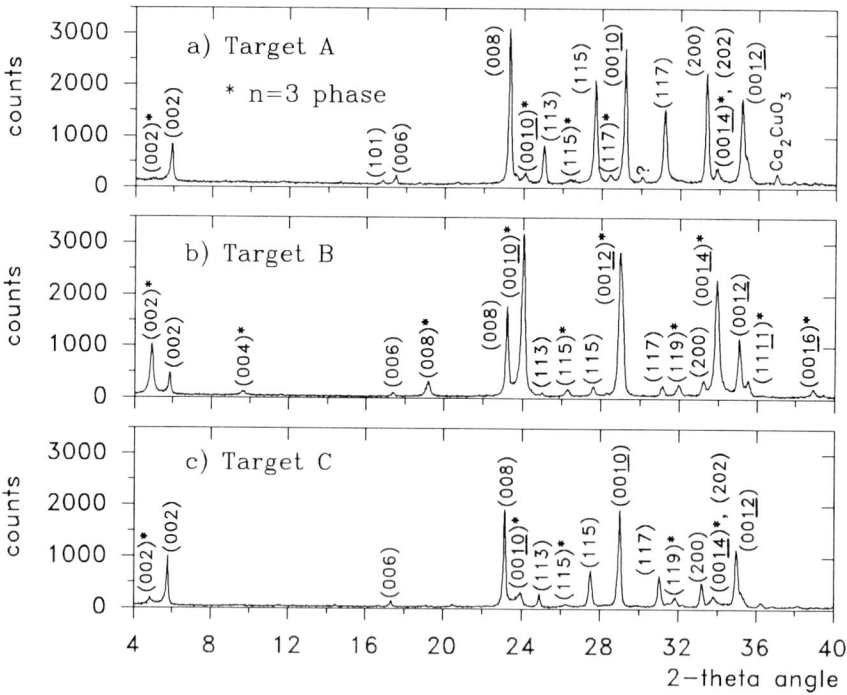

Figure 3. XRD patterns for targets A, B, and C.

place during deposition. Norton et al. [19] found that the growth on heated substrates of $YBa_2Cu_3O_{7-\delta}$ films by laser–ablated deposition took place via the formation of islands, following nucleation on substrate features. However, with Pb–doped Bi–Sr–Ca–Cu–O, it is difficult to produce superconducting films with good microstructures by using *in situ* annealing because of the volatility of Bi and Pb. The best results have so far been obtained when films, deposited by any technique, have been post–annealed in the presence of a source of these volatile components, such as a Pb–doped bulk pellet [20].

Macroscopic particles about 0.5 μm across were observed by SEM on the surface of a Si substrate after only 5 laser pulses had been fired at the target. After 600 pulses, the number per unit area and the sizes of the particles had increased and there was a visibly detectable thin film with a thickness of about 200 nm. This implied that for 5 pulses the film would have been only 1 ~ 2 nm thick. It therefore seems unlikely that the 0.5 μm particle could have grown from the surrounding film during such a short deposition time.

The LIMA spectrum for positive ions (Fig. 4) was produced with a single laser pulse. Under the defocussed analysis conditions used, the intensity of the secondary ions was generally low, with prominent peaks corresponding to the major constituents occurring at lower masses. At higher masses, weak signals close to the mass value for a single unit cell (~ 888 amu) of the dominant (Bi, Pb)$_2$Sr$_2$CaCu$_2$O$_{8\pm\delta}$ phase in target C were detected. These clusters may have condensed out of the plume [21] or they may have been detached from the target surface, but considering that a large proportion of the plume constituents are neutrals, which the LIMA system cannot detect, this clearly supports the transfer of high molecular mass clusters during the ablation process.

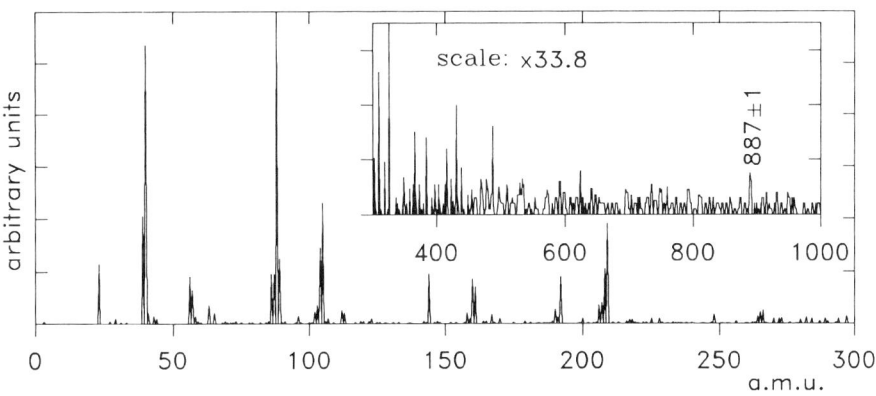

Figure 4. LIMA spectrum of positive ions for target C.

4. CONCLUSIONS

The T_c of Pb–doped Bi–Sr–Ca–Cu–O films produced by laser ablation appeared to be influenced by the phase composition of the target. This may be due to the deposition of molecular clusters containing unit cells of superconducting phases on to the film–growing surface and the subsequent nucleation and ordering of the film around these "seeds" during the annealing stage. Signals have been detected during the laser ionisation mass analyses of the laser ablation plume which could indicate the presence of such clusters.

These results suggest that it may be important to use a target with a high relative content of the 110 K phase to produce a film with a T_c of 110 K, or to use a target which has only been calcined, so that there are no phases of $(Bi, Pb)_2Sr_2Ca_{n-1}Cu_nO_x$ present to influence the film growth. Alternatively, cluster and particle emission could probably be minimised by adjusting the laser energy density and laser wavelength [16]. Particles on the surface must, in any case, be minimised if devices are to be produced on commercial scales.

5. ACKNOWLEDGEMENTS

This work was carried out with the support of the Science and Engineering Research Council (SERC).

6. REFERENCES

1. H. Maeda, Y. Tanaka, M. Fukutomi, T. Asano, Jpn. J. Appl. Phys. 27 (1988) L209.
2. J.M. Tarascon, W.R. McKinnon, P. Barboux, D.M. Huang, B.G. Bagley, L.H. Greene, G.W. Hull, Y. Le Page, N. Stoffel, M. Giroud, Phys. Rev. B 38 (1988) 8885.
3. S.A. Sunshine, T. Siegrist, L.F. Schneemeyer, D.W. Murphy, R.J. Cava, B. Batlogg, R.B. van Dover, R.M. Fleming, S.H. Glarum, S. Nakahara, R. Farrow, J.J. Krajewski, S.M. Zahurak, J.W. Waszczak, J.H. Marshall, P. Marsh, L.W. Rupp Jr., W.F. Peck, Phys. Rev. B 38 (1988) 893
4. R. Kelly, J.J. Cuomo, P.A. Leary, J.E. Rothenberg, B.E. Braren, C.F. Aliotta, Nucl. Instr. and Meth. B 9 (1985) 329.
5. R. Kelly, J.E. Rothenberg, Nucl. Instr. and Meth. B 7/8 (1985) 755.
6. J.E. Rothenberg, R. Kelly, Nucl. Instr. and Meth. B1 (1984) 291.
7. H. Sankur, J.T. Cheung, Appl. Phys. A 47 (1988) 284.
8. O. Eryu, K. Murakami, K. Masuda, A. Kusuya, Y. Nishima, Appl. Phys. Lett. 54 (1989) 2716.
9. R.K. Singh, N. Biunno, J. Narayan, Appl. Phys. Lett. 53 (1988) 1013.
10. T. Venkatesan, X.D. Wu, A. Inam, J.B. Wachtman, Appl. Phys. Lett. 52 (1988) 1193.
11. B. Roas, L.Schultz, G. Endres, Appl. Phys. Lett. 53 (1988) 1557.
12. H. Tabata, T. Kawai, M. Kanai, O. Murata, S. Kawai, Jpn. J. Appl. Phys. 28 (1989) L430.
13. J. Levoska, T. Murtonieni, S. Leppävuori, J. Less–Common Metals 164/165 (1990) 710.
14. A. Sajjadi, K. Kuen-Lau, F. Saba, F. Beech, I.W. Boyd, Appl. Surf. Sci. 46 (1990) 84.
15. Y. Iwai, M. Takata, T. Yamashita, M. Ishii, H. Koinuma, Jpn. J. Appl. Phys. 28 (1989) L1518.
16. W. Kautek, B. Roas, L. Schultz, Thin Solid Films 191 (1990) 317.
17. J. Perrière, G. Hauchecorne, F. Kerhervé, F. Rochet, R.M. Defourneau, C. Simon, I. Rosenman, J.P. Enard, A. Laurent, E. Fogarassy, C. Fuchs J. Mater. Res. 5 (1990) 258.
18. A. Briggs, B.A. Bellamy, I.E. Denton, J.M. Perks, J. Less–Common Metals 164/165 (1990) 559.
19. M.G. Norton, C.B. Carter, B.H. Moeckly, S.E. Russek, R.A. Buhrman, in Science and Technology of Thin–Film Superconductors 2, (R.D. McConnell and R. Noufi (SERI) (eds.)), Plenum, N.Y., 1990 (in press).
20. S. Labdi, H. Raffy, S. Megtert, A. Vaures, P. Tremblay, J. Less–Common Metals 164/165 (1990) 687.
21. C.H. Becker, J.B. Pallix, J. Appl. Phys. 64 (1988) 5152.

High T_c Superconductor Thin Films
L. Correra (Editor)
© 1992 Elsevier Science Publishers B.V. All rights reserved.

The Properties of Sputtered and Laser Ablated YBa$_2$Cu$_3$O$_{7-x}$ Thin Films for Microwave Applications

D Jedamzik, S J Zammattio, N C Vicars, C Dineen, K A Gehring, W A Phillips; GEC Marconi Ltd., Hirst Research Centre, East Lane, Wembley HA9 7PP, United Kingdom.

M Adams, B F Nicholson; GEC Marconi Ltd., Marconi Research Centre, West Hanningfield Road, Great Baddow, Chelmsford, Essex CM2 8HN, United Kingdom.

K Scott, A Mackenzie; Interdisciplinary Research Centre in Superconductivity, University of Cambridge, Madingley Road, Cambridge CB3 OHE, United Kingdom.

Abstract

Y-Ba-Cu-O thin films have been deposited on SrTiO$_3$, and MgO both by laser ablation and by off-axis DC magnetron sputtering, under a range of conditions. Transition temperatures of >90K and critical currents of > 10^6 Acm^{-2} at 77K have been achieved. A range of physical measurements have been used to characterize the samples, and include: I-V characteristics on ion-beam etched constrictions as a function of temperature, magnetization as a function of both magnetic field and temperature and surface resistance at 50GHz as a function of temperature. Structural examination using a range of X-ray diffraction techniques have been used to determine phase purity, crystallite size, lattice parameters and degree of epitaxy. The composition of the thin films was analysed using a wavelength-dispersive X-ray microanalytic technique.

Introduction

Although the transition temperature of YBa$_2$Cu$_3$O$_7$ is approximately 30K below that of the highest known thallium material, the higher available sample quality and low toxicity of YBCO mean that it is being developed much more extensively for microwave applications. This is in spite of the fact that the widespread use of superconducting components will require thin films that can operate at 77K or above. In satellite applications, for example, passive (radiation) cooling would be ideal, but it is unlikely that this method could achieve temperatures significantly below 80K even with extensive redesign of the satellite system. It is important, therefore, not only to understand the factors that control the surface resistance R$_S$, but also to ensure that the transition temperature T$_C$ of deposited films is as high as possible.

Sample Preparation

Laser ablated samples were prepared using a Sopra S520 laser (XeCl 308nm), operating at a pulse energy density of 1-2Jcm^{-2} and a repetition rate of 3Hz. During deposition the substrate temperature was varied over the range 660-760°C in 20°C steps (as monitored by a thermocouple in the heating block). Thermal contact between the MgO or SrTiO$_3$

and the heater was provided by silver paste. The deposition was carried out in pure oxygen at a pressure of 0.3mbar. The deposition rate was approximately 0.2nm/pulse and the sample thickness was varied from 80-380 nm. After the deposition the chamber was vented with pure oxygen to atmosphere and the sample was cooled to 450°C, held for 5 minutes, and then allowed to cool to room temperature. No further post deposition anneal processes were necessary.

Sputtered samples were prepared using an off-axis DC sputtering technique with a cylindrical stoichiometric YBCO target [1]. A range of substrate temperatures, total pressures, oxygen partial pressures , Ar/O_2 ratios and in-situ soak conditions were explored. Typically, the total gas pressure was 4×10^{-1}mbar, with an oxygen/argon ratio of 3:1, and a sputtering power 100 W which gave a deposition rate of ~4nm/min. The thickness of the thin films were varied between 120-350nm. The MgO substrates were clamped onto the heater with an intermediate gold foil (50μm thick). The temperature was monitored by two thermocouples, one in the heater block and the other in a hole drilled in a dummy substrate. The temperature difference was less than 30K. After deposition the sample was cooled in 1400mbar of pure oxygen. Although a number of deposition parameters were investigated (see above) we will limit the discussion of sputtered films in this paper to a range of deposition temperatures 760-800°C (in 10°C steps measured inside the dummy substrate).

Stoichiometry and Structure

Film stoichiometry was determined on the laser ablated samples by electron-probe microanalysis (EPMA) using a Caneca SX-50 equipped with three wavelength-dispersive spectrometers. An YBCO single crystal with Tc=90 was used as a standard and an oxygen content of 6.8 was assumed. Reliable data requires a sufficiently low electron energy to keep the beam within the film, which in turn requires a film thickness greater than 250nm. Typically the accuracy is 2%. No significant variation of composition with substrate temperature could be detected in laser-ablated samples on MgO or $SrTiO_3$, and the mean composition was close to 123. Oxygen content was measured directly using this technique [2], and found to have a mean value of 6.85, with no dependence on substrate temperature.

X-ray patterns were obtained using a standard powder diffractometer with nickel filtered Cu Kα radiation. The patterns show only sharp (OOℓ) lines corresponding to the 123 phase for both sputtered and laser ablated films at their optimum deposition temperature. For laser ablated films deposited at lower temperatures the peaks broaden, as shown for the (009) peak in fig. 1, and move to slightly lower angle, corresponding to an increased c-axis spacing. This shift was not found for the sputtered films. However it should be noted that the laser-ablated samples were deposited over a wider range of substrate temperatures. Despite the shift of the 009 peak for the laser samples they show a slightly better resolution of the $\alpha1/\alpha2$ satellite peak compared to the best sputtered film.

The increased c-axis in the laser ablated films was found not to be related to oxygen concentration (as measured by EPMA), and presumably arises from increased disorder, with larger Ba atoms occupying Y sites. For substrate temperatures below 700°C a new peak emerges at higher angle, and examination of the complete X-ray pattern indicates that this corresponds to the (003) peak for films with the a-axis perpendicular to the substrate.

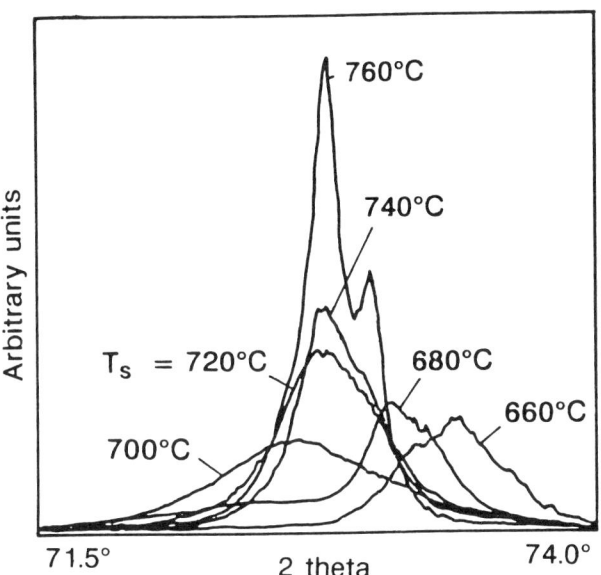

Figure. 1: Aθ/2θ x-ray scan showing the (009) peak for a laser-ablated film as a function of substrate temperature.

The thickness dependence between 0.1 and 0.35μm was measured for the laser ablated samples on unpolished substrates. X-ray patterns showed a small increase in peak width with increasing film thickness above 0.25μm.

No change in c-axis parameter or mixed c- and a-axis orientation was detected in the sputtered films over the temperature range discussed here.

Electrical Properties

Transition temperatures were measured inductively and resistively. The inductance was measured using a surface coil placed against the film, to which a magnetic field, supplied by a drive coil, was applied perpendicular to the substrate. The mutual inductance between the coils was used to detect the transition to superconductivity and by performing the experiment at different drive currents an estimate of the critical current Jc could be obtained. Resistance was measured by a standard four-terminal method on both unpatterned and patterned films (25μm track width of different lengths). Patterning, using positive photoresist soft contact printing (UV 405nm) and Ar ion-beam etching, did not appear to affect the superconducting properties.

The transition temperatures varied consistently with Ts, showing a maximum when the film quality was highest, as judged from the narrowness of the (009) peak in the X-ray pattern. This maximum T$_c$ corresponded to the minimum transition width, but there was no obvious correlation with the resistivity, which was typically $100\mu\Omega$cm at 100K.

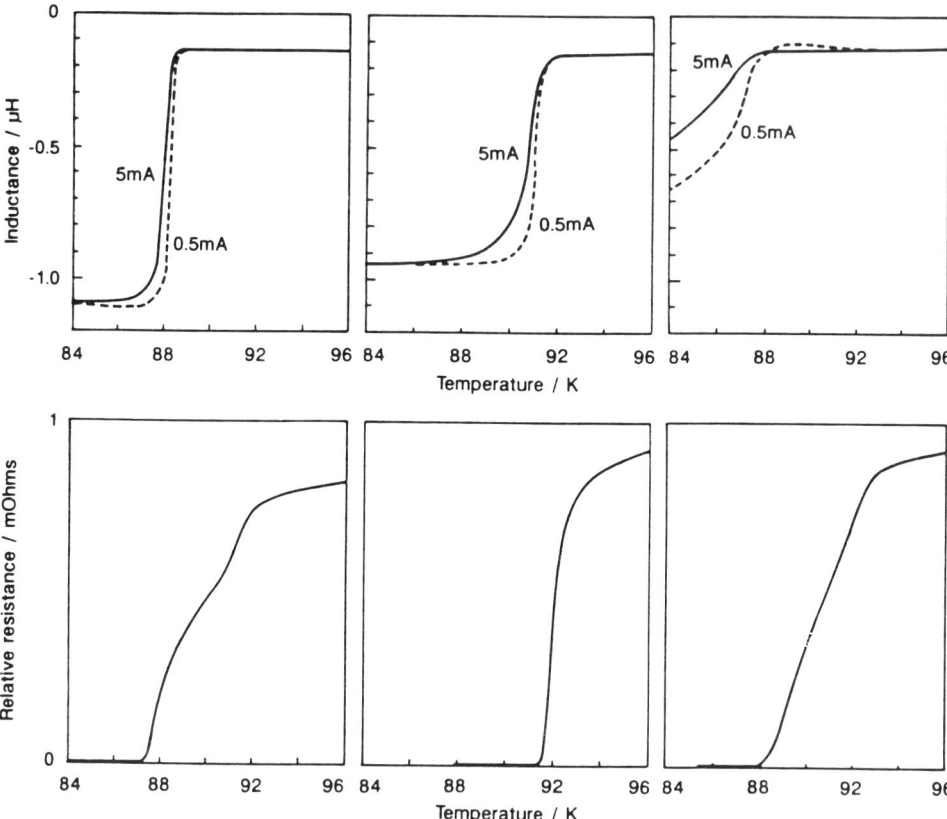

Figure 2: The inductance traces a), b) and c) (top row, left to right) together with the resistance d), e) and f) for three sputtered thin films on an expanded temperature scale close to the transition. The substrate temperatures are 760°C (a and d), 780°C (b and e) and 800°C (c and f).

An interesting feature of the inductive and resistive transitions is shown in fig. 2a-2f for sputtered films. These transitions can be categorised into three groups. The first group which occurs for samples deposited at substrate temperatures of 760°C (below the optimum substrate temperature) shows a sharp inductive transition (for the two drive currents) with $T_c \sim 88$K (fig. 2a) but a two-step resistive transition. The first step occurs at approximately 92-91K and the second at 88K with $T_c(zero) = 87$K (fig. 2d). Films deposited at a substrate temperature of 800°C (above the optimum temperature) show a broadened inductive transition below 88K (fig. 2c) and a single-step resistive transition with a broad transition width and $T_c(zero) = 87.4$K (fig. 2f). For the optimum substrate temperature of 780°C the inductive and resistive transitions (figs. 2b + 2e) are sharp and show $T_c(zero) = 91.2$K. Similar 2-step transitions are seen in the laser ablated films, which have a maximum inductive T_c of 87K. However higher T_cs were not measured on the laser ablated samples, probably because the maximum substrate temperature was too low.

A possible explanation for the broadened resistive transition in the sputtered films deposited at a substrate temperatures of 760°C is misorientation of the grains in the plane of the substrate. There is no evidence from XRD measurements that the films contain regions with a-axis and c-axis orientation. Instead it is proposed that the a- and b-axes of the grains are randomly orientated so that the grains are weakly coupled with high-angle grain boundaries which may be deficient in oxygen effectively forming a second phase. In the films deposited at 780°C the a- and b-axes of the grains are better correlated so that the grains are strongly coupled and no two-phase behaviour is observed. If the sample is heated above 780°C it is possible that a reaction with the MgO substrate occurs leading to contaminated grain boundaries.

Critical currents, measured on patterned sputtered films varied from $J_c(77K) = 1.2 \times 10^6$Acm^{-2} for the film with the highest transition temperature shown in figs. 2b+2e to $J_c(77K) = 5 \times 10^5$Acm^{-2} for the film shown in figs. 2a+2d.

Surface Resistance

The surface resistance R$_s$ was measured using a cavity end-wall replacement technique [3], as a function of temperature between 20 and 120K. Results are shown in fig 3 for a laser-ablated film of thickness $t = 0.35\mu$m and T$_c$ = 87K at a frequency f = 50GHz, and for a sputtered film of thickness t = 0.22μm and T$_c$ = 91K at 60GHz (where T$_c$ is measured inductively). The thicknesses are small compared to the skin depth in the normal state (approximately 2μm) and comparable to the expected zero-temperature penetration depth in the superconducting state (0.15-2μm). In this limit the resistivity in the normal state is given by R$_s$t, approximately 85$\mu\Omega$cm in both films at 100K.

Data for the two films are very similar. Scaling the 60GHz data as f^2 to 50GHz, and applying a correction factor to take into account microwave penetration of the film gives R$_s$ values approaching 10mΩ for both films which compares well with [6]. This thickness correction in the superconducting state can give an effective R$_s$ lower by a factor which is typically 2 at 77K, but which depends on the way in which the sample is mounted, the substrate, and on the value assumed for the penetration depth. In general thinner films are of higher quality and, after correction, show smaller values of R$_s$, but the predicted performance cannot be supported by measurements on thicker films because of degradation in film quality. Our view is that for practical purposes a comparison based on actual values is the most significant and useful test, taking 77K as a reference temperature. On this basis the data shown in fig. 3 (scaled to 50GHz) gives an R$_s$ of approximately 20mΩ for both films.

A comparison with theory does involve a correction for film thickness, and is important in order to assess the scope for further reductions in R$_s$. Calculations (not fitting procedures) of R$_s$ using the two-fluid or full BCS theories [5,6] give values in the range 1-5mΩ at 50GHz and 77K, suggesting that relatively little improvement in performance is possible in this technology important frequency and temperature range.

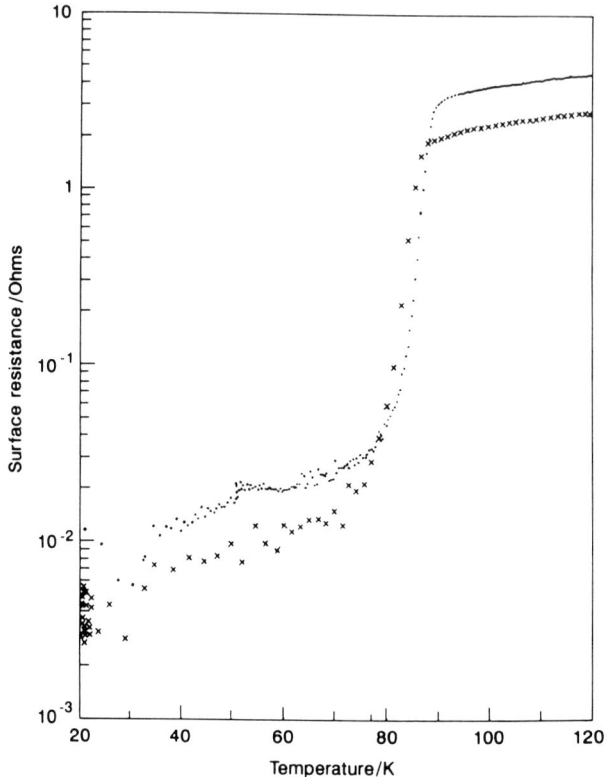

Figure 3. The surface resistance R_s of a laser ablated film (crosses, 0.35μm) at 50GHz and a sputtered film (0.22μm) at 60GHz.

References

[1]   X.X. Xi, G. Linker, O. Meyer, E. Nold, B. Obst, F. Ratzel, R. Smithey, B. Strehlau, F. Weschenfelder, and J. Geerk; Z. Phys. B Condensed Matter, 74, 1, p.13-19, 1989.

[2]   M. Takauo, J. Takada, K. Oda, H. Kitaguchi, Y. Miura, Y. Ikeda, Y. Tomii and H. Mazaki; Jpn. J. Appl. Phys 27 L1041 (1988).

[3]   M.J. Adams, S.J. Hedges, N.G. Chew; Surface Resistance Measurement of High Temperature Superconductors at Microwave Frequencies; Extended abstracts of Third ISEC 1991, Glasgow.

[4]   N. Klein, H. Chaloupka, G. Muller, S. Orbach and H. Piel; J. Applied Phys. 67(11), 1 June 1990.

[5]   S.S. Laderman, R.C. Taber, D.C. Jacowitz, J.L. Mull, C.B. Eom, T.L. Hylton, A.F. Marshall, T.H. Geballe and M.R. Beasley
Phys Rev B43 2922 (1991).

[6]   J.J. Chang and D.J. Scalapino; Phys. Rev B40 4299 (1989).

High T$_c$ Superconductor Thin Films
L. Correra (Editor)

Improved Surface Resistance Properties of YBa$_2$Cu$_3$O$_{7-x}$ Thin Films produced by Pulsed Laser Deposition on MgO Substrates

Y. Ariel[a], M. Mahrizi[a] ,M. Schieber[b,c] ,B. L. Zhou[b], J. A. Virtanen[b] , S. C. Han[c], G. Deutscher[d], D. Racah[d], A. Raizman[e] and S. Rotter[e]

[a]Xsirius Superconductivity Materials Ltd, POB 23566, Jerusalem, Israel 91030

[b]Xsirius Superconductivity Inc, 1110 N Globe Road, Arlington, VA 22201, USA

[c] Graduate School Appl. Sci. & Technol. Hebrew University of Jerusalem, Jerusalem , 91904, Israel

[d]School of Physics and Astronomy, Tel Aviv University, Ramat Aviv, Israel

[e]Solid State Physics Dept, Soreq Nuclear Center, Yavne 70600, Israel

Abstract

YBa$_2$Cu$_3$O$_{7-X}$ thin films with low Surface resistance (R$_s$) films of have been reproducibly deposited on MgO by pulsed laser deposition(PLD). The R$_s$ on MgO values are less than 20 mW at 50.9 GHz at 77 K. The T$_C$ (R=0) is 90K and the J$_C$ is $> 10^6$ A/cm^2 at 77 K. The Rs data is correlated to the surface quality of the substrates as determined by Scanning Tunnel Microscopy (STM) and X-ray diffraction (XRD) rocking curves. Comparing STM images of different YBCO films, we find that films with lower R$_S$ values generally have lower particulate densities. The structural and crystalline quality analysis of the epilayers was carried out using x-ray diffractometry. The Q/2Q diffractogram shows a predominantly aligned epilayer while the rocking curve width of the (005) diffraction varies between 0.42^o-0.55^o .

1. INTRODUCTION

A number of reports on YBCO thin films deposited on (200) MgO and YSZ/(1102) Al_2O_3 using Pulsed Laser Deposition (PLD) and Organo-Metallic Chemical Vapor Deposition (OMCVD) have recently been published.[1-3] The present paper is focused on improved microwave surface resistance (R_s) of YBCO/MgO films and its correlation with its structural perfection. Among the methods mostly used, PLD and in-situ off-axis sputtering have produced low surface resistance films of YBCO. on $LaAlO_3$.[4,5] Although $LaAlO_3$ is closely lattice matched to YBCO and possesses acceptable dielectric properties, it still is plagued by problems of twinning, availability and cost, criteria which are particularly crucial for the application of YBCO thin films to passive microwave devices. In this respect, MgO is a better substrate. Besides lack of twinning, wide availability and low cost, it has much lower dielectric loses than the AlO_3 in both the microwave and millimeter wave regions.[6] Despite a relatively large lattice mismatch of 9%, high quality YBCO films have been grown on MgO.[7]

In this paper we report on our progress of achieving reproducible YBCO thin films with low R_s on MgO by PLD. In addition, we have observed some correlation between R_s and surface morphology as observed by STM and XRD.

2. EXPERIMENTAL RESULTS AND DISCUSSION

The MgO substrates were cut and polished at Xsirius Superconductivity Materials Ltd., Jerusalem. PLD was performed using a Lambda Physik LPX 300 Excimer Laser operating at a wavelength of 248 nm. The base pressure was 10^{-5} Torr and the chamber was backfilled to a pressure of 250 mTorr of oxygen. Deposition was performed at a pulse rate of 2 Hz and an energy density of 4 J/cm². The substrate temperature was 740ºC. The films were cooled in almost one atm oxygen and were given an in-situ postanneal at 400ºC for one half hour. The film thickness was 200nm. The substrate size was 1x1 cm.

Resistivity as a function of temperature was measured using the four probe Van der Pauw technique with Ag paint contacts. Surface resistance measurements were made at 50.9 GHz using the end wall replacement technique which has been described in detail elsewhere.[8] The limit of sensitivity of the present measurements is 20 mΩ. STM imaging was performed in a Nanoscope II STM equipped with a W tip at a bias of 1.5V.

The YBCO on MgO films produced by the PLD process were black and shiny and single phased as revealed by x-ray diffraction. No a- or b- axis oriented material has been observed within the XRD sensitivity range. A more complete description of the XRD characterization of one of the low R_s films will be given later in this paper.

Typical critical temperatures are > 88 K with a transition width less than 2K. The critical current density J_c measured on patterned bridge of 20 μm width is $> 10^6$ A/cm² at 77K.

For this paper, the R$_s$ of four samples of YBCO on MgO were measured. Three samples showed R$_s$ < 20 mΩ at 50.9 GHz and 77K. Typical result is shown in Fig. 1. Assuming f^2 dependence of R$_s$, this translates to a value < 0.77 mΩ at 10GHz and 77 K. This result compares favorably to results of other laboratories using LaAlO$_3$.[4,5]

The fourth sample of this batch had R$_s$ = 96 mΩ at 50.9 GHz. This is higher than the other samples by approximately a factor of five. Since the deposition conditions and substrate preparation were identical for the four samples, there is no obvious reason for this high value and it is necessary to look for other aspects of the PLD process in order to understand this result.

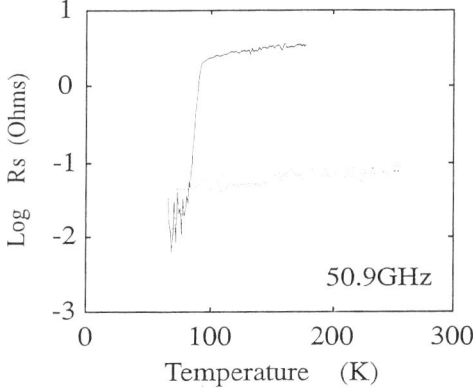

Figure 1. Temperature dependence of Rs upper curve YBCO, lower curve of Hc copper.

Since R$_s$ may be a function of the surface morphology, STM scans were performed on the four samples. A typical scan for the three low R$_s$ samples is shown in Fig. 2. The surface consists of a smooth background interrupted with particulates as typically observed in PLD films.

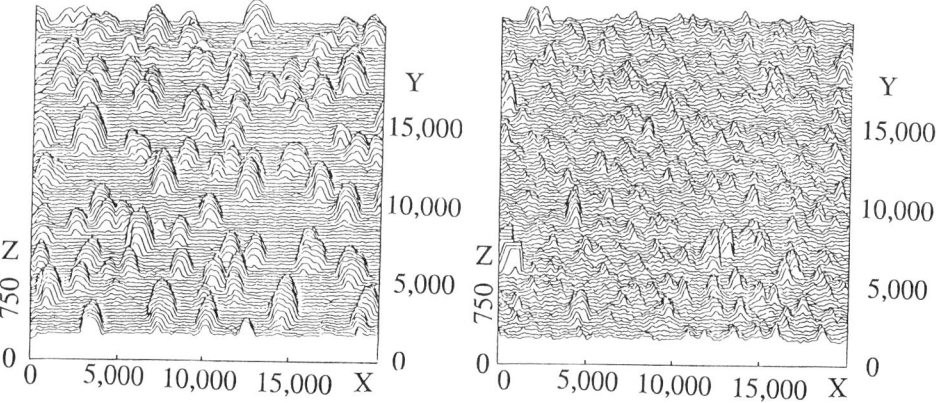

Figure 2. STM of High Rs YBCO
All numbers are in nm units.

Figure 3. STM of a low Rs YBCO
All numbers are in nm units.

In contrast to this morphology, the STM scan of the higher R_s film shows a much rougher background(Fig. 3). Although the particulates size in this film is smaller than those in the lower R_s films, the background is much rougher with a higher particle density. It appears that this roughness contributes to the poor R_s result.

One of the better samples whose STM is shown in Fig. 2 was thoroughly examined by XRD. A $\Theta-2\Theta$ diffractometer with a graphite monochromator and a double crystal diffractometer (DCD) aligned to the non-dispersive setting, both with a copper tube source were utilized.

The diffractogram given in Fig. 4 shows a series of peaks corresponding to the (001) reflections. The (002) reflection of the MgO underlying substrate is also seen in the figure. This is an indication to a predominantly aligned epilayer whose c-axis is perpendicular to the substrate's (001) surface. The full width at half maximum (FWHM) of the (005) symmetric rocking curve ($\Theta = 19.25°$) varies between $0.42°$-$0.55°$. The value obtained in this work is fairly comparable to the results of Moeckly et al[9] They found FWHM values of their YBCO films, grown on MgO, to be in the range of 0.3-$1.02°$ depending on the substrate preparation method. Please note that although FWHM of 0.5 is not as good as those found for off-axis sputterd films, it is considered fairly good for PLD films. Similar FWHM values were also obtained for YBCO layers grown on Si[10] and SrTiO$_3$[11] substrates.

Perhaps one of the most interesting results is the fact that the relative roughness of the film shown by the STM images seems to have a stronger correlation with the R_s value than the XRD rocking curve width. It is well known that one of the problems associated with the PLD process is the formation of particulates on the surface.

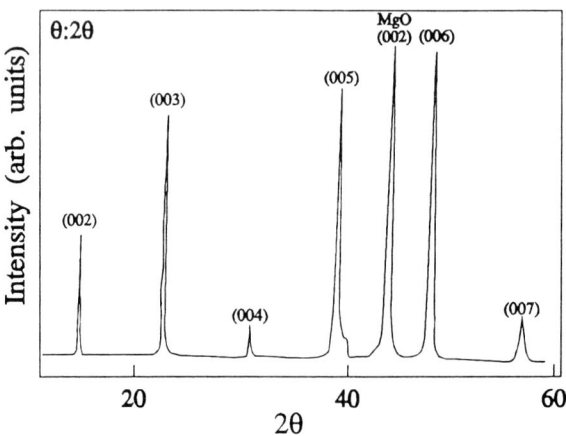

Fig 4. $\Theta-2\Theta$ x-ray diffraction pattern of YBCO thin film grown on (001) MgO.

The source of the difference in surface morphologies between low and high R$_s$ samples reported here may be in the target. It is known that the condition (i.e. density, pitting, bubbles, etc.) of the target is a source of particles.[12] A deterioration of the target condition during the PLD process certainly affects the surface morphology of the films. This is an important issue which needs to be considered to make the PLD process adaptable to continuous production at an industrial scale. Indeed, preliminary SEM studies do show a large decrease of the number of particles in two different targets prepared in different ways. These results will be shown elsewhere.

3. CONCLUSIONS

We have shown that low R$_s$ YBCO films can be prepared reproducibly on MgO. This is an important result as it shows that MgO is a suitable substrate for HTSC films for passive microwave device application. In addition, we have shown a possible correlation between R$_s$ and the surface morphology of these films. We have also shown that the XRD rocking curves do not show a clear correlation with the R$_s$ values.

4. ACKNOWLEDGEMENT

We wish to acknowledge Prof. G. Koren of the Physics Dept.,Technion - Israel Institute of Technology for his assistance with the PLD process and general discussions.

5. REFERENCES

1.   M. Schieber, J. of Cryst. Growth 109, (1991), 401.

2.   M. Schieber, Y. Ariel, M. Schwartz, M. Levinsky, S. Chokrun, M. Maharizi, B. L. Zhou, and S. C. Han, Supercond. Sci. Technology 4,(1991) 5268 .

3.   M. Schieber, M. Schwartz, G. Koren, and E. Aharoni, Appl. Phys. Lett. 58,(1991) 301 .

4.   A. Inam, X. D. Wu, L. Nazar, M. S. Hegde, C. T. Rogers, T. Venkatesan, R. W. Simon, K. Daly, H. Padamsee, J. Kirchgessner, D. Moffat, D. Rubin, Q. S. Shu, D. Kalokitis, A. Fathy, V. Pendrick, R. Brown, B. Brycki, E. Belohoubek, L. Drabeck, G. Gruner, R. Hammond, F. Gamble, B. M. Lairson, and J. C. Bravman, Appl. Phys. Lett. 56, (1990)1178 .

5. N. Newman, K. Char, S. M. Garrison, R. W. Barton, R. C. Taber, C. B. Eom, T. H. Geballe, and B. Wilkins, Appl. Phys. Lett. <u>57</u>, (1990) 520 .

6. D. Grischowsky and S. Keiding, Appl. Phys. Lett. <u>57</u>, (1990) 1055 .

7. Q. Li, O. Meyer, X. X. Xi, J. Geerk, and G. Linker, Appl. Phys. Lett. <u>55</u>,(1989) 310 .

8. B. L. Zhou and S. C. Han, IEEE Mag. <u>Mag-27(2)</u>, (1991), 1268.

9. B. H. Moeckly, S. E. Russek, D. Lathrop, R. A. Buhrman, Jian Li, and J. W. Mayer, Appl. Phys. Lett. <u>57</u>, (1990) 1687.

10. D. K. Fork, D. B. Fenner, R. W. Barton, J. M. Philips, G. A. N. Connel, J. B. Boyce, and T. H. Geballe, Appl. Phys. Lett. <u>57</u>, (1990) 1161.

11. B. Roas, G. Endres, and L. Shultz, Appl. Phys. Lett. <u>53</u>, (1988) 1557 .

12. G. Koren, R. J. Baseman, A. Gupta, M. I. Lutwyche, and R. B. Laibowitz, Appl. Phys. Lett. <u>56</u>, (1990) 2144 .

High T$_c$ Superconductor Thin Films
L. Correra (Editor)
1992 Elsevier Science Publishers B.V.

EVAPORATED YBa$_2$Cu$_3$O$_7$ THIN FILMS AND DEVICE TECHNOLOGY

R.G.Humphreys, J.S.Satchell, N.G.Chew, J.A.Edwards, S.W.Goodyear and M.N.Keene

DRA Electronics Division (RSRE), Malvern, Worcs. WR14 3PS, UK

S.J.Hedges

Marconi Research Centre, Great Baddow, Essex CM2 8HN, UK

Abstract
Epitaxial thin films of YBa$_2$Cu$_3$O$_7$ have been grown by evaporation of the metals from e-guns in the presence of atomic oxygen. The films have been characterised structurally by X-ray diffraction and electrically by measurements of DC magnetisation and flux creep. The progress in device technology is discussed. Results are presented on Josephson junctions, and a superconducting microwave filter with the ground plane grown on the back.

1. INTRODUCTION

High temperature superconductor thin films have potential as the basis for a wide range of devices. We reviewed the film technology fairly extensively about 18 months ago[1]. At that time, the key parameters being studied were the critical temperatures and critical current densities of single layer films. Since then, the focus of work on YBa$_2$Cu$_3$O$_7$ thin films has shifted markedly towards multilayers and device structures, although the control of thin film growth has continued to advance. Good progress has been made in device fabrication technology, which has turned out to be easier than expected overall, but with much work still to be done. It is especially true of superconductors that growth and processing need to be considered as a complete technology rather than distinct activities. In this paper, we address film growth, characterisation, device processing and device results.

2. GROWTH BY EVAPORATION

An important aspect of thin film growth is the choice of substrate. We have chosen to work almost exclusively with MgO because it is readily available in sizes up to 25mm square, gives good film quality, and has low dielectric constant and loss[2]. In common with most groups, we use polished substrates, the surfaces of which may be expected to be heavily damaged. Fig.1a shows an atomic force microscope image of a polished MgO surface. It is clearly very rough, and does not look promising as a substrate for epitaxy. Fig.1b shows a scan (on rather different scales) of a similar substrate after furnace annealing at 1050°C. Crystallographically related steps are clearly visible. These are not single unit cell steps, but are of the order of 1nm high. Obviously the surface of MgO reorders rather readily. A similar but less striking improvement was seen for films heated in the evaporator, as normally occurs before growth. This correlates with RHEED observations on polished substrates: the diffraction maxima are initially diffuse, but sharpen over a period of minutes after the temperature is raised before growth. These results are less surprising if one knows that MgO can be grown epitaxially at temperatures down to 140K[3]. This remarkable surface mobility of MgO seems to be general for ionic materials. We suggest

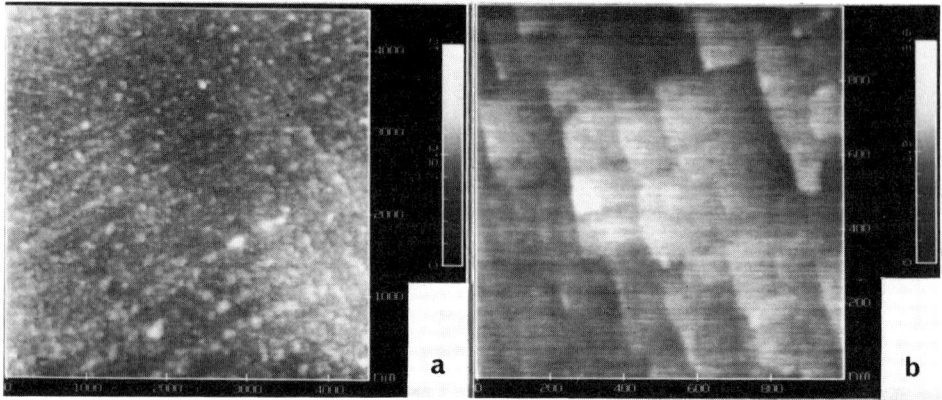

Fig.1 Atomic force microscope images of (a) a polished MgO surface and (b) a similar surface after annealing at 1050°C.

below that high temperature superconductors also have high surface mobilities, and that this makes a major contribution to minimising the effects of processing damage in device manufacture.

Our evaporation system has been described in detail elsewhere[1,4], and is still configured and used in basically the same way. The metals are evaporated from three electron beam heated hearths. The evaporation rates are sensed by a mass spectrometer whose output is fed back at 12Hz to keep the rates constant. The rates are independently monitored by quartz oscillator film thickness monitors with an integration time of 60s per source to maintain the precise run to run reproducibility required. The substrate is clipped to a stainless steel platen, good thermal contact being achieved by placing a well−annealed copper shim between substrate and platen. The platen is heated radiatively from behind, and is rotated at 0.5Hz during growth. The large separation (530mm) between the sources and the substrate together with the rotation ensure excellent uniformity of composition. The substrate is partially enclosed in a silica box into which oxygen is introduced. About 10% of the gas by volume is atomic oxygen formed in a microwave discharge outside the chamber[5]. The oxygen pressure during growth is estimated to be about 10^{-4}mbar, but the actual atomic fraction depends on the recombination probability for atoms on the walls of the hot silica box. This is known to be low when the box is cold, but our atomic oxygen sensor[1] cannot be used at the growth temperature. The silica box also serves as a partial radiation shield for the platen. This is desirable because the platen emissivity changes as evaporation proceeds, and the substrate temperature tends to fall during growth. With the silica box, this amounts to 20°C, which is easily compensated for by ramping the set point as growth proceeds. Without it, this change was over 100°C[1]. The substrate temperature routinely used is 670°C. Films are grown at a rate of 0.12 nm/s on polished (001) oriented MgO crystals. They are cooled in ~10^{-2} mbar of oxygen, containing 10% atomic. The standard film thickness grown is 0.35um, corresponding to a growth time of 50 min. Thicker films can be grown without any noticeable degradation in properties[6]. In fact, scanning electron microscope images of 0.7um thick films suggest that there is a tendency for the surface to become smoother as growth proceeds.

Evaporation has not been used as widely as sputtering or laser ablation for growing high temperature superconductors. There are two main reasons for this. With evaporation, a considerable expenditure of effort is required to achieve accurate composition control, requiring feed−back electronics from rate sensors to stabilise the evaporation rates from the individual sources. The other techniques largely eliminate this problem by yielding a reproducible composition fairly close to that of the target, but which may change by a small amount as other growth parameters are varied. The groups using these techniques which have been most successful have sought a regime where the superconducting properties and/or surface morphology are the required ones. In evaporation, on the other hand, once the required control of composition is achieved, it is possible to explore the effect it has on film properties independently of other growth parameters. We have studied this in some detail[4]. The structure, morphology and critical current density (J_c) were all found to be sensitive to small changes in composition. Similar conclusions have been reached by the Stanford group[7]. We have space here only to summarise the results briefly. Cu deficiency was found to be very deleterious to film properties. The morphology was rough, with deep holes in the surface, and the critical current density was low. Excess Cu leads to the segregation of precipitates, mainly copper oxide. Their density and size scale with growth temperature: high temperature gives relatively few, large precipitates, as would be expected if their density were controlled by surface diffusion. Since excess Cu segregates into relatively small numbers of precipitates, leaving an otherwise well connected film of (presumably) good stoichiometry, one would expect that the other properties are relatively little affected by it, and this is indeed found to be the case. We believe that stoichiometric Cu content corresponds to the disappearance of the precipitates. Excess Ba leads to a granular morphology, a–oriented grains, and sharply falling critical current density. Excess Y gives smooth films with good electrical properties, and is the preferred growth regime for many practical purposes. We are uncertain of the calibration of the Ba/Y ratio. We now believe that our original calibration in ref.4 was in error, as suggested by the Stanford group[7], but think that the correction they suggested was too large. It seems most likely to us that the sharp drop in critical current density associated with excess Ba marks the change from one region of the phase diagram to another, and is close to the stoichiometric composition.

A second reason for the less widespread use of evaporation is that the oxygen pressures compatible with stable and reproducible growth tend to be rather low. This has been thought to be a serious problem[8] in growing YBa₂Cu₃O₇. Fig. 2 shows a graph of oxygen pressure aginst temperature, with lines from two different sources[9,10] plotted to indicate the edge of the thermodynamic stability region of the YBa₂Cu₃Oₓ structure. We have superimposed on this the points corresponding to a number of growth experiments both with and without atomic oxygen present (for more details, see ref. 11). The conclusion is that our growth conditions correspond to a region in which YBa₂Cu₃Oₓ is thermodynamically unstable, but grows for kinetic reasons. This conclusion is supported by the observation[7] of YCuO₂ as a second phase in thin films grown in a similar regime. YCuO₂ was not found to be a contact phase in the thorough investigation of the equilibrium phase diagram by Beyers and Ahn[10]. Because growth is possible beyond the region of thermodynamic stability, the range of temperatures over which high

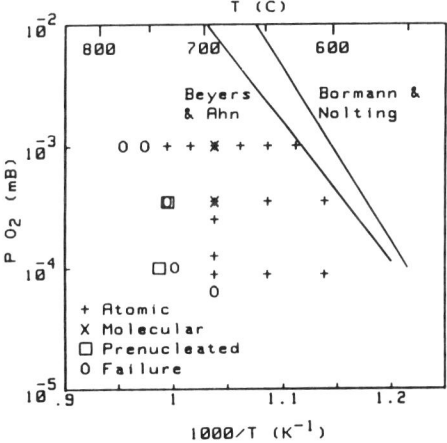

Fig.2   Temperature–pressure diagram showing that our growth conditions lie outside the region of thermodynamic stability, extrapolated from the data of refs. 9 and 10. "Prenucleated" means that growth was initiated at 670°C and successfully continued at the higher temperature with atomic oxygen. "Failure" means $T_c < 80K$.

quality material can be grown by evaporation is very wide. This is a growth regime which does not seem to be accessible with other techniques, and would be expected to lead to different defects, and hence give a different perspective on which film properties are intrinsic and which extrinsic.

It is worth pointing out that it is true of all growth techniques that the films are not fully oxygenated under the growth conditions, so an oxygenation step is an essential part of film growth. Usually this takes place during cooling "in situ", and is not discussed, but it is an important part of the process of making and controlling high quality thin films. Systematic experiments in this area require some care, as the film morphology has a significant effect on the diffusion rate of oxygen[12]. We have found that for our growth process it is necessary to have atomic O present during cooling[11] in order to achieve the best properties. If it is not present, no amount of subsequent annealing in 1 bar molecular oxygen seems to restore the film properties. This is still not understood, but is consistent with the observations of other groups[7]. The most likely explanations are either that some non-chain oxygen sites are incompletely filled or that the cations are disordered in some way. Understanding this point is particularly important for making multilayers for device manufacture. We do not yet have systematic data in this area, but it seems to be difficult to achieve full oxygenation of the bottom layer in a thick superconductor–insulator–superconductor structure (with Y_2O_3 as insulator) if the insulator is low–leakage, i.e. pinhole free.

3. CHARACTERISATION

In research aimed at making devices, it is essential to be able to characterise material quickly, routinely and non–destructively. We use X–ray diffraction and DC magnetisation as our two standard techniques which fulfil these requirements. In this section, the samples used to illustrate the techniques are the same ones as were used for microwave surface resistance measurements, the results of which are presented in another paper at this conference[6].

Structural assessment of the films is routinely carried out by X–ray diffraction. This ranges from a standard lattice parameter measurement from the planes parallel to the substrate

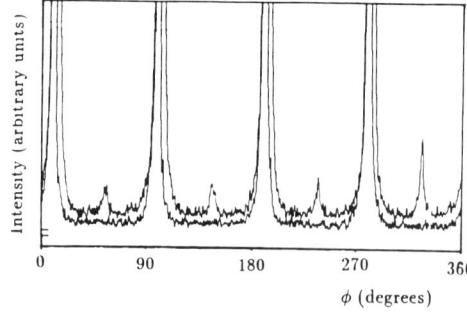

Fig.3 Scanning electron microscope image of sample 477. The long thin rectangles are a–oriented grains, and the background morphology is in the form of approximately square grains. This morphology is suggestive of a slight excess of Ba.

Fig.4 X–ray diffractometer Φ–scans for two films, 363 and 364. The former shows ~0.2% 45° rotated grains, while the latter does not. The zeros of the vertical scales have been offset for clarity.

(Bragg–Brentano geometry) to rocking curves and Φ–scans. The standard lattice parameter measurement is so widely used as to need no further explanation. Apart from the lattice parameter, Bragg–Brentano geometry also gives an estimate of the amount of a–oriented material in a film. For our growth conditions this is primarily controlled by composition[4]. Fig. 3 shows a scanning electron microscope image of a sample with an estimated 2.7% a–oriented material determined by X–ray diffraction. The long thin rectangles are a–oriented grains, which have the c–axis parallel to their short edges. It is evident from this picture that although 2.7% is a relatively small volume fraction, when distributed in this way it could have a very significant effect on a microwave measurement, for example. The sample illustrated actually had $R_s <$ 3 mohm at 40K and 87GHz, and J_c = 4 10^6 A/cm^2 at 77K. Since the microwave loss is very low[6], we conclude that the presence of quite large fractions of a–oriented grains of unfavourable geometry has not had a seriously deleterious effect on the electrical properties. This is consistent with the work of the Wisconsin group on 90° grain boundaries[13] and with the observations of Laderman et al.[14].

When polished substrates are used, there is inevitably some slight misorientation of the substrate with respect to the crystallographic axes. This is presumably the origin of the steps in fig.1b. YBa₂Cu₃O₇ films deposited on MgO tend to have their c–axes normal to the substrate rather than aligned with respect to the crystallographic axes[15,16]. This is also true for large misorientations[17]. This distinguishes MgO from the perovskite substrates on which the crystallographic axes of the film are parallel to those of the substrate. This has consequences for the properties of step junctions like those discussed in section 5. On a perovskite substrate, two 90° grain boundaries are expected at each step, while on MgO the step angle should affect the angle of the grain boundaries.

Φ–scans have been used by several groups to search for rotationally misoriented grains. These have been associated with excess microwave loss, for example[14]. Fig.4 shows data from two films containing different amounts of misoriented material to illustrate the technique. We deliberately use low angular resolution in these measurements to make them more quantitative for determining the misoriented fraction, and to improve the sensitivity to small quantities of misoriented material.

As our primary electrical characterisation technique, we use DC magnetisation[18]. This is fast and non–destructive, and yields measurements of J_c and flux creep rate (S = –d(logM)/ d(log t)) as functions of temperature. The heart of this equipment is an InSb Hall effect sensor and low noise electronics. The film is firmly held down on top of a "pancake" coil. When a pulse of current is applied to the coil, the film is swept round half a hysteresis loop. After the pulse is switched off, the film is left in the critical state, and the field associated with the persistent current monitored by the Hall effect sensor 6mm away. The process is then repeated with the sense of the field reversed, and the difference between the two transients taken as a singleset of data. This eliminates the effect of static fields and ensures that the film is taken round complete hysteresis loops. The measured flux density is converted to a current density assuming that the film carries a uniform current everywhere. The decay transient of the persistent current is measured between 15 and 500ms after the field pulse, and fitted to a straight line when plotted against log (time). The mean value of the current density is taken to be the critical current density. The temperature is varied by lowering this assembly into a helium storage dewar under computer control using a linear drive. Because the transient is recorded on such a fast timescale, the temperature stability required during a measurement is not difficult to achieve, and data can be collected rapidly. A complete routine experiment takes less than an hour.

It is not obvious that data obtained on an intact film are representative of the material and free of spurious geometric effects, since it is never possible to drive the centre of a film into the critical state. Some results have therefore been measured using films patterned into circular discs, and then removing their centres to form rings. An example is shown in fig.5. These data were measured with the sample immersed in liquid nitrogen in order to obtain the necessary temperature stability for the long transients shown to be recorded. The results show that the measured flux creep rate is indeed independent of geometry, while the data from the unpatterned film underestimate the value of J_c by about 6%. If this is due to the central region of the film not reaching the critical state, then this region would have to be 2.5mm in diameter. The results of fig.5 illustrate that even when transients are measured over 6 decades in time, they are only very slightly non-linear on a log(j) vs log(t) plot.

At present, measurements are limited to temperatures over 40K or more, as the large field pulse required at the lower temperatures has frequently been found to cause catastrophic failure of the films in the form of radial cracks. An example is shown in fig.6. In the scanning electron microscope, the cracks show evidence of local melting. We presume that they are caused by the energy dissipated as the sample is swept too rapidly round a large hysteresis loop, but they could also indicate that the experimental conditions are approaching a region of inherent instability[19]. A typical film in this experiment sustains a circulating current of some hundreds of amps at low temperatures.

To illustrate the DC magnetisation technique, the critical current density and flux creep rate are plotted as functions of temperature in fig.7. The samples shown are those whose microwave properties are reported in ref.6. The critical current densities at 77K cover a wide range up to $6 \ 10^6$ A/cm^2. It is worth noting that the low temperature critical current density does not necessary correlate with T_c; in fact there is some evidence for an inverse correlation. It is not clear to us whether this is a statistically significant trend when a large number of samples is considered.

The flux creep data are fairly typical of the range of curves we have observed. From a few degrees below T_c, values of S are in the range 0.015 – 0.035 for nearly all samples. The creep rate has the nearly temperature independent character which is so surprising for a thermally activated process. A divergence in creep rate is observed near T_c. Although the magnetic relaxation is more difficult to measure in this region because the signal is small, we have observed this in a nearly all the 300 or more samples studied. It is not surprising that the creep rate should diverge near T_c, as the flux pinning energy must tend to zero at T_c.

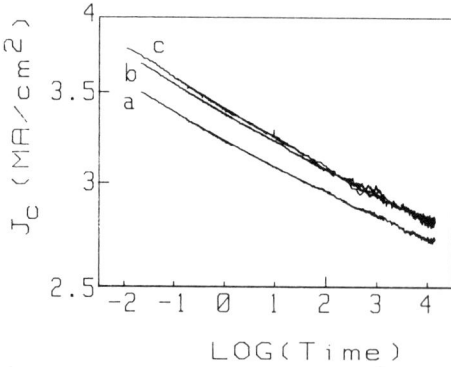

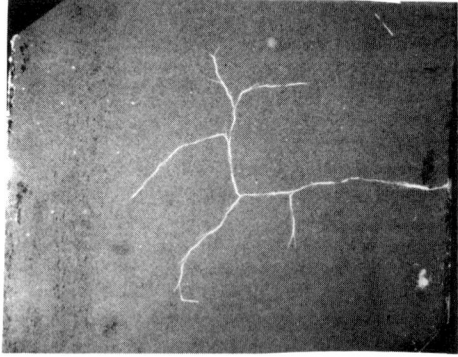

Fig.5 Magnetisation transients recorded in zero applied field for the same 8mm diameter sample as: (a) a disc, (b) a ring with a 4mm central hole and (c) a ring with a 6mm central hole. Note that both scales are logarithmic.

Fig.6 Optical transmission photograph of a film damaged by the fast field pulse in the DC magnetometer. The film was 10mm square, and two of its edges are visible in the picture. Nearly crossed polarisers were used in taking the photograph to enhance the contrast from the cracks, which were of submicron dimensions.

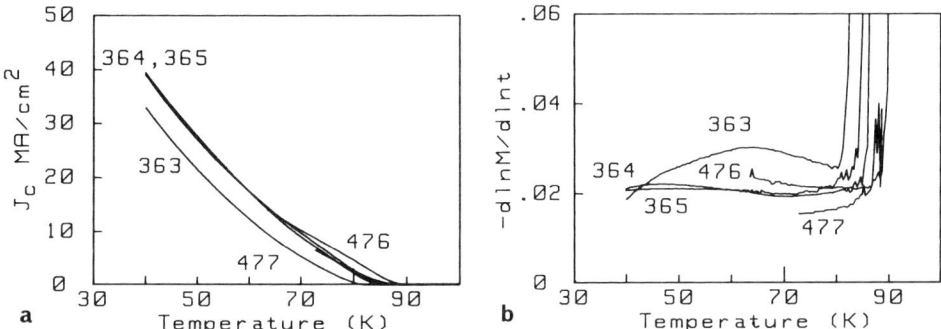

Fig.7 Critical current density (a) and flux creep rate (b) as functions of temperature for the samples used in ref. 6.

4. DEVICE PROCESSING

The main reason for the extensive world–wide effort in developing high temperature superconductor thin films is to make devices, and attention in the last year or two has increasingly been shifting from the growth of single layer films to multilayers, and to the technology necessary for devices. Fortunately, many of the necessary processes can be taken over, often without modification, from well known semiconductor technology. Early on, there was some speculation that the sensitivity of thin films to water would make standard (positive) photolithography inapplicable, but modern high quality films are undamaged even by extensive exposure to water provided acid solutions are avoided. Films can still be degraded by condensation arising from careless cryogenic practice. Presumably this is due to atmospheric CO_2 making the condensate acidic.

For any useful superconductor technology, it is essential to be able to interleave growth and patterning steps, something which is never attempted in semiconductor device manufacture because of the impurities introduced. Fortunately, superconductors are many orders of magnitude less sensitive to impurities than semiconductors, so there is no reason a priori why this should not work. Quite a number of growth and patterning processes need to be mastered to achieve a complete technology. One needs to be able to grow all–epitaxial multilayers, selectively etch through them and stop at the desired depth, and then continue growth on the various exposed surfaces. A complete list of the separate processes is quite long. However, we have found that nearly all of them work without optimisation. We believe that there are two main reasons for this. Firstly, as noted above, damage anneals out easily under the conditions of growth. Secondly, the effect of the atmosphere or non–acid processing chemicals on the surface of a superconductor layer seems to be at worst to create a thin passive hydroxide layer, which presumably decomposes again at the growth temperature. We have tried taking a film after several months storage in a conventional dessicator cabinet and putting it back into the evaporator. On heating to the growth temperature, a clear RHEED pattern was observed, albeit with some superstructure.

Contacts to the films do not appear at present to be a serious problem. We use evaporated Ag or Au annealed in at 500C in oxygen. Apart from a tendency of silver to ball up[20] if it is too thin, such contacts seem to work satisfactorily both at dc and at microwave frequencies. The metallisation is sufficiently adherent to allow reliable use of conventional gold wire ball bonding.
We use standard photolithography (AZ1518 resist) and 500eV Ar^+ ion milling for virtually all our processing. The film is rotated during milling to improve depth uniformity, and a

secondary ion mass spectrometry (SIMS) system detects the composition of the surface being milled. This gives a clear indication when an end point is reached. The milled edges are found to be deoxygenated to a depth of the order of microns, and need to be annealed in oxygen to recover their properties. This is at first sight surprising, as it appears to require a significant diffusion coefficient for oxygen near room temperature. However, there might well be a large gradient of chemical potential near an edge which is being milled, and in such cases, effective diffusion rates can be enhanced by orders of magnitude. Cooling the sample to below room temperature would be expected to minimise loss of oxygen. We note with interest that no deoxygenation apparently occurred in the very narrow track experiments recently reported[21].

We describe below some results on Josephson junctions made by an edge junction technique, the geometry of which is shown in the inset to fig. 11. This is a good example of a device manufacturing sequence. It requires the growth of a superconductor layer coated with insulator, milling at 45° to create an edge, growth of a further layer of superconductor, patterning again, and contacting. Any deliberately introduced barrier layer is an extra step. We have used epitaxial Y_2O_3 as the insulator. This is the least satisfactory part of the process, as the Y_2O_3 gives poor isolation due to cracks (although adequate for this purpose), and also inhibits the oxygenation of the base electrode, reducing its T_c. This seems to be a problem with our technology, as Hirata et al.[22] have achieved good isolation using Y_2O_3. End point detection is now used while milling the edge, so that any step in the substrate is kept small. When this is done, the "junction" is a superconducting short. The critical current density as a function of temperature of one such structure is shown in fig.8. It is comparable to that of an undamaged film, and insensitive to the application of a weak magnetic field. This is a very useful thing to be able to make, and is strong evidence that milling damage is not a serious issue, at least in this geometry. The junctions whose properties are described in section 5 were made before the SIMS end point detection was added to the ion miller. A scanning electron microscope image of one of them is shown in fig.9. This suggests that there was a significant step in the substrate at the junction, so that what we have in fact made was probably a step junction similar to those described by TRW[23] and Julich[24]. It differs (as remarked in section 3) in that they use a perovskite substrate ($LaAlO_3$) and would expect to have 90° grain boundaries, while we are growing on MgO which is expected to give grain boundaries which depend on the milling angle.

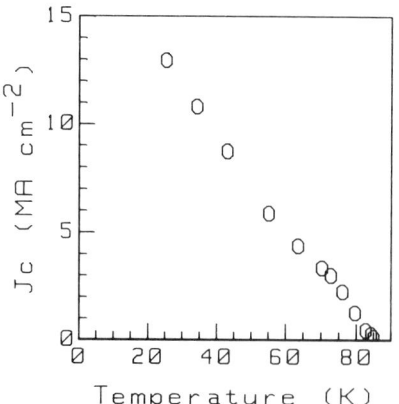

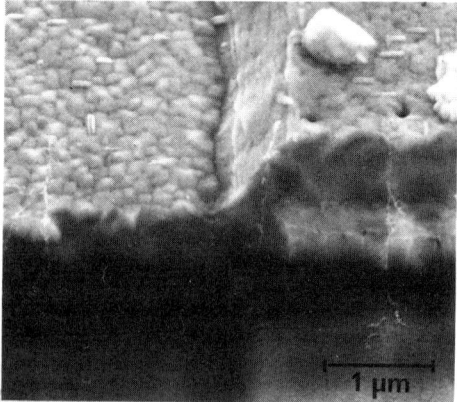

Fig.8 Critical current density as a function of temperature for a superconducting short between layers deposited in different growth runs.

Fig.9 Scanning electron microscope image of a junction made on a 45° step as described in the text. The junction behaviour is believed to be due to the material near the junction being crystallographically tilted.

We identify two problem areas in which more work is needed. One of these is referred to above: it is achieving full oxygenation of the bottom layer of a superconductor–insulator–superconductor trilayer while at the same time having a good pin–hole free insulator. Good progress in solving this problem has been made by by the Berkeley group[25], using fast heating to minimise oxygen loss, but it is not clear that this will be possible in all situations. Although the leakage resistance obtained was quite high, they are looking for further improvement[26]. The other unsolved problem is, of course, that of making a low leakage tunnel junction. Tunneling barriers are so thin that it is difficult to predict how to achieve layers with sufficient integrity. Historically, success in this area has depended on either finding a process which naturally favours such a barrier being formed, or a good deal of hard work, or both.

5. DEVICES

Planar microwave filters patterned from thin film HTS on low loss substrates can offer waveguide–like performance in a fraction of the mass and volume. Several groups have reported the growth and performance of narrow band filters[27]. All have used films on separate substrates to serve as the patterned filter and its ground plane or a normal metal ground plane, to avoid the difficulty of depositing epitaxial films on both sides of the substrate. All who have tried it have found this arrangement unsatisfactory because of the irreproducible air gap between them. We have also experimented with this approach, and come to the same conclusion.

Fig.10 shows the design and performance of a narrow band filter[28] in microstrip in which both the ground plane and the filter layer were grown "in situ" on opposite sides of a 0.75mm thick MgO 19x22mm² substrate in two growth runs. The films were 0.35um thick and were chosen to be slightly Cu rich. The filter is a four section bandpass filter with a 4% bandwidth centred on 6.25GHz. Folded resonators were used for compactness. The minimum insertion loss was measured to be 0.24 dB at 77K. For comparison, similar filters in which the ground plane was on a separate substrate had minimum insertion losses at 77K

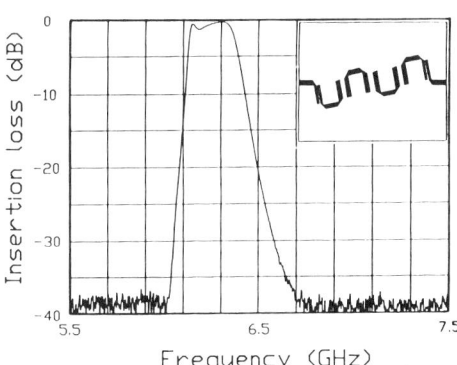

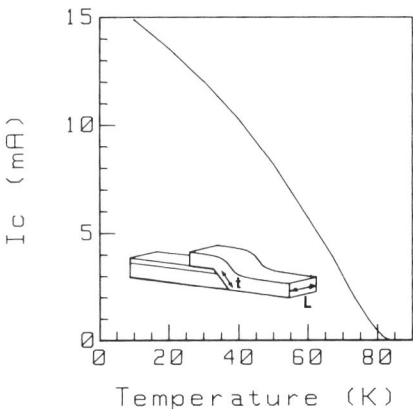

Fig.10 Insertion loss as a function of frequency for a microwave filter made from two films grown either side of the same substrate. The inset shows the filter pattern.

Fig.11 The critical current of a junction as a function of temperature. These data were taken from the maxima of field sweeps, and were measured with a voltage criterion of 0.4uV. The inset shows the idealised geometry of the junction.

of 0.37 dB when the filter was made in $YBa_2Cu_3O_7$, and 1.15 dB when it was made in Au. All these results include connector losses, estimated at 0.13dB at room temperature. There is clearly a large performance improvement compared to the gold reference structure. The performance of the double sided filter was significantly better than the equivalent made from films on separate substrates, and the centre frequency was closer to the design value. It is evident from fig.10 that the filter is slightly mistuned. This is believed to be because the design had been optimised to compensate for an air gap. Up to now, only TRW have reported such double sided growth (using MOCVD) demonstrated in a meander line resonator. The present results suggest that the narrowband (<1%) designs required for communications links will have good performance when realised in superconductor microstrip.

In the previous section, our technology for making Josephson junctions was described. Here we give experimental results for a junction made using a 0.35um thick base electrode and a 0.1um thick insulator[29]. The counterelectrode was 0.35um thick, and the junction was a nominal 12um wide. A schematic of the device is shown in the inset of fig.11, and a photograph of a similar device in fig.9. The devices were measured in a mumetal shielded continuous flow cryostat, in which all the wiring to the sample passed through cold RF filtering.

The electrode T_c in these devices was 84K, and all showed critical currents >5uA at temperatures over 83K. The critical currents rose approximately linearly with temperature on cooling (fig.11). The critical currents shown are a factor 30–100 lower than measured on a track of similar geometry patterned in a single film, but are still very large currents for a junction. For example, at 70K, J_c is about $5 \ 10^4$ A/cm^2 averaged over the junction area. If we naively identify this with the Josephson current density, we derive a Josephson penetration depth of 1um. This is much smaller than the junction size, so a self consistent large junction treatment is essential.

The critical current was found to depend strongly on applied magnetic field normal to the plane of the substrate, in contrast to what we find on a single film. Fig.12a shows data obtained at 82.1K, where several subsidiary maxima and minima are clearly visible, suggesting that the current distribution in the junction was moderately uniform, even so close to T_c. Fig.12b shows a similar set of data at 70K. The smooth curve has been replaced by well defined triangular dependence on field. Similar behaviour was predicted for long junctions by Owen and Scalapino[30] and observed in long low T_c junctions by Matisoo[31]. Our geometry is much less ideal than that considered in ref. 30, so that a quantitative interpretation of the data is not possible without extending the theory of long junctions. However, in qualitative terms, the results are consistent with what might be expected. The critical current of a small junction is the product of the Josephson current density (j_1) and its area. In a long junction, current is only carried in a region λ_j (the Josephson penetration depth) in size at its ends, and the critical current in zero field is $4j_1\lambda_jt$, where t is the junction thickness. An applied field causes a screening current to flow whose magnitude is proportional to the field. This subtracts directly from the critical current, causing the peak to be triangular in shape. The subsidiary peaks can be accounted for in a similar way, but with one or more fluxons in the junction.

The differential resistance of this device above about 1mV was temperature and bias independent. If we take this as the normal state resistance, the I_cR_N product was 0.27mV at 77K and 4mV at low temperature. The critical current is independent of junction length for a long junction, so higher values of I_cR_N are expected for shorter junctions. These devices show long junction behaviour at all temperatures more than a few degrees below T_c. They are heavily damped, but at least some of the long junction devices (e.g. flux flow transistors) could be implemented. We estimate[29] the fluxon velocity in the junction to be ~3 10^5m/s, which indicates that such devices can be fast. It is worth emphasising that although strongly coupled junctions like these are electrically long, they are physically quite small, so quite dense circuits can be contemplated.

More recent devices made by a similar process, but with better control of the height of the substrate step via end point detection show broadly similar characteristics. Fig.13 shows the voltage–field characteristics of a SQUID loop made using a pair of somewhat weaker junctions ~2um wide separated by a loop ~3um square.

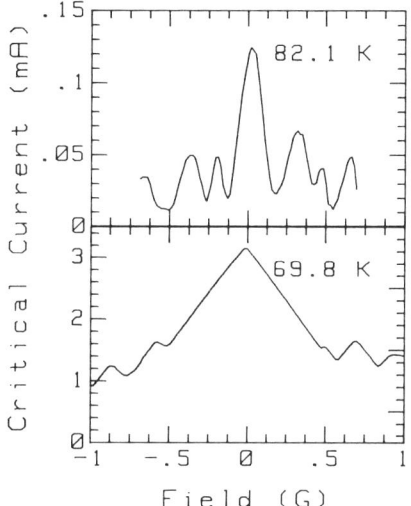

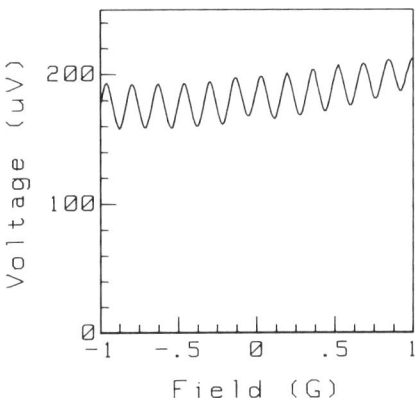

Fig.12 The critical current of a junction as a function of applied magnetic field at 82.1K and 69.8K. The actual field at the junction is believed to be about seven times larger than the applied field due to flux focussing effects from the electrodes. Some hysteresis was observed, due presumably to flux trapping in the electrodes, and so the zero of field is arbitrary.

Fig.13 Voltage field characteristic of a SQUID at 60K. The bias current was 0.3mA.

6. CONCLUSION

In this paper, we have attempted to cover the whole range of thin film superconductor technology from film growth to recent progress in devices. It has been a pleasant surprise to find how easily many of the processes necessary to make devices can be made to work, often with little or no optimisation. Although low leakage tunnel junctions are also desirable, junctions of the type described above will allow many many promising device concepts to be explored. For example, it has been persuasively argued by Likharev and Semenov[32] that the form of logic best suited to a high T_c realisation is single flux quantum logic, and this requires non-hysteretic junctions.

Note Added: We have just received a preprint by T.B.Lindemeyer et al. (to be published in Physica C) which suggests that the stability line for YBa₂Cu₃Oₓ is significantly to the left of the Beyers and Ahn line in fig.2, and runs through the centre of the our data points. The conclusion that all our films have been grown in a thermodynamically unstable regime may therefore be incorrect.

REFERENCES

1. R.G.Humphreys, J.S.Satchell, N.G.Chew, J.A.Edwards, S.W.Goodyear, S.E.Blenkinsop, O.D.Dosser, and A.G.Cullis Superconductor Sci. Technol. 3, 38 (1990)
2. M.Morisue, S.Furusawa, J.Asahina,and A.Kanasugi IEEE Trans. Magnetics 27, 2805 (1991)
3. S.Yadavalli, M.H.Yang and C.P.Flynn Phys. Rev. B41, 7961 (1990)
4. N.G.Chew, S.W.Goodyear, J.A.Edwards, J.S.Satchell, S.E.Blenkinsop and R.G.Humphreys App. Phys. Lett. 57, 2016 (1990)
5. N.G.Chew, R.G.Humphreys and J.S.Satchell UK patent application 9009319 (1990)
6. M. Hein, S.Hensen, G.Muller, S.Orbach, H.Piel,M.Strupp, N.G.Chew, J.A.Edwards, S.W.Goodyear, J.S.Satchell and R.G.Humphreys Paper A1-IV.2 This conference.
7. V.Matijasevic, P.Rosenthal, K.Shinohara, A.F.Marshall, R.H.Hammond and M.R.Beasley J.Mater. Res. 6, 682 (1991)
8. R.H.Hammond and R.Bormann Physica C 162-4, 703 (1989)
9. R.Bormann and J.Nolting App. Phys. Lett. 54, 2148 (1989)
10. R.Beyers and B.T.Ahn to be published in Annual Review of Materials Science (1991)
11.   R.G.Humphreys,   N.G.Chew,   J.S.Satchell,   S.W.Goodyear,   J.A.Edwards   and S.E.Blenkinsop IEEE Trans Magnetics 27, 1357 (1991).
12. K. Yamamoto, B.M.Lairson, J.C.Bravman and T.H.Geballe App. Phys. Lett. 57, 1936 (1990) and J. App. Phys. To be published
13.S.E.Babcock, X.Y.Cai, D.L.Kaiser and D.C.Larbalestier Nature 347, 167 (1990)
14. S.S.Laderman, R.C.Taber, R.D.Jacowitz, J.L.Moll, C.B.Eom, T.L.Hylton, A.F.Marshall, T.H.Geballe and M.R.Beasley Phys. Rev. B43, 2922 (1991)
15. S.E.Russek, B.Jeanneret, D.A.Rudman and J.W.Ekin IEEE Trans Magnetics 27, 931 (1991)
16. S.K.Streiffer, B.M.Lairson and J.C.Bravman App Phys. Lett. 57, 2501 (1990)
17.   J.A.Edwards,   N.G.Chew,   S.W.Goodyear,   J.S.Satchell,   S.E.Blenkinsop   and R.G.Humphreys. J. Less Common Metals 164-5, 414 (1990)
18. S.W.Goodyear, J.S.Satchell, R.G.Humphreys, N.G.Chew and J.A.Edwards to be published
19. M.I.Flik and C.L.Tien J. Heat Transfer 112, 11 (1990)
20. A.Roshko, R.H.Ono, J.A.Beall and J.Moreland IEEE Trans. Magnetics 27, 1616 (1991)
21. H.Jiang, Y.Huang, H.How, S.Zhang, C.Vittoria, A.Widom, D.B.Chrisey, J.S.Horwitz and R.Lee Phys. Rev. Lett. 66, 1785, (1991)
22. K.Hirata, K.Yamamoto, K.Iijima, J.Takada, T.Terashima, Y.Bando and H.Mazaki App. Phys. Lett. 56, 683 (1990)
23. K.P.Daly, W.D.Dozier, J.F.Burch, S.B.Coons, R.Hu, C.E.Platt and R.W.Simon App. Phys. Lett. 58, 543 (1991)
24. C.L.Jia, B.Kabius, K.Urban, K.Herrman, C.J.Cui, J.Schubert, W.Zander, A.I.Braginski and C.Heiden To be published in Physica C
25. J.J.Kingston, F.C.Wellstood, P.Lerch, A.H.Miklich and J.Clarke App. Phys.Lett. 56, 189 (1990)
26. J.J.Kingston, F.C.Wellstood, D. Quan and J. Clarke IEEE Trans Magnetics 27, 974 (1991)
27. See for example IEEE Trans Magnetics 27, 2533-2553 (1991)
28. S.J.Hedges, N.G.Chew and R.G.Humphreys to be published
29. J.S.Satchell, R.G.Humphreys, N.G.Chew, S.W.Goodyear, J.A.Edwards and M.N.Keene to be published
30. C.S.Owen and D.J.Scalapino Phys. Rev. 164, 538 (1967)
31. J.Matisoo J. App. Phys. 40, 1813 (1969)
32. K.K.Likharev and V.K.Semonov IEEE Trans Applied Superconductivity 1, 3 (1991)

High T$_c$ Superconductor Thin Films
L. Correra (Editor)
© 1992 Elsevier Science Publishers B.V. All rights reserved.

SUPERCONDUCTING ARTIFICIAL LATTICES GROWN BY LASER MBE

Tomoji KAWAI, Masaki KANAI, Takuya MATSUMOTO, Hitoshi TABATA, Ken HORIUCHI and Shichio KAWAI

The Institute of Scientific and Industrial Research, Osaka University, Mihogaoka, Ibaraki, 567 Japan

ABSTRACT

The laser molecular beam epitaxial technique to form superconducting thin films and artificial lattices is presented. Tunnel junctions, superlattices and tailored thin films have been formed by the layer-by-layer method using this technique. In the junction of $Au/Bi_2Sr_2CuO_6/Bi_2Sr_2CaCu_2O_8$, reproducible tunnel spectra have been obtained. Strain effects are observed in $YBa_2Cu_3O_7/La_2CuO_4$ and La_2CuO_4/Sm_2CuO_4 superlattices to induce Tc changes and this effect is discussed based on the pressure effect of high Tc cuprates. The growth mechanism of the $Bi_2Sr_2Ca_{n-1}Cu_nO_{2n+4}$ (n=1 to 8) films in the order of unit cell, sub-unit cell and atomic layer has been revealed.

[I] Atomic layer and sub-unitcell layer growth of $(Ca,Sr)CuO_2$ and $Bi_2Sr_2Ca_{n-1}Cu_nO_{2n+4}$ thin films

Thin films of $(Ca,Sr)CuO_2$, the parent material of high-Tc cuprate superconductors, have been formed by laser ablation method under molecular beam epitaxial condition, and the growth mechanism has been investigated with RHEED and AES.[1] (see Fig.1) Analyses of RHEED patterns and intensity oscillations show this material grows with two dimensional layer growth. When all the metal elements are supplied simultaneously in NO_2 atmosphere, the layer growth occurs with the unit-cell layer of $(Ca,Sr)CuO_2$. Furthermore, it has become evident that the growth unit can be separated into Ca (Sr) atomic layer and CuO_2 atomic layer by monitoring the RHEED intensity oscillation. The successive supply of each metal element leads to one atomic layer growth of this metal oxide material.[1](Fig.2)

The $Bi_2Sr_2Ca_{n-1}Cu_nO_{2n+4}$ with even n=5 to 8 [2,3] are formed by the successive supply of Bi, Sr, CaCuO and SrCuO elements and by monitoring the surface with reflection high energy electron diffraction.(Fig.3) The diffraction pattern shows that these materials can be formed with layer-by-layer growth. The monitoring of the changes of the diffraction intensity as well as the analysis of the total diffraction pattern makes it

possible to control the growth of the atomic layer or the sub-
unit layers in this $Bi_2Sr_2Ca_{n-1}Cu_nO_{2n+4}$ system.(Fig.4)

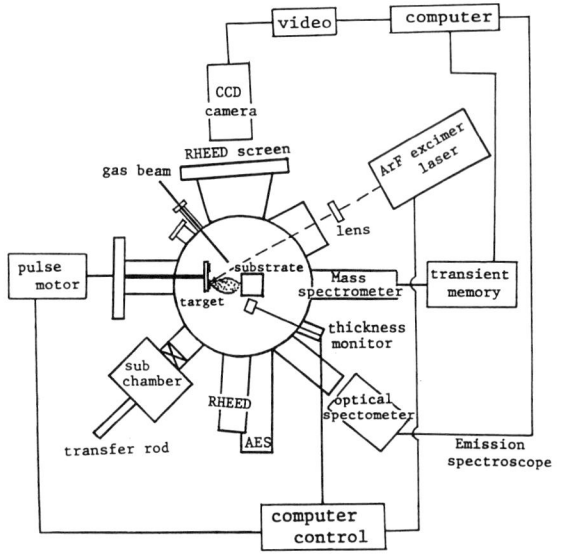

Fig.1. Apparatus

for Laser MBE

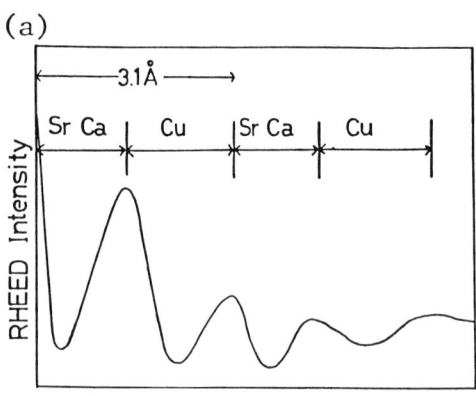

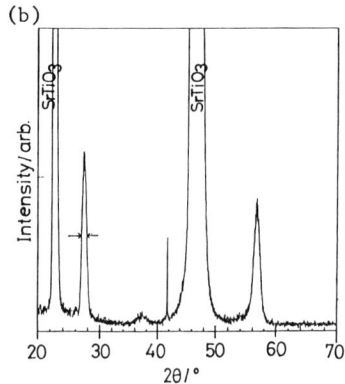

Fig.2. (a) Changes of RHEED intensity during the alternate
deposition of Ca(Sr) and Cu in NO_2. (b) X-ray pattern of of
the film indicating the formation of layered Ca(Sr)CuO_2.

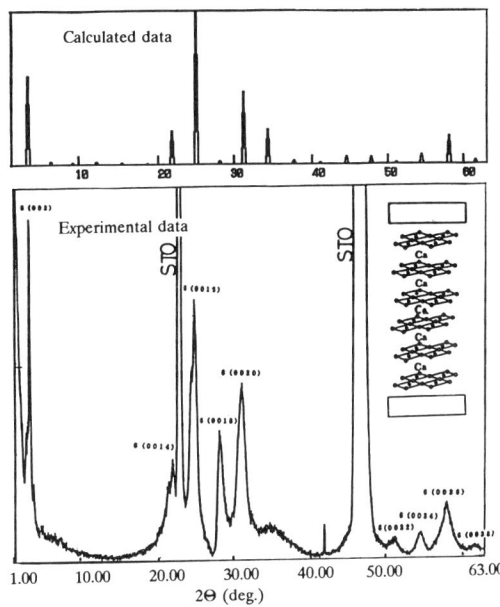

Fig.3. n=6 compound of $Bi_2Sr_2Ca_{n-1}Cu_nO_{2n+4}$

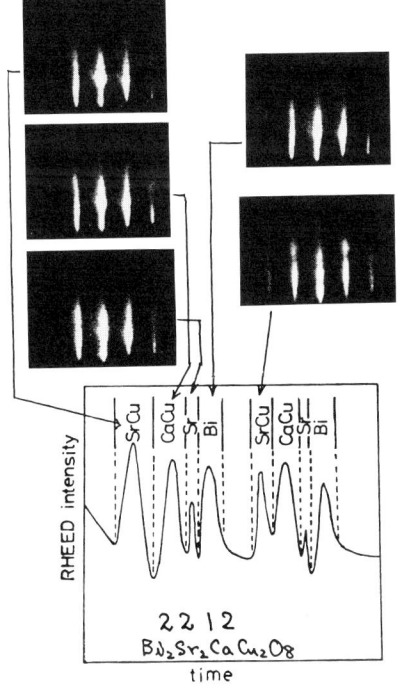

Fig.4 RHEED pattern during the growth of $Bi_2Sr_2CaCu_2O_8$ thin film

[II]   Unit cell layer growth of $Bi_2Sr_2CuO_6$ and the tunneling spectra of $Au/Bi_2Sr_2CuO_6/Bi_2Sr_2CaCu_2O_8$

The very short coherence length in high Tc cuprates require a sharp interface for a tunnel junction and device applications.   We have fabricated junctions of $Au/Bi_2Sr_2CuO_6$(2201) thin film / $Bi_2Sr_2CaCu_2O_8$ (2212) single crystal taking advantage of the Laser MBE technique.   The 2212 single crystal is as large as 5x5x50 mm with high quality.   In this system, the growth of the 2201 film on the 2212 single crystal is pseudo-homoepitaxial because both lattices have only a difference of c-axis length.   The RHEED patterns are streak lines all through and the RHEED intensity oscillates as the unit cell of the 2201 layer grows.[4](Fig.5)   These results show that the 2201 phase grows as a flat film with layer-by-layer growth mechanism.   As a result, we have obtained the well defined interface of the 2201/2212 junction.(Fig.6) Using this junction, we have measured tunneling spectra and have obtained reproducible tunneling spectra.   The gap energy derived from the peak top separation is 40meV along the direction of c axis.

RHEED patterns

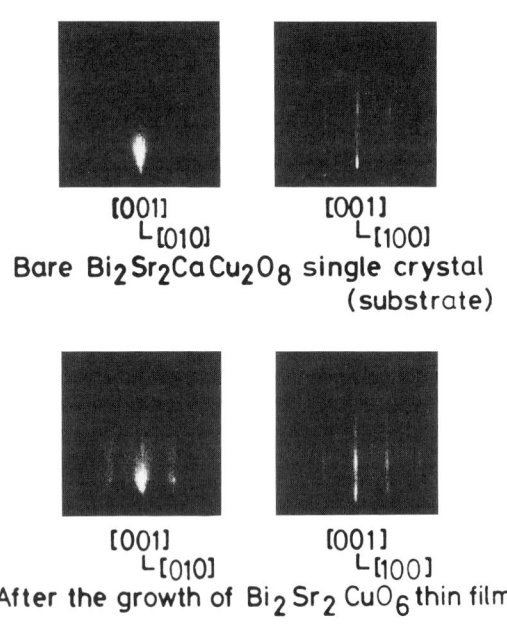

[001]
└[010]

[001]
└[100]

Bare $Bi_2Sr_2CaCu_2O_8$ single crystal
(substrate)

[001]
└[010]

[001]
└[100]

After the growth of $Bi_2Sr_2CuO_6$ thin film

Fig.5.   RHEED pattern for $Bi_2Sr_2CuO_6$ film on $Bi_2Sr_2CaCu_2O_8$ single crystal

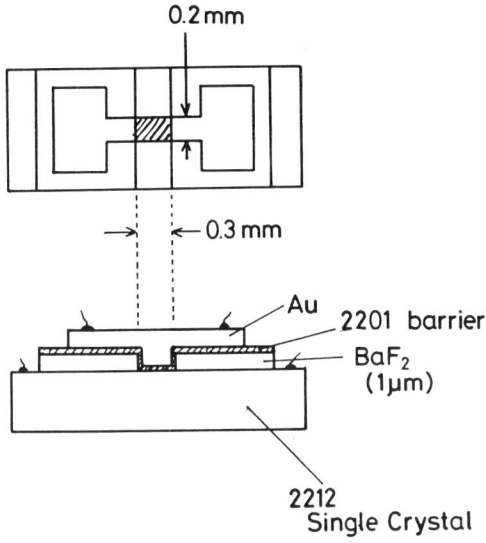

Fig.6. Scheme of the tunneling measurement for $Au/Bi_2Sr_2CuO_6/Bi_2Sr_2CaCu_2O_8$

[III] Superconducting superlattices of $YBa_2Cu_3O_7/La_2CuO_4$ and $(La,Sr)_2CuO_4/Sm_2CuO_4$

The control of atomic layers in the high Tc cuprates has become possible by laser MBE method, and variety of superconducting superlattices and tailored thin films are formed to study physical properties. The superlattices have been fabricated for $Bi_2Sr_2CaCu_2O_8/Bi_2Sr_2CuO_6$ and $Bi_2Sr_2(Ca_{1-x}Y_x)Cu_2O_8$ x=0.5/0.15.[5] The two dimensional nature of $Bi_2Sr_2CaCu_2O_8$ together with the three dimensional interaction between CuO_2 layers has been proved in these superlattices utilizing superconductor/semiconductor interface.

This laser MBE has further been applied to the unit cell layer growth of $YBa_2Cu_3O_7$(YBCO) and La_2CuO_4(LSCO). In the $YBa_2Cu_3O_7/La_2CuO_4$ superlattice, the ab plane of YBCO is compressed and the c axis is elongated due to the LSCO layer which has smaller lattice constants a than that of the YBCO layer. When the thickness of the YBCO layer is reduced, superconducting transition temperature(Tc_{zero}) of YBCO gradually decreases. The deterioration of Tc, however, is much smaller in this system than that in the $YBa_2Cu_3O_y/PrBa_2Cu_3O_7$ and $YBa_2Cu_3O_7/(Nd,Ce)_2CuO_4$ superlattices.(Fig.7) The Tc_{zero} for only two unit cells(24A) of the YBCO layers are as high as 72K. The correlation between the Tc and the in plane strain along b-axes suggests that the pressure effect to change the Cu-O bond length along CuO_2 plane of superconducting YBCO layer is important to raise the Tc value.

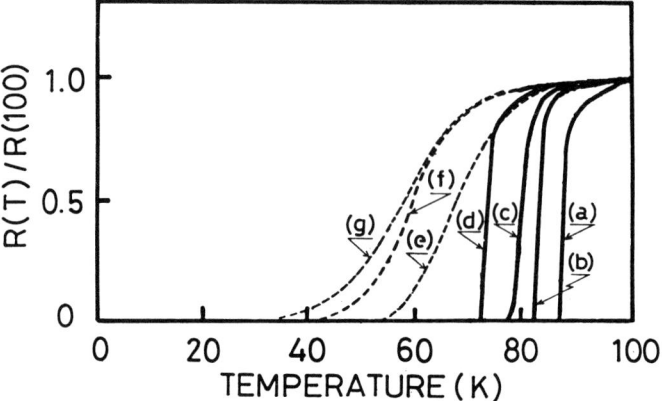

Fig.7-1.Resistance-temperature curves of YBCO/LSCO and YBCO/PBCO

:a)   YBCO/LSCO(=150A/300A);   b)   YBCO/LSCO(=80A/300A);   c)

YBCO/LSCO(=36A/300A);   d)   YBCO/LSCO(=24A/300A);   e)

YBCO/PBCO(=24A/100A) (Q.Li et al.); f) YBCO/PBCO(=24A/144A)

(Triscone et al.); g) ultra thin film for YBCO(=30A) (Terashima

et al.). The thickness of LSCO is fixed at 300A and only those

of YBCO layer are varied.

The superlattice of $(La,Sr)_2CuO_4/Sm_2CuO_4$ can be
classified into three regions.[6](Fig.8) For the short
periodicity of the stacking of each layer up to 4/4 unit
combinations, new structures are constructed having a large
unit cell made of a CuO_5 pyramid, CuO_6 octahedral , and CuO_4
sheet. For the large periodicity more than 60/60, the
superlattices exhibit the same superconducting behavior as the
standard LSCO. The superlattices of the intermediate
periodicity, from 60/60 to 15/15 on the other hand, show
changes in the critical temperature as the variation of the
stacking periodicity. This phenomenon is explained to be
derived from a pressure effect caused by a lattice mismatch
between LSCO and SCO. Thus the systematic correlation between
the compression and the expansion of ab plane and the Tc are
observed in $(La,Sr)_2CuO_4/Sm_2CuO_4$ and $YBa_2Cu_3O_7/La_2CuO_4$
superlattices, indicating that the changes of the in plane Cu-
O bond length is effective to control the Tc vallue .

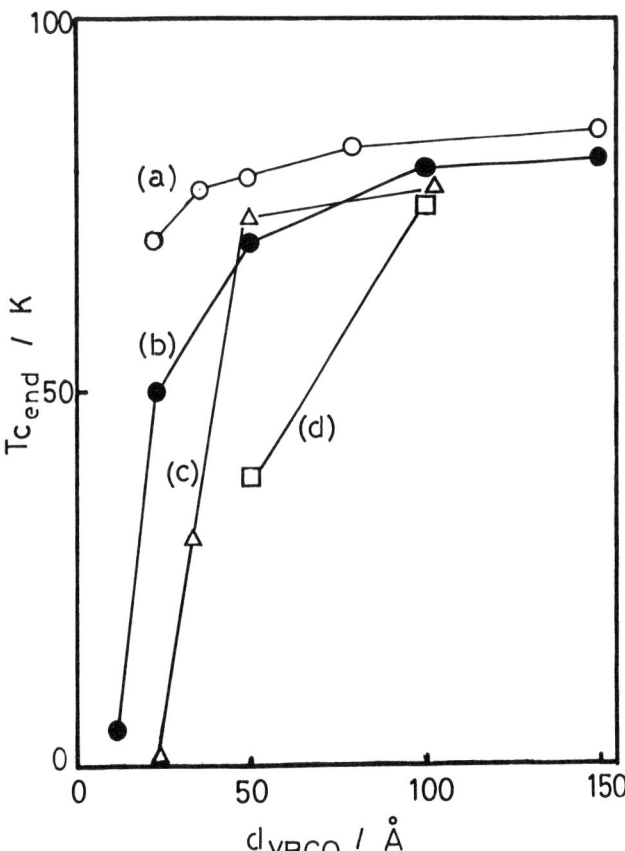

Fig. 7-2. The Tc$_{zero}$ vs. the thickness of YBCO layer of various or ultra thin Y-Ba-Cu-O films: a) YBCO/LSCO (open circles); b) YBCO/PBCO (Q.Li et al.) (filled circles); c) YBCO ultra thin films on MgO(100) substrate (Terashima et al.) (open triangles); d) YBCO/NCCO (Gupta et al.)(open squares).

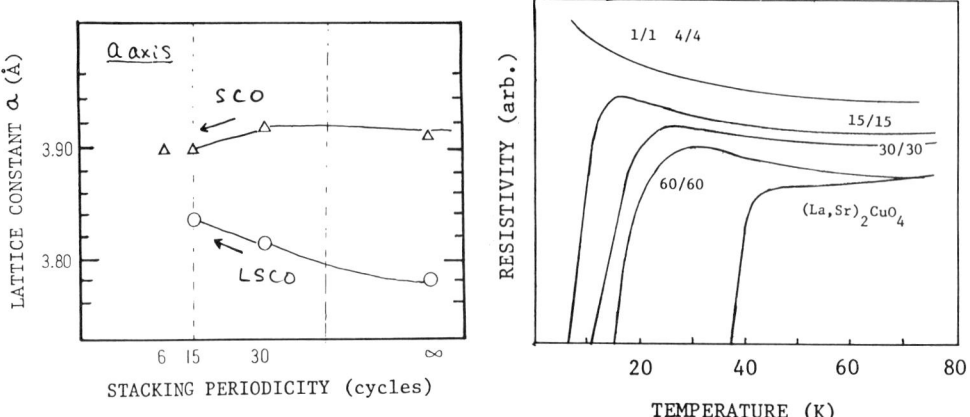

Fig.8.   Changes   of   Tc   and   lattice   constant   c   of
(La,Sr)$_2$CuO$_4$/Sm$_2$CuO$_4$ superlattice versus stacking periodicity.

REFERENCES

1. M.Kanai, T.Kawai, S.Kawai; Appl.Phys.Lett., 58, 771 (1991).
2. M.Kanai et al: Appl.Phys.Lett. 42, (1989) 1802.
3. T.Kawai et al: Nature, 349, (1991) 200.
4. T.Matsumoto, T.Kawai, K.Kitahama, S.Kawai,I.Shigaki and
    Y.Kawate, Appl.Phys.Lett.,58, May 1st (1991).
5. M.Kanai, T.Kawai, S.Kawai: Appl.Phys.Lett., 57, 198 (1990).
6. H.Tabata, T.Kawai,S.Kawai;Appl.Phys.Lett.,58,1443 (1991).

High T_c Superconductor Thin Films
L. Correra (Editor)
© 1992 Elsevier Science Publishers B.V. All rights reserved.

LASER GROWN THIN FILMS OF (RE)Ba$_2$Cu$_3$O$_7$: SURFACE AND IN-PLANE STRUCTURAL DETERMINATION BY RHEED AND WEISSENBERG TECHNIQUES.

M.G. Karkut, M.Guilloux-Viry*, A.Perrin, C.Thivet, J. Padiou, O. Pena, and M. Sergent

Université de Rennes I, Laboratoire de Chimie Minérale B, URA CNRS 254, Avenue du Général Leclerc - 35042 Rennes Cédex, France

* Also C.N.E.T. Lannion B, Division OCM, BP 40, 22301 Lannion, France

ABSTRACT

In addition to the more standard characterisation techniques of θ–2θ x-ray diffractometry, θ-scans (rocking curves), scanning electron microoscopy (SEM), resistance and ac susceptibility measurements, we also use reflection high energy electron diffraction (RHEED) and oscillating crystal x-ray diffraction using a Weissenberg camera to obtain information on the surface and in-plane structure of in-situ laser ablated films of (RE)Ba$_2$Cu$_3$O$_7$ (RE = Y, Pr, Gd, Ho). We obtain RHEED streaks, characteristic of atomically smooth surfaces and epitaxial order, for films grown on (100) surfaces. The films grown on (110) SrTiO$_3$ substrates produce well defined arrays of sharp diffraction spots. Weissenberg photography allows us to determine the in-plane structure of some films on MgO that display anomalous surface features as seen by RHEED and SEM.

1. INTRODUCTION

We have grown by laser ablation thin films of (RE)Ba$_2$Cu$_3$O$_7$ (RE = Y, Pr, Gd, Ho). Part of the motivation for growing other rare earth "123" thin films has been due to the results we obtained on GdBa2Cu3O7 by dc sputtering [1]. We found that the crystallisation of the GdBCO thin films was clearly superior to that of the dc sputtered YBCO films in that both the widths of the θ–2θ reflections and the widths of the rocking curves for GdBCO were clearly and consistently narrower than those of YBCO films. We wanted to see if these results could be duplicated with laser ablated thin films and also where the other rare earth "123" thin films fall in this crystallization characterization scheme. We use the following characterization tools: θ–2θ x-ray diffractometry, rocking curves, scanning electron microscopy, resistance and susceptibility measurements, reflection high energy electron diffraction, and oscillating crystal x-ray diffraction. The latter two methods are interesting in that they give structural information in the plane of the film, thus allowing us to establish the overall structural relation of the thin film to the substrate. We present results on (RE)BCO films grown onto (100)MgO, (100) and (110) SrTiO$_3$.

2. DEPOSITION , SUPERCONDUCTIVITY, AND DIFFRACTOMETRY

The films were grown in-situ by laser ablation[2]. Briefly, incident laser pulses of 40 ns duration and wavelength $\lambda = 308$ nm are focused onto a rotating ceramic target which is at 45° with respect to the pulses. The substrate is glued with silver paint onto a stainless steel holder which is heated to about 700°C during deposition. The substrate directly faces the target and its distance from the target is adjustable and is usually kept at 45 mm. The base pressure of the system before deposition is $\sim 5 \times 10^{-7}$ mbar and during deposition there is a flowing O_2 pressure of ~ 0.5 mbar. We operate the laser at a frequency of 3 Hz. The energy of each pulse is between 80 and 200 mJ and is focused onto a square $\sim 2 \times 2$ mm^2. After deposition the films are cooled to room temperature in an oxygen atmosphere and no further processing is performed.

Before discussing the in-plane methods, we will first briefly describe the film characterization by the more standard methods of θ-2θ diffractometry, rocking curves, resistance and susceptibility measurements. The superconducting properties are the same for YBCO, GdBCO, and HoBCO thin films (PrBCO is, of course, semiconducting): the 87-90K resistive transitions are sharp and the ac susceptibility measurements imply good film homogeneity. This can be seen in Figure 1 in which we show a typical resistance curve for a GdBCO film and and an ac susceptibility transition of a HoBCO film with both the in-phase χ' and the quadrature χ'' components [3]. X-ray diffractograms show the films to be solely c-axis oriented. The rocking curves about the 005 film reflection for the different rare earth "123" films are narrow . These are presented in Figure 2. Rocking curves provide information on the mosaicity of the material and the full-width-half-maximum FWHM is a measure of the film's crystalline quality. We can see that, as with dc sputtered films, GdBCO has a rocking curve narrower than the other (RE)BCO films. Whether the superior crystallization properties of GdBCO are intrinsic or not has yet to be determined. We have shown it to be a result that is reproducible both by laser ablation and by dc sputtering.

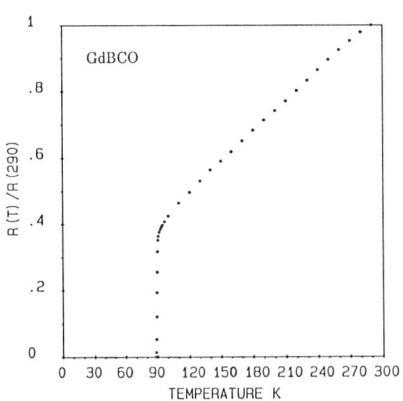 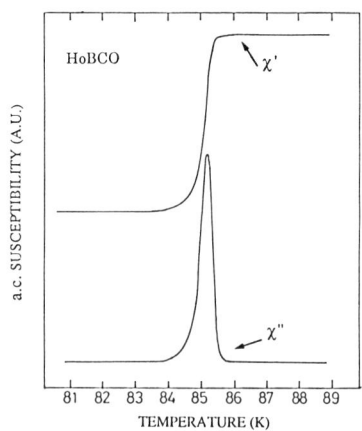

Figure 1. (a) Normalized resistance vs temperature for a GdBCO film grown on (100) SrTiO$_3$. (b) ac susceptibility for a HoBCO film on SrTiO$_3$.

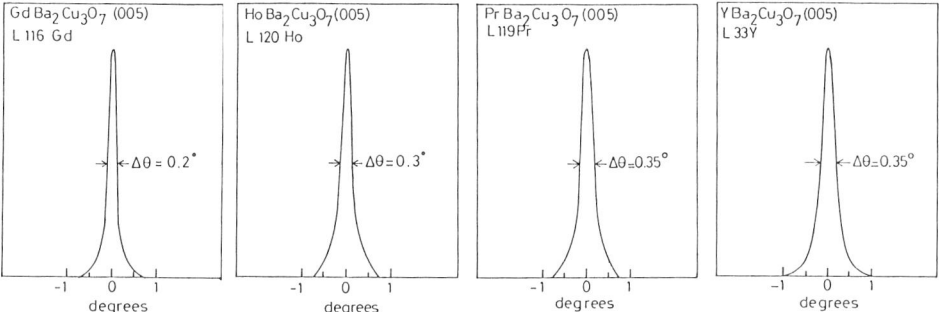

Figure 2. Rocking curves (θ-scans) about the 005 reflection for different (RE)BCO thin films.

3. IN-PLANE CHARACTERIZATION

We use a RIBER model 603 10 keV RHEED system. The RHEED is useful not only for studying the surface structure of the films but also for evaluating the quality of the substrates. RHEED streaks are considered indicative of an atomically smooth surface, and the distance between the streaks is related to the distance between the surface rows of atoms which diffract the electron beam. This is because the reciprocal lattice of an ordered surface layer of atoms is an array of rods perpendicular to the surface and the condition for diffraction is met when there is intersection of a rod or rods with the Ewald sphere. Since the radius of the Ewald sphere is very large, the first row of rods will touch a relatively large section of the sphere, and these are seen as streaks on the phosphor detecting screen. The presence of ordered diffraction spots suggests that there is scattering from the bulk of the material i.e. the surface is no longer smooth. Figure 3a shows RHEED photographs taken of the SrTiO3 surface. We emphasize here that we do **not** use the RHEED during the deposition since the e-gun cannot support the high O₂ pressures necessary to produce the film. The streaks and Kikuchi lines (resulting from the diffraction of inelastically scattered electrons) in Fig. 3a demonstrate the high quality of the substrate surface. The bright circular spot in the center of the film is only an artifact and is the light from the filament of the e-gun. We have found that the substrate must present RHEED streaks before deposition. If a substrate has a RHEED pattern composed only of oriented spots then the film on which it is deposited is more than likely to be of inferior quality ,i.e., not epitaxial and not superconducting down to 77K. The diffraction patterns of Figure 3 b, c, and d are produced by films of GdBCO, PrBCO, and YBCO grown on SrTiO3, respectively. Since the photographs are taken with the same azimuth as their SrTiO3 substrates, these diffraction patterns demonstrate the epitaxial relation of the thin film to the in-plane substrate axes as well as the atomically smooth surfaces of these films. The sharpest and best defined streaks are those produced by GdBCO thin films thus reinforcing the other structural information indicating the superior crystalline quality of the GdBa2Cu3O7 over the other "123" films in this study.

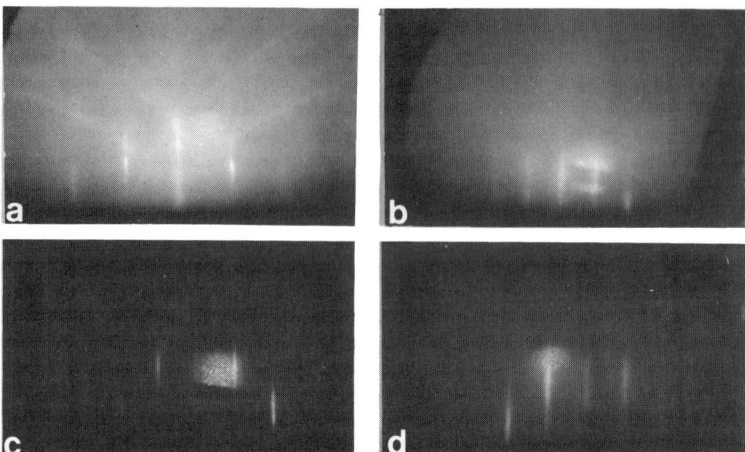

Figure 3. RHEED patterns of (a) a SrTiO₃ substrate, (b) a PrBCO film, (c) a GdBCO film, and (d) aYBCO film. The RHEED patterns of the films were taken after deposition.

We have also grown YBCO and GdBCO films on (110)SrTiO₃ substrates. We present the the RHEED patterns for the substrates and the films in Figure 4. We have not been able to grow films on (110)SrTiO3 that do not have diffraction spots i.e. the surfaces, even though epitaxially oriented, are not as smooth as the surfaces of films grown on (100) substrates. There can be at least two possible reasons for this. One is that even though streaks are observed for the substrates, they are not as sharp as those streaks produced by (100) SrTiO3 nor do they show well defined Kikuchi lines as do the (100) substrates. This could be disadvantageous to the growth of smooth films. The second reason is that some type facetting is produced when the copper oxide planes are not aligned parallel to the substrate surface. The fact that these are (103) oriented films means that the copper oxide planes make an angle of 45° with respect to the substrate surface. Since crystal growth is considerably faster parallel to these planes , one could imagine a sawtooth-like structure resulting from this orientation.

We now turn to oscillating crystal diffraction using a Weissenberg camera. We have recently shown that this is a very useful tool in evaluating the in-plane structure of thin films [4]. Briefly, a single crystal substrate axis is aligned parallel to the camera axis. A monochromatic x-ray beam impinges onto the oscillating substrate. A row of reflections will be recorded which correspond to the scattering vector perpendicular to the plane of the substrate and also parallel rows to the left and right of this center row if the substrate is single crystal. If now a thin film is grown on the substrate, additional reflections will be recorded due to the film. If this film is epitaxially grown onto

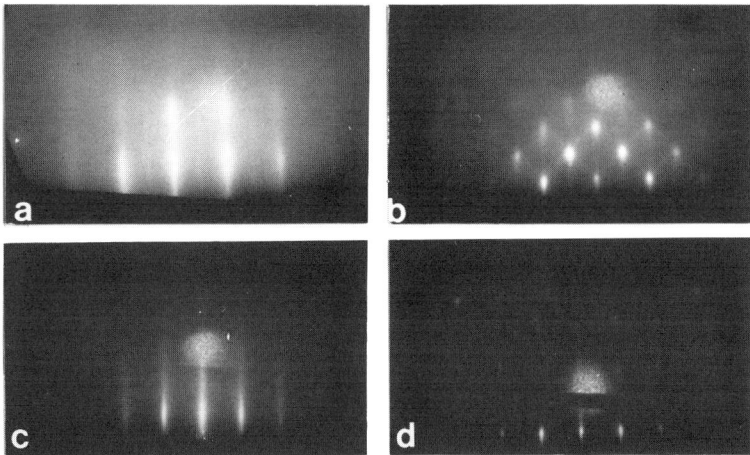

Figure 4. RHEED patterns of (a) a (110) SrTiO$_3$ substrate, and (b) the (103) GdBCO film. (c) and (d) are for the (110) SrTiO$_3$ substrate and the (103) YBCO film grown on it.

the substrate, then additional rows of spots will appear and will be parallel to the substrate axes. With this Weissenberg method the in-plane lattice parameter a of the substrate or the thin film can can be determined from the formula: a = nλ/ sin(arctan(d / 2R)). Here n is the order of the row of streaks (for the central row n = 0), R is the radius of the camera (for this work 2R = 57.295 mm), d is the distance between two rows symmetric about the central row, and λ is the wavelength of the x-radiation (λ = 1.542 Å).

In Figure 5 we show the x-ray diffraction patterns from a laser ablated film of YBCO thin film grown on (100) MgO. In (a) the substrate has been oscillated about its <100> axis while in (b) the substrate has been oscillated about its <110> axis. We denote the substrate reflections by the horizontal arrows and the rows of spots due to the YBCO thin film by the vertical arrows. The well-aligned rows of spots imply the in-plane epitaxy of the film and the interplanar spacing determined from the formula agrees with the expected film values. In contrast to our sputtered films some of our laser ablated films on MgO exhibit abnormal surfaces of needle shaped crystals oriented at 45° to the MgO axes as seen in SEM. Their Weissenberg pattern[2] , when oscillated about either the <100> or <110> MgO axis, is the superposition of the two diagrams of Figure (4a) and (4b). This means that these films consist of two sets of "123" crystals: one with its a/b axes parallel to the <100> MgO axes and the other with its a/b axes parallel to the <110> MgO axes. Thus both types of these crystallites are epitaxially oriented.

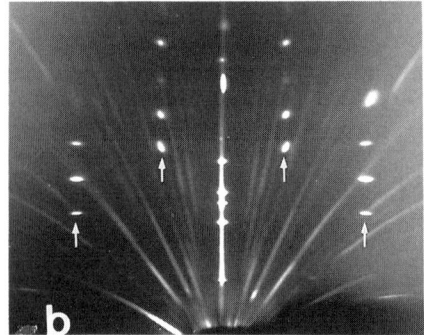

Figure 5. Weissenberg photographs taken of a YBCO film on a MgO substrate. In (a) the rotation axis is <100> MgO and in (b) the rotation axis is <110> MgO. The vertical arrows mark the rows of reflections due to the thin film.

4. CONCLUSION

We have laser ablated high quality $(RE)Ba_2Cu_3O_7$ films on substrates of (100)MgO, (100) $SrTiO_3$ and (110)$SrTiO_3$. Films on (100) substrates are c-axis oriented and have narrow (FWHM $\geq 0.2°$ on $SrTiO_3$) rocking curves. The superconducting properties are good with T_{co}'s > 87 K and Jc (77 K) ~ 10^6 A/cm^2. In addition, we have demonstrated the epitaxial order of these films by RHEED and Weissenberg x-ray photography. Films grown on (100) substrates can display well-defined RHEED streaks which is indicative of atomic smoothness. We observe only well ordered diffraction spots for films grown on (110) $SrTiO_3$. Weissenberg photographs show that the films are aligned with the substrate axes even in the case of the "anomalous" 45° crystal growth. Finally we have again observed the cystalline superiority of $GdBa_2Cu_3O_7$.

This work was supported in part by the Centre National d'Etudes des Telecommunications, CNET Lannion B, under contract No. 89 8B054, by the Ministère de la Recherche et de la Technologie under contract No. 89 H0556 , and by the Fondation Langlois.

5. REFERENCES

1 M. G. Karkut, M. Guilloux-Viry, A. Perrin, O. Pena, J. Padiou, and
   M. Sergent, J. Less-Common Met. 164&165 (1990) 336.
2 M. G. Karkut, M. Guilloux-Viry, A. Perrin, J. Padiou, O. Pena, and
   M. Sergent, submitted to Physica C.
3 O. Pena, Measurement Science and Technology, in press;
   M. Guilloux-Viry, M. G. Karkut, A. Perrin, O. Pena, J. Padiou, and
   M.Sergent, Physica C166 (1990) 105.
4 A. Perrin, M. G. Karkut, M. Guilloux-Viry, and M. Sergent,
   Appl. Phys. Lett. 58 (1991) 412.

High T$_c$ Superconductor Thin Films
L. Correra (Editor)

Chemical vapor deposition of high-Tc superconducting oxide thin films

H. Yamane and T. Hirai

Institute for Materials Research, Tohoku University
2-1-1 Katahira, Aoba-ku, Sendai 980, Japan

Abstract

Recent work on chemical vapor deposition (CVD) of high-temperature superconducting thin films was reviewed. Y-Ba-Cu-O (YBCO) thin films having zero-resistance transition temperatures (Tc(R=0)) of about 90 K and critical current densities (Jc) over 10^6 A/cm^2 at 77 K and 0 T were made on oxide single-crystal substrates by CVD using β-diketonate metal-organic (MO) precursors. Bi-Sr-Ca-Cu-O (BSCCO) superconducting thin films were prepared by halide CVD and MOCVD. The BSCCO film composed of the high-Tc phase, Bi$_2$Sr$_2$Ca$_2$Cu$_3$O$_x$, showed a Tc of 93 K. Tl-Ba-Ca-Cu-O(TBCCO) films with Tc(R=0) over 100 K and Jc of 10^4 A/cm^2 at 90 K and 0 T were fabricated by the heat treatment of MOCVD Ba-Ca-Cu-O films in Tl vapor.

1. INTRODUCTION

Since the discovery of a superconductor with zero-resistance transition temperature (Tc(R=0)) above the boiling point of liquid nitrogen (77.3 K), many techniques have been studied for the preparation of high-Tc superconducting thin films [1]. Chemical vapor deposition (CVD) techniques are characterized by greater compatibility with various shaped substrates and greater flexibility for large-scale deposition in comparison with physical vapor deposition (PVD) techniques such as sputtering and vacuum evaporation.

Dated March 2, 1988, two papers reported successful synthesis of Y-Ba-Cu-O (YBCO) superconducting films by CVD using metal-organic (MO) compounds of β-diketone metal chelates as sources: one was by Berry et al. [2] and the other by the present authors [3]. Since then, chemical vapor deposition of superconducting oxide films, including Bi-Sr-Ca-Cu-O (BSCCO) and Tl-Ba-Ca-Cu-O (TBCCO) superconducting films, has been pursued by many investigators [4].

The film preparation process that Berry et al. reported was a post-deposition annealing process [2]. They prepared an amorphous YBCO film. The film was then taken out of the CVD reactor and crystallized by annealing at 920°C. Our process was in-situ growth of crystalline YBCO films at 800-900°C and succeeding in-situ oxygen treatment [3,5,6]. Recently, much work has been reported on in-situ preparation of YBCO and BSCCO films by CVD at lower deposition temperatures.

This paper reviews work on in-situ preparation of YBCO and BSCCO films and on fabrication of TBCCO films by CVD.

2. MOCVD-YBCO THIN FILMS

2.1. Precursors
    The sources used for in-situ deposition of YBCO films were usually β-diketonate of Y, Ba and Cu such as Y(DPM)3, Ba(DPM)2 and Cu(DPM)2 (DPM: dipivaloylmethanate, 2,2,6,6-tetramethyl-3,5-heptanedionate: C11H19O2). The structure, sublimation or evaporation behavior and thermal stability of the DPM chelates have been reported [7-10]. In order to vaporize Ba(thd)2 at a steady condition, dipivaloylmethane (H(DPM)) was added to the carrier gas [11,12]. Matsuno et al. reported that addition of tetrahydrofuran (THF) to the carrier gas stabilized the vaporization of Ba(thd)2 and increased the volatility [13,14]. Some β-diketone chelates with fluorocarbon ligands, which have higher volatility than chelates with hydrocarbon ligands, were used for MOCVD sources [15-17]. However, it has been reported that fluorine was contained in as-deposited films as fluoride and that heat treatment was necessary in order to obtain superconductivity.

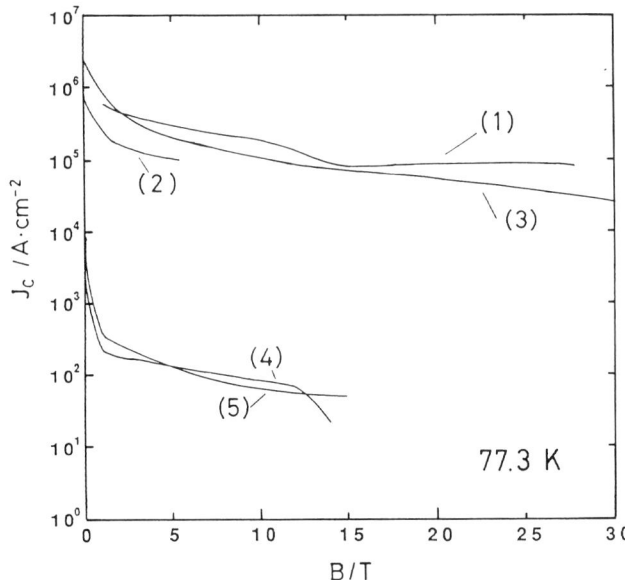

Fig. 1. Critical current densities of MOCVD-YBCO films prepared on SrTiO3(100) (1)[19], (2)[12] and (3)[20], on MgO(100) (4)[21] and on a Y2O3 stabilized ZrO2 polycrystalline substrate (5)[24] as a function of applied magnetic field parallel to the substrate plane and perpendicular to current.

2.2. High-Jc films

YBCO films prepared on SrTiO3(100) single-crystal substrates by MOCVD were mainly composed of YBa2Cu3Ox with c-axis orientation perpendicular to the substrate plane. The films exhibited Jc over 10^6 A/cm^2 at 77.3 K and 0 T [13,18]. High Jc was also measured in magnetic fields. Figure 1 shows the dependence of Jc on magnetic fields at the temperature of 77.3 K. We measured a Jc of 6.5x10^4 A/cm^2 at 77.3 K and in a field of 27 T for the 1 μm thick film prepared at the deposition temperature of 850°C [19]. The value of Jc over 10^4 A/cm^2 at 77.3 K and 30.2 T was measured for an MOCVD-YBCO film with a thickness of 0.25 μm by Matsuno et al. [9,20].

These high values of Jc were measured under the fields applied parallel to the substrate plane and perpendicular to the transport current. When the direction of the fields was perpendicular to the substrate plane, Jc decreased rapidly with increasing field strength [21]. Large anisotropy of upper critical fields as well as the anisotropy of Jc was reported for the MOCVD-YBCO films [22,23].

YBCO films prepared on MgO(100) single-crystal substrates and Y2O3 stabilized ZrO2 polycrystal substrates were also mainly composed of c-axis oriented YBa2Cu3Ox. However, Jc of these films was two orders of magnitude lower than that of the films prepared on SrTiO3 [21,24]. The low Jc was probably caused by the many boundaries between the YBCO grains with different orientations in the substrate plane, which were observed by scanning electron microscopy and transmission electron microscopy [24,25].

Table 1
Tc and Jc of MOCVD-YBCO films prepared at various conditions: deposition temperature (Tdep), partial pressure of reactant gas (P(r.g.))

	Tdep (°C)	reactant gas	P(r.g.) (Torr)	Tc(R=0) (K)	Jc(77 K, 0 T)(A/cm^2)	ref.
thermal MOCVD	900	O2	1.1	>90	8.5x10^5	[12]
	850	O2	3.6	92	2.0x10^6	[18]
	850	O2	7.4	92	6.3x10^6	[13]
	700	O2	0.036	89	2.2x10^6	[38]
	650	O2	0.036	85	1.3x10^4	[38]
	650	O2+O3	1.7	80	--	[28]
	730	N2O	2	89	2.3x10^6	[27]
	650	N2O	27	79	--	[26]
plasma MOCVD	580	O2	0.035	85	1.0x10^5	[29]
	515	O2	0.035	60	--	[29]
	730	N2O	2	88	5.0x10^5	[31]
	670	N2O	2	90	1.0x10^6	[32]
	570	N2O	5-10	72	--	[30]

2.3. Low-temperature growth

Low-temperature preparation of high-Tc superconducting oxide thin films is necessary for their wide application. In order to lower the deposition temperature, active reactant gas and plasma enhancement have been introduced into the MOCVD process, as well as into the PVD process. Recently, low temperature growth of YBCO films by conventional thermal CVD has also been studied under low oxygen partial pressure (P(O2)). Table 1 lists superconducting properties of YBCO films prepared by various MOCVD techniques.

A. Active reactant gas

Tsuruoka et al. first reported CVD of YBCO films using N2O as an active reactant gas [26]. The YBCO film they prepared at 650°C showed a Tc(R=0) of 79 K. Li et al. also used N2O as the reactant gas and prepared a YBCO film on LaAlO3(100) at 730°C [27]. The Tc(R=0) and Jc at 77.3 K and 0 T of the film were 89 K and 2.3×10^6 A/cm^2, respectively. The film having Tc(R=0) = 80 K was deposited at 650°C by addition of O3 as an active reactant gas [28].

B. Plasma-enhancement

Kanehori et al. were successful in the preparation of a YBCO film at a deposition temperature below 600°C by microwave plasma-enhanced (PE) MOCVD [29]. The Tc(R=0) and Jc(77.3 K, 0 T) of the film obtained at 580°C were 85 K and 10^5 A/cm^2, respectively. Zhao et al. prepared a YBCO film with Tc(R=0)=72 K at 570°C by introducing N2O into a PE-MOCVD process [14,30]. The YBCO films they prepared on LaAlO3 at 670°C showed Tc(R=0) = 90 K and Jc (77.3 K, 0 T) = 1.0×10^6 A/cm^2 [31,32]. YBCO films having Tc(R=0) = 85 and 82 K were prepared by PE-MOCVD on Ag substrates and Al2O3 substrates, respectively [33,34]. The effect of plasma enhancement on the CVD of YBCO films was studied spectroscopically [35,36].

C. Oxygen partial pressure

In our previous studies and other studies [12,13,18,37], high quality YBCO films were obtained around 850°C by a conventional thermal CVD technique. Recently, we succeeded in preparing a high-Jc YBCO film at 700°C by thermal CVD under a low oxygen partial pressure (P(O2)) of 0.036 Torr which was a hundredth of the P(O2) in the previous preparation at 850°C [38]. A YBCO film with Tc(R=0) = 85 K was also obtained at 650°C and P(O2) = 0.036 Torr [39].

Figure 2 shows the deposition temperature (Tdep) and oxygen partial pressure P(O2) at which high-Jc YBCO films were prepared by thermal CVD, together with the CuO-Cu2O-O2 equilibrium line and the YBa2Cu3Ox stability line [40-42]. The conditions of Tdep and P(O2) were plotted in close proximity to the CuO-Cu2O-O2 line where high quality YBCO films were usually prepared in-situ by PVD [43].

Vahlas et al. investigated MOCVD conditions of YBCO films by thermodynamic calculation. They showed that the MOCVD parameters reported for the preparation of YBCO films were in the region of the formation of YBa2Cu3Ox as a stable phase [44].

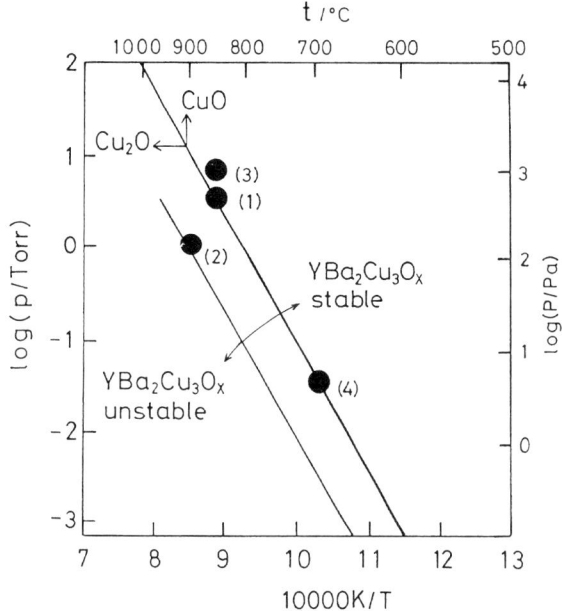

Fig. 2. Oxygen partial pressure vs. temperature plot showing the critical stability line for YBa2Cu3Ox [41,42] and the CuO-Cu2O-O2 equilibrium line [40] together with the conditions at which high-Jc YBCO films were obtained by thermal MOCVD. The Jc(77.3 K and 0 T) of the films obtained at (1), (2), (3) and (4) were 2.0x10^6 [18], 8.5x10^5 [12], 6.3x10^6 [13] and 2.2x10^6 A/cm^2 [38], respectively.

2.4. 124 phase

YBa2Cu4Ox (124) is another high-Tc superconducting phase in the YBCO system. Hayashi et al. reported the in-situ formation of a film composed of YBa2Cu4Ox and YBa2Cu3Ox (123) at deposition temperatures from 750 to 800°C and P(O2) from 7.5 to 34 Torr [45]. The films deposited at 850°C consisted of the 123 phase and CuO. The volume ratio of 124/123 increased with film thickness. They concluded from an annealing experiment that the 124 phase was not formed by solid state reaction. The 123 and 124 phases were co-deposited during the early stage of film growth and the 124 phase was the main phase near the film surface [46].

3. CVD-BSCCO THIN FILMS

3.1. Halide CVD

The first successful preparation of BSCCO superconducting films by CVD was conducted by Ihara et al. using halides, $BaCl_3$, SrI_2, CaI_2 and CuI, as sources [47]. They prepared a film composed of $Bi_2Sr_2CaCu_2O_x$ (2212, low T_c phase) and $Bi_2Sr_2Ca_2Cu_3O_x$ (2223, high T_c phase) at 850°C under 12% oxygen in an open tube reactor [48]. The film exhibited $T_c(R=0)$ at 98 K and a J_c of 1.7×10^4 A/cm^2 at 4.2 K and 0 T. A J_c of 1.0×10^6 A/cm^2 was measured at 50 K and 0 T for the BSCCO films prepared on MgO(100) at 775°C [49].

Transmission electron microscope observation revealed that the films consisted of c-axis oriented domains with diameters of 15-100 μm [50]. Transport properties of the boundaries between these grains were investigated [51,52].

3.2. Plasma-enhanced halide CVD

Kimura et al. developed a plasma-enhanced (PE) halide CVD process for in-situ formation of BSCCO films at low deposition temperatures [53,54]. A film of the 2212 phase with $T_c(R=0)$ = 70 K and $J_c(10$ K, 0 T) = 2.5×10^6 A/cm^2 was prepared at 580°C on sapphire ($1\bar{1}02$). A film mainly composed of the 2223 phase, having $T_c(R=0)$ = 55 K, was also obtained.

The heterostructure of a BSCCO superconducting thin film and a Bi-Sr-Cu-O metallic thin film were constructed by halide CVD with the open tube and by PE-halide CVD [49].

3.3. MOCVD

The sources used for in-situ preparation of MOCVD-BSCCO films were usually triphenylbismuth ($Bi(C_6H_5)_3$) and β-diketone metal chelates such as $Sr(DPM)_2$, $Ca(DPM)_2$ and $Cu(DPM)_2$. Purification, vaporization behavior and thermal stability of these sources were investigated [7,55-57].

The deposition temperatures reported for the in-situ growth of high-T_c BSCCO films by MOCVD were around 800°C in the oxygen partial pressure range from 0.9 to 23 Torr. Table 2 lists growth rates, thickness and $T_c(R=0)$ of the BSCCO films prepared at 800°C on MgO(100) substrates. The films obtained at a growth rate of 0.5-1 μm/h consisted of the 2212 phase or mixtures of the 2212 and 2223 phases. The $T_c(R=0)$ of these films was about 72-82 K [58,59].

Endo et al. reported the preparation of a BSCCO film made up of only the 2223 phase [60]. However, this single-phase film exhibited a $T_c(R=0)$ of 73 K. Recently, they were successful in the MOCVD of a 2223 film with $T_c(R=0)$ = 93 K by lowering the deposition rate from 50 nm/h to 6 nm/h [61]. The thickness of the film was 180 nm. Elongation of the c-axis length was observed for the films with thicknesses less than 100 nm. $T_c(R=0)$ of the films decreased with increasing c-axis length. Endo et al. considered that the strain caused by the misfit of in-plane lattice constants between the BSCCO and MgO caused the expansion of the c-axis length for the thinner films.

An ultrathin BSCCO film of the 2212 phase, having an average thickness of 3.5 nm and $T_c(R=0)$ = 64 K, was prepared by MOCVD [62]. The films having thicknesses over 5.5 nm mainly consisted of the 2223 phase with a trace amount of the 2212 phase and exhibited a $T_c(R=0)$ around 80 K. The c-axis lengths of the 2212 and 2223 phases in both the thinner and thicker films coincided with the lengths of the bulk samples. Sugimoto et al. presumed

Table 2
Deposition rate and film thickness of BSCCO films prepared
at 800°C and various partial pressures (P(O2)) by MOCVD
and their Tc

phase	P(O2) (Torr)	deposition rate(μm/h)	thickness (μm)	Tc(R=0) (K)	ref.
2212>2223	0.9	1	1	80	[58]
2212	10	0.5-1	1	72	[59]
2223	20	0.05	0.25	73	[60]
2223	23	0.006	0.18	93	[61]
2212	8.8	0.02	0.0035	64	[62]
2223>2212	8.8	0.02	0.0055	80	[62]

that the stress caused by the large lattice mismatch between BSCCO and MgO
was released by formation of the misfit dislocations observed in a cross-
sectional transmission electron micrograph of a thicker film [63,64]. They
considered that the 2212 phase was grown on the MgO substrate and that the
2223 phase was formed by the diffusion of CaO and CuO into the 2212 phase
from the gas phase or the solid phase.

4. TBCCO THIN FILMS

Only Zhang et al. have used four metal sources of Tl, Ba, Ca and Cu
(Tl(DPM), Ba(DPM)2, Ca(DPM)2 and Cu(DPM)2) during film deposition [65]. In
order to obtain a superconducting phase in the film, the as-deposited TBCCO
film was annealed at 850°C in the presence of a Tl2Ca2Ba2Cu3Oy pellet. The
films showed a Tc(R=0) of 94 K after further heat treatment at 500°C in
oxygen.
   The preparation of TBCCO films by CVD in other studies was achieved by
the use of two steps probably because of the high volatility of Tl [66-68].
Ba-Ca-Cu-O films were first prepared by MOCVD using β-diketone metal
chelates as sources. Thallium was next incorporated in these films either
by vapor diffusion at 800-900°C using a bulk of TBCCO [66] or pellets of
TBCCO powder [67,68], or by MOCVD using Tl(C5H5) [66].
   Hamaguchi et al. obtained a c-axis oriented Tl2Ba2CaCu2Ox film with
Tc(R=0) = 108 K. The Jc of the film was as high as 10^4 A/cm^2 at 90 K and 0
T [67]. The films composed predominantly of Tl2Ba2Ca2Cu3Ox or
Tl2Ba2CaCu2Ox were prepared using films with approximate Ba-Ca-Cu
stoichiometries of 2:2:3 or 2:1:2. The TBCCO films exhibited Tc(R=0) of 93
and 85 K, respectively [68].

5. CONCLUSION

Low temperature growth of high quality superconducting oxide films have recently been carried out by CVD using active reactant gases and plasma-enhancement (PE). YBCO films having J_c over 10^6 A/cm^2 at 77.3 K and 0 T have already been prepared at a deposition temperature around 700°C. Further study will lead to the reduction of deposition temperature and open the way to the development of high-T_c superconducting films on semiconductor substrates such as Si. In the system of BSCCO, a heterostructure has been fabricated by CVD. A multilayered heterostructure may also be constructed by CVD in the system of YBCO in the near future. Optimum conditions of Tl incorporation into the films of Ba-Ca-Cu-O will raise the critical current density of the TBCCO films.

6. REFERENCES

1   M. Leskela, J.K. Truman, C.H. Mueller and P.H. Holloway, J. Vac. Sci. Technol., A7(1989)3147.
2   A.D. Berry, D.K. Gaskill, R.T. Holm, E.J. Cukauskas, R. Kaplan and R.L. Henry, Appl. Phys. Lett., 52(1988)1743.
3   H. Yamane, H. Kurosawa and T. Hirai, Chem. Lett., (1988)939.
4   T. Hirai and Y. Yamane, Oyo Buturi, 59(1990)134.
5   H. Yamane, H. Kurosawa, H. Iwasaki, H. Masumoto, T. Hirai, N. Kobayashi and Y. Muto, Jpn. J. Appl. Phys., 27(1988)L1275.
6   H. Yamane, H. Masumoto, T. Hirai, H. Iwasaki, K. Watanabe, N. Kobayashi, Y. Muto, and H. Kurosawa, Appl. Phys. Lett., 53(1988)1548.
7   S. Yuhya, K. Kikuchi, M. Yoshida, K. Sugawara and Y. Shinohara, Mol. Cryst. Liq. Cryst. 184(1990)231.
8   S.B. Turnipseed, R.M. Barkley and R.E. Sievers, submitted to Inorg. Chem., (1990).
9   S. Matsuno, F. Uchikawa, S. Utsunomiya and M. Tanaka, Proceedings of the 3rd International Symposium on Superconductivity, Sendai 1990, K. Kajimura and H. Hayakawa (eds.), Advances in Superconductivity III, Springer-Verlag, Tokyo, 1991 (to be published).
10 K. Higashiyama, T. Ushida, H. Higa, I. Hirabayashi and S. Tanaka, Jpn. J. Appl. Phys., (1991) (to be published).
11 P.H. Dickinson, T.H. Geballe, A. Sanjurjo, D. Hildenbrand, G. Craig, M. Zinsk, J. Collman, S.A. Banning, and R.E. Sievers, J. Appl. Phys., 66(1989)444.
12 F. Schmaderer and G. Wahl, J. Phys., C5(1989)119.
13 S. Matsuno, F. Uchikawa and K. Yoshizaki, Jpn. J. Appl. Phys., 29(1990)L947.
14 J. Zhao, C.S. Chern, Y.Q. Li, D.W. Noh, P.E. Norris, P. Zawadzki, B. Kear and B. Gallois, J. Cryst. Growth., 107(1991)699.
15 K. Shinohara, F. Munakata and M. Yamanaka, Jpn. J. Appl. Phys., 27 (1988)L1683.
16 J. Zhao, K.-H. Dahmen, H.O. Marcy, L.M. Tonge, T.J. Marks, B.W. Wessels, and C.R. Kannewurf, Appl. Phys. Lett., 53(1988)1750.
17 A.J. Panson, R.G. Charles, D.N. Schmidt, J.R. Szedon, G.J. Machiko and A.I. Braginski, Appl. Phys. Lett., 53(1988)1756.
18 H. Yamane, H. Kurosawa, T. Hirai, K. Watanabe, H. Iwasaki, N. Kobayashi and Y. Muto, Supercond. Sci. Technol., 2(1989)115.

19 K. Watanabe, H. Yamane, H. Kurosawa, T. Hirai, N. Kobayashi, H.
    Iwasaki, K. Noto and Y. Muto, Appl. Phys. Lett., 54(1989)575.
20 S. Matsuno, F. Uchikawa, K. Yoshizaki, N. Kobayashi, K. Watanabe,
    Y. Muto and M. Tanaka, Proceedings of the 1990 Applied
    Superconductivity Conference, Colorado 1990 (to be published).
21 H. Yamane, T. Hirai, H. Kurosawa, A. Suhara, K. Watanabe, N. Kobayashi,
    H. Iwasaki, E. Aoyagi, K. Hiraga and Y. Muto, Proceedings of the 2nd
    International Symposium on Superconductivity, Tsukuba, 1989,
    T. Ishiguro and K. Kajimura (Eds.), Advances in Superconductivity II,
    Springer-Verlag, Tokyo, 1990, p.767.
22 K. Watanabe, S. Awaji, N. Kobayashi, H. Yamane, T. Hirai and Y. Muto,
    J. Appl. Phys., 69(1991)1543.
23 N. Kobayashi, H. Kawabe, K. Watanabe, S. Awaji, H. Yamane, H. Kurosawa,
    T. Hirai and Y. Muto, Supercond. Sci. Technol., 4(1991)S328.
24 H. Kurosawa, H. Yamane, T. Hirai, K. Watanabe, S. Awaji, N. Kobayashi
    and Y. Muto, Supercond. Sci. Technol. (1991) (to be published).
25 H. Suzuki, H. Kurosawa, K. Miyagawa, Y. Hirotsu, M. Era, T. Yamashita,
    H. Yamane and T. Hirai, Jpn. J. Appl. Phys., 29(1990)L1648.
26 T. Tsuruoka, R. Kawasaki and H. Abe, Jpn. J. Appl. Phys., 28(1989)
    L1800.
27 Y.Q. Li, J. Zhao, C.S. Chern, W. Huang, G.A. Kulesha, P. Lu, B.
    Gallois, P. Norris, B. Kear and F. Cosandey, Appl. Phys. Lett.,
    58(1991)648.
28 H. Ohnishi, H. Harima, Y. Kusakabe, M. Kobayashi, S. Hoshinouchi and
    K. Tachibana, Jpn. J. Appl. Phys., 29(1990)L2041.
29 K. Kanehori, N. Sugii and K. Miyauchi, Proc. Mat. Res. Soc. 196(1989)
    585.
30 J. Zhao, D.W. Noh, C. Chern, Y.Q. Li, P. Norris, B. Gallois and B.
    Kear, Appl. Phys. Lett., 56(1990)2342.
31 C.S. Chern, J. Zhao, Y.Q. Li, P. Norris, B. Kear, B. Gallois and Z.
    Kalman, Appl. Phys. Lett., 58(1991)185.
32 J. Zhao, Y.Q. Li, C.S. Chern, W. Huang, P. Norris, B. Gallois, B. Kear,
    P. Lu and F. Cosandey, Proceedings of the 3rd International Symposium
    on Superconductivity, Sendai 1990, K. Kajimura and H. Hayakawa (eds.),
    Advances in Superconductivity III, Springer-Verlag, Tokyo, 1991 (to be
    published).
33 J. Zhao, Y.Q. Li, C.S. Chern, P. Norris, B. Gallois, B. Kear and
    B.W. Wessels, Appl. Phys. Lett., 58(1991)89.
34 C.S. Chern, J. Zhao, Y.Q. Li, P. Norris, B. Kear and B. Gallois,
    Appl. Phys. Lett., 57(1990)721.
35 H. Harima, H. Ohnishi, K. Hanaoka, K. Tachibana, M. Kobayashi and
    S. Hoshinouchi, Jpn. J. Appl. Phys., 29(1990)1932.
36 N. Sugii, K. Imagawa, S. Saito and K. Kanehori, Jpn. J. Appl. Phys.,
    30(1991)48.
37 H. Yamane, H. Kurosawa, T. Hirai, K. Watanabe, H. Iwasaki, N. Kobayashi
    and Y. Muto, J. Cryst. Growth., 98(1989)860.
38 H. Yamane, T. Hirai, K. Watanabe, N. Kobayashi, Y. Muto, M. Hasei and
    H. Kurosawa, J. Appl. Phys. 69(1991) (to be published).
39 H. Yamane, M. Hasei, H. Kurosawa and T. Hirai, Jpn. J. Appl. Phys.,
    (1991) (to be published).
40 O. Kubachewski and C.B. Alcock, Metallugical Thermochemistry, Pergamon
    Press, Oxford, 1979.
41 T.B. Lindemer, J.F. Hunley, J.E. Gates, A.L. Sutton, Jr., J. Brynestad,
    C.R. Hubbard and P.K. Gallagher, J. Am. Ceram. Soc., 72(1989)1175.

42 T.B. Lindemer, F.A. Washburn, C.S. MacDougall and O.B. Cavin, Physica C
    174(1991)135.
43 R.H. Hammond and R. Bormann, Physica C 162-164(1989)703.
44 C. Vahlas and T.M. Besmann, Proceedings of the 11th International
    Conference on Chemical Vapor Deposition, Seattle, K.E. Spear and G.W.
    Cullen (eds.) (Electrochemical Society, Pennington, 1990), p.188.
45 H. Hayashi, Y. Yamada, K. Sugawara, Y. Shiohara and S. Tanaka,
    Jpn. J. Appl. Phys., 30(1991)L352.
46 H. Hayashi, Y. Yamada, T. Sugimoto, Y. Shiohara and S. Tanaka,
    Jpn. J. Appl. Phys., 30(1991)L725.
47 M. Ihara and T. Kimura, Extended Abstracts of 5th International
    Workshop on Future Electron Devices, Miyagi-Zao, (1988), p.137.
48 T. Kimura, M. Ihara, H. Yamawaki, K. Ikeda and M. Ozeki, Proceedings of
    the 1st International Symposium on Superconductivity, Nagoya 1988,
    K. Kitazawa and H. Ishiguro (eds.), Advances in Superconductivity,
    Springer-Verlag, Tokyo, 1989, 495.
49 M. Ihara, H. Nakao, H. Yamawaki and T. Kimura, Proceedings of 3rd FED
    Workshop on High-Temperature Superconducting Electron Devices, R & D
    Association for Future Electron Devices, Kumamoto 1991 (to be
    published).
50 O. Ueda, T. Kimura, H. Yamada, H. Yamawaki, K. Ikeda, M. Ihara and
    M. Ozeki, J. Cryst. Growth, 99(1990)958.
51 C. Tanaka, T. Nakamura, H. Yamawaki, T. Kimura and M. Ihara, Physica
    B, 165&166(1990)1421.
52 C. Tanaka, T. Nakamura, H. Yamawaki, T. Kimura, O. Ueda and M. Ihara,
    Proceedings of the 3rd International Symposium on Superconductivity,
    Sendai 1990, K. Kajimura and H. Hayakawa (eds.), Advances in
    Superconductivity III, Springer-Verlag, Tokyo, 1991 (to be published).
53 T. Kimura, H. Nakao, H. Yamawaki, M. Ihara and M. Ozeki, IEEE Trans.
    Magn., 27(1991)1211.
54 H. Nakao, T. Kimura, H. Yamawaki, M. Ihara and M. Ozeki, Proceedings of
    the 3rd International Symposium on Superconductivity, Sendai 1990,
    K. Kajimura and H. Hayakawa (eds.), Advances in Superconductivity III,
    Springer-Verlag, Tokyo, 1991 (to be published).
55 T. Hashimoto, K. Kitazawa, Y. Suemune, T. Yamamoto and H. Koinuma,
    Jpn. J. Appl. Phys., 29(1990)L2215.
56 T. Sugimoto, S. Yuhya, Y. Yamada, H. Hayashi, K. Kikuchi, M. Yoshida,
    K. Sugawara and Y. Shiohara, Proceedings of the 2nd International
    Symposium on Superconductivity, Tsukuba, 1989, T. Ishiguro and K.
    Kajimura (Eds.), Advances in Superconductivity II, Springer-Verlag,
    Tokyo, 1990, p. 907.
57 A.D. Berry, R.T. Holm, M. Fatemi and D.K. Gaskill, J. Mater. Res.,
    5(1990)1169.
58 K. Natori, S. Yoshizawa, J. Yoshino and H. Kukimoto, Jpn. J.
    Appl. Phys., 28(1989)L1578.
59 K. Saikusa, T. Sugihara and T. Takeshita, Proceedings of the 2nd
    International Symposium on Superconductivity, Tsukuba, 1989,
    T. Ishiguro and K. Kajimura (Eds.), Advances in Superconductivity II,
    Springer-Verlag, Tokyo, 1990, p.899.
60 K. Endo, S. Hayashida, J. Ishiai, Y. Matsuki, Y. Ikedo, S. Misawa and
    S. Yoshida, Jpn. J. Appl. Phys., 29(1990)L294.
61 K. Endo, K. Nakatsuka, S. Misawa and S. Yoshida, Proceedings of the 3rd
    International Symposium on Superconductivity, Sendai 1990, K. Kajimura
    and H. Hayakawa (eds.), Advances in Superconductivity III, Springer-

Verlag, Tokyo, 1991 (to be published).

62 T. Sugimoto, M. Yoshida, K. Sugawara, Y. Shiohara and S. Tanaka, Appl. Phys. Lett., 58(1991)1103.

63 T. Sugimoto, M. Yoshida, K. Yamaguchi, K. Sugawara, Y. Shiohara and S. Tanaka, Appl. Phys. Lett., 57(1990)928.

64 T. Sugimoto, M. Yoshida, K. Yamaguchi, Y. Yamada, K. Sugawara, Y. Shiohara and S. Tanaka, J. Cryst. Growth, 107(1991)692.

65 K. Zhang, E.P. Boyd, B.S. Kwak, A.C. Wright and A. Erbil, Appl. Phys. Lett., 55(1989)1258.

66 D.S. Richeson, L.M. Tonge, J. Zhao, J. Zhang, H.O. Marcy, T.J. Marks, B.W. Wessels and C.R. Kannewurf, Appl. Phys. Lett., 54(1989)2154.

67 N. Hamaguchi, R. Gardiner, P.S. Kirlin, R. Dye, K.M. Hubbard and R.E. Muenchausen, Appl. Phys. Lett., 57(1990)2136.

68 G. Malandrino, D.S. Richeson, T.J. Marks, D.C. DeGroot, J.L. Schindler and C.R. Kannewurf, Appl. Phys. Lett., 58(1991)182.

High T$_c$ Superconductor Thin Films
L. Correra (Editor)
1992 Elsevier Science Publishers B.V.

Computer Modeling of Y-Ba-Cu-O Thin Film Deposition and Growth

C.P. Burmester[1,2], L.T. Wille[3], and R. Gronsky[2,4]

[1]Materials Science Division, Lawrence Berkeley Laboratory, 1 Cyclotron Road, Berkeley, CA 94720, USA.
[2]Department of Materials Science and Mineral Engineering, University of California, Berkeley, CA 94720, USA.
[3]Department of Physics, Florida Atlantic University, Boca Raton, FL 33431, USA.
[4]National Center for Electron Microscopy, Lawrence Berkeley Laboratory, 1 Cyclotron Road, Berkeley, CA 94720, USA.

The deposition and growth of epitaxial thin films of $YBa_2Cu_3O_7$ are modeled by means of Monte Carlo simulations of the deposition and diffusion of Y, Ba, and Cu oxide particles. This complements existing experimental characterization techniques to allow the study of kinetic phenomena expected to play a dominant role in the inherently non-equilibrium thin film deposition process. Surface morphologies and defect structures obtained in the simulated films are found to closely resemble those observed experimentally. A systematic study of the effects of deposition rate and substrate temperature during in-situ film fabrication reveals that the kinetics of film growth can readily dominate the structural formation of the thin film.

INTRODUCTION

Bulk studies of the $YBa_2Cu_3O_z$ high T$_c$ oxide superconductor have made significant progress in the last few years in improving engineering properties. However, bulk polycrystalline materials still have less than adequate properties to motivate commercial application, in part due to weak superconducting links between grains.[1] In contrast, thin films of this material allow the production of highly oriented and textured material as well as providing a greater degree of control over the interior microstructure. Thus, the first viable applications promise to be in the area of microelectronic devices utilizing thin films of the YBaCuO superconductor.[2] Thin film deposition, however, is an inherently non-equilibrium process and thus a true understanding and, more importantly, control of the structural characteristics and properties of YBaCuO thin film will be best pursued through an understanding of the kinetic processes dominant during deposition. The evolution of non-equilibrium systems is expected to be strongly influenced by impurities and by surface inhomogeneities.[3] As a result, it is difficult to interpret experimental kinetic data as attributable to a specific macroscopic parameter as microscopic effects can easily dominate. In addition, most characterization techniques lack either sufficient time or spatial resolution necessary to allow study of the kinetic evolution of thin film structures. In many other systems, the problem of characterizing non-equilibrium evolution has been addressed by "enhancing" the resolution of experimental techniques with the effectively "infinite" time and spatial resolution of atomistic computer simulation. Computer simulation allows both the development and evaluation of microscopic theories of kinetic evolution and a degree of control over the "experimental" conditions inaccessible by experimental methods. Thus a technique for the computer modeling of thin film deposition and growth is introduced and its theoretical and experimental basis briefly described. The ability of this technique to model established trends is examined and results of the effects of deposition techniques on film characteristics are reported.

THEORETICAL BASIS

The simulation must be based on the predominant microscopic mechanisms operating during thin film deposition in order to mirror the rich range of fabrication conditions afforded by deposition techniques. It is based on an extension of the Monte Carlo technique to model deposition, surface redistribution, and bulk realignment of deposited species. As the YBaCuO system can be described by a ternary phase diagram of yttrium, barium, and copper oxides,[4] the simulation "deposits" variable ratios of these oxides particles. Within a unit cell of a YBaCuO phase, each atom can be assigned a type and an orientation with respect to the substrate which reflects the crystallographic orientation of the domain it participates in within the thin film. In addition to the three oxide types, two additional particle types are used to represent the substrate and "unoccupied" sites. These unoccupied sites correspond to voids within the film and unoccupied sites above the current surface of the film. Amongst the yttrium, barium, and copper oxide particles, the pair interactions are taken over the six nearest neighbor sites in the lattice: four "in-plane" (001) sites and two "out-of-plane" [001] sites. These interactions are dependent on both type and direction and their magnitudes are chosen to reflect the in-plane clustering and out-of-plane layer ordering experimentally established for the layered YBaCuO superconductors. Oxide-substrate interactions vary in the same manner allowing the modeling of different substrate types and orientations. Thus the Hamiltonian describing the configurational energy of microstates arising under the simulated deposition process is composed of a generalized Ising model reflecting surface, vacancy/surface, and oxide interactions, coupled with a Heisenberg spin interaction reflecting the variation of interactions with particle misorientation. This Hamiltonian can be written in the compact form:

$$H = -\sum_{<i,j>} V_n(n_i) n_i n_j - \mu_i \sum_{<i>} n_i - \sum_{<i,j>} V_s(\theta_{ij})(s_i \cdot s_j)$$

where $n_i = 1,2,..,5$ denotes the type at site i, and $|s_i| = 1$. The chemical potential terms, μ, in the Ising model component of the Hamiltonian, while taken to be zero in this study, can be used to describe different deposition atmospheres above the film with respect to the different metallic species in the film. This is particularly useful in modeling the transformations that occur in the pseudo-binary subcomponent of the overall ternary system described by $Y_2Ba_4Cu_{6+x}O_{14+x}$.[5-8] In the current study, particles interact via symmetrical pair interactions $V_n V_s$ only. Thus a logical enhancement of this model might include multi-spin interactions.

The simulation employs three excitation mechanisms which model deposition, surface redistribution, and bulk realignment of oxide particles. The deposition step, an inherently open process, is carried out in the grand canonical scheme with Kawasaki dynamics.[3] The surface redistribution step, involving the exchange of nearest neighbor surface particles and vacancies, conserves the total number of particles in the film and is thus carried out in the canonical ensemble with Glauber dynamics.[3] Finally, the bulk realignment step involves the attempted orientation change of an oxide particle within the film. As the "orientation" property is not conserved under this step, orientation changes are calculated in the grand canonical scheme with Kawasaki dynamics.

The deposition rate for the simulation is defined as the number of deposition steps attempted divided by the total number of simulation steps attempted (deposition, surface redistribution, and bulk realignment steps). Thus, a deposition rate equal to 1 corresponds to pure deposition and no "annealing." In this way, in-situ vs. ex-situ thin film deposition schemes are modeled. During in-situ deposition, deposition and "annealing" (surface redistribution and bulk realignment) are carried out simultaneously in a ratio corresponding to the deposition rate selected for the current simulation. By contrast, during ex-situ deposition, all the deposition steps are carried out first, often producing amorphous films, after which the annealing steps are performed.

In this study, the substrate is chosen to be atomically smooth, defect free, and its interaction with the deposited material constant, attractive, and independent of crystallographic orientation so as to effectively eliminate the influence of the substrate from the study of the effects of other parameters. Deposited material is also taken to be fully oxygenated. In addition, the orientation vector of the deposited particles is constrained to the <100> unit cell directions. Studies are in progress which remove these constraints in the study of substrate variations and the development of the full range of grain boundary and surface orientations.

SIMULATION RESULTS

Many hundreds of simulated films were produced under various deposition conditions. All films studied examined near substrate growth with typical lattice sizes consisting of 5×10^5 $YBa_2Cu_3O_7$ unit cells. For a simulation lattice with toroidal boundary conditions in the [100] and [010] directions created such that its lateral dimensions are an order of magnitude greater than its vertical extent, this produced films approximately 15 nm thick. All films reported here were deposited with a cation ratio of 1:2:3 (Y:Ba:Cu) with a slight shortage of barium in order to produce stoichiometric $YBa_2Cu_3O_7$ thin films.

As the orientations of particles is constrained to the cube directions, individual domains within the film are found to fall into two general classes: domains whose unit cell [001] 'c' axis was oriented perpendicular to the substrate, or 'c' type, and domains whose 'a' or 'b' axis was oriented perpendicular to the substrate, or 'a' type.

By varying the conditions under which the films were produced, a broad range of orientation types and surface morphologies were observed. Ex-situ films of the 'a' type show surface morphologies composed of a network of [100] and [010] oriented lenticular shaped interlocking "bars" of YBCO material. The surface of such a film is shown in Figure 1a. The size of such domains was seen to coarsen upon increasing annealing times. Overall, ex-situ 'a' type film surfaces were found to be very "rough" and characterized by rectangular shaped protrusions of 'a' type material in which the height of the film surface from the substrate varies markedly from point to point. Domains were found to be commonly subdivided by pronounced anti-phase boundaries. In contrast, ex-situ films of the 'c' type (see Figure 1b) displayed a much more consistent surface height. The overall shape of the 'c' type domains was more circular in nature than the interlocked 'a' type domains, producing a "pebbly" morphology. Domains of anti-phase relation were similarly separated by anti-phase boundaries in the film surface. Surfaces of in-situ grown films displayed a much less dramatic surface structure. Surfaces were generally atomically smooth but domain and anti-phase boundaries were decorated with surface defects and depressions. These decoration patterns were often sufficient to identify the surface type as the pattern inscribed would be orthogonal and rectangular in nature for 'a' type surfaces and more curvelinear in nature for 'c' type surfaces. Numerous thin film studies describe surfaces or include surface micrographs of ex-situ[9-12] or in-situ[13,14] deposited films which display completely analogous features as those produced by the simulation.

Figure 1a: Surface structure of ex-situ grown thin film displaying 'a' type domains.

Figure 1b: Surface structure of ex-situ grown thin film displaying 'c' type domains.

Under specific conditions, adjacent domains of 'c' type and 'a' type material can be observed to grow. The growth rate anisotropy between the 'a' and 'c' directions of the YBCO unit cell is apparent in the surface characteristics of such films. The greater growth rate of the 'a' type domains leads to regions of the film surface that are notably greater in vertical extent than adjacent 'c' type domains.

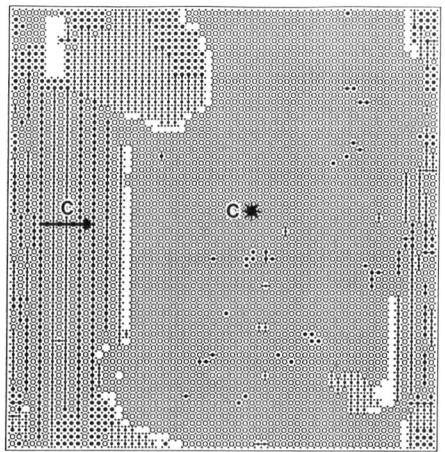

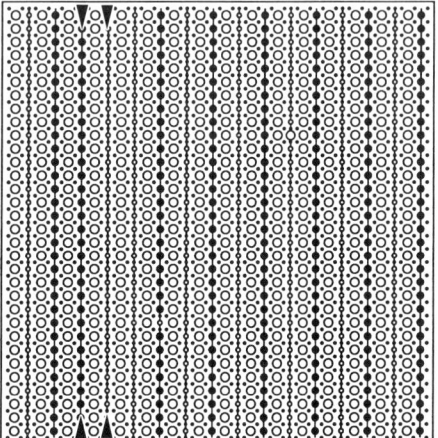

Figure 2: Plane (a) and cross-section (b) view of ex-situ grown film displaying both 'a' and 'c' type growth and an assortment of defects.

Figure 3: Plane section of an nearly perfect in-situ grown film illustrating a growth layer defect. Large filled circles represent barium, medium open circles yttrium, small open circles oxygen in the CuO basal plane, and small filled dots, copper. The lines included delineate the [100] or [0 1 0]   d i r e c t i o n s .

Many different types of defects were found to occur in the simulated thin films as shown in the cross-sectional and planar slices of Figures 2 and 3. Growth defects include stacking faults in which two layers were observed to switch position in a unit cell, for example, the exchange of a CuO and Y layer, may be observed in the nearly perfect in-situ grown film depicted in Figure 3. Additional defects of this type are illustrated in both plane and cross-section in Figure 2. An anti-phase boundary delineates a small circular subdomain in the upper left hand portion of Figure 2a within the broad central 'c' type domain. Within the subdomain, a layer defect is visable where a Y layer changes to a CuO layer. The domain boundary between the 'c' and 'a' type regions of this film is sharply delineated by a surface groove when running along the <100> type directions, but inclined boundaries of the <110> type seem more easily accommodated and do not retain vacancies at the boundary. In general, growth layer defects, as opposed to anti-phase defects and orthogonal grain boundaries, were more easily accommodated by the film. Through the film thickness, domain boundaries between orthogonal domains are a common occurrence and are often found to retain deviations from stoichiometry, as observed in the cross-section of Figure 2b. The surface variation of the cross-section of Figure 2 is similar to those reported by cross-sectional TEM studies.[15] Point defect inclusions of oxide particles are typically observed on growth surfaces. In general, in-situ films grown with lower deposition rates and ex-situ films annealed for longer times produced films of higher crystallographic homogeneity, as expected.

To investigate the effects of deposition rate and substrate temperature on the growth and orientation of in-situ grown films, a systematic study was performed under conditions where 'c' type growth is energetically preferred. In these studies, the effects of the substrate and the deposition atmosphere were held constant. In order to quantify the kinetic processes operating during deposition, the surface mobilities of particles on 'a' and 'c' type surfaces within the film were monitored. The results of these studies are depicted graphically in Figure 4. Figure 4a depicts the variation of the fraction of the film sites having a 'c' type orientation vs. deposition rate at constant simulation temperature of 0.8. At high deposition rates, the distribution of sites in 'c' and 'a' type orientations are nearly equal, reflecting the nearly amorphous nature of films produced under these conditions. As the deposition rate is decreased, the fraction of 'c' type growth gradually increases. In this range, the growth of films is nucleation limited and thus consist primarily of many small domains of 'c' type material. Note that the surface mobilities remain essentially equal throughout this region reflecting that a nucleation and not growth mechanism is predominant. However, as the deposition rate is decreased below 0.01, a sudden transformation to 'a' type growth occurs. This transformation to 'a' type growth occurs with a corresponding a drop in the surface mobility of 'a' type domains, and an increase in the mobilities of 'c' type domains. This bifurcation of the surface mobilities indicates a transition to growth limited kinetics and the lower 'a' surface mobilities reflect that more particles are adsorbing to 'a' type surfaces.

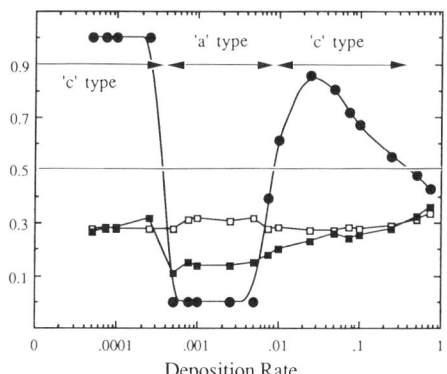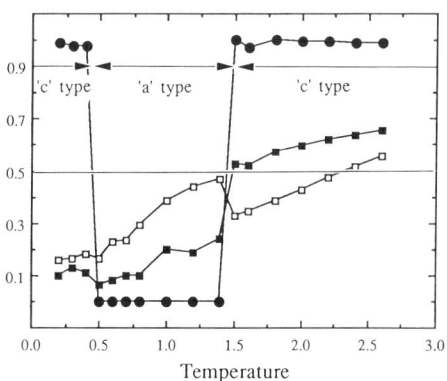

Figure 4a: A plot of film orientation and surface mobilities vs. deposition rate at constant temperature equal to 0.8. In both figures, filled circles represent fraction of 'c' type sites, filled squares 'a' surface mobilites, and open squares 'c' mobilities.

Figure 4b: A similar plot to 4a but varying temperature at constant deposition rate equal to 0.001.

While equilibrium considerations indicate that 'c' type growth should still prevail, the growth anisotropy between the 'a' and 'c' direction of the YBaCuO material allows provides 'a' type growth with a kinetically more favorable growth mechanism. It is felt that most thin film experiments are carried out in this regime. Substrate effects lead to the nucleation of 'c' type domains, but the growth limited kinetics quickly lead to the formation of 'a' type growth thus producing the type of 'c' to 'a' type transformation through thickness in thin films which is commonly reported[9,10,13-16] and also observed in Figure 2b. Under very low deposition rates, the trend is again seen to reverse. This regime approaches equilibrium growth conditions under which the equilibrium free energy of formation of the 'c' type phase is great enough to overcome the kinetic path afforded to 'a' type domains by the growth rate anisotropy and only 'c' type grains nucleate and grow. Note that the 'a' and 'c' surface mobility approach a constant value. A similar trend is seen in the variation of the film orientation with temperature at a constant deposition rate of 0.001. High temperatures are again nucleation limited, and polycrystalline films of 'c' type domains prevail. As the temperature decreases, the growth anisotropy intercedes and 'a' type films become dominant. Note the corresponding crossover in surface mobilities. At very low temperature, the growth anisotropy is overcome by the equilibrium free energy of formation for 'c' type material, and 'c' type grains prevail. Preliminary results for conditions under which 'a' type growth is energetically preferred do not show the 'a' to 'c' to 'a' transformation, as here, 'a' type is both the kinetically and energetically preferred phase.

The films used in these studies were very thin and on the order of 3.2 nm thick. Thus the resulting films tend to be completely homogeneous through thickness which results in the sharp transitions from 'a' to 'c' type films. It is expected that these sharp transitions will "round" out for greater film thicknesses, reflecting substrate retained material of a subordinate orientation type in a given deposition range. In many ways the modified plots of Figures 4a and 4b could then be thought as isothermal and isochronal sections taken through a classical TTT (Time-Temperature-Transformation) where time is replaced by deposition rate and the transformation altered to indicate the degree of 'a' or 'c' type film.

ACKNOWLEDGEMENTS

C.P.B. and R.G. are supported by a University of Houston subcontract under DARPA Grant No. MDA972-88-J-1002, using facilities at Lawrence Berkeley Laboratory funded by the Director, Office of Energy Research, Office of Basic Energy Sciences, Materials Sciences Division of the U.S. Department of Energy under Contract Number DE-AC03-76SF00098. L.T.W. is supported by an Internal Research Grant from the Division of Sponsored Research at Florida Atlantic University.

REFERENCES

1   J. Evetts, Phys. World, **3** (1990) 24.
2   R. G. Humphreys, J. S. Satchell, N. G. Chew, J. A. Edwards, S. W. Goodyear, S. E. Blenkisop, O. D. Dosser, and A. G. Cullis, Supercond. Sci. Technol., **3** (1990) 38.
3   O. G. Mouritsen, *Computer Studies of Phase Transistions and Critical Phenomena*, Springer Series in Computational Physics, Springer-Verlag, Berlin Heidelberg, 1984.
4   K. Char, M. R. Hahn, T. L. Hylton, M. R. Beasley, T. H. Geballe, and A. Kapitulnik, IEEE Trans. Magnetics, **25** (1989) 2422.
5   M. Fendorf, C.P. Burmester, L.T. Wille, and R. Gronsky, J. Less-Common Metals **164-165**, 84 (1990).
6   M. Fendorf, C.P. Burmester, L. T. Wille, and R. Gronsky, Appl. Phys. Lett. **57**, 2481 (1990).
7   J. Karpinski, S. Rusiecki, E. Kaldis, and E. Jilek, in *High T_c Superconductor Materials;* edited by H.U. Habermeier, E. Kaldis, and J. Schoenes (North Holland, Amsterdam, 1990).
8   M. Fendorf, M. E. Tidjani, C. P. Burmester, L. T. Wille, and R. Gronsky, *these proceedings*.
9   D. X. Li, X. K. Wang, D. Q. Li, R. P. H. Chang, and J. B. Ketterson, J. Appl. Phys., **66**:11 (1989) 5505.
10   R. L. Sandstrom, W. J. Gallagher, T. R. Dinger, R. H. Koch, R. B. Laibowitz, A. W. Kleinsasser, R. J. Gambino, B. Bumble, and M. F. Chisholm, Appl. Phys. Lett., **53**:5 (1988) 444.
11   J. R. Phillips, J. W. Mayer, J. A. Martin, and M. Nastasi, Appl. Phys. Lett., **56**:14 (1990) 1374.
12   X. K. Wang, K. C. Sheng, S. J. Lee, Y. H. Shen, S. N. Song, D. X. Li, R. P. H. Chang, and J. B. Ketterson, Appl. Phys. Lett., **54**:16 (1989) 1573.
13   R. Ramesh, C. C. Chang, T. S. Ravi, D. M. Hwang, A. Inam, X. X. Xi, Q. Li, X. D. Wu, and T. Venkatesan, Appl. Phy. Lett., **57**:10 (1990) 1064.
14   C. B. Eom, J. Z. Sun, B. M. Lairson, S. K. Streiffer, A. F. Marshall, K. Yamamoto, S. M. Anlage, J. C. Bravman, and T. H. Geballe, Physica C, **171** (1990) 354.
15   B. M. Clemens, C. W. Nieh, J. A. Kittl, W. L. Johnson, J. Y. Josefowicz, and A. T. Hunter, Appl. Phys. Lett., **53**:19 (1988) 1871.
16   C. W. Nieh and L. Anthony, Appl. Phys. Lett., **56**:21 (1990) 2138.

High T$_c$ Superconductor Thin Films
L. Correra (Editor)
319

Effect of oxygen partial pressure on the chemical vapor deposition of Y-Ba-Cu-O superconducting films

H. Yamane, M. Hasei, H. Kurosawa, T. Hirai, K. Watanabe, N. Kobayashi and Y. Muto

Institute for Materials Research, Tohoku University
2-1-1 Katahira, Aoba-ku, Sendai 980, Japan

Riken Co.,
810 Kumagaya, Kumagaya 360, Japan

Abstract
    The relationship between oxygen partial pressure (P(O$_2$)) and deposition temperature (T$_{dep}$) was studied for the preparation of Y-Ba-Cu-O superconducting oxide films by thermal chemical vapor deposition using β-diketone metal chelates as sources. The films were deposited on SrTiO$_3$ (100) single-crystal substrates heated at 600-900°C in a hot-wall type reactor. Superconducting transition temperature defined by zero resistivity (T$_c$(R=0)) at and above 90 K and critical current density (J$_c$) over 10^6 A/cm^2 at 77.3 K and 0 T were measured for the films prepared at P(O$_2$)=3.6 Torr and T$_{dep}$ = 850°C and at P(O$_2$) = 0.036 Torr and T$_{dep}$ = 700-720°C. Those conditions of P(O$_2$) and T$_{dep}$ were on and around the CuO-Cu$_2$O-O$_2$ equilibrium line.

1. INTRODUCTION

    The in-situ preparation of Y-Ba-Cu-O (YBCO) superconducting films has been performed by various physical vapor deposition (PVD) techniques such as sputtering, laser ablation and vacuum evaporation [1,2] and by chemical vapor deposition (CVD) techniques [1,3,4]. The YBCO films epitaxially grown on single-crystal substrates by these techniques exhibited resistivity zero above 90 K and high critical current density (J$_c$) over 10^6 A/cm^2 at 77.3 K and 0 T.
    Previously, YBCO films having high superconducting transition temperature (T$_c$) and high J$_c$ were prepared at 850°C by thermal CVD [5,6]. Recently, we were successful in thermal CVD of a high-J$_c$ YBCO film at 700°C by reducing oxygen partial pressure during film deposition [7].
    The present paper describes the effect of oxygen partial pressure on the crystallographic orientation and crystallinity of YBCO films prepared at various deposition temperatures under two different oxygen partial pressures. The conditions of the deposition of high-T$_c$ and high-J$_c$ YBCO films by thermal CVD are reported.

2. EXPERIMENTAL

The reactor used for CVD was a vertical hot-wall type described in detail
in our previous report [8]. The sources used were 2,2,6,6-tetramethyl-
3,5-heptanedionate (thd) of Y, Ba and Cu. Each source was evaporated and
the resulting vapor was introduced into the reactor by Ar gas at a flow
rate of 150 ml/min. The evaporation temperatures of the sources were 126-
133°C for Y(thd)3, 240-250°C for Ba(thd)2 and 106-133°C for Cu(thd)2.
Either pure oxygen gas or 1% O2 gas balanced with Ar gas was introduced
into the reactor at a rate of 250 ml/min. Total gas flow rate of the gases
was 750 ml/min. Total gas pressure was kept at 10 Torr during film
deposition. Oxygen partial pressures in the total gas (P(O2)) introduced
into the reactor were 3.6 and 0.036 Torr in the cases of the pure oxygen
gas and the 1% oxygen gas, respectively.

The substrates used were SrTiO3(100) single crystals (10x5x1 mm^3) with
mirror surfaces. Deposition temperature (Tdep) was from 600 to 900°C. The
conditions of the deposition temperature and oxygen partial pressure are
plotted in Figure 1. After deposition for 30-60 min, the films were cooled
in the reactor to about 150°C at a rate of 10-15°C/min under 760 Torr of
oxygen (in-situ oxygen treatment).

Crystal structure and crystallographic orientation of the films were
characterized by the standard $2\theta-\theta$ x-ray diffraction (XRD) method with
the CuKα line. The thickness of the films was measured with a stylus
instrument. The deposition rate of the films was from 0.6 to 1.0 μm/h.
The resistivity and Jc of the films were measured by a DC four-probe method
with Au electrodes sputtered on the films.

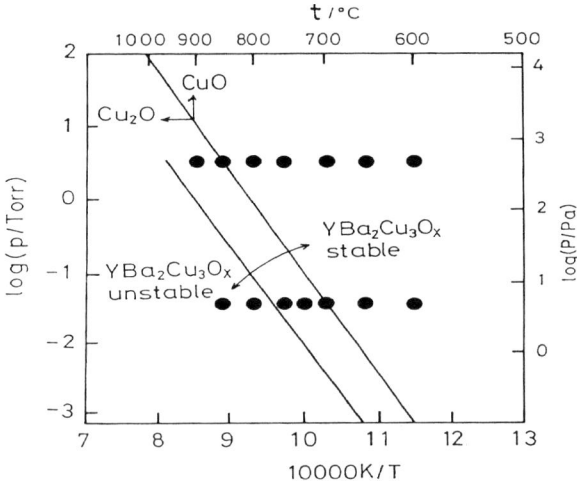

Figure 1. Log oxygen partial pressure versus reciprocal temperature plot
showing CVD conditions together with the CuO-Cu2O-O2 equilibrium line
[9,10] and the critical stability line for YBa2Cu3Ox [10,11].

3. RESULTS AND DISCUSSION

3.1. XRD analysis

Figure 2 shows the XRD patterns of the films deposited at 850, 700 and 650°C under the P(O_2) of 3.6 Torr. The peaks of YBa2Cu3Ox (001) (l = 1, 2, 3....) in the pattern of the film deposited at 850°C (Figure 2(a)) represent the orientation of the c-axis perpendicular to the substrate plane. A similar XRD pattern was obtained for the film deposited at 900°C. The large relative intensities of (100) and (200) peaks in Figure 2(b) indicate that the film deposited at 700°C was mainly composed of a-axis oriented YBa2Cu3Ox. As shown in Figure 2(c), broad and weak peaks of (110), (103) and (013) were seen in the XRD pattern of the films deposited at 650°C.

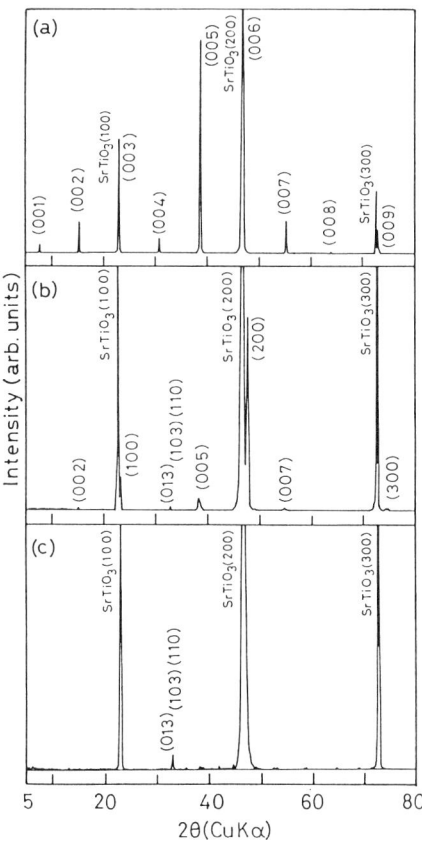

Figure 2. X-ray diffraction patterns of YBCO films prepared at P(O_2) = 3.6 Torr and T_{dep} = 850 (a), 700 (b) and 650°C (c).

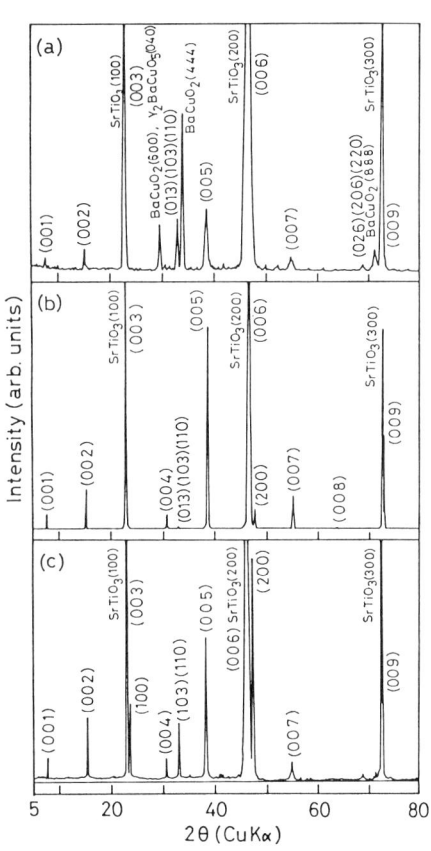

Figure 3. X-ray diffraction patterns of YBCO films prepared at P(O_2) = 0.036 Torr and T_{dep} = 850 (a), 700 (b) and 650°C (c).

The XRD patterns of the YBCO films prepared at 850, 700 and 650°C under the P(O2) of 0.036 Torr are illustrated in Figure 3(a), (b) and (c), respectively. Peaks having large relative intensities were observed at 2θ = 29.4° (d = 3.04 Å) and 33.9° (d = 2.64 Å), in addition to the peaks from (001), (103), (013) of YBa2Cu3Ox in the XRD pattern of the films deposited at 850°C. The peak position at 2θ = 29.4° was in agreement with the peak positions of BaCuO2 (600) [12] and Y2BaCuO5 (004) [13]. The peak at 33.9° can be assigned to the diffraction of BaCuO2 (444).

Prominent (001) peaks of YBa2Cu3Ox were seen with small peaks (100), (200), (103), (013) and (110) in the XRD patterns of the film deposited at 700°C (Fig.3 (b)). This means that most of the grains in the films were c-axis oriented YBa2Cu3Ox. Large relative intensities of (100) and (200) were seen in the XRD pattern of the films deposited at 650°C (Fig. 3(c)). Main peaks in the XRD pattern of the films prepared at 600°C were (110), (103) and (103) of YBa2Cu3Ox with small intensities.

The films obtained under the two different oxygen partial pressures showed that the direction of the crystallographic orientation normal to the substrate surface changed from the c axis to the a axis with the lowering of deposition temperature. This change was observed at a lower temperature for the films prepared at the lower P(O2) (0.036 Torr).

The films having a high degree of crystallinity and c-axis orientation were prepared around the CuO-Cu2O-O2 equilibrium line shown in Figure 1 [9]. The condition of P(O2) = 0.036 Torr and Tdep = 850°C is in the unstable region of YBa2Cu3Ox [10,11]. Thus, Y2BaCuO5, BaCuO2 and Cu2O were probably deposited at this condition and YBa2Cu3Ox should be produced during the in-situ oxygen treatment.

3.2. Tc and Jc

Figure 4(a) shows the relationship between deposition temperature and Tc(R=0) of the films prepared under the oxygen partial pressure of 3.6 Torr. Tc(R=0) above 90 K was measured for the films obtained at 850°C. The Tc of the films obtained at 900°C were lowered to 84 K. The film prepared at 700°C did not exhibit zero resistivity above 30 K.

Under the condition of the lower P(O2) of 0.036 Torr, films having Tc(R=0) = 90-91 K were deposited at 700-750°C (Figure 4(b)). The films deposited above 750°C decreased in Tc(R=0). The Tc(R=0) of 85 K was measured for the film deposited at 650°C. The films prepared at 600°C did not exhibit zero resistivity above 30 K.

Jc at 77.3 K and 0 T is plotted against the deposition temperatures in Figure 5. The Jc over 10^6 A/cm^2 was measured for the films prepared at 850°C and P(O2) = 3.6 Torr (Figure 5(a)) and at 700-720°C and P(O2) = 0.036 Torr (Figure 5(b)).

Hammond and Bohmann noticed that the successful in-situ growth of high quality YBCO films by PVD was achieved at the deposition parameters in close proximity to the equilibrium line of YBa2Cu3Ox decomposition to Y2BaCuO2, BaCuO2 and Cu2O, reported previously by Bohmann and Nolting [14,15]. According to the study of Lindemer et al. [10,11], this equilibrium line was in accord with the CuO-Cu2O-O2 equilibrium line and the YBa2Cu3Ox thermodynamic stability line was situated at lower oxygen partial pressure as shown in Figure 1.

The conditions of P(O2) and Tdep at which high-Tc and high-Jc YBCO films were prepared by the thermal CVD were in close proximity to the CuO-Cu2O-O2 equilibrium line. The YBCO films obtained at the conditions apart from

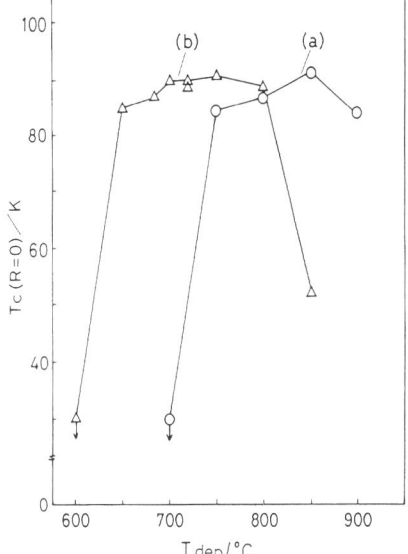

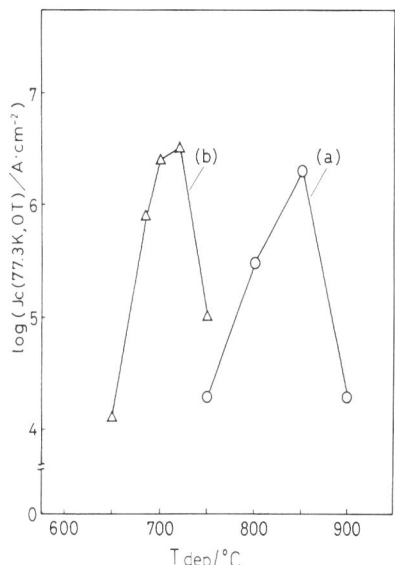

Figure 4.   Tc(R=0) of the films
prepared at various deposition
temperatures and at P(O2) = 3.6 (a)
and 0.036 Torr (b).

Figure 5.   Jc(77.3 K and 0 T) of the
films prepared at various deposition
temperatures and at P(O2) = 3.6 (a)
and 0.036 Torr (b).

this line exhibited lower Tc and Jc.

The relationship between the superconducting properties and the
deposition conditions of P(O2) and Tdep for the thermal CVD of YBCO films
were the same as those reported for the PVD.   However, the in-situ growth
of high quality YBCO films could be carried out by thermal CVD at the
deposition temperature around 700°C under total gas pressure in the CVD
reactor over 100 times higher than that in the deposition chambers of PVD.

4. SUMMARY

The deposition temperature of the high-Tc and high-Jc YBCO films
prepared by thermal CVD strongly depended on the oxygen partial pressure.
The YBCO films having Jc over 10^6 A/cm^2 at 77.3 K and 0 T were obtained at
the conditions on and near the CuO-Cu2O-O2 equilibrium line.

ACKNOWLEDGEMENTS

This work was supported in part by a Grant-in-Aid for Developmental Scientific Research under Contract No. 0255146 and a Grant-in-Aid for Scientific Research on Chemistry of New Superconductors under Contract No. 0227201 from the Ministry of Education, Science and Culture, Japan.

REFERENCES

1 M. Leskela, J.K. Truman, C.H. Mueller and P.H. Holloway, J. Vac. Sci. Technol., A7(1989)3147.
2 R.G. Humphreys, J.S. Satchell, N.G. Chew, J.A. Edwards, S.W. Gookyear, S.E. Blenkinsop, O.D. Dosser and A.G. Cullis, Supercond. Sci. Technol., 3(1990)38.
3 T. Hirai and H. Yamane, Oyo Buturi, 59(1990)134.
4 T. Hirai and H. Yamane, J. Cryst. Growth, 107(1991)683.
5 H. Yamane, H. Kurosawa, T. Hirai, K. Watanabe, H. Iwasaki, N. Kobayashi and Y. Muto, Supercond. Sci. Technol., 2(1989)115.
6 S. Matsuno, F. Uchikawa and K. Yoshizaki, Jpn. J. Appl. Phys., 29(1990) L947.
7 H. Yamane, T. Hirai, K. Watanabe, N. Kobayashi, Y. Muto, M. Hasei and H. Kurosawa, J. Appl. Phys., 69(1991), (to be published).
8 H. Yamane, H. Kurosawa, T. Hirai, K. Watanabe, H. Iwasaki, N. Kobayashi and Y. Muto, J. Cryst. Growth, 98(1989)860.
9 I. Barin and O. Knacke, Thermochemical Properties of Inorganic Substances, Springer-Verlag, 1973.
10 T.B. Lindemer, J.F. Hunley, J.E. Gates, A.L. Sutton, Jr., J. Brynestad, C.R. Hubbard and P.K. Gallagher, J. Am. Ceram. Soc., 72 (1989)1775.
11 T.B. Lindermer, F.A. Washburn, C.S. MacDougall and O.B. Cavin, Physica C, 174(1991)135.
12 Powder Diffraction File, JCPDS International Center for Diffraction Data, File No. 38-1402.
13 Powder Diffraction File, JCPDS International Center for Diffraction Data, File No. 38-1434.
14 R.H. Hammond and R. Bormann, Physica C, 162-164(1989)703.
15 R. Bormann and J. Nolting, Appl. Phys. Lett., 54(1989)2148.

High T$_c$ Superconductor Thin Films
L. Correra (Editor)
© 1992 Elsevier Science Publishers B.V. All rights reserved.

A new MO-CVD technique for the preparation of $YBa_2Cu_3O_{7-\delta}$ thin films

B. Schulte[a,b], M. Maul[a,b], W. Becker[a], S. Elschner[a], E.G. Schlosser[a], P. Häussler[a], and H. Adrian[b]

[a] Hoechst AG, Central Research, 6230 Frankfurt 80, Germany

[b] Institute of Solid State Physics, TH Darmstadt, 6100 Darmstadt, Germany

Abstract

We report on a new metalorganic chemical vapour-deposition technique (MO-CVD) for the *in situ* preparation of thin films of $YBa_2Cu_3O_{7-\delta}$ (YBCO). The new technique, which utilizes a very compact apparatus, works *without* carrier gases. Both a high deposition rate (1-$10\,\mu$m/h) and a high yield of the precursors (2,2,6,6-tetramethyl-3,5-heptanedionates (thd)) have been achieved. Thin films of YBCO with smooth surfaces and thicknesses $d=0.4$-$0.6\,\mu$m were prepared on single crystalline (100) $SrTiO_3$ and (100) MgO. At temperatures $T=800°$C the films grow with a high degree of $\vec{c}$-axis orientation.

The samples were characterized by X-ray diffraction, four-probe-resistance measurements, ac-susceptometry, transport critical-current measurements, and scanning-electron microscopy. Critical data of $T_c^{\mathrm{mid}}=92.3\,$K, $j_c^{77\,\mathrm{K}}=1.3{\cdot}10^6\,$A/cm^2 on $SrTiO_3$ and $T_c^{\mathrm{mid}}=91.8\,$K, $j_c^{77\,\mathrm{K}}=3{\cdot}10^5\,$A/cm^2 on MgO, respectively, were achieved.

1. INTRODUCTION

From the numerous techniques used for the in situ growth of thin films of high-temperature superconductors (HTSC) as thermal evaporation, sputtering, laser ablation and all their modifications, the chemical vapour deposition has become more and more important. Like in the microelectronic industry for thin-film applications, metalorganic chemical vapour deposition can play a major role due to its easy and relatively cheap realization and high throughout. Further advantages include the protective role, played by the organic ligands, to the highly reactive Y, Ba, and Cu atoms from their oxygenation on the path from the source to the substrate; which contrasts to any thermal evaporation technique of the pure metals where efficient pressure stages are required. Due to the high working pressure, MO-CVD has a high potential for the 3-dimensional deposition of HTSC materials onto multifilaments capable for power lines or magnets. Several authors have shown by (conventional) MO-CVD, that thin films of YBCO with excellent crystallographic orientation and superconducting properties can be achieved on single crystalline substrates [1,2]. Critical temperatures $T_c \geq 90$ K and critical current densities $j_c \geq 10^6$ A/cm^2 at $T = 77$ K and $B = 0$ T are routinely obtained.

As source materials 2,2,6,6-tetramethyl-3,5-heptanedionates (thd) of Y, Ba, and Cu are used. Both Y(thd)$_3$ and Cu(thd)$_2$ are volatile at relatively low temperatures without appreciable thermal decomposition, resulting in constant evaporation rates for hours. Ba(thd)$_2$ behaves quite differently since it crystallizes, depending on the preparation conditions, with 0.5-1.6 molecules of water co-ordinated to each Ba [3]. If improperly prepared, its volatility, subsequently, varies from batch to batch [3]. Further, Ba(thd)$_2$ partially decomposes during vaporizing [4], which causes its vapour-phase concentration to drop significantly as a function of time at a constant source temperature T_{Ba} [5]. Mass spectra show fragments corresponding to dimers, trimers, and tetramers indicating that the precursor exists in oligomeric form with a varying range of its properties [6].

Thus the stability of the Ba-precursor needs further improvements prior to any large-scale application. Additionally, the deposition rate is still too low. A scale-up of the vapour source is limited by the loss of easy temperature control. Higher evaporation rates or a higher content of the metalorganics in the vapour phase, on the other hand, are somehow contrasted to the Ba(thd)$_2$ instabilities at elevated temperatures. Accordingly, the precursors are produced with high reproducibiliy in our own laboratories, and a modified MO-CVD technique which functions *without* a carrier gas, thus leading to a higher density of the metalorganics in the vapour phase, has been developed. The new technique, additionally, prevents the loss and decomposition of the precursors on route from the source to the substrate [3]. Moreover the new technique allows the design of a compact, low-cost apparatus with improved simplicity and hence easier process control.

2. EXPERIMENTAL SETUP AND PROCEDURES

The apparatus (shown in Fig. 1) consists of a vertical reactor with three individual aluminum crucibles, one for each precursor, inside the same vacuum chamber. The crucibles are individually temperature controlled by thermocouples located inside the sources. The vapours are mixed together and guided to the substrate inside the chimney over a short distance. The short distance and the heating of the chimney avoids loss and premature condensation. The major modification to conventional CVD consists in the lack of a carrier gas for the transport of the precursors. Controlled by mass flow, O$_2$ is coaxially supplied close to the substrate only for the growth of the HTSC-oxide [7]. During preheating, a shutter prevents the substrate from beeing coated. Single crystalline substrates of (100) SrTiO$_3$ and (100) MgO are attached to the sample holder by silver paste.

Typical process data are $T_Y = 140°$C, $T_{Cu} = 130°$C, and $T_{Ba} = 264°$C, respectively, for the individual sources. The temperature of the chimney T_{ch} is chosen to be a few degrees higher than the highest evaporation temperature ($T_{ch} = 270°$C) and the temperature of the sample holder is chosen to be $T_s = 800°$C. The background pressure is 5 Pa, the partial pressure of oxygen $p_{O_2} = 45$ Pa, measured by a capacitive vacuum gauge. The flow rate of O$_2$ is 200 cm^3/min (standard conditions, sccm).

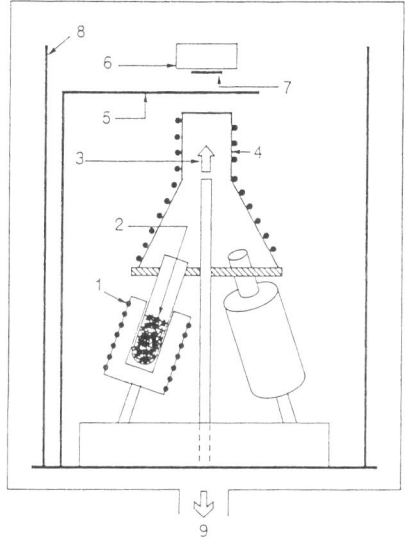

Figure 1:
Scheme of the new MO-CVD apparatues [3]. 1 vapour sources, **2** precursor, **3** oxygen inlet, 4 chimney, **5** shutter, **6** sample holder, **7** substrate, 8 quartz cylinder in order to direct the mass flow to the substrate, **9** to pump.

The precursors are thermally decomposed in close proximity to the hot substrate to form the HTSC film under the presence of oxygen. A deposition rate of about $2\,\mu$m/h is easily achieved. The films do not undergo postprocessing except for a low-temperature anneal (500°C, 1 atm O_2) for 40 min to ensure full oxidation. From 800°C to 500°C cooling takes place within 5 min at 1 atm O_2, from 500°C to room temperature within 10 min. The thickness is measured by a stylus monitor.

3. RESULTS

The films are characterized by X-ray diffraction, four-probe-resistance measurements $R(T)$, ac-susceptometry, transport critical-current measurements $j_c(T, B)$, and scanning-electron microscopy (SEM). Fig. 2 shows the X-ray diffraction pattern of an $YBa_2Cu_3O_x$ thin film on a (100) $SrTiO_3$ single-crystalline substrate. The large and narrow (00ℓ)-peaks indicate a large volume fraction of $\vec{c}$-axis orientation. Θ-scans give further information on the angular distribution of the $\vec{c}$-axis. The full width at half maximum (FWHM) of the rocking curve of thin films on (100) $SrTiO_3$ as well as (100) MgO both give 0.42° indicating epitaxial growth. X-ray patterns show that $\vec{c}$-axis lattice parameters in the range of c_o=1.1680 nm are routinely obtained, corresponding to an oxygen content of x=6.9±0.1 [8].

The temperature dependence of the electrical resistance was determined by the usual four-probe technique. At zero magnetic field we obtain transition temperatures $T_c(R = 50\%)$=92.3 K with a width of $\Delta T_c(90\% - 10\%)$=0.8 K on $SrTiO_3$ (Fig. 3) and T_c=91.8 K with ΔT_c=1.3 K on MgO, respectively. The linear extrapolation of the normal state intersects the vertical axis close to zero. Measurements of the ac-susceptibility performed with

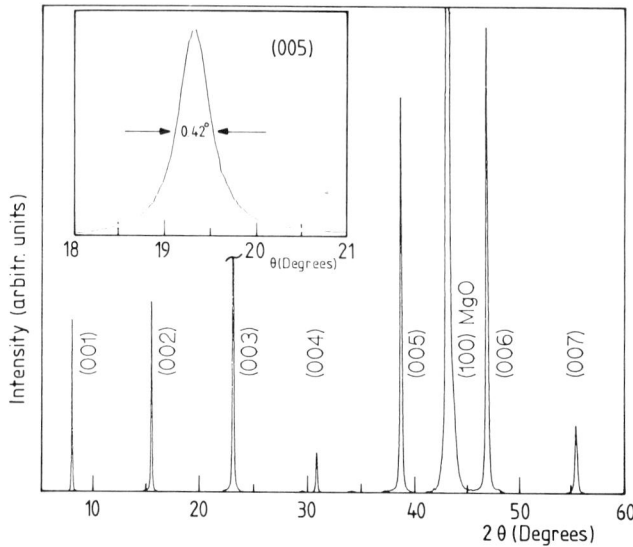

Figure 2: X-ray diffraction pattern of YBa$_2$Cu$_3$O$_{7-\delta}$ on (100) MgO. The inset shows the rocking curve of the (005) peak with the FWHM indicated.

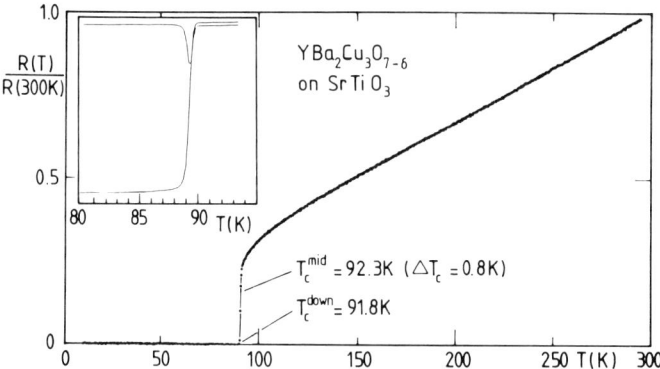

Figure 3: Normalized resistance of YBCO versus temperature. ΔT_c is defined as $T(R=90\%)$-$T(R=10\%)$. The inset shows the corresponding in-phase and out-of-phase ac-signal versus temperature.

a driving field $B_a \leq 1\,\mu$T at frequency $\nu=70$Hz give sharp transitions with $T_c=89.5$ K and $\Delta T_c=0.7$ K (inset of Fig. 3). The narrow out-of-phase signal indicates high homogeneity.

For measurements of the transport critical-current density j_c, the films were patterned by both direct laser writing or wet chemical etching. Microbridges, respectively, 200 μm long and 20 μm wide (laser writing) or 100 μm long and 10 μm wide (wet etching) were prepared. The films were contacted by copper leads, which are connected via silver paste onto gold contacts which themselves were sputtered onto the film surface. Since the current-voltage curves are very sharp the particular choice of the critical-current criterion is not critical. We used a criterion of $1\,\mu V$. At $T=77.3\,\mathrm{K}$ and $B=0\,\mathrm{T}$ we obtain critical current densities j_c up to $1.3\cdot10^6\,\mathrm{A/cm^2}$ on (100) $SrTiO_3$ and $j_c=3\cdot10^5\,\mathrm{A/cm^2}$ on (100) MgO.

Besides the intrinsic properties of HTSC thin films, experimentally determined critical-current densities strongly depend on the correct determination of the effective, current carrying cross section of the sample. In Fig. 4 scanning electron micrographs of four films with different ratios of Y:Ba:Cu are shown. Whereas a) - c) shows the topography

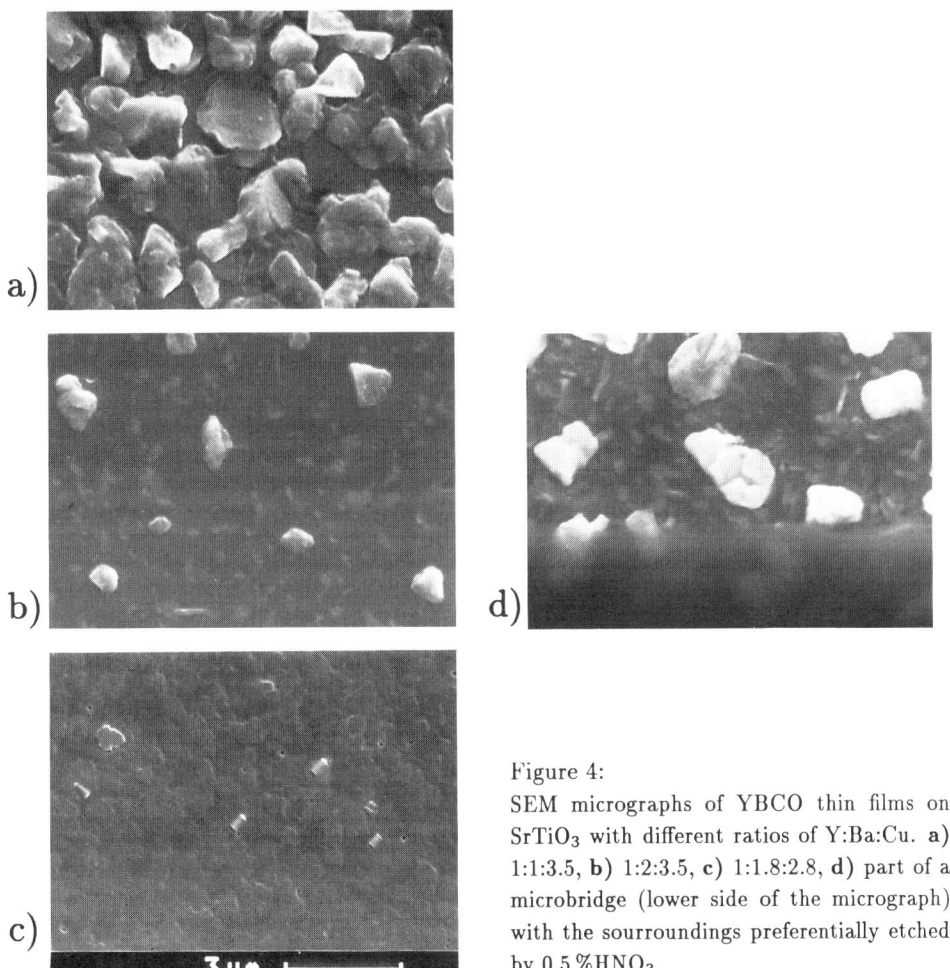

a)

b)

d)

c)

Figure 4:
SEM micrographs of YBCO thin films on $SrTiO_3$ with different ratios of Y:Ba:Cu. **a)** 1:1:3.5, **b)** 1:2:3.5, **c)** 1:1.8:2.8, **d)** part of a microbridge (lower side of the micrograph) with the sourroundings preferentially etched by 0.5 % HNO_3.

as-deposited, in d) part of the superconducting film is preferentially etched away by 0.5 %HNO_3 leaving the precipitates and a microbridge for critical-current measurements to stand alone at the substrate. Part of the microbridge is seen at the lower side of Fig. 4d. Inductively-coupled plasma atomic-emission spectroscopy (ICP-AES) is used to analyse films a) -c) entirely and the resolved part of film d). Fig. 4a shows the surface of a film with a composition ratio Y:Ba:Cu of 1:1:3.5. Due to the non-stoichiometry large precipitates exist. Using EDX-analysis, two types of precipitates are distinguished. One is very rich in Cu (presumably CuO) and the other one has a higher content of Cu and Y compared to the areas between the precipitates. The number of precipitates decreases with increasing Ba ratio, as can be seen in Fig. 4b. Here essentially CuO-precipitates seem to remain. A smooth shiny surface can be obtained if the composition is very close to 1:2:3 as shown in Fig. 4c.

Nearly independent of the amount of precipitates, all films of the type a) - d) show $T_c^{\text{down}} \simeq 91$ K and metallic behaviour above T_c. Accordingly, we assume that using the CVD technique, Ba forms stoichiometric $YBa_2Cu_3O_{7-\delta}$ leaving the excess Y and Cu to form precipitates. ICP-AES analysis of that part of the film having been removed by the acid (shown in Fig. 4d), infact shows, within experimental resolution, the exact ratio Y:Ba:Cu = 1:2:3. The precipitates grow from the substrate through the complete film. By properly taking into account the corresponding reduction of the effective cross section by the non-superconducting precipitates, critical current-densities of films of the type a) come close to those of type c).

In conclusion we have shown that a new simplified MO-CVD technique without any carrier gas allows the reproducible production of high-quality epitaxially grown supercon-ducting thin films of $YBa_2Cu_3O_{7-\delta}$ on (100) $SrTiO_3$ and (100) MgO.

4. REFERENCES

1 F. Schmaderer, G. Wahl: Journal de Psysique C5 sup. no 5, Tome **50** (1989).

2 H. Yamane, H. Masumoto, T. Hirai, H. Iwasaki, K. Watanabe, N. Kobayashi, and Y. Muto: Appl. Phys. Lett. **53**, 1548 (1988).

3 B. Schulte, M. Maul, W. Becker, S. Elschner, E.G. Schlosser, P. Häussler, and H. Adrian: Submitted to Appl. Phys. Lett.

4 J.E. Schwarzenberg, R.E Sievers, and R.W. Moshier: Anal. Chem. **42**, 1828 (1970); G.S. Hammond, D.C. Nonhebel, and C.S. Wu: Inorg. Chem. **2**, 73 (1962).

5 E. Fitzer, H. Oetzmann, F. Schmaderer, G. Wahl: private communication; will be published in Proc. of EURO CVD 8, Sept. 91, Glasgow.

6 P.H. Dickinson, T.H. Geballe, S. Sanjurjo, D. Hildenbrand, G. Graig, M. Zisk, J. Collman, S.A. Banning, and R.E. Sievers: J. Appl. Phys. **66(1)**, 444 (1989).

7 R.H. Hammond, R. Bormann: Physica **C162-164**, 703 (1989).

8 S. Rusiecki, B. Bucher, E. Kaldis, E. Jilek, J. Karpinski, C. Rossel, B. Pümpin, H. Keller, W. Kündig, G. van Tendeloo, T. Krekels, and C. Amelinckx: J. Less-Comm. Met., **164-165**, 31 (1990).

High T$_c$ Superconductor Thin Films
L. Correra (Editor)
1992 Elsevier Science Publishers B.V.

Properties of Low Temperature, Low Oxygen Pressure Post-annealed YBa$_2$Cu$_3$O$_{7-x}$ Thin Films.

R. Feenstra, D. K. Christen, J. D. Budai, S. J. Pennycook,
D. P. Norton, D. H. Lowndes, C. E. Klabunde, and M. D. Galloway

Solid State Division, Oak Ridge National Laboratory,
P.O. Box 2008, Oak Ridge, Tennessee 37831-6057

Abstract
The application and effects of low oxygen pressures for the growth of YBa$_2$Cu$_3$O$_{7-x}$ films by post deposition annealing are reviewed and related to available thermodynamic data at subatmospheric oxygen pressures. Special emphasis is given to the correlation between epitaxial growth properties and high temperature annealing conditions and the effect of variable low temperature oxidation. The application of low oxygen pressures is extended to the recrystallization of ion implanted, amorphized YBa$_2$Cu$_3$O$_{7-x}$ layers for comparison with epitaxial growth properties during initial post deposition annealing.

1. INTRODUCTION

During the last few years remarkable progress has been made in the processing of YBa$_2$Cu$_3$O$_{7-x}$ (YBCO), made possible for a large part by the increasing availability of systematic thermodynamic data concerning the Y-Ba-Cu-O phase diagram [1] and effects related to the oxygen solubility in YBCO [2,3]. Of interest for the processing of films is the high temperature phase diagram of YBCO at subatmospheric oxygen pressures $p(O_2) \leq 1.0$ atm, for cation compositions close to the ideal 1:2:3 stoichiometry. This pressure range correlates with the two basically different methods developed for YBCO film growth, i.e. by *ex situ* thermal processing of predeposited oxide layers, or more directly, by deposition controlled, epitaxial growth in a so-called *in situ* process. The oxygen pressure plays an important role in this distinction, since most *in situ* synthesis methods utilize vacuum deposition techniques, requiring substantially subatmospheric oxygen (partial) pressures, whereas, "historically," for post-annealing the oxygen pressure was set arbitrarily at 1.0 atm. For *in situ* film growth, typically $0.1 < p(O_2) < 1000$ mTorr. With both methods, as for all YBCO, the films require low temperature annealing at oxygen pressures $p(O_2) \simeq 1.0$ atm for transformation into the orthorhombic, superconducting phase of YBCO with transition temperature $T_c \simeq 90$ K.

Recently, we reported on the use of substantially reduced oxygen partial pressures (down to 0.0001 atm $\sim$ 100 mTorr) for the high temperature growth of coevaporated Y, BaF$_2$, and Cu precursor films by post deposition annealing, bridging the gap in oxygen pressure between *ex situ* and *in situ* processing [4]. Guided by recent information concerning the depression of melting temperature with decreasing oxygen pressure [5] and the stability limit of YBCO at low $p(O_2)$ [6], not only was compatibility of *ex situ* processing demonstrated (overcoming conjecture related to possible YBCO metastability during *in situ* film growth), but low oxygen pressures, in fact, proved to be beneficial for enhanced c-axis epitaxial growth and improved structural quality of the films. A mutual dependence of growth properties on both oxygen pressure and temperature was observed, allowing for compensation of lower annealing temperatures by lower oxygen pressures. Elimination of the oxygen pressure gap substantially diminishes the distinction between *in situ* and *ex situ* processing temperatures, previously thought to be inherent for the difference in growth mechanisms.

In this paper, the observed systematics in growth properties are reviewed and the resulting superconducting properties of films annealed at low oxygen pressures are evaluated in comparison with films annealed at higher oxygen pressures and films produced by *in situ* laser ablation. Annealing at low oxygen pressures induces a lower initial oxygen content in the films during growth and it is shown that this may lead to unique behavior upon low temperature oxidation. Lastly, the general applicability of low oxygen pressure thermal processing is tested for *c*-axis YBCO films, previously amorphized by high dose oxygen implantation. While the results confirm our previous conclusions regarding YBCO stability at low $p(O_2)$, the implanted films exhibited less tendency for $c \perp$ epitaxial (re)growth than the fluoride containing precursor films under the same conditions. This suggests a previously unanticipated role of the BaF_2 in the films and its dissociation in the high temperature growth process. A possible mechanism and cause for this discrepancy will be discussed.

2. FILM DEPOSITION AND ANNEALING

Details of the two-step preparation method involving low $p(O_2)$ annealing of amorphous precursor films have been described in [4]. Briefly, the precursor films were formed by low temperature codepositions of Y, BaF_2, and Cu from three rate-controlled e-beam guns in a low background pressure of 1-5 μTorr. The films were annealed in flowing mixtures of argon and oxygen, adjusted to the desired oxygen partial pressure by flow control. The $p(O_2)$ value was monitored continuously with a zirconia-cell-based oxygen meter at the furnace exit. Small amounts of water vapor were added to the high temperature annealing ambient to enhance the dissociation of BaF_2 (water partial pressure 0.025 atm). The films were cooled in dry $p(O_2)$ ambient to 550°C, at which temperature the oxygen pressure was increased to 1.0 atm for an intermediate 0.5 h soak before further cooling to room temperature.

For comparison, *in situ* laser ablated films were grown in an oxygen background pressure of 200 mTorr at a substrate temperature of 730°C (estimated actual temperature). A KrF excimer laser beam (~ 350 mJ, 38 ns pulse duration) was focussed to a horizontal line on a 25 mm diameter, rotating, high density YBCO target [7]. The focussed energy density was 2.5-3.0 J/cm^2. After deposition, the substrate heater was turned off, the oxygen pressure increased to 600 Torr, and the samples were cooled at this pressure without further oxidation annealing at a constant temperature.

For the study of the effect of residual oxygen deficiency x, the films were reannealed at 550°C for 1 h at $1.0 \geq p(O_2) \geq 0.01$ atm, followed by slow cooling at the same oxygen pressure. This resulted in reversible changes in residual oxygen deficiency within the "90 K plateau". Electrical transport properties were measured by standard four probe resistance measurements after patterning of 40-100 μm wide measuring bridges by standard photolithography and wet etching.

3. CORRELATION OF GROWTH PROPERTIES WITH OXYGEN PRESSURE AND ANNEALING TEMPERATURE : $p(O_2) - 1/T$ DIAGRAM

A $p(O_2)$-$1/T$ diagram, similar to that previously [8] introduced by Hammond and Bormann (H&B), comparing *in situ* and *ex situ* film processing conditions and thermodynamic properties at oxygen pressures below 1.0 atm, is presented in Fig. 1.

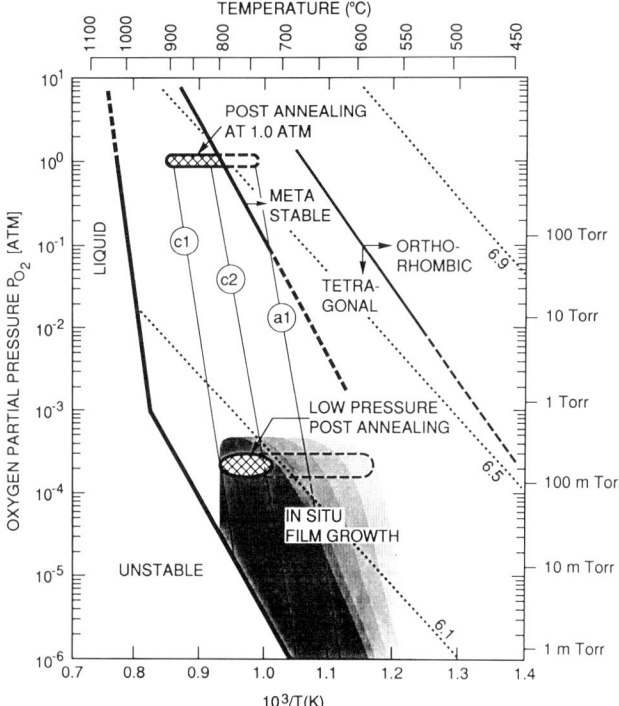

Figure 1. Oxygen partial pressure versus temperature diagram showing $YBa_2Cu_3O_{7-x}$ phase stability limits [6, 9], lines of constant oxygen stoichiometry [2], tetragonal to orthorhombic transition [10], and thin film processing conditions as indicated. The observed systematics for *ex situ* annealing are represented by straight lines denoted $c1$, $c2$, and $a1$, respectively. Predominantly c-axis films were obtained between $c1$ and $c2$ and mixed c- and a-axis films between $c2$ and $a1$.

A wide range of processing conditions may be employed for the synthesis of YBCO films, bounded by stability limits both at high oxygen pressures-low temperatures [9] or at high temperatures-low oxygen pressures. In the original H&B diagram the low $p(O_2)$ decomposition line for YBCO practically coincided with the CuO-Cu_2O reduction line, for example, indicating phase stability at $p(O_2) = 200$ mTorr up to only 750°C or at $p(O_2) = 1$ mTorr up to 600°C. A close correlation with this line was observed for *in situ* T, $p(O_2)$ combinations compiled from literature as employed with various vacuum deposition techniques. More recent studies indicate a significantly expanded stability region for YBCO, reaching up to temperatures well above, or oxygen pressures well below the CuO-Cu_2O-O_2 equilibrium. The stability line included in Fig. 1 was derived from a recent study by Lindemer *et al.* [6] and even surpasses recently updated pressure limits reported by Beyers and Ahn [1] and König *et al* [10]. Many low $p(O_2)$ *in situ* growth conditions previously thought to have a kinetic origin still fall within the stability region defined by this line. For a discussion of decomposition reactions and outstanding uncertainties, the reader is referred to [1, 4, 6].

Growth conditions employed for *in situ* film growth are indicated as a shaded region with (arbitrary) cutoffs at 800°C and 500 mTorr for vacuum deposition methods. Predominantly *c*-axis films (*c*-axis perpendicular to the substrate) are grown in the region with darker shading, basically in accordance with the original H&B diagram [8]. The lighter shading indicates a diminished tendency for *c*-axis epitaxy and overall decreasing film quality with decreasing substrate temperature. Processing conditions for post-annealing at high and low oxygen pressures, respectively, are indicated by the cross-hatched regions. The region at 1.0 atm represents the traditional annealing interval at temperatures between 800 and 900°C and clearly exemplifies the early oxygen pressure and temperature gaps between *in situ* and *ex situ* processing. The indicated low $p(O_2)$ annealing region, on the other hand, is well within the range of deposition conditions applied for *in situ* film growth.

Evidently, processing conditions for post annealing are not restricted by the two regions indicated in Fig. 1 and an important part of our previous study [4] concerned the identification of a mutual correlation between oxygen pressure and annealing temperature. Precursor films were deposited on (100) SrTiO₃ as described in Section 2 and each film was given a different heat treatment, in all according to a grid-like array in the $p(O_2)$-$1/T$ diagram, extending downwards in oxygen pressure and temperature from annealing conditions at 1.0 atm. The resulting systematics in surface morphology are illustrated by scanning electron micrographs (SEM) of representative films in Fig. 2.

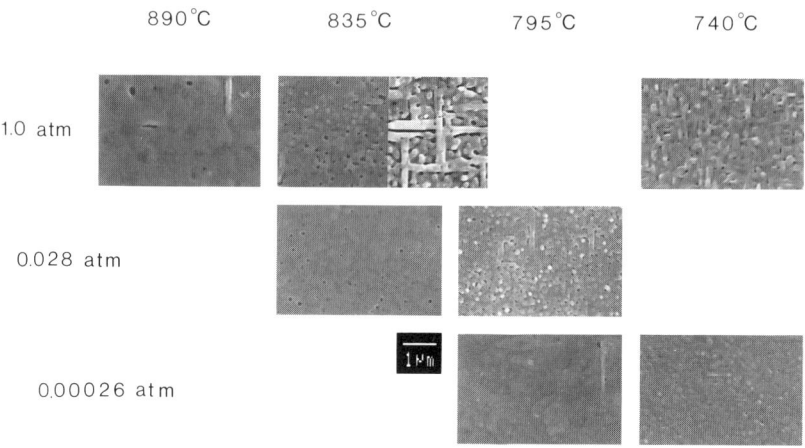

Figure 2. Scanning electron micrographs of YBa₂Cu₃O₇₋ₓ films on (100) SrTiO₃ annealed at different temperatures and oxygen pressures.

Basically two trends may be distinguished. First, at each oxygen pressure, the surface morphology becomes denser and more platelike, indicating larger lateral grain sizes, with increasing temperature. This morphology is typical of post-annealed *c*-axis YBCO films. Decreasing the temperature to 740°C at 1.0 atm changes the predominant epitaxy to *a*-oriented, giving rise to small, perpendicular grains at the film surface. At 0.00026 atm a similar epitaxial change-over occurred at 680°C (not shown in Fig. 2).

The annealing condition at 835°C, 1.0 atm clearly shows intermediate character with enhanced tendency for a-axis outgrowths and sensitively to extraneous factors such as film thickness (Section 4), or substrate surface preparation, as illustrated by the two micrographs included for this temperature.

Secondly, at constant annealing temperature, a similar densification and tendency for c-axis growth as that occurring with increasing temperature at fixed $p(O_2)$, results upon lowering of the oxygen pressure. Together, this leads to a staggered occurrence of comparable structural features upon adjustment of both temperature and oxygen pressure. With the identification of $c1$ for c-axis films with a dense, platelike surface morphology, $c2$ for c-axis films with a more granular morphology and increased sensitivity for a-axis growth, and $a1$ for predominantly a-axis films, the observed systematics are summarized in Fig. 1 by corresponding annealing lines. Predominantly $c \perp$ films are grown between $c1$ and $c2$ and mixed $c \perp + a \perp$ films between $c2$ and $a1$.

Similar systematics were observed with other characterization methods such as x-ray diffraction, Rutherford backscattering spectroscopy-ion channeling (RBS-II), and electrical transport measurements. In the RBS-II measurements, a systematic decrease in minimum channeling yield χ_{min}, indicating improved crystalline quality, resulted both with increasing temperature or decreasing oxygen pressure. A χ_{min} value as low as 0.04 was recorded for the film shown in Fig. 2 annealed at 835°C, 0.028 atm. χ_{min} values less than 0.1 are routinely obtained for films annealed at 730°C, 0.00026 atm.

As indicated by the annealing lines, *reduced oxygen pressures may compensate for lower annealing temperatures*. This observation is basic to the low temperature growth of c-axis YBCO films at reduced oxygen pressures. Although a complete explanation for this effect at present is still elusive, it is worthwhile noting some interesting correlations with other thermodynamic properties. For example, the annealing lines are nearly parallel to the high temperature phase stability line, representing the depression of melting temperature with $p(O_2)$ [5, 6]. This suggests that the *epitaxial ordering is mostly determined by the atomic mobilities* of participating cations. At constant temperature, these mobilities apparently increase with decreasing oxygen pressure, or, as indicated by the oxygen solubility lines, with increasing oxygen vacancies in YBCO [12]. Accordingly, an enhancement of the Cu self-diffusion in YBCO with decreasing oxygen pressure recently was reported by Routbort *et al* [13]. Similar oxygen vacancy induced enhancements of cation diffusion coefficients have been observed in other nonstoichiometric oxides [14].

As shown in Fig. 1, the oxygen content in YBCO during low $p(O_2)$ annealing of c-axis films may reach equilibrium values less than $z \simeq 6.1$, whereas $z \simeq 6.45$ for anneals at 835°C, 1.0 atm. Similarly large oxygen deficiencies may also occur during the *in situ* growth of c-axis films. Since lower processing temperatures apparently correlate with lower oxygen pressures, this leads to the somewhat paradoxical conclusion that, for practical reasons for the growth of YBCO films, it is preferable to initially form $YBa_2Cu_3O_{6.1}$ rather than a more oxygen rich compound, even though the desired final composition should contain as few as possible oxygen vacancies. Evidently, this emphasizes the role of low temperature oxidation in the total synthesis process. The effects of initial and residual oxygen deficiencies on the superconducting properties are studied in Section 5.

4. ELECTRICAL TRANSPORT PROPERTIES: THICKNESS DEPENDENCE

Important information concerning the growth properties may be derived from the dependence of film properties on thickness. For YBCO films, because of the lattice anisotropy, electrical transport properties depend sensitively on epitaxial orientation.

A series of precursor films with thicknesses between 73 nm and 1 μm on (100) $SrTiO_3$ was formed in a number of evaporation runs. Because of limited source supply, two depositions were needed to produce a total thickness of 1 μm. The films were annealed at 730°C and $p(O_2) = 0.00026$ atm to facilitate direct comparison with a similar series of films grown by *in situ* laser ablation under nominally the same applied thermodynamic conditions. With increasing film thickness, the annealing duration was extended from 1 to 4 h to ensure complete conversion of the BaF$_2$. The water partial pressure was kept at 0.025 atm.

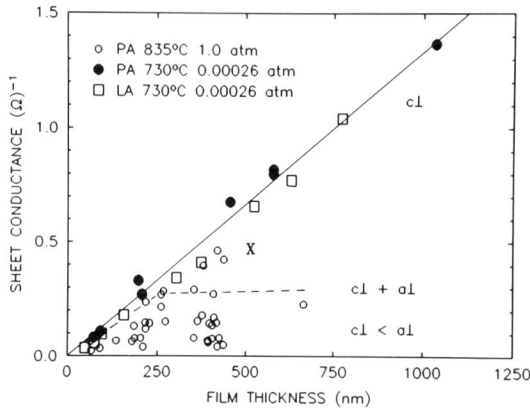

Figure 3. Sheet conductance at 100 K of post-annealed (PA) and *in situ* laser ablated (LA) YBa$_2$Cu$_3$O$_{7-x}$ films on (100) SrTiO$_3$ as a function of film thickness. Growth conditions and predominant epitaxial orientations are indicated in the Figure. The data marked **X** were taken from films containing the YBa$_2$Cu$_4$O$_8$ phase. The straight line indicates a resistivity $\rho(100\ \text{K}) = 75\ \mu\Omega.\text{cm}$

The results are shown in Fig. 3 as plots of the sheet conductance $\sigma_s = (\rho/d)^{-1}$ at 100 K as a function of film thickness d. For uniform deposits σ_s should scale linearly with d, indicating a thickness independent resistivity ρ. This behavior indeed is followed by both low $p(O_2)$ annealed and laser ablated films, with relatively little scatter over the entire thickness range $d > 100$ nm. $\rho(100\ \text{K}) \simeq 75\ \mu\Omega.\text{cm}$ as indicated by the straight line. Each of these films was characterized to be c-axis epitaxial by x-ray diffraction and only very few a-axis grains were observed in SEM micrographs. Additional diffraction intensities due to intergrowths of the Y124 phase, which would tend to lower ρ, were not observed [15]. Deviations mainly occur for $d < 100$ nm, where ρ increases by $\sim 20\%$ for a 73 nm thick low $p(O_2)$ annealed film and by $\sim 60\%$ for a 50 nm thick laser ablated film.

By contrast, for films annealed at 835°C, 1.0 atm, the sheet conductance levels off for thicknesses greater than ~ 300 nm, indicating a resistivity which increases with increasing film thickness. Electron microscopy and x-ray diffraction show that this increase should be attributed to the formation of $a\perp$ grains, growing out from a $c\perp$ layer adjacent to the substrate. As reported by others [16], the maximal thickness of this highly conductive layer that can be formed by post annealing at $p(O_2) = 1.0$ atm appears to be limited to only 250-400 nm, thus leading to a leveling-off of σ_s at greater thicknesses. The relatively high σ_s-values marked X in Fig. 3 for ~ 400 nm thick films annealed at $p(O_2) = 1.0$ atm were influenced by c-axis intergrowths of the Y124 phase, as detected by x-ray diffraction. In the absence of this phase, the average value of

$\rho(100$ K$)$ for c-axis films with thicknesses less than 300 nm annealed at $p(O_2) = 1.0$ atm amounted ~ 100 $\mu\Omega$.cm.

Additionally, Fig. 3 shows a wide scatter of data for films annealed at $p(O_2) = 1.0$ atm. This reflects the sensitivity of epitaxial growth properties to extraneous factors such as substrate surface finish, gas flow conditions during the high temperature anneal, compositional homogeneity of the precursor films as determined by the stability of individual evaporation sources, and presumably also the Y:Ba:Cu composition itself. Although some influence of these factors may also be expected for annealing at low oxygen pressures, thus far this sensitivity appears to be less pronounced than at $p(O_2) = 1.0$ atm. Indeed, the enhanced thickness range for c-axis growth indicates a more orderly growth mechanism at low $p(O_2)$, producing a more uniform microstructure, consistent with the improved crystalline quality observed in RBS-II and overall lower resistivities observed for low $p(O_2)$ annealed films.

Consistent with the c-axis epitaxy and low resistivities, the low $p(O_2)$ annealed films exhibited high critical current densities for all thicknesses. J_c typically was $\sim 50\%$ higher than the 2-3 MA/cm^2 measured for films annealed at 1.0 atm, even though T_c usually was 1-2 K lower. For the 1 μm thick film, $J_c(H=0,$ 77 K$) = 1.9$ MA/cm^2, showing the usual anisotropy with respect to orientation in applied magnetic fields H [17]. For fields parallel to the film surface $(H\|ab)$, $J_c(8$ T, 77 K$) = 60$ kA/cm^2. These high J_c values evidence a well-connected, c-axis film morphology, despite the fact that two depositions were needed to arrive at a thickness of 1 μm. This clearly illustrates the flexibility of the BaF$_2$ preparation method, especially in combination with low $p(O_2)$ annealing.

5. PROCESSING DEPENDENT T_c AND J_c CHARACTERISTICS UPON LOW TEMPERATURE OXIDATION

As noted in Section 3, the initial oxygen content z in films post-annealed at low oxygen pressures may reach very low values, quite possibly as low as $z \leq 6.1$ if complete equilibrium with the oxygen ambient is reached. This obviously increases the likelihood for a larger residual oxygen deficiency after low temperature oxidation. On the other hand, oxygen vacancies might provide an additional source of pinning defects, giving rise to larger critical current densities J_c [18].

To examine whether the somewhat suppressed T_c values observed for low $p(O_2)$ reacted films might relate to a systematically larger residual oxygen deficiency, the low temperature oxygen content was varied through successive reanneals at 550°C under des(as)cending oxygen pressures in the range $1.0 \geq p(O_2) \geq 0.01$ atm. J_c measurements were performed concurrently to investigate the role of oxygen vacancies in flux pinning.

The typical response to such anneals for a c-axis film reacted at $p(O_2) = 0.00026$ atm is shown in Fig. 4, together with results for a 1.0 atm reacted film and an *in situ* laser ablated film. Synthesis conditions and properties in the fully oxygenated starting condition for these films are listed at the bottom of Table 1. Surprisingly, the T_c response is not the same for all films and depends on initial processing conditions. Whereas a semi-flat, plateau-like T_c dependence, as expected for oxygen deficient YBCO [19], is observed for the film post-annealed at 835°C, 1.0 atm and also for the laser ablated film, a non-monotonic T_c variation resulted for the low $p(O_2)$ reacted film, reaching its maximal value $T_c \simeq 92$ K only after oxidation in a slightly reducing oxygen ambient with $p(O_2) \simeq 0.05$ atm. By contrast, J_c decreases monotonically with increasing oxygen deficiency for all films. This effect becomes more prominent at temperatures below 77 K. As will be discussed elsewhere [20] in more detail, this decline may be attributed to diminished pinning energies upon oxygen removal and implies that oxygen vacancies do not represent effective pinning centers in high current

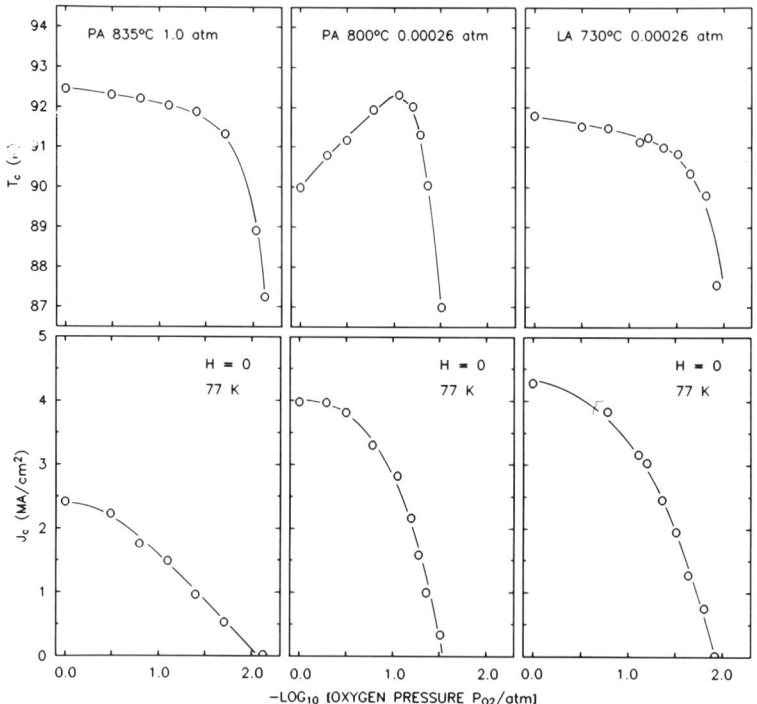

Figure 4. Variation of midpoint transition temperature T_c and critical current density J_c at 77 K in self field for post-annealed (PA) and *in situ* laser ablated (LA) YBa$_2$Cu$_3$O$_{7-x}$ films on (100) SrTiO$_3$ with oxygen partial pressure during low temperature oxidation at 550°C. Initial high temperature growth conditions are indicated in the Figure.

density films.

Thus, upon oxygen desorption in the low $p(O_2)$ reacted films, T_c initially increases, whereas J_c decreases (at temperatures sufficiently below T_c). These changes were reversible upon ascending or descending oxygen pressures during the low temperature anneals. Clearly, the somewhat suppressed T_c values after oxidation in 1.0 atm of oxygen do not simply result from a larger residual oxygen deficiency. A more oxygen deficient composition with similarly suppressed T_c would yield significantly higher normal state resistivities [21] and, as shown in Fig. 4, 3-4 times lower J_c values.

The peaked T_c variation exhibited by the low $p(O_2)$ reacted films resembles the more generally observed T_c dependence on charge carrier density in high-T_c superconducting oxides, in the present case induced by variation of the chain site oxygen occupancy [19]. Since oxygen donates holes to the electronic system, the lower T_c values upon oxygenation in 1.0 atm oxygen indicate overdoping due to an additional source of holes, introduced either during the initial high temperature-low oxygen pressure anneal or during the initial oxidation from $z \simeq 6.1$ to $z \simeq 7.0$.

Alternatively, one might interpret the peaked T_c variation as the absence of a flat "90 K-plateau," indicating, by analogy with the 60 K-plateau, a difference in oxygen vacancy ordering [3]. Assuming a similar doping relation between T_c and oxygen ordering indeed exists for the 90 K-plateau, the peaked T_c variation could result from a less ordered chain site oxygen vacancy distribution, due to, for example, residual *cation* disorder. A positive correlation with oxidation route in the $p(O_2)$-$1/T$ diagram was not observed.

In a recent collaboration with the Stanford group, a similar peaked T_c variation was observed for YBCO films prepared either by *in situ* coevaporation or off-axis sputtering at very low oxygen pressures (< 1 mTorr). Based on observed correlations with the film's cation stoichiometry, especially with respect to Y excess, Matijasevic *et al* [21] proposed that the observed large c-axis lengths and suppressed T_c values after complete low temperature oxidation might result from partial substitution of Ba for Y. Indeed, the peaked T_c variation observed in Fig. 4 is very similar to that observed for the extrinsically doped system $(Y_{1-y}Ca_y)Ba_2Cu_3O_{7-x}$ [22], where divalent Ca substitutes for trivalent Y. It was argued that formation of this type of lattice disorder might be thermodynamically driven under conditions close to the low $p(O_2)$ stability limit. Because of the similarities it is reasonable to assume that a similar effect might occur in the low $p(O_2)$ post-annealed films. However, the fact that laser ablated films grown at the same oxygen pressure as the low $p(O_2)$ annealed films did not exhibit such anomaly suggests that kinetic factors in the growth process may also be important. Moreover, neither the laser ablated films, nor the low $p(O_2)$ annealed films exhibited expanded c-axes. Clearly, more research is needed before a final conclusion may be reached on this topic.

6. THERMAL PROCESSING OF OXYGEN IMPLANTED $YBa_2Cu_3O_{7-x}$ LAYERS

An area where the use of reduced oxygen pressures also might yield improved results is in the thermal processing of YBCO layers previously amorphized by ion implantation. The superconducting and structural properties of YBCO react strongly to the interaction with incident particles (ions, neutrons, electrons), leading successively to enhanced flux pinning (mostly in crystals), loss of superconductivity, metal to insulator transition, and ultimately, to loss of long range crystalline order (amorphization). Because of the structural likeness, the recrystallization of previously amorphized near-surface layers is expected to be very similar to the post annealing of amorphous precursor films.

Fig. 6 compares the effects of thermal processing either at $p(O_2) = 1.0$ atm or at 0.00026 atm, as monitored by RBS-II measurements, in c-axis YBCO films partially amorphized by oxygen implantation. A post-annealed film ($835°C$, 1.0 atm) was used for the experiment at $p(O_2) = 1.0$ atm and a laser ablated film for the low $p(O_2)$ recovery anneal. The significantly lower channeling yield ($\chi_{min} = 0.027$) in the (virgin) starting condition for the latter should not represent an overriding factor in the regrowth properties after amorphization.

The films were implanted at 77 K with 40 keV O^+ ions at near normal incidence to a dose of 10^{16} cm^{-2}. As indicated by the corresponding channeling spectra reaching up to the random spectra, this dose was sufficient to turn a near-surface layer of approximately 100 nm thick amorphous. The amorphous nature of this layer was confirmed by x-ray diffraction and cross-sectional transmission electron microscopy (TEM). Subsequently, the post-annealed film was broken into two pieces which were given separate anneals. The first piece was annealed in atmospheric oxygen at $650°C$ for 3 h and $750°C$ for 1 h. The second piece was annealed at $650°C$ and $850°C$, both for 1 h. Although some ordering occurred at the crystalline-amorphous interface at $750°C$, epitaxial regrowth was limited to this region only and a higher annealing temperature was needed for epitaxial regrowth of the entire layer at $p(O_2) = 1.0$ atm. The χ_{min} value after regrowth at $850°C$ was 0.61. In contrast, nearly complete regrowth of the implanted layer in the laser ablated film resulted after a 10 min anneal at $740°C$,

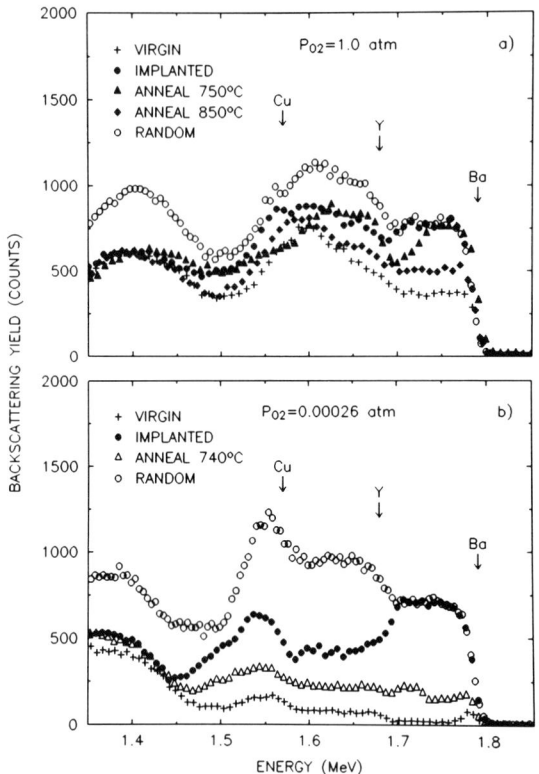

Figure 5. Aligned and random Rutherford backscattering spectra of c-axis YBa$_2$Cu$_3$O$_{7-x}$ films on (100) SrTiO$_3$ before and after implantation with 40 keV oxygen ions (dose 10^{16} cm^{-2}) and after thermal processing at indicated temperatures and oxygen pressures. The spectra were measured with 2.0 MeV He$^+$ ions at near-normal incidence, detected at a 160° scattering angle. Y, Ba, and Cu surface energies are indicated by arrows.

0.00026 atm, yielding χ_{min}=0.20. A second anneal for 20 min under the same conditions did not produce any further improvement in channeling yield.

Cross-sectional TEM images of related samples suggest that the absence of channeling in the near-surface region after annealing at 750°C, 1.0 atm should be attributed to polycrystalline regrowth originating at the film surface. On the other hand, a substantial fraction of the amorphized layer regrown at 740°C, 0.00026 atm, proved to be oriented with the c-axis parallel to the substrate. In other oxygen implanted films, nearly complete conversion of epitaxy from $c \perp$ to $a \perp$ occurred upon annealing at 760°C, 0,00026 atm and even at 820°C predominantly a-axis regrowth prevailed. Thus, while the use of low oxygen pressures clearly enhances the recrystallization process, the preferred epitaxial orientation upon regrowth apparently deviates from the systematics established for the postannealing of amorphous Y, BaF$_2$, Cu precursor films (Fig. 1).

A relatively simple explanation for this qualitative difference may be given by considering that the epitaxial growth during conversion of precursor films with BaF_2 is governed by two principally inequivalent process rates: (i) the net dissociation rate of the BaF_2 and (ii) the intrinsic growth rate of YBCO at the applied annealing temperature and oxygen pressure. If the latter rate is faster, epitaxial growth is controlled by the dissociation of the BaF_2 (dependent on temperature, water partial pressure, and presumably also fluorine and oxygen transport through the unreacted layer). Because of the high chemical stability of BaF_2, random nucleation and growth of YBCO or other phases away from the substrate or moving c-axis YBCO front may be suppressed, resulting in an orderly growth mechanism, dosed by the availability of dissociated Ba for the formation of YBCO.

Conversely, if the dissociation rate is faster, BaF_2 may be removed from the entire precursor film before the epitaxial layer growing up from the substrate has reached the film surface. In this case, random nucleation and growth away from the substrate is not inhibited by the presence of BaF_2 and may lead to a more disordered morphology. The change-over from c-axis to a-axis epitaxy in thick films annealed at $p(O_2) = 1.0$ atm could well reflect this situation. At low $p(O_2)$, because of enhanced cation mobilities, the *balance between YBCO growth rate and BaF_2 conversion rate* apparently is more favorable, giving rise to more orderly growth and allowing for thicker c-axis films to be formed. During the thermal processing of oxygen implanted, amorphous layers this type of dosing mechanism evidently is absent and the final structure with enhanced a-axis growth may result from competing effects taking place at the crystalline-amorphous interface and other regions of the film.

Introduced originally to avoid decomposition of the precursor films [23], this explanation assumes a significantly expanded role of the BaF_2, which appears to be consistent with observations and may be further tested with simple experiments. For example, preliminary results indicate that this balance of rates may be shifted by changing the water partial pressure. Increasing the water content of the annealing ambient indeed tends to result in a more chaotic surface morphology, whereas a lower water content occasionally has been observed to have the opposite effect. Likewise, the recrystallization of fluorine implanted layers may deviate from that of oxygen implanted layers. Further research is planned to resolve these issues.

7. SUMMARY

In summary, the use of low oxygen partial pressures has pronounced advantages for the post processing of amorphous films, both in the initial synthesis of high current density, epitaxial YBCO films from Y, BaF_2, and Cu precursors, and in the recrystallization of oxygen implanted near-surface layers. In either application, lower processing temperatures may be used at lower oxygen pressures, while thicker c-axis films may be synthesized. By further optimization of the annealing ambient, we anticipate that annealing temperatures for high current density films may be reduced to $\sim 700°C$.

Proper low temperature oxidation was observed to be essential for obtaining films with the highest current densities J_c. However, depending on initial growth conditions, the low temperature oxygenation treatments for maximal T_c and J_c are not necessarily the same. A peaked T_c dependence on residual oxygen deficiency was observed for the low $p(O_2)$ annealed films, giving rise to slightly suppressed transition temperatures after complete low temperature oxygenation in 1.0 atm oxygen. Qualitatively, this anomalous behavior can be explained by hole doping due to partial Ba substitution on the Y site, indicating a new type of lattice defect in YBCO formed preferentially under low $p(O_2)$ conditions.

ACKNOWLEDGEMENTS

The authors wish thank T. B. Lindemer and D. M. Kroeger (Oak Ridge National Laboratory), R. Beyers (IBM Almaden), R. Bormann (Forschungszentrum Geesthacht, Germany), and V. Matijasevic (Stanford University) for stimulating discussions and making available experimental data prior to publication. This research was sponsored by the Division of Materials Sciences, U. S. Department of Energy under contract DE-AC05-84OR21400 with Martin Marietta Energy Systems, Inc.

REFERENCES

1. R. Beyers and B. T. Ahn, in *Annual Review of Materials Science*, Vol. 21, 1991 (in press)
2. T. B. Lindemer, J. F. Hunley, J. E. Gates, A. L. Sutton Jr., J. Brynestad, C. R. Hubbard, and P. K. Gallagher, *J. Am. Cer. Soc.* **72**, 1775 (1989)
3. B. W. Veal, A. P. Paulikas, H. You, H. Shi, Y. Fang, and J. W. Downey, *Phys. Rev B* **42**, 6305 (1990)
4. R. Feenstra, T. B. Lindemer, J. D. Budai, and M. D. Galloway, *J. Appl. Phys.* 1991 (in press)
5. K. W. Lay and G. M. Renlund, *J. Am. Cer. Soc.* **73**, 102 (1990)
6. T. B. Lindemer, F. A. Washburn, C. S. MacDougall, R. Feenstra, and O. B. Cavin, *Physica C*, 1991 (in press)
7. D. H. Lowndes, D. P. Norton, J. W. McCamy, R. Feenstra, J. D. Budai, D. K. Christen, and D. B. Poker, in *High-Temperature Superconductors: Fundamental Properties and Novel Materials Processing*, Mater. Res. Soc. Symp. Proc. 169, edited by D. K. Christen, L. Schneemeyer, and J. Narayan (Mater. Res. Soc., Pittsburgh, PA, 1990), p.431
8. R. H. Hammond and R. Bormann, *Physica C* **162-164**, 703 (1989)
9. R. K. Williams, K. B. Alexander, J. Brynestad, T. J. Henson, D. M. Kroeger, T. B. Lindemer, G. C. Marsh, and J. O. Scarbrough, *J. Appl. Phys.* **67**, 6934 (1990)
10. E. D. Specht, C. J. Sparks, A. G. Dhere, J. Brynestad, O. B. Cavin, and D. M. Kroeger, *Phys. Rev. B* **37**, 7426 (1988)
11. P. König, R. Bormann, and J. Nölting, internal report, University Göttingen, 1990
12. N. Chen, D. Shi, and K. C. Goreatta, *Appl. Phys. Lett.* **66**, 2485 (1989)
13. J. L. Routbort, S. J. Rothman, N. Chen, and J. N. Mundy, *Phys. Rev. B* **43**, 5489 (1991)
14. P. Kofstad, in *Nonstoichiometry, Diffusion, and Electrical Conductivity in Binary Metal Oxides*, (Wiley, New York, 1972)
15. K. Char, M. Lee, R. W. Barton, A. F. Marshall, I. Bozovic, R, H. Hammond, M. R. Beasley, T. H. Geballe, and A. Kapitulnik, *Phys. Rev. B* **38**, 834, (1988)
16. A. Mogro-Campero, L. G. Turner, E. L. Hall, and N. Lewis, in *High-Temperature Superconductors: Fundamental Properties and Novel Material Processing*, Mater. Res. Soc. Symp. Proc. 169, edited by D. K. Christen, L. Schneemeyer, and J. Narayan (Mater. Res. Soc., Pittsburgh, PA, 1990), p. 703
17. D. K. Christen, C. E. Klabunde, R. Feenstra, D. H. Lowndes, D. P. Norton, J. D. Budai, H. R. Kerchner, J. R. Thompson, L. A. Boatner, J. Narayan, and R. Singh, *Physica B* **165-166**, 1415 (1990)
18. M. Daeumling, J. M. Seuntjes, and D. C. Larbalestier, *Nature* **346**, 332 (1990)
19. R. J. Cava, A. W. Hewat, B. Batlogg, M. Marezio, K. M. Rabe, J. J. Krajewski, W. F. Peck, and L. W. Rupp, *Physica C* **165**, 419 (1990)
20. R. Feenstra, D. K. Christen, C. E. Klabunde, and J. D. Budai, submitted to *Phys. Rev.*
21. V. Matijasevic, P. Rosenthal, A. F. Marshall, R. H. Hammond, and M. R. Beasley, *J. Mater. Res.* **6**, 682 (1991)
22. Y. Tokura, J. B. Torrance, T. C. Huang, and A. I. Nazal, *Phys. Rev. B38*, 7156 1988)
23. P. M. Mankiewich, J. H. Scofield, W. J. Skopcol, R. E. Howard, A. H. Dayem, and E. Good, *Appl. Phys. Lett.* **51**, 1753 (1987)

High T$_c$ Superconductor Thin Films
L. Correra (Editor)

Preparation and properties of YBCO thin films with the c– axis aligned in the film plane

H.–U. Habermeier[a], A.A.C.S. Lourenco[a,c], B. Leibold[a], J. Kircher[a], B. Friedl[a] and G. Lu[b]

a.) Max–Planck–Institut für Festkörperforschung, Heisenbergstr. 1 D 7000 Stuttgart–80, FRG

b.) Max– Planck–Institut für Metallforschung, Institut für Physik, Heisenbergstr. 1, D 7000 Stuttgart–80

c.) Laboratorio de Fisica, Universidade do Porto, PL 4000 Porto, Portugal

Abstract

High temperature superconductor thin films with the CuO_2 planes oriented perpendicular to the substrate open a possible way to fabricate planar Josephson junctions, because unlike in c– axis oriented films the superconducting coherence length of the films is relatively large perpendicular to the surface. In our experiments we demonstrate that the use of (110) oriented $SrTiO_3$ substrates in conjunction with a combination of heteroepitaxial growth at low substrate temperatures and homoepitaxial growth at high substrate temperatures results in thin films with the c– axis aligned along the [100] direction of the substrate surface over the whole specimen and critical temperatures up to 88 K. Mandatory for the convincing analysis of the orientation and the homogeneity of the films is a combination of X– ray diffractometry, Raman spectroscopy and ellipsometry. Cross–sectional TEM gives the direct proof of the film orientation and reveals some details of the growth close to the substrate/ film interface.

1. INTRODUCTION

Progress in high temperature superconductor [HTS] thin film technology has advanced rapidly in the past two years especially in preparing thin films of $Y_1Ba_2Cu_3O_{7-x}$ [YBCO] with critical temperatures, T_c, above 90 K and critical currents, j_c, above 10^6 A/cm^2 at liquid nitrogen temperatures. The standard thin film preparation process consists in the deposition of a crystalline film with the correct stoi– chiometry of the metallic constituents at substrate temperatures, T_d, around 750°C at an oxygen partial pressure, p_{ox}, of the order of 100 Pa. Using (100) oriented substrates like $SrTiO_3$, MgO or $LaAlO_3$ such films grow epitaxially with the c— axis perpendicular to the film plane. A review covering the description of different techniques for the HTS thin film deposition technology is given by Leskelä et. al. [1]. Due to the anisotropy of the crystal structure and consequently the anisotropy of the superconducting parameters single phase YBCO films with the c— axis aligned either perpendicular or parallel to the film plane are of importance, both, for fundamental research as well as for for applications. Using the anisotropy of electrical transport properties such as j_c or the anisotropy of intrinsic parameters like the superconducting coherence length, ξ, [the coherence length along the c— axis $\xi_c = 0.2$ nm, along the a,b plane $\xi_{ab} = 1.5$ nm] new types of superconducting devices are feasible or already existing device types like Josephson junctions can be fabricated as planar devices. The reliable fabrication of planar Josephson junctions e.g. is facilitated, if the a,b planes are perpendicular to the film plane and the coherence length with the comparably large value is in this direction, too. In Fig. 1a,b the orientations of the CuO_2 planes with respect to the substrate surface are shown.

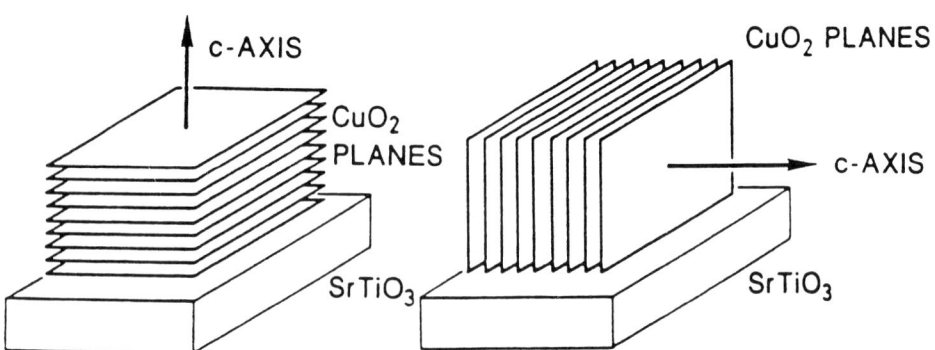

Fig1:   Schematic representation of the orientation of the CuO_2 YBCO planes with respect to the film plane a.) for the c—axis perpendicular and b.) for the c— axis in the film plane.

The desired configuration in Fig. 1b can be realized if the film has a (100) or (110) orientation. In a number of papers methods are described to prepare single phase (110) oriented YBCO thin films on (110) $SrTiO_3$ substrates [1,2,3]. Enomoto et al. [1] have successfully grown films on (110) $SrTiO_3$ substrates and reported anisotropic critical currents and H_{c2} values. Their films, however, were only analyzed with respect to the occurrence of other phases by X—ray diffraction and no further experimental evidence was given to discriminate the observed X—ray diffraction peaks ascribed to the (110) reflections from the (103) peaks located at similar positions in the diffraction pattern [the lattice spacings $d_{110} = 0.27254$ nm and $d_{103} = 0.27280$ nm are too close to be discriminated in an usual X—ray diffraction pattern]. Terashima et al. [2]

reported a film growth procedure at rather low deposition temperatures around 530° C using an activated evaporation process. The orientations of the films were determined by RHEED patterns only and no further information about superconducting transport properties were given. Linker et al. [3] used the hollow cathode sputtering technique and studied the epitaxial growth of YBCO thin films grown on (110) oriented substrates as a function of the deposition temperature. In this paper the film growth on (110) SrTiO$_3$ is found to be either in the (110) or in the mixed (110)/(103) direction depending on the substrate temperature. Transport measurements revealed a sharp transition at 90 K even for the mixed phase. The authors demonstrate clearly the dependence of the film growth on the substrate temperature.

In this paper we elaborate the preparation of single phase (110) − oriented YBCO thin films on (110) SrTiO$_3$ substrates as described previously [4]. The films are analyzed with respect to their electrical transport properties and their structure as revealed by X− ray diffraction and cross−sectional TEM. Raman spectroscopy is used as a nondestructive tool to determine the orientation, especially to discriminate between the competing (110) and (103) orientations. Ellipsometry is applied to test the overall homogeneity of the orientation of the films.

2. FILM DEPOSITION PROCESS

For the film deposition process we used the standard pulsed laser deposition technique [PLD] as introduced by Venkatesan [5]. The description of the experimental setup and the details of the in−situ growth are given elsewhere [6]. The standard conditions for epitaxial film growth combine a deposition temperature, T_d, of 780° C and an oxygen background pressure, p_{ox}, of 100 Pa during deposition. Films prepared using these conditions are single phase c− axis oriented if (100) SrTiO$_3$, LaAlO$_3$ or ZrO$_2$ substrates are used. The high reliability of the PLD technique allows not only to grow thin films with high T_c and j_c but also to study the film growth process as a function of the deposition parameters e.g. the substrate temperature, oxygen pressure and deposition rate. Especially the deposition temperature has been demonstrated to be of decisive importance for the growth mode of the films [2,3]. The general trend is that films grow crystalline at deposition temperatures exceeding 550° C with the c− axis in the plane [a − axis orientation] whereas with increasing substrate temperature the c− axis growth is preferred. Mixed orientations are observed between the two extreme orientations (h00) and (00l) for deposition temperatures between 700°C and 780° C.

3. CRYSTALLOGRAPHIC REQUIREMENTS FOR EPITAXIAL GROWTH

The unit cell of YBCO is composed of two primitive perovskite BaCuO$_{3-x}$ cells with a primitive cell of YCuO$_{3-y}$ sandwiched between them. Thus, the YBCO unit cell is tripled in the c− direction with respect to the a and b direction. The lattice parameters are a= 0.3827 nm, b= 0,3877 nm and c= 1.1708 nm, respectively. The orthorhombic distortion is of the [100] type leading to a ≠ b with an orthorhombicity (b–a)/a ≅ 1.3%. Consequently a substrate with a good lattice match for a and b shows from the crystallographic point of view an equally favorable epitaxy for the c − in plane orientation. In the case of $SrTiO_3$ as substrate material with the primitive cubic perovskite structure and a lattice constant a= 0.3905 nm the crystallographic lattice mismatch $(a_{SrTiO_3} - b_{YBCO}) / a_{SrTiO_3}$ is less than 1% and good epitaxy is expect–ed. Using (001) oriented substrates, however, the epitaxy relation for c− axis growth holds simultaneously for both in plane < 100 > directions and an epitaxial growth will lead to films with grains with their c− axis along one of the two perpendicular < 100 > directions. In order to prepare films with the c− axis aligned in the film plane the symmetry of the crystallographic epitaxy relation has to be violated. The step from the highly symmetrical (001) substrate to the (110) substrate leaves in the surface plane of the substrate instead of two perpendicular oriented < 100 > directions only one [001] direction and perpendicular to that a [110] direction. Since along the [001] direction the lattice mismatch between film and substrate

$$(3a_{SrTiO_3} - c_{YBCO}) / c_{YBCO} = 1.3 . 10^{-3}$$

is rather small a preferential growth with the c − axis macroscopically aligned is gener–ated. Fig. 2 shows the orientations of the crystal axes in the substrates of different cuts schematically.

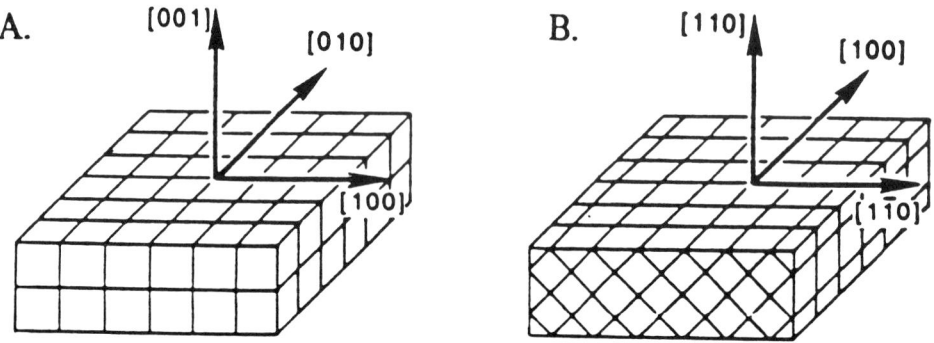

Fig. 2: Crystal orientations in a (100) cut [a] and in a (110) cut [b].

4.   YBCO THIN FILM GROWTH ON (110) $SrTiO_3$ AT HIGH SUBSTRATE

TEMPERATURES [T_d = 750° C]

If the standard deposition process is applied using (110) oriented substrates we obtain films with a T_c of 91 K (c.f. Fig. 3). The X− ray diffraction patterns show only

2 sets of double peaks centered at $2\theta = 33°$ and $2\theta = 68°$ (see Fig. 4). The high intensity peaks are attributed to the $SrTiO_3$ (110) and (220) reflections, respectively,

and the neighbouring peaks are due to the YBCO thin film.

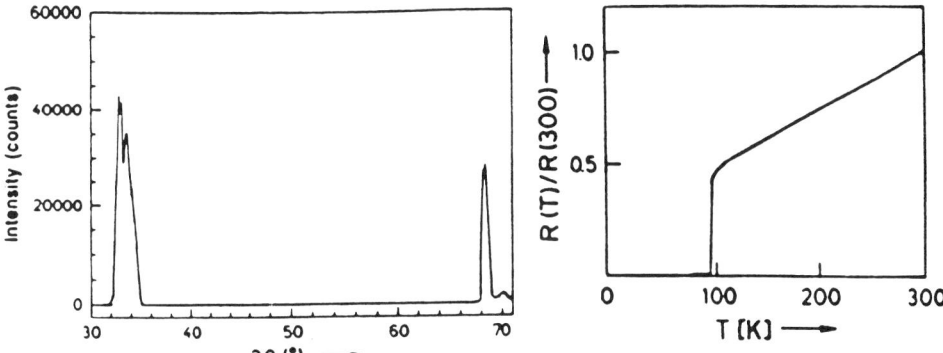

Fig. 3: X ray diffraction pattern of a
YBCO film on (110) strontiumtitanate
$T_d = 780^\circ C.$

Fig. 4 Transition curve of
the film in Fig.3.

In an X– ray diffraction pattern with higher angular resolution the substrate peaks appear to be broadened with a shoulder at the high angle side which can be due to some interfacial stress or the occurrence of a mixture of (110), (103) and (013) phases. For a nondestructive optical analysis Raman spectroscopy is applied which has been proven to be a valuable tool for the characterization of HTS superconductors [7]. Based on the fact that the relative magnitudes and signs of the Raman tensor (α_{ij}) for a given phonon mode are rather different for the incident light polarized parallel or perpendicular to the CuO_2 planes Raman scattering in an easy and fast method to determine the orientation of single crystalline films [8]. In Fig. 5 the Raman spectra of an YBCO thin film deposited on a (110) $SrTiO_3$ substrate at $750^\circ C$ are represented with the polarization vector of incident light beam and reflected light oriented parallel to the c– axis (top) and parallel to the (110) axis (middle). In the geometry with the incident and scattered light polarized parallel to the [100] direction the four A_g modes at 116, 150, 438 and 501 cm^{-1} are observed corresponding to the vibrations of the Ba, planar Cu(2), planar O(II) – O (III) [in phase] and O(IV) atoms. The peak at 339 cm^{-1} corresponds to the out of phase vibration of the planar O (II) – O (III). For this mode $\alpha_{xx} \approx - \alpha_{yy}$ and $\alpha_{zz} \approx 0$. The fact that this B_{1g} like mode is observed in the polarization along the c– axis indicates that the film is at least not single phase with the c–axis aligned in the film plane. The spectra for the polarisation parallel to the substrate (110) direction shows a low intensity B_{1g}–like mode which is suppressed compared to the spectra taken in the xx or yy geometry [8]. From these spectra we conclude that the film is a mixture of (110) and (103) orientation and not a single phase (110) material. Compared to the X– ray analysis the Raman result gives the more detailled information concerning the orientation of the film.

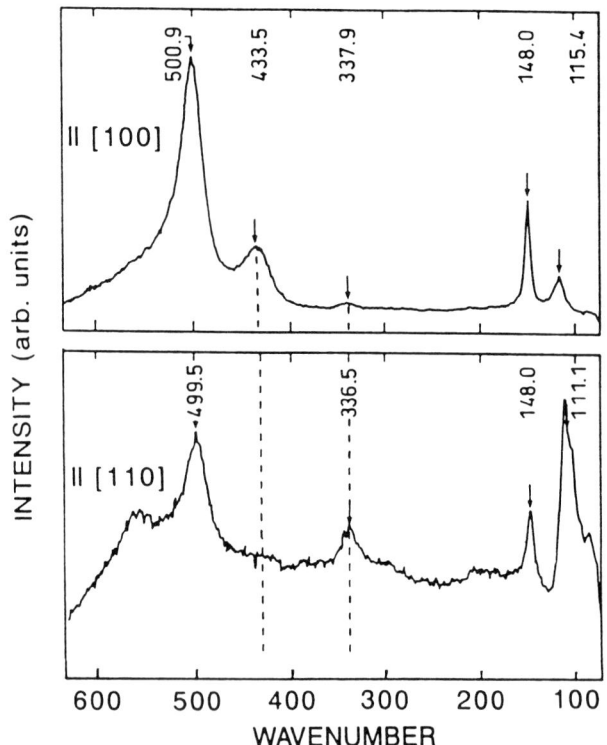

Fig. 5 : Raman spectra of an YBCO thin film on (110) $SrTiO_3$ deposited at 750° C for the different orientations of the polarization vector for incident and scattered light.

5.   YBCO THIN FILMS GROWN ON (110) $SrTiO_3$ SUBSTRATES WITH THE C– AXIS ALIGNED IN THE FILM PLANE.

The technique used for the growth of YBCO films with the c– axis aligned in the film plane combines heteroepitaxial growth at low substrate temperatures and homo–epitaxial growth at higher substrate temperatures. In Fig.6 the temperature / time profile for the growth conditions is given. At $T_d = 663^o$ C a 60 nm thick film is grown heteroepitaxially, the starting temperature was choosen in such a way that good crystallinity could be expected while the c– axis is still oriented in the plane. The substrate temperature is continuously increased to 720° C without interruption of the growth process. The total thickness of the film is approximately 350 nm. Films grown according to this process show a sharp transition to superconductivity between 83K and 87K [c.f. the insert of Fig. 7]. The X– ray diffraction patterns revealed almost single phase (110) oriented films with the dominant diffraction peaks ascribed to the (110) and (220) reflections adjacent to the corresponding substrate peaks as shown in Fig. 7.

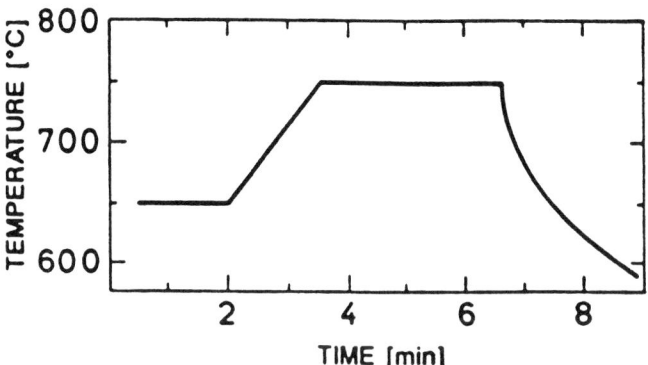

Fig. 6: Temperature / time profile for the epitaxy of (110) oriented YBCO thin films.

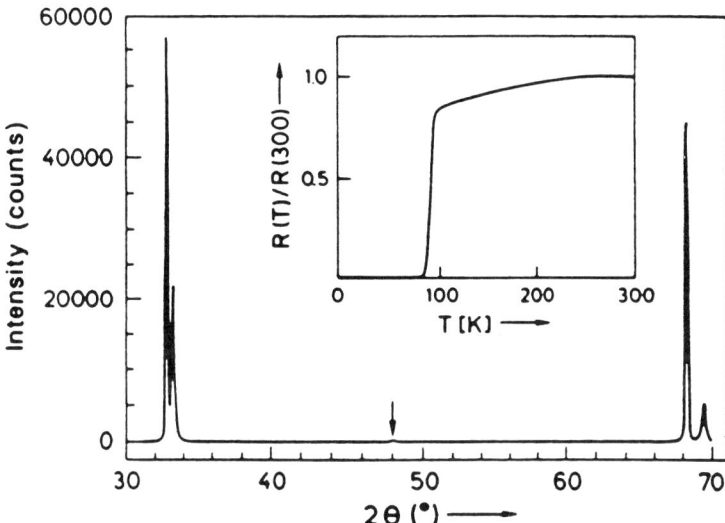

Fig. 7: X − ray diffraction pattern of an (110) oriented YBCO thin film [the insert shows the electrical transition to superconductivity].

The volume fraction of the misoriented material is estimated to be less than 1/ 1000. A Raman spectrum of this film taken for different polarizations is given in Fig. 8. In contrast to the spectra shown in Fig.5 this spectrum does not show any indications of a B_{1g} like mode for the polarization parallel

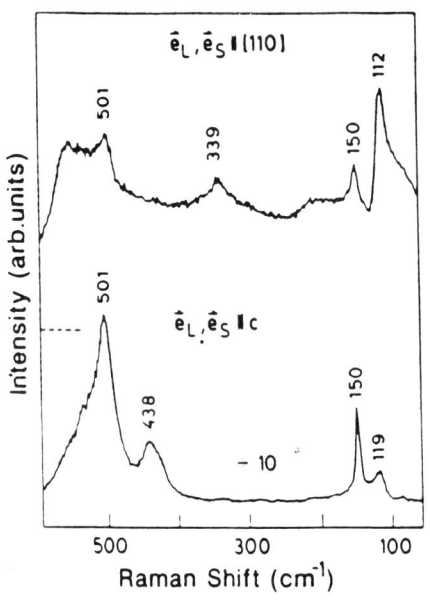

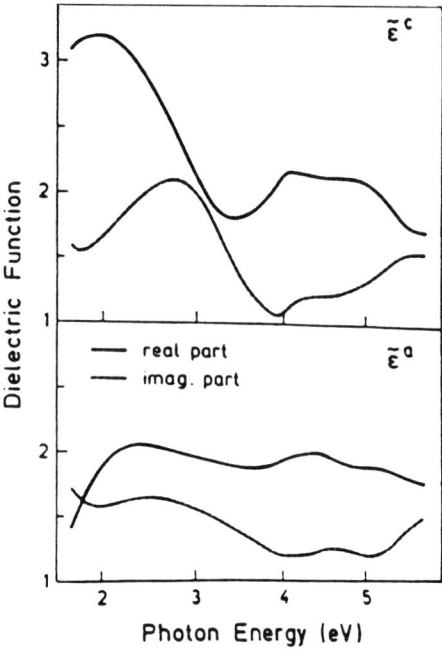

Fig. 8 Raman spectra of a (110) YBCO
thin film for polarization of incident
and scattered light parallel [110] (top)
and parallel [100] (bottom)

Fig. 9: Ellipsometric mea—
surements: upper part:
pseudodielectric function
for c parallel to the plane
of incidence. Lower part:
pseudodielectric   function
for c perpendicular to the
film plane

to c (lower part of the figure). This is a proof that the c— axis is really in the plane and
not only a projection of the c— axis on to the plane is seen. Furthermore, it should be
pointed out that in the Raman spectra no signal of spurious phases like $BaCuO_2$ or
Y_2BaCuO_5 could be observed which indicates the high quality of the films investigated.

To check the overall homogeneity of the film quality, especially of the orientation
ellipsometric measurements have been performed on these films. Ellipsometry is a
powerful tool for the characterization of oxygen stoichiometry and the orientation of
high T_c materials [9]. The method probes a large area of the sample surface (in this

case the whole sample of 5 x 5 mm^2) and thus delivers information about the lateral
homogeneity of the material. For moderate anisotropies the pseudodielectric function is
measured which is a good approximation to the components of the dielectric tensor
along the line of intersection between the plane of incidence and the sample surface [10].

The measured spectra are shown in Fig. 9 for the pseudodielectric function parallel to the c— axis (upper part) and perpendicular to the c— axis (lower part). The presence of the strong 2.7 eV peak assigned to the plane—to—chain transition for c parallel to the plane of incidence and the absence of this peak for the direction perpendicular to the c — axis is a strong indication that the c — axis is aligned over the whole sample. Structural analysis of these samples by cross— sectional TEM show clearly that the c— axis is aligned parallel to the film plane. It is noticed in Fig.10 that in a layer with an average thickness of 20 nm along the interface the ordering fringes along the c— direction of the YBCO structure are not preserved i.e. the film is here of not so perfect crystalline quality compared to areas further apart from the interface. This finding certainly resembles the growth process in which the first couple 10 of nm are deposited under conditions not optimized to obtain the best crystalline film quality.

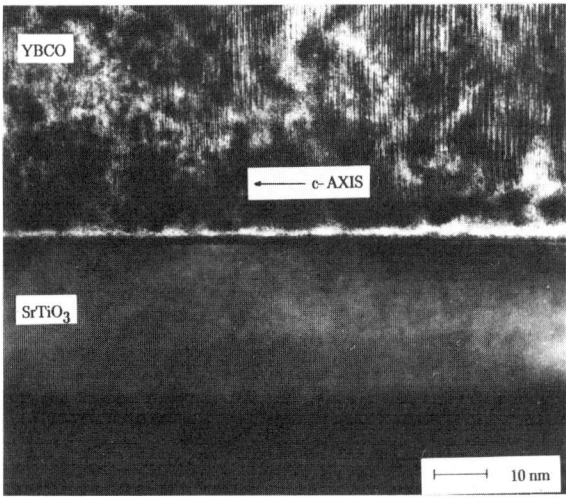

Fig. 10.: Cross sectional TEM micrograph of a YBCO (110) thin film.

6. CONCLUSIONS

We have grown high quality (110) oriented superconducting YBCO thin films with the c— axis aligned over the whole film surface along the [100] direction. Our method is characterized by the film growth starting at low substrate temperatures, and a subsequent increase of the deposition temperature enables us to overcome the problems of the mediocre superconducting properties usually inherent with film growth at low deposition temperatures. Furthermore, the combination of optical methods, Raman spectroscopy and ellipsometry, is of essence to avoid incorrect conclusions about the film orientation which can be drawn from X — ray analysis alone. The precise interpretation of the X — ray diffraction patterns is further hampered by the fact that film / substrate interactions can either lead to interfacial layers with somewhat changed lattice constants or chemical reactions of film and substrate material which gives rise to diffuse scattering. These effects lead to a substrate peak splitting and/ or broadening which can mask or simulate the presence of diffraction peaks. Consequently, the application of other nondestructive optical methods which make use of the anisotropic structure of the material is necessary. The more detailed investigations of the epitaxial growth using cross –sectional TEM gives the direct proof of a homogeneous crystallization of the fim in the desired c in plane orientation. The detailled analysis of the changing deposition temperature during the epitaxial process and its consequences for the perfection of the crystallinity remains as a task for the future.

In summary, single phase (110) oriented YBCO thin films with excellent superconducting parameters can be deposited using a continuous increase of the substrate temperature during film growth. The films are macroscopically oriented with the c– axis aligned along the [100] direction of the substrate surface. The development of a reliable deposition technique for this type of films is seen as a basic prerequisite for the fabrication of planar Josephson junctions.

Acknowledgments

We would like to thank Dr. J. Köhler and L. Viczian for the X– ray diffraction measurements, Dipl. phys. G. Wagner for encouraging discussions and assistance in specimen analysis.

REFERENCES

[1]   Y. Enomoto, T. Murakami, M. Suzuki, and K. Moriwaki, Jap. J. Appl. Phys. 26 (1987) L 1248

[2]   T. Terashima, Y. Bando, K. Iijima, K.Yamamoto, and K. Hirata, Appl. Phys. Lett. 53 (1988) 2232

[3]   G. Linker, X.X.Xi, O. Meyer,Q.Li, and J. Geerk, Solid Stata Comm. 69 (1989) 249

[4]   H.–U. Habermeier, A.A.C.S. Lourenco, B. Friedl, J. Kircher, and J. Köhler, Solid State Comm. 77 (1991) 683

[5]   D. Dijkamp, T. Venkatesan, and X.D. Wu, Appl. Phys. Lett. 51 (1987) 619

[6]   H.–U. Habermeier, Eur. J. Solid State Inorg. Chem. 28 (1991) 619

[7]   C. Thomsen and M. Cardona,in Physical Properties of High Temperature Superconductors, ed. by D. M. Ginsberg, (World Scientific, Singapore, 1989), p.409

[8]   K.F. McCarty, J.Z. Liu, R.N. Shelton, and H.B. Radousky, Phus. Rev. B 41 (1990) 8792

[9]   J. Kircher, M. K. Kelly, S. Rashkeev, M. Alouani, D. Fuchs, and M. Cardona, Phys. Rev. B. 1991 (in print)

[10]   D. E. Aspnes, J. Opt. Soc. Am. 70 (1980) 1275

High T$_c$ Superconductor Thin Films
L. Correra (Editor)

Evidence for Cation Disorder in *In-Situ* Grown YBaCuO Superconducting Films

V.C. Matijasevic,[a,b] R.H. Hammond,[a] P. Rosenthal,[a] K. Shinohara,[a,c] A.F. Marshall,[a] M.R. Beasley,[a] and R. Feenstra[d]

[a] Department of Applied Physics, Stanford University, Stanford, CA 94305,USA

[b] Applied Physics, TU Delft, Lorentzweg 1, 2628 CJ Delft, The Netherlands

[c] Central Eng. Labs, Nissan Motor Co., 1, Natsushima, Kanagawa 237, Japan

[d] Oak Ridge National Lab, Oak Ridge, TN 37831, USA

Abstract

Some of the still unanswered materials questions regarding *in-situ* films of cuprate superconductors are reviewed. As a case study we present $YBa_2Cu_3O_y$ thin films grown by evaporation in low oxygen pressure. Several anomalous results are found. For O_2 pressure ≤ 10 mTorr, films with average composition substantially off the 1:2:3 stoichiometry have more bulk-like properties, including higher T$_c$'s, compared to films made on-stoichiometry. Films made at lower oxygen pressure also have c-axis lattice constant expanded compared to the films made at higher pressures (> 100 mTorr). The films show evidence that they are hole-doped compared to the ideal material. Altogether, the results strongly suggest that metal-atom point-like defects are quenched into the films. A model based on the presence of Ba-for-Y substitution is discussed and found to be consistent with the experimental results. We suggest that these defects might be important for superconducting properties of *in-situ* films in general.

1. INTRODUCTION

It is by now well established that *in-situ* thin film synthesis of cuprate superconductors provides good quality materials for both basic research as well as their applications. A number of deposition techniques are especially well suited for providing films of a standard high quality. In this paper we will discuss, in particular, the cuprate superconductor $YBa_2Cu_3O_y$. What is not yet understood from a materials' standpoint, however, are the important differences between the materials in this form and other forms of the same material (such as single crystals). Generally speaking, several qualities of the *in-situ* films are still mysteries, especially the structural origins of these properties. These are, for example, the relatively high critical currents that are obtained for most good quality *in-situ* films, the occasionally occurring depressed T$_c$'s (by a few degrees), and also occasionally reported expanded c-axis lattice parameters. These films have high critical currents, within an order of magnitude of the depairing critical currents. Such critical currents imply a high density of pinning centers.

At the same time, a number of recent *in-situ* efforts has focused on producing cuprate superconductor thin films at even lower pressures of oxygen. This is important in order to achieve lower growth temperatures. The thermodynamic conditions for growth of the cuprate superconductors require such a correlation between temperature and pressure, see for example Ref. 1. In addition, lower oxygen pressures would allow various deposition techniques to be more manageable, e.g. in case of e-beam evaporation, and might allow for mbe-type growth. Figure 1 shows regions of the p-T phase space where *in-situ* efforts have been successfully attempted (a list of references is in Ref. 2). The solid lines (and the dashed extrapolations) in the figure represent the stability lines for the $YBa_2Cu_3O_y$ compound.[1] The wavy vertical line is the substrate interaction temperature which limits growth for the usual *in-situ* substrates, such as MgO. The top dashed line represents roughly the pressure below which *in-situ* methods have more anomalous results, as will be discussed below.

However, growth at lower oxygen pressures, and correspondingly lower temperatures, does not usually produce as good quality films as does growth at higher pressures and temperatures. This appears largely to be due to kinetic constraints. One solution is to use activated species of oxygen (e.g. atomic oxygen or ozone) in order to enhance kinetics. A substantial research effort has been devoted to such growth. For pressures below 1 mTorr (0.2 Pa) activated species have almost always yielded improvements in film quality.

We have tried to address the issues surrounding *in-situ* growth by examining in detail the films made at the minimum O_2 pressure required for YBaCuO growth. We have determined this minimum pressure to be about 1 mTorr. By looking for clues in the films made at these marginal conditions, one can try to learn something more general about the *in-situ* films. Our growth conditions are shown by the region outlined in bold in Fig. 1.

2. RESULTS OF LOW PRESSURE REACTIVE COEVAPORATION

We present here the summary of results of an extensive study, presented in more detail in Ref. 3. The technique used for thin film synthesis is reactive coevaporation. Reference 3 also explains most of the technical details. We will just mention here that this technique allows for independent variation of the metal fluxes during growth and independent control over a number of other growth variables.

The properties of these evaporated films are discussed in Refs. 2 and 3. Here we focus only on some of the more anomalous results. These are: growth in the unstable part of the phase diagram, depressed superconducting transition temperatures, off-stoichiometry growth yielding higher T_c's, expanded c-axes, and the hole-doped nature of the films. We will then try to discuss these results within a single interpretation.

2.1. Growth in the thermodynamically unstable region

Successful growth can vary over several hundred degrees in temperature and several orders of magnitude in pressure. However, what is intriguing is that these conditions extend beyond the stability line;[1] in Fig.1 our growth

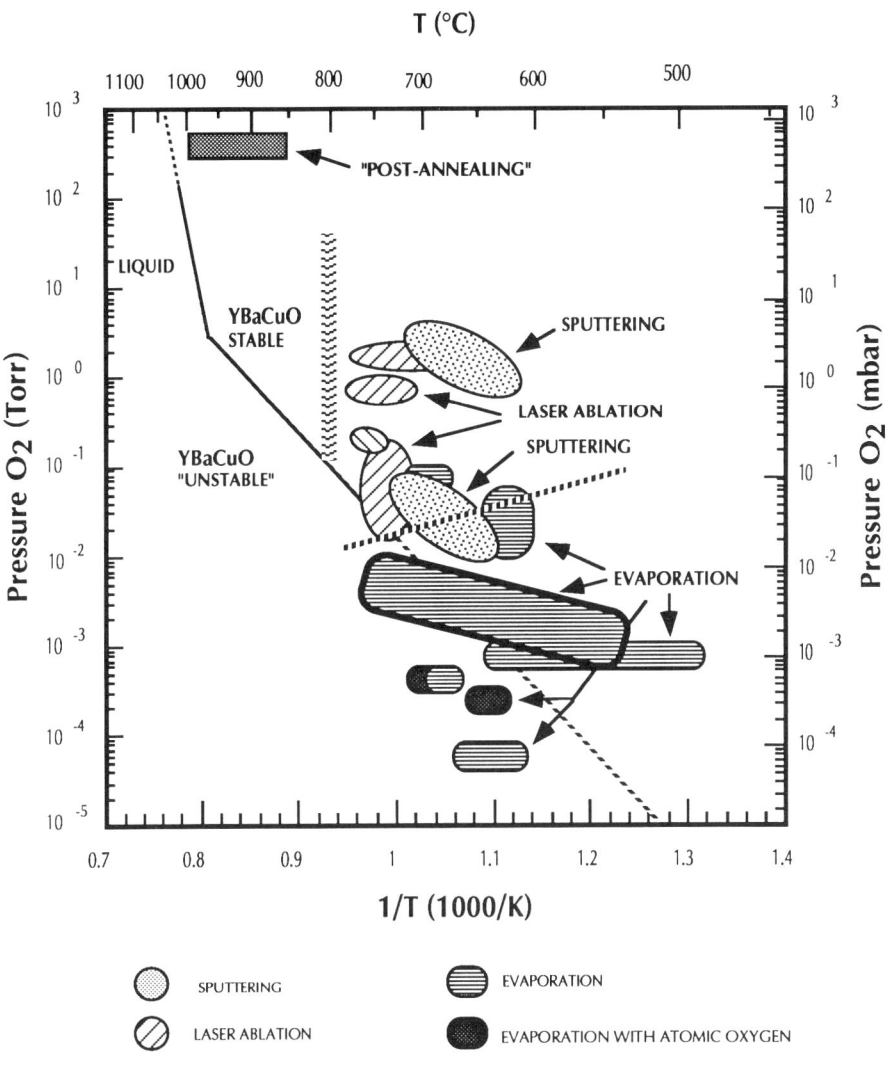

Figure 1. Synthesis efforts for YBaCuO thin films. The various shaded regions represent regions of the p-T phase space where efforts in synthesizing superconducting films were successful.

conditions, outlined in bold, cross the stability line. In other words, we do not observe any sharp decomposition crossover in the regimes where we have worked. Superconducting YBaCuO films were made up to 80°C beyond (to the left of) the extrapolated stability line.

A possible explanation is that a metastable YBaCuO may be formed, especially at low temperature: for a small free energy driving force (close to the line), the kinetics could favor YBaCuO, rather than nucleating and growing three separate phases.[1] However, there are other explanations as will be discussed below.

2.2 Ba-deficient films

Another striking observation for these low pressure films is that films which are made off-composition, and Ba-deficient, have higher T_c's and c-axis lattice parameters (c_0) than do films which are made at the same p-T conditions, but on 1:2:3 stoichiometry. For pressure < 10 mTorr, the highest T_c and lowest c_0 (close to 11.70Å), i.e. most bulk-like, are obtained for a Ba/Y deposition ratio ≤1.4. This is in spite of the fact that such off-stoichiometry films have greater amounts of second phases present in them. Such behavior suggests a more complicated phase diagram than is usually presented.

2.3. Expanded c-axis

In-situ films have often been reported to have slightly dilated c-axes. Figure 2 shows c_0 as a function of the oxygen pressure during growth for our evaporated films as well as for films made by laser ablation and sputtering. The values of c_0 were determined by x-ray diffraction analysis. All of these films were superconducting (with varying transition temperatures), except for the films which made at 0.2 mTorr, which were not superconducting but had the $YBa_2Cu_3O_y$ structure (so-called "123-like"). Although there is an inhomogeneous component to the c_0 expansion, this dilation is largely homogeneous, i.e. the broadening of the x-ray peaks is only slight with c_0 expansion. This means that the structural causes of crystal dilation are homogeneously distributed in the material.

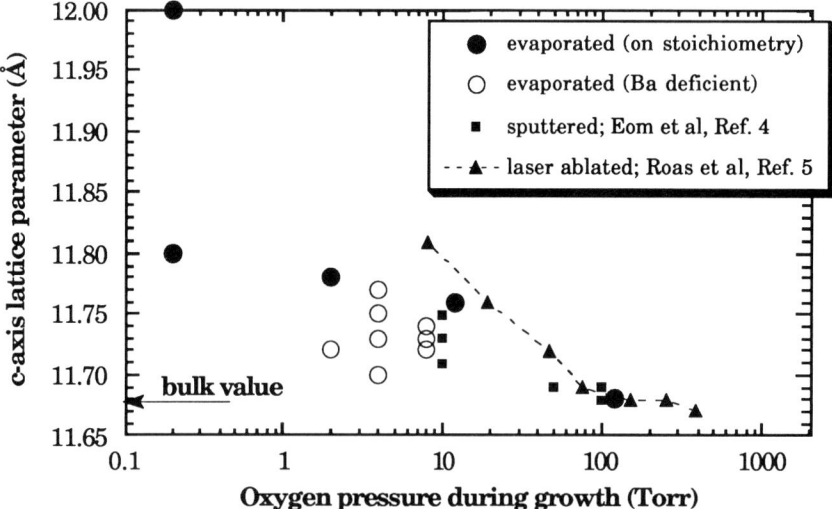

Figure 2. c-axis lattice parameter vs. oxygen pressure during growth.

On the other hand, this c-axis expansion appears stable to low temperature oxygen anneals. T_c of these films cannot be increased, nor can c_0 be made shorter, by a low temperature (<700°C) anneal in an atmosphere of oxygen. A higher temperature anneal will shorten c_0. This suggests that these structural defects are metal-atom related, since metal atoms require the higher temperatures for bulk mobility, whereas oxygen is mobile even at lower temperatures.

2.4. Hole-doped nature of films

A further intriguing feature of these films is that T_c, which is usually depressed from the bulk value (92K) by 2-10 degrees, can actually be increased by a small reduction in oxygen content. For bulk $YBa_2Cu_3O_y$ it is normally reported[6] that maximum oxygen content corresponds to the highest T_c, and the superconducting transition is depressed as oxygen is removed from the compound. For in-situ films, however, this is not quite the case. Maximum oxygen content corresponds to a few degrees lower T_c than the maximum value; see paper by R. Feenstra in this volume. Figure 3 shows T_c for one of the evaporated films as a function of annealing in successively lower pressures of oxygen. The x-axis of Fig. 3 is the decrease in conductivity, or equivalently the reduction of oxygen content. There is a slight increase in T_c before it goes down. Such behavior implies that the material is slightly overdoped with holes over the maximum T_c, and is observed, for example, when Ca is intentionally doped into $YBa_2Cu_3O_y$ where it substitutes for Y (Ca is 2+ and Y 3+).[3]

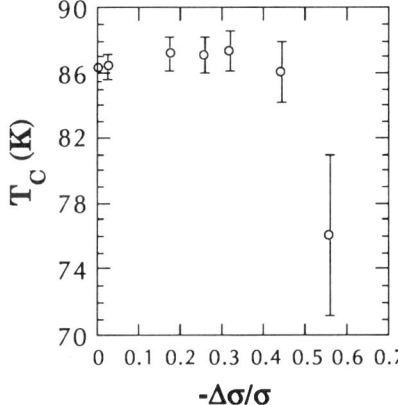

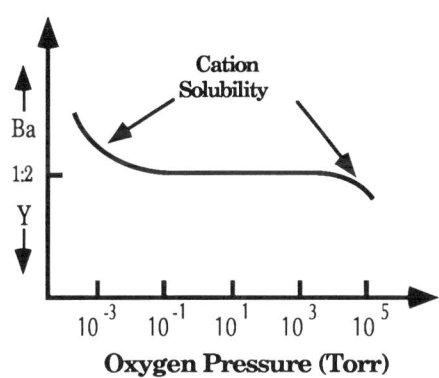

Figure 3. T_c as a function of the reduction in conductivity

Figure 4. Proposed model for cation solubility as a function of oxygen pressure

3. DISCUSSION AND PROPOSED MODEL

Largely based on the homogeneous nature of the c-axis dilation, it has been argued in Ref. 3 that metal-atom point-like defects are most likely present in these films. The likely candidate for this point defect was suggested as Ba substituting on the Y site. (Alternately a proposal by Hellman[7] is that Y could substitute Cu on the chain site.) Further evidence now from the hole-doped nature of these films corroborates the Ba-for-Y model. Thus, Ba^{2+} would behave similarly to Ca^{2+} substituting on the Y^{3+} site, as is in fact observed.

Furthermore, a plausible mechanism for such a substitution can be argued based on charge compensation: Ba^{2+} could be favored over Y^{3+} for a lower pressure of oxygen, or equivalently a lower chemical potential of oxygen. This is shown schematically in Fig. 4. At lower oxygen pressures there could be extended cation solubility (not normally observed at standard pressure) in the Y layer, with Ba increasingly substituting Y. Note the similarity of this proposal to the data of Fig. 2. In principle, the converse could then also be possible, whereby Y could substitute for Ba at higher pressures of oxygen. There exists mixed evidence whether something like the latter actually exists.

4. CONCLUSIONS AND IMPLICATIONS

We have argued from the c-axis expansion of films made at low pressures of oxygen, that metal-atom point-like defects are quenched in during synthesis. A likely scenario appears to be Ba preferentially substituting on the Y site.

Such a model has other implications. It can explain the slight T_c reductions, as argued in Ref. 3. More importantly, if these defects can act as pinning centers, then these defects would be in sufficient quantity to give rise to the high critical currents that are observed for these films. In fact even a much lower density of defects would be sufficient to give rise to these high critical currents.[2]

Growth in the "unstable" part of the phase diagram can then be explained by the fact that this is actually another phase with a different phase stability boundary, and because it contains cation disorder it would be entropy stabilized over the ideal $YBa_2Cu_3O_y$ phase.

The authors wish to acknowledge discussions with T. Geballe, T. Hylton, C.B. Eom, D. Mitzi, E. Hellman, T. Siegrist, I. Bozovic as well as many others in the KGB group at Stanford. This work was supported by the AFOSR under contract F49620-89-C-0001.

5. REFERENCES

1 R. Beyers and B.T. Ahn, to appear in Vol. 21 of the Annual Review of Material Science (1991).
2 V. Matijasevic, PhD Thesis, Stanford University (1991).
3 V. Matijasevic, P. Rosenthal, K. Shinohara, A.F. Marshall, R.H. Hammond, and M.R. Beasley, J. Mater. Res. **6**, 682 (1991).
4 C.B. Eom, J.Z. Sun, S.K. Streiffer, A.F. Marshall, B.M. Lairson, K. Yamamoto, S.M. Anlage, J.C. Bravman, T.H. Geballe, S.S Laderman, and R.C. Taber, Physica C **171**, 354 (1990).
5 B. Roas, B. Hensel, G.Endres, L. Schultz, S. Klaumünzer, and G. Saemann-Ischenko Physica C **162-164**, 135 (1989).
6 J.D. Jorgensen, B.W. Veal, A.P. Paulikas, L.J. Nowicki, G.W. Crabtree, H. Claus, and W.K. Kwok, Phys. Rev. B **41**, 1863 (1990).
7 E.S. Hellman, private communication.

High T_c Superconductor Thin Films
L. Correra (Editor)
© 1992 Elsevier Science Publishers B.V. All rights reserved.

Growth of a-axis oriented YBaCuO and EuBaCuO thin films

G. Linker, E. Brecht, J. Geerk, O. Meyer, B. Rauschenbach, J. Reiner*, J. Remmel, Ch. Ritschel, and R.L. Wang[+]

Kernforschungszentrum Karlsruhe, INFP, ITP*, P.O. Box 3640, W-7500 Karlsruhe, FRG

[+] Chinese Academy of Science, Beijing, PR China

Abstract

Thin HTSC EuBaCuO and YBaCuO films have been deposited onto (100) $LaAlO_3$, MgO, $SrTiO_3$ and $Zr(Y)O_2$ substrates at different substrate temperatures, T_s, by magnetron sputtering. A reduction of T_s by 100-150°C reverses the growth direction from c-axis to a-axis growth. The growth quality of the a-axis films is characterized by X_{min} values down to 4% in channeling experiments and narrow mosaic spreads below 0.2° determined in x-ray diffraction measurements. The surfaces of the films are very smooth. Two perpendicular in plane orientations of the c-axis have been observed in TEM measurements. The deposition of a thin template film at reduced T_s and growth of the bulk of the films at elevated T_s preserves a-axis growth and yields T_c values up to 84 K.

1. INTRODUCTION

The layered perovskite like structure of the high temperature superconductors (HTSC) is reflected in many properties which reveal a large unisotropy especially if determined in or perpendicular to the copper-oxygen planes. Because the availability of thin films is an important precondition for various spectroscopic investigations and thin films with respect to their properties often are superior to bulk material, the growth of thin films of different orientation is of great interest if effects depending on orientation have to be studied. This equally well applies to applications where films of different orientation or film packages with alternating orientation will be needed for the development of three-dimensional device structures. Finally, from the viewpoint of epitaxial growth, the growth conditions for differently oriented films appear as an interesting problem in itself.

Most of the high quality HTSC thin films prepared by many groups with diverse methods were such with the „1-2-3"-structure. The growth of the films depends on many parameters which may be divided into such equally important for all preparation techniques and such specific for a special technique with possible synergistic effects between the parameters influencing their quantity. The optimization of the deposition parameters in most cases has been performed to grow high quality c-axis films; less attention has been paid to other orientations. We have detected earlier, applying the sputtering technique, that the substrate temperature during deposition, T_s, appears as the most important parameter controlling the growth direction of a film [1]. Especially, the reduction of T_s by about 100°C changes the growth direction, e.g., from c-axis to a-axis growth, with the nomenclature referring to the axis perpendicular to the substrate surface. Now, we have studied this behavior in more detail. Some of the results, especially those of the growth on $SrTiO_3$ substrates, have been reported recently [2]. In this contribution we include further data of the temperature dependent growth of EuBaCuO and YBaCuO films with „1-2-3"-structure on (100) oriented substrates like $LaAlO_3$, MgO and Zr(Y)O_2.

2. EXPERIMENTAL DETAILS

We have employed the inverted cylindrical magnetron (ICM) sputtering technique for film deposition. The geometrical arrangement of the target and the relatively high total pressure of 6×10^{-1} Torr during sputtering largely avoids detrimental negative ion bombardment of a growing film. The film preparation was performed in two steps, i.e., deposition at elevated T_S and subsequent cooling in oxygen atmosphere for oxygen incorporation. Details of the ICM-sputtering technique, which due to its reliability and reproducibility allows systematic studies as a function of various parameters, are described elsewhere [3]. In our experiments we have kept all parameters constant except T_S which was the variable parameter under investigation. The deposition rate which may influence the growth direction, as we conlcude from preliminary measurements, was about 0.3 nm/s.

We have characterized the growth quality and the growth direction of the films by different methods like X-ray diffraction, ion beam channeling, SEM and TEM, with emphasis on diffraction and channeling which have been routinely applied. Especially, the FWHM of rocking curves, Δ (mosaic spread), and the ratio of random and aligned yields, X_{min}, from backscattering-channeling spectra were considered as measures of the growth quality of the films. The optimization of a-axis oriented growth was performed with respect to a minimization of these two parameters and of the content of c-axis oriented grains. Since the properties of a-axis films grown at one reduced T_S are degraded we have employed a special procedure of growth at different temperatures where a rather thin template film is deposited at low T_S and the bulk of the film is grown in situ at elevatred T_S (template procedure). We have varied the deposition times for the template film growth and temperature increase. The thickness of the template film was in the range of 10-100 nm and the total film thickness arose to about 500 to 600 nm.

3. RESULTS

We have observed that from the intrinsic deposition parameters T_S is the essential parameter determining the growth direction of HTSC films with „1-2-3"-structure. The growth direction, however, also depends on the substrate orientation. For example, on (110) $SrTiO_3$ YBaCuO films also grow (110) oriented irrespective of T_S [1]. On (100) oriented substrates - our experiments were performed on $SrTiO_3$, $LaAlO_3$, MgO, and Zr(Y)O_2 - the films grow c-axis oriented for T_S values optimized for best superconducting properties and the growth direction reverses to a-orientation for T_S-values of 100-150°C lower depending on the special substrate material under consideration. The change of the growth direction occurs continuously, i.e. at intermediate temperatures we observe the presence of c- and a-axis oriented grains simultaneously, with their quantity ratio changing as a function of T_S.

To illustrate the gradual change of the growth direction we show X-ray and channeling data in Figs. 1 and 2, respectively. X-ray diffraction diagrams of YBaCuO films deposited on Zr(Y)O_2 substrates at different T_S (we quote the heater block temperature measured with a thermocouple) are displayed in Fig. 1. At

Figure 1

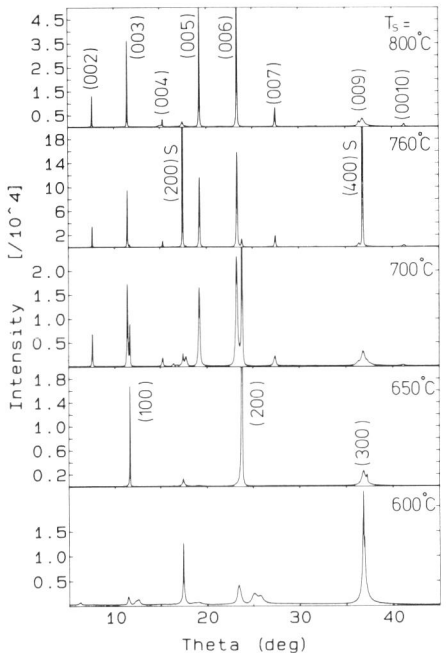

X-ray diffraction diagrams of YBaCuO films deposited on (100) Zr(Y)O$_2$ substrates at different T$_s$ demonstrating the growth reversal from c-axis to a-axis orientation.

$T_s = 800°C$ only lines of the (00ℓ)-type are present demonstrating purely c-axis textured growth. With decreasing T_s we observe mixtures of (00ℓ) and (h00) lines with growing intensity of the (h00) peaks. At $T_s = 650°C$ merely (h00) peaks are observed; this manifests pure a-axis oriented growth. At even lower T_s the appearance of additional peaks indicates growth deterioration. The gradual change of the growth direction is also reflected in the channeling data. In Fig. 2 we have plotted X_{min} values of EuBaCuO films deposited on LaAlO$_3$ substrates vs deposition temperature. Two distinct minima at T_s values close to 710° and 810°C correspond to pure a-axis and c-axis orientation, respectively, and the increase of X_{min} at intermediate T_s values is due to the presence of a mixture of a-axis and c-axis oriented grains. It is interesting to note that pure a-axis growth on LaAlO$_3$ appears at T_s values very close to those observed for SrTiO$_3$ [1,2] while on Zr(Y)O$_2$ as shown in Fig. 1 and on MgO [3] it is shifted to lower T_s by at least 50°C. This probably is due to different lattice matching as will be discussed later. The growth quality of a-axis films in terms of the mosaic spread Δ and X_{min} in comparison to c-axis films is not largely degraded. Typical Δ values are below 0.4° and best films reach values below 0.2°. With regard to X_{min} the EuBaCuO films of Fig. 2 for a-axis orientation are superior to those of c-axis orientation, though our best X_{min} values below 3% were observed in c-axis oriented films. For EuBaCuO on SrTiO$_3$ best X_{min} values were close to 5% which is somewhat above the observations for c-axis films. In general, no significant differences were observed in the growth of YBaCuO and EuBaCuO films and therefore the findings reported here refer to both compounds.

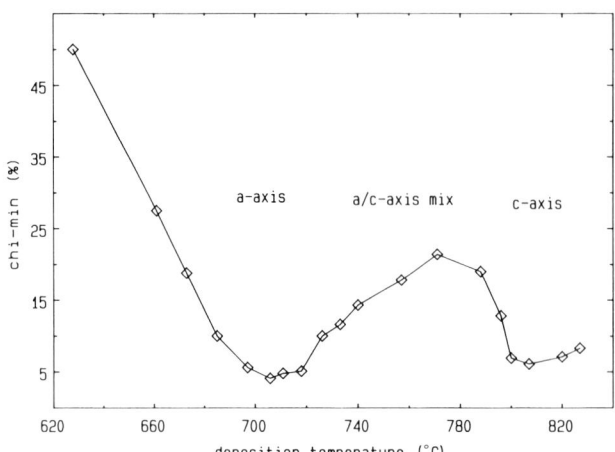

Figure 2

X$_{min}$ values from channeling measurements of
EuBaCuO films depo-
sited on LaAlO$_3$ at
different T$_s$. The two
minima correspond to
pure a-axis and c-axis
orientation of the films.

The growth of „1-2-3"-films at low T$_s$ with the a-axis perpendicular to the
substrate surface and correspondingly with the b- and c-axes in the film plane in
general appears unique. This is demonstrated in Fig. 1 where no splitting of the
(h00) lines is oberved, i.e. (0k0) lines are absent. On SrTiO$_3$ substrates (0k0) lines
from the film would overlap with (h00) lines of the substrate and therefore are
difficult to detect. With increasing lattice parameter at very low T$_s$, which has
been observed for the a-axis [2], a similar behavior of the b-axis would lead to a
separation of the (0k0) film line from the substrate peaks. This effect, however, has
not been observed. But in special cases also b-axis growth may occur. This
behavior is shown in Fig. 3 where sections of X-ray diagrams of EuBaCuO films on
LaAlO$_3$ are plotted at two deposition temperatures. While at lower T$_s$ besides the
substrate line there is only the (400) peak of the a-axis film visible, a clearsplitting
occurs at the 50°C higher T$_s$ indicating the presence of both, a-axis and b-axis
oriented grains.

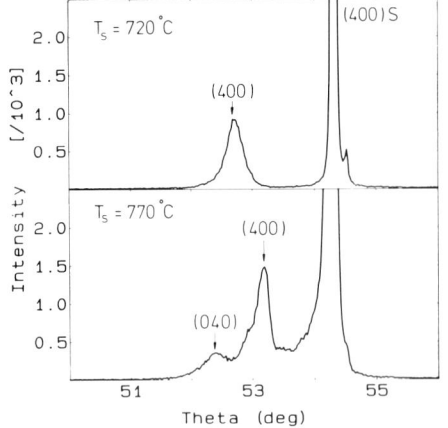

Figure 3

Sections of X-ray diagrams of
EuBaCuO films on LaAlO$_3$ showing
pure a-axis and mixed a- and b-axis
growth at the lower and higher T$_s$
values, respectively.

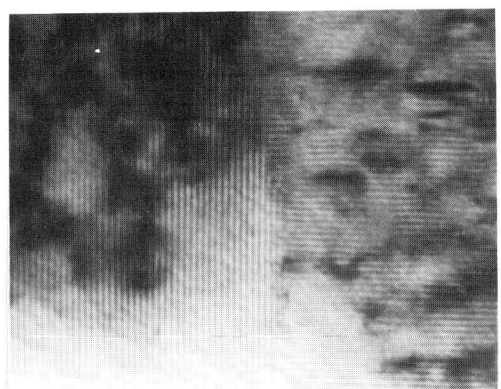

Figure 4

TEM micrograph of a EuBaCuO film on $SrTiO_3$ revealing two perpendicular in plane growth directions of the c-axis. The diameter of the in plane regions exceeds 100 nm.

Despite the high growth quality of the a-axis films in terms of narrow mosaic distributions and small X_{min} values in channeling experiments, their properties like T_c or resistivity are degraded, especially the transitions to superconductivity are broadened with depressed zero resistance values. This probably is due to the formation of a defect structure on atomic scale which inhibits a proper oxygen incorporation in the second preparation step. Annealing tests in oxygen atmosphere at elevated temperature for property improvement were not successful. We therefore applied the template procedure for property improvement. This procedure was first used by Inam et al. [4] in the preparation of laser ablated films with PrBaCuO as a template layer. In the second step of this process we applied T_s values close to those optimized for c-axis growth. In the template procedure the a-axis growth of the films is preserved and the growth quality is even improved. Our best X_{min} values were 4% and best Δ values close to 0.1°, however, with some change in the total shape of the particle distribution. Here, the distribution of the grains in a-axis films deposited in one step at reduced T_s appeared steeper. The presence of a minute (005) peak in Θ-2Θ X-ray diagrams indicated that the films contained some c-axis oriented grains. For the concentration determination the intensities should be taken from rocking curves rather than from Θ-2Θ scans because the

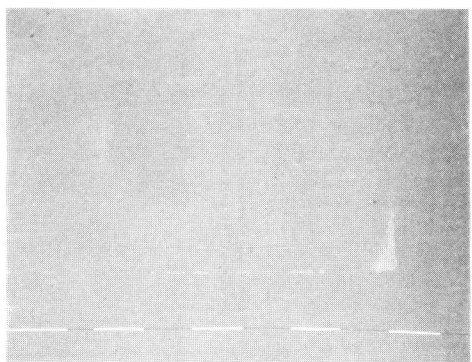

Figure 5

SEM micrograph revealing the smoothness of a-axis films. The length of a marker corresponds to 10 μm.

widths of the distributions may be different for different orientations. In particular, a broader distribution has been observed for c-grains in a-axis films [2]. Generally, the concentration of c-axis grains is distinctly below 1%.

The properties of the films grown in the template procedure are improved. The transitions to superconductivity are sharp with T_c values above 80 K and best values of 84 K in resistive and inductive measurements. The films reveal a metallic behavior with linear R vs. T relationships. These curves, however, do not extrapolate to zero like in good c-axis films and the resistivities at 100 K are rather high with values in the range of 800-1000 $\mu\Omega$cm. These high values mainly are due to the in plane growth of the films with two perpendicular directions of the c-axis which are equally probable on cubic substrates. Such growth can be directly seen in TEM measurements as displayed in Fig. 4 and has also been reported by [5]. In such films the current has to pass high resistivity grains perpendicular to the Cu-O planes or the path length through low resistance directions along the chains or planes, [6] is increased in both cases enlarging the average resistivity of a film. The surfaces of the a-axis films are extremely smooth and shiny, an important property for layered growth and applications. A SEM picture is displayed in Fig. 5 as a demonstration.

4. CONCLUSIONS

The substrate temperature, T_s, during sputter deposition of HTSC films is one of the most important parameters controlling the growth quality and growth direction of films with „1-2-3" structure. On various (100) oriented substrates a reduction of T_s by 100-150°C is sufficient to reverse the growth direction from c-axis to a-axis orientation. It is thought that the change of the preference axis is due to a reduced surface mobility of the atoms at lower T_s. Thus lattice matching which is more favorable for a-axis growth dominates layered growth which is preferred if the mobility of the atoms is sufficiently high. The shift of the a-axis growth mode to lower T_s on substrates with less adjusted lattice parameters like for MgO or Zr(Y)O$_2$ gives experimental support for this argumentation.

The growth quality of the a-axis films with X_{min} values down to 4% and mosaic spreads below 0.2° is similar and with respect to the surface smoothness even superior to c-axis films. This may extend the potentials for applications. The improvement of properties with growth direction preservation by deposition at different temperatures in the template procedure is feasible. The a-axis films reveal metallic behavior with T_c values up to 84 K.

5. REFERENCES

1   G. Linker, X.X. Xi, O. Meyer, Q. Li, J. Geerk, Solid State Commun. 69 (1989) 249.
2   R.L. Wang, J. Reiner, J. Remmel, E. Brecht, J. Geerk, O. Meyer, G. Linker, E-MRS Int. Workshop HTCS Thin Films, Properties and Applications, Rome, Apr. 91, Physica C, to be published.
3   J. Geerk, G. Linker, O. Meyer, Mat. Sci. Rep. 4 (1989) 193.
4   A. Inam, C.T. Rogers, R. Ramesh, K. Remschnig, L. Farrow, D. Hart, T. Venkatesan, B. Wilkens, Appl. Phys. Lett. 57 (1990) 2484.
5   R. Ramesh, A. Inam, D.L. Hart, C.T. Rogers, Physica C 170 (1990) 325.
6   T.A. Friedmann, M.W. Rabin, J. Giapintzakis, J.P. Rice, D.M. Ginsberg, Phys. Rev. B 42 (1990) 6217.

High T$_c$ Superconductor Thin Films
L. Correra (Editor)
365

Smooth YBCO Films On MgO With Good Superconducting Properties

F. Baudenbacher, K. Hirata, P. Berberich, H. Kinder and W. Assmann*

Fakultät für Physik der Technischen Universität München, D-8046 Garching

* Sektion Physik, Ludwig-Maximilians-Universität München, D-8046 Garching

Abstract

We studied the film growth mechanism and the surface morphology of YBa$_2$Cu$_3$O$_{7-x}$ on (001) MgO substrates as a function of chemical composition by reactive thermal co-evaporation at substrate temperatures of 650° C. The evaporation rates are stabilised using a single multiplexed cross beam mass spectrometer. The absolute composition was determined by heavy-ion RBS measurements with an accuracy of 2%. Extremely smooth films with good superconducting properties (T$_c$ = 90 K, ΔT = 0.8 K and J$_c$ = 3.4 *10^6 A/cm^2 at 77K) can be prepared close to ideal composition. The high quality of the thin film is reflected in the low level of the minimum aligned backscattering yield for Ba of 4.4% in 1.66-MeV He channeling experiments. RHEED observations during the growth demonstrate the existence of well defined composition regions with different growth mechanisms resulting in a characteristic surface morphology. Smooth films show sharp streaks with indications of Kikuchi lines during all stages of the growth.

1. Introduction

HTSC films with perfect crystallinity and no grains or outgrowth are required for ultra fine structuring, multilayer applications, Josephson junctions, and probably also to achieve the lowest possible RF resistance. The perfect surface morphology must be achieved without degradation in the superconducting properties. Many groups have reported on the experience that both requirements are contradicting each other. We have succeeded to grow films which have perfect morphology and extremely good superconducting properties at the same time. The key was selecting substrates by their RHEED patterns on one hand, and a close control of the composition of the YBCO films on the other.

2. Film preparation

Films were deposited in a high vacuum evaporator using three resistively heated boats for Y, Ba and Cu metals. The evaporation rate from each boat is monitored by a single multiplexed quadrupole mass spectrometer. The quadrupole signals are incorporated into a feedback system which maintains a constant evaporation rate. A high accuracy quartz crystal film thickness monitor at the substrate position is used before each evaporation run to calibrate the quadrupole reading. The growth rate was approximately 2 Å/s, and the film thickness 1200 Å in all cases.

The substrate, situated in a small oven surrounding it, was heated by radiation. This ensures a uniform temperature distribution and a reliable temperature calibration. The substrate temperature was 650° C prior to growth and drops 5 - 10° C during growth. Molecular oxygen was confined near the substrate by guiding it through a ring of small nozzles inside the oven onto the substrate surface. This results in a pressure of about $2 \cdot 10^{-3}$ mbar at substrate position and a background chamber pressure of $2 \cdot 10^{-5}$ mbar. After the evaporation the pressure was raised to 20 mbar and the heater was switched off, taking about 20 min to cooldown to 100° C. Details of the experimental setup are described elsewhere. [1]

3. Heavy Ion RBS and Channeling

The stoichiometric ratios of the metals in YBCO films were measured using Rutherford backscattering spectroscopy (RBS) with 25 MeV ^{16}O ions. The backscattered ions are detected with PIN-diodes at a scattering angle of 155°. Due to the heavy ion beam and the finite film thickness the signals from the different components are well separated in the energy spectrum of the backscattered particles (see Fig. 1). Backscattering from the MgO substrate is below the threshold energy of the detector. Pure Rutherford cross sections can be assumed at this beam energy, the screening effect corrections are less than 0.4%.
This allowed the determination of the Y:Ba:Cu ratio within the statistical error of the peak areas. The statistical error is reduced with increasing measurement time, which is chosen to determine the individual concentration of each element with an accuracy of better than 2%. The following table shows the results of systematic variations in composition and represents 11 typical examples out of 30 YBCO films. Some of them were prepared under identical conditions to determine the error in calibration of the feedbacksystem to be 3%.

Sample No.	at % Yttrium	at % Barium	at % Copper	T_c	$J_c \cdot 10^6$ A/cm^2
1	15.2 ± 0.3	31.4 ± 0.4	53.4 ± 0.5	83.1 K	-
2	16.4 ± 0.3	32.7 ± 0.5	50.9 ± 0.5	90.1 K	3.4 [77 K]
3	16.6 ± 0.3	33.4 ± 0.4	49.9 ± 0.5	84.0 K	-
4	14.9± 0.3	35.0 ± 0.4	50.1± 0.5	89.0 K	-
5	17.1 ± 0.3	34.3 ± 0.4	48.6 ± 0.4	90.1 K	2.7 [77 K]
6	16.8 ± 0.3	35.7 ± 0.5	47.6 ± 0.5	58.8 K	-
7	16.9± 0.3	37.8 ± 0.4	45.5± 0.5	47.7 K	-
8	18.8± 0.4	33.7 ± 0.5	47.5 ± 0.6	60.0 K	-
9	18.1± 0.4	33.9 ± 0.4	48.0 ± 0.4	77.0 K	-
10	19.5 ± 0.3	31.5 ± 0.4	49.1 ± 0.4	64.0 K	-
11	17.7± 0.4	31.2 ± 0.5	51.1 ± 0.5	90.1 K	2.8 [77 K]

We have also performed 1.7 MeV He RBS in axial channeling mode which is sensitive to long-range crystalline order. In Fig. 2 we compare an aligned channeling spectrum and an unaligned spectrum for the nominally stoichiometric film #3. The high degree of order and lack of mosaic spread over the entire beam area of 1 mm^2 is reflected in the low level of the aligned yield. The estimated minimum aligned backscattering yield for Ba in a perfect crystal would be 3.2 %. N.G. Stoffel et al. reported a value of 3.5% for single crystals. [2] The observed minimum yield of 4.4% below the surface peak of Ba indicates nearly ideal channeling and reveals the stability of the rates during the evaporation run. The increase in the aligned spectrum towards the low energy edge may be partly attributed to the large lattice misfit of 10% and the formation of misfit dislocations near the substrate interface.

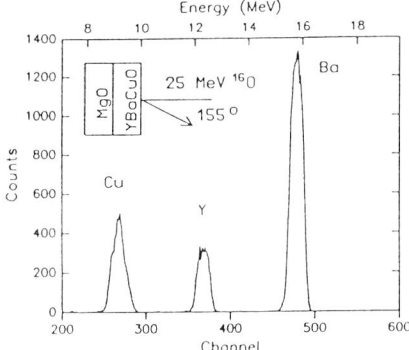

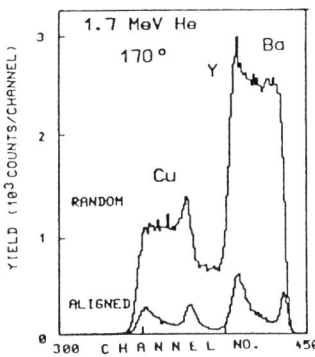

FIG. 1. 25 MeV ^{16}O backscattering spectra from a YBCO film (#3) on (001) MgO.

FIG. 2. Random and [001]-aligned RBS spectra of a YBCO film (#3) with ideal composition.

4. Growth mechanism and surface morphology

Scanning electron microscopy (SEM) was performed on a field emission Hitachi S 4000 system with a nominal SEM spatial resolution of 1.5 nm and all micrographs were taken with the sample tilted 40° from the electron beam axis. The reflection high energy diffraction (RHEED) was done in-situ after closing the shutter for a short period at certain film thicknesses. The energy of the incident electrons was 20 keV and the grazing angle less than 1°, making RHEED very sensitive to changes in surface morphology.

High quality mechano-chemically polished (001) oriented single crystal MgO substrates were used to investigate the growth mechanism and the resulting surface morphology. The substrate surface was characterised by RHEED and only those which showed a sharp streaky pattern were selected. In-situ RHEED observations showed that all the films on these substrates were epitaxial, c-axis oriented. They also demonstrated that there were systematic differences in the growth mechanism in different ranges of composition on a fine scale. A strong tendency towards island growth was observed for films with a Ba rich and Cu poor composition. The RHEED pictures during different growth stages of film #6 are shown in Fig. 3. The first picture was taken at a thickness of 10 Å. The streaks have a pronounced spotty appearance, thus indicating the scattering from three-dimensional objects, due to island formation. This growth behaviour remains unchanged up to 120 Å. The spots disappeared almost and a streaky pattern indicates a two dimensional growth up to the final thickness. A similar series of diffraction patterns was observed for film #7 where the diffration spots on the streaks remained up to a thickness of 500 Å.

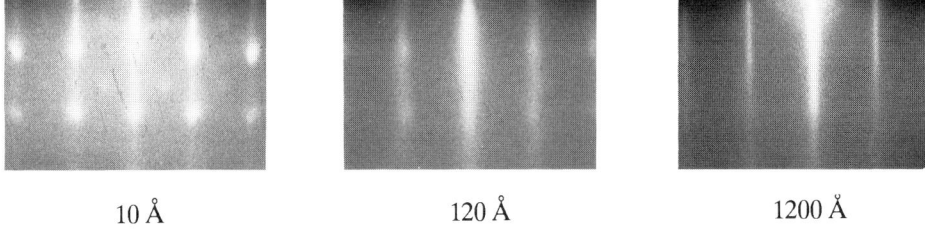

| 10 Å | 120 Å | 1200 Å |

Fig. 3. In-situ RHEED patterns demonstrating island growth.

SEM micrographs of these films (see Fig. 6) have pinholes prior to coalescence of the nucleated islands.

For Cu rich compositions (sample #1 and #4) and Ba/Y > 2, layer by layer growth dominated only during the early stages and was then followed by three dimensional growth. The corresponding RHEED patterns are shown in Fig. 4. SEM pictures (see Fig. 6) of the surface structure showed a pronounced grain structure and lens type outgrowths. As the excess of Cu is decreased a lower density of these precipitates is observed.

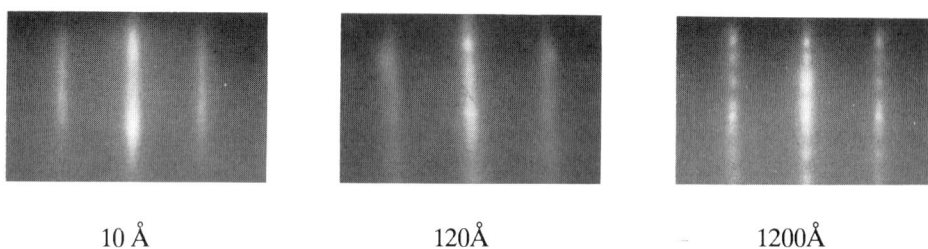

| 10 Å | 120Å | 1200Å |

Fig. 4. Layer by layer growth with transition to 3 dimensional growth at 120 Å.

The RHEED patterns (see Fig. 5) of films with the Ba/Y ratios ≤ 2 and films which are close to ideal composition show sharp streaks during all stages of growth indicating a two dimensional growth. This region seems to extend more to the Cu poor side.

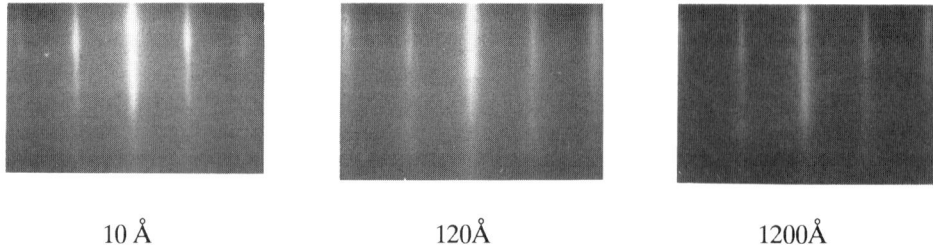

| 10 Å | 120Å | 1200Å |

Fig. 5. Layer by layer growth during all stages (sample #5).

SEM observations reveal that surfaces of these films (see Fig. 6) are structured on a much finer scale. On a 10 μm scale there were no structures or particles visible. Typically for a quite large Y content on the Cu poor side is the evidence of about 100nm non-aligned particles. Some rectangular, crystallographically aligned a-axis oriented grains can be found on some samples with very different composition, which is consistent with the substrate induced growth of a-axis oriented YBCO at surface steps. [3]

From SEM observations (see Fig. 6) it can be concluded that the surface morphology is very sensitive to film composition. Similar results using thermal evaporation, at a higher growth temperature of 690° C and a lower oxygen pressure (molecular with 10% atomic) were published. [4] A pitted morphology on all Cu deficient samples was reported. We observed also a pitted surface at 690° C for Cu deficient films. So the substrate temperature is another important growth parameter which has to be chosen carefully. Some films were prepared at a growth rate of 1 Å/s, but no change in surface morphology could be found.

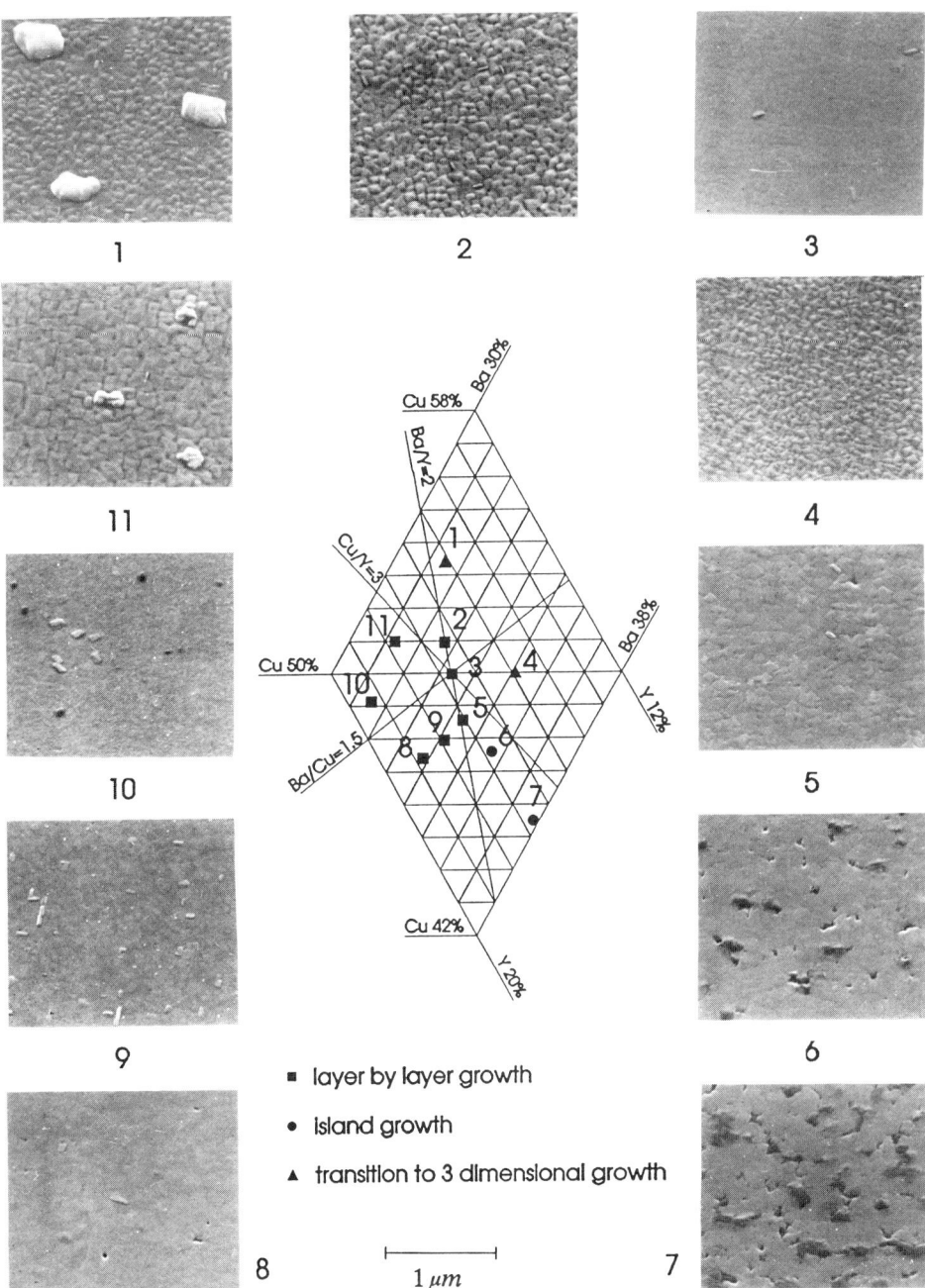

Fig.6. Surface morphology and growth mechanism of YBCO thin films as a function of chemical composition.

4. Transport properties

Values of T_c ($R = 0$) and J_c were obtained on 50 μm wide and 50 μm long strips prepared by photolithography and wet etching. The voltage criterion for J_c was 1 μV. J_c measurements as a function of temperature of one film prepared close to ideal stoichiometry were also done on 30 μm long and 3 μm wide strips and are shown in Fig (8).

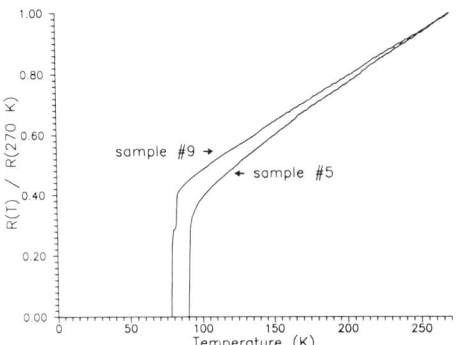

Fig. 7. R(T) measurements of sample #5 and # 9.

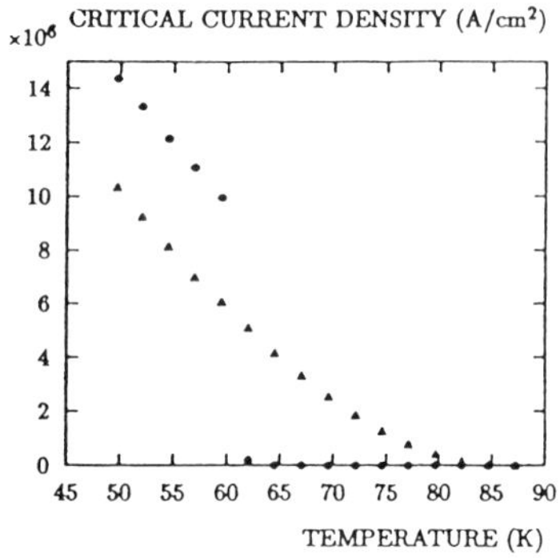

Fig. 8. J_c (T) measurements on a 30 μm long and 3 μm wide microbridge.

T_c is relatively insensitive to a quite large excess of Cu. The T_c and J_c remain high even for slightly Cu poor compositions only for exactly stoichiometric Y/Ba ratios, but it becomes very sensitive to deviations from this ratio. The transition curves of two films on the Cu poor side are shown in Fig. 7. Sample #5 with a Ba/Y = 2 and sample #9 with a Ba/Y = 1.9. The broad transition and the steplike behaviour of film #9 may be caused by the formation of insulating phases at the grain boundaries. The systematically lower T_c in comparison with the values reported by N. G. Chew et al [4] may be attributed to the use of molecular oxygen and the missing post annealing process.

5. Conclusion

There are well defined regions in composition with different growth mechanisms resulting in a typical surface morphology. Extremely smooth films with good superconducting properties can be prepared at ideal composition which are the key for microstructuring, microwave and multilayer applications.

Acknowledgements

We wish to thank R. Grötzschel, Rossendorf, for his assistance and providing us with the facilities for RBS measurements. The J_c measurements on the 3 μm wide microbridge were done by B. Roas, Siemens Erlangen, and is greatly acknowledged. This work was supported by BMFT under contract #TK 0338/2.

REFERENCES

1. F. Baudenbacher, H. Karl, P. Berberich, H. Kinder, J.L.Com. Met. 164 & 165 (1990) 269
2. N. G. Stoffel, P. A. Morris, W. A. Bonner, B. J. Wilkens, Phys. Rev. B 37, (1988) 2297
3. Hiromi Takahashi, Yuji Aoki, Toshio Usui, Rainer Fromknecht, Tadataka Morishita and Shoji Tanaka, Physica C 175 (1991) 381-385
4. N. G. Chew, S. W. Goodyear, J. A. Edwards, S. E. Blenkinsop, and R. G. Humphreys, Appl. Phys. Lett. 57 (19), (1990), 2016

High T$_c$ Superconductor Thin Films
L. Correra (Editor)
371

In-situ sputter deposition of very smooth large area YBa$_2$Cu$_3$O$_{7-\delta}$ thin films

G. Wagner and H.-U. Habermeier

Max-Planck-Institut für Festkörperforschung, P.O. Box 800665,
W-7000 Stuttgart 80, Germany

Abstract

One important requirement for large scale application of high T$_c$ superconducting films is the successful deposition of high quality large area films. For that purpose not only the electrical properties but also the surface quality are decisive. The utilization of a RF magnetron sputtering system, especially designed for preparing film sizes up to two inches in diameter, allows to fulfill these conditions. Three planar magnetron sputtering guns, each equipped with a superconducting YBa$_2$Cu$_3$O$_{7-\delta}$-target, are mounted into a flange so that the central axis intersect in a common focal point. The substrate is not placed in the focus whereby the result is an off-axis geometry with deposition from three directions thus getting a very homogeneous film. During deposition, the substrate is heated up by quartz lamps. Sputtering gases are argon and oxygen with an accurate controlled oxygen partial pressure using a quadrupole mass spectrometer. The films grown in-situ show high critical temperatures and high critical current densities. The films' surfaces are very smooth without contamination of any particles. In this paper, we describe the deposition technique and discuss the correlation of the film parameters as to electrical, optical, and structural properties of the films.

1 Introduction

The development of advanced preparation techniques for high temperature superconductor thin films during the last two years allows to think about their practical application in microelectronics or microwave technologies. Beside excellent electrical properties two other conditions have to be fulfilled to give high T$_c$ superconductors a real chance for practical implementation in the field mentioned above. First, good surface quality without any contamination with particles which is a basic requirement for multilayer structures and microelectronic devices, has to be achieved. Furthermore, a large area deposition technique is essential for most industrial applications of high T$_c$ films.

Although the laser deposition is the most common way of preparing high T$_c$ thin films, the problem of particle contamination is still unsolved. As investigated by several authors, all films produced by pulsed laser deposition show particulates in larger or smaller numbers whatever deposition conditions like target density, different pressures, laser powers or laser wavelengths are used [1,2,3,4]. Moreover, it is not a very suitable technique for homogeneous deposition of large areas. This is due to the small ablation area and the narrow angle where the stoichiometry is transferred correctly to the substrate [5,6]. In contrast, MOCVD [7] and MBE [8] are favourable to produce large and smooth films but controlling the stoichiometry during deposition from three different sources is rather unreliable. Furthermore, MBE has the disadvantage of very low operating pressures. In-situ growth of $YBa_2Cu_3O_{7-\delta}$ (YBCO) thin films is merely possible using activated oxygen sources [9].

At sputtering, the possibility of operating with a single superconducting target allows to reproduce the stoichiometry in a much easier way. Problems arising from bombardement by negativeley charged ions can be solved utilizing either high pressures [10] or off-axis geometries [11,12,13]. In this work, we describe a special off-axis sputtering technique with three superconducting targets to get smooth and homogeneous films on a large area.

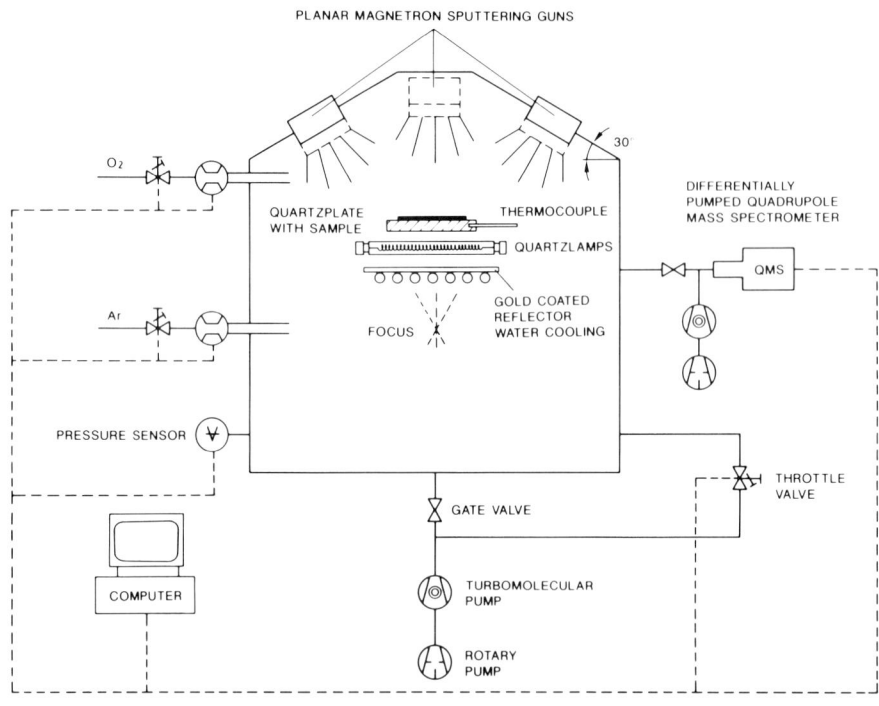

Figure 1: Schematic diagram of the sputtering system

2 Experimental Setup

A schematic diagram of the most important parts of our sputtering system is shown in Figure 1. We are operating with three planar magnetron sputtering guns which are mounted into the top flange of the deposition chamber with an angle of 30° in respect to the baseplane. The central axis of all three guns intersect in a common focal point, about 15 cm away from the targets' surface. The substrate lies not in the focus but about 5 cm closer to the targets. This arrangement is a special way of an off-axis geometry to avoid resputtering effects. The deposition from three different directions leads to a much better homogeneity of the films as in usual 90° off-axis sputtering machines. Each gun, equipped with a three inch YBa$_2$Cu$_3$O$_{7-\delta}$-target, is driven by a RF power of 75 Watt.

Substrate heating is performed by using four rod-shaped quartz lamps mounted between an even mirror and a quartz plate carrying the substrate which is heated up only by infrared radiation. For temperature measurement we have fitted a thermocouple inside the quartz plate. The values obtained this way are rather uncertain, because the temperature readout corresponds only with the radiation absorbed by the thermocouple.

A computer controlled gas inlet and pumping system guarantees an accurately defined pressure and composition of the atmosphere during sputtering. The gas flow rates of argon and oxygen into the chamber are regulated by mass flow controllers. A capacitive pressure transducer measures the total pressure which is kept up on a constant level by regulating a throttle valve in the pumping main. The gas composition is analysed by a differentially pumped quadrupole mass spectrometer. The computer calculates the partial pressure of each gas and adjusts the setpoint of the oxygen flow to maintain the oxygen partial pressure at the desired value.

3 Film Deposition

We developed a sputtering process to produce high quality thin films of YBa$_2$Cu$_3$O$_{7-\delta}$ on SrTiO$_3$ or LaAlO$_3$ substrates using the equipment just described. The parameters which are decisive for this process are position of the substrate, RF power, substrate temperature, oxygen partial pressure, and cooling procedure. The position of the substrate is chosen in a way that the normal from any point of the targets' surface does not meet the substrate whatever substrate size is taken up to two inches in diameter. A shift of the substrate towards the focal point would change the off-axis to an on-axis geometry resulting in a decrease of the deposition rate and in a deviation of the film homogeneity and stoichiometry arising from resputtering effects. The alteration of the RF power is limited to a narrow range around 75 W which is the value we have chosen in all our runs. A lower power decreases the deposition rate below about 5Å/min depending on the pressure, whereas a higher one leads to an inappropriate thermal stress of the targets. The substrate temperature, T$_D$, and the oxygen partial pressure, p$_{O_2}$, are primarily subject to thermodynamic requirements.

As shown by Bormann et al. [14], there is a critical oxygen partial pressure for every temperature T$_D$ below which the perovskite structure YBa$_2$Cu$_3$O$_{6+x}$ becomes unstable and coexists with Cu$_2$O, BaCuO$_2$, and Y$_2$BaCuO$_5$. On the other hand, kinetic constraints necessiate a high substrate temperature for crystallization of YBa$_2$Cu$_3$O$_{6+x}$. We obtained best results with p$_{O_2}$ = 20Pa (total pressure p$_0$ = 25Pa) and T$_D$ ≈ 800°C.

Note that the real substrate temperature is propably lower due to a higher IR absorption coefficient of the metallic thermocouple in comparision to the substrate. Any decrease of p_{O_2} or T_D respectively, leads to a lower critical temperature T_c and a broader transition to superconductivity. After deposition, the chamber is vented with oxygen to atmospheric pressure. The subsequent cooling to room temperature takes about 20 minutes. Meanwhile, the tetragonal phase is oxidized to orthorombic $YBa_2Cu_3O_{7-\delta}$.

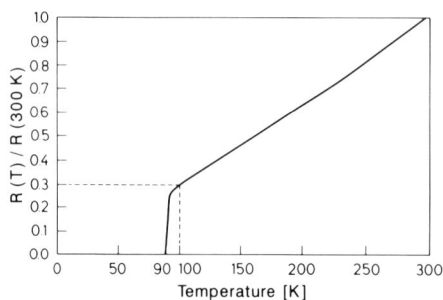

Figure 2: Four probe DC resistance measurement of a film grown on $SrTiO_3$

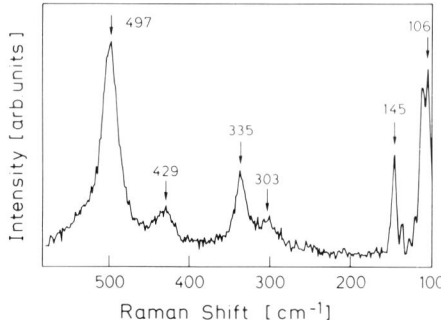

Figure 3: Raman spectrum of a sputtered film (thickness 3000Å) on $SrTiO_3$

4   Results and Discussion

A typical result of a four probe DC resistance measurement of our samples on $SrTiO_3$ as well as on $LaAlO_3$ substrates is plotted in Figure 2. The film shows metallic behaviour of resistance vs. temperature with a resistance ratio $\Gamma = R(300K)/R(100K) > 3$ which extrapolates to vanishing resistance at a temperature near absolute zero indicating an epitaxial growth of the films. The critical temperatures $T_c(R = 0)$ are about 90K and the critical current densities in zero magnetic field are roughly $1 MA/cm^2$ at 77K as measured on patterned films using conventional photo lithography and wet chemical etching. Moreover, inductive measurements of T_c also show a narrow transition to superconductivity.

Our films are single phase and have a high oxygen content as shown by Raman spectroscopy (Figure 3). The sharp decrease of the Raman intensity at wavenumbers above $500cm^{-1}$ as well as the absence of any peak around $220cm^{-1}$ point out a very well oxygen ordering [15]. The film is not completely textured. The intensity ratio of the peaks at $497cm^{-1}$ and $335cm^{-1}$ indicates the presence of grains having the c-axis parallel to the plane [16]. These results are confirmed by X-ray diffraction analysis. Beside the (00l) peaks two small peaks are visible which can be identified as the (100) and (200) diffraction peaks.

One important advantage of the sputtering technology is the film deposition without any contamination by particulates. Nevertheless, the film surface is not perfectly smooth

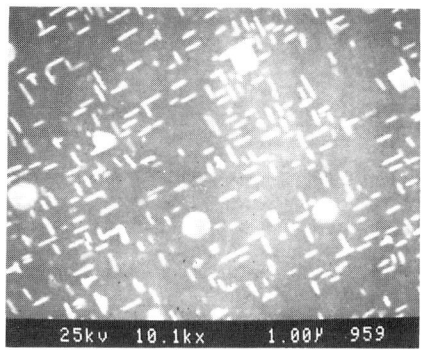

Figure 4: SEM photograph of a film grown at $p_{O_2} = 20$Pa and $T_D \approx 780°C$

Figure 5: SEM photograph of a film grown at $p_{O_2} = 20$Pa and $T_D \approx 815°C$

in principle but the surface morphology depends on the growing conditions. Figure 4 shows a SEM micrograph of the surface of a YBCO film produced by sputtering. The deposition parameters are $p_{O_2} = 20$Pa and $T_D \approx 780°C$ resulting in a perfectly smooth film also having a T_c of 91 K and a high critical current density. The surface roughness is below 5nm as investigated by scanning tunneling microscopy (STM). In contrast, the film shown in Figure 5 is grown using an increased deposition temperature of 815°C whereas all the other parameters are just the same as in the film mentioned before. The electrical properties do not vary significantly between both films but the surface morphology looks completely different. The second film clearly shows two types of outgrowths: (i) needle shaped outgrowths which are orientated in two perpendicular directions, and (ii) lens shaped outgrowths. Although the details of the outgrowth formation are not clear so far it is obviously that outgrowth formation depends on growing kinetics of the film [17] and must not be confound with surface roughness resulting from particulate deposition. It is important to emphasize that we have found a narrow range of deposition temperature which results in both, a perfectly smooth surface and very good electrical properties.

For checking the homogeneity of our films over a large area, we produce four small samples in the same run. Each substrate has a different but well defined position spread over an area of two inches in diameter. Both, the film thickness, and the electrical properties do not differ significantly from one position to another. Therefore, we have also grown superconducting large area YBCO films on LaAlO₃-wafers of two inches in diameter. Detailed investigations about the microwave absorption of these films in the 100GHz range are currently underway and will be published in a future paper.

5   Conclusions

In this report we pointed out that single target off-axis sputtering is a very suitable technique to produce high quality YBCO thin films. Our special arrangement using three sputtering cathodes allows depositing homogeneous films over an area of at least two inches in diameter. The fims have excellent surface properties without any contamination with particulates. The off-axis sputtering technique just described has the potential to accomplish multilayer deposition and preparation of epitaxial heterostructures just as line patterning in the μm range. Therefore, it is a favourable technique especially for application orientated research and development.

Acknowledgements

The authors would like to thank J. Buhn, J. Chi, D. Eißler, B. Friedel, Y. Kershaw, G. Kölz, B. Leibold, F. Razavi, F. Schartner, I. Skupin, S. Tippmann, L. Viczian, and M. Wurster for encouraging technical assistance in film preparation and characterization.

References

[1] W. Kautek, B. Roas, and L. Schultz. *Thin Solid Film*, 191:317, 1990.

[2] G. Koren, A. Gupta, R.J. Baseman, M.I. Lutwyche, and R.B. Laibowitz. *J. Appl. Phys. Let.*, 55:2450, 1989.

[3] H.-U. Habermeier, G. Beddies, B. Leibold, G. Lu, and G. Wagner. *Physica C*, in press.

[4] H.-U. Habermeier. *Eur. J. Solid State Inorg. Chem.*, 28:619, 1991.

[5] T. Venkatesan, X.D. Wu, A. Inam, and J.B. Wachtman. *J. Appl. Phys. Let.*, 52:1193, 1988.

[6] R.K. Singh and J. Narayan. B 41:8843, 1990.

[7] P.A. Zawadzki, G.S. Tompa, P.E. Norris, C.S. Chern, R. Caracciolo, B.H. Kear, D.W. Noh, and B. Gallois. *J. Electronic Materials*, 1989. 4th Biennial Workshop on OMVPE.

[8] J. Kwo, D.J. Hong, R.M. Trevor, R.M. Fleming, A.E. White, R.C. Farrow, A.R. Kortan, and K.T. Short. *J. Appl. Phys. Let.*, 53:2683, 1988.

[9] J Kwo. 1990. private communication.

[10] U. Poppe, J. Schubert, R.R. Arons, W. Evers, C.H. Freiburg, W. Reichert, K. Schmidt, W. Sybertz, and K. Urban. *Solid State Comm.*, 66:661, 1988.

[11] X.X. Xi, J. Geerk, Q. Linker, and O. Meyer. *J. Appl. Phys. Let.*, 54:2367, 1989.

[12] N. Newman, K. Char, S.M. Garrison, R.W. Barton, R.C. Taber, C.B. Eom, T.H. Geballe, and B. Wilkens. *J. Appl. Phys. Let.*, 57:520, 1990.

[13] A.M. Kadin, P.H. Ballentine, D.S. Mallory, and J.P. Allen. *Proc. LT-19 Cambridge England*, in press, 1990.

[14] R. Bormann and J. Nölting. *J. Appl. Phys. Let.*, 54:2148, 1989.

[15] Christian Thomson and Manuel Cardona. Raman scattering in high-tc superconductors. In *Physical Properties of High-Temperature Superconductors*, Ginsberg, D.M., 1990.

[16] R. Feile, U. Schmitt, P. Leiderer, and U. Poppe. *Z. Phys. B -Condensed Matter*, 72:161, 1988.

[17] C.C. Chang, X.D. Wu, R. Ramesh, X. X. Xi, T.S. Ravi, T. Venkatesan, D.M. Hwang, R.E. Muenchhausen, S. Foltyn, and N.S. Nogar. *J. Appl. Phys. Let.*, 57:1814, 1990.

High T$_c$ Superconductor Thin Films
L. Correra (Editor)
377

Growth and patterning of Y$_1$Ba$_2$Cu$_3$O$_7$ films

P. K. Srivastava, Ph. Flückiger, Ch. Leemann and P. Martinoli

Institut de Physique, Université de Neuchâtel, 2000 Neuchâtel, Switzerland.

Abstract

Highly c-axis oriented films of Y$_1$Ba$_2$Cu$_3$O$_7$ have been prepared by dc inverted cylindrical magnetron sputtering onto single crystal SrTiO$_3$ (100) substrates. These films have a superconducting transition temperature T$_c$ of 90K, a very narrow transition width (1K), a resistivity ratio of $\rho(300K)/\rho(90K) = 3$ and critical current densities of $10^6 A/cm^2$ at 77K. Triangular and square wire networks consisting of 10^6 nodes with a periodicity of $5\mu m$ and line width of $1\mu m$ have been prepared on these films. The details of the deposition parameters and the fabrication of the wire networks are reported. Also, we will report the measurements of the complex ac impedance Z with a two coil mutual inductance technique from which the in-plane penetration depth was determined. Near the transition temperature, where vortex pinning is relatively weak, both the real and imaginary parts of Z show oscillations in a perpendicular magnetic field, resulting from flux quantization in the cells of the network.

1. Introduction

In order to be able to use superconducting oxide films in device fabrication it is essential for the films to have the right stoichiometry, right crystal structure, smooth surface, sharp interface between film and substrate, high superconducting transition temperature, small transition width and high critical current density at 77K. Among the different techniques which have been used for the deposition of superconducting oxide films, sputtering is the most common one. Both dc and rf sputtering using single or multiple targets in the planar magnetron mode have been employed. It has been observed that the film composition often deviates from the target composition due to the different sputtering rates of the constituents of the target and due to the bombardmemt of the negative ions emerging from the target resulting in selective resputtering. Many workers have overcome this problem by suitably adjusting the target composition or by using unconventional sputtering geometries[1-3]. A simple way to circumvent this problem, as used by Geerk [4], is to make films in a high sputtering gas pressure using an inverted cylindrical magnetron with a single composite target of Y$_1$Ba$_2$Cu$_3$O$_7$. We have used this technique to make superconducting oxide films of Y$_1$Ba$_2$Cu$_3$O$_7$ and have used these films for making superconducting networks with contact photolithography and ion milling. In this report we present fabrication and characterization of our films and measurements of flux quantization phenomena in Y$_1$Ba$_2$Cu$_3$O$_7$ networks.

2. Experimental details

A high vacuum chamber fitted with a turbo-molecular pump was used for the deposition of $Y_1Ba_2Cu_3O_7$ films on $SrTiO_3$ (100) substrates. A single stoichiometric target of composition Y:Ba:Cu in a 1:2:3 ratio was used. The target was of hollow cylindrical form having 3.0 cm inner diameter, 3.0 cm height and 0.5 cm wall thickness. It was mounted vertically, with the substrate on its heater at a distance of 2.0 cm below the target. Before each run the substrate heater was cleaned and outgassed at 980C for 2 hours in 10^{-6} Torr. The substrate was then mounted and cleaned at 900C for 1 hour in pure oxygen. Again the system was evacuated to the 10^{-6} Torr range, then the turbo-molecular pump was switched off and the rotary pump directly connected to the system. Two flowmeters, a mass-flow ratio controller and a pressure controller were used to maintain the desired sputtering pressure and an oxygen to argon ratio of 1:6. The substrate temperature was controlled by a programmable temperature controller and measured with a 0.5mm thick chromel-alumel thermocouple mounted in the substrate holder 0.2mm below the substrate. Due to the positioning of the thermocouple in the heater we believe that the real temperature of our substrate could be 50C to 100C lower than the measured value. After attaining the desired substrate temperature, the dc power supply was switched on for the deposition of the film. The various sputtering parameters used are: nominal substrate temperature between 800C and 900C, sputtering current 0.2A which developed a dc voltage 125V. After the deposition, the system was filled with pure oxygen and the films cooled to 450C in 15 sec. They were annealed for 5 minutes and cooled further to room temperature in 2 minutes. Typical film thicknesses as measured by Rutherford backscattering spectroscopy (RBS) are 200Å - 2000Å. The films as produced are black and shiny are microscopically smooth exhibiting a featureless microstucture as observed by optical microscopy. Observation with a scanning electron microscope (SEM) reveals a slightly grainy structure on a scale of 1000Å, as shown in figure 1.

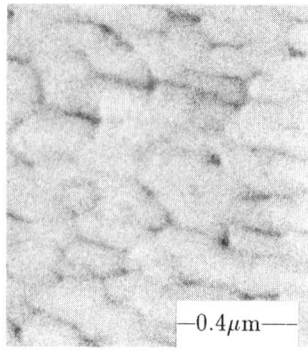

Fig. 1: SEM micrograph of a c-axis oriented film showing a smooth surface with some structure on a scale of 1000Å.

3. Results and Discussions

We have observed that along with other sputtering parameters, substrate temperature, Ar/O_2 ratio, power applied to the target and the cooling rate after the growth of the films play an important role in determining the crystal structure and the superconducting

properties of the films. X-ray diffraction of
a film produced as described, with the substrate at 890C (figure 2(a)) shows the presence of a number of $(00l)$ peaks suggesting
that the major phase consists of grains with
the c-axis perpendicular to the film surface.
For another film, prepared with an oxygen
to argon ratio of 1:1 at a substrate temperature of 830C and slowly cooled from 830C
to 450C (about 30 minutes), the presence of
only (100) and (200) diffraction peaks with
relatively large intensities (figure 2(b)) suggests a preferred grain growth with the a-
axis perpendicular to the film surface. The
parameters determining the good overall quality of our films, namely correct composition
and c-axis orientation, appear to be substrate temperature, oxygen to argon ratio
and a fast cooling rate.

The RBS spectrum of a c-axis oriented
film along with a numerically calculated spectrum for a film with a thickness of $1600\mathring{A}$ is
shown in figure 3. From the step heights we
calculate the composition of the film which
in this case turns out to be Y:Ba:Cu in a ratio 1.6:2.0:3.0 The important feature of this
data is the sharpness of the spectrum which
suggests a sharp interface, without excessive
intermixing or diffusion of film material into
the substrate material.

The resistivity vs temperature curve for
a good quality sample, as measured by a
four probe method using indium as a contact material, is shown in figure 4. The
$\rho(300K)/\rho(90K)$ ratio for the film is 3.0 and
its resistivity at 95K is $110\mu\Omega$cm. The transition temperature is 90K and the transition width about 1K. For the critical current density measurements we have fabricated 40μm wide and 500μm long strips with
wet chemical etching. A voltage criterium of
0.2μV was used for the measurement of the
critical current. As shown in figure 5, this
film has a critical current density at 77K in
excess of 10^6 A/cm^2.

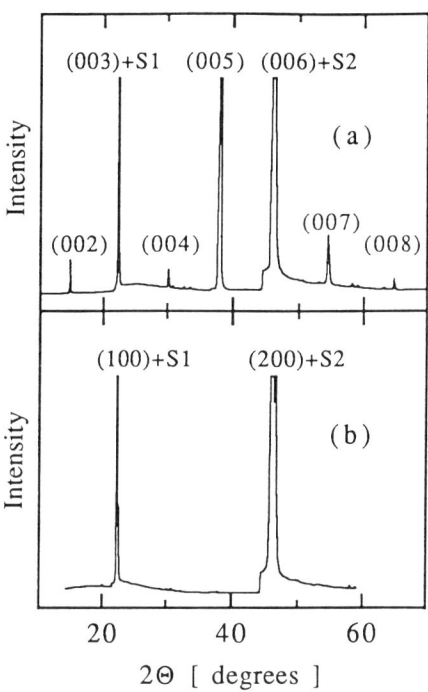

Fig. 2: X-ray diffraction pattern (a) of c-
axis oriented and (b) a-axis oriented films.
S1 and S2 are the (100) and (200) lines
of (100) oriented $SrTiO_3$ substrate.

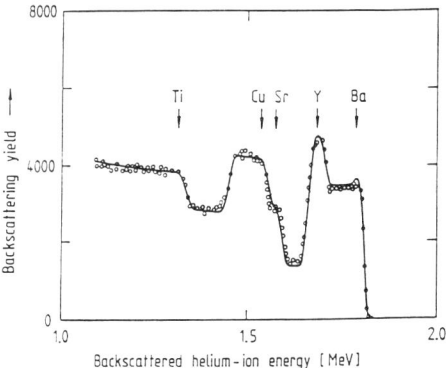

Fig. 3: RBS spectrum of a c-axis oriented film. Circles are the measured spectrum, full line is the spectrum calculated
for a $Y_{1.6}Ba_{2.0}Cu_{3.0}O_{6.0}$ film of thickness
$1600\mathring{A}$.

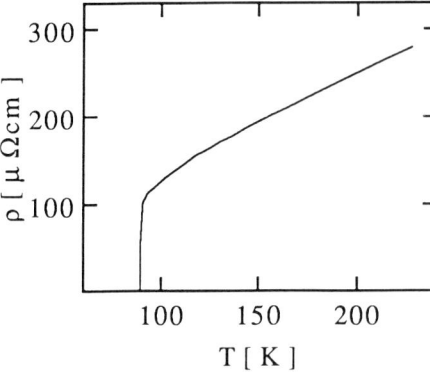

Fig. 4: Temperature dependence of the dc resistivity of good quality film showing metallic behaviour above the transition temperature and a narrow transition width at 90K.

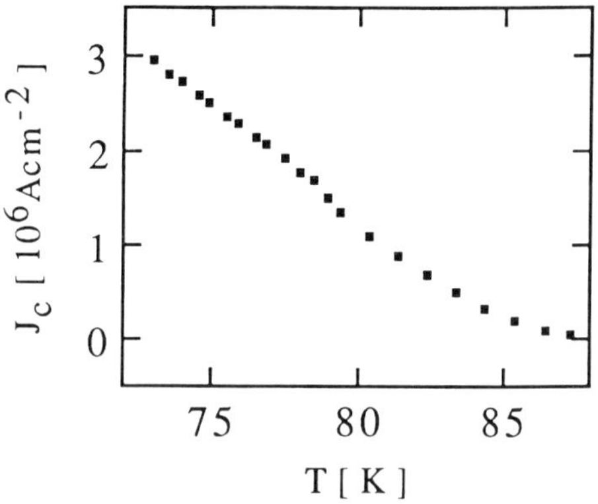

Fig. 5: Temperature dependence of the critical current density of a c-axis oriented thin film.

The penetration depth of the films was deduced from inductive conductance data as measured with a two coil mutual inductance method described in earlier papers [5-8]. Basically it is an eddy current technique. The sample is positioned directly under a coil assembly consisting of an excitation coil and a concentric gradiometer detection coil. A current flowing in the excitation coil induces sreening currents in the film which in turn induce a signal (δV) in the detection coil which can be phase senstively detected. The variation of the in phase Re(δV) and quadrature component Im(δV) of the signal with

temperature for a c-axis film is shown in figure 6. With a numerical inversion procedure [8] we can extract from the signal (δV) the complex impedance of the film Z= R + iωL_k , where R is the resistance and L_k the sheet kinetic inductance. The penetration depth λ [8] is directly connected to the L_k by the relation $L_k = \mu_0 \lambda^2/d$ where d is the thickness of the film. From a fit to the London expression for λ, with the mean field transition temperature and $\lambda(0)$, the penetration depth at zero temperature, as adjustable parameters, we find, for our best films, $\lambda(0)$=2500Å, not far from 1400Å obtained for single crystal thin films [9].

Triangular and square networks with lattice parameter a = 5μm and wire width w = 1μm were fabricated on c-axis oriented

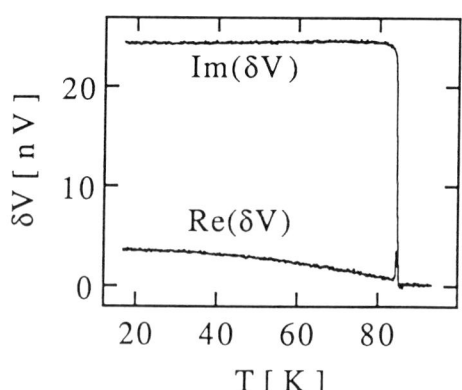

Fig. 6: Temperature dependence of the measured signal Im(δV) and Re(δV) from which the thin film penetration depth can be extracted.

films with contact photolithography and argon ion milling. First photoresist 2.5μm thick was spinned onto the film and exposed through a contact mask fabricated by electron beam lithography. After developing the photoresist, the unwanted $Y_1Ba_2Cu_3O_7$ was removed by ion milling with 2kV argon ions. An SEM mircograph of a resulting network, consisting of $\sim 10^6$ nodes, is shown in figure 7. After the patterning process the superconducting transition temperature of the network was found to be 83.5K, this degration is due to the ion beam etching.

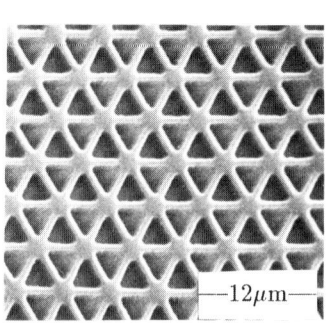

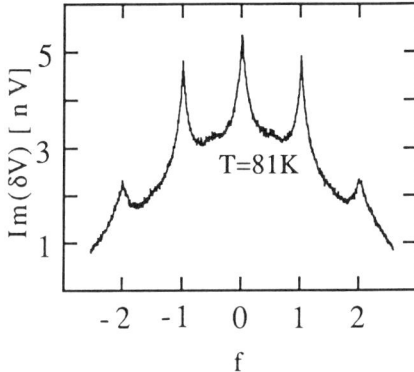

Fig. 7: SEM micrograph of a triangular network.

Fig. 8: Magnetoconductance oscillations of the network shown in figure 7 as a fuction of magnetic field

In the presence of a perpendicular magnetic field B the imaginary and real parts of the conductance of the networks show oscillations which are visible in a very narrow region of about 0.4K below the transition temperature. These oscillations are the result of flux quantization in the loops of the networks. In figure 8, $Im(\delta V)$ of a triangular network measured at 6kHz is shown as a function of the frustration parameter $f = \Phi/\Phi_o$, expressing the applied flux per unit cell, $\Phi = \sqrt{3}a^2B/4$ in units of the superconducting flux quantum Φ_o. The oscillations observed correspond to integer numbers of flux quanta per unit cell of the network.

4. Conclusions

High temperature superconducting oxide films of $Y_1Ba_2Cu_3O_7$ have been prepared by dc cylindrical magnetron sputtering by utilizing high gas pressure and insitu annealing. By optimizing the substrate temperature, the cooling cycle and the oxygen to argon ratio in the sputtering gas, we have obtained good quality a-axis and c-axis oriented films. Further we have optimized the lithographic process to obtain triangular and square networks without substantially degrading the superconducting properties and have shown that these superconducting networks exhibit magnetoconductance oscillations in a perpendicular magnetic field.

Our work was supported by the Swiss National Science Foundation.

5. References

1. H. C. Li, G. Linker, F. Ratzel, R. Smithey and J. Geerk, Appl. Phys. Lett., 52 (1988) 1098.

2. R. L. Sandstrom, W. J. Gallagher, T. R. Dinger, R. H. Koch, R. B. Laibowitz, A. W. Kleinsasser, R. J. Gambino, B. Bumble and M. F. Chisholm, Appl. Phys. Lett., 53 (1988) 444.

3. H. Koinuma, M. Kawasaki, S. Nagata, K. Takeuchi and K. Fueki, J. Appl. Phys., 27 (1988) L376.

4. X. X. Xi, G. Linker, O. Mayer, E. Nold, B. Obst, F. Ratzel, R. Smithey, B. Strehlau, F. Weschenfelder and J. Geerk, Z. Phys. B, 74 (1989) 13.

5. P. K. Srivastava, P.Debély, H. E. Hintermann, Ch. Leemann, Ph. Flückiger, O. Caccivio, J. L. Gavilano, J. Weber and P. Martinoli, IEEE Trans. Magn., MAG-25 (1989) 2575.

6. P. K. Srivastava, P.Debély, H. E. Hintermann, Ph. Flückiger, J. L. Gavilano, Ch. Leemann, J. Weber and P. Martinoli, in*Surface Modification Technologies III*, edited by T. S. Sudarshan and D. G. Bhat (The Minerals, Metals and Material Society, Warrendale, PA, 1990) p. 219

7. A. F. Hebard and A. T. Fiory, Appl. Phys. Lett. 52 (1988) 2165.

8. B. Jeanneret, J. L. Gavilano, G. -A. Racine, Ch. Leemann and P. Martinoli, Appl. Phys. Lett., 55 (1989) 2336.

9. D. R. Harshman, L. F. Schneemeyer, J. V. Waszczak, G. Aeppli, R. J. Cava, B. Batlogg, L. W. Rupp, E. J.Ansaldo and D. Li, Phys. Rev., B39 (1989) 851.

High T$_c$ Superconductor Thin Films
L. Correra (Editor)
© 1992 Elsevier Science Publishers B.V. All rights reserved.

In-situ growth of YBa$_2$Cu$_3$O$_y$ superconducting films by reactive coevaporation at low pressure using ozone

H.M. Appelboom, A.W. Fortuin, G. Rietveld, V.C. Matijasevic, P. Hadley, D. van der Marel, and J.E. Mooij.

Department of Applied Physics, TU Delft, P.O. Box 5046, 2600 GA Delft, The Netherlands.

Abstract

YBa$_2$Cu$_3$O$_y$ thin films are grown in-situ on SrTiO$_3$ and MgO substrates using MBE techniques and ozone. Yttrium and copper are evaporated from electron guns and barium is evaporated from a Knudsen cell. Pressure and substrate temperature have been varied to study growth conditions. The results are compared to thermodynamic stability conditions for YBa$_2$Cu$_3$O$_y$ and oxygen. The use of ozone enables growth of in situ superconducting films at pressures substantially below the YBa$_2$Cu$_3$O$_y$ stability line for oxygen. The films are analyzed with R(T) and x-ray diffraction measurements. The best film so far has been grown on SrTiO$_3$ and has a T$_{c,0}$ of 83 K.

1. INTRODUCTION

High temperature superconductor thin films can be made by several fabrication methods, such as sputtering, laser ablation, MOCVD and evaporation techniques. When the superconductor is crystallized during the deposition lower temperatures can be used compared to typical post annealing temperatures. This lower temperature will result in fewer reactions at the superconductor-substrate interface and in less reactions with artificial barriers. Several succesful attempts to fabricate in-situ superconducting thin films at considerably lower growth temperatures as compared to post-annealing processes have been reported by other groups using various techniques[1,2,3,4]. The evaporation method requires the use of multiple sources for the fabrication of YBa$_2$Cu$_3$O$_y$ thin films. This enables one to study the dependance of film properties on compostion[5,6] or to grow artificial structures, either by using layer by layer growth[7], or by varying the deposition rates during growth.

The most severe problem of using evaporation techniques for in situ growth is the presence of sufficient amounts of oxygen during growth. Other fabrication techniques such as sputtering and laser-ablation can use high enough oxygen partial pressures during growth to form YBa$_2$Cu$_3$O$_y$ within the region of the pressure-temperature diagram where YBa$_2$Cu$_3$O$_y$ is thermodynamically stable according to the data of Beyers and Ahn[8]. Evaporation techniques limit the overall pressure to roughly 5×10^{-4} mbar, which is not sufficient. Solutions to this problem are to make locally a very high oxygen pressure near the substrate, as has been demonstrated by K.Shinohara et.al [6], or to use some form of activated oxygen, like ozone[4], or atomic oxygen[1,5,6]. The advantage of using ozone or atomic oxygen with respect to a high differential oxygen pressure is that it also allows to coat large areas and that it is easier to obtain a homogenous pressure distribution.

2. FABRICATION EQUIPMENT

The films are fabricated in a VG MBE system which consists of a growth chamber for evaporation, an analysis chamber for XPS/Auger measurements, an annealing station, and aditionally a laser ablation system. Samples can be transported from one chamber to another without breaking the vacuum. The growth chamber contains two Airco Temescal electron guns and one VG Knudsen cell. It is pumped by a turbo molcular pump and a Ti sublimation pump. Base pressure of the system is below 10^{-10} mbar. Yttrium and copper are evaporated from the e-guns and the barium is evaporated from the Knudsen cell. The filament of the Knudsen cell has been modified to deliver sufficient power at the end of the crucible to prevent condensation of material at the lip. This improves the long term drift of the barium evaporation rate from 10-20% per hour to less than 1 % per hour.   The yttrium gun contains a tantalum liner to improve the stability of the melt. The evaporation flux from the electron guns is controlled by a feedback system which uses one differentially pumped mass spectrometer for each gun and a high frequency sweep for high frequency control. The bandwidth of the feedback system is larger than 100 Hz, and the long term drift is on the order of 1-2% per hour.[9] However, the sensitivity of the mass spectrometers is strongly dependent on pressure above 10^{-5} mbar. In spite of the differential pumping of the mass spectrometers this can still result in a 20-30% drift of the yttrium and copper evaporation rates during a deposition run at the highest pressure we use, because the pressure distribution in the growth chamber can change over time. The pressure dependence of the sensitivity of the mass spectrometers also limits the bandwidth of the feedback loop considerably by the gettering effect of the evaporated metals. The total pressure in the system is measured with an ionization vacuum gauge which is located near the sample holder. The fluxes from the various sources are adjusted to the desired ratios with a single quartz crystal monitor which can be moved into the sample position to provide a tooling factor of 100 %.

The ozone is produced in a commercial "silent" discharge generator.[10] This generator yields a 5% ozone/oxygen mixture at a flow rate of 50 l/h. We can produce enough ozone in 15 minutes for a deposition run of several hours. The ozone/oxygen mixture is pumped through a glass chamber which is kept at a temperature of 140 Kelvin. At this temperature mainly ozone condenses. After purification the ozone can be lead into the growth chamber through a leak valve and a stainless steel tube. The tube is directed towards the substrate to ensure that the ozone does not decompose on the chamber walls before reaching the substrate. The angle between the tube and the substrate plane is 20° and the distance between the end of the tube and the substrate is 8 centimeters. The pressure in the deposition chamber, which is proportional to the vapour pressure in the liquid ozone vessel, can be adjusted by heating the glass chamber relative to the liquid nitrogen bath. Based on geometrical and pumping speed considerations we estimate that a pressure of 1.0×10^{-4} mbar in the growth chamber corresponds to a flux of 1.5×10^{-8} mol cm^{-2} s^{-1} of ozone on the sample position, assuming all the ozone is delivered. This flux corresponds to a partial pressure of 3.4×10^{-5} mbar. We verified the ozone flux at the sample postion with a silver coated quartz crystal[11] and estimate it to be at least 1.6×10^{-9} mol cm^{-2} s^{-1} at a chamber pressure of 1.0×10^{-4} mbar.

The substrates are glued with silver paste to a platinum transport plate which is clamped to a tantalum block. The tantalum block is radiatively heated from the back by a 250 W quartz lamp. Because we cannot measure the temperature of the sample directly we have to deduce the temperature of the sample from the temperature of the tantalum plate. Therefore a thermocouple has been fitted in a hole in the tantalum block. The relation between the temperature of the platinum transport plate and the temperature of the tantalum block has been calibrated with a pyrometer and a thermocouple several times over a period of one year. We estimate the reproducibility of the substrate temperature measurement to be smaller than 30 °C.

3. FABRICATION PROCEDURE

The fluxes from the various sources are switched on first and are brought into their consecutive feedback loops. Then the ozone flow is turned on by heating the still. After adjustment of the ozone flow to the desired value the evaporation rates of the respective sources are set to give the desired composition. A typical total evaporation flux is 0.2 nm/s. When this calibration procedure has been completed the quartz crystal is removed and the substrate is brought into position. The tantalum block is heated to the required temperature and the shutters are opened until a film with the required thcikness is grown. The shutters are closed and sample holder is allowed to cool down in about half an hour to room temperature. The evaporation sources are switched off immediately after closing the shutters to prevent deposition of material during the cooling by scattering. During the deposition and the cool down the temperature of the ozone still is held constant, which implies that the ozone flux into the growth chamber is also constant. The sample is removed from the holder and transported to the load-lock when the temperature has dropped below 100 °C.

4. RESULTS

The results obtained on samples grown at 2×10^{-4} mbar and at 3×10^{-5} mbar chamber pressure for various temperatures are summarized in table 1. Only samples that were grown at chamber pressures of 2×10^{-4} mbar and with maximum purity of ozone flux are superconducting. Figure 1 shows the position of the growth conditions of the superconducting and the non-superconducting samples in the p,T diagram in combination with the extrapolated stability line for YBa$_2$Cu$_3$O$_y$ and O$_2$. The superconducting samples are clearly grown in the "unstable" region. The position of the superconducting and the nonsuperconducting samples in the p,T diagram is comparable to the positions found of films grown with atomic oxygen by Edwards et al.[12] The reason why the samples can be grown in the "unstable" region is still a question to be answered: either the stability line does shift due to the presence of the ozone, or the presence of ozone enhances the formation of YBa$_2$Cu$_3$O$_y$, while the decomposition of YBa$_2$Cu$_3$O$_y$ is limited by kinetics. Figure 2 shows the R(T) curves of the superconducting films. The minimum growth temperature required to obtain a superconducting film is 600 °C

Table 1. Properties of samples grown under various growth conditions.

Sample	Substrate	d [nm]	p [mbar]	ozone [%]	T [°C]	T$_c$ [K]	XRD	c-axis [Å]
BH1a	MgO	500	2.0×10^{-4}	100	577	--	YCuO$_2$	--
BH1b	SrTiO$_3$+Nb$^{a)}$	500	2.0×10^{-4}	100	577	--	--	--
BG1	SrTiO$_3$	91	1.8×10^{-4}	100	599	50-4.2	--	--
BI1a	MgO	500	1.8×10^{-4}	100	627	80-60	poly YBa$_2$Cu$_3$O$_y$	--
BI1b	SrTiO$_3$+Nb$^{a)}$	500	1.8×10^{-4}	100	627	80-68	001..006 YBa$_2$Cu$_3$O$_y$	11.86
BJ1a	MgO	500	2.9×10^{-4}	100	652	87-47	poly YBa$_2$Cu$_3$O$_y$	--
BJ1b	SrTiO$_3$+Nb$^{a)}$	500	2.9×10^{-4}	100	652	88-42	poly YBa$_2$Cu$_3$O$_y$	--
BM1b	SrTiO$_3$+Nb$^{a)}$	192	2.3×10^{-4}	100	686	87-83	001..007 YBa$_2$Cu$_3$O$_y$	11.72
BP1a	SrTiO$_3$	239	6.8×10^{-5}	100	631	--	--	--
BP1b	Si	239	6.8×10^{-5}	100	631	--	--	--
BO1a	SrTiO$_3$	61	3.6×10^{-5}	100	652	--	CuO	--
BO1b	Si	61	3.6×10^{-5}	100	652	--	--	--
BN1	SrTiO$_3$	500	3.0×10^{-5}	100	672	--	YBa$_2$Cu$_3$O$_y$ (a and c axis mixed), BaCu$_2$O$_2$, Ba$_2$CuO$_3$, YCuO$_2$, Y$_2$Cu$_2$O$_5$ Y$_2$O$_3$, Cu$_2$O YBa$_3$Cu$_2$O$_y$, Y$_2$BaCuO$_5$	--
BK1	MgO	500	2.5×10^{-4}	5	656	--	Y$_2$BaCuO$_5$ CuO	-- --
BL1a	MgO	500	1.7×10^{-4}	5	646	--	BaCu$_2$O$_2$	--
BL1b	SrTiO$_3$+Nb$^{a)}$	500	1.7×10^{-4}	5	646	--	BaCu$_2$O$_2$	--
BE1	SrTiO$_3$	200	$>4.0 \times 10^{-4}$	0	661	--	--	--

a) Some of the SrTiO$_3$ substrates were doped with 200 ppm Nb.

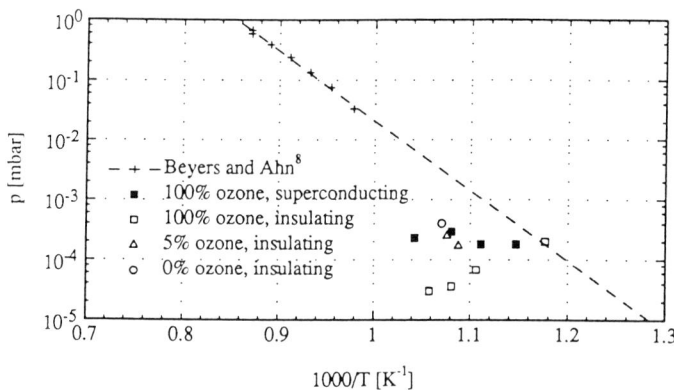

Figure 1. Position of the samples listed in table 1 in the pressure versus inverse temperature diagram as compared to the extrapolation of the stability line of Beyers and Ahn.[8]

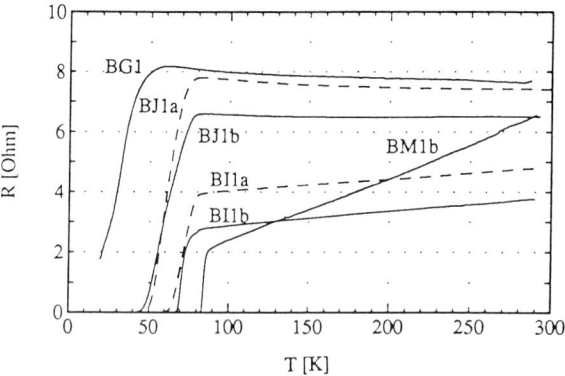

Figure 2. Resistance vs. temperature of the superconducting samples listed in table 1.

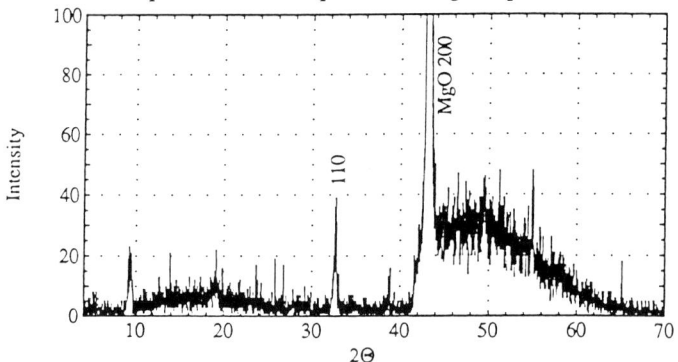

Figure 3. XRD scan of sample BJ1a. The scan shows only the (110) peak of $YBa_2Cu_3O_y$ and the MgO 200 peak, indicating that the sample is polycrystalline. The sample was grown at a temperature of 652 °C and a pressure of 2.9×10^{-4} mbar, using the maximum purity ozone flux. The sample has a transition of 87-47 K.

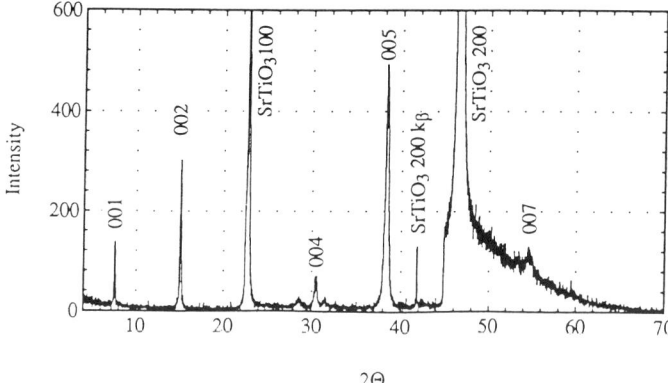

Figure 4. XRD scan of sample BM1b. This sample is predominantly c-axis oriented. It was grown at a temperature of 686 °C and a pressure of 2.3×10^{-4} mbar, using the maximum purity ozone flux. The sample has a transition of 87-83 K.

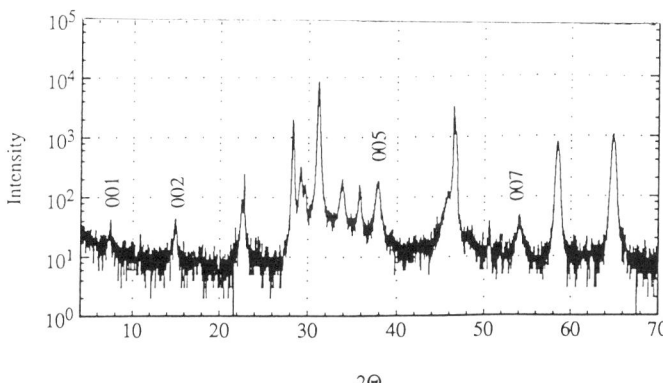

Figure 5. XRD scan of the non superconducting sample BN1. The strongest peaks can be assigned to $YCuO_2$ (or $YBa_3Cu_2O_y$), Ba_2CuO_3, $BaCu_2O_2$, and $YCuO_2$ (or Y_2BaCuO_5). Note that a small amount of $YBa_2Cu_3O_y$ is also present. The sample was grown at a temperature of 672 °C and a pressure of 3.0×10^{-5} mbar, using the maximum purity ozone flux.

(sample BG1). Figure 3 shows the XRD scan a sample BJ1a which was grown at 652 °C. This scan shows only the 110 peak of $YBa_2Cu_3O_y$, which indicates that the film is polycrystalline. A well oriented film with a linear R,T behaviour (like sample BM1b) requires a growth temperature of at least 680 °C. Figure 4 shows the XRD scan of sample BM1b: this sample is predominantly c-axis oriented. A 400 °C post anneal in ambient oxygen did not change the R(T) curve of this sample, indicating that the sample is fully loaded with oxygen during the cooling in the evaporator.

Table 1 also contains data on samples grown with a 5 % ozone/oxygen mixture (i.e. directly from the generator) or with pure oxygen at conditions comparable to the growth conditions of the superconducting samples. None of these samples are superconducting. Figure 5 shows a XRD scan of one of the non-superconducting samples, BN1, which was grown at 672°C and at 3×10^{-5} mbar. The strongest XRD peaks can be assigned to $YCuO_2$ (or $YBa_3Cu_2O_y$), Ba_2CuO_3, $BaCu_2O_2$, and $YCuO_2$ (or Y_2BaCuO_5). Note that a small amount of $YBa_2Cu_3O_y$ is also present.

5. CONCLUSIONS

We have grown in situ superconducting $YBa_2Cu_3O_y$ films by reactive coevaporation at low pressures using ozone. The films are grown below the extrapolated $YBa_2Cu_3O_y$ stability line for oxygen. The minimum temperature to grow a superconducting film at a chamber pressure of 1.8×10^{-4} mbar and maximum purity ozone flux was determined to be 600 °C. The growth of a well crystallized c-axis oriented film with a $T_{c,0}$ of 83 K requires a temperature of at least 680 °C at a pressure of 2.3×10^{-4} mbar and maximum purity ozone flux. Superconducting films grown at lower temperatures are polycrystalline according to XRD scans. Films grown in the temperature region 631-672 °C and at pressures below 10^{-4} mbar with maximum purity ozone flux (or with low purity ozone flux at 2×10^{-4} mbar) are not superconducting.

6. REFERENCES

1   R.M.Silver, A.B.Berezin, M.Wendman, and A.L.de Lozanne, Appl.Phys.Lett 52, 2174
    (1989).
2   C.B.Eom, J.Z.Sun, K.Yamamoto, A.F.Marshall, K.E.Luther, T.H.Geballe, and
    S.S.Laderman, Appl. Phys. Lett. 55, 595 (1989).
3   T. Venkatesan, X.D.Wu, B.Dutta, A.Inam, M.S.Hegde, D.M.Hwang, C.C.Chang,
    L.Nazar, and B.Wilkens, Appl. Phys. Lett. 54, 581 (1989).
4   D.D.Berkley, B.R.Johnson, N.Anand, K.M.Beauchamp, L.E.Conroy, A.M.Goldman,
    J.Maps, K.Mauersberger, M.L.Mecartney, J.Morton, M.Tuominen, and Y-J.Zhang,
    Appl. Phys. Lett. 53, 1973 (1988).
5   N.G.Chew, S.W.Goodyear, J.A.Edwards, J.S.Satchell, S.E.Blenkinsop, and
    R.G.Humphreys, Appl.Phys.Lett. 57, 2016 (1990).
6   K.Shinohara, V.Matijasevic, P.A.Rosenthal, A.F.Marshall, R.H.Hammond, and
    M.R.Beasley, Appl.Phys.Lett.58, 756 (1991).
7   D.G.Schlom, A.F.Marshall, J.T.Sizemore, Z.J.Chen, J.N.Eckstein, I.Bozovic, K.E.von
    Dessonneck, J.S.Harris jr., and J.C.Bravman, submitted to Journal of Crystal Growth.
8   R.Beyers, and B.T.Ahn, to appear in Annual Review of Materials Science, Vol.21,
    (1991).
9   H.M.Appelboom, P.Hadley, D.van der Marel, and J.E.Mooij, IEEE Trans. Magn.
    Vol.27, No.2, 1467 (1991).
10 M.Horváth, L.Bilitzky, and L.Hüttner, Ozone (Elsevier Science Publishers, Amsterdam ,
    1985).
11 V.Matijasevic, E.L.Garwin, and R.H.Hammond, Rev.Sci.Instrum. 61, 1747 (1990).
12 J.A.Edwards, N.G.Chew, S.W.Goodyear, J.S.Satchell, S.E.Blenkinsop, and
    R.G.Humphreys, J.Less Comm. Mat.164&165, 414 (1990).

High T_c Superconductor Thin Films
L. Correra (Editor)

DC Sputtered Bi (2212) Cuprate films on MgO

M. Asfaj[a], Mamidanna S. R. Rao[b], R. Suryanarayanan[b] and O. Gorochov[b].

[a]Laboratoire de Physique des Matériaux, Faculté des Sciences, B. P. 1014, Rabat, Morocco

[b]Laboratoire de Physique des solides de Bellevue, CNRS, F-92195 Meudon, France

Abstract

Preparation, X ray diffraction, SIMS ,EDX, resistivity (ρ), and ac susceptibility (χ', χ'') of DC sputtered Bi(2212) films onto MgO substrates are described. Optimization of post annealing treatments resulted in the following properties $\rho = 0.7$ m Ω cm at 300K, $T_c = 85$K, diamagnetic onset at 84 K. Data obtained on ρ with H parallel and perpendicular to the film which was annealed at 870 C , cooled to 650 C and quenched in air indicated an anisotropy ratio of 7.5 .

1. INTRODUCTION

Thin films of high T_c superconductors are of current interest beacause of possible applications and contributions to fundamental studies. Among the well known CuO based superconductors, different methods of preparation and various properties of thin films of $YBa_2Cu_3O_{6+z}$ have been extensively documented (1). The properties of $YBa_2Cu_3O_{6+z}$ are strongly affected by the oxygen content z and hence could pose some problems in film processing. The Tl superconductors though have the highest T_c of 125 K require special precautions for handling because of high toxicity. The Bi superconductors form a closely knit family usually denoted as $Bi_2Sr_2Ca_nCu_{n+1}O_{6+2n}$ (n=0,1,2) with T_c ranging from 10 to 110 K. The high T_c phase with n=2 is difficult to obtain in bulk form without the incorporation of Pb though there have been reports on the preparation of films having a high T_c phase without Pb (2). On the other hand the n=1 member usually denoted as Bi(2212) is more easily obtained with T_c ranging from 70 to 85 K depending on the heat treatment which does not normally require long time oxygen annealing unlike in the case of

$YBa_2Cu_3O_{6+z}$ .However, we have noticed that there have been relatively few reports, especially on the magnetic properties, of the Bi(2212) films. This motivated us to examine this system. Among the different techniques available to prepare thin films, DC sputtering is a relatively simple and versatile technique. We report here on the preparation, X-ray diffraction (XRD), energy dispersive X-ray analyses (EDX), secondary mass ion spectroscopy (SIMS), resistivity (ρ), magneto resistance, real (χ') and imaginary (χ'') part of the ac susceptibility of the Bi (2212) DC sputtered films.

2. EXPERIMENTAL TECHNIQUES

Several pellets with nominal composition of Bi(2212) were made by solid state sintering technique starting from the respective oxides or carbonates. The pellets were sintered at 860 C in air for a period of 12-18 h. These pellets (about 12 of them) were fixed by silver paint to a Cu disc which was held 2 to 3 cms above a metal plate on which several polished (100) oriented MgO substrates were kept. This was housed inside a classical vacuum system. Pure argon was used as the sputtering gas. The residual pressure was of the order of 10^{-2} Torr and the dc voltage applied was of the order of 500 V and the current varied between 15 to 25 mA. The sputtering rate was of the order of 6 - 8 A/min and the thickness of the films studied ranged between 0.5 to 2.5 microns.Electrical contacts were made on the film using indium metal and the resistivity was evaluated by the van der Pauw method by an ac technique. χ' and χ'' were simultaneously measured in a ac field of 0.22 Oe at a frequency of 1500 Hz using an ac mutual inductance bridge model ATNE PMS 02.

3. RESULTS AND DISCUSSION.

The as sputtered films showed very broad X-ray pattern and were insulating. The properties changed remarkably after annealing them in air. However, the annealing conditions were found to be very critical. To find out the optimal conditions, we measured the ac susceptibilty of the films which in addition to being a contactless method is also a rapid, sensitive and convenient technique to monitor the evoluation of superconducting properties. Further, any percolation path present in a film due to incomplete formation of the superconducting phase would show a resistive transition but often not a diamagnetic one. The heat treatment schedule was as follows: the films were heated in air to T_a, kept there for t_a hours, cooled to a temperature T_b in 60 to 80 minutes and then quenched to 300 K . Some typical values are given in table I along with some relevant properties.The x-ray diffraction pattern of a typical annealed film is shown in figure 1. In general, irrespective of the values of T_c, the films were always oriented with c-axis perpendicular to the substrate. In particular,only (00l) peaks were seen. The c-lattice parameter calculated from (00 10) peak is 30.7 A which is the known value for the (2212) phase.We also see some unidentified impurity peaks of very small intensity. The crystallites are needle shaped with 10 to 50 microns in length. The depth profile obtained by SIMS on one of the samples is shown in figure 2. The diffusion of Mg into the

Table I
Properties of Bi(2212) films as a function of heat treatment

Sample No	T_a C	t_a h	T_b C	ρ (300K) mΩcm	T_c^* K	ΔT_c^* K	T^d K	ΔT (χ'') K
1	870	1	650	0.68	88	7	82.5	12
2	870	1	620	2.10	85	7.7	80	very broad
3	870	2	650	2.3	88	8	83	8
4	870	3	650	0.78	85	9.4	83.5	2.9

T^d diamagnetic onset; ΔT full width at half maximum; * from ρ data.

film is clearly seen. Further, Sr and Ca diffuse into the substrate. The constant intensity of Cu and Bi show that they do not diffuse into the substrate. EDX analyses of sample 3 is shown in figure 3 . Since the Sr content was found to be fairly constant at different regions of the sample, we have plotted the Bi, Ca and Cu contents relative to Sr. Whereas, there was hardly any variation in the Ca and Cu concentration, that of Bi was found to vary at different regions of the sample. This was also earlier pointed out by others in the case of films obtained by single target dc sputtering (3).

We now present and discuss the resistivity (fig.4)and the magnetic properties (fig.5) which show the effect of annealing. ρ does not reach zero if the films are slowly cooled or oxygen annealed or quenched from 870 C.As an example figure 4 shows the variation of ρ as a function of T. The oxygen content in the film is influenced by the residual oxygen content in the sputtering chamber during sputtering and the heat treatment. This in turn affects the hole density in Cu-O$_2$ planes which is crucial for observing superconductivity . Thus we found that when the film was quenched from 620 C, though there was a resistive transition , the χ'' peak was very broad (sample 2). On annealing the film at 870 C for t_a= 1h in air and cooling it to T_b =650 C followed by quenching, the width of the χ'' peak reduced to 12 K in the case of sample 1 which also showed a resistive transition . This width further reduced to 8 K when t_a was increased to 2h and then to 3 K for t_a= 3h in the case of sample 4.

χ' shows a saturation for T< 75 K for the sample 4 whereas in the case of the sample 1 it shows a saturation below 55 K indicating the presence of weak links. Both the samples 1 and 4 show a linear decrease in ρ as a function of T. However, the residual ρ at T=0 K of the sample 4 is 0.058 mohm cm which is 2.2 times smaller than that observed in the case of the sample 1. This value is still

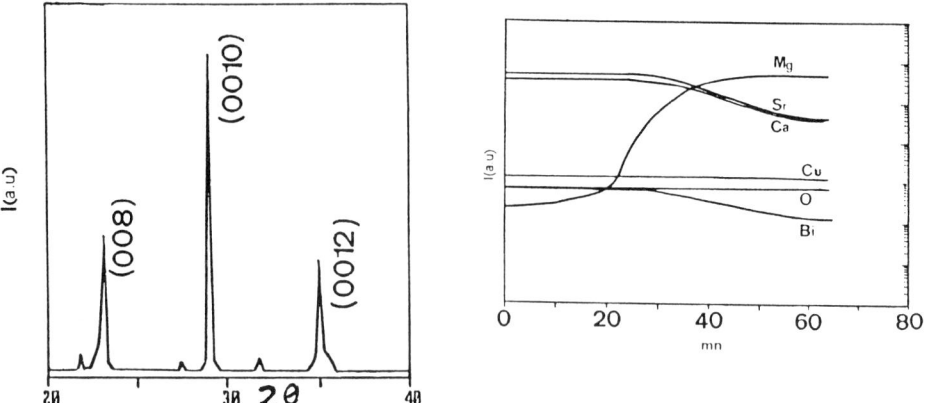

Figure 1(left). XRD pattern of sample 3 taken with Cu Kα.
Figure 2 (right). SIMS analyses of sample 2 as a function of sputtering time.

higher comapared to epitaxial films of YBaCuO, possibly because of not achieving the optimal annealing condictions and partly due to small diffusion of Mg which could substitute for Cu.

Next, we discuss ρ as a function of T with H (11.1 kOe) applied parallel (H// ab) and perpendicular (H I ab) to the film surface but the current (1mA) always being perpendicular to H . The shape of the resistive transition in oxide superconductors in magnetic fields has been discussed by several authors (4 - 8). The long tails in ρ (T,H) curves have been observed not only in sintered polycrystalline bulk but also in single crystals and c axis oriented films (4 - 8) and were ascribed to the magnetic flux creep (4 - 8). The limited amount of data obtained by us here do not allow us to discuss the diffrent models, but help us to investigate the influence of annealing conditions on the slope dH/dT. As an example, in the case of sample 4 we find from fig.6 the slope (taken at 50% of transition) to be 4.1tesla/K for H//ab and 0.55 tesla/K for H I ab. On the other hand, these slopes are found to be respectively 1.2 tesla/K and 0.4 tesla/K in the case of sample 3. The determination of H_{c2} is not unambigous. However, assuming a linear variation of the slope with T, one could estimate (for sample 4) a value of 250 tesla for $H_{c2//ab}$ which is of the right order (9). Finally we note that the slope evaluated for sample 4 is 15 to 20 % smaller than that reported by Fukami et al in the case of laser evaporataed films (8). By taking the ratio of the two slopes given above, the anisotropy is calculated to be 7.5 in the case of sample 4 and 2.5 times smaller in the case of sample 3 which again indicates the necessity of annealing for 3h at 870 C. The sample 4 which shows a narrow width of χ'' peak compared to that of sample 3 is expected to have a higher intergranular critical current (J_c) .Indeed preliminary measurements of χ''as a function of T at different externally apllied dc magnetic field show that J_c of sample 4 at 70 K is about 75% higher .

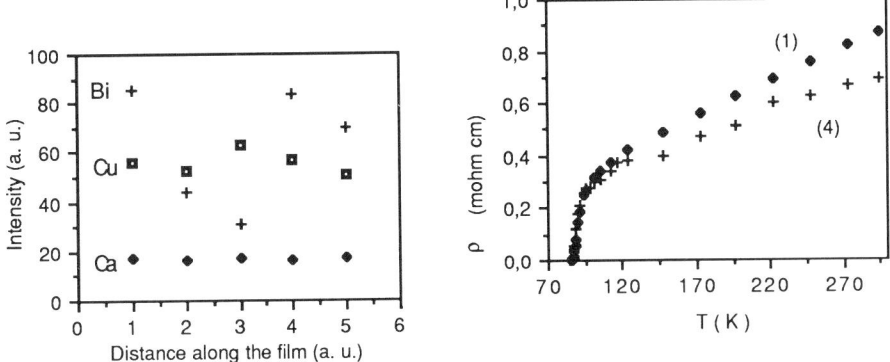

Figure 3. (left) EDX analyses of sample 3. The intensity is in arbitrary units with respect to Sr.
Figure 4. (right) Resistivity as a function of temperature of samples 1 and 4.

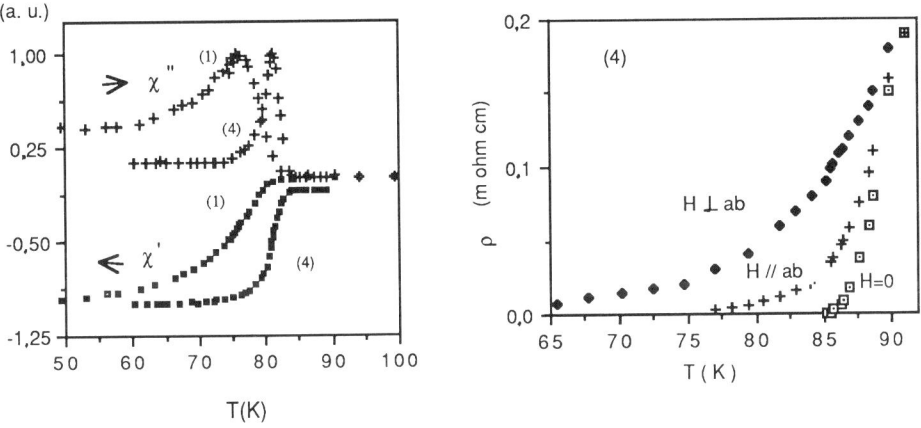

Figure 5. (left) Real (χ ') and imaginary (χ ") part of ac susceptibility of samples 1 and 4 as a function of temperature.
Figure 6. Resistivity as a function of temperature of sample 4 for H // ab and H ⊥ ab , current ⊥ H. (H=11.1 kG).

4. CONCLUSIONS

Bi(2212) films have been obtained by dc sputtering on MgO substrates. The films were annealed at 870 C for 3 h in air , cooled to 650 C and then quenched. ac susceptibility was shown to be a powerful technique to determine the optimum annealing conditions. The best films with c axis perpendicular to the film surface showed a T_c of 85 K with a residual resistivty of 0.058 m ohm cm and a slope (measured at 50% transition) $dH_{//ab}/dT$ of 4.1 tesla/K and $dH_{//c}/dT$ of 0.55 tesla/K. The anisotropy was evaluated to be 7.5. SIMS analyses show Mg diffusion into the film and that Sr and Ca diffuse into the substrate.

5. ACKNOWLEDGMENTS

The authors would like to thank C. Grattepain and G. Zribi for SIMS and EDX analyses. The present research was partially funded by the Commission of the European Communities under Grant No. CI 1- 0337 - F.

6. REFERENCES

1. For a recent review, see, M. Schieber, J. Crystal Growth, 109 (1991) 401.
2. T. Yoshitake, T. Satoh, Y. Kubo and H. Igarashi, Jpn. J. Appl. Phys., 27 (1988) L 1089; M. Makaida, K. Kuroda, T. Tazoh and S. Miyazawa, in Advances in Superconductivity, (eds.),T. Kitazawa and S. Ishiguro, Springer Verlag, Tokyo, 1989 p 609; K. Wasa, H. Adachi, Y. Ichihawa, K. Setsune and K. Hirochi in Rev. Solid State Science (eds.), A. K. Gupta and S. K. Joshi, World Scientific, Singapore, 1988, Vol.2 (453).
3. See for example, H. Raffy, A. Vaures, J. Arabeski, S. Megtert, F. Rochet and J. Perriere, Solid State Comm., 68 (1988) 235.
4. T. T. M. Palstra, B. Batlogg, L.F. Schneemeyer and J. V. Wasczak, Phys. Rev. Lett., 61 (1988) 1662.
5. H. Hidaka, M. Oda, M. Suzuki, A. Katsui, T. Murakami, N. Kobayashi and Y. Muto, Physica B, 148 (1987) 329.
6. Y. Iye, T. Tamegi and H. Talei, Jpn. J. Appl. Phys., 26 (1987) L 1057.
7. K. Kitazawa, S. Kambe, M. Naito, I. Tanaka and H. Kojima, Jpn. J. Appl. Phys., 28 (1988) L 555.
8. T. Fukami, T. Kamura, T. Yamamuto and S. MAse, Physica C, 160 (1990) 391.
9. S. Mase in Studies of High Temperature Superconductors, (ed.), A. Narlikar, Nova Science Publisher, New York, 1990, Vol 6, p 81.

High T$_c$ Superconductor Thin Films
L. Correra (Editor)

PROPERTIES OF HIGHLY ORIENTED BSCCO THIN FILMS

C. Attanasio, G. Balestrino, L. Maritato, A. Nigro, S. Prishepa[a], R. Scafuro, and R. Vaglio

Dipartimento di Fisica, Università degli Studi di Salerno, I–84081, Baronissi (Sa), Italy

[a]Permanent address: Radio Engineering Institute, Minsk, 220600, USSR

Abstract

By completely electron beam evaporating weighted amounts of BSCCO pellets on MgO substrates, we produced good quality BSCCO thin films. The structural and superconducting properties shown by these films along with the repeatibility of the simple deposition technique is very promising in view of further applications. As starting pellets we used both Pb doped and undoped BSCCO pellets. In the case of undoped pellets the stoichiometry was $Bi_{1.6}Sr_{1.8}Ca_{1.05}Cu_{2.4}O_x$ which allowed to obtain film stoichiometries very close to 2:2:1:2. After an annealing in air at 880 °C, for times depending on the final stoichiometry of the samples, the films showed a $T_c(R = 0)$ higher than 80 K. The films were characterized by $\Theta - 2\Theta$ X–Ray diffraction and EDS analysis and by surface impedance, paraconductivity, (I–V) characteristics and critical current measurements.

1. INTRODUCTION

Good superconducting BSCCO and YBCO thin films have been obtained by using different deposition techniques[1]. The control during the deposition of various parameters (the film stoichiometry, the substrate temperature, and the oxygen partial pressure) is the major factor limiting the repeatibility of the process[2]. We have realized BSCCO superconducting thin films with $T_c(R=0)$ higher than 80 K using an e–beam evaporation process in which weighted amounts of BSCCO pellets were deposited onto MgO (100) single crystals [3]. The films were superconducting after an ex-situ annealing in air at temperature higher than 875 °C for times longer than 1 hour. This simple method, because of the very few deposition parameters to control, allows a good reproducibility from a deposition to another.

2. SAMPLE REALIZATION

We realized the starting pellets by the usual sintering process[4–5]. Both Pb–doped and undoped pellets have been realized. In particular we have been able to fabricate bulk superconducting pellets with $T_c \sim 105$ K when doped with Pb. The $\Theta - 2\Theta$ X–Ray diffraction analysis performed on these pellets always showed the expected spectra related with the presence of the superconducting BSSCO phases[6]. During the deposition the pellets were situated in a water cooled copper crucible in which it was possible to deflect a 15 kV electron beam with adjustable focus and X–Y position. The substrates were not heated and we did not introduce any Oxygen in the vacuum chamber during the process. Small amounts of the pellets were always present in the crucible at the end of the deposition with a typical weight less than   5 % of the starting pellets. By correcting the initial stoichiometry, we successed in realizing, with good repeatibility, BSCCO superconducting thin films with atomic ratio very close to the desired one. The ex–situ annealing process was held in air using a muffle furnace. The annealing times and temperatures producing superconducting thin films were strictly related to the stoichiometry of the starting pellets. For films obtained by Pb–doped pellets the annealing was performed around 840 °C for 1–2 hours, while for films realized from undoped pellets it was done around 880 °C for 10–12 hours. After the annealing process the samples were cooled down to 500 °C in 2–3 hours.

3. SAMPLE CHARACTERIZATION

The structural properties of our films have been studied by means of EDS and $\Theta - 2\Theta$ X–Ray diffraction analysis. In the case of Pb doped films the stoichiometry of the starting pellets producing films with an atomic ratio very close to the 2:2:1:2 (Bi:Sr:Ca:Cu) phase ($T_c(R = 0) \sim 85$ K)[7] was $Pb_{0.96}Bi_{1.62}Sr_{1.87}Ca_2Cu_{4.45}O_x$. The EDS microprobe analysis showed slight disomogeneity on zones with radius of about 5 μm^2 while where larger areas were sampled the EDS measurements were highly homogeneous. The granularity of the films was also confirmed by scanning microscope pictures. The $\Theta - 2\Theta$ X–Ray analysis of these films showed peaks related with both the 2:2:1:2 and the 2:2:2:3 phase ($T_c(R = 0) \sim 110$ K)[6]. The X–Ray spectra also showed a slight preferential orientation of the grains with the c–axis perpendicular to the plane of the substrate. Typical transition curves for films annealed at 850°C show a large drop in resistance at temperatures around 110 K[5], clearly indicating the presence of the 2:2:2:3 phase in the film [8], but $T_c(R=0)$ is about 30 K. Because the overall stoichiometry of the film is very close to 2:2:1:2, this low T_c value can be related to the formation, with longer annealing times, of both the Copper rich 2:2:2:3 phase and the Copper poor 2:2:0:1

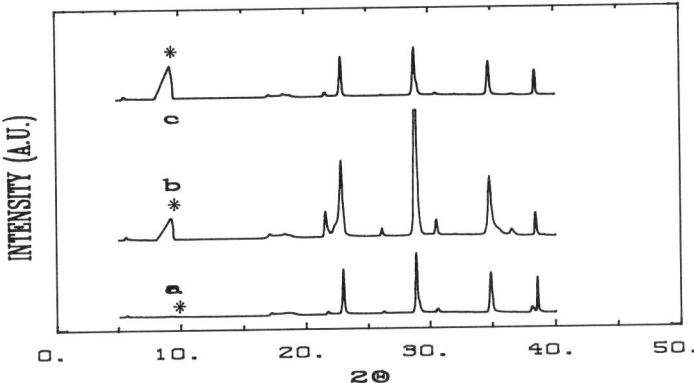

Figure 1. $\Theta - 2\Theta$ X–Ray spectra of films having an atomic ratio $Bi_{1.85}Sr_{1.88}Ca_{1.03}Cu_{2.22}O_x$ realized by the undoped pellet quoted in the text and annealed at 880 °C for a) 3 hours, b) 6 hours, c) 12 hours. The 2:2:2:3 peaks are labeled by (*).

($T_c \sim 10$ K) phase[9]. This hypothesis has been confirmed by $\Theta - 2\Theta$ X–Ray spectra showing peaks related to the three different phases[5]. When the annealing procedure was done at 835 °C for 1 hour superconducting thin films with $T_c(R = 0) = 72$ K have been obtained. In the $\Theta - 2\Theta$ X–Ray spectrum of this film the peaks related to the 2:2:0:1 phase were completely absent. However, even though the transition width is sharp and the $R(300)/R(100)$ ratio is close to 2, the $T_c(R=0)$ value is lower than the one expected for the pure 2:2:1:2 phase.

For undoped films we obtained the 2:2:1:2 stoichiometry starting from pellets with atomic ratio $Bi_{1.6}Sr_{1.8}Ca_{1.05}Cu_{2.4}O_x$. Large uniform areas with platelike structures, separated by fractures about 5 μm wide, were shown by scanning microscope pictures. EDS analysis confirmed the   scanning microscope pictures showed a larger homogeneity of these films pointing out that in the fractured zones the thickness of the film was largely reduced. The $\Theta - 2\Theta$ spectra show a strong preferential orientation with the c-axis perpendicular to the plane of the substrate. Peaks related to 2:2:1:2 and 2:2:0:1 or to the 2:2:1:2 and 2:2:2:3 phases were present in the spectra, depending on the final stoichiometry of the samples. The amplitude of the 2:2:0:1 and 2:2:2:3 peaks was strongly dependent on the annealing regime. For samples with a slight Ca rich stoichiometry an increase in $T_c(R = 0)$ was obtained by annealing the film at the same temperature for a shorter time. In this case the amplitude of the peaks related to the 2:2:2:3 phase decreases with the decreasing annealing time, as it is shown in figure 1[5]. The best results have been obtained with annealing time of 3 hours and showed a $T_c(R=0) = 80$ K. The density of the bias current was usually not more than $2 \cdot A/cm^2$.

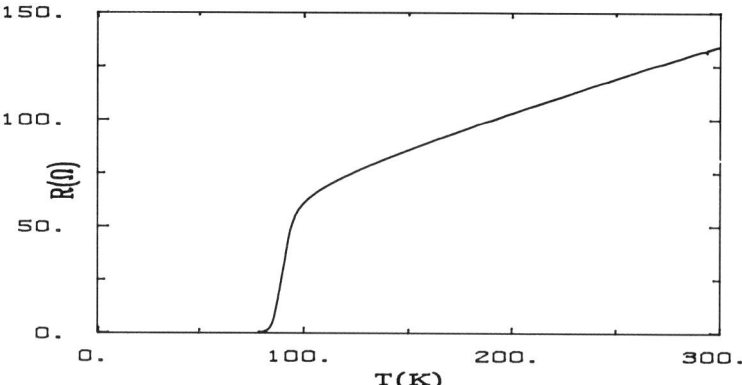

Figure 2. R versus T curve of a patterned film.

Our samples have also been characterized studying different superconducting properties. In particular surface resistance, paraconductivity and critical current measurements have been realized. The microwave properties of our sample have been tested by
measuring the quality factor of ring microstrip resonators. A dependence of the surface
resistance from the applied rf power has been observed[10]. Measurements of the excess
conductivity above the superconducting transition have also been performed and the
data have been well interpreted in terms of Aslamazov–Larkin theory of thermodynamic
fluctuactions for a two-dimensional superconductor[5].

In this paper much more attention is paid to critical current measurements. To
perform this kind of measurements we have designed a four probe geometry on our
films by using a photolithografic procedure in which we etched the BSCCO films in a
dilute solution (4%) of nitric acid. The width of the final strip was 50 μm.
In figure 2 is shown the R (electrical resistance) versus T (temperature) curve of a
patterned film evaporated from an undoped pellet and annealed at 880 °C for 12 h.
$T_c(R=0)$ is 78 K. The transport critical current density J_c was measured using a 5
μV/cm criterion. Typical values for the critical current density were about $5 \cdot 10^4$ A/cm^2
at the temperature of 4.2 K.
In figure 3a and 3b are respectively shown the J_c versus B and J_c versus T curves for
the same film in fig.2. In figure 3a the external magnetic field is applied perpendicular
to both the substrate plane and bias current flowing direction and the curve is taken
at 4.2 K. The solid lines in the figures are best fits to the experimental data obtained
using the formula:

$$J_c(B, T) = J_c(B, 0)(1 - \alpha t - \beta t^2)$$ (1)

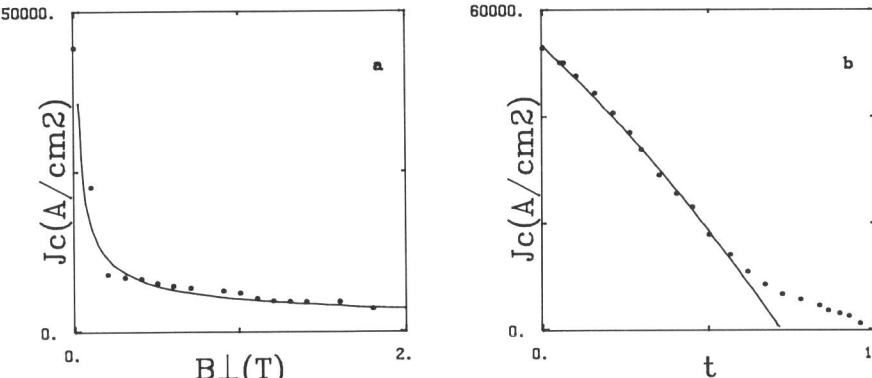

Figure 3. a) J_c versus B curve. The solid line is the best fit obtained using the equation (3) in the text. b) J_c versus T curve. The solid line is the best fit obtained using the equation (1) in the text.

with $t = (T/T_c)$, valid in the presence of flux creep related phenomena. For $T < 50$ K the best fit of the data in figure 3b is obtained for $\alpha = 1.1$ and $\beta = 0.41$. The flux creep theory in low magnetics field[11] gives for α the expression:

$$\alpha = \frac{k_B T_c}{U(B,0)} \ln\left(\frac{a_o B \Omega}{E_{min}}\right) \tag{2}$$

$U(B,0)$ is the $T = 0$ free energy difference between pinned and unpinned flux quanta, a_o is the average hopping distance of the flux quanta, Ω is the attempt frequency for escape and E_{min} is the electric field criterion defining J_c.

Assuming $a_o = 0.1$ μm, $\beta = 10$ Oe and $\Omega = 10^{10}$ Hz we calculated $U(B,0) \simeq 30$ meV which is in agreement with other values obtained on BSCCO, using different methods[12–13]. The flux creep theory also gives for $J_c(B,0)$ the expression:

$$J_c(B,0) = \frac{N_p U(B,0)}{1.07 \, (\phi_o B)^{1/2}} \tag{3}$$

where N_p is the density of the pinning sites. The solid line in figure 3a is obtained by equation (3). Inserting in this formula a constant value for $U(B,0)$ of about 30 meV we obtained from the experimental data in low magnetic fields a value of $N_p \sim 4 \cdot 10^{20} \mathrm{m}^{-3}$

4. REFERENCES

1   See for example "Science and Technology of Thin Films Superconductors", edited by R.D. McConnell and S.A. Wolf, Plenum Press, 1989.

2   L. Maritato and C.M. Falco, Modern Phys. Lett. B, 4 (1990), 639.

3   M.S. Osofsky, P. Lubitz, M.Z. Harford, A.K. Singh, S.B. Quadri, E.F. Skelton, W.T. Elam, R.J. Soulen Jr., W.L. Lechter, and S.A. Wolf, Appl. Phys. Lett., 53 (1988), 1663.

4   H. Maeda, Y. Tanaka, M. Fukutomi, and T. Asano, Jpn. J. Appl. Phys., 27 (1988), 209.

5   C. Attanasio, L. Maritato, A. Nigro, S. Prishepa, and R. Scafuro, submitted to Physica C.

6   M. Okhubo, Jpn. J. Appl. Phys., 27 (1988), 1271.

7   R.M. Hazen, C.T. Prewitt, R.J. Angel, N.L. Ross, L.W. Finger, C.G. Hadidiacos, D.R. Veblen, P.J. Heaney, P.H. Hor, R.L. Meng, Y.Y. Sun, Y.Q. Wang, Y.Y. Xue, Z.J. Huang, L. Gao, J. Bechtold, and C.W. Chu, Phys. Rev. Lett., 60 (1988), 1174.

8   K. Kuroda, M. Mukaida, M. Yamamoto, and S. Miyazawa, Jpn. J. Appl. Phys., 27 (1988), 625.

9   C. Michel, M. Hervieu, M.M. Borel, A. Grandin, F. Deslandes, J. Provost, and B. Raveau, Z. Phys. B, 68 (1987), 421.

10 A. Andreone, C. Attanasio, A. Di Chiara, L. Maritato, A. Nigro, G. Peluso, and R. Vaglio, to be published on Physica C.

11 M. Tinkham, Helv. Phys. Acta, 61 (1988), 443.

12 T.T.M. Palstra, B. Batlogg, L.F. Schneemeyer, and J.V. Waszczak, Phys. Rev. Lett., 61 (1988), 1662.

13 Y. Yeshurun, A.P. Malozemoff, T.K. Worthington, R.M. Yandrofski, L. Krusin-Elbaum, F.H. Holtzberg, T.R. Dinger, and G.V. Chandrashekhar, Cryogenics, 29 (1989), 258.

High T$_c$ Superconductor Thin Films
L. Correra (Editor)

Characterisation of YBa$_2$Cu$_3$O$_{7-\delta}$ films prepared by laser ablation and DC-sputtering

M. Brakmann, M. Schnell, R. Scholtes, F. Stellmach, S. Ewert

2. Physikalisches Institut, RWTH Aachen, D-5100 Aachen, FRG

W. Albrecht, A. Krings

Institut für Halbleitertechnik II, RWTH Aachen, D-5100 Aachen, FRG

J. Kissel, A. Comberg

Philips GmbH, Forschungslaboratorium Aachen, D-5100 Aachen, FRG

Abstract
Superconducting YBa$_2$Cu$_3$O$_{7-\delta}$ films were prepared by pulsed laser ablation and DC-sputtering on random oriented ZrO$_2$ and single crystal SrTiO$_3$ substrates. X-ray diffraction pattern show the films to be highly oriented with the c-axis perpendicular to the substrate surface. We obtained values for the critical current density j$_c$ in films on SrTiO$_3$ of 5×10^6 A/cm^2 at 77 K. AC-susceptibility measurements give hints that preferably microbridge-type weak links occur in our films. The appearance of particulates in the films is discussed.

1. INTRODUCTION

Since the discovery of high-T$_c$ superconductivity many methods have been used to prepare superconducting thin films [1-7]. Beside economical aspects some general features, such as critical current density or homogeneity of films on different substrates are of considerable interest for technical applications. Among various deposition techniques available for superconducting thin film fabrication, laser ablation and DC-sputtering of a bulk stoichiometric target are two of the simplest and most promising techniques. The characterisation of films prepared by these two different in-situ processes with the same methods is useful to investigate and compare their physical properties.

2. EXPERIMENTAL

For the deposition of thin films by laser ablation two different types of excimer laser were used: An ArF laser (EMG 201MSC, Lambda Physik) with a wavelength $\lambda = 193$ nm, maximum pulse energy of 150 mJ and a repetition rate of 10 Hz and a XeCl laser (XP2020, Siemens/KWU) with $\lambda = 308$ nm, pulse energy 2 J (200 mJ behind an aperture) and repetition rate of 5 Hz. The laser beam was focused on a rotating sintered YBa$_2$Cu$_3$O$_{7-\delta}$ pellet in a vacuum chamber with a base pressure of 10^{-5} mbar. The spot sizes of the laser beams on the target were about $8 \, mm^2$ (EMG 201MSC) and $3 \times 3 \, mm^2$ (XP2020). The O_2 pressure during deposition was about 0.3 mbar. The substrates were glued with silver epoxy to a heater which was situated opposite the target at distances between 30 and 60 mm. The temperature of the substrate was determined with a thermocouple. At the end of the deposition process the chamber was filled with O_2 up to 1 bar, whereas the substrate temperature was reduced to 550 °C for 30 minutes.

The sputtered films were prepared in a planar DC-sputtering configuration with oxygen pressures of 3-5 mbar. Sputtering was performed with a constant current power supply using currents of about 150-180 mA at 200-300 V. The target was a superconducting $YBa_2Cu_3O_{7-\delta}$ disk with a diameter of 28 mm and a thickness of 3 mm. A typical target to substrate distance was 15 mm. The substrate temperatures were about 740-780 °C, although some superconducting films ($T_c^{onset} = 88 \, K$, transition width 10 K) on ZrO_2 could be prepared at 680 °C. The substrate temperature was measured with a thermocouple, too. There is a significant temperature difference (up to 200 °C) between heater and substrate. This difference depends mainly on pressure but changes with sputtering current and target-substrate distance, too, which should be considered when comparing films prepared under different conditions. Oxygen was pumped with LN_2 cooled sorption pumps. The gas flow was about 6 l·mbar/s. The pressure could be kept constant within $50 \, \mu bar$. After deposition the chamber was vented and films were cooled down to room temperature within 5 min .

With both techniques $YBa_2Cu_3O_{7-\delta}$ films were deposited in-situ on single crystal (100) $SrTiO_3$ and random oriented ZrO_2 substrates. The thickness of the films ranged from 100 - 1000 nm (laser ablation) and from 60 - 400 nm (DC-sputtering). X-ray diffraction pattern have been obtained with a Debye-Scherrer diffractometer using a Cu K_α - source. AC-susceptibility measurements were made with a Hartshorn bridge. Critical current density measurements were performed on microstructured films by pulse current method.

3. RESULTS AND DISCUSSION

A study of the surface morphology of the target with SEM shows different surface structures generated by the two lasers. Behind an aperture the beam of the XP2020 laser is very homogeneous and yields a constant roughness of the target surface over the whole spot area, whereas the varying intensity of the EMG 201MSC laser causes areas with different forms of troughs and cones on the target surface. Power densities lower than 40 MW/cm^2 or higher than 100 MW/cm^2 diminish the quality of deposited films, because a non-stoichiometric ablation of the target sets in. Other groups obtained similar results [8,9].

Figures 1 and 2 show the X-ray diffraction patterns of a laser-ablated and a sputtered $YBa_2Cu_3O_{7-\delta}$ film on $SrTiO_3$ substrates. The spectra show the high orientation of the films perpendicular to the c-axis. A weak (200) diffraction peak of the laser-ablated film indicates that small parts of the film have grown in a-axis orientation.

AC-susceptibility measurements are useful for determining superconducting phase transition temperature and detecting different superconducting phases. The frequency applied was 195 Hz and the exciting field amplitude ranged from 5.3 to 1590 μT. Figures 3-5 show the imaginary part of magnetic susceptibility χ" of films deposited by laser-ablation

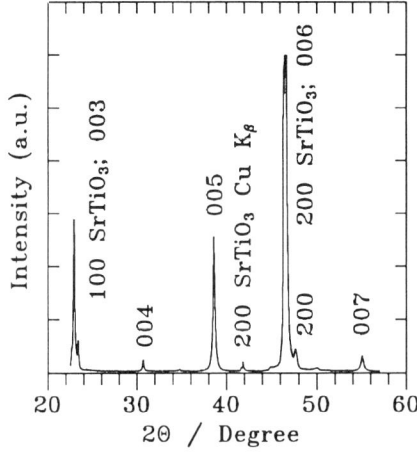

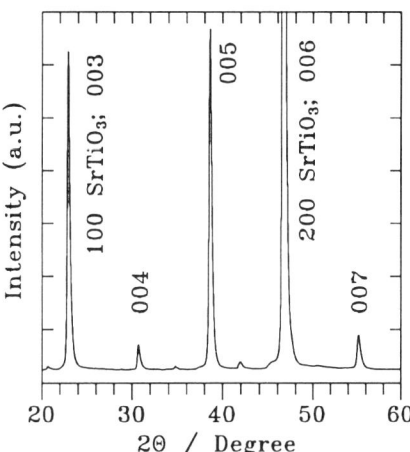

Figure 1. X-ray pattern of a laser-ablated film on a SrTiO₃ substrate (film R5).

Figure 2. X-ray pattern of a DC-sputtered film on a SrTiO₃ substrate (film M51).

on SrTiO₃ and ZrO₂ and by DC-sputtering on SrTiO₃ respectively. The curves are normalized and shifted.

We obtained values for $T_c^{midpoint}$ of about 85 K (laser ablation) and 88 K (DC-sputtering). Resistive measurements always showed $T_c^{midpoint}$ to be a few degrees higher. The highest values for the superconducting phase transition temperature we obtained for films on SrTiO₃ substrates.

The occurence of only one peak in each curve indicates that the superconducting properties are homogeneous over the whole films. For all measured amplitudes of the exciting field the imaginary part $\chi"$ shows a peak, whose maximum is located approximately at the point of inflection of the real part χ'. The $\chi"$ peak becomes broad with increasing field amplitude, while the superconductivity onset temperature T_c^{onset} is independent of field amplitude. The width of the curves depends on the critical current density [10]. This behaviour of the complex AC susceptibility has been observed on YBa₂Cu₃O₇₋δ bulk samples and films before [11,12] and can be understood in terms of a model by Ishida and Mazaki [13]. The authors pointed out that depending on the type of junction, either Josephson tunnel or microbridge-type weak links, a symmetric or asymmetric curve of $\chi"(T)$ and $\chi'(T)$ with respect to its midpoint is expected. From this theory we deduce that in these YBa₂Cu₃O₇₋δ films mainly microbridge-type weak links occur.

The $\chi"$ peaks of the films on ZrO₂ are broader compared to those on SrTiO₃ corresponding to a lower critical current density j_c on ZrO₂ . In figures 6-8 the critical current densities j_c for films deposited by laser ablation on SrTiO₃ and ZrO₂ and sputtering on SrTiO₃ are shown respectively. The absolute value and the temperature dependence of the critical current are affected by the homogeneity of the films, the presence of grains, and the nature of contacts between the grains. We obtained values for j_c in films on SrTiO₃ of about 5×10^6 A/cm² at 77 K, whereas in films on ZrO₂ j_c did not exceed 3×10^4 A/cm² at this temperature.

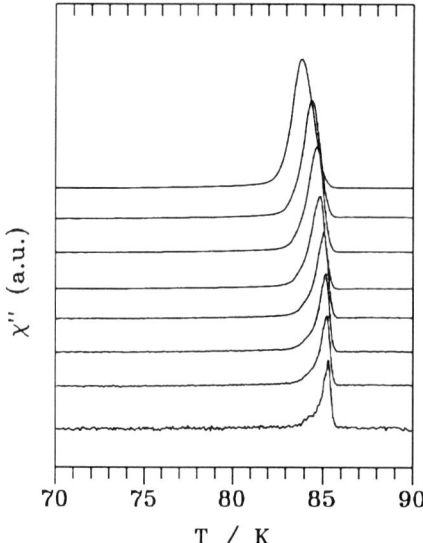

Figure 3. Imaginary part χ'' of magnetic susceptibility of a laser ablated film on SrTiO$_3$ versus temperature (film R5), exciting field amplitudes were 5.3, 15.9, 26.5, 53, 159, 265, 530 and 1590 μT.

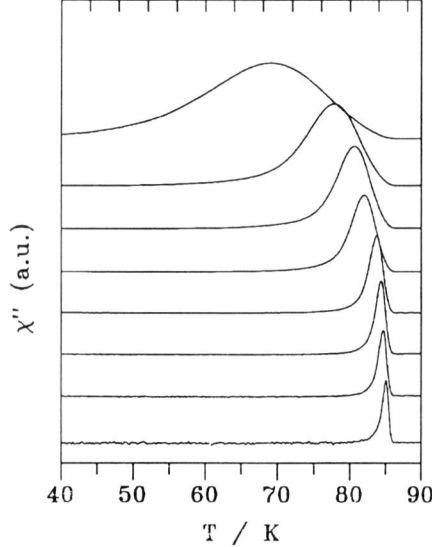

Figure 4. Temperature dependence of χ'' of a laser-ablated film on ZrO$_2$ (film R2), exciting field amplitudes were the same as in figure 3.

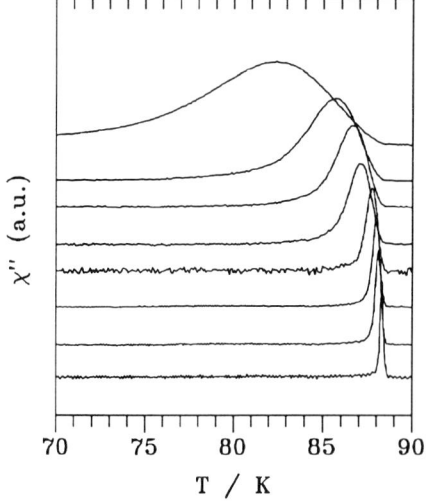

Figure 5. χ'' versus temperature for a DC-sputtered film on SrTiO$_3$ (film M51), exciting field amplitudes were the same as in figure 3.

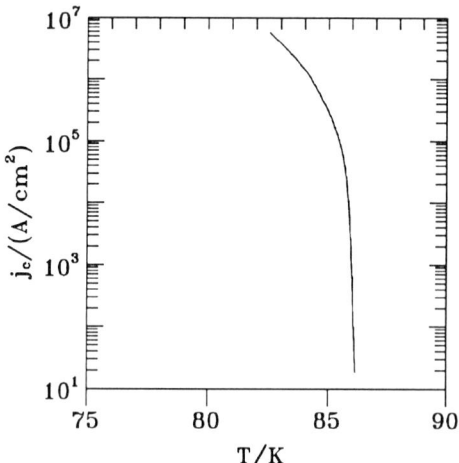

Figure 6. Critical current density j_c of a laser-ablated film on SrTiO$_3$ versus temperature (film R102).

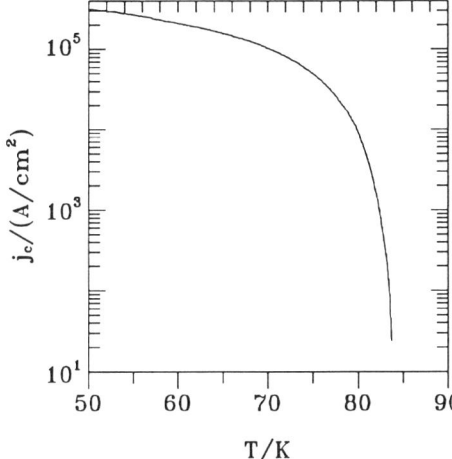
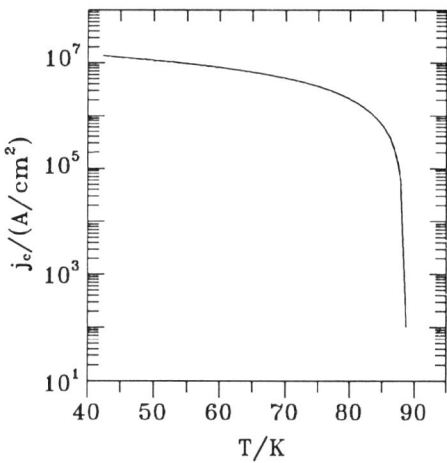

Figure 7. j$_c$ as function of temperature of a laser-ablated film on ZrO$_2$ (film R107).

Figure 8. Temperature dependence of j$_c$ of a DC-sputtered film on SrTiO$_3$ (film M51).

A fundamental problem is the appearance of particulates in the films. These particulates with diameters less than 1 μm restrain the use of films in technology. Fig. 9 shows a scanning electron micrograph of a film deposited by laser on a SrTiO$_3$ substrate. Particulates were found in larger or smaller numbers in all laser-ablated films. Compared to films on ZrO$_2$ substrates films on SrTiO$_3$ substrates show a diminished number of particulates. We obtained a further reduction of number of particulates with decreasing film thickness, which should be lower than 300 nm for good films. A photo made with an optical microscope (fig. 10) shows that the particulates appear preferably in scratches on the substrate. All this indicates that the particulates accumulate during the film deposition by laser ablation. In figure 11 a scanning electron micrograph of a sputtered film on SrTiO$_3$ is shown. The form of the particulates is different from those in laser-ablated films and the interface is smoother between them.

Acknowledgement

We wish to thank Prof. G. Herziger and Dr. E. Kreutz for placing the XP2020 laser at our disposal.

References

[1]   J. Schubert, U. Poppe, W. Sybertz, J. Less. Comm. Met. 151 (1989) 277

[2]   M. Hong, S.H. Liou, J. Kwo, B.A. Davidson, Appl. Phys. Lett. 51 (1987) 694

[3]   B. Roas, L. Schultz, G. Endres, Appl. Phys. Lett. 53 (1988) 1557

[4]   J. Fröhlingsdorf, W. Zander, B. Stritzker, Solid State Commun. 67 (1988) 965

[5]   C. Webb, S.L. Weng, I.N. Eckstein, N. Missert, K. Char, D.G. Schlom, E. Hellmann, M.R. Beasley, A. Kapitulnik, J.S. Harris, Appl. Phys. Lett. 51 (1987) 1191

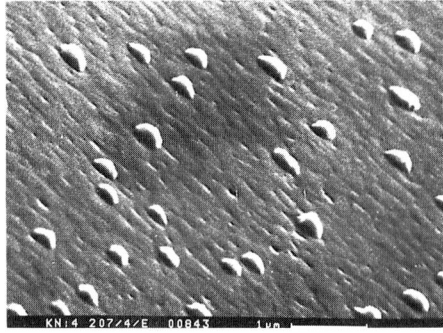

Figure 9. Scanning electron micrograph of a laser deposited film on SrTiO₃ (film R157) [——— 1 μm] .

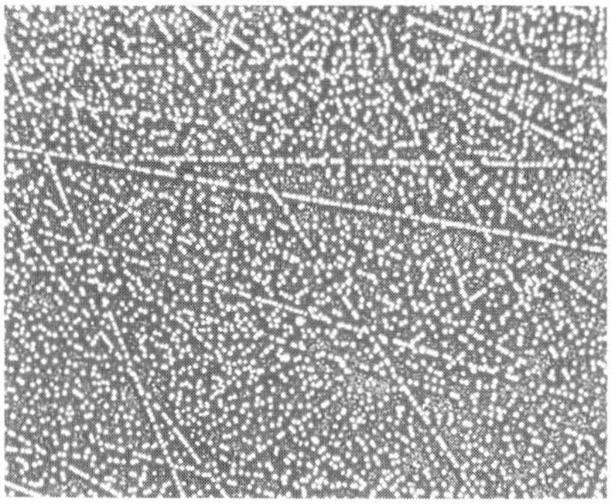

Figure 10. Photo of a laser-ablated film on SrTiO₃ (film R161) made with an optical microscope [——— 30 μm].

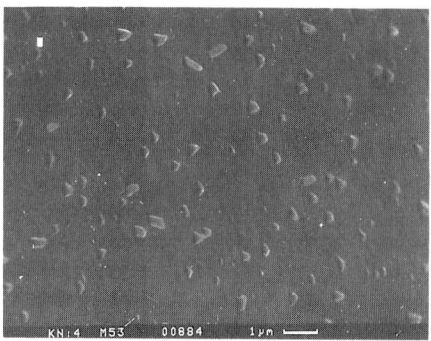

Figure 11. Scanning electron micrograph of a sputtered film on SrTiO₃ (film M53) [── 1 μm].

[6]   J. Tate, P. Berberich, W. Dietschke, H. Kinder, J. Less Comm. Met. 151 (1989) 311
[7]   H. Yamane, T. Hirai, H. Iwasaki, K. Watanabe, N. Kobayashi, Y. Muto, Appl. Phys. Lett. 53 (1988) 1548
[8]   W. Kautek, B. Roas, L. Schultz, Thin Film Solids 191 (1990) 317
[9]   U. Sudarsan, N.W. Cody, M.J. Bozack, R. Solanki, J. Mater. Res. 3 (1988) 825
[10]   T. Ishida, R.B. Goldfarb, Phys. Rev. B 41 (1990) 8937
[11]   J. Garcia, C. Rillo, F. Lera, J. Bartolomé, R. Navarro, D.H.A. Blank, J. Flokstra, J. Magn. Magn. Mat. 69 (1987) L225
[12]   J.W.C. de Vries, G.M. Stollman, M.A.M. Gijs, Physica C 157 (1989) 406
[13]   T. Ishida, H. Mazaki, J Appl. Phys. 52 (1981) 6798

High T$_c$ Superconductor Thin Films
L. Correra (Editor)

Growth and properties of YBCO thin films grown on MgO substrates

D. Chambonnet, C. Fages, D. Sinobad, C. Belouet[a]

A. Cheenne, J. Perrière[b]

C. Vignolle, A. Gervais[c]

[a] Alcatel Alsthom Recherche, Route de Nozay, 91460 Marcoussis, France

[b] G.P.S., Univ. Paris VI et VII, Tour 23, 2 Place Jussieu, 75251 Paris Cedex 05

[c] Laboratoire de Minéralogie Cristallographie, Université Paris VI, Tour 16, 4 Place Jussieu, 75251 Paris Cedex 05

Summary

YBa$_2$Cu$_3$O$_{7-x}$ superconducting thin films were prepared using the Pulsed Laser Deposition (PLD) technique. The films were formed in situ on MgO (001) single crystalline substrates at a temperature of 750 °C in a 0.1 mbar oxygen pressure. An XeCl laser was used for the ablation of the YBa$_2$Cu$_3$O$_{7-x}$ stoichiometric target.

These films have a characteristic granular texture with a large fraction of the grains having their **a** or **b** axis preferentially aligned with the [100] and [110] directions of the substrate and their **c** axis perpendicular to the substrate surface. Best films show a sharp superconducting transition at a temperature, Tc (R = 0), up to 92 K with a width below 1 K and critical current densities, Jc (77 K, 0T), up to 10^6 A/cm^2.

INTRODUCTION

Numerous papers to date have shown the potential applicability of thin films of oxide based high temperature superconductors to passive microwave devices. Thus, studies of patterned resonators have already demonstrated that high quality YBa$_2$Cu$_3$O$_{7-x}$ (YBCO) films can have surface resistances significantly lower than those of normal metals as gold and copper currently used in microave applications [1].

Despite these encouraging results, the microwave performances of the films is ₅till strongly influenced by important factors of technical relevance for applications such as the film thickness and surface area and the nature of the substrate. The general approach followed by material scientists in order to meet film quality requirements consists in the development of in-situ, low temperature growth protocols on pertinent substrates selected for their microwave characteristics.

Among the various preparation routes, the Pulsed Laser Deposition (PLD) technique has received considerable attention and substrates as MgO, LaAlO$_3$ and more recently Al$_2$O$_3$ with a buffer layer have been widely used [2 to 4]. The research presented here falls into the latter approach ; it is reported on the results of a study on YBCO films grown at 750 °C on MgO substrates by the PLD technique. The paper describes in succession the main technological aspects of film growth and patterning and the characteristics of the films, namely : texture, chemical composition and superconducting properties (Tc and Jc (T)).

EXPERIMENTAL

Thin films were deposited on (001) MgO substrates using the Pulsed Laser Deposition (PLD) technique as described by Venkatesan et al. [5]. In this study, a SOPRA type 510 Excimer laser (λ = 308 nm) was employed to evaporate a high density (90 to 95 %) YBa$_2$Cu$_3$O$_{x-7}$ target ; this laser delivers a pulse energy of typically 80 mJ at a repetition rate of 1 to 2 Hz (the pulse duration is about 30 ns). A laser fluence of 2-3 J/cm^2 was used to transfer the target material onto the substrates placed at the vertical of the target at a distance of 5 to 7 cm. The laser beam, focused on the target surface by means of an optical lens at a 45° impingement angle, was rastered along a segment across the target surface by means of an external mirror ; the target itself was rotated about its vertical axis so that the ablation protocol could be monitored for each size of the laser impact by adjusting the target rotation rate and the laser beam scanning length and rate. Computer simulation was used to optimize the ablation protocol on the basis of a minimal overlapping of laser impacts during a growth run.

The MgO substrates used, 1 cm x 1 cm in size, were mechano-chemically polished and cleaned in isopropanol vapour prior to deposition. They were mounted on a stainless steel holder and pressed on the Inconel bed plate of a resistance furnace. A conducting foil placed between the substrate backside and the furnace bed plate yielded excellent control of the temperature of the substrate deposition surface, estimated to be in a 5 to 15 °C range below that recorded by a thermocouple tightly bound to the bed plate.

The deposition protocol was started as the residual pressure in the growth chamber reached 10^{-7} mbar. During the heat-up and deposition steps, the ambient gas was pure oxygen at a 10^{-1} mbar pressure, the oxygen gas being injected through a nozzle directed towards the substrates.

Current runs were made at a pulse frequency of 2 Hz, which yielded an averaged deposition rate of around 0.1 nm/s. As deposition was interrupted, the cooling sequence was started at a rate of 10 °C/min and the oxygen pressure was set at 300 mbar. Films thus made were not submitted to any ex-situ post-annealing.

The chemical composition of the films was measured by means of Induced Coupled Plasma (ICP) and Rutherford Back Scattering (RBS) analysis for cations ; the oxygen content was estimated on the basis of c-lattice parameter measurements [6] and of Raman Spectra [7].The texture, the microstructure and the orientation relationship of the films were determined from Transmission Electron Microscopy (TEM) investigations mainly.

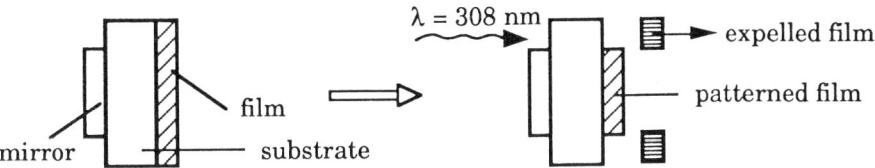

Figure 1. Sketch of the laser patterning technique.

The superconducting transition temperature, Tc (R = 0), was obtained from R (T) curves using the classical four-probe technique. These measurements were made either on the whole sample or on patterned channels also used for the measurement of the critical current density, Jc. These channels were made with the laser patterning technique shown in fig. 1. This technique allows the preparation of superconducting patterns which are the one-to-one replica of the metallic mirror inserted on the path of a high power laser beam ; the laser-beam expels the irradiated part of the film whereas its complement, in the shadow of the mirror, is untouched and left with a virgin surface. The laser used in this task was the Excimer laser described in the above. Jc measurements were made on bridges 200 μm x 50 μm up to 4000 μm x 40 μm in size, the latter bridges being aimed at testing the long-distance homogeneity of the films.

RESULTS

Morphology of the layers

X-ray diffraction patterns of as grown films deposited on MgO substrates at 750 °C show (00l) and (0k0) diffraction peaks with relatively large and small intensities respectively, thus indicating a preferred orientation of the grains with their **c** axis perpendicular to the substrate surface.

However, it was found that the value of Tc (R = 0) of these films decreased as their thickness increased for otherwise identical preparation conditions, with the exception of the deposition time. Thus, Tc (R = 0) decreased from 85-83 K to 75 K on a series of films as their thickness increased from 100 to 300 nm.These films were carefully studied by TEM in order to investigate on their microscopic texture (TEM studies of more recent films 150 nm thick with Tc (R = 0) at 92 K are in progress). The overall results of this first TEM study, detailed in refs. 8, 9, are briefly described in the following.

Films about 100 nm thick essentially consist of a mosaic of two families of grains typically 500 nm in edge having their **c** axis perpendicular to the substrate surface and their **a** (or **b**) axis along the [100] and [110] axes of MgO respectively, fig. 2. As the film thickness increases, competing grain orientations with the **c** axis parallel to the [100] or [010] directions of the substrate become increasingly important. Needle shaped grains, typically one tenth of a micrometer in width and several micrometers in length nucleate on top of the two above families, although a fraction of them do nucleate at the substrate surface. The needles build up a network of grains at right angles, which eventually completely isolate islands of the regular mosaic described above (fig. 3). The surface area covered by needles grows to above 20 % of the film surface as the thickness reaches 300 nm.

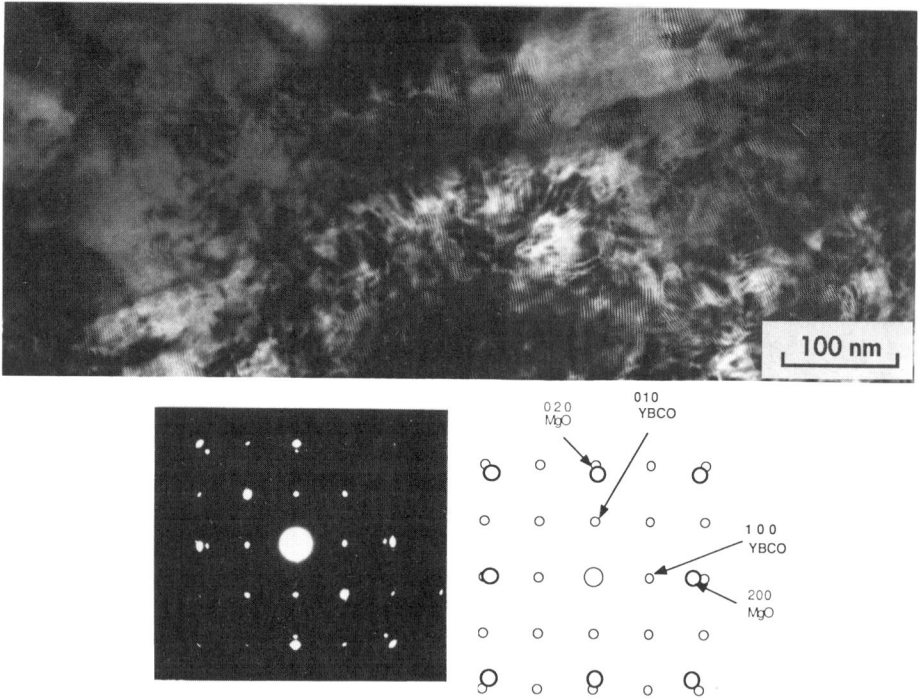

Figure 2. TEM plan view. Moiré fringes testify to the homogeneity of the layer and to the orientation relationship : [100] YBCO // [100] MgO.

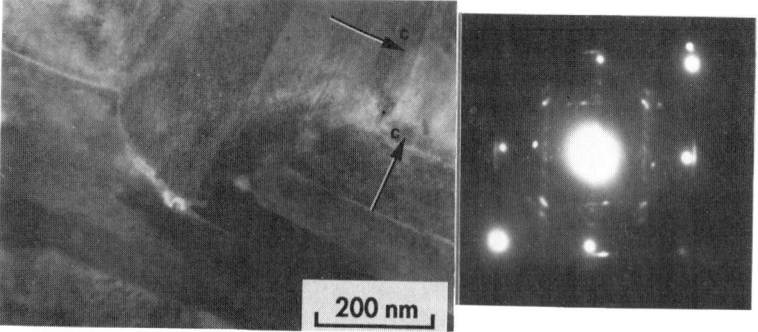

Figure 3. TEM plan view. Note two needles (**c** //) at right angles.

In one sample 200 nm thick, square shaped grains 100 nm in edge and identified as YBCO polytypes, were present within the regular mosaic structure.

As the thickness of the films is increased, more complicated structures develop together with a high density of twins. Grains with peculiar orientations of their **a** (or **b**) axis about the [100] direction of the (001) MgO substrate and similar to those reported in other works have been observed [10].

Finally, electron diffraction patterns reveal the occurrence of Pmmm and Ammm structures in all the samples regardless of their thickness.

The origin of the complicated textures observed has not been unambiguously determined. The deposition temperature (not discussed here), the preparation of the substrate surface and the oxygen partial pressure are certainly important factors which influence the film texture. However, studies in progress strongly suggest that minor deviations from the perfect cationic stoichiometric composition may have a dramatic impact on the film texture and its superconducting properties as well.

Chemical composition of the layers

The angular distribution of cationic species in the plume was studied in a series of dedicated experiments where PLD was made at room temperature on silicon wafers. The target and silicon wafer were mounted parallel, coaxial and 6 cm apart. Four identical runs were conducted and the films thus deposited were analyzed by both RBS and ICP techniques. It was first found that largest compostion variations occurred along the substrate diameter in the incidence plane (Y direction). The overall results of the analysis in the Y direction are given in fig. 4. The accuracy on the Ba/Y and Cu/Y ratios is estimated to be better than 4 %. The asymmetry of composition, which appears in fig. 4, may be due to the inclination of the laser plume towards the laser beam in the incidence plane, at the time when this study was conducted. These experiments clearly show that the 1:2:3 composition, within our accuracy, is obtained in a narrow region in the Y direction (ca. 1 cm for the operating conditions). This phenomenon presently limits the formation of stoichiometric films of large surface area. It was also found that the target wear did influence the superconducting properties (T_c (R = 0)), probably due to an evolution of the cationic composition with the impingement number.

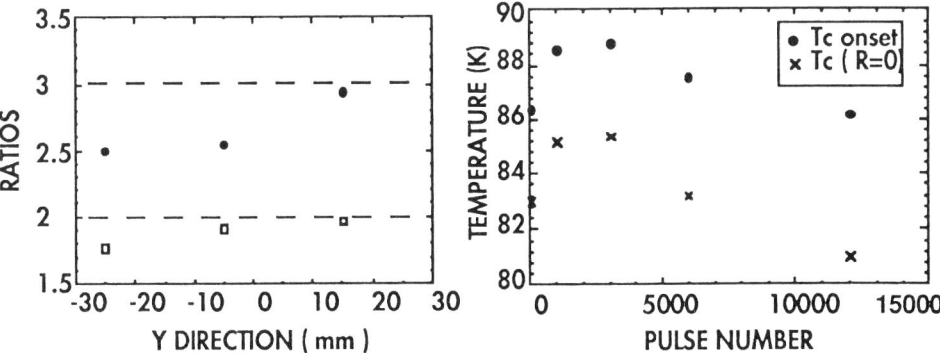

Figure 4. Ba/Y (□) and Cu/Y (•) vs. Y direction.

Figure 5. Target wear. Evolution of T_c vs. target ablation (pulse number) prior to deposition.

The best operating conditions require the use of freshly polished targets submitted to a moderate ablation prior to film deposition, fig. 5.

Electrical measurements

$T_c(R = 0)$ (noted T_c) values measured on 9 x 9 mm^2 films were repeatedly obtained above 90 K; best results were T_c = 92 K over a 2 cm^2 surface area (fig. 6). R(T) measurements made on 40 x 4000 µm^2 bridges patterned by the laser technique in fig. 1 yielded T_c values up to 90 K. J_c measurements made on small (40 x 200 µm^2) bridges yielded J_c values of 10^6 A/cm^2 (77 K, 0 T, fig. 7); values of 5.10^5 A/cm^2 are currently obtained on large (40 x 4000 µm^2) bridges. However, it has been observed that imperfections such as substrate defects - scratches, inclusions - or larger defects, as needle networks cited above - resulted in large drops of J_c values. The overall results are reported in table 1. Surface impedance measurements at the microwave frequencies have recently been initiated ; early results are given in ref. 11.

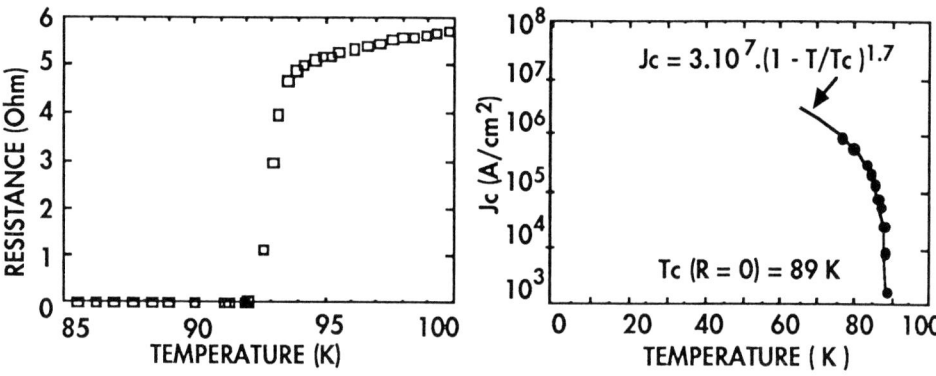

Figure 6. R(T) curve typical of best films, $T_c(R = 0)$ = 92 K.

Figure 7. J_c(T) curve typical of the best 40 x 200 µm^2 bridges.

Table 1
Typical characteristics of the best YBCO films grown on MgO substrates at 750°C

Film	Overall	40 x 200 µm^2	40 x 4000 µm^2
T_c (R = 0) (K)	92	89	90
J_c (77 K, 0T) (A/cm^2)	-	10^6	5.10^5
$\Delta T_{10\% - 90\%}$ (K)	0.6	1	2
Field criterion	-	5 µV/cm	0.25 µV/cm

Concluding remarks

The superconducting properties, T_c and J_c mainly, of YBCO films grown on MgO substrates have been continuously improved so that to date they approach those of films grown on $SrTiO_3$ and they permit the fabrication of prototype microwave passive devices. The quality of the films is most probably adversely affected by the occurrence of undesirable textures with **c**// grains and the family of **c** $\perp$ grains where the **a** (or **b**) axis of YBCO are not aligned with the [100] direction of MgO. The latter family has been associated with large drops in $J_c(T)$ values [12] ; it may partly account for the scatter of J_c values across the films and the generally observed degradation of J_c as the bridge length is increased up to the millimeter dimension. Further progress in film quality appears to be bound, among other things, to the control of i) the film texture and ii) the film composition in large surface areas.

Acknowledgments

The authors whish to thank contributing members of Alcatel Alsthom Recherche, in particular the Analysis Group of J-Y. Barraud for ICP and XRD studies, A. Wicker for target fabrication and S. Prévot for J_c measurements.

REFERENCES

1　H. Piel and G. Müller, IEEE Trans. Mag 27-2 (1991), 854.

2　T. Nishino, H. Nakane, Y. Tarutani, M. Hirano, T. Aida, S. Kominami and U. Kawabe, Jpn. J. Appl. Phys. 26-8 (1987), L1320.

3　R.W. Simon, C.E. Patt, A.E. Lee, G.S. Lee, K.P. Daly, M.S. Wire and J.A. Luine, Appl. Phys. Lett. 53-26 (1988), 2677.

4　J.C. Villegier, H. Moriceau, H. Boucher, R. Chicault, J. Di Cioccio, A. Jager, M. Schwerdtfeger, M. Vabre and C. Villard, IEEE Trans. Mag, 27-2 (1991), 1552.

5　D. Dijkkamp, T. Venkatesan, X.D. Wu, S.A. Shaheen, N. Jisrawi, Y.H. Min-Lee, W.C. Mc Lean and M. Croft, Appl. Phys. Lett. 51 (1987), 619.

6　C.B. Eom, J.Z. Sun, K. Yamamoto, A.F. Marshall, K.E. Luther, T.H. Geballe and S.S. Lederman, Appl. Phys. Lett. 55-6 (1989), 595.

7　G.A. Kourouklis, A. Jayaraman, B. Batlogg, R.J. Cava, M. Stavola, D.M. Krol, E.A. Ritman and L.F. Scheneemeyer, Phys. Rev. B-36 (1987), 8320.

8　C. Vignolle, A. Gervais, D. Chambonnet and C. Belouet, Mat. Res. Bull 26 (1991), 171.

9　C. Vignolle and A. Gervais, paper sub. for publ. to Phys. Status Sol.

10　T.S. Ravi, D.M. Hwang, R. Ramish, Siu Wai Chan, L. Nazar, C.Y. Chen, A. Inam and T. Venkatesan, Phys. Rev. B-42-16 (1990), 42.

11　M. Pyée, P. Meisse, M. Chaubet, D. Chambonnet, D. Sinobad, this conference.

12　D.K. Lathrop, B.H. Moeckly, S.E. Russek and R.A. Buhrman, Appl. Phys. Lett. 58-10 (1991), 1095.

High T$_c$ Superconductor Thin Films
L. Correra (Editor)
© 1992 Elsevier Science Publishers B.V. All rights reserved.

YBa$_2$Cu$_3$O$_{7-\delta}$ Thin films grown in-situ by laser ablation

N.Y.Chen, K. van Dijk, L.W.Lander, H.M.Appelboom, D. van der Marel, P.Hadley
and J.E.Mooij

Delft University of Technology. P.O.Box 5046, 2600GA Delft,
The Netherlands.

Abstract

Superconducting thin films of Y$_1$Ba$_2$Cu$_3$O$_{7-\delta}$ have been prepared in-situ on MgO and SrTiO$_3$ <001> substrates by laser ablation, using a XeCl excimer laser. The influences of substrate temperature from 700 °C to 770 °C during ablation and the method of cooling, immediately after the ablation, to room temperature have been studied. We have found that the films prepared at higher substrate temperature and cooled down during a longer period, up to 1 hour, have higher T$_c$ (92 K) and a sharper transition (2 K), indicating a better homogeneity in the films. X-ray diffraction and Fourier Transform Infrared measurements showed that the films prepared at substrate temperatures higher than 720°C are highly aligned, with c-axis oriented perpendicular to the substrate surface. The thickness of our films is uniform in an area with a diameter as large as 5.4 mm.

1. Introduction

The ability to make reproducible high quality high-T$_c$ superconducting films determines their future applications in industry. Furthermore, the ability to control the characteristics of the films is of fundamental importance for the research of the physics of their superconducting behavior. Among a number of successful techniques employed to date, laser ablation has come out to be the most popular one due to its unique advantages. This technique is relatively inexpensive, produces films of reproducible characteristics easily and has excellent control over its stoichiometry. Because a relatively low substrate temperature (<800 °C) can be used for this process, problems of diffusion between a number of substrates and the film can be largely avoided.

In the first stage of our recently started laser ablation project, our attention is mainly focused on the influences of the substrate temperature during the ablation process and the duration of cooling in an oxygen environment on the properties of our films.

2. Experimental details

The target of YBa$_2$Cu$_3$O$_{7-\delta}$ (YBCO) for laser ablation was prepared by citrate synthesis and pyrolysis method [1], and pressed into a pellet of 2.5 mm thick. It has a diameter of 15 mm. X-ray diffraction measurement showed that this target consisted of single-phase orthorhombic material. T$_c$ of this target is 93 K with 3 K transition width. It was glued onto a rotatable stainless steel plate mounted into the ablation chamber and irradiated during the deposition with a focused XeCl laser beam (308 nm) at about 45° to its surface plane. The spot size of the laser on the target surface is about 2.5x0.5 mm^2. The pulsed laser was fired at a repetition rate of 5 Hz. The energy of each pulse was 125 mJ.

Single-crystalline <001> SrTiO$_3$ and MgO of 10x10x1mm^3 were placed at a distance of about 30 mm from the target surface, with its center right opposite to the center of the laser spot. They were glued onto a 2.5 mm thick stainless steel plate with silver paint, and were heated by a quartz lamp placed on the other side of this plate. The highest temperature attainable at the substrate surface was 800 °C.

The films described in this paper were all made at an oxygen pressure of 1 mbar with 4×10^{-4} mbar base pressure. It should be mentioned that films were also made with base pressure lower than 10^{-6} mbar recently, with no improvement in its superconducting properties.

After ablating at substrate temperatures ranging from 700 to 770 °C, the ablation chamber was vented with oxygen and the heater switched off. Both the sequence of the last two steps and the duration of cooling turn out to affect the superconducting properties of our films significantly. The resistance as a function of temperature was determined by a four-probe method in a flowcryostat. During this measurement the temperature in the flowcryostat was changed slowly, typically 1 K every 3 minutes. After etching part of the film away by wet-etching, the thickness of our film was measured with an Alfa-step.

3. Results and discussion

As mentioned earlier in this paper, the sequence of venting the ablation chamber with oxygen and cooling after the ablation has great influence on the superconducting properties of our films. Apart from adopting different cooling procedures, the films shown in figure 1a and 1b were all made under the same conditions on MgO substrates at substrate temperature of 730 °C. Their thicknesses were 250 nm.

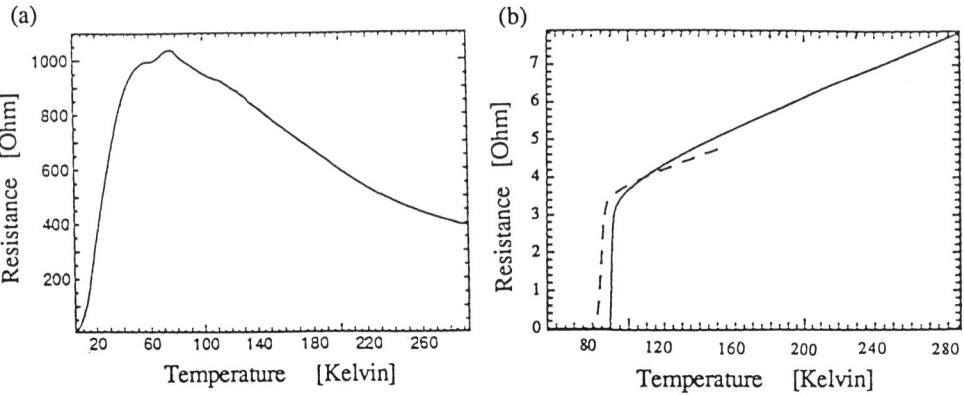

Figure 1. Resistance-temperature characteristics of $YBa_2Cu_3O_{7-\delta}$ films cooled after the ablation a) by switching off the heater first and venting the ablation chamber afterwards, and b) by venting the chamber first to 1 atmosphere and then starting to cool the films respectively in 60 minutes (——) and in 30 minutes (— — —).

The film in figure 1a is cooled by switching off the heater immediately after the ablation and starting to vent the ablation chamber about 30 seconds afterward. After the heater was switched off, the substrate temperature dropped sharply to around 620 °C in 30 seconds. By the time that the chamber was vented with oxygen to 1 atmosphere, the substrate temperature was already lower than 500 °C. It has a high resistance at room temperature and shows initially semiconducting behavior upon cooling, although it finally becomes superconducting at a very low temperature. In contrast, the films in figure 1b were made by first venting the chamber with oxygen to 1 atmosphere and turning off the heater afterwards. The film represented by a broken line in this figure was cooled in

30 minutes to 60 °C before being taken out of the ablation chamber, whereas the one represented by a solid line was cooled in 60 minutes. Their resistances at room temperature were a factor of 50 lower than that shown in figure 1a. and show metallic behavior upon cooling to their respective transision temperatures. One can see quite obviously that the film cooled during a longer period in oxygen has a higher T_c of 90 K and a narrower transition range of about 3 K.

We studied the influences of substrate temperature ranging from 700 °C to 770 °C on the properties of our films using $SrTiO_3$ substrates. All films made at substrate temperatures lower than 720 °C did not have shiny surfaces. Their T_c's were all lower than 80 K. In figure 2, resistance - temperature characteristics of 3 films on $SrTiO_3$ made at substrate temperatures of respectively 750, 760 and 770 °C are given.

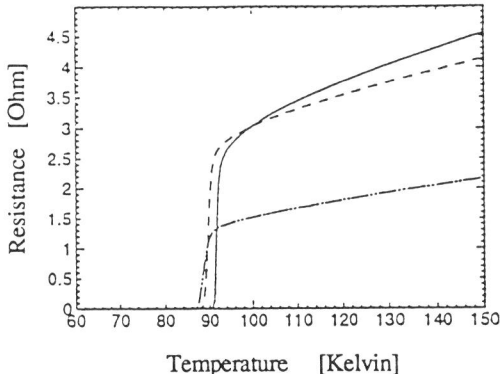

Figure 2. Resistance-temperature characteristic of films grown at substrate temperature of respectively 750 °C (‒··‒··‒), 760 °C (‒ ‒ ‒) and 770 °C (——).

It shows that when a higher substrate temperature is chosen during the ablation, the films have a higher T_c (92 K) and a narrower transition range (2 K). We have found no relationship between the substrate temperature used during the ablation and the room temperature resistance of the film.

X-ray diffraction of our films shows the presence of only (*001*) reflection lines of orthorhombic YBCO crystal, except for the reflection lines of the substrates. This indicates that our films are highly aligned, with their c-axis directed normal to the substrate surface. One of the results from films grown on MgO substrate is shown in figure 3.

The thickness profile of one of the films can be seen in figure 4. The area with constant thickness has a diameter larger than 5 mm.

In figure 5 we display an infrared reflectivity spectrum of one of our films measured at room temperature. The strong feature below 100 cm^{-1} and the peak at 550 cm^{-1} correspond to phonons of the $SrTiO_3$ substrate material. For a sufficiently thick film with a high reflectivity these features are shielded by the $YBa_2Cu_3O_7$ top layer and are no longer visible [2]. Phonons of the $YBa_2Cu_3O_7$ are strongly screened by the Drude-like electronic response, except in ab-axis oriented grains and in secondary phases. The presence of ab-axis oriented grains would for example result in a phonon-peak at 315 cm^{-1} which, however, we do not observe. Hence there is no indication for the presence of secondary phases or differentorientations from figure 5. The relatively low reflectivity of our film should therefore be taken as an indication of a reduced electronic mobility.

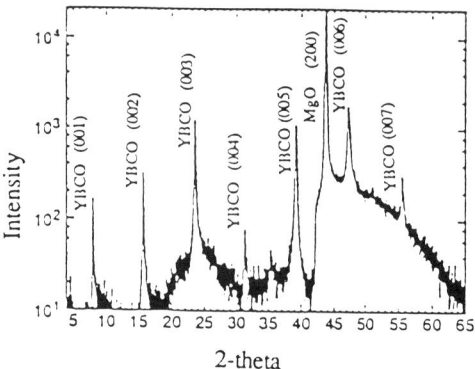

Figure 3. Cu Kα X-ray diffractogram of YBa$_2$Cu$_3$O$_{7-\delta}$ grown on MgO substrate.

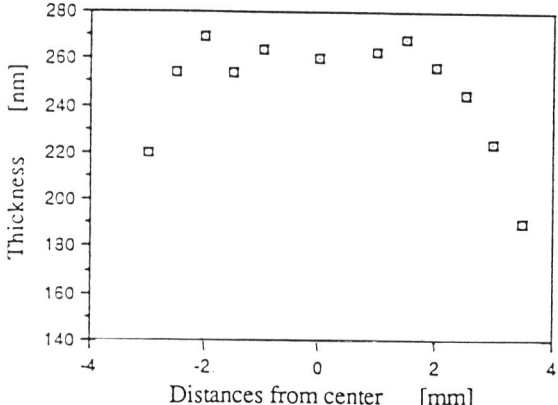

Figure 4. Typical thickness profile of our films.

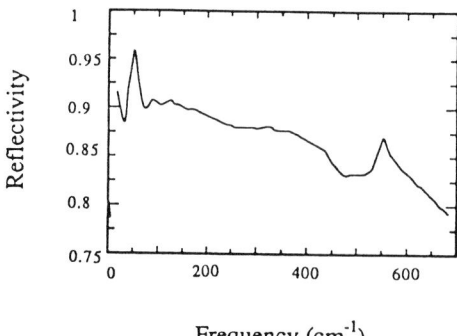

Figure 5. Infrared reflectivity of YBa$_2$Cu$_3$O$_{7-\delta}$ on SrTiO$_3$ measured at room temperature.

4. Conclusion

Our experiments show clearly that $YBa_2Cu_3O_{7-\delta}$ films with c-axis perpendicular to the substrate surface can be easily made using substrate temperature higher than 720 °C. Better superconducting properties were achieved by choosing higher substrate temperature up to 770 °C. Films made at even higher substrate temperature will be produced and fully examined in the near future.

The fact that our film properties strongly depend on the cooling duration in oxygen shows that the process of oxygen incorporation into the films was not completed during the ablation, as can be expected from the thermodynamic phase diagram. Oxygen incorporation can only be fully achieved at lower temperature given a sufficient period of time. This is in agreement with Kishio *et al* [3] have found. They have shown that films annealed at 350 °C in 1 atmosphere of oxygen in a long enough period of time have the highest T_c and the lowest oxygen deficiency.

Acknowledgment

We wish to thank the group of Dr. H.-U.Habermeier of MPI Festkörperforschung in Stuttgart for sharing with us their invaluable experiences in laser ablation and etching techniques during the starting period of our project.

References

1) D. H. Blank, H. Kruidhof and J. Flokstra, J. Phys. D: Appl. Phys. 21 (1988) 226-227.
2) D. van der Marel, H.-U Habermeier, D. Heitmann, W. König, and A. Wittlin, Physica C 176, (1991).1-18.
3) K. Kishio, J. Shimoyama, T. Hasegawa, K. Kitazawa and K. Fueki, Jap. J. Appl. Phys. 26 (1987) L1521.

High T$_c$ Superconductor Thin Films
L. Correra (Editor)
© 1992 Elsevier Science Publishers B.V. All rights reserved.

Very thin YBaCuO films made by coevaporation

M. Danerud, P.Å. Nilsson, and J. Alarco

Department of Physics, Chalmers University of Technology,
S-412 96 Göteborg, Sweden.

Abstract
Superconducting $YBa_2Cu_3O_7$ thin films with various thicknesses from 100 Å to 5000 Å have been deposited on (100) $SrTiO_3$ substrates with a standard BaF_2 coevaporation process. The films had critical temperatures (T_C) of up to 93 K. The best critical currents were 1×10^6 A/cm^2 at 77 K and 3×10^7 A/cm^2 at 4.2 K. The critical current was generally higher for thinner films. The thinner films had smoother surfaces and a more pronounced c-axis orientation than the thicker films.

Two different etching methods were used to pattern the films for jc measurements: argon ion etching and EDTA wet etching. The wet etching was found to work well for thicker films (>1000 Å). For the thinner films, the ion etching process was preferred because of the reduced film surface degradation.

1. INTRODUCTION

High T$_c$ thin films can be made with several methods including laser deposition [1], sputtering [2] and coevaporation [3]. All these deposition methods have advantages and limitations. The in situ laser deposition process is easy to implement, and gives films with high T$_c$ and j$_c$. Disadvantages are e.g. the formation of precipitates on the surface and the small substrate areas that can easily be covered from one target position (usually less then one cm^2).

Sputtering has a possibility to give smoother films than the laser process. For on-axis sputtering there is a problem with negative ions accelerated towards the substrate. This problem is avoided for off-axis sputtering. However, the deposition areas are limited in this case and the sputtering rate is low.

The coevaporation process gives a possibility to cover large areas with high T$_c$ thin films. This can be important for e.g. microwave applications and high speed chip interconnect wires. The in situ process is difficult to implement due to the high oxygen pressure which is needed to oxygenate the films during deposition. Here, we have chosen to investigate a post-anneal process which is easy to implement and gives reproducible results [4].

Very thin films may be used for various types of studies of critical parameters. They may also be exploited in different sorts of applications such as IR-detectors with films thinner than the optical penetration depth ($\approx$1200 Å). In this way, the possibility of one normal and one superconducting region of the film is avoided when the film is irradiated.

2. FILM FABRICATION

The films were made by coevaporation of Y, BaF$_2$ and Cu in a diffusion pumped evaporation chamber using a method similar to the one described by Mankiewich et al [5] . Y and Cu metal were evaporated from electron beam heated guns, and BaF_2 was evaporated from a resistively heated boat. The evaporation rates were 0.7 Å/s, 0.8 Å/s and 2.5 Å/s respectively. Good films were made with $SrTiO_3$ as substrate material. Tests with MgO and sapphire did not produce films with T_cs in the 90 K range.

To get the right evaporation rates, energy dispersive x-ray spectroscopy (EDX) was used to examine the films. Since Sr is emitting a high energy when the substrate is reached by the x-rays that creates extra excitations in Cu, Al_2O_3 substrates instead of $SrTiO_3$ substrates, were needed for these measurements.

The substrates were heated to 400° C during the evaporation. A stream of oxygen, which gave a chamber pressure of 5×10^{-5} mbar, was directed onto the substrates. After an evaporation cycle, the pump line was closed and the oxygen pressure was raised to 1 mbar in the chamber before cool down.

This procedure resulted in amorphous, insulating and shiny films with a yellow-brownish colour. These "raw materials" for high T_c films were resistant to air and could be stored for long times without any degradation. This is due to the fact that BaF_2 does not dissociate during the evaporation and that Cu and Y were oxidized during the evaporation.

In order to make the films superconducting, they were annealed at a high temperature. They were put in a cold furnace and heated in 1000 mbar of dry oxygen to 890°C. The heating time was one hour. At this temperature, oxygen was bubbled through water and fed to the furnace for 15 min to remove the fluorine from the BaF_2 [5]. Then the films were furnace cooled in a stream of dry O_2. A cool down took about 12 hours.

This procedure resulted in black, superconducting films. Films were made with seven different thicknesses from 100 Å up to 5000 Å.

3. PATTERNING

Two patterning methods, wet and ion etching, were tried to make thin lines suitable for transport j_c measurements on the films. EDTA was used for wet etching [6]. Photoresist (Shipley S1813 [7]) was spun onto the films at 4000 rpm, which resulted in a 1.3 μm thick resist layer. This layer was soft baked at 105 °C for 45 seconds. The pattern was exposed for 8 seconds in UV light and developed for 30 seconds in a 1:1 mixture of Microposit developer and water. The pattern consisted of five 16 μm wide bridges with corresponding contact pads for four-point measurements (Fig.1). After these steps, the resist was hard baked for 30 minutes in 120 °C.

According to Venkatesan et. al. [6] EDTA should etch YBCO at 24 Å/s for the bulk of the film and at 70 Å/s for a thin surface layer. The etch worked well for our thicker films, but for thinner samples (< 1000 Å) the etching did not work properly. The film was not removed by the etch. Later, a SEM examination of a wet etched, 1800 Å film showed a remaining bottom layer which had not been etched away. This bottom layer was not superconducting. It is at the present stage difficult to establish, whether this layer was due to film-substrate interactions which changed the chemical properties of the films close to the substrates or to problems with the wet etching method.

After removal of the photo-resist, 800 Å thick silver contacts were evaporated onto the films and defined with an ordinary lift off process.

Ion etching was also used to pattern the films. First, the films were covered with a silver layer (800 Å). This layer protected the films from being contaminated by carbon during the ion etching. A photoresist pattern was then applied with the same method as in the wet etching process. The films were etched in an ion etcher in pure argon at 0.2 mA/cm² at 500 V. These etching conditions gave an etch rate of 72 Å/min for YBCO [8].

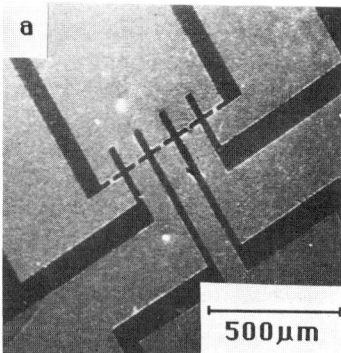 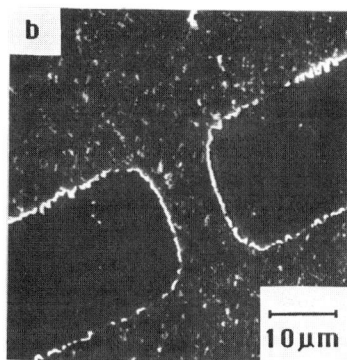

Figure 1. SEM picture of a) the pattern used for critical current measurements, ion etched out of a 1800 Å thick film and b) one of the microbridges of this film at a higher magnification.

The dry etching resulted in a YBCO pattern covered with silver. The same resist steps as for the wet etching method was utilized to make a contact pattern on the silver. Then ion etching was used to remove the unwanted silver. It was stopped when the silver layer was etched through. This was possible because of the large difference in etch rates of Ag and YBCO (360 Å/min compared to 72 Å/min) and the large optical differences of the two materials.

To conclude the comparison of the two etching techniques, we found that the wet etching method is better for large structures in thick films, and the ion etching method is better for the making of small structures in thin films.

4. MEASUREMENTS AND RESULTS

4.1. Critical temperatures

Resistance as a function of temperature for the different films has been recorded by using resistive four-point measurements. The four contacts on the film were made by silver paint.

The results are summarized in figure 2. For the five thickest films (5000 Å) the normal resistance decreases with a factor three from room temperature down to 100 K, the critical temperatures at zero resistance vary from 93 K to 88 K and the widths of the transitions are below 2 K. For the two thinnest films, however, a certain degradation could be noticed. The 225 Å film has a critical temperature of 83 K and the 100 Å film has a T_c as low as 47 K. At the same time the slopes of the normal resistances are smaller and the transitions are wider.

4.2. Critical currents

The pattern with 16 μm wide bridges described above was used for four-point measurements of the critical currents. The wet etched films as well as the ion etched films were measured for comparison.

Figure 3 shows the results for the three thickest films. The critical current densities increased clearly with smaller film thicknesses. The results for the two different etching methods were similar. The highest critical current density measured at 77 K was 1×10^6 A/cm^2 for an 1800 Å thick film. The thinner films were heavily degraded after etching and critical currents could not be measured for higher temperatures. Nevertheless, for the 450 Å wet etched film a j_c of 3×10^7 A/cm^2 was measured at 4.2 K. The two thinnest films had almost completely lost their superconductivity after etching.

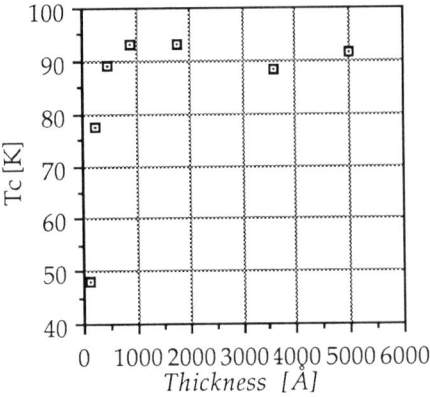
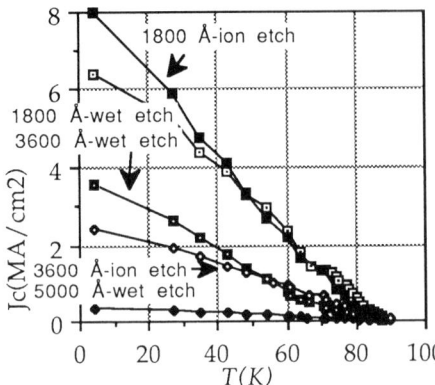

Figure 2. T_c (R=0) versus
thickness for the seven films with
different thicknesses.

Figure 3. Critical current versus temperature for
the three thickest films with a comparison between
the etching methods for two of them.

4.3. Surface topography

All films were examined by scanning electron microscopy (SEM). The thicker films showed
a trellis-like lattice of rods of about 1 μm which is typical for co-evaporated thin films with
subsequent annealing [8], but for the 900 Å film and the thinner films this pattern had
apparently not been developed (Fig.4). A much smoother surface with grains smaller than 100
nm was present.

4.4 Structure

X-ray diffractograms were made for each of the films. As could be seen from the
diffractogram of the 5000 Å film (Fig.5) there was a c-axis orientation. However, a- and b-axis
peaks were also visible. This is typical for co-evaporated thin films with subsequent annealing
and it has been related to the surface structure [9]. The a- and b-axis peaks became relatively
weaker for the thinner films. For the thinnest films only the substrate peaks were visible.

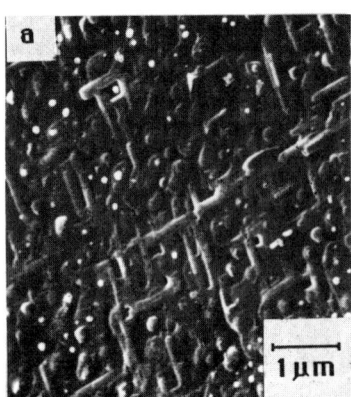
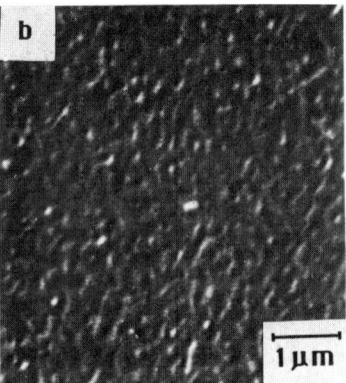

Figure 4. SEM images of a) a 1800 Å thick film and b) a 900 Å thick film. Notice the
difference in surface topography. For the thinner films the typical trellis-like lattice of rods has
not developed.

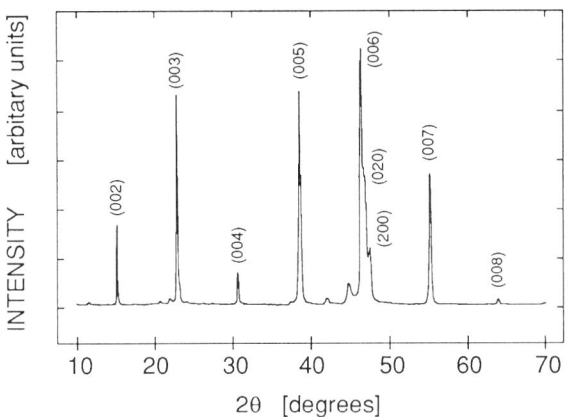

Figure 5. X-ray diffractogram for a 5000 Å thick film. The a- and b- axis peaks were relatively less pronounced for thinner films.

5. CONCLUSIONS

We have made coevaporated $YBa_2Cu_3O_7$ films from Y, Cu and BaF_2 sources with a subsequent heat treatment. Superconducting films as thin as 100 Å were made.

SEM pictures revealed a smoother surface topography for the thinner films where the typical trellis-like lattice could not be seen any more. X-ray diffraction showed less a-axis orientation for the thinner films which is consistent with the SEM results. j_c increased for thinner films.

6. ACKNOWLEDGEMENTS

We would like to thank our supervisor Tord Claeson for his help during this work. Our work would have been impossible without the assistance from our mechanical engineer Staffan Pehrsson. We would also like to thank Dag Winkler, Zdravko Ivanov and Göran Brorsson for fruitful discussions.

7. REFERENCES

1. Venkatesan, T., Wu X.D., Dutta, B., Inam, A., Hegde, M.S., Wang, D.M., Chang, C.C., Nazar, L., and Wilkens, B., Appl. Phys. Lett. **54,** (1989) 581.

2. Sandstrom, R.L., Gallagher, W.J., Dinger, T.R., Koch, R.H., Laibowitz, R.B., Kleinsasser, A.W., Gambino, R.J., Bumble, B., and Chisholm, M.F.,   Appl. Phys. Lett. **53,** (1988) 444-446.

3. Matijasevic, V., Rosenthal, P., Shinohara, K., Marshall, A.F., Hammond, R.H., and Beasley, M.R., submitted to Journal of Materials   Research.

4. Alarco, J.A., Brorsson, G., Claeson, T., Danerud, M., Engström, U., Ivanov, Z.G., Nilsson, P.Å., Olin, H., Winkler, D., to be published in Physica Scripta.

5. Mankiewich, P.M., Scofield, J.H., Scocpol, W.J., Howard, R.E., Dayem, A.H., and Good, E., Appl. Phys. Lett. **51,** (1987) 1753.

6. Shokoohi, F.K., Schiavone, L.M., Rogers, C.T., Inam, A., Wu, X.D., Nazar, L., and Venkatesan, T., Appl. Phys. Lett. **55,** (1989) 2661-2663.

7. Shipley Company, Inc., 2300 Washington Street, Newton, MA 02162-1469, USA.

8. Ivanov, Z.G., Nilsson, P.Å., Andersson, E-K, and Claeson, T., Supercond. Sci. Technol. **4,** (1991) 112-114.

9. Hebard, A.F., Fleming, R.M., Short, K.T., White, A.E., Rice, C.E., Levi, A.F.J., and Eick, R.H., Appl. Phys. Lett. **55,** (1989) 1915-1917.

High T$_c$ Superconductor Thin Films
L. Correra (Editor)
© 1992 Elsevier Science Publishers B.V. All rights reserved.

Preparation of Bi-Sr-Ca-Cu-O superconducting films showing Tc(zero) of 97K without post-annealing by MOCVD

K. Endo[a], S. Misawa[a], H. Yamasaki[a] Y. Moriyasu[b] and S. Yoshida[a]

[a]Electrotechnical Laboratory, Tsukuba, Ibaraki, 305 Japan

[b]Asahi Chemical Industry Co., Fuji, Shizuoka, 416 Japan

Abstract

Bi-Sr-Ca-Cu-O superconducting films were grown on MgO(100) and LaAlO$_3$(100) substrates by MOCVD using triphenyl bismuth and β− diketonates of Sr, Ca and Cu as source materials. The film thickness and growth rate dependences of the zero resistance temperature Tc(zero) were studied. A high-Tc single phase film showing Tc(zero) of 97 K was successfully obtained without post-annealing, which is the highest Tc(zero) among as-deposited films reported so far.

1. INTRODUCTION

The Bi-Sr-Ca-Cu-O system attracts much attention for its higher superconducting transition temperature Tc. In this system, however, there are many superconducting phases showing various Tc(zero), *e.g.*, the high-Tc phase (Bi$_2$Sr$_2$Ca$_2$Cu$_3$O$_x$) with Tc around 110 K and the low-Tc phase (Bi$_2$Sr$_2$CaCu$_2$O$_y$) with Tc around 80 K. It is difficult to obtain the films with the high-Tc single phase. Much effort has been devoted to preparing as-deposited films with the high-Tc single phase, especially for microelectronics.

In the previous study [1], for the first time, we succeeded in preparing *high-Tc single phase films* on MgO substrates *without post-annealing* by metalorganic chemical vapor deposition (MOCVD). However, the zero resistance temperature Tc(zero) of the films was as low as 74 K. Also, we tried *in-situ* Pb-doping during the growth of Bi-Sr-Ca-Cu-O films by MOCVD using Pb(DPM)$_2$ [2]. However, there were little effects of Pb-doping on the rise of Tc(zero), whereas the high-Tc single phase films doped with Pb were formed.

In order to obtain *as-deposited* Bi-Sr-Ca-Cu-O films showing the higher Tc(zero), we have studied the dependences of Tc(zero) on film thicknesses and growth rates. As a result, we have succeeded in growing high-Tc single phase Bi-Sr-Ca-Cu-O films having *Tc(zero) of 97K without post-annealing* by MOCVD, which is the highest Tc(zero) among as-deposited films reported so far.

2. EXPERIMENTAL

Figure 1 is a schematic drawing of a cold-wall type MOCVD apparatus with a horizontal reaction tube used in this study. The source materials were triphenyl bismuth $Bi(C_6H_5)_3$, bis (-2, 2, 6, 6-tetramethyl-3, 5-heptanediono)-strontium $Sr(DPM)_2$, -calcium $Ca(DPM)_2$ and -copper $Cu(DPM)_2$. Each source material was loaded into an individual vaporizer and was heated to an appropriate temperature. These source materials were carried into a quartz reactor with Ar gas. Total flow rate of Ar gas was 770 cc/min. The gas line was heated to the temperatures higher than those of the vaporizers to prevent the condensation of source materials. Oxygen gas was separately introduced into the reactor at a flow rate of 675 cc/min. $MgO(100)$ and $LaAlO_3(100)$ single crystals, $5 \times 10 \times 0.5$ mm^3 in size, were used as substrates. They were put on an inconel susceptor and then heated to 800 °C, inductively. The pressure in the reactor was kept at 50 Torr during the film growth. Oxygen partial pressure was 23 Torr.

We have grown Bi-Sr-Ca-Cu-O films on MgO(100) substrates with thicknesses from 200 to 5200 Å at a growth rate around 100 Å/h. We have also grown films at different growth rates between 10 and 500 Å/h. Growth rates were varied by changing the temperature of the vaporizers.

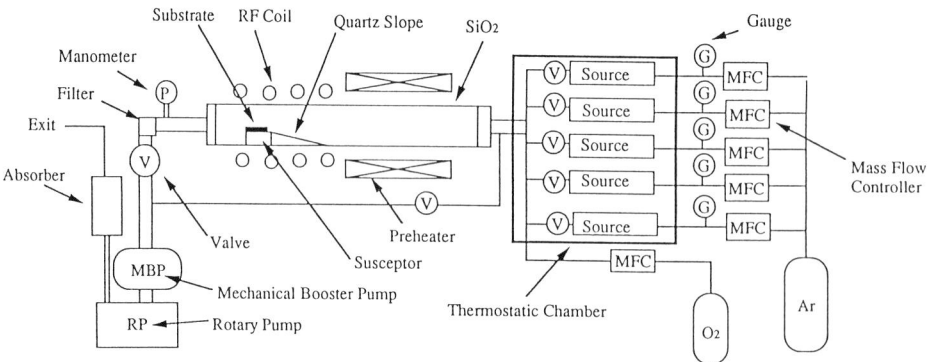

Fig.1 Schematic drawing of MOCVD apparatus.

The compositions and the crystal structures of the films were examined using inductively coupled plasma spectroscopy (ICP) and X-ray diffraction (XRD), respectively. Resistance measurements were carried out by a standard four-probe DC method.

3. RESULTS AND DISCUSSION

Figure 2 shows the thickness dependence of Tc(zero) for the as-deposited films on MgO substrates. It should be noted that Tc(zero) increases steeply with the increase of film thickness, and then looks saturated in the large thickness range. The XRD patterns of these films are classified into two groups. The thinner film (sample A, 200 Å thick) consists of the high-Tc single phase, and the thicker one (sample B, 5200 Å thick) consists of the high-Tc and low-Tc mixed phases. Figures 3 and 4 show the XRD patterns of these films. All the films show the crystal orientation with the c-axis normal to the substrate surface.

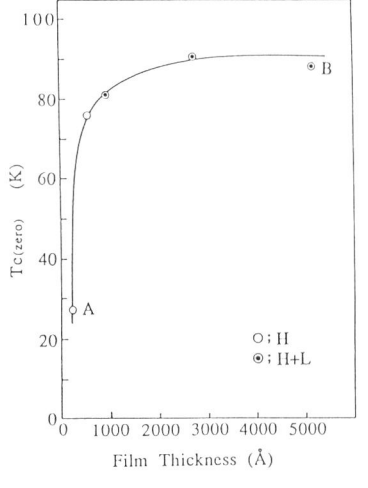

Fig.2 Dependence of Tc(zero) on film thickness. Growth rate is around 100 Å/h;
○ : the film with high-Tc singie phase,
◉ : the film with high-Tc and low-Tc mixed phases.

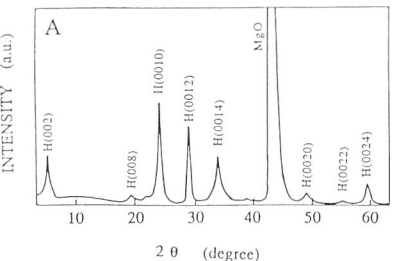

Fig.3 XRD pattern of sample A, H(00n)'s denote the peaks corresponding to the high-Tc phase.

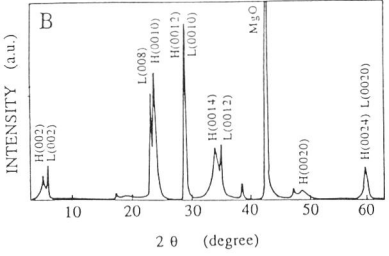

Fig.4 XRD pattern of sample B. H(00n)'s and L(00n)'s denote the peaks corresponding to the high-Tc and low-Tc phase, respectively.

We have estimated c-axis length from 2θ-values of the XRD peaks indexed as H(00n) of the high-Tc phase. Figure 5 shows the dependence of c-axis length on film thickness. Clearly, c-axis length decreases with the increase of film thickness. Such a dependence is considered to be a result of the lattice strain caused by the misfit of in-plane lattice constants between the Bi-Sr-Ca-Cu-O film and the MgO substrate. Namely, the misfit at the interface between the film and the substrate introduces the compressive stress into the film along the interface, because the in-plane lattice constant of the film (a-axis length) is twenty percent larger than that of MgO, although the greater part of the misfit is relaxed by the introduction of so-called misfit dislocation. The origin of the expansion of the c-axis in the films can be attributed to such a compressive stress along the interface. It is presumed that the increase of the film thickness brings about the reduction of the compressive stress, which results in the decrease of c-axis length with the increase of thickness.

Figure 6 shows the dependence of Tc(zero) on c-axis length. Tc(zero) increases with the decrease of c-axis length. This result suggests that the use of the substrates with small misfit is favorable to increase Tc(zero), in particular, for the thinner films.

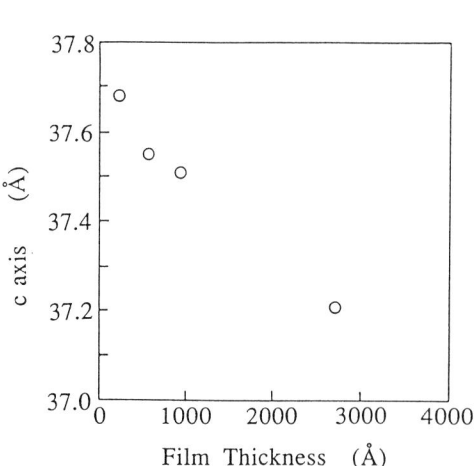

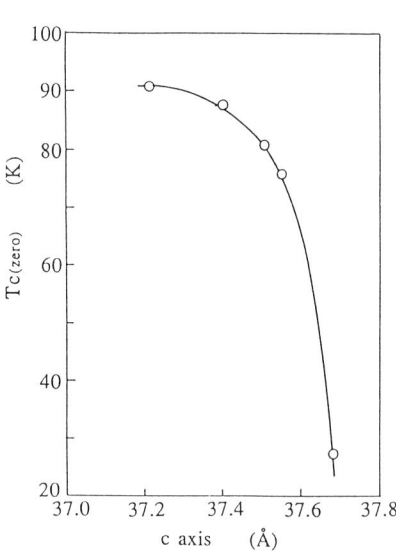

Fig.5 Dependence of c-axis length on film thickness. Growth rate is around   100Å/h.

Fig.6 Dependence of Tc(zero) on c-axis length. Growth rate is around 100Å/h.

We have also examined the dependence of Tc(zero) on the growth rate of the films. The result for the films with the thickness of 1800 Å is shown in Fig. 7. It is to be noted that Tc(zero) increases with the decrease of growth rate.

On the basis of these results, we have grown the films on $LaAlO_3(100)$ substrates at a growth rate around 12 Å/h. Figure 8 shows the temperature dependence of the resistivity for the as-deposited film with the thickness of 960 Å. Tc(zero) was as high as 97 K, which is the highest Tc(zero) as reported so far. XRD study revealed that all the peaks can be indexed as (00n) of the high-Tc phase, *i.e.* the film is apparently *a single phase of the high-Tc phase* with the c-axis oriented normal to the substrate.

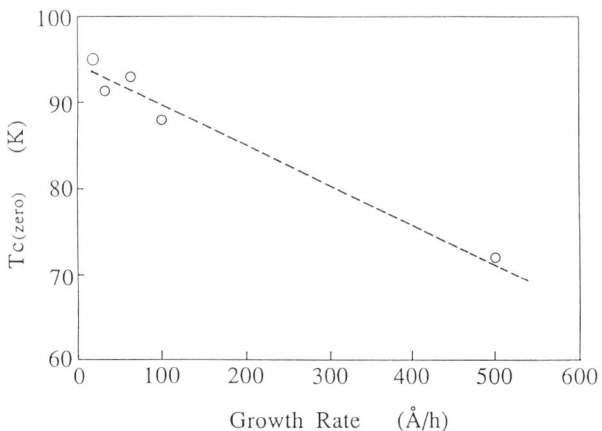

Fig.7   Dependence of Tc(zero) on growth rate. The film thicknesses are around 1800Å.

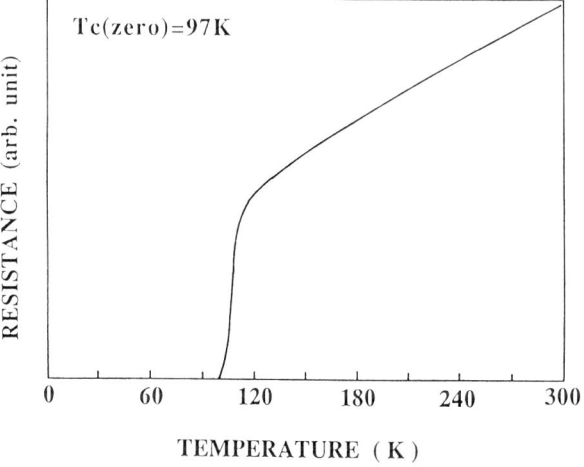

Fig.8   Temperature dependence of the resistance of BSCCO film showing the highest Tc(zero) of 97K.

4. CONCLUSION

We have grown Bi-Sr-Ca-Cu-O superconducting films on MgO(100) and LaAlO$_3$(100) substrates by MOCVD. It was found that Tc(zero) increases with the increase of film thickness and with the decrease of c-length and the growth rate. On the basis of these results, we have succeeded in obtaining the Bi-Sr-Ca-Cu-O film showing Tc(zero)=97K without post-annealing by MOCVD, which is the highest Tc(zero) among as-deposited films reported so far.

5. REFERENCES

1. K. Endo, S. Hayashida, J. Ishiai, Y. Matsuki, Y. Ikedo, S. Misawa and S. Yoshida, Jpn. J. Appl. Phys. 29 (1990) 294
2. K. Endo, S. Hayashida, K. Nakatsuka, J. Ishiai, Y. Ikedo, S. Misawa and S. Yoshida, in: Advances in Superconductiviy II, Ed. T. Ishiguro and K. Kajimura (Springer-Verlag, Tokyo, 1990) p. 911

High T$_c$ Superconductor Thin Films
L. Correra (Editor)
© 1992 Elsevier Science Publishers B.V. All rights reserved.

CHARACTERIZATION OF THIN SUPERCONDUCTING YBa$_2$Cu$_3$O$_{7-x}$ FILMS PREPARED BY MO CVD

K Fröhlich, J. Šouc, D. Machajdík and I. Vávra

Institute of Electrical Engineering, SAS, 842 39 Bratislava, Czech and Slovak Federal Republic

Abstract

YBa$_2$Cu$_3$O$_{7-x}$ superconducting thin films prepared by MO CVD on single crystalline MgO substrate were studied by means of scanning electron microscopy, transmission electron microscopy as well as by X-ray diffraction analysis. Investigations show very similar microstructure for the samples with different Cu content; Cu rich precipitates on the surface, strong c-axis orientation and no preferred orientation in a-b plane.

1. INTRODUCTION

High-T$_c$ thin superconducting YBa$_2$Cu$_3$O$_{7-x}$ films have been grown by a number of different techniques; sputtering, coevaporation, laser ablation, chemical vapour deposition. Metalorganic chemical vapour deposition (MO CVD) allows the in situ growth of highly oriented YBa$_2$Cu$_3$O$_{7-x}$ films with excellent superconducting properties [1,2]. The films with the best properties were deposited on closely-matched substrates such as SrTiO$_3$.

However, highly oriented c-axis YBa$_2$Cu$_3$O$_{7-x}$ films were prepared also on non-perovskite substrate - (100) single crystalline MgO although a large lattice mismatch(about 9%) exists between the substrate and the film. Superconducting properties are partially deteriorated for the sample deposited on MgO [3]. In this paper we present the study of the microstructure of MO CVD YBa$_2$Cu$_3$O$_{7-x}$ films prepared on single crystal MgO substrate.

2. EXPERIMENTAL

$YBa_2Cu_3O_{7-x}$ films were prepared in a horizontal quartz- tube reactor having three separate parallel glass-inlet tubes for the introduction of the precursors [4]. Resistance furnaces were used for heating of four zones. The source materials tetramethylheptadionates of Y, Ba and acetylacetonate of Cu were sublimated from aluminium boats inserted into zones heated at 120, 234, and 150 ^{O}C, respectively. The substrates were placed in the fourth furnace on stainless steel sample holder and were heated to 800 ^{O}C. The total Ar flow was 30 l/h, O_2 flow was 10 l/h, and working pressure was 10 torr. The growth rate was approximately 1μm/h and the time of deposition was 30 min. After the deposition the samples were cooled down to 500 ^{O}C over a period of 30 min in 1 atmosphere of O_2 and held at this temperature for 60 min prior to cooling to the room temperature.

The temperature dependence of resistance was measured by a conventional dc four-probe method using In contacts. The thickness was estimated from the weight increase. A sample for TEM observations was prepared by ion-milling first from the rear side in order to remove the substrate. Then, the sample was milled for short time from both sides so that the microstructure near the substrate as well as near the top of the film could be investigated. X-ray diffraction spectra were measured by X-ray powder diffractometer in Bragg-Brentano focusing geometry using CuKα radiation. Slow scanning speed 1/16 ^{O} ϑ/min, external Si powder standard and full width at half maximum (FWHM) of Kβ 200 MgO diffraction was used for the determination of c parameter and for the determination of the coherent crystalline region in the c direction (D_{001}). We have used ω-scan procedure with fixed detector at the position of specified 001 diffraction in order to determine the quality of the texture.

3. Results and discussion

Properties of two samples prepared on (100) MgO single crystals in the same run are summarized in Table 1. The temperature dependence of the relative resistance for the samples is shown in Fig. 1. The samples were placed in the deposition chamber in row at different position from the

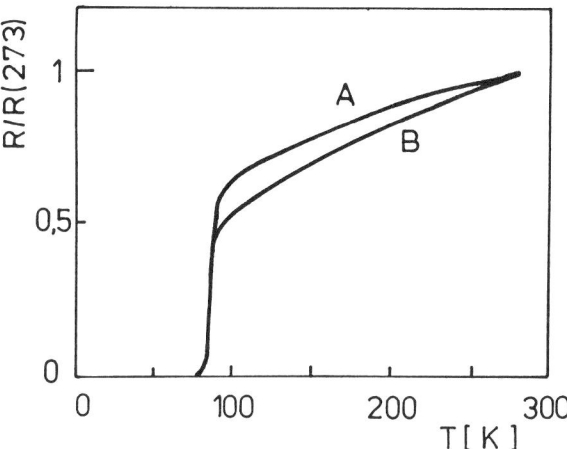

Figure 1. Temperature dependence of relative resistance for samples from Table 1.

Table 1

Properties of the samples

Sample	h	T_c(R=0)	Cu content[a]	c parameter	D_{001}	FWHM
	[μm]	[K]		[Å]	[Å]	[°ψ]
A	0.7	80.1	0.84	11.682	1350	0.8
B	0.4	80.7	0.64	11.683	1250	0.9

a) atomic fraction of the metals Y, Ba, Cu

FWHM - full width at half maximum obtained from ω-scan

inlet nozzle. The distance between the samples was 20 mm. Rutherford backscattering analysis of the samples revealed excess of Cu if compared with the stoichiometric composition. A gradual decrease of Cu content along the deposition chamber was observed. The Cu content for sample B is very close to those of MO CVD prepared samples with very high current critical density, ($J_c = 10^4$ A/cm^2 at T = 77 K and B = 30 T [2]).

SEM micrograph for sample B is shown in Figure 2. The film has numerous 1 μm size particles visible on the surface. Electron-probe microanalysis

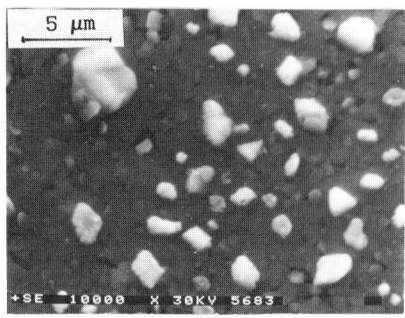

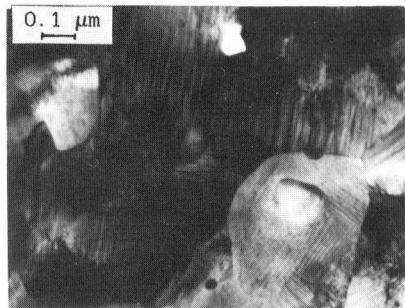

Figure 2   Figure 3

Figure 2. Scanning electron-microscopy image of surface

Figure 3. Bright- field transmission electron - microscopy image of
         $YBa_2Cu_3O_{7-x}$ layer near the substrate

revealed that these particles are Cu rich. The matrix of the film consist
of grains with 0.5 μm diameter. At some areas pits in the film are
observed. The sample A has very similar surface morphology, except that the
density of Cu-rich surface precipitates is higher.

TEM investigations of sample A revealed polycrystalline nature of
$YBa_2Cu_3O_{7-x}$ film with pits between some grains. Pits observed in Figure 2
by SEM in some cases go throughout the whole thickness of the film. The
particles on the surface have polycrystalline character and in some areas
the $YBa_2Cu_3O_{7-x}$ polycrystalline layer could be observed bellow. Some
unknown precipitates were seen using TEM in the layer close to the surface
of the film. TEM micrograph showing the microstructure of an $YBa_2Cu_3O_{7-x}$
layer is displayed in Figure 3. Grains are haevily twinned in two principal
directions. They exhibit preferred orientation with c-axis perpendicular to
the substrate. No preferred orientation in a-b plane was observed.

XRD analysis revealed series of 001 $YBa_2Cu_3O_{7-x}$ diffractions. This
showed preferred orientation of 1-2-3 phase with the c- axis perpendicular
to the substrate surface. Diffractions (-111, 002, 111, 200, -202) of
monoclinic CuO phase are also present in the diffraction pattern. However,

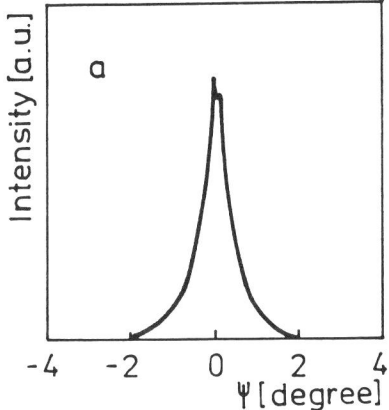
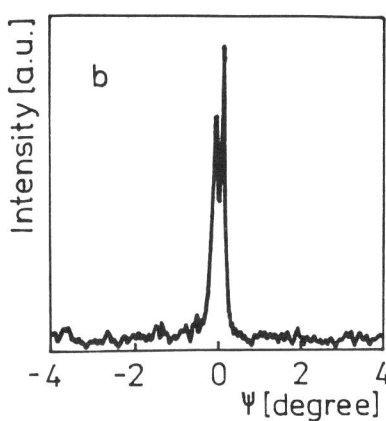

Figure 4. Rocking curves (ω-scan) for a) 006 line of YBa$_2$Cu$_3$O$_{7-x}$ film
b) Kβ 200 line of MgO substrate film

their intensities compared with those for 1-2-3 phase are very weak and their Miller indexes imply random orientation. We suggest that they are caused by the precipates shown in the SEM micrograph. We used the 006 diffraction for the determination of the lattice parameter c and for the determination of the coherent crystalline regions in c direction. The average crystalline size in c direction was 0.135 and 0.125 nm for samples A and B, respectively.

Figure 4 shows rocking curve (ω-scan) for the sample A as well as for the substrate. The FWHM value measured for several diffractions yields 0.8 for sample A and 0.9 for sample B. Measurements of the ω-scan for several MgO single crystalline substrates using Kβ 200 diffraction show a splitting of the rocking curve to several overlapped maxima. This indicates the division of the single crystal to several blocks disoriented to 0.1 - 0.3 degree. Consequently, the FWHM value for YBa$_2$Cu$_3$O$_{7-x}$ layer is distorted and does not represent the quality of the layer texture itself. Values related to the texture are expected to be up to two times smaller.

4. CONCLUSIONS

The properties of YBa$_2$Cu$_3$O$_{7-x}$ samples prepared on single crystalline MgO substrate were studied. The samples have different content of Cu, however the values of $T_c(R=0)$ and c-axis parameter are very close to each other. The surface morphology including the Cu-rich precipitates observed for our MO CVD prepared $YBa_2Cu_3O_{7-x}$ films is typical for Cu-rich samples prepared by different techniques [2,5,6]. It seems, that at high temperatures of deposition needed for growing films with the best transport properties, formation of Cu rich surface precipitations (mostly determined as CuO) is a fundamental problem for c-axis oriented $YBa_2Cu_3O_{7-x}$ films. With regard to the disorientation of the substrate, the values of FWHM determined by XRD ω-scan imply a high degree of c-axis orientation. However, no preferred orientation was indicated in a-b plane by TEM investigations.

As it was shown elsewhere [7], annealing of the single crystalline MgO substrate at 1100 - 1200 ^{o}C results in a growth of $YBa_2Cu_3O_{7-x}$ with well aligned grains in a-b plane. Thus, for the growth of high quality $YBa_2Cu_3O_{7-x}$ films on single crystal MgO a special attention must be paid to the choice and thermal treatment of the substrate.

REFERENCES

1   K. Watanabe, T. Matsushita, N. Kobayashi, H. Kawabe, E. Aoyagi, K. Hiraga, H. Yamane, H. Kurosawa, T. Hirai, and T. Muto, Appl. Phys. Lett., 56 (1990) 1490

2   S. Matsuno, F. Uchikawa, K. Yoshizaki, N. Kobayashi, K. Watanabe, Y. Muto, and M. Tanaka, IEEE Trans. on Magn., 27 (1991) 1398

3   O. Thomas, A. Pisch, E. Mossang, F. Weiss, R. Madar, and J.P. Senateur, J. Less-Common Metals, 164 & 165 (1990) 444

4   J. Šouc, K. Fröhlich, V. Bukovenová, D. Machajdík, V. Šmatko, Š. Chromik, and V. Kliment, Proc. 2nd Czechoslovak - Italian Symposium on Superconductivity, Smolenice, CSFR, May 1991, to be published

5   R. G. Humphreys, N. G. Chew, J. S. Satchell, S. W. Goodyear, J. A. Edwards, and S. E. Blenkinsop, IEEE Trans. on Magn., 27 (1991) 1357

6   T. Kobayashi and M. Iyori, Proc. of Int. Conf. on Superconductivity, S. K. Joshi, C. N. R. Rao and S. V. Subramanyam (eds), World Scientific Publishing, (1990) 1

7   B. H. Moeckly, S. E. Russek, D. K. Lathrop, R. A. Buhrman, Jian Li, and J.W. Mayer, Appl. Phys. Lett., 57 (1990) 1687

High T$_c$ Superconductor Thin Films
L. Correra (Editor)
439

Y Ba Cu O thin films deposited by ion beam sputtering assisted by ion implantation

R.J. Gaboriaud, S. Benayoun and E. Salmon

Laboratoire de Métallurgie Physique, URA 131 - CNRS, Faculté des Sciences, 40, Av. du recteur Pineau, 86022 Poitiers France

Abstract

Thin films of Y Ba Cu O are deposited by ion beam sputtering (Kaufman source) assisted by energetic ion implantation (200 KV ion implanter). Influence of the atomic mass of the implanted ions (O, Ne, Ar, Kr at 150 Kev) on the composition and densification of the as-deposited amorphous layer are studied by RBS and grazing X-Ray reflectometry. The results which are obtained show a densification effect with heavy implanted ions and a dramatic effect of loss of matter as a function of ion mass and temperature.

1. INTRODUCTION

Among the enormous amount of activities generated by the high critical temperature (Tc) superconductors since their discovery, there has been a high degree of interest in deposition of copper oxide superconductors thin films by a variety of physical and chemical techniques (1). On a technical point of view, applications involving superconducting thin films seem to be closer than applications of bulk material. Nevertheless, preparation of such thin films as 1-2-3 phase with high Tc and high critical current density (Jc) is more complicated than expected.

All the deposition techniques require either a high temperature (> 800°C) post deposition annealing of the films or growth at elevated substrate temperature (> 600°C) to obtain superconducting films (2). Sputtering (3), electron beam evaporation (4), laser ablation (5), Molecular beam epitaxy (6) and other deposition methods are now commonly used. Among the most serious obstacles encountered with all these methods are reproducible control of film stoichiometry, phase purity and stability, good cristallinity. Furthermore, associated with high temperature post-annealing are undesirable interactions between film and substrate (7) which lead to severe restrictions on the choice of substrate available for film deposition. Most of these films are deposited on $SrTiO_3$, MgO, ZrO_2, $LaGaO_3$ etc... and obviously, for microelectronic application, it would be more desirable to deposit on other important technical substrate as Si substrate.

There is therefore a considerable interest in the deposition at lower temperature (T < 600°C) and the general trend is the in-situ growth technique which try to eliminate the as deposited amorphous stage of the process.

One of the possibilities is the energetic ion beam assisted sputtering deposition (8) which offers interesting avantages. The most important is probably the possibility of a lower substrate temperature, as the energy of the sputtered species coupled with the ion assisted implantation allows irradiation enhanced cristallization, densification and might produce irradiation induced epitaxy. On an other hand, bombardment of the film surface by energetic ions causes composition change either by preferential sputtering or by irradiation enhanced surface diffusion and sometime surface faceting (8). All the deposition conditions of superconducting thin films including ion bombardment of the growing film are an interesting field of research which is not very well developped yet.

Therefore this work gives some preliminary results obtained with YBaCuO thin films obtained by ion beam sputtering assisted by ion implantation of different atomic species. This study is focused on the very first step of the deposition. Influence of the implanted atom on the composition and on the density of the as deposited film is carried out by RBS and X-ray technique respectively.

2. EXPRIMENTAL PROCEDURE

Thin films (1000-2000 Å) of YBaCuO were deposited in an ion sputtering chamber in line with a 200 KV ion implanter (Fig.1).

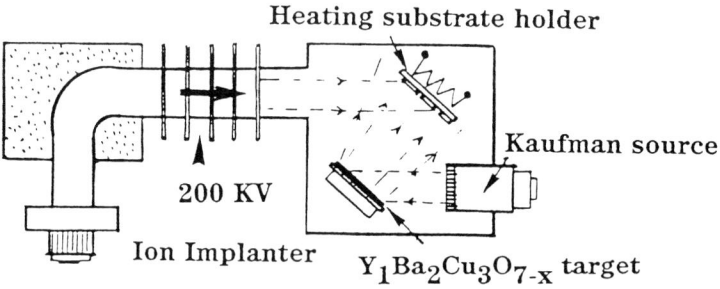

Figure 1. YBaCuO thin film sputtering system

The sputtering system includes a 7 cm diameter Kaufman source which uses a beam of argon whose accelerating voltage is 1200 eV and intensity around 60 mA. A 10 cm diameter, water cooled, sputtering target of stoichiometric superconducting $YBa_2Cu_3O_{7-x}$ is used at 15 cm from the Kaufman source. The substrate holder is mounted at 20 cm from the target. The sputtering deposition chamber work at a base pressure of 10^{-4} torr. Deposition rate which is measured by a water cooled quartz crystal monitor is around 0,7 Å/s. Ions implanted during the film growth were O, Ne, Ar, Kr with an energy of 150 KeV and a current density less than 1 µA/cm². Total ion doses received by the thin films at the end of the deposition were approximately 10^{14} ion/cm². Temperatures of the substrate holder were room temperature and 400°C. Substrates used were Si, SiO_2, MgO, $SrTiO_3$ and $LaGaO_3$. The as deposited films were analysed by :

1) Rutherford backscattering spectrometry (RBS) (9, 10) using a 2 MeV helium beam with a scattering angle of 165°. Experimental spectra were fitted using the RUMP program package (11)

2) Grazing X-ray reflectometry (12).

3. EXPERIMENTAL RESULTS

The results described in this work concern 7 different as-deposited thin films realized in the following conditions :

- sputtering deposition, without assisted ion beam, at room temperature
- oxygen, neon, argon, krypton energetic beam assisted sputtering deposition at room temperature
- oxygen and argon energetic beam assisted sputtering deposition at 400°C.

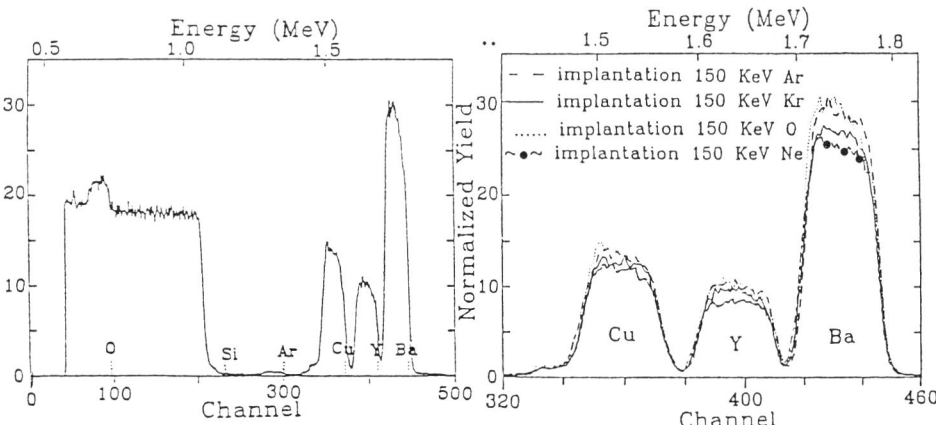

Figure 2. General aspect of RBS spectra obtained in this study

Figure 3. RBS spectra corresponding to implantation assisted deposition.

All the samples are amorphous, as shown by TEM and X-ray.
For the sake of clarity only RBS spectra corresponding to thin films of
YBaCuO deposited on silicon are displayed in this paper. Results are
similar for all the substrates which have been used. General aspect of
the RBS spectra obtained in the analysis of the thin films, is shown in
figure 2. Thickness of the samples has been choosen in order to give
separate signals for the different elements which are in the layer. This
thickness corresponds to a number of atoms per unit area of
approximately $0.75 \ 10^{18}$ at.cm^{-2}. The same following features appear on
each spectra of this work :
    - A large oxygen peak
    - Well defined Cu, Y, Ba signals
    - Argon signal coming from the sputtering process
    - Fe small signal at the back side of the copper signal coming very
likely from sputtering of the chamber walls.
    When the sputtering deposition is assisted by O, Ne, Ar and Kr,
implantation at room temperature, the RBS spectra obtained are shown
in figure 3. The thickness is roughly the same in the four cases,
nevertheless variations in the composition appear.
    Sputtering depositions assisted by O and Ar implantation with a
substrate temperature of 400°C lead to the RBS spectra shown in
figure 4. Contrarily to the results obtained at room temperature,
dramatic effect appears with the sputtering assisted by the argon
implantation.
    All the samples mentioned in this work were studied by grazing X-
ray reflectometry (11) which allows an experimental determination of
the volumic density. The figure 5 exhibits the caracteristic asymmetric
shape of the reflectometry curve obtained in this study.

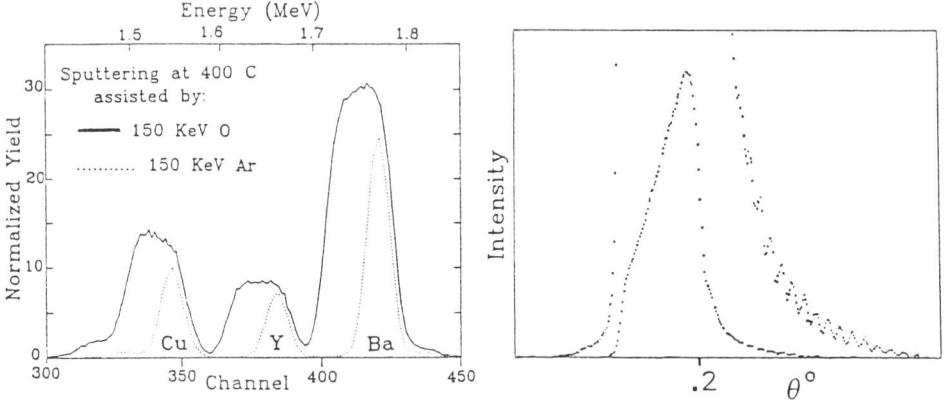

Figure 4. RBS spectra of the thin   Figure   5.   Grazing   X-ray
film deposited at 400°C.   reflectometry   curve   used   to
                                          determine the density of the
                                          thin film.

Salient features determined by RBS and X-ray are summarized in
Table 1 and figure 6.

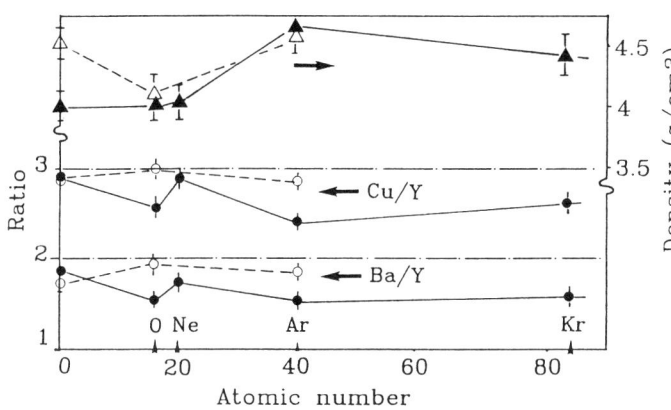

Fig.6. Results obtained with :
- RBS for the element ratio (righ scale)
- X-ray reflectometry for the density (left scale)
- full marks : RT
- open marks : 400°C

Table 1
Summary of the results obtained by RBS and X-ray reflectometry on the as sputtered thin film of YBaCuO.

	Samples	Implanted 150 KeV ion	X-ray density g/cm^3	$\dfrac{Ba}{Y}$ ±0.05	$\dfrac{Cu}{Y}$ ±0.05	$\dfrac{0}{Y}$
RT	No impl.		4 ± 0.2	1.87	2.87	10
	Impl.	O_{16}	4	1.52	2.55	7
	Impl.	Ne_{20}	4.1	1.73	2.93	12
	Impl.	Ar_{40}	4.7	1.52	2.45	6
	Impl.	Kr_{84}	4.4	1.53	2.59	-
400°C	No impl.	-	4.6	1.75	2.90	-
	Impl.	O_{16}	4.1	1.97	3	7
	Impl.	Ar_{40}	4.4	1.84	2.8	9

4. DISCUSSION OF THE RESULTS

It is emphasized that this study on YBaCuO thin films obtained from the ion beam sputtering of a unique stoichiometric target is focussed on the very first step of the elaboration process. In all the cases described above, the as-elaborated thin layers are amorphous. Parameters under investigation are composition and volumic density as a function of ion mass implanted during the film growth. Those preliminary results show that : (table 1 - Figure 6).

a) At room temperature :
- without implantation the stoichiometry is close to the 1,2,3 ratio and the density is 62% of the theoretical density (6.38 g/cm^3)
- depositions assisted by ion implantation result in a deficiency of Cu and Ba. A densification effect appears only with the heavy ions Ar and Kr which leads to a density of approximately 72% of the theoretical density.

- The difference in atomic ratio obtained between oxygen and neon implantation is surprising and not well understood yet.
b) Deposition at 400°C.
At 400°C deposited thin films of YBaCuO with and without assisted implantation are still amorphous. But the effect of implantation are much more pronounced.
- without implantation it is worthy of mention that the stoichiometry is close to the good values of 1-2-3 compound (Table 1)
- depositions assisted by oxygen implantation give very good stoichiometry with the same thickness as the non-implanted sample. With argon implantation, a dramatic effect of loss of matter appears. A small deficiency in Ba and Cu is noticed but the thickness is approximately divided by 3. This strong difference cannot be explained in terms of classic sputtering theory. This effect needs to be clarified. Implantations with other ions, like Cu and Xe, are in progress in order to determine the re-sputtering and sticking coefficient influence on the thin film as a function of temperature.
In table I the oxygen ratio is also mentioned. It is worth noting that the value of O/Y is obtained from the RBS spectra assuming a Rutherford regime. Therefore those value must be considered with care. Anyhow the oxygen peak in all the RBS spectra of this study is always large.

ACKNOWLEDGMENTS

The authors wish to thank Dr J. Perriere from GPS Paris and Dr A. Naudon from CNRS-Poitiers for assistance with regard to the analysis by RBS and X-ray respectively.

REFERENCES

1   M. Leskelä, J.K. Truman, C.H. Mueller and P.H. Holloway, J. Vac. Technol. A7 (1989) 3147.
2   H. Adachi, K. Setsune, K. Hiroghi, T. Kamada and K. Wasa, Physica C 153-155 (1988) 1696.
3   J. Gao, Y.Z. Zhang, R.R. Zaho, P. Out, C.W. Yvan and L. Li, Appl. Phys. Lett. 53 (1988) 2675.
4   R.B. Laibowitz, R.H. Koch, P. Chaudari and R.J. Gambino, Phys. Rev. B 35 (1987) 8821.
5   D. Dijkkamp, T. Venkatesan, X.D. Wu, S.A. Shaheen, N. Jisrawi, Y.M. Min-Lee, W.L. Lean, and M. Croft, Appl. Phys. Lett. 51 (1987) 619.
6   J.K. Wo, T.C. Hsien, R.M. Flemming, M. Hong, S.H. Liou, B.A. Davidson and L.C. Feldman, Phys. Rev. B 36 (1987) 4039.
7   R.L. Sandstrom, W.J. Gallagher, T.R. Dinger and R.M. Koch, Appl. Phys. Lett. 53 (1988) 444.
8   F.A. Smidt, Inst. Mater. Rev. 35-2 (1990) 61.
9   W.K. Chu, J.W. Mayer, M.A. Nicolet, Backskattering spectrometry, Academic press. W.Y. (1978).
10 J. Keinonen, J. Raisanen and E. Rankala, Studies of high temperature superconductors, vol.4, Nova Science Publ. N.Y. (1990) 239.
11 L.R. Doolittle, Nucl. Instrum. Methods B9 (1985) 344.
12 A. Naudon, J. Chihab, P. Goudeau and J. Mimault, J. Appl. Cryst. 22 (1989) 460.

High T$_c$ Superconductor Thin Films
L. Correra (Editor)
© 1992 Elsevier Science Publishers B.V. All rights reserved.

Thin YBaCuO superconductor film deposited by high power and high repetition rate excimer lasers

M. Gerri[a], W. I. Marine[a], N. Merk[a], D. Pailharey[a], Y. Mathey[a], M. Sentis[b], P. Delaporte[b], B. Forestier[b] and B. Fontaine[b]

[a]URA CNRS 783 and [b]I.M.F.M. UM 34 CNRS Faculté des Sciences de Luminy, Case 901, 13288 Marseille Cedex 9, France

Abstract

Thin YBaCuO films on MgO substrates were prepared using pulsed laser ablation method. The film quality is mainly controlled by the deposition rate, the laser fluence and the substrate temperature. The best quality films are produced when the deposition parameters are low enough to allow well-oriented grain growth with c-axis perpendicular to the substrate surface. Higher deposition parameters result in higher crystal nucleation rates leading to a gradual loss of epitaxy.

1. INTRODUCTION

Thin film deposition by pulsed laser ablation is now one of the most promising methods to produce thin Y-Ba-Cu-O films in an oxygene ambient. High power and high repetition rate lasers open up a new way of producing single phase superconductors over large substrate areas [1]. However both the morphology and the properties of these films are strongly dependent on the deposition parameters such as the substrate temperature, the deposition rate and the laser fluence. In this paper we present the microstructural evolution of Y-Ba-Cu-O thin films deposited on MgO as a function of these parameters.

2. EXPERIMENTAL TECHNIQUES

Deposition was carried out using an X-rays pre-ionized discharged pumped excimer XeCl laser (λ=308 nm). The laser was operating with 40 ns output pulses, at repetition rates of 5 to 200 Hz. The pulse energy was 1 to 8 J/cm^2. The target was a 13 mm diameter and a 1.5 mm thick disk processed with the usual procedure of mixing, calcining, regrinding and sintering. The laser beam was focused at 40° on a 300 μm (FWHM) spot on the target (rotated constantly during the deposition). The films were deposited on (100) MgO single crystals of 10x10 mm in size mounted on a heater block at 3 cm from the target. Oxygen was introduced at a distance of about 2-3 mm from the substrate with an equilibrium pressure of 160 mTorr.

The films prepared at high laser fluence were deposited by the same number of laser shots. The films thickness (of about 750nm) was not dependant on the laser repetition rate. The change in the films properties was related only to effective deposition rate change.

The observation and analysis were carried out by means of scanning, transmission and high resolution electron microscopy (SEM, TEM, HREM), energy dispersive spectroscopy (EDS) and X-ray diffractometry (XRD). The superconducting properties were caracterized by resistivity measurement using the conventional four point probe technique.

3. RESULTS AND DISCUSSION

The films were produced at different substrate temperatures (620-700°C) and different oxygen pressures (50-600 mTorr), and then rapidly cooled down to the ambient temperature at high oxygen pressure ($2.6 * 10^{+4}$ Torr). The overall composition of the YBaCuO films was confirmed to be 123 phase by energy dispersive spectroscopy.

The best quality superconducting films on (100) MgO are produced for a substrate temperature of 650°C for a deposition rate of 5 Hz under a low laser fluence (2 J/cm^2). The characteristics typical of such 123 films are presented in Fig. 1. The sharp resistive transition (Fig.1.a), in agreement with the textured XRD pattern (Fig.1.b) suggests a fairly well oriented film i.e. a good epitaxy at the interface with the c-axis of YBCO grains perpendicular to the MgO substrate surface. Accordingly, besides the commonly observed 123 droplets detached from the target during the explosive laser pulse, the surface microstructure of the film (Fig.1.c) is smooth and featureless. TEM observations confirm that the films are almost entirely c-oriented. A typical conventional TEM micrograph in a planar view together with the corresponding diffraction pattern obtained from two large grains misoriented through an angle of 32.5° is presented in Fig. 1.d. The grains contain a number of nanocrystalline (10 nm size) inclusions distinguished by their moiré fringes of well defined directions suggesting a preferred orientation of the precipitates in relation with the 123 matrix (formation of different variants). The small size of the inclusions in comparison to the surrounding matrix does not allow a proper compositional variation determination upon EDS analysis. However, the measurements of the lattice image spacing in addition to a careful indexing of the very weak diffraction spots corresponding to the oriented inclusion suggest a cubic type structure with a=0.407 nm. In agreement with previous observations [2], the inclusions may be identified as the YBa$_3$Cu$_2$O$_{7-x}$ equilbrium phase which has a distorted perovskite structure with a=0.407 nm and c=0.401 nm.

One of the almost perfect epitaxial relationships of the thin film to the substrate is illustrated in the high resolution planar view of Fig.2. Indeed, under 2g condition, where the (200)$_{123}$ diffraction vector interferes with the (200)$_{MgO}$, the predominantly parallel moiré pattern suggest a good epitaxy.

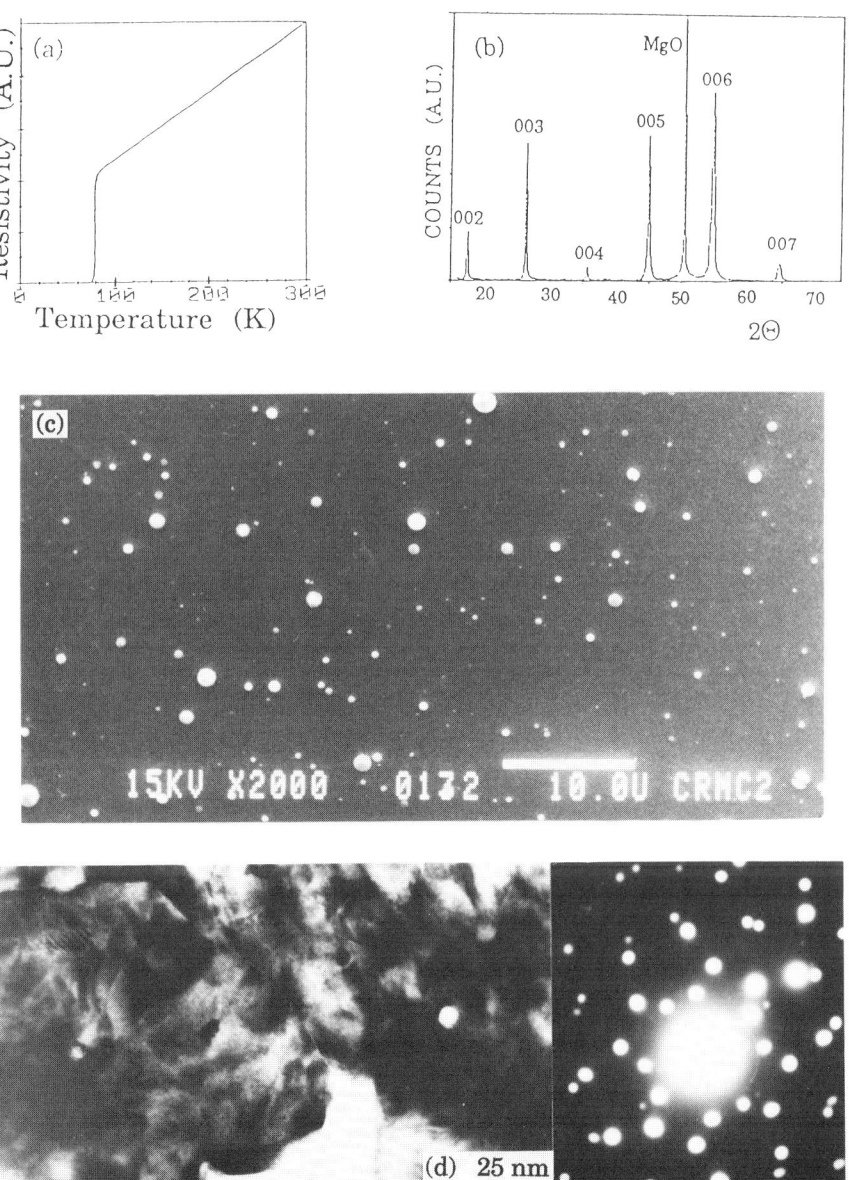

Figure 1. a) Resistivity curve, b) XRD pattern, X-rays wavelength was λ=0.1788nm (Co Kα), c) SEM and d) TEM micrographs typical of the best quality films produced under $2J/cm^2$ fluence with a repetition rate of 5 Hz.

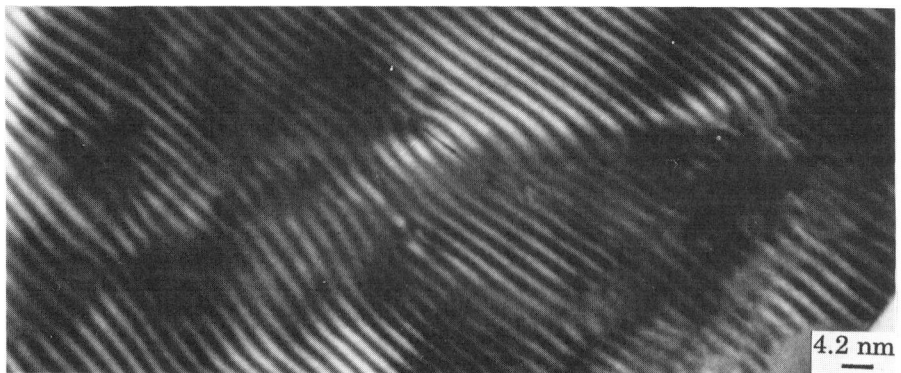

Figure 2. Typical high resolution planar view along [001] direction of the 123 film on MgO resulting in a predominantly parallel moiré pattern which intensifies the misfit dislocations.

The moiré fringes which have an expected periodicity of D= 2.1 nm intensify strongly the crystalline defects i.e. the misfit dislocations present due to the lattice mismatch and the surface atomic steps of the cleaved MgO.
    At higher fluences (5-6J/cm^2), the film quality degrades. Although the Tc onset is about 84-90 K (see Fig.3.a), the transition width is rather large (ΔT>23K). The resistivity ratio at 300K and 100K of about 1 was only observed for the films deposited at 25 Hz laser repetition rate. The slight emprovement of the electrical behaviour of these films can be explained by the formation of a surface microstructure denser than the one observed with films deposited at 10 Hz. However the effective deposition rate increase (100-200Hz laser repetition rate) leads to a strong transition width enlargement. The X-ray diffraction patterns of these films (see Fig 3.b) exhibit the presence of (00l) Bragg peaks but also contain the (110) and the (h00) reflections with an increasing peak intensity upon frequency suggesting a gradual loss of epitaxy.
    The evolution of the 123 surface microstructure observed by SEM as a function of the repetition rate, under a laser fluence of 5-6 J/cm^2 is presented in Fig.4. The smooth and featureless surface of the best films (Fig.1) transforms to a rougher surface containing crystal grain outgrowth with elongated rectangular morpholgy. In agreement with previous observations [3] the rectangular crystals are 123 grains with c-axis parallel to the substrate surface. The higher the repetition rate the larger is the crystal outgrowth density. Furthermore as the repetition rate increases, the growth axis directions in the plane become more and more random. Finally for a very high frequency (200Hz) a much higher rate of crystal nucleation leads to the formation of nanocrocrystalline films. TEM planar view micrographs presented in Fig. 5 confirm the presence of crystals with c-axis parallel to the MgO surface for medium frequencies and the nanocrystalline nature of the films produced at 200 Hz. The increase in grain boundary surface area leads to an increase in current leaking which explains the resistivity behaviour.

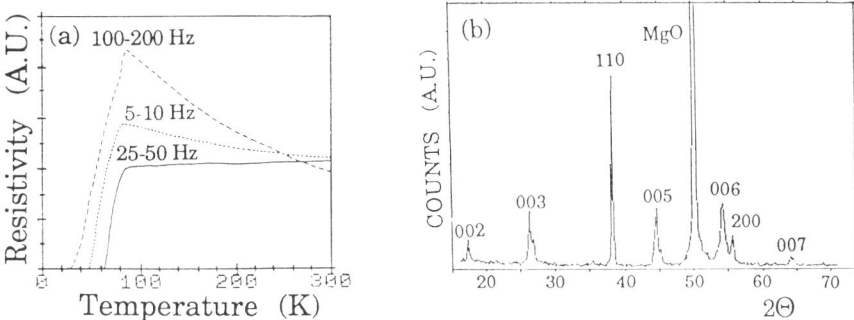

Figure 3. a) Resistivity curves of thin films produced respectively at 5-10 Hz, 25-50 Hz and 100-200 Hz and b) XRD pattern corresponding to a 50 Hz film. X-rays wavelength was λ=0.1788nm (Co Kα).

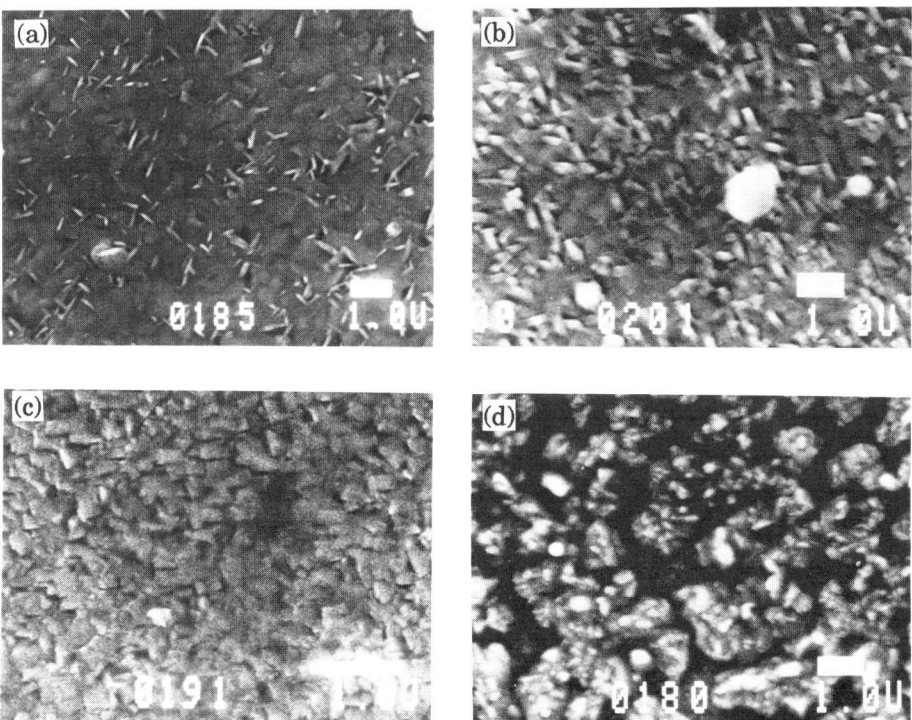

Figure 4. SEM micrographs obtained from thin films produced under laser fluence 6J/cm^2 and at a) 10 Hz, b) 50 Hz, c) 100 Hz and d) 200 Hz showing the gradual increase in density of crystal outgrowths with frequency. Substrate temperature was 650°.

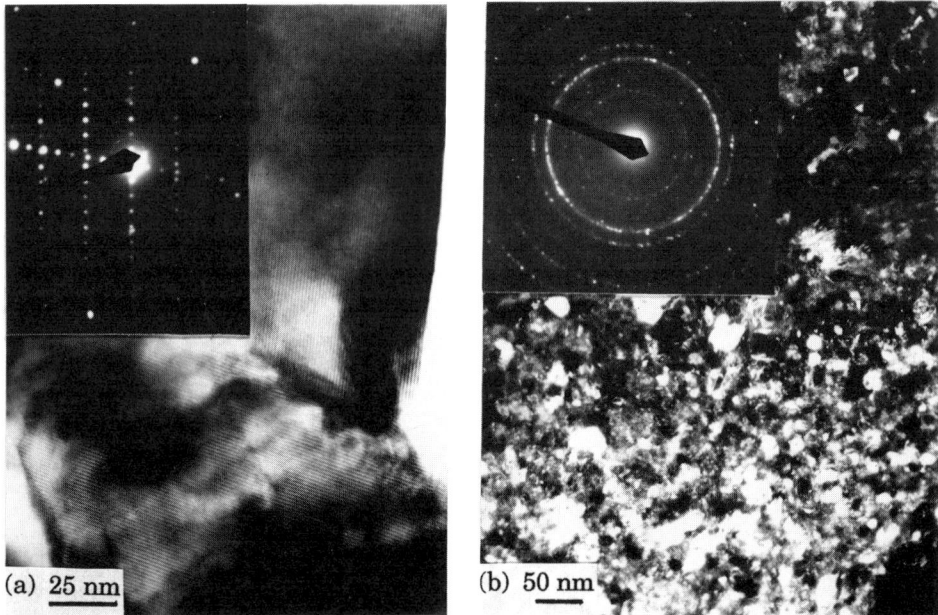

Figure 5. TEM micrographs and the corresponding selected area diffraction patterns of a) grains oriented with c-axis parallel to the MgO surface for a thin film produced at 25 Hz and b) nanocrystalline microstructure of thin films produced at 200 Hz.

4. CONCLUSIONS

During laser ablation, low repetition rate and low laser fluence give rise to the formation of perfectly c-oriented 123 thin films. Higher repetition rate and/or higher laser fluence lead to a gradual loss of epitaxy up to the formation of randomly oriented nanocrystalline films which affect the resistivity temperature and the width of transition due to the increase of grain boundaries.

5. REFERENCES

1   M. Gerri, W. Marine, M. Sentis and P. Delaporte, to be published in Revue de Métallurgie 88, 321 (1991).
2   R. Ramesh, A. Inam, D. M. Hwang, T .D. Sand, C. C. Chang and D. L. Hart, Appl. Phys. Lett., 58 (1991) 1557.
3   A. H. Carim, S. N. Basu and R. E. Muenchausen, Appl. Phys. Lett., 58 (1991) 871.

High T$_c$ Superconductor Thin Films
L. Correra (Editor)
© 1992 Elsevier Science Publishers B.V. All rights reserved.

LASER DEPOSITION OF THIN FILMS AND CHARACTERIZATION OF SUPERCONDUCTING BiSrCaCuO AND YBaCuO

Anna Giardini Guidoni, Marianna Ambrico, Antonio Morone, Luigi Maritato+, Rosa Scafuro+, Elio Desimoni°, Anna Maria Salvi°, Veronica Marotta

Istituto Materiali Speciali, C.N.R., Tito Scalo (PZ) Italy

+Phys. Dept. Università di Salerno, Salerno Italy

°Chem. Dept. Università della Basilicata, Potenza, Italy

Abstract

An investigation on "in situ" laser induced deposition of superconductor $Bi_2Sr_2CaCu_2O_x$ (BSCCO) and $YBa_2Cu_3O_x$ (YBCO) material is here reported. The laser irradiation of the solid sample was performed by a frequency doubled Nd-YAG laser. The thin films deposited on $SrTiO_3$ substrate were annealed in a high temperature oven. Resistance measurements showed the presence of superconductive transition above 80 K for BSCCO. XPS analysis of "as deposited" and annealed films appears to indicate that in the annealed sample the oxygen and copper valence are restored. Laser ionization mass analysis studies, performed on BSCCO sample showed the presence of cluster ions.

1. INTRODUCTION

The study of advanced materials, their preparation and their examination has undergone an explosive growth in the last few years [1]. Since superconductivity with an onset temperature above 100 K was discovered in the BSCCO system, the investigation of copper-based oxides and the different methods of deposition have undergone a considerable development. Laser ablation of BSCCO and YBCO is a very promising technique in the preparation of the films [2].

The study of the evaporation process of the ablated species, ions and neutrals, is important in understanding fundamental aspects of the chemistry and physics of deposition [3]. This work presents the analysis of BSCCO solid target made from mixtures of simple oxides, the volume density of the target (BSCCO) was generally about seventy % of the theoretical value. The production of superconducting thin films will be discussed in relation to previous studies on laser formation of ion clusters and annealing effects on the deposited films. Some preliminary results on YBCO will be also reported.

2. EXPERIMENTAL

The ablative photodecomposition of BSCCO and YBCO pellets has been performed using a frequency doubled ($\lambda = 532$ nm) or quadrupled ($\lambda = 266$ nm) Nd-YAG Quantel 581 laser. The experimental system has been described in detail previously [4] and here will be only summarized.

The six arms stainless steel deposition chamber is equipped with a quartz window such that the rotable target pellet can be irradiated by the laser light at normal incidence. Irradiation is accomplished using beam energies of 2.5 - 3 $J \cdot cm^{-2}$ .Before irradiation the chamber was pumped down to 1.5 x IO^{-6} mbar. The ablated material originates a luminous plume about 2,5 cm long. An $SrTiO_3$ substrate glued by silver paint to a mechanical holder was positioned at 2,5 cm distance from the target and tilted by 45° with respect to the laser. A deposition rate on the substrate of 0,18 nm $\cdot$ s $^{-1}$ was measured. The deposited film thickness was determined by an a step profilometer to be about 1 µm after 2 hours of irradiation. The produced films of BSCCO, deposited in high vacuum condition, were annealed in an air oven at 880° for 2 hours. The YBCO deposits were annealed "in situ" in presence of oxygen at 450°.Analysis of standard pellets and of the produced thin film either "as deposited" and after annealing was performed by SEM and XPS. XPS spectra were collected by an LH XI Leybold spectrometer using unmonocromatized Mg K_α radiation (1253.6 eV) for BSCCO and Al K_α radiation (1486.6 eV) for YBCO. The experimental conditions were similar to those previously discussed [4,7]. Annealed films were also characterized by X-Ray diffraction (XRD) and energy dispersive spectrum (EDS). Resistance measurements of the annealed samples were performed with a standard four-probe technique.

3. RESULTS AND DISCUSSION

YBCO. The stoichiometry of the deposited films analyzed by INA3 technique [5], give a ratio of 1:2:3 for Y:Ba:Cu respectively .

It was also previously shown [7] that Copper valence is restored after in situ annealing at 450° at an oxygen pressure of 3.8 mbar.

However, the resistance of the annealed film was not zero at 4.2 K and showed an onset with a fast decrease at about 90 K.

BSCCO. Fig. 1 shows the resistance (R) versus temperature (T) curve for an annealed BSCCO film. It can be seen that the Tc (R=O) value is 73 K and the ratio R(300 K)/R(100 K) is about 1.2. The Θ - 2Θ X-ray diffraction pattern is reported in fig. 2; it clearly shows that the film is polycristalline with only a slight preferential c-axis orientation, of the grains. The observed peaks are related to the 2:2:1:2 (Bi:Sr:Ca:Cu) phase (Tc 85 K).

The EDS spectrum, not reported here, indicates that the overall stoichiometry of the film is 2:2:1:2 .

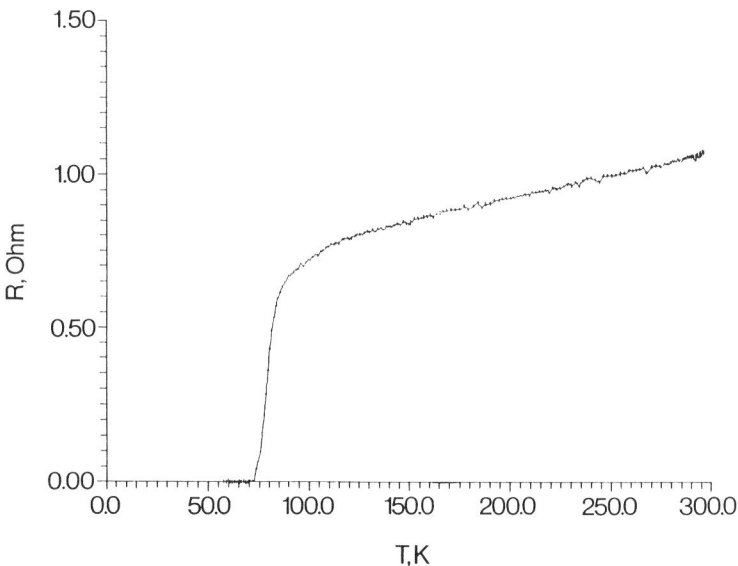

Figure 1. Resistance as a function of temperature of an annealed BSCCO thin film produced by laser deposition.

Characterization of the deposits has been also performed by XPS surface analysis. In fig.3 the XPS wide spectra of the BSCCO target and of the BSCCO films deposited on SrTiO$_3$ substrates before and after annealing treatment show that all the constituent elements of the sample are observed.

Detailed spectra of all these elements have been performed in order to carry out speciation analysis [6]. Fig. 4 compares the O$_{1s}$ region of the target and deposited film before and after annealing.

The structure found in the O$_{1s}$ region displays two contributing peaks. As reported in a review paper [6], the peak at higher Binding Energy (BE) can be due to a variety of hydroxydes and carbonates, while the other can be related to oxygen in the superconducting phase. From the analysis of the curve fitting it appears that the O$_{1s}$ peak of the film depositedat 300 K, curve C of figure 4, is mostly due to the higher BE component.

The lower BE component, already present in the target (curve A), becomes preminent after annealing (curve B). This confirms the opportunity of the annealing treatment after the film deposition.

The usefulness of the annealing treatement appears evident from the analysis of the Cu$_{2p}$ region of the same samples as reported in fig. 5. The comparison of spectra B and C, indicates that annealing allows to increase the Cu(II)/Cu(I) ratio. Spectrum B is quite similar to spectrum A relevant to an as-received BSCCO target. This is in agreement with previous findings [4,7] which have shown that the copper valence is reduced by ablating pure copper (11) oxide.

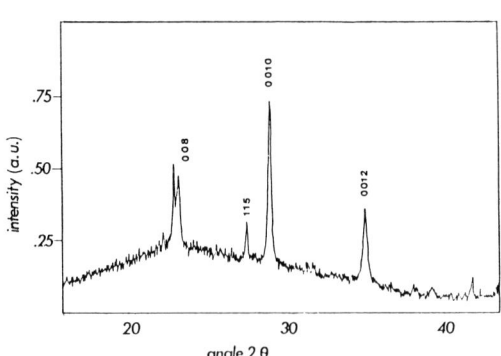

Figure 2 X-ray diffraction pattern of the annealed BSCCO film.

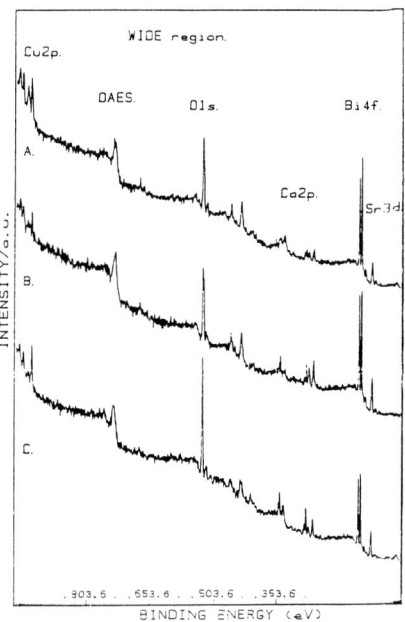

Figure 3. XPS wide spectra of BSCCO samples: A)BSCCO target B) annealed BSCCO film, C)as deposited BSCCO film.

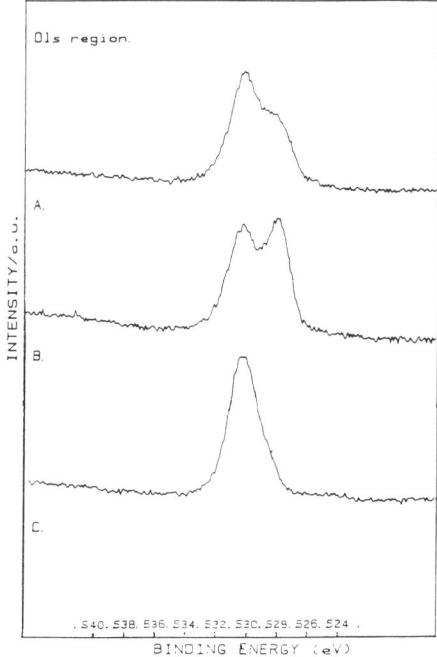

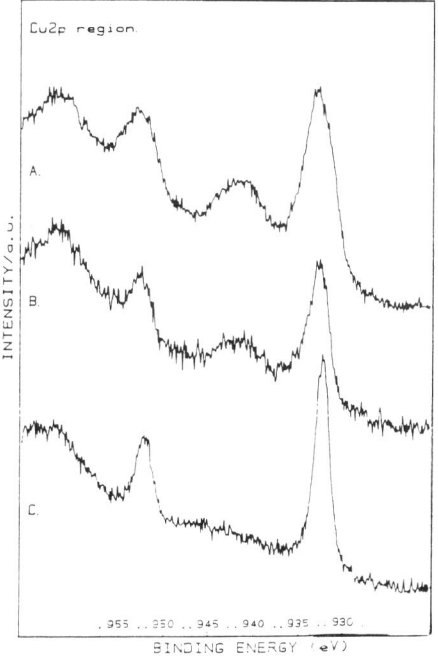

Figure 4. XPS spectra of O_{1s} region in BSCCO samples: A) BSCCO target, B) annealed BSCCO film, C) as deposited BSCCO film.

Figure 5. XPS spectra of Cu_{2p} region of BSCCO sample A) BSCCO target, B) annealed BSCCO film, C) as deposited BSCCO film.

In order to ascertain the role played by the transient species present in the plume and which can give information on the aggregation process in plume region, laser ionizzation mass analysis (LAMMA) has been performed and is reported in fig. 6.

It can be noted that masses of mixed positive cluster ions as $Sr_2Ca_2O_2^+$, $Sr_2CaO_3^+$ and others, containing only two of the four metal oxides of the BSCCO material are present. This feature is characteristic also of other metal oxide mixtures and of oxide mixtures of superconductor precursors [3].

At the present the mass spectrometric plume analysis has been, performed only in the region near to the laser hit surface. Aggregates of larger dimensions could be formed at longer distance from the target surface. Further work is in progress in our laboratory for a better understanding of deposition process dynamics.

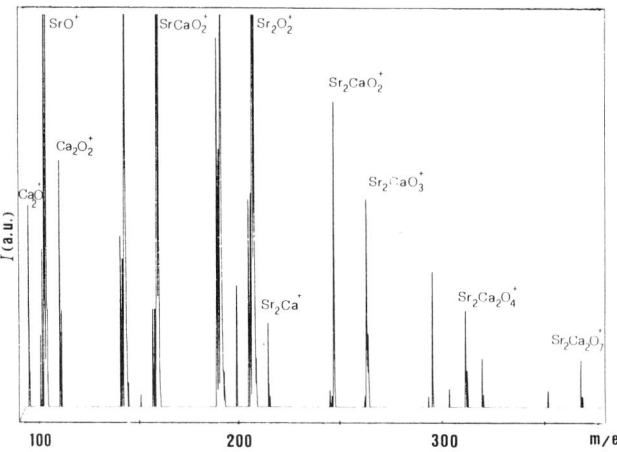

Figure 6. Laser Ionization mass spectrum of BSCCO target.

AKNOWLEDGMENTS

This work bas been partially supported by CNR "Progetto Finalizzato Tecnologie Elettroottiche" and "Progetto Finalizzato Chimica Fine.

REFERENCES

1. Deposition technologies for films and coatings (Noyes New Jersey 1982).

2. D.Bäuerle, Laser Processing and Diagnostics (Springer-Verlag, Berlin, 166 (1984)). T. Venkatesan, Appl. Phys. Lett. 53 (1988) 1431.

3. A. Mele, A. Giardini-Guidoni, D. Consalvo, D. Stranges, R. Teghil, Int. J. Mass. Spectr. Ion. Proc. 91 (1989) 319.

4. A. Giardini-Guidoni, A. Morone, M. Snels, E. Desimoni, A.M. Salvi, R. Fantoni, W.C.M. Berden, M. Giorgi, Appl. Surf. Sci. 46 (1990) 321.

5. A. Giardini-Guidoni, R. Teghil, A. Mele, Spectrochimica Acta 46A 503,1990

6F.A. Shama, J.C. Fuggle, Physica C, 169 (1990), 325.

7. A. Giardini-Guidoni, E. Desimoni, A.M. Salvi, R. Teghil, M. Ambrico, A. Morone, S. Piccirillo, M. Snels, Proc. of Intern. Conf. "Laser 90", San Diego, USA 1990, in Press.

High T$_c$ Superconductor Thin Films
L. Correra (Editor)

The properties of YBa$_2$Cu$_3$O$_{7-x}$ thin films prepared by MOCVD

O.Yu.Gorbenko[a], A.R.Kaul[a], S.V.Pozigun[b], Yu.D.Tretyakov[a]

[a]Chemistry Department, Moscow State University, Leninskie Gory, Moscow, 119899, USSR

[b]Department of Atomic Energy Institute, Moscow, USSR

Abstract

YBa$_2$Cu$_3$O$_{7-x}$ thin films were obtained by MOCVD using especially developed set-up allowing to try different thermal and flow regimes. Influence of process conditions on morphological properties is discussed. It is shown that small composition deviations may lead to striking difference of the morphology as well as magnetic and transport properties.

1. Introduction

Film deposition from vapours of metallorganic compounds (MOCVD) belongs to a number of HTSC film fabrication technologies with bright outlook. Providing all technological advantages of chemical techniques (availability of large area covering, possibility of multilayer structure preparing, etc.) MOCVD is a real alternative to physical deposition techniques (PVD) as far as morphological perfection and electrophysical properties of films are concerned [1]. In the paper the influence of thermal conditions and gas-dynamics of MOC vapour transport on composition and morphology of films is discussed. Then there is a comparison of properties of YBa$_2$Cu$_3$O$_{7-x}$ films prepared by MOCVD which demonstrates influence of small deviations of film composition.

2. Experimental

2.1 Set-up

Preparation of films by MOCVD technique was carried out *in vacuo* using especially developed set-up including 5 gas-lines (4 for career-gas - Ar, 1 for oxygen), 3 vapourizers of MOC with individual thermal control, manifold with 3 thermal zones, quartz reactor. The set-up included electronic controllers of flows and temperature.

Set-up used contained horizontal hot-wall reactor allowing to try different flow regimes.

2.2 Volatile compounds

These were dipivaloylmethanates of Y, Ba and Cu : $Y(DPM)_3$, $Ba(DPM)_2$, $Cu(DPM)_2$ [DPM = $C_{11}H_{19}O_2^-$]. Barium dipivaloylmethanate was dihydrate $Ba(DPM)_2 \times 2H_2O$ obtained by reaction of saturated aqueous solution of $Ba(OH)_2$ with dipivaloylmethane $C_{11}H_{20}O_2$. Volatile complexes of Cu and Y were purified by vacuum sublimation.

2.3 Process conditions

Vapourizer temperature constant during individual experiment varied as follows : $Ba(DPM)_2$ - 240 - 270 °C, $Cu(DPM)_2$ and $Y(DPM)_3$ - 120-150 °C. Manifold zone temperature was maintained 10 °C above temperature of corresponding vapourizer (fig.1a). Argon flows through all vapourizers were the same (100 ml/min at s.s.) and constant during experiment. Oxygen feeding was accomplished direct to reactor inlet using inner coaxial tube in the manifold. Ratio of flows was $\Sigma V_{Ar}/V_{O2} = 2:1$ at overall gas flow 600 ml/min. Overall gas pressure in reactor was 40-80 torr. Monocrystalline MgO (100) was used as a substrate material.

At *ex-situ* deposition substrate temperature varied in range 400-600 °C then high temperature annealing took place in separate furnace at 900-950 °C (10-30 min) in oxygen flow followed by cooling during 1 h down to 400 °C, staying for 1 h and cooling with furnace.

If *in-situ* process was carried out direct after deposition at 750-850°C cooling with furnace took place down to 400°C, then staying for 1 h and cooling with furnace down to room temperature.

2.4 Diagnostics of films

Films obtained were characterized by SEM with EPMA, XRD patterns and resistivity measurements. Magnetic susceptibility was measured by AC susceptometer using mutual inductance balancing bridge technique with exciting magnetic field frequency 640 Hz and amplitude h_o from 8 mOe to 8 Oe. The magnetic properties of films in external magnetic field H_{ext} normal to the film surface were investigated at constant values of H_{ext} while temperature was scanning from 4.2K to 100K. Magnetization loops M(H) were measured at 4.2K.

Problem of approximation to desirable cation stoichiometry was solved using diagnostic cycle : calculation of vapour composition (by MOC evaporation rate) → determination of film composition → correction of evaporation rate by means of vapourizers temperature alterations.

3. Results and Discussion

Fig. 1a demonstrates results of experiments in film deposition using MOC vapour with ratio $Y(DPM)_3/Ba(DPM)_2/Cu(DPM)_2 \approx 1:2:3$. It shows that deposition regimes used lead to deviation between cation composition of film and that of vapour. Composition deviations were not occasional but on the contrary were determined basically by thermal and gas dynamic conditions. For instance deposition at 450°C resulted in good correspondence of cation compositions of film and vapour (fig. 1a, points 1,1'). At the same time composition variation along film surface were negligible and comparable with the accuracy of EPMA. Deposition temperature growth led to effect of barium loss in film composition (fig.1a, points 2,2′,3,3′).

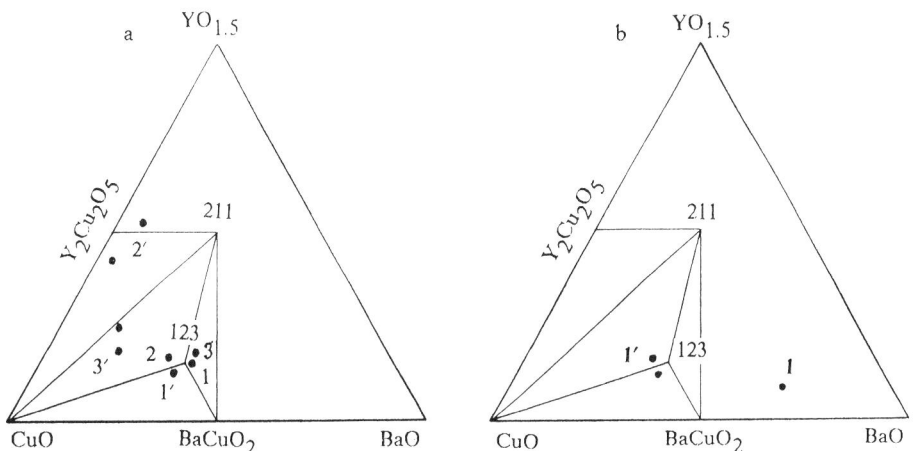

Figure 1. Comparison of film and vapour compositions.
(**a**): initial vapour composition for ***ex-situ*** (1), ***in-situ*** with laminar flow (2) and stagnation flow (3) regimes; 1′,2′,3′ - final film compositions correspondently. (**b**): 1,1′ - vapour and film compositions accounting barium loss during ***in-situ*** process.

The results differ from Schmaderer's data [2] concerning film deposition in cold-wall reactor where composition deviation from stoichiometry took place at the temperature $\leq 600°C$, but at higher temperature cation composition of film reproduced that of vapour. We believe the discrepancy results from reactor type difference. Barium loss observed in hot-wall reactor evidently was a consequence of interaction between Ba-containing vapour and reactor material (quartz) with the formation of boundary layer containing barium silicates.

Near stoichiometric $YBa_2Cu_3O_{7-x}$ composition was achieved in modified conditions of ***in-situ*** process using mixture of MOC vapours with intentional excess of barium (mol. fraction of $Ba(DPM)_2 > 0.50$, fig. 1b). Nevertheless films with average $YBa_2Cu_3O_{7-x}$ stoichiometry showed some nonuniformity of composition according to EPMA. We tried to use the fact for the investigation of small composition deviations influence on film properties. One film (13×13 mm) was divided into smaller shares, two corner parts (3×3 mm) were investigated more carefully. The parts had close but not the same composition. They belonged to neighbouring triangles of phase diagram Y_2O_3-CuO-BaO : 123-CuO-$BaCuO_2$ (sample 1), 123-CuO-211 (sample 2). Compositions were 53, 32, 15 and 52, 30, 18 % mol. CuO, BaO, $YO_{1.5}$ correspondently.

Different types of microstructure were observed in dependence of approximation to stoichiometric composition and deposition conditions. Films deposited ***ex-situ*** always had chaotic arangement of shapeless grains with average size about 0.5 mμ.

There was a great variety of microstructure types for films deposited ***in-situ***. Ordinary in conditions of stagnation flow regime monotonous dense structures with more or less regular arrangement of grains were observed. More developed crystallinity differed such grains from ones prepared ***ex-situ***. Two types of film growth were observed. The first was "one-dimension" growth resulting in smooth matrix of $YBa_2Cu_3O_{7-x}$ with inclusions of additional phases. Sample 1 (fig.2a) belonged to the type like majority of films with composition corresponding phase triangle 123-$BaCuO_2$-CuO in vicinity of 123. The second was "three-dimension" growth leading to dense package of grains with rectangular habitus. Sample 2 (fig.2b) was one of the type. No wonder that according to XRD patterns the most prominent texture (00l) was observed for films like shown on fig.2a. On the contrary films prepared ***ex-situ*** had the least prominent texture.

 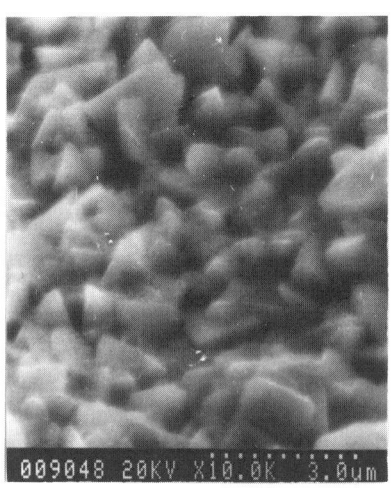

Figure 2. SEM photographs of sample 1 (**a**) and 2 (**b**) belonging to neighbouring phase triangles 123-CuO-$BaCuO_2$, 123-CuO-211 correspondently.

We consider the differences in morphology mentioned result from the influence of additional phases. The presence of such phases leads to different cation diffusion mobility during the film growth. It can be proposed this influence arises from different Tamman's temperature in possible phase assemblies. The good evidence of this may be the lowest temperature of liquid phase appearance (890°C in air [3]) in Y123-CuO-$BaCuO_2$ coexistence in comparison with other equilibrium phases assemblies of Y_2O_3-BaO-CuO system.

Diffusion in ***ex-situ*** prepared films may be slowed down by $BaCO_3$ arising during high-temperature annealing of the precursor film. This kinetic difficulty should prevent texture development.

The curves $x'(T)$ representing the measurement of real part of AC susceptibility versus temperature under exciting field for sample 1 are shown on fig.3a. The critical temperature for sample 1 (calculated as onset of superconductive magnetic

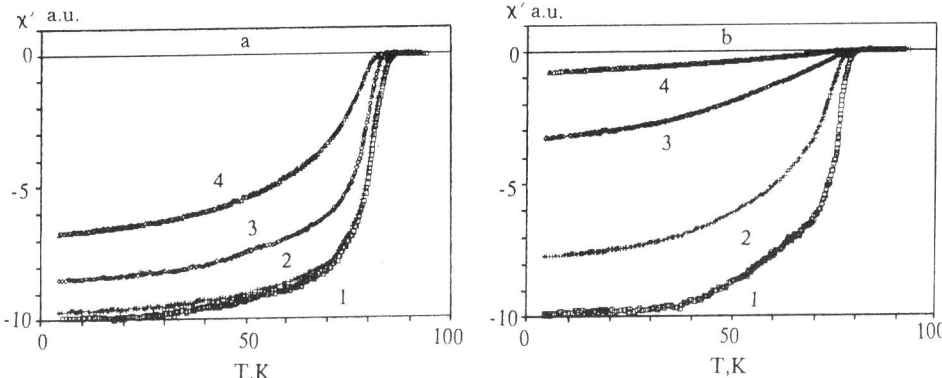

Figure 3. (**a**): χ′(T) curves for sample 1 under various exiting fields h_o (1- 8mOe, 2- 80 mOe, 3- 0.8 Oe, 4- 8 Oe). (**b**): anologous curves for sample 2.

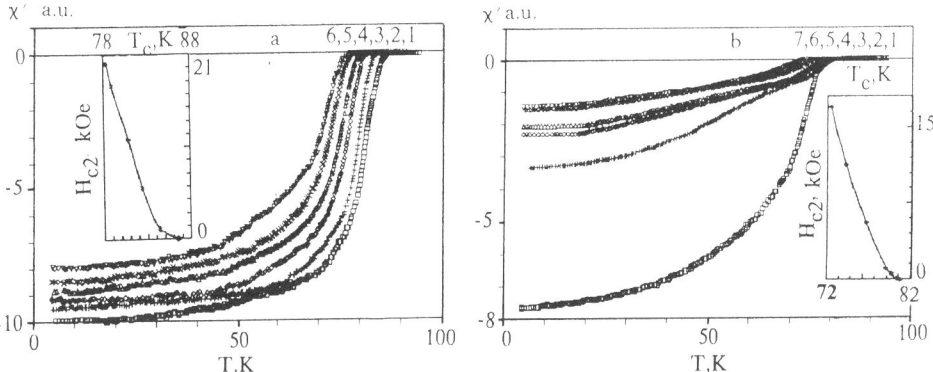

Figure 4. (**a**): series of χ′(T) curves for sample 1 under various external fields H_{ext} (1- 0kOe, 2- 1.0kOe, 3- 5.0kOe, 4- 9.8kOe, 5- 15.0kOe, 6- 17.2kOe) with corresponding $H_{c2}(T_c)$ diagram; (**b**): anologous data for sample 2 under H_{ext} (1- 0Oe, 2- 99Oe, 3- 504Oe, 4- 1.0kOe, 5- 5.0kOe, 6- 10.0kOe, 7- 15.0kOe).

transition under h_o = 8 mOe) was 86.5 K. With increasing h_o the superconductive transition width grew monotonously and critical temperature reduced. The analogous series of the χ′(T) curves for sample 2 at the same values of h_o is shown on fig. 3b (critical temperature was 81.5 K). Comparison of 3a and 3b points out the higher field stability of sample 1 than that of sample 2. Although the contents of the Meissner phase was comparable for both samples it reduced faster for sample 2 when exciting field increased.

Fig. 4a represents series of the $\chi'(T)$ curves under various H_{ext} for sample 1 (h_o = 8 mOe). For all H_{ext} values up to the highest one (17,6 kOe) the superconductive transition was one-phase with width monotonously growing and educing T_c as well as Meissner phase contents. The $H_{c2}(T_c)$ phase diagram derived from the curves is shown in corner of fig. 4a. Sample 2 demonstrated two-phase behaviour at $H_{ext} > 0$ and T_c reducing when H_{ext} grew. Corresponding $H_{c2}(T_c)$ diagram is inserted in fig. 4b.

Such great difference of magnetic properties of samples undoubtedly is connected with their morphology. Comparatively weak field stability of sample 2 and two-phase superconducting transition nature at $H_{ext} > 0$ point to field penetration into inter-grain medium. The sample's morphological features (espesially the degree of texture [4]) strongly affect on transport properties too. Using Bean's formula from M(H) loops for samples 1,2 J_c values were estimated as 1.6×10^6 A/cm^2 for sample 1 and 7×10^4 A/cm^2 for sample 2. So separate grains structure and weaker texture resulting from small deviation of film composition are responsible for lower values of j_c and T_c obtained from magnetic measurement for sample 2.

4. Conclusions

Thus *in-situ* MOCVD of $YBa_2Cu_3O_{7-x}$ leads to obtaining films with higher morphological perfection then *ex-situ* deposition. At the same time small deviations of final film composition result in great difference of critical magnetic properties. The best values of j_c and T_c may be achieved for composition belonging to Y123-CuO-BaCuO$_2$ assembly in vicinity of 123-stoichiometry.

5. References

1   R.Simon, Solid.St.Techn., 32 (1989) 141.
2   F.Schmaderer et al., Appl.Surf.Sci., 46 (1990) 53.
3   T.Aselage, K.Keefer, J.Mater.Res., 3 (1989) 1279.
4   T.Hirai, H.Yamane, Oyo Buturi (Appl.Phys.), 59 (1990) 134.

High T$_c$ Superconductor Thin Films
L. Correra (Editor)
463

Preparation and Properties of superconducting thin Bi-Sr-Ca-Cu-O
films from metalorganic complexes

H. Gruber[a], E. Krautz[a], H.P. Fritzer[b] and A. Popitsch[c]

[a]Institut für Festkörperphysik, Graz University of Technology,
Petersgasse 16, A-8010 Graz, Austria

[b]Institut für Physikalische u. Theoretische Chemie, Graz
University of Technology, Rechbauerstraße 12, A-8010 Graz, Austria

[c]Institut für Anorganische Chemie, University of Graz,
Universitätsplatz 1, A-8010 Graz, Austria

Abstract
   Thin films of the 2212 and 2223 phase in the Bi-Sr-Ca-Cu-O
system have been produced by decomposition of organic precursor
compounds containing different metallo-organic complexes on
different substrates such as MgO-single-crystals, Ag-ribbons
and Au-ribbons. SEM, EDX and X-ray diffraction patterns have
been used for the detection of the microstructure and composition of the prepared films.

1. INTRODUCTION

   Thin films of high-T$_c$-metal-oxide-superconductors allow applications to the communication, computation, sensor technology
and transmission technology. For all these different applications the superconducting materials have to be fabricated in
suitable geometric forms. For the synthesis of high superconductors including Y-Ba-Cu-O, Bi-Sr-Ca-Cu-O and Pb-Bi-Sr-Ca-Cu-O
compounds polymer-metal-complex (PMC) precursor processes have
now been reported [1-7]. In this contribution a special process
for preparation of superconducting layers in the system
Bi-Sr-Ca-Cu-O has been investigated.

2. PREPARATION

   As starting materials are used metal nitrates as commercially
suprapure available powders and ethylenediamine or N,N,N',N'-
tetramethylene and pyridine. The complexes are prepared from

mixtures of metal-nitrates of the nominal composition exactly
for the 2212 and 2223 phases with organic solvents. This precursor paste is then used for the depositions on substrates
such as (100)-MgO single crystals or polycrystalline silver or
gold ribbons which are then dried. After that the films are
heated at 850°C, in the case of the (100)-MgO substrate for 12 h
at 840°C, in case of Ag- and Au-ribbons for 12 h at 850°C for
the 2212-phase and repeated 14 h in atmosphere consisting of
90% Ar and 10% O_2 for the 2223-phase. The samples with 2223-
phase were also heated for 27 h at 850°C in the same gas mixture. The composition and the phases of the different samples
have been controlled by X-ray diffraction patterns and EDX. SEM
and EDX have been used for the detection of the microstructure
of the films. Microsectioning has been applied for the determination of the thickness of the films. The first 10 measured
samples had a thickness close to 10 μm. The application
of our paste allows, however, to prepare thinner as well as
thicker HT$_c$-films.
   The resistivity has been measured with a standard four point
dc method. The electrical contacts were attached to the superconducting film with silver paint.

3. RESULTS AND DISCUSSION

3.1 Films of the 2212-phase
   The SEM micrograph of the produced Bi-Sr-Ca-Cu-O films on a
MgO-(100)-single crystal by the new method is shown in Figure 1.
The SEM picture shows clearly the layered structure of the film
on the MgO substrate. The X-ray diffraction patterns for the
2212-phase on different substrates is shown in Figure 2 and 3.

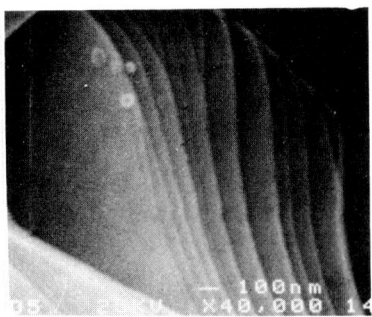

Figure 1. SEM micrograph of a single phase
2212 film of Bi-Sr-Ca-Cu-O on a MgO-single
crystal plane (100)

The FWHM value of the rocking curve of the (0010) reflection in
Figure 2 is 0.6°. This value is nearly the same as the value
for Bi-Sr-Ca-Cu-O films with epitaxial growth. The 2212 phase
film on the MgO-single crystal substrate was obtained by
heating 12 h at 840°C in a flowing gas mixture of 90% Ar and
10% O_2. Figure 4 shows the same phase on a Ag-ribbon.

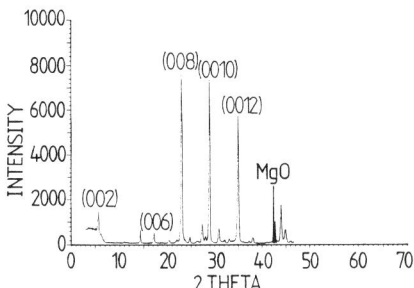

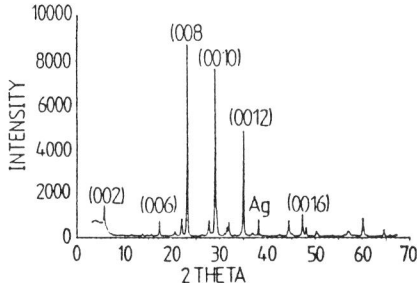

Figure 2. X-ray diffraction
patterns of the 2212 phase
film on a MgO-single crystal
plane (100).

Figure 3. X-ray diffraction
patterns of a single 2212
phase film on a polycrystalline
Ag-ribbon.

Already after a heating time of two minutes the 2212 phase is
formed. With an increased heating time at 850°C in a gas mixture
of 90% Ar and 10% O_2 an increased amount of the 2223-phase
appears.
  The single phase 2212 film on a polycrystalline Ag-ribbon is
got by heating 12 h at 850°C in a flowing gas mixture of 90% Ar
and 10% O_2. The X-ray diffraction patterns (Figure 3) indicate
that for a Ag-ribbon substrate the c-axis is also preferentially oriented perpendicular to the surfaces. Within the
resolution of EDX analysis only the 2212-phase could be detected. Because the Ag- or Au-ribbons were polycrystalline the
superconducting Bi-Sr-Ca-Cu-O films contain platelike grains
which are predominantly oriented perpendicular to the c-axis
of the substrate. From the SEM pictures (Figure 4) we conclude
that the orientation of the individual crystallites in the film
is nearly parallel with only small angle deviation in orientation to each other.
  The results of the resistivity measurements for 2212 phase
films are shown in Figure 5. The temperature dependence of the
resistivity shows two transitions when the film is heated 27
hours in a flowing gas mixture of 90% Ar and 10% O_2 on the
MgO (100) substrate crystal. The transition with an onset at
90 K belongs to the 2212 phase, the transition with an onset at
110 K belongs to the minor content of the 2223 phase. The zero

resistance of this sample is 85 K.

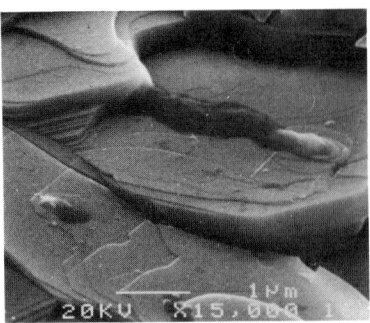

Figure 4. SEM micrograph of the single phase 2212 film
of Bi-Sr-Ca-Cu-O on a polycrystalline Ag-ribbon.

The temperature dependence of the resistivity for the thin
films on Ag- or Au-ribbons shows only one sharp transition with
a zero resistance at 79 K which belongs to the 2212 phase. The
X-ray diffraction patterns indicate only the dominating 2212-
phase in these samples.

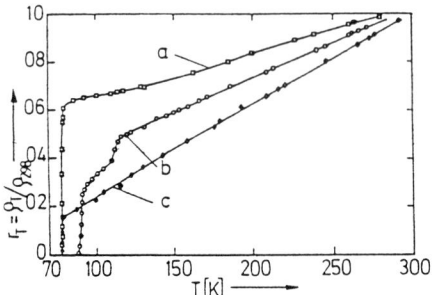

Figure 5. Temperature dependence of the resistivity
ratio r= ρ_T/ρ_{298} of Bi-Sr-Ca-Cu-O films on different
substrates.
(a) polycrystalline Au-ribbon
(b) MgO-single crystal plane (100)
(c) polycrystalline Ag-ribbon

3.2 Films of the 2223-phase

Films with mainly 2223-phase could be obtained on the poly-crystalline Ag- or Au-ribbon substrates. The corresponding SEM-micrograph of such a film on Ag-ribbon substrate is shown in Figure 6.

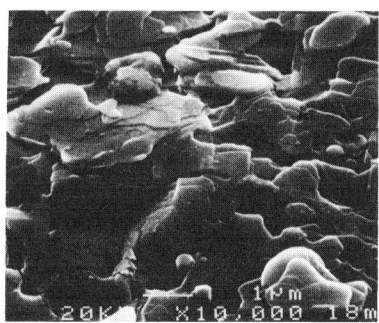

Figure 6. SEM micrograph of the 2223 phase film of Bi-Sr-Ca-Cu-O on a polycrystalline Au-ribbon.

The SEM image of the sample surface shows clearly platelike crystal grains which are also predominantly oriented with the c-axis perpendicularly to the substrate surface. Within the resolution of EDX analysis only the 2223 phase could be detec-ted.

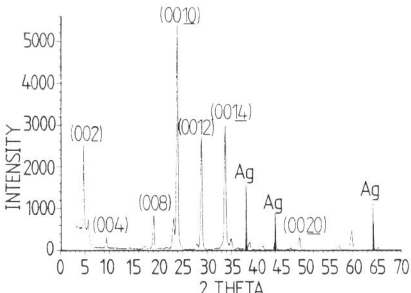

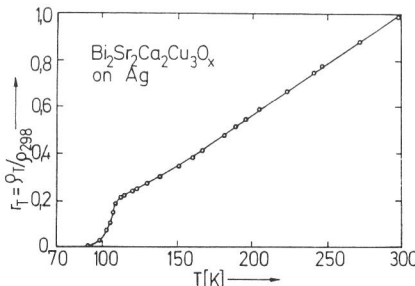

Figure 7. X-ray diffraction patterns of a 2223 phase film on a polycrystalline Ag-ribbon

Figure 8. Temperature dependence of the resistivity ratio $r = \rho_T / \rho_{298}$ of 2223 phase film

The X-ray diffraction patterns of Figure 7 confirm the presence of the 2223-phase. The X-ray diffraction patterns indicate that for the Ag-ribbon substrate the c-axis is also preferentially oriented perpendicular to the surface of the substrate.

The results of our electrical resistivity measurements are shown in Figure 8. The 2223-phase sample on Ag-ribbon shows a broad transition with the onset at 110 K and zero resistance at 85 K.

4. CONCLUSION

A new method is developed to prepare films for superconducting Bi-Sr-Ca-Cu-O phases by using a paste consisting of a mixture of metal nitrates with ethylenediamine or N,N,N',N'-tetramethylenediamine and pyridine. On the polycrystalline Ag- and Au-ribbons both the 2212 and the 2223-phase were obtained. On the (100)-MgO-single crystal plane the 2212-phase is dominating. In all films the c-axis is preferentially oriented perpendicular to the substrate surface. With this method not only thin films can be produced but also thick films.

5. REFERENCES

1   J.A. Agostinelli, G.R. Paz-Pujalt and A.K. Mehrotra, physica C, 154 (1988) 208.
2   C. Chiang, C.Y. Shei, Y.T. Huang, W.H. Lee and P.T. Wu, physica C, 170 (1990) 383.
3   H. Zhuang, H. Kozuka and S. Sakka, Jap. J. Appl. Phys., 28 (1989) L1805.
4   W.H. Wang, C.M. Wang, H.C.I. Kao, D.C. Ling, H.C. Ku and K.H. Lu, Jap. J. Appl. Phys., 28, (1989) L1505.
5   G. Kordas and M.R. Teepe, Appl. Phys. Lett., 27 (1990) 1461.
6   T. Kobayashi, K. Nomura, F. Uchikawa, T. Masumi and Y. Uehara, Jap. J. Appl. Phys., 27 (1988) L1880.
7   H. Jaeger, M. Aslan, K. Schulze and G. Petzow, J. Cryst. Growth, 96 (1989) 459.

High T$_c$ Superconductor Thin Films
L. Correra (Editor)
© 1992 Elsevier Science Publishers B.V. All rights reserved.

Change in the electronic structure of the high Tc superconductor YBa$_2$Cu$_3$O$_{7-\delta}$ induced by oxygen deficiency.

H. Guyot[a], J. Devenyi[a] and O. Thomas[b]

[a]Laboratoire d'Etudes des Propriétés Electroniques des Solides - CNRS, BP 166 X, 38042 Grenoble Cedex, FRANCE.

[b]Laboratoire des Matériaux et du Génie Physique - ENSPG / INPG, 38402 Saint Martin d'Hères, FRANCE.

Abstract

The electronic structure of the cuprate is analyzed by XPS in different states of oxygenation, realized by in- and out-diffusion of oxygen through the surface. The changes of the Cu$_{2p}$, O$_{1s}$ and Ba$_{4d}$ core level lineshapes are found to be completely reversible, within the limits imposed by the texture of the samples. A variable low binding energy component of intrinsic origin is detected in the O$_{1s}$ and Ba$_{4d}$ core level lines and correlated to the presence of oxygen vacancies.

1. INTRODUCTION

In the cuprate YBa$_2$Cu$_3$O$_{7-\delta}$, the oxygen content acts as a tunable parameter which governs the superconducting transition and decreases the transition temperature and the density of carriers at the Fermi level when δ increases from zero to 0.65 [1-2]. For this value, a metallic-semiconducting transition occurs and is accompanied by a orthorhombic-tetragonal structural transition. The decrease of oxygen content in the orthorhombic structure results from the appearance of oxygen vacancies on the O-1 site, which affects particularly the environment of the Cu-1, Ba and O-4 neighboring sites [2-3].

Previous XPS studies performed on samples of various textures have reported the complexity of the Cu$_{2p3/2}$, O$_{1s}$, Ba$_{4d}$ or Ba$_{3d}$ core level lines [4-6]. The O-line and Ba-lines have been found multiple and generally an intense high binding energy (HBE) component has been attributed to spoiling phases. Nevertheless, on clean surface of single phase samples, such a HBE component has also been detected [7-9]. Up to now the origins of the different main components are not completely well established and the present study contributes to the determination of the nature of these components.

Since the existence of a vacancy in the vicinity of an atom should modify its XPS signature, as a result of a possible variation of its valence or of an eventual modification of the relaxation process of the XPS final state, we have studied in details the evolutions of the Cu-, O- and Ba-lines produced by oxidizing or reducing treatments on slightly dirty and clean surfaces of several samples. Thin films are the best candidates for changing the oxygen content by in- or out-diffusion through the surface: the reported results essentially concern thin films realized by MOCVD, but the discussion which underlines the influence of the sample texture, refers also to results obtained with sintered pellets and single crystals.

2. EXPERIMENTS.

The XPS experiments were carried out in a VSW ESCA system, at room temperature, under a pressure of 3 x 10^{-10} mbar, with a $Mg_{K\alpha}$ radiation. All spectra were recorded in the configuration of normal detection, except the 'Ox-50' ones for which the sample was tilted by 50 degrees. The oxygen in- and out-diffusion was realized in-situ, in an adjacent chamber where the pressure of oxygen was controlled between 3 x 10^{-9} and 10^3 mbar. Samples were annealed in ultra high vacuum (UHV) at the lowest pressure, at a temperature ranging from 200° C and 650° C, generally during 20 min. They were oxygenated at 50 mbar of pure oxygen at 550° C (spectra labeled Ox-1) or at 1 bar and 450° C (spectra labeled Ox-2 and S-Ox). They were also etched (label Et) with argon ions of 600 V.

Argon ion etches and anneals in UHV are well known to chemically reduce $YBa_2Cu_3O_{7-\delta}$ [7-8]. The duration of the treatments were adjusted to take into account the variation of the relaxation time of the oxygen diffusion [10-11]. For 1 mm thick samples like sintered pellets, it has been experimentally established that the XPS spectra become reproducible after successive anneals in UHV at 520° C, when the cumulative duration of the treatments passes over 60 min. This stability, which is obtained for a shorter time with thin films, indicates that the oxygen concentration is homogeneous in the entire region investigated by XPS, but not necessarily in the whole sample. Magnetic measurements realized on annealed samples show a broad superconducting transition, which is attributed to the existence of a gradient of oxygen concentration in the samples.

3. RESULTS AND DISCUSSION.

In figure 1 are reported the evolutions of the $Cu_{2p3/2}$, O_{1s} and Ba_{4d} core level lines of a superconducting film untreated and submitted to two sets of anneals in UHV at increasing temperatures. The fist set (upper part of fig. 1) is ended by an oxygenation and the second set (lower part of fig. 1) is followed by an argon ion etch, an oxygenation and a final anneal.

The *$Cu_{2p3/2}$ line* is characterized by a complex structure, consisting in a main peak close to 933 eV and a satellite peak at higher binding energy. The intensity of the satellite, the width and the position of the main peak allow to follow the valence fluctuation of copper atoms, varying between a mono- and a divalency [4,7-9]. The relative intensity of the satellite (ratio of the intensities of the two peaks I_s/I_m) is presented in figure 2, as a function of the anneal temperature. In the untreated state, the satellite is weak and does not significantly change during the first set of anneals. It is slightly decreased by the last anneal at 550° C. This confirms that the untreated film is covered by a poor oxygenated overlayer, sometimes insulating, which is progressively removed by the anneals, acting as cleaning process [7-8]. To this first effect is added an oxygen out-diffusion, which becomes easily detectable above 450° C. The oxygen in-diffusion, resulting from the oxygenation (Ox-1), enhances the satellite intensity and broadens the main peak, where a detection of a contribution at $\approx$ 934 eV reveals the presence of screened trivalent copper atoms, according to [4,7]. The relatively high value of I_s/I_m observed after the first oxygenation, in comparison with the 35 % measured by several authors and in this present study in the ' Ox-50' conditions, is attributed to an overoxygenation of the surface due to an oxygen physical adsorption, which relaxes within 12 hours to the standard state.

The second set of spectra characterizes the cleaned surface of the sample, free of external spoiling phases. When the sample is annealed at a temperature above $\approx$ 500° C in UHV or in oxygen (Ox-2), the satellite intensity reversibly varies (figure 2). This variation, which expresses the valence fluctuation of copper, reflects in fact the oxygen out- and in-diffusion. After the anneal at 605° C, the sample is found to be semiconducting: the orthorhombic-tetragonal transition occurred, that allows to correlate the diminution of the satellite intensity to the increase of the number of oxygen vacancies.

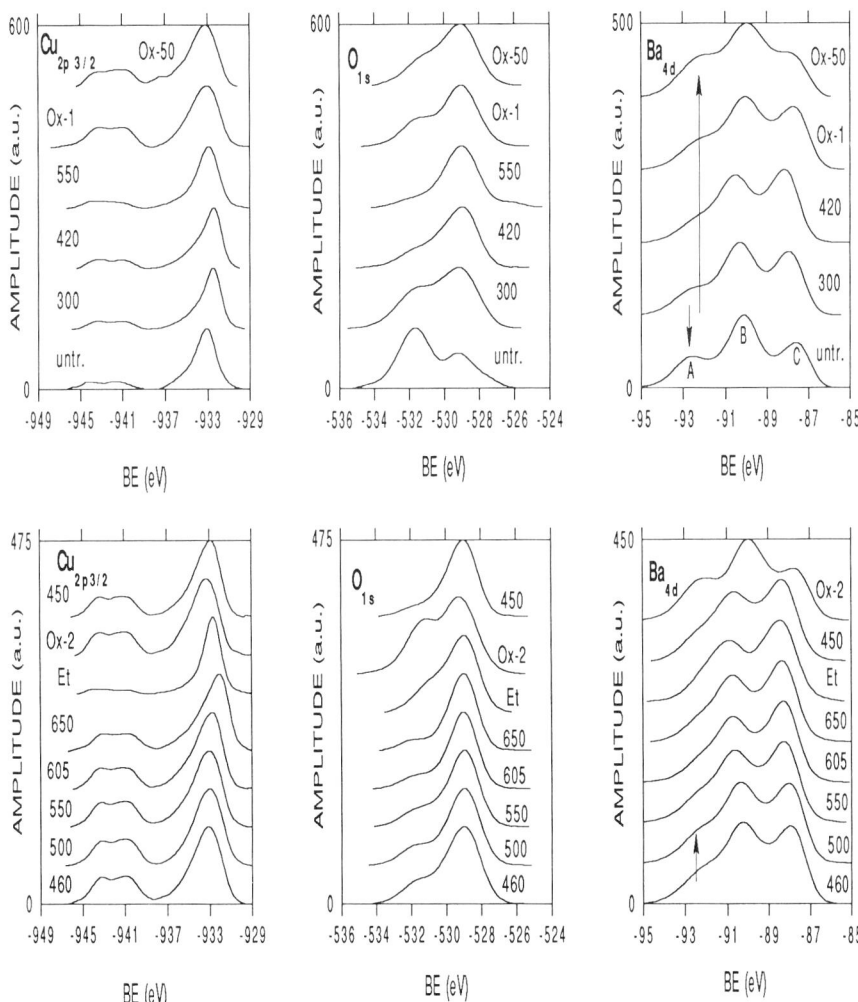

Figure 1. Evolutions of the Cu$_{2p3/2}$, O$_{1s}$, Ba$_{4d}$ core level spectra characterizing the surface of a thin film, untreated (untr.), after annealing in ultra high vacuum, after oxygenation (Ox-1, Ox-2) and after Ar$^+$ etching (Et). Label Ox-50 refers to a detection at 50 degrees off the normal to the film. Numbers indicate the annealing temperatures in degrees. Successive treatments are ordered from bottom to top.

The Ar$^+$ etching decreases the satellite intensity down to 10 %. The concomitant appearance of an additional HBE O-component indicates that in addition to a simple chemical reduction of copper, a drastic change in the cuprate structure occurs, attributed to a strong distortion of the surface. Nevertheless, one should note that an oxygenation at 500° C followed by an annealing at moderate temperature completely cancels the damage produced by the Ar$^+$ etching.

The modifications of the O_{1s} *core level line* shown in figure 1 result from the variations of several components detectable on each spectrum:

- a ≈ 529 eV line, specific from $YBa_2Cu_3O_{7-\delta}$, independent of the metallic or semiconducting state of the cuprate. This line is the most intense, except in the 'untreated' spectrum.

- a HBE line with position ranging from ≈ 531.1 eV to 531.8 eV. The relative weight of this line in the total O-line and the energy shift between the two components are presented in figure 2 as a function of the various treatments.

These two parameters decrease with increasing annealing temperature in the first set of treatments, while in the second one, the intensity of the HBE O-line still decreases down to 17 % but the energy shift remains roughly constant. These evolutions are understandable if it is assumed that the HBE O-line originates from two independents contributions. The first one has an extrinsic origin and comes from the spoiling phases of the overlayer or of an internal pollution. For the thin film, the cleaning of the surface during the first set of anneals is consistent with the decrease of the intensity of this HBE component. The second contribution is intrinsic to the cuprate and is detected on clean surface at ≈ 2.3 eV above the main line. The change of the intensity of this component produced by anneals at increasing temperatures or by oxygenations is found to be completely reversible, confirming the role played by the oxygen out- and in-diffusion through the sample surface. The similarity of the evolutions of the Cu-satellite and the HBE O-line intensities leads to correlate those changes to the variation of the number of oxygen vacancies. In addition, the intrinsic HBE O-line is clearly distinguishable from the extrinsic one observed on the 'untreated' spectrum, since both lines are separated by ≈ 0.4 eV.

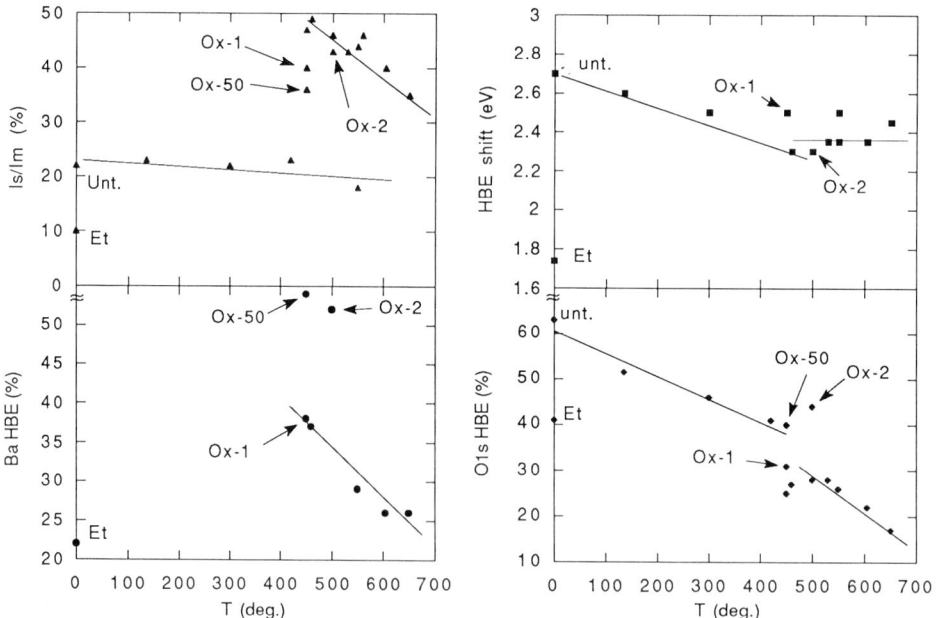

Figure 2. Relative intensity of the Cu- satellite peak (I_s/I_m), relative weight of the HBE Ba-component, energy shift between the two O-components and relative weight of the HBE O-component as a function of temperature of anneals in UHV or in oxygen (labels Ox-). The parameters related to the untreated (unt.) and etched (Et.) surfaces are plotted with an arbitrary zero abscissa. Straight lines underline the general evolution.

The Ar$^+$ etch produces a large modification of the parameters of the HBE O-component: the HBE line at ≈ 2.3 eV is replaced by an intense (40 %) HBE line located at 1.7 eV from the main line. The unexpected values of these parameters reflect the roughness of the surface resulting from the etching. Since the etched surface appears oxygen poor, the shifted HBE O-line may also express the presence in the cuprate of oxygen vacancies located at positions differing from the O-1 site, in the basal plane for instance.

The lineshape of the *Ba$_{4d}$ core level line* is either ternary (peaks A,B,C) or binary with weak additional structures (figure 1). These modifications result from the presence of two main doublets of variable intensities. The decomposition of the Ba-line in two doublets with a spin-orbit splitting of 2.6 eV, the evolutions of their positions and their relative weights are very similar to the results concerning the O-line. Referring each doublet to the Ba$_{4d5/2}$ component position, the cuprate is characterized by a Ba-line at ≈ 88.3 eV. The 'untreated' spectrum exhibits a HBE component at 90.1 eV, while the spectra related to the clean surface show an intrinsic HBE component at 89.7 eV. The variation of the intensity of this doublet is presented in figure 2. This figure also shows that the 'Ox-50' and 'Ox-2' spectra have intense HBE Ba- and O-components. This particularity should be related to the adsorbed oxygen present at the sample surface after a very efficient oxygenation. This kind of pollution enhances the HBE components, by adding an extrinsic contribution.

The similarity of behavior of the HBE O-line and the HBE Ba-component as resulting from annealings, etching or oxygenations should be underlined and confirms the influence of the oxygen vacancies in the evolutions of the XPS structures.

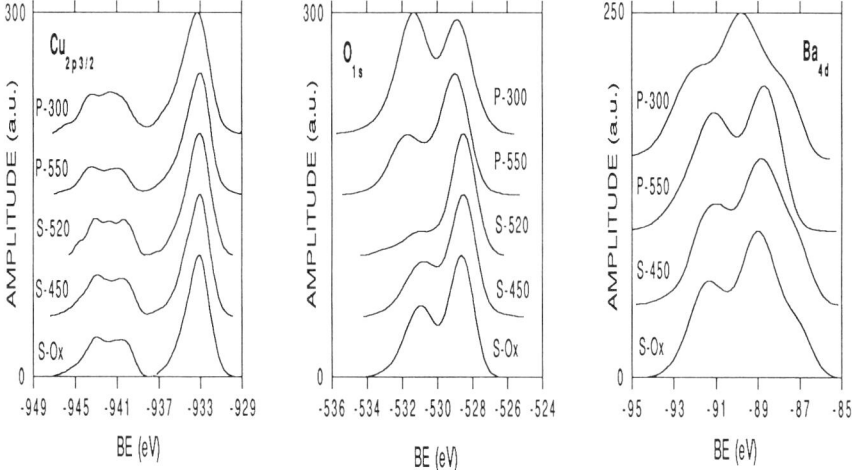

Figure 3. Comparison of the Cu$_{2p3/2}$, O$_{1s}$, Ba $_{4d}$ core level spectra characterizing the surface of a single crystal (S) and a sintered pellet (P) after oxygenation (Ox) or annealing in UHV. Numbers indicate the anneal temperatures in degrees.

The shapes of the Cu-, O- and Ba-lines and their evolutions by reducing or oxidizing treatments are found to be weakly dependent on the nature of investigated samples. For three kinds of samples - thin film, sintered pellet and single crystal - the HBE Ba- and O-components are always detected, as shown on figure 3. These results are in good agreement with the most recent publication devoted to the XPS on $YBa_2Cu_3O_{7-\delta}$, and differ only about the Ba-line that is considered as a single doublet in single crystal [12]. Nevertheless, the texture of the sample has a direct influence on the limits of the variations of the parameters characterizing each XPS lines. For instance, while the HBE O-line is considerably reduced by an annealing in UHV at 550 -600° C for thin films or single crystal, this line remains at ≈ 40 % of the total O-line for a sintered pellet treated in the same conditions. This difference is attributed to the presence in this last sample of a large amount of grain boundaries, which generates an important extrinsic HBE O-line and superimposes a high extrinsic signal to the intrinsic HBE O-line.

4. CONCLUSION

The analysis of the O_{1s} and Ba_{4d} core level lines reveals two intrinsic HBE components, that can be enhanced by other contributions of extrinsic origins. The importance of these contributions are strongly related to the sample texture. The variations of the parameters of the intrinsic HBE components are driven by the in- and out-diffusion of oxygen and are related to the presence of oxygen vacancies in the cuprate.

5. ACKNOWLEDGMENTS

The authors wish to thank G. Collin (University of Paris-Orsay) for providing single crystals and R. Suryanarayanan (CNRS, Meudon) for sintered samples. They are grateful to S.K. Agarwal for magnetic characterizations. This work was supported in part by the EEC Contract SC1-0038C.

6. REFERENCES

1   I.V. Aleksandrov, A.F. Goncharov, I.N. Makarenko, A.J. Shapiro, S.M. Stishov, I.P. Zibrov et al., Physica C 162-164 (1989) 1057.
2   J.D. Jorgensen, B.W. Veal, A.P. Paulikas, L.J. Nowicki, G.W. Cabtree, H. Claus and W.K. Kwok, Phys. Rev.B 41 (1990) 1863.
3   S. Jantsch, J. Ihringer, J.K. Maichle, W. Prandl, S. Kemmler-Sack, R. Kiemel, S. Lösch, W. Schäfer, M. Schlichenmaier and A. W. Hewat, J. Less Comm. Met. 150 (1989) 167.
4   P. Steiner, S. Hüfner, V. Kissinger, I. Sander, B. Siegwart, H. Schmitt, R. Schulz, S. Junk, G. Schwitzgebel, A. Gold, C. Pilotis, H. P. Müller, R. Hoppe, S. Kemmler-Sack and C. Kunz , Z. Phys. B 69 (1988) 449.
5   P.C. Healy, S. Myhra and A.M. Stewart, Philos. Mag. B 58 (1988) 257.
6   P.A.P. Lindberg, Z.X. Shen, W.E. Spicer and I. Lindau, Surf. Science Rep. 11 (1990).
7   H. Guyot, L. Schmidt, R. Cinti, P.L. Reydet, J. Marcus, Physica C 162-164 (1989) 1305.
8   H. Guyot, J. Schubert, W. Zander, B. Stritzker, J. Less Comm. Met. 164-165 (1990) 1369.
9   A. Balzarotti, M. De Crescenzi, N. Motta, F. Patella and A. Sgarlata Phys. Rev. B 38 (1988) 6461.
10 X.M. Xie, T.G. Chen and Z.L. Wu, Phys. Rev. B 40 (1989) 4549.
11 N.H. Andersen, B. Lebech and H.F. Poulsen, J. Less Comm. Met. 164-165 (1990) 124.
12 A. Balzarotti, F. Patella, M. De Crescenzi, N. Motta, A. Sgarlata and F. Licci, Phys. Rev. B 43 (1991) 351.

High T_c Superconductor Thin Films
L. Correra (Editor)
© 1992 Elsevier Science Publishers B.V. All rights reserved.

IN SITU LASER ABLATION DEPOSITION OF THIN FILM SUPERCONDUCTING $Bi_2Sr_2CaCu_2O_x$.

J. F. Lawler, T. P. O'Brien, J. G. Lunney and J. M. D. Coey

Department of Pure and Applied Physics, Trinity College, Dublin 2, Ireland.

Abstract

Laser ablation deposition is used to prepare thin films of $Bi_2Sr_2CaCu_2O_x$ onto single crystal MgO (100) substrates. This is achieved using stoichiometric superconducting ceramic targets which are ablated using an excimer laser (KrF 248 nm). In situ preparation is achieved using a heated substrate (710-750 °C) in an oxygen pressure of 0.1-1.0 mbar. X-ray diffraction, resistivity, electron microscopy and energy dispersive x-ray analyses are presented showing the effect of varying the laser fluence on both deposited films and target itself. The formation of surface structure on the target at fluences below 3 J/cm^2 produces time dependant variation of both deposition rate and stoichiometry of the film. There is increased particulate contamination of the film at higher fluence.

1. INTRODUCTION

Pulsed laser deposition has proved to be one of the most successful techniques for preparation of thin film superconductors and is now widely used. As with the bulk material much of the work to date has concentrated on $YBa_2Cu_3O_7$ leading to practically optimised deposition conditions for the production good quality films exhibiting high critical currents [1,2]. There has though been less success with in situ films of the $Bi_2Sr_2Ca_{n-1}Cu_nO_x$ (n=1,2 or 3) series of compounds, the n=3 material only being produced after post annealing [3,4] and the n = 1 and 2 species show reduced transition temperatures when made without post annealing . Deposition studies now concentrate on problems of lowering the substrate temperature, particulate contamination and growth mechanisms of the film [5,6]. Apart from some theoretical modelling [7] of the ablation process the laser-target interaction has recieved little attention to date. Previously [8] we reported on parameters affecting production of films of $YBa_2Cu_3O_x$. Here we present our work on the *in situ* deposition of $Bi_2Sr_2CaCu_2O_x$ and studies of the parameters affecting ablation of this material. Laser ablation deposition is generally considered to preserve stoichiometry, we find this to be fluence dependant.

2. EXPERIMENT

The laser ablation arrangement is similar to that used by many others. A Lambda Physik EMG 102 MSC excimer laser operating with ArF at 248 nm, produces appoximately 250 mJ/pulse with a 23 ns (FWHM) pulse duration. A 14×6 mm^2 aperture was used to provide a uniform beam profile. The aperture is imaged onto the target resulting in incident fluences of between 0.1-10J/cm^2 after the the geometrical and interface optical reflections at the lens and window of the vacuum chamber have been accounted for. The image area is also measured by burn marks at the target position. The beam energy is measured at the imaging lens and in the target position. Both he target and the lens can be rotated, the latter eccentrically, allowing the image to scan over an area of 0.5 cm^2 on the target. The target material is a polycrystalline ceramic pellet of $Bi_2Sr_2CaCu_2O_x$ formed by the solid state reaction of oxides and carbonates with a metal ion ratio of 2:2:1:2, pressed at 15 MPa and fired at 860 °C in air with intermediate grindings. The pellets consist of thin flat crystallites which are typically 10-20 μm wide and 0.5 μm thick. The deposition rate is monitored by a quartz crystal oscillator. The single crystal MgO substrates are fixed to a resistance heater using silver paint. The silver paint producing a good thermal contact between the front of the heater and the substrate. The paint is cured in air at 250 °C to reduce outgassing in the chamber. Temperature measurement is by a thermocouple placed beside the substrate and also by an optical pyrometer. Deposition takes place in a high vacuum chamber which has a base pressure of 10^{-7} mbar. Typical target substrate distance is 4 cm and deposition takes place in an oxygen pressure of 0.1-1 mbar and the laser is incident on the target at an angle of 30° from the normal.

3. RESULTS AND DISCUSSION

It was found that films of Bi$_2$Sr$_2$CaCu$_2$O$_x$ could be formed *in situ* at substrate temperatures of 710 - 750 °C. Films prepared at higher temperatures contained a mixture of the n = 1 and 2 members of the homologous series and films prepared below 710 °C were only partially crystalline. X-ray diffraction patterns of films formed close to the upper stability line for the Bi$_2$Sr$_2$CaCu$_2$O$_x$ phase exhibited the narrowest line widths. The effect of deposition in different oxygen pressures show narrowest line widths at 0.15 mbar with increased broadening above and below this pressure. The x-ray diffraction pattern in Fig. 1 shows the presence of only 001 peaks indicating the preferential c-axis orientation. Resistivity measurements on these films shown in Fig. 2 display a broad transition starting at 85 K.

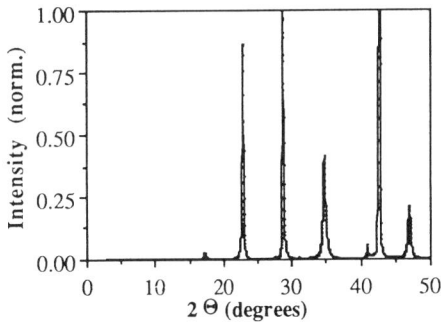
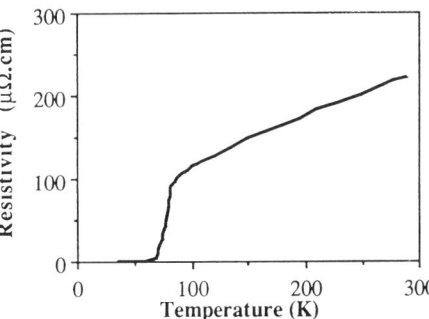

Fig. 1 X-ray diffraction (CuK$_\alpha$) pattern of Bi$_2$Sr$_2$CaCu$_2$O$_x$ deposited at 740 °C in 0.15 mbar O$_2$.

Fig. 2 Resistivity as a function of temperature for Bi$_2$Sr$_2$CaCu$_2$O$_x$. T$_{c0}$ = 70 K

Other workers have reported that particulate contamination is minimised when ablation is carried out just above threshold. Films deposited at these fluences when viewed under the electron microscope show very little particulate contamination and exhibit no apparent grain structure. The number of particles deposited with the film increases with the fluence above threshold. The deposition rate as a function of incident fluence in 0.1mbar of O$_2$ (Fig. 3) is measured using a quartz crystal oscillator. By keeping the pulse energy constant the size of the imaged area is varied thus altering the incident fluence. The deposition rates are normalised to a 1 cm^2 area of irradiation. The deposition rate is low below 0.5 J/cm^2 at which point there is a sharp rise up to about 1 J/cm^2, above this value the rate is again slowly increasing. Some of the increased mass measured by the thickness monitor at these higher fluences is due to the particles deposited with the film material. For fluences below 3 J/cm^2 it was found that there was a marked decrease in deposition rate with time when only the lens or the target were rotated. This was found to be associated with the formation of cone-like surface structures on the target.

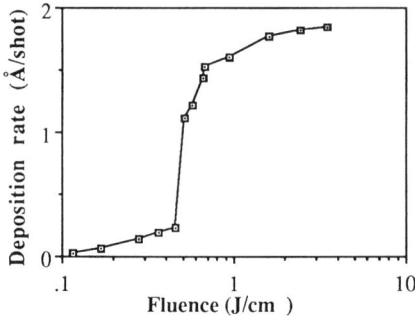

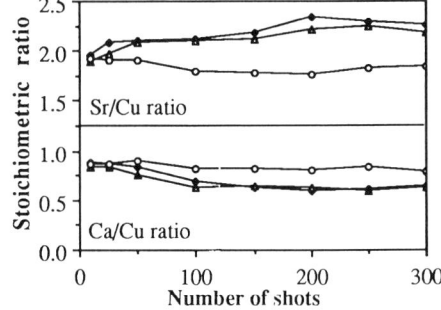

Fig. 3 Deposition rate as a function of fluence (normalised to an irradiated area of 1 cm^2), for a ceramic target of Bi$_2$Sr$_2$CaCu$_2$O$_x$.

Fig. 4 Variation of target surface stoichiometry as a function of incident shot number. —♦— 1.0 J/cm^2, —Δ— 2.0 J/cm^2 and —o— 3.2 J/cm^2.

The effect on the target stoichiometry was measured by energy dispersive x-ray analysis of target material after irradiation with different number of shots. The relative stoichiometry of Sr and Ca as a function of shot number for three different fluences 1.0, 1.9 and 3.2 J/cm^2 is shown in Fig. 4. The ratio of Bi and Cu varied little with time but the relative concentration of Sr shows an increase and Ca a decrease with progressive exposure for the two lower fluences. The variation in stoichiometry at 3.2 J/cm^2 is much less and corresponds to a smooth surface on the target material. After about 200 shots the Ca and Sr ratios vary only slightly and coincide with the completion of cone growth on the target.

Figure 5 shows the mature target surface, the cones have their axes aligned parallel to the laser beam. During cone formation the variation in the target stoichiometry is most evident. The formation of these cones is due to the fact that any surface which is not perpendicular to the incident laser beam recieves a lower effective fluence. Ablation from an uneven target will be greater from those areas whose surface is perpendicular to the beam and this process of preferential etching will suitably roughen the surface to the point where the effective fluence is reduced to the threshold fluence. Higher fluences produce more acuetly angled cones and hence a larger surface area.

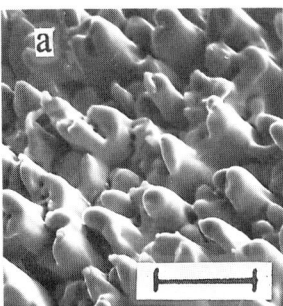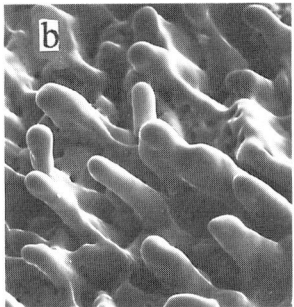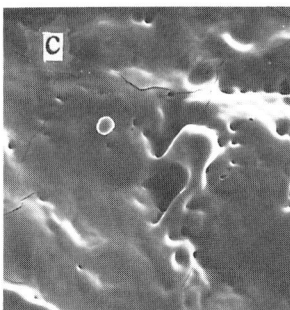

Fig. 5 Electron micrographs of target surface after 200 shots a) 1.0 J/cm^2, b) 2.0 J/cm^2 and c) 3.2 J/cm^2, bar represents 25 µm.

Electron micrographs of the edge of broken surfaces show the melt depth to be of the order of a few µm, much smaller than the dimensions of the cones themselves. Above 3 J/cm^2 cones appear to be unsustainable and the target has a flat smooth surface, this produces deposition close to the target stoichiometry but the films are very rough due to the particles on the surface. The surface morphology of Bi$_2$Sr$_2$CaCu$_2$O$_x$ targets shows similar behaviour to our and other previously reported results on YBa$_2$Cu$_3$O$_x$ [8,9]. These structure are not unique to ceramic superconductors, similar features have been reported in polyamides [10]. In order to try and limit cone formation both the target and the imaging lens were rotated (the imaging lens being rotated eccentrically), this had the effect of irradiating the target surface from different directions and produced a constant deposition rate. Another solution which would give more even coverage of the target would be to raster the beam while rotating the target.

4. CONCLUSIONS

We have deposited smooth *in situ* thin films of Bi$_2$Sr$_2$CaCu$_2$O$_x$ requiring no post annealing. The effect of deposition at fluences between 3 J/cm^2 and

0.5 J/cm^2 (the threshold for ablation) have been studied. This shows a strong dependance of stoichiometry on shot number, this is due to the formation of cone like surface structures on the target. Below threshold the deposition rate is low and non stoichiometric. Above 3 J/cm^2 the target surface becomes smooth and films though stoichiometric are subject to increased particulate contamination. Stoichiometry of the target can be preserved in the film when ablation takes place above the threshold fluence and the formation of surface structure is avoided.

5. ACKNOWLEDGEMENTS

We are grateful to C. G. Reid and the Electron Microscope Unit T.C.D. for their assistance with the microprobe analysis of our samples. This research is supported by the European Commission under SCIENCE contract SCI*0104 C and Brite/Euram 0201 (P-3560).

6. REFERENCES

1. B. Roas, L. Schultz and G. Endres, Appl. Phys. Lett. **53** (1988) 1557.
2. T. Venkatesen, X. D. Wu, B. Dutta, A. Inam, M. S. Hegde, D. M. Hwang, C. C. Chang, L. Nazar and B. Wilkens, Appl. Phys. Lett. **54** (1989) 581.
3. J. Levoska, T. Murtoniemi and S. Leppavuori, J. Less Comm. Met. **164-165** (1990) 710
4. M. Viret, K. Donnelly, J. G. Lunney and J. M. D. Coey, J. Appl. Phys. (1990) **69** 2423 .
5. R. Remesh, A. Inam, D. M. Hwang, T. D. Sands, C. C. Chang and D. L. Hart, Appl. Phys. Lett. **58** (1991) 1557.
6. C. C. Chang, W. D. Wu, R. Ramesh, X. X. Xi, T. S. Ravi, T. Venkatesan, D. M. Hwang, R. E. Muenchausen, S. Foltyn and N. S. Nogar, Appl. Phys. Lett. **57** (1990) 1814.
7. A. N. Jette and W. J. Green, J. Appl. Phys. **68** 5273 (1990).
8. T. P. O'Brien, J. F. Lawler, J. G. Lunney and W. J. Blau, Eng. Mat. B. (to be published)
9. O. Auciello, A. R. Krauss, J. Santiago-Aviles, A. F. Schreiner and D. M. Gruen, Appl. Phys. Lett. **52** (1988) 239.
10. P. E. Dyer, S. D. Jenkins and J. Sidhu, Appl. Phys. Lett. **49** (1986) 453.

High T$_c$ Superconductor Thin Films
L. Correra (Editor)

High T$_c$ superconducting Li-doped Bi-Sr-Ca-Cu-O thin films deposited by laser ablation

J. Levoska, T. Murtoniemi and S. Leppävuori

Microelectronics and Material Physics Laboratories, University of Oulu, SF-90570 Oulu, Finland

Abstract

Li doped Bi-Sr-Ca-Cu-O thin films were deposited on MgO(100) substrates at room temperature using a Nd:YAG laser (λ=1064 nm). The target materials had compositions near $Bi_2Sr_2Ca_1Cu_{2-x}Li_xO_y$ (x<0.7), $Bi_2Sr_2Ca_1Cu_2LiO_y$ and $Bi_2Sr_2Ca_2Cu_3Li_{1.5}O_y$. The structure of the films was studied by x-ray diffraction and SEM/EDS. After annealing at 840 °C - 850 °C, the films consisted mainly of the 2212 phase. These films had T_{c0} at 83 K - 86 K, 3 to 6 K higher than undoped films prepared under similar conditions. Large amounts of the 2212 phase were obtained even at 740 °C, but these films had T_{c0} below 80 K. The best films had critical currents exceeding 10^4 A/cm^2 at 77 K. The superconducting phases in the annealed films were strongly oriented with the Cu-O planes parallel with the substrate surface but with no in-plane epitaxy.

1. INTRODUCTION

Undoped high T$_c$ superconducting Bi-Sr-Ca-Cu-O 2212 phase films usually have zero resistivity temperatures, T_{c0}, of about 80 K or lower. For applications at 77 K, a larger temperature margin is needed. The 2223 phase has a T_{c0} of about 110 K, but it is difficult to obtain the pure 2223 phase, or even a high fraction of it, in undoped films. It is possible by Pb doping to increase the fraction of the 2223 phase, both in bulk material [1] and in films [2,3]. However, the critical currents in Pb doped films at 77 K or lower temperatures are not considerably higher than in 2212 phase films.

The effect of Li doping on the critical temperature of the 2212 phase, bulk material [4,5,6] and whiskers [7] have been studied. It was found that by Li doping a T_{c0} of 91 K was obtained and the formation of the 2212 phase was achieved at much lower temperatures than in undoped material [5].

In this study, high T_c superconducting Bi-Sr-Ca-Cu-O 2212 thin films doped with Li were deposited by laser ablation. Their structural and electrical properties were characterised. Room temperature deposition and post annealing were used because of the high volatility of Li (and even Bi) during high temperature deposition.

2. EXPERIMENTAL DETAILS

2.1. Preparation of the films

Targets for the laser ablation deposition were made by the mixed oxide/carbonate route. Appropriate amounts of Bi_2O_3, $SrCO_3$, $CaCO_3$, CuO and Li_2CO_3 powders were mixed, ground and calcined in air at 600 °C for 5 h. The calcined powder was ground, pelletised and sintered for 20 h at 740 °C in a covered crucible. The compositions of the target materials were $Bi_2Sr_2Ca_1Cu_{1.45}Li_{0.7}O_x$ (221CuLi2), $Bi_2Sr_2Ca_1Cu_2Li_1O_x$ (2212Li) and $Bi_2Sr_2Ca_2Cu_3Li_{1.5}O_x$ (2223Li1.5).

The laser ablation deposition process was carried out in a vacuum chamber with a base pressure of about 10^{-5} mbar using a focused beam from a Q-switched Nd:YAG laser at 1064 nm. The beam was scanned over the target at an angle of 45 degrees. The energy density was above 10 J/cm^2, the repetition rate 1000 Hz and the target-to-substrate distance was 3 cm. The films were deposited on MgO(100) single crystal substrates at room temperature in order to inhibit evaporation of Li.

The films were annealed in an air furnace. They were placed in loosely covered crucibles together with powder of corresponding target material. After annealing, they were slowly cooled to 560 °C and then quenched in air.

Films were annealed at 740 °C for 14 h (temperature at which enhancement of T_c has been observed [5]) and at 840 °C for 14 h. 2223Li1.5 films were also annealed at 850 °C for 60 h (as Pb doped 2223 composition films in recent studies [2,3]). Some loss of Li occurred during the annealing. It was not possible to analyse the amount of Li in the film by the methods used.

2.2 Characterisation of the films

The structural properties of the films were characterised by x-ray diffraction and scanning electron microscopy (SEM/EDS). The electrical measurements were carried out using the standard four probe method. The electrical contacts were attached to the films using silver paint. The resistivity curves were measured using current densities of 0.2 to 1 A/cm^2. Critical currents were measured from laser patterned stripes (width 30-80 μm).

3. RESULTS AND DISCUSSION

3.1 Microstructure

X-ray diffraction patterns of are shown in Figure 1. The 221CuLi2 films annealed at 740 °C consisted mainly of the 2212 phase and a smaller amount of the 2201 phase. In the 2212Li and 2223Li1.5 films, the fraction of the 2201 phase was larger, and the 2212 phase was a minority phase. As the main reflections were of the (00l) type, the crystals were mainly c-axis oriented. After annealing at 840 °C, the peaks of the 2201 phase almost disappeared in all the films, and the orientation became stronger. The 2223Li1.5 films annealed for 60 h at 850 °C showed strong (00l) reflections of the 2212 phase. Some amount of non-superconducting Sr-Ca cuprates was also found. No reflections of the 2223 phase were observed in contrast to the behaviour of undoped and Pb doped films annealed at the same temperature [2,3]

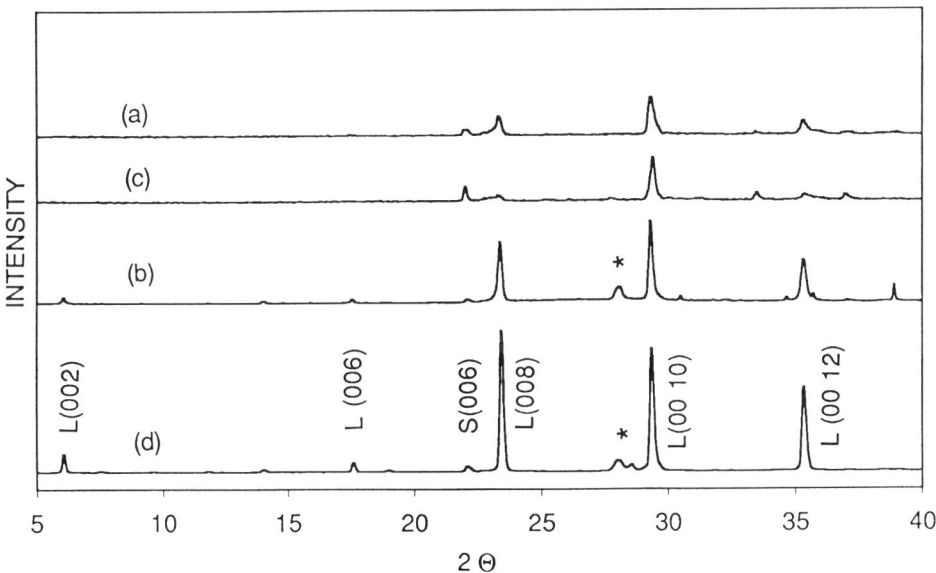

Figure 1. X-ray diffraction patterns (CuKa) of films doped with Li (a) target composition $Bi_2Sr_2Ca_1Cu_{1.45}Li_{0.7}O_x$ annealed at 740 °C for 14 h and (b) at 840 °C for 14 h, (c) target composition $Bi_2Sr_2Ca_1Cu_2Li_1O_x$, film annealed at 840 °C for 14 h, and (d) target composition $Bi_2Sr_2Ca_2Cu_3Li_{1.5}O_x$, film annealed at 850 °C for 60 h. (00l) type reflections are indicated. L: 2212, S: 2201, ✲: Ca-Sr-cuprates.

SEM studies (Figure 2.) revealed that the films annealed at 740 °C consisted of platelike grains mainly oriented parallel with the substrate surface and having a grain size of several microns. The films annealed at 840 °C - 850 °C had a larger grain size (from tens to hundreds of μm) and good orientation along the substrate surface, but there were also some holes in the films.

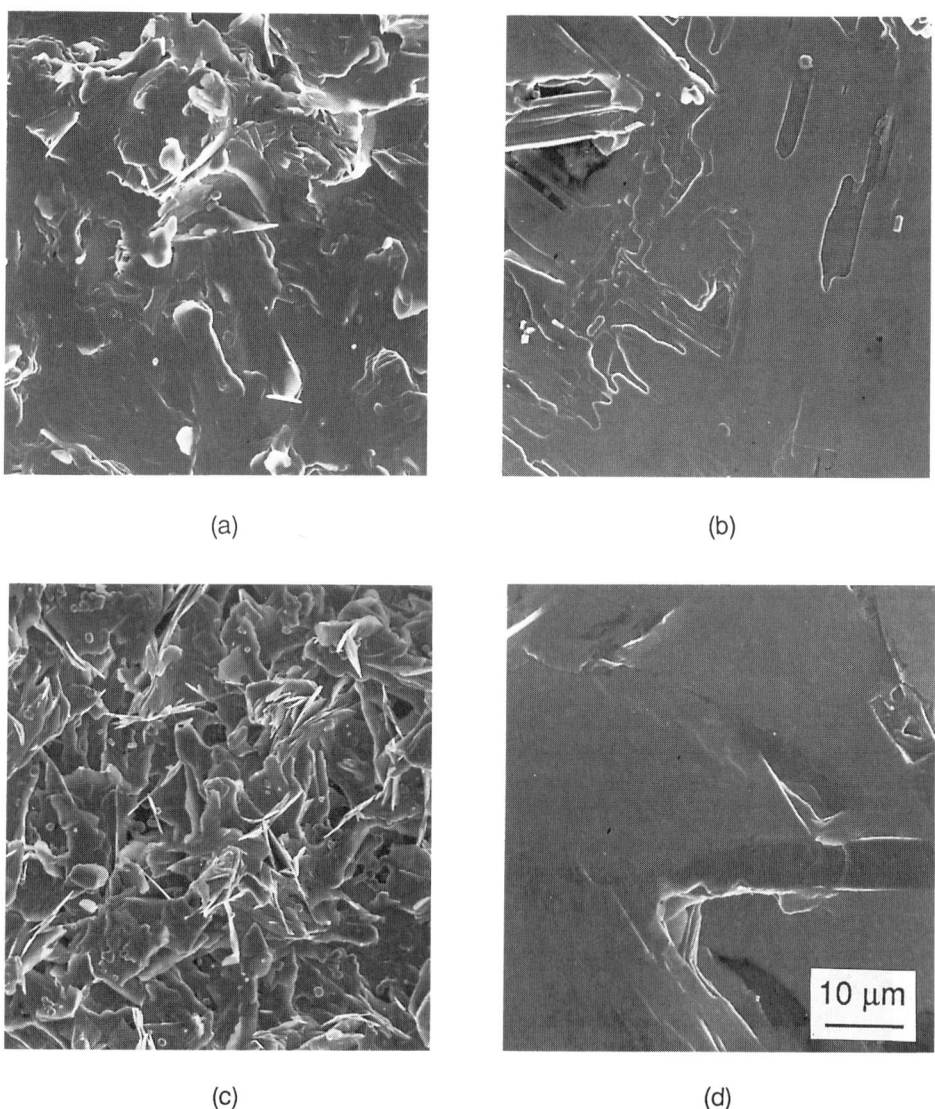

(a)   (b)

(c)   (d)

Figure 2. SEM secondary electron images of the films of Figure 1 (a) - (d).

3.2 Electrical properties

Resistivity curves of Li doped films are presented in Figure 3. The zero resistivities were in the range 73 K - 77 K and 82 K - 84 K for the films annealed at 740 °C and at 840 °C, respectively, and 85 K for the 2223Li1.5 film annealed at 850 °C for 60 h. A transition near 110K in the 2223Li1.5 film (Fig. 3b) indicates the presence of a small amount of the 2223 phase not observed by x-ray diffraction. The T_{c0} temperatures for the films annealed at 840 °C - 850 °C were 3 to 6 K higher than those obtained for undoped films under similar conditions.

The films annealed at 840 °C - 850 °C had the largest grain size, the highest degree of orientation and also the highest critical currents, exceeding 10^4 A/cm² at 77 K. Figure 4 shows the critical current of a 2223Li1.5 film annealed at 850 °C for 60 h, measured from a 30 μm wide laser patterned strip.

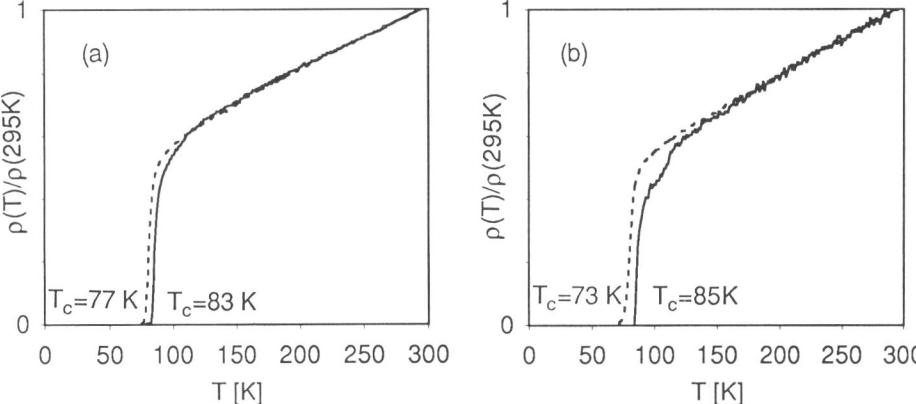

Figure. 3. Resistivity vs temperature plots of films doped with Li: (a) target composition $Bi_2Sr_2Ca_1Cu_{1.45}Li_{0.7}O_x$) annealed at 740 °C for 14 h (broken line) and at 840 °C for 14 h (solid line), (b) target composition $Bi_2Sr_2Ca_1Cu_2LiO_x$) annealed at 840 °C for 14 h (broken line) target composition $Bi_2Sr_2Ca_2Cu_3Li_{1.5}O_x$, film annealed at 850 °C for 60 h (solid line).

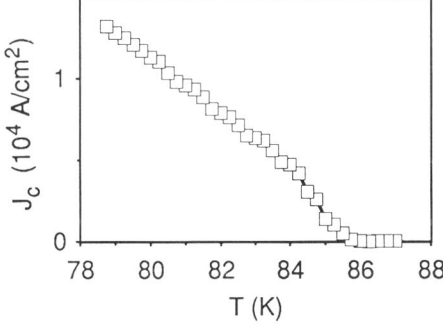

Figure 4. Critical current vs temperature for a film with target composition $Bi_2Sr_2Ca_2Cu_3Li_{1.5}O_x$, annealed at 850 °C for 60 h .

3.3. Effect of Li doping

There is some evidence [5,6] that Li substitutes Cu in the lattice and reduces the hole concentration of the Cu-O planes of the 2212 phase, thus enhancing T_{c0}. Li could also produce a flux phase. That would explain the low formation temperature for the 2212 phase. The flux could also change the oxygen intake of the film during cooling and make the oxygen concentration and the hole density of the 2212 phase more ideal. However, no Li rich (flux) phases were observed by x-ray diffraction or SEM/EDS.

It is possible to conclude that enhancement of 3 to 6 K of the T_{c0} of the 2212 phase was obtained by Li doping. The formation of the 2223 phase was inhibited or reduced considerably and the formation of the 2212 phase occurred at lower temperatures.

4. ACKNOWLEDGMENT

The authors gratefully acknowledge the assistance of FPRI J. Matthan in editing the text.

5. REFERENCES

1   S. A. Sunshine, T. Siegrist, L. F. Schneemeyer, D. W. Murphy, R.J. Cava, B. Batlogg, R. B. van Dover, R. M. Fleming, S. H. Glarum, S. Nakahara, R. Farrow, J. J. Krajewski, S. M Zahurak, J. V. Waszczak, J. H. Marshall, P. Marsh, L. W. Rupp, Jr. and W.F. Peck, Phys. Rev. B 38 (1988) 893.

2   J. Levoska, J. Hagberg, P. Pusa, A. Uusimäki, S. Leppävuori, High Temperature Superconductors: Fundamental Properties and Novel Materials Processing, MRS Symposium Proceedings Series, vol. 169, (1990) 523.

3   J. Levoska, T. Murtoniemi, S. Leppävuori, J. Less-Common Metals, 164 & 165 (1990) 710.

4   T. Kawai, T. Horiuchi, K. Mitsui, K. Ogura, S. Takagi and S. Kawai, Physica C, 161 (1989) 561.

5   T. Horiuchi, T. Kawai, K. Mitsui, K. Ogura and S. Kawai, Physica C, 168 (1990) 309

6   P. Strobel, B. Bonnet, C. Mouget and B. Souletie, Physica C, 172 (1990) 193.

7   I. Matsubara, H. Tanigava, T. Ogura, H. Yamashita, M. Kinoshita and T. Kawai, Physica C, 167 (1990) 503.

High T$_c$ Superconductor Thin Films
L. Correra (Editor)
© 1992 Elsevier Science Publishers B.V. All rights reserved.

In situ growth of superconducting Bi-Sr-Ca-Cu-O thin films by RF magnetron sputtering.

Z.Z. Li, S. Labdi, A. Vaurès, S. Megtert et H. Raffy.

Laboratoire de Physique des Solides , Bât. 510, Université Paris-Sud, 91405 Orsay, France

Abstract

Superconducting BiSrCaCuO thin films with $T_c(R=0)$ up to 76K were prepared in situ on heated MgO (100) substrates by reactive RF magnetron sputtering under high sputtering gas pressure.The influence of deposition parameters on film properties are described.

1. INTRODUCTION

Since Maeda et al [1] discovered superconductivity in the Bi-Sr-Ca-Cu-O system, high quality thin films of this compound have been sucessfully obtained by a two-step process : thin films are deposited in the amorphous state and are subsequently annealed at high temperature T_a (>800°C). We have regularly used this process [2,3]. However such a high temperature T_a is unsuitable for fabricating multilayer films or heterostructures and also the surface of annealed thin films may be rough. Furthermore in the case of Bi-Sr-Ca-Cu-O it appeared to be very difficult to obtain single phase films of the 2223 phase [3]. Therefore, we attempted to produce superconducting thin films in situ. This process is widely used for YBaCuO films, giving films with $T_c(R=0)$ comparable to that of bulk YBaCuO. By comparison, there have been few reports on the in-situ growth of thin films of Bi-Sr-Ca-CuO with $T_c(R=0)$ up to liquid nitrogen temperature [4-7].

In this paper, we report the in-situ growth by reactive sputtering of Bi-Sr-Ca-Cu-O superconducting thin films with $T_c(R=0)=76K$ and with a very smooth and bright surface. First we will indicate the experimental techniques used for their preparation and characterization. Then we will describe the results of a systematic study of the relationships between several important deposition parameters and thin film composition, structure, and superconductivity. These parameters include the total process gas pressure, the percentage of O_2, the temperature of the substrates and the cooling conditions.

2. EXPERIMENTAL

The thin films were deposited on heated MgO (100) substrates by reactive RF magnetron sputtering with a single target (sputtering up configuration). We used a target 50mm in diameter with a nominal cation composition of (Bi,Pb) : Sr : Ca :Cu equal to (2.5, 1.1):2:2:3. The target-substrate separation was 30mm. The RF power applied to the electrode was 50W. The substrates were glued with Ag paint on a temperature - controlled substrate holder plate facing the target in the so-called "on-axis geometry". The sample holder temperature T_S was varied from 675°C to 775°C. During the deposition process, a mixture of Ar and O_2 was used for the sputtering gas. In different experiments, the total pressure was varied from 100 mTorr to 400 mTorr and the percentage of O_2 from 10% to 60%. The deposition rate was between 0.5 and 0.8Å/s and the sample thickness between 4000 and 7000Å. Once the deposition was finished, the samples were immediately cooled down to room temperature under a given pressure of O_2 : generally under 150-180 Torr of O_2 down to 400°C and under 760 Torr of O_2 from 400°C to room temperature. It takes about 10 minutes down to 400°C and 30 minutes for the whole cooling process.

X-ray diffraction (XRD) studies for phase identification were carried out with a diffractomer using CuKα radiation. Chemical composition of the films were determined by energy dipersive X-ray spectroscopy (EDX) and the surface morphology was examined by scanning electron microscopy (SEM). The superconducting transition was measured with a conventional dc resistive four probe method.

3.RESULTS   AND   DISCUSSION

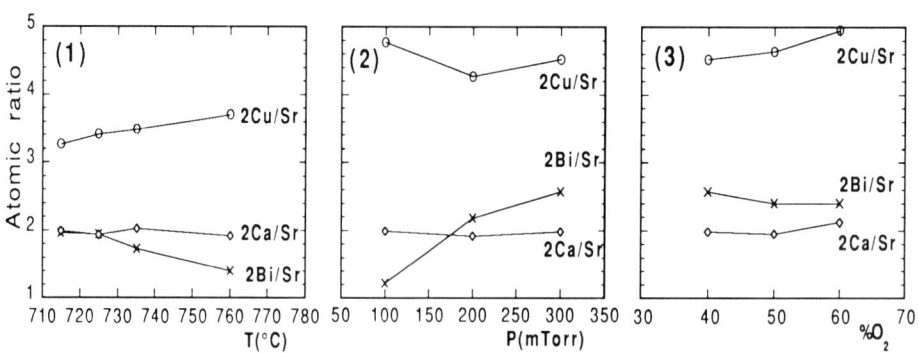

Figure 1-3. Relationships between chemical compositions and different deposition parameters :
(1) - substrate holder temperature T_S (p=100mTorr, %O_2 =20%)
(2) - sputtering gas pressure p (T_S=735°C, %O_2 =40%)
(3) - percentage of O_2 in the process gas (T_S=735°C, p=300mTorr)

Figures 1, 2, 3 display the *composition* of the thin films, respectively, as a function of the substrate temperature, the total pressure and the percentage of O_2 in the sputtering gas. The compositions of Bi, Ca and Cu were normalized to Sr. From these figures, it was found that the Bi content tends to decrease with increasing the substrate temperature or increasing the percentage of O_2 and it increases with increasing total pressure. The influence of above mentioned deposition parameters on the Cu was opposite to that on Bi.

X-ray diffraction spectra of the thin films with various substrate temperatures are shown in figure 4. They indicate that in the thin films deposited at 715°C, 735°C, and 775°C with a pressure of 100mTorr and 20% O_2, the following phases were formed : 2223, 2223+2212, 2212 , respectively. All the diffraction lines of these patterns are indexed as (00l) lines. This means that the films have a strongly preferred orientation with the c-axis perpendicular to the film plane. These results indicate that the crystal structure is dependent on the substrate temperature. So the various phases could be obtained by controlling the temperature of the substrates.

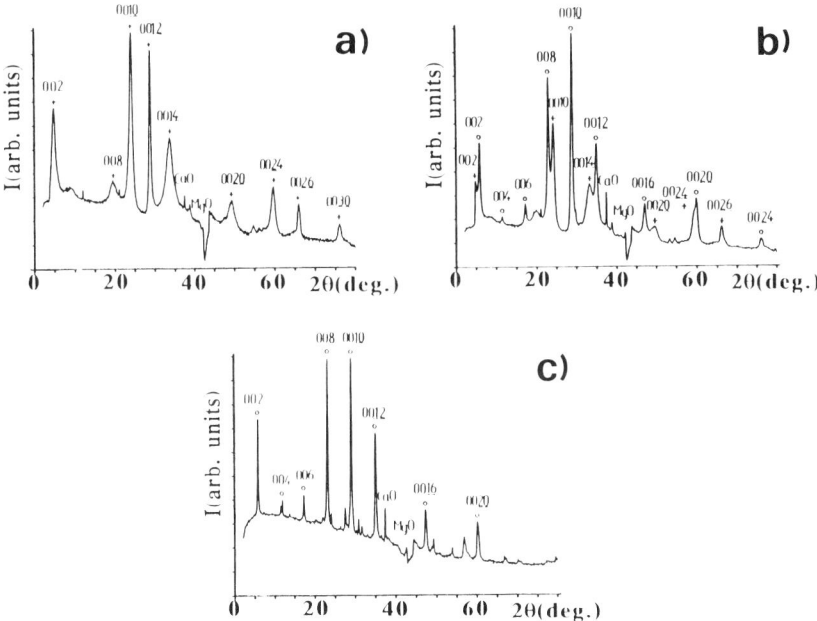

Figure 4. X-Ray diffraction patterns of in- situ films deposited at three different substrate temperatures : a) 715°C, b) 735°C, c)775°C. In each case , we had : p=100mTorr, %O_2=20% . (o : 2212 phase; + : 2223 phase)

Although films deposited at 715-775°C with a pressure of 100mTorr had a nearly stoichiometric composition 2223, their superconducting transition temperature were rather low. The T_c(R=0) were in the range of 20-60K for T_s = 715°C-775°C. It might be caused by the imperfections of the films, such as stacking faults or oxygen deficiency .

Figure 5 shows the temperature dependence of the normalized resistance for films prepared at 675-735°C with a process pressure of 300mTorr and 50%O_2. All the films displayed metallic linear temperature dependence of the resistivity in the normal state having R(300K)/R(150K)=1.59-1.79. The superconducting properties were generally improved with increasing substrate temperature. The best film was obtained at 735°C. It shows an onset of transition at 90K and a zero resistance at 76K.

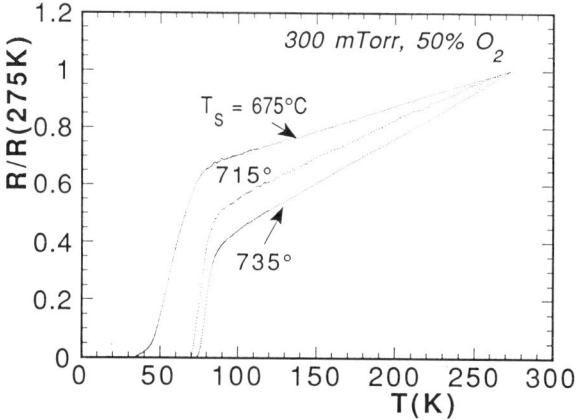

Figure 5. Temperature dependence of the normalized resistance for in- situ films deposited at three different substrate temperatures : a) 675°C, b) 715°C, c)735°C. In each case , we had : p=300mTorr, %O_2=50%.

Figure 6 shows X-ray diffraction patterns of the films deposited at 675-735°C with a process pressure of 300mTorr and 50% O_2. The peaks of these films display a strongly preferred (00l) orientation of the phase 2212 .

Comparing with the films prepared under 100mTorr, 20% O_2, the superconducting properties of the films were improved by increasing the pressure and the percentage of O_2 of the sputtering gas. The latter improved the T_c(R=0) of sputtered films and decreased the crystallization temperature of the films. In order to keep the film with a stoichiometric composition (especially for Bi), the total pressure was increased simultaneously. In our experiments, the films with highest T_c(R=0) were obtained with the following relationship between total pressure and percentage of O_2 : 100 mTorr-20% O_2, 150mT-30% O_2, 200mTorr-40% O_2 and 300mTorr-50% O_2. Another reason for increasing the total pressure, also indicated in [7], is that during the sputtering process, the high energetic particles causing irradiation damage or back-sputtering effect, are thermalized by increasing the total pressure. On the other hand, as shown in Figure 7, the surface morphology of the film prepared under 300mTorr, 50% was much smoother than the films prepared under 100mTorr 20% O_2, although there are some particles on their surfaces due to off-stoichiometry.

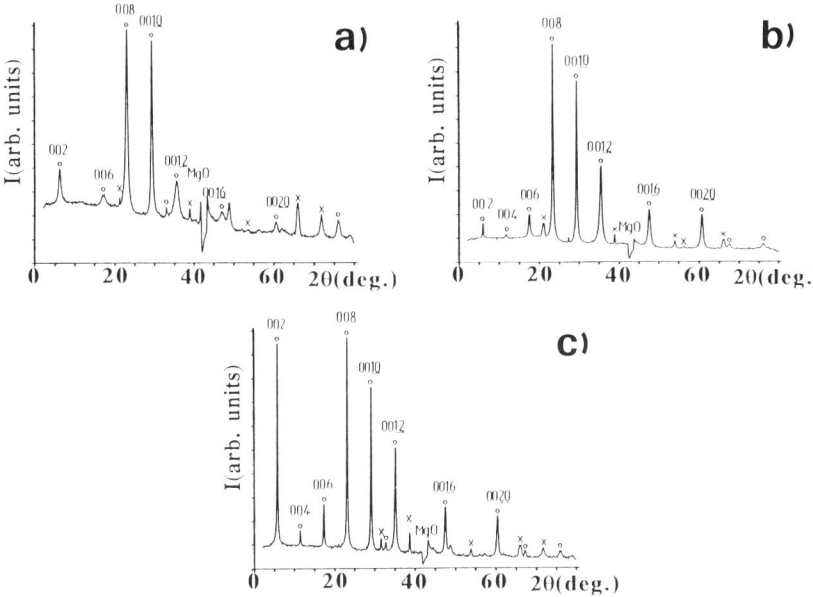

Figure 6. X-Ray diffraction patterns of the same in situ films as in Fig.5 (o :2212 phase; x : mixed 2212/2223 phase as already reported [8] (orthorombic c ~33.92Å)

Different *cooling conditions*, such as cooling in a mixture of N_2-7% O_2 or 100% O_2 at 760mTorr were examined. There was no significant effect on T_c compared to our currently used cooling conditions.

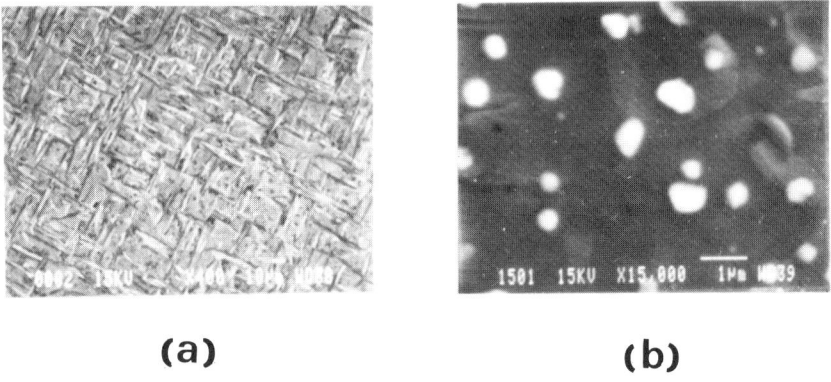

(a)   (b)

Figure 7. SEM photographs for two in- situ thin films :
    a) same film as in Fig.4c
    b) same film as in Fig.5c

4. CONCLUSION

In conclusion, superconducting Bi-Sr-Ca-Cu-O films have been prepared in-situ by a RF magnetron sputtering method on heated MgO (100) substrates. The effect of several deposition parameters on the composition of thin films was studied. Films with superconducting temperature $T_c(R=0)=75$-76K and R(300K)/R(150K) =1.79 have been obtained. We indicate that the T_c value of our BSSCO films could be improved by increasing together the total pressure and the percentage of O_2 in the process gas

Acknowledgements : We thank R. Laval (Institut d'Electronique Fondamentale, Université Paris-Sud) for his help in SEM and EDX analyses ; J.Arabski for the preparation of the targets; J.G. Zhao and J.Sanchez for technical assistance; Dr H. Hurdequint for his contribution in the definition and setting up of the sputtering equipment and his advice for the use of this machine.

4.REFERENCES

1 H. Maeda,Y. Tanaka, M. Fukutomi and T. Asano, Japan. J. Appl. Phys., 27(1988) 209.
2 H. Raffy, A. Vaurès, J. Arabski, S. Megtert, J. Perrière , Solid State Comm. 68 (1988) 235.
3 S. Labdi, H. Raffy, A. Vaurès, J. Arabski, S. Megtert, P. Tremblay, J. Less.Comm. Met. 164-165 (1990) 687.
4 K. Ohbayashi, H. Anma, Y. Takai, and H. Hayakawa , Japan J. Appl. 29 (1990) L2049.
5 T. Matsushima; Y. Ichikawa, H. Adachi, K. Setsune and K. Wasa , Solid State Comm. 76 (1990) 1201.
6 M.Kanai, T.Kawai, and S.Kawai, Appl.Phys.Lett., 54(1990)1802
7 T. Matsushima, Thesis, Osaka University (1991).
8 D.G.Schlom, A.F.Marshall, J.T.Sizemore, Z.J. Chen, J.N. Eckstein, I. Bozovic, K.E. Von Dessonneck, J.S.Harris, J.C. Bravman, J. Cryst. Growth 102 (1990) 361

High T$_c$ Superconductor Thin Films
L. Correra (Editor)
493

STRUCTURAL CHARACTERIZATION OF YBaCuO SUPERCONDUCTING SCREEN-PRINTED FILMS ON SILICON DIOXIDE AND NITRIDE.

J. MARFAING*, X.Y.SONG*, A. BETTAHI*, Y. WANG*, C. BOULESTEIX*, P. ODIER+, J.P. SORBIER++.

*Lab. de Micros. Electr. Appl., URA 797, Fac. Sci. St. Jérôme, B 61, 13397 Marseille Cedex 13.F.

+Centre de Recherc. sur la Physique des Hautes Températures/CNRS, 45071 Orléans Cédex 2.F.

++Lab. d'Electronique, URA 784, Fac. Sci. St Jérôme, D 22, 13397 Marseille Cedex13.F.

Abstract

Superconductive YBaCuO films (10 to 40 μm thick) have been screen-printed on SiO$_2$/Si (obtained by wet and dry way) and Si$_3$N$_4$/Si from different powders obtained via an atomisation method. After optimization of the sintering conditions for each powder the resistance measurements, the X-ray diffraction spectra and the mechanical and microstructural properties of the samples were studied and found to be sensitive to the initial morphology of the powders. All the samples present a zero resistance above 77K with a ΔT$_c$ varying between 3 and 14K.

Analysis of the results shows the good quality of the screen-printed films on both SiO$_2$/Si or Si$_3$N$_4$/Si substrates; the superconducting character is not affected by silicon.

1. INTRODUCTION

Among the various processes of the superconducting films synthesis [1-3], the screen-printing technique is certainly the lowest cost one and does not require sophisticated or vacuum apparatus. Moreover, this technique is well controlled through its large use in the semiconductor technology and can be easily adapted in the case of superconducting films elaboration, for example for the YBaCuO compound which properties start to be well known. The major problem comes from the substrate which frequently interferes with the material during the process and degrades the properties.

Until now, many works were devoted to the study of screen-printing films on substrates such as alumina, YSZ and MgO [4-6] as these kinds of substrates minimize the distortions of the lattice which otherwise weaken the superconducting properties. However, these substrates cannot reasonably be used for industrial applications.

A few experiments were performed on pure silicon [1,3,7]; however, diffusion of silicon into the YBaCuO breaks down the structure as secondary phases are formed and destroys the superconducting properties [8-9]. The silicon diffusion problem can be solved using a buffer film of dioxide or nitride because of their great stability up to 1500 and 1900°C respectively.

In this paper, we report results on the physical properties of YBaCuO films, screen-printed on SiO_2/Si, and Si_3N_4/Si substrates, in relationship with morphology of the initial YBaCuO powders.

The structural characterizations of the elaborated films are with reference to their resistive, crystalline, morphological and mechanical properties. In all cases, the screen-printed films have presented a superconducting character with a zero resistance of at least 77K. Analysis of the results has shown the influence of both nature of the substrate and morphology of the initial powders on the films properties.

2. EXPERIMENTAL PROCEDURE

2.1. Powders and screen-printing method. Subtrates.

The powder with nominal composition of $YBa_2Cu_3O_7$ was prepared by an atomisation method of nitrates precursors at high temperature which has been previously described [10]. Powders A (PA) and B (PB) were annealed by a flash treatment at 700 and 860°C, repectively, while powder C (PC) was annealed at increasing temperatures in oxygen, that is up to 860°C in order to accomplish the solid state reaction between the constituants. Between each thermal treatment, the powder was deagglomerated and structurally characterized by X-ray diffraction. In the case of the PC powder, $YBa_2Cu_3O_7$ was almost pure while a mixture of oxides homogeneously distributed among the small spheres for PA and PB. The particle sizes were controlled by scanning electron microscopy (SEM) and are composed of spherical empty voids less than 1 and 0.5 μm for PA and PB respectively, and platelets of about 0.8-1 μm large for PC. Adequate viscosity was first obtained with an organic solvent, and then screen-printing was achieved through a 300-mesh stainless screen on the different substrates.

The typical thermal cycle consists in a preliminar drying (100°C for one hour) of the samples (thickness varying from 10 to 40 μm) then of a sintering under Argon (up to temperature 850-900°C depending on the substrates) at 18°C/min, followed by an annealing in an oxygen atmosphere (450°C) during two hours. The cooling was done at 1-2°C/min. down to 200°C.

The substrates of Si_{100} were coated either with a Si_3N_4 film, 200nm thick, or with a SiO_2 film, 170nm thick, obtained in a wet way or 200nm grown in a dry way.

2.2. Resistance and X-rays measurements. Adhesion.

The ac resistance measurements were carried out in a classical helium cryostat by the four probe method and silver paste contacts; the alternative current values can vary from 1 μA to 100 μA in order to have a typical lock-in voltage value of about 10 μV. In some cases, the onset of a voltage near 1μV was used as a measurement of the critical current I_c. An underestimation of the critical current density, J_c, can be obtained by dividing I_c values by the maximum cross-sectional area of the films taking into account the contact points dimensions and the film thickness.

The X-ray diffraction measurements were made on a 2Θ−goniometer equipped with a Cu Kα source operating at 30kV and 15mA.

Adhesion characterization was done by the "stripe-test". Calibrated forces, applied on a diamond tip moved on the surface film, create stripes for which the width and other effects were measured. For medium adhesion, the stripe induces scratches of the screen-printed material, while for good adhesion the pressure induces a consecutive and evolutive deformation of grains, depending on the applied forces value on the substrate.

3. RESULTS AND DISCUSSION

Table I summarizes all the results, i.e. the onset temperature T_c^{on} (defined when the resistance significatively decreases), the zero resistance temperature T_c^0, the transition width ΔT_c (90–10% T_c), and the adhesive behaviour.

Table 1
Values of the critical onset temperature T_c^{on}, the zero resistance temperature T_c^0, the transition width ΔT_c and the adhesive behavior for the samples screen-printed on SiO_2 [prepared by a dry way (dw) or a wet way (ww)] and Si_3N_4 with three YBaCuO powders (PA, PB and PC).

Sample	Powder	Subs.	Cycle	Adhesion	Tc(onset)	Tc(0)	ΔTc
a	PA	SiO2 (dw)	880°C-2' 860°C-5' 450°C-2h	medium	95K	77K	6K
b	PB	SiO2 (dw)	880°C-3'5 860°C-5' 450°C-2h	good	96K	63K	14K
c	PC	SiO2 (dw)	875°C-3'5 860°C-5' 450°C-2h	excellent	96K	80K	6K
d	PA	SiO2 (dw)	850°C-5' 450°C-2h	medium	92K	72K	13K
e	PB	SiO2 (ww)	870°C-3'5 860°C-5' 450°C-2h	good	96K	78K	7K
f	PC	Si3N4	890°C-2' 860°C-5' 450°C-2h	excellent	96K	80K	8K
g	PA	Si3N4	890°C-2'5 860°C-5' 450°C-2h	good	95K	76K	5K
h	PC	Si3N4	885°C-5' 860°C-5' 450°C-2h	excellent	96K	84K	3K

Zero resistance above 77K was observed in all samples and reproducible results were obtained. Figure 1 exhibits representative curves obtained on the SiO_2/Si (a,c,e) and Si_3N_4/Si (f) substrates with the three used powders PA (a) and PB (e) and PC (c and f).

In all cases the superconductive transition presents a sharp slope and a tail which extends to T_c^0. The existence of this tail suggests that grain boundaries or other phases can be present in the film. From these resistance measurements, no significant difference in the transition width can be attributed to the substrate type or to the powders.

When measured, the J_c are of about 10-70A/cm^2 at 77K which are comparable with the values given in the literature for screen-printed films[4,11]. However, in the case of screen-printed films, the notion of critical current density is questionable as the samples are porous and present many grain boundaries.

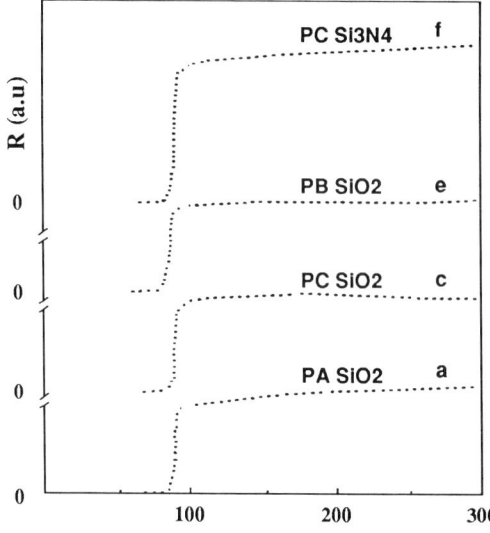

Figure 1. Representative temperature resistance curves obtained for "YBaCuO" samples sreen-printed on SiO2 [prepared by a dry way (a and c) or a wet way (e)] and on Si3N4 (f) with the three powders [PA (a), PB (e) and PC (c,f)].

Another characteristic of these films is adhesion on the substrates; this property varies depending on the subtrate and qualitative measurements, i.e. width and height of the stripe in the film, can inform on the influence of thermal treatment, used substrates or starting powders. Figure 2 shows the mechanical differences of two films screen-printed on SiO₂/Si substrate

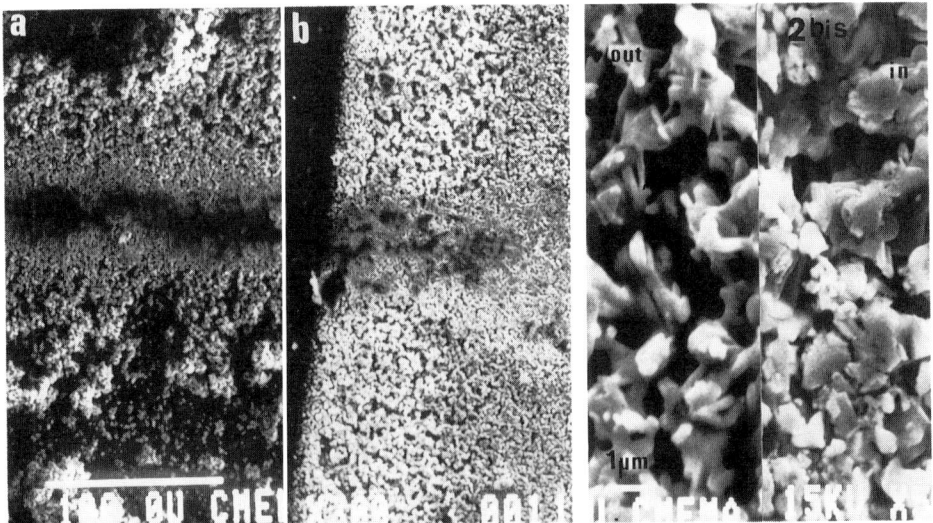

Figure 2 Effect of the diamond stripe test, on YBaCuO films screen-printed on SiO₂/Si substrate with PA (a) and PB (b). The applied force was defined as 30g. Sample a was partially destroyed while sample b shows best mechanical resistance even if grains are deformed under pressure as fig 2 b shows with the same magnification.

using the two powders PA and PB (samples a and b respectively with corresponding thicknesses 22 and 25µm). Sample a has been scratched up to the substrate by the diamond tip while sample b has resisted with a consequent deformation of the crystals as shown on the magnification on fig2b: under pressure, crystals were pressed together with a change in shape and dimensions. The diamond- stripe test gives evidence of the best adhesive and mechanical properties of the films screen-printed on SiO_2 substrates with powder PB.

Peaks of spectra obtained by X-ray diffraction measurements have shown a majority of the superconductive YBaCuO phase "123" with a random orientation of the grains. The peaks are well-defined and sharp. In a reproducible manner, the Y_2BaCuO_5 insulating "green phase" and CuO were formed, especially in samples screen-printed on SiO_2. However, in any case, the presence of small amount of secondary phases does not significantly affect the T_c values, as can be seen from Table I comparing the respective values for the samples a and c.

All films on these insulating substrates were polycrystalline, with crystals of about 1-2 µm size as observed by transmission electron microscopy.

From the resistance and X-ray measurements, it is clear that YBaCuO films of the superconducting phase can be screen-printed successfully upon SiO_2/Si or Si_3N_4/Si substrates with a zero resistance above 77K. This result is very important for technological applications even though it appears that the SiO_2/Si substrate may favour the formation of impurity phases.

The initial morphology of the powders does not significantly affect the superconducting behavior, while induces a great influence on their mechanical and adhesive properties. Systematic observations by SEM have exhibited a striking feature in the micrographs, i.e. that small changes in the thermal cycle (5-8°C) of a powder have a relatively large effect on the scale of the microstructure of the film. As seen on figure 3, the surface topology of a film screen-printed on SiO_2 with PA (a) shows large cracks that extend from the surface into the volume with dimensions of many grain sizes in length and width of about 3µm.

The presence of these cracks can be generally attributed to volume changes during the sintering schedule (layer constrained densification, evaporation of the ink vehicle and/or by cooling from the sintering temperature). However such cracks are also present on bulk specimens for which role of grain size and oxygenation schedules were explored [12].

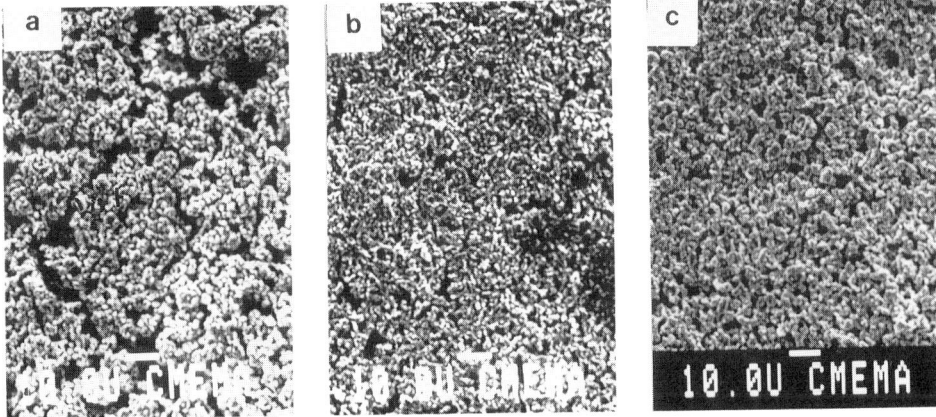

Figure 3 Surface topology of films screen-printed on SiO2/Si substrate with PA (a), PB (b) and PC (c).

Checking on the different powders, these cracks have a tendency to be absent on the samples prepared with PB and are not visible for films screen-printed with PC, as Figs. 3b and c respectively show. The phenomenon is complex and seems to be due to a form factor of the powders. Porosity increases with the initial oxide sphere size, while densification of the material occurs with platelets of PC. A more realistic film density can be given from both cross-sectional and surface SEM patterns and varies from a factor of 77% with PA to 82% with PB.

4. CONCLUSIONS

In conclusion, the study clearly shows that, despite the simplicity of the technique, films with good superconductive behavior could be screen-printed on silicon dioxide and nitride. All the films have presented a zero resistance above 77K and almost same value of ΔT_c. To our knowledge, these results are new. The X-ray diffraction spectra confirm that the majority phase is the orthorhombic phase on the two kinds of substrates.

Finally, the microstructure of the initial powders has a crucial influence on the adhesive behavior of the films, and the best results were obtained with initial powders constituted of plate-like grains. This latter microstructure seems to avoid the crack formation in the screen-printed films and the study of this parameter is in progress.

We can expect that improvement of the results, concerning, in particular, densification of these films and their critical current density values, will contribute to a better understanding of the substrate/material interaction in relationship with the microstructure of the initial powders.

We greatly acknowledge M. Legeleux for facilities for using screen-printed apparatus, and F. Floret and C.Viguier (Ion Beam Service) for supplying substrates for this study.

4. REFERENCES

1   S. WITANACHCHI, S. PATEL ,H.S. KWOK and D.T. SHAW, Appl. Phys. Lett.54 (1989) 578

2   R. SINGH, S. SINHA, N.J. HSU and P. CHOU, J. Appl. Phys. 67,(1990) 3764.

3   G. KOREN, E. POLTURAK, B. FISHER and D. COHEN, G. KIMEL, Appl.Phys. Lett. 63 (1988) 2330.

4   D. BHATTACHARYA, C.K. MAITI, P. PRAMANIK, T.K DEY, S.K.GHATAK and K.L. CHOPRA, J.C. KASHYAP, D.K. PANDYA and B.GOGOI, Thin Solid Films 164 (1988) 155.

5   D.K. FORK, A. BARRERA and T.H. GEBALLE, A.M. VIANO and D.B. FENNER, Appl. Phys. Lett. 57 (1990) 2504

6   M.G. NORTON, S. MCKERMAN and C.B. CARTER, Phil. Mag. Lett. 62 (1990) 77.

7   T.VENKATESAN, E.W. CHASE, X.D. WU, A. INAM, C.C. CHANG and F.K. SHOKOOHI, Appl. Phys. Lett. 53 (1988) 243.

8   A. MOGRO-CAMPERO, B.D. HUNT, L.G. TURNER, M.C. BURRELL and W.E. BALZ, Appl. Phys. Lett. 52 (1988) 584.

9   Q.Y. MA, X.WU, M.T. SCHMIDT, E.S. YANG and CHIN-AN CHANG, Physica C 162-164 (1989) 637.

1 0 P.ODIER, B. DUBOIS, C. CLINARD, H. STROUMBOS and P. MONOD, "Ceramic Powder Science vol III", Ceramic Transactions vol 12, ed. G.L. MESSING, S. HIRANO, and H. HAUSNER, The American Ceramic Society, Westenville, Ohio 1990

11 A. TAFFIN, T. BROUSSE, B. MERCEY, M. HERVIEU, C. GUNTHER and Y. MONTFORT, Physica C 170 (1990) 59.

1 2 D.R. CLARKE, T.M. SHAW and D. DIMOS, J. Am. Ceram. Soc. 72 (1989) 1103.

High T$_c$ Superconductor Thin Films
L. Correra (Editor)
499

Highly textured Bi-Sr-Ca-Cu-O thin films, produced ex-situ by sequential thermal evaporation

F J Müller[a,b], J C Gallop[a,b], M Stewart[a], F M Saba[b] and A D Caplin[b]

[a]National Physical Laboratory, Teddington, Middlesex TW11 OLW, United Kingdom

[b]Centre for High Temperature Superconductivity, The Blackett Laboratory,
Imperial College, London SW7 2BZ, United Kingdom

Abstract
   Sequential thermal evaporation has been employed for the ex-situ production of 2212 phase Bi-Sr-Ca-Cu-O thin films of varying thickness (0.2 to 1.54 µm). The high vacuum deposition onto MgO single crystal substrates is thereby accomplished by resistively heating the boats containing the respective constituents.
   The transition temperatures T$_c$(zero) for most of these films were recorded between 75 and 80 Kelvin. The microstructure of these polycrystalline thin films was found to depend crucially on the post-anneal conditions. The film coverage of the substrate was dramatically improved by increasing the thickness of the as-deposited film, underlining the impact of the initial grain-size-to-film-thickness ratio D/t.

1. Introduction

   Sequential thermal evaporation is an elementary deposition technique which does not require sophisticated equipment for composition control or an intricate set of different evaporation sources. Nevertheless, it readily facilitates the ex-situ production of 2212 phase Bi-Sr-Ca-Cu-O (BISCCO) polycrystalline thin films of respectable quality. These features make this deposition technique a candidate for industrial mass production of HTS (High Temperature Superconductivity) thin films. However, we encountered several severe problems in terms of both microstructure and coverage until we were finally able to produce good quality films. This development is shown here pictorially, with the display of SEM (Scanning Electron Microscopy) photographs.

2. Experimental: Thin Film Production

   Prior to the post-anneal, the constituents of our Bi-Sr-Ca-Cu-O (BISCCO) films were deposited onto 8 by 8 mm or 5 by 5 mm MgO single crystal <100> substrates at room temperature. Several vacuum deposition cycles were used, with the pressure not higher than 10^{-5} torr. During each cycle the constituents were, in the desired ratio, sequentially evaporated from four resistive boats, each containing one of the constituents: Bi metal, SrF$_2$, CaF$_2$ and Cu metal. The use of a fifth resistive boat containing Pb metal for doping was optional.

   An XTC crystal monitor determined the thickness of individual layers. This device naturally monitors the overall amount of an individual constituent bombarding and sticking to the substrate. To obtain an idea about the relative sticking coefficients for the respective constituents, we arranged a chemical analysis of some of our as-deposited BISCCO films. This analysis was performed on a Perkin Elmer Sciex Elan 500 ICP-MS (Inductively Coupled Plasma Mass

Spectrometer) after dissolution of the sample in nitric acid. Reference 1 contains a description of the instrument and the (standard) technique used. For our purposes, the detection of Ca constituted a problem, since the plasma is propagated in argon with its amu of 40. The most abundant Ca isotope equally possesses an amu of 40 and is thus rendered not usable for ICP-MS. The only other isotope of Ca, amu 44, has a low abundance and hence dictated a low sensitivity. However, the relative sticking coefficients of the other constituents were all calculated to be high. They will generally depend on the deposition rate; the following figures refer to two films simultaneously produced with deposition rates of around 1.0 to 2.0 Å/s; Cu, the first material evaporated in each cycle, scored 82.1 %. The first Cu layer, 320 Å thick, probably provided a buffer layer to enhance the adherence of subsequent layers. For Bi and SrF_2, the third and fourth constituent in this sequence, the sticking coefficients were estimated at around 98.3% and 99.1%, respectively.

The post-annealing was accomplished in a wet atmosphere, in order to allow the fluoride in the as-deposited film to react with hydrogen and evaporate away. The recipe for the post-anneal normally includes a temperature $T(1)= 810°C$ at which the film is held for 30 minutes for the HF to evaporate away. It is then heated up, at a rate of 2 °C/min, to temperature $T(2)$, where it is kept for varying lengths of time. $T(2)$ typically lies between 840°C and 855°C.

The partial pressure of oxygen during the post-anneal seemed to have a crucial impact on the microstructure, as we shall show pictorially in the next section, and was varied between 0.1 and 1 atm.

3. Microstructure and Coverage

This paragraph will first describe two unsuccessful attempts to produce a thin film with both decent microstructure and coverage, and then confront this with a series of successful ones. The BISCCO sample which we shall refer to as film 1 had an as-deposited thickness of around 0.57 µm, according to the XTC crystal monitor. Taking relative sticking coefficients into account, the molar ratio Bi:Sr:Ca:Cu was estimated as 2.0:2.0:2.3:3.1. Film 1 was lead-doped, with a high Pb:Bi ratio of 1.25. It was post-annealed in pure oxygen with pressure 1.0 atm, for 120 minutes at $T(2)=845°C$. Its microstructure, along with many other films annealed in O_2, exhibits a network of dendrites, occasionally connecting isolated patches of material. The dendrites, photographed in Fig.1 on the next page, were typically up to 200 µm long and tens of µm thick. An analysis with Energy Dispersive X-rays (EDAX) revealed these dendrites to carry the 2212 phase and thus be responsible for the superconducting transition with a T_c(zero) of around 70 Kelvin. Despite the molar ratio of the as-deposited film, the 2223 phase was not detected by X-ray diffraction (XRD). For a film produced to a similar recipe but with a greater as-deposited thickness the dendritic jungle was becoming denser, albeit still leaving the substrate exposed in between the dendrites. With a lower as-deposited thickness, the uncovered substrate area increased.

Reducing the partial pressure of oxygen and the lead content helped to transform this network of dendrites into a plate-like microstructure as shwon in Fig.2a. The molar ratio of the as-deposited film was estimated as 2.0:2.0:2.0:2.5, with its overall thickness 0.284 µm and a Pb:Bi ratio of around 0.5. It was post-annealed in an Ar/O_2 atmosphere, with partial O_2 pressure at around 0.1 atm and partial Ar pressure of around 0.9 atm. The onset of the superconducting transition was at around 90 K, with T_c(zero) at around 75 K. X-ray diffraction showed a very high degree of c-axis oriented texturing for this film, more than 95 %; the corresponding XRD graph is shown in Fig.2b. The only shortcoming of this polycrystalline film remained the patchy coverage of the substrate.

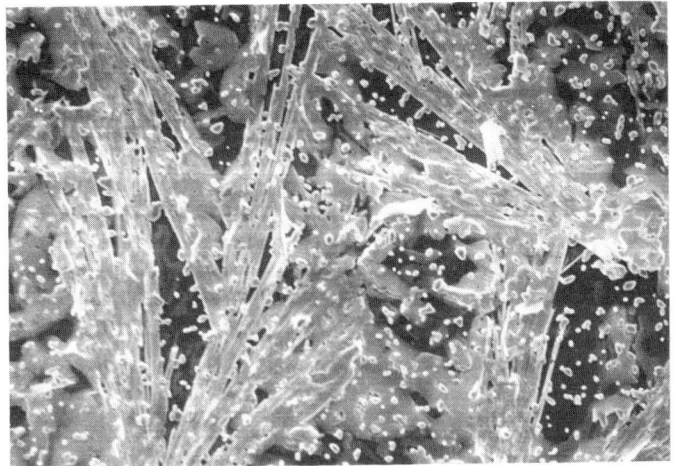

Fig.1: *SEM photograph for film 1 shows a jungle of dendrites. This thin film microstructure is unsuited for electronic applications.*

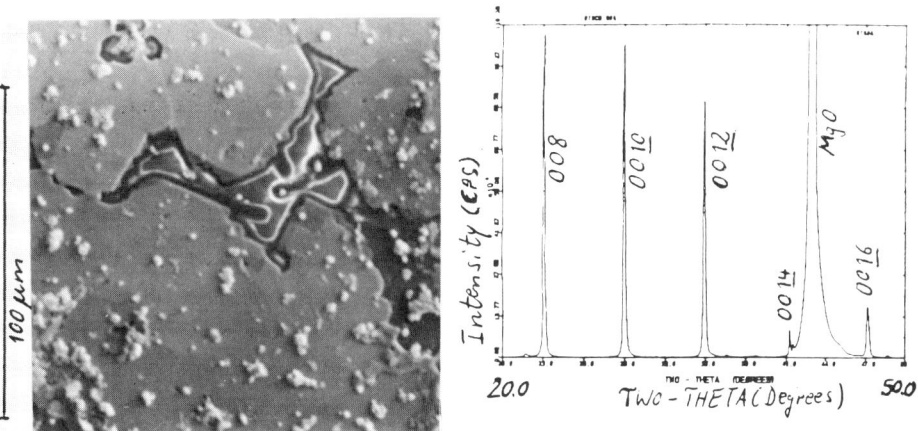

Fig.2a and 2b: *SEM photograph for film 2. The XRD graph on the right (intensity vs two-theta, from 20 to 50 degrees) shows the 2212 phase with a very high degree of c-axis oriented texturing (more than 95 %). Coverage of the substrate, though, is still unsatisfactory for film 2, which had an as-deposited thickness of around 0.284 μm.*

Finally, Fig.3 starts a series of SEM evidence for both good coverage and microstructure. All films in the series were simultaneously deposited in the molar ratio 2.0:2.0:1.0:1.7. The overall thickness of the films is 1.54 μm and seems to take prime responsibility for the good results. We note in passing that we shall also investigate more systematically to what extent the as-deposited molar ratio (which for the films in this series is closer to '2212') and the lead doping influences coverage. No lead was deposited for this series. Film 3 in Fig.3 was post-annealed at $T(2)= 845°C$ for 120 minutes in wet air at atmospheric pressure. Fig.3 shows the typical grain dimension in its polycrystalline microstructure to be around 3 to 5 μm.

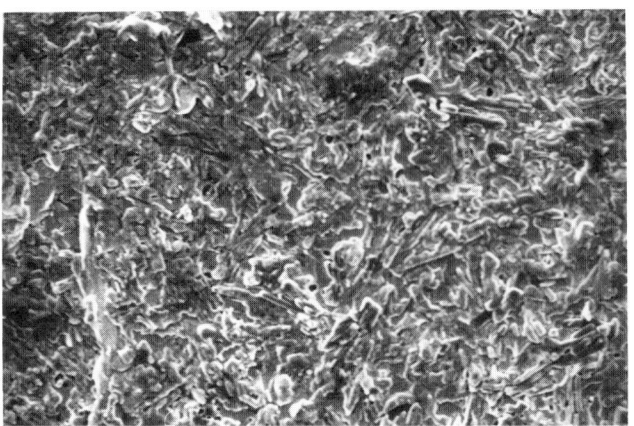

50 μm

Fig.3: *SEM photograph of film 3: finally, the problem of coverage is resolved. The grainy surface structure is suitable for wet chemical etching [2] , but we would still like a smooth surface.*

Film 4 in Fig.4 was post-annealed at T(2)= 845°C for 250 minutes in wet air and in the proximity of a Pb doped BISCCO pellet of the 2223 stoichiometry. The dimension of most granular plates has increased to tens of microns. The white coloured blobs littered across the surface consist predominantly of lead, according to an EDAX scan. T_c(zero) for both film 3 and film 4 lies between 79 and 80 K, with the transition onset at around 90 K.

50 μm

Fig.4: *SEM photograph of film 4. In contrast to film 3, film 4 was placed in the proximity of a BISCCO pellet of the 2223 stoichiometry and the anneal time was increased, from 120 to 250 minutes. The grains in film 4 are significantly larger than in film 3, and their surface is much smoother.*

Film 5 (not pictured) was post-annealed at T(2)= 840°C for 960 minutes (16 hours) in wet air, in the proximity of the same pellet as film 4. The density of the lead rich blobs increased. T_c(zero), surprisingly, dropped by around 5 Kelvin to 74 K. On the other hand, a sudden decrease (around 5 percent) in resistivity was now visible at 110 K, suggesting the presence of grains of the 2223 phase. However, the abundance of 2223 phase grains, if at all present, lies below the detection limit for our X-ray diffractometer.

X-ray diffraction shows that the degree of c-axis oriented texturing for films 3, 4 and 5 is almost identical, but lower than for the thinner film 2. The corresponding XRD graph of film 3 in Fig.5a also reveals the presence of the 20 K chemical phase, which is not detected in film 2. Finally, film 6 stems from the same batch of 1.54 μm as-deposited films. It was annealed simply in wet air at atmospheric pressure for 16 hours, but at the higher T(2) of 855°C. As a result, coverage has deteriorated, a certain fraction of the material seems to have evaporated away and macroscopic cracks are visible in the centre of the film surface. Its coverage thus resembles more that of the thinner as-deposited film 2 than that of films 3,4 or 5. However, the degree of c-axis oriented texturing has improved markedly, as established in Fig.5b, reminescent more of film 2 than of the other films of equal as-deposited thickness. Remnants of the 20 K 2201 phase, however, are still visible.

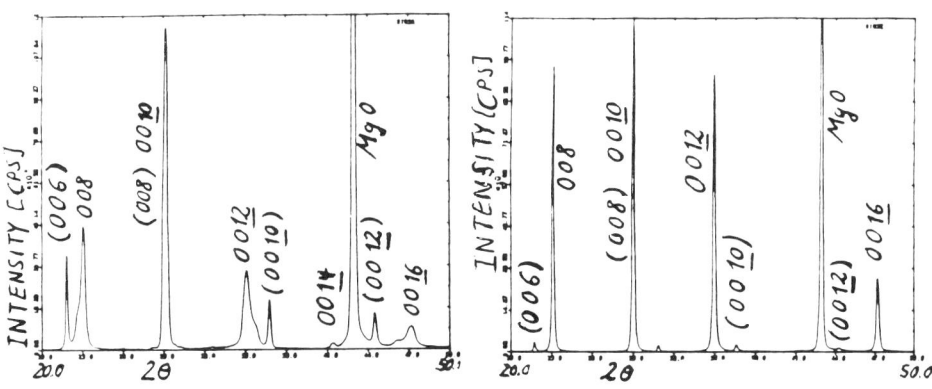

Fig.5a (left): *XRD graph for film 3 (intensity vs two-theta, from 20 to 50 degrees) which is almost identical to the XRD graphs for film 4 and 5. The films are predominantly of the 2212 phase, with around 80 to 85 % c-axis oriented texturing. The as-deposited thickness of these films is around 1.54 μm. Notice the peaks belonging to the 2201 (20 K) chemical phase, which are labelled in brackets.*
Fig.5b (right): *XRD graph for film 6, with the same as-deposited thickness, but a higher anneal temperature (855°C). The degree of c-axis oriented texturing has increased markedly, while remnants of the 2201 K phase are still visible.*

4. Conclusion

Thermal sequential evaporation in a vacuum from resistive boats is a suitable technique for the production of polycrystalline thin Bi-Sr-Ca-Cu-O films of the 2212 phase. As demonstrated in this paper we encountered severe problems in terms of microstructure and coverage on our way to arrive at good quality films. Apropos microstructure: annealing in pure O_2 produces dendrites which are deleterious for electronic applications of these films. Dendritic films are very difficult to pattern, e.g. by means of chemical etching. Annealing in air (i.e. partial pressure of O_2 at 0.2

atm) or in an Ar/O$_2$ gaseous mixture with reduced oxygen pressure can eliminate dendrites. Apropos coverage: increasing the thickness of the as-deposited films prevents the breaking up of the film into separate islands during the post-anneal.

In this context we refer to a study of Y$_2$O$_3$ doped ZrO$_2$ polycrystalline thin films by Miller, Lange and Marshall [3]. Their free-energy calculations predict that "when the grain-size-to-film-thickness ratio D/t is greater than a critical value, the grains will lower the free energy of the sysytem by uncovering the substrate." The authors claim that, according to their calculations and supported by their experimental observations, films with an initial D/t ratio *less* than 8/π will always retain a boundary between the grains. This conjecture is at least not falsified by our own experimental results. For film 1 the final D/t ratio after the post-anneal is of the order of at least 100; for film 2 the same ratio is still well above 10. Both films have broken up into islands, dendritic in the first case, plate-like in the second. For the good quality film 3, however, the final D/t ratio lies between 1 and 2, i.e. below the critical value 8/π. The initial grain dimension at the start of the annealing cycle will most probably be lower than after the anneal, the situation illustrated in Fig.3. Films 4 and 5 belong to the same batch of as-deposited films as film 3, were annealed at the same temperature, and their coverage is equally good. Film 6, again from the same batch but with a deteriorated coverage, is an indication of a temperature threshold at which the surface energy equations force the film to break up. We shall test the D/t ratio argument by depositing films of sub-micron thickness onto a cold substrate, which is in thermal contact with liquid nitrogen, to decrease the initial grain size.

There is a possible price to be paid for the increase in film thickness and the corresponding dramatic improvement in substrate coverage: the transition temperature T$_c$(zero) can become slightly depressed. A film with approximately the same as-deposited molar ratio as films 3, 4 and 5 but with only about 20 percent of their thickness (and a very patchy coverage) exhibited zero resistance at around 85 K. This might be due to the limited diffusion length of the constituent molecules during the anneal. This postulate can be tested by increasing the number of deposition cycles for a given deposition thickness, i.e. reducing the thickness of individual layers. We have already put this into practise on previous occasions, while studying properties of dendritic thin films. The results seemed to back our postulate.

Incidentally, the film with T$_c$(zero) at 85 K, showed an XRD pattern similar to that of films 3, 4 and 5. Increasing the thickness does hence not necessarily reduce the degree of texturing, as long as the thickness of the individual deposited layers is kept comparable. Again, the argument of the limited diffusion length of constituent atoms applies here. Note that the 1.54 μm thick film 6 has in fact a very high degree of texturing but shows a deterioration in coverage. Is there a trade-off between the two ?

It should be noted that numerous other groups have produced good coverage and a decent microstructure of thinner as-deposited films by other more complicated ex-situ production methods. It could perhaps be argued that our evaporation technique is less energetic than most of its competitors and thus needs thicker as-deposited films. Some of these methods deposit the material in tiny lumps of already stoichiometric structure, which overcomes the problem of limited diffussion lengths during the post-anneal.

The 1.54 μm thin films are easily patterned by means of wet chemical etching in EDTA.[2] The lead rich blobs on the surface of films 4 and 5, however, proved disadvantageous for etching purposes.

5. REFERENCES

1 J. Luck and U. Siewers, Fresenius Z.Anal.Chem., 331 (1988) 129
2 F.J. Müller et al - see same conference proceedings Strasbourg 1991
3 K.T. Miller, F.F. Lange and D.B. Marshall, J.Mater.Res., 5 (1990) 151

High T$_c$ Superconductor Thin Films
L. Correra (Editor)
© 1992 Elsevier Science Publishers B.V. All rights reserved.

Patterning of YBa$_2$Cu$_3$O$_7$ films using laser induced oxygen diffusion

Y.Q. Shen, T. Freltoft, and P. Vase

NKT Research Center, Sognevej 11, DK-2605 Brøndby, Denmark

Abstract

High resolution patterning (4μm) has been achieved on epitaxial thin films of the high temperature superconductor YBa$_2$Cu$_3$O$_{7-\delta}$ using laser writing. A focused laser beam is applied to write semiconducting patterns on superconducting films in a vacuum or in a nitrogen atmosphere. The semiconducting patterns are shown to be formed by a reduction of the oxygen content due to local heating caused by the laser beam. The process does not cause any structural damage and does not change the surface morphology. When reapplying the same laser beam and rewriting the film in an oxygen atmosphere, it is possible to restore the superconductive properties. The application of this method for production of superconducting microelectronic circuits is discussed.

1. INTRODUCTION

It is well established that the superconducting and electrical transport properties of the high-T$_c$ superconductor YBa$_2$Cu$_3$O$_{7-\delta}$ depend strongly on the oxygen depletion δ [1]. In the orthorhombic phase (i.e. $0.5 > \delta > 0$), the compound behaves like a traditional superconducting metal concerning the electronic transport properties and below 92K, it becomes superconducting. The tetragonal phase of the compound (i.e. $0.5 < \delta < 1$) has an activated ground state, i.e. it behaves like a semiconductor. By heat processing in a controlled atmosphere, the oxygen content can be changed resulting in oxygen diffusion in or out of the material.

A method for modifying the properties of YBa$_2$Cu$_3$O$_{7-\delta}$ by laser writing was suggested by Ib Johannsen [2]. The laser writing method prescribes application of a focused laser beam in order to locally heat the YBa$_2$Cu$_3$O$_{7-\delta}$ and thereby causing oxygen to diffuse out of the superconductor. The direction of the oxygen diffusion (i.e. in or out) may be controlled through the application of either a reducing or an oxidizing atmosphere. In this way, modified patterns may be written on the YBa$_2$Cu$_3$O$_{7-\delta}$ materials by scanning a focused laser beam (see Fig.1).

Independent investigations by R.R. Krchnavek et al. [3] and R.C. Dye et al. [4] have proven that by laser writing in vacuum, the YBa$_2$Cu$_3$O$_{7-\delta}$ can be modified into superconductors at lower transition temperatures or insulators/semiconductors [3], and writing in oxygen can generate superconducting patterns on YBa$_2$Cu$_3$O$_{6+x}$ films [4].

In the following, the term "writing" applies to a laser writing process enabling the oxygen

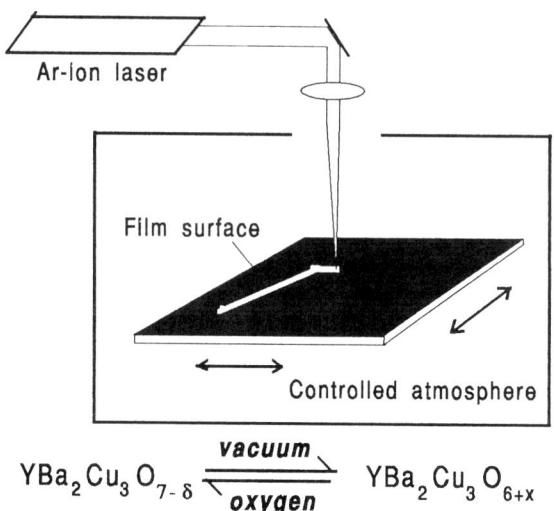

Figure 1 Schematic diagram of the laser writing process.

to diffuse *out* of the $YBa_2Cu_3O_{7-\delta}$ film, while the term "rewriting" applies to a laser writing process *recovering* the lost oxygen when writing in an oxygen atmosphere.

2. EXPERIMENTAL

In this experiment, the superconducting films were prepared by laser ablation of $YBa_2Cu_3O_7$ on 1 cm^2 MgO substrates [5]. The epitaxial films have typical transition temperatures around 88 K, and critical current densities at 77 K measured on 10 μm or 50 μm microbridges up to $4{\cdot}10^6$ A/cm^2, and typically above 10^6 A/cm^2. The crystal mosaicity was measured to $\approx$ 1.3° by X-ray diffraction analysis which also showed that only *one* crystal orientation (and its twin) was present in the film with *a* and *b* axes along the MgO (100) and (010) directions [6].

The contacts for 4-probe measurements were laser ablated Ag contacts, and the thickness of the film was $\approx$ 500 nm. Prior to laser writing, each film was patterned into a 500 μm wide strip by laser ablation lithography [7]. Ablation patterning was performed using the same Q-switched Nd-YAG laser as was used for the film deposition, but with the laser pulses applied directly to the film surface. A thin metal mask was placed in contact with the thin film defining the desired pattern. In this way, micro bridges down to 10 μm width may be produced [7].

The film sample was mounted on a set of motor controlled translation stages and electrically connected to a 4-probe Ohm meter. The laser beam was provided by a multi-wavelength (488/515.5 nm) Ar-ion laser and focused at a 4 μmdiameter spot by a

20x Ø0.35 microscope lens. The translation stages and the lens were sealed in a vacuum chamber, and scanning of the beam was carried out by displacing the translation stages at a speed of 5 μm/s. In order to focus the laser beam at the film surface, and to control the absolute position of the beam spot, a beam splitter was placed in front of the microscope lens. Through the beam splitter, the irradiated area and the beam spot may be viewed.

It is well known that $YBa_2Cu_3O_7$ changes its optical properties when losing oxygen [3,8]. This phenomenon was used to optimize the intensity of the focused beam. It was found that after laser writing at intensities between 3.5 mW/μm^2 and 10 mW/μm^2 in vacuum (between 3.5 mW/μm^2 and 15 mW/μm^2 in nitrogen), a bright trace was left on the exposed part of the film. The trace was much more transparent than $YBa_2Cu_3O_7$ and did not show any physical damage. When a higher degree of intensity was applied, the exposed part showed signs of melting.

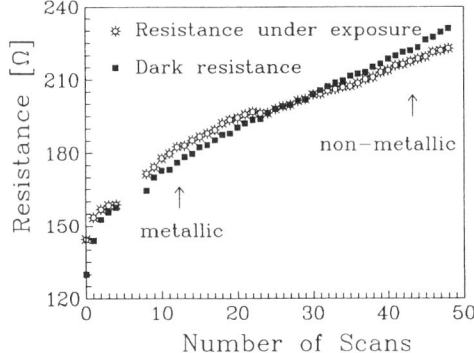

Figure 2 Dark resistance (■) and maximum resistance under exposure (✿) during laser writing in nitrogen atmosphere.

During the laser writing process, the vacuum chamber was pumped to 10^{-3} Torr or filled with 1 atm. nitrogen. The laser intensity was set to a fixed value according to the optimization. The motor controlled translation stage was applied so that the focused beam could be scanned perpendicularly to the strip. By *in situ* four probe resistance measurements, the dark resistance and the resistance under the exposure were monitored. During scanning, the resistance of the strip increased due to the loss of oxygen of the exposed part. The *in situ* resistance measurement also reflects the metal to non-metal transition which is caused by the oxygen depletion. As long as the metallic nature of the film is dominating, the resistance measured during the exposure was higher than the dark resistance due to the elevated temperature. After passing the metallic to non-metallic transition, the dark resistance became higher because photoconduction reduces the resistance of the non-metallic phase. Figure 2 shows the resistance change of a strip recorded during the laser writing process indicating the metal to non-metal transition. During this laser writing process, the intensity of the focused beam was 6mW/μm^2, and the speed of scanning was 5μm/s. Typically, the transition was observed after 30 scans. After roughly 60 scans, the superconducting strip was cut by a 4μm wide, bright path, and the resistance of the path had increased about 500 times.

In the rewriting process, the chamber was filled with 1 atm oxygen. The translation stage directed the focused beam to scan parallel to the strip in order to reconnect the strip by recovering the oxygen content in the semiconducting path. The intensity of the focused beam was below $15\text{mW}/\mu\text{m}^2$ to avoid melting of the film. By applying the same scanning speed and intensity as was used in the writing process, the rewriting process was much faster than writing. After only 1 or 2 scans, the resistance decreased strongly, and the photoconduction effect disappeared. A more accurate control of the rewriting process may be obtained by increasing the scanning speed or by decreasing the intensity.

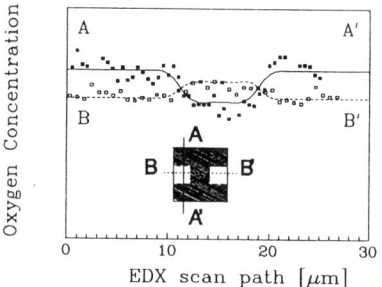

Figure 3 Optical reflection micrograph of a YBCO film after laser writing and rewriting. The bright line is the trace left after writing in nitrogen atmosphere and the to bridges across the line have been recovered by rewriting in 1 atm. oxygen.

Figure 4 Profile of oxygen distribution determined by EDX, over the same area as shown in fig. 3. The oxygen EDX count rate was normalized by the EDX copper count rate before plotting.

3. RESULTS AND DISCUSSION

Figure 3 shows an optical micrograph of the traces left by writing and by rewriting on a $YBa_2Cu_3O_7$ film. The semiconducting path caused by the writing is the bright line, and the bridge across the line is the rewritten area. The profile of the oxygen distribution across the semiconducting line and across the bridge was determined by EDX (window less Energy Dispersive X-ray analyses) and is shown in Fig. 4. Due to background oxygen signals from the MgO substrate, an absolute calibration of the oxygen content in the film was not possible. In the figure, the ratio between the oxygen and copper count rates is plotted. As may be seen from Fig. 4, there is a clear reduction of the oxygen signal of profile A when crossing the semiconducting line. The opposite behaviour is observed in profile B along the semiconducting line when crossing the restored area.

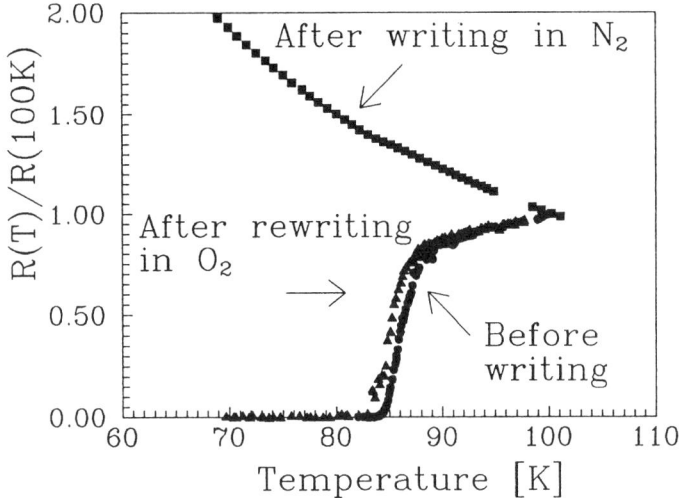

Figure 5 Resistance versus temperature of a YBCO strip before writing, after disconnection by writing, and after recovering by rewriting. R(100K) for the three data sets were 62Ω, 1850Ω and 85Ω, respectively.

The temperature dependent resistance of a $500\,\mu$m strip before and after disconnecting it by laser writing is plotted in Fig. 5 as is the resistance of the strip after it is reconnected by rewriting. Fig. 5 clearly shows the semiconducting nature of the laser writing path as well as the metallic nature and superconducting transition of the recovered area. For the recovered part, a T_c of 83K was measured, whereas the original strip showed a superconducting transition with $T_c = 84$K. Laser writing on this strip was carried out in 1 atm. nitrogen scanning the focused beam 48 times, and using the intensity 6 mW/μm^2. The same intensity was applied in the rewriting process carried out in 1 atm. oxygen where the number of scans was only two.

Laser writing and rewriting were also carried out on 50 μm wet etched YBa₂Cu₃O₇ strips. However, it was found that both processes went much slower on the wet etched strips than on the strips etched by ablation lithography. Furthermore, it was not possible to fully recover the semiconducting part created by the writing process to a condition comparable to the original state of the superconductor. The reason for the lower oxygen diffusion rates could be that remaining photo resist changed the absorbtivity of the film, prevented the diffusion of oxygen, or even destroyed the film during the process.

4. CONCLUSIONS

Our investigations have shown that the application of a focused laser beam for writing on YBa₂Cu₃O₇ films is a reversible process capable of modifying films of this material from a superconductor to a semiconductor and vice versa with a resolution of at least 4 μm. The modifications are solely due to a change of the oxygen content, δ.

The structures formed by laser writing are likely to be less prone to environmental degradation than e.g. physically etched structures since the diffusion of H_2O and CO_2 into the film is restricted along the c axis.

Laser writing is simple and may be useful for the patterning of $YBa_2Cu_3O_7$ films for the manufacture of weak links e.g. in SQUIDs, and for modification (tuning) of the critical current of weak links. Moreover, the unchanged epitaxy and surface morphology of the patterned film may enable continued epitaxial growth of other materials after applying the laser writing technology, and in this way integrated superconductor - semiconductor devices may be manufactured.

5. ACKNOWLEDGEMENTS

We like to thank Dr. Bilde Sørensen at Risø National Laboratory for EDX measurements, and Dr. J. Bindslev Hansen at the Technical University of Denmark and Dr. I. Johannsen for stimulating discussions and guidance. The work was supported in part by the Danish National Program for High Temperature Superconductors, the power distribution company, ELSAM, and the Danish Academy of Technical Sciences (ATV).

6. REFERENCES

1   J.-M. Tarascon, W.R. Mckinnon, L.H. Greene, G.W. Hull, and E.M. Vogel, Phys.Rev., B36 (1978) 226.
2   I. Johannsen, DK Patent application No 4980/87 (1987).
3   R.R. Krchnavek, S.J. Allen, S.W. Chan, F. De Rosa, M.K. Kelly, S. Sampere, C.T. Rogers and P.F.
4   Miceli, SPIE 1187 (1989) 261.R.C. Dye, R.E. Muenchausen, N.S. Nogar, A. Mukherjee ans S.R. Brueck, Appl. Phys. Lett., 57 (1990) 1149.
5   T. Freltoft and P. Vase, in *Electronic Properties of High-T$_c$ Superconductors and Related Compounds, Springer Series in Solid-State Sciences, 99*, edited by H. Kuzmany, M. Mehring and J. Fink (Springer-Verlag, Heidelberg 1990) p.19.
6   X-ray texture analysis was performed by R. Kromann, Risø National Laboratory, Denmark.
7   P. Vase, Y.Q. Shen and T. Freltoft, Applied Surface Science 46 (1990) 61.
8   M.K. Kelly, P. Barboux, J.-M. Tarascon and D.E. Aspnes, Phys.Rev., B40 (1989) 6797.

High T$_c$ Superconductor Thin Films
L. Correra (Editor)
511

Role of oxygen in the incorporation of bismuth in Layer By Layer deposited BiSrCaCuO thin films

Laurent Ranno*, Pierre Bernstein**, Jean Pierre Contour**, Laurent Force**, MichelViret*, Xiang Zhen Xu*, Michel Laguës*

* Laboratoire de physique quantique UA 421 CNRS, ESPCI, 10 rue Vauquelin, Paris
** Laboratoire de physique des solides, ESPCI, 10 rue Vauquelin, Paris

Abstract

BiSrCaCuO thin films (about 40 nm thick) were deposited at low temperatures (500°C) on MgO (100) and Si (100) by a Layer By Layer process. An atomic oxygen plasma source was used during the deposition. Since this source is presently very directional, its impact spot is well localised on the substrate. Therefore the influence of the local oxygen pressure (10^{-4} torr) on the atomic concentrations obtained in the films was studied using RBS and EDX analyses. In this range of pressures and temperatures the incorporation of bismuth is enhanced by the presence of the plasma. We also found strong correlations between the incorporations of some of the elements.

One important application of HTc superconducting thin films is the design of junctions or multilayers. In these structures sharp interfaces must be achieved. That is why MBE (ref 1,2,3) or laser ablation (ref 4) methods are widely used. Since the crystallisation must be obtained in-situ, the substrate is always heated but its temperature should not be too high (ref 5) in order to prevent diffusion and deterioration of interfaces. Moreover correct oxydisation of films must be performed during the growth and only low temperature post-annealing to introduce extra oxygen are allowed. Although some groups use N$_2$O (ref 5), NO$_2$ (ref 3) or O$_3$ (ref 1,2,4) as oxydising agents, our group uses atomic oxygen in a MBE system. We present results on the influence of oxygen on the incorporation of bismuth and on the importance of the sequence of evaporation when several elements are co-evaporated.

1 EXPERIMENTAL

1.1 Equipment

An introduction chamber is coupled with an ultra-high-vacuum growth chamber where the deposition is made (chamber pressure 10^{-6} torr during the process, 10^{-10} torr otherwise). The substrates are MgO (100) or Si (100). Bismuth, calcium and strontium are evaporated from three Knudsen cells at 500°C, 450°C and 440°C respectively. Copper is evaporated using an electron gun and monitored by a thickness recorder. We control the fluxes of these four elements using a computer which drives the shutters. Oxygen is introduced near the substrate (distance source-substrate 7 to 25 mm) by an oxygen DC plasma source. The atomic to molecular oxygen ratio is 10% (ref 6). The local pressure of molecular oxygen is 10^{-4} torr. RHEED (Reflection High Energy Electron Diffraction) equipment provides information about the state of the substrate surface and the film surface during deposition.

1.2 Process

The substrate is introduced and outgassed for two hours at 700°C and for MgO substrates the oxygen source is turned on. Thus the RHEED patterns show an improvement in the quality of the MgO surface. Then we use a Layer By Layer sequence, each layer consisting of the evaporation of one or several elements simultaneously. The evapororation of the elements during this period follows the crystal bidimensional structure $Bi_2Sr_2Ca_{n-1}Cu_nO_{2n+4}$. The temperature of the substrate is set between 500°C and 600°C.

1.3 Analyses

After the deposition, Energy Dispersive X-ray (EDX) and Rutherford Backscattering (RBS) analyses are performed in order to determine the composition of the films and X-ray Diffraction is performed to assess their cristallisation. We have motorised the substrate holder of the EDX analyser, so that the electron beam can be focused on the film and then swept along a line, 0.1mm wide and several centimeters long. During each sweep the signal corresponding to one element is measured and after the fourth one we have the four compositions along the selected line instead of having the composition averaged on a 2 mm spot when using RBS.

2 RESULTS

2.1 Incorporation of bismuth

Since the oxygen source is very directional, there exists a natural oxygen pressure gradient on the surface of the substrate during the deposition. Thus the films show a radial oxydisation around the centre of the oxygen spot, which is determined by the oxygen pressure. Before deciding on a new oxygen source, which is characterised by a wider spot and a higher pressure, we studied the influence of oxygen on the incorporation of bismuth.

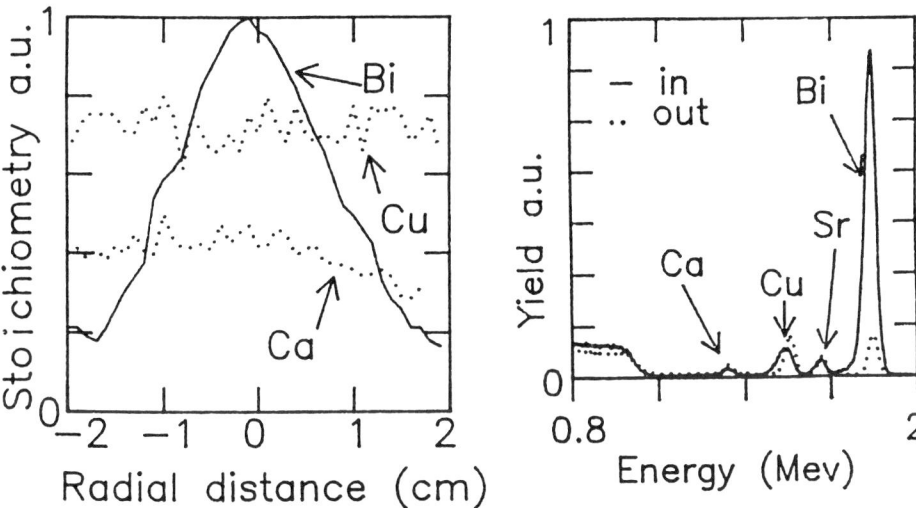

Figure 1 : Composition of a BiSrCaCuO film deposited on Si as a function of distance from the center of the most oxydised zone.

Figure 2 : RBS spectrum of a BiSrCaCuO film deposited on MgO in and outside the oxydised zone.

Figure 1 is an EDX graph which shows the variations of the concentrations of bismuth, calcium and copper along a diameter of the plasma oxydised spot on a BiSrCaCuO film deposited on Si at 500°C. Strontium cannot be analysed because of the superimposition of its main X-ray line on one of silicon's. The fluxes of the four elements are homogeneous in this area and the temperature of the substrate is uniform on this scale. The only non uniform parameter is the pressure of molecular and atomic oxygen. We observe a strong variation in the incorporation of bismuth as a function of the oxygen pressure on the substrate surface. The stoichiometry of Bi is 5 times larger in the center of the spot than in the non oxydised area. On the contrary the quantities of copper and calcium are far less affected by the presence of oxygen. This result has already been observed by Ogihara et al.(ref 7) who measured a higher level of incorporation of bismuth under an ozone beam than in the region not reached by ozone. This result is related to the low vapour pressures of oxydes. Figure 2 presents RBS spectra of the concentrations of the four elements on a BiSrCaCuO film deposited on MgO with the same experimental conditions as the one deposited on silicon. The first spectrum (solid line) was taken at the centre of the most oxydised part of the film, the other one (dotted line) was taken on the outside zone. One can notice that the shapes of the peaks are different because of the difference of surface conductivity of the two regions. The quantitative analysis does not reveal any significant variation of concentration for Cu, Sr and Ca, but the amount of bismuth is 10 times larger under the oxygen beam than outside the beam spot. We also did not notice any major difference in the incorporation of bismuth in films deposited on MgO or Si substrates. Even if the bismuth incorporation is higher with only molecular oxygen (ref 8), we need atomic oxygen to oxidise copper (ref 6) and it is more convenient to keep the oxygen plasma on during the whole deposition if the pressure threshold for incorporating bismuth is to be reached.

2.2 Correlations between elements

Table 1 : sequences for the evaporation of films 1, 2 and 3

Depositions on Si (100) at 500°C. O_2 pressure 10^{-4} torr, O pressure 10^{-5} torr

opened shutters	film 1	film 2	film 3
Bi+Sr+Ca+Cu+O	50 sec	50 sec	50 sec
Bi+Sr+O	-	-	50 sec
Bi+O	-	100 sec	-

We made three depositions with the same conditions of fluxes, substrate temperature (500°C on Si) and oxygen pressure (O_2 10^{-4} torr, atomic oxygen 10^{-5} torr). The only difference was the sequence chosen for the shutters movements. Table 1 shows the sequences of these three experiments and figure 3 the composition of the three films obtained.

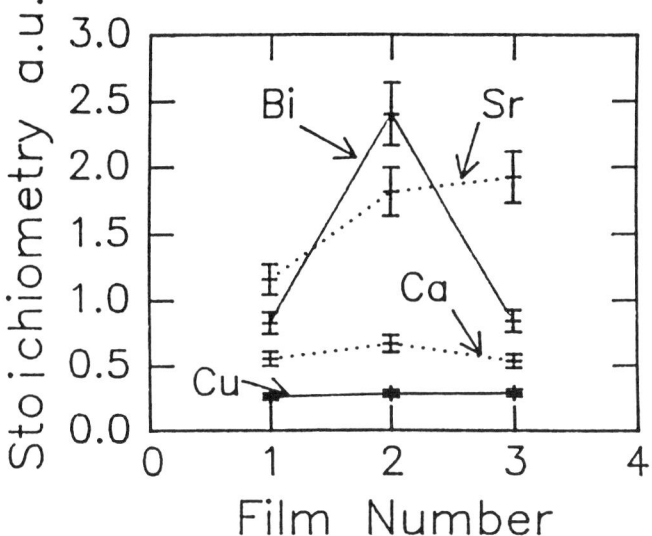

Figure 3 : Composition of films 1, 2 and 3 (RBS)

The oxygen plasma was always on, and the oxygen beam permanently reached the surface. The holder was rotating in order to have a more homogeneous centreWe observed an influence of strontium on the incorporation of bismuth. When we triple the time Bi is evaporated, the amount of incorporated Bi increases of 190±20 % but when we double the time Bi and Sr are evaporated together, the amount of incorporated Bi does not vary significantly (13±20%). Therefore strontium seems to prevent bismuth from incorporating. The strontium incorporation itself depends strongly on the elements which are evaporated at the same time.

When copper is evaporated with bismuth, its sticking coefficient is 4 times smaller than when it is evaporated alone. This result has already been observed by P.Luzeau et al.(ref 9).

3 Conclusion

Our films of BiSrCaCuO deposited under MBE conditions on MgO or Si substrates at low temperatures (about 500°C) presented strong inhomogeneities in bismuth concentration. We showed that the problem of bismuth incorporation is strongly correlated to a gradient of oxygen pressure on the surface of the substrate. The three other elements do not present such a dependence on oxygen pressure. We are now developping a new oxygen source which, coupled with the rotation of the sample holder, is expected to provide a larger homogeneously oxidised area on the substrate.

Acknowledgments

We would like to thank Dr Jacques Perrière, Annie Cheenne and Rafael Perez for RBS analyses and fruitful discussions.

References

(1) J.N Eckstein, I.Bozovic, K.E. von Dessonneck, D.G.Schlom, J.S.Harris, S.M. Baumann, Appl.Phys.Lett. 57 (9) (1990)

(2) Y.Nakayama, I.Tsukada, A. Maeda and K.Uchinokura Jpn.J.Appl.Phys. 28 (10) pp L1809-L1811 (1989)

(3) S.Watanabe, M.Kawai and T.Hanada Jpn.J.Appl.Phys. 29 (7) pp L1111-L1113 (1990)

(4) M.Kanai, T.Kawai and S. Kawai Appl.Phys.Lett. 58 (7) (1991)

(5) M.Kanai, T.Kawai and S.Kawai Appl.Phys.Lett. 54 (18) (1989)

(6) P.Luzeau, X.Z.Xu, M.Lagues, N.Hess, J.P.Contour, M.Nanot, F.Queyroux, M.Touzeau and D.Pagnon J. Vac.Sci.Tech. A8, 3938 (1990)

(7) M.Ogihara, T.Yamada,T.Tsurdoka and H.Abe, Abstract of the 22nd Conference of solid state devices and materials, Sendai (1990) pp581

(8) P.Luzeau,X.Z.Xu, M.Laguës, M.Nanot and F.Queyroux, J. Less Comm.Met. 164-165, 695 (1990)

(9) P.Luzeau, Thèse de l'université Paris XI (1990)

High T$_c$ Superconductor Thin Films
L. Correra (Editor)
517

Preparation and Characterization of Thin Films of (Bi,Pb)$_2$Sr$_2$Ca$_2$Cu$_3$O$_{10+\delta}$ Using the Pulsed Laser Deposition Technique

F.S. Razavi[†] and H.-U. Habermeier

Max-Planck-Institut für Festkörperforschung , Heisenbergstraße 1, 7000 Stuttgart 80, F.R. Germany

Abstract

(Bi,Pb)$_2$Sr$_2$Ca$_2$Cu$_3$O$_{10+\delta}$ (2223-Bi) thin films, were fabricated on [100] oriented MgO single crystal substrates using the pulsed laser deposition technique. For preparing the films, two different routes were followed. First, insulating amorphous films of Bi-2223 were deposited on the substrates at room temperature. After the annealing procedure of these films at 865 °C, they showed a T$_{c,zero}$ as high as 105K. Second, films were prepared at substrate's heater temperature about 835 °C. The SEM and XRD results indicated the c-axis oriented films. The resistivity results showed the T$_{c,zero}$ between 60K to 70 K on these films, after annealing at 865 °C, the onset of superconducting transition increased to 117 K and T$_{c,zero}$ to 105 K. The annealing procedure and the effect of annealing on the surface quality of the films are discussed.

1   Introduction

There have been many reports on preparation of the thin film of high T$_c$ Pb doped Bi-2223 compound [1-5]. Generally, these films were produced either by pulsed laser deposition technique (PLD) or by means of DC magnetron sputtering technique. The films synthesized by PLD technique initially showed T$_{c,zero}$ below 50 K and required a long time annealing (more than 15 hours) for obtaining T$_{c,zero}$ between 95 K to 100 K [1,2]. Using DC magnetron sputtering technique, Lin et al. [4] reported the T$_{c,zero}$ at 106 K following the annealing of the film in presence of the Bi-2223 pellet at 840 °C for 4 hours. The films were not single Bi-2223 phase, but the Bi-2212 phase were also present in the film. The "in-situ" film of Bi-2223 deposited on MgO single crystal substrate at 720 °C was reported by Hakuraku et al. [5]. This film showed the onset of superconductivity at about 100 K and T$_{c,zero}$ near 70 K.

We used the PLD technique on nearly stoichiometric target of Bi-2223 for "in-situ" fabrication of the thin films of this high T$_c$ oxide. The effect of oxygen pressure during deposition, the substrate temperature and post annealing on the T$_{c,zero}$ of these films were investigated.

[†]permanent address : Department of Physics , Brock University, St.Catharines, Ontario, Canada L2S 3A1

2   Experimental

Several 1.3 cm diameter pellets of $Bi_{1.7}Pb_{.3}Sr_2Ca_2Cu_3O_{10+\delta}$ were prepared from the powder of the same chemical constitution (purchased commercially) [6]. The powder was pressed into several pellets and sintered at 860 °C for 24 hours. The resistivity measurement on a pellet showed the onset of superconductivity at 115 K and $T_{c,zero}$ at 102 K. The dc-magnetization measurement also confirmed the above result with the diamagnetic moment saturate at 84 K. This indicates that our pellets are multi-phase consisting of Bi-2223 and Bi-2212 phases. A KrF excimer laser (λ=248 nm, energy density about 1.5 J/cm^2) with the repetition rate of 4 Hz and an incident angle of 15° on the rotating target were employed to deposit the thin film of Bi-2223 on [100] oriented MgO single crystal. The films were deposited at the rate between .2 to 1 nm/s and the oxygen pressure in the chamber was varied between .3 to 6 mbar during different runs. The substrate was heated using a flat heater with its temperature monitored by Pt-10%Rh thermocouple. The films were investigated by x-ray diffraction (XRD), scanning electron microscope (SEM), four probe electrical resistivity and ac-magnetization measurement. The critical current density ($J_c(0)$) of the films were measured at 4.2 K.

3   Results and Discussion

We carried out two separate studies on the post annealing of the thin films of Bi-2223. First, amorphous films were produced on the substrate at a temperature below 400°C. Second, films were grown epitaxially on the substrate at a temperature of the substrate heater between 740°C to 860°C.

Thin films of Bi-2223 were prepared by annealing the amorphous films in the deposition chamber after each deposition process. These films were deposited in the presence of .4 mbar O_2 gas. Two samples were prepared at substrate temperature of 300 °C and 350 °C with the thickness of about 100 nm and 300 nm, respectively. The annealing process of these two samples is characterized by a temperature rise to 860°C in 8 minutes, and a 6 minute halt at this temperature; finally, a rapid cooling down to room temperature in 15 minutes. The resistivity measurement on the 100 nm sample showed a semiconducting behaviour with the drop in the resistivity at about 80 K, whereas, the 300 nm sample showed a metallic behaviour down to 120 K and below this temperature it showed a double transition, one between 120 K to 110 K and the second between 95 K and 82 K (Fig. 1).

For insuring the zero loss of Pb during the deposition, we have deposited several films with the substrate at room temperature. The above annealing procedure was carried out on one of the films with the thickness of 400 nm, the resistivity measurement on this film showed the onset of superconductivity $T_{c,onset}$ at 82 K and $T_{c,zero}$ at 55 K. Subsequent external annealing on this film at 860 °C for 65 hours and quenching to room temperature afterwards, increased the $T_{c,onset}$ to 120 K, $T_{c,zero}$ to 92 K and $J_c(0)$ increased by a factor of 50 to about $.18 \times 10^6$ A/cm^2. In this annealing procedure, the substrate was laid on a quartz boat and a pellet of Bi-2223 was put on top of the film without the contact to the surface of the film and the whole assembly was wrapped in gold foil.

This external method of annealing was improved by finally chosing the right thickness of

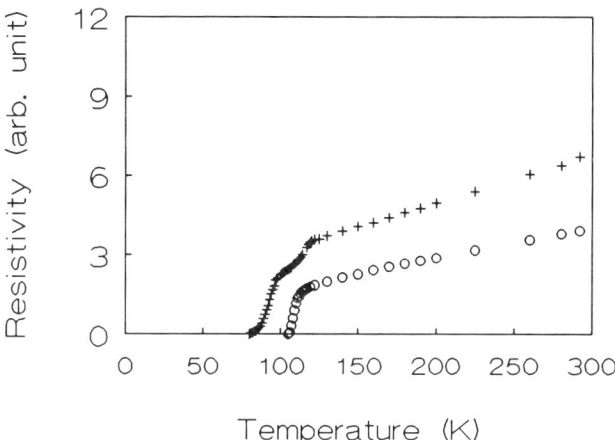

Fig. 1. Resistivity vs temperature of Bi-2223 film deposited at room temperature. (+) After "in-situ" annealing for 6 minutes at 860°C, (○) after external annealing at 860°C for 24 hours.

the sample, at about 250 nm, and the annealing time (24 hours) and slow cooling to room temperature . An example of the resistivity measurement after this annealing procedure is given in Fig. 1. The $T_{c,onset}$ is about 122 K and the $T_{c,zero}$ is at 105 K. The critical current density also improved slightly to about $.48 \times 10^6$ A/cm² at 4.2 K. The XRD measurement on these films showed that the film is mainly c-axis oriented with c about 3.65 nm. However, other peaks due to different planes were also observed. The SEM pictures on these films indicated that these films consist of many platelets with the surface parallel to the surface of the substrate. The slanted platelets were also observed in the SEM picture which was confirmed by the XRD result.

As we have reported previously [7] it is possible to fabricate "in-situ" a thin film of Bi-2212 epitaxially by PLD. To obtain an ideal condition for the "in-situ" fabrication of Bi-2223 thin films, we produced and analysed series of films of Bi-2223 by varing the oxygen pressure, the substrate temperature and the distance between the target and the substrate. The best Bi-2223 films were obtained by chosing 3 mbar oxygen pressure, 2.45 cm distance between the substrate to the target and the substrate heater temperature between 740 °C to 865 °C. The resistivity measurement on all these films revealed metallic behaviour where the film resistance decreases as the substrate temperature increases. The increase in the substrate heater also increases the $T_{c,onset}$ (varied between 80 K to 120 K), as well as the $T_{c,zero}$ (varied between 5K to 60K) and the $J_c(0)$.

The best result was obtained for the substrate heater temperature at 863° which indicated the $T_{c,onset}$ of 120 K, $T_{c,zero}$ at 65 K and the $J_c(0)$ about $.33 \times 10^6$ A/cm² (Fig. 2). The SEM studies on this film showed smooth films plus a large number of impurity phases on the surface of the film. Tabata et al. [2] attributed these impurity phases to the Ca_2PbO_2. The peaks due to this phase can be seen in the XRD with the 2Θ at about 17.8° and 32° which are near to the peaks of (008) and (0014) planes of Bi-2223. The XRD result of revealed (00l) reflections with the c=3.75 nm which is slightly larger than the reported c for the bulk crystal of Bi-2223 [7].

These films were annealed similar to the procedure described above. After annealing, all these films showed $T_{c,zero}$ above 90 K with the $T_{c,onset}$ varying between 115 K to 122 K. The best $T_{c,zero}$=104 K was obtained for films deposited at the substrate heater temperature between 830°C and 840°C in which prior to the annealing procedure they have indicated $T_{c,zero}$ about 46 K. As seen in Fig. 2 the resistivity remains the same before and after annealing at room temperature, but the slope of resistivity versus temperature increases by a factor of 2 and the $J_c(0)$ increases to about $.4 \times 10^6$ A/cm². The SEM pictures on these films showed a slight improvement of the surface quality of these films and the reduction of impurity phases from the film's surface. It is possible that, during the annealing procedure the amount of Pb required for the growth of the Bi-2223 phase is partially supplied by the Ca_2PbO_2 thus, it causes the reduction of the surface impurities.

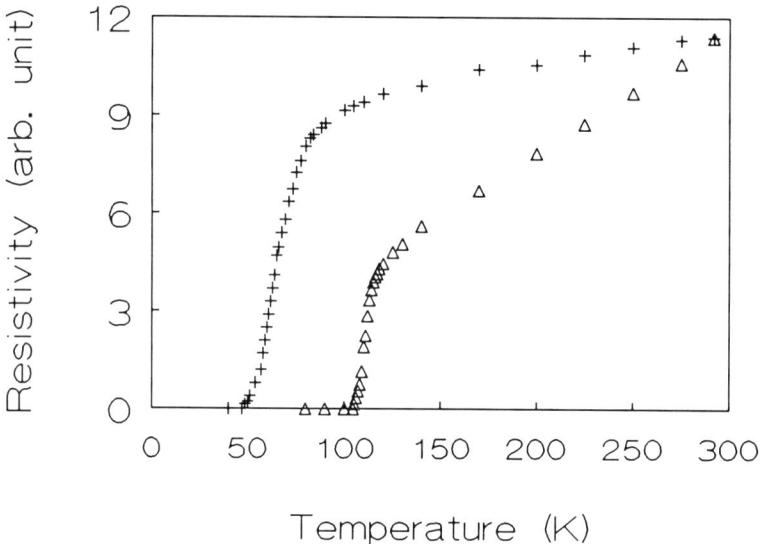

Fig. 2. Resistivity vs temperature of Bi-2223 films. (+) for the as-grown film on the substrate at 830°C, (△) after annealing at 860°C for 24 hours.

The above "in-situ" process although it produced crystalographically the c-axis oriented thin film of the Bi-2223, but the lack of right oxygen stoichiometry and the loss of Pb in the films, prevent the stablization of the high T_c phase. We should also note here that the above annealing process is long. The following procedure was carried out for improving the annealing procedure in production of the thin film of Bi-2223. After the deposition of the film at 840 °C in 3 mbar O_2 pressure the film was rapidly cooled down to 700°C, then, one atmosphere

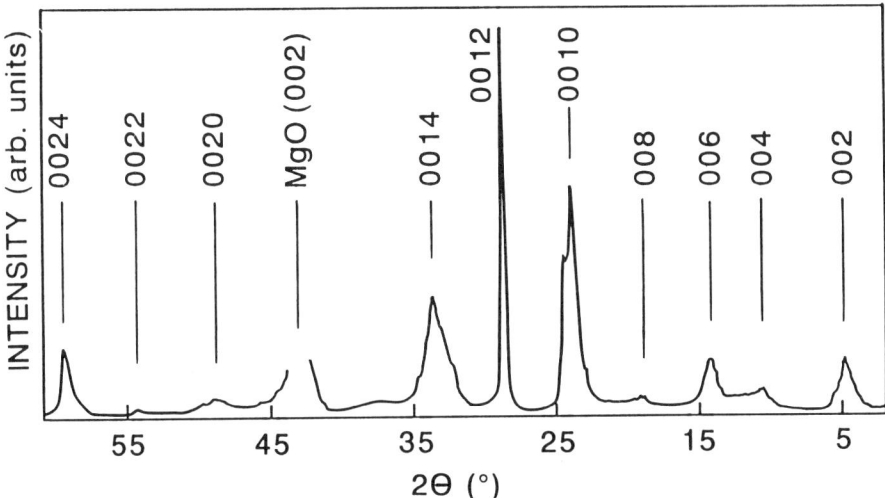

Fig. 3. The XRD results on the Bi-2223 film deposited at 830°C and annealed for two hours at 860°C .

of oxygen was introduced into the chamber and finally the sample was cooled down to room temperature in 14 minutes. Finally, the film was externally annealed (in the presence of the Bi-2223 pallet similar to the above) by rapidly heating the film to 860°C, holding at this temperature for two hours and finally quenching the film to room temperature. This film showed the $T_{c,onset}$ about 115 K and $T_{c,zero}$ at 104 K and the $J_c(0)=.9\times10^6$ which is the highest among our Bi-2223 thin films. The XRD result revealed all peaks due to (00l) plane with the c=3.68 nm (Fig. 3). The SEM result on this film showed a smooth film but some unknown crystalline outgrowth on the surface of the film can clearly be seen (Fig. 4).

The initial ac-susceptibility on this film showed a very broad superconducting transition with the $T_{c,onset}$ starting at 122 K. This indicates that our films are multi-phase superconductors and the phases can be due to the lack or the excess of oxygen during the annealing procedure.

4   Conclusion

Using PLD technique, we have shown the possibility of the "in-situ" growth of Bi-2223 on MgO substrate at temperatures above 740°C. The c parameter of these films is slightly larger than the one of the Bi-2223 oxide indicating a partial CuO stacking of more than three in these films. A two hours external annealing in the presence of saturated Pb atmosphere is required for obtaining the Bi-2223 phase. Although, these films showed $T_{c,zero}$ at about 104 K, the ac-susceptibility result revealed multi phase superconductivity . This indicates a further improvement in the annealing process may be required for obtaining single phase Bi-2223 type superconductor.

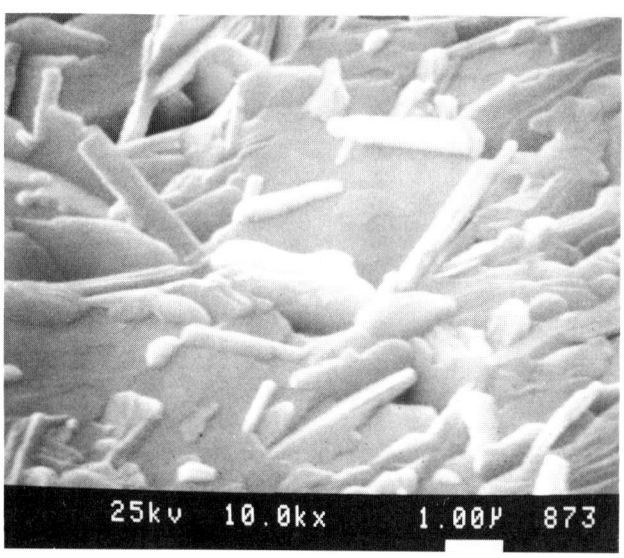

Fig. 4. The SEM micrograph on the Bi-2223 film deposited at 830°C. and annealed for two hours at 860°C .

References

[1] Y. Nakayama, I. Tsukada, A. Maeda and K. Uchinokura, Jpn. J. Appl. Phys., Vol. 28 (1989) L1089.

[2] H. Tabata, T. Kawai, M. Kanai, O. Murata and S. Kawai, Jpn. J. Appl. Phys., Vol. 28 (1989) L430.

[3] J. Levoska, T. Murtoniemi and S. Leppävuori, High T_c Superconducting $YBA_2Cu_3O_{7-d}$ and Bi-Pb-Sr-Ca-Cu-O Thin Films Deposited by Laser Ablation, in: High-T_c Super-conductor Materials, eds. H-U. Habermeier, E. Kaldis and J. Schoenes (North-Holland, Amsterdam, 1990) pp. 710-715.

[4] W.J. Lin, L.C. Wang, H.B. Lu, I.N. Lin, C.C. Young, W.Y. Lin, S.J. Yang and S.E. Hsu, Jpn. J. Appl. Phys., Vol. 29 (1990) L846.

[5] Y. Hakuraka, D. Miyagi, S. Higo andT. Ogushi Jpn. J. Appl. Phys., Vol. 29 (1990) L926.

[6] The $Bi_{1.7}Pb_{.3}Sr_2Ca_2Cu_3O_y$ was supplied by Kalichemie, Hannover , FR Germany.

[7] F.S. Razavi, H-U. Habermeier and P. Majewski, to be published in Mater. Sci. Eng. B.

[8] J.M. Tarascon, W.R. McKinnon, P. Barboux, D. M. Hwang, B.G. Bagley, L.H. Greene, G. Hull, Y. Lepage, N. Stoffel and M. Giroud, Phys. Rev. B 38 (1988) 8885.

High T_c Superconductor Thin Films
L. Correra (Editor)

THE EFFECT OF THE SUBSTRATE TEMPERATURE ON THE GROWTH OF YBa$_2$Cu$_3$O$_{7-d}$ ON LaAlO$_3$ USING RF MAGNETRON SPUTTERING.

E.M.C.M. Reuvekamp, W.A.M. Aarnink, W.H. Minkjan, G.J. Gerritsma, H. Rogalla.

University of Twente, Faculty of Applied Physics, Low Temperature Division, P.O.B. 217, 7500 AE Enschede, The Netherlands.

Abstract

A series of YBa$_2$Cu$_3$O$_{7-d}$ thin films have been in-situ RF magnetron sputtered at different substrate temperatures using a direct facing target. At the highest substrate-temperatures, which were about 700 °C, the films showed $T_{c,zero}$ of 85 to 87 K. Superconducting onset-temperatures higher than 93 K could be obtained. Optical- and electron-microscopy showed a very pronounced surface structure, consisting of needle-like crystallites, which were oriented perpendicularly to one another in the plane of the film. The substrate-temperature 'window' for obtaining a $T_{c,zero}$ of around 85 K was found to be 20-30 K. At a substrate-temperature of approximately 110 °C below the high-temperature window, another temperature range (width ~ 20 K) was found in which very good oriented films could be sputtered. Although $T_{c,zero}$ was drastically reduced below 77 K, the onset-temperatures were almost as high as in the high substrate-temperature regime. These films were smoother than the c-axis oriented films and the pronounced needle-like crystallites at the surface were absent. The films were found to be semiconductive out of the two temperature ranges, without any sign of a superconducting transition temperature. For characterization of the films we combined a number of techniques. To determine the superconducting properties, we applied critical temperature T_c measurements. X-ray diffraction (XRD), Rutherford backscattering spectrometry (RBS) and scanning Auger measurements (SAM) were used as analysis techniques. Results of these techniques are presented here.

1. INTRODUCTION

A series of YBa$_2$Cu$_3$O$_{7-d}$ thin films have been in-situ RF magnetron sputtered on LaAlO$_3$ twinned substrates (AKZO) at different substrate temperatures. The goal of this research is to deposit good superconducting films with controlled orientation on LaAlO$_3$. Films with the c-axis parallel to the substrate are of special interest, because of their expected good superconducting properties perpendicular to the substrate. It is obvious that this is of great interest for planar type junctions. It has been shown by other authors [1,2,3] that the substrate temperature during deposition is a critical parameter for both the growth of the right phase and the orientation of YBaCuO. In this paper we extend this for RF magnetron sputtering in a top-down configuration.

2. FILM PREPARATION

The films were in-situ RF magnetron sputtered using a top-down configuration. A sintered stoichiometric target with a diameter of 50 mm has been used. The used sputtergas was a mixture of argon and oxygen, consisting of 60% Ar and 40% O_2, with a total sputter pressure of 56 Pa. The deposition rate was of the order of 4 Å/min. The thicknesses of the films were between 40 and 100 nm. The sputter-power was 80 W, which equals a power density of ~ 4 W/cm^2 on the target. The films were sputtered with substrate temperatures between about 600 °C and 740 °C. To avoid contamination in the sputter chamber a vacuum load-lock system has been developed and used. After film deposition the sample was cooled down to room temperature within 30 to 45 minutes in a oxygen atmosphere of at least 2000 Pa.. No other anneal procedures were performed. The $LaAlO_3$ substrates were twinned, as is normal for this material, and had a stepped surface with step heigths up to 20 nm.

3. TRANSITION TEMPERATURES

At the highest substrate temperatures (720 °C) the films showed a $T_{c,zero}$ of 85 to 87 K. For optimal settings superconducting onset-temperatures slightly higher than 93 K could be obtained. The substrate-temperature 'window' for obtaining an optimal $T_{c,zero}$ of around 85 K was found to be 20-30 K. The slope of the resistance-curve was about 2.6, which was defined as the resistance at room temperature divided by the value at 100 K. The $T_{c,zero}$ for the lower substrate-temperature films (610 °C) was drastically reduced to about 55 K. The onset-temperatures were however almost as high as in the high substrate-temperature regime. The slope of the resistance-curve for these 'cold-sputtered' films was about 1.2. The width of the substrate-temperature 'window' for growing optimal films with $T_{c,zero}$~ 50 K was about 20 K. The YBaCuO films on $LaAlO_3$ showed a semiconductor-like behaviour when a substrate temperature outside the two 'windows' was used. In those cases there was no possibility to measure any superconducting onset-temperature. In figure 1 two examples of the temperature dependent resistance is given for an optimized $T_{c,zero}$ within the two substrate temperature windows.

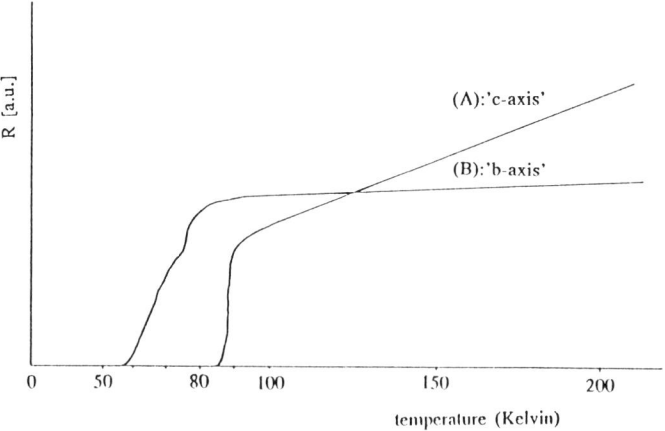

Figure 1: Temperature dependent resistance of two optimized YBaCuO films on $LaAlO_3$; (A) at T_{substr}= 720 °C ('high-temperature window') and (B) at T_{substr}= 610 °C ('low-temperature window').

4. X-RAY DIFFRACTION

X-Ray Diffraction (XRD)-measurements, using a Debeye-Scherrer diffractometer with a Cu-Kα source ($\lambda = 1.5418$ Å) were performed on both the high and low substrate-temperature films, as can be observed in figure 2. For the high substrate-temperature films, X-ray diffractograms always reveal high orientation with the c-axis perpendicular to the substrate. This conclusion can be made because only *(00l)* peaks can be observed, as regularly found for YBCO layers using comparable substrate temperatures. Using a Philips texture goniometer, and applying the rocking-curve method, it was measured that the full half-width angle of the *(005)* reflection was approximately 1°. So we can conclude that these films are reasonably good textured. The XRD pattern for the low substrate-temperature films showed (relative) very high intensity reflections, which could be indexed as the *(003)/(006)* and/or *(010)/(020)* YBa₂Cu₃O₇₋x reflections, in addition to the reflections due to the LaAlO₃ substrate material. These intense reflections due to the sputtered layer are related to lattice parameters of 3.89 and 1.95 Å respectively. These values are very closely related to the bulk values of YBa₂Cu₃O₇, which equals 11.68 Å and 3.89 Å for c- and b-axis respectively [4]. The diffractogram did not show, however, the *(002)*, *(005)* and *(007)* reflections. Therefore the highest intensity reflections will most probably be due to *(010)* and *(020)* reflections only. This should mean that the b-axis is oriented normal to the substrate, in contrast to other publications that mention only a-axis oriented films. The other reflections for the 'cold-sputtered' films, identified with a unit-cell lenght varying between 11.84 to 11.93 Å from sample to sample, could not yet be identified for sure. These unidentified reflections are certainly not 123-reflections, but could be due to the tetragonal (YBa₂Cu₃O₆), cubic (YBa₃Cu₂Oₓ) or cation-disordered [1] phases of the YBaCuO compound.

5. FILM SURFACE

Optical- and electron-microscopy showed a very pronounced surface structure, for the c-axis, 'hot-sputtered', films. The surface consisted of needle-like crystallites, which were oriented perpendicularly to one another in the plane of the film, with sizes up to several micron. The b-axis films were smoother and the needle-like crystallites at the surface were absent, but some different shaped sub-micron precipitates are present, as can be seen in figure 3.

6. RUTHERFORD BACKSCATTERING

Rutherford Backscattering (RBS), using 2 MeV He^{+}-ions, was performed at the AMOLF-institute in Amsterdam. RBS experiments on the 'hot- and cold-sputtered' LaAlO₃/YBCO samples are shown in fig. 4. Since the lattice constants of YBCO and LaAlO₃ (pseudo-cubic, a$\cong$ 3.79 Å) are very similar, LaAlO₃ is interesting for growing YBCO thin films nearly free of strain. The random spectra are simulated using the RUMP computer code [5] and assuming a stoichiometric composition. Using the simulations, the YBCO layer thicknesses were found to be 43 nm and 66 nm for the c- and b-axis oriented films respectively. Auger sputter profiling (see below) confirmed these thicknesses. For the c-axis, 'hot-sputtered', film the deviations between the simulated and experimental values near channel 340 are ascribed to some interface reaction taking place at the LaAlO₃/YBCO interface. SAM indicated that some Y and Ba diffused into the LaAlO₃ over a distance of about 5 nm, leaving a Cu-rich interface layer with also a thickness of about 5 nm. The RBS-spectrum of the b-axis, 'cold-sputtered', film showed no interdiffusion at the interface. The simulation fits very well to the random spectrum.

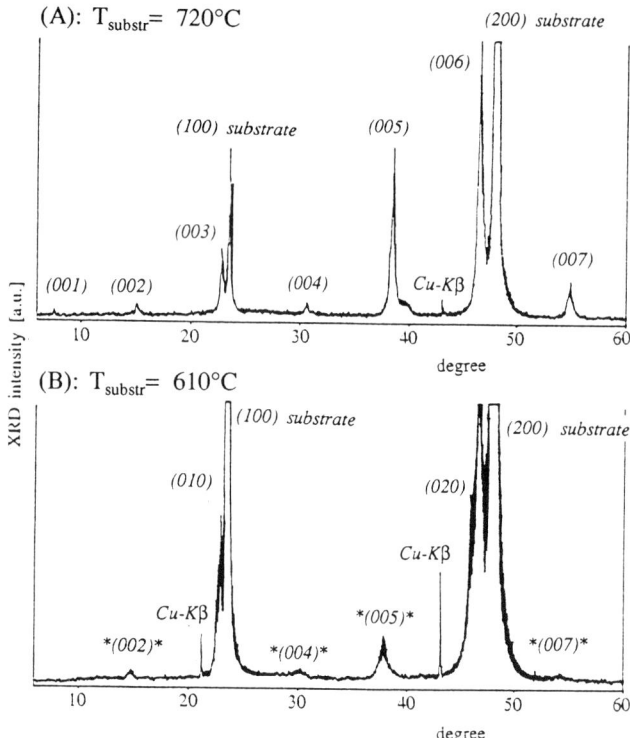

Figure 2: X-ray diffraction patterns of YBCO grown on *(100)* LaAlO₃ substrates: (A) at T$_{substr}$= 720 °C ('high-temperature window') and (B) at T$_{substr}$= 610 °C ('low-temperature window'), the extra (unidentified) peaks are marked with an asterix.

(A): T$_{substr}$= 720°C   (B): T$_{substr}$= 610°C

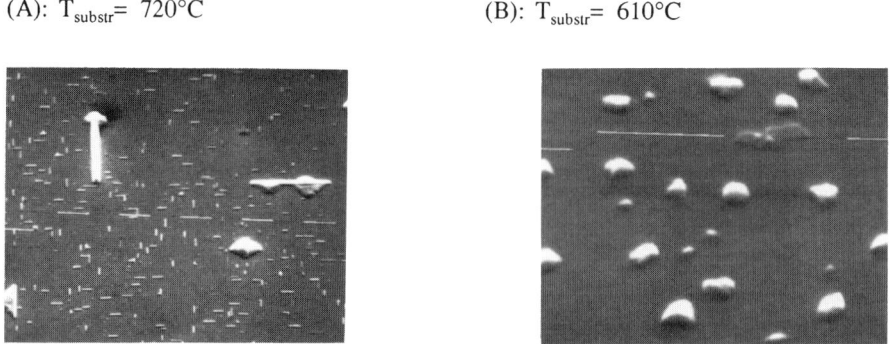

Figure 3: SEM pictures of the surface of the sputtered films. (A) at T$_{substr}$= 720 °C ('high-temperature window') and (B) at T$_{substr}$= 610 °C ('low-temperature window'). For both pictures is the length of a scale-marker equal to 1 μm.

In figure 4, also channeling experiments along the (*001*) direction are shown. For the minimum backscattered yield χ_{min} we found 7% and 12%, for the c-axis, and b-axis, 'cold-sputtered', YBCO films respectively. These are very low values when compared to YBCO films with comparable thickness sputtered on substrates of YSZ. From this we may conclude that that in the YBCO thin film on the LaAlO₃ substrate dechanneling effects are low and, therefore, that the YBCO thin film contains very little strain. Note that in the channeled RBS spectrum, de- channeling effects of the LaAlO₃/YBCO and YBCO/vacuum interfaces can be observed clearly for both types sputtered films.

(A): T_{substr} = 720°C   (B): T_{substr} = 610°C

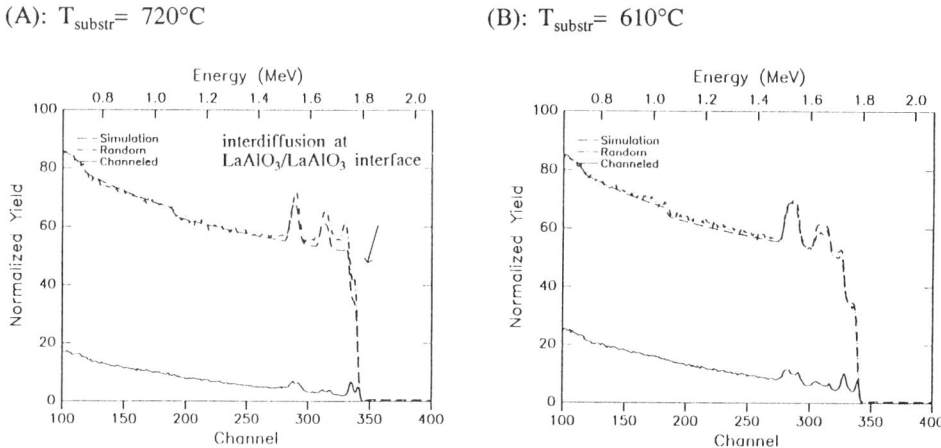

Figure 4: The RBS and He⁺ion channeling spectroscopy of two LaAlO₃/YBCO samples: (A) at T_{substr} = 720 °C ('high-temperature window') with thickness of 42 nm and (B) at T_{substr} = 610 °C ('low-temperature window') with thickness of 66 nm.

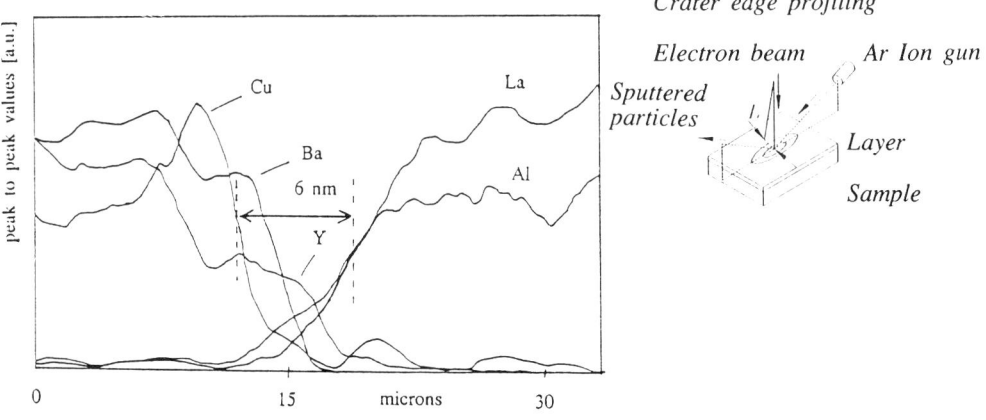

Figure 5: The Auger depth profiling using a crater edge technique of a LaAlO₃/YBCO sample sputtered at T_{substr} = 720 °C ('high-temperature window') with thickness of 42 nm.

7. AUGER ELECTRON SPECTROSCOPY

For the interface studies on the c-axis orientated films, Auger depth analysis (on PHI SAM 600, Perkin-Elmer) was performed , after having formed a crater edge by ion milling (3.5 keV Ar^+-ions), see figure 5 inset. The primary electrons had an energy of 10 keV. Calculations showed that the angle between the normal to the surface of the edge and the normal to the sample surface equals about $0.06°$. Therefore, interfaces are stretched by a factor of about 1×10^4 and can be studied in detail [6]. However, the resolution of this technique is limited to ≈ 7 nm as in the case of RBS, Auger depth profiling. This may be due to effects of intermixing during Ar^+-ion bombardment.

SAM indicated for the 'hot-sputtered' c-axis oriented film that some Y and Ba diffused into the $LaAlO_3$ over a distance of about 5 nm, leaving a Cu-rich interface layer on the superconductor side of the interface, with also a thickness of about 5 nm, as can be observed in figure 5. Interface contamination by a non-scanned element (probably carbon) might be the reason for the very low total intensity measured at the substrate side of the interface.

8. CONCLUSIONS

It is has been shown that 2 narrow temperature windows exist for the sputtering of thin superconducting YBCO films on $LaAlO_3$, using in-situ RF-magnetron sputtering. Outside these windows only semiconducting films can be obtained. Films sputtered at high substrate temperatures were found to be c-axis oriented and had a $T_{c,zero}$ of 85-87 K. Due to the high substrate temperature, some interface reaction took place at the $LaAlO_3$/YBCO interface. SAM showed that some Y and Ba diffused into the $LaAlO_3$ over a distance of about 5 nm, leaving a Cu rich interface layer with also a thickness of about 5 nm. The surfaces of these films show a very pronounced 'rough' structure. Films sputtered at low substrate temperatures are most probable b-axis orientated. The highest $T_{c,zero}$ obtained yet is 55 K. No interdiffusion at the interface is observed, while the surfaces of these films are smoother. In both types of the YBCO thin films on the $LaAlO_3$ substrate dechanneling effects are low. The YBCO thin films contain very little strain and very few defects and are very well textured. Modified off-axis sputter procedure is currently under research to improve the $T_{c,zero}$ of the films to ~ 90 K and to make smoother surfaces.

9. REFERENCES

1 V. Matijasevic et al, J. Mater. Res., Vol. 6, No. 4 (1991) 682.
2 C.B. Eom et al, Science 249 (1991) 1549.
3 R. Ramesh et al, Appl. Phys. Lett. 57 (1990) 1064.
4 R.J. Cava et al, Phys. Rev. B. 36 (1987) 5719.
5 L.R. Doolittle, Nucl. Instr. Meth. b 9, (1985) 344.
6 W.A.M. Aarnink et al, to be published in Appl. Surf. Sci.

High T$_c$ Superconductor Thin Films
L. Correra (Editor)

Reactively co–sputtered in situ annealed superconducting $Y_1Ba_2Cu_3O_{7-x}$ thin films

J. Salmi, J. Saarilahti and I. Suni

Technical Research Centre of Finland, Semiconductor Laboratory, Otakaari 7B, SF–02150 Espoo, Finland

Abstract

Thin films of $Y_1Ba_2Cu_3O_{7-x}$ have been deposited using reactive co–sputtering from Y, BaCu and Cu targets and combined with in situ postannealing. Zero resistance temperatures of 89 K on cubic yttrium stabilized zirconia substrates and 82 K on sapphire substrates with ZrO_2 buffer layer have been achieved. Effects of deposition and postannealing parameters on thin film stoichiometry, morphology and superconducting properties are discussed.

INTRODUCTION

Among the methods used for thin film deposition of high temperature superconductors co–sputtering offers many advantages. Compared to sputtering from a single compound target these include avoiding long presputtering times needed to reach steady state, deviation of stoichiometry from target composition and negative oxygen ion effect. Film composition can also be easily changed. Co–sputtering has been studied for the deposition of Y–Ba–Cu–O based [1–8] materials usually from metallic targets. We have examined possibilities to grow in situ superconducting films in the low partial pressure regime of reactive gases such that oxidation of the metallic targets can be avoided.

EXPERIMENTAL

Schematic diagram of the growth chamber is shown in Figure 1. The cryopumped deposition system is equipped with four planar dc–magnetron sources surrounded by Ar injection rings for easier low–pressure operation. A substrate, with upto 100 mm diameter, can be positioned to an adjustable 300–400 mm distance from the targets of 3 inch in diameter through a gate valve between a loading chamber and the deposition chamber. A turbomolecular pump mounted on the loading recipient is used to provide a differential pumping action on the oxidizing gas constituent injected at the substrate. Radiating quartz lamp heaters are used to raise the substrate temperature up to 950 °C.

$Y_1Ba_2Cu_3O_{7-x}$ films have been co–deposited from metallic Y, BaCu 1:1 alloy and Cu targets. BaCu has been substituted for elemental Ba to reduce the reactivity of the target material in oxidizing ambient. Cu shields have also been used at the Y target to suppress sparking on the oxidizing and slowly eroding part of the surface. The sparks are found to render the deposition process unstable and to degrade the film stoichiometry. The film composition is controlled using precalibrated deposition rates. The magnetron power supplies are maintained at a constant output power during the deposition. Typical power levels used are 180 W, 205 W and 45 W for Y, BaCu and Cu respectively. A silent discharge ozone generator producing approximately 5 % O_3 in O_2 has been used to provide a highly efficient oxidation ambient. Stable operation is achieved using O_2/O_3 partial pressure of $4 \cdot 10^{-5}$ Torr, as measured in the deposition chamber. The pressure near the substrate surface is estimated to be about one order of magnitude higher. The total pressure has been 2 mTorr. The effective deposition rate is approximately 10 nm/min. After film deposition in situ postannealing in the loading chamber at a higher temperature has been carried out in a O_2/O_3 pressure of upto 50 Torr. Alternatively, ex situ postannealing has been carried out in flowing oxygen at atmospheric pressure for samples which were not in situ annealed above the growth temperature.

High energy ion backscattering and X-ray diffraction with Cu Kα radiation have been used for composition analysis and crystalline phase identification. Resonant scattering of α-particles from oxygen at 3.045 MeV has been used to determine the oxygen content in the films. The film resistivity as a function of temperature has been measured in a cryocooler using the van der Pauw method. Film morphology and lateral homogeneity have been examined using SEM and SIMS element mapping.

Single crystalline (100) YSZ, ($1\bar{1}02$) Al_2O_3, (100) MgO or (100) $SrTiO_3$ were used as substrates. In this study, the substrate size has been limited to 8 by 12 mm for YSZ and 10 by 10 mm for other materials, respectively. HfO_2 and ZrO_2 have been studied as buffer layers. HfO_2 is reactively rf–sputtered at room temperature from metallic target in a separate system. ZrO_2 is thermally oxidized at growth temperature with standard O_2/O_3 flow from e–gun evaporated Zr. The thicknesses for HfO_2 and ZrO_2 are 25 – 75 nm and 30 nm respectively.

RESULTS AND DISCUSSION

The influence of reactive gases

Use of O_2 or N_2O was first tried for the reactive deposition of $Y_1Ba_2Cu_3O_y$ at 600 °C to 700 °C. Their partial pressure during sputtering was $2 \cdot 10^{-4}$ Torr. With O_2 the oxygen content in the as deposited films was in the range y = 4.4 – 5.1. Using N_2O instead of O_2 results in oxygen content of y = 4.7 – 5.7. N_2O has an advantage over O_2 due to less tendency for target oxidation. The room temperature resistivity of the films is ρ_{rt} ~500 Ωcm. The X-ray diffraction analysis reveals an almost amorphous like structure with vanishingly small signals from $Y_1Ba_2Cu_3O_{7-x}$ and Y_2BaCuO_5 phases. Substituting O_3 for O_2 improves the crystallinity and increases the oxygen fraction above y = 6.3. Films, however, do not show superconducting transition.

Substrate temperature

The temperature control is found to be critical for reproducible film deposition. Molybdenum disk with stainless steel springs for sample clamping was first used. Large variation especially in the Cu content was observed from run to run. Mounting the samples on the Mo holder with silver paste or indium was found necessary for reliable thermal contact between sample and Mo. Effect of growth temperature on the metal composition of the film is shown in Fig. 2. The amount of Cu in the film is found to decrease with increasing temperature.

Annealing procedures

Ex situ furnace annealing in flowing oxygen was tried for films deposited at 580 – 720 °C. Peak annealing temperatures varied between 600 °C and 925 °C. Considerable interdiffusion results in films deposited on ($1\bar{1}02$) Al_2O_3 above 700 °C and on $Si/SiO_2/HfO_2$ above 670 °C. Best results were achieved loading the samples into the furnace at 850 °C, maintaining that temperature for 30 min and cooling down to 100 °C within 3 h before unloading. All films displayed a temperature dependence of the

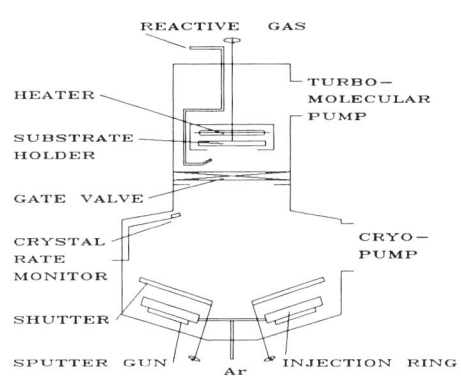

Fig. 1. The schematic structure of the co–sputtering system.

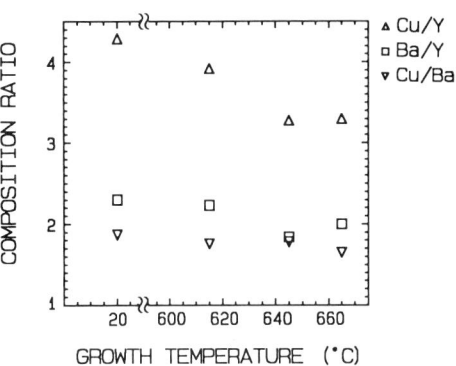

Fig. 2. Composition ratios of the metallic elements at various growth temperatures. Films were grown on Al_2O_3 at a $2 \cdot 10^{-4}$ Torr partial pressure of O_2/O_3 with the target power set at 220 W, 220 W and 60 W for Y, BaCu and Cu respectively.

resistivity characteristic to a semiconductor. The onset of the superconducting transition T_{co} was 90 – 93 K for the samples which were cooled down at 20 Torr O_2/O_3 after deposition maintained at 650 °C. For samples with 1 Torr O_2/O_3 or 20 Torr O_2 during cooldown T_{co} of 72 – 87 K was obtained. On (100) YSZ, (100) MgO and (100) SrTiO$_3$ substrates the zero resistance was reached at T_{cz} of 52 K, 40 K and 35 K respectively. Resistive change with a wide transition range for a 440 nm thick film on YSZ is shown in Fig. 3. According to XRD the amount of 1-2-3 phase in the film increases with increasing annealing temperature or time. Y_2BaCuO_5 is, however, always seen in the annealed films even when a film with $Y_1Ba_2Cu_3O_7$ composition was annealed. The lattice parameters along c-axis for two films grown on YSZ with T_{cz} of 47 K and 52 K were calculated from (001) peak positions and were found to be 11.70 Å and 11.69 Å. These are close to the bulk value of 11.680 Å, which supports the assumption that the resistive tail is caused by impurity phases and not oxygen deficiency in the films.

For in situ postannealing, direct radiation heating on the backside of the substrates was used. In this way the heat–up times were considerably shortened favouring less substrate reactions. Drawback of this scheme is that the temperature calibration depends on substrate material and surface properties. The samples were maintained at the peak temperature for 5 min and then cooled down at 20 Torr of O_2/O_3.

The film morphology on YSZ and on ZrO_2 buffered Al_2O_3 changes from smooth to whisker type with increasing in situ annealing temperature. The highest T_{cz} have been achieved at an intermediate temperature range of 800 °C. Smooth films exhibit T_{co} 85 – 89 K, T_{cz} $\leq$ 57 K, ρ_{rt} 1.0 –1.5 mΩcm and no preferential alignment. A typical resistive transition for a film with smooth surface morphology is shown in Fig. 4. For the whisker type growth $\rho_{rt} \geq 2 \cdot 10^3$ Ωcm is characteristic. The temperature dependence of resistivity changes from metallic to semiconductor type when increasing annealing temperature.

The best superconducting films are preferentially oriented with the c-axis normal to the substrate. A plate and whisker type morphology prevails in these films. On YSZ the plates are aligned in angle of about 60° from the substrate (100) direction. This angle matches with the 63° inclination of the line between the opposite corners of two neighboring unit cells of YSZ with a = 5.13 Å. The length of this line is 11.471 Å, which is close to three unit cells of $Y_1Ba_2Cu_3O_7$ along basal plane directions. Cu segregates, confirmed by SIMS, are typically seen on film surface. Resistive transition of a 205 nm thick film of this type is shown in Fig. 5.

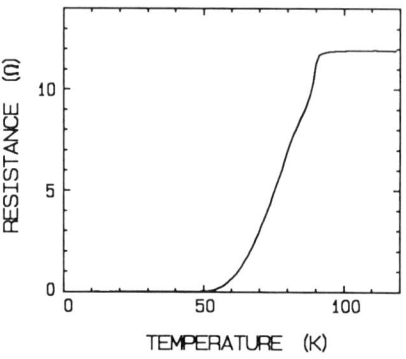

Fig. 3. Resistance vs temperature of an ex situ annealed 440 nm thick film on YSZ.

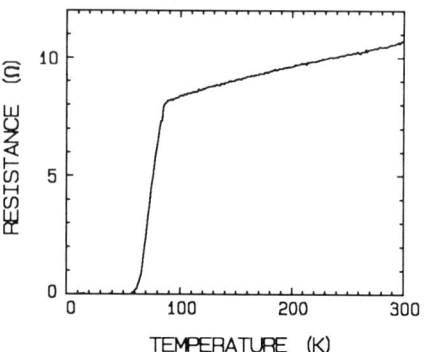

Fig. 4 Resistance vs temperature of a 205 nm thick film on Al_2O_3/ZrO_2. The film has been deposited at 580 °C and annealed in situ at 820 °C. T_{cz} is 57 K.

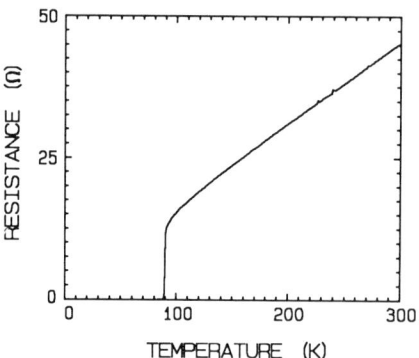

Fig. 5. Resistance vs temperature of a 205 nm thick film on YSZ. The film has been deposited at 620 °C and annealed in situ at 800 °C. T_{cz} is 89 K.

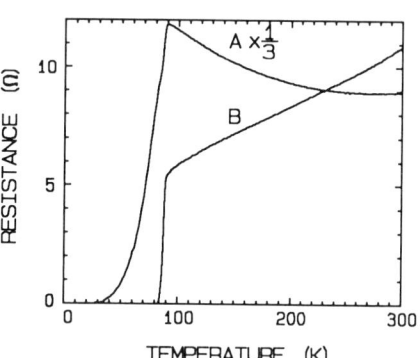

Fig. 6. Resistance vs temperature of 205 nm thick films on Al_2O_3/ZrO_2 with two different Cu deposition rates (power settings): A) $P(Cu) = 45$ W, $T_{cz} = 27$ K, B) $P(Cu) = 55$ W, $T_{cz} = 82$ K. Films have been deposited at 630 °C and in situ annealed at 800 °C.

Fig. 6. shows the resistive transitions of two 205 nm thick films with slightly differing deposition rates for Cu. Temperatures for deposition and annealing were 630 °C and 800 °C respectively. Both films have the same T_{co} of 91 K. Excess Cu changes the temperature dependence from semiconductor type to metallic and increases T_{cz}. However, Cu segregation on the film surface is also enhanced.

CONCLUSIONS

Reactive co-sputtering and in situ annealing in oxidizing ambient have been used to fabricate superconducting $Y_1Ba_2Cu_3O_{7-x}$ thin films. The superconducting properties are found to depend critically on the substrate material, deposition temperature, oxidizing gas pressure, annealing temperature and the ratio of the constituent metal elements. A transition to the zero resistance state is observed at 89 K for films deposited on yttrium stabilized zirconia at 620 °C and annealed at 800 °C. The deposition and the annealing have been carried out using ozone as the oxidizing gas. The films show plate and whisker type morphology and they are preferentially oriented with c-axis perpendicular to the substrate surface. The composition is slightly on the copper rich side to ensure the formation of the correct superconducting phase. Promising results have also been obtained using more readily available substrate materials such as alumina and silicon with thin buffer layers.

ACKNOWLEDGEMENT

M. Karppinen and H. Mölsä at the Laboratory of the Analytical and Inorganic Chemistry of the Technical University of Helsinki are acknowledged for making the XRD measurements. J. Likonen and R. Zilliacus at the Reactor Laboratory of the Technical Research Centre of Finland are acknowledged for the SIMS analysis. This work has been partially sponsored by the Technology Development Centre (TEKES).

REFERENCES

1 K. Char, M. R. Hahn, T. L. Hylton, M. R. Beasley, T. H. Geballe and A. Kapitulnik, Appl. Phys. Lett. 51(1987)1370
2 K. Char, M. R. Hahn, T. L. Hylton, M. R. Beasley, T. H. Geballe and A. Kapitulnik, IEEE Trans. Magn., 25(1989)2422
3 N. Akutsu, M. Fukutomi, K. Katoh, H. Takahara, Y. Tanaka, T. Asano and H. Maeda, Jpn. J. Appl. Phys., 29(1990)604
4 M. Sagoi, Y. Terashima, K. Kubo, Y. Mizutani, T. Miura, J. Yoshida and K. Mizushima, Jpn. J. Appl. Phys., 28(1989)444
5 M. Scheuermann, C. C. Chi, C. C. Tsuei, D. S. Yee, J. J. Cuomo, R. B. Laibowitz, R. H. Koch, B. Braren, R. Srinivasan and M. M. Plechaty, Appl. Phys. Lett., 51(1987)1951
6 Y. Shirakawa and M. Kobayashi, Jpn. J. Appl. Phys., 28(1989)1405
7 R. W. Simon, C. E. Platt, K. P. Daly, A. E. Lee and M. K. Wagner, IEEE Trans. Magn., 25(1989)2433
8 H. Wiesmann, D. H. Chen, R. L. Sabatini, J. Hurst, J. Ochab and M. W. Ruckman, J. Appl. Phys., 65(1989)1644

High T_c Superconductor Thin Films
L. Correra (Editor)

Laserdeposition of $YBa_2Cu_3O_7$ films with low droplet densities

M. Schilling, F. Goerke and U. Merkt

Universität Hamburg, Institut für Angewandte Physik, Jungiusstraße 11,
D-2000 Hamburg 36, F.R.G.

Abstract

Pulsed laser deposition is a very successful method of preparing thin films of high-temperature superconductors with optimum electrical properties. For multilayer systems in superconductor electronics very smooth surfaces are required. In order to achieve such films that are free of pinholes, droplets and outgrowths, the surface morphology on (100) MgO and (100) $SrTiO_3$ is studied with electron microscopy. We report on the dependence of the droplet density on film thickness and growth parameters. Epitaxial films with standard superconducting parameters and droplet densities in the range of 10^2 - 10^3 cm^{-2} could be deposited with our KrF excimer laser.

1. INTRODUCTION

For multilevel thin film devices, such as superconducting quantum interference devices (SQUIDs), very smooth films of $YBa_2Cu_3O_7$ are required to avoid short circuits between adjacent layers. With sputtering techniques the preparation of very smooth films has been reported,[1] while in the laser deposition process a higher surface roughness is usually observed. For the origin of the particles on the film surface various mechanisms were proposed recently.[2-3] It is difficult to compare these results, because many different particles are involved, differing in shape and size as well as in stoichiometry. Here we report on the four most common species illustrated in Figure 1 (a)-(d). First there are a-axis oriented $YBa_2Cu_3O_x$ needles. The number of needles can be substantially reduced using the optimum substrate temperature, because it is well known that substrate temperatures that are too low lead to mixed a/c-orientation of the film normals.[4] No needles were observed on the films studied later on.

(a)   (b)

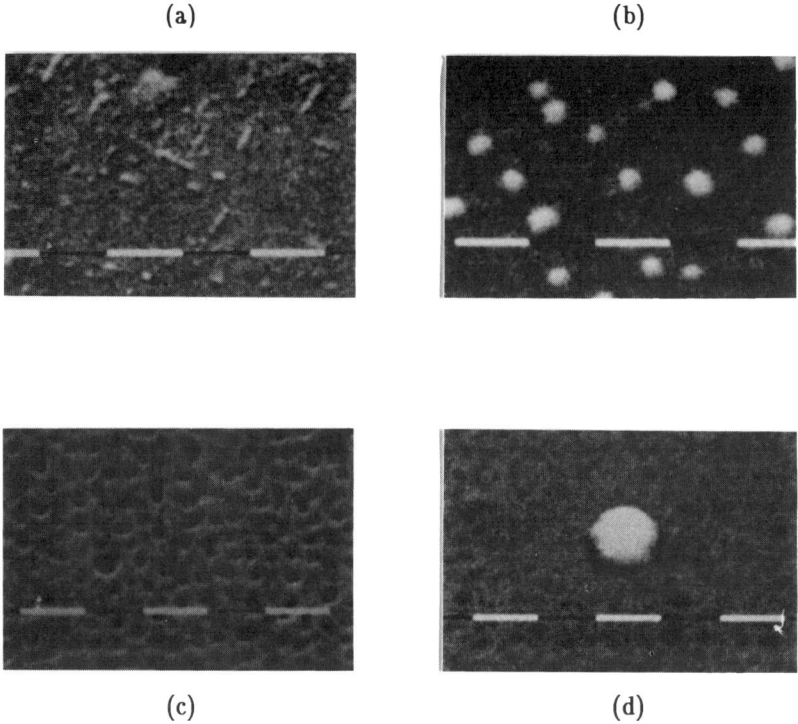

(c)   (d)

Figure 1. Most common species observed in the morphology of laser deposited YBa$_2$Cu$_3$O$_x$ films. (a) a-axis oriented needles, (b) outgrowth particles of differing stoichiometry, (c) inherent surface roughness due to screw dislocations, and (d) spherical droplets. The markers are 1 μm long.

However, we often find irregular shaped outgrowth particles of similiar size but of wrong stoichiometry. Also, there is surface roughness on a smaller scale (0.5 μm) due to screw dislocations most clearly visible in Figure 1(c).[5] This roughness is also observed in YBa$_2$Cu$_3$O$_7$ films prepared by other methods and does not seem to be typical of laser deposited films but rather reflects the inherent growth mechanism.[5] Last we consider spherical droplets with typical diameters of 1 μm. These droplets are typical of the laser deposition process and it was previously shown[6] that their density increases during film growth. Here we investigate the dependence of the droplet density on two additional parameters. First the energy density of the laser beam on the target and second the distance between target and substrate is varied.

2. EXPERIMENTAL SET-UP

Targets for the laser ablation process are prepared by standard calcination methods in order to obtain polycrystalline ceramic pellets of densities ranging from 70% to 90% of the theoretical density of the YBa$_2$Cu$_3$O$_7$ single crystal. For the ablation we employ a KrF excimer laser LPX 301 iCC (248 nm wavelength, 25 ns pulse length), which is normally used at an energy density of 2 J/cm^2 on the target and at a pulse frequency of 5 Hz. The ablated material is deposited in an oxygen atmosphere of 0.4 mbar on substrates, which are mounted on a heater of 785 ^{0}C temperature. On (100) SrTiO$_3$ substrates epitaxial orthorhombic YBa$_2$Cu$_3$O$_7$ films are obtained after our standard cooling procedure.[7-8] With the same set of parameters thin YBa$_2$Cu$_3$O$_x$ films are prepared on (100) MgO where the lattice mismatch is larger and the superconducting properties are slightly degraded. Still we routinely achieve critical temperatures T$_{c0}$ above 85 K.

3. VARIATION OF ENERGY DENSITY AND TARGET-SUBSTRATE DISTANCE

To achieve uniform illumination of the ablated area on the target, a lens forms an image of an aperture on the target. For an accurate determination of the energy density the transmissions of the recipient window and the lens are measured each time and the angle of incidence on the target is taken into account. Here the energy density is varied from 0.5 to 2.5 J/cm^2. For higher values large deviations from the proper stoichiometry are observed,[8] for lower values the YBa$_2$Cu$_3$O$_x$ ablation threshold of 0.1 J/cm^2 is a limit. In Figure 2 the surface morphologies of four films on (100) MgO substrates are shown. At 0.5 J/cm^2 very many outgrowth particles are observed, indicating the wrong stoichiometry of the deposited material. Increase of the energy density leads to a reduction of their number and between 2.0 and 2.5 J/cm^2 the lowest density is obtained. All micrographs show nearly the same underlying surface roughness. This supports the conclusion, that this kind of roughness is independent of stoichiometry and reflects the inherent growth mechanism of YBa$_2$Cu$_3$O$_x$.[5]

The distance between target and substrate was varied between 30 and 55 mm, limited by the present geometry of our chamber, at the optimum energy density of 2.0 J/cm^2. At 30 mm the substrates lie within the plasma plume, resulting in a clearly visible spot on such films. Inside this spot we find almost no outgrowth particles, while outside their density is much higher. The spot becomes larger and fainter at 55 mm distance, where a small increase of outgrowth density is observed, but there no longer is a difference between inside and outside. No significant dependence of the

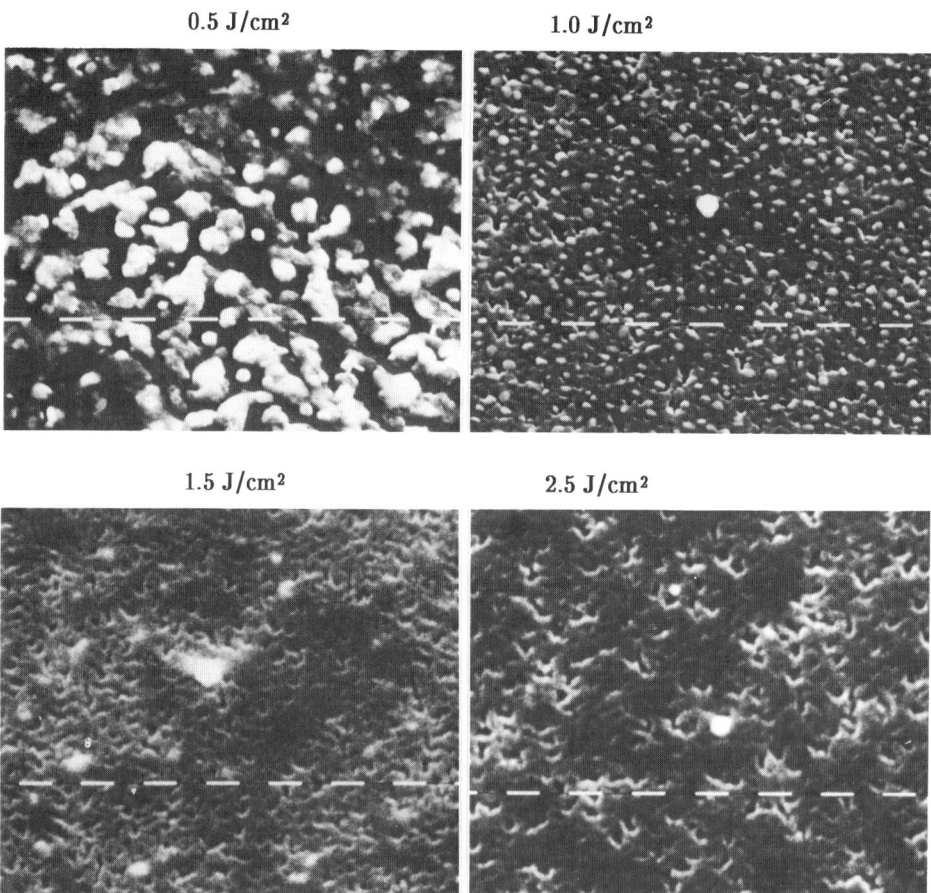

Figure 2. Micrographs of $YBa_2Cu_3O_x$ films on (100) MgO prepared at different energy densities indicated near the photos. All films were deposited at a target-substrate distance of 46 mm, a substrate temperature of 785 ^{0}C, and an oxygen pressure of 0.4 mbar. The films are about 200 nm thick. The markers are 1 μm long.

droplet density on the distance is found. In Figure 3 the dependence of the superconducting transition temperature T_{c0} and the resistance ratio R(300K)/R(100K) on the distance are shown, both, for SrTiO$_3$ and MgO substrates. On MgO no influence on the transition temperature is observed and the resistance ratio exhibits a degraded value only at 30 mm. On SrTiO$_3$ better results are obtained. The lowest droplet densities we found at the smallest distance of 30 mm and best thickness homogeneities at 55 mm.

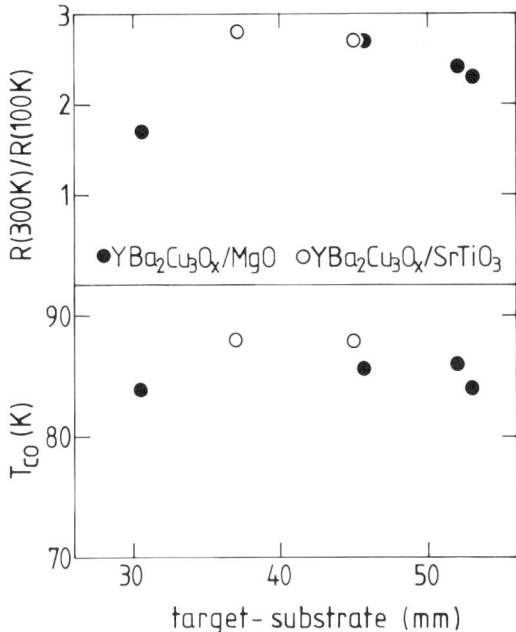

Figure 3. Dependence of transition temperature T$_{c0}$ and resistance ratio on target-substrate distance. The energy density for all films was 2.0 J/cm^2, the substrate temperature 785 ^{0}C and the oxygen pressure 0.4 mbar. The films are about 200 nm thick.

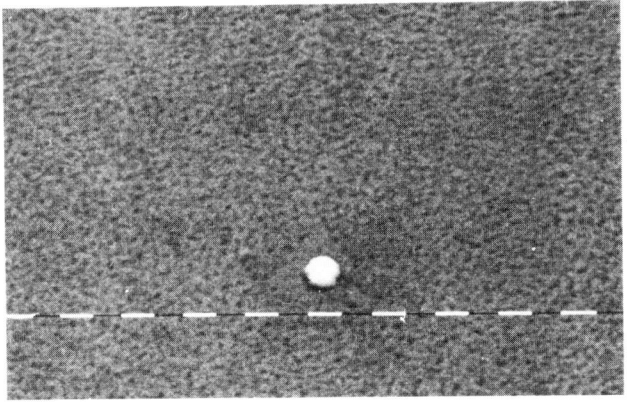

Figure 4. YBa$_2$Cu$_3$O$_7$ film on (100) SrTiO$_3$ with a droplet density around 100 cm^{-2}. The marker is 1 μm long.

In Figure 4 a $YBa_2Cu_3O_7$ film on (100) $SrTiO_3$ with a droplet density around 100 cm^{-2} is shown. It was grown at 2.0 J/cm^2 and 34 mm target-substrate distance. On this film the inherent surface roughness also is reduced.

4. CONCLUSIONS

We previously found that the number of particles on the surface of $YBa_2Cu_3O_x$ films increases with the film thickness.[6] The energy density of the laser on the target surface determines the stoichiometry of the ablated material. We obtain stoichiometric, smooth films that are essentially free of outgrowth particles for energy densities between 2.0 and 2.5 J/cm^2 and for a target-substrate distance of about 45 mm. The droplet density does not depend on the distance between target and substrate. The best homogeneity of the film thickness is found for the 55 mm distance.

We acknowledge financial support from the Bundesministerium für Forschung und Technologie, Federal Republic of Germany.

5. REFERENCES

1) B. Stritzker, J. Schubert, U. Poppe, W. Zander, U. Krüger, A. Lubig and Ch. Buchal, J. Less Common Metals **164 & 165**, 279 (1990).

2) N.G. Chew, S.W. Goodyear, J.A. Edwards, J.S. Satchell, S.E. Blenkinsop and R.G. Humphreys, Appl. Phys. Lett. **57**, 2016 (1990).

3) C.C. Chang, X.D. Wu, R. Ramesh, X.X. Xi, T.S. Ravi, T. Venkatesan, D.M. Hwang, R.E. Muenchausen, S. Foltyn and N.S. Nogar, Appl. Phys. Lett. **57**, 1814 (1990).

4) A. Inam, R. Ramesh, C.T. Rogers, B. Wilkens, K. Remschnig, D. Hart and J. Barner, IEEE Trans. Magn. **27**, 1603 (1991).

5) M. Hawley, I.D. Raistrick, J.G. Beery and R.J.Houlton, Science **251**, 1587 (1991).

6) M. Schilling, F. Goerke, D. Reimer and U. Merkt, Proceedings of the BMFT-Statusseminar, Baden-Baden, p. 323 (1991).

7) M. Schilling, R. Handt, N.H. Hansen, A. Bock, T. Zettler and U. Merkt, Proceedings of the German Soviet Bilateral Seminar, Karlsruhe, p. 424 (1990).

8) M. Schilling, K.D. Laue and U. Merkt, J. Less Common Metals **164 & 165**, 400 (1990).

High T$_c$ Superconductor Thin Films
L. Correra (Editor)
© 1992 Elsevier Science Publishers B.V. All rights reserved.

Preparation of YBa$_2$Cu$_3$O$_{7-x}$ superconducting thin films on silicon nitride by plasma-assisted laser ablation

P. Tejedor, M. Cagigal[a], F. Briones, J.L. Vicent[ab]

Centro Nacional de Microelectrónica (CSIC)
Serrano 144, 28006 Madrid

[a]Departamento de Física de Materiales
Universidad Complutense, 28040 Madrid

[b]Instituto de Ciencia de Materiales (CSIC)
Serrano 144, 28006 Madrid

Abstract
   High-quality films of YBa$_2$Cu$_3$O$_{7-x}$ were deposited on (100) MgO and Si$_3$N$_4$/MgO substrates by plasma-assisted laser ablation at 600^{0}C in a 200-350 mTorr oxygen discharge (-300 V). Characterization by SEM and XRD showed that the films are polycrystalline (0.1-0.8 μm grain size) and grow preferentially with the c-axis perpendicular to the substrate surface. Resistance measurements indicate that these films have metallic behavior in the normal state and have sharp superconducting transitions with zero resistance temperatures T$_{c0}$ of 82 and 86 K for films grown on Si$_3$N$_4$/MgO and MgO, respectively. The J$_c$ value for laser patterned films measured at 77 K was up to 10^3 A/cm^2.

1. INTRODUCTION

   The development of in-situ low temperature techniques to grow epitaxial superconducting films that avoid undesirable film-substrate reactions and can provide high-quality multilayer structures is of great importance to their applicability to microelectronics, microwave, and novel superconductor/semiconductor device technology. The utilization of buffer layers compatible with integrated-circuit technology, such as SiO$_2$ and Si$_3$N$_4$, seems to solve very efficiently the substrate-film interdiffusion problem [1, 2]. However, the as-deposited YBa$_2$Cu$_3$O$_{7-x}$ films show poor superconducting properties and require high temperature (900-940°C) post-annealing treatments. After annealing, the films exhibited zero resistance temperatures between 50 and 75 K [2-4]. During the last few years, a number of research groups have prepared high-quality thin films of YBa$_2$Cu$_3$O$_{7-x}$ at temperatures between 650 and 750°C by the in-situ laser deposition technique [5-8]. Other authors have shown that the deposition temperature can be further reduced by incorporating an oxygen plasma [9,10].
   We report here on the preparation of high-quality textured thin

films of YBa$_2$Cu$_3$O$_{7-x}$ on Si$_3$N$_4$ buffer layers deposited on (100) MgO
substrates by plasma-assisted laser deposition at 600°C, in a low
pressure oxygen dicharge, followed by an in-situ low temperature
(400-600°C) annealing treatment. The films obtained show improved
superconducting properties with respect to previously reported
films on Si$_3$N$_4$, having zero resistance temperatures as high as 82 K.

2. EXPERIMENTAL

The experimental setup used for deposition of Y-Ba-Cu-O thin
films was similar to one described elsewhere [11], with a few
modifications. A schematic diagram of the deposition chamber is
shown in Figure 1. The 10^{-5} Torr cylindrical vacuum chamber is made
of stainless steel and has a UV quartz window for laser
irradiation. The rotating target is a 3 mm thick 50 mm diameter
disk of superconducting YBa$_2$Cu$_3$O$_{7-x}$ material. The substrate is placed
at a distance of 3 cm away from the target and is heated by a
quartz lamp. The deposition temperature is measured by a type K
thermocouple mechanically clamped on the sample. An ArF excimer
laser (Lambda Physik LPX 200i) emitting pulses of 17 ns duration at
193 nm and 50 Hz repetition rate was focused to a 1x3 mm^2 area on
the edge of the rotating target using a 200 mm focal length lens to
give a fluence of 2-3 J/cm^2 on the target. The target is biased
negatively (-300 V) with respect to the chamber flange, while the
substrate is held at a floating potential. This originates a flow
of O$_2$+ ions directed towards the sample, which is known to enhance
the oxygen content of the film and improve its superconducting
properties [9].

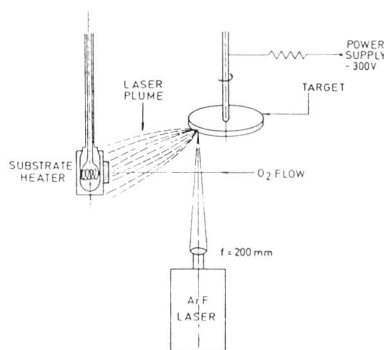

Figure 1. Schematic diagram of the plasma-assisted laser deposition
system

Optically polished (100) MgO crystals and 1500-2000 Å thick Si$_3$N$_4$
buffer layers grown at 300°C by PECVD on (100) MgO crystals were
used as deposition substrates. Prior to each experiment, the
substrates were ultrasonically cleaned in trichloroethylene,
acetone, methanol, and de-ionized water and further heated inside

the chamber at 500°C for 1/2 hr in 200 mTorr of oxygen. Subsequent deposition of YBa$_2$Cu$_3$O$_{7-x}$ was carried out at 600°C in 200-350 mTorr oxygen partial pressure over a period of 1/2-2 hr to give 0.4-1.8 μm thick films. An in-situ annealing step in 1 Torr of oxygen at 600°C for 1/2 hr and 425°C for 1 1/2-2 hr was followed by slow cooling of the film to room temperature over a period of 1/2 hr.

Film thickness measurements were performed using a mechanical stylus. X-ray diffraction patterns of the films were recorded on a Philips PW-1730/10 diffractometer using Cu K$_\alpha$ radiation. SEM micrographs of the films surface morphology and cross-sections were taken in a Hitachi S-800 field emission scanning electron microscope. The dependence of the films resistance on temperature was determined by standard four-contact dc measurements. A constant current of 100 μA was used. The critical current density was determined in films patterned by laser ablation to form two blanket film pads connected by a 20 μm wide and 100 μm long stripe for the four-contact dc measurements. Laser ablation was carried out in air at room temperature by focusing the output of the ArF laser operating at 193 nm and 6 Hz with a 15x (N.A.=0.28) reflective objective and scanning it over the film masked with a 20 μm diameter gold wire. The laser energy density on the film was tipically 5 J/cm^2.

3. RESULTS AND DISCUSSION

Under the experimental conditions described above, we obtained black shiny films of YBa$_2$Cu$_3$O$_{7-x}$ with smooth surface morphology at typical deposition rates of 0.8-1.1 μm/hr.

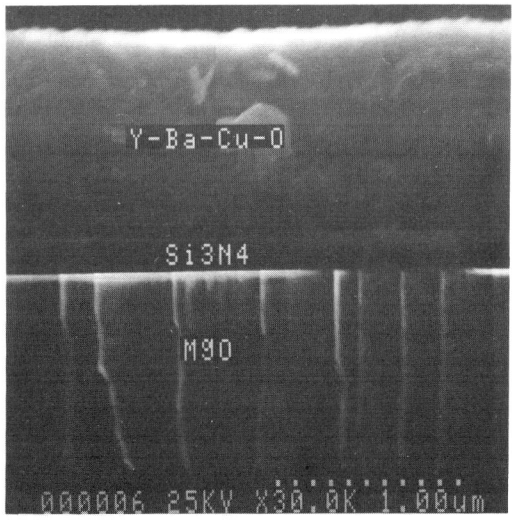

Figure 2. SEM micrograph of the cross-section of a superconducting film deposited on Si$_3$N$_4$/MgO.

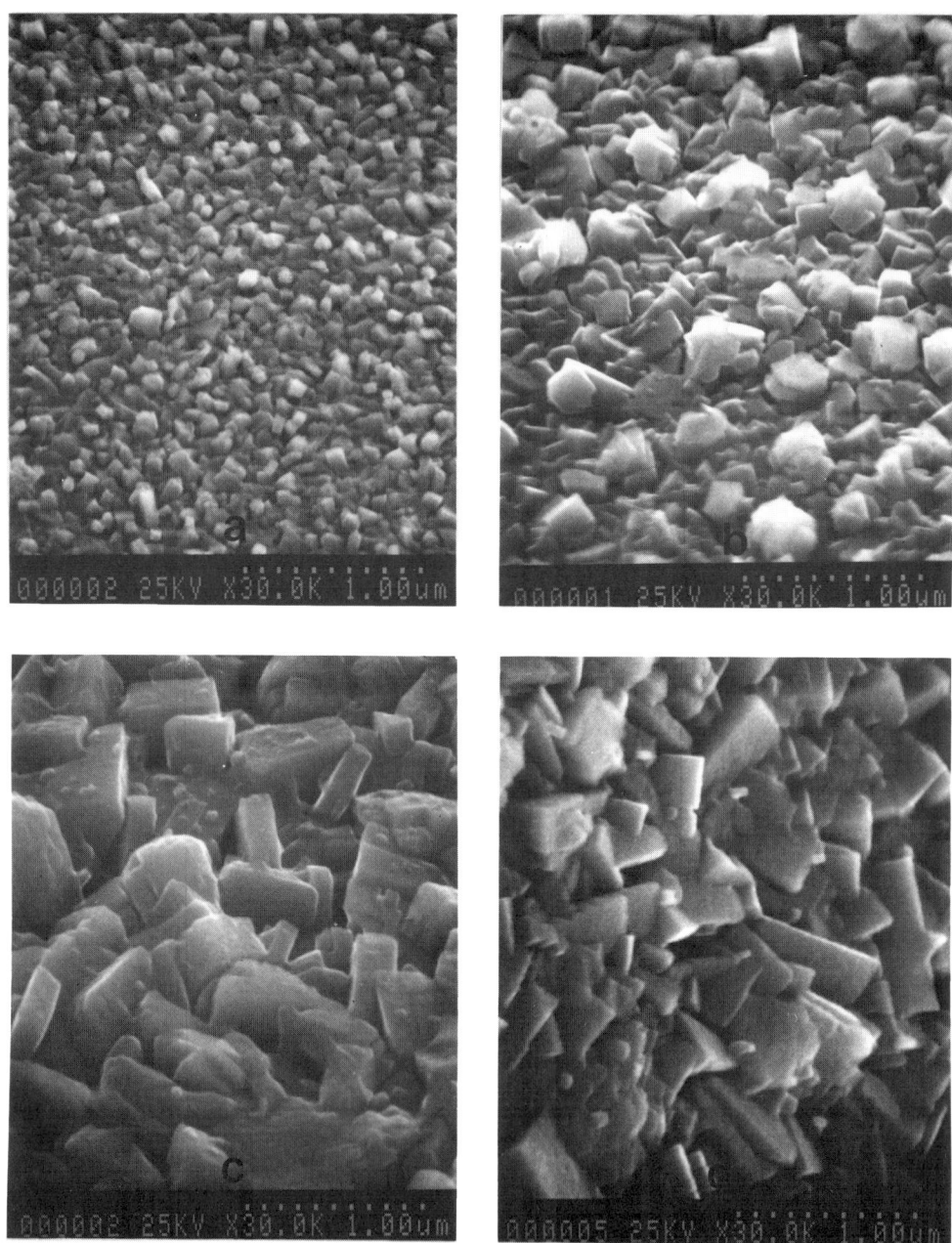

Figure 3. Surface morphology of $YBa_2Cu_3O_{7-x}$ films deposited on Si_3N_4/MgO over periods of a) 10 min, b) 30 min, and c) 120 min, and a film grown on MgO over a period of 120 min.

The samples have sharp interfaces and uniform thickness, as depicted in Figure 2, where the cross-section of a 1.1 μm thick superconducting film deposited on a MgO substrate coated with a 1900 Å thick Si_3N_4 buffer layer is shown. Figures 3a,b, and c show the surface morphology of $YBa_2Cu_3O_{7-x}$ films of different thicknesses grown on Si_3N_4/MgO over periods of 10 min (0.2 μm), 30 min (0.4 μm), and 120 min (1.5 μm), respectively. In Fig 3d we present the surface morphology of a film grown on (100) MgO simultaneously with the film of Figure 3c, for comparison. We observe a polycrystalline grain structure in all cases. The grain size increases with film thickness, ranging from 0.1 to 0.8 μm. The presence of the Si_3N_4 buffer layer, on the other hand, does not influence appreciably the grain size of the superconducting film.

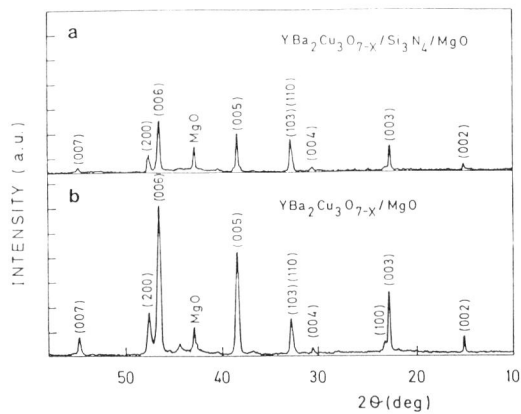

Figure 4. X-ray diffraction patterns of $YBa_2Cu_3O_{7-x}$ fims grown on MgO substrates a) with and b) without a Si_3N_4 buffer layer.

Figure 4 shows the x-ray diffraction patterns corresponding to 1,5 μm thick films of $YBa_2Cu_3O_{7-x}$ grown on Si_3N_4/MgO and MgO, respectively. A well-crystallized single orthorhombic phase is formed in both cases, with somehow poorer crystallinity when the film is deposited on Si_3N_4, as indicated by the lower intensity of the reflection peaks. This is presumably due to the amorphous structure of the buffer layer. Besides the peak at 43° corresponding to the (002) reflection of the MgO substrate, we observe primarily the (001) reflections of the superconducting film and some weak (h00) peaks. This indicates that the films are mainly preferentially oriented with the c-axis normal to the substrate surface, although a minority domain with the a-axis normal to the surface is also present. The lattice parameters calculated from the (200) and (002) reflections were a=3,82 Å, c=11,66 Å for the film grown on Si_3N_4/MgO and a=3,82 Å, c=11,69 Å for the film grown on MgO. The crystalline texture described here is consistent with the general observation reported by other authors [5, 8, 10] for optimally annealed films deposited at low temperatures.

The temperature dependence of the resistance corresponding to the
1.5 μm thick superconducting films grown on Si_3N_4/MgO and MgO
substrates are shown in Figure 5. These films exhibit metallic
behavior in the normal state and sharp superconducting transitions
at 87 K and 91 K, respectively.

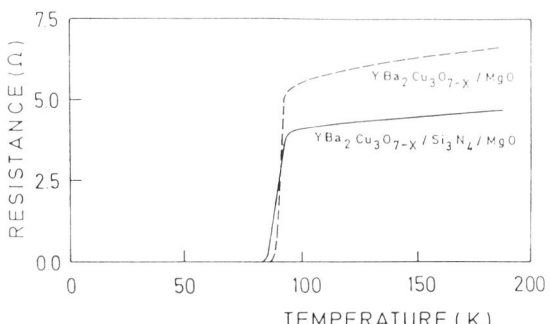

Figure 5. Resistence versus temperature plot for $YBa_2Cu_3Cu_{7-x}$ films
deposited on a) MgO and b) Si_3N_4 substrates.

As can be seen in the figure, the transition temperature onset is
95 K for both films, while zero resistance is reached at 82 K for
the film grown on Si_3N_4/MgO and at a slightly higher temperature,
86 K, for the film grown directly on MgO. Preliminary critical
current density measurements at liquid nitrogen temperature (77 K)
have rendered values of up to 2×10^3 A/cm^2.

4. CONCLUSIONS

We have deposited high-quality superconducting films of $YBa_2Cu_3O_{7-x}$
on MgO and Si_3N_4/MgO substrates by plasma-assisted laser ablation at
temperatures as low as 600^0C, in the presence of a low pressure
(200-350 mTorr) oxygen discharge, without further high temperature
post-annealing treatments. The films are polycrystalline and grow
preferentially oriented with the c-axis perpendicular to the
substrate. We obtained zero resistance temperature values (T_{c0}) as
high as 86 K and 82 K for films deposited directly on MgO and on
Si_3N_4/MgO, respectively. Finally, preliminary measurements of the
critical current density have given a value of 2×10^3 A/cm^2.

5. ACKNOWLEDGMENTS

The authors wish to thank P.S. Domínguez for the scanning
electron micrographs and M. Garriga for help in building the
deposition chamber. This work was partially supported by CICYT
(grant MAT88-0250-C02-02) and program MIDAS (grant 90/639).

6. REFERENCES

1. M. Scheib, H. Hofmann, B. Lengeler, H. Oechsner, and G. Zorn, Thin Solid Films,174 (1989) 5.
2. M. Aslam, R. E. Soltis, E. M. Logothetis, R. E. Chase, L. E. Wenger, and J. T. Chen, IEEE Trans. Electron Devices, 36 (1989) 2693.
3. B. Mercey, G. Poullain, E. Bouteloup, H. Murray, and B. Raveau, Vide Couches Minces, 43 (1988) 17.
4. G. Poullain, B. Mercey, H. Murray, and B. Raveau, Mod. Phys. Lett. B, 2 (1988) 523.
5. G. Koren, A. Gupta, R. J. Baseman, M. I. Lutwyche, and R. B. Laibowitz, Appl. Phys. Lett.,55 (1989) 2450.
6. G. Koren, A. Gupta, E. A. Giess, A. Segmüller, and R. B. Laibowitz, Appl. Phys. Lett., 54 (1989) 1054.
7. T. Venkatesan, E. W. Chase, X. D. Wu, A. Inam, C.C. Chang, and F. K. Shokoohi, Appl. Phys. Lett., 53 (1988) 243.
8. A. Kumar, L. Ganapathi, S. M. Kanetkar, and J. Narayan, Appl. Phys. Lett., 57 (1990) 2594.
9. S. Witanachchi, H. S. Kwok, X. W. Wang, and D. T. Shaw, Appl. Phys. Lett., 53 (1988) 234.
10. R. K. Singh, L. Ganapathi, P. Tiwari, and J. Narayan, Appl. Phys. Lett., 55 (1989) 2351.
11. P. Tejedor and F. Briones, Materials Research Society Symposium Proceedings, edited by H. A. Atwater (MRS, Boston 1990) in press.

High T$_c$ Superconductor Thin Films
L. Correra (Editor)

High Tc Films by Molecular Beam Epitaxy

H.S.Wang,W.Dietsche,D.Eissler,Y.Kershaw,K.Ploog,A.Fisher and
O.Buresch

Max-Planck-Institut für Festkörperforschung,Heisenbergstr.1,
W 7000 Stuttgart 80, Germany

Abstract
   A mixture of ozone and molecular oxygen has been successfully as the oxidant in a MBE system to grow HTSC films.High
quality DyBa$_2$Cu$_3$O$_y$ films with Tcon of 90-93K and Tco of 80-85K
have been prepared in situ in spite of some contamination with
impurities.Some results of superconductivity,structure and
morphology of the films are presented and discussed.

1.INTRODUCTION

   Because of the strong anistropy and very short coherence
length[1] of the high Tc oxide superconductors,the fabrication
of some superconducting electronic devices made of these
materials needs atomic scale controlled growth processes.From
this point of view,molecular beam epitaxy(MBE)is a promising
method, as evidenced widely in semiconductor processing.The
particular features of MBE are the very good starting
vacuum,the avaliability of reliable effusion cells to evaporate
the respective materials, and the well established technology
for sample manipulation and analysis.Despite these
advantages,MBE has rarely been used to prepare HTSC films
compared to other techniques such as laser ablation and
sputtering.This is mainly because of the very low oxygen
pressure required for MBE.Recently,several efforts have been
made to use active oxidants for the growth of the HTSC films by
MBE and other high vacuum systems.These include pure ozone(O$_3$)
[2],atomic oxygen[3],excited molecular oxygen[4],NO$_2$[5] and so
on.Among them,pure ozone has been proved [2,6,7] to be one of
the most efficient oxidizing agents.However,particular
precaution must be taken when pure ozone is used because of the
high possibility of explosion[2,7].In addition,the continuous
operating time is limited.
   In this article,we report for the first time on the
successful use of a mixture of (O$_3$+O$_2$) with O$_3$/O$_2$=6% for the in
situ preparation of high Tc DyBa$_2$Cu$_3$O$_y$ films using a modified
MBE system.Results of superconductivity, structure and
morphology of the films will be presented.Some particular
points concerning the impurity problems related to the MBE
system and their effects on superconductivity will be
discussed.

2.FILM PREPARATION

The MBE machine used is a modified Riber vacuum system (Fig.1) with the base pressure of $1*10^{-9}$ torr. Dysprosium,barium and copper are evaporated from three water cooled Knudsen cells using pyrolytic boron nitride(BN) crucibles.The Knudsen cell temperatures are controlled by Eurotherm 820 regulators with minimum temperature interval of 1°C, corresponding to a composition change of Ba,Dy and Cu of less than 5%, in the evaporating temperature range. Deposition rates are monitored by a quadrupole mass spectrometer(QMS). The correct partial pressures of the evaporated metals were found by chemical analysis of test films.

The mixture of (O_3+O_2) is produced continuously by a commercial ozonisator with the highest O_3/O_2 of 6%,using pure oxygen as the feed gas.A special gas inlet consisting of a narrow and very short pinhole, along which a rapid stream of fresh ozone flows on the high pressure side,minimises the decay rate of ozone before it enters into the chamber.A Titanium sublimation pump was operating continuously during deposition to enhance the pumping speed of oxygen.

The sample holder was originally made of molybdenum which has caused contamination problems when used in the ozone atmosphere as will be discussed in Section 4.The substrate is heated by a tantalum wire. The substrate temperature(Ts) is detected with a W-Re thermocouple positioned just behind the stainless steel enclosure on which the substrate is clamped.Polished (100) MgO substrates with sizes of 10mm*10mm*0.5-1mm were used.

The typical experimental parameters are:Ts=600-800°C, $P(O_3+O_2)<3*10^{-5}$mbar during deposition and $<1*10^{-3}$mbar during cooling.The growth rate is 0.3-0.6 A°/S.Film thickness is between 150-2000 A°,but typically 1000A°.After deposition,the film is cooled for between 30 minutes to 2 hours to form the orthorhombic superconducting phase.

Fig.1,the diagram of the MBE system.

3.CHARACTERIZATION AND PROPERTIES

The DC resistance vs temperature curves R(T) are measured using the standard four terminal method.A Pt-Resistor is used as the temperature monitor.The composition of the films is determined by chemical analysis and EDX.The structures are studied by x-ray diffraction(XRD) and Raman scattering.STM and AFM are used to investigate the morphology of the films with a Nanoscope-II type microscope operated in air.

The film properties were researched by varying the deposition parameters, such as Ts and $P(O_3+O_2)$.By optimizing the conditions as described in the last section,we are able to reproducibly prepare $DyBa_2Cu_3O_y$ films with Tcon of 90-93K and Tco of 80-85K.The films are very smooth and uniform.However,the transition widths of the R(T) curves are relatively large-typically 4-7K.This, we assume,is due to the impurity contamination as will be discussed in the next section.

Table 1 presents the superconductivity,the resistance and the structural properties of a few films prepared at different Ts.A few interesting points are seen:(1),when Ts is between 740-780°C,the films are grown with c axis normal to the substrate(c-orientation),the resistances are lower and Tc's are higher and they are not sensitive to Ts,(2);when Ts is lower than 740°C,the b-orientation is enhanced and Tc is reduced drasticly with Ts,(3);the c values of the c-oriented films are comparable to those of $YBa_2Cu_3O_y$ films prepared in much higher oxygen pressure[8] and of $DyBa_2Cu_3O_y$ films made with pure ozone[6].This means that the oxygen content of our films prepared with the mixture of (O_3+O_2) is as high as that of the good films when high oxygen pressure or pure ozone is applied.

Table 1
Superconductivity,resistance and structural properties of $DyBa_2Cu_3O_y$ films prepared at various substrate temperatures.

No.	Ts/°C	t/A°	c/A°	b/A°	R100K/	Rroom/R100K	Tc/K	Tc/K
M-2-27	780	1150	11.670	–	6.0	2.17	87	4.0
M-2-28	760	930	11.750	–	15.9	1.42	83	8.0
M-3-1	740	950	11.686	–	22.0	1.72	84	9.5
M-3-2	720	1100	11.847	3.915	68.0	.84	45	15
M-3-3	700	1000	–	3.869	190.0	.69	30	–
M-3-4	650	1000	–	3.894	227.0	.57	–	–
M-3-5	600	700	–	–	570.0	.4	–	–

The dependence of the superconductivity of the films on the oxygen pressure was also studied when other conditions were fixed.We found that as the total pressure $P(O_3+O_2)$ is reduced from $3*10^{-5}$ mbar to $6*10^{-6}$ mbar during deposition,Tcon is drops from 90K to 45K.One particularly interesting point is demonstrated in Fig.2 and Fig.3.Film M-1-34(b) was prepared in the similar conditions as for other good films except that the O_3/O_2 ratio was zero.The room temperature resistance is by one order of magnitude higher than others and no superconducting transition was detected above 4.2K.The XRD spectra of this film (Fig.3(b)) shows perfect c axis oriented single phase structure

with c value of 11.863 which corresponds to y=6.25.After
annealing this film at 500°C in flowing oxygen for 5 hours,it's
resistance was reduced by one order of magnitude and it shows
a relatively sharp transition.The value of c was decreased to
11.756,corresponding   to   y=6.65.These   results   are strong
evidence that ozone is crucial for the formation of the
orthorhombic superconducting phase in our system,especially
during the cooling process.

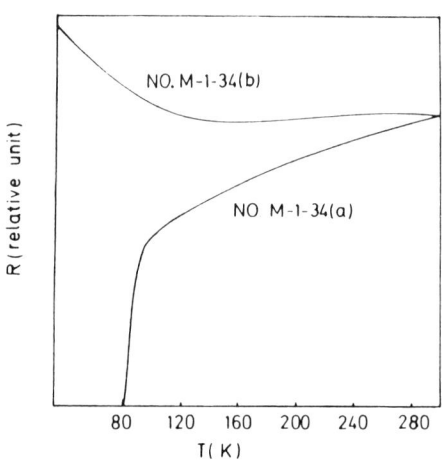

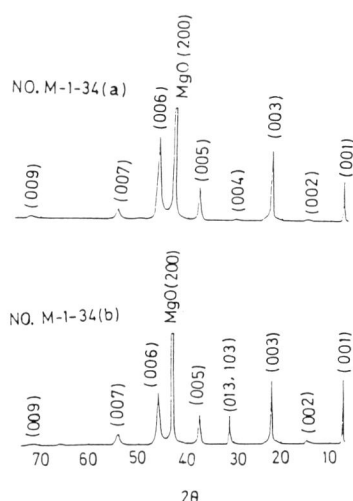

Fig.2,R(T) curves of sample No.
M-1-34 prepared without ozone.
(b),before annealing;(a),after
annealing(see text).

Fig.3,XRD spectra of samples
corresponding to figue 2.

The STM and AFM study convinced us that most of our films are
relatively smooth with vertical roughness of the order of 1-15
nm.The detailed film structure,however,varied from sample to
sample,depending on the preparation conditions.An example is
shown in Fig.4(a).This film has been grown at high Ts(780°C).
It shows grains with a typical(horizontal) size of about
100nm.The   flat   areas   of   the   grains   are   particularly
remarkable.From XRD and EDX data we conclude that this film is
c-oriented and chemically homogeneous and the grains are due to
different crystallographic growth.In Fig.4(b) we present a AFM
image of another film grown at lower Ts(film M-3-1 in table
1).It shows a mixture of orientations:the very flat background
surface are c-oriented crystals,whereas the sharp peaks are
assumed to be b-oriented grains.These grains protrude from the
surface because the growth rate of the a-b plane is faster than
that along the c direction.

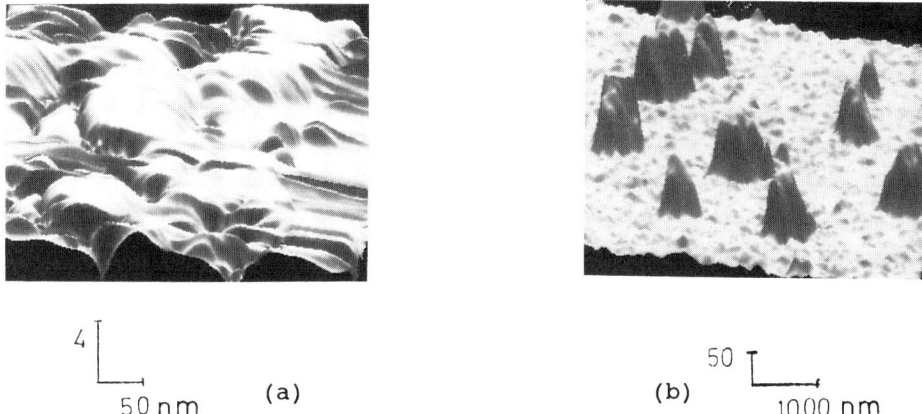

Fig.4,(a)STM image of sample M-2-21 with c-orientation prepared
at Ts=780°C;(b)AFM image of sample M-3-1 with some b-orientated
grains on the c-orientated background,prepared at Ts=740°C.

4.DISCUSSION

As is already mentioned in the last section,although the Tcon
values of our films are comparable to that of the bulk
samples,the transitions are relatively broad.We assume this is
mainly caused from the impurities in the film.According to the
chemical analysis,the impurities include Mo and As.As comes
from the vacuum parts which had been used to grow GaAs
before.Although the vapour pressure of As has been reduced by
more than 4 orders of magnitude in the chamber after a few
months of various cleaning processes,As can still be detected
by QMS as the vacuum chamber becomes hot during evaporation.The
typical As content in the film is 3-10 percent to Dy.Mo is
found to be at least 5 percent to Dy.It comes from the
molybdenum parts of the sample holder.As we know molybdenum is
widely used in commercial MBE machines from the consideration
of purity.At present,the problem appears when ozone is applied
because ozone reacts strongly with Molybdenum to form Mo-oxides
with very high vapour pressures at high temperature[6].Johnson
et al.could not make any superconducting film when a Molybdenum
enclosure was used[6].In our case it seems to be much
better,which may be because our ozone partial pressure is lower
than Johnson et al.so that the reaction is not as strong as
theirs. Nevertheless,the impurity content in our films is high
enough to reduce Tc drastically when compared with many other
doping experiments such as Zn,Co and Ni[9].Furthermore, we
could not find any impurity peaks in our XRD spectra(as seen in
Fig.3).This implies that Mo and/or As atoms go, at least
partially,to the metal sites of the (123) structure.If so,it is
most likely they go into the copper sites because the atomic

radii of Mo(0.68) and As(0.69) are relatively close to that of copper(0.96).Raman scattering experiments for two films have shown strong disorder of the Cu(I)-O(I) chains and a significant suppression of the intensity of the Cu model as well as a large shift of the O(IV) model to higher frequency.These facts suggest a partial substitution of the Cu(I) site by Mo and/or As.It is widely believed that[10] superconductivity is less sensitive to Cu(I) site substitution than to Cu(II) in the (123) phase.This might provide a clue to explaining why our films are so good considering they have so high a impurity content.

We believe that the transition width of our films can be much sharper and Tco can be increased further with the same preparation conditions if the impurities can be reduced to a tolerable level.Recently we have made new substrate holders from copper and stainless steel.Further experiment is under way.

In conclusion,we have succeeded in preparing high Tc $DyBa_2Cu_3O_y$ thin films in situ by MBE using a mixture of (O_3+O_2) as the oxidizing agent.Superconductivity,structure and morphology of the films prepared under various conditions were studied.Some impurity problems orignated from the MBE system and their effects on superconductivity were discussed.

We thank L.Viczian for XRD, and C.Thomsen for Raman scattering experiments and stimulating discussions.

REFERENCES

1.M.B.Salamon,"Phys.Properties High Tem.Supercon."I
  Chap.2,(1989) 39 Edited by D.M.Ginsberg.
2.D.D.Berkley,A.M.Goldman,B.R.Johnson,J.Morton,and T.Wang
  Rev.Sci.Instrum.60(12),(1989) 3769.
3.N.Missert,R.Hammond,J.E.Mooij,V.Matijasevic,P.Rosenthal,
  T.H.Geballe,A.Kapultnik,M.R.Beasley,S.S.Laderman,C.Lu,E.
  Garwin,and R.Barton..1989 IEEE Trans.Magn. MAG-25 2418;
  J.Kwo,M.Hong,D.J.Trevor,R.M.Fleming,A.E.White,
  J.P.Mannaerts,R.C.Farrow,A.R.Kortran,and K.T.Short.1989 Int.
  Conf.Materials and Mechanism of Superconductivity(Stanford).
4.R.M.Silver,A.B.Berezin,M.Wendman,and A.L.de Lozanne,
  Appl.Phys Lett.52(1988) 2174.
5.Hidehiko Nonaka,Takashi Shimizu,and KazuoArai,Appl.Phys.Lett.
  57(26),(1990) 2850.
6.B.R.Johnson,K.M.Beauchamp,T.Wang,J-X.Liu,K.A.McGreer,J-C.Wan,
  M.Tuominen,Y-J.Zhang,M.L.Macartney,and A.M.Goldman.
  Appl Phys.Lett.56(19),(1990) 1911.
7.H.M.Appelboom,J.P.Adriaanse,H.I.de Groot,G.Rietveld,
  D.van der Marel,and J.E.Mooij,Physica B.165 &166 (1990) 1497
8.V.Matijasevic,P.Rosenthal,K.Shinohara,A.F.Marshall,R.H.Hammod
  and M.R:Beasley,J.Mater.Res.,Vol.6(1991) 682.
9.J.T.Markert,Y.Dalichaouch and M.B.Maple,"Phys.Properties
  High Tem.Supercon."I Chap.6,(1989) 265,Edited by
  D.M.Ginsberg.

High T$_c$ Superconductor Thin Films
L. Correra (Editor)
555

Preparation and Characterization of Epitaxially Grown Thin Films of YBa$_2$Cu$_3$O$_y$

R. Wördenweber, J. Schneider, T. Göddenhenrich, and U. Krüger

Forschungszentrum Jülich, Institut für Schicht- und Ionentechnik, Postfach 1913, 5170 Jülich, FRG)

Abstract

Thin films of YBa$_2$Cu$_3$O$_y$ are grown on (100) oriented MgO and SrTiO$_3$ substrates via planar dc-magnetron sputtering at high pressure. At optimized deposition parameters we obtained epitaxially grown, c-axes oriented films of high quality. Structure and orientation are determined by x-ray and channeling experiments, stoichiometry by RBS measurements. The surfaces of the films are very smooth. Force microscopy reveals a surface roughness less than 10 nm. The morphology of the surface is described by an arrangement of hillocks of an elliptic form, which might result from a growth mechanism on the base of screw dislocations. The films were structured by ion beam etching. Bridges with widths down to 1 μm were formed without detectable degradation of the superconducting properties.

1. INTRODUCTION

The preparation of layers or multilayers of high quality high temperature superconductor (HTS) in connection with compatible normal-, semiconduction or isolating buffer layers is one of the important issues of the thin film research of these days. For potential applications in the field of cryoelectronics different classes of oxide ceramics are examined, e.g. ReBa$_2$Cu$_3$O$_y$ (Re=Y, rare earths) with a transition temperature T$_c$ up to 92 K, Bi-Sr-Ca-Cu-O compounds (T$_c$ up to 125 K) and Tl-Ba-Ca-Cu-O (T$_c$ up to 125 K). In spite of the slightly lower transition temperature ReBa$_2$Cu$_3$O$_y$ offers a number of advantages for thin film preparation. In this paper we report on our work on the preparation, characterization and patterning of epitaxially grown, c-axis oriented YBa$_2$Cu$_3$O$_y$ (YBCO) films.

2. SAMPLE PREPARATION

For the deposition of the thin films a planar on-axis dc magnetron sputtering technique at high pressure was chosen /1/.
There are a large number of preparational parameters, which do
have influence on the quality of the resulting film. The most
important parameter is the substrate temperature. Others are
gas pressure, consistence of the sputter gas, gas flow during
deposition (the mean free path of a particle at 1 mbar is
about 0.5 mm), target properties, target diameter, target-substrate spacing, substrate material and thickness, sputter voltage, the electrical field distribution and many more. The
optimization of these parameters is a very complicated process.
Most of the parameters are related to each other and can not be
optimized seperately.

  Our aim was the deposition of thin films, suitable for
patterning down to at least $1\mu m$ without degradation or aging
effects taking place during or after patterning. Therefore, an
exact stoichiometry and a smooth surface without any defects
and precipitates is required. A sputter concept with a
relatively low sputter rate and deposition temperature was
chosen. Force microscopy measurements showed, that in this case
the surface roughness can be reduced drastically /2/. A
disadvantage of this concept is the slightly reduced transition
temperature of $T_{c,mid} \approx$ 87-88 K with ΔT_c < 1K on $SrTiO_3$ and
$T_{c,mid} \approx$ 85-87K with ΔT_c < 1.5K on MgO. This might result from
chemical interdiffusion at the interface between wafer and film
or from oxygen deficiency in the deposited layer.

  For our deposition device the following parameters are used.
The sputtering gas consisted of a 2:1 mixture of Ar and O_2 at a
total pressure of 100 Pa. During deposition the substrate was
heated to about 660°C. Applying a dc-power of 3.1 W/cm^2 on a
stoichiometric target (diameter 5 cm) a growth rate of approximately 2-3 nm/min was obtained at a substrate-target spacing of
2.8 cm. After deposition the substrate was cooled down to room
temperature in pure O_2 at atmospheric pressure. The substrate
temperature had to be slightly adjusted for each new target.

3. SAMPLE CHARACTERIZATION

X-ray analysis of the films confirms the perfect structure and
epitaxy of these films. The powder diffraction pattern reveals
the absence of diffraction lines different from (00n). The rocking curves are symmetrical with a full width at half maximum
of 0.1° for the (111) reflection of the MgO substrate and 0.4°
for the (113) reflection of the YBCO. The accuracy of the
rocking curve data is restricted to the resolution of the slit-

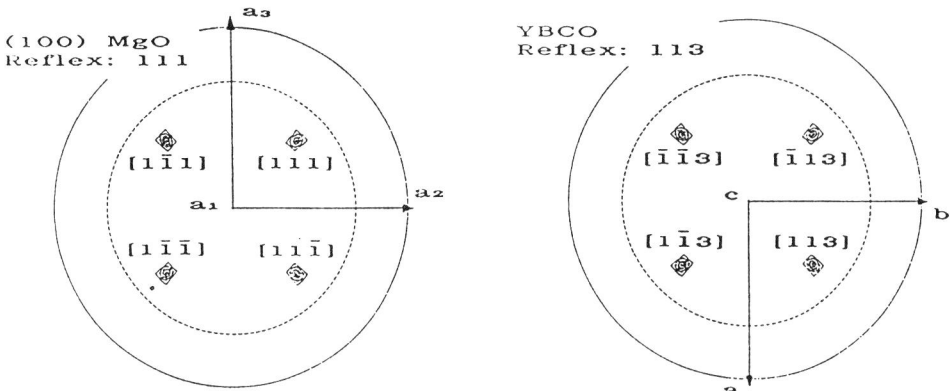

Fig.1: Set of pole figures for MgO substrate and YBCO film in standard stereographic projection.

colimated x-ray beam geometry. Therefore, the actual alignment might even be better. Pole figure measurements (Fig.1) demonstrate the perfect epitaxial orientation of the YBCO film. The film grows with it's c axis perpendicular to the substrate surface, i.e. along the (100) direction of the substrate. The a and b axes of the YBCO are aligned along the (010) and (001) directions of the substrate.

The quality of the orientation can be obtained from channeling experiments. Fig. 2 shows the RBS and channeling data of

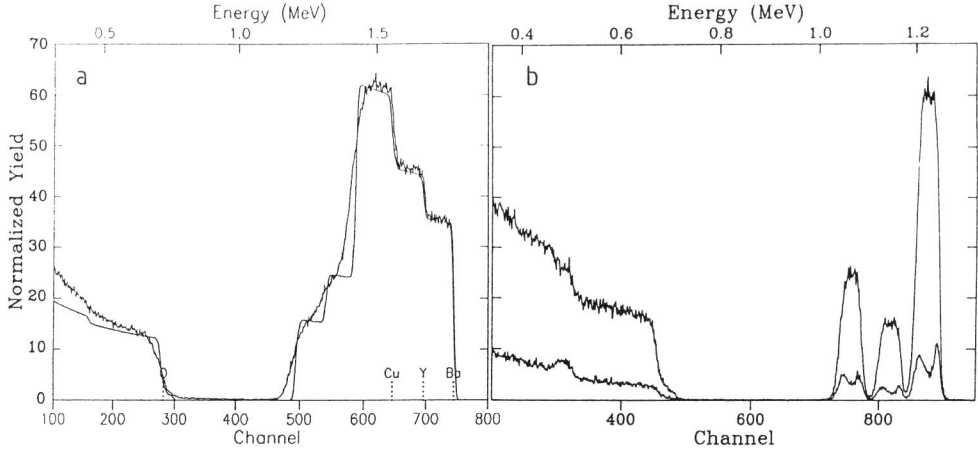

Fig.2: RBS and Channeling data of two YBCO films on MgO. The fit in a) reflects the perfect 123-stoichiometry.

a 50 nm thick YBCO film on MgO. A minimum yield of 5-7% is
obtained for the normal direction. However, a relatively large
interface contribution in the channeling spectrum can be seen.
The large lattice mismatch of 9% between the cubic MgO (a=0.421
nm) and YBCO leads to strain at the interface. This might be
the reason why films with thicknesses less than 50 nm deposited
on MgO show aging effects especially when they are exposed to
the patterning processes (see Fig. 5). Films on $SrTiO_3$ (lattice
mismatch 1.2 %) show less pronounced interface peaks in the
channeling data.

The stoichiometry can be obtained via RBS measurements. Fig.
2 shows the comparison between the experimental data and the
fit with the 123-stoichiometry and the thickness of the film as
parameters. Experiment and fit are in excellent agreement.

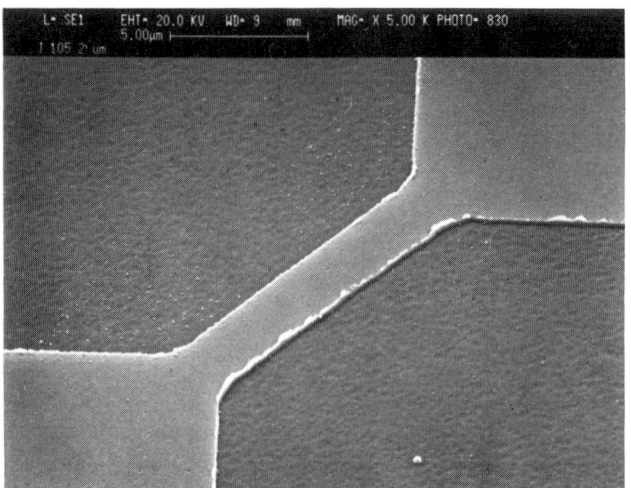

Fig.3: SEM photo of a patterned YBCO film on MgO.

The surfaces of the films are extremely smooth. Fig. 3 shows
a SEM-photo of a structured layer. No cracks or signs of grain
boundaries are visible, except for defects, which are caused by
damages of the underlying substrate surface. The presence of
large-angle grain boundaries is unlikely as can be seen from
the values of the critical current densities up to 2×10^7 A/cm^2
at the temperature of liquid nitrogen and 1 T /6/. Force
microscopy photos reveal a surface roughness of less than 10 nm
(Fig.4). The surface morphology of the films can be described
by an arrangement of hillocks of elliptical forms. The
extensions of the hillocks are given in the Fig. 4b for two
samples, which have been deposited at different substrate
temperatures. The sample, which has been deposited at a
slightly higher substrate

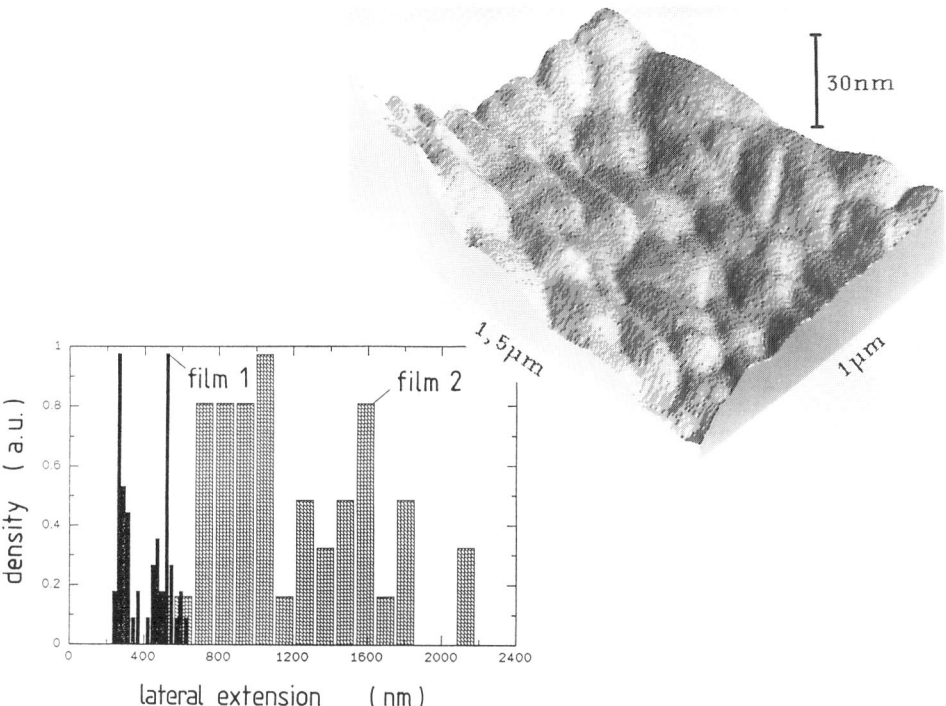

Fig.4: Image of the surface of a YBCO film on MgO obtained by force microscopy. The plot represents the lateral extensions of the hillocks of films sputtered at about 660°C (film 1) and about 20°C higher temperature (film 2).

temperature, possesses typical lateral extensions of 1000 and 1600 nm (100-150 nm in c-direction). The film, deposited at the lower temperature, is characterized by smaller lateral extensions , i.e. 300 and 500 nm (5-15 nm in c-direction). This corresponds to a density of hillocks of 0.63×10^8 cm^{-2} and 6.7×10^8 cm^{-2}. Similar features are obtained by scanning tunneling microscopic experiments on YBCO-films on SrTiO$_3$ /3,4/. It is suggested that screw dislocations nucleate during the growth of the film at defects, e.g. stacking faults, or during island coalescence. Due to the anisotropic growth properties of YBCO these screw dislocations might serve as growth centres for the film. The presence of growth spirals and their interactions are consistent with the theory of crystal growth /5/. The plateaus observed in our films might be the result of such a growths mechanism. The difference in the extensions can be explained in the difference in the diffusion constants during the growth at different substrate temperatures.

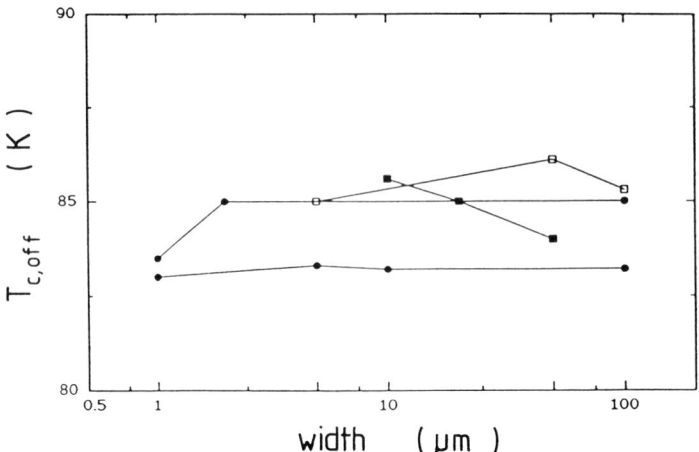

Fig.5: $T_{c,off}$ of sets of bridges as a function of widths.

The films proved to be ideal for patterning. Narrow strips with widths ranging from 100 to 1 μm were formed applying standard optical lithography and ion beam etching with 500 eV Ar-ions. No degradation of the superconducting properties could be observed . The transition temperatures were as high as before patterning. The values obtained for the critical current densities increased with decreasing width of the brigde /6/. A detailed report on the critical properties of the films is given in Ref. 6.

We like to acknowlege the helpful discussions with C. Heiden and A. Braginski, and the technical support of R. Kutzner, R. Lehmann, and F. Schroeteler.

4. REFERENCES

1   A. Höhler, H. Neeb, and C. Heiden, J. of the Less-Comm. Met., 151, 341 (1989); A. Höhler, D. Guggi, H. Neeb, and C. Heiden, Appl. Phys. Lett. 54, 1066 (1989).
2   To be published by the authors.
3   H.P. Lang, T. Frey, and H.-J. Güntherodt, "Atomic resolution and nanostructure of $YBa_2Cu_3O_{7-x}$ laser-ablated thin films studied by scanning tunneling microscopy (STM)", submitted to Europhysics Letters.
4   C. Gerber, D. Anselmetti, J. G. Bednorz, J. Mannhardt, and D.G. Schlom, Nature 350, (1991) p. 279.
5   W.K. Burton, N. Cabrera, and F.C. Frank, Phil. Trans. R. Soc. A243, (1951) P. 299.
6   R. Wördenweber, J. Schneider, M.O. Abd-El-Hamed, R. Lehmann, and D. Guggi, proceedings of ICMC'90, Garmisch-Partenkirchen, FRG (1990), to be published.

High T$_c$ Superconductor Thin Films
L. Correra (Editor)

Superconducting Y–Ba–Cu–O tapes fabricated by excimer laser ablation

N.Yoshida[a],M.Kubota[a],S.Takano[a],M.Nagata[a],T.Hara[b],H.Ishii[b],T.Yamamoto[b],K.Maekawa[b]

[a] Osaka Research Laboratories,Sumitomo Electric Industries,1–1–3,Shimaya,Konohana–ku, Osaka,554 Japan

[b] Engineering Research Center,Tokyo Electric Power Company,2–4–1, Nishi–Tsutsujigaoka,Chofu city,Tokyo,182 Japan

Abstract

We report on high rate deposition and continuous deposition as essential techniques for fabricating superconducting Y–Ba–Cu–O tapes by excimer laser ablation. Using MgO(100) substrates, a Y–Ba–Cu–O film with J_c(77.3K,0T) of 1.2×10^6A/cm^2 has been obtained at the deposition rate of 3.7μm/min, which is the highest rate reported for Y–Ba–Cu–O films. With continuous deposition, we have successfully deposited long Y–Ba–Cu–O films on sintered yittria–stabilized zirconia tapes. The best J_c(77.3K,0T) over 40mm voltage tap spacing is 1.6×10^4A/cm^2(the critical current;1.7A). Stable continuous deposition has been achieved.

1.INTRODUCTION

In recent years the excimer laser ablation technique has been widely used for fabricating high quality films of Y–Ba–Cu–O[1–3]. We previously reported that an as–grown Y–Ba–Cu–O film with a high J_c(77.3K,0T) of 8.0×10^6A/cm^2 was obtained on MgO(100)[4]. An outstanding advantage of this method is the high rate of deposition which maintains the high quality of the film. Wu et al. have demonstrated that increasing the deposition rate up to 14.5nm/min(0.87μm/min) does not degrade the critical current density[5]. The laser ablation technique seems applicable to the large scale deposition of film such as tapes and wires.

In this paper we will first discuss our successful high rate deposition of Y–Ba–Cu–O on MgO(100). Films with J_c(77.3K,0T) exceeding 1×10^6A/cm^2 have been obtained even at the rate of 3.7μm/min. This rate is the highest reported at least for high quality superconducting films. Secondly, we will describe a technique for depositing a long tape substrate. Stable continuous deposition has been successfully performed.

These two thechniques are essential to large scale deposition. We are now working on a high–current long superconductor with high producibility using these techniques.

2.EXPERIMENTAL

The fundamental setup for the deposition was previously described[6]. Briefly a rotating Y$_1$Ba$_2$Cu$_3$O$_x$ target was irradiated by a focused KrF excimer laser with 248nm output. During deposition the substrates were held at 700–750°C in a few hundred–mTorr oxygen ambient.

After deposition the chamber was back-filled with oxygen up to 1 atm and the sample was then slowly cooled down.

In the experiment of high rate deposition on MgO(100), the laser was operated at up to 250 pulses per second. The effects of the repetition rates and the oxygen gas were investigated. The substrate was 8mm×15mm.

In order to fabricate a long superconducting tape, a sintered yittria-stabilized zirconia(YSZ) tape was employed as the substrate. The tapes were 100 to 200 mm long, 10 mm wide and 0.1–0.2 mm thick. As is well known, YSZ tapes of about this thickness can be easily bent elastically, which allows flexibility of the superconducting tape. In the tape deposition, a mechanism for moving the tape substrate was added to the deposition apparatus. As illustrated in Figure 1, the tape was sliding

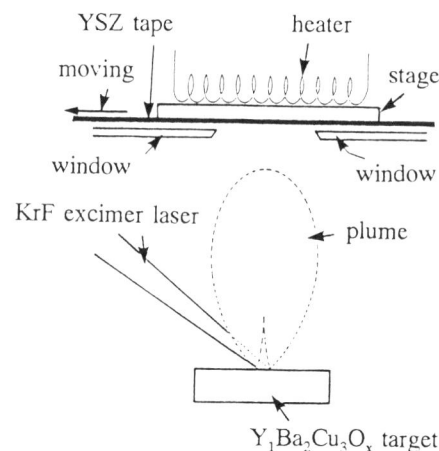

Figure 1. Experimental setup for Y–Ba–Cu–O tape depositions

on the heater stage during deposition. A deposition rate of around 0.1μm/min was chosen. The running speed was between 1 and 2mm/min for a film thickness of 1 to 2μm. A stainless steel window was fixed just above the tape substrate. This enabled the film deposition only within the region where process conditions such as the substrate temperature and the ablation plume were proper for formation of high quality films.

Structural analyses were performed using electron microscopy and X-ray diffraction measurement. The conventional four probe technique was used for determining the T_c and the J_c.

3.RESULTS AND DISCUSSION

3.1.High rate deposition on MgO(100)

The relationship between the deposition rates and the superconducting properties is summarized in Figure 2(a). The c-axis lattice parameter and the full width at half maximum(FWHM) derived from the (00$\underline{10}$) Bragg reflection are shown in Figure 2(b). In these cases only the repetition rate of laser pulses was changed to increase the deposition rate. All the other process conditions were kept the same. The oxygen gas pressure was fixed at 200 mTorr. The deposition rates were found to increase almost in proportion to the repetition rates.

For deposition rates of up to 1.3μm/min the T_c's were constant within 86 to 88K and the J_c's exceeded 1×10^6A/cm^2. At the deposition rate of 1.4μm/min, a rapid drop in the T_c was observed. This behavior corresponds well to the rapid increases in the c-axis lattice parameter and the FWHM of the Bragg peak of (00$\underline{10}$) in Figure 2(b), implying the occurrence of oxygen deficiency and lattice disorder respectively in the films.

One of the causes of this rapid degradation is possibly related to the condition of the oxygen gas over the surface of the growing film. The oxygen gas over the surface is believed to be consumed during one pulse of deposition, resulting in a local pressure decrease near the film surface. A finite time will be required for the recovery of this pressure after one pulse. As the repetition rate is increased, return to the initial condition of the oxygen gas will be incomplete. As a result, the local condition of the oxygen gas near the film surface for high repetitions will be different from that for low repetitions. Assuming that this phenomenon is critical, the sudden occurrence of the oxygen deficiency and the lattice disorder shown in Figure 2(b) can be explained.

If this assumption is valid, film growth will be sensitive to the oxygen pressure in the chamber. In fact, surprizingly, both the growth rate and the film quality are greatly enhanced by increasing the oxygen pressure up to 500 mTorr. In Table 1 the improved superconducting properties and the enhanced deposition rates are summarized. These films are highly c–axis oriented. For the film with a 3.7μm/min rate, the c–axis lattice parameter is 11.68Å and the FWHM of (00$\underline{10}$) is 0.29 deg.

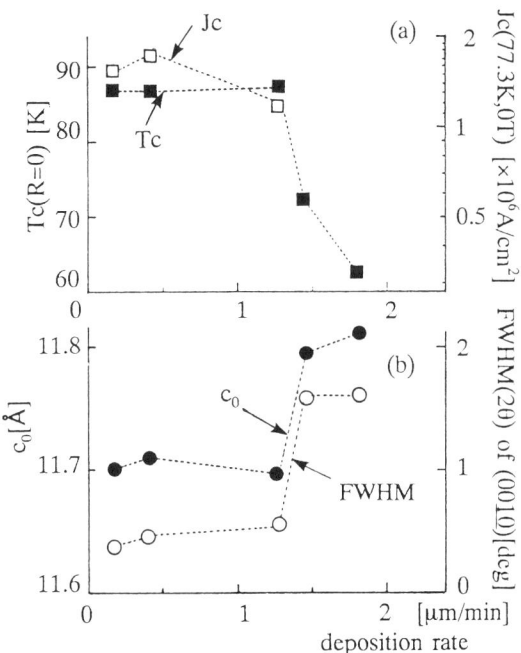

Figure 2. Critical current density, critical temperature(upper), c-axis lattice parameter(c_0) and FWHM of (00$\underline{10}$) peak(lower) vs deposition rate, where all the process conditions are kept the same. Pressure of oxygen gas is 200mTorr.

Table 1. Improved superconducting properties at enhanced deposition rates[*]

$T_c(R=0)$	87.1	K	88.1	K
$J_c(77.3K,0T)$	2.0×10^6A/cm^2		1.2×10^6A/cm^2	
deposition rate	1.5	μm/min	3.7	μm/min
(deposition time)	20	sec	10	sec
repetition rate	100	pps	250	pps

[*] Oxygen pressure of 500mTorr was used

3.2. Y–Ba–Cu–O tape deposition

Figure 3 shows an X–ray diffraction pattern for a small part of the as–grown Y–Ba–Cu–O tape. This film is highly oriented with its c–axis perpendicular to the tape substrate. This high orientation was found to be possible only when mirror–polished YSZ tapes are used. There is little difference in the c–axis orientation over the full length of the tape. The distribution of the c–axis lattice parameter calculated from the Bragg angle of (00$\underline{10}$) reflection is shown in Figure 4. The front part of the as–grown Y–Ba–Cu–O tape was found to have a c–axis lattice parameter longer than 11.68Å, which is expected for fully oxygenated Y–Ba–Cu–O. This fact corresponds to the existence of oxygen deficiencies in the film and originates from the cooling of the front part of the tape in several hundred–mTorr oxygen ambient during deposition. The as–grown Y–Ba–Cu–O tapes were annealed at 400–500°C for 1hour in 1 atm of oxygen. These annealed Y–Ba–Cu–O tapes was found to have 11.68Å of the c–axis lattice parameter over the full length of the tape.

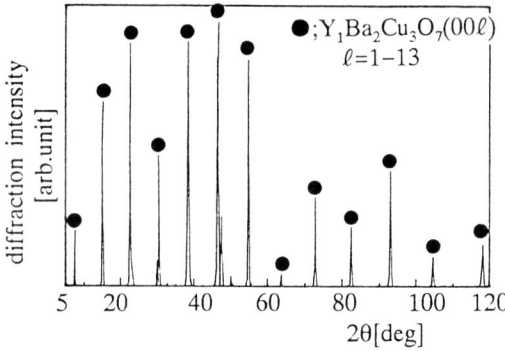

Figure 3. X–ray diffraction pattern for as–grown Y–Ba–Cu–O tape

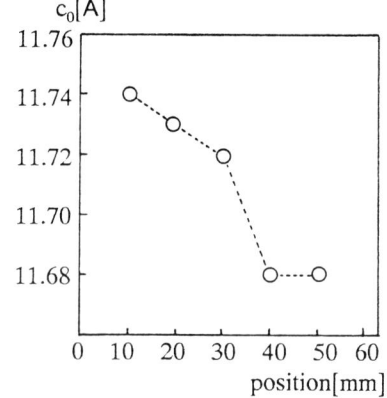

Figure 4. Longitudinal distribution of c–axis lattice parameter(c_0) of as–grown Y–Ba–Cu–O tape

Figure 5 shows a set of I–V curves for a 50mm long sample of our superconducting Y–Ba–Cu–O tapes. The inset explains the configuration of the current and the voltage taps. These taps are gold thin films which are deposited to reduce the contact resistance[7]. Each voltage tap was located at 10mm intervals. $V_0, V_1, \cdots, V_4$ denote the voltage tap spacings as shown in the figure. Every I–V curve shows smooth and sharp transitions. The uniformities of the critical currents, the Jc's and the thicknesses are summarized in Figure 6. In each case, 1μV criterion was used to determine the data. The J_c value for 40mm voltage tap spacing(V_0) is $1.6 \times 10^4 A/cm^2$, which is comparable to the data measured across a small bridge(0.1mm×0.2mm) for the films deposited on the YSZ substrates[8]. The critical current obtained over the same spacing is 1.7A. This value is almost the same as the lowest(V_4) of the local critical current values($V_1, \cdots V_4$).

Figure 7 is a comparative result of the film surface morphology of the Y–Ba–Cu–O tape along the longitudinal direction. The SEM micrographs were taken at positions of 10, 30 and 50mm from the front end of the tape. There is little difference among the morphological features

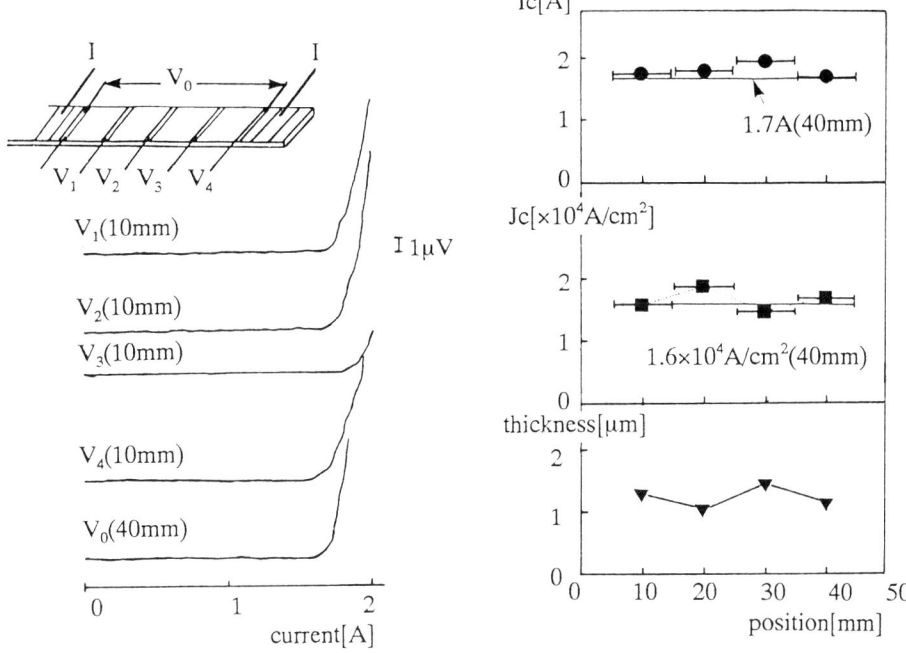

Figure 5. I–V curves for a 50mm long Y–Ba–Cu–O tape

Figure 6. Longitudinal distribution of J_c's, T_c's and thicknesses of the Y–Ba–Cu–O tape

in these photos. A densely packed morphology is commonly observed. This means that stable moving and heating of the tape is achieved during deposition.

Here we need to refer to the two orders of magnitude difference between the J_c's of the film on the YSZ tape and that on MgO(100). Figure 8 is a plan view TEM photograph of the Y–Ba–Cu–O film on the YSZ tape. A polycrystalline structure composed of submicron grains with random in–plane orientations is observed. This feature differs remarkably from the in–plane structure of the film on MgO(100), which has entirely orthogonal twin boundaries[4]. The dictinct grain boundaries seen in the photo seem to behave as weak links of the supercurrent, resulting in the low J_c. To improve the J_c value of this Y–Ba–Cu–O tape further, some locking mechanism to align the grains of the film with each other may be necessary.

4.CONCLUSIONS

We reported on high rate deposition and continuous tape deposition by excimer laser ablation. For high rate deposition, at the deposition rate of 3.7μm/min a Y–Ba–Cu–O film with J_c(77.3K,0T) of 1.2×10^6A/cm^2 was obtained on MgO(100) substrate. For continuous tape deposition, we successfully deposited long Y–Ba–Cu–O films on sintered yittria–stabilized zirconia tapes. A uniform J_c distribution was achieved. The best J_c(77.3K,0T) over 40mm voltage tap spacing is 1.6×10^4A/cm^2(the critical current;1.7A). These two deposition techniques are essential to large scale depositions such as tapes and wire.

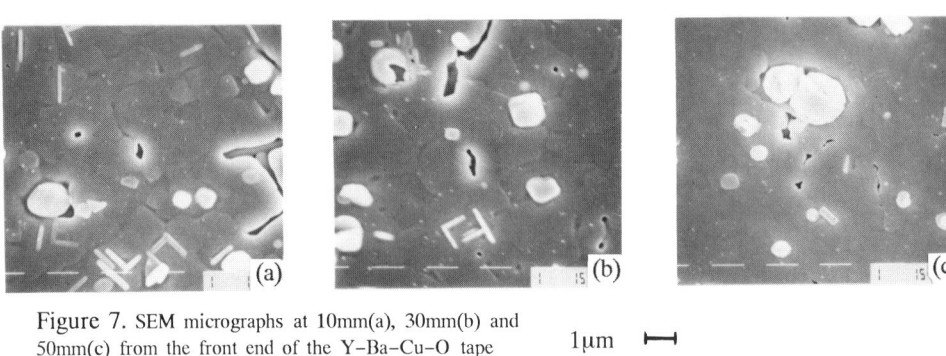

Figure 7. SEM micrographs at 10mm(a), 30mm(b) and
50mm(c) from the front end of the Y–Ba–Cu–O tape   1μm   ⊢──┤

1μm   ⊢──────┤

Figure 8. Plan view TEM micrograph
for the Y–Ba–Cu–O tape

The J_c value of the film on the YSZ tape is two orders lower than that on MgO(100). This seems to be due to the polycrystalline structure of the film, which was composed of submicron grains with random in–plane orientations. To improve the J_c value of this Y–Ba–Cu–O tape further, some locking mechanism to align the grains of the film with each other may be necessary.

REFERENCES

1 R.Roas, L.Schulz, G.Endres :Appl.Phys.Lett.,**53**,(1988)p1557.

2 T.Venkatesan, X.D.Wu, B.Dutta, A.Inam, M.S.Hedge, D.M.Hwang, C.C.Chang, L.Nazar and B.Wilkens :Appl.Phys.Lett.,**54**,(1989)p581.

3 R.K.Singh, J.Narayan, A.K.Singh and J.Krishnaswamy :Appl.Phys.Lett.,**54**,(1989)p2271.

4 S.Okuda, N.Yoshida, M.Kubota, S.Takano, K.Sato, H.Hitotsuyanagi, M.Kawashima, T.Hara, K.Okaniwa and T.Yamamoto :in Proc. ICMC Topical Conf., Garmisch–Partenkirchen, (1990) in press.

5 X.D.Wu, R.E.Muenchausen, S.Foltyn, R.C.Estler, R.C.Dye, A.R.Garcia, N.S.Nogar, P.England, R.Ramesh, D.M.Hwang, T.S.Ravi, C.C.Chang, T.Venkatesan, X.X.Xi, Q.Li and A.Inam :Appl.Phys.Lett.,**57**,(1990)p523.

6 N.Yoshida, M.Kubota, S.Takano, K.Sato, H.HItotsuyanagi, M.Kawashima, T.Hara, K.Okaniwa and T.Yamamoto :in Proc. the 2nd Int. Symp. on Superconductivity, Tsukuba,(1989)p891.

7 J.W.Ekin, A.I.Braginski, A.J.Panson, M.A.Janocko, D.W.Kapone II, N.J.Zaluzec, B.Flandermeyer, O.F.de Lima, M.Hong, J.Kwo and S.H.Lion :J.Appl.Phys.,62,(1987)p4821.

8 N.Yoshida, M.Kubota, S.Okuda, S.Takano, M.Nagata, T.Hara, K.Okaniwa, T.Yamamoto: in Proc. the 3rd Int. Symp. on Superconductivity, Sendai,(1990) in press.

High T$_c$ Superconductor Thin Films
L. Correra (Editor)
© 1992 Elsevier Science Publishers B.V. All rights reserved.

Properties of laser deposited BiSrCaCuO thin films

A. Chéenne[a], R.M.Défourneau[a], C. Le Fiblec[a], J. Perrière[a], A. Raboutou[b], P. Peyral[b], C. Lebeau[b], J. Rosenblatt[b], J.P. Burin[b]

[a]Groupe de Physique des Solides, Université Paris VII et VI, URA 17 du CNRS, Tour 23, 2, Place Jussieu, 75251 Paris Cedex 05, France.

[b]Laboratoire de Physique des Solides, URA 786 du CNRS, INSA, 20, Avenue des Buttes de Coësmes, 35043 Rennes Cedex, France.

Abstract

The resistive transition of thin superconducting $Bi_2Sr_2Ca_1Cu_2O_8$ films has been analyzed in the frame of a *granular model.*. In addition to demonstrate granular aspects in the films, this analysis has allowed to access to relevant parameters of the intragranular material like T_{CS} the critical temperature of the grains and the *width of the critical region*. In particular the *incorporation and/or the distribution of the oxygen* in the grains has been related to the critical temperature of the grains.

INTRODUCTION

Superconducting thin films of the Bi-compound have been prepared by a two step process : deposition by pulsed laser evaporation technique followed by oxygen annealing at high temperature. Depending upon the post deposition thermal treatment, the films present a variable broadening of the resistive transition which may be characteristic of granular effects. We have tried to explain the resistive transition measured on our films, following a model developped for granular superconductors [1]. The superconducting transition of an assembly of weakly superconducting grains can be described as a double transition : the intragranular superconducting transition of the grains at T_{CS} followed by the intergranular coherence transition at $T_C < T_{CS}$ induced by Josephson coupling. The successful application of this model to our "2212" Bi-films demonstrated the granular properties of the films. Moreover, using this model we could have access to important parameters characterizing our films such as T_{CS} and the width of the intragranular critical region. Finally, the critical temperature of the grains (T_{CS}) has been related to the incorporation and/or distribution of the oxygen content incorporated during the thermal treatment. Thus, the quality of the grains, and therefore the quality of the films, can be improved by an adjustement of the oxygen concentration.

MODEL

In this model, the sample is considered as formed by superconducting homogeneous grains separated by insulating barriers. The overall resistivity of the film can be expressed by considering the contribution of the barriers and that of the grains by setting :

$$\rho(T) = \rho_G(T) + \rho_b(T) \tag{1}$$

with: $\rho(T)$ = Resistivity of the material
        $\rho_G(T)$ = Global resistivity of the grains (= 0 for T < T_{CS})
        $\rho b(T)$ = Global resistivity of the barriers

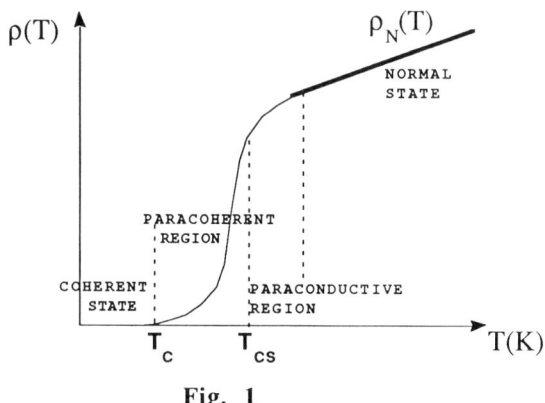

Fig. 1

As shown in Fig. 1, the grains first become individually superconducting at T_{CS} and the global phase coherence is established at $T_C < T_{CS}$. The transition to phase coherence is analogous to the paramagnetic-ferromagnetic transition of ferromagnets with 2D spins (XY model), the Josephson coupling playing the role of the exchange integral [2]. The thermodynamic fluctuations of the order parameter existing around second-order phase transitions lead to consider two important regions of the $\rho(T)$ curve : the paraconductive region for $T>T_{CS}$ (Gaussian fluctuations) and the critical paracoherent region for $T_C < T < T_{CS}$. An estimate of the width of the critical regions for the intragranular and the intergranular transitions is obtained from the Ginzburg criterion [3] which defines a temperature interval (respectively ε_{S0} and ε_0) outside which a mean-field theory is valid [4].

Paraconductive Region (T > T_{CS})
        Above T_{CS}, the rounding of the curve is due to thermal fluctuations of the *Amplitude* of the order parameters of the grains. Physically, this corresponds to the appearance of limited lifetime superconducting pairs leading to an excess conductivity in the grains (σ') with respect to the normal conductivity of the grains (σ_{GN}) :

$\sigma' = \sigma_G - \sigma_{GN}$ with σ_G, the measured conductivity of the grains   (2)

σ' has been determined by Lawrence and Doniach [5] in the case of a material composed of weakly coupled superconducting planes.

$$\sigma'_{LD} = g[d^2\varepsilon_S^2 + 4\xi^2_c(0)\varepsilon_S]^{-1/2} \tag{3}$$

with : $g = e^2/16h$
        d: distance between superconducting planes
        $\xi_C(0)$: Coherence length in the direction perpendicular to the planes, extrapolated to 0 T
        $\varepsilon_S = (T - T_{CS})/T_{CS}$

Paracoherent Region ($T_C < T < T_{CS}$)

Fluctuations of the phase of the order parameter of the superconducting grains lead to an incipient Josephson coupling resulting in an excess of conductivity. The analogy between the coherence transition and the XY ferromagnets implies that in the critical region near T_C, the physical quantities should scale with $t = (T-T_C)/T_C$, which allows to write:

$$\sigma_{XY}(T) = 1/\rho_b(T) - 1/\rho_{bN}(T) \sim ((T-T_C)/T_C)^{-\gamma} \tag{4}$$

with : $\gamma = 2.7$
   $\rho_b(T) = \rho(T)$ (since $\rho_G(T) = 0$ for $T < T_{CS}$)
   $\rho_{bN}(T) = \rho_N(T).\rho_b(T_{CS}).[\rho_N(T_{CS})]^{-1}$

EXPERIMENTS

Superconducting thin films of the Bi-compound were prepared by a two-step process : pulsed laser deposition followed by a thermal treatment. The deposition takes place in a vacuum chamber by irradiation of a $Bi_2Sr_2Ca_1Cu_2O_8$ target using a doubled frequency Nd:YAG laser (532 nm) delivering pulses with a 5ns width, at a power density around 10^8 W/cm^2. The material ejected from the target was collected onto non-heated MgO(100) subtrates. After deposition, the films were amorphous and insulating and needed a high temperature thermal treatment (872°C) under oxygen to become superconducting. After the annealing, the cationic composition of the films determined by RBS is found close to 2212. As shown in Fig. 2, X-Ray patterns exhibit only (00l) peaks of the "2212" superconducting phase. First, this indicates that the films are highly textured (with their c-axis normal to the substrate plane). This last point is in agreement with SEM observations, which show a very lamellar material [5]. These RBS and X-ray measurements show that the films are single phased ("2212") without measurable quantities of spurious phases.

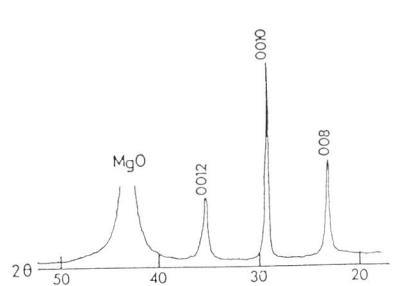

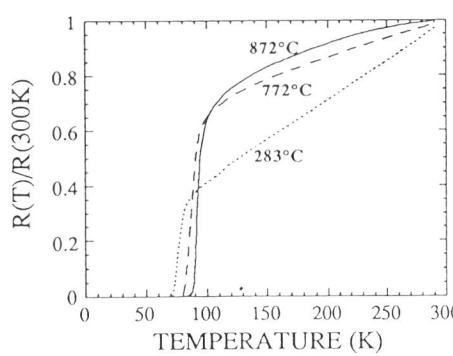

Fig. 2   Fig.3

Resistivity measurements show that the critical temperature (T_C) can vary from a film to the other, depending upon the thermal treatment. For example, Fig.3 presents the resistive curves of three samples annealed at 872°C during 90 minutes, then quenched from three different temperatures (872°C, 772°C, 283°C) As shown in Fig. 3, T_C decreases with the temperature of the quench.

In order to correlate this variation of T_C with the intergranular and/or the intragranular features of the films, we have applied the model previously described. T_{CS} and $\xi_C(0)$ were first determined from the treatment of the paraconductive region, which has allowed then to estimate ε_0 by the describing the paracoherent region.

Paraconductive region treatment

Rewriting eq(3) we obtain : $g^2[\sigma'^2_{LD}\varepsilon_S]^{-1} = d^2\varepsilon_S + 4\xi^2_{C(0)}$. Thus, outside the critical zone near T_{CS}, in the region where the Lawrence-Doniach description is applicable, the representation of $g^2[\sigma'^2_{EXP}\varepsilon_S]^{-1} = f(\varepsilon_S)$ is linear with a d^2 slope and an extrapolated value at the origin of $4\xi^2_{C}(0)$. The excess conductivity of the grain (σ'_{EXP}) is related to the measured excess conductivity of the material $\Delta\sigma_{EXP}$ by the following relation :

$$\Delta\sigma_{EXP} = 1/\rho - 1/\rho_N = \sigma'_{EXP}(1 - x)^2[1 + \sigma'_{EXP}\rho_b(1 - x)]^{-1} \qquad (5)$$

where $x = \rho_b(T)/\rho_N(T) = \rho_b(T_{CS})/\rho_N(T_{CS})$, which estimates the granularity of the material is assumed to be constant. For the fits (Fig. 4), T_{CS} was taken as adjustable parameter with *d fixed at 0.32 nm*. This value corresponds to the distance between the copper oxides planes located in the same unit cell [7]. This choice is justified by the very anisotropic lamellar character of the Bi-compound, and by the particularly important role of the copper oxide planes for superconducting properties in these new materials. In the fit presented in Fig. 4, $\xi_C(0)$ is found close to zero, which has been confirmed by the other samples studied in this work $(\xi_C(0) < 0.06$ nm). This demonstrates the quasi-bidimensional behaviour of $Bi_2Sr_2Ca_1Cu_2O_8$. Moreover, as shown in Fig. 5, the size of the critical region around T_{CS} (ε_{SO}) appears to be very small $(< 3\ 10^{-3})$ since σ'_{EXP} is perfectly fitted by σ_{LD} beyond this value

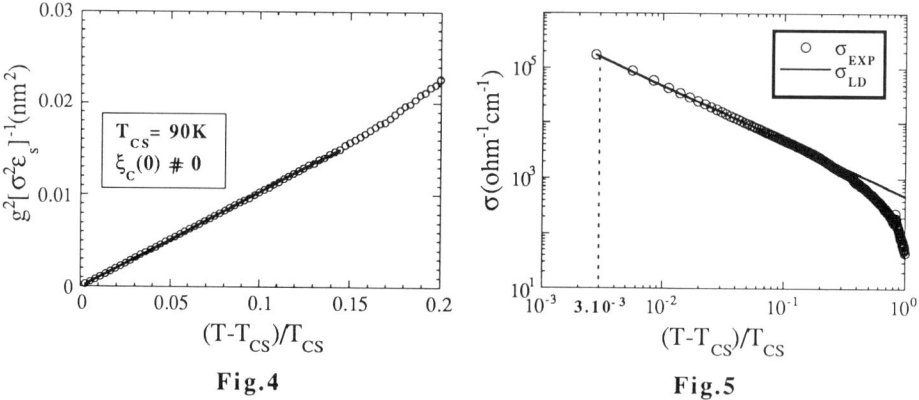

Fig.4   Fig.5

Paracoherent region treatment ($T_C<T<T_{CS}$)

The determination of T_{CS} from the paraconductive region treatment allows to calculate $x = \rho_b(T_{CS})/\rho_N(T_{CS})$ as also σ_{XY} (the excess conductivity in the paracoherent region) from Eq(4). In the critical region, the representation of $\sigma_{XY}(T)^{-1/\gamma}$ versus T should be linear. Thus, T_C is given by the intersection of the linear fit with the x-axis. The plotting of $\sigma_{XY}^{-1/\gamma}$ versus $(T - T_C)/T_C$ in Fig. 6 allows to estimate the temperature range inside which the scaling law can be applied. As shown in Fig. 6, the size of the critical region (ε_0) vary with the samples.

According to the Ginzburg criterion [3], ε_0 and ε_{S0} should be related by the following law :

$$\varepsilon_0 = \varepsilon_{S0} \, [T_C/T_{CS}]^2 [1 - x]^{-3} \qquad \text{with } x = \rho_b(T_{CS})/\rho_N(T_{CS}) \qquad (6)$$

Fig 7 presents the plotting of $\varepsilon_0[T_{CS}/T_C]^2$ versus $(1-x)$ for different single phase samples of $Bi_2Sr_2Ca_1Cu_2O_8$, which allows the experimental checking of the Ginzburg criterion through Eq(6). Moreover, the size of the critical region around T_{CS} (ε_{S0}), has been estimated at $1.48 \; 10^{-3}$.

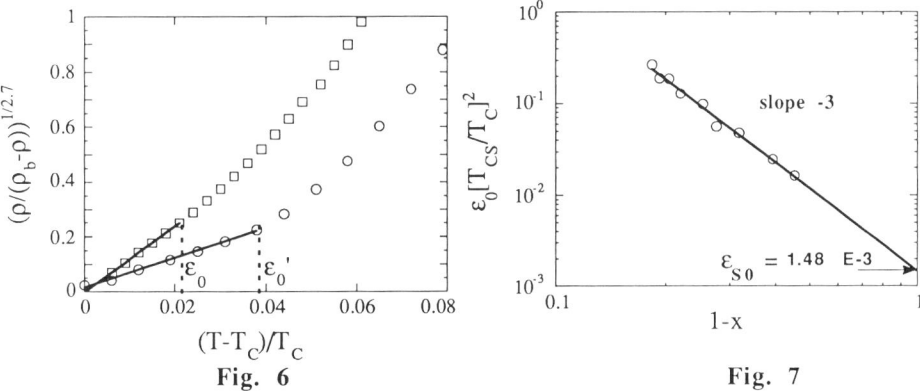

Fig. 6   Fig. 7

The treatment of the samples quenched from various temperatures, whom resistive transitions are presented in Fig. 3 leads, to the results reported in table 1.

Table 1

$T_q(^\circ C)$	$T_C(K)$	$T_{CS}(K)$
872	80.3	105.5
772	70.5	97.5
287	68.5	94.85

The critical temperature of the grains (T_{CS}) follows the variations of T_C by decreasing with the temperature of the quench (T_q). It has been demonstrated [9] that the reduction of the oxygen content in the $Bi_2Sr_2Ca_1Cu_2O_8$ films changes the concentration of carriers and therefore, the critical temperature. That is the reason why the limitation of the oxygen content induced by a quench from high temperature improves T_{CS}. As a result of this study, it seems reasonable to think that T_{CS} (and T_C) can be degraded or improved by incorporing or removing oxygen. This assumption has been checked experimentally as shown in Fig. 8. Oxygen has been removed from a superconducting thin film by a low temperature thermal treatment under hydrogen (350°C). After this step, the film was completely insulating. Then an annealing under oxygen at the same temperature has been performed. As seen in Fig 8, this has not only allowed to recover superconducting properties but also to improve the T_{CS} and therefore T_C by adjusting the oxygen content in the grains.

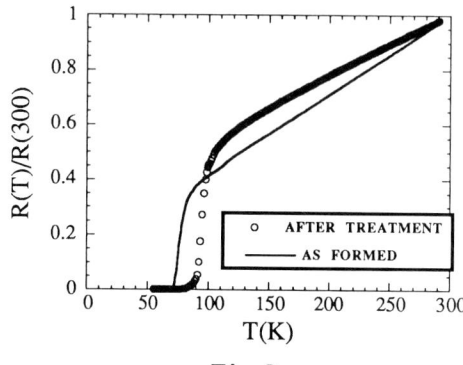

Fig.8

SUMMARY

The granular behaviour of $Bi_2Sr_2Ca_1Cu_2O_8$ thin films has been demonstrated in this paper. The definition of the granularity appears to be complex in the new HT_C superconductors : in addition to the image of grains and barriers, localized inhomogeneity at a microscopic scale might induce the same effects. The application of this granular model has allowed to determine the size of the critical region near T_{CS} for this material (1.48 E^{-3}) and the Ginzburg criterion has been checked experimentally. The choice of d = 0,32 nm has appeared to be justified, which confirms the importance of the copper oxide planes in the HT_C superconductors. The determination of $\xi_C(0)$ close to 0 has demonstrated the quasi-bidimensionnal feature of the Bi-compound. Finally, the precise determination of T_{CS} via this treatment has allowed to prove the strong correlation between the quality of the grains and the oxygen concentration. Moreover the fact that T_{CS} can be easily changed at low temperature opens a way for the enhancement of the quality of the BiSrCaCuO thin films, in particular for the in-situ growth.

REFERENCES

1   J. Rosenblatt, A. Raboutou, P. Peyral, C. Lebeau, Revue Phys. Appl. 25 (1990) 73
2   J. Rosenblatt, A. Raboutou and P. Pellan, Low Temperature Physics, LT 14, M.Krusius and M. Viurio Eds, (American Elsevier, New York, 1975), Vol 2, p. 361
3   V.L. Ginzburg, Fiz. Tverd. Tela 2, 2031 (1960)(Sov. Phys. Solid St. 2, 1824 (1960))
4   A. Raboutou, P. Peyral, C. Lebeau, J.Rosenblatt and J.P. Burin, J. Less Common Met. 164, 1400 (1990); J. Rosenblatt, J.P. Burin, A. Raboutou, P. Peyral and C. Lebeau, Proc. Colloque Transphase 3, Djerba 1990, to be published in phase transition.
5   W.E. Lawrence and S. Doniach, Proc. of the 12th International Conference on low temperature Physics, Kyoto 1970, edited by E. Kanda (Kergaku, tokyo, 1970), p.361
6   J. Perrière,G. Hauchecorne, F. Kerhervé, F.Rochet, R.M. Defourneau, C. Simon, I. Rosenmann, J.P. Enard, A. Laurent, J. Mat. Res., Vol. 5, No. 2, Feb. 1990
7   H.W. Zandbergen, W.A. Groen, F.C. Mijlhoff, G. van Tendeloo, S. Amelinckx, Physica C 156 (1988) 325-354
8   M.A. Denia, O. Penã, C. Perrin, M. Sergent, Sol. State Comm., Vol 73, p.715, 1990

High T$_c$ Superconductor Thin Films
L. Correra (Editor)
573

FORMATION OF WIRE-TYPE STRONGLY ORIENTED STRUCTURES IN HTSC FILMS

I.N.Chukanova, N.V.Mitina, Zh.I.Klitzova, P.V.Mateichenko, G.H.Rozenberg, E.N.Rudenko and A.I.Usoskin

Institute for Single Crystals, Kharkov, 310141, USSR

Abstract

Thin wire-type structural formations growing during the laser pulse deposition of HTSC films have been observed and investigated for the first time. The thin "wires" with a length up to 0.05 mm and with transverse dimensions of about 100 nm were found in $YBa_2Cu_3O_x$ films on $SrTiO_3$ substrates. The increase of the temperature of superconducting transition as well as its narrowing have been observed for films with long "wires".

Most of the research activity in the field of HTSC films development has focused on the uniform isotropic samples. However the tendency to anisotropic crystallite formation which has been observed for HTSC film under certain growth conditions [1] seems to be important not only from the viewpoint of study of a new growing mechanisms but also for technical applications which obviously can acquire HTSC film with anisotropic superconducting properties.

This paper is devoted to the description of thin wire-type formations in $YBa_2Cu_3O_x$ films which have been observed and investigated for the first time.

Wire-type formations have been found in the $YBa_2Cu_3O_x$ films deposited on the $SrTiO_3$ substrates under ~ 800° temperature and ~ 0.1 Torr partial oxygen pressure. IAG laser with 1.06 μm wavelength, 0.4 J pulse energy and ~ 10 ns pulse duration has been used for the deposition procedure.

The length of observed wires reaches 50 μm while their characteristic transversal dimensions over the cross-section vary from 0.2 x 0.2 m to 1 x 1 μm (see Fig. 1).

It's important that the wires spontaneously acquire two strongly perpendicular to each other orientations. According to the results of X-ray analysis these orientations coincide with

projections of crystal axes [010] and [001] onto the plane of
the substrate SrTiO$_3$ the orientation of which is close to
(100).

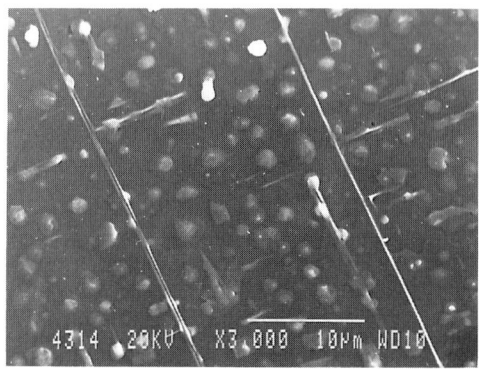

Figure 1. Scanning electron microscopical view of wire-type
formations in YBa$_2$Cu$_3$O$_x$ film on SrTiO$_3$ substrate (3000 X).

Scanning electron microscopical analysis has shown that
intermediate parts of surface between wires are covered by
uniform YBa$_2$Cu$_3$O$_x$ layer with a thickness much less than the
wire one. Wire-type structural formations have been also
observed at the initial stage of film growth when the uniform
HTSC film practically can not be distinguished.
    A great number of structural types with various
characteristics of wire ordering has been found (see Fig. 2)
for the different conditions of film preparation and different
substrate quality. The conditions which can lead to the wire
formation include:
    - availability of the temperature gradient along the
      substrate during film deposition (see Fig 2 c),
    - presence of disoriented crystallites within the
      substrate (see Fig. 3).
Considerable difficulties arise when it is necessary to
determine wire composition since the wires possess a fine
dimensions as well as high adhesion to the substrate due to
which transmittance electron microdiffractometry becomes unfit
technique for such examination. However the estimations
obtained with the help of electron probe microanalysis allowed
to determine that the composition of a certain part of wires
differs from YBa$_2$Cu$_3$O$_x$ composition and possess a deficiency of
Y (see Table).
    As it follows from X-ray analysis films containing the
wire-type formations as a rule possess good crystalline
perfection corresponding to mosaic single crystal. Angular

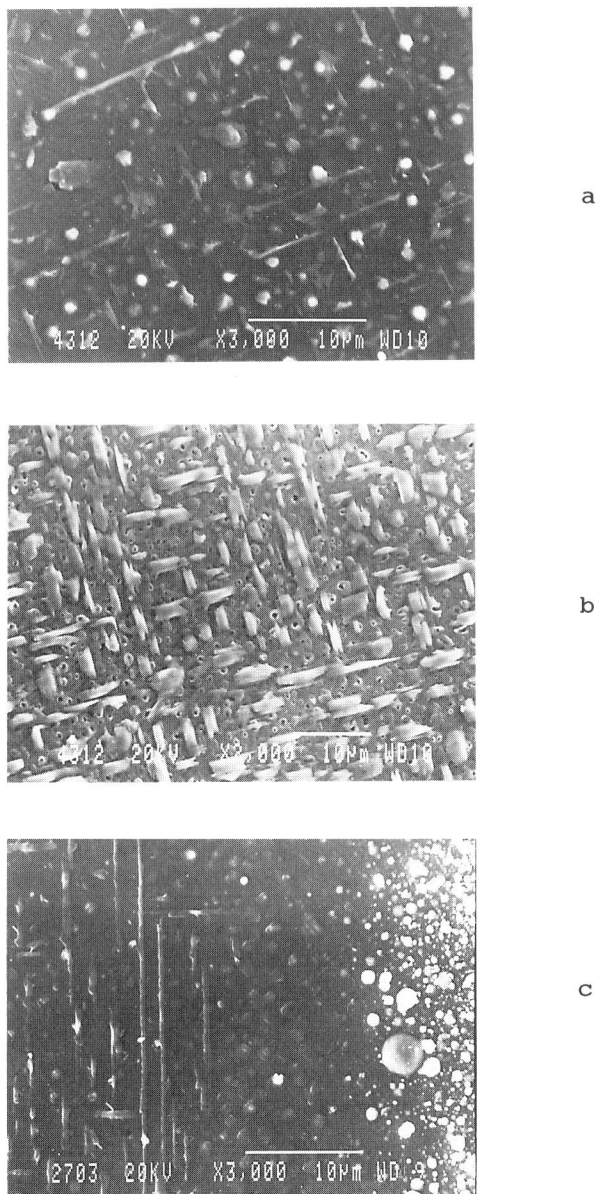

Figure 2. Various types of wire ordering in $YBa_2Cu_3O_x$ films.
c - film was obtained under temperature gradient along the substrate; right side of photo corresponds to the higher temperature. a, b - 3000 X, c - 2000 X.

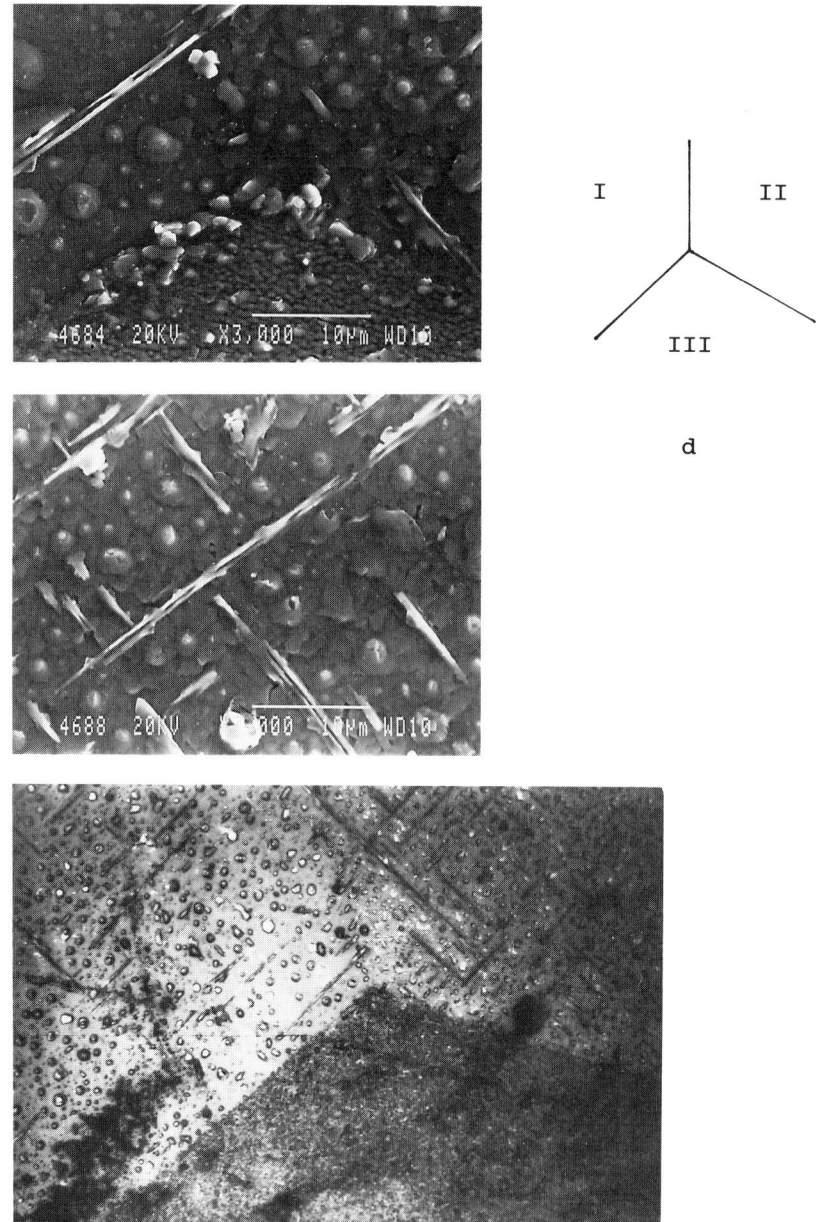

Figure 3. Scanning electron (a, b) and optical microscopical (c) views of $YBa_2Cu_3O_x$ film on $SrTiO_3$ substrate containing three adjacent crystallites (scheme d) with the disorientation of $< 1^o$ (I), 6.5^o (II) and 13.5^o (III) relatively to the (100) plane of $SrTiO_3$. a, c - correspond to the middle part of scheme d, b - to crystallite II. a, b - 3000 X, c - 600 X.

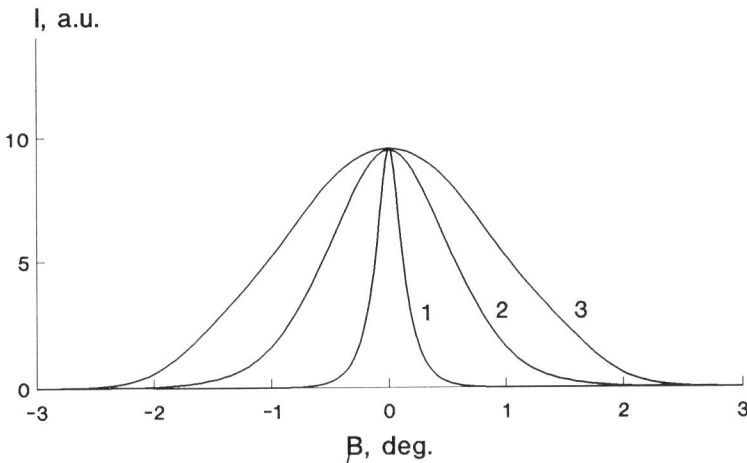

Figure 4. Ω -scanning curves for X-ray reflections corresponding to various phases in $YBa_2Cu_3O_x$ with wire-type structure. 1 - $YBa_2Cu_3O_x$, 2 - phase I, 3 - phase II.

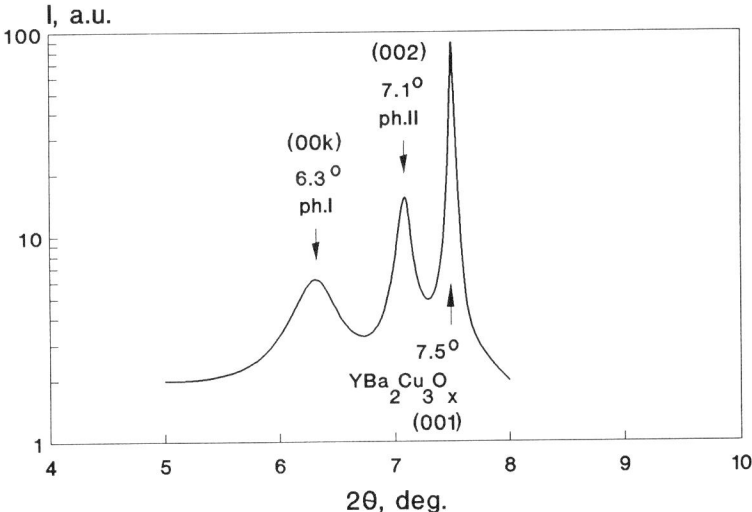

Figure 5. X-ray diffraction (θ - 2θ scanning) for $YBa_2Cu_3O_x$ film containing thin wires (see Fig. 1).

Table
Weight content of elements in wires (crude estimation), %

	Theoretical values	Experimental values	
	$YBa_2Cu_3O_x$	Wires	Film
Y	13	6	12
Ba	41	37	43
Cu	29	41	29
O	16	16	16

width of — scanning curve corresponds to $0.25-0.4^\circ$ (see Fig. 4) while the angular deviation of film crystallites orientation along $\vec{a}$ and $\vec{b}$ axes does not exceed 0.5°. At the same time these films contain a certain portions of supplementary phases. Composition of one of them (phase I) could not be identified however its lattice parameter c is equal to 14.76 Å or 29.52 Å, and the composition of another phase (phase II) was identified as $Y_2Ba_4Cu_7O_x$ with c = 25.2 Å (see Fig. 5). It was found that c-axis orientation of phase I is normal to the substrate surface while c-axis of phase II is normal to (100) crystallographic plane of $SrTiO_3$. The width of —curve for phases I and II (see Fig. 4) corresponds to 1.1° and 2° accordingly.

The availability of these supplementary phases with a content ~ 3% can not completely explain the origin of the wire-type structure since the estimation of volume concentration of wires gives the value of 6 - 8 %.

Temperature dependencies for real part of diamagnetic susceptibility (T) for the film containing wire-type structure and for the film without this structure obtained on the same substrate are shown in Fig. 6. As follows from this figure the appearance of wire-type formations is accompanied by considerable increase of the superconducting transition temperature as well as by decrease of the transition width. At the same time the deterioration of superconducting properties has been observed for the films with relatively short (< 10 m) wires.

Thus we conclude that the formation of wire-type structures considerable changes superconducting properties of $YBa_2Cu_3O_x$ films. We hope that in prospect the development of the

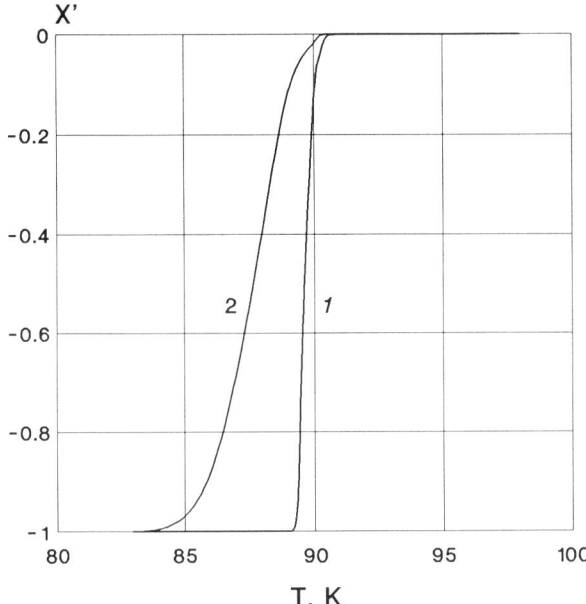

Figure. 6. Temperature dependence of real part of diamagnetic susceptibility for the film containing wire-type structure (1) and for the film without this structure (2).

technique for single oriented wires growing will allow to obtain the planar high temperature superconducting structures with anisotropic properties and perhaps quasi-one-dimensional superconductors, which can be applied in certain fundamental investigations and also might find technical applications.

1   H.Wang, B.Zhao, Y.Lu et al. Chinese Science Bulletin, 34, No. 5 (1989) 370.

High T$_c$ Superconductor Thin Films
L. Correra (Editor)

Comparison of superconducting parameters of variously prepared and treated thin films

S. Leppävuori[a], I. Kirschner[b], A. Uusimäki[a], T. Porjesz[b],
Gy. Kovács[b], I. Dódony[c], R. Laiho[d] and G. Zsolt[b]

[a]Microelectronics Laboratory, University of Oulu, Oulu, Finland

[b]Department for Low Temperature Physics and [c]Mineralogical Department,
 Eötvös University, Budapest, Hungary

[d]Wihuri Physical Laboratory, University of Turku, Turku, Finland

Abstract
 The interaction of Y-Ba-Cu-O and Pr-Ba-Cu-O or Y-Pr-Ba-Cu-O double layers has been studied. The investigation of the electrical properties has been served for the characterization of the quality of samples. The experimental results of a.c. susceptibility and d.c. magnetization investigations can be evaluated on the basis of the proximity effect.

1. INTRODUCTION

The task of this paper, presented was to investigate the interaction of the conductivity electrons between two different layers, which can be superconducting and normal or two superconducting of different critical parameters.

The laser deposition technique [1,2] is suitable to produce thin films or thin film arrangements for the study of fine superconducting phenomena, namely quantum size effects, tunneling and proximity effects.

The effect of preparation process on the properties of films and the interaction between high temperature superconducting and normal (semiconducting) layers or superconducting films of different T$_c$-s were examined.

The results have been interpreted as the appearance of a possible proximity effect, causing in one the hand, superconductor induced superconductivity, or in the other hand, a deviation in the superconducting properties.

2. SAMPLE PREPARATION

Sample films with different layer structures were deposited using the conventional laser ablation method. Details of the sample configurations are given below. The ablation process, used [4] has been reported in detail elsewhere . In this process, the pulses of a XeCl excimer laser were applied to evaporate the Y- and Pr-based compositions onto heated (100) oriented $SrTiO_3$ and MgO single crystal substrates.

Two types of double layer samples were made which had different heat treatments. The first type consisted of two variations and consisted of a basic $YBa_2Cu_3O_{7-d}$ layer of ~200 nm thickness with either a ~20 nm thick $PrBa_2Cu_3O_{7-d}$ layer (**A**) or a $(Pr_{0.25}Y_{0.75})Ba_2Cu_3O_{7-d}$ layer (**B**) partially covering (by ~60 %) it. At the second type, a ~200 nm thick $YBa_2Cu_3O_{7-d}$ layer was completely covered by $(Pr_{0.25}Y_{0.75})Ba_2Cu_3O_{7-d}$ films, of thicknesses of either 20 nm (**C**) or 30 nm (**D**). Sample **E** was the same as the specimen **C** except of its substrate which was MgO. It was prepared for the purpose of structural comparison. In the case of the first type, the samples had to be cooled between the two ablation processes as the mask had to be moved. In the case of the second type, the samples were at a constant temperature throughout the ablation.

For the control measurements, reference layers, made of Y-Ba-Cu-O, Pr-Ba-Cu-O or Y-Pr-Ba-Cu-O were also simultaneously deposited by laser ablation.

3. ANALYTICAL INVESTIGATIONS

For the X-ray studies (Siemens Kristalloflex D-5000) the crystallographic orientation of the layer deposited onto $SrTiO_3$ substrates was different from that deposited onto MgO. In Figure 1.a, the lower curve shows the diffraction pattern of the deposit on the MgO substrate. The [00l] peaks of the Y-123 phase are dominant, but there are also some hkl peaks of Y-123 and also $BaCuO_2$ peaks. The ab, i.e. (001) plane, of Y-123 is mostly parallel to the (100) surface of the MgO. The upper pattern shown in Figure 1.a demonstrates the orientation of the double layer on $SrTiO_3$ and differs from that of the lower one. The [200] reflection of $SrTiO_3$ was used as the inner standard but this peak was cut in

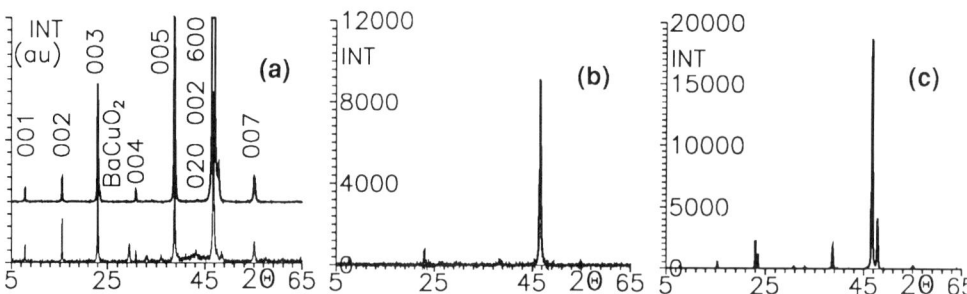

Figure 1. XRD patterns of samples on substrates (a) MgO (upper) and $SrTiO_3$ (lower), the most intensive [006] peaks are cut, (b) on MgO and (c) on $SrTiO_3$

Fig. 1.a to be able to show the other peaks too. In Fig. 1.b and c the same patterns plotted without cut. If the reflexions are normalized to the same intensity, e.g. $I[005]_b=I[005]_c$, then the differences are clearly visible except the [001] ones, since the [003] and [006] of Y-123 coincide with [100] and [200] of $SrTiO_3$, respectively. The $I[200]/I[100]\sim 4$ for $SrTiO_3$. Substracting the normalized intensities of the two layers the resulted intensity ratio was $I[006]/I[003]\sim 8$. It has not been caused by anything else but the [200] oriented Y-123.The morphological observations and the measurement of chemical composition were carried out on a JEOL JSM 6400 scanning and on a JEOL JEM 100CXII analytical TEM-SCAN electron microscope. The chemical homogeneity was measured on the BEI and by a semi-quantitative evaluation of the EDS spectra obtained by TEM-SCAN.

There were no traces of the Pr-diffusion into the Y-123 layers. In the cases of $SrTiO_3$ substrates, the lath shaped crystallites of $Y_{0.75}Pr_{0.25}Ba_2Cu_3O_x$ were oriented with Y-123 (Figure 2a). At a high level of contrast, the BEI micrographs show the chemical inhomogeneity (Figure 2b). As the BEI-TOPO micrograph (Figure 2c) represents, the upper layer is Pr rich. The pale grey spots are probably due to the $BaCuO_2$ in the Y-123 host material. The surface of the double layered film consisted of some spherical clusters (Figure 2d) as well. The Pr-poor part of the same sample is shown in Figure 2e. The Y-123 had a tweed-like texture with uncovered $SrTiO_3$ areas. Figure 2f is the BEI-COMPO image of the same area as in the previous image.

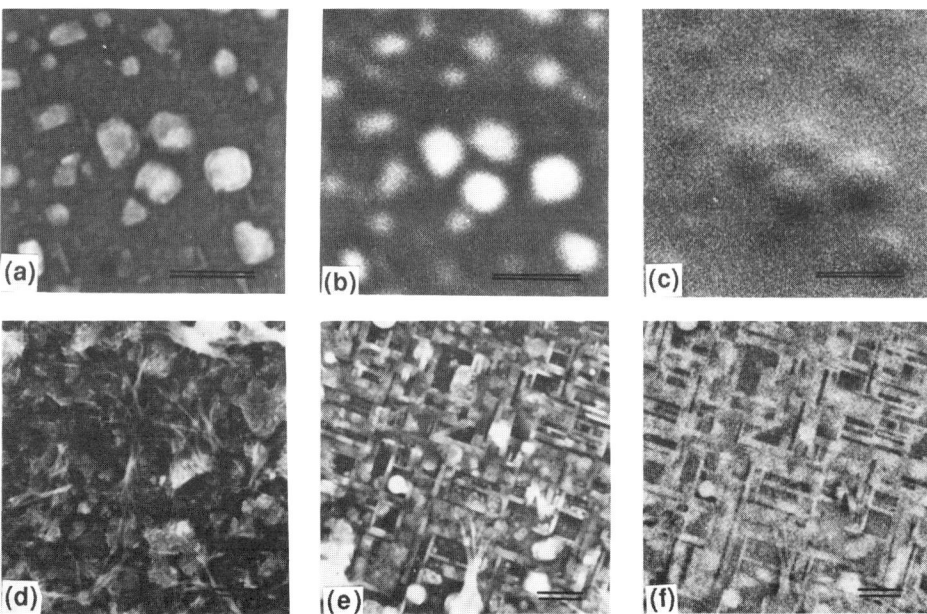

Figure 2. SEM (a), BEI-COMPO (b) and BEI-TOPO (c) micrographs of the same area of a double layer of **A**, SEM image of **B** (d), SEM (e) and BEI-COMPO (f) image of a Pr-poor area of **B**.

4. ELECTRICAL AND MAGNETIC MEASUREMENTS

The electrical resistivity measurements shows a metallic character and a rather narrow superconducting transition for the samples **A** in which the basic Y-Ba-Cu-O layer was dominating (Fig. 3).

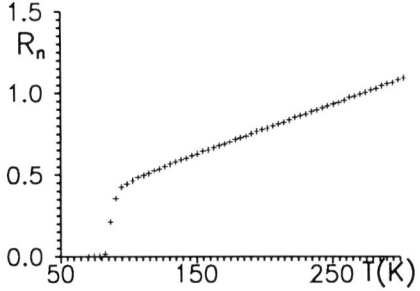

Figure 3. Resistivity curve of **A**.

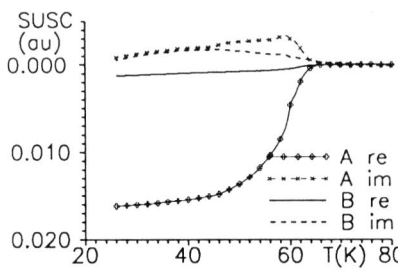

Figure 4. AC susceptibility of **A** and **B**.

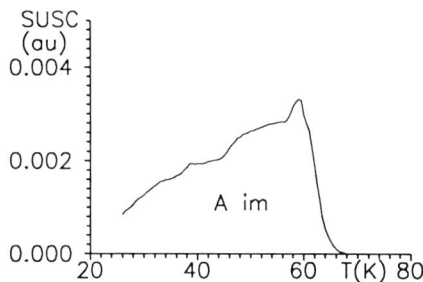

Figure 5. Magnified imaginary part of susceptibility of **A**.

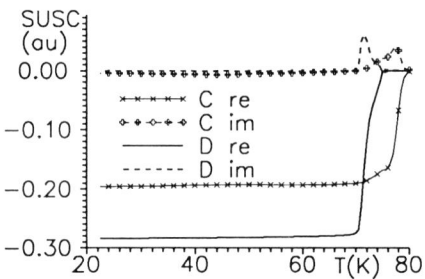

Figure 6. AC susceptibility of samples **C** and **D**.

The magnetic investigation, namely a.c. susceptibility and d.c. magnetization measurements seemed to be the very informative in the case of a probably small interaction between layers. The a.c. susceptibility was measured using opposed secondary coils in series and a lock-in amplifier (EG&G PARC 5210) while the d.c. magnetization measurement was carried out by a SQuID (Biomagnetic Technology). The a.c. field, with a frequency of 190 Hz, was 0.4 Oe rms. The amplitude and the phase shift were measured in order to calculate the real and imaginary part of the susceptibility as a function of temperature. The d.c. field was one order higher. The magnetization was normalized to 1 mm^2 surface area.

The real and imaginary parts of the susceptibility of **A** and **B** are plotted in Figure 4. Both of them show a critical temperature of ~74 K, close to that of pure YBCO layer (Figure 3) and have a characteristic knee at about 45 K in the imaginary part of the susceptibility as is seen in the magnified curve of **A** in Figure 5. In Fig. 8. the normalized d.c. magnetization curves of **A** and its reference counterpart plotted. It can be noted, that nearly at the "knee" of the a.c. suscepteptibility of Fig. 5, the d.c. magnetization of the double layer starts to become larger than that of the reference one.

The real and imaginary parts of the susceptibility of **C** and **D** are plotted in Figure 6. It can be concluded that the higher signal and the lower T_c of **D** were caused by the thicker upper layer (30 nm) compared to **C** which had a thinner second layer (20 nm). **B** and **C** had lower T_c-s relative to the reference sample (see Figure 7). The area and the thickness of the reference sample were the same as that of the basic layers of **C** and **D**.

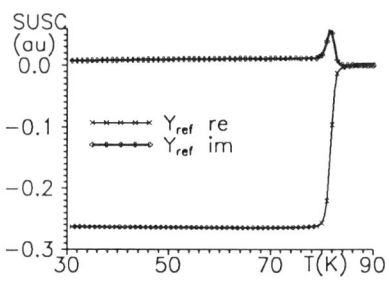

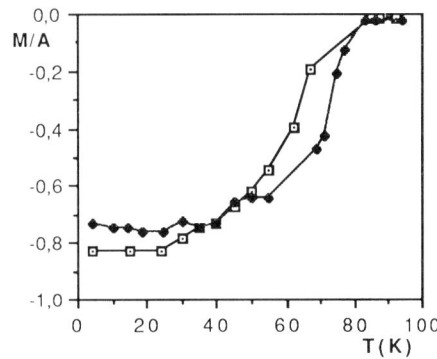

Figure 7. AC susceptibility of an Y-Ba-Cu-O reference sample

Figure 8. DC magnetization of **A** and its reference (normalized to the surface area)

5. DISCUSSION

Since the observed effects are similar to those of the proximity ones obtained for conventional superconductors [4,5], it influences us to treat our results to originate from the proximity interactions of high-T_c materials. Comparing our results to some new observations [6,7] the possibility of the proximity effect seems to be supported.

The most probable structure to get the proximity effect in high-T_c materials, as was expected by Deutscher and Simon [5], was to make a sandwich of a host high temperature superconducting layer (e.g. Y-Ba-Cu-O) and to deposit onto it a normal (semiconducting) layer of possibly similar structure (like Pr-Ba-Cu-O or Y-Pr-Ba-Cu-O). In this case, the ratio of the density of superconducting states to the normal one is large enough (~10) to compensate for the short coherence length of the high-T_c superconducting materials. Furthermore it fulfils the chemical, thermal and structural compatibility of the materials in question.

Taking into consideration all the results obtained, the observed phenomena can be explained on the basis of the proximity interaction of the conductivity electrons of different layers deposited on to each other. As the experiments show, the superconducting behaviour of samples **A** and **B** is characterized by critical temperatures to be at about 45 K as it can be seen from the a.c.- and d.c. measurements coincidently. A decrease of the critical temperature of them could be caused either by the deterioration of the film by subsequent heating or also by some interaction of the conduction electrons. Contrary, in specimens **C** and **D**, only the latter effect was detectable, probably due to the different heat treatment and larger area.

6. ACKNOWLEDGEMENTS

The authors thank J. Levoska and T. Murtoniemi for the preparation of the samples and J. Matthan for editing the text.

7. REFERENCES

1   X. D. Wu, A. Inam, T. Venkatesan, C. C. Chang, P. Barboux, J. M. Tarascon and B. Wilkens, Appl. Phys. Lett. 52 (1988) 754.
2   J. Fröhlingsdorf, W. Zander, and B. Stritzker, Solid State. Commun. 67 (1988) 965.
3   H. -U. Krebs, M. Kehlenbeck, M. Steins and V. Kupcik, J. Appl. Phys. 69 (1991) 2405.
4   J. Levoska, T. Murtoniemi and S. Leppävuori, J. Less-Common Metals 164 & 165 (1990) 710.
5   G. Deutscher and R. W. Simon, J. Appl. Phys. 69 (1991) 4137.
6   V. Kresin, Advances in Superconductivity II, p. 971 (eds. T. Ishiguro and K. Kajimura) 1989, Springer-Verlag Tokyo, Berlin, Heidelberg, New York, London, Paris, Hong Kong.
7. H. Akoh, Advances in Superconductivity II, p. 975 (eds. T. Ishiguro and K. Kajimura) 1989, Springer-Verlag Tokyo, Berlin, Heidelberg, New York, London, Paris, Hong Kong.

High T$_c$ Superconductor Thin Films
L. Correra (Editor)
© 1992 Elsevier Science Publishers B.V. All rights reserved.

Patterning of Bi-Sr-Ca-Cu-O thin films by wet chemical etching in EDTA

F J Müller[a,b], J C Gallop[a,b], J R Laverty[b], M A Angadi[b], A D Caplin[b], S Labdi[c] and H Raffy[c]

[a]National Physical Laboratory, Teddington, Middlesex TW11 0LW, United Kingdom

[b]Centre for High Temperature Superconductivity, The Blackett Laboratory, Imperial College, London SW7 2BZ, United Kingdom

[c]Laboratoire de Physique des Solides, Bât. 510, Université Paris XI, 91405 Orsay, France

Abstract

The wet chemical etchant EDTA (*ethylenediaminetetraacetic acid*) in water has been applied successfully to impose dc SQUID patterns onto superconducting Bi-Sr-Ca-Cu-O (BISCCO) thin films of varying thickness. Sub-micron resolution was achieved for the pattern.

The transition temperature was monitored both resistively and inductively before and after the wet etching process. No depression of T$_c$ was observed within experimental accuracy.

Using a Vibrating Sample Magnetometer (VSM), the critical current density J$_c$ before and after patterning was measured. Again, no depression has been seen. One film even possessed improved T$_c$ and J$_c$ values after wet chemical etching in EDTA.

1. Introduction

So far, dry etching is used by most groups to pattern HTS (High Temperature Superconductivity) thin films, with an exhaustive list of examples like reactive ion etching [1] and pulsed laser evaporation [2]. Although these processes are promising for producing micron-sized patterns, there is a chance that they degrade the transition temperature [3] and/or damage the substrate [4]. Wet chemical etching presents a viable alternative. Various acid etchants such as solutions of phosphoric acid, nitric acid, and hydrochloric acid have been used recently. These etchants provide rapid removal rates larger than 0.5 μm per minute for very low concentrations [4], and are thus not suitable for thin film patterning where a controllable rate providing smooth sidewalls is desired. This is the rationale for employing EDTA in this study, an organic solvent which provides a relatively slow, controllable rate of wet etching; Shokoohi et al [4] have previously applied EDTA on laser deposited Y-Ba-Cu-O thin films.

We believe it does not suffice to simply monitor the transition temperature before and after patterning! After all, a drop in T$_c$(zero) of a few Kelvin may well correspond to a loss in critical current density by orders of magnitude. Monitoring J$_c$ before and after etching is hence crucial. However, a transport measurement of J$_c$ usually requires a patterned bridge in the first place. This is the first reason to opt instead for a non-destructive VSM magnetisation J$_c$ measurement, as described in depth in [5]. The second reason in favour of the VSM technique is the (several orders of magnitude) higher sensitivity of the irreversible magnetisation to small residual dissipation.

Equally, an inductive portrait of the superconducting transition is warranted before and after etching. It provides information about the homogeneity of a particular sample, besides allowing a more subtle determination of T$_c$(zero). We have shown this in [6] with the aid of a percolative computer simulation.

Our inductive characterization set-up is a modified version of an ac susceptibility apparatus designed by Xenikos and Lemberger [7], which we have described in more detail in [6]. The sample is laid on top of a flat spiral coil which forms part of a resonant LC circuit and induces changes in the coil inductance during the superconducting transition. Our inductive system has allowed us to determine the transition temperature of a variety of samples, even that of a tiny single crystal, weighing just 0.1 mg. Generally speaking, the inductive transition will extend below T_c(zero) for any sample containing individual grains with a lower transition temperature than those grains that form the first percolative path across the sample.

The BISCCO thin films under investigation here come from two sources: Some were produced ex-situ at the National Physical Laboratory, U.K., by sequential thermal evaporation.[8] Others are ex-situ films produced by dc triode single target sputtering at the Université Paris XI at Orsay, France.[9]

2. Experimental: Patterning BISCCO thin films

We generate dc SQUID (Superconducting **Q**uantum Interference **D**evice) patterns using several layers of positive photoresist, AZ 1512. A small quantity of microposit ensures a smooth uniform distribution of the AZ 1512 over the whole film surface, which is subsequently spun (on a rotating vacuum chuck) at 4000 rpm for 30 seconds per cycle. After a "soft-bake" at 70 °C for 25 minutes, the sample is carefully positioned under a well-focussed mask aligner, supplying UV light (wavelength 320 nm) with an intensity of $5mW/cm^2$ during exposure. Empirically, an exposure time of 600 seconds has proved ideal. The photoresist layer is subsequently developed for 10 minutes, before being "hard-baked" at 120°C for a further 10 minutes.

The mask pattern we designed for this purpose contains 4 dc SQUIDs with different parameters, to make optimum use of an individual film. Each dc SQUID pattern consists of a round washer design with two symmetrically arranged constrictions, aiming to exploit the grain boundaries in polycrystalline films as naturally occurring Josephson junctions.

Finally, we immerse the sample into a saturated solution of EDTA/H_2O. Excess amounts of EDTA crystals are present in the solution to eliminate concentration degradation with time, and the EDTA solution is repeatedly stirred during the etching procedure. Shokoohi *et al* [4] have measured the etch rate to depend linearly on time and exponentially on the solution temperature. In accordance with another of their observations, we found that the removal rates are reduced if etching is interrupted and the samples are exposed to atmosphere, e.g. in order to look at them under an optical microscope.

The overall etching time is determined by the microstructure of the sample. Thick dendrites are thereby particularly difficult and time consuming to etch away. Conversely, platelets are easily removed. Fig.1 and Fig.2 below display typical results of our wet etching technique.

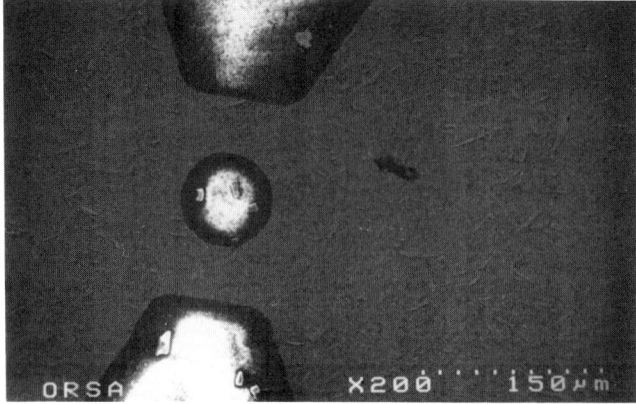

Fig.1: *A dc SQUID pattern (round washer design with two constrictions) imposed on a Bi-Sr-Ca-Cu-O thin film ("film 1", 0.8 μm thick) produced ex-situ at Orsay by dc triode single target*

sputtering [9]. *The diameter of the SQUID hole is around 75* μm, *the two constrictions on either side are approximately 45* μm *wide. This clear-cut pattern was obtained after only 5 minutes immersion in the EDTA solution, corresponding to an etching rate of around 27 Å/s.*

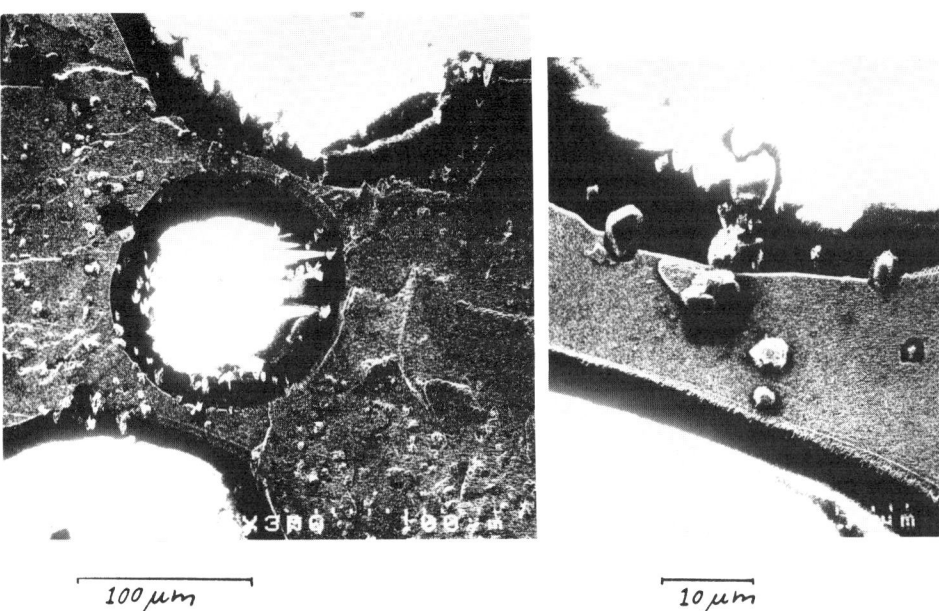

$\overline{\quad 100\,\mu m \quad}$ $\overline{\quad 10\,\mu m \quad}$

Fig.2a (left): *A similar dc SQUID pattern as in Fig.1, this time imposed on a BISCCO thin film ("film 3", 1.54* μm *as-deposited thickness), produced at the National Physical Laboratory in Teddington by sequential thermal evaporation.*[8] *The pattern was obtained after immersion in the EDTA solution for 25 minutes.*

Fig.2b *on the right shows a detail of the lower constriction as displayed on the left. The resolution for the smooth plate-like grains can here be inspected to be around 0.6* μm. *Their etching rate lies around 10.3 Å/s, while that of the lead rich blobs littered across their surfaces is significantly higher.*

3. Experimental: Film production and T_c before and after patterning

We double-checked a possible change in the superconducting transition temperatures both resistively (a standard four-terminal van der Pauw measurement) and inductively, as described in the introduction and in [6].

Film 1 (see Fig.1) is around 0.8 μm thick and was produced ex-situ by dc triode single target (2223, 30% lead doped) sputtering. It was post-annealed in air at 873°C for 14.5 hours, in the proximity of a lead-doped BISCCO pellet of 2223 stoichiometry. Its T_c(zero) before patterning was around 107 K, with the resistive onset of the superconducting transition, T_c(onset), at around 120 K. The inductive transition extended down to about 90 K.

Film 2 had an as-deposited thickness of 1.54 μm and was produced ex-situ by sequential thermal evaporation from resistive boats. Its molar ratio for deposition was 2.0:2.0:1.0:1.7. It was post-annealed in wet air at 845°C for 120 minutes. Its T_c(zero) before patterning was around 79 K, with T_c(onset) at around 92 K. Its inductive transition extended down to about 74 K.

Film 3 (see Fig.2a and Fig.2b), finally, stems from the same deposition batch as film 2. However, it was annealed differently, for 16 hours at 840°C in wet air, in the proximity of a lead-doped BISCCO pellet. Its T_c(zero) was around 5 K lower than that of film 2, i.e. around 74 K, with T_c(onset) at around 80 K. Its inductive transition before and after etching is displayed in Fig.3.

No change in T_c(zero) was observed for film 1 and 2, within experimental accuracy. Of equal importance is the fact that the inductive measurement curves before and after patterning are virtually identical, except for the step height of the inductive transition. The latter, however, is a result of the loss of (i.e. the etching away of) superconducting film area. The (In)homogeneity of the films, i.e. their distribution of different grains, has seemingly not changed.

On the other hand, T_c(zero) of film 3 has been increased slightly to around 75 K (See Fig.3 below). This increase is in line with an improvement in critical current density, discussed in the next section. A similar phenomenon has been observed on laser ablated ex-situ BISCCO films, produced by Sajjadi et al [10] and EDTA etched by us. This is probably due to the removal of "impure" areas containing grains of lower T_c(zero).

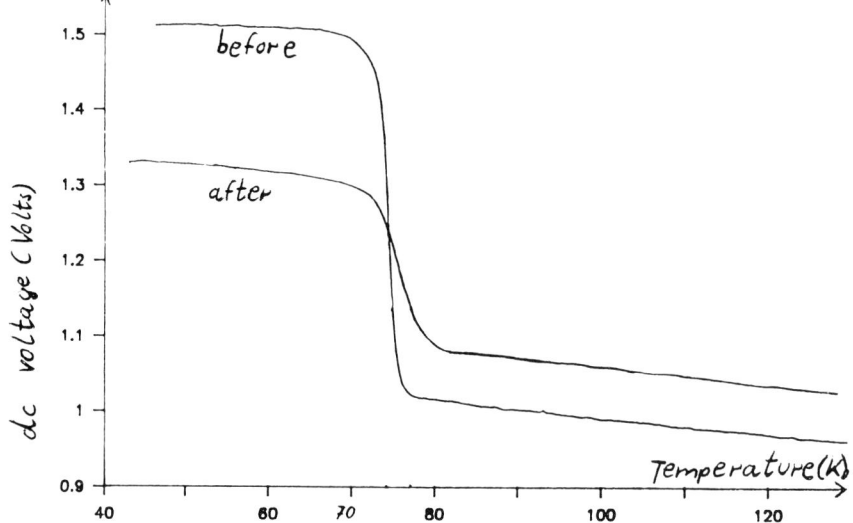

Fig.3: *Inductive measurement curve for film 3 before and after wet etching in EDTA. As shown in reference [6], T_c(zero) can be found in the region with the highest differential voltage increase within the inductive transition. The dc voltage step is proportional to the change in inductance of the flat spiral coil onto which the sample is laid.*

4. J_c before and after patterning

The critical current density in film 2 and 3 was measured by Laverty and Angadi before and after etching in EDTA, employing the VSM magnetization technique described in detail in [5]. For film 2, no significant changes were observed at 10 K in a 0.5 Tesla field, while film 3 showed an improvement.

Λ denotes the current carrying length scale in our superconducting films, i.e. the diameter of the intergranular screening currents. In film 2, Λ was approximately 6.2 mm before and 1.4 mm after etching. This decrease can be entirely attributed to the loss in superconducting area. Both values indicate that the film is fully connected before and after patterning. J_c(10 K, 0.5 T) was measured as 1.1×10^5 Amps/cm^2 before etching in EDTA and as 1.4×10^5 Amps/cm^2 afterwards. Due to the limited accuracy to which we were able to determine the film area before and after patterning, we do not believe that this represents a genuine increase in critical current density.

On the other hand, J_c(10 K, 0.5 T) for film 3 was measured as 4.0×10^5 Amps/cm^2 before etching and as 5.6 x 10^5 Amps/cm^2 afterwards. This 40 % increase indicates a genuine improvement. Λ for film 3 was measured as 4.8 mm before and 2.1 mm after patterning. In unison with the results for film 2, both values show full connectivity.

5. Conclusion

An inductive characterization of the superconducting transition, as well as a J_c measurement before and after patterning should be recommended as a standard litmus test for the suitability of patterning techniques for HTS films. EDTA has passed this test well, and we can warmly recommend this wet chemical etchant.

6. References

1 S. Matsui, N. Takado, H. Tsuge and K. Asakawa, Appl.Phys.Lett.,52 (1988) 69
2 A. Inam, X.D. Wu, T. Venkatesan, S.B. Ogale, C.C. Chang and D. Dijkamp, Appl.Phys.Lett., 51 (1987) 1112
3 D.W. Face, J.M. Graybeal, T.P. Orlando and D.A. Rudman, Appl.Phys.Lett., 56 (1990) 1493
4 F.K. Shokoohi, L.M. Schiavone, C.T. Rogers, A. Inam, X.D. Wu, L. Nazar and T. Venkatesan, Appl.Phys.Lett., 55 (1989) 2661
5 M.A. Angadi, A.D. Caplin, J.R. Laverty and Z.X. Shen, submitted to Physica C
6 F.J. Müller, J.C. Gallop and A.D. Caplin - to be published as an extended abstract for ISEC 1991 conference in Glasgow
7 D.G. Xenikos and T.R. Lemberger, Rev.Sci.Instrum., 60 (1989) 831
8 F.J. Müller, J.C. Gallop, M. Stewart, F.M. Saba and A.D. Caplin - see this conference's proceedings ERM Strasbourg 1991
9 H. Raffy, S. Labdi, A. Vaures, J. Arabski, S. Megtert, Physica C, 162-164 (1989) 613
10 A. Sajjadi, F.J. Müller, A.D. Caplin, F.M. Saba, J.C. Gallop, F. Beech and I.W. Boyd, to be published

High T$_c$ Superconductor Thin Films
L. Correra (Editor)

PREPARATION, PATTERNING AND PROPERTIES OF YBCO FILMS

R.SCHNEIDER[a], G.SCHULZ[a], G.GRABE[a], D.MAY[b] and T.KUHLEMANN[c]

[a] Daimler-Benz Forschungsinstitut, Frankfurt, Germany

[b] Physikalisch-Technische Bundesanstalt, Braunschweig, Germany

[c] Institut für Hochfrequenztechnik, TU Braunschweig, Germany

Abstract

YBCO films were prepared on various single-crystalline substrates, (100)-oriented LaAlO$_3$, SrTiO$_3$, MgO, randomly oriented Zr(Y)O$_2$ and r-plane sapphire, by Inverted Cylindrical Magnetron Sputtering (ICMS). The surface morphology was investigated by scanning electron microscopy (SEM). The inductively measured T$_c$ values of the films "as prepared" ranged from 87 to 90K. The j$_c$ values at 77K of films on (100) LaAlO$_3$ and SrTiO$_3$ were $5 \cdot 10^6$ and $4 \cdot 10^5$ A/cm^2, respectively. j$_c$ of films on r-plane sapphire was $5 \cdot 10^4$ A/cm^2. The surface resistance R$_s$ was measured using a 67 GHz resonator made of copper. R$_s$ of a 0.4µm thick YBCO film on LaAlO$_3$ was 17mΩ at 77K. All the films were highly c-axis textured with small admixtures of a-axis oriented grains. Microstrips of a width down to 2µm were patterned by means of conventional UV-photolithography. The etch step was performed in diluted phosphoric acid or EDTA. Measuring T$_c$ of the microstrips as a function of their width, we found that the T$_c$-degradation starts at 8µm using phosphoric acid but not above 2µm using EDTA.

1. INTRODUCTION

It is likely that microwave devices will be first applications of YBCO films. Many passive high-frequency components have been already presented [1]. For their production a reliable and reproducible thin film deposition technique is needed. One of the most successful techniques is the Inverted Cylindrical Magnetron Sputtering (ICM) which we have used for our YBCO film preparation. The employment of the YBCO films in the microwave technique requires c-axis oriented films with small surface resistance on substrates with low HF losses, e.g. sapphire and LaAlO$_3$. In addition the film surfaces have to be so smooth that reproducible patterning to device structures is possible. With respect to these standards we report on our investigations of YBCO films on various single-crystalline substrates.

2. EXPERIMENTAL

For the deposition of the YBCO films we used the inverted cylindrical magnetron (ICM) sputter technique. The films were grown in two steps, i.e. in situ deposition at elevated substrate temperature T_s and subsequent oxygen take-up in the cooling process to room temperature. Details of the ICM sputter technique are described elsewhere [2,3].

The quality of the film surfaces was investigated by SEM. The transition temperature to superconductivity T_c of the films "as prepared" was measured inductively. The film resistivity was measured by a standard dc four-probe arrangement after patterning to microstrips of various dimensions. From measurements of the I-V characteristics at various fixed temperatures in zero magnetic field, the temperature dependence of the critical current density j_c (T) was deduced with the criterion of 1μV voltage drop over the microstrip. The film growth direction and growth quality were characterized by X-ray diffraction measurements. The growth direction was deduced from θ-2θ diffraction patterns and the FWHM of X-ray rocking curves served as a measure of growth quality. The surface resistance was measured using a copper cavity at 67GHz in the temperature range from 20 to 90K. Details of measurement principle and system were published earlier [4].

Microbridges of various dimensions were structured by means of conventional UV photolithography using Shipley AZ 1350 J or Hoechst AZ 8112 positive photoresist followed by wet chemical etching. The etchants used were diluted phosphoric acid (1.5%) or a PS 80-04 solution, which is a mixture of phosphoric acid, nitric acid and water in the ratio of 20:4:1. The films were also etched using EDTA dissolved in water at 50°C. More details concerning wet chemical etching of YBCO films are reported elsewhere [5].

3. RESULTS

Our SEM investigations can be summarized as follows: on the most film surfaces we observe particles less than 1 micron in size with a variety of morphologies. Many of the "outgrowths" have a lenticular morphology that is typical of a,b-axis orientation [6]. Other features of many surfaces are small pinholes of flat rectangular needles which form an interconnected network. The size and density of the lenticular outgrowths is correlated to the deposition conditions, especially the deposition rate and the substrate temperature. At the lower deposition temperature the outgrowths are smaller and their density is reduced. Films deposited on (100) LaAlO$_3$ substrates at high deposition rate (0.3nm/s) are much smoother than films deposited at lower rates (down to 0.03nm/s). Although the substrate temperature and the deposition rate are well controllable growth parameters, the surface morphologies of films prepared under the same conditions can differ considerably. This indicates that other parameters, e.g. the substrate quality, also control the surface morphology.

The superconducting transition temperature T_c is routinely measured after the film preparation by an inductive method. The T_c values of films on various single-crystalline substrates, (100) LaAlO$_3$, SrTiO$_3$, MgO, randomly oriented Zr(Y)O$_2$ and r-plane sapphire, are usually between 87 and 90K.

The resistive transition to superconductivity is measured on wet chemically etched bridges of widths from 10 to 30µm and lengths from 100 to 500µm. The T_c(R=0) values range from 87 to 90K with the exception of films on r-plane sapphire where T_c(R=0) is not larger than 84K even in case of a 30µm wide bridge. Typical normal conducting properties of YBCO measured on bridges of a width of 10µm and a length of 500µm on (100) SrTiO$_3$ are resistivities less than 100µΩcm at 100K and resistance ratios R(300K)/R(100K) near 3. In Fig. 1 we show the critical current densities j_c versus temperature in zero magnetic field of YBCO films on (100) LaAlO$_3$, SrTiO$_3$ and randomly oriented Zr(Y)O$_2$.

The growth conditions, especially the substrate temperature, were optimized with respect to c-axis oriented films. The growth direction was determined on a two axes goniometer. θ-2θ diffraction patterns of films on (100) LaAlO$_3$, SrTiO$_3$ and MgO revealed strong (001) and much weaker (h00) lines indicating that the films are highly c-axis textured with small admixtures of a-axis oriented grains. The minor growth quality of films on sapphire becomes pronounced in the rocking curves for the (005) line. The FWHM of the curves, also called mosaic spread, typically amounts to 2.5° whereas the mosaic spreads of films on the above mentioned substrates are typically 0.3-0.4°.

Fig. 2 shows the surface resistance R_s of 0.3-0.4µm thick films on (100) LaAlO$_3$, Zr(Y)O$_2$ and r-plane sapphire versus temperature at a frequency of 67GHz. For comparison, R_s of copper is plotted. At 77K only R_s of the film on (100) LaAlO$_3$ is smaller than R_s of copper.

To pattern the YBCO films we used a conventional photolithographic process followed by wet chemical etching which provides high etching rates. With the intent of determining the geometrical limits of the wet chemical etching process we have structured microbridges of various dimensions on a film area of 1cm^2. The results are shown in Fig. 3, where T_c(R=0) of two films on randomly oriented Zr(Y)O$_2$ is plotted as a function of bridge width w. One film was patterned using PS80-04 solution and the other using EDTA. The T_c(R=0) of the unstructured films was 88.3K and 87.2K respectively with R(300K)/R(100K)=2.4. Both films had a thickness of 300nm. The T_c-degradation starts at 8µm using PS80-04 solution but not above 2µm using EDTA. No superconductivity was observed for w<2µm. It should be noticed that the film is not completely removed from the substrate by these wet-etching methods. We observed the formation of a residual insulating film which weakly adheres to the substrate but is very resistant to the used etchants.

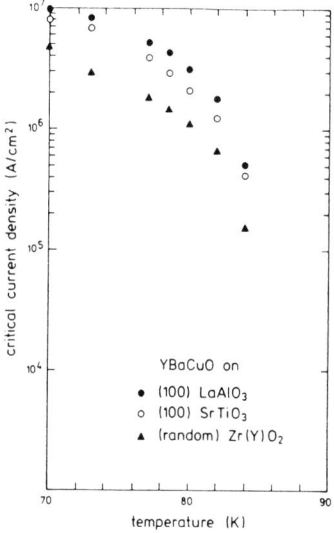

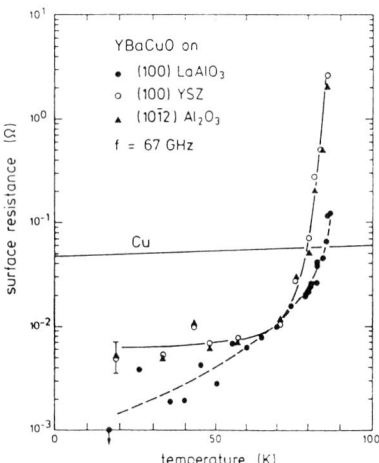

Figure 1. j$_c$ versus temperature in zero magnetic field
of YBCO films on (100) LaAlO$_3$,
SrTiO$_3$ and randomly oriented
Zr(Y)O$_2$.

Figure 2. Surface resistance
R$_s$ of YBCO films on (100)
LaAlO$_3$, Zr(Y)O$_2$ and r-plane
sapphire versus temperature
at a frequency of 67GHz.

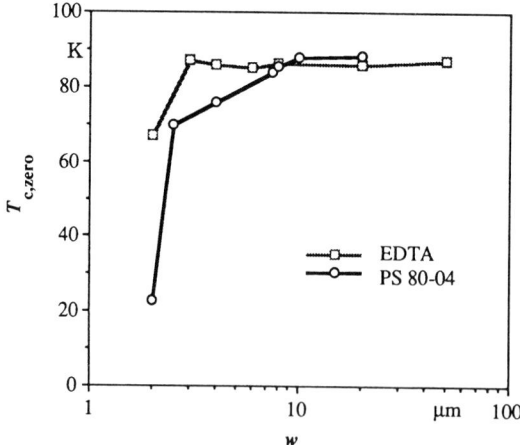

Figure 3. Variation of T$_c$(R=0) as a function of microbridge
width w for two YBCO films on randomly oriented Zr(Y)O$_2$ etched
with PS80-04 and EDTA.

4. SUMMARY

We have presented our results on c-axis oriented YBCO films grown by inverted cylindrical magnetron sputtering. The most remarkable surface irregularity is "outgrowths". We have demonstrated that their density and size depends on the growth conditions, especially the substrate temperature and the deposition rate. Despite of this feature we reproducibly patterned microbridges with EDTA to 3μm without any degradation of the superconducting properties.

Growth quality and transport properties of films on r-plane sapphire which is of interest for RF applications are poor so that suited buffer layers are needed. Due to its low microwave losses and the excellent YBCO film quality, (100) oriented $LaAlO_3$ is the favoured substrate material for microwave applications.

ACKNOWLEDGEMENT

We wish to thank G.Linker (Kernforschungszentrum Karlsruhe, Institut für Nukleare Festkörperphysik) for support in X-ray diffraction measurements. – This work was supported by the German Bundesministerium für Forschung und Technologie under the program High-Temperature Superconductors for Micro/Cryoelectronics.

REFERENCES

1 HTS Microwave Devices, in IEEE Trans. Magn.27(2)(1991)

2 J.Geerk, G.Linker, O.Meyer, Mat.Sci.Rep.4(5,6) (1989) 193

3 R.Schneider, G.Grabe, M.Hartlieb, Proc.Int.Cryogenic Materials Conf., ICMC 90, Garmisch-Partenkirchen, Mai 1990 (in print)

4 T.Kuhlemann and J.H.Hinken, IEEE Trans.Magn.27(2) (1991) 872

5 N.D.Kataria, D.May, H.Wolf, R.Schneider, J.Niemeyer, HTSC Thin Films: Prop.and Appl., Rome, April 1991 (to be published in Physica C)

6 R.Ramesh, A.Inam, D.M.Hwang, T.D.Sands, C.C.Chang and D.L.Hart, Appl.Phys.Lett. 58(14) (1991) 1557

High T$_c$ Superconductor Thin Films
L. Correra (Editor)

Chemistry of high T$_C$ superconductors by metallo-organic decomposition

Gustavo R. Paz-Pujalt

Corporate Research Laboratories, Eastman Kodak Company, Rochester, New York 14650-2011, USA

ABSTRACT

Metallo-organic decomposition (MOD) is a convenient non-vacuum technique for the deposition of high T$_c$ superconductor thin films, and other materials. It consists of coating a precursor solution, in this case metal carboxylates, containing the desired cations in the desired proportions onto a substrate followed by solvent removal and thermal decomposition. After the decomposition, an inorganic film consisting of metal oxides/carbonates is obtained. Upon thermal treatment this film undergoes solid state reactions to form the superconductor. Schemes for precursor preparation are presented and thermal decomposition products are discussed in terms of the cation's position in the periodic table. The solid state reactions preceding the formation of Y-Ba-Cu and Bi-Sr-Ca-Cu superconductors are contrasted.

1. INTRODUCTION

Various approaches have been used for the preparation of supercon-ductor thin films, each with its advantages and disadvantages. Metallo-organic decomposition (MOD) is a simple technique with low equipment requirements that permits excellent control of overall stoichiometry, high uniformity of thickness and composition, and the ability to coat irregular substrate shapes. Some of the disadvantages, however, are that it is a high temperature process and it appears to be limited to a thickness below 1 μm/layer.

MOD precursor solutions of metallo-organic compounds are mixed at the desired stoichiometry and the rheology is adjusted to fit the chosen deposition technique (spraying, dip or spin coating). The precursor solution is then deposited onto a substrate followed by thermal decomposition and thermal treatment to obtain the desired phase.

The precursors of choice are classified as metallo-organic because the metal, is linked to the organic groups by an atom other than carbon; most of the time oxygen. Examples of metallo-organic compounds are the metal alkoxides, metal carboxylates, and ß-diketonates.

Metal carboxylates are ideal precursors, because of their stability toward air and water, ease of preparation, low toxicity, high metal content, and amphiphilic nature. They consist of a nonpolar organic "tail" and a polar metal-oxygen "head". The organic tail imparts solubility in organic solvents that have a wide range of properties. This is very important for adjusting rheology. Because polar and nonpolar groups are present within the same molecule, metal carboxylates have special properties that make them particularly suitable for film formation.

2. PRECURSOR PREPARATION

The synthesis and chemistry of metal carboxylates has been extensively covered by Mehrotra, and Bohra [1]. The preparative methods may be grouped into two general categories.

2.1. Fusion

The carboxylic acid of choice reacts with an appropriate salt, oxide, or hydroxide of the metal under reflux conditions with solvent.

$$MX + 2RCOOH \longrightarrow M(OOCR)_2 + H_2X \tag{1}$$

where M is a divalent metal, X a suitable ligand (oxide, hydroxide, halide, carbonate, etc.), RCOOH is a carboxylic acid, and HX a reaction product.

2.2. Metathesis

In metathesis a lighter ligand is exchanged from a metal carboxylate or alkoxide, for example, by the desired carboxylic group. The liberated ligand is generally driven off thermally.

$$M(Oac)_2 + 2RCOOH \longrightarrow M(OOCR)_2 + 2HOac \tag{2}$$

In this example Oac represents the acetate group, and HOac acetic acid.

2.3. Experimental

Precursors were prepared for the systems Y-Ba-Cu and Bi-Sr-Ca-Cu.
The yttrium precursor was prepared by reacting yttrium carbonate with excess 2-ethylhexanoic acid (EHNA), and the barium precursor by the reaction of barium hydroxide with EHNA. Both reactions were carried out in toluene while refluxing continuously to drive off carbon dioxide and water, respectively. The copper precursor was obtained by the transcarboxylation of Cu(II) acetate with a sufficient amount of EHNA. The bismuth precursor was prepared by the reaction of bismuth oxide with excess EHNA. The calcium precursor was prepared by reacting a slurry of calcium carbonate with EHNA. Both reactions were carried out in toluene. The strontium precursor was commercial cyclohexane-butyrate from Eastman Chemicals.

All the precursors were filtered and analyzed by thermogravimetric (TGA) and inductively coupled plasma analyses for cation contents. The precursors were mixed in the desired ratios and rosin was added as an aid to film formation.

3. THERMAL DECOMPOSITION

Metal carboxylates generally decompose into the corresponding ketone and metal carbonate. The ketone and carbonate then decompose following their own thermodynamic characteristics.

$$M(OOCR)_2 \longrightarrow MCO_3 + R\overset{\displaystyle O}{\overset{\displaystyle \|}{C}}R \tag{3}$$

Lighter ketones tend to volatilize while heavier ones pyrolize into smaller fragments or undergo complete oxidation.

The stability of metal carbonates is determined by the basicity or electropositivity of the cation, the more electropositive the metal the more stable the carbonate. Thus alkali metals, alkaline earths, and lanthanides (including scandium, yttrium, and lanthanum) form stable carbonates that decompose at elevated temperatures. Transition metal carboxylates go through a carbonate stage that readily decomposes into the metal oxide and carbon dioxide. Noble and "noble-like" metal carboxylates decompose into the metallic form of the element.

4. RESULTS

Portions of the mixed precursors were decomposed in a hot plate and the resulting powders were used to study the solid state reactions preceding the formation of the superconductors.

4.1. Y-Ba-Cu system

Initially the powder consisted of yttrium and barium carbonates, which appeared to be in a solid solution, and copper oxide. After heating to 600°C the yttrium carbonate decomposed into the oxide and carbon dioxide.

$$\left.\begin{array}{l}\text{Y-precursor}\\\text{Ba-precursor}\\\text{Cu-precursor}\end{array}\right\} \text{---> } 1/2Y_2O_3 + 2BaCO_3 + 3CuO \tag{4}$$

Differential thermal analysis (DTA) of the oxide/carbonate mixture showed several endothermic peaks (Figure 1, Table 1). No intermediate compounds were detected by x-ray diffraction (XRD) as a function of the processing temperature. The formation of the 123-perovskite may be represented as a one step process accompanied by the evolution of carbon dioxide.

$$1/2Y_2O_3 + 2BaCO_3 + 3CuO \text{ ---> } YBa_2Cu_3O_{7-\delta} + 2CO_2 \tag{5}$$

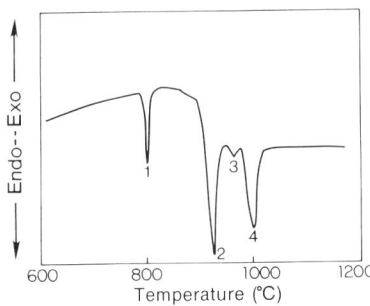

Figure 1. DTA profile of Y-Ba-Cu oxide/carbonate mixture derived by precursor decomposition.

Table 1
Peak assignments for DTA profile of Y-Ba-Cu system

Process	Peak #	T(°C)	Ref.
$BaCO_3$(ortho)---> $BaCO_3$(hex)	1	820°C	2
Formation of 123 perovskite	2	930°C	2
$BaCO_3$(hex) --> $BaCO_3$(cubic)	3	969°C	2
Melting of 123-perovskite	4	1000°C	2

4.2. Bi-Sr-Ca-Cu system

The powders obtained after decomposition of a $Bi_2Sr_2CaCu_2O_8$ (2212) precursor mixture consisted of a solid solution of strontium calcium carbonate, bismuth and copper oxides.

$$\left.\begin{array}{l}\text{Sr-precursor}\\\text{Ca-precursor}\\\text{Bi-precursor}\\\text{Cu-precursor}\end{array}\right\} Sr_{1-x}Ca_xCO_3 + CuO + Bi_2O_3 \tag{6}$$

Copper and bismuth oxides form an bimetallic oxide, at ~500°C, with a melting point of 840°C, whereas the double carbonate separates into $SrCO_3$, CaO and CO_2. This is followed by the reaction of $SrCO_3$ with the Cu-Bi-oxide intermediate to form a second intermediate

>700°C. This second intermediate reacts then with CaO and CuO to form the super-conductor composition 2212.

$$CuO + Bi_2O_3 \longrightarrow CuBi_2O_4 \tag{7}$$

$$Sr_{1-x}CaCO_3 \longrightarrow SrCO_3 + CaO + CO_2 \tag{8}$$

$$CuBi_2O_4 + 2SrCO_3 \longrightarrow Bi_2Sr_2CuO_6 + 2CO_2 \tag{9}$$

$$Bi_2Sr_2CuO_6 + CaO + CuO \longrightarrow Bi_2Sr_2CaCu_2O_8 \tag{10}$$

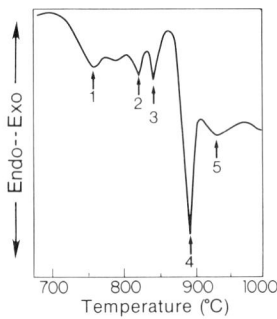

Figure 2. DTA profiles of Bi-Sr-Ca-Cu-oxide/carbonate powders derived by precursor decomposition.

The formation of the Cu-Bi- and Bi-Sr-Cu-oxide intermediates is independent of the starting stoichiometries. They appear to be the building blocks in the Bi-Sr-Ca-Cu-oxide family of superconductors. Figure 2 shows a DTA profile for this system.

Table 2
DTA assignments for the Bi-Sr-Ca-Cu system

Process	Peak #	T(°C)	Ref.
$CaCO_3 \longrightarrow CaO + CO_2$	Broad endotherm	~500°C	3
$\alpha\text{-}Bi_2O_3 \longrightarrow \gamma\text{-}Bi_2O_3$	1	770°C	3
Bi_2O_3 melting	2	817°C	3
$SrCO_3 \longrightarrow SrO + CO_2$	Broad endotherm	~750°C	3
$CuBi_2O_4$ melting	3	840°C	3
$Bi_2Sr_2CaCu_2O_8$ melting	4	890°C	3
$SrCO_3 \longrightarrow SrCO_3$ ortho --> hexagonal	5	925°C	3

5. THIN FILMS

Thin films have been deposited by MOD onto a variety of substrates with good results [4-8]. Patterning of MOD-derived films has also been reported [9]. XRD indicates single phase highly oriented films, and Rutherford Backscattering shows correct compositional profiles and channeling. Scanning electron microscopy and x-ray energy dispersive spectroscopy (EDS) showed uniform microstructure and composition. Transmission electron microscopy and electron diffraction measure-ments indicated an epitaxial relationship between film and substrate [10]. Figure 3 shows a resistivity versus temperature plot for a $Bi_2Sr_2CaCu_2O_8$ film on (100) $SrTiO_3$.

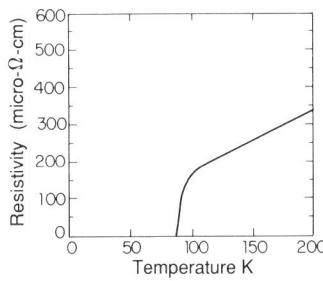

Figure 3. $Bi_2Sr_2CaCu_2O_8$ thin film on (100) $SrTiO_3$. Film thickness is 2 µm.

6. CONCLUSIONS

No intermediates are detected in the formation of 123-perovskite whereas 2 intermediates are identified in the formation of Bi-Sr-Ca-Cu oxides. The formation of Bi-Sr-Ca-Cu oxides appears to be aided by liquid phase transport at 840°C when the Bi-Cu-oxide intermediate melts. No oxygen annealing is necessary for Bi-Sr-Ca-Cu-type super-conductors whereas it is essential for the 123-perovskite. The formation of carbonate solid solutions is a clear indication of the "atomic level" mixing of cations attainable by the precursor decomposition method.

MOD is a versatile technique that produces adequate films for certain applications, and it lends itself well for coating HTcSC thin films on flexible substrates.

7. ACKNOWLEDGMENTS

This paper is dedicated to Professor Melvin E. Gleiter, of the University of Wisconsin Eau-Claire Chemistry Department, on occasion of his retirement. The author gratefully acknowledges contributions of Drs. John Agostinelli, G. Lubberts, L.S. Hung, and the members of the Ceramics Science Laboratory at Eastman Kodak Company.

8. REFERENCES

1   R.C. Mehrotra and R. Bohra, Metal Carboxylates (Academic Press, New York, (1983).
2   G.R. Paz-Pujalt, A.K. Mehrotra, S.A. Ferranti and J.A. Agostinelli, Solid State Ionics 32/33 (1989) 1179.
3   G.R. Paz-Pujalt, Physica C 166 (1990) 177
4   T. Kumagai, H. Yakota, K. Kawaguchi, W. Kondo and S. Mizuta, Chem. Lett. (Chem. Soc. Jpn.) (1987) 1645.
5   Jose M. Mir, John A. Agostinelli, D. L. Peterson, G. R. Paz-Pujalt, Brian J. Higberg, and G. Rajeswaran, Metalorganic Deposition Process for Preparing Superconducting Oxide Films, US Patent No. 4 880 770 (1989).
6   J.A. Agostinelli, G.R. Paz-Pujalt and A.K. Mehrotra, Physica C 156 (1988) 208.

7 L.S. Hung, J.A. Agostinelli, G.R. Paz-Pujalt, and Jose M. Mir, Appl. Phys. Lett. 53(54) (1988) 2450.
8 J.A. Agostinelli, G.R. Paz-Pujalt, A.K. Mehrotra, L.S. Hung, Metalorganic Deposition Process for the Deposition of Heavy Pnictide Superconducting Oxide Films, US Patent No. 4 950 643 (1990).
9 G. Lubberts, J. Appl. Phys. 68(2) (1990) 688.
10 S.-Tong Lee, S. Chen, L.S. Hung, and G. Braunstein, Appl. Phys. Lett. 55(3) (1989) 286.

Part III
Substrates & Multilayers

High T$_c$ Superconductor Thin Films
L. Correra (Editor)
© 1992 Elsevier Science Publishers B.V. All rights reserved.

Buffer layers on silicon for growth of YBCO films

D. K. Fork[a,b], G.A.N. Connell[a], F.A. Ponce[a], J.B. Boyce[a], and T.H. Geballe[b]

[a]Xerox Palo Alto Research Center, 3333 Coyote Hill Rd., Palo Alto, CA 94304, USA

[b]Dept. of Applied Physics, Stanford University, Stanford, CA 94305, USA

Abstract

Pulsed laser deposition is a powerful technique for the epitaxial growth of superconducting oxide films on numerous oxide substrates. Until recently, epitaxy on silicon substrates was made impossible by the chemical reactivity and thermal expansion mismatch of silicon and superconducting oxides. By interleaving an epitaxial buffer layer of yttria-stabilized zirconia in a fully *in-situ* laser deposition process, we have addressed both of these problems to grow c-axis oriented, epitaxial $Y_1Ba_2Cu_3O_{7-\delta}$ (YBCO) films on silicon. X-ray diffraction indicates an in-plane, epitaxial alignment within 2°. The films not only have high superconducting transition temperatures of 88 K but also have critical current densities of 2.2x10^6 A/cm^2 at 77 K. Transition widths are 1 K and normal state resistivities are 0.28 mOhm-cm at 300 K. Additional studies of YBCO on silicon-on-sapphire show that the limit on thickness of about 50 nm, set by thermal strain, for oxide films on bulk silicon can be overcome, and films with thicknesses far beyond this critical thickness have been grown. The surface resistance of these films at 4.2K and 11.8GHz is 72 µΩ cm and the critical current at 77 K is 4.6x10^6 A/cm^2. The values of the above parameters are now high enough to enable the fabrication of micro-electronic devices with superior performance.

1. OUTLINE

In this extended abstract, we review our work on the epitaxial growth of YBCO on silicon and silicon-on-sapphire using YSZ as an epitaxial buffer layer.

2. PULSED LASER DEPOSITION

All growths in this study were performed by pulsed laser deposition using a rotating polygon target system ('PolyGun') designed to accommodate ten different materials during a single deposition run. By triggering a 308 nm XeCl excimer laser to ablate a particular target with each revolution of the polygon, interfacial monolayers, films formed by atomic level mixing of numerous target materials, and multilayer stacks can all readily be deposited under computer control. We have used all of these growth techniques in the work reported here.

3. WHY CHOOSE YSZ AS A BUFFER LAYER ?

YSZ is an attractive buffer layer for numerous reasons: it is chemically inert both with silicon and YBCO at the growth temperature of 750 °C [1]; its cubic fluorite structure is related to the silicon structure and has a lattice constant that may be controlled by varying the yttria content [2]; and it provides a good silicon diffusion barrier when more than 20 nm thick, such that thicknesses well below the critical thickness for crack propagation from thermally induced strain can be used [3]. But previous work on the deposition of YBCO on single crystal YSZ points to a potential problem: the in-plane epitaxy of YBCO on YSZ is tristable with the YBCO a-axis either parallel to the YSZ<100>, parallel to the YSZ <110>, or broadly distributed about ±9° to the YSZ <100> [4]. Such grain boundaries will severely restrict the critical current attainable in the YBCO film [5]. We have shown that the occurrence of grains other than those with the a-axis parallel to the YSZ <110> is minimized by first growing a homoepitaxial YSZ layer on the YSZ and initiating the YBCO growth with a monolayer of CuO or Y_2O_3 [6]. We therefore have an effective process for growing high quality YBCO films on single crystal YSZ and epitaxial YSZ films on silicon.

4. EPITAXIAL GROWTH AND STRUCTURE OF YSZ AND YBCO ON SILICON

An essential step for the epitaxial growth of YSZ on silicon is the preparation of the silicon surface. In our work, the silicon wafers are spin cleaned and etched in flowing nitrogen first using high purity alcohol and then a 1:10:1 mixture of HF, ethanol, and water. This process produces an oxide-free, passivated wafer surface, covered by one monolayer of hydrogen, that remains uncontaminated during the passage of the wafer through the nitrogen-purged load lock into the deposition chamber [7]. Epitaxial growth of a 50 nm thick YSZ film is then achieved using an excimer laser energy density of 1-2 J/cm², a substrate temperature of 750 °C, and background oxygen pressure of 5x10⁻⁴ Torr. The epitaxial growth of YBCO follows immediately using the same laser energy density and substrate temperature, but with a background oxygen pressure of 200 mTorr [8].

The structural orientation and in-plane texturing of the YSZ and YBCO films have been examined by x-ray diffraction. The YSZ films have [001] perpendicular to the surface to within 0.8° and [100] parallel to silicon [100] to within 1.1°. This cube-on-cube epitaxy occurs with a lattice mismatch to silicon of almost 6%, but the ratio of the Rutherford backscattered yield along <100> to that in a random direction (X_{min}) of 5.3% is little greater than the 3% observed in high quality YSZ single crystals. There is strong dechannelling at the YSZ-silicon interface attributable to the large mismatch in the thermal expansion coefficients [9].

The YBCO films have their c-axes oriented perpendicular to the surface to within 0.8° and [110] parallel to YSZ [100] to within 2°. Thin films are placed under strain by the great difference in the thermal expansion coefficients of silicon and YBCO - the average ab-distance is expanded by 0.41% while the c-distance is contracted by a similar amount in 13 nm thick films. In films thicker than about 50 nm, the strain energy is relaxed by cracks occuring at about 1 μm spacings along <100> directions [10]. Cracking in thicker films can be overcome by using silicon-on-sapphire as the substrate [11]. In this case, the closer match of the thermal expansion coefficients of sapphire and YBCO allow films with thicknesses of at least 400 nm to be grown before strain relaxation by cracking occurs.

5. ELECTRICAL PROPERTIES OF YBCO ON SILICON AND SILICON-ON-SAPPHIRE

YBCO films grown in this study have a zero resistance at 88 K and a transistion width of less than 1 K. For films on silicon-on-sapphire, the critical current at 77 K is 4.6×10^6 A/cm^2 and the surface resistance at 11.8 GHz and 4.2 K is 72$\mu\Omega$. For films on silicon, the electrical properties depend strongly on film thickness. When cracking occurs, the critical current at 77 K drops precipitously to 1×10^5 A/cm^2 from a value of 2×10^6 A/cm^2 in thin films.

6. CONCLUSION

Using a straightforward *in-situ* deposition process, we believe that we have taken the first steps in developing the materials technology needed to design electronic circuits in which superconducting and semiconducting elements are integrated. We now plan to address the challenging materials issues associated with circuit delineation, cross-over and via fabrication, and active device design.

7. REFERENCES

1   D.B. Fenner, A.M. Viano, D.K. Fork, G.A.N. Connell, J.B. Boyce, F.A. Ponce, and J.C. Tramontana, J. Appl. Phys., 69 (1991) 2176.
2   I. Golecki, H.M. Manasevit, L.A. Moudy, J.J. Yang, and J.E. Mee, Appl. Phys. Lett., 42 (1983) 501.
3   M.S. Hu, M.D. Thouless, and A.G. Evans, Acta Metall., 36 (1988) 1301.
4   D.K. Fork, A. Barrera, T.H. Geballe, A. M. Viano, and D.B. Fenner, Appl. Phys. Lett., 57 (1990) 2504.
5   D. Dimos, P. Chaudhari, J. Mannhart, and F.K. Legouse, Phys. Rev., B41 (1990) 4038.
6   D.K. Fork, S.M. Garrison, F.A. Ponce, Marlyn Hawley, D.M. Wehner, and T.H. Geballe, J. Mats. Res. (to be published).
7   D.B. Fenner, D.K. Biegelsen, and R.D. Bringans, J. Appl. Phys., 66 (1989) 419.
8   D.K. Fork, D.B. Fenner, A. Barrera, J.M. Phillips, T.H. Geballe, G.A.N. Connell, and J.B. Boyce, IEEE Trans. Appl. Supercond., 1 (1991) 67.
9   L.J. Schowalter, R.W. Fathauer, R.P. Goehner, L.G. Turner, R.W. DeBlois, S. Hashimoto, J.L. Peng, W.M. Gibson, and J.P. Krusius, J. Appl. Phys., 58 (1985) 203.
10 D.K. Fork, D.B. Fenner, R.W. Barton, J.M. Phillips, G.A.N. Connell, J.B. Boyce, and T.H. Geballe, Appl. Phys. Lett., 57 (1990) 1161.
11 D.K. Fork, F.A. Ponce, J.C. Tramontana, N. Newman, J.M. Phillips, and T.H. Geballe, Appl. Phys. Lett., (to be published May 27, 1991).

High T_c Superconductor Thin Films
L. Correra (Editor)
© 1992 Elsevier Science Publishers B.V. All rights reserved.

Epitaxial growth of yttria-stabilized zirconia thin films on R-plane sapphire as buffer layers for HTSC films

F. Wang*, M. Siegel, R. Smithey, J. Geerk, G. Linker, and O. Meyer

Kernforschungszentrum Karlsruhe, Institut für Nukleare Festkörperphysik, P.O.B. 3640, W-7500 Karlsruhe, FRG

*On leave from Peking University, Beijing, China

Abstract

Yttria-stabilized zirconia (YSZ) films have been deposited on R-plane ($\bar{1}012$) sapphire substrates by inverted cylindrical rf magnetron sputtering from sintered targets with different yttria contents of 0.033 and 0.1, respectively. X-ray diffraction observations revealed epitaxial growth of the YSZ films in a substrate temperature range of 800-1100°C and under an oxygen pressure ranging from 1×10^{-4} to 6×10^{-2} mbar. The $(ZrO_2)_{0.967}(Y_2O_3)_{0.033}$ films were formed in a tetragonal phase while the $(ZrO_2)_{0.9}(Y_2O_3)_{0.1}$ films exhibited a cubic structure with the growth orientation of $YSZ(001)//Al_2O_3(\bar{1}012)$. The crystalline quality of the as-grown films could be significantly improved by post-annealing in air at 1300°C. The mosaic spread of the post-annealed films typically was 0.15° and the channeling minimum yield, X_{min}, in the Y and Zr sublattice reached values down to 4%. Epitaxial $GdBa_2Cu_3O_{7-x}$ thin films deposited on sapphire with YSZ buffer layers had $T_c(R=0)$ of 92.5 K and best $J_c(77\ K)$ values of 3×10^6 A/cm^2.

1. INTRODUCTION

Successful growth of epitaxial high-T_c superconducting films has attracted research interest directed towards applications in passive microwave devices such as filters, high-Q resonators, infrared bolometers, delay lines, etc. Sapphire is a highly desirable substrate for microwave device applications since it has a small dielectric constant and lowest rf losses of all the applicable substrates for high-T_c films [1]. Usually, however, "123" films directly grown on sapphire have poor T_c and J_c values [2] due to interdiffusion between film and substrate at the temperature required to form a good superconducting phase. Therefore, a suitable buffer layer on sapphire is necessary. Apart from the advantage in microwave applications, sapphire with an epitaxial buffer layer may be exploited for cheap substrates for high-T_c superconductors because, unlike $SrTiO_3$ and $LaAlO_3$, sapphire is available in large sizes at low cost.

Yttria-stabilized zirconia (YSZ) is one of the most promising substrates for the growth of high-T_c superconducting materials [3-5] and has been shown to be an excellent barrier material for the deposition of $YBa_2Cu_3O_{7-x}$ on Si [6,7] and Al_2O_3 [8]. The epitaxial growth of ZrO_2 and YSZ on Si and sapphire has been achieved by e-beam evaporation [6], ion sputter depostion [9], pulsed laser ablation [7] and rf sputtering [10]. Most recently, Wu et al. [8] have reported the successful growth of YBCO films on $YSZ/Al_2O_3(\bar{1}102)$ with T_c and J_c of up to 89 K and 1×10^6 A/cm^2 at

77 K, respectively. In this letter, we discuss the development of epitaxial tetragonal and cubic YSZ buffer layers on (1̄012) sapphire (R-plane) and the properties
of "123" films deposited on them.

2. BUFFER LAYER FORMATION

The growth of YSZ films was carried out by inverted cylindrical rf magnetron
sputtering using a 1×10^{-1} mbar mixture of argon and oxygen gases. The
deposition system has been described elsewhere [11]. Two $(ZrO_2)_{1-m}(Y_2O_3)_m$
targets containing traces of HfO_2 were used in the experiment with Y_2O_3 contents
of m=0.033 and 0.1, respectively. The substrates were heated during depositon
and the substrate temperature, T_s, was monitored by a thermocouple. The
sputtering rate was 1.0-1.3 Å/s. Typical film thicknesses were around 2000 Å. The
films were characterized by X-ray diffraction (2-axis goniometers and Seemann-
Bohlin geometry) and Rutherford backscattering spectrometry.

Prior to deposition, the Al_2O_3 substrates were cleaned in ethanol, and then
heated to 1200°C in air for 3 hours. This predeposition annealing was found to be
the most important step in obtaining good epitaxial YSZ films. It resulted in a
significant improvement in the crystalline quality of the Al_2O_3 substrate, as
indicated by RBS channeling results. After annealing, the surface peak area
decreased dramatically and the minimum yield, X_{min}, behind the surface peak for
the Al sublattice decreased from 40% to 1.5%, the latter value corresponding to an
essentially perfect single crystal.

Fig. 1 shows a typical diffraction pattern of a $(ZrO_2)_{0.967}(Y_2O_3)_{0.033}$ film grown
at 950°C under 1×10^{-2} mbar oxygen. It can be seen that only the (00ℓ) and $(\ell00)$
type peaks appeared in the figure, indicating the presence of the tetragonal phase
and the highly textured growth with either the c- or a-axis perpendicular to the

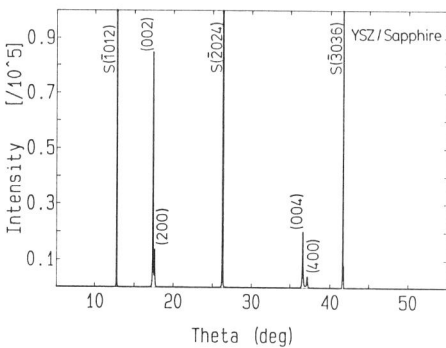 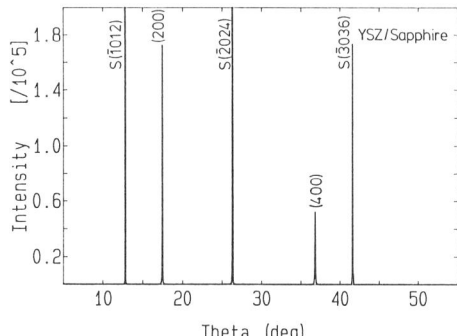

Figure 1.   X-ray diffraction pattern
of a $(ZrO_2)_{0.967}(Y_2O_3)_{0.033}$ film grown
at 950°C on R-plane sapphire under
1×10^{-2} mbar oxygen. The lattice
parameter a is 5.103 Å, b is 5.167 Å.

Figure 2.   X-ray diffraction pattern of a
$(ZrO_2)_{0.9}(Y_2O_3)_{0.1}$ film grown at 950°C
on R-plane sapphire under 1×10^{-2} mbar
oxygen. The lattice parameter a is
5.14 Å.

substrate surface. Although the $(ZrO_2)_{0.967}(Y_2O_3)_{0.033}$ films grew highly textured, the rocking curve widths of the YSZ(002) and (200) peaks, which characterize the spread of the (002) and (200) grains, respectively, were still large ranging from 1.2° to 0.6° for the as-deposited films. In addition, we have found that the spread of the (200) grains was always larger than the corresponding values of the (002) grains. This implies that the epitaxial imperfection was largely caused by preferentially oriented tetragonal YSZ(100) grains. Thus an improvement in the crystalline quality of the YSZ films has been expected if the films are in the cubic phase.

We therefore have deposited YSZ films from a target with a higher Y_2O_3 content. Fig. 2 shows typical X-ray diffraction spectra for a $(Y_2O_3)_{0.1}(ZrO_2)_{0.9}$ film grown at T_s of 950°C under an oxygen pressure of 1×10^{-2} mbar. In contrast to Fig. 1, no splitting of the {001} type peaks could be observed indicating that the film was formed in the cubic phase as anticipated. The oxygen partial pressure dependence of the epitaxial cubic YSZ formation on T_s is similar to the case of tetragonal YSZ. The number of the (111) oriented YSZ grains decreased with increasing T_s and pure (001) oriented epitaxial growth could be obtained in the temperature range of 800-1100°C and O_2 pressures ranging from 1×10^{-4} to 6×10^{-2} mbar. The mosaic spread for the (002) peak of the as-deposited cubic films of typically 0.7° was still large but could be highly improved by a post-annealing procedure as discussed below.

Fig. 3 shows the channeling minimum yield values of several (001) oriented cubic YSZ samples before and after a 1h, 1300°C anneal plotted as a function of the original growth temperature. In all cases, the X_{min} values of the annealed films were improved, but the improvement was most striking for films grown under non-ideal conditions. We have found that epitaxial YSZ films with good crystalline quality could be obtained through such annealing even if the as-grown films contained (111) oriented grains. The ability to improve films with initially poor epitaxy relaxes the stringent growth conditions during deposition. The mosaic

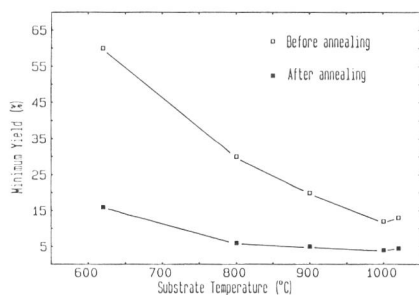

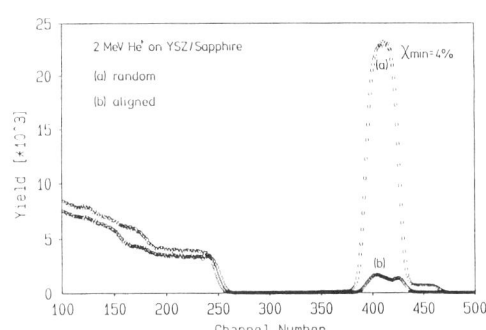

Figure 3. RBS minimum yield of $(ZrO_2)_{0.9}(Y_2O_3)_{0.1}/Al_2O_3$ before and after a 1h, 1300°C annealing as a function of substrate temperature during deposition.

Figure 4. Random and aligned backscattering spectra with 2 MeV $^4He^+$, of a post-annealed $(ZrO_2)_{0.9}(Y_2O_3)_{0.1}$ film grown on R-plane sapphire.

spread determined at the YSZ(002) peak for the post-annealed films was typically 0.15°, showing the highly oriented nature of the films. Rutherford backscattering spectrometry (Fig. 4) also indicated a high degree of crystalline perfection with a channeling minimum yield in the Zr(Y) sublattice as low as 4%, being the best value so far reported for the YSZ/Al$_2$O$_3$ system.

3. GdBa$_2$Cu$_3$O$_{7-x}$ FILMS

GdBa$_2$Cu$_3$O$_{7-x}$ films have been deposited on Al$_2$O$_3$ with epitaxial YSZ buffer layers using inverted cylindrical dc magnetron sputtering. The deposition procedure has been described elsewhere [11]. The GdBa$_2$Cu$_3$O$_{7-x}$ films on cubic YSZ buffer layers were of higher quality than those on tetragonal buffer layers. The best films had a zero resistance temperature of 92.5 K and a resistivity of 70 µΩcm at 100 K. The critical current densities of the best films were 3×10^6 A/cm^2 at 77 K which is superior to results for films which we have grown directly on YSZ (100) single-crystal substrates [4]. Fig. 5 shows the X-ray diffraction pattern of a GdBa$_2$Cu$_3$O$_{7-x}$ film grown on Al$_2$O$_3$ with a cubic (ZrO$_2$)$_{0.9}$(Y$_2$O$_3$)$_{0.1}$ buffer layer. One observes only c-axis GdBa$_2$Cu$_3$O$_{7-x}$ peaks and the YSZ(00ℓ) and R-plane Al$_2$O$_3$ peaks. The full width at half maximum of the ω-scan across the GdBa$_2$Cu$_3$O$_{7-x}$ (005) peak was 0.5°, indicating the highly oriented nature of the films. Fig. 6 shows aligned and random backscattering spectra for a GdBa$_2$Cu$_3$O$_{7-x}$ film grown on a cubic YSZ buffer layer. The X_{min} is 17%. More details of the growth of superconducting thin films on buffer layers are given in the contribution of M. Siegel et al., this conference proceedings.

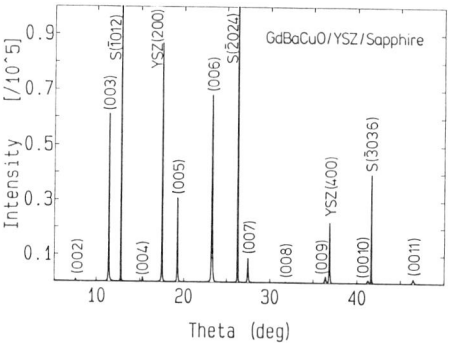

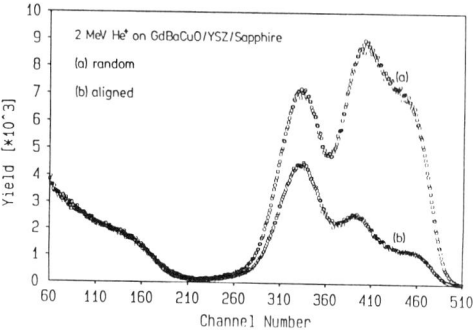

Figure 5. Diffraction pattern from a GdBa$_2$Cu$_3$O$_{7-x}$ film deposited on Al$_2$O$_3$($\bar{1}$012) with an epitaxial (ZrO$_2$)$_{0.9}$(ZrO$_2$)$_{0.1}$ buffer layer. The GdBa$_2$Cu$_3$O$_{7-x}$ film shows pure c-axis oriented growth.

Figure 6. Energy spectra of <001> aligned and random 2 MeV ^{4}He$^+$ ions backscattered from a GdBa$_2$Cu$_3$O$_{7-x}$ film with a (ZrO$_2$)$_{0.9}$(Y$_2$O$_3$)$_{0.1}$ buffer layer on sapphire. The channeling minimum yield for Gd is 17%.

4. SUMMARY

We have grown tetragonal and cubic YSZ films on R-plane sapphire substrates using inverted cylindrical rf magnetron sputtering. X-ray diffraction observations revealed the epitaxial growth of tetragonal $(ZrO_2)_{0.967}(Y_2O_3)_{0.033}$ and of cubic $(ZrO_2)_{0.9}(Y_2O_3)_{0.1}$ films on sapphire with preference of the YSZ(001) // $Al_2O_3(\bar{1}012)$ orientation. The quality of the epitaxial YSZ could be improved by post-annealing in air at 1300°C. RBS channeling measurements indicated a high degree of crystalline perfection with a channeling minimum yield for the Zr(Y) sublattice as low as 7% and 4% for the tetragonal and cubic phases, respectively. $GdBa_2Cu_3O_{7-x}$ films grown on YSZ buffer layers had properties comparable to those deposited on YSZ (100) single-crystal substrates.

5. REFERENCES

1   R.W. Simon, in "Proceedings of SPIE, Processing of Films for High-T_c Superconducting Electronics", edited by T. Venkatesan (SPIE, Bellingham, WA, 1989), SPIE Vol. 1187, p. 1.

2   A. Stamper, D.W. Greve, D. Wong, and T.E. Schlesinger, Appl. Phys. Lett. 52 (1988) 1746.

3   G. Koren, E. Polturak, B. Fisher, D. Cohen, and G. Kimel, Appl. Phys. Lett. 53 (1988) 2330.

4   Q. Li, O. Meyer, X.X. Xi, J. Geerk, and G. Linker, Appl. Phys. Lett. 55 (1989) 2351.

5   R.K. Singh, L. Ganapathi, P. Tiwari, and J. Narayan, Appl. Phys. Lett. 55 (1989) 2351.

6   H. Myoren, Y. Nishiyama, H. Fukumoto, H. Nasu, and Y. Osaka, Jpn. J. Appl. Phys. 28 (1989) 351.

7   D.K. Fork, D.B. Fenner, R.W. Barton, Julia M. Philips, G.A.N. Connel, J.B. Boyce, T.H. Geballe, Appl. Phys. Lett. 57 (1990) 1161.

8   X.D. Wu, R.E. Muenchausen, N.S. Nogar, A. Pique, R. Edwards, B. Wilkens, T.S. Ravi, D.M. Hwang, and C.Y. Chen, Appl. Phys. Lett. 58 (1991) 304.

9   P. Legagneux, G. Garry, D. Dieumegard, C. Schwebel, C. Pellet, G. Gantherin, and J. Siejka, Appl. Phys. Lett. 53 (1988) 1506.

10   F. Konushi, T. Doi, H. Matsunaga, Y. Kakihara, M. Koba, K. Awane, and I. Nakamura, Mat. Res. Soc. Symp. Proc. 56 (1986) 259.

11   X.X. Xi, G. Linker, O. Meyer, E. Nold, B. Obst, F. Ratzel, R. Smithey, B. Strehlau, F. Weschenfelder, and J. Geerk, Z. Phys. B., Condensed Matter 74 (1989) 13.

High T$_c$ Superconductor Thin Films
L. Correra (Editor)
© 1992 Elsevier Science Publishers B.V. All rights reserved.

Epitaxial YBa$_2$Cu$_3$O$_7$ - Films on Silicon with Epitaxial Buffers of Yttrium-Stabilized Zirconia

W. Prusseit, S.Corsépius, F.Baudenbacher, K.Hirata, P.Berberich and H.Kinder

Physik-Department E10, Technischen Universität München, James-Franck-Straße, D-8046 Garching, Germany

Abstract

For the preparation of very smooth epitaxial YBa$_2$Cu$_3$O$_7$ - films on silicon substrates, Yttrium-stabilized zirconia (YSZ) buffer layers turn out to serve as effective barriers against interdiffusion and show an epitaxial relationship with the silicon (100) single-crystal as well as the superconductor. To accomplish good epitaxial growth of YSZ on silicon the amorphous native oxide has to be removed by a careful chemical etching prodecure. To prevent the reactive surface from contamination it was passivated by a thin oxide layer which decomposes just when the process temperature of 800 °C is approached. YSZ - films evaporated by electron gun on so prepared substrates showed an excellently smooth and crystalline surface.

These buffers served as the basis for the subsequent YBa$_2$Cu$_3$O$_7$ deposition by reactive thermal co-evaporation. YSZ as well as YBa$_2$Cu$_3$O$_7$ - films were characterized by RHEED, XRD, SEM and TEM studies. Besides their good superconducting properties (Tc $\approx$ 86 K) our YBa$_2$Cu$_3$O$_7$ - films exhibit an extraordinarily smooth surface which is necessary for applications in microelectronic devices.

1. INTRODUCTION

Although thin films of the high temperature superconductor YBa$_2$Cu$_3$O$_7$ with excellent electronic properties can be grown on a variety of substrates, e.g. SrTiO$_3$, MgO, LaAlO$_3$ etc., it would be of superior technological interest to use silicon as basic substrate material. On the one hand there exists a highly developed technology to produce high quality crystals on the other hand future applications will probably require HTSC material integrated next to microelectronic devices on the same chip. Unfortunately the combination of cuprate superconductors like YBa$_2$Cu$_3$O$_7$ and silicon bears difficulties related with the high chemical reactivity of silicon and copper at typical process temperatures. Silicon in YBa$_2$Cu$_3$O$_7$ severely degrades the superconducting properties, and, conversely, copper diffuses rapidly in silicon and affects its semiconductor properties. At present it is widely accepted that only an appropriate buffer layer which effectively prevents interdiffusion may solve the problem. Furthermore this barrier should be grown

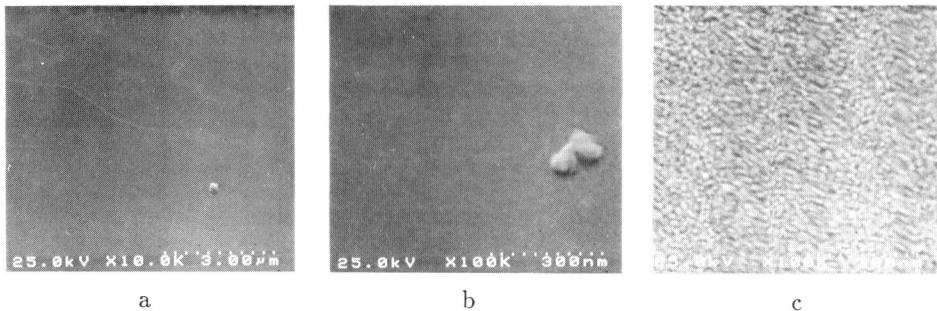

Figure 1. SEM pictures of two YSZ films of different quality. Smooth film at a magnification of 10^4 (a), smooth film at 10^5 (b) and grainy film at 10^5 (c).

epitaxially to allow epitaxial growth of the $YBa_2Cu_3O_7$ as well. Yttrium-stabilized zirconia (YSZ) appears to be a candidate of good choice because it seems to meet all these requirements [1—3].

2. PREPARATION AND GROWTH OF YSZ BUFFER LAYERS

To achieve epitaxial growth of YSZ on silicon it is absolutely necessary to remove the relatively thick (15 Å) and contaminated amorphous native oxide from the surface. Therefore the silicon substrate is prepared by a beforehand two-step chemical treatment. For details of this process see Ref. 4. In the first step the native oxide is removed by spin etching with several drops of a dilute alcoholic HF solution. The second step consists of a dip-oxidation in a concentrated H_2SO_4-H_2O_2 mixture and provides a very thin and clean oxygen passivation which is stable enough to persist up to temperatures of about 700 °C. Above this temperature the dioxide transforms to volatile SiO [5] leaving a proper silicon surface for the subsequent evaporation. Although other groups have reported on successful YSZ deposition without any surface treatment we made the experience that no smooth YSZ films were achievable without spin etching.

At 800 °C substrate temperature the YSZ buffer layer was evaporated by an electron gun in an oxygen background pressure of $2 \cdot 10^{-4}$ mbar. Best results were obtained at deposition rates of 0.5 to 1 Å/s. Typically the thickness of the films was lying in the range of 500 to 800 Å. To prove the excellent crystallinity and smoothness the films were characterized by RHEED, SEM, XRD and TEM studies [4]. In Figure 1 the surface morphology of a high and a low quality YSZ film is compared. The pictures were taken with a high resolution SEM (field emission tip) under a tilt angle of 40°. The photographs in Figure 1a und b show parts of a smooth film on different scales. Except from a focussing object no surface structure could be resolved. In contrast the film of poor quality (Figure 1c) exhibits a distinct grainy morphology. The YSZ is epitaxial and [100]-aligned with the silicon substrate as shown in the lower trace of Figure 2.

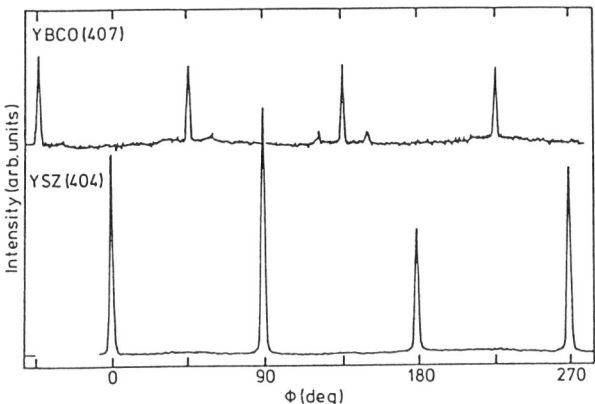

Figure 2. ϕ-scan of the YSZ (404)-peak (lower trace) and the YBa$_2$Cu$_3$O$_7$ (407)-peak (upper trace) indicating good epitaxy and the relative alignment. The small side-peaks in the upper trace correspond to silicon (533) index planes.

3. YTTRIA - AND YBa$_2$Cu$_3$O$_7$ - DEPOSITION

After characterization of the YSZ buffer layers, the samples were mounted in the chamber for thermal co-evaporation of YBa$_2$Cu$_3$O$_7$ [6]. As a first step we deposited a very thin layer of Y$_2$O$_3$ ($\approx$ 50 Å) [7] by using only the yttrium-boat under the same conditions as described previously for the YBa$_2$Cu$_3$O$_7$ - films. Then the usual YBa$_2$Cu$_3$O$_7$ - process was carried out. The growth of both, yttria and YBa$_2$Cu$_3$O$_7$, was observed by RHEED. The sequence in Figure 3 gives an impression of how the epitaxy proceeds starting from the YSZ buffer and ending at the final thickness of YBa$_2$Cu$_3$O$_7$. Although the RHEED pattern of the YSZ buffer layer appears to be streaky, it is not yet atomically smooth (a). After deposition of about 50 Å of Y$_2$O$_3$ the pattern has become very spotty (b). We attribute this to a reconstruction of the surface where yttria forms steps of 10.6 Å height (1 unit cell). These steps in turn probably serve as nucleation centers for a rapid lateral YBa$_2$Cu$_3$O$_7$ - growth which finds its manifestation in the streaky appearance even after deposition of only 10 Å of YBa$_2$Cu$_3$O$_7$ (c). After 60 Å we can see a well developed layer by layer growth of the superconductor (d). After 120 Å streaks have become even sharper and initial spots have totally disappeared (e). From this stage of growth no further change could be observed until the final thickness of 1200 Å was reached (f). Both RHEED and XRD (see upper trace of Figure 2) indicate that YBa$_2$Cu$_3$O$_7$ is rotated by 45° with respect to the underlying silicon , YSZ and Y$_2$O$_3$.

In contrast to these sharp and streaky patterns of YBa$_2$Cu$_3$O$_7$ on yttria intermediate layers the growth of YBa$_2$Cu$_3$O$_7$ directly on the YSZ buffer can be described as rather disordered. Spotty and partially ring-shaped RHEED patterns of those films indicate a high degree of misorientation. This behaviour may be due to a degradation of the YSZ surface under atmospheric conditions [8]. The resulting YBa$_2$Cu$_3$O$_7$ - surfaces are compared in the SEM pictures of Figure 4. Without yttria the surface is strongly

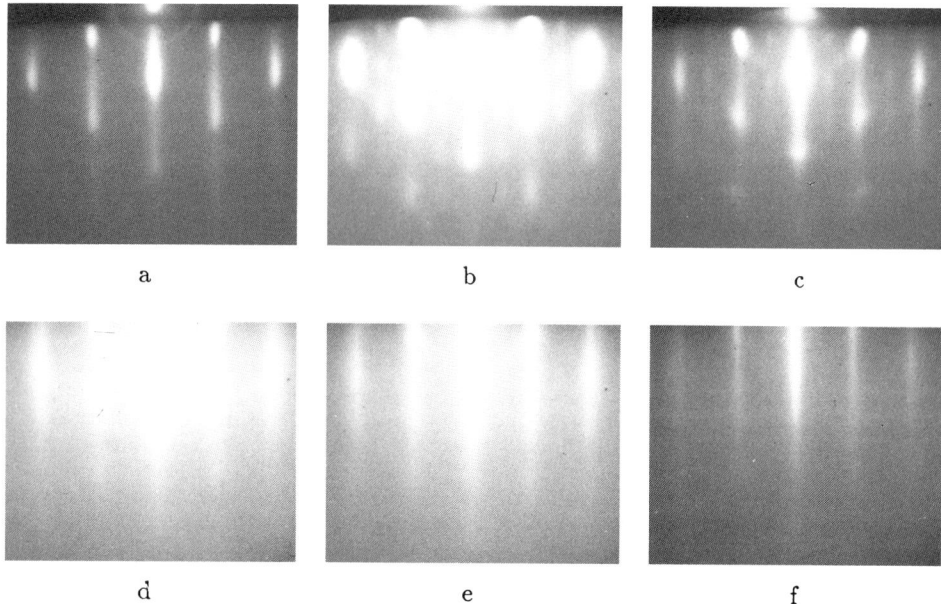

Figure 3. RHEED patterns of different stages of growth: YSZ buffer layer in [110]-direction before evaporation (a), after deposition of 50 Å Y_2O_3 (b), after $YBa_2Cu_3O_7$ deposition of 10 Å (c), 60 Å (d), 120 Å (e) and 1200 Å (f). See text for explanation.

disordered (a). In contrast, the $YBa_2Cu_3O_7$ - film with the intermediate yttria layer appears to be very smooth on a large scale (b) as well as on a small scale (c), *without any outgrowth* and *without any cracks*. These films exhibit good electronic properties. T_c (zero resistance) is typically 86 K with an onset at 91 K and the resistive transition curve extrapolates well to zero. The specific resistance at 100 K is about $\rho_{DC}(100K)$ $\approx 100\mu\Omega$cm. Critical currents of smooth films could not yet been measured because of technical problems with our structuring. We have only a result, $5 \cdot 10^5$ A/cm^2 at 77 K, for a film made by a less optimized process. On this film we observed microcracks by FE-SEM, indeed.

4. CONCLUSIONS

For the fabrication of microelectronic devices incorporating HTSC material two basic requirements have to be met: good superconducting properties on a cheap and common substrate like silicon and a well defined smooth surface of the superconductor to avoid shorts. We were able to satisfy both conditions by employing a YSZ buffer layer grown on carefully etched and passivated silicon substrates. Degradation problems which may arise when the buffer is exposed to room air could be solved by evaporation of some Angstøms of yttria immediately before the $YBa_2Cu_3O_7$ deposition. Thereby

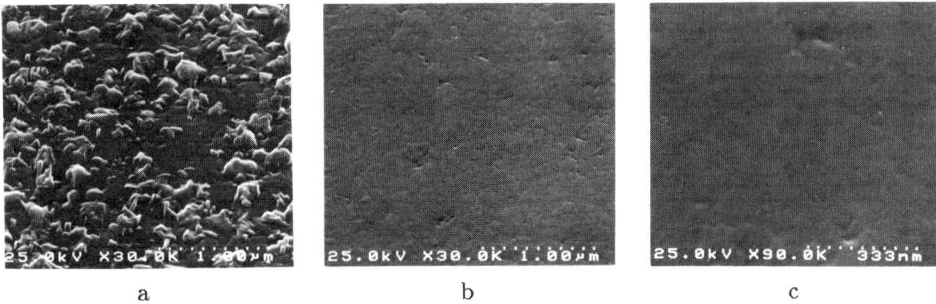

Figure 4. SEM pictures of YBa$_2$Cu$_3$O$_7$ deposited directly on YSZ with a magnification of $3 \cdot 10^4$ (a) and with an intermediate yttria layer magnified by a factor of $3 \cdot 10^4$ (b) and $9 \cdot 10^4$ (c).

we observed the astonishing feature that YBa$_2$Cu$_3$O$_7$ tends to compensate the initial roughness and we arrived at nearly atomically smooth YBa$_2$Cu$_3$O$_7$ - films without segregations or cracks. This fact and their electronic properties make them comparable to high quality films we obtained on MgO.

ACKNOWLEDGEMENTS

We wish to thank M.Zwerger and O.Eibl for providing the TEM studies on our films. The work was supported by the Bundesminister für Forschung und Technologie under contract No. TK0338 .

REFERENCES

1. H.Fukumoto, T.Imura, Y.Osaka, Jpn.J.Appl.Phys. **27** (1988) L1404
2. H.Myoren, Y.Nishiyama, H.Fukumoto, H.Nasu, Y. Osaka, Jpn.J.Appl.Phys. **28** (1989) 351
3. D.K.Fork, D.B.Fenner, G.A.N.Connell, J.M.Phillips, T.H.Geballe Appl.Phys.Lett. **57** (1990) 1161;   D.K.Fork, D.B.Fenner, A.Barrera, J.M.Phillips, T.H.Geballe, G.A.N.Connell, J.B.Boyce, IEEE Trans.Appl.Supercond. **1** (1991) 67
4. To be publ. in Physica C: Proc.Int. Workshop on HTSC thin films, Rome, 1991
5. Y.Kobayashi, Y.Shinoda, K.Sugi, Jpn.J. Appl.Phys. **29** (1990) 1004
6. F.Baudenbacher, H.Karl, P.Berberich, H.Kinder, J.Less Common Met. **164 & 165** (1990) 269
7. H.Myoren, Y.Nishiyama, N.Miyamoto, Y.Kai, Y.Yamanaka, Y. Osaka, F.Nishiyama, Jpn.J.Appl.Phys. **29** (1990) L955
8. D.K.Fork, A.Barrera, T.H.Geballe, A.M.Viano, D.B.Fenner Appl.Phys.Lett. **57** (1990) 2504

High T$_c$ Superconductor Thin Films
L. Correra (Editor)
623

Epitaxial yttria stabilized zirconia on Si(100) by evaporation - an XPS and LEED study

H. Behner, J. Wecker, B. Heines

Siemens AG, Corporate Research Laboratories, Paul Gossen Str. 100, D-8520 Erlangen, FRG

Abstract

The growth of thin films of yttria stabilized zirconia (YZO) on Si(100) surfaces by e-beam evaporation was studied by means of XPS and LEED. The occurrence of metallic Zr in the films as well as the formation of a reaction zone at the interface Si/YZO depending on the oxygen partial pressure during deposition is shown. As a result of the high formation enthalpy of ZrO_2 it was possible to obtain an epitaxial growth of the YZO on Si(100) samples without removing the native oxide before the deposition process. Single crystalline YBCO films could be deposited with laser ablation onto Si/YZO(40 nm)/Y_2O_3(40 nm) multilayer structures.

1. INTRODUCTION

With the application of high temperature superconductor (HTSC) thin films a variety of liquid nitrogen cooled electronic devices with new properties can be established. Passive devices such as antennas and resonators for HF-applications, inter- and intrachip connections of integrated circuits and active devices such as SQUID's or Josephson logic circuits are examples.

In order to combine high quality HTSC thin films with the silicon technology it is necessary to overcome the well known reaction and diffusion problems [1] which occur at the elevated substrate temperatures by the direct deposition of $YBa_2Cu_3O_{7-\delta}$ (YBCO) films onto Si surfaces and which cause a severe degradation of the superconducting properties. Only with the help of buffer layers, e. g. epitaxially grown films of yttria stabilized zirconia (YZO) it has been shown that high current densities in the order of $j_c > 10^6$ A/cm^2 at 77K in the YBCO layer can be achieved [2-3].

In this work we investigated the deposition of YZO thin films on Si(100) by means of X-ray photoelectron spectroscopy (XPS) and low energy electron diffraction (LEED). We will show results concerning the properties of the YZO films and the YZO/Si-interface in dependence on various deposition parameters with a special focus on the pretreatment of the silicon surface. In particular we will demonstrate that it is possible to grow epitaxial YZO films without removing the amorphous oxide from the Si substrates before the deposition process.

2. EXPERIMENTAL

The experiments were carried out in an UHV system, which consists of two sputter chambers and one evaporation chamber connected by a transfer tube with a Kratos XSAM 800 surface analysis system with facilities for XPS, LEED and scanning auger microscopy (SAM). Sample transfer can be performed at a base pressure of $p < 5 * 10^{-9}$ mbar.

Buffer layers of YZO $((ZrO_2)_{91}(Y_2O_3)_9)$ and Y_2O_3 were grown in a Balzers UMS 630 system by e-beam evaporation from sintered targets. The base pressure was typically $< 10^{-8}$ mbar and the growth of the oxide layers was studied at oxygen partial pressures up to $5 * 10^{-5}$ mbar. The oxygen was introduced through a tube near the substrate. The substrate temperature and the evaporation rates were kept constant at 730 °C and 0.1 Å/s, respectively.

The Si(100) substrates were either cleaned by sputtering with a fast atom beam (Ion Tech FAB 114) source (denoted as precleaned substrates) or were simply rinsed in acetone and ethanol before loading (denoted as uncleaned substrates).

XP-spectra were obtained by using Mg K_α-radiation with the hemispherical electron energy analyzer operated in the high magnification mode at a constant pass energy of 20 eV, which results in a resolution of the Ag $3d_{5/2}$ line of 0.9 eV full width at half maximum (FWHM). The binding energies are referred to the Au $4f_{7/2}$ emission at 83.8 eV and the "adventitious" C 1s emission of adsorbed hydrocarbons at 285.0 eV.

For characterizing the interface reaction and the first stages of epitaxy of the YZO layer by means of XPS and LEED we deposited films of a thickness between 1 and 10 nm at various oxygen partial pressures. Under these conditions and with an inelastic mean free path of 1 keV photoelectrons (e.g. Si 2p, Zr 3d) of about 2.5 nm [4] the sampling depth of XPS is large enough to get information of the YZO film, the interface region and the substrate as well.

3. RESULTS AND DISCUSSION

3.1. Precleaned Si(100) substrates

In order to get a well ordered and oxide free Si surface for epitaxy the substrates were cleaned by means of a FAB treatment for 30 min at an Ar pressure of $5 * 10^{-3}$ mbar, a discharge voltage of 1.5 keV and a current density equivalent of about 70 $\mu A/cm^2$. This procedure resulted in a heavily damaged surface showing no LEED pattern. After heating the substrates to 730 °C, the LEED experiment exhibited a sharp pattern of a reconstructed Si(100) 2x1 surface and no contamination could be detected with XPS. The precondition to obtain this well defined Si surface was a carefully baked system with a base pressure with *hot* substrate heater of $p < 2 * 10^{-8}$ mbar, otherwise a drastic recontamination of the substrate (XPS revealed the formation of oxidic and carbidic spezies on the Si surface) was observed.

A typical feature during e-beam evaporation of YZO and Y_2O_3 is the dissociation of the oxides into species with lower oxidation states, e.g. ZrO and metallic Zr. The metallic components can even be detected in the film if it is grown under low oxygen partial pressures. Figure 1 compares the Zr 3d emission after deposition of 1 nm YZO on a precleaned Si surface at base

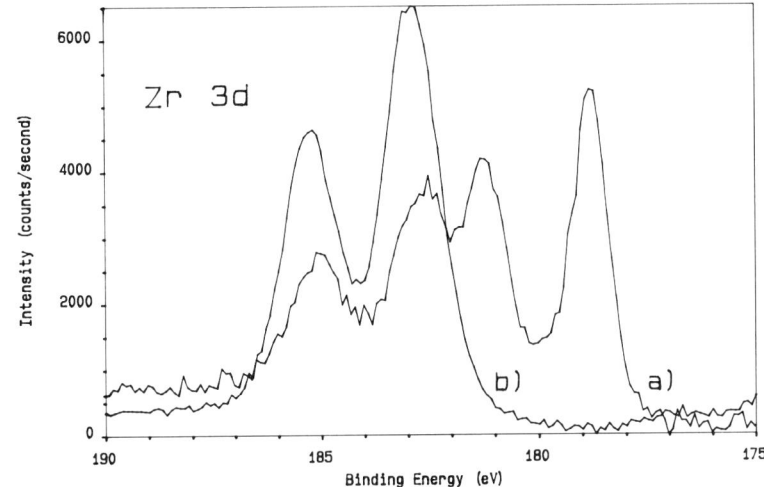

Figure 1. Zr 3d spectra of 1 nm YZO on precleaned Si(100). a) deposited at base pressure, b) deposited at $3*10^{-6}$ mbar O_2.

pressure (curve a)) with the Zr 3d signal of a 1 nm YZO film deposited at an oxygen pressure of $3*10^{-6}$ mbar (curve b)). In curve a) the existence of metallic Zr ($3d_{5/2}$ at 178.8 eV) besides oxidized Zr ($3d_{5/2}$ at 182.7 eV) is evident. As long as metallic Zr was present, a 1x1 overstructure with increased background intensity could be observed in the LEED experiment. With increasing oxygen pressure during deposition the amount of the metallic component was reduced and at an oxygen pressure of $p > 3*10^6$ mbar only the signal of ZrO_2 ($3d_{5/2}$ at 183.0 eV) could be detected.

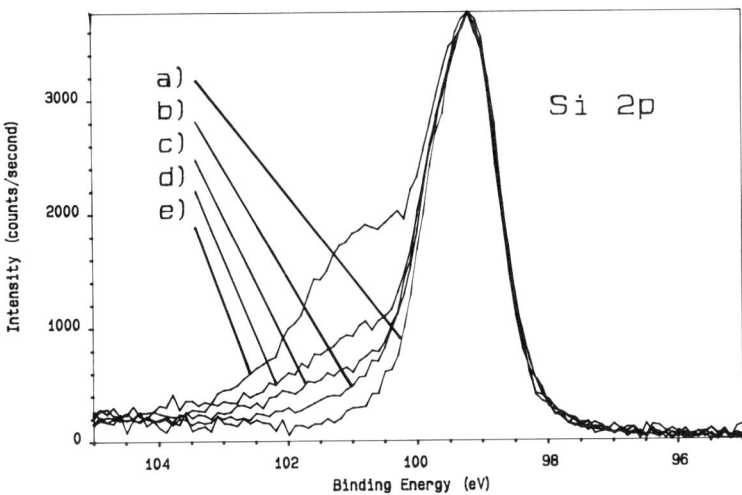

Figure 2. Si 2p spectra of a) clean Si(100) and after deposition of 1 nm YZO at: b) base pressure, c) $5*10^{-7}$ mbar O_2, d) $1*10^{-6}$ mbar O_2, e) $3*10^{-6}$ mbar O_2

On the other hand the admission of oxygen during the first stages of the film growth led to the formation of a reaction zone between the silicon surface and the YZO layer. This can be seen in Figure 2, where a comparison of the Si 2p emission of a clean Si surface (curve a)) with the Si 2p signals after deposition of 1 nm YZO at increasing oxygen pressures (curves b)-e)) is shown. With increasing oxygen pressure during growth an additional Si 2p line at a binding energy of 101.5 eV shows up. This line is assigned to the formation of a silicate at the Si/YZO interface according to the reaction : $Si + O_2 + ZrO_2 \rightarrow SiZrO_4$.

Too high oxygen partial pressures during the deposition of the first nanometers as well as growing conditions under which more than 10 % metallic Zr were observed always led to nonepitaxial growth of the YZO films. Epitaxially grown YZO layers on precleaned substrates could only be obtained at a low oxygen pressure (typically $5*10^{-7}$ mbar) during the growth of the first 2 nm followed by a deposition at pressures above $5*10^{-6}$ mbar.

3.2. Uncleaned Si(100) substrates

Besides a carbon contamination our Si samples had a native oxide layer of 1.7 nm thickness as calculated from the intensity ratio of the Si 2p lines. Increasing the substrate temperature to 730 °C (deposition temperature) left the native oxide layer unchanged (Figure 3, curve a) shows the Si 2p emission), the carbon signal was no longer detectable and no LEED pattern could be observed. After the deposition of 1 nm YZO at base pressure (Figure 3, curve b)) and at an oxygen partial pressure of $1*10^{-6}$ mbar (Figure 3, curve c)) the emission line at 103.2 eV corresponding to the native oxide vanished. The surfaces b) and c) exhibited the LEED pattern of a Si(100) 1x1 overstructure.

Simultanuously, the amount of metallic Zr detected in the YZO films deposited at base pressure was significantly lower than compared with films deposited at the same parameters on precleaned substrates (compare Figure 1, curve a) with Figure 4, curve a)).

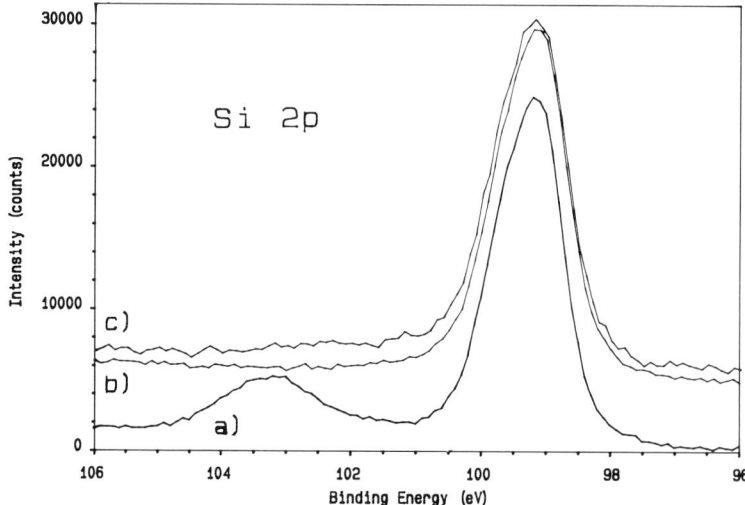

Figure 3. Si 2p spectra of a) Si(100) with native oxide and after deposition of 1 nm YZO at : b) base pressure, c) $1*10^{-6}$ mbar O_2

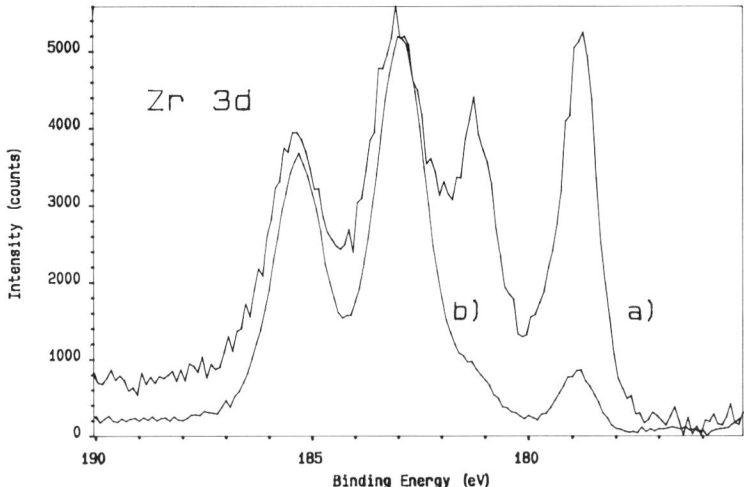

Figure 4. Zr 3d spectra of 1 nm YZO on Si(100) with native oxide. a) deposited at base pressure, b) deposited at $1*10^{-6}$ mbar O_2

Obviously, the metallic Zr component has reduced the native Si oxide, which is due to the higher formation enthalpy of ZrO_2 as compared to SiO_2 at 730 °C ($ZrO_2: \Delta G$ = -26 kJ/mol, $SiO_2: \Delta G$ = -21 kJ/mol). In the same way as for the precleaned samples the deposition of the first atomic layers at a low oxygen partial pressure resulted in a nearly complete depression of the amount of metallic Zr as can be seen in Figure 4, curve b).

In order to grow epitaxial YZO films it is necessary to carefully balance the amount of metallic Zr in the film against the formation of the silicate at the interface. This was done by using low oxygen pressures (typically $1*10^{-6}$ mbar) at the beginning of the deposition process and higher oxygen pressures (typically $5*10^{-6}$ mbar) during the further growth.

YZO films deposited under these conditions were characterized by X-ray diffraction (XRD) and Rutherford backscattering (RBS) channeling. The films grew in (h00) orientation with FWHM values of the rocking curves (YZO 200 reflex) between 0.9 ° and 1.1 °. The epitaxy was verified by Φ-scans, which showed the a-axis of the YZO (5.12 Å) to be aligned with the Si a-axis (5.43 Å). RBS channeling revealed a x_{min} = 22 % on a sample having a FWHM of the YZO 200 rocking curve of 1.2 °. The epitaxial structure of the YZO layers was also proved by transmission electron microscopy, which additionally revealed the regrowth of a 2 nm thin reaction layer at the Si/YZO interface. These experiments gave no hint for the existence of a difference between buffer layers grown on uncleaned or precleaned substrates, respectively.

To judge the quality of our growth process YBCO layers were deposited on Si/YZO(40 nm)/Y_2O_3(40 nm) by laser ablation [5]. Figure 5 shows the XRD pattern of a typical sample and, as an insert, the rocking curve of the YZO 200 reflex. The deposition was performed with the usual set of parameters [6] at substrate temperatures of 800 °C. To avoid a deterioration of the super-

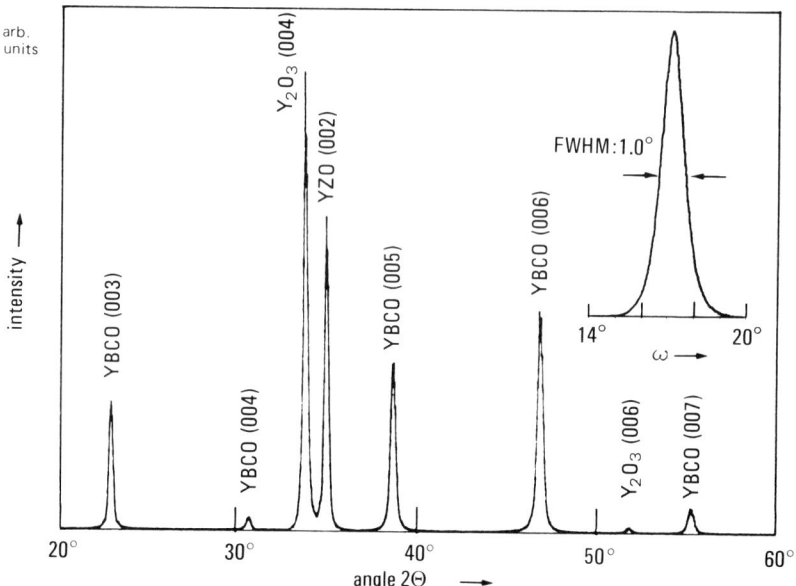

Figure 5. XRD pattern and rocking curve of the YZO 200 reflex (insert) of a Si/YZO/Y$_2$O$_3$/YBCO multilayer

conducting properties by interdiffusion only thin (40 nm) YBCO films were grown. So far, the best films showed a zero resistivity at 87 K and j$_c$ of 0.8 *10^6 A/cm^2 at 77K.

5. REFERENCES

1   T. Venkatesan, E. W. Chase, X. D. Wu, A. Inam, C. C. Chang and F. K. Shokoohi, Appl. Phys. Lett., 53 (1988) 243.
2   D. K. Fork, D. B. Fenner, R. W. Barton, Julia M. Phillips, G. A. N. Connell, J. B. Boyce and T. H. Geballe, Appl. Phys. Lett., 57 (1990) 1161.
3   H. Myoren, Y. Nishiyama, N. Miyamato, Y. Kai, Y. Yamanaka, Y. Osaka and F. Nichiyama, Jap. J. Appl. Phys., 29 (1990) L955.
4   S. Tanuma, C. J. Powell and D. R. Penn, Surf. Interface Anal., 11 (1988) 577.
5   G. Friedl, T. Matthee and J. Wecker, to be published.
6   B. Roas, L. Schultz, G. Endres, Appl. Phys. Lett., 16 (1988) 1557.

Acknowledgements
   The authors are grateful to W. Hösler and O. Eibl for the channeling and TEM work, to Y. Uzel for the rocking curve data and to T. Matthee and M. Said for technical assistance. They acknowledge the fruitful discussions with the members of the Siemens thin film group. This work was partly supported by the German Ministry of Research and Technology.

High T$_c$ Superconductor Thin Films
L. Correra (Editor)

Properties of Epitaxial Layers of YBaCuO/PrBaCuO on Sapphire

B.B.G. Klopman, A.E. Reitsma, J. Gao, W.A.M. Aarnink, G.J. Gerritsma and H. Rogalla

University of Twente, Faculty of Applied Physics, P.O.Box 217, 7500 AE Eschede, The Netherlands

Abstract

We have investigated the surface impedance of coplanar waveguide resonators (CPW) of YBa$_2$Cu$_3$O$_{7-x}$ (YBCO) thin films deposited on sapphire for different temperatures and microwave power levels. The surface resistance is obtained from the experimental data for the quality factor. The influence of the surface reactance on the resonance frequency is used to determine the penetration depth. The experimental results for the quality factor and resonance frequency are processed with the use of a spectral domain approach, modified by complex resistive boundary conditions [1].

1. INTRODUCTION

The surface impedance of superconductors is characterized by a low resistance and a relatively high reactance. This makes superconductors suitable for applications, which are impossible if normal conductors are used. As a result of the low surface resistance, the power losses in superconductors are low. This offers the possibility of the fabrication of high-Q filters, for instance, which are characterized by a sharp frequency response with low losses. The surface reactance represents the stored energy in the superconductor. This energy can be relatively high if the thickness t of a superconducting film is chosen to be smaller than the magnetic penetration depth λ. The increased stored energy has the effect of slowing the electromagnetic wave. In this way very compact microwave devices, such as filters and delay lines, can be fabricated.

Besides the applicational point of view, measurements of the surface impedance of superconductors as a function of frequency, temperature and magnetic field are of scientific interest because they yield information on different physical parameters, such as the penetration depth and the energy gap.

2. THEORY

At sufficiently high magnetic fields ($H > H_{c1}$) the magnetic field dependence of the surface impedance of type-II superconductors is a consequence of the production of vortices. In this flux-flow regime the electrodynamical properties can be analyzed by a phenomenological approach, based on the London equations and on the vortex motion equation. Within this framework the current density is given by [2]:

$$\mathbf{J} = \sigma_n \mathbf{E} - j\,\sigma_s\,(\mathbf{E} - \mathbf{v} \times f\mathbf{B}) \tag{1}$$

Here $\mathbf{E}$ is the electric field, $\mathbf{v}$ the vortex velocity, $\mathbf{B}$ the magnetic induction and f the fraction of the weakly pinned vortices. According to the two-fluid model, the conductivities $\sigma_n = 2/\omega\mu_0\delta^2$ and $\sigma_s = 1/\omega\mu_0\lambda^2$ are related to the normal and superconducting electrons respectively. The classical skin-depth is denoted by δ, the magnetic penetration depth by λ. Expressions for the Meißner state ($H < H_{c1}$) are obtained by setting $\mathbf{B} = 0$.

At microwave frequencies the motion of vortices is usually approximately purely viscous, meaning that the vortex mass and pinning force can be neglected [3]:

$$\eta\,\mathbf{v} = \mathbf{J} \times \phi_0 \tag{2}$$

where η is the coefficient of viscosity and ϕ_0 the flux quantum.

From (1) and (2) the effective conductivity is evaluated as:

$$\sigma = \mathbf{J}/\mathbf{E} = \frac{\sigma_n - j\sigma_s}{1 - j\sigma_s\phi_0 f\mathbf{B}/\eta} \tag{3}$$

For a film with thickness t much less than λ the surface impedance is, with the neglect of the displacement current:

$$Z_s = 1/\sigma t = 1/\sigma_s t\,[\sigma_n/\sigma_s + \sigma_s\phi_0 f\mathbf{B}/\eta + j\,(1 - \sigma_n\phi_0 f\mathbf{B}/\eta)] \tag{4}$$

Thus the surface resistance $R_s = \mathrm{Re}\{Z_s\}$ is proportional with B if λ (i.e. σ_s) is a much slower varying quantity. The surface reactance $X_s = \mathrm{Im}\{Z_s\}$ depends on the field through λ; the second term of X_s can be neglected in most cases.

In the Meißner state ($B = 0$) the field dependence of Z_s is determined by the variation of λ with the field. On thermodynamic grounds λ is expected to depend quadratically on the field, due to the pair-breaking effect on the gap.

With respect to propagating waves R_s has its main effect on the attenuation constant α, while the propagation constant β depends mainly on X_s. For quasi TEM-modes the exact dependence of α and β on Z_s can be evaluated numerically (section 4).

3. EXPERIMENTAL

The $YBa_2Cu_3O_{7-x}$ (YBCO) c-axis oriented thin films are deposited on sapphire substrates using an off-axis RF-magnetron sputtering technique [4]. Sapphire is a very suitable substrate for microwave applications because of its low dielectric loss-tangent. A $PrBa_2Cu_3O_{7-x}$ (PBCO) buffer layer is used, which has been proved to be an effective block for interdiffusion [5], while it has similar lattice constants and the same perovskite structure as YBCO.

The critical temperature T_c of the films range from 85 to 87 K, while the typical critical current density j_c at 77 K equals $2 \cdot 10^5$ A/cm^2. The surface roughness is about 50 nm.

The thin films are patterned to $\lambda/2$ coplanar waveguide (CPW) resonators and are surface mounted on a alumina motherboard, as described in [6]. Results for the unloaded quality factor Q_0 and the resonance frequency f_0 are obtained by measuring the resonance peak in reflection. From this, data for R_s and λ are deduced with the aid of a numerical analysis of the propagation characteristics of the CPW transmission line.

4. NUMERICAL

Transmission lines with very thin superconducting films ($t \ll \lambda$) can be simulated by a spectral domain approach, modified by complex resistive boundary conditions [1, 7]. The change of the normalized propagation constant β/k_0 is proportional to $R = 1/\sigma_s t = \omega\mu_0\lambda^2/t$ for lower values, as is to be expected on the basis of a transmission line model for CPW [8]. Deviations from the linear dependence occur for higher values of R, because the current distribution in the strips changes with R.

With the same numerical method it is also possible to calculate the power losses. If the losses are low, α is proportional to R_s, which means that the quantity $Q_0 R_s$ is independent of α.

The results for β/k_0 will be used to determine the penetration depth λ from the resonance frequency f_0, while the surface resistance R_s is calculated from the quality factor Q_0 by means of the quantity $Q_0 R_s$.

5. RESULTS AND DISCUSSION

We report on the results of films with an YBCO layer and a PBCO bufferlayer, both layers having a thickness of 50 nm. First we studied the properties of the resonator ($f_0 \cong 11$ GHz) in the temperature range 4.2-80 K at excitation levels for which the data are independent of applied microwave power (-30 dBm).

The normalized propagation constant β/k_0 can be calculated from the resonance frequency f_0: $\beta/k_0 = c/2lf_0$. Here c is the phase velocity in free space and l is the length of the resonator (5.4 mm). Open-end effects are estimated to increase the effective length of the resonator to 5.70 mm. With the results of the numerical analysis we obtain λ for different temperatures. Figure 1 shows λ as a function of $y = [1-(T/T_c)^4]^{-1/2}$. Assuming the two-fluid model is valid, we obtain the penetration depth at zero temperature λ_0 from the slope of this plot. The best fit is represented by the drawn line and corresponds to $\lambda_0 = 230$ nm and $T_c = 86$ K. Indeed the very thin film limit is appropriate: $\lambda_0 = 4.6t$. T_c is consistent with dc-resistance measurements. Especially at lower temperatures the fit is not very good. This is in agreement with accurate measurements of λ [9]. It is well-known that λ is sensitive to impurities and defects in the material. As a result, λ is substrate and deposition dependent. Our value of λ_0 for YBCO on sapphire is somewhat higher than $\lambda_0 = 140$ nm [9], which is a result for very good films on MgO.

From the experimental data for Q_0, the surface resistance R_s is determined with the aid of the numerical results. To account for the small thickness of the film the R_s is corrected by a factor $(2\lambda/t)^{-1}$. Then we obtain the value of R_s for a thick film (or bulk material) with the same properties (fig. 2). As usual R_s shows a sharp transition near T_c and approaches a residual value well down T_c. With respect to this residual value it should be mentioned that the results are not corrected for dielectric losses and conductor losses in the box, enclosing the resonator. In any event, the value $R_s = 80$ μΩ at lower temperatures is comparable to the best results obtained so far on sapphire.

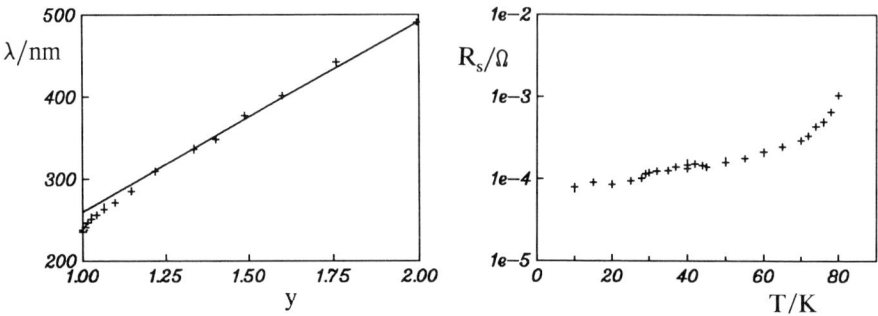

Figure 1. The penetration depth λ as a function of $y=[1-(T/T_c)^4]^{-1/2}$.

Figure 2. The surface resistance R_s for different temperatures.

To analyze the rf magnetic field dependence of R_s and λ, we note that the maximum amplitude of the current I in the central strip of the resonator is given by:

$$I^2 = 8|a|(1-|a|)Q_0P_{in}/n\pi Z_0 \tag{5}$$

Here a is the minimum reflection ratio, Q_0 the unloaded quality factor, P_{in} the incident microwave power, n the resonant mode number (here n = 1) and Z_0 the characteristic impedance of the transmission line, defined on the power-current basis. Although the current density and the rf magnetic field H_{rf} are not uniform but peaked at the edges of the strips, an indication of the magnitude of H_{rf} is obtained from I = $2H_{rf}s$, where s is the width of the central strip (0.5 mm). The factor 2 is a result of the existence of H_{rf} at both sides of the film.

The H_{rf} dependence of R_s and λ is shown in fig. 3 and 4 for different temperatures. At T = 4.2 and 30 K R_s is approximately proportional to H_{rf} for $H_{rf} > 0.1$ G. At T = 50 K R_s saturates at the higher values of H_{rf}, which can be attributed to the lower critical currents at this temperature. At the lowest fields the data seem to indicate a field dependence of R_s through λ in the Meißner state.

The change of λ with H_{rf} is shown in fig. 4 for T = 4.2 and 30 K. Again two different regimes can be distinguished, with a cross-over at about 0.1 G. Above 0.1 G the magnetic field dependence of λ is weaker.

A lower critical field of the order of 0.1 G seems to be very small. However, at the edges of the strips the field will be considerably higher. Also it is well-known that defects like twin- or grain-boundaries can lower the critical fields significantly. The non-ideal values of T_c (86 K), j_c at 77 K $(2 \cdot 10^5 A/cm^2)$ and a surface roughness of about 50 nm validate the presumption that defects are present in the films.

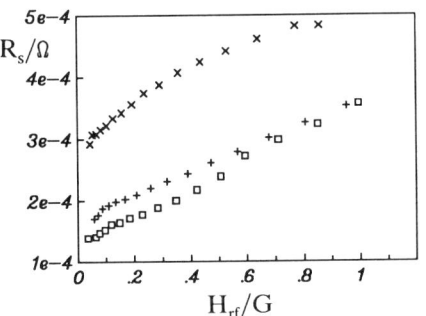
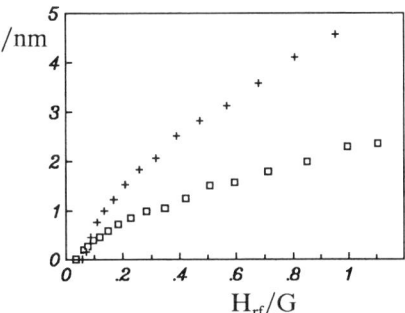

Figure 3. The surface resistance R_s as a function of rf magnetic field H_{rf} for T = 4.2 K (□), T = 30 K (+) and T = 50 K (×).

Figure 4. The change of the penetration depth $\Delta\lambda$ with rf magnetic field H_{rf} for T = 4.2 K (□) and T = 30 K (+).

6. CONCLUSIONS

We have measured the surface resistance R_s and the penetration depth λ of YBCO thin films on sapphire as a function of temperature T and rf magnetic field H_{rf} at 11 GHz. The penetration depth at zero temperature is evaluated as $\lambda_0 = 230$ nm. The surface resistance R_s is 80 $\mu\Omega$ at the lowest temperatures. The result for λ_0 is somewhat higher than the best results obtained so far for other substrates, R_s is comparable to good results with sapphire as substrate.

At least qualitively the magnetic field dependence of R_s and λ can be understood in terms of the Meißner state for low fields and a flux-flow regime for higher fields.

7. REFERENCES

1   J.M. Pond, C.M. Krowne, and W.L. Carter, IEEE Trans. Microwave Theory Tech. MTT-37, 181 (1989)
2   A.M. Portis, K.W. Blazey, K.A. Müller, and J.G. Bednorz, EuroPhys. Lett. 5, 467 (1988)
3   H. Suhl, Phys. Rev. Lett. 14, 226 (1965)
4   J. Gao, B. Häuser, and H. Rogalla, J. Appl. Phys. 67, 2512 (1990)
5   W.A.M. Aarnink, A. van Silfhout, and H. Rogalla, presented at the MRS 91 spring meeting

6 B.B.G. Klopman, H.W. Weijers, J. Gao, G.J. gerritsma, and H. Rogalla, IEEE
   Trans. Magn. MAG-27, 2821 (1991)
7 D. Kinowski, F. Huret, P. Pribetich, and P. Kennis, Ann. Telecommun. 45, 334
   (1990)
8 D. Kinowski, C. Seguinot, P. Pribetich, and P. Kennis, Electronics Lett. 26,
   148 (1990)
9 S.M. Anlage, B.W. Langley, G. Deutscher, J. Halbritter, and M.R. Beasley, to
   be published

High T$_c$ Superconductor Thin Films
L. Correra (Editor)
© 1992 Elsevier Science Publishers B.V. All rights reserved.

Superconducting Y-Ba-Cu-O thin films on Si and Al₂O₃ substrates

X. Queralt, M. Varela, M. V. García-Cuenca and J. L. Morenza

Universitat de Barcelona, Departament de Física Aplicada i Electrònica,
Avda. Diagonal 647, E-08028 Barcelona, Spain.

Abstract

YBaCuO superconducting thin films have been prepared by sequential evaporation of Cu, Y_2O_3 and BaF_2 layers on Si[100] and polycrystalline Al_2O_3 substrates. One, ten and thirty-sequences were evaporated on Si[100] with a 0.6 μm ZrO_2 buffer layer. The effects of different buffer layers were studied for Al_2O_3 substrates by means of the deposition of ZrO_2, Cu, Ag and $BaTiO_3$ layers with thicknesses of 20, 100 and 500 nm. The film properties were studied using electrical measurements, scanning electron microscopy, x-ray diffractometry, electron microprobe analysis and secondary ion mass spectrometry. The best properties were found in the film obtained on ZrO_2/Si[100] with one-sequence. The studies of the different buffers show a strongly dependence of the film properties on the nature and thickness of the buffer, with the best results for films deposited on an Ag layer.

1. INTRODUCTION

Superconducting YBaCuO thin films have been successfully prepared by sequential evaporation on oxide single-crystal substrates such as MgO, ZrO_2 and $SrTiO_3$ [1,2], however, for technological applications it is interesting to form the superconducting films on substrates such as Si, GaAs and Al_2O_3 [3-5]. The preparation of superconducting thin films on these substrates is limited by the considerable chemical interaction between the film and the substrate owing to the high temperature required for crystalline growth [6,7].

In this paper, we report the preparation of superconducting YBaCuO thin films on silicon and alumina substrates by means of sequential evaporation of Cu, Y_2O_3 and BaF_2, and subsequent heat treatments. In order to minimize the detrimental annealing effects two approaches have been used, the deposition of different buffer layers between the films and the substrates, and the deposition of several groups of three layers with the corresponding decrease of the individual layer thickness. Here we report the preparation of superconducting films from one, ten and thirty groups of three layers on Si [100] with a ZrO_2 buffer layer. We present also the results obtained from thirty groups of three layers on polycrystalline Al_2O_3 coated with ZrO_2, Cu, Ag and $BaTiO_3$ layers.

2. EXPERIMENTAL

The films were prepared using sequential evaporation as previously described [8]. The buffer layers were obtained in the same chamber that the Cu-Y_2O_3-BaF_2 layed thin films, but in different evaporation conditions. ZrO_2 and $BaTiO_3$ were evaporated by electron gun under an oxygen pressure of 10^{-4} mbar, at 0.10 - 0.15 $nm \cdot s^{-1}$ deposition rate. Cu was evaporated with W boat under the same oxygen pressure and the deposition rate was between 0.20 - 0.30 $nm \cdot s^{-1}$. Ag was evaporated in W boat too, but without oxygen pressure, and the deposition rate was about 0.50 $nm \cdot s^{-1}$.

Table 1
Preparation conditions and surface resistance

Sample[#]	B.M.[+]	Th.[Δ] (nm)	N.S.[*]	T_1 (°C)	t_1 (min)	R_1 (kΩ)	T_2 (°C)	t_2 (min)	R_2 (kΩ)
S11	ZrO_2	500	1	800	35	4.50	400	15	5.00
S12	ZrO_2	500	1	825	12	1.10	400	8	0.50
S13	ZrO_2	500	1	850	7	1.30	400	8	1.80
S21	ZrO_2	500	10	825	6	2.20	400	10	1.10
S22	ZrO_2	500	10	825	6	1.40	400	10	1.10
S31	ZrO_2	500	30	800	10	1.00	400	10	1.10
S32	ZrO_2	500	30	825	10	1.30	400	10	0.50
S33	ZrO_2	500	30	850	11	4.20	400	10	1.20
A11	ZrO_2	20	30	850	42	1.70	400	30	0.30
A12	ZrO_2	100	30	850	41	0.60	400	30	0.20
A13	ZrO_2	500	30	850	41	0.50	400	30	0.20
A21	Cu	20	30	850	57	1.30	400	30	0.50
A22	Cu	100	30	850	40	∞	400	--	--
A31	Ag	20	30	850	10	∞	400	--	--
A32	Ag	100	30	850	78	0.20	400	20	0.02
A33	Ag	500	30	850	16	0.04	400	30	0.01
A41	$BaTiO_3$	20	30	850	39	1.30	400	30	0.40
A42	$BaTiO_3$	100	30	850	60	15.0	400	60	4.00

[#] S for samples on Si; A for samples on Al_2O_3
[+] B.M. = Buffer material.
[Δ] Th. = Buffer thickness.
[*] N.S. = Number of sequences.

The thickness of the ZrO_2 buffer layer on Si [100] was 0.5 μm, and for ZrO_2, Cu, Ag and $BaTiO_3$ on Al_2O_3 was 20, 100 and 500 nm. The temperature of the substrates was maintained at 400 °C for ZrO_2, at 300 °C for Cu and $BaTiO_3$, and at 200 °C for Ag.

After the deposition of the Cu-Y_2O_3-BaF_2 multilayer, 0.7 μm thick, a two steps post-annealing process was necessary to achieve the superconducting YBaCuO phase. In the first one, the layered films were annealed in flowing oxygen, wet for films on $ZrO_2/Si[100]$ substrates and dry for films on Al_2O_3, at 800-850 °C. In the second step, a dry oxygen flow was used at 400 °C for all samples. In Table I are collected the annealing conditions of the films.

The resistance versus temperature dependence was measured by the four-contacts method, in the 10-300 K range. The structure of the films was studied using scaning electron microscopy (SEM) and x-ray diffraction analysis (XRD). The film composition was determined by electron microprobe analysis (EDX), and the composition profile was obtained by secondary ion mass spectrometry (SIMS).

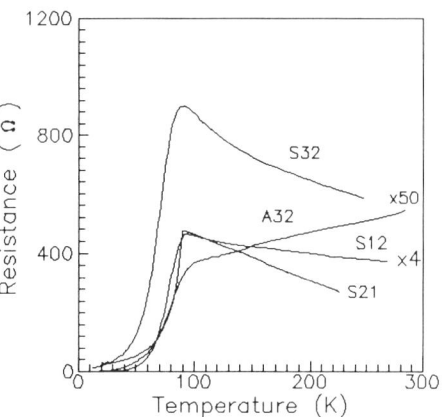

Figure 1 Resistance versus temperature for some samples.

3. RESULTS AND DISCUSSION

3.1 Thin films on $ZrO_2/Si[100]$

In Figure 1 resistance versus temperature is plotted for some representative superconducting films. The best film was obtained after a 825 °C, 12 min high temperature annealing and a 400 °C, 8 min low temperature annealing for an one-sequence film (sample S12), with a $T_{c,onset} \approx 92$ K and a $T_{c,0} \approx 25$ K. All the films display a nonmetallic normal-state behaviour, with a $T_{c,onset}$ around 90 K and long tails near zero resistance.

The x-ray diffraction spectra of some films are shown in Figure 2. The YBaCuO superconducting phase has been observed in all samples. The films do not show important preferential orientation, however the intensity ratios between [005] and [103] peaks indicate a little [005] preferential orientation which is higher for the thirty-sequences films than for the ten and one-sequence ones. The spectra show also the presence of compounds other than YBaCuO, such as BaF_2, $BaZrO_3$, Y_2BaCuO_5 and CuF_2. The residual resistance observed at low temperature could be attributed to the presence of these compounds between YBaCuO grains.

The SEM micrographs of the films indicate a granular surface morphology with globular and elongated grains (Figure 3). It is also noticeable the presence of microcracks, attributed to the different thermal expansion coefficient between films and substrates.

The films show compositions near to the 1:2:3 stoichiometry ratios, with small inhomogeneities along the sample.

The composition profiles of a ten-sequences thin film as deposited and after annealing are represented in Figure 4. The as deposited film profile shows the oscillations corresponding to the Cu-Y_2O_3-BaF_2 layers. After the annealing the profile presents an Y, Ba, Cu and O composition quite constant near the surface with a low signal of Zr. The ZrO_2 layer seems to be effective as diffusion barrier for silicon, copper and yttrium, however a high barium signal is observed in the barrier. This behaviour could explain the $BaZrO_3$ peaks in the XRD spectra.

3.2 Thin films on Al_2O_3 with different buffer layers

The x-ray diffraction spectra of the films deposited on Al_2O_3 with different buffer layers indicate the presence of the orthorombic phase of YBaCuO and other binary and ternary compounds. The formation of compounds other than YBaCuO is strongly dependent on the buffer layer material. However, the spectra of the films grown on Ag/Al_2O_3 only present the YBaCuO peaks for the 100 nm thick buffer layer.

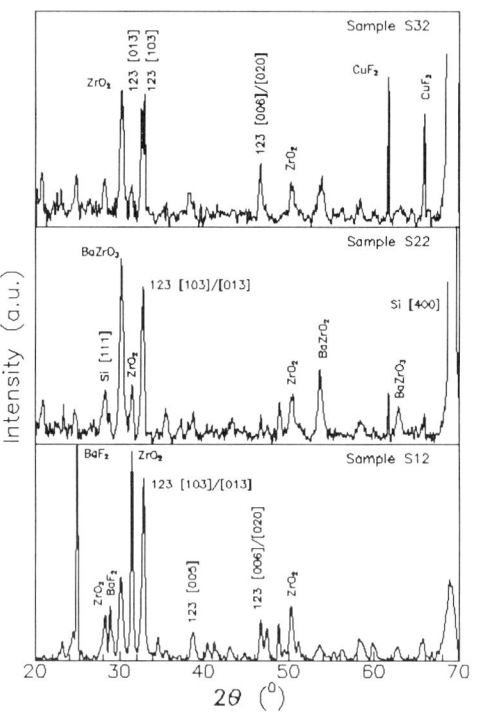

Figure 2 X-ray spectra for one, ten and thirty-sequences samples.

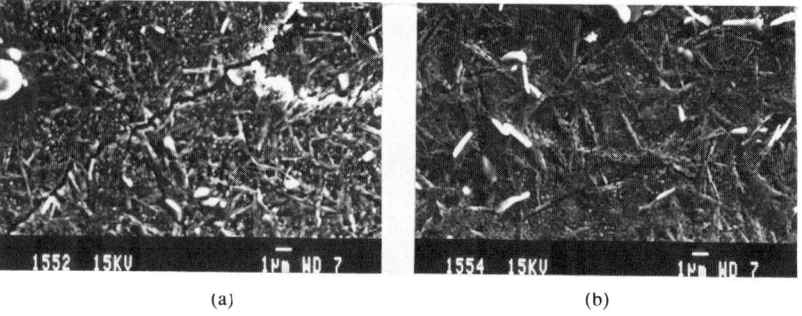

(a)   (b)

Figure 3 SEM micrographs of two films on Si. (a) Sample S21 and (b) sample S32

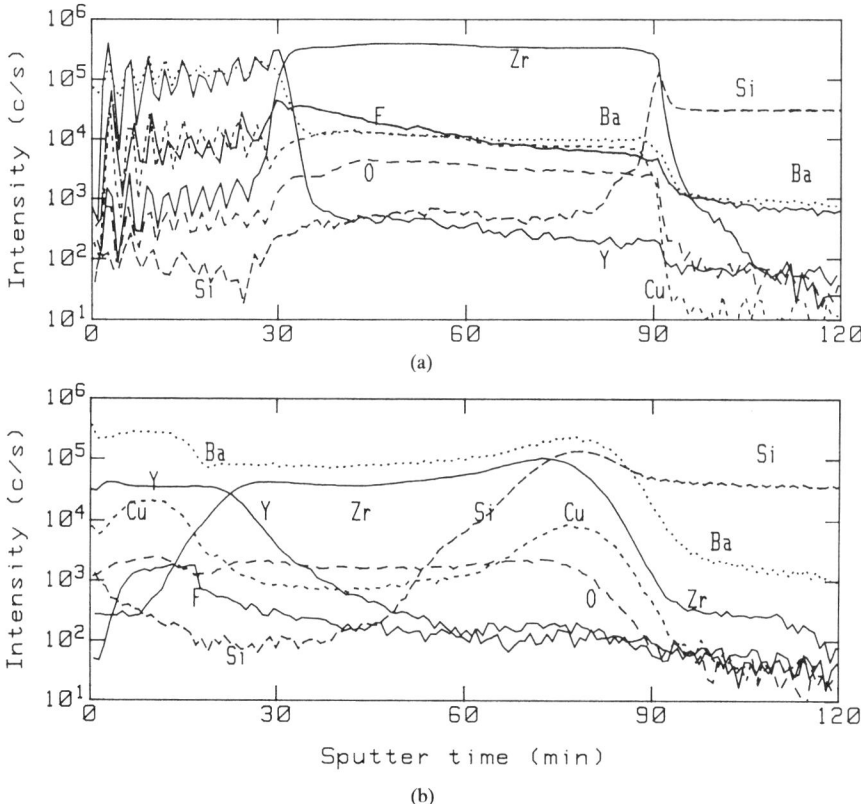

Figure 4 SIMS profiles of a sample on Si. (a) As deposited and (b) after annealing.

The films analyzed by SEM show globular and elongated grains. The elongated grains are predominant in the films grown on ZrO_2 buffer layers, and in the other cases are only present in the films deposited on a 100 nm Ag buffer layer, and on a 20 nm $BaTiO_3$ layer. The films on Cu/Al_2O_3 show a poor globular structure and the presence of microcraks.

The SIMS profiles indicate the diffusion of Ba, Cu, Y and F toward the substrate. The presence of the buffer layers is only clear with barrier thicknesses of 500 nm. For the films with Ag buffers, the presence of Ag in the film is only crearly observed when the barrier has a thickness of 500 nm.

The diffusion of Ba and F toward the substrate explains the presence of $BaAlF_5$ in the x-ray diffraction spectra. The formation of compounds other than YBaCuO in the films with thick (500 nm) Ag buffer layer can be attributed to the Ag diffusion toward de film. The absence of YBaCuO in the films with thin (20 nm) Ag buffer layer can be due to the low barrier efficiency of this thickness.

The measurements of the resistance versus temperature show the best properties for the film obtained with an Ag buffer layer of 100 nm (Figure 1). This film displays a metal normal-state behaviour, with an onset temperature of 92 K and zero resistance around 30 K. The film grown on a 500 nm Ag buffer layer shows metallic behaviour but without superconducing transition. The film obtained on a 20 nm Ag buffer is insulating. The superconducting transition is also observed in the films deposited on ZrO_2/Al_2O_3, $BaTiO_3/Al_2O_3$ and Cu/Al_2O_3, with nonmetallic normal-state behaviour. The film grown on a 20 nm Cu buffer layer is insulating.

4. CONCLUSION

YBaCuO superconducting films have been obtained by sequential evaporation on silicon and alumina substrates. The deposition of several sequences of three Cu, Y_2O_3, BaF_2 layers does not improve the properties of the films on Si[100]. The films obtained from thirty sequences on Al_2O_3 show the best properties when an Ag 100 nm buffer layer is used.

ACKNOWLEDGEMENTS

The authors acknowledge the collaboration of the Secondary Ion Mass Spectrometry, the Electron Microscopy and the Spectroscopy Services of the University of Barcelona.

This work is a part of a research programme financed by CICYT of the Spanish Government (project MAT 89-0253) and Red Eléctrica de España, S.A./UNESA (MIDAS programme).

REFERENCES

1 N. Hess, L.R. Tessler, U. Dai and G. Deutstcher, Appl. Phys. Lett. 53 (1988) 698.
2 B.-Y. Tsaur, M.S. Dilorio and A.J. Strauss, Appl. Phys. Lett. (1987) 858.
3 C.X. Qui and I. Shih, Appl. Phys. Lett. 52 (1988) 587.
4 M. Migluolo, A.K. Stamper, D.W. Greve and T.E. Schlesinger, Appl. Phys. Lett. 54 (1989) 859.
5 Q.Y. Ma, E.S. Yang and Ch.-A. Chang, J. Appl. Phys. 66 (1989) 1866.
6 A. Mogro-Campero, B.D. Hunt, L.G. Turner, M.C. Burrell and W.E. Balz, Appl. Phys. Lett. 52 (1987) 584.
7 Ch.-A. Chang, C.C. Tsuer, C.C. Chi and T.R. McGuire, Appl. Phys. Lett. 52 (1988) 72.
8 X. Queralt, M. Varela, M.V. García-Cuenca and J.L. Morenza, J. of Less-Common Metals, 164&165 (1990) 430.

High T$_c$ Superconductor Thin Films
L. Correra (Editor)

Deposition and patterning of epitaxial Y–Ba–Cu–O thin films on silicon with ZrO$_2$ buffer layers

G.Beddies, B. Leibold, H.–U. Habermeier, and G. Lu

Max–Planck–Institut für Festkörperforschung, Heisenbergstr.1, D 7000 Stuttgart 80

Abstract

   Superconducting Y–Ba–Cu–O thin films are deposited on Si (100) substrates where an intermediate buffer layer of yttria stabilized zirconia is used to prevent interdiffusion and interfacial reactions. Both layers are grown epitaxially using the pulsed laser deposition technique. The films are patterned by two different schemes. One is the conventional photoresist masking combined with ion milling, the other is a type of selective epitaxy of the YBCO thin films on prepatterned YSZ layers on silicon substrates combined with the use of Si–Y–Ba–Cu–O intermixing during the in situ film growth process. In both cases the properties of the patterned YBCO films are nearly identical showing critical temperatures as high as 90K and critical transport currents exceeding 10^6 A/cm^2 at 77K.

1. INTRODUCTION

   High quality superconducting thin films of Y$_1$Ba$_2$Cu$_3$O$_{7-x}$ [YBCO] with transition temperatures, T$_c$, as high as those of bulk materials and critical current densities, j$_c$, larger than 10^6A/cm^2 at 77 K can be deposited on oxide substrates with good lattice match to the orthorhombic unit cell of YBCO [a= 0.3823 nm, b= 0.3886 nm, c= 11.681 nm] by various methods including sputtering [1,2], reactive evaporation [3,4], pulsed laser deposition [5,6] and chemical vapor deposition [7]. Single crystal SrTiO$_3$, MgO, LaAlO$_3$ and ZrO$_2$ are commonly used substrate materials in accordance with this requirement. Application driven research activities directed towards the use of high temperature superconductor thin films in microelectronic devices demand cheap substrates of large size which are capable of combining semiconductor device tehnology with superconducting electronics. Silicon single crystals would be the preferred choice due to their availability and the well established device technology developed on silicon wafers. Although the direct deposition of YBCO thin films on bare silicon has already be accomplished with some success [8,9], the strong chemical interaction of silicon with the constituents of YBCO, the interdiffusion at the interface and the mismatch of thermal expansion coefficients lead to the formation of rough film surfaces and microcracks in the films resulting in poor film quality especially with regard to the critical currents [10]. These problems can be overcome using an intermediate buffer layer between the silicon and the YBCO thin film. The buffer layer has to be optimized with respect to three different properties: (i) efficiency as diffusion barrier, (ii) relaxation of thermal stress and (iii) possibility of YBCO epitaxy, since only epitaxially grown YBCO thin films give superior values for the critical current. The most successful buffer layer used to date has been yttria stabilized zirconia [YSZ] with an yttria content of 8 mol% [11].

Micropatterning of such YBCO thin films on buffered silicon has been done so far by means of different schemes including direct laser patterning [12] as well as photolithographic techniques in conjunction with either wet chemical etching [13] or ion milling [14]. The disadvantages of these techniques are either the restriction to pattern dimensions in the order of 5 μm in the case of laser patterning and the still not clear chemical interaction of the YBCO surface with photoresist which can, but not necessarily must degrade the superconducting properties of the top layer of the YBCO thin film.

In this paper we report the in–situ preparation of superconducting YBCO thin films on (100)–oriented silicon covered with an epitaxially grown thin film of cubic YSZ and the micropatterning of these multilayers by two different methods. One is the conventional lithography combined with ion milling and the other is a selective epitaxy based on the use of Si–Y–Ba–Cu intermixing as introduced by Ma et al [15] and Hatani et al. [16]. The films are characterized by their electrical transport properties, their structure using X– ray as well as TEM analysis and their surface morphology as revealed by SEM.

2. EXPERIMENTAL DETAILS

The substrates, cut from single crystal (100) –oriented silicon wafers to a dimension of 7x 7x .5 mm^3 were RCA cleaned, transferred to the growth chamber in buffered HF, rinsed in ultrapure water, dried with nitrogen and inserted into the specimen holder of the deposition system. The deposition of the YSZ films as well as the YBCO films are done in a standard pulsed laser deposition [PLD] chamber equipped with a KrF excimer laser as energy source. Details of the PLD system are given elsewhere [6]. The processing parameters for the deposition of the YSZ and the YBCO differ only in the oxygen pressure, p_{ox}, during deposition [$p_{ox} = 2.10^{-2}$ Pa for YSZ and $p_{ox} = 100$ Pa for YBCO] and the substrate temperature, T_d [$T_d = 800^o$ C for YSZ and $T_d = 780^o$ for YBCO].

The application of etching masks with a Hall bar geometry is done using Shipley 1400–17 photoresist in the usual processing scheme; for the ion milling a VEECO microetch system is applied equipped with a Kaufman type Ar$^+$ ion source. To prevent a degradation of the etching mask during the ion milling process the specimen is mounted onto a watercooled rotating sampleholder. Typical parameters for the etching process are:

Base pressure	$< 2. 10^{-4}$ Pa
Argon pressure	$2. 10^{-2}$ Pa
Beam Energy	500 eV
Ar$^+$ current density	0.2 mA/cm^2

Using these conditions the etching rates were approximately 5 nm/min for all materials involved in the multilayer. After ion milling the remaining photoresist was stripped in a parallel plate RIE system using an oxygen rf plasma [o.1 mbar, 5o W input power]. This process has the further advantage of to restore the full oxygen—ation of the YBCO thin film which can be detoriated during the processing [17].

3. RESULTS AND DISCUSSION

To optimize the crystal quality of the YSZ thin film the oxygen background pressure as well as the deposition temperature were varied systematically according to the process parameters published by Fork et al. [11].The optimized values for the oxygen pressure and depostition temperature in our experiments were quite comparable to the published data. At $T_d = 800^{o}C$ and $p_{ox} = 2.10^{-2}$ Pa during deposition the YSZ films grow in a single phase cubic lattice and show a smooth surface in the SEM. In Fig. 1a the X—ray diffraction pattern of a YSZ thin film on Si is given, indicating the presence of the (h00) reflection peaks, only. The epitaxial growth of these YSZ films on Si could also be demonstrated by cross— sectional TEM [18]. Auger depth profiling analysis shows a sharp interface and no appreciable interdiffusion could be observed. For the following YBCO deposition the standard parameters as given above are used. The thickness of the YSZ layers have been varied between 30 nm and 150 nm, however, no influence on the properties of the

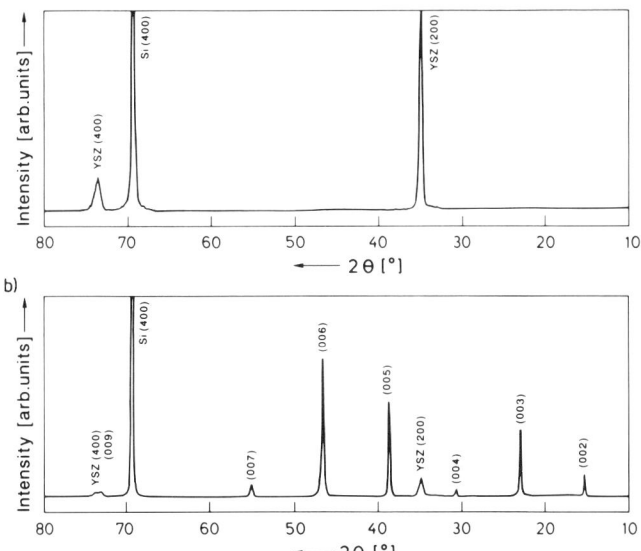

Fig. 1: X ray diffraction pattern of a YSZ thin film on silicon [a] and a YBCO thin film on YSZ buffered silicon [b].

superconducting thin film was observed. X— ray analysis as well as cross—sectional TEM show the epitaxial growth of the YBCO film on YSZ. The X— ray diffraction pattern in Fig. 1b indicates that single phase YBCO is growing exclusively with the c—axis perpendicular to the film plane. The processing steps for the two different

patterning schemes are given in Fig. 2. The conventional sequence [right hand side]
consists in the photoresist masking and ion milling followed by the oxygen plasma
stripping/ oxygenation process. As shown in Fig. 3a these YBCO films have a me—
tallic temperature characteristic above 100 K and a sharp transition to super—
conductivity

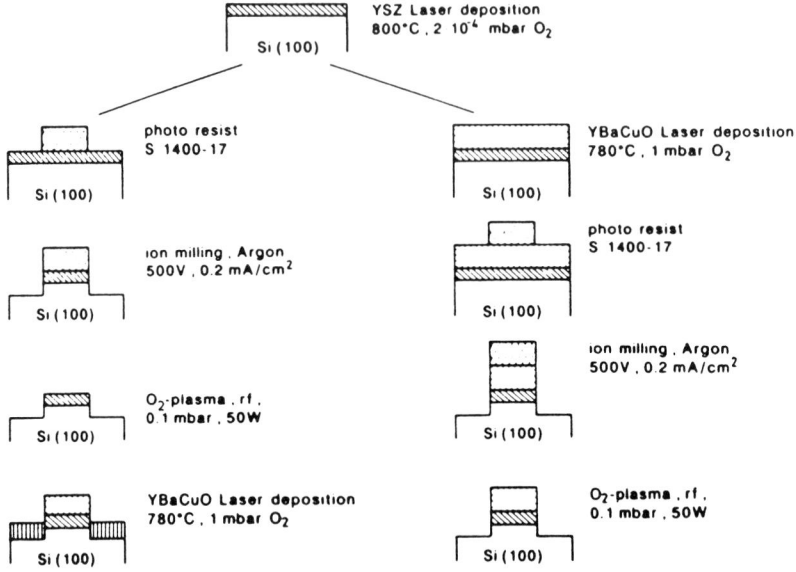

Fig. 2:   Patterning scheme for conventional lithography [right hand side] and
selective epitaxy [left hand side].

around 90 K. The transport critical current at 77 K exceeds 10^6 A/ cm^2.
Alternatively a second technique for micropatterning we tested the selective epitaxy
as described schematically in Fig. 2 [left hand side]. This method is based on

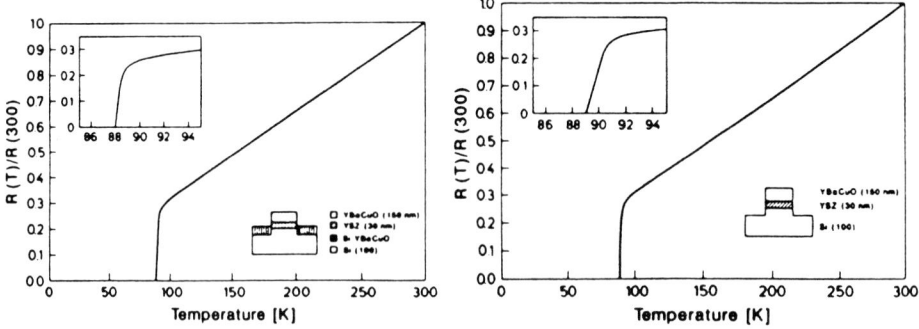

Fig. 3:   Electrical transition of a patterned YBCO film on YSZ buffered silicon by
conventional microlithography [right] and selective epitaxy [left].

a mesa–type micro patterning of the YSZ buffer layer on the silicon substrate and a subsequent homogeneous deposition of the whole device with YBCO using the in situ PLD technique. Due to the high substrate temperatures required for the in situ YBCO process the superconductor grows epitaxially on the YSZ whereas on the other regions of the device an intermixing at the interface occurs giving rise to a nonsuper–conducting Si–Y–Ba–Cu–O mixtured. Using this technique of selective epitaxy the total thickness of the YBCO film is restricted to 100nm– 200 nm, for much thicker films the extension of the layer in which the intermixing occurs will not reach the surface of the YBCO and the top layer will be superconducting [9]. The electrical transition to superconductivity of films patterned by selective epitaxy [c.f. Fig. 3b] is quite comparable to that of the film patterned directly. The critical currents of patterns prepared in the different sequences are comparable high. The surface morphology of the patterns prepared by both different methods are given in Fig. 4 a and 4b. covering an area including substrate and the patterned YBCO with underlying YSZ. The appearance of the surfac in the SEM of the films on YSZ are comparable smooth with some surface granularity. The YBCO film on bare silicon shows a rather inhomogeneous morphology with individual grains somehow similar to those observed in 5oo nm thick films on bare silicon. [9].

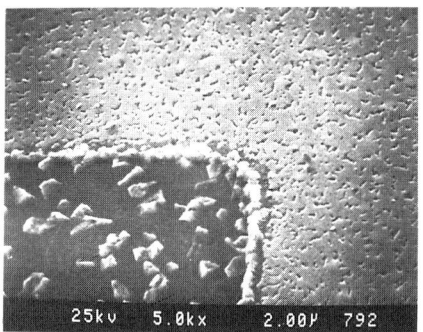

Fig. 4: Surface morphology of an YBCO thin film pattern on buffered silicon prepared by conventional processing [right hand side] and selective epitaxy [left hand side].

4. CONCLUSIONS

Superconducting YBCO thin films were deposited on an epitaxially grown YSZ buffer layer on Si (100) –oriented substrates by pulsed laser deposition. For film patterning we have used two different methods, conventional Ar^+ ion milling and a new scheme of selective epitaxy of the YBCO film on a patterned YSZ layer. In both cases the resulting films have comparable electrical transport data with critical curent densities at 77 K close to those measured on epitaxially grown YBCO films on $SrTiO_3$ substrates. The technique to prepare YBCO micropatterns on buffered silicon by selective epitaxy opens a simple path to avoid degradation effects associated with conventional patterning schemes especially if structures in the sub– μm range are envisaged.

Acknowledgments

We would like to thank Dipl. phys. G. Wagner for encouraging discussions, L. Viczian for the X– ray diffraction measurements and Dr. W.–E. Seibt from the KFK Karlsruhe for the AES investigations.

REFERENCES

[1]   H. Adachi, K. Setsune, T. Mitsuyu, K. Hirochi, Y. Ichikawa, Japn. J. Appl. Phys. 26 (1987) L 709

[2]   H.C. Li,G. Linker,F. Ratzel, R. Smithey,and J. Geerk, Appl. Phys. Lett. 52 (1988) 1098

[3]   P. Chaudhari, R.H. Koch, R.B. Laibowitz an R.J. Gambino, Phys. Rev. Lett. 58 (1987) 2684

[4]   B. Oh, M. Saito, S. Arnason, P. Rosenthal, R. Barton Appl. phys. Lett. 51 (1987) 852

[5]   T. Venkatesan, X.D. Wu, A. Inam, and J.B. Wachtman, Appl. Phys. Lett. 52 (1988) 1193.

[6]   H.–U. Habermeier, Eur. J. Solid State Inorg. Chem. 28 (1991) 619

[7]   A.D. Berry, D.K. Gaskill, R.T. Holm, E.J. Cukauskas, R. Kaplan and R.L.Henry, Appl. phys. Lett. 52 (1988) 1743

[8]   P. Berberich, J. Tate, W. Dietsche, and H. Kinder, Appl.Phys, Lett 53 (1988) 925.

[9]   H.–U. Habermeier and G. Mertens, Physica C 153–155 (1988) 1429

[10]   H.–U. Habermeier, G. Mertens and G. Wagner, Vacuum 4 (1990) 859

[11]   D.K. Fork, D.B. Fenner, G.A.N. Connell, J.M. Philips and T.H. Geballe, Appl. Phys. Lett. 57 (1990) 1137.

[12]   A. Inam, X.D. Wu, T. Venkatesan,S.B. Ogale, C.C. Chang, and D. Dijkamp, Appl. Phys. Lett. 51 (1987) 1112

[13]   F.K. Shokoohi, L.M. Schiavone, C.T. Rogers, A. Inam, X.D. Wu, L. Nazar, and T. Venkatesan, Appl. Phys. Lett. 55 (1989) 2661

[14]   G.C. Hilton, E.B. Harris and D.J. van Harlingen, Appl. Phus. Lett. 53 (1988) 1107

[15]   Q.Y. Ma, E.S. Yang, G.V.Treyz and C.–A. Chang, Appl. Phys. Lett. 55 (1989) 896

[16]   T. Hatano, A. Fujimaki, Y. Takai and H. Hayakawa, Jap. J. Appl. Phys. 29 (1990) 1076

[17]   H.–U. Habermeier, W. Ebert, S. Kalt, G. Wagner, and G. Mertens, Thin Solid Films 174 (1989) 2659

[18]   G.Lu, R. Philipp, H.–U. Habermeier and B. Leibold, private communication, to be published.

High T$_c$ Superconductor Thin Films
L. Correra (Editor)
647

Growth and properties of NdGaO3/YBa2Cu3O7-δ multilayer structures

Yu. Boikov [a], G. Brorsson, T. Claeson, and Z. G. Ivanov

Department of Physics, Chalmers University of Technology
S-412 96 Gothenburg, Sweden

[a] permanent address: Ioffe Physico-Technical Institute, Academy of Sciences of USSR, Leningrad, USSR.

Abstract

We have used laser deposition to make two types of NdGaO$_3$/YBa$_2$Cu$_3$O$_{7-\delta}$ multilayer structures. The growth of NdGaO$_3$ thin films on different substrates was studied and it was shown that the films grow epitaxially only on well lattice matched substrates, such as SrTiO$_3$ (100) and c-axis oriented YBa$_2$Cu$_3$O$_{7-\delta}$ films. YBa$_2$Cu$_3$O$_{7-\delta}$/NdGaO$_3$/YBa$_2$Cu$_3$O$_{7-\delta}$ trilayers were epitaxially grown and a T$_c$ of 86 K was measured for the top and bottom layers. The resistivity of a 300 nm thick NdGaO$_3$ interlayer was as high as 10^8 Ωcm at 300 K. NdGaO$_3$/YBa$_2$Cu$_3$O$_{7-\delta}$ double layers were grown and tested as a complex buffer layer for the YBa$_2$Cu$_3$O$_{7-\delta}$ growth on sapphire.

1. INTRODUCTION

A number of devices, such as SQUIDs, microwave transmission lines, and filters have been developed using a single layer of a high T$_c$ superconductor (HTS) [1,2,3]. Further development depends to a large extent on the availability of trilayer structures, S/I/S, where S is an HTS layer and I is an insulating intermediate layer with a typical thickness of a few tenths of a micrometer. Such a structure is the key element for the construction of multilevel SQUIDs with integrated input coils and transformers, and of microwave integrated circuits. A general requirement for such structures is to have two YBa$_2$Cu$_3$O$_{7-\delta}$ (YBCO) layers (bottom and top) with critical temperature T$_c \approx$ 90 K and critical current density j$_c \approx$ 10^6 A/cm^2 at 77 K. To satisfy these demands, the insulating layer should allow epitaxial growth of the top layer and should have minimal diffusion into the YBCO layers. Preferably all three layers should be grown in the same deposition cycle. For microwave devices, insulators with comparatively low dielectric constants, $\varepsilon \approx$ 10-30, and low microwave losses, tgδ < 10^{-3}, are needed.

Several insulating materials are good candidates for being used as intermediate layers. Experiments have been performed with SrTiO$_3$, and in this case both the bottom and top YBCO layers showed high T$_c$ and j$_c$ [4,5]. SrTiO$_3$ looks promising as an intermediate layer in SQUID applications, however, its high tgδ and ε make it useless for microwave applications. Recently, LaAlO$_3$, which has comparatively good microwave properties, was used as an intermediate layer in S/I/S structures made by sputtering, however problems with pinholes were reported [6]. Two other interesting materials for microwave applications are the rare earth gallates $LaGaO_3$ and $NdGaO_3$. Single crystalline substrates of these materials have been used to grow high quality YBCO thin films [7]. The gallates have $\varepsilon \approx 25$ and $tg\delta \approx 1.8 \times 10^{-3}$ at room temperature ($tg\delta \approx 1.6 \times 10^{-6}$ at 4.2 K) [8] , i.e. at least two orders of magnitude lower than for $SrTiO_3$. Most promising of these two is the $NdGaO_3$ (NGO). It has a small difference in the thermal expansion coefficient as compared to YBCO in a broad temperature interval [9], good matching between the YBCO a-b plane and the NGO oxygen sublattice (mismatch less than 0.27% [8]), and the possibility of growing twin-free single crystals [7]. Because of these properties NGO is attractive to be used as a buffer layer on sapphire substrates.

The aim of this work was to investigate the conditions for epitaxial growth of YBCO/NGO/YBCO trilayer structures by pulsed laser deposition. As no data on the epitaxial growth of thin NGO films was available, we first studied the NGO growth on various single crystal substrates, such as $SrTiO_3$, MgO, and sapphire. The growth of NGO films as a buffer layer on sapphire was studied too.

2. TRILAYER YBCO/NGO/YBCO STRUCTURE

The NGO and YBCO films were grown in a diffusion pumped vacuum system (details of the process are given elsewhere [10]) using a KrF excimer laser, (λ=248 nm, pulse length $\tau \approx 30$ ns). The NGO and YBCO films were deposited at 675-765 °C in oxygen at 0.2 mbar and cooled down to 100 °C in oxygen at atmospheric pressure. For both materials, a laser beam spot size of 1x3 mm^2 and a laser energy density of 1.5 J/cm^2 were used, which, for a target-to-substrate distance of 3 cm, gave deposition rates of 1.5 and 1 Å/pulse for NGO and YBCO respectively. The laser was operated at 5 Hz.

When depositing single NGO layers on (100) $SrTiO_3$ substrates, the NGO films grew epitaxially at temperatures of 675-765 °C. A diffractogram of a 300 nm thick film made at 745 °C is shown in Fig. 1, and one can see that there is a well pronounced orientation (001) NGO parallel to (100) $SrTiO_3$. The full width at the

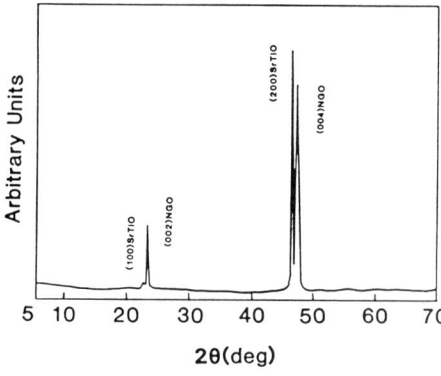

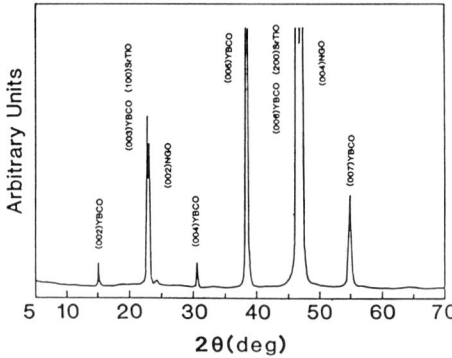

Figure 1. X-ray diffractogram for a 300 nm NGO film grown at 745 °C on a (100) $SrTiO_3$ substrate. The film shows strong c-axis orientation with only (00n) peaks.

Figure 2. X-ray diffractogram for an YBCO/NGO/YBCO trilayer structure on a $SrTiO_3$ substrate. All three layers had thicknesses of 300 nm. All layers are strongly c-axis oriented.

half maximum (FWHM) value of the (004) peak was 0.5°. In the case of MgO and sapphire substrates, the films were amorphous at condensation temperatures up to 765 °C. This shows the importance of having a good lattice match to achieve epitaxial growth of NGO films. The structure of the NGO also depended on laser beam energy density, and the best results were obtained at an energy density of 1.5 J/cm^2. Increasing the energy density up to 4-5 J/cm^2 created a polycrystalline and misaligned phase, and peaks corresponding to reflections from (200) and (131) were observed. A possible reason for this could be an increasing amount of micro-droplets and clusters in the ejected material. They have low mobility on the substrate and the low substrate temperatures did not allow an effective recrystallization. YBCO films deposited on SrTiO$_3$ substrates, usually at 730-760°C, were fully c-axis oriented and typically had a T$_c$ of 87 K under the deposition conditions used here.

The S/I/S structures were deposited with and without breaking the vacuum between the deposition steps. In both cases similar results were obtained. A diffractogram of such a structure where all layers have a thickness of 300 nm is shown in Fig. 2. The structure was grown epitaxially and only the NGO (002) and (004) peaks and the YBCO (00n) peaks, were observed. The FWHM value for the YBCO (005) peak was measured to be 0.25°.

The temperature dependences of the resistances of the top and bottom layers of a trilayer with 100 nm NGO are shown in Fig. 3. The resistance R as a function of the temperature T for the bottom layer was measured after etching the top S and I layers. Both S layers showed good superconducting properties: T$_c$ of 86.5 K and 86 K, resistivity of 2x10^{-3} Ωcm and 1x10^{-3} Ωcm at room temperature and R$_{300}$/R$_{100}$ ratios of 2.2 and 1.8, for the bottom and top layers respectively. The higher resistivity of the bottom layer might be an indication of diffusion from the NGO layer. In one case, part of the substrate was shielded off by a metal mask after depositing the first layer, thus making it possible to measure the resistance between top and bottom layers. Over an area of 9 mm^2 with an intermediate layer thickness of 300 nm, we measured a resistance of 500 kΩ at 300 K, giving a resistivity of >10^8 Ωcm.

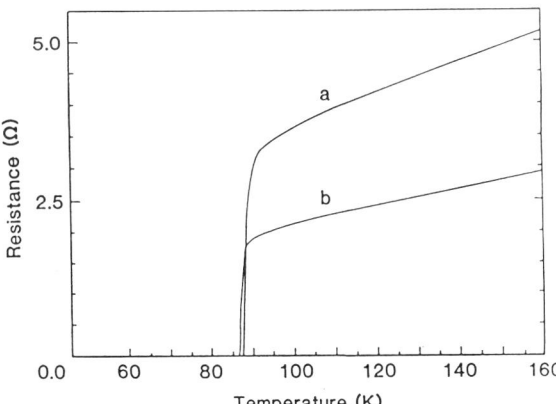

Figure 3. Resistance vs. temperature for the bottom (curve a) and top (curve b) layers of a trilayer structure. Top and bottom layers both had a thickness of 300 nm and the intermediate NGO layer had a thickness of 100 nm. The top layer was measured before patterning and the bottom layer after.

Under cooling, the resistance of the NGO layer showed an insulating behaviour and was more than 2 MΩ at 77 K. It should be pointed out that several atempts to make S/I/S structures, using insulators with thicknesses less than 100 nm, gave shortages between the S layers.

3. COMPLEX NGO/YBCO BUFFER LAYERS

In this section we report the results of our study of the properties of NGO as a buffer layer on sapphire substrates. The latter was chosen as it is an attractive material for microwave applications, because of its low $\varepsilon \approx 10$ and $tg\delta \approx 10^{-6}$ at 77 K. Although it is possible to obtain YBCO films on sapphire with sufficiently high T_c ≈ 90 K, the reaction between the YBCO film and the substrate resulted in very low $j_c \approx 10^2$ A/cm^2 at 77 K [11] and a buffer layer is required.

As was mentioned above, the NGO growth mechanism is very sensitive to the lattice match to the surface on which it is growing. The direct growth of NGO on sapphire substrates resulted in an amorphous film. To create conditions for epitaxial growth of NGO on sapphire there is a need to introduce, between the NGO and the substrate, a sublayer of some material with a low lattice mismatch as compared to NGO. Furthermore, this material should grow epitaxially on sapphire. One of the main requirements for a highly oriented growth is that a particular orientation has a lower free energy of formation. In the case of YBCO this requirement is fullfilled for the [00l] direction [12]. This is the reason for the two-dimensional growth mechanism of YBCO (ledge mechanism [12]). TheYBCO grows highly c-axis oriented at the initial stage of nucleation, even on polycrystalline substrates with large lattice mismath [13]. We suggest that a thin sublayer of YBCO could serve as a structural "catalyst" to the epitaxial growth of NGO.

The growth of NGO was directly related to the orientation and quality of the YBCO sublayer. We studied the orientation of NGO as a function of thickness of the YBCO. At 50-80 nm the YBCO sublayer was smooth and well c-axis oriented. At lower thicknesses the density of pinholes increased. At thicknesses of 150 nm and above, YBCO (110) and (103) peaks appear in the diffractograms. This can be explained by the fact that with increase of the film thickness, an increase of the energy of tension occurs, which results in a break of the oriented growth and the appearence of misoriented phases. In both latter cases, the NGO grows misoriented and a (200) NGO peak appears in the diffractogram.

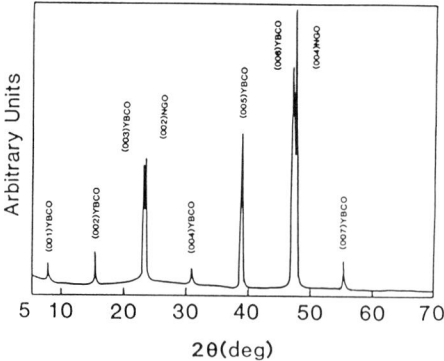

Figure 4. X-ray diffractogram for a 300 nm thick YBCO film deposited on a sapphire substrate with buffer layer. The buffer layer consists of 80 nm YBCO deposited on sapphire and 200 nm NGO deposited above it.

Figure 5. X-ray diffraction pattern (low incidence angle) for YBCO film grown on sapphire with complex buffer layer NGO/YBCO, where the thickness of the YBCO sublayer varies: (a) 80 nm and (b) 20 nm.

An 80 nm thick YBCO sublayer, followed by a 200 nm thick NGO layer and a 300 nm thick YBCO layer, were grown according to the deposition process described above. All layers were c-axis oriented and only [00n] peaks were observed in the diffractogram (Fig. 4). Low angle diffraction was performed to obtain information about the azimuth orientation of the subcrystals. According to the photos (Fig. 5a) well defined texture exist in the a-b plane.

The R vs. T dependences of YBCO films deposited on bare sapphire and sapphire with a complex buffer NGO/YBCO layer are shown in Fig.6. In the case of YBCO on bare sapphire, a typical semiconductor behaviour was observed with a T$_c$ of 46 K. The YBCO film deposited on the complex buffer layer was significantly better with a T$_c$ of 81 K and metallic behaviour of R(T).

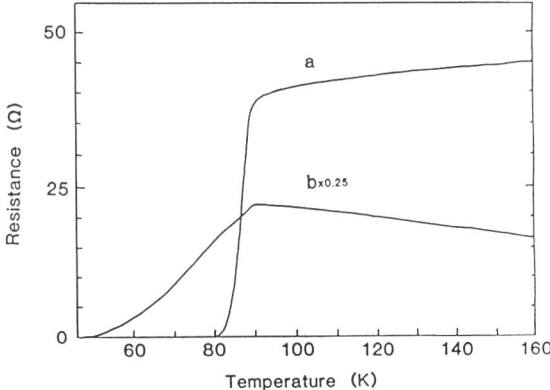

Figure 6. Resistance vs. temperature for the 300 nm thick YBCO film deposited on a sapphire substrate covered with buffer layer (curve a) and on a bare sapphire (curve b). The buffer layer consists of 80 nm YBCO deposited on sapphire and 200 nm NGO deposited above it.

The diffusion processes between different layers in the multilayer structure described above were studied by Auger depth profile analysis. The Al diffuses through the YBCO sublayer. However the NGO layer stopped further diffusion and preserved the upper YBCO layer from Al diffusion. The intensity of Ga diffusion from the NGO buffer layer into YBCO was depending upon its azimuth disorientation. The length of the diffusion region was about 600 Å in the case of YBCO film disoriented in a-b plane. The low T_c as compared to the one of YBCO films grown on NGO single crystal substrates can be explained with the disorientation in a-b plane of the NGO buffer layer.

In summary, we have studied the epitaxial growth of multilayer structures made by pulsed laser deposition, using NGO as an intermediate or buffer layer. In the case of the S/I/S trilayer structures, the bottom and the top YBCO layers were not affected by the NGO and showed a T_c of 86 K. A resistivity of $>10^8$ Ωcm was measured for a 300 nm thick NGO film at 300 K. The microwave properties of the NGO promise a big potential for use of such S/I/S structures in various microwave applications. The growth of the NGO films showed very high sensitivity to the lattice mismatch to the substrate. A double NGO/YBCO layer was developed and used as a buffer for growing YBCO on sapphire. The study showed that the disorientation in the a-b plane of the NGO layer in this case, was the reason for the comparatively low T_c of 81 K and further optimization of the buffer layer is required.

We gratefully acknowledge the assistance of G. Johansson in preparing the $NdGaO_3$ targets, J. Ramos for supplying YBCO targets, and D. Winkler for stimulating discussions. The work was funded by the Swedish Board for Technical Development and the Swedish Industrial Superconductive Initiative.

4. REFERENCES

1 R. Gross, P. Chaudhari, M. Kawasaki, M B. Ketchen, and A. Gupta, Appl. Phys. Lett. 57 (1990) 727.

2 J. Gao, W.A.M. Aarnink, G.J.Gerritsma, and H. Rogalla, Physica C, 171 (1990) 126.

3 W. G. Lyons, and R. S. Withers, Microwave Jour. vol **33**, No 11, 85, (1990).

4 F.C. Wellstood, J.J. Kingston, and J. Clarke, Appl. Phys. Lett., 56 (1990) 2336.

5 J.J. Kingston, F.C. Wellstood, P. Lerch, A.H. Miklich, and J. Clarke, Appl. Phys. Lett., 56 (1990) 189.

6 A.E. Lee, C.E. Platt, J.F. Burch, R.W. Simon, J.P. Goral, and M.M. Al-Jassim, Appl. Phys. Lett., 57 (1990) , 2019.

7 G. Koren, A. Gupta, E.A. Giess, A. Segmüller, and R.B. Laibowitz, Appl. Phys. Lett.,54 (1989) 1054.

8 M. Sasaura, S. Miyazawa, and M. Mukaida, J. Appl. Phys., 68 (1990) , 3643.

9 R.L. Sandstrom, E.A. Giess, W.J. Gallagher, A. Segmüller, E.I. Cooper, M.F. Chisholm, A. Gupta, S. Shinde, and R.B. Laibowitz, Appl. Phys. Lett., 53 (1988) 1874.

10 G. Brorsson, Z. Ivanov, and P.Å. Nilsson, Science and Technology of Thin Film Superconductors 2, edited by R. D. McConnell and R. Noufi.

11 Z. Ivanov, Mod. Phys. Lett. B, 2 (1988) 805.

12 R. Ramesh, C.C. Chang, T.S. Ravi,D.M. Hwang, A. Inam, X.X. Xi, Q. Li, X.D. Wu, and T. Vankatesan, Appl. Phy. Lett., 57 (1990) 1064.

13 D.P. Norton, D.H. Lowndes, J.D. Budai, D.K. Christen, E.C. Jones, J.W. McCamy, T.D. Ketcham, D.St. Julein, K.W. Lay, J.E. Tkaczuk, J. Appl. Phys., 68 (1990) 223.

High T$_c$ Superconductor Thin Films
L. Correra (Editor)
653

Plasma-Enhanced CVD of Yttria Stabilized Zirconia

H. Holzschuh and H. Suhr

University of Tübingen, Department of Organic Chemistry
Auf der Morgenstelle 18, D-7400 Tübingen, FRG

Abstract

By using plasma enhanced chemical vapour deposition (PECVD) (100) textured yttria stabilized zirconia (YSZ) films were deposited onto quartz, silicon, nickel, and stainless steels. Substrate temperatures were T_S= 673-873 K. The cubic YSZ was observed within the yttria range of 3.5 to 80 mol% while the lattice parameters increased in the range from 0.508 to 0.527 nm (at T_S=773 K). Smooth and crackfree films could be deposited on the metals and on silicon. Interdiffusion and chemical reactions were investigated between metal substrates and films after annealing at 1173 K for 15 min. Films on nickel showed a layered structure of Ni/NiO/YSZ while on steel deep penetration of Fe and Ni into the oxide layer and destruction of the surface was found.

1. INTRODUCTION

High-quality high-T_c-superconducting thin films have been deposited on single crystal substrates (SrTiO$_3$, MgO and LaAlO$_3$) and on polycrystalline YSZ, but it has been difficult to extend these superior results to substrates used in electronics or other technologies. Hence there is a great demand for buffer layers to reduce chemical reaction, interdiffusion, and thermal stress. Yttria stabilized zirconia (YSZ) buffer layers are commonly used [1-2].

We report here the preparation of (Y$_2$O$_3$)$_m$(ZrO$_2$)$_{1-m}$ thin films on different substrates by plasma-enhanced chemical vapour deposition (PECVD). We investigated the influence of film composition on the cubic lattice parameter and the crystal phase. Film-substrate interactions of YSZ films on metal substrates are demonstrated after annealing at 1173 K for 15 min.

2. EXPERIMENTAL

To prepare $(Y_2O_3)_m(ZrO_2)_{1-m}$ thin films the 2,2,6,6-tetramethyl-3,5-heptan-dionates of yttrium and zirconium ($Y(thd)_3$ and $Zr(thd)_4$) were vapourized from two separate reservoirs and carried into a parallel-plate reactor by a gas stream. The precursors have been synthesized according to published procedures [3] and were characterized by NMR and masspectra. Evaporation temperatures were 373-413 K for $Y(thd)_3$ and 403-473 K for $Zr(thd)_4$.

The parallel-plate reactor was constructed of stainless steel (ø: 100 mm, h: 260 mm) and equipped with two electrodes (ø: 50 mm) with a spacing of 2 cm. The upper, insulated electrode was connected to the radio frequency power supply (13.56 MHz) and was provided with a shielding to avoid discharge towards the reactor walls. The substrates were attached to the upper electrode which could be heated up to 873 K. The precursors were transported into the reactor vessel through the lower grounded electrode which was heated to 483 K. Mass flow controllers were used to regulate the gas flow. The substrates (quartz, silicon wafers ((100) and (111)), nickel and the stainless steels (V2A and Hastelloy)) were degreased by solvents. For the removal or reduction of surface oxides on silicon, nickel, and the steels, the substrates were subjected to a cleaning plasma prior to the deposition experiments (15 min Ar plasma at 5 W/cm^2, 737 K, 20 Pa). Films were deposited at different substrate temperatures (673-873 K) and vapourizer temperatures while all other parameters were kept constant (rf power density: 3 W/cm^2, deposition pressure 20 Pa, gas flow 10 cc/min Ar and 60 cc/min O_2).

The microstructure and composition of the deposits were studied by scanning electron microscopy (SEM) coupled with a wavelength dispersive X-ray detector (WDX). The composition of the cross-section was investigated by backscattered electron imaging in composition contrast (BEI). X-ray diffraction (XRD) spectra were obtained using CuKα radiation. The lattice constants were calculated from the 2 ϑ value of the (200) reflection from cubic YSZ films. The film thickness was determined by a stylus profilometer (Tencor Instruments alpha-step 200). In some cases electron diffraction for chemical analysis (ESCA) was used as an independent confirmation of the composition of the films.

3. RESULTS AND DISCUSSION

Transparent and crystalline $(Y_2O_3)_m(ZrO_2)_{1-m}$ thin films were obtained at substrate temperatures Ts $\geq$ 637 K and deposition rates ranging from 0.3 to 0.8 µm/h.

The single YSZ phase was formed at yttria contents of 3.5 to 80 mol%. Below 3.5 mol% the films consist of a multiphase of cubic YSZ and monoclinic ZrO_2, whereas above 80 mol% they consist of cubic YSZ and cubic Y_2O_3 multiphase. The range in which the single cubic YSZ was obtained by PECVD far exceeds the range predicted from the phase diagram for the bulk yttria-zirconia system [4] (Fig.1). The plasma deposition probably produced a metastable phase. The stability of this phase was examined by XRD after annealing the YSZ films on quartz and Si(100) (yttria content: 50-80 mol%) for 8 h at 1248 K. No phase changes from cubic YSZ to the higher ordered $Zr_3Y_4O_{12}$ was observed.

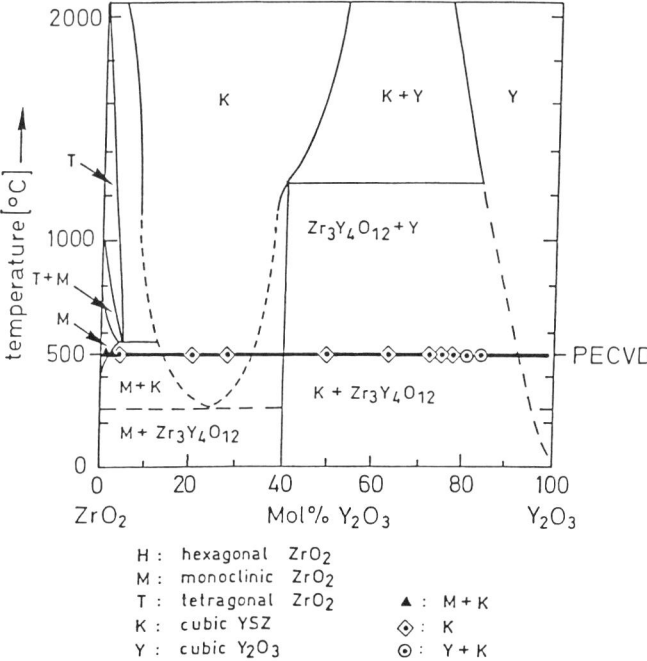

Figure 1. The binary yttria-zirconia phase diagram [4], with the thin film data presented here

An increase in the yttria content from 3.5 to 80 mol% in the film causes the cubic lattice parameter of the fully stabilized zirconia to increase from 0.508 to 0.527 nm (T_s= 773 K). This greater range of composition and consequently larger variation in lattice parameters than reported by other authors [5,6] is certainly due to the low deposition temperature used by PECVD.

Cubic YSZ films thicker than 1 μm were found to be completely (100) textured on all examined substrates and showed smooth and crackfree surfaces on the metals and on silicon (Fig. 2), while films on quartz cracked, which we attribute to high thermal stresses (thermal expansion coefficients α in 10^{-6}/K are: YSZ (10 mol%): 11.4, quartz: 0.5, Ni: 13, V2A: 17.3, Hastelloy: 11.3, Si: 3.8 [7-9]).

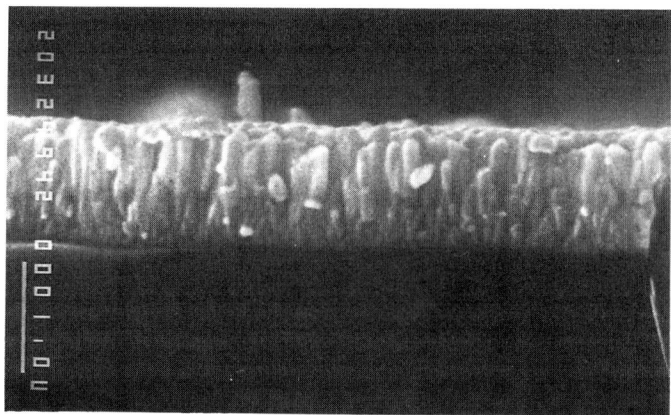

Figure 2. Fracture cross-section of (100) textured YSZ film on Si(100)

Annealing of YSZ films deposited on Ni and the stainless steels at 1173 K for 15 min resulted in cracks and blisters of the films and chemical reactions as demonstrated by the XRD spectra and cross-section BEI pictures in figure 3 and 4. Films on Ni showed a graded zone consisting of NiO and penetration of small amounts of Ni into the YSZ film. On the Hastelloy stainless steel a deep penetration of iron, chromium and nickel into the oxide layer was found. BEI pictures showed the total destruction of the surface after annealing the YSZ on steel due to an intensive reaction between the substrate and the oxide layer. In addition XRD spectra showed an oxidation of the metals at the interface.

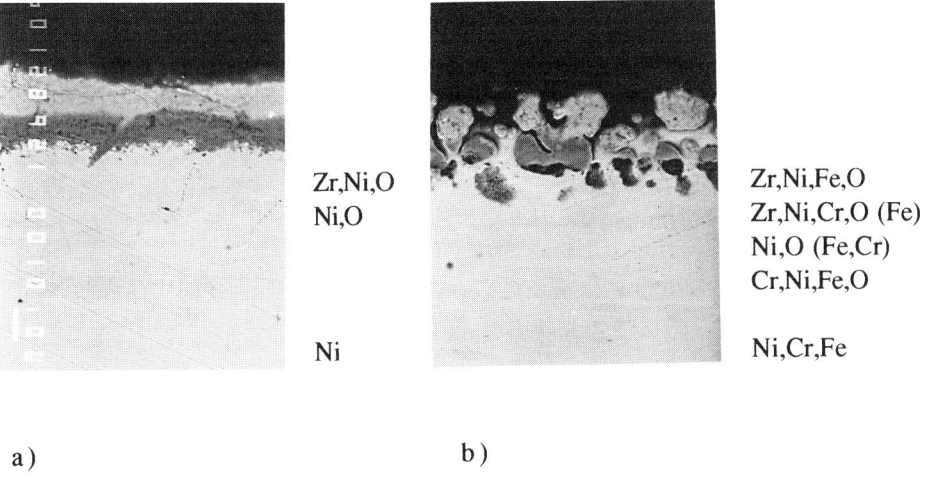

a)   b)

Figure 3. BEI pictures of annealed YSZ films on a) Ni and b) steel (Hastelloy)

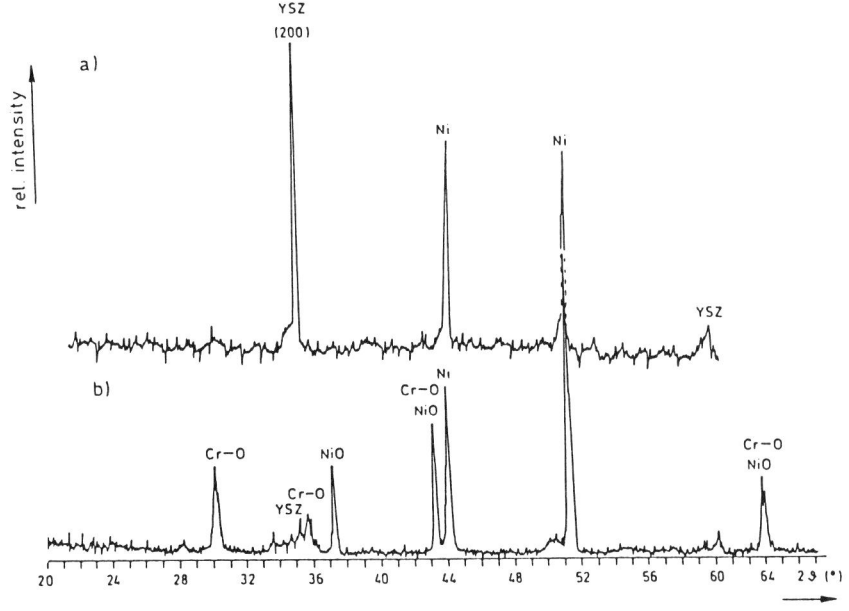

Figure 4. XRD-spectra of YSZ films on steel a) as deposited b) annealed

4. CONCLUSION

Cubic (100) textured YSZ films have been deposited at low deposition temperatures (T_S: 737 K) by plasma-enhanced chemical vapour deposition. The fully stabilized zirconia was obtained over a wide range of yttria contents (3.5 to 80 mol%). Simultaneously the cubic lattice parameters increased with the yttria concentration from 0.508 to 0.527 nm. YSZ thin films were deposited on quartz, silicon, nickel, and steel substrates. Films on quartz cracked while films on the other substrates were smooth and crackfree after deposition. Annealing at 1173 K for 15 min of YSZ on metal substrates resulted in film-substrate reactions. For the stainless steels these resulted in the total destruction of the film surface. On Ni substrates, however, a layered structure Ni/NiO/YSZ was formed after annealing.

5. ACKNOWLEDGEMENTS

This work has been supported by the Bundesministerium für Forschung und Technologie (grant 13N56750).

6. REFERENCES

1 D. K. Fork, A. Barrera, T. H. Geballe, A. M. Viano, and D. B. Fenner, Appl. Phys. Lett., 57 (1990) 2504-2506.

2 S. B. Ogale, R. D. Vispule, and R. R. Rao, Appl. Phys. Lett., 57 (1990) 1805-1807.

3 R. C. Mehrotra, R. Bohra, and D. P. Gaur (eds.), Metal ß-Diketonates And Allied Derivatives, Academic Press, London-NewYork-San Fransisko, 1978.

4 V. S. Stubican, R. C. Hink, and S. P. Ray, J. Amer. Ceram. Soc., 61 (1978) 17-21.

5 H. Yamane and T. Hirai, J. Cryst. Growth, 94 (1989) 880-884.

6 R. P. Ingel and D. Lewis III, J. Amer. Ceram. Soc., 69 (1986) 325-332.

7 R. C. Weast, M. J. Astle, and W. H. Beyer (eds.), CRC Handbook of Chemistry and Physics, 66th edition, Boca Raton, Florida, 1985-1986.

8 W. Foerst (ed.), Ullmanns Enzyklopädie der technischen Chemie, 3rd edition, Urban&Schwarzenberg, München-Berlin, 1960.

9 D. R. Fork, D. B. Fenner, R. W. Barton, J. M. Philips, G. A. Conell, J. B. Boyse, and T. H. Geballe, Appl. Phys., 57 (1990) 1161-1163.

High T$_c$ Superconductor Thin Films
L. Correra (Editor)
© 1992 Elsevier Science Publishers B.V. All rights reserved.
659

Preparation and characterization of YBa$_2$Cu$_3$O$_{7-x}$ high T$_c$ superconducting thin films on Al$_2$O$_3$ (1$\bar{1}$02) with (ZrO$_2$)$_{1-x}$ (Y$_2$O$_3$)$_x$ buffer layers

K. Hradil[a,b], H.Schmidt[a], W.Hösler[a], W.Wersing[a], F.Frey[b]

[a]   Siemens AG, Research Laboratories, P.O.Box 830953, 8000 München 83,FRG

[b]   Institut für Kristallographie der LMU München, Theresienstr. 41, 8000 München 2, FRG

Abstract
YSZ was investigated as an intermediate buffer layer for the deposition of epitaxial YBa$_2$Cu$_3$O$_{7-x}$ (YBCO) thin films on sapphire . Both layers were deposited by a sputtering process. X- ray θ -2θ patterns indicated highly (h00) textured YSZ films and (001) textured YBCO films. Ion channeling showed χ_{min} values of 6% to 9% for the YSZ layers and χ_{min} values between 16% and 19% for the YBCO layers, confirming the epitaxial growth of both layers. X- ray pole figure measurements showed a sharp (001) pole next to other orientations for the YSZ and tilts for the YBCO [001] axis up to 5° for the YBCO layer investigated in this paper. The electrical characterisation of this YBCO layer resulted in a T$_c$ of 85˙K and a J$_c$ of 5x10^5 A/cm^2 at 77K which is attributed to a poorer degree of texturing and stoichiometric deviations in the YBCO layer. The best results obtained for YBCO thin films on the here investigated YSZ/sapphire were a T$_c$= 90 K and J$_c$= 1.2x10^6 A/cm^2 at 77K.[1]

1. INTRODUCTION

For microwave applications high T$_c$ superconducting YBCO thin films must be deposited on substrates with small dielectric losses and close lattice match to the superconductor to obtain good superconducting properties. These requirements are well fulfilled by sapphire single crystal substrates cut in (1$\bar{1}$02). The in situ crystalline deposition of YBCO thin films on sapphire above 700°C, however, lead to interdiffusion between the substrate and the deposited film, causing a deterioration of the superconducting properties.[2] Therefore, an intermediate buffer layer is required to prevent the detrimental reaction. This buffer layer has to be closely matched to the lattice constants of both, the sapphire substrate and the YBCO film. Various possible buffer layers have been investigated for this substrate such as $SrTiO_3$[3], YSZ[4] and MgO[5]. Within this report we present results using in situ rf- magnetron sputtered YSZ buffer layers on the (1$\bar{1}$02) plane of sapphire substrates.

2. EXPERIMENTAL

The YSZ and YBCO thin films were deposited by rf- magnetron sputtering. For the YSZ sputtering process a ceramic target with a Y_2O_3 content of 10-15 mol% was used to ensure the presence of the cubic phase in the YSZ film. For the deposition of the YBCO films a Cu- enriched $YBa_2Cu_{3.3}O_x$ target was used (because previously deposited films were found slightly poorer in their copper content compared to the target composition[6]). Table 1 shows the deposition parameters for the YSZ- and the YBCO films.

Table 1
Parameters for the deposition of the YSZ- and the YBCO films. The basis pressure of the sputtering chamber was 1×10^{-3} Pa.

	YSZ	YBCO
substrate temperature [°C]	780	740
gas pressure [Pa]	1.3	70
O_2 content [%]	10	33
rf- power [W]	165	260
deposition rate [Å/s]	0.1- 0.2	~0.5
film thickness [nm]	40- 50	200- 250
cooling parameters	in O_2- atmosphere p_{O2}= 100 Pa, 30 min	in O_2- atmosphere p_{O2}= 100 Pa, 60 min,

The sapphire substrates were (1$\bar{1}$02) oriented and had dimensions of 10x10x0.5 mm.

The deposited films were characterized by X- ray diffractometry (θ-2θ scan), Rutherford backscattering spectrometry, scanning electron microscopy and X- ray pole figure measurements. The last method is very useful to characterize the texture of films. Fig. 1 shows the geometry of this method. During the measurement the diffraction angle θ is fixed. The tilting angle α of the specimen is varied between 0° and 90°. Simultaneously, the sample is rotated 360° arround an axis perpendicular to its surface. In this way the resulting pole figure is able to indicate dominant misorientations and the degree of texturing.

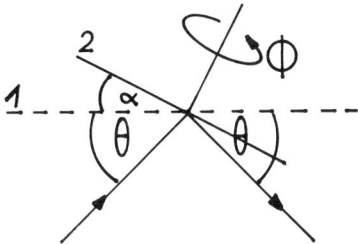

Figure 1. Geometry of the X- ray pole figure measurement. 1: diffraction plane; 2: crystal surface; θ: diffraction angle; α: tilt angle; Θ: rotation angle

3.RESULTS AND DISCUSSIONS

3.1 YSZ layer on sapphire

The X- ray diffraction θ -2 θ scan of YSZ film deposited on a $(1\bar{1}02)$ sapphire plane showed the YSZ layer to consist of a single phase and to be (001) oriented.[1] The surfaces examined by scanning electron microscopy (SEM) were very smooth (Fig. 2). The lattice parameter calculated with an internal Si standard from the (400) peak resulted in a value of $\underline{a}=0.516$ nm, which is 0.4% higher than that of $Y_{0.15}Zr_{0.85}O_{1.93}$ (JCPDS 30- 1468).

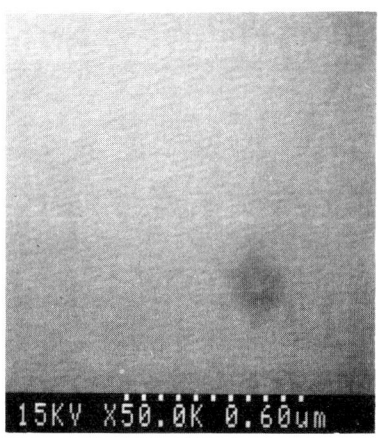

Figure2. SEM image from the surface of an YSZ layer.

Figure 3 shows a RBS channeling spectrum for random and channeling incidence of 2.0 MeV He^{++} ions for a 40 nm thick YSZ film. A minimum of the backscattered yield (χ_{min}) for the YSZ layer is observed for an ion incidence perpendicular to the substrate surface, which can be attributed to ion channeling along the [001] axis of the YSZ. The χ_{min} at the Al$_2$O$_3$/YSZ interface is 0.11, just below the surface 0.06 and at the surface of the YSZ 0.14. The increased χ_{min} of 0.11 at the interface is due to defects and mechanical strains caused by the difference of the in- plane anisotropic lattice mismatch between the sapphire and the YSZ. Strain fields at this interface were also observed by transmission electron microscopy[7].

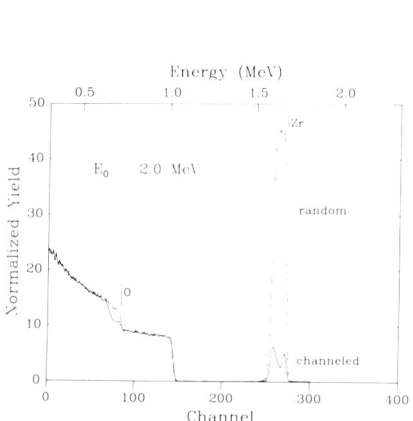

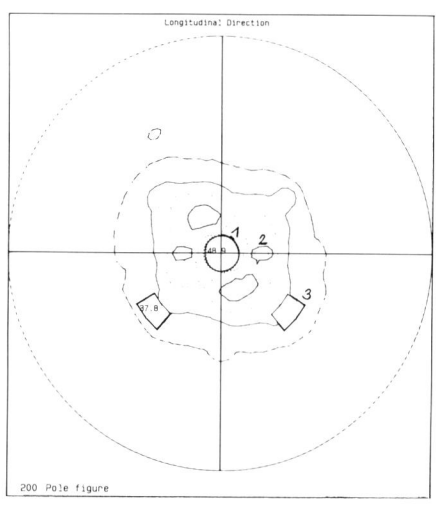

Figure 3. RBS- spectra from an epitaxial YSZ film

Figure4. X- ray pole figure of an epitaxial YSZ layer. 1: (200) pole ; 2: (311) pole of the YSZ; 3: (1$\bar{1}$02) pole of the sapphire

Figure 4 shows an X- ray (200) pole figure of an YSZ layer obtained with the measuring method described in **2**. A sharp maximum of the (200) pole and no peak splitting is observed in the middle of the pole figure, indicating a highly (001) textured YSZ film. In addition to the (200) pole two other peaks are detected and can be identified from the angle between these peaks and the (200) pole. In this way one pole was determined as the (311) pole for YSZ, while the other pole originates from the (1$\bar{1}$02) pole of the sapphire substrate (possibly superimposed by the (220) pole of the YSZ). The presence of the (311) pole indicates that also areas of the layer exhibit a (311) texture. An estimation of the peak intensity resulted in $I_{311}/I_{200}= 0.03$.

3.2 YBCO layer on YSZ/sapphire (1$\bar{1}$02)

YBCO layers were deposited in situ upon the YSZ/sapphire at substrate temperatures between 720 °C and 740 °C. The best results achieved up to now are YBCO films with a T_c of 90 K and a J_c of 1.2x10^6 A/cm^2 at 77K.[1] In this work an YBCO film with a T_c of 85 K and a J_c of 5x10^5 A/cm^2 at 77K was investigated. For both layers the X-ray diffraction θ-2 θ scan showed in addition to the (001) reflection from the YSZ and the substrate peaks the (001) peaks of the YBCO film, indicating the growth of the (001) plane parallel to the YSZ layer[1]. RBS channeling measurements of a (001) oriented film with reduced T_c, J_c- values resulted in an χ_{min} of 0.19 using 2.0 MeV He^{++} ions, while the film with higher T_c, J_c- values showed an χ_{min} of 0.16. The variation of the ion energy from 3.3 MeV to 1.0 MeV showed a variation in the χ_{min} between 0.26 and 0.11 which can be attributed to a mosaic spread in the YBCO layer[1]. The texture of the YBCO layer was also examined by an X-ray pole figure measurement. The (005) plane was used for this purpous resulting pole figure as shown in Fig.5. In the middle of the pole figure two sharp maxima are detected with an angle of 5° between the two poles. This corresponds to two dominant orientations of the YBCO grains, different by 5°. In addition to the sharp (005) pole three other peaks of low intensity are apparent in the pole figure, identified as the (0kl)or (h0l) poles of the YBCO. A correct distinction between the (0kl)/ (h0l) planes cannot be achieved, due to the resolution of the goniometer. The lower crystalline quality of this YBCO layer seen from the pole figure and the stoichiometric deviations (Y$_1$Ba$_{2.3}$Cu$_{3.2}$O$_{8.8}$) calculated from the RBS measurements[1], are responsible for the reduction of T_c to 85K and J_c to 5x10^{5}A/cm^2 (at 77K).

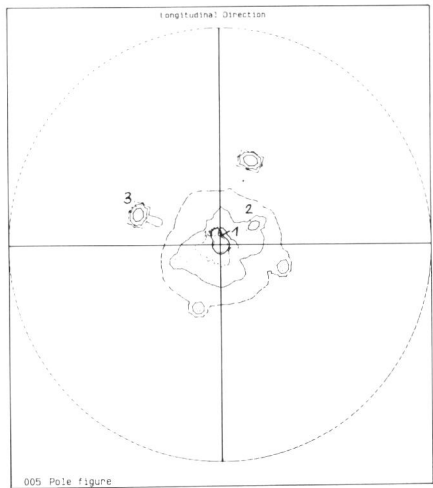

Figure 5: X- ray pole figure of the YBCO (005) pole. 1: (005) pole of the YBCO with a splitting of 5°. 2,3: (0kl)/(h0l) poles of the YBCO.

4. CONCLUSION

Epitaxial YBCO thin films have been deposited on sapphire using YSZ as an intermediate bufferlayer. The best results recieved up to now are a T_c of 90 K and a J_c of 1.2×10^6 A/cm^2 at 77K. Reduced T_c and J_c values have been correlated to crystalline defects and deviations from stoichiometry.

5. ACKNOWLEDGMENT

This work was supported by the " Bundesministerium für Forschung und Technologie" under the file number 13N5812.

6. REFERENCES

1   H. Schmidt, K.Hradil, W.Hösler, W.Wersing, G.Gieres and R.J.Seeböck, to be published in APL

2   K.Char, D.F.Fork, T.H.Geballe, S.S.Laderman, R.C.Tabe, R.D. Jacowitz, F.Bridges, G.A.N.Connell and J.B.Boyce, Appl. Phys. Lett., 56(8), (1990),785.

3   K.Char, N.Newman, S.M. Garrison, R.W. Barton, R.C.Taber, S.S.Laderman and R.D.Jacowitz, Appl. Phys. Lett., 57(4), (1990), 409.

4   X.D.Wu, R.E.Muenchausen, N.S.Nogar, A.Pique, R.Edwards, B.Wilkens, T.S.Ravi, D.M.Hwang and C.Y.Chen, Appl. Phys. Lett., 58(3), (1991), 304.

5   A.B.Berezin, C.W. Yuan, and A.L.de Lozanne, Appl.Phys.Lett., 57(1), (1990), 90.

6   H.Schmidt, O.Eibl and B.Jobst, Proceed. ICMC 90, Topical Conf. High Temperature Superconductors, 9- 11.5., 1990, Garmisch- Partenkirchen.

7   O.Eibl, K.Hradil and H.Schmidt, to be published in Physica C.

High T$_c$ Superconductor Thin Films
L. Correra (Editor)
665

SHIFT OF THE PHASE TRANSITION OF LaGaO$_3$ ABOVE THE
PREPARATION TEMPERATURE OF SUPERCONDUCTOR FILMS

B.Jobst and T.Berthold

Siemens AG, Corporate Research and Development,
Otto-Hahn-Ring 6, D-8000 München 83

ABSTRACT

Substitution of La by Gd in LaGaO$_3$ shifts the temperature
of the orthorhombic/rhombohedral phase transition linearly to
750°C at the composition (La$_{.9}$Gd$_{.1}$)GaO$_3$. The lattice
constants decrease with increasing Gd concentration, while a$_0$
and b$_0$ become very similar. At more than 10 mole% Gd the garnet phase Ga$_5$Gd$_3$O$_{12}$ is found in samples reacted above 1200°C.
At substitution with rare earth elements from Tb to Lu and
with Y, the garnet phase already occurred below 5 mole %.

1. INTRODUCTION

LaGaO$_3$ single crystals were among those discussed as the
first substrates for the epitaxy of superconducting YBa$_2$Cu$_3$O$_7$
layers, due to their perovskite structure, the low misfit of
their lattice constants and their interesting dielectric properties [1,2]. But the critical currents obtained with epita-
xial films of YBa$_2$Cu$_3$O$_{7-x}$ and Bi$_2$Sr$_2$CaCu$_2$O$_{8+x}$ on LaGaO$_3$
were much lower than on other substrates [3,4]. This is at-
tributed to the phase transition of LaGaO$_3$ at 150 $^{\circ}$C [5], at
which the films are damaged by a roughening of the substrate
surface. We recently observed that a substitution of La by Nd
raises the transition temperature [6]. To reach temperatures
above the normally applied deposition temperature of superconductor films, 30 mole% Nd or more are necessary. As such
high concentrations cause problems in the growth of single
crystals, we looked in this work for other elements which
would shift the transition temperature at low concentrations.

2. EXPERIMENTAL

We prepared $(La_{1-x}, RE_x)GaO_3$ with $x \leqslant 15$ mole % (RE = rare earth elements from Sm to Er, Yb, Lu and Y) from the oxides with 4N purity. The oxides were dried at 800 °C for 10 hours before weighing since they adsorb a lot of water, and were then mixed in an agate mortar. The powders were pressed into pellets using an organic binder and fired in air at 1400 °C for 10 hours, in some cases at 1600 or 1250 °C. Afterwards the pelletswere crushed, mixed, pressed and fired again under the same conditions to obtain a complete reaction.

The pellets were ground and examined for crystalline phases with an X-ray powder diffractometer at room temperature. If the powders consisted only of the orthorhombic perovskite pha- se we measured the phase transition temperature with a high temperature diffractometer [7]. We observed the transition of the reflections of the low temperature phase into those of the rhombohedral phase with changing temperature (Fig.1). In a first cycle we heated the powder in steps of 50 °C. After

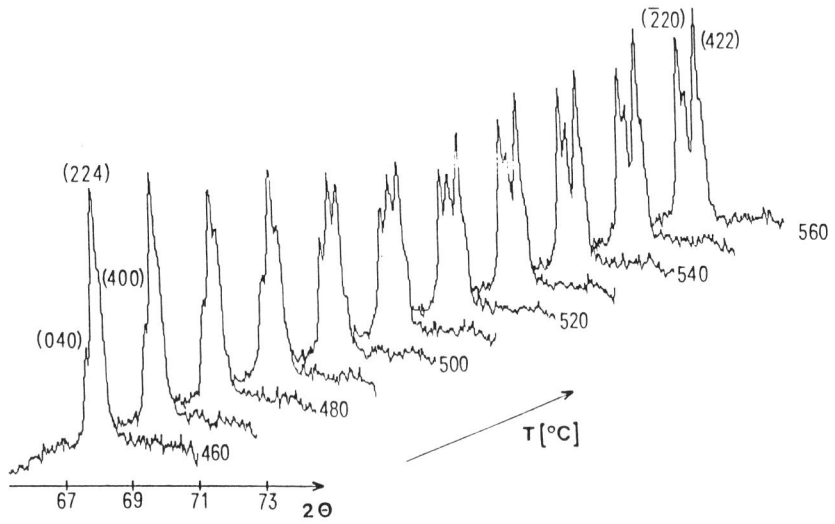

Fig.1 The group of the (040), (224) and (400) reflections of the orthorhombic and the ($\bar{2}$20) and (422) reflections of the rhombohedral phase at different temperatures.

this, in a second cycle the powder was heated and cooled down
again in steps of 5 °C over the 50 ° step in which the transition was found during the first cycle. In all cases the temperature of the phase transition was determined to within ± 20°.

3. RESULTS

Specimens with Gd concentrations between 0 and 10 mole% were
shown to be single phase orthorhombic at room temperature. The
orthorhombic/rhombohedral transition temperature increased
linearly with increasing substitution up to 750 °C at 10 mole%
Gd. At higher Gd concentrations, additional reflections of the
$Ga_5Gd_3O_{12}$ garnet phase were also observed in samples synthesized at 1600 °C and at 1250 °C. In samples substituted with Dy,
Er and Y, the corresponding garnet was already formed at concentrations above 3 mole %. Samarium-substituted samples were
single-phase orthorhombic at least up to 5 mole %.

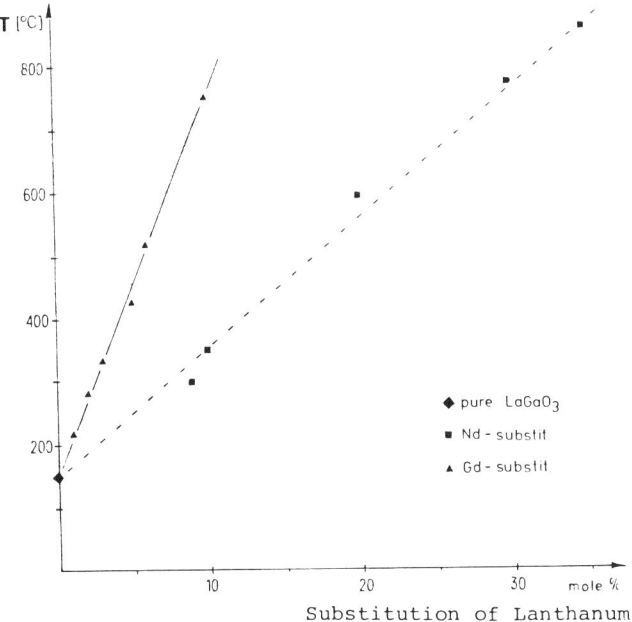

Fig.2 Temperature of the orthorhombic/rhombohedral phase transition of LaGaO₃ at different substitution rates

In some samples with Gd substitution the transition was smea-
red over a larger temperature range (up to $20°$) than at Nd substitution ($5°$). This is due to small inhomogeneities of the Gd
concentrations in the specimens, which are more clearly visible
because of the steeper shift of the transition temperature due
to Gd.

With increasing Gd concentration the length of the lattice
constants slightly decreased. The characteristic reflection
groups of orthorhombic perovskites merged into broadened single
peaks (Fig.3). The lattice could be called pseudocubic with
tetragonal superstructure at room temperature. We also observed
this behavior with the Nd substitution, but with Gd we could no
longer distinguish between a_0 and b_0 even at the lowest Gd concentrations. A similar behavior was observed in the high temperature phase.

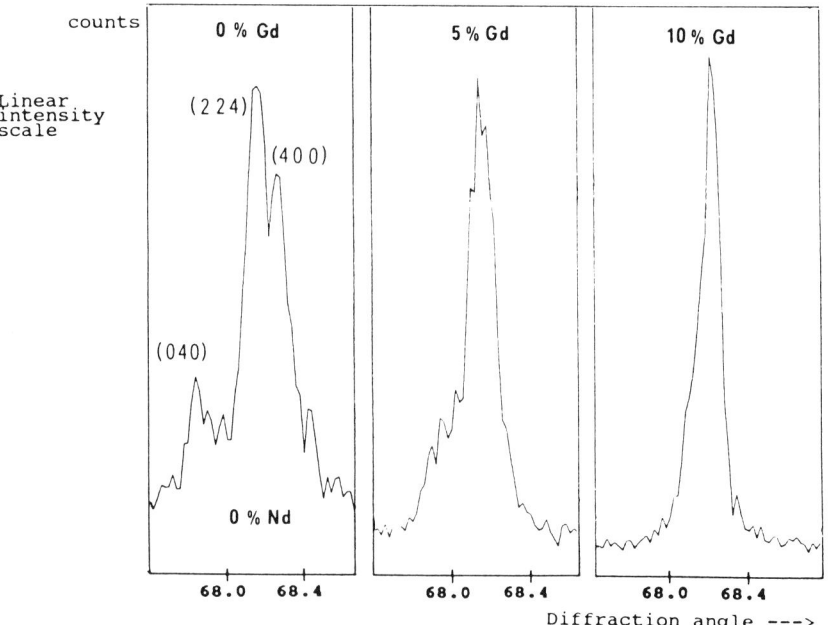

Fig.3   The (040)(224)(400) reflection group of pure and substi-
        tuted $LaGaO_3$ at room temperature

4. DISCUSSION

As we pointed out earlier, the phase transition behavior of substituted $LaGaO_3$ is on the whole governed by the average size of the atoms on the A-sites ($\bar{r}_A$) together with a non-negligible influence of the chemical properties of the elements which replace the lanthanum [6]. This is confirmed by our observations in the case of Gd substitution. The same mechanism seems to determine the solubility range of rare eart elements in $LaGaO_3$. Transition shift and solubility end up at $\bar{r}_A \approx 1.008$ Å (r_{La} 1.016 Å), which may be the deviation limit for $\bar{r}_A$ in the $LaGaO_3$ lattice. The reduction of $\bar{r}_A$ results in slightly different shifts of the transition temperature for Nd and Gd substitution (see Fig. 4). This is probably due to the different chemical properties of Nd and Gd. Another reason could be that the actual sizes of the A-atoms differ from the sizes used for evaluating the averages [8].

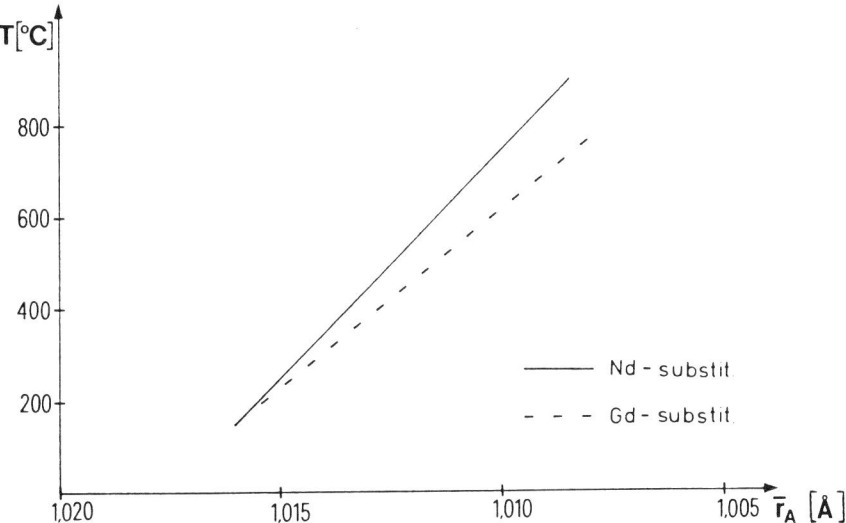

Fig.4 Transition temperature versus $\bar{r}_A$ for Nd and Gd substitution. The difference may be due to the different chemical properties of the rare earth elements.

The different results show that only a coarse estimation of the transition temperature from $\bar{r}_A$ is possible for substitution with any rare earth element. More experiments will be necessary for tailoring low-substituted $LaGaO_3$ with a transition temperature above the deposition temperatures of superconductor films.

The authors wish to thank J. Stielfried for her assistance in preparing the samples. The work was supported by the German Ministry of Research and Technology, support No. 13N5688.

REFERENCES

[1] R.L.Sandstrom, E.A.Giess, W.J.Gallagher, A.Segmüller, E.I.Cooper, M.F.Chisholm, A.Gupta, S.Shinde, R.B.Laibowitz: Appl.Phys.Lett. 53 (1988) 1874 – 1876

[2] D.W.Cooke, E.R.Gray, R.J.Houlton, B.Rusnak, E.A.Meyer, J.G.Beery, D.R.Brown, F.H.Garzon, I.D.Raistrick, A.Rollet, R.Bolmaro: Appl.Phys.Lett. 55 (1989) 914 – 916

[3] G.Koren, A.Gupta, E.A.Giess, A.Segmüller, R.B.Laibowitz: Appl.Phys.Lett. 54 (1989) 1054 – 1056

[4] G.Balestrino, V.Foglietti, M.Marinelli, E.Milani, A.Paoletti, P.Paroli: Appl.Phys.Lett.57 (1990) 2359-2361

[5] S.Miyazawa: Appl.Phys.Lett. 55 (1989) 2230 – 2232

[6] T.Berthold, B.Jobst: To be published in: Mater. Sci. Forum (1991)

[7] G.Zorn, E.Hellstern, H.Göbel, L.Schultz: Advances in X-ray Analysis, Vol.30 (1987) 483 – 491

[8] R.C.Weast: Handbook of Chemistry and Physics CRC Press, Cleveland, Ohio, 56.ed. (1976) F209 – F210

High T$_c$ Superconductor Thin Films
L. Correra (Editor)
© 1992 Elsevier Science Publishers B.V. All rights reserved.   671

BaRO$_3$ (R = Zr, Hf) as possible buffer layer materials for HTSC-films

I.E Korsakov[a], V.N Fuflygin[a], A.R. Kaul[a], A.V.Permyakov[a], I.V.Zubov[b], S.V.Red'ko[b], B.P.Gorshunov[c], A.G.Pimenov[c]

[a]Chemistry Department, Moscow State University, Moscow, 119899, USSR

[b]Department of Physics, Moscow State University, Moscow, 119899, USSR

[c]Institute of General Physics , USSR Academy of Sciences, Moscow, USSR

Abstract
BaR$_{0.8}$Y$_{0.2}$O$_{3-y}$ (R =Zr,Hf) proposed in the previous work of authors as possible substrate material with reduced chemical reactivity to YBa$_2$Cu$_3$O$_{7-y}$ are characterized by high temperature X-ray diffraction and microwave spectrometry. Thermal expansion coefficients of these solid solutions are calculated and dielectric constants are given. 0.25-1 mkm buffer layers of these compositions are obtained by alcoxide-sol-gel method on YSZ and Al$_2$O$_3$ substrates. Kinetics of "123" film-substrate interaction is determined by high temperature resistivity measurements.

Introduction
Low chemical activity, minimal mismatch in lattice parameters, low dielectric losses at high frequencies, absence of phase transitions, accordance of thermal expansion coefficients (TEC) are the criteria for choice of substrate materials.

All substrate materials in use possess some drawback, for instance great value dielectric constant of SrTiO$_3$, phase transition of LaGaO$_3$ at 418 K, lattice mismatch of YSZ striking the epitaxy, strong chemical activity of sapphire and silicon to YBa$_2$Cu$_3$O$_{7-y}$. Thus now search of new substrate materials for HTSC-films is problem to dissolve. Materials with low reactivity are of particular importance when low temperature *in situ* deposition is impossible for some reason or other and step of high temperature annealing is necessary.

In our previous work chemical reactivity of perovskite-like solid solutions BaR$_{1-x}$Y$_x$O$_{3-y}$ (R = Ce, Zr, Hf, Th) was studied and two composition BaR$_{0.8}$Y$_{0.2}$O$_{3-y}$ (R = Zr, Hf) were found to be most perspective. In the present work we communicate thermal expansion characteristic, dielectric properties, sol-gel preparation of buffer layers from the compounds mentioned above and kinetics of their interaction with YBa$_2$Cu$_3$O$_{7-y}$.

Thermal Expansion Coefficients
The crystal structure, values of unit cell parameter a and the thermal expansion at high temperatures (in air) were obtained from the X-ray diffraction data taken

with the use of the diffractometer with the CuK $_\alpha$ monochromatic radiation. The function a(T) in the temperature range 290K < T < 1200 K was determined from the temperature shift of (211) and (321) reflections. To minimize experimental errors the dependence of a(ϑ) was extrapolated to the Bragg angle ϑ = 90°. All samples were identified in the structure of cubic perovskite at all temperatures. As can be seen from Fig.1 the thermal expansion of the $BaHf_{1-x}Y_xO_{3-y}$ compounds is rather linear in the temperature range studied, the same is true for the $BaHf_{1-x}Y_xO_{3-y}$ compounds. The calculated parameters and linear thermal expansion coefficients are given in the Table 1 in comparison with the available data for other substrate materials.

Table 1
Unit cell parameters (a), linear thermal expansion coefficients (TEC) and dielectric properties (ε' and tg δ) of some substrate materials

Composition	a,Å [a]	TEC*10^6 [b]	ε' [a,c]	tgδ [a,c]
		Our data		
$BaHfO_3$	4.177(1)	7.7(2)	6	0.017
$BaHf_{0.9}Y_{0.1}O_{3-y}$	4.180(1)	7.6(1)	9	0.026
$BaHf_{0.8}Y_{0.2}O_{3-y}$	4.184(1)	8.3(2)	10	0.057
$BaZrO_3$	4.191(1)	-	9	0.012
$BaZr_{0.9}Y_{0.1}O_{3-y}$	4.195(2)	8.4(6)	19	0.013
$BaZr_{0.8}Y_{0.2}O_{3-y}$	4.201(2)	8.8(2)	27	0.016
		Literature data		
$SrTiO_3$	3.905	11.1	310	0.09
MgO	4.213	13.0	13	0.01
$ZrO_2(Y_2O_3)$	5.15	10.3	10.3	0.015
$\alpha-Al_2O_3$	hexagon.	7.5	7.5	0.001
SiO_2	amorph.	0.01	0.01	0.0001
$YBa_2Cu_3O_{7-y}$	a = 3.824	16.9		
	b = 3.889			
	c = 11.169			
$LaGaO_3$	a = 5.485	7-13		
	b = 5.521			
	c = 7.771			
$NdGaO_3$	a = 5.417	9-10		
	b = 5.499			
	c = 7.717			

[a] measurements at room temperature
[b] measurements up to 1000°C
[c] frequency range 200-500 GHz

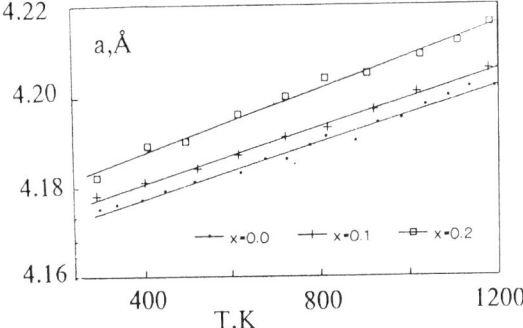

Figure 1. Temperature dependence of $BaHf_{0.8}Y_{0.2}O_{3-y}$ unit cell parameters.

One can see from the Table 1 that TEC's of the substances studied are similar to each other, but there is the tendency of their increase with Y-content increasing as well as with the change of Hf for Zr. The values of TEC's are appreciably less than those of $YBa_2Cu_3O_{7-y}$ and most substrate materials excluding Al_2O_3, but it is known that films with good morphology were obtained on Al_2O_3-substrate also.

Dielectric properties

The backward-wave-oscillator transmission spectroscopy was performed on the laboratory-made spectrometer "Epsilon" [2] in the frequency range 200-500 GHz at room temperature. Specimens were ~50 mkm thick polished ceramic pellets with the relative density ~80%.

Results of measurements are given in the Table 1. It is seen that dielectric constant ε' and dielectric loss tgδ of the studied compounds $BaR_{1-x}Y_xO_{3-y}$ are considerably less than those of $SrTiO_3$ and seem to be similar to dielectric characteristics of YSZ, MgO and Al_2O_3. Taking into consideration that measurements were performed on ceramic specimens one may expect that loss of $BaR_{1-x}Y_xO_{3-y}$ single crystals could be less. Two tendencies may be noticed: higher dielectric losses of Hf-based solid solutions compared with Zr-based ones and augmentation of dielectric losses with Y-content increasing.

Buffer layers: synthesis and characterization

The attempts to prepare single crystals of the investigated substances by Chokhralsky method and by spontaneous crystallization of melt in cold-wall crucible were undertaken. The first method was unsuccessful because of great volatility of BaO at temperature near the $BaZrO_3$ melting point 2800 K, appreciable loss of BaO from $BaZrO_3$ was observed even at 1500 K in course of the ceramic sintering. Spontaneous crystallization of the melt gave very dense polycrystalline product with some crystals several mm in size.

That's why instead of bulk substrates we obtained buffer layers of $BaR_{1-x}Y_xO_{3-y}$ on polished single crystal substrates of $ZrO_2(Y_2O_3)$ 3×4 cm^2 and sapphire 76 mm in diameter.

Buffer layers were prepared by spin-on technique using the solutions of Ba, Zr, Hf, Y -β-methoxy-ethylates in absolute β-methoxy-ethanol (concentration was 7-9 weight% of metals). To obtain the needed thickness 0.25-1 mkm from 5 to 20 cycles of dripping were performed. After final heat treatment of precursors at 1073-1273 K very dense, smooth and transparent layers were obtained. X-ray diffraction showed the layers to be single phase and polycrystalline. SEM revealed very uniform morphology of all the layer which consists of very fine grains invisible even at magnification up to 10000^x (Fig.2). The layers were practically free of cracks and pores except the narrow band (1 mm of width) along the substrate edge. Tallistep profilograph investigation of buffer layers showed surface roughness amplitude to be no more than 400 Å and thickness fluctuations to be less than 2% per 1 cm^2.

Thickness of the layer on $ZrO_2(Y_2O_3)$ after heat treatment depends linearly on the number of drippings. The tangent is 400-600 Å per dripping depending on concentration of the solution. It is interesting to note, that tangent also depends on the type of the substrate to coat and if solution contained Zr or Hf.

According to SEM data, single dripping gave entire precursor layers. However, after the precursor was annealed the separated oriented particles of thickness more than expected 400-600 Å were formed (Fig.3). Evidently, the rupture and the local thickening are caused by the tendency to minimize surface energy. Thicker precursors (2 and more drippings) gave entire layers after high temperature annealing.

There are several reasons for defects formation in layers:
1) gel-like particles of hydrolyzed alcoholates make worse morphology of layer;
2) heating (or cooling) rate more than 40 K/min leads to extensive cracking during 2heat treatment;
3) increase of the precursor thickness $> \sim 0.8$ mkm gives cracking even before heat treatment.

It is interesting to note that the alcoholates solutions used are more stable for hydrolysis than those with the same Zr:Y ratio 4:1 but not including Ba. It results in more perfect morphology of $BaHf_{0.8}Y_{0.2}O_{3-y}$ and $BaZr_{0.8}Y_{0.2}O_{3-y}$ layers than that of YSZ ($Zr_{0.8}Y_{0.2}O_{2-y}$) layers produced under the same conditions (Fig.4).

Kinetics of the $YBa_2Cu_3O_{7-y}$ film - substrate interaction

YSZ substrates with buffer layers obtained as described in previous section were coated by 2-3 mkm thick films of $YBa_2Cu_3O_{7-y}$ using spray-pyrolysis technique.

Kinetics of film - substrate interaction was studied by isothermal resistance mesurements at 1173 K and 1213 K after heating the sample with the rate of 20 K/min in oxygen flow. Decreasing of film resistance is caused by the formation and sintering of $YBa_2Cu_3O_{7-y}$ - phase while increasing of resistance reflects degradation in consequence of film - substrate chemical interaction.

One can see from the Fig. 6a that "123" film reacts strongly with non protected YSZ-substrate while the 0.25 mkm thick $BaZr_{0.8}Y_{0.2}O_{3-y}$ layer provides very significant protecting action. $BaHf_{0.8}Y_{0.2}O_{3-y}$ layer of the same thickness demonstrates stronger protecting behavior.

Figure 2.
0.5 mkm $BaZr_{0.8}Y_{0.2}O_{3-y}$ on YSZ
heat treated at 1273 K for 1 hr.

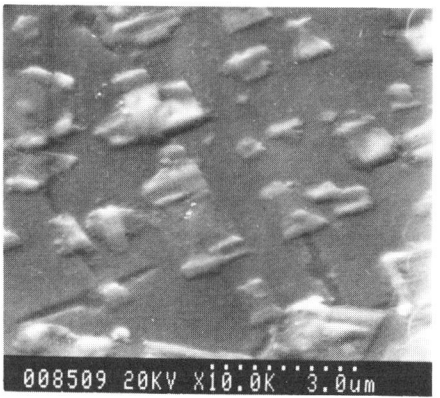

Figure 3.
0.5 mkm $BaZr_{0.8}Y_{0.2}O_{3-y}$ on YSZ
single dripping layer after
annealing at 1273 K for 1 hr.

Figure 4.
Bad morphology of YSZ layer on
YSZ substrate,1273 K, 1 hr.

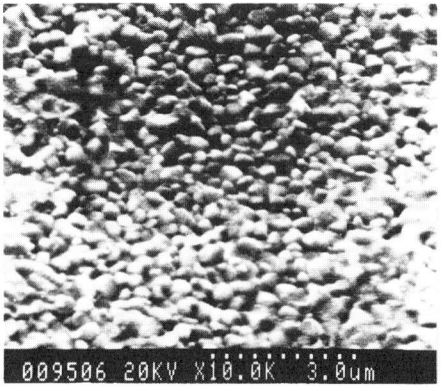

Figure 5.
0.5 mkm $BaZr_{0.8}Y_{0.2}O_{3-y}$ on YSZ,
e⁻-beam annealing.

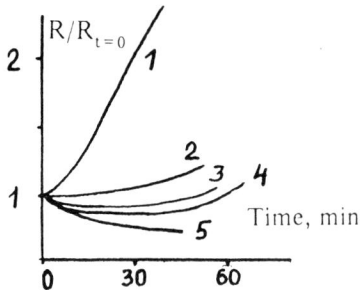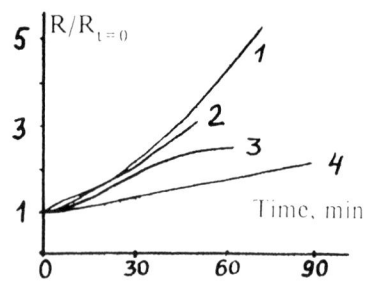

Figure 6. Kinetics of YBa_2CuO_{7-y} film - substrate interaction at 1213 K:

a) 1-YSZ single crystal
    2-0.25 mkm $BaZr_{0.8}Y_{0.2}O_{3-y}$
    on YSZ
    3-0.25 mkm $BaHf_{0.8}Y_{0.2}O_{3-y}$
    on YSZ
    4-1 mkm $BaZr_{0.8}Y_{0.2}O_{3-y}$ on
    YSZ
    5-$BaZrO_3$ fused ceramics

b) 1-0.5 mkm $BaZr_{0.8}Y_{0.2}O_{3-y}$
    on Al_2O_3, e^--beam annealied
    2-0.5 mkm $BaZr_{0.8}Y_{0.2}O_{3-y}$
    on Al_2O_3, 1 hr 1273 K
    3-0.5 mkm $BaZr_{0.8}Y_{0.2}O_{3-y}$
    on YSZ, 1 hr 1273 K
    4-0.5 mkm $BaZr_{0.8}Y_{0.2}O_{3-y}$
    on YSZ, e^--beam annealing.

Increasing of buffer layer thickness prevents chemical interaction and, as a limit, "123" film does not react with $BaZrO_3$ fused ceramic substrate during the time limited by experiment. XRD-analysis showed that YBaCuO films on buffer layers, while being annealed at 1213 K for a long time, interact with YSZ substrate, Y_2BaCuO_5 being formed. Apparently, chemical interaction of "123" with YSZ takes place despite of protecting layer presence: $2YBa_2Cu_3O_{6.5} + 3ZrO_2 = 3BaZrO_3 + Y_2BaCuO_5 + 5CuO$, "BaO" diffuses trough the $BaR_{1-x}Y_xO_{3-y}$ coating which grows at the layer - substrate interface.

To reduce the diffusion permeability of highly disordered fine-crystalline buffer-layers we tried to perform their crystallization at tough conditioned: the samples were heated up to 1550 - 1800 K and cooled down during 2 minutes by electron beam (E = 1MeV). Very fast crystallization took place (see Fig.5), this resulted in reducing of the "123" film - substrate reactivity (Fig.6b) in the case of YSZ substrate, but no effect was observed in the case of very reactive Al_2O_3 substrate.

Acknowledgments

Authors are very grateful to Dr.B.A.Melekh for providing $BaZrO_3$ fused ceramic sample, Dr. V.R.Sokolovsky for performing e-beam annealing of layers and Dr. Yu.A.Sorokin for the assistance in X-ray experiments.

References

1   A.R.Kaul, I.E.Korsakov, A.V.Permjakov, in: Science and Technology of Thin Films Superconductors 2 / Proceedings of the second Conference on the Science and Technology of Thin-Film Superconductors, April 30 - May 4, 1990, Denver, Colorado. Plenum Press, New-York (1991) 403.

2   A.A.Volkov, Y.G.Goncharov et al., Infrared Phys., 25 (1985) 369.

High T$_c$ Superconductor Thin Films
L. Correra (Editor)
© 1992 Elsevier Science Publishers B.V. All rights reserved.

Epitaxial high T_c thin film on (100) Si with BaF_2/CaF_2 buffer layers

Ph. Lerch[α], F. Marcenat, M. Dénoreaz, J. Weber and P. Martinoli and A.N. Tiwari[ω], S. Blunier and H. Zogg

[α] Institut de Physique, Université de Neuchâtel, A.L. Breguet 1, 2000 Neuchâtel, Switzerland

[ω]AFIF, ETH-Hönggerberg, 8083 Zürich, Switzerland

Abstract

IIa-fluorides have been used as buffers to grow $YBa_2Cu_3O_{7-\delta}$ (YBCO) films on Si. The buffers are either molecular beam epitaxially grown stacks of 10 nm thick CaF_2 and 240 nm thick BaF_2, or a single layer of 80 nm CaF_2 on (100) Si. Growth of YBCO films is obtained using an in-situ laser ablation process. On both types of buffer the YBCO films are epitaxial with the c-axis perpendicular to the (001) surface. High transition temperature (84 K) and metallic behavior in the normal state resistance are observed in thick (500 - 700 nm) YBCO films on BaF_2/CaF_2 buffers. However, surface cracks, probably due to strain release in the films, degrade their transport properties. Thin 60 nm films show much less cracking, but also very poor superconducting properties. On 80 nm thick CaF_2 buffer, crack free epitaxial 60 nm thick YBCO layers are obtained and show 69 K superconducting transition temperature.

1 Introduction

The trend to solve the problem of growing of $YBa_2Cu_3O_{7-\delta}$ (YBCO) onto Si substrates is to use buffer layers. The buffer should act as an interdiffusion barrier, be nonreacting, absorb all the strain induced by the thermal expansion differences between YBCO and Si, be single crystal and provide a template for the growth of YBCO. Up to now, no buffer layer is known which fulfils **all** these requirements. To date best results on Si have been achieved by using oxide layers like yttrium stabilized ZrO_2 (YSZ) [1]. Fluorides like BaF_2, CaF_2, or SrF_2 were used as substrates [2] or as buffer layers for growth of polycrystalline YBCO [3], or for growth of $TlCa_2Ba_2Cu_3O_x$ [4] on MgO . Additional motivation for using the IIa - fluorides can be sketched as follows:

- A wide range of lattice constants can be covered by using CaF_2, SrF_2, BaF_2 or mixtures thereof. Stacks like CaF_2/BaF_2 or $CaF_2/SrF_2/BaF_2$ have been used to overcome lattice mismatches as large as 20 % [5,6] to grow II-VI, IV-VI, and III-V semiconductors.

- These buffers were used in combination with materials exhibiting thermal expansion coefficients up to 7 times that of Si [7].

- Their thermal expansion coefficient is higher than that of Si (but also higher than that in the a-b plane of YBCO).

One possible reason why insulating IIa-fluorides work well as lattice and thermal expansion mismatch buffers is their ionicity. The surface field in ionic compounds decreases exponentially, leading to an interface in which the bonds across the interface are not too strong [8]. This allows to overcome lattice mismatch and, probably, also to release the strain due to mismatch of thermal expansion coefficient. In this paper we report on the growth of epitaxial YBCO films on epitaxial stacks of BaF_2/CaF_2 buffers and on single CaF_2 buffer on Si.

2   Experimental

For the growth of CaF_2/BaF_2 stacks a 10 nm thick CaF_2 layer is grown by molecular beam epitaxy (MBE) on Si(100) at about 550 C, immediately followed by 240 nm of BaF_2. The lattice mismatch of Si with CaF_2 is 0.6 % at room temperature, while that of BaF_2 is 13 %. Without CaF_2 as a starting layer, highly twinned (111)-oriented BaF_2 with unsuitable quality forms. If deposition is continued at 550 C for the whole BaF_2 layer, 3-d island growth prevails leading to slightly uneven surfaces consisting of pyramids with basal width of the order of 20 nm and (111)- oriented side faces. (The (111) surface of the fluorides has the lowest free surface energy). A smoother, true (100) terminated surface is obtained when the growth temperature is increased to 720 C after formation of 10-20 nm of the layer, and 2-dimensional growth is observed by RHEED. Typical total layer thickness is 250 nm. Rutherford backscattering spectroscopy (RBS) channelling minimum yields are less than 10 %, and the layers exhibit a tensile strain of about 0.3 %, which is much lower than the 1.4% expected if no plastic relaxation would occur on cool down from growth (720 C) to room temperature. More details of the growth procedure may be found elsewhere [5,6]. We also used samples with an only 80 nm thick layer of CaF_2.

We use a KrF excimer laser ablation system similar to those described in the literature [9]. An optical pyrometer (1-2 μm window) is used to measure the actual surface temperature, and care is taken to take into account the changes of emissivity during the deposition.

Parameters of a typical run are: Starting vacuum 10^{-5} mbar, working oxygen pressure between 0.1 mbar and 0.3 mbar, sample surface temperature between 680 C and 720 C, laser repetition rate 4 Hz, beamsize apertured to 1 x 2 cm^2, laser fluence on target ranging between 1.7 and 2.5 J/cm^2 and sample to target distance 4 cm. After deposition, the pump is shut off, oxygen is flushed into the chamber and the heater power is decreased. Temperature is kept around 500 C for 5 minutes in a pressure of 1 bar of oxygen.

3   Results

In figure 1 we show an x-ray spectrum $(\theta - 2\theta)$ scan with symmetrical reflection of a typical 700 nm thick film of YBCO (sample A-12). Only planes lying parallel to the (001)-surface of single crystal samples are revealed in such a scan. Clearly, only (00 *l*) reflections are visible besides the peaks due to the buffer layer and the substrate. A slight YBCO(110) or YBCO(103) reflection at 33 degrees is due to some misoriented parts, but these grains must be small in volume in light of the very weak intensity of the lines. The layer is therefore almost completely c-axis oriented.

Epitaxial growth can be proven from x-ray diffraction scans of out-of-surface peaks. The c-axis of YBCO is about 3 times longer than the a- or b-axis. Si(101) and YBCO (103) lines are thus accessible if the surface normal of the sample is inclined by 45 degrees with respect to the x-ray reflection plane, and the symmetry can be tested by rotating the sample around a direction normal to its (001)-surface and around an axis which is inclined 45 degrees from the reflecting plane. Figure 2a shows a $(\theta - 2\theta)$ spectrum with the sample position tilted to obtain the Si(101) lines and figure 2b shows the spectrum of the sample in the same position but after a rotation of 45 degress around its normal axis. Clearly, (103)-YBCO is visible in the second spectrum only. In the first spectrum almost no peak appears at the YBCO position, but strong peaks are visible from the (220)-Si and (220)-BaF_2. A four fold rotational symmetry about a direction normal to the sample surface is experimentally observed. This shows that the a- and b-axis of the YBCO layer are rotated by 45 degrees with respect to the Si (100)- and (010)-axis and that most of the YBCO layer has therefore grown epitaxially. The epitaxial relationship with respect to the Si (or BaF_2) is (001)YBCO / (001)Si and (010)YBCO / (110)Si.

The line widths of the YBCO x-ray lines of the planes parallel to the surface are slightly larger than the Si-line (0.4 degrees) obtained with the resolution in the present measurements (without x-ray monochromator). The width of the inclined (103)-YBCO line is 1.7 degree, i.e. larger than the (00 *l*) lines. This indicates better structural perfection along the c-axis than for inclined orientations. This may be related to some mosaicity and, most probably, heavy a/b twinning of the layers.

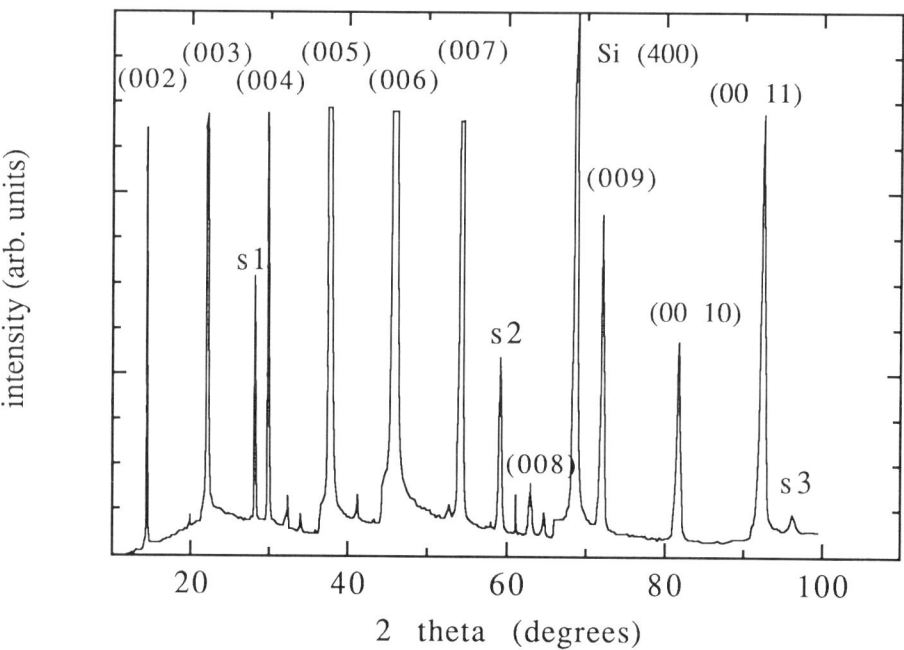

Figure 1: X-ray spectrum $(\theta - 2\theta)$ scan of a Si substrate covered with 10 nm of CaF_2 and 240 nm of BaF_2 and after deposition of 700 nm of YBCO (sample A-12). C-axis (00 *l*) lines with $l = 2$ up to 11 are visible. Note some spurious beta-lines, the substrate and BaF_2 buffer layer peaks as well as a weak YBCO peak near 33 degrees which is due to (103) and/or (110) reflections.

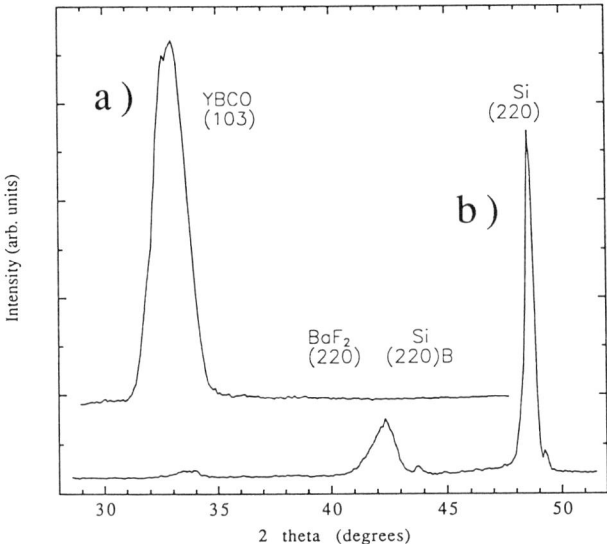

Figure 2: X-ray spectrum ($\theta - 2\theta$ scans) of a) planes inclined by 45 degrees with respect to the (001) surface with the sample (100)-Si axis in the reflection plane and b) with the sample rotated 45 degrees around the (001) surface normal. The near absence of YBCO peaks in spectrum (b) demonstrates the almost perfect epitaxy with 45 degrees rotation of the YBCO lattice around the [001] surface normal compared to the Si-substrate and the single crystal BaF_2 buffer layer.

In figure 3 we present the normalized 4 point resistance as a function of temperature for sample A-12 grown on a BaF_2/CaF_2 buffer and for sample STO-6 grown on $SrTiO_3$ (STO). For both samples we observe metallic behavior followed by a transition to the superconducting state, occurring around 84 K for the sample on Si. The ratio $RRR = R(300K)/R(90K) = 2.6$, compares well with a good quality film obtained on STO. However, careful observation reveals that the surface of the film grown on Si is heavily cracked. This is probably due to the release of thermally induced strain between the YBCO film and the Si substrate. These cracks are detrimental to transport properties of the layer. The fact that we observe a relatively high transition temperature and reasonable RRR value shows that, intrinsically, the YBCO material is of good quality. It seems obvious that the transport capability of such a film is limited by the ability to carry supercurrent of the weakest region in the percolating system formed by the "islands" and the cracks in the film.

We find that there is a strong reaction between the fluoride and oxygen during growth, in contrast with other reports [10]. RBS shows that the fluorine of the BaF_2 partially disappears and is replaced by oxygen. X-ray data, taken on a substrate (with its buffer but without YBCO deposition) after heat treatment in oxygen, show that the amplitude of the BaF_2 lines is reduced, whereas samples heated in vacuum survive that kind of treatment. In order to avoid this deterioration we modified our deposition procedure; the laser deposition is started in vacuum and the oxygen pressure is established after a few seconds of deposition. This enables us to keep the buffer layer almost unreacted. In order to prevent relaxation of thermally induced strain, the thickness of YBCO is reduced, resulting indeed in crack free films. However, 30 to

60 nm thick YBCO layers are either polycristalline (strong(103) line and some c-axis lines) or the films peel off at room temperature. The origin of this reaction between the YBCO and the BaF_2 remains unknown.

The same laser deposition procedure has been applied to 80 nm of CaF_2 buffer layers. In this case we obtain epitaxial, crack-free, 30 to 60 nm thick YBCO films. Typical superconducting properties are shown in figure 3 where resistance (right hand scale) is plotted against temperature. Again, metallic behavior is observed, followed by a transition 5 K wide. On the left hand scale we show the resistive $Re(\delta V)$ and inductive $Im(\delta V)$ part of the ac response of the film driven by a small ac magnetic field. This detection system consists of a small cylindrical coil, which applies an ac field perpendicular to the film. Inside this drive coil, an astatic pair of detection coils is sensitive to the screening response of the sample [11]. The inductive part of the signal clearly demonstrates the ability of the sample to shield the magnetic field. It also provides a more accurate description of the superconducting transition. The sharp peak in the resistive part is an indication of the dissipation occurring at the transition and of the homogeneity of the superconducting phase present in the sample. In bad quality films we often observe a double peak (correspondingly a double step appears in the inductive part) which reveals two superconducting phases with different transition temperatures. Clearly, this is not the case here. To our knowledge, this is the best result obtained thus far for YBCO films on Si with fluoride buffers.

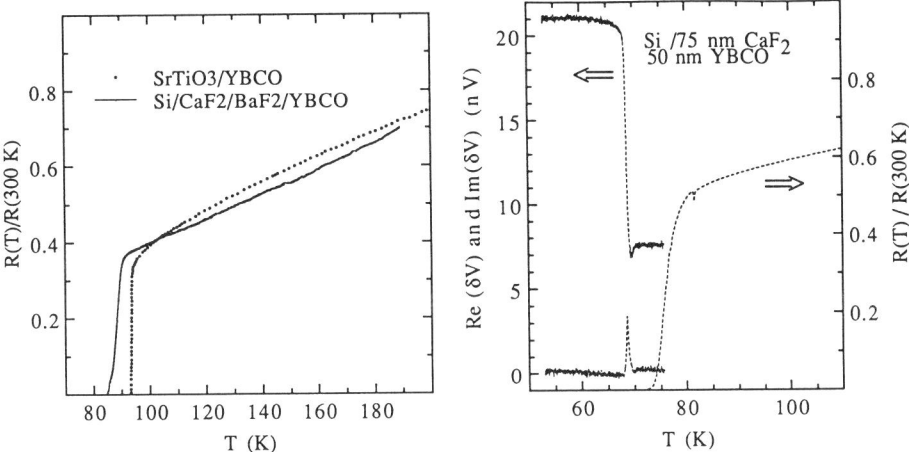

Figure 3: Normalized 4 point resistance versus temperature of 700 nm of YBCO (sample A-12, full line) grown on a BaF_2/CaF_2 buffer and of 300 nm of YBCO grown on STO (sample STO-6, dots). Measuring current is 10 μA, voltage resolution is $1\mu V$. The critical temperatures are 84 K and 91 K for the samples grown on Si and STO respectively. RRR values are 2.6 and 3 respectively.

Figure 4: On the right hand scale, normalized 4 point resistance of a 60 nm sample of YBCO on a 80 nm thick $CaF2$ buffer. RRR= 2.0, transition width is 5 K. On the left hand scale, real and imaginary part of the ac response of the film. T_c is around 69 K. Note the different values of T_c obtained with the two measurments.

4 Conclusions

On BaF_2, we are able to obtain thick, epitaxial YBCO films, with critical temperature values around 84 K, but cracked surfaces. We have no explanation why thin YBCO films on BaF_2 do not survive mechanically or why they grow polycrystalline (strong (103) lines). We can only speculate that the higher reactivity of this material may be at the origin of this problem. For samples grown without oxygen at the very beginning of the laser deposition, RBS data indicate that the YBCO/BaF_2 interface region does not extend over distances larger than 10 to 20 nm.

On CaF_2 buffer we obtain crack-free, epitaxial YBCO films, with best T_c values around 69 K. From RBS analysis we see that the YBCO/CaF_2 interface does not extend over distances larger than 10 to 20 nm. However, the CaF_2 has a modified stoichiometry and is, over a distance of 30 nm, close to 1 Ca, 1 F, and $2O$. We speculate that high oxygen mobility through thin YBCO precludes the total absence of reaction of the buffer with oxygen, and leads, in addition to a moderate reaction with the buffer, to an overall oxygen deficient YBCO layer.

Careful reading of the manuscript by Chris Leemann is kindly aknowledged, as well as his efforts at the very early stage of this project. This work is supported by the Swiss National Science Foundation.

References

[1] see for example D.K. Fork, D.B. Fenner, A. Barrera, J.M. Phillips, and T.H. Geballe, G.A.N. Connell, J.B. Boyce, IEEE Trans. Appl. Supercond. 1, 67 (1991) and refs. therein.

[2] S-W. Chan, E.W. Chase, B.J. Wilkens, D.L. Hart, Appl. Phys. Lett. 54, 2032, (1989).

[3] F. Radpour, R. Singh, S. Sinha, A.M. Tulpule, P. Chou, R.P.S. Thakur, M. Rahmati, N.J. Hsu, A. Kumar, Appl. Phys. Lett. 54, 2479, (1989).

[4] K.M. Hubbard, P.N. Arendt, D.R. Brown, D.W. Cook, N.E. Elliott, J.D. Farr, M. Nastasi, K.C. Ott, E.J. Peterson, and G.A. Reeves, preprint (1989).

[5] S. Blunier, H. Zogg, H. Weibel, Appl. Phys. Lett. 53, 1512, (1988).

[6] S. Blunier, H. Zogg, A. Regge, Thin Solid Films 184, 387, (1990).

[7] H. Zogg, G. Masek, C. Maissen, T, Hoshino, S. Blunier, Mat. Res. Soc. Symp. Proc. 216, (1991) in print.

[8] M.H. Yang, C.P. Flynn, Phys. Rev. B41, 8500, (1990).

[9] see for example D. Bäuerle, Appl. Phys. A 48, 527 (1989) and refs. therein.

[10] T. Komatsu, O. Tanaka, K. Matusita, T. Yamashita, Jpn. J. Appl. Phys. 27, L1686, (1988).

[11] B. Jeanneret, J.L. Gavilano, G.A. Racine, Ch. Leemann, and P. Martinoli, Appl. Phys. Lett. 55, 2336 (1989).

High T$_c$ Superconductor Thin Films
L. Correra (Editor)

PLASMA SPRAYED YBa$_2$Cu$_3$O$_{7-x}$ ON NICKEL-RICH SUBSTRATES: XPS ANALYSIS AND INTERFACE INTERACTION

W. Lisowski[1,3], H. Hemmes[1], D. Jäger[2], A. van Silfhout[1], D. Stöver[2] and L.J.M. van de Klundert[1]

[1]Department of Applied Physics, University of Twente, P.O. Box 217, NL-7500 AE Enschede, The Netherlands

[2]Institut für Angewandte Werkstofforschung, Forschungszentrum Jülich, Postfach 1913, D-5170 Jülich, Federal Republic of Germany

[3]Permanent address: Institute of Physical Chemistry, Polish Academy of Sciences, ul. Kasprzaka 44/52, 01-224 Warszawa, Poland

Abstract

YBa$_2$Cu$_3$O$_{7-x}$ layers deposited on nickel-rich substrates by atmospheric plasma spraying have been investigated by XPS. Both the top of the layer and the interface region have been studied in terms of chemical composition and peak shapes. The effects of Ar$^+$ sputtering and air-scraping of both the surface and the interface region of the YBa$_2$Cu$_3$O$_{7-x}$ layer have been considered. Air-scraping was found to be not sufficiently effective in cleaning the sample surface. Sputtering removes effectively the surface contaminants but results in a structural and electronic rearrangement with an apparent partial reduction of the three- and di-valent states of copper to a mono-valent state and the disappearance of low binding energy component of Ba 4d. The preferential sputtering of oxygen is also observed. Finally evidence is found for migration of chromium from the substrate into the YBa$_2$Cu$_3$O$_{7-x}$ layer. The sputtering of the interface results in a reduction of the six-valent state of chromium to a state with a valency of three or four.

1. INTRODUCTION

In the application of bulk high-T$_c$ materials one of the limiting factors is the small critical current density J$_c$, although not all applications call for a high J$_c$ [1]. Another problem is obtaining the desired shape. For this a potentially useful technique is plasma spraying. With this technique virtually any shape can be produced at a high deposition rate. Although reasonable results have been obtained on Ni-rich substrates [2,3], prolonged annealing at high temperatures, which is often used to improve J$_c$ in bulk material, destroys the superconducting properties.

X-ray Photoelectron Spectroscopy (XPS) is a widely used technique for investigation of the surface region of superconducting YBa$_2$Cu$_3$O$_{7-x}$ materials [4-8]. The results revealed always more than one chemical state for Ba, Cu and O [4-10]. Significant variations of the core levels indicate an important role of the preparation procedure [7]. Recent publications have also shown a significant influence of argon sputtering [5,8] and scraping [7,8] on the surface region core levels. These methods are routinely used to prepare the clean surfaces for study in UHV conditions.

In this paper we present the results of XPS analysis of YBa$_2$Cu$_3$O$_{7-x}$ layers on

nickel-rich substrates, prepared by Atmospheric Plasma Spraying (APS). The influence of
argon sputtering and scraping are analyzed in terms of the elemental XPS line shapes and
the surface composition. For one sample, the $YBa_2Cu_3O_{7-x}$ layer was separated from the
substrate which gave us the opportunity to study the interface region with XPS.

2. EXPERIMENTAL

The samples are ~100 μm thick $YBa_2Cu_3O_{7-x}$ layers deposited by APS on nickel-rich
substrates. Samples with two types of substrates have been analyzed: nickel covered with
a sprayed $Ni_{80}Cr_{20}$ layer and Nimonic75 (containing 19% of Cr), denoted by MNN and MM
respectively. After deposition the samples received a two step heat treatment: 10 h. at
930 °C followed by 10 h. at 430 °C, all in 1 atm. of flowing oxygen.

The XPS measurements were carried out on a KRATOS 800 instrument, controlled by a
PDP-11 computer. The excitation source was Mg Kα radiation. The spectrometer was
calibrated using the Ag $3d_{5/2}$ peak and the X-ray induced Ag MNN peak on a clean
sputtered silver sample. The position of the C 1s peak, relative to its normal position
at 285.1 eV, is used to correct the measured
binding energies (BE) for electrostatic
charging of the sample. The spectra were
taken and analyzed using a DS 800 software
package. For the peak synthesis 100%
Gaussians were used. Compositions were
calculated using empirical sensitivity
factors for the pure elements [11]. The
influence of the matrix was not taken into
account. Although the absolute values for the
compositions may not be correct, changes in
composition can be determined.

The samples were studied under three
conditions: 'as-received', after 30 min of
argon-ion sputtering (2.5 kV Ar^+ at a
pressure of $2x10^{-7}$Torr) and after scraping.
As our system has no in-situ facilities all
scraping was done in air, after which the
sample was mounted as fast as possible in the
vacuum chamber. The time that the scraped
surface was exposed to the atmosphere,
including the pumping down of the vacuum
chamber, was typically 2-3 minutes.

3. RESULTS AND DISCUSSION

3.1. XPS measurements of the surface

The core level spectra of the surface of
sample MNN in the 'as-received' state are
represented by line 1 in Figures 1, 2a and
3a. The results after Ar^+ sputtering and
scraping are shown as lines 2 and 3
respectively.

The Y 3d spectra, shown in Figure 1,
suggest the presence of two components both

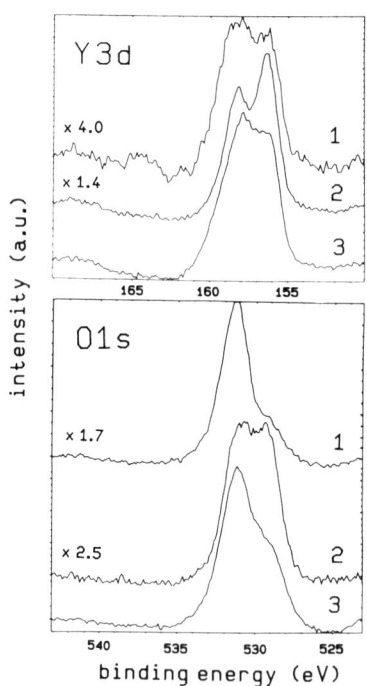

Figure 1. Y 3d and O 1s XPS spectra
of the surface of sample MNN:
'as-received' (1), after sputtering
(2) and after scraping (3).

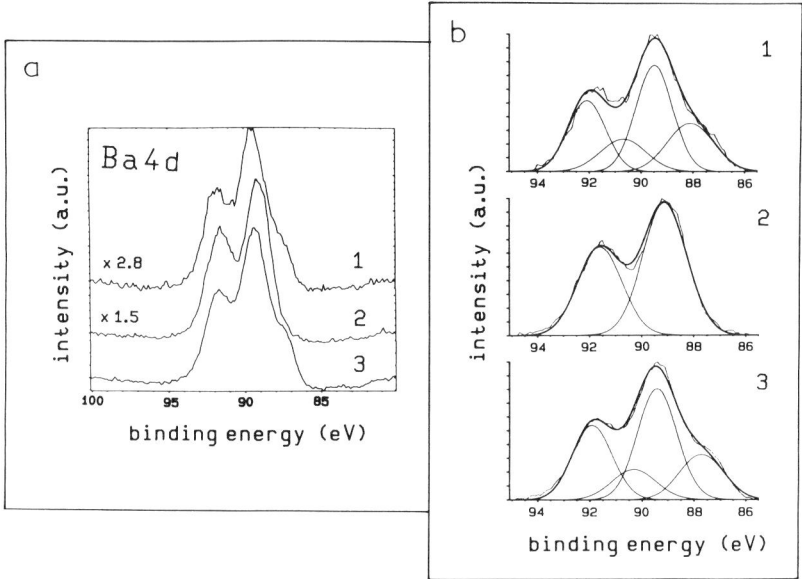

Figure 2. (a) Ba 4d XPS spectra of the surface of sample MNN: 'as-received' (1), after sputtering (2) and after scraping (3). (b) Deconvolution of the spectra in (a).

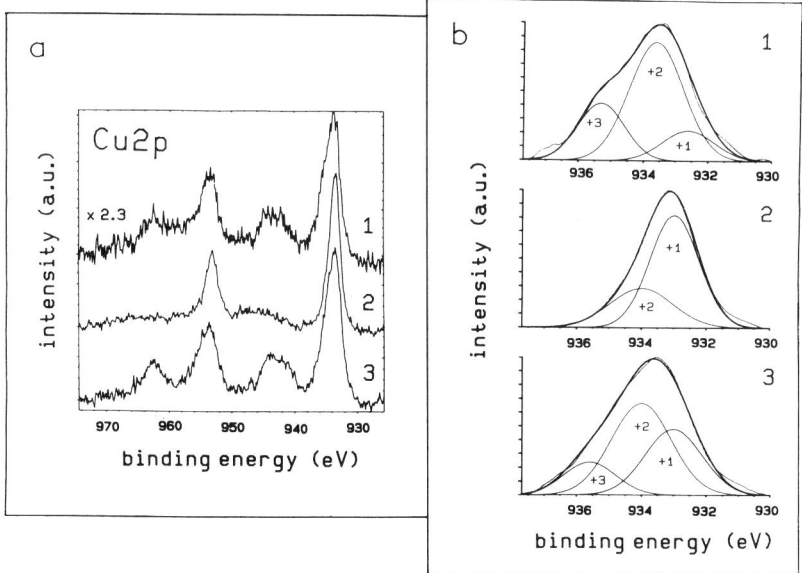

Figure 3. (a) Cu 2p XPS spectra of the surface of sample MNN: 'as-received' (1), after sputtering (2) and after scraping (3). (b) Deconvolution of the spectra in (a).

Table 1. Atomic concentration ratios from the XPS spectra.

Sample:	Treatment	Atomic concentration ratios	
		Y : Ba : Cu : O	C/(Y+Ba+Cu+O)
MM:	'as-received'	1:1.5:0.9:12.1	0.4
MNN:	'as-received'	1:2.3:1.5:18.5	0.6
	sputtered	1:1.8:1.0: 7.7	0.2
	air-scraped	1:1.7:1.1:10.1	0.2
POWDER:	'as-received'	1:1.2:2.8:19.2	0.3

before sputtering and after scraping. After sputtering there is evidence for only one component.

The O 1s spectrum (Figure 1) shows the existence of two peaks at 531.4 and 528.9 eV. The ratio of the high and low BE component areas are 4.5, 1.2 and 2.0 respectively for the 'as-received', sputtered and scraped state. The high BE contribution is reduced as a result of the sputtering and scraping. The same phenomenon has been observed earlier [5,7,8]. Also the amount of oxygen at the surface of the sample is reduced, relative to the other elements, as is shown in Table 1.

A detailed analysis of the Ba 4d spectra is presented in Figure 2. Two doublets with their main peaks at 89.5 and 88.0 eV can be distinguished (Figure 2b-1). The latter one is attributed to the superconducting material [12] and disappears after sputtering (Figure 2b-2). After scraping, however, the Ba 4d is found in its former chemical state (Figure 2b-3).

The Cu 2p spectra in Figure 3a for the 'as-received' and scraped state show prominent shake-up satellite structures next to a broad low-BE main peak. After sputtering the spectrum exhibits a narrower main peak and a much smaller satellite contribution. The ratio of the intensity of the satellite peak (I_{sat}) and the main peak (I_{main}) is 0.26, 0.11 and 0.22 respectively for the 'as-received', sputtered and scraped samples. This suggests that sputtering reduces Cu to a state of lower valency [5,8]. A comparison of the I_{sat}/I_{main} ratio reported for CuO (~0.33) [10] with that for the

Table 2. Relative abundance, binding energy and half-width (FWHM) of the peak synthesis components of the Cu 2p spectra for the surface of sample MNN in the 'as-received', sputtered and air-scraped state.

Treatment	Relative abundance [%]			Binding energy [eV] (FWHM)		
	Cu^{1+}	Cu^{2+}	Cu^{3+}	Cu^{1+}	Cu^{2+}	Cu^{3+}
'as-received'	14.5	60.7	24.8	932.6 (1.9)	933.6 (2.0)	935.6 (1.7)
sputtered	69.3	30.7	—	933.0 (1.8)	934.0 (2.3)	—
air-scraped	34.4	50.2	15.4	933.0 (2.1)	934.0 (2.2)	935.6 (1.9)

YBa₂Cu₃O₇₋ₓ sample in the 'as-received' state (0.26) suggests the coexistence of the dominant Cu^{2+} state with the Cu^{1+} and Cu^{3+} states which have a low intensity satellite structure [10]. This was also found in other investigations [4,6]. The detailed analysis of the low BE component of the Cu 2p spectra shown in Figure 3b, and the characteristic parameters of the deconvoluted Cu peaks presented in Table 2, confirm this. The BE corresponding to Cu^{1+}, Cu^{2+} and Cu^{3+} peaks, found from the deconvolution, are in good agreement with those, reported in the literature [10,13,14].

Sputtering transforms Cu^{3+} and Cu^{2+} to the more stable Cu^{1+} state [15]. This is accompanied by a relatively high loss of oxygen (Table 1). It is well known that the most mobile oxygen atoms in YBa₂Cu₃O₇₋ₓ are close to the copper atoms in the square CuO_4 cluster [6]. The de-oxygenation as a result of sputtering reduces the Cu in these clusters to the di- or mono-valent state.

The relative abundance of oxygen and carbon in different samples (Table 1) can be interpreted in terms of surface degradation products like $BaCO_3$ and $Ba(OH)_2$ [7].

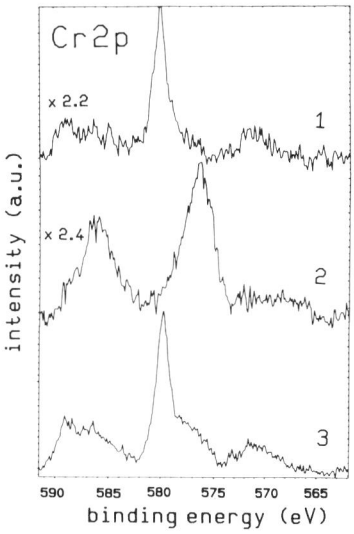

Figure 4. Cr 2p XPS spectra of the interface of sample MM: 'as-received' (1), after sputtering (2) and after scraping (3).

3.2. XPS measurements of the interface region

The interface area of sample MM was green at the edges and black at the center. Both areas were analyzed, revealing the presence of a substantial amount of Cr (but no Ni) together with Y, Ba, Cu, O and C. This indicates a considerable amount of migration of Cr into the YBa₂Cu₃O₇₋ₓ layer. The different relative amounts of Y, Ba, Cu, O and Cr, measured on both the green and black part, show that the stoichiometry in the interface region is not homogeneous.

The Ba 4d and Cu 2p spectra are comparable to those measured on sample MNN. Sputtering changes the ratio of I_{sat}/I_{main} of Cu 2p spectra from 0.25 to 0.10, indicating a reduction to Cu^{1+}. It also reduces strongly the amount of O and C.

A significant change of the Cr 2p spectrum can be observed in Figure 4, as a result of the sputtering. In the 'as-received' state the major Cr 2p peak was found at 579.8 eV. This indicates the presence of Cr^{6+}, which suggests the formation of $BaCrO_4$ [16]. After sputtering the peak is shifted to 576.5 eV, which is typical for Cr^{3+} and Cr^{4+} and can be attributed to Cr_2O_3 [17,18] and CrO_2 [19] respectively. So, the sputtering changes the oxidation state of chromium from Cr^{6+} to Cr^{4+} or Cr^{3+}. After scraping, the Cr 2p peak shifs back to 579.8 eV, although a substantial amount of Cr^{3+} and/or Cu^{4+} is still present. A detailed peak-synthesis analysis of the Cr 2p peak is difficult because of an overlap with the Ba MNN Auger peak.

4. CONCLUSIONS

The composition of both the surface and interface of a plasma sprayed YBa₂Cu₃O₇₋ₓ on a nickel-rich substrate is not homogeneous and depends strongly on the preparation.

Argon sputtering reduces strongly the amount of contaminants, but results in a structural and electronic rearrangement: oxygen is preferentially removed, the valency of Cu and Cr is reduced and the low-BE component of the Ba 4d spectrum disappears. Air-scraping on the other hand has much less influence on the peak shapes but is not sufficiently effective in cleaning the $YBa_2Cu_3O_{7-x}$ surface. The scraping should therefore be done in-situ. Apart from the dominant Cu^{2+} state, copper is also found as Cu^{1+} and Cu^{3+}. As the Cu^{3+} state could play a role in the mechanism of superconductivity for these materials the reason for the presence of Cu^{3+} in some XPS studies and the absence in others should be clarified in the future. Cr is detected in the interface region both after sputtering and scraping. However, no nickel is found. This shows that there is an extensive interaction between Cr and the $YBa_2Cu_3O_{7-x}$ layer. This explains why the superconductivity disappears after prolonged annealing at high temperatures.

ACKNOWLEDGEMENTS
This work was supported by the European Community under contract no. BREU-0124, within the BRITE/EURAM program.

REFERENCES
1 P.E. Chandler, in *Proceedings of 'The 12th International Conference on THERMAL SPRAYING'*, London 4-9 June 1989, preprints vol. 2, ed. I.A. Bucklow, (1989) p93-1.
2 K. Tachikawa, I. Watanabe, S. Kosuge, M. Kabasawa, T. Suzuke, Y. Matsuda and Y. Shinbo, *Appl. Phys. Lett. 52* (1988) 1011
3 J. Lacombe, J. Danroc and G. Kurka, *J. Less Common Met. 164&165* (1990) 509.
4 P. Steiner, S. Hüfner, V. Kinsinger, I. Sander, B. Siegwart, H. Schmitt, R. Schulz, S. Junk, G. Schwitzgebel, A. Gold, C. Politis, H.P. Müller, R. Hoppe, S. Kemmler-Sack and C. Kunz, *Z. Phys. B-Condensed Matter, 69* (1988) 449, and references therein.
5 P.C. Healy, S. Myhra and A.M. Stewart, *Philos. Mag., B, 58* (1988) 257.
6 F. Werfel, M. Heinonen and E. Suoninen, *Z. Phys. B, 70* (1988) 317.
7 S.L.T. Andersson and J.C. Otamiri, *Appl. Surface Sci., 45* (1990) 1.
8 H.M. Meijer III, D.M. Hill, T.J. Wagener, Y. Gao, J.H. Weaver, D.W. Capone II and K.C. Goretta, *Phys. Rev., B, 38* (1988) 6500, and references therein.
9 Y. Fukuda, M. Nagoshi, T. Suzuki, Y. Namba, Y. Syono and M. Tachiki, *Phys. Rev., B, 39* (1989) 11494.
10 P. Steiner, V. Kinsinger, I. Sander, B. Siegwart, S. Hüfner, C. Politis, R. Hoppe and H.P. Müller, *Z. Phys. B-Condensed Matter, 67* (1987) 497.
11 D. Briggs and M.P. Seah, *Practical Surface Analysis by Auger and X-ray Photo-electron Spectroscopy*, Wiley, Chichester, 1983.
12 W.A.M. Aarnink, J. Gao, H. Rogalla and A. van Silfhout, *J. Less-Common Met. 164-165* (1990) 321.
13 G. Shoen, *Surface Sci., 35* (1973) 96.
14 N.S. McIntyre, S. Sunder, D.W. Shoesmith and F.W. Stanchell, *J. Vac. Sci. Technol., 18* (1981) 714.
15 G. Panzner, B. Egert and H.P. Schmidt, *Surface Sci., 151* (1985) 400.
16 G.C. Allen, P.M. Tucker, *Inorg. Chim. Acta, 16* (1976) 41.
17 C. Battistoni, J.L. Dormann, D. Fiorani, E. Paparazzo and S. Viticoli, *Solid State Commun., 39* (1981) 581.
18 G.C. Allen, P.M. Tucker and R.K. Wild, *J. Chem. Soc. Faraday Trans. II, 74* (1978) 1126.
19 I. Ikemoto, K. Ishii, S. Kinoshita, H. Kuroda, M.A.A. Franco and J.M. Thomas, *J. Solid State Chem., 17* (1976) 425.

High T$_c$ Superconductor Thin Films
L. Correra (Editor)

Interface structures of YBCO thin films on different substrates

G. Lu[a,c], F. Phillipp[a], B. Leibold[b], A.A.C.S. Lourenco[b] and H.-U. Habermeier[b]

[a]Max-Planck-Institut für Metallforschung, Institut für Physik,
Heisenbergstr. 1, 7000 Stuttgart 80, FRG

[b]Max-Planck-Institut für Festkörperforschung,
Heisenbergstr. 1, 7000 Stuttgart 80, FRG

[c]On leave from Physics Department, Wuhan University, P.R. China

Abstract

The interface structures of YBa$_2$Cu$_3$O$_{7-\delta}$ (YBCO) thin films, as prepared by pulsed laser deposition, on substrates of different materials have been investigated by transmission electron microscopy (TEM). It is found that the interfaces on SrTiO$_3$ substrate are in general sharper than those on LaAlO$_3$, where extended defects such as misoriented grains and continuous layers of misoriented material occur. When the substrate temperature is lower during the deposition process, the film can have different orientation. Films with c-axis oriented in the interfacial plane can also grow epitaxially with a sharp interface.

1. INTRODUCTION

The *in situ* pulsed laser deposition (PLD) has been established to produce YBCO thin films, which normally show T$_c$ above 86 K and j$_c$ of the order of 10^5 to 10^6 A/cm at 77 K [1,2]. In studying the structures, defects and interfaces in the thin films TEM is a powerful tool, which can image these structural features directly and thus provide information about the film quality. This is of physical interest in understanding the formation of the superconducting thin films in the PLD process, and is practically important because the TEM results provide a direct estimate of the structural perfection of the film, which may influence the superconducting properties and gives hints for optimization of preparation conditions for materials for devices such as optical multilayers and Josephson junctions. TEM is particularly useful and unique in observing the detailed interface structure and thus provide information on the early stages of the film growth. In this paper we report a TEM investigation of the interfaces between YBCO thin films and different substrates under various deposition conditions.

2. EXPERIMENT

YBCO thin films were evaporated on substrates of LaAlO$_3$ and SrTiO$_3$ by pulsed laser deposition of bulk YBCO pellets. The details of the technique are described elsewhere [2]. Majority of the samples were deposited on substrates of [001] orientation at a substrate temperature measured as 780°C at oxygen flow of pressure of 1 mbar. In this case the films are mostly textured with the c-axis perpendicular to the interface (c$_\perp$). Substrates of ⟨110⟩ orientation were also used as well as deposition temperatures lower than 780°C to produce films with the c-axis parallel to the interface (c$_\parallel$).

Most of the films show a T$_c$ above 86 K and j$_c$ of the order of 10^6 A/cm^2. After measuring the electrical properties, the films were prepared for TEM cross-section study in the conventional way that is cutting, sticking face to face, grinding, dimpling and ion milling. In the process of the specimen preparation water moisture was avoided and contact with solutions was diminished to reduce structural deterioration. During ion milling a low voltage (3 - 4.5 keV) and liquid nitrogen cooling were used. The cross-section specimens were examined in a JEM 4000FX machine. The superconducting 1-2-3 phase and its orientation were determined by the ordering fringes of 1.17nm spacing along the c-axis as observed in high resolution electron microscopy (HREM) mode.

3. RESULTS AND DISCUSSION

In the films grown on SrTiO$_3$ substrates, the interface is normally clean and sharp. Fig. 1a shows a typical example of interface of this kind. The cleanness and smoothness of the substrate surface are important for the interface structure. However, even on SrTiO$_3$ substrate where there are surface steps of several unit cell height, a reasonably sharp and direct growth of the film is observed without forming extended defects (fig. 1b).

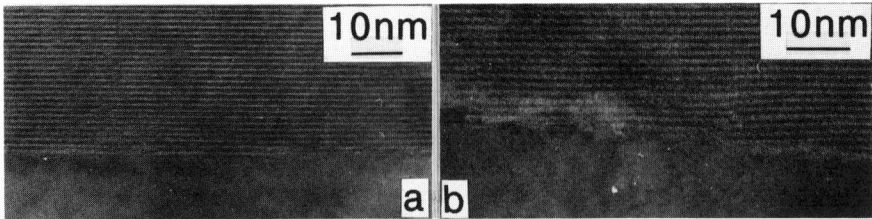

Fig. 1 Interface structure of YBCO film on SrTiO$_3$. Clean and sharp interface on flat substrate (a), and on substrate with deep surface steps (b).

The YBCO thin films on LaAlO$_3$ substrates are on the whole epitaxial, and the interface sometimes can be as sharp as that on SrTiO$_3$, such an example is shown in fig. 2a.

However, extended defects are often found in the interface region of the film on LaAlO$_3$, one of which is seen on the left in fig. 2a. This defects is not identified and is seen to associate with a dislocation in the substrate coming up to the surface. Some of these extended defects are determined to be misoriented 1-2-3 phase as tiled about ⟨100⟩ axis and show the characteristic fringes inclined to the interface. In some cases, these defects are observed to extended parallel to the interface to form a layer up to 50nm thick along the interface (fig. 2c). In one sample the misoriented grains were observed to be extended perpendicular to the interface and to reach the top surface of the film (fig. 3a). These through grown defects appear in TEM amorphous like, apart from the region close to the

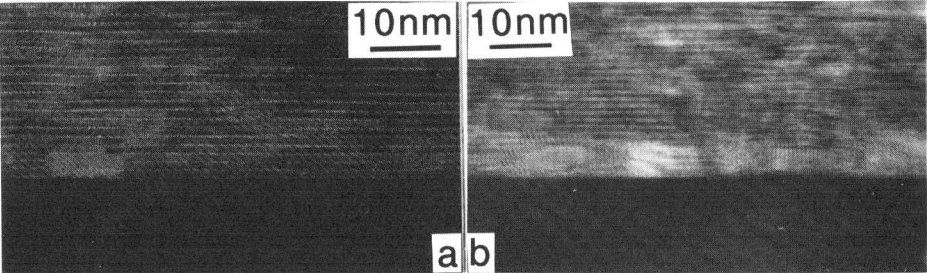

Fig. 2 HREM images of interfacial structures of YBCO on LaAlO$_3$, showing a sharp interface, but with an extended defect on the right hand side (a), and a continuous layer of misoriented material along the interface (b).

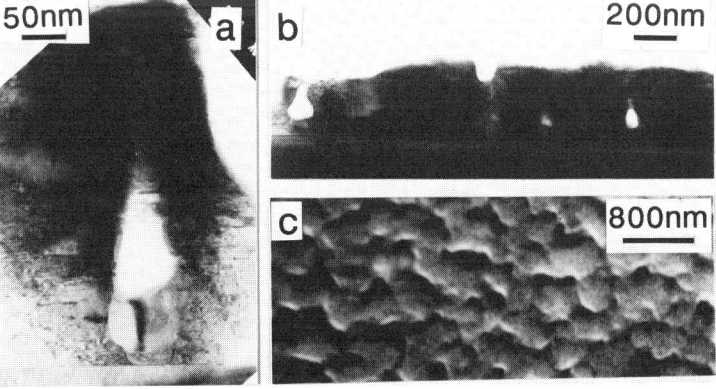

Fig. 3 Images of misoriented grains extended perpendicular to the interface, (a) HREM cross-section image of such a grain, (b) low magnification cross-section image showing distribution of the extended defects, (c) SEM image showing the surface roughness which can be correlated to the defects in (b).

interface. When reaching the top surface of the sample, they result in surface roughness. Fig. 3b and c show these defects imaged with low magnification cross-section TEM and surface SEM respectively. A comparison correlates these defects in fig. 2b to the surface roughness in fig. 2c convincingly. This sample shows a sharp transition to superconductivity at 88 K. But the critical current density of the film is rather low.

As reported in the literature (e.g., [3,4]), we also observed epitaxial growth of the YBCO films and sharp interfaces on LaAlO₃ as shown in majority part of fig. 3a. However, our experimental results shown that these interfaces contain much more defects when compared with the films grown on SrTiO₃. Note that the two kinds of substrates were cleaned in the same way prior to the deposition and the films were grown on these substrates under the same conditions. Also the surface smoothness of the LaAlO₃ substrates is not worse than that of the SrTiO₃ substrates and the dislocation density in the former substrate is even lower than in the latter as shown by TEM . Thus, the observed difference in sharpness of interfaces on these two kinds of substrates is likely to be intrinsic when used as a substrate for YBCO growth. It is probably the interdiffusion and the reactivity with YBCO that causes the different appearance of the interfaces. This direct observation of the interface qualities of the films on these two substrates support the infrared reflection results reported previously [5]. The appearance of substrate phonons in the infrared reflectivity curves of the films on LaAlO₃ is probably due to some bending of the CuO₂ planes. In the case of the film on SrTiO₃ with a much better interface structure, this bending is missing and no substrate phonons are observed.

On the SrTiO₃ substrate of {110} surface, YBCO films with the c-axis in the interface plane (c∥) were grown by controlling the substrate temperature first at 663°C, allowing the film to grow for about 60nm and then increasing continuously to 720°C for the film to grow up to 350nm. In this case the film is also epitaxial and the interface is sharp (fig. 4a). Whereas in x-ray diffraction it is difficult to discriminate between ⟨110⟩ and ⟨103⟩ orientation [6], the TEM results show clearly that film is ⟨110⟩ oriented. One may notice in fig. 4a that within a layer of an average thickness of 20nm along the interface the ordering fringes in the c-direction of the YBCO 1-2-3 structure are not preserved. Furthermore, in some areas of the sample a layer of another phase with a thickness up to 60nm exists (fig. 4b). Under the particular diffraction condition used in fig. 4b this layer does not show any lattice fringes, however it is not amorphous as indicated by the microdiffraction pattern in the insert of fig. 4b. The formation of this layer may mainly be due to ion damage during ion milling since in a specimen prepared with 6 keV Ar⁺ ions this kind of interface layer exists all over the thin area, while after 20 minutes polishing with 3.5 keV Ar⁺ ions most interface regions show structure as in fig. 4a. With the evidence of radiation damage due to Ar⁺ ions during the milling, one may consider the loss of the 1-2-3 ordering in fig. 4a as the early stage of the radiation damage. On the other hand, this observation indicates that there exists some structure difference in the interface region of the as grown film, which is responsible for the higher sensitivity to the ion damage than the normal 1-2-3 phase. The damaged layer as shown in fig. 4b is about 60nm thick, which is the thickness estimated to be grown at the early stage of the deposition at lower temperature. Therefore this may be attributed to the consequence of the deposition conditions.

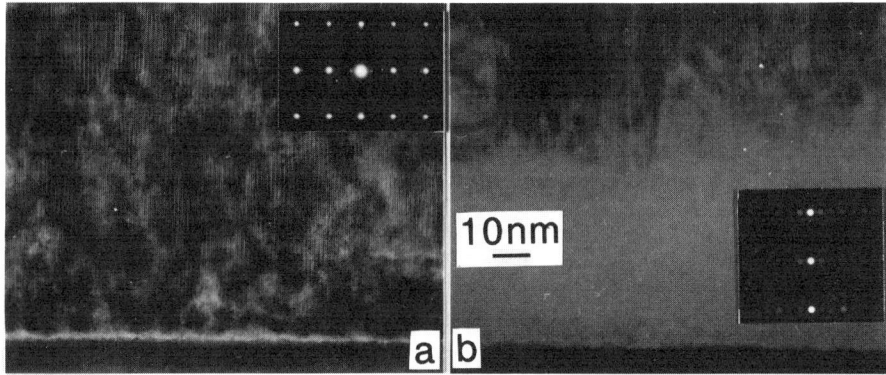

Fig. 4 A $c_\parallel$ film on a $\langle 011 \rangle$ substrate. The interface is quite sharp in (a), but note the loss of ordering of the 1-2-3 phase in the region near the interface. A layer of a different phase appears along the interface due to Ar^+ damage (b). The inserted microdiffraction pattern in (b) is taken from this layer.

Films with $c_\parallel$ orientation can also be grown on $\langle 001 \rangle$ $SrTiO_3$ substrate under the same conditions as for the $\langle 110 \rangle$ film described above. However, the film in this case tends to grow with $c_\perp$ orientation immediately next to the substrate for about 40nm and then show a region of 30nm thickness with a mixture of $c_\perp$ and $c_\parallel$ before it turns to $c_\parallel$ completely (fig. 5). The reason for this growth mode is not yet clear.

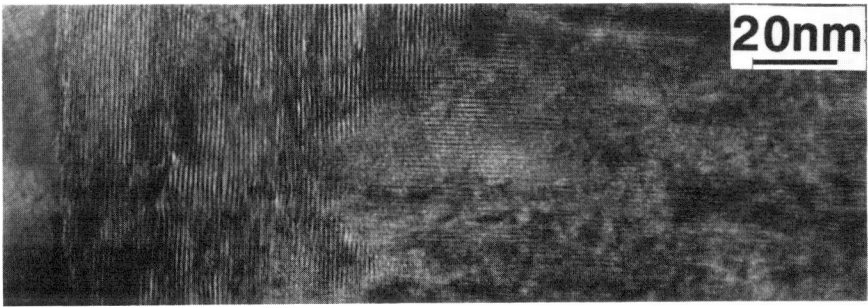

Fig. 5 A $c_\parallel$ film grown on $\langle 001 \rangle$ $SrTiO_3$. Note the $c_\perp$ orientation near the interface and the change to $c_\parallel$.

4. CONCLUSION

The TEM analysis of YBCO thin films deposited on $LaAlO_3$ and $SrTiO_3$ shows that films of higher structural perfection, especially with sharper interfaces are grown on $SrTiO_3$ substrates. Whether the diffusion of Al into the superconducting films is responsible for the production of the extended defects in films on LaALO$_3$ is not clear. However, the absence of the extended defects in films prepared on $SrTiO_3$ gives some indication correlation between the interdiffusion and the defect formation.

The preparation of $c_{\parallel}$ YBCO films using template technique give some interesting structural properties, especially in the case of films deposited on $\langle 001 \rangle$ substrates, where a change from $c_{\perp}$ growth to $c_{\parallel}$ growth occurs close to the interface.

TEM studies of the interface turn out to be more delicate than expected, especially due to the fact that ion milling is causing damage or even amorphization of those parts of the films close to the interface. Further investigations are required to analyze whether this mechanical softening is due to some interdifussion of the metal elements.

5. REFERENCES

1. D. Dijkamp, T. Venkatesan, X.D. Wu, S.A. Shaheen, N. Jisrawi, Y.H. Min-Lee, W.L. McLean and M. Croft, Appl. Phys. Lett., 51 (1987) 619.

2. Hanns-Ulrich Habermeier, Gunter Beddies, Bernd Leibold, Ganghua Lu and Gunter Wagner, Physica C, in the press.

3. Y.Q. Li, J. Zhao, C.S. Chern, W. Huang, G.A. Kulesha, P. Lu, B. Gallois, P. Norris, B. Kear and F. Cosandey, Appl. Phys., Lett., 58 (1991) 648.

4. A.E. Lee, C.E. Platt, J.F. Burch and R.W. Simon, Appl. Phys. lett., 57 (1990) 2019.

5. Von der Marel and H.-U. Habermeier, Physica C, in the press.

6. H.-U. Habermeier, A.A.C.S. Lourenco, B. Friedl, J. Kircher and J. Köhler, Solid State Communications, 77 (1991) 683.

High T$_c$ Superconductor Thin Films
L. Correra (Editor)

METAL SUBSTRATES FOR HTSC-FILMS PREPARATION

T.E.Os'kina and Y.D.Tret'yakov

Moscow State University, Dept. of Chemistry, Moscow V-234, USSR

Abstract

The possibility to use Ag, Au, Ni, Fe (stainless steel), Cu, V, Ta, Mo and Ag buffer layers for the preparation of superconducting films is evaluated. The choice of metals depends on the type of film and on the annealing conditions.

1. INTRODUCTION

A number of successful attempts has been reported to obtain HTSC-films with the use of MgO, ZrO$_2$, Al$_2$O$_3$ as substrates and also of other high temperature materials as monocrystals and ceramics. As a rule it concerns highly expensive and tiresome technologies. The use of metallic substrates can be promising especially when dealing with easily available metals. This allows to modify technological means and to use the methods of HTSC-covers processing employed only for metals (pressing, rolling, bending, etc). Moreover in some high current devices (energy storages, superconducting magnets) metal substrates and metal buffer layers are indispensable.

2. Ag SUBSTRATES

The properties of silver make it the most convenient metal for practical use, as demonstrated by the intensive research in the preparation of silver tapes based on YBCO superconductors [1,2], BSCCO superconductors [3-5] and silver containing composites [6,7]. The specific form of interaction of different types of superconductors with silver has been reported. In TBCCO materials only small amounts of silver (about 2 wt.%) do not deteriorate superconducting properties, and with the increase of Ag concentration the amount of superconducting phase obtained decreases and the volume resistivity of samples rises [6]. In Y-Ba-Cu system silver does not appreciably interact with the stoichiometric 123-phase but Ba admixture containing phases, usually present in superconductors (BaCO$_3$, BaCuO$_2$), interacts with silver facilitating its melting [7]. The optimum Ag content in the system is probably 10-20 wt% and, according to some data, up to 40 wt%. In BSCCO systems the properties of superconducting phases are not deteriorated by silver if annealing is carried out at low oxygen partial pressure, but T_c and J_c decrease after heating in air [4]. In our experiments we prepared silver substrates by melting granules in He at 1050°C and rolling of the plates to the thickness of 100-150 μm. The superconducting phase with the composition $Bi_{1.8}Pb_{0.3}Sr_{1.9}Ca_{2.0}Cu_{3.0}O_x$ was prepared by the dissolution of respective oxides-carbonates in nitric acid and the decomposition of nitrates followed by firing of the composition at 850°C in air for 170 hours. The pellets with 3.3g/cm^{-3} density consisted of homologue-phases mixture, the content of 2223-phase being 65-75%. They were ground to powder, suspended in acetone and applied to the cleaned surface of Ag plates. The films were compacted by cold rolling and subjected to cyclic annealing at 830°C ± 2°C in air-argon mixture at $P(O_2)$ = 0.1-0.5 atm in closed system in the presence of Pb-containing BSCCO powders. Figure 1 shows the film resistivity after every "heating-rolling" cycle. The T_c value increases with the number of cycles, reaching the value of T_c = 101 K after 200 hours annealing. Further heating resulted in the deterioration of the film properties. Coating thickness was varied in the range of 5-100 μm by changing the thickness of the starting coating and the rolling intensity. As follows from EPMA data after 200 hours annealing the films 20 μm thick lost up to 50% of the initial Pb content. The XRD spectra revealed the presence of 2212, 2223 phases and small amounts of Ca_2PbO_4 and CuO.

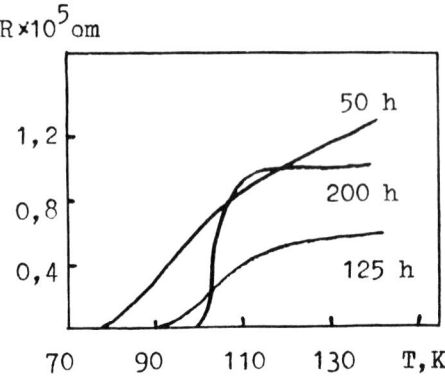

Figure 1. R(T) for the BSCCO films on the Ag substrates.

A short heating to 880°C ("thermal shock") before the first rolling increases the degree of texturing, but the phase composition shifts towards the increase of 2212 phase content. The same effect is observed when the annealing temperature is increased. On annealing in air at 830°C the T_c values dropped to 68 K. The nature of the interaction between BSCCO and silver is still somewhat obscure, as far as annealing media are concerned. The deterioration of the superconducting properties after heating in air or in oxygen could be explained by eutectics formation [8], but this model does not account for the fact that in Pb absence this dependence is still observed. Figures 2 and 3 show the emanation-thermal analysis spectra, illustrating the characteristics of BSCCO on Ag after heating in air and in argon. An exponential dependence on the temperature is usually observed for solids if transitions in the solid do not occur. The appearance of liquid regions during the heating of the superconductor induces an abrupt decrease of the internal surface of the particles, resulting in the decrease of the emanation level. Thus the spectrum of emanation acquires the shape of a peak, its occurrence coinciding with the very beginning of liquid phases appearing in the system under investigation. For BSCCO powder in air the appearance of the liquid phase occurs at 840°C, and in the case of contact with silver at 830°C. In argon atmosphere these values decrease to 815° and 800°C respectively. In conclusion, the heating in argon lowers the temperature of the liquid phase appearance by about 25-30°C both for BSCCO and BSCCO on silver.

I, imp/s

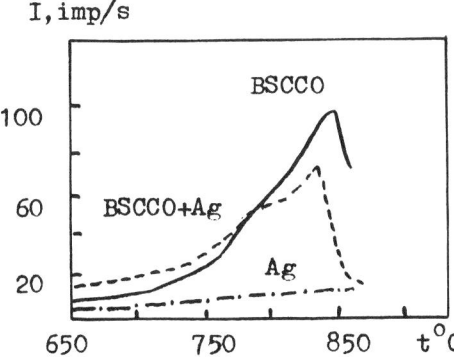 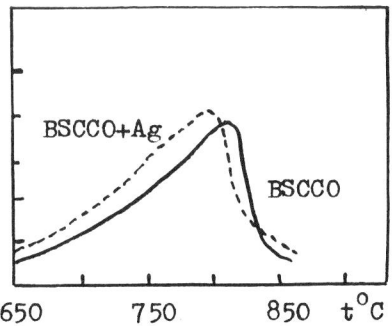

Figure 2. ETA curves in air.   Figure 3. ETA curves in argon.

The lowering of the melting point can be critical for homologue-phases ratio in the BSCCO system and for the superconducting film properties. The study of Ag sheathed BSCCO tapes prepared by folding film prior to rolling, demonstrates a number of advantages of this method. The superconducting properties of the film can be achieved easier. Thus, after 125 h annealing the films obtained by routine procedure (Fig.1) had T_c (initial) = 100 K and T_c=90.2 K, whereas for silver-sheathed tapes T_c = 112 K and T_c = 99.6 K were obtained. Moreover, the silver sheathed tapes are not so sensitive to possible contamination of superconducting layer during rolling and annealing, have more stable cation composition due to the prevention of volatilization, can be rolled to very low thickness, are not susceptible to curvature due to the difference in thermal expansion coefficients of silver and superconducting layers, and the film is less exposed to degradation due to environmental factors.

3. Au SUBSTRATES

It has been shown that YBCO interact with gold forming a solid solution, where gold atoms substitute the position of Cu(1) sites. This results in axial expansion along C axis, which corresponds to an increase of T_c from 91 to 93 K [9]. This interaction is confirmed by photoemission data [10]. In the case of BSCCO the interaction with gold was not observed [8,10]. Our results show that

BSCCO films can be successfully obtained on Au substrates by the "heating-rolling" cycles method (Fig.4). In the case of Au substrates the T_c value resulted to be lower than that measured for Ag substrates (92 and 101 K respectively). Rather than to possible poisoning effects of gold, that is inert, the higher T_c value in the case of silver should be attributed to the fact that contact melting on silver substrates improves the properties of the film.

4. Cu, Fe (STAINLESS STEEL) AND Ni SUBSTRATES

If superconducting films can be formed on widely used metals, the range of their utilization can be substantially increased. YBCO films on copper substrates are liable to degradation with time. Copper is easily oxidized on heating. BSCCO films with T_c = 83 K have been obtained by reaction of Bi_2SrCaO and CuO/Cu substrate during a short heating. YBCO films with T_c = 84 K were obtained on stainless steel substrates. Nickel is a more thermal stable material. The information on the possibility to prepare thin YBCO layers on Ni substrates seem to be reliable. BSCCO superconductors interact with Ni even at room temperature [13] and in this particular case films without buffer layers cannot be prepared.

5. METAL SUBSTRATES WITH Ag SUBLAYER

Many authors have reported Ag as the best buffer layer. The calculation of the optimal buffer layer thickness is of great practical importance. For Ni substrates the following dependence was found. The diffusion coefficient of Ni into silver can be calculated from the equation:

$$D = D_0 \exp(-Q/RT) = 21.9 \exp(-54800/RT) \text{ cm}^2/\text{s} \qquad (1)$$

For a temperature of 830°C, D = 2.96×10^{-10} cm^2/s, is obtained, which is a relatively large value. For this reason during the annealing a 5 μm thick Ag layer is contaminated by the Ni in only 3 minutes. The time to reach equilibrium concentration of Ni in Ag does not exceed 12 min. Thus the Ag layer only decreases the Ni concentration at the superconductor boundary to a value corresponding to the level of Ni solubility in Ag (at the considered temperature it is about 0.15 at.%), and at Ag/film interface the decrease is 670-fold. The thickness of Ag layer is not important under these circumstances and can be very low. The figure 4 shows that the use of Ag buffer layer on Ni substrates allows to obtain BSCCO films with T_c = 99 K, that is a temperature only 2 K lower than that obtained for pure silver substrates. It can be expected that the use of Ag buffer layers on stainless steel substrates can be very effective since Ag layer decreases the concentration of iron atoms on the film surface by a factor of 1.2×10^4.

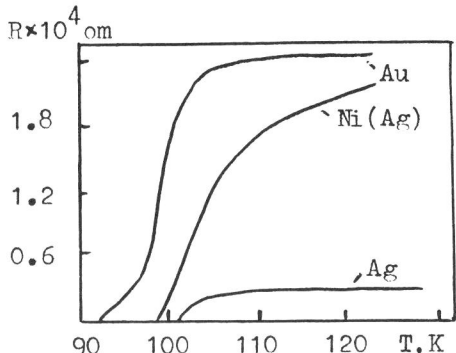

Figure 4. R(T) for the BSCCO films after 200 h annealing.

The first attempts to use stainless steel with Ag sublayer to prepare BSCCO films were not successful, because after 20 hours heating at 830°C the Fe/Ag boundary was oxidized and the Ag/film layer peeled off. V, Nb, Ta, Ti and Mo are nearly insoluble in silver, and some of them (V, Ta and Mo) became fragile on heating. For these reasons different methods of films preparation must be developed, employing low temperature annealings.

6. REFERENCES

1   C.T.Chu and B.Dunn, Appl.Phys.Lett., 55 (1989) 492
2   S.Witanachi, S.Patel, D.T.Shaw and H.S.Kwok, Appl.Phys.Lett., 55 (1989) 295
3   Y.Yamoda, K.Jikihara, T.Hasebe et al., Japan J.Appl.Phys., 29 (1990) L 456
4   S.X.Dou, H.K.Liu, M.H.Apperley et al., Supercond.Sci.Technol., 3 (1990) 138
5   K.Osamura, Soo Oh Sang and O.Shojiro, Supercond.Sci.Technol., 3 (1990) 143
6   D.March, F.Arammash, J.Bennet et al., Appl.Phys.Communications, 9 (1989-90) 245
7   F.Deslandes, B.Raveau, P.Dubots and D.Legat, Solid State Communications, 71 (1989) 407
8   H.K.Liu, S.X.Dou, K.H.Song et al., Supercond.Sci.Technol., 3 (1990) 210
9   M.Z.Cieplak, G.Xiao, C.L.Chien et al., Appl.Phys.Lett., 57 (1990) 934
10   D.S.Dessau, Z-X.Shen, B.O.Wells et al., Appl.Phys.Lett. 57 (1990) 307
11   T-H.Sung, Z.Nakagawa, M.Yoshimoto et al., Japan J.Appl.Phys., 29 (1990) 949
12   R.E.Russo, R.P.Reade, J.M.McMillan and B.L.Olsen, J.Appl.Phys., 68 (1990) 1354
13   P.S.Asoka Kumar, S.Mahumini, P.Kulkarni et al., J.Appl.Phys., 67 (1990) 3184

High T$_c$ Superconductor Thin Films
L. Correra (Editor)
© 1992 Elsevier Science Publishers B.V. All rights reserved.

Epitaxial Growth of GdBa$_2$Cu$_3$O$_{7-x}$ Thin Films on Yttria-Stabilized Zirconia Buffer Layers

M. Siegel[a], F. Wang[b], R. Smithey, U. Kaufmann[c], G. Linker, O. Meyer, and J.Geerk

Kernforschunszentrum Karlsruhe, Institut für Nukleare Festkörperphysik, P. O. Box 3640, W-7500 Karlsruhe, Germany

Abstract

Yttria-stabilized zirconia (YSZ) buffer layers have been deposited on R-plane sapphire by inverted cylindrical magnetron sputtering. The buffer layers grow epitaxially in a cubic phase with a mosaic spread of 0.15° and a χ_{min} of 4%.

Epitaxial GdBa$_2$Cu$_3$O$_{7-x}$-films were grown on these buffer layers at different substrate temperatures, T$_S$. A transition from c-axis to a-axis oriented growth was obtained by decreasing substrate temperature. The c-axis oriented GdBa$_2$Cu$_3$O$_{7-x}$-films exhibit a mosaic spread of 0.5°, a χ_{min} of 17%, a normal state resistivity at 100 K of 66 $\mu\Omega$cm, a zero resistance transition temperature of 92.5 K, and a critical current density of 2.6·10^6 A/cm^2.

1. Introduction

Deposition of high quality epitaxial thin films of high T$_C$ superconductors on sapphire is considered to be an important step towards the development of superconducting electronics and combination of superconducting and integrated circuit devices. One of the potential applications for the high T$_C$ superconductors is the area of passive microwave devices such as filters, resonators, delay lines, and antennas. Sapphire is a preferred substrate for microwave device applications due to its small dielectric constant (ϵ=9 at 9GHz) and low losses (tanδ is less than 10^{-4} at 300 K). Sapphire has good mechanical strength, and is less expensive than the other substrates used for the deposition of HTSC thin films. However, HTSC thin films grown on pure sapphire have low j$_C$ values and depressed T$_C$ values [1] due to interdiffusion between film and substrate at the substrate temperature required to form

---

[a] On leave from University Jena, Jena, FRG
[b] On leave from Peking University, Beijing, P.R. China
[c] Dornier Systems, Friedrichshafen, FRG

a good superconducting film. Despite the reasonably high T_C values up to 88 K the critical current density is still very low [2]. One method to prevent substrate-film reactions is to cover the sapphire substrates with an intermediate layer such as YSZ. In addition the buffer layer has to be lattice matched with both sapphire and HTSC thin film. The epitaxial growth of YSZ on Si and sapphire has been already achieved by e-beam evaporation [3], ion sputter deposition [4], pulse laser deposition [5], and rf sputtering [6]. Wu et. al. [7] have reported the growth of $YBa_2Cu_3O_{7-x}$-films on YSZ-buffer layers on R-plane sapphire with T_C and j_C of up to 89 K and $1 \cdot 10^6$ A/cm^2 at 77 K, respectively.

In this paper, we report on the epitaxial growth of $GdBa_2Cu_3O_{7-x}$-thin films deposited on R-plane sapphire substrates covered with cubic YSZ-buffer layers. The films were characterized for their electrical and structural properties using four point probe dc resistance measurements, X-ray diffraction, and ion beam channeling.

2. Experimental details

The YSZ-buffer layer were prepared by inverted cylindrical magnetron rf- sputtering using $1 \cdot 10^{-1}$ mbar mixture of argon and oxygen. A target with the composition of $(Y_2O_3)_{0.1}(ZrO_2)_{0.9}$ was used. The sputtering rate was 1.0-1.3 Å/s. Before deposition, the sapphire was cleaned using ethanol, and heated up to 1200°C in air for 3 hours.

The $GdBa_2Cu_3O_{7-x}$-thin films were deposited by inverted cylindrical magnetron dc-sputtering using 0.7 mbar mixture of argon and oxygen (Ar:O_2=2:1). This method, which yields highly reliable and reproducible results, has been described in detail previously [8, 9]. The substrates were fixed onto the heater block using silver paste, and the substrate temperature given in the text was measured with a thermocouple attached to the heater block. The superconducting transitions of the films were measured resistively. The transport current density was determined at 77 K on microbridges with dimensions of 10 μm x 100 μm. The microbridges were patterned either by amorphisation with 300 keV Ar$^+$ ions or by inhibit layer patterning. X-ray diffraction diagrams were taken in the Bragg-Brentano focusing geometry. Ion channeling and Rutherford backscattering measurements were made using 2 MeV He$^+$ ions.

3. Results and discussion

In the deposition experiments of the YSZ-buffer layer on R-plane sapphire substrates we found, that growth in the desired cubic phase starts already at comparably low substrate temperatures near 600°C, but in order to obtain films with high quality crystalline properties substrate temperatures near 950°C are necessary. An additional improvement provides a post-annealing of the YSZ-buffer layer at 1300°C in air. The epitaxial buffer layer reveals in the X-ray diffraction spectra only (h00)-peaks, and has mosaic spreads of the (200)-peak of 0.15°. The buffer layers show pronounced axial channeling. After the post-annealing procedure the minimum backscattering yield χ_{min} was 4%. In more detail, we report the growth of the YSZ-buffer layer on R-plane sapphire substrates in the contribution by F. Wang et. al. at this conference.

GdBa₂Cu₃O₇₋ₓ-thin films were deposited by inverted cylindrical magnetron dc-sputtering on the epitaxial cubic YSZ-buffer layers. The epitaxial growth was studied as a function of the substrate temperature during deposition, T_S. X-ray diffraction diagrams of films deposited at different temperatures T_S are shown in Figure 1. At a temperature of $T_S=790°C$ the GdBa₂Cu₃O₇₋ₓ-film grow c-axis oriented. With decreasing substrate temperatures mixtures of c-axis and a-axis oriented grains were observed. At a temperature of $T_S=640°C$ the most part of the film contained a-axis oriented grains. However, a small (013)-peak appeared in the diagramm at lower substrate temperatures.

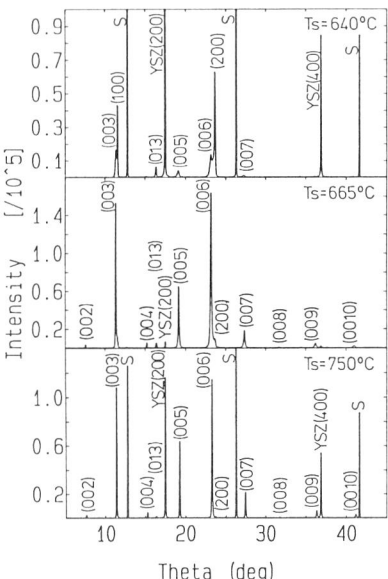

Figure 1.   X-ray diffraction diagramm in the Bragg-Brentano geometry of GdBa₂Cu₃O₇₋ₓ -films on R-plane sapphire with YSZ-buffer layer at different temperatures.

The X-ray diffraction diagramm of an c-axis oriented GdBa₂Cu₃O₇₋ₓ-film deposited at optimized conditions is shown in Figure 2. The Θ-2Θ-diagramm shows only (h00)-peaks of the YSZ-buffer layer, the substrate peaks, and the (00l)-peaks of the GdBa₂Cu₃O₇₋ₓ-film demonstrating the highly textured growth. Figure 3 shows the rocking curve of this film. The mosaic spread determined for the (005)-peak was 0.5° indicating a highly oriented growth of the GdBa₂Cu₃O₇₋ₓ-film on the YSZ-buffer layer.

Figure 4 shows the random and aligned backscattering spectra of a (001)-oriented GdBa₂Cu₃O₇₋ₓ-film on sapphire with YSZ-buffer layer. The χ_{min} value is 17%.

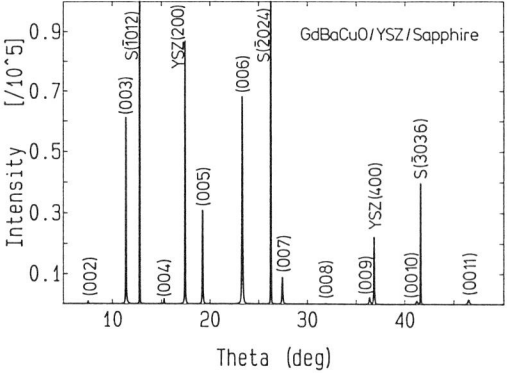

Figure 2. X-ray diffraction diagramm of a c-axis oriented GdBa$_2$Cu$_3$O$_{7-x}$-film on YSZ-buffer layer.

Figure 3. Rocking curve of the GdBa$_2$Cu$_3$O$_{7-x}$-film shown in Figure 2.

This confirms the epitaxial growth of the (001)-oriented GdBa$_2$Cu$_3$O$_{7-x}$-films on YSZ-buffer layers.

The high quality of the GdBa$_2$Cu$_3$O$_{7-x}$-films on YSZ-buffer layers is demonstrated by their superconducting properties. The optimal grown c-axis oriented GdBa$_2$Cu$_3$O$_{7-x}$-films on YSZ-buffer layers had a zero resistance temperature of 92.5 K, and a transition with of 1 K. Figure 5 shows the temperature dependence of the resistivity of a GdBa$_2$Cu$_3$O$_{7-x}$-film patterned with dimensions of 10 μm x 100 μm. The normal state resistivity of the GdBa$_2$Cu$_3$O$_{7-x}$-film was 66 $\mu\Omega$cm at 100 K. The critical current density was measured on this

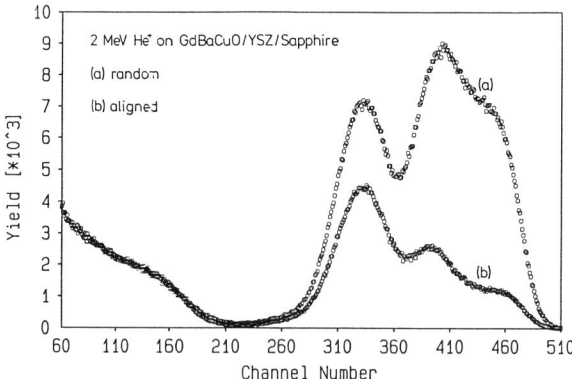

Figure 4.   Random and aligned backscattering spectra of a GdBa$_2$Cu$_3$O$_{7-x}$-film on sapphire with YSZ-buffer layer.

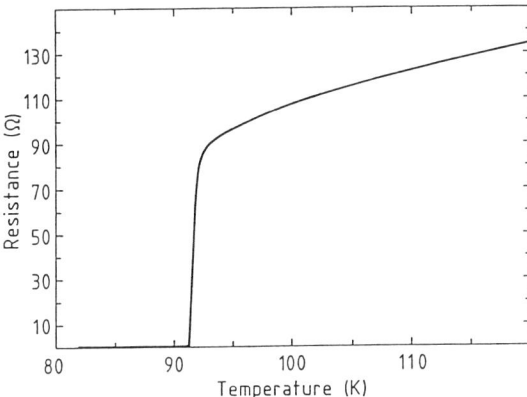

Figure 5. Resistance vs. temperature of a GdBa$_2$Cu$_3$O$_{7-x}$-film on sapphire with YSZ-buffer layer. The dimensions of the microbridge were 10 μm x 100 μm.

microbridge with a voltage criterion of 10 μV/cm. At 77 K we obtained maximum critical current densities of 2.6·10^6 A/cm^2.

It should be mentioned, however, that on different films and even on different microbridges on the same film quite a large scatter has been observed in the critical current density values ranging from 10^5 A/cm^2 to the maximum values quoted above. This probably is due to the different in plane orientation of the grains for various microbridges as has been discussed recently by Rowell [10]. We have not controlled the in plane growth so far in our experiments.

With decreasing deposition temperatures the superconducting properties of the films usually degrade. This degradation structurely is accompanied by the change of film orientation from c-axis to a-axis oriented growth as has been reported for YBa$_2$Cu$_3$O$_{7-x}$-films on SrTiO$_3$ substrates [11]. We observed a similar behavior for the GdBa$_2$Cu$_3$O$_{7-x}$-films on the YSZ/Al$_2$O$_3$ substrates. It is interesting to note, however, that the onset temperature of the degradation in comparison to films on SrTiO$_3$ is shifted by at least 50°C to lower temperatures. This confirms our earlier findings for YBa$_2$Cu$_3$O$_{7-x}$-films deposited on (100)Zr(Y)O$_2$ single crystalline substrates [9]. The higher T$_c$ values at lower T$_s$ coincide with the preservation of c-axis growth to lower temperatures. We believe that this effect is to due the reduced lattice matching on Zr(Y)O$_2$ as compared, e.g., to SrTiO$_3$, because we observed this behavior also on other substrates where we intentionally deteriorated lattice matching.

4. Conclusion

Epitaxial growth of $GdBa_2Cu_3O_{7-x}$-thin films on R-plane sapphire with yttria-stabilized zirconia buffer layer with inverted cylindrical magnetron sputtering has been demonstrated. X-ray diffraction and ion channeling indicated the high structural quality of the YSZ-buffer layers and the $GdBa_2Cu_3O_{7-x}$-films. We have found the optimal substrate temperature for c-axis oriented growth of the $GdBa_2Cu_3O_{7-x}$-film on YSZ-buffer layer. The $GdBa_2Cu_3O_{7-x}$-films on the YSZ-buffer layer had a resistivity of 66 $\mu\Omega$cm at 100 K, a zero resistance temperature of 92.5 K, and a transport critical current density of $2.6 \cdot 10^6$ A/cm^2 at 77 K.

In comparison, we have grown c-axis oriented $GdBa_2Cu_3O_{7-x}$-films on $(100)SrTiO_3$-substrates in the same sputtering system. These films had normal state resistivity of 60 $\mu\Omega$cm, zero resistance temperature of 93.8 K, transport critical current densities of $5 \cdot 10^6$ A/cm^2 at 77 K, mosaic spreads of 0.1°, and χ_{min} values of 4%. These data demonstrate the high quality of the YSZ-buffer layer and the quality of the $GdBa_2Cu_3O_{7-x}$-films grown on these buffer layers.

Acknowledgements
The authors thank Mr. Massing for technical assistance in the X-ray analysis and Mr. Strehlau for the ion implantation.

5. References

1   A. Stamper, D.W. Greve, D. Wong, and T.E. Schlesinger, Appl. Phys. Lett. **52** (1988) 1476

2   K. Char, D.K. Fork, T.H. Geballe, S.S. Ladermann, R.C. Taber, R.D. Jacowitz, F. Bridges, G.A.N. Connell, and C.B. Boyce, Appl. Phys. Lett. **56** (1990) 785

3   H. Myoren, Y. Nishiyama, H. Fukomoto, H. Nasu, and Y. Osaka, Japn. Journ. Appl. Phys. **28** (1989) 351

4   P. Legagneux, G. Garry, D. Dieumegard, C. Schwebel, C. Pellet, G. Gautherin, and J. Siejka, Appl. Phys. Lett. **53** (1988) 1506

5   D.K. Fork, D.B. Fenner, R.W. Barton, Julia M. Philips, G.A.N. Connel, J.B. Boyce, and T.H. Geballe, Appl. Phys. Lett. **57** (1990) 1161

6   F. Konushi, T. Doi, H. Matsunaga, Y. Kakihara, M. Koba, K. Awane, and I. Nakamura, Mat. Res. Soc. Symp. Proc. **56** (1986) 259

7   X.D. Wu, R.E. Muenchhausen, N.S. Nogar, A. Pique, R. Edwards, B. Wilkens, T.S. Ravi, D.M. Hwang, and C.Y. Chen, Appl. Phys. Lett. **58** (1991) 304

8   X.X. Xi, G. Linker, O. Meyer, E. Nold, B. Obst, F. Ratzel, R. Smithey, B. Strehlau, F. Weschenfelder, and J. Geerk, Z. Phys. B, Cond. Matter **74** (1989) 13

9   J. Geerk, G. Linker, O. Meyer, Mat. Sci. Rep. vol.4(5,6) (1989)

10 J.M. Rowell, 3rd FED Workshop on HTSC Electron Devices, Kumamoto, Japan, May 1991

11 G. Linker, X.X. Xi, O. Meyer, Q. Li, J. Geerk, Solid State Comm. **69** (1989) 249

High T$_c$ Superconductor Thin Films
L. Correra (Editor)
© 1992 Elsevier Science Publishers B.V. All rights reserved.   707

Passivation Layers on YBa$_2$Cu$_3$O$_{7-x}$ Thin Films

Ch. Ziegler, G. Frank, and W. Göpel

Institute of Physical and Theoretical Chemistry, University of Tübingen, Morgenstelle 8,
W-7400 Tübingen, FRG

Abstract

The passivation of YBa$_2$Cu$_3$O$_{7-x}$ epitaxial thin films was tested against water after preparing SiO$_2$, ZrO$_2$(Y$_2$O$_3$) (YSZ), and CaF$_2$ protection layers and subsequent annealing the as-passivated superconductor in 100% humid air at elevated temperatures.

We first characterize the clean and covered surfaces of epitaxial thin YBa$_2$Cu$_3$O$_{7-x}$ films by means of photoemission spectroscopies (XPS and UPS). All passivation layers have been deposited at room temperature with negligible interface reaction with the superconductor. At elevated temperatures the formation of silicates, zirconates, and fluorides is observed.

The passivated films are stable up to 500K in 100% humid air. Above this temperature the superconductor is reduced chemically. Non-passivated films, however, are stable only at temperatures below 450K and are destroyed totally in humid air at higher temperatures.

1. Introduction

For practical applications of high temperature superconductors, long-term stability is highly desired. In contrast to the Bi-based superconductors, YBa$_2$Cu$_3$O$_{7-x}$ is not stable against humidity /1-9/. Water may easily react at grain boundaries, which are present even in epitaxial thin films. This reaction was studied in detail mainly on polycrystalline material by means of X-Ray diffraction /1/, electron microscopy /1-5/, pH-measurements /1/, T$_c$-measurements /2,3,5-7/, X-ray photoemission (XPS) /2,4,8,9/, and UV photoemission spectroscopy (UPS) /9/. These investigations show that YBa$_2$Cu$_3$O$_{7-x}$ is destroyed according to the overall reaction scheme

$$2\ YBa_2Cu_3O_7 + 3\ H_2O \rightleftharpoons 5\ CuO + 3\ Ba(OH)_2 + Y_2BaCuO_5 + 1/2\ O_2.$$

We now report on formation and stability of thin (3 nm) protection layers of SiO$_2$, YSZ, and CaF$_2$ on YBa$_2$Cu$_3$O$_{7-x}$ thin film surfaces. In this context we first studied the thin film growth of these materials during their deposition at room temperature with XPS.

Subsequently the reactivity was investigated of the passivation layer/superconductor interface at elevated temperatures.

In a second set of experiments we compared the stability of non-passivated and passivated thin $YBa_2Cu_3O_{7-x}$ films against 100% humid air as it shows up in changed resistance versus temperature behavior.

2. Experimental

The samples were epitaxial thin films of $YBa_2Cu_3O_{7-x}$ with a thickness of 200 nm prepared on $SrTiO_3(100)$ or $MgO(100)$ and with $T_{c,0} > 87K$ and $J_c(77K) > 10^6$ A/cm^2. The c-axis of $YBa_2Cu_3O_{7-x}$ was oriented perpendicular to the surface.

Prior to the deposition of passivation layers on the $YBa_2Cu_3O_{7-x}$ films their surface was cleaned carefully by gentle heating in UHV until the total amount of carbon was less than 1.5 at% as checked by XPS /10/.

SiO_2 and yttria stabilized zirconia (YSZ) were evaporated in-situ with an electron beam gun and CaF_2 with a Knudsen cell at $T_{substrate} = 298K$ with layer thicknesses between 3-50 nm. These sandwich systems were subsequently heat-treated at elevated temperatures and then investigated in-situ with XPS.

The XPS studies were performed in a combined high pressure-ultrahigh vacuum system (for details see, e.g., /11/) with a VSW Ha 150 analyzer equipped with a MgK_α-source. The pass energy was set to 25 or 50 eV in different experiments with an overall resolution of 1.0 and 1.45 eV, respectively. All spectra were taken at room temperature. The base pressure during analysis was less than $3x10^{-8}$ Pa with the same base pressure in the separate sample preparation chamber.

The XPS binding energies are referred to the Fermi level of platinum. The energy scale was calculated from a linear regression for 17 reference levels covering the energy range up to 1000 eV. The decomposition of the lines was performed using Gaussian-shaped peaks.

The passivation measurements were performed in a separate UHV/high pressure system consisting of two separate chambers for high and low temperature resistivity measurements. Temperatures down to 25 K (monitored by a Ni/Ni(Cr) thermocouple attached to the sam   ple surface) were adjusted by a Leybold Heraeus RGD 510 cryo cooler at a base pressure of 10^{-9} mbar. In the high temperature chamber temperatures up to 800 K (monitored by a Pt/Pt(Rh) thermocouple) can be adjusted in the pressure range between 10^{-9} mbar and 1 bar. The four point resistivity measurements were performed according to the method of van der Pauw /12/.

The samples were subsequently exposed to 1 bar 100% humid synthetic air. The temperature was raised from 300K in steps of 50 K and kept at constant temperature for 1 hour. During this heating cycle the resistivity was measured in-situ and after each temperature step the low-temperature resistivity behavior was monitored.

3. Passivation Layer Formation and Stability

3.1 SiO₂ Layers

Photoemitted electrons were collected of the core levels Cu $2p_{3/2}$, Ba $3d_{5/2}$, O 1s, Y $3p_{3/2}$, C 1s, Y 3d, Si 2s, Si 2p, and Ba 4d before and after deposition of SiO_2 and after each heat treatment step (see Sec.2) (Fig.1).

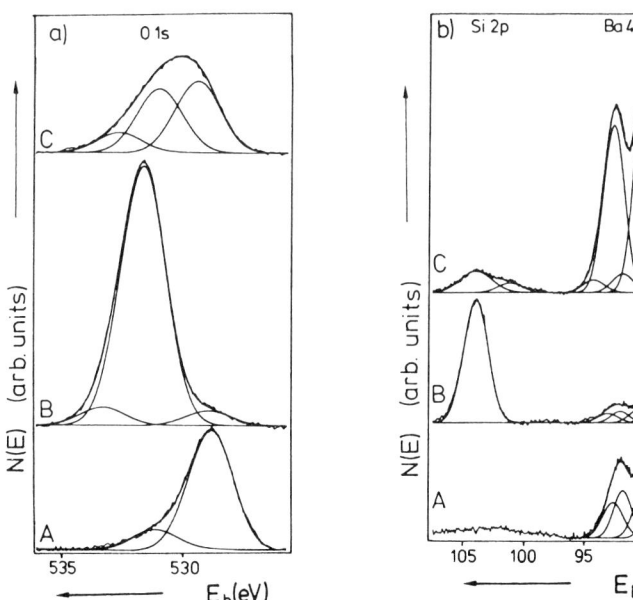

Figure 1. XPS spectra of a clean $YBa_2Cu_3O_{7-x}$ surface (A), after evaporation of 28 nm SiO_2 (B), and after heat treatment at T = 750K in $p(O_2)$ = 10^{-4} mbar (C) with typcal O 1s (a) and Si 2p + Ba 4d (b) emissions

The clean $YBa_2Cu_3O_{7-x}$ surface shows XPS features well known from earlier measurements (see /10,13/ and references therein). The higher binding energy components of the Ba and O core levels are associated with a surface component attributed to relaxed surface bonds /10,13/.

There is no interface reaction during the SiO_2 deposition at 300K (Fig.1A,B). The $YBa_2Cu_3O_{7-x}$ substrate signals decrease exponentially during SiO_2 evaporation which is characteristic for a random cluster growth /14/. Negligible interface reactivity can also be deduced from UPS measurements where no shift in the zero cut-off energies is seen.

The system is stable up to 750 K in $p(O_2)$ = 10^{-4} mbar. Higher oxygen partial pressure leads to reduced reactivity at the same temperature. At higher temperatures the formation of new interface features can be detected:

The O 1s peak shows three new contributions at 534.0, 532.5, and 530.0 eV instead of two at 529.1 (bulk) and 532.1 eV (surface) before the heat treatment.

The Ba $3d_{5/2}$ peak consists of two components at 780.3 (surface) and 778.6 eV (bulk) before and of one main component at 780 eV with a small contribution at 782.2 eV (carbonate) after the heat treatment. Comparable results are deduced from the Ba $4d_{5/2}$ emission (Fig.1b).

The Y $3d_{5/2}$ peak shows a slight shift of 0.4 eV from 156.4 to 156.8 eV after the heat treatment.

The core level spectrum of copper shows the chemical reduction from Cu^{2+} before the heat treatment (with a main peak at 933.7 eV and a satellite structure around 942 eV) to Cu^{1+} or Cu^{0} afterwards (with one feature at 932.6 eV).

In the Si 2p region no shift of the main peak at 103.6 eV (SiO_2) is observed after the heat treatment. A new peak appears at 101 eV (Si_2O) which is not observed if the heat treatment is performed in 1 bar O_2.

The new oxygen features can be assigned to SiO_2 and/or carbonates (534 eV), silicates (532.5 eV) /15/, and oxidic or peroxidic species (530.0 eV). As the binding energy of the Ba $3d_{5/2}$ core levels is not identical with that of the pure oxide (778.9 or 779.4 eV) /16/ the formation of Ba silicates is more probable.

For the yttrium compounds we deduce the formation of Y_2O_3 with E_b(Y $3d_{5/2}$) = 156.8 eV and E_b(O 1s) = 529.9 eV /2,17/. For the copper compounds, formation of Cu_2O is thermodynamically most probable which also determines the oxygen feature at 530.0 eV /18/.

These findings are in line with those obtained on Si-covered $YBa_2Cu_3O_{7-x}$ surfaces where the same reaction products are observed at lower temperatures /19/.

3.2 Yttria Stabilized Zirconia (YSZ) Layers

The substrate XPS peaks of $YBa_2Cu_3O_{7-x}$ were investigated as a function of YSZ overlayer thickness. The YSZ layer grows on $YBa_2Cu_3O_{7-x}$ in a layer-by-cluster growth mode (Stransky Krastanov mode) as it can be deduced from a linear decrease of the $YBa_2Cu_3O_{7-x}$ substrate emission up to one monolayer, followed by an exponential decrease above one monolayer.

No interface reaction occurs during the overlayer deposition at room temperature. However, the superconductor is chemically reduced because of an oxygen loss of YSZ during the electron beam evaporation under UHV conditions. This oxygen loss is compensated with oxygen atoms from the superconductor surface. This leads to a reduced work function of the layer system for thin YSZ coatings as deduced from the zero energy onset in UPS measurements.

The YSZ/$YBa_2Cu_3O_{7-x}$ interface is stable up to 500 K in 10^{-5} mbar oxygen. Above this temperature a diffusion of the superconductor into the YSZ layer occurs. At temperatures above 800 K barium zirconate, Y_2O_3, and CuO are formed. In contrast to SiO_2 the reactivity of YSZ increases with increasing oxygen partial pressure.

3.3 CaF$_2$ Layers

CaF$_2$ also growths in a layer-by-cluster mode with negligible reaction with YBa$_2$Cu$_3$O$_{7-x}$ during the deposition. The reactivity of the CaF$_2$/YBa$_2$Cu$_3$O$_{7-x}$ interface at elevated temperatures is under investigation now. From our experience with CaF$_2$ buffer layers between Si and YBa$_2$Cu$_3$O$_{7-x}$, a reduced stability is expected if compared with SiO$_2$ and YSZ /20/.

4. Passivation measurements

In Fig.2 the typical R(T)-behavior of a non-passivated YBa$_2$Cu$_3$O$_{7-x}$ film is shown.

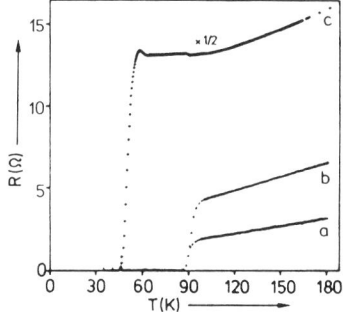

Figure 2. R(T)-behavior of a non-passivated YBa$_2$Cu$_3$O$_{7-x}$ film before (a) and after heat treatment in 100% humidity at 400K (b) and 450K (c)

We observe no change before and after humidity treatment at room temperature. After heating at 400K an increase in R(100K) can be observed with unchanged T$_{c,0}$. This effect is associated with changed grain boundary resistances /21/. Evidently the decomposition starts at grain boundaries but leads the bulk material unaffected. After heating at 450K, a pronounced increase in R(100K) and a strong decrease of T$_{c,0}$ is found. This is due to the additional decomposition of bulk material most probably along the reaction scheme mentioned in Sec.1.

Fig.3 shows the typical R(T)-behavior of a passivated YBa$_2$Cu$_3$O$_{7-x}$ film. No differences could be observed for different passivation materials or layer thicknesses. The increase in R(100K) is negligible in all experiments. Above 500K, however, a drastic change occurs which is characterized by semiconductor behavior. The latter is known to occur in YBa$_2$Cu$_3$O$_{7-x}$ after oxygen loss with x > 0.5. The latter follows independently from XPS results which indicate chemical reduction in the Cu 2p$_{3/2}$ peak (compare Sec.3.1) with all other core level peaks remaining unaffected. Formation of carbonates or hydroxides in particular can be ruled out.

This chemical reduction at 500K can only be explained by reactions with water because at this temperature any oxygen loss is negligible in YBa$_2$Cu$_3$O$_{7-x}$ even in UHV. This is mainly due to a kinetically hindered oxygen transport through the surface /22/. We therefore conclude a catalytical process to occur in which water reduces the activation barrier for bulk oxygen loss in $YBa_2Cu_3O_{7-x}$.

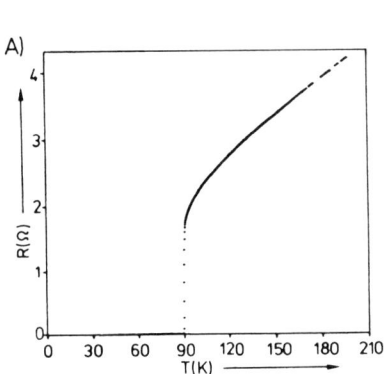 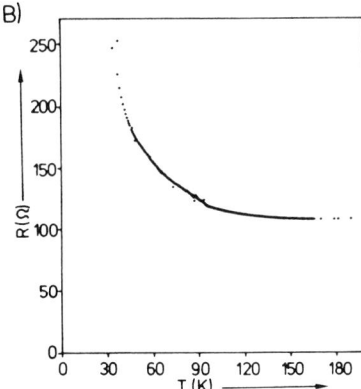

Figure 3. R(T)-behavior of a passivated $YBa_2Cu_3O_{7-x}$ film before and after all treatments up to 500K (A) and after treatment at 550K (B)

5. Conclusions

SiO_2, YSZ, and CaF_2 are suitable materials for passivation layers on epitaxial thin $YBa_2Cu_3O_{7-x}$ films up to 500K in 100% humidity. Above this temperature bulk decomposition of $YBa_2Cu_3O_{7-x}$ can still be avoided. Only a chemical reduction due to oxygen loss is observed.

The chemical stability of the passivation layer/$YBa_2Cu_3O_{7-x}$ interface can be investigated in detail by XPS data as a function of overlayer thickness and annealing temperature. From this we conclude that SiO_2 is stable up to 750K, YSZ up to 800K, whereas CaF_2 may react at lower temperatures. As $YBa_2Cu_3O_{7-x}$ grows epitaxially on thin YSZ layers /23/, we expect that YSZ may also grow epitaxially on $YBa_2Cu_3O_{7-x}$ at higher deposition temperatures. This would then lead to even denser and hence better passivation layers.

6. Acknowledgments

We greatfully acknowledge P. Berberich and H. Kinder, TU Munich, for $YBa_2Cu_3O_{7-x}$ sample preparation and characterization, W. Neu for excellent technical assistance during the performance of the experiments, and B. Reusch for performing part of the experiments. This work was supported by the Bundesminister für Forschung und Technologie, No. 13N5482 and the Fonds der Chemischen Industrie.

7. References

/1/ N.P. Bansal, A.L. Sandkuhl, Appl. Phys. Lett. **52** (1988) 323

/2/ S. Myhra, P.R. Chalker, P.T. Moseley, J.C. Riviere, Physica C **165** (1990) 270

/3/ Z. Dexin, X. Mingshan, Z. Ziqing, Y. Shubin, Z. Huansui, S. Shuxia, Sol. State Comm. **65** (1988) 339

/4/ J.H. Thomas III, M.E. Labib, in: G. Lucovski (Ed.); AIP Conf. Proc. **165** American Institute of Physics, New York 1988

/5/ M.F. Yan, R.L. Barns, H.M. O'Brian Jr., P.L. Gallagher, R.C. Sherwood, S. Jin, Appl. Phys. Lett. **51** (1987) 532

/6/ J. Dominec, L. Smrcka, P. Vasek, S Geurten, O. Smrckova, D. Sykorova, B. Hajek, Sol. State Comm. **65** (1988) 373

/7/ R.L. Barns, R.A. Laudise, Appl. Phys. Lett. **51** (1987) 1373

/8/ V.I.Nefedov, A.N.Sokolov, M.A.Tyzykov, N.N.Oleinikov,Y.A. Yeremina, M.A. Kolotyrkina, J. Electron Spectrosc. Relat. Phenom. **49** (1989) 47

/9/ S.L. Qiu, M.W. Ruckman, N.B. Brookes, P.D. Johnson, J. Chen, C.L. Lin, M. Strongin, B. Sinkovic, J.E. Crow, C.-S. Jee, Phys. Rev. B **37** (1988) 3747

/10/ Ch. Ziegler, G. Frank, W. Göpel, Z. Phys. B **81** (1990) 349

/11/ W. Göpel, Sens. Act. **16** (1989) 167

/12/ J. van der Pauw, Philips Res. Rep. **13** (1958) 1

/13/ G. Frank, Ch. Ziegler, W. Göpel, Phys. Rev. B **43** (1991) 2828

/14/ J.A. Venables, G.D.T. Spiller, M. Hanbücken, Rep. Progr. Phys. **47** (1984) 339

/15/ B. Carriere, J.P. Deville, D. Brion, J. Escard, J.Electron Spectrosc.Relat.Phenom. **10** (1977) 85

/16/ H. van Doveren, J.A.T. Verhoeven, J.Electron Spectrosc.Relat.Phenom. **21** (1980) 265

/17/ Y. Uwamino, T. Ishizuka, H. Yamatera, J.Electron Spectr.Relat.Phenom. **34** (1984) 67

/18/ J. Ghijsen, L.H. Tjeng, J. van Elp, H. Eskes, J.Westerink, G.A. Sawatzky, M.T. Czyzyk, Phys. Rev. B **38**(1988) 11322

/19/ Ch. Ziegler, F. Baudenbacher, H. Karl, H. Kinder, W.Göpel, Fres. Z. Analyt. Chem., in press

/20/ Ch. Ziegler, Ph.D. Thesis, Tübingen (FRG) 1991

/21/ J. Halbritter, Int. J. Mod. Phys. B **3** (1989) 719

/22/ K.N. Tu, S.I. Park, C.C. Tsuei, Appl. Phys. Lett. **51**(1987) 2158

/23/ G. Frank, Ph.D. Thesis, Tübingen (FRG) 1991

Preparation of Thin Films of CeO$_2$ by MOCVD

Klaus-Hermann Dahmen*, Michael Becht, Tobias Gerfin

Lab. für Anorganische Chemie, ETH Zürich, Universitätsstrasse 6, CH-8092 Zürich, Schweiz

Abstract

Thin films of CeO$_2$ were grown by thermal MOCVD using Ce(dpm)$_4$ as precursor. Additional information about this precursor received by thermal analysis showed that the complex decomposes at 265°C in air. Films were grown on quartz, sapphire, MgO[100] and Si[100] using helium, helium/Hdpm, helium/ammonia or helium/isopropanol as carrier gas. No stabilization of the complex by ammonia or isopropanol were found, whereas the ligand Hdpm stabilizes the complex. The cristallinity of the films were analyzed by XRD and the thicknesses were measured by ellipsometry.

1. Introduction

Cerium dioxide films have long been considered as useful material for optical coatings due to their high refractive index. The use of CeO$_2$ films as insulating layers on silicon substrates and its applicability to high quality silicon on insulator (SOI) structures were discussed by Inoue et al. [1,2]. Yoshimoto et al. [3] have proposed its use as buffer layer for epitaxial growth of high T$_c$-superconducting films on Si. Thin films of CeO$_2$ were prepared by vacuum evaporation [4], electron beam evaporation [1,2,5], laser ablation [3] and by chemical vapor deposition from CeCl$_3$ [6].

We report here for the first time the preparation of CeO$_2$ films deposited by metalorganic chemical vapor deposition (MOCVD) on different substrates. We would like to study the growth of thin films and its use as a buffer layer. Furthermore Ce is one of the elements of the n-type superconductor (Nd,Ce)$_2$CuO$_{4-\delta}$.

2. Experiments

All films were prepared from the precursor Ce(dpm)$_4$. This complex was prepared as described in reference [7]. The compound was studied by thermal analysis (DTA/TG, Mettler TA-2000C Thermoanalyzer).

Thin films of CeO_2 were grown on quartz, sapphire, MgO[100] and Si[100] in a horizontal quartz reactor. The substrates were located on a graphite susceptor heated by an infrared lamp. In all experiments helium was used as carrier gas transporting the precursor vapor to the substrate. In an attempt to stabilize the precursor, the carrier gas was at first passed through a saturator filled with the ligand (Hdpm 20°C, isopropanol 20°C or NH_3 -78°C) and than introduced into the apparatus. Typical growth parameters are summarized in Table I.

Table I Typical growth parameters

substrate	quartz, sapphire, Si[100], MgO[100]
substrate temperature	350 - 600° C
sublimation temperature	175 - 205° C
He flow	4 l/h
O_2 flow	0 - 6.4 l/h
pressure	30 mbar

The deposited films were analyzed by XRD (Phillips PW1710), EMPA (CAMECA SX50), profilometry (TENCOR alpha-step 200) and ellipsometry (PLASMOS SD2300).

3. Results and Discussion

In order to determine the use of the volatile $Ce(dpm)_4$ as a MOCVD precursor the thermal stability was measured by thermal analysis. The results of this analysis are shown in figure 1. The thermogravity (TG) experiment in air shows a residue of 13 percent and XRD analysis of the residue identified CeO_2. For this reason the exothermic peak at 265°C in the differential thermal analysis (DTA) experiment (fig. 1b) can be explained by thermal decomposition

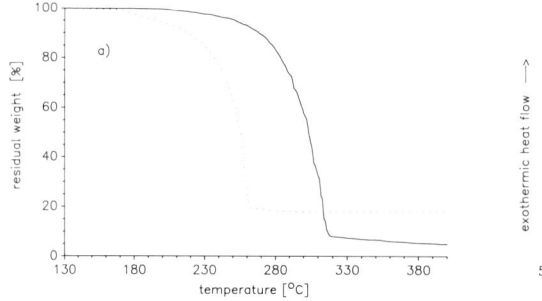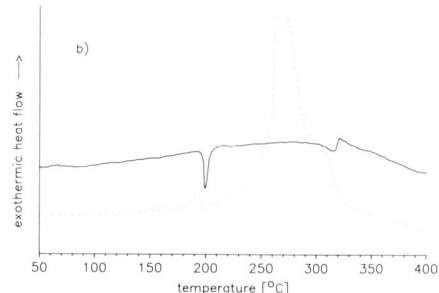

Figure 1. Thermogravity (TG) (a) and differential thermal analysis (DTA) (b) measurements of $Ce(dpm)_4$ in air (dotted line) and in Argon (solid line) both at 1 atm

of Ce(dpm)$_4$ and most of the weight loss is due to evaporation of decomposition products. The results of the experiments with Argon show no exothermic peak, a residue of only 2-3 percent and the weight loss is probably due to evaporation of the whole precursor Ce(dpm)$_4$. The melting point is identified as the endothermic peak at 200°C. Therefore, we can conclude that the thermal stability of Ce(dpm)$_4$ is not very high in air.

Figure 2 shows two XRD patterns of CeO$_2$ film depositions on MgO[100]. The films can be grown either with introduction of O$_2$ into the reaction chamber (2a) or without O$_2$ (2b). From the XRD pattern of 2b a cubic lattice constant of a = 5.402 Å and a FWHM of 20' can be calculated. The XRD pattern of 2b shows that the CeO$_2$ film has a preferential orientation in [100] direction, whereas films grown in the presence of O$_2$ do not show this preferential orientation. The orientation of CeO$_2$[100]-films on sapphire, quartz, Si[100] and MgO[100] is different than CeO$_2$ deposited films reported by other authors [3]. Therefore, we believe that the mechanism of the CeO$_2$ deposition in MOCVD process is different from the mechanism of deposition by other physical vapor deposition methods.

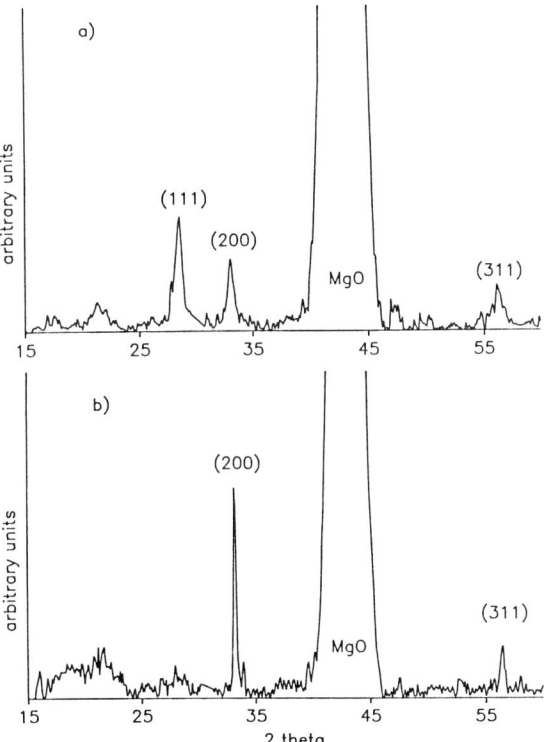

Figure 2 XRD of CeO$_2$ deposited films on MgO [100] a) in the presence of O$_2$ b) in the absence of O$_2$.

The thickness of the thin films on Si [100] were determined by ellipsometry at a wavelength of 632.8 nm. The film thicknesses were found in the range of 500 - 3000 Å.

Figure 3 shows the relationship between the substrate temperature and deposition rate in different carrier gas mixtures and under the same deposition conditions. Below 450°C the deposition rate increased linearly with the substrate temperature, so we conclude a kinetically controlled growth. Above 450°C for He a drastic drop of the deposition rate can be noticed. This is probably due to additional decomposition of Ce(dpm)$_4$ in the reaction chamber; therefore the concentration of volatile Ce-species at the substrate becomes smaller. The XRD pattern shows that above 490°C no cristalline reflection can be found anymore. In order to avoid the decomposition of Ce(dpm)$_4$ isopropanol, ammonia or Hdpm vapor were introduced with the He carrier gas flow into the apparatus as was proposed for Ba(dpm)$_2$ by Dickinson et al. [8] and Barron [9]. For NH$_3$ and isopropanol no stabilization of Ce(dpm)$_4$ could be demonstrated. However, for Hdpm a high deposition rate above 500° C could be observed. Though the origin of the unusual temperature dependence of the deposition rate is not clear at the moment, we can conclude a stabilization of the complex by Hdpm above 500° C.

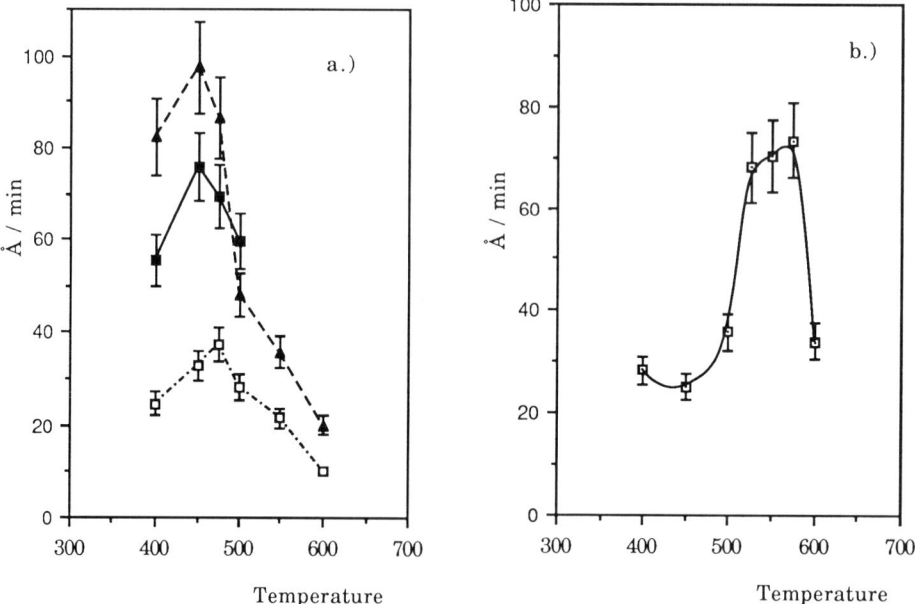

Figure 3 Deposition rate of CeO$_2$ on Si[100] with a flow of 6.4 l/h of O$_2$ a) in He (triangle), in He/isopropanol (full square), in He/NH$_3$ (empty square) and b) in He/Hdpm (circle)

Whereas the increase of oxygen concentration in pure He leads to smaller preferential orientation, as mentioned earlier, the employment of a Helium/ Hdpm mixture showed in the XRD a clear CeO$_2$ [100] preferential orientation for different oxygen concentrations. The reason for this phenomenon and the special feature of the deposition rate of CeO$_2$ in the presence of Hdpm is at present under investigation.

Figure 4 illustrates the relationship between the effect of oxygen flow rate and the deposition rate in He carrier gas. The result is comparable to the investigation of Tominaga et al. on the growth of La$_2$O$_3$ [10].

The EMPA studies showed a very smooth, homogeneous deposition of CeO$_2$ films. Cerium was distributed very homogeneously on the substrate.

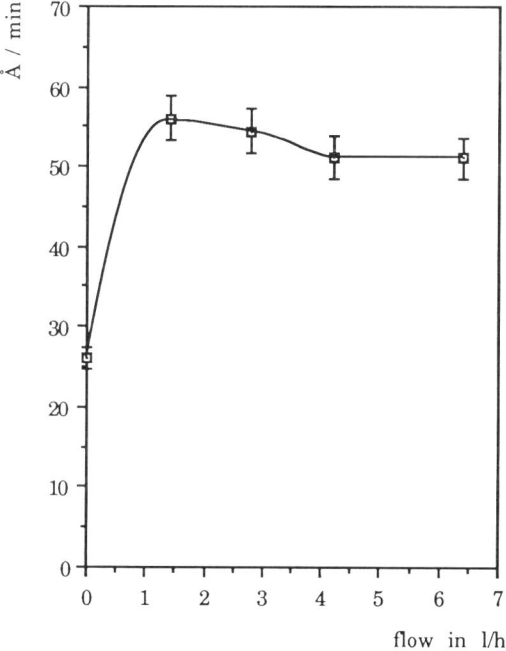

Figure 4 Deposition rate of CeO$_2$ on Si[100] with He as carrier gas, a substrate temperature of 450° C as a function of the oxygen flow

4. Conclusion

Cerium dioxide films were prepared by MOCVD on quartz, sapphire, MgO[100] and Si[100]. For the first time CeO$_2$ films with a preferential orientation in

[100] direction were grown. These films may be useful buffer layers on Si for the growth of cuprate superconductors. It was possible to stabilize the thermally not very stable precursor by using a carrier gas mixture of helium and Hdpm. Even if $Ce(dpm)_4$ decomposes at 265° C, it can be used as a precursor to grow films of the n-type superconducting system of Nd-Ce-Cu-O by MOCVD.

Acknowledgment

The research was supported by Kredite für Unterricht und Forschung, ETH Zürich (T.Gerfin, K.-H. Dahmen) and the Swiss National Research Foundation (M.Becht, K.-H.Dahmen). We would like to express our thanks to Dr. Marek Maciejewski for the DTA/TG measurements.

References

1) T.Inoue, Y.Yamamoto, S.Koyama, S.Suzuki, Y.Ueda, Appl. Phys. Lett., 56 (1990), 1332
2) T.Inoue, M.Osonoe, H.Tohda, M.Hiramatsu, Y.Yamamoto, A.Yama-naka, T.Nakayama, Rep. Res. Ion Beam Technol., 9 (1990), 59
3) M.Yoshimoto, H.Nagata, T.Tsukahara, H.Koinuma, Jpn. J. Appl. Phys., 29 (1990), L1199
4) G.Hass, J.B.Ramsey, R.Thun, J. Opt. Soc. Am., 48 (1958), 324
5) R.P.Netterfield, W.G.Sainty, P.J.Martin, S.H.Sie, Appl. Opt., 24 (1985), 2267
6) H.L.Taylor, J.D.Trotter, Proc. Int. Conf. Chem. Vap. Deposition, 3rd, 1972, 475
7) E.Uhlemann, F.Dietze; Z. anorg. allg. Chemie, 386 (1971) 329-334
8) P.H.Dickinson, T.H.Geballe, A.Sanjurio, D.Hildenbrand, G.Craig, M.Zisk, J.Collman, S.A.Banning, R.E.Sievers, J. Appl. Phys., 66 (1989), 444
9) A.R.Barron, STREM Chemiker, 13 (1990), 1
10) K.Tominaga, M.Miyajima, Y.Sakashita, H.Segawa, M.Okada, Jpn. J. Appl. Phys., 29 (1990), L1874

High T$_c$ Superconductor Thin Films
L. Correra (Editor)
© 1992 Elsevier Science Publishers B.V. All rights reserved.

YBaCuO thin films on Yttria-stabilized Zirconia: growth and properties

J.A. Alarco, G. Brorsson, T. Claeson, Z.G. Ivanov, M. Löfgren[a], and P.Å. Nilsson

Department of Physics, Chalmers University of Technology
S-412 96 Gothenburg, Sweden

[a]Department of Applied Electron Physics, Chalmers University of Technology
S-412 96 Gothenburg, Sweden

Abstract

YBaCuO thin films were grown epitaxially on single crystalline Yttria-stabilized Zirconia substrates by laser deposition. The films were investigated by X-ray diffractometry analysis, SEM and TEM. Critical temperatures, microwave surface resistances and critical current densities were also measured for the films, which were deposited at substrate temperatures (T_{sub}) in the range 710-765 °C. The surface morphology of the films depended on T_{sub}, and the smoothest films had a comparatively smooth background with a low density of small additional outgrowths. The superconducting transition temperatures of the films did not depend strongly on T_{sub} and all the measured values were in the range from 89.5 K to 92.0 K. The microwave surface resistance and the critical current however, showed strong T_{sub} dependences. A surface resistance as low as 0.37 mΩ at 4.2 K and 21.5 GHz, and a critical current of 5×10^6 A/cm^2 at 77 K, were measured, showing the excellent quality of the best of these films. The TEM studies showed that the differences in critical current and microwave surface resistance between the films could be related to the microstructural properties of the films; at T_{sub} below ~750 °C the films were single-crystalline, and for T_{sub} above ~750 °C, the films consisted of grains connected by low angle grain boundaries.

1. INTRODUCTION

Our latest work on artificial grain boundary weak links, grown on bicrystal Yttria-stabilized Zirconia (YSZ) substrates, required high quality YBa$_2$Cu$_3$O$_{7-d}$ (YBCO) thin films [1]. The YBCO films are easy to grow epitaxially on well lattice matched substrates, such as SrTiO$_3$ [2] and LaAlO$_3$ [3], when the lattice mismatch is less than 1%. Growth on YSZ substrates however, differs significantly because of the large lattice mismatch of -6.2%. It has been shown that YBCO films are growing oriented with c-axis perpendicular to the substrate and with lattice matching to the oxygen sublattice of the YSZ [2,4]. The objective of this work was to make high quality films on YSZ substrates and to study the film properties as function of the substrate temperature (T_{sub}). In particular, we wanted to relate the variations of the electrical transport properties to the film microstructure.

2. EXPERIMENTS

The films were deposited in a conventional laser deposition system, consisting of a KrF excimer laser and a vacuum chamber (details of the deposition system are given elsewhere [5]). The energy density at the target was ~1.5 J/cm^2, the substrate-to-target distance 5 cm, the O_2 pressure 0.2 mbar, and the laser repetition rate 10 Hz. After deposition, the O_2 pressure was increased to 1 atm and the temperature ramped down to 100 °C at 20 °C/min. The substrates were YSZ (12 % Y_2O_3) with (001) orientation .

Five films were grown at temperatures (T_{sub}) ranging from 710 to 765 °C. All films were investigated by X-ray diffractometry and Scanning Electron Microscopy (SEM) and their critical temperature (T_c) and microwave surface resistances (R_s) were measured. Three films were taken through an additional patterning procedure to measure the critical current density (j_c). Two of this films were studied by Transmission Electron Microscopy (TEM).

The temperature dependence of the resistance was measured by a four-point ac method, and T_c (R=0) was determined.

The surface resistance, R_s, was measured at 4.2 K by a parallel plate resonator method [6,7] at a frequency of 21.5 GHz. The resonator was formed with a thin dielectric spacer placed between two superconducting films; the film to be measured and a Nb reference film.

To measure j_c, the films were patterned using photolithography and ion beam milling [8], and 6 μm wide bridges were defined. The j_c values of the bridges were determined using a 1 μV criterion. An evaporated Ag layer was used for contacts.

TEM specimens for plane views were polished from the substrate side, down to about 30 μm, and thinned to electron transparency using an Ar ion beam. The beam was incident at an angle of 15°, and a gun voltage of 3 kV and a gun current of 0.5 mA were used. The samples were fixed to a cold stage, and the final thinning done at 2.5 kV.

3. RESULTS

The X-ray diffractometry showed that the films were highly c-axis oriented with sharp (00n) peaks. As can be seen in Table 1, the Full Width at Half Maximum (FWHM) value of the (005) peak, was less than or equal to 0.25° for all the films. The ratio between the intensity of the (200) peak and the intensity of the (005) peak, was taken as a measure of the amount of a-axis oriented grains in the films. The two films deposited at lower T_{sub}, displayed traces of a-axis oriented grains. These results were consistent with the results from the SEM and TEM investigations.

Table 1.
Column 1 and 2 contain the substrate holder temperatures and the thicknesses of the films. In column 3 the the Full Width at Half Maximum values of the (005) peaks in the X-ray diffractograms are listed, and in column 4, the ratios between the (200) and (005) peaks.

Tsub [°C]	Thickness [nm]	FWHM (005)[deg]	I(200)/ I(005)
710	300	0.25	0.02
730	300	0.25	0.007
740	260	0.20	0.00
755	300	0.25	0.00
765	300	0.25	0.00

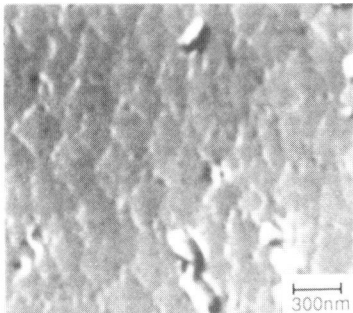

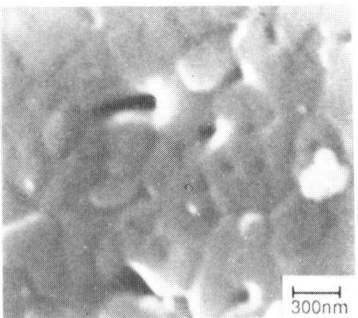

Figure 1 a. SEM micrograph of the film made at 710 °C. The magnification was x40 000.

Figure 1 b. SEM micrograph of the film made at 765 °C. The magnification was x40 000.

The SEM pictures showed the films to consist of a textured background film with a distribution of particles and outgrowths. Figures 1a, b show two micrographs of the films made at 710 and 765 °C. The films made at lower temperatures showed a rectangular-like texture aligned preferentially along two perpendicular directions. The two films which showed traces of a-axis oriented grains in the X-ray diffractograms, both displayed rectangularly shaped outgrowths on the surface.

These outgrowths were also directed along the two perpendicular directions identified in the texture. The film with the strongest a-axis signal, had more rectangularly shaped outgrowths. The films made at 755 and 765 °C had a more irregular texture with rounded domains and pinholes.

TEM investigations were performed on the films made at 710 and 765 °C. A TEM picture of the 710 °C film is shown in Fig.2 a. The film was heavily twinned on the two perpendicular planes (110) and (1$\bar{1}$0), but it was otherwise single crystalline. The twin planes often showed discontinuities at defects. Rectangular a-axis oriented grains, corresponding in shape and size to the rectangular outgrowths observed in the SEM studies, were observed. The normal vectors of the twin planes of the epitaxial film were approximately parallel to the (100) and (010) directions of the substrate, or equivalently, the a- and b-axis of the film were oriented along the (110) and (1$\bar{1}$0) directions of the substrate.

A TEM picture of the film made at 765 °C is shown in Fig. 2 b. This film was polycrystalline. It consisted of twinned c-axis oriented grains, a few hundred nm in size, generally interconnected by low angle grain boundaries.

Resistance vs. temperature measurements of the films showed that all films had R(300)/R(100) ratios close to 3, and extrapolations of the linear high temperature part of the curves, gave intersections with the resistance axis close to the origin.

In Figs. 3a and 3b, the T_{sub} dependence of T_c, j_c, and R_s are shown. The T_c values of the films were fairly constant within the investigated T_{sub} interval; all films had a T_c between 89.5 and 92 K. R_s, on the other hand, showed a large T_{sub} dependence. At the higher T_{sub} values, R_s was more than an order of magnitude higher than at T_{sub}=740 °C. The j_c values of the three measured films correlated very well to the respective R_s values; a high j_c corresponded to a low R_s, and vice versa.

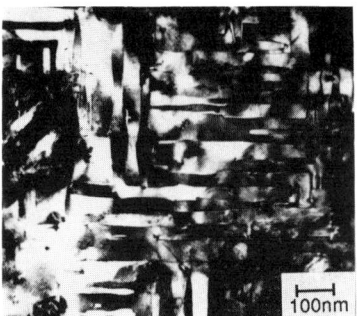

Figure 2 a. TEM micrograph of the film made at 710 °C. The magnification was x68 000.

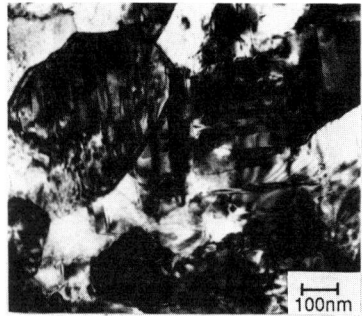

Figure 2 b. TEM micrograph of the film made at 765 °C. The magnification was x68 000.

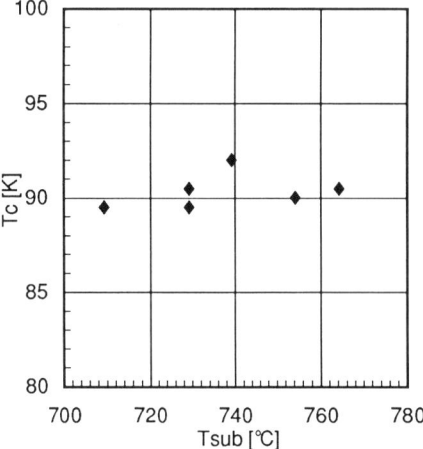

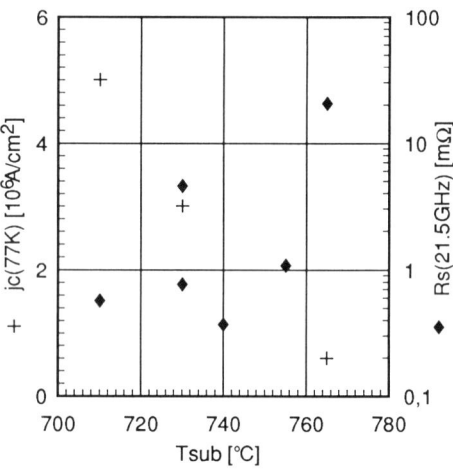

Figure 3 a. T_c (R=0) vs. T_{sub} for the five films.

Figure 3 b. j_c (at 77 K) and R_s (at 4.2 K and 21.5 GHz) vs. T_{sub} for the five films.

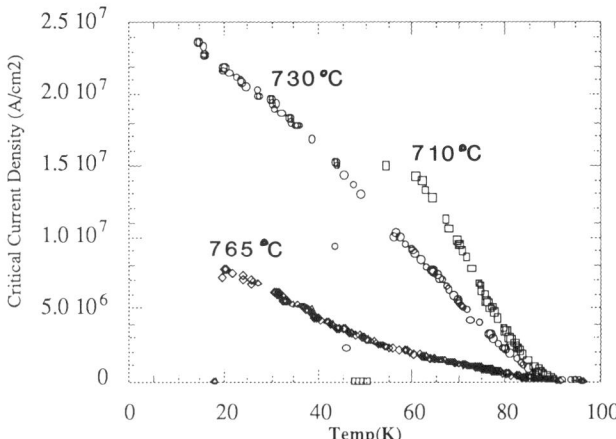

Figure 4. j_c vs. temperature for the films made at 710, 730 and 765 °C.

The temperature dependences of j_c for the three patterned films are shown in Fig. 4. For the two best films (made at 710 and 730 °C) the dependence was approximately linear with a slight bending close to T_c. The film made at 765 °C showed a more pronounced bending, extending down to lower temperatures.

4. DISCUSSION AND CONCLUSIONS

By varying the substrate temperature and keeping the energy density and the O_2 pressure constant, at 1.5 J/cm^2 and 0.2 mbar respectively, we were able to find conditions for making high quality YBCO films on YSZ substrates. The films were highly c-axis oriented with narrow X-ray diffraction peaks. At lower substrate temperatures, a-axis oriented rectangularly shaped outgrowths were present. The a-axis orientation of these outgrowths was inferred from X-ray diffraction, SEM and TEM investigations.

Determining the T_c of a film by a resistive method, gives limited information of the film quality; as soon as there is a superconducting path through the film, the resistance will be zero. We showed that properties such as j_c and R_s, can show large differences between films with similar T_c values of around 90 K.

All films had a T_c between 89.5 and 92 K, but from the results of the other measurements, the films could be divided into two groups; the films made between 710-740 °C and the films made at 755 and 765 °C. The films of the first group generally had a high j_c and a low R_s, while the films of the second group had low j_c and high R_s values. The films of the two groups also showed morphological and microstructural differences. The differences are illustrated by the two films discussed in section 3.2. The films of the first group were heavily twinned, but otherwise single crystalline. The films of the second group had grains which were misoriented in the a-b plane, and were characterized by the presence of low angle grain boundaries. We conclude that under the given deposition conditions, a change in the growth process occured at a substrate holder temperature of about 750 °C. This change resulted in c-axis oriented granular films, instead of single crystalline films.

Our recent results on artificially created grain boundaries in YBCO films deposited on YSZ bicrystal substrates [1], showed that the j_c and the resistance of the grain boundaries depended on the misorientation angle between the grains. In the case of grain boundaries with misorientation angles less than 10°, the j_c was depressed an order of magnitude at 77 K. Those results support the increase of R_s and the decrease of j_c that we observe here in films deposited at higher substrate temperatures, displaying low angle grain boundaries.

REFERENCES

1   Z. G. Ivanov, P-Å Nilsson, D. Winkler J. Alarco, G. Brorsson, T. Claesson, E. A. Stepantsov, and A. Ya Tzalenchuk, To be presented at the International Superconducting Electronics Conference, Glasgow, 24-27 June, 1991.
2   D. H. Lowndes, D. P. Norton, J. W. McCamy, R. Feenstra, J. D. Budai, D. K. Christen, and D. B. Poker, Mater. Res. Soc. Symp. Proc. 169 (1990).
3   A. H. Carim, S. N. Basu, and R. E. Muenchausen, Appl. Phys. Lett. 58, 871 (1990).
4   R. K. Singh, J. Narayan, and A. K. Singh, J. Appl. Phys. 67, 3452 (1990).
5   G. Brorsson, Z. Ivanov, and P-Å Nilsson, Science and Technology of Thin Film Superconductors 2, edited by R. D. McConnell and R. Noufi.
6   R. C. Taber, Rev. Sci. Instrum. 61, 2200 (1990).
7   M. Löfgren and B. Todorov, To be presented at the International Superconductive Electronics Conference, Glasgow, 24-27 June, 1991.
8   Z. G. Ivanov, P-Å Nilsson, E-K Andersson, and T. Claeson, in High Temperature Superconductivity, edited by J. Evetts.

Part IV
Structural Characterization

High T$_c$ Superconductor Thin Films
L. Correra (Editor)
729

High−resolution and analytical TEM of epitaxial YBa$_2$Cu$_3$O$_{7-x}$ thin films and YBa$_2$Cu$_3$O$_7$/PrBa$_2$Cu$_3$O$_7$ superlattices

O.Eibl

Siemens Research Laboratories, Otto Hahn Ring 6, D-8000 Munich 83, F.R.G.

Abstract

TEM studies of epitaxial YBa$_2$Cu$_3$O$_{7-x}$ thin films and YBa$_2$Cu$_3$O$_7$/PrBa$_2$Cu$_3$O$_7$ superlattices will be summarized. High-resolution imaging of cross sections and plan view samples, EDX microanalysis in the TEM and EELS in the TEM are the methods applied. In the first section results on YBa$_2$Cu$_3$O$_7$-x thin films with varying oxygen stoichiometry deposited on SrTiO$_3$ are discussed. Then, YBa$_2$Cu$_3$O$_7$/PrBa$_2$Cu$_3$O$_7$ superlattices deposited on SrTiO$_3$ and MgO are investigated. Finally, an interface analysis of high-quality YBa$_2$Cu$_3$O$_{7-x}$ thin films deposited on sapphire with YSZ buffer layers is presented.

1. Introduction

Single orientation (but twinned), c-axis oriented, high-T$_c$ superconducting YBa$_2$Cu$_3$O$_7$ thin films exhibit electrical proper-ties (high critical current densities and low ac/microwave surface resistivity) which make them appear as first candidates for larger scale device appilcation. Strong efforts have been undertaken in depositing YBa$_2$Cu$_3$O$_7$ thin films on substrates with perovskite-related structures (SrTiO$_3$, LaAlO$_3$ etc.) and MgO by various techniques, e.g. laser ablation, DC and rf (magnetron) sputtering, CVD and, more recently, MBE techniques. A review covering most of the deposition techniques for the preparation of high-quality films is given in [1]. Single oriented YBa$_2$Cu$_3$O$_7$ films permitted to study the bulk (rather than grain boundary) electrical transport properties of YBa$_2$Cu$_3$O$_7$ in the normal and superconducting states [2,3]. Evi-dence could be obtained for the potential of the material of carrying large superconducting currents (densities $> 10^6$ A/cm^2 at 77 K) and showing only a moderate drop when magnetic fields of several Teslas were applied.

Besides   growing single-oriented c-axis YBa$_2$Cu$_3$O$_7$ films, Poppe
et al.   [4] first succeded in depositing   single   orientation
YB$_2$Cu$_3$O$_7$/PrBa$_2$Cu$_3$O$_7$ (YBCO/PrBCO) superlattices. YBCO/PrBCO superlattices with   superlattice   periodicities as small as 2.3
nm, i.e. a sequence of one unit cell YBCO and one unit cell
PrBCO, were first   grown by Fischer et al. [5]. For the films
with ulra thin layers,   x-ray diffraction clearly revealed the
presence of satellite reflections and showed that 87% of Y and
Pr were built into the structure in an ordered way. By controlling the YBCO and PrBCO layer thicknesses independently,
the   superlattices allow to study   the   amount of electronic
coupling between superconducting CuO$_2$ layers in YBCO, separated by semiconducting PrBCO layers. Systematic studies of the
resistivity vs. temperature behaviour with an applied magnetic
field for YBCO/PrBCO superlattices with different superconducting and semiconducting layer thicknesses were carried out
[6]. The scaling of the activation energy of the magnetic flux
creep with the YBCO layer thickness could be determined [6].
More   recently, a breakthrough in depositing high-quality
YBa$_2$Cu$_3$O$_7$ thin films on technologically relevant substrates as
e.g. sapphire [7,8] and Si [9,10] has been achieved. Because
of interdiffusion, deposition on sapphire and Si requires buffer layers which act as diffusion barriers but still yield a
single-oriented YBa$_2$Cu$_3$O$_{7-x}$ layer on top of it.
Transmission   electron microscopy is an indispensable tool for
an improved understanding (giving way to an improved control)
of the growth behaviour, the structure and microstructure of
the deposited YBa$_2$Cu$_3$O$_{7-x}$ films and the structural analysis of
interfaces which appear in superlattices and films deposited
on buffer layers. Results obtained on YBa$_2$Cu$_3$O$_7$ thin films deposited on different substrates will be presented in this paper and show that besides the high-resolution imaging capabilities available with 400 kV TEMs also analytical TEM (EDX,
EELS) yields valuable information on intrinsic properties
(chemical composition, electronic structure) of the films with
significant accuracy and high lateral resolution.

2. Results and discussion

2.1 YBa$_2$Cu$_3$O$_{7-x}$ thin films deposited on SrTiO$_3$ with varying
     oxygen stoichiometry

The microstructure of laser ablated YBa$_2$Cu$_3$O$_7$ thin films yielding a T$_c$> 90 K and critical current densities > 10^6 A/cm^2 at
77 K was analysed by TEM [11]. A typical cross section of such
a film showing the YBa$_2$Cu$_3$O$_7$-SrTiO$_3$ interface under two beam
conditions is presented in fig.1 . A highly defected region
occurs up to 50 nm from the SrTiO$_3$ interface. From the
interface dislocations are threading towards the surface of

the film. The dislocations can be identified by a dark line contrast and their density in the film is approximately 10^9/cm^2. The films aslo contain a high density of (110) twins with twin widths of approximately 30 - 50 nm. Due to the presence of twins, splitted reflections appear in [001] diffraction patterns (fig.2). From the amount of spot splitting the lattice parameter ratio (b-a)/b can be determined with significant accuracy. The lattice parameter ratio (b-a)/b depends sensitively on the oxygen content and permits its determination for x < 0.6. The (b-a)/b ratio was 1.6% for the high-j$_c$ film (fig.2a) indicating a composition of YBa$_2$Cu$_3$O$_{6.8}$.

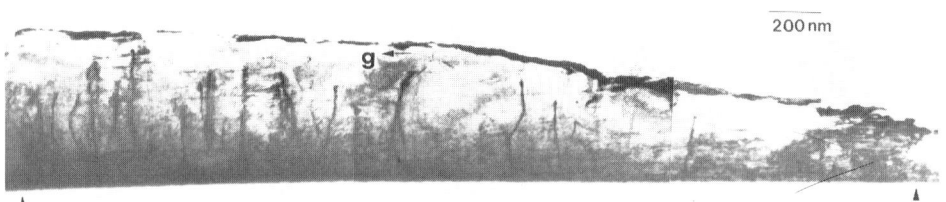

Figure 1: Bright-field image of an YBa$_2$Cu$_3$O$_{6.8}$-SrTiO$_3$ cross section.

A systematic study of the microstructure of films deposited at high (0.3 mbar) and low (10^{-2} mbar) oxygen partial pressure was carried out and the effect of cooling (slow cooling in a 1 bar oxygen atmosphere vs. quenching in the atmosphere present during deposition) was investigated. Table 1 summarizes the preparation conditions for the four samples which were studied.

Table 1:

Specimen	Oxygen partial pressure For deposition	For cooling	Cooling rate	T$_c$(on)
1	0.3 mbar	1 bar	10-100 K/min	90 K
2	0.3 mbar	0.3 mbar	>10^3 K/min	61 K
3	10^{-2} mbar	1 bar	10-100 K/min	50 K
4	10^{-2} mbar	10^{-2} mbar	>10^3 K/min	not sc.

The [001] diffraction patterns for sample 1 and 2 are shown in fig.2a and b. In the quenched film (sample 2) oxygen is removed from the chain sites [12] yielding a (b-a)/b latice paramter ratio of < 0.5%. The oxygen content of this film is assumed to be close to YBa$_2$Cu$_3$O$_{6.4}$.

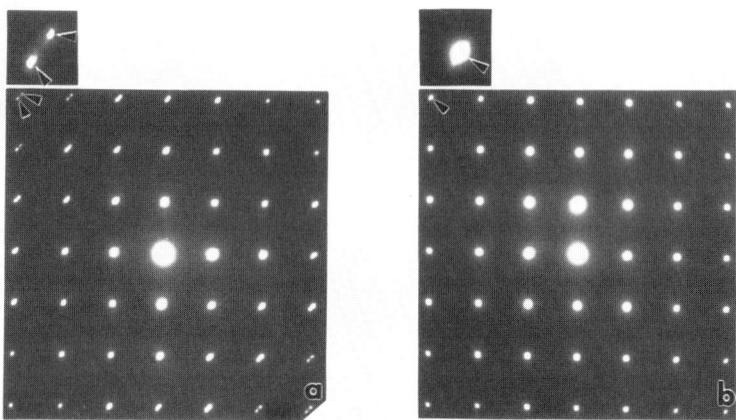

Figure 2: Electron diffraction pattern in the {001} pole for specimen 1 (a) and specimen 2 (b).

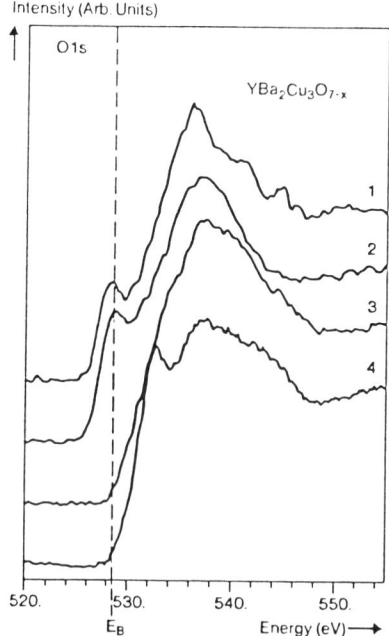

Figure 3: EEL spectra of the oxygen K edge for sample 1-4.

Electron energy loss spectroscopy (EELS) has shown to be a very powerful tool to analyse the elctronic structure of $YBa_2Cu_3O_{7-x}$ [13]. In particular, the O 1s ionization edge shows charcteristic differences for $YBa_2Cu_3O_{7-x}$ with different oxygen content. EELS in the TEM can be performed with an energy resolution of better than 1 eV. For higher energy-losses, eg. O 1s, the momentum transfer resolution is, however, very limited (several ten mrad) because of signal to noise problems. Under favourable conditions (which are explained in more detail in ref.[14]), however, i.e. operating the TEM in image mode (i.e. diffraction coupling to the spectrometer), having the electron beam parallel [001] and inserting no apertures in the backfocal plane, spectra are obtained which show the characteristic features similar to spectra obtained with very good momentum transfer resolution with q parallel to the a or b axis. Spectra obtained under conditions described above are shown for the O 1s core-loss for the four different films (table 1) in fig.3. The electron beam was parallel to the [001] direction and momentum transfer is predominantly in the a,b plane. For spectrum 1 and 2 the characteristic prepeak at the oxygen K ionization edge appears (fig.3, at 528 eV) which is due to an O 1s - O 2p electron transition. Since the occupancy of oxygen in the chain cite differs considerably between sample 1 (close to 1) and sample 2 ($\approx$ 0.25), the hole states giving rise to the prepeak both in spectrum 1 and 2 cannot be

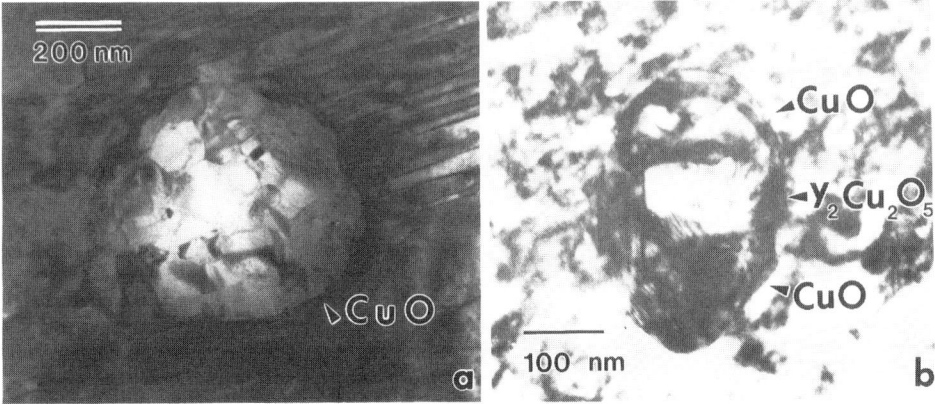

Figure 4: Second phases in specimens 1 and 2. A region of CuO grains in (a) and CuO and $Y_2Cu_2O_5$ in (b).

due to oxygen sites in the chains but must be linked either to
the oxygen in the Ba plane (O4) or the oxygen in the CuO_2 layers (O2). The spectrum of sample 1 (prepeak feature) agrees
nicely with results in ref. [13] for material with composition
$YBa_2Cu_3O_7$. The thin film of sample 4 was deposited at low oxygen partial pressure and subsequently quenched to room temperature. At the ionization edge the spectrum shows a characteristic peak and subsequently a local minimum. The energy losses at the ionization onset (fig.3, 528 eV), the first maximum
(532.5 eV) and the subsequent minimum (534 eV) fit very well
to data obtained in ref. [13] for composition $YBa_2Cu_3O_6$.
Laser ablated $YBa_2Cu_3O_{7-x}$ thin films contain a significant
amount of second phases, which are typically 100 nm in size
and are located at or close to the surface of the film. The
cation composition of these second phase particles could be
identified by EDX in the TEM (Fig.4). These were either CuO or
$Y_2Cu_2O_5$ particles the latter were found adjacent to CuO par-
tilces. It was therefore concluded that $Y_2Cu_2O_5$ forms on the
surface of the deposited film by a reaction of CuO and Y_2O_3.
The presence of the observed second phase particles is due to
a nominal excess of Cu and Y during deposition.

2.2 $YBa_2Cu_3O_7$/$PrBa_2Cu_3O_7$ superlattices

$YBa_2Cu_3O_7$/$PrBa_2Cu_3O_7$ superlattices investigated in this paper
were grown by DC planar magnetron sputtering [15]. TEM provided the proof that the deposited superlattices consisted indeed of well defined layers and sharp interfaces between them
[16]. Fig.5 shows a high-resolution image of a cross section
of a superlattice with 10.5 nm layer thicknesses. $YBa_2Cu_3O_7$
layers appear brighter than the $PrBa_2Cu_3O_7$ layers and the
(001) lattice fringe contrast allows to determine the
interface sharpness more precisely. The high-resolution image
of fig.5 was digitized and an intensity line scan was plotted
across the interface (fig.6). The background level shifts from
one position ($PrBa_2Cu_3O_7$) to the other ($YBa_2Cu_3O_7$) within one
(001) lattice plane spacing. The (001) lattice fringes run
from top to bottom and the arrow (bottom in fig.6) points towards the surface of the film. The position of the interface
is indicated by a double arrow.
The $SrTiO_3$-substrate thin film interface of this particular
film showed a significant density of steps and the reaction of
the deposited film to these steps was studied by high-resolution microscopy. Fig.7 shows the $SrTiO_3$-$YBa_2Cu_3O_{7-x}$ interface
which contains steps indicated by double arrows. A highly defected region is found in the first 3 nm of the film, however,
6 nm away from the interface (at A) the (001) lattice planes
are already continous and bent so that they align parallel to
average surface of the $SrTiO_3$ substrate.

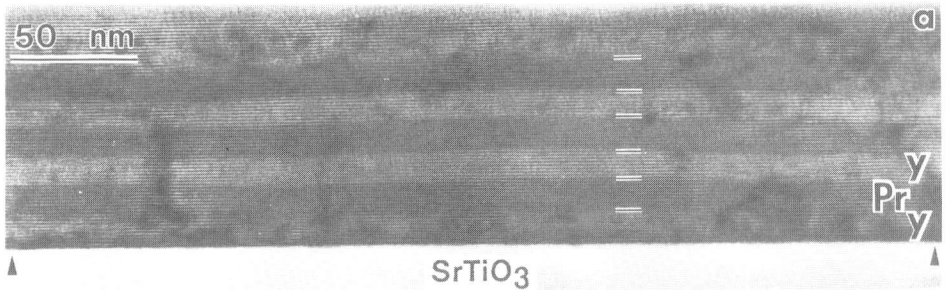

Figure 5: High-resolution, low-magnification image of the YBa₂Cu₃O₇/PrBa₂Cu₃O₇ superlattice (cross section).

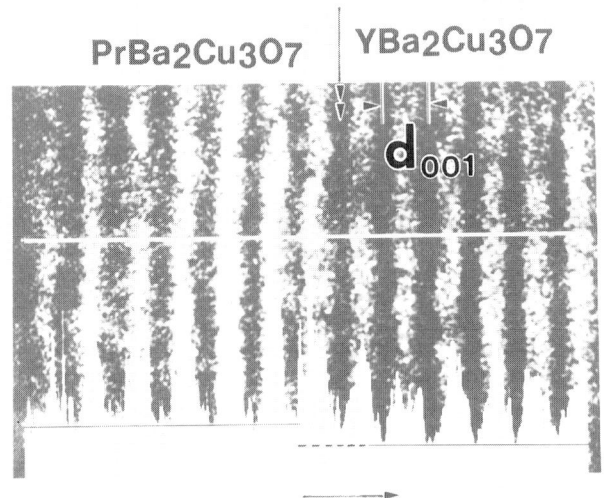

Figure 6: Intensity scan across a YBa₂Cu₃O₇/PrBa₂Cu₃O₇ interface. The (001) lattice fringe spacing is 1.17 nm.

The misfit in the a,b-plane between YBa₂Cu₃O₇ and MgO is considerably larger (8%) than between YBa₂Cu₃O₇ and SrTiO₃ (< 3%). Still, high-quality, single orientation, c-oriented YBCO/PrBCO superlattices can be grown on MgO [17]. For a study of extended defects by TEM [18], however, a mixed a- and c-oriented YBCO/PrBCO superlattice with 3.5 nm layer thickness was

Figure 7: High-resolution image of the $SrTiO_3-YBa_2Cu_3O_7$ interface (cross section). A number of steps are observed at the substrate surface and the position of the interface is indicated by arrows. Note the bending of (001) lattice planes in the right part of the image.

deposited. With the rather small superlattice periodicity and the reduced deposition temperature (T= 750°C, usually for single orientation films T= 790°C), this film seemed to be a good candidate for studying defects introduced during the deposition process. A cross section of this film is shown in fig. 8a and details of the microstructure are explained in the schematic drawing, fig.8b. The film started to grow with the c-axis perpendicular to the substrate surface. Only at points "A" and "B" a-oriented areas approximately 25 nm in diameter are observed at the interface. At "B" surface steps appear at the boundaries which separate the a and c-oriented part of the film. Up to a distance of 11 nm from the MgO interface the size of the a-oriented grains parallel to the substrate remains roughly constant and the interface between a- and c-oriented parts of the film is a (001) plane of the a-oriented crystallite. Beyond 11 nm from the MgO interface both a-oriented crystallites in fig.8 grew rapidly in the lateral direction. The contrast between the various YBCO/PrBCO layers is best visible in the a-oriented regions of the film and is not visible in the c-oriented regions. In the a-oriented regions the YBCO/PrBCO layers do not appear flat but bent and have to be understood as growth contours. Due to the different growth rates in the a- and c-oriented regions, the growth contour reaches a

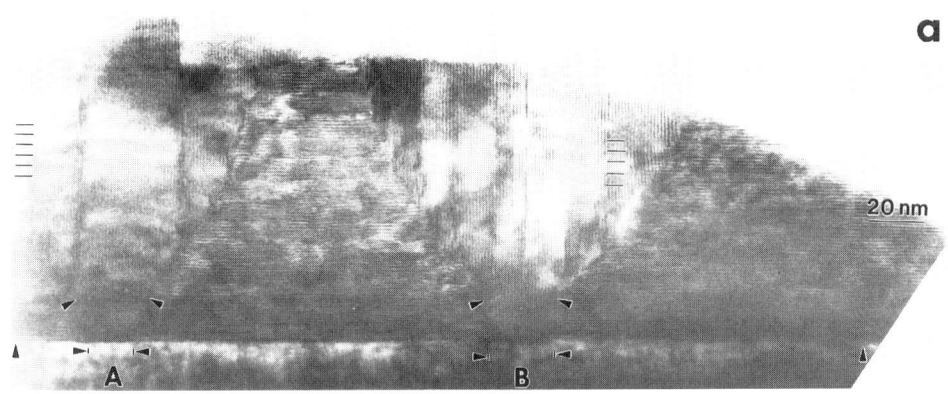

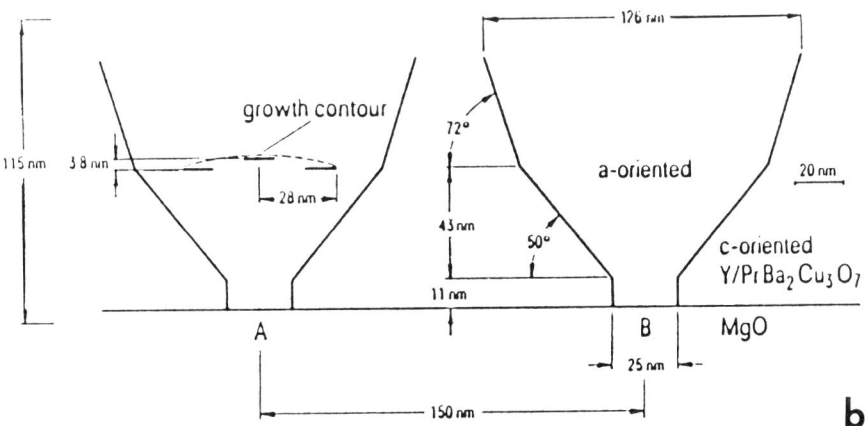

Figure 8: (a) High-resolution, low-magnification image of the superlattice-MgO cross-section. "A" and "B" indicate areas of a-oriented nuclei. The Y/PrBA$_2$Cu$_3$O$_7$ layers are indicated by horizontal bars. (b) Schematic drawing of the grain boundaries and interfaces in (a).

maximum in the center of the a-oriented part of the film. It is remarkable, that although the layers are curved the interfaces between the YBCO and PrBCO layers appear to be very sharp, of the order of one to two times the lattice parameter in a-direction, i.e. 0.4 – 0.8 nm.

2.3 YBa$_2$Cu$_3$O$_{7-x}$ thin films deposited on sapphire with ZrO$_2$ (YSZ) buffer layers

The deposition of YBa$_2$Cu$_3$O$_{7-x}$ thin films on sapphire requires buffer layers which act as effective diffusion barrieres at temperatures required for c-oriented, single crystalline growth. More recently, deposition conditions for depositing single crystalline, epitactic Y stabilized ZrO_2 buffer layers on sapphire were established [8]. On top of the buffer layers the $YBa_2Cu_3O_{7-x}$ films were deposited and critical transport currents of up to 1.2 10^6 A/cm^2 at 77 K in zero magnetic field were obtained [8,19].

Figure 9: Low-magnification image of the Al_2O_3-YSZ-$BaZrO_3$-$YBa_2Cu_3O_7$ layer structure (cross section).

A TEM cross section of a film with j_c= 5.10^5 A/cm^2 at 77 K is shown in fig.9. The 80 nm thick YSZ layer reveals a strong diffraction contrast due to mechanical strain fields in the buffer layer. The $YBa_2Cu_3O_7$/YSZ interface was analysed by high-resolution in more detail (fig.10).
The approximately 3 nm thick intermediate layer at the interface was identified as cubic $BaZrO_3$. It shows characteristic properties, e.g. (i) the (100) lattice planes are bent with respect to the (200) YSZ lattice planes, (ii) the structure appears severely strained, especially close to the YSZ

Figure 10: High-resolution image of the $YBa_2Cu_3O_{7-x}$-YSZ interface showing a $BaZrO_3$ intermediate layer.

layer, (iii) the $BaZrO_3$/YSZ interface is rough and (iv) in contrast, the $YBa_2Cu_3O_7$/$BaZrO_3$ interface is plane. The presence of the $BaZrO_3$ layer has to be linked to the out diffusion of Ba from the $YBa_2Cu_3O_7$ film [19]. Since no interdiffusion between $YBa_2Cu_3O_7$ and $BaZrO_3$ was detected up to 950°C [20] it is assumed that the formation of the intermediate layer occurs at the beginning of the $YBa_2Cu_3O_7$ deposition and the layer inhibits further Ba out diffusion. On top of the layer the $YBa_2Cu_3O_7$ film grows with the (001) planes parallel to the surface plane.

Acknowledgement

The author gratefully appreciates collaboration with all thin film growers envolved, in particular B.Roas (Siemens, laser ablated films), O.Fischer, J.M.Triscone and all other members of the Geneva thin film group working on deposition and characterization of superlattices and K.Hradil and H.Schmidt (Siemens, films deposited on sapphire). Financial support by the Federal Department of Science and Technology (BMFT) is gratefully acknowledged.

References

[1] M.Leskelä, J.K.Truman, C.H.Mueller and P.Holloway; J.Vac.Sci.Technol. $\underline{A7}$ (1989) 3147

[2] B.Roas, L.Schultz, G.Endres; Appl.Phys.Lett. $\underline{53}$ (1988) 1557

[3] B.Roas, L.Schultz, G.Saemann-Ischneko; Phys.Rev.Lett. $\underline{64}$ (1990) 479

[4] U.Poppe, P.Prieto, J.Schubert, H.Soltner, K.Urban; Sol.State Comm. $\underline{71}$ (1989) 569

[5] J.M.Triscone, $\emptyset$.Fischer, O.Brunner, L.Antognazza, A.D.Kent, M.G.Karkut; Phys.Rev.Let. $\underline{63}$ (1990) 1016

[6] O.Brunner, L.Antognazza, J.M.Triscone, L.Mieville and $\emptyset$.Fischer; "Thermally activated flux motion in artificially grown $YBa_2Cu_3O_7/PrBa_2Cu_3O_7$ superlattices" (1991), submitted to Phys.Rev.Lett.

[7] K.Char, N.Mewman, S.M.Garrison, R.W. Barton, R.C.Taber, S.S.Laderman and R.D.Jacowitz, Appl.Phys.Lett. $\underline{57}$ (1990) 409

[8] H.Schmidt, K.Hradil, W.Hösler, W.Wersing, G.Gieres, R.J.Seeböck; Appl.Phys.Lett. July 8th 1991

[9] H.Myoren, Y.Nishiyama, N.Miyamoto, Y.Kai, Y.Yamanaka, Y.Osaka and F.Nishiyama; Jap.J.Appl.Phys. $\underline{29}$ (1990) 955

[10] D.K.Fork, D.B.Fenner, R.W.Barton, J.M.Phillips, G.A.N.Connell, J.B.Boyce and T.H.Geballe; Appl.Phys.Lett. $\underline{57}$ (1990) 1161

[11] O.Eibl and B.Roas; J.Mater.Res. $\underline{5}$ (1990) 2620

[12] J.D.Jorgensen, B.W.Veal, W.K.Kwok, G.W.Crabtree, A.Umezawa, L.J.Nowicki and A.P.Paulikas; Phys.Rev.B $\underline{36}$ (1987) 5731

[13] J.Fink, N.Nücker, H.Romberg, S.Nakai; "High-T_c superconductors – electronic structure", edts. A.Bianconi and A.Marcelli; Pergamon Press (1989) p. 293

[14] O.Eibl, P.van Aken, W.F.Müller; "EELS in the TEM – a high-spatial resolution method for analysing the electronic structure of high-T_c $YBa_2Cu_3O_{7-x}$ thin films"; (1991), submitted to Physica C

[15] J.M.Triscone, M.G.Karkut, O.Brunner, L.Antognazza, M.Decroux and $\emptyset$.Fischer; Physica C $\underline{158}$ (1990) 293

[16] O.Eibl, H.E.Hoenig, J.M.Triscone, O.Fischer, L.Antognazza and O.Brunner; Physica C $\underline{172}$ (1990) 365

[17] J.M.Triscone, $\emptyset$.Fischer, O.Brunner, L.Antognazza, A.D.Kent, and M.G.Karkut; Phys.Rev.Lett. $\underline{64}$ (1990) 804

[18] O.Eibl, H.E.Hoenig, J.M.Triscone, $\emptyset$.Fischer, L.Antognazza, O.Brunner; Physica C $\underline{172}$ (1990) 373

[19] O.Eibl, K.Hradil, H.Schmidt; Physica C (1991) in print

[20] M.J.Cima, J.S.Schneider, S.C.Peterson and W.Coblenz; Appl.Phys.Lett. $\underline{53}$ (1988) 710

High T$_c$ Superconductor Thin Films
L. Correra (Editor)

Evidence of two-dimensional epitaxial growth of YBa2Cu3Ox thin films on lattice-matched NdGaO3 substrates

S.Miyazawa, M.Sasaura, and M.Mukaida

NTT LSI Laboratories, 3-1, Morinosato Wakamiya, Atsugi-shi, Kanagawa 243-01, Japan

Abstract

The nucleation-and-growth stages of superconducting YBa$_2$Cu$_3$O$_x$ thin films on lattice-matched NdGaO$_3$ substrates were studied using an Atomic Force Microscope (AFM) and field-emission type high resolution scanning electron microscope (FE-SEM). The films were grown by ArF pulsed laser ablation on (110) just-cut and 1.7° off-cut NdGaO$_3$ substrates. AFM revealed that the films were grown two-dimensionally with 12Å-high islands through "epitaxy", and FE-SEM observations of the vicinal surface clearly showed that the films were grown through two-dimensional step growth mode. Growth mechanism of YBa$_2$Cu$_3$O$_x$ thin films will be discussed briefly.

1. INTRODUCTION

The thin film microstructure of high Tc superconducting YBa$_2$Cu$_3$O$_x$ (abbreviated hereafter as YBCO) controls electrical properties and superconducting behaviors, and also often influences surface morphology. Final surface morphology is very important to the application of electronic devices. Surface morphology tightly correlated to final microstructures of the films depends on the nucleation-and-growth stages of film growth. The most important issue in thin film growth is the substrate material on which YBCO thin films are deposited either "epitaxially" or "preferentially".

The preparation and properties of thin films have been widely discussed, but the mechanism of thin film growth have received less attention. More recently, it has been reported that pulsed laser ablated YBCO thin films were grown by an island formation mode on MgO substrates [1], where a mismatch between the substrate and YBCO resulted in a misaligned, mosaic structure.

This study reports on our latest results of direct observation into the early stages of nucleation-and-growth of YBCO ultrathin films on well lattice-matched $NdGaO_3$ substrates. Lattice-matching of $NdGaO_3$ to YBCO at deposition temperatures around 700°C is only 0.3% [2], while that of $SrTiO_3$ is 1.03%. Regarding the "epitaxy", the use of lattice-matched substrates is inevitable for deeply considering film growth of good morphology through epitactic nature.

2. EXPERIMENTS

The substrates used in this study were just-cut and 1.7° off-axis-cut (110) oriented $NdGaO_3$ substrates. Ultrathin YBCO films were deposited onto the substrates by pulsed ArF excimer laser ablation from a stoichiometric YBCO ceramics target (Tc = 92K). Though the details of film preparation have been reported elsewhere [3], they are briefly described as follows. ArF excimer laser (~1joule/pulse, 1Hz repetition, 20ns pulse duration) was irradiated on the target under an oxygen pressure of 0.2~0.4Torr. Substrate temperature during deposition was maintained at 700~750°C. The deposition procedure to attain ordinary film growth provided c-axis oriented thin films with a zero-resistance temperature (Tc) as high as 90K. Film thickness on the order of 1Å per pulse was expected on the substrates, as determined by extrapolation from our standard deposition procedure. After deposition, samples were cooled down under the atmospheric pressure of oxygen.

Surface microstructure of the films was examined with a standard atomic force microscope (AFM; Digital Instr.) in ambient atmosphere and a field-emission type, high resolution scanning electron microscope (FE-SEM).

3. RESULTS

From a film thickness dependence of Tc for YBCO thin films on $NdGaO_3$ and $SrTiO_3$ substrates, films on lattice-matched $NdGaO_3$ substrates exhibited Tc's of 80 and 63K for thickness of 80 and 40Å, respectively [4]. Venkatesan et al. earlier reported that less than 50Å thick films on $SrTiO_3$ substrates did not show any superconducting transition down to 10K, while 100Å thick film exhibited a Tc as high as 82K [5]. Reactive co-evaporated ~100Å-thick YBCO films on $SrTiO_3$ also showed a Tc of 82K [6]. The film thickness dependence of Tc seems identical, although film deposition methods were different. It was apparent that the thinner the film, the more effective the influence of substrates, presumably indicating that the film on $NdGaO_3$ substrates must

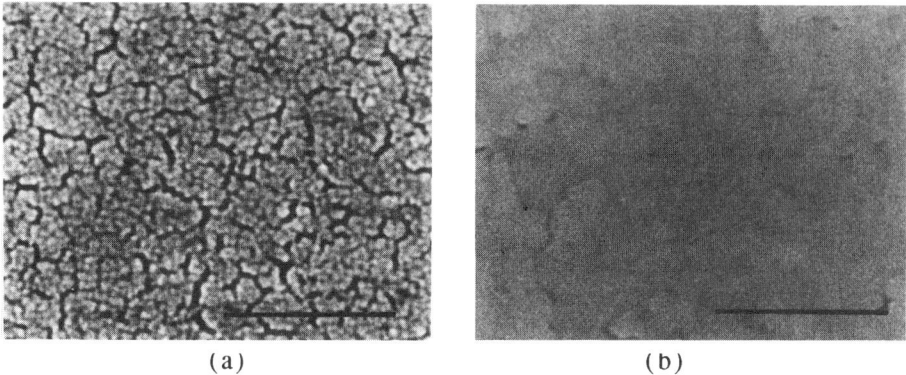

(a)   (b)

Figure 1   Field-emission-type SEM pictures of YBCO thin film surface on (a) SrTiO$_3$ and (b) on NdGaO$_3$ substrates. Bar is 3000Å.

interconnect more preferentially than that on SrTiO$_3$ substrates. This means the enhancement of epitactic growth nature on the NdGaO$_3$.

Another characteristic feature of the films on two different substrates appeared on surface morphology, as shown in Fig.1. Figure 1(a) demonstrates the surface texture of a film on SrTiO$_3$, where irregular mosaic structure was evidently detected. This mosaic morphology seems to be reflected of three-dimensional granular growth of films. On a NdGaO$_3$ substrate, the surface appeared quite smooth, as shown in (b). The smooth surface gives proof that the film grows two-dimensionally to form interconnecting thin films even when it is very thin.

To verify the results described above, early stages of film growth were examined using an AFM [7]. The samples were ultrathin films about 30Å or less in thickness deposited simultaneously on just-cut NdGaO$_3$ and SrTiO$_3$ substrates. Figure 2 shows typical topographs (surface mode) for (a) on SrTiO$_3$ and (b) on NdGaO$_3$. A film surface on SrTiO$_3$ demonstrates a stalactite-like granular morphology, in which grains do not seem to be interconnected. This surface topograph explains the appearance of mosaic surface structure shown in Fig.1(a). In contrast, a very smooth feature was obtained on NdGaO$_3$ substrate, as shown in (b).

Note that rectangular-shaped hillocks or islands are distributed on a relatively smooth background, except for large stalactites that are presumably the origin of particulates peculier to pulsed-laser ablated thin films. The height of most islands was readily measured to be essentially 12Å or multiples of 4Å. Figure 3 shows examples of cross-sectional profiles of islands, where numerals give a height from averaged surroundings. The observed value is equal to a unit length of

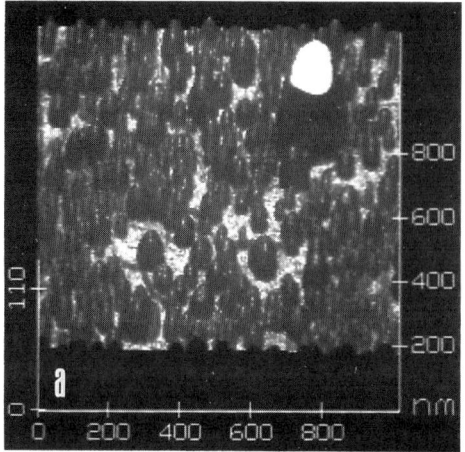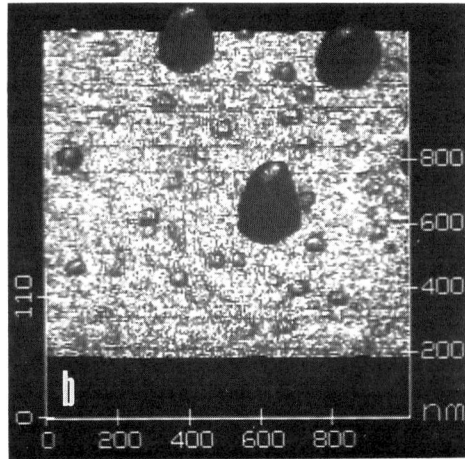

Figure 2   AFM pictures of YBCO ultrathin films on (a) SrTiO₃ and (b) NdGaO₃ substrates.

the c-axis or a treble of the basic lattice constant of the c-axis. Terashima et al., [8] reported that a RHEED oscillation rapidly degraded during deposition, which indicates that the surface becomes atomically rough.   Frequent nucleation of hillocks or islands must be the origin of this phenomenon.   A careful observation on AFM topographs revealed pits or dimples which were also essentially 12Å or multiples of 4Å in depth.

It is presumed that YBCO film grows by the formation of two-dimensional hillocks with a c-axis unit length (nucleation) with their lateral spreading mode on lattice-matched substrates to coalesce with each other (film growth), while it grows in the form of three-dimensional granules on lattice-mismatched substrates.   Norton et al., found that pulsed-laser ablated YBCO on lattice-mismatched MgO substrates were grown by the formation of discrete nuclei of ~200Å in diameter, resulting in a rotationally misaligned, polycrystalline, mosaic structure [1].

A NdGaO₃ substrate inclined at 1.7° with respect to a substrate surface provides surface facets and steps.   Then, it is expected that film nucleation occurs preferentially at steps, allowing a lateral spreading of steps on terraces through "epitaxy".   About 170Å thick YBCO thin film was deposited on the off-cut NdGaO₃ substrate by the same manner. The resultant film exhibited a Tc of 87K.   The as-deposited film surface was examined with FE-SEM, as demonstrated in Fig.4.   Very ragged step-like fronts are clearly   observed   together with very small particles

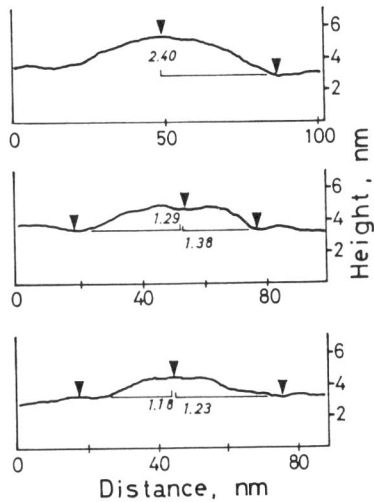

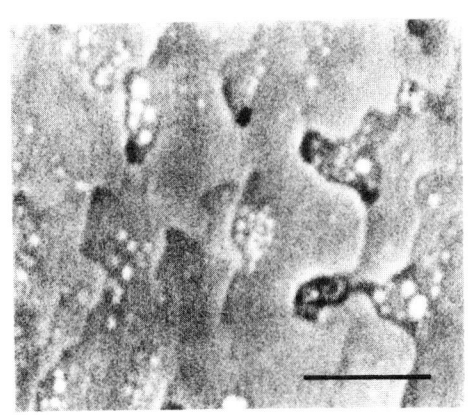

Figure 3   Cross-sectional profiles of growth-islands on a selected area in Fig.2(b).

Figure 4   Field-emission-type SEM picture of a film grown on a vicinal surface of NdGaO$_3$ substrate.   Bar is 3000Å.

on terraces.   These small particles have been established to be excess CuO, but they were not examined in this study.   A more significant feature is that step-growth seems to be interrupted by these particles. A possible mechanism of film growth is that nucleation occurs preferentially at step edges and steps grow laterally on preceding terraces successively., i.e., two-dimensional epitaxy.   Figure 4 allows us to image this plausible mechanism.

The width between neighboring steps was directly measured to be 770 ± 60Å on the average, and step height was estimated to be 20~30Å from a carbon replica examination by electron microscope.   A geometrical feature on a 1.7° vicinal surface numerically gives ~23Å-high surface steps with 770Å-wide surface facets.   Surprisingly, the estimated step height closely agrees with twice the c-axis unit length of YBCO.   This examination leads us to conclude that the film grew two-dimensionally on terraces by a lateral motion of steps through "epitaxy", even though nominal film thickness was about 170Å.   The interruption of lateral step-growth at particles is evidence of two-dimensional growth, sometimes called the "step bunching effect".

As was examined above, a vicinal surface substrate lattice-matched with YBCO film demonstrated a two-dimensional nucleation-and-growth mechanism in layer-by-layer fashion.   The growth unit was essentially ~12Å high, equal to the c-axis unit length.

4. SUMMARY AND CONCLUSION

The early growth stages of nucleation-and-growth of superconducting $YBa_2Cu_3O_x$ thin films were examined with regards to both film thickness dependence of Tc on two different susbtrates, lattice-mismatched $SrTiO_3$ and lattice-matched $NdGaO_3$, and their surface morphologies. An atomic force microscope revealed a significant difference in film growth mode between two substrates, indicating that the film on lattice-matched $NdGaO_3$ grows epitaxially by the formation of two-dimensional hillocks or islands with a height of ~12Å or multiples of ~4Å. The AFM pictures taken in this study show that film growth on the $NdGaO_3$ substrate is presumably in Stranski-Krastanov mode, while that on $SrTiO_3$ substrate in Volmer-Weber mode.

On a 1.7° off-cut vicinal surface of $NdGaO_3$ substrate, two-dimensional step-growth takes place on surface facets. Height was twice the c-axis unit length. This indicates a film growth mode to be epitaxial on a lattice-matched substrate.

The formation of multilayer structures such as SIS and SNS with good transport properties require atomically smooth surfaces and/or interfaces for each layer. They should be realized from the perspective of lattice-match, and much attention must be paid to film growth mechanism proper to each film deposition technique of high Tc superconductors.

The authors express their thanks to Dr.Y.Tazoh and J.Kobayashi for their useful discussions and comments.

REFERENCES

1   M.G.Norton and C.B.Carter, J.Cryst.Growth, 110 (1991) 641

2   M.Sasaura, S.Miyazawa, and M.Mukaida, J.Appl.Phys., 68 (1990) 3643

3   M.Mukaida, S.Miyazawa, M.Sasaura, and K.Kuroda, Jpn.J.Appl.Phys., 29 (1990) L936

4   M.Mukaida, S.Miyazawa, and M.Sasaura, Proceedings of Int.Conf.on Superconductors (ISS'90, Sendai 1991), to be published.

5   T.Venkatesan, V.D.Wu, B.Dutta, A.Inam, M.S.Hegde, D.M.Hwang, C.C.Chang, L.Nazar, and B.Wilkens, Appl.Phys.Lett., 54 (1989) 581

6   Y.Bando, T.Terashima, K.Iijima, K.Yamamoto, and H.Mazaki, Physica C, 153-155 (1988) 810

7   S.Miyazawa, M.sasaura, and M.Mukaida, unpublished work

8   T.Terashima, Y.Bando, K.Iijima, K.Uamamoto, K.Hirata, K.Hayashi, K.Kamogaki, and H.Terauchi, Phys.Rev.Lett., 65 (1990) 2684

High T$_c$ Superconductor Thin Films
L. Correra (Editor)
747

HREM study of epitaxial CuO and DyBa$_2$Cu$_3$O$_7$ films grown with layer-by-layer epitaxy

A. CATANA, J.-P. LOCQUET and R. BROOM

IBM Research Division, Zurich Research Laboratory, 8803 Rüschlikon, Switzerland

Abstract
The structure of molecular beam epitaxy-grown CuO and DyBa$_2$Cu$_3$O$_7$ thin films on MgO and SrTiO$_3$ substrates has been investigated with high-resolution electron microscopy. It has been verified by X-ray and electron diffraction that CuO grows with the (111) planes parallel to the (001) MgO planes. Owing to the 4-fold symmetry of MgO, several CuO domains contribute to the continuous film. The matching conditions imposed by the substrate induce distorted regions in the CuO film. The DyBa$_2$Cu$_3$O$_7$ films were grown on SrTiO$_3$ using sequential layer deposition. They show predominantly **c**-axis oriented domains, with a small fraction of **a**-axis regions. In addition, small **c**-axis domains are visible on top of the **a**-axis grains, which we relate to the layer-by-layer growth process. Finally, we report on the interface chemistry. Comparisons between experimental and calculated images favor an interface configuration where the first layer to be deposited is a CuO plane as imposed by the deposition conditions.

1. Introduction

Presently, considerably effort is being devoted to the preparation of artificially layered structures or ultrathin films in which superconducting layers with one unit cell or more alternate with one or more insulating layers [1,2,3]. It is therefore of great importance to be able to identify the starting and ending layers of the as-grown unit cells. Using layer-by-layer epitaxy should enable one to monitor the nature of the starting layer. Bando et al. [4] recently suggested that the terminating layer in a co-evaporation experiment at 680 °C is the CuO$_2$ plane, while Streiffer et al., using off-axis single target sputtering at 650-670 °C, favored the BaO layer to be the terminating layer [3]. Furthermore reflection high-energy electron diffraction (RHEED) oscillations recently observed during co-evaporation indicated that growth proceeds by the sequential stacking of complete unit cells [2]. This would indicate that the growth does not proceed by sequential stacking of monolayers.

Previously, it has been reported that CuO grows epitaxially on MgO [5]. Here we report on the morphology and structure of the CuO film and its interface with MgO and on structural features related with a layer-by-layer growth mode. In particular, we focus on the structure of DyBa$_2$Cu$_3$O$_7$ films and on the atomic configuration at DyBa$_2$Cu$_3$O$_7$/SrTiO$_3$ interfaces. Using image calculations we show that it is possible to distinguish between different interfacial atomic configurations. Finally, careful comparison between experimental and calculated images allows us to assess locally the interfacial atomic configuration. The results show that the starting evaporation sequence CuO-BaO-CuO$_2$ is preserved after completion of the film growth.

2. Experimental

A description of the experimental details has been given elsewhere [5]. The epitaxial layers were prepared in a custom-designed molecular beam epitaxy (MBE) system consisting of four effusion cells, a substrate heater and a sample transfer module. The system is pumped with a 400 l/s ion pump and a 2200 l/s turbo pump. Cu, Ba and Dy are evaporated from standard effusion cells in an oxygen background pressure generated by an rf plasma source inside the vacuum system which delivers a flow of activated oxgyen onto the substrate. This system has recently led to superconducting thin films with a sharp transition ($\Delta T = 1$ K) and $T_{c0} = 86.5$ K. The DyBa$_2$Cu$_3$O$_7$ film which has been used for the HREM study was grown on SrTiO$_3$ at 660 °C in an oxygen pressure of 2×10^{-5} Torr. A layer-by-layer deposition was used with a CuO-BaO-CuO$_2$ starting sequence. The HREM study was performed on a JEOL 2010 microscope operating at 200 kV. Thin specimens were prepared for cross-sectional observations by conventional thinning and ion milling.

3. Results and Discussion

CuO/MgO

Copper oxide (CuO) is an insulator which is, chemically, very close to the cuprate superconductors. This makes it an interesting candidate to be used as buffer or interdiffusion layer. In addition CuO is often present as an impurity phase in cuprate superconductors and is difficult to detect by standard analysis tools like x-ray diffraction.

Cross-sectional TEM observations indicate that the CuO layer is polycrystalline and that both interface and top surface of the film are rough. Typical grain sizes are 20-40 nm. Selected area diffraction (SAD) patterns and HREM recorded on different domains show that CuO grows preferentially with its close-packed (111) planes parallel to the (001) MgO planes in agreement with earlier x-ray diffraction measurements [5]. In these (111) planes, two main orientation relationships have been observed locally: [$\bar{1}$10] CuO ∥ [110] MgO and [10$\bar{1}$] CuO ∥ [100] MgO. The HREM micrograph in Fig. 1 shows small CuO domains oriented with [10$\bar{1}$] CuO ∥ [100] MgO and (111) CuO ∥ (001) MgO. The corresponding SAD (inset in Fig. 1) clearly shows that the (111) CuO planes are parallel to the (001) MgO planes. This result is confirmed for the first 4-6 monolayers by RHEED observations. HREM results show that further CuO growth preserves the parallellism between the (111) CuO and (001) MgO planes. However, as one approaches the top surface of the film, the CuO looses its exact [10$\bar{1}$] orientation. Close to the interface, highly distorted regions are present which may be related to the lattice mismatch and/or to the occurrence of strain during the film growth.

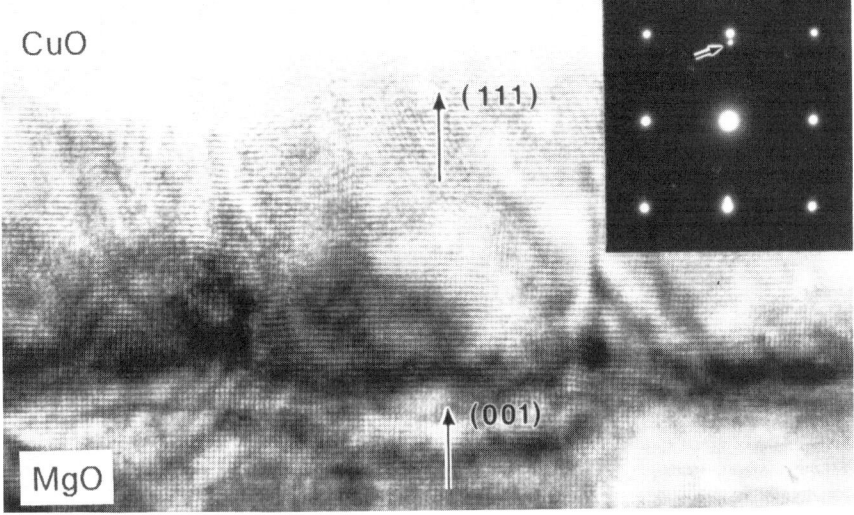

Figure 1. HREM and SAD of the CuO/MgO interface showing (111) CuO ∥ (001) MgO. Small domains with [10$\bar{1}$] CuO ∥ [100] MgO are locally observed. Contrast modulations close to the interface indicate the presence of distorted regions.

DyBa$_2$Cu$_3$O$_7$/SrTiO$_3$

The layer-by-layer deposition of DyBa$_2$Cu$_3$O$_7$ at 660 °C results in the formation of dominant **c**-axis regions. In addition to these well-ordered domains the film also displays **a,b**-type domains adjacent to inclusions of both crystalline and amorphous impurity phases. Figure 2 shows a crystalline precipitate grown epitaxially onto a **c**-axis region at a **a,b/c** boundary. In the case of sputtered and laser-deposited films, it has been shown by TEM on planar sections that **a,b**-axis regions nucleate at second phase regions with the 132 structure [6]. Our observations, recorded on cross sections, show that in the case of the layer-by-layer deposition, well-ordered 123 regions grow epitaxially on top of the inclusions. This result indicates that second phase inclusions related to **a,b**-axis domains do not necessarily grow up to the film surface.

The growth of epitaxial second phases as well as the overgrowth of well-ordered 123 domains on top of both crystalline and amorphous precipitates may well be a specific feature related to the layer-by-layer growth process. Additionally, **c**-axis growth can be promoted on top of **a,b**-axis regions as shown by the arrow in Fig. 2. To our knowledge, this feature has not been observed on samples prepared by other techniques. Although we have not yet identified the mechanism responsible for this behavior, one can speculate that it is related to the way atoms are supplied to the growing front. The low temperature that has been used may also impede the 'instant' transformation of the **c**-axis top layer into an **a,b**-axis region.

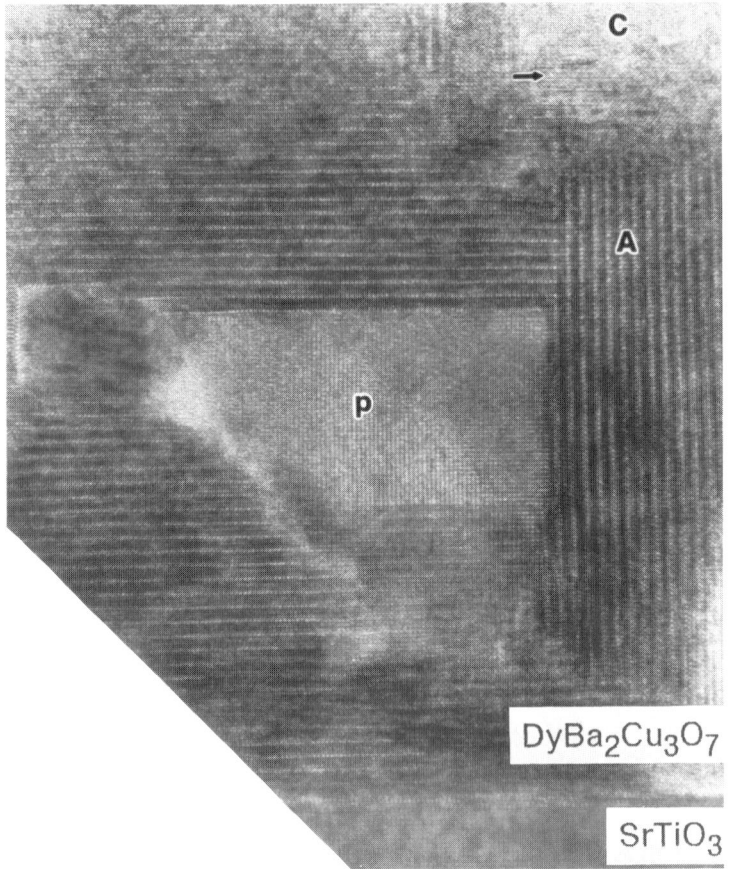

Figure 2. HREM showing an epitaxial precipitate (**P**) adjacent to an **a,b**-axis domain (**A**). Arrows indicate atomically flat **c**-axis/**P** and **a,b**-axis/**P** interfaces.

Figure 3 shows an HREM micrograph of a well-defined 123/SrTiO$_3$ interface. The electron beam is parallel to the [100] SrTiO$_3$ direction. In order to determine the starting 123 layer, image calculations have been performed using a multislice algorithm [7]. Two interfacial models have been considered: in the first model the interface joins the CuO chain to the terminating TiO plane whereas in the second, the interface joins the CuO$_2$ plane to the TiO plane. By careful analysis of the image contrast in the interfacial region one can easily distinguish between the two models (Fig. 4). The main intensity contribution to the image (bright dots) is provided by the Cu atoms of the CuO$_2$ planes. Pairs of such planes separated by Dy chains are clearly visible in both calculations. In the case where CuO chains are adjacent to the SrTiO$_3$ surface, the first pair of CuO$_2$ planes is imaged as equally bright spots, which is not the case for the second configuration (Fig. 4b).

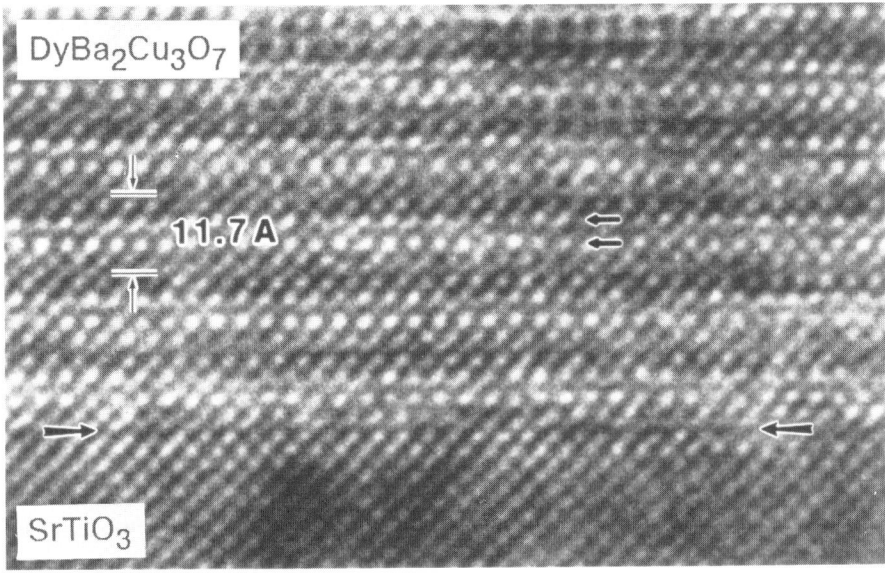

Figure 3. HREM of a well-defined interface projected along [100] SrTiO$_3$. Bright pairs of spots (see arrows) indicate the position of the CuO$_2$ planes.

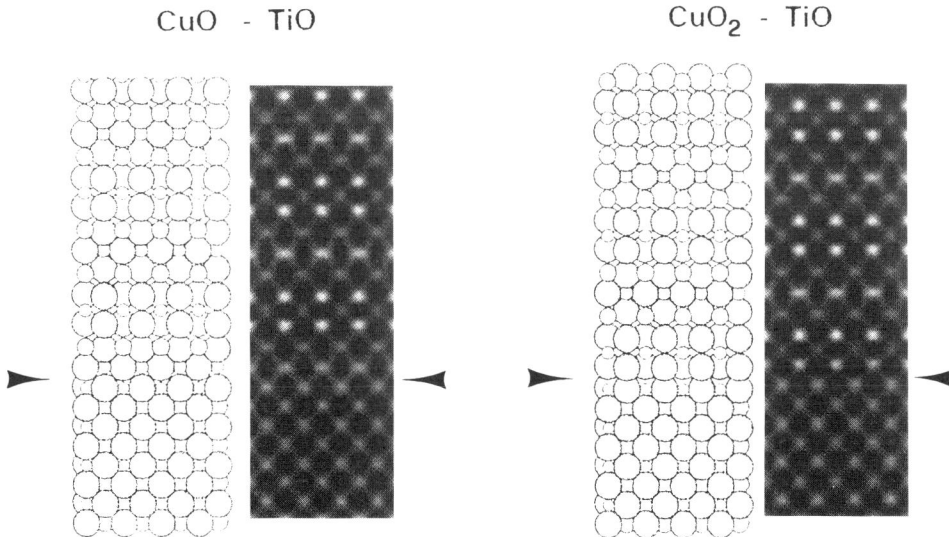

Figure 4. Image calculations (defocus = 52 nm^{-1}, sample thickness = 4.6 nm) and corresponding supercells for two interfacial configurations: (a) CuO/TiO and (b) CuO$_2$/TiO.

A. Catana, J.-P. Locquet and R. Broom

The best match between the experimental and calculated images is obtained at a defocus of 52 nm. The sample thickness is estimated at 4.6 nm. Under these conditions, equally bright pairs of CuO_2 planes are clearly observed in the experimental image. The periodic repetition of these image features up to the interface as well as the two-layer spacing between the first pair of CuO_2 planes and the interface indicates that in this case the starting layer is the CuO plane. This plane was chosen as starting plane in the deposition sequence. It is thus confirmed that under these preparation conditions the chosen starting layer is preserved. It is interesting to note that different interfacial layers have been proposed in the case of co-evaporated [2] and sputtered [3] films.

4. Conclusion

The growth of CuO on MgO is shown to be polycrystalline and strongly textured. Using HREM and SAD we show that CuO aligns its (111) planes parallel to the (001) MgO planes. In spite of the large crystallographic difference between both crystals, small epitaxial domains are observed. However these domains lose their exact orientation towards the film surface.

In the layer-by-layer growth process, we find that the growth of a-axis domains is associated with the presence of both crystalline and amorphous impurities. The precipitates do not inhibit the further epitaxial growth of well-ordered c-axis regions. The resulting interfaces between the precipitate and both the a-axis and the overgrown c-axis domains are atomically sharp.

Finally, using image calculations, we have investigated the local interface chemistry at an atomic scale. It is shown that the chosen starting deposition sequence (CuO-BaO-CuO_2) is preserved at the $DyBa_2Cu_3O_7/SrTiO_3$ interface, indicating that the layer-by-layer deposition technique allows the interfacial chemistry to be tuned. This result opens interesting perspectives in the field of ultrathin film and raises the question as to *which sequence of atoms is necessary for a unit cell to be superconducting.*

We are grateful to Erich Mächler for the excellent assistance he provided.

References

[1] J.M. Triscone, O. Fisher, O. Brunner, L. Antognazza, A. D. Kent and M. G. Karkut, Phys. Rev. Lett. **64** (1990) 804.
[2] T. Terashima, Y. Bando, K. Iijima, K. Yamamoto, K. Kirata, K. Hayashi, K. Kamigaki and H. Terauchi, Phys. Rev. Lett. **65** (1990) 2684.
[3] S. K. Streiffer, B. M. Lairson, C. B. Eom, B. M. Clemens, T. H. Geballe and J. C. Bravman, submitted to Phys. Rev. B.
[4] Y. Bando, T. Terashima, K. Shimura, T. Sato, Y. Matsuda, S. Komiyama, K. Kamigaki and H. Terauchi, Proc. Int'l Workshop on HTSC Thin Films, Properties and Applications, Rome, Italy, April 1991 (special issue of Physica C).
[5] J.-P. Locquet, J. Less Common Metals **164-165** (1990) 300.
[6] R. Ramesh, A. Inam, D. M. Hwang, T. D. Sands, C. C Chang and D. L. Hart, Appl. Phys. Lett. **58** (1991) 1557.
[7] P. Stadelmann, Ultramicroscopy **21** (1987) 131.

High T$_c$ Superconductor Thin Films
L. Correra (Editor)
© 1992 Elsevier Science Publishers B.V. All rights reserved.

TEM observation of twin domain boundaries in Y$_1$Ba$_2$Cu$_3$O$_{7-x}$ thin films

M. Cantoni, H.-U. Nissen

Laboratorium für Festkörperphysik, ETH Zürich, CH-8093 Zürich, Switzerland

Abstract

The twinning texture of Y$_1$Ba$_2$Cu$_3$O$_{7-x}$ superconducting thin films grown on SrTiO$_3$ substrates has been investigated by electron microscopy. The orientation of the twin lamellae is constant within domains having an average size of 1μm and changes at the boundaries of these domains. The average thickness of the polysynthetic twin lamellae is 20-30 nm; this average width is about one third of that found in bulk specimens. At the boundary of the domains two sets of twin lamellae merge nearly at right angles, and the lamellae have bottleneck-shaped as well as T-shaped endings. Although selected area diffraction patterns reveal four different lattice orientations near the domain boundaries, one set of twin lamellae merges coherently into a single lamella of the adjacent domain, without any boundary. The lattice planes in the junction are found to be continuously bent over a distance of about 2 nm. This deformation is consistent with the presence of four diffraction spots in SAD patterns observed for regions near domain boundaries. The films are uniformly covered by coherent crystalline inclusions or possibly voids having an overall density of the magnitude 10^{11}cm^{-2}.

1. Introduction

We report transmission electron microscopic (TEM) results on Y$_1$Ba$_2$Cu$_3$O$_{7-x}$ superconducting thin films grown epitactically on (100)-SrTiO$_3$ substrates. The critical current densities j$_c$ of these films exceed those of bulk samples or even of single crystals. A major limiting factor for high critical currents in bulk samples is the occurrence of grain boundaries which act as weak links. The process limiting the critical current in single crystals and in thin films is the depinning of magnetic flux lines. The detailed study of the microstructure is therefore important to obtain superconductors for industrial application. Here we present a description of the twin textures and a genetic growth model for this texture.

2. Experimental

Thin films of Y$_1$Ba$_2$Cu$_3$O$_{7-x}$ of 200 nm thickness were produced at the Kernforschungszentrum Karlsruhe by W. Schauer et al. [1], using inverted cylindrical magnetron sputtering. A characterization of the physical properties of these films has been published elsewhere [1]. The substrate material and the deposition temperature of about 800°C have been chosen so as to obtain a pure c-axis oriented growth and, consequently, highest critical current density. The films have critical temperatures of 90 - 92 K and critical current densities measured by the four point method of more than 10^6 A/cm² at liquid nitrogen temperature.

The preparation of the samples for the TEM observations had to be done absolutely water-free. Even exposure to normal air for several hours resulted in a decrease of the critical temperature and the critical current density. The samples were kept under vacuum in order to prevent changes in the film structure and in physical properties due to water vapor. For observations along the direction normal to the film, specimens were prepared by breaking the substrate. Wedge-shaped grains containing sufficiently large thin regions were glued onto a copper grid. The planar surface of the grains, together with the superconducting film, was oriented parallel to the grid. The c-axis in the film was therefore oriented almost parallel to the electron beam, and therefore the required tilting angle usually was very small. A Philips CM 30 electron microscope (300 kV) was used for selected area diffraction work as well as for high resolution electron microscopy.

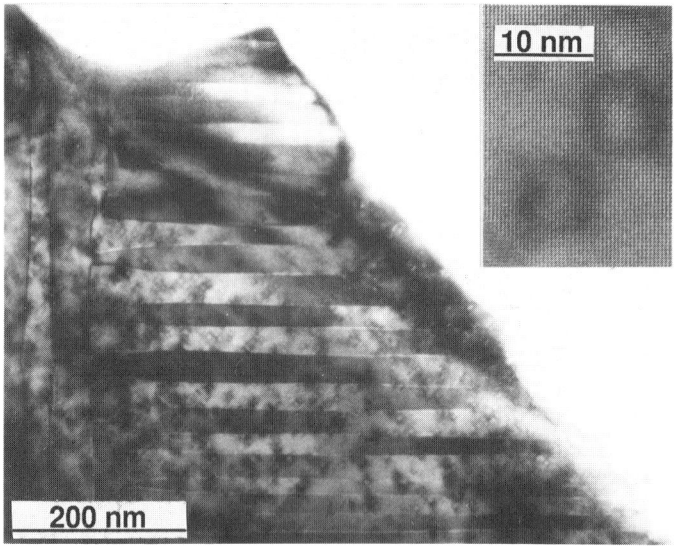

Fig.1 Bright-field image; notice size and shapes of twin lamellae. Enlarged detail on the right shows the shape of the dark spots.

3. Results and discussion

Bright field images show a textured c-axis growth, with the a- and b-axes parallel to the surface of the substrate. The film contains domains (about 1μm wide) characterized by differently oriented polysynthetic lamellar twins with widths between 20 and 40 nm, similar to the twin lamellae found in bulk $Ho_1Ba_2Cu_3O_{7-x}$ specimens [2]. This is much smaller than the widths reported for single crystals (~75 nm [3]) and polycrystalline ceramics (approx. 100-500 nm [2,3,4]). No crossing twin lamellae as reported for bulk samples [5] were observed (Fig. 1).

Near the domain boundaries the twin lamellae show characteristic shapes: T-shape terminations alternate with bottleneck terminations, similar to the texture described in [2]. This indicates that these domain boundary regions are highly strained. The spotty contrast visible in Fig.1 is due to homogeneously distributed defects of a special type localized near the surface of the film. These defects have square shape with rounded corners, and the edges are parallel to the a- and b- axes of the surrounding material. These outlines can be recognized in the magnified detail inserted in Fig. 1.

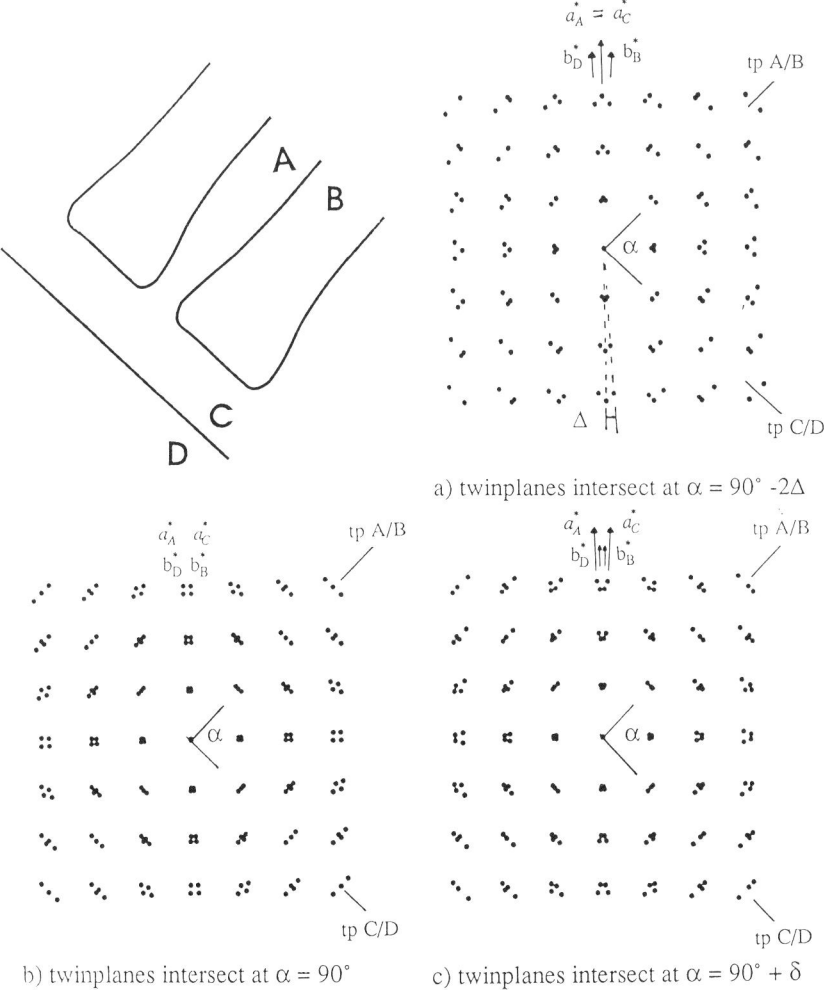

a) twinplanes intersect at $\alpha = 90° -2\Delta$

b) twinplanes intersect at $\alpha = 90°$ c) twinplanes intersect at $\alpha = 90° + \delta$

Fig. 2 Schematic drawing of the diffraction patterns expected for the different models. The ratio a/b was chosen as 0.95 (instead of 0.984) for clarity. The directions of the a^* and b^* axes are labeled corresponding to the spot positions.

The upper left portion of Fig. 2 presents a schematic drawing of the domain boundary. In the junction between the lamellae A and C the width of lamella A decreases from about 30 nm to 5 nm, and in the junction no boundary can be observed. From this observation the existence of three different orthorhombic lattices with an orientation relation as in Fig. 2a could be concluded, since A and C are assumed to have one common lattice orientation only. With this assumption the idealized SAD pattern, Fig. 2a was calculated on the basis of one orthorhombic lattice, twinned once with (110) as twin plane, corresponding to lamellae C and D, and twinned a second time with (1$\overline{1}$0) as twin plane, corresponding to A and B. The orientation of the lamellae A and C would in this case be the same, and consequently no boundary would be observed at the junction. This calculated SAD pattern shows a spot-splitting into three distinct diffraction spots, corresponding to the three different orientations.

By contrast, the observed SAD pattern shows a spot-splitting into four instead of three subspots (Fig. 3). This indicates that there are four different orthorhombic lattice orientations. The calculated diffraction pattern in Fig. 2b was generated on the assumption that the twin planes of A/B and C/D intersect at exactly 90°. In an additional, alternative model (Fig. 2c) the intersection angle was increased by a small angle δ, in accordance with the angle as measured in Fig. 3. The enlarged details in Fig. 3 show four different orthorhombic lattice orientations. The intersection angle of the corresponding twin planes in this figure is 90°+δ, where the angle δ has been measured as $\delta = 0.3° \pm 0.05°$.

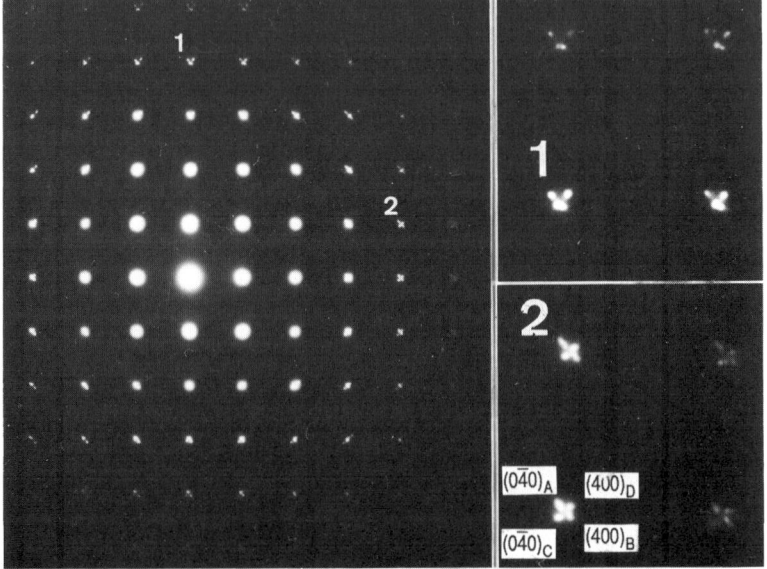

Fig. 3 Observed diffraction pattern, enlarged details on the right show spot-splitting due to twinning.

High resolution images were taken in order to investigate the structure of the junction between lamellae A and C, as well as to identify the four lattice orientations A,B,C and D. Fig. 4 shows the region in which lamella C merges into lamella A. Looking at glancing angle, the difference between the orientation of the lattice planes in A and in B can be recognized. This angle Δ is the angle between two orthorhombic lattices twinned with twin plane (100). This angle was measured as $0.9° \pm 0.1°$. The same angle was observed between C and D. At the junction lamella $\overline{C}$ is coherently connected with lamella A. This coherent transition between the two lattice orientations is possible because of a continuous bending of the lattice at the junction. The orientation of lattice A differs from that of C by the angle $2\Delta-\delta = 1.5°\pm0.2°$.

In order to understand the implications of the models 2b and 2c, it is necessary to consider the geometry of the lattice of the substrate material and the process of formation of the orthorhombic structure in the superconducting thin film.

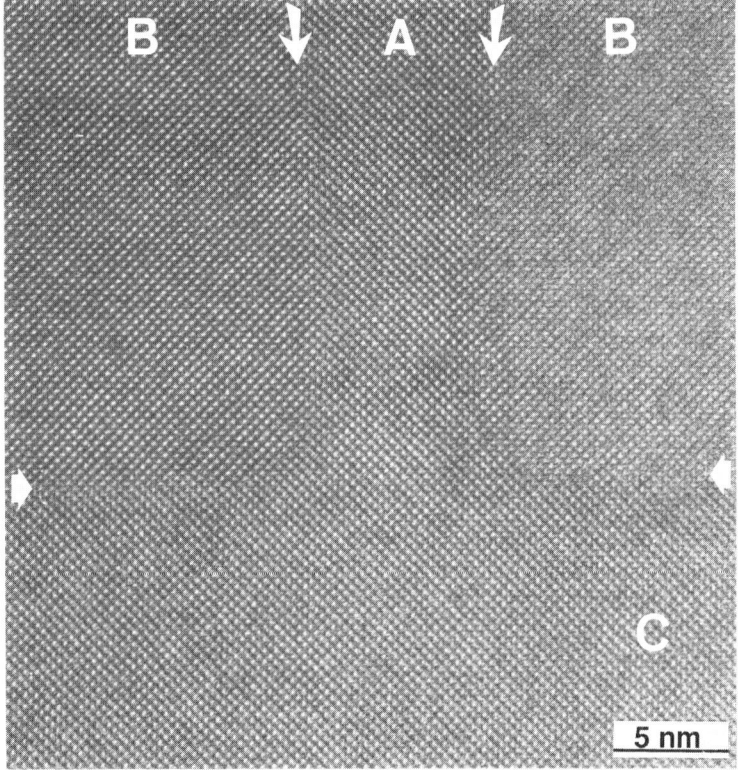

Fig. 4 High resolution image of a junction. Bent lines can be recognized looking along the vertical direction. Arrows indicate the domain boundary.

The epitactic growth of the tetragonal phase $Y_1Ba_2Cu_3O_6$ onto the cubic substrate lattice during film deposition [1] would favour model 2b (with exactly perpendicular intersection of the twin planes). This model could also explain the regular spacing of the twin lamellae. However, the boundary between B and C would then be an incoherent grain boundary with dislocations, implying insertion of material on a macroscopic scale during the tetragonal-orthorhombic transition. The additional rotation in model 2c of one twin plane against the other by the angle δ increases the orientational mismatch between lattices A and C, but eliminates the necessity of dislocations in the grain boundaries between lamella C and lamella B, since the <110> lattice planes in C and B are more nearly parallel. This model does not require macroscopic insertion of material. The orientational difference of $2\Delta-\delta$ between the lattice planes in lamellae A and C causes the observed elastic deformation of the lattice in the junction and the reduction of the width of the lamellae A at that junction. This bottleneck shape may be the result of a minimization of the strained area.

4. Conclusions

Thin films of $Y_1Ba_2Cu_3O_{7-x}$ of 200 nm thickness show a growth texture with c-axis orientation. The texture consists of domains approximately $1\mu m$ in diameter which differ with regard to the orientation of polysynthetic twin lamellae. The equivalent <110>-planes are the twin planes, and the spacing of the twins varies between 20 and 40 nm. At the domain boundaries one set of twin lamellae merges coherently into a single lamella of the adjacent domain. In spite of the fact that no boundary between the merging lamellae can be recognized, the SAD patterns of these regions indicate the existence of four separate lattice orientations. The observed orientation relation at the domain boundaries can be explained by a genetic process, which avoids the assumption of a macroscopic insertion of material during the transition from tetragonal to orthorhombic. The observed specific terminations of the twins at the domain boundaries are the result of a minimization of the strained area. The lattice orientation mismatch of $1.5°\pm0.2°$ in the bottleneck-shaped junctions of the coherently connected lamellae is found to be related to a continuous lattice bending. Homogeneously distributed crystalline inclusions (or possibly voids) with square cross-section and rounded edges do not interrupt the twin texture. Since the overall density of these inclusions (or voids) does not change with specimen thickness, it is concluded that they are located near the film surface.

The authors are very grateful to Dr. W. Schauer for the thin film specimens and they thank him as well as Mr. C. Beeli for critical reading of this manuscript.

References

[1]   W. Schauer, X.X. Xi, V. Windte, O. Meyer, G. Linker, Q. Li, J. Geerk,
       Cryogenics **30** (1990) 586
[2]   Y. Kawamata, H.-U. Nissen, C.Beeli, F. Hulliger, H.R. Ott, Z. Phys.B,
       Cond. Mat. **72** (1988) 345
[3]   C.H. Chen, D.J. Werder, S.H. Liou, J. Kwo, M. Hong,
       Phys. Rev. B **35**(1987) 8767
[4]   H.W. Zandbergen, G. Thomas, Phys. stat. sol. **107** (1988) 825
[5]   W.F. Müller, Neues Jahrb. Miner., Abh. **161**,(1989) 41

High T$_c$ Superconductor Thin Films
L. Correra (Editor)
© 1992 Elsevier Science Publishers B.V. All rights reserved.   759

T.E.M. characterization of high T$_c$ superconductive multilayered thin films

A. Alimoussa[a], M.J. Casanove[a], C. Roucau[a], C. Villard[b], M. Schwerdtfeger[b], L. di Cioccio[b], H. Moriceau[b] and J.C. Villegier[b]

[a]Centre d'Elaboration de Matériaux et d'Etudes Structurales / Laboratoire d'Optique Electronique (CEMES / LOE), B.P. 4347, 31055 Toulouse Cedex, France.

[b]DTA / LETI / Département Optronique, 85 X, 38041 Grenoble Cedex, France.

Abstract

Superconductive multilayered thin films have been deposited by excimer laser ablation and d.c. magnetron sputtering. The laser deposited films were grown on (100) SrTiO$_3$ substrates and consist of alternating layers of La$_{2-x}$Sr$_x$CuO$_4$ and YBa$_2$Cu$_3$O$_{7-\delta}$ while the sputtered films, deposited on (100) MgO substrates, consist of a 3 nm MgO layer sandwiched in 100 nm YBCO layers.

Cross-sectional T.E.M. specimens have been prepared so as to investigate the different interfaces and the local structure of the films. The results show that the laser deposited films are epitaxially grown with the c-axis normal to the surface. The interfaces are rather sharp but some roughness of the YBCO over LSCO interface, possibly related to the presence of shear defects in the LSCO layers, is evidenced. The observations of the sputtered films reveal the presence of some a-axis oriented YBCO grains and the partial crystallization of the amorphous deposited MgO layer. Some a-axis oriented grains are also observed in the laser deposited films but only in the upper layers. The different types of structural defects as well as the different interfaces, characterized in high resolution electron microscopy (HREM) will be reported.

1. INTRODUCTION

Technological applications of high T$_c$ superconducting films in junction-devices require the deposition of reproducible, high-quality, multilayered films. The electrical properties of these films depend strongly on the sharpness of the different interfaces as well as on the size and orientation of the high T$_c$ grains. Heteroepitaxial growth of the multilayered films has therefore to be achieved.

Two different hetero-structures were investigated. Multilayered films consisting of alternating La$_{1.77}$Sr$_{0.23}$CuO$_4$ and YBa$_2$Cu$_3$O$_{7-\delta}$ (with $\delta \simeq 0$) layers were deposited by excimer laser ablation on (100) $SrTiO_3$ substrate. Different superlattices involving YBCO layers, as Y-Ba-Cu-O / Dy-Ba-Cu-O, Y-Ba-Cu-O / Pr-Ba-Cu-O or Y-Ba-Cu-O / Bi-Sr-Cu-O, have already been reported in the literature, [1-4]. The YBCO / LSCO sandwich-type layers present the particular interest to produce different types of devices depending on the Sr content in the LSCO layers. The $La_{2-x}Sr_xCuO_4$ compound is indeed known to present different types of electrical behaviour for $x < 0.25$ (insulator, normal metal, superconductor), [5].

The second type of films studied in this paper consists of YBCO/MgO/YBCO layers deposited by d.c. magnetron sputtering on (100) MgO substrate.

The films were investigated by plan view and cross-sectional T.E.M. experiments. We report in this paper the microstructures of both types of multilayered films. The heteroepitaxial growth of the LSCO/YBCO superlattices is evidenced, as well as sharp interfaces between the layers. The amorphous MgO tunnel barrier, present in the sputtered YBCO films, is shown to partially crystallize during the deposition process.

2. ELABORATION

2.1. LSCO/YBCO superlattices

The multilayered structure was obtained on $SrTiO_3$ substrates by the pulsed laser deposition technique using a 248 nm (KrF) excimer laser. Both materials (LSCO and YBCO) have been deposited under the same conditions ; i.e. a laser energy density of 2.5 J / cm^2 at 2 Hz The substrate holder temperature was maintained at 730°C throughout the deposition while the N_2O pressure was hold at 0.5 Torr. This multilayer consists of five bi-layers with a modulation wavelength of $\Lambda = 48$ nm (26 nm $YBa_2Cu_3O_{7-\delta}$ + 22 nm $La_{1.77}Sr_{0.23}CuO_4$). Results of X-Ray, SIMS and electrical characterization will be reported elsewhere.

2.2. YBCO/MgO/YBCO tunnel junctions

In order to study the growth of Superconductor / Insulator / Superconductor (SIS) structures using a MgO ultra thin film (3nm) as a tunnel barrier, we chose an alternate approach to the YBCO / MgO / YBCO heteroepitaxial growth as reported in [6-7] and yielding a shorted and discontinuous barrier. A thin (about 75 nm) base electrode was deposited by d.c. magnetron sputtering using a hollow target (HITEC inc.), at about 750°C in a mixture of O_2 / Ar (0.2 / 0.4 mbar). The sample was then cooled down, in a 300 mbar O_2 atmosphere, to 50°C where a 3nm thick amorphous and continuous MgO layer was deposited by RF sputtering. The films were then heated again up to 650°C where the YBCO counterelectrode (about 100 nm thick) was deposited in the same conditions as the base electrode. After cooling down under oxygen, a thicker (10 nm) amorphous MgO layer is deposited as a passivation layer before patterning and junction processing, described elsewhere [8].

3. T.E.M. CHARACTERIZATIONS

3.1. YBCO / LSCO multilayers

The mean a-parameter of the LSCO/ YBCO / STO stacking sequence (first bi-layer over substrate) is about 3.83 Å (with a_{LSCO} = 3.76 Å for x = 0.23, $[a, b]_{YBCO}$ = 3.82, 3.89 Å for δ = 0, and a_{STO} = 3.9 Å). Considering the lattice matching of the 3 different unit cells, heteroepitaxial growth in the c-direction can be expected. Cross-sectional specimens were prepared according to the method already described in [9]. Figure 1 presents the first three bi-layers of the film. The c-axis orientation is effectively obtained, as evidenced by the (001) planes lying parallel to the substrate surface (see the fringes in figure 1) and the alignment of the reflections due to the 3 different diffraction patterns. The experimental SAD pattern enclosed in figure 1 corresponds to a region centred on the first YBCO layer. Strong SrTiO$_3$ reflections are also present in the pattern while the LSCO layer reflections are weak.

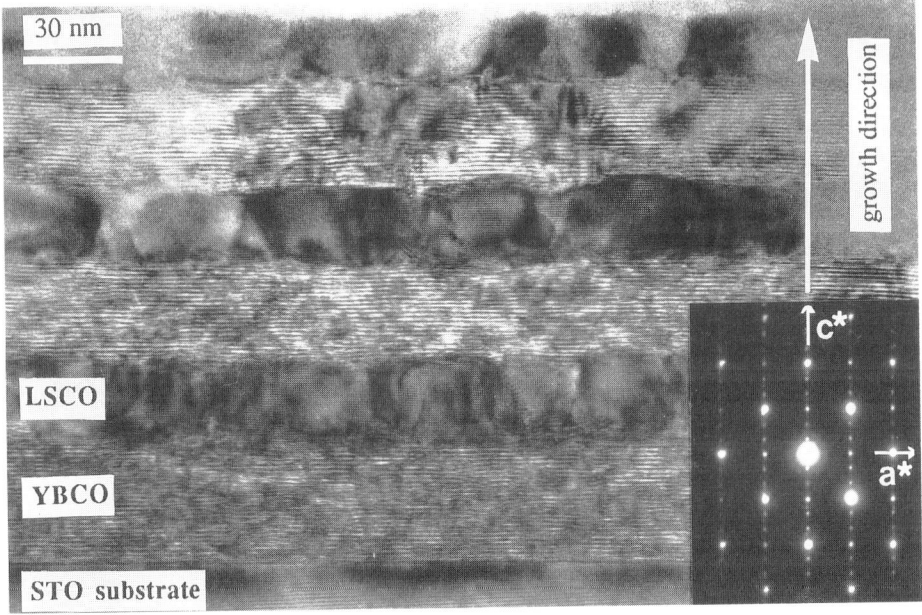

Figure 1. High resolution, low magnification image of the first three YBCO/ LSCO bi-layers showing the c-axis orientation of the film. A [010] diffraction pattern centered on the first YBCO layer is inserted.

Some a- or b-axis oriented grains appear at the fourth YBCO layer. These misoriented grains are displayed in figure 2 where the LSCO layers are shown to keep their c-axis orientation. Plan view specimens having the YBCO layer at their surface have also been investigated. Although the majority of the grains have the c-axis orientation, the observations reveal the presence of a- or b- axis oriented grains.

The different interfaces can be observed in figures 1 and 2. All the interfaces appear sharp but a difference in their smoothness can be observed. In particular, we have to make a distinction between the YBCO over LSCO and the LSCO over YBCO interfaces. The upper planes of the YBCO layers are nearly always flat even if some a-axis oriented grains are present in the layer (see figure 2). The different steps that can be observed in the HREM mode are not higher than a single YBCO unit cell, and rarely exceed 0.66nm (half c_{LSCO}), see the step arrowed in figure 2. The upper planes of the LSCO layers are more rough and local extensions of YBCO in the LSCO layer can be observed at these interfaces (see for instance the lower interface displayed in figure 2). Some steps about one or two YBCO unit cells high are also observed at the YBCO over LSCO interfaces.

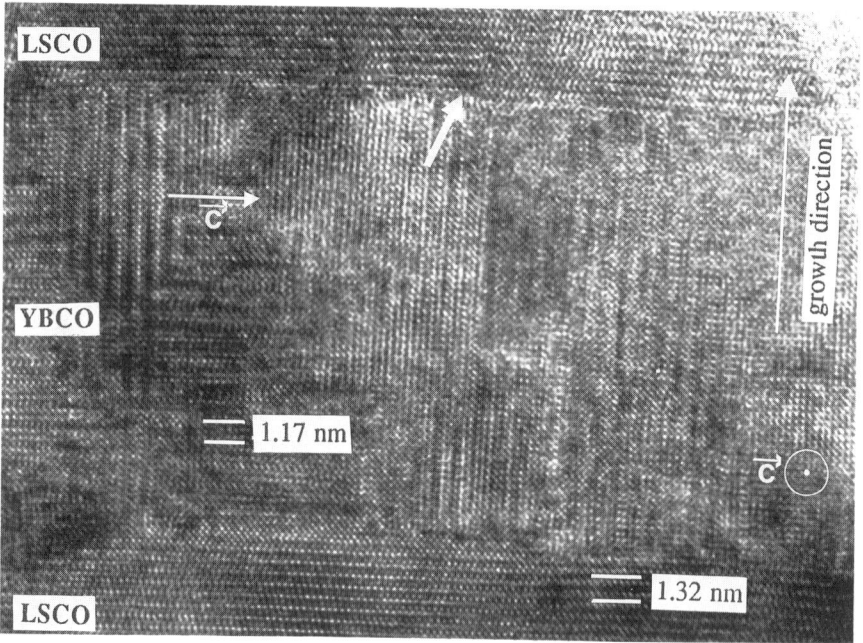

Figure 2. High-resolution micrograph of a region of the fourth YBCO layer presenting a- and b- oriented grains.

The roughness of the LSCO layers surfaces may be attributed to the presence of the defects imaged at the YBCO / LSCO interface in figure 3. These defects, present all across the LSCO layers, have been identified as shear defects, the shear plane lying parallel to the (101) planes. A high resolution micrograph of this defect, which is still being analysed, is displayed in figure 4.

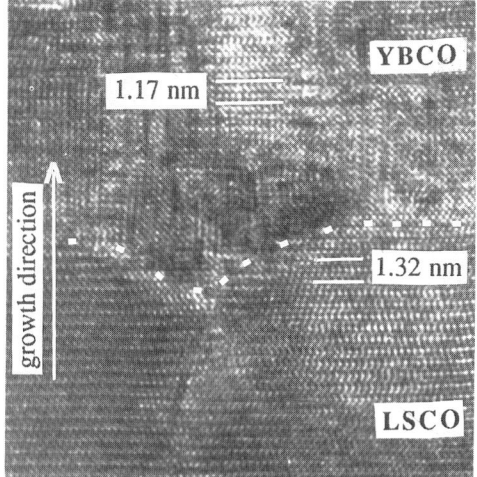

Figure 3. Local extension of YBCO in the LSCO layer. The dotted line corresponds to the interface

Figure 4. HRTEM image of one of the shear defects present in the LSCO layers.

3.2. YBCO / MgO / YBCO multilayers

The first T.E.M. experiments on these sputtered films reveal the presence of some a-axis oriented grains close to the film / substrate interface. These misorientations were already reported by Eibl *et al.* in [2]. A cross-sectional general view of the film is displayed in figure 5. An a-axis oriented grain can be observed on this high resolution micrograph where the 3 nm thick MgO tunnel barrier is seen as a chain of small grains.

4. CONCLUSION

The cross-sectional TEM experiments have provided useful information on the microstructure of the films studied. The sharpness of the different interfaces of YBCO / LSCO superlattices has been evidenced. It has been found that the interfaces were rather flat except for the YBCO over LSCO interface which presents a non negligible roughness. It has been suggested that the numerous shear defects present the LSCO layers may play a part in the bending of the LSCO surfaces. These defects may be attributed to an oxygen deficiency in the LSCO layers, as reported in [10]. Electrical and micro-structural studies on multilayered films having different Sr contents and different layer thicknesses are just beginning.

The amorphous deposited MgO tunnel barrier embedded in the YBCO sputtered films has been seen to crystallize partially, forming small islands. We have not observed any disruption of this layer, but complementary studies in HREM are still necessary.

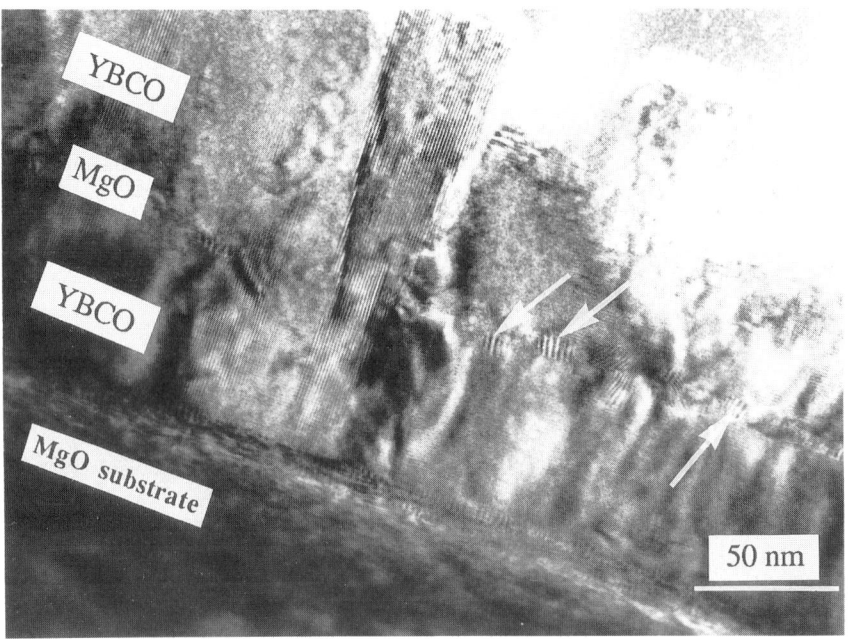

Figure 5. High resolution low magnification image of a YBCO/MgO/YBCO multilayered film. Note the partial crystallization in small islands (arrowed) of the amorphously deposited MgO layer.

REFERENCES

1 J.M. Triscone, M.G. Karkut, L. Antognazza, O. Brunner and Ø. Fischer, Phys. Rev. Letters, 63 n°9 (1989) 1016

2 O. Eibl, H.E. Hoenig, J.M. Triscone, Ø. Fischer, L. Antognazza and O. Brunner, Physica C, 172 (1990) 365 and 373.

3 T. Venkatesan, A. Inam, B. Dutta, R. Ramesh, M.S. Hedge, X.D. Wu, L. Nazar,, C.C. Chang, J.B. Barner, D.M. Hwang and C.T. Rogers, Appl. Phys. Letters, 56 (1990) 391.

4 K. Harada, S. Tanaka, H. Itozaki and S. Yazu, Jpn. J. Appl. Phys.,29 n°7 (1990) L1114.

5 J.B. Torrance, A. Bezinge, A.I. Nazzal, T.C. Huang, S.S.P. Parkin, D.T. Keance, S.J. La Placa, P.M. Horn, G.A. Held, Phys. Rev. B, 40 (1989) 8872.

6 K. Sakuta, M. Iyori, T. Kobayashi, M. Matsui and N. Nakajima, IEEE Trans.on Magnetics, 27 n°2 (1991) 1361.

7 S. Tanaka, H. Nakanishi, T. Matsuura, K. Higaki and S. Yazu, IEEE Trans.on Magnetics, 27 n°2 (1991) 1607.

8 J.C. Villegier, H. Moriceau, H. Boucher, R. Chicault, L. di Cioccio, A. Jager, M. Schwerdtfeger, M. Vabre and C. Villard, IEEE Trans.on Magnetics, 27 n°2 (1991) 1552.

9 M.J. Casanove, A. Alimoussa, C. Roucau, C. Escribe-Filippini, P.L. Reydet and P. Marcus, Physica C , 175 (1991) 285.

10 P.L. Gai, E.M. McCarron III, Science, 247 (1990) 553.

High T$_c$ Superconductor Thin Films
L. Correra (Editor)

765

The influence of the native BaAl$_2$O$_4$ boundary layer on microstructure and properties of YBa$_2$Cu$_3$O$_{7-x}$ thin films grown on sapphire

K. Dovidenko, S. Oktyabrsky, D. Tokarchuk, A. Michaltsov and A. Ivanov

P.N. Lebedev Physical Institute, 53, Leninsky prospect, Moscow, 117924, USSR

Abstract

The microstructure and the interface of YBa$_2$Cu$_3$O$_{7-X}$ high-temperature superconducting (HTSC) thin films grown on ($\bar{1}$012) α-Al$_2$O$_3$ by laser deposition and magnetron sputtering have been investigated by TEM and x-ray microanalysis. We observed the formation of hexagonal BaAl$_2$O$_4$ boundary layer at the film/substrate interface. Depending on the interlayer origin polycrystalline textured or epitaxially oriented film appears to grow that results in different superconductive properties. Additionally the BaAl$_2$O$_4$ layer acts as a diffusion buffer for Al.

1. INTRODUCTION

Among different substrates for high-temperature superconductors single-crystalline sapphire seems to be highly suitable due to its excellent mechanical and dielectric properties. However, the quality of these films [1-5] is still not as high as on SrTiO$_3$, MgO and YSZ [6]. Nevertheless, the giant advantages for practical applications of sapphire (and also Si) substrates stimulate further investigations. Among the common problems of HTSC thin film technology such as stoichiometry, oxygen content etc. there are two typical for sapphire substrates. The first is the Al diffusion from the substrate [7,8] as an Al doping with a concentration higher than 1 at% (relatively to Cu) causes strong degradation of critical parameters (temperature, current, etc.) [9]. The second is a microstructure of the film determined by the substrate, in particular, the film's crystallinity, orientation of domains, their dimensions, type of grain boundaries.

We present a study of the microstructure of YBa$_2$Cu$_3$O$_{7-x}$ films as well as the film/α-Al$_2$O$_3$ interface by high resolution analytical electron microscopy. The point of special interest is the influence of the native intermediate layer on the structure and properties of the film.

2. EXPERIMENTAL

We have investigated the YBa$_2$Cu$_3$O$_{7-x}$ films of 300-400 nm thickness grown *in situ* by both pulsed laser deposition and magnetron sputtering herein-after referred as L# and M# samples respectively. Table 1 contains an overview of investigated films grown at various substrate temperatures T$_s$.

The deposition rate in all the cases was 0.1-0.5 nm/sec. The laser deposition was carried out at air pressure 1 Torr from ceramic target of stoichiometrical composition. The D.C. magnetron sputtering was produced from a ceramic target at total $Ar+O_2$ pressure 7 mTorr. These methods were also used for HTSC film deposition on $SrTiO_3$ and gave quite similar results - transition temperature as high as 92 K with the width down to 0.8 K and $J_c(T=77\ K) \geq 10^6\ A/cm^2$.

The resistive measurements were carried out using four probe method with silver paint contacts. The microstructural investigations of the specimens prepared by ordinary planer and cross-sectional techniques were performed on a Philips CM30 electron microscope equipped with x-ray dispersive analyzer.

Table 1
The overview of investigated films

Sample	T_s, K	T_c(R=0), K	Epitaxial or Textured **	Thickness of interm. layer, nm	Typical grain size μm
L#1	600	43	Textured	100-300	0.3
L#2	640	80	Textured	100-300	0.4
L#3	660	71	Textured	100-300	0.5
M#1*	600	40	Epitaxial	–	0.3
M#2	600	83	Epitaxial	30-80	0.3
M#3	680	84	Epitaxial	30-80	1.0
M#4	720	87	Epitaxial	30-80	1.5

* DC magnetron film deposited from segmented metallic target in an activated oxygen flow with pressure 4 mTorr.
** The main volume of the film.

3. RESULTS AND DISCUSSION

3.1. RESISTANCE MEASUREMENTS
Fig.1 presents R(T) normalized to R(130K) curves for four selected samples. Besides the well known improvement of HTSC properties with increasing growth temperature in this range which is due to more equilibrium crystallization [6] one can notice a rather delayed offset in laser films as compared with magnetron ones. Furthermore, laser deposited films have rather high temperature onset which does not vary noticeably with T_s.

On the contrary, the R(T) curve for magnetron films shifts as a whole with T_s without appreciable change in the transition width. We believe that this difference is defined by film's microstructure.

3.2. MICROSTRUCTURE
The planar TEM observations show that all the films investigated consist of single-crystalline domains of quite different shape and size with the c-axis mostly perpendicular to the substrate surface (c-axis oriented films). The typical microdiffraction patterns along [$\bar{2}021$] direction of sapphire, i.e. perpendicular to the substrate surface, (Fig.2) were taken from rather large regions (~ 4 μm in diameter) of the two films comprising

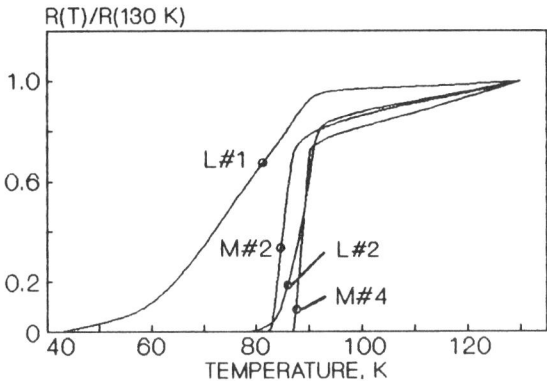

Figure 1. Normallized resistivity vs. tempreture curves of YBaCuO films on sapphire. Samples as in Table 1.

several adjacent domains. The patterns show that there is an essential distinction in microstructure.

The above means that laser films are somewhat more inhomogeneous than magnetron films provided the other conditions are more or less equal.

The laser films have polycrystalline c-axis textured structure with no any special orientations of a or b axes of HTSC blocks. Even the largest grains of 1.5 µm in size do not follow the substrate orientation. Moreover, at lower temperature some low angle (~5°) c-axis tilting of grains occurs. Hence such films contain a high density of high angle grain boundaries. Acting as weak links [10] they deteriorate noticeably the superconducting properties. It is worth noting that polycrystalline films usually display rather high T_c onset ~ 92 K with delayed transition [7,3]. It means that there are regions with perfect HTSC properties but as a whole the films are strongly inhomogeneous.

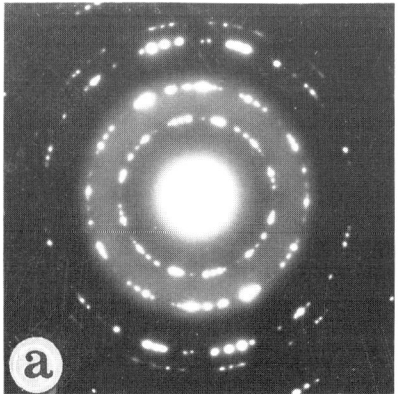 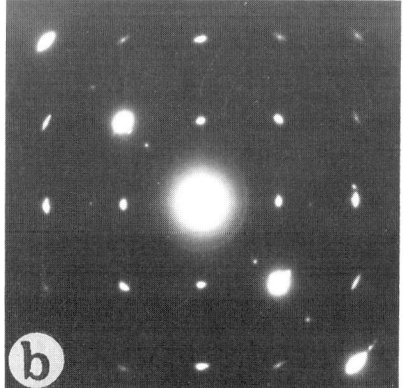

Figure 2. Electron diffraction patterns of the L#2 laser (a) and M#4 magnetron (b) films along the ~[2̄021] zone axis of sapphire.

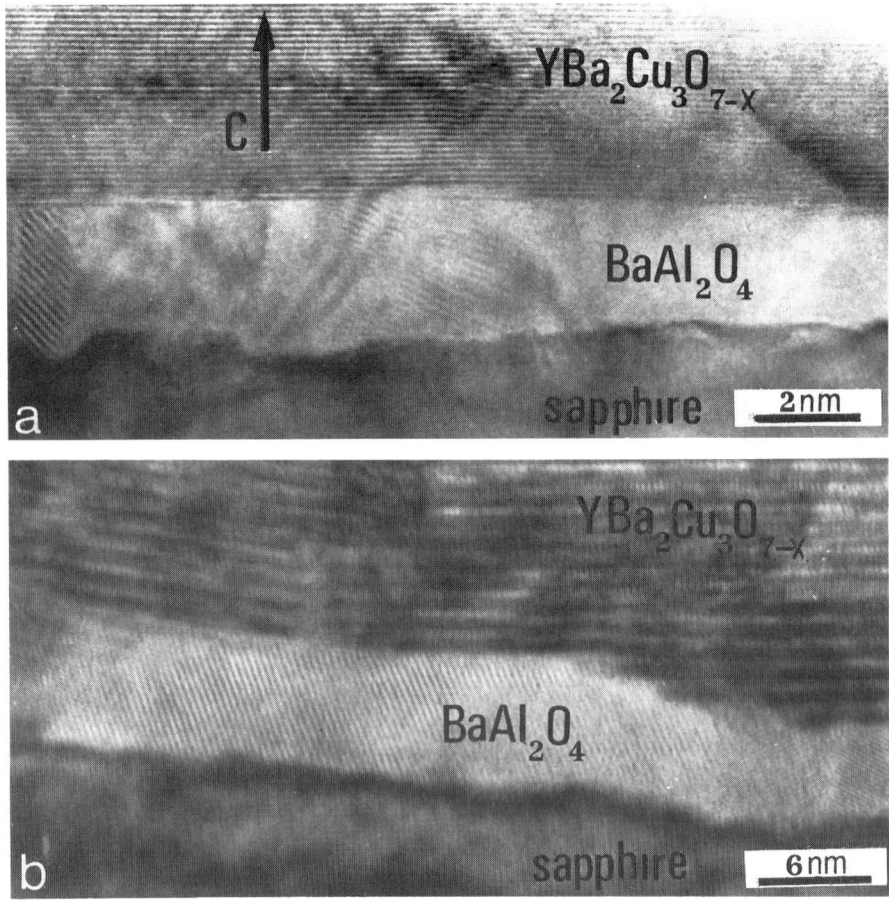

Figure 3. Typical cross-sectional images of laser (a) and magnetron (b) films. The intermediate layer of ~20 nm and ~7 nm thikness is visible between sapphire and YBaCuO for laser and magnetron film consequently. Pay attention to the different scales of the images.

On the contrary, the magnetron films (fig.2b) appear to fit the epitaxial relations with the substrate. However, in general it does not mean that the film is single-crystalline. Moreover, as a rule our films contain a certain amount of misoriented grains, but from the view of percolation the current transport medium is mostly epitaxial. Thus, the most of the grain boundaries are symmetrical or low angle ones and do not influence appreciably the critical properties.

3.3. EPITAXIAL RELATIONS

It is well known that in sapphire there are two systems of nearly equivalent planes with the interplanar distance 0.347 nm almost perpendi-

cular to the substrate surface which form the quasisquare lattice in the ($\bar{1}012$) plane. These are planes ($01\bar{1}2$) and ($1\bar{1}02$) with the angle ~86° between them. So there is not only lattice parameter misfit of 9% but also the angle misfit of about 4°. Consequently, the crystallographic planes of the epitaxial grain and the substrate can fit in at least two possible ways:
- only in one certain direction, i.e. one of the three perovskite planes of the YBaCuO fitting to either ($1\bar{1}02$) or ($01\bar{1}2$) planes of sapphire. This situation was dominant in M#1 sample. The 4°-rotation of crystal lattices of the adjacent HTSC domains caused the formation of noncoherent boundaries. We found them to be a surfaces where the second phase containing some impurities segregates.
- The angle misfit is divided almost equally between the two equivalent directions normal to the planes ($1\bar{1}02$) and ($01\bar{1}2$) of the substrate. The latter relationship was observed in [11] and in our M#2-4 films. The boundaries between superconductor grains were found to be free of any signs of second phase.

Rather striking difference in the film's microstructure is suggested to result from the formation of interfacial boundary layer.

3.4. INTERFACIAL BOUNDARY LAYER

The cross-sectional images of the HTSC film are is presented in Fig. 3(a,b). Some intermediate layer is visible between the substrate and the film. The thickness of the layer varies from 10 to 30 nm for laser films and from 3 to 8 nm for magnetron films. The film M#1 displays no boundary layer at the interface. EDX microanalysis carried out on the interfaces proved that the boundary layer consisted of Ba and Al oxide compounds. The diffraction analysis and its comparison with the crystallographic data [12] for many possible structural modifications of Ba-Al-O compounds proved this layer to be BaAl$_2$O$_4$ hexagonal phase with a=0.522nm and c=0.875nm [13].

Our laser and magnetron deposited films displayed not only different intermediate layer thickness but also different shapes of interfaces. In laser samples the boundary between YBaCuO and the layer is quite flat as compared to the layer-substrate boundary (Fig. 3a). So the original α-Al$_2$O$_3$ surface is now the boundary between HTSC and the intermediate layer. We consider this layer to be formed due to diffusion of Ba into the substrate.

The magnetron samples contain much more thinner boundary layers with interfaces of quite similar shapes (Fig. 3b). Consequently, the layer must be formed by interdiffusion of both Ba and Al atoms.

Furthermore, in both cases the thickness of the boundary layer does not depend on deposition temperature. Therefore, the interlayers of two kinds are suggested to result from different formation processes.

The laser layer originates during the initial stage of deposition before the coalescence of 3-dimensional islands that results in suppression of epitaxy. The relatively large thickness of the interlayers is probably due to higher ion energies in laser than in magnetron plasma. The energetic ions from the tail of laser plasma distribution seem to facilitate the reaction or even cause Ba implantation in the substrate with BaAl$_2$O$_4$ formation.

The magnetron interlayer formation is likely the result of a diffusion limited solid state reaction at the beginning of the actual HTSC growth.

Rather high thermal stability of the thickness of the interlayer indicates that BaAl$_2$O$_4$ layer acts as a diffusion buffer for both Ba and Al. The latter property appears to be highly advantageous for HTSC films growing.

The origin of the intermediate layer is not yet quite clear. For instance, the results of other authors [1,2] have brought us to a conclusion that "magnetron" interlayer can be formed in laser ablation process. It was also established [14,15] that the interlayer was always formed during high temperature annealing (850-900°C) that defined a polycrystalline structure of such films. Thus, the peculiarities of real technological conditions actually play a significant role in the interlayer formation process.

4. SUMMARY

In conclusion, we believe that the main feature of YBaCuO deposition on sapphire is the formation of an intermediate boundary layer, hexagonal $BaAl_2O_4$, which determines whether polycrystalline textured or epitaxially oriented film grows. The boundary layer formed at the initials stages of deposition prevents the HTSC from epitaxial growth causing the arbitrary in-plane orientation. Additionally, the $BaAl_2O_4$ layer considers acting as a diffusion buffer for Al making it possible to increase the growth temperature that improves the crystallinity of the HTSC layer.

5. REFERENCES

1   K. Char, D. K. Fork, T. H. Geballe, S. S. Laderman, R. C. Jacowitz, F. Bridges, G. A. N. Connell and J. B. Bouce. Appl. Phys. Lett., 56 (1990) 785.
2   YU. D. Varlamov, M. R. Predtechenskii, A. N. Smal, A. V. Turbin, S. N. Smirnov. Soviet Superconductivity : Physics, Chemistry, Technique, 3 (1990) 461. (The X-ray diffraction study of the films is reported in [11])
3   S. Chromik, F. Hanic, R. Adam, M. Jergel, J. Liday, and S. Benacka, Appl. Phys. Lett., 56 (1990) 2237.
4   X. X. Xi, G. Linker, O. Meyer, E. Nold, B. Obst, F. Ratzel, R. Smithey, B. Strehlau, F. Weschenfelder and J. Geerk. Z. Phys. B, 74 (1989) 13
5   S. Witanachchi, S. Patel, D. T. Shaw and H. S. Kwok. Appl. Phys. Lett., 55 (1989) 295.
6   Malkolm R. Beasley. Proc. IEEE, 77 (1989) 1155.
7   T. Venkatesan, C. C. Chang, D. Dijkkamp, S. B. Ogale, E. W. Chase, L. A. Farrow, D. M. Hwang, P. F. Miceli, S. A. Schwarz, J. M. Tarascon, X. D. Wu and A. Inam. J. Appl. Phys., 69 (1988) 4591.
8   E. J. Tomlinson, Z. H. Barber, G. M. Morris, R. E. Somekh and J. E. Evetts. IEEE Trans. on Magn., 25 (1989) 2530.
9   R. Suryanarayanan, J. I. Bhandage and M. Rakan. J. Less-Common Metals, 1989, 151, 109.
10 D. Dimos, P. Chaudhari, J. Mannhart, and F. K. LeGoues. Phys. Rev. Lett., 61 (1988) 219.
11 V. P, Martovitsky, V. V. Rodin and V. F. Vratskich Soviet Physics - Lebedev Institute reports, 3 (1990) 15.
12 Structure reports, 29-51 (1964-1984).
13 Strukturbericht, 5 (1937) 97.
14 P. Madakson, J. J. Cuomo, D. S. Yee, R. A. Roy, and G. Scilla. J. Appl. Phys., 63 (1988) 2046-2053.
15 Y. Ichikawa, H. Adachi, T. Mitsuyu and K. Wasa. Jap. J. Appl. Phys., 27 (1988) L381

High T$_c$ Superconductor Thin Films
L. Correra (Editor)
1992 Elsevier Science Publishers B.V.

NON-STOICHIOMETRIC DEFECTS IN YBaCuO THIN FILMS

M. Fendorf[1,2], M.E. Tidjani[1,2], C.P. Burmester[1,2], L.T. Wille[3], and R. Gronsky[2,4]

[1]Materials Science Division, Lawrence Berkeley Laboratory
1 Cyclotron Road, Berkeley, CA 94720 USA

[2]Department of Materials Science and Mineral Engineering, University of California
Berkeley, CA 94720 USA

[3]Department of Physics, Florida Atlantic University, Boca Raton, FL 33431 USA

[4]National Center for Electron Microscopy, Lawrence Berkeley Laboratory
1 Cyclotron Road, Berkeley, CA 94720 USA

Defects in superconducting YBaCuO thin films deposited by laser ablation are investigated by high resolution transmission electron microscopy. Micrographs reveal numerous defects in the YBaCuO film, falling into four basic classes. One of these is an interesting, non-stoichiometric helical defect structure which possibly corresponds to a growth-related screw dislocation. In general, defects in the YBaCuO film are associated both with growth geometry and with local deviations in stoichiometry. Substrate surface geometry is seen to have a profound effect on the number and type of defects produced. Simulation of annealing transformations using a static lattice, three dimensional, Monte Carlo technique is carried out to gain further insight into specific defect formation mechanisms. The results of these studies suggest preparation conditions that are expected to lead to films with improved critical current densities.

1. INTRODUCTION

Because the growth of YBaCuO thin films is typically carried out under non-equilibrium conditions, most films are highly defected. However, YBaCuO films generally show transport properties superior to more perfectly crystalline bulk material.[1] Understanding the relation between crystallographic defects and electrical properties in superconducting YBaCuO therefore has important practical implications, as low critical current density values presently stand as one of the important obstacles in the way of more widespread application of high-T$_c$ oxide superconductors.

In addition, detailed characterization of defects may lead to insights on film growth processes. For example, it has been found that extrinsic stacking faults in the CuO layers of the YBaCuO structure can be introduced during the YBa$_2$Cu$_4$O$_8$ to YBa$_2$Cu$_3$O$_7$ phase transformation, when local deviation from ideal, single-phase stoichiometry occurs.[2] A dramatic increase in intragrain critical current density has recently been reported for YBa$_2$Cu$_3$O$_7$ containing a fine dispersion of these double CuO layer structures.[3] Therefore, control of this type of defect (either producing or preventing it) requires a thorough understanding of the kinetics of phase transformations in the YBaCuO system. Many other defects also form in such a way as to accommodate local deviations in stoichiometry, and attention is focused on these in the present study, since film composition during deposition is an area where improvements in process control can be made.

2. EXPERIMENTAL PROCEDURE

In the present study, high resolution transmission electron microscopy (TEM) images are used to study thin films of superconducting YBaCuO. The films were nominally deposited as c-oriented YBa$_2$Cu$_3$O$_7$ on

(100)-oriented magnesium oxide (MgO) and ($1\bar{1}02$)-oriented alumina (Al_2O_3) substrates. Deposition was carried out using laser ablation at 200mTorr oxygen pressure and a substrate temperature of 750°C. A buffer layer of approximately 100Å $CaTiO_3$ was deposited on the alumina substrates before growth of the YBaCuO film was initiated. The YBaCuO layers were typically 300nm thick and had critical current densities on the order of 1 to 2×10^6 A/cm^2 at 74 K (as measured by the ac mutual inductance response of the films).[4]

Cross-sectional specimens for TEM were prepared by non-reactive ion milling at 5.0 to 5.5 kV, followed by a low-angle final thinning at approximately 4 kV. Subsequent TEM imaging was carried out using a JEOL *JEM200CX* operating at 200kV and the Berkeley Atomic Resolution Microscope operating at 800kV. The NCEMSS software package[5] developed at the National Center for Electron Microscopy was used for simulation of high-resolution TEM images.

3. EXPERIMENTAL RESULTS

Computer simulation of high-resolution TEM images shows that, at optimum defocus and proper crystal orientation, the CuO planes in the $YBa_2Cu_3O_7$ structure produce bright white contrast and the heavy cations (Y and BaO layers) appear as prominent black dots, while the CuO_2 planes show almost no contrast. The actual TEM images show a proliferation of defects of many types throughout the films examined. The films deposited on Al_2O_3 with $CaTiO_3$ buffer layer are notably poorer in quality than those deposited on MgO substrates, exhibiting extremely high defect densities, including pockets of amorphous material in the film. This can be attributed in part to the nature of the buffer layer, which is polycrystalline with some surface irregularity.

At many locations in the specimens, double CuO layers ($YBa_2Cu_4O_8$ structure) appear as isolated stacking defects (with displacement $\mathbf{R}=1/6[031]$[6]), typically extending a short distance (50~100 Å) within a matrix of material that is predominantly found to have the single CuO layer ($YBa_2Cu_3O_7$) structure. An example of this is shown in Figure 1, where no less than six CuO-related stacking faults can be seen. An unusual occurrence also appears in this figure, where a step in the interface is just the correct height to accommodate one additional YBaCuO unit cell. As a result, registry of all lattice planes is maintained in the film below this point; no defects are introduced in the YBaCuO, although a roughly semicircular strain field is visible as a dark band of contrast. An additional example of the presence of double CuO layers is shown in Figure 2. Here the copper and oxygen content are locally higher than was the case in Figure 1, and the average stoichiometry approaches $YBa_2Cu_{3.5}O_{7.5}$ (or $Y_2Ba_4Cu_7O_{15}$, the "2-4-7" phase[7-9]) in the region near the edge of the specimen.

Another stacking defect observed in one instance involves the heavy cations rather than the CuO planes. Since the stacking of heavy cations in $YBa_2Cu_3O_7$ follows the sequence Ba-Y-Ba, a high-resolution image will show three black dots in each unit cell of the 1-2-3 structure. However, in Figure 3, there is a region where a YBaCuO unit cell can be seen which is wider than those of the $YBa_2Cu_3O_7$ material surrounding it and contains four rows of heavy cations, indicating the presence of an extra Y or BaO layer.

In general, irregularities in the substrate surface were found to generate a series of defects in the YBaCuO films, appearing as a type of antiphase boundary between two mis-matched regions of the crystal. As shown in Figure 4, there are many instances where lattice planes bifurcate at such boundaries, and thus these defects become sites for the accommodation of non-stoichiometry in the film. Crystallographic defects, especially dislocations, in the substrate material have been observed to result in this same type of boundary in the YBaCuO film.[10]

A very interesting defect structure of a different type can be seen in Figure 5. One possible interpretation of this image is that a type of screw dislocation, helical in shape, intersects the surface of the specimen at the point where the defect is visible. It can be seen that over part of its extent, the defect has produced a displacement whose component in the plane of the figure is 1/3[001]. In addition, the pitch of the dislocation "helix" is not constant, indicating that the defect is a non-conservative one (i.e., that it incorporates changes in stoichiometry).

4. COMPUTER MODELING

Of the many non-stoichiometric defects that arise in the YBaCuO system, one of the most significant with

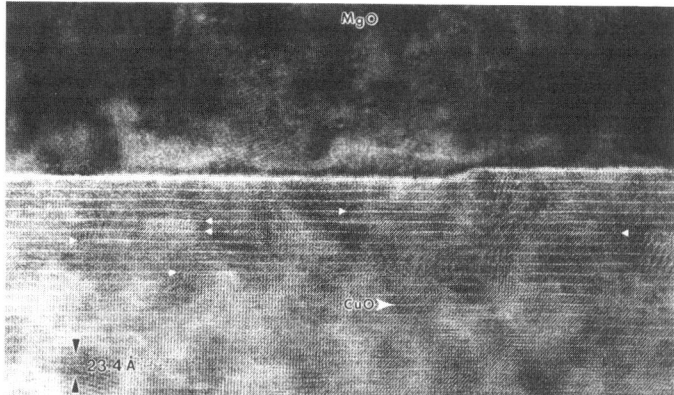

Figure 1:   High resolution TEM image of YBaCuO on MgO, showing CuO stacking faults (small arrowheads) and an interface step which accommodates exactly one additional YBaCuO unit cell.

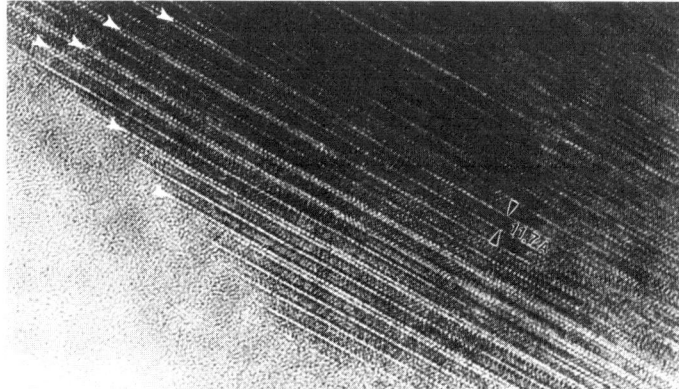

Figure 2:   Additional CuO double layers (white arrows), more extended than those in Figure 1. The overall stoichiometry deviates appriciably from $YBa_2Cu_3O_7$ in this region.

respect to the superconducting properties is the CuO stacking faults which occur as a result of deviations in the amount of CuO near the basal plane. Therefore, in order to better understand how to control the number, density, and distribution of such defects as a function of deposition parameters and post-deposition annealing, phase transformations leading to their formation are studied by developing a static lattice, three dimensional Monte Carlo simulation. The simulation is based on an intercalation scheme, and the details of this are reported elsewhere.[11,12]

Changes in CuO content in the YBaCuO system generate a series of layered structures, $Y_2Ba_4Cu_{6+x}O_{14+x}$, where x=0 corresponds to the well known $YBa_2Cu_3O_7$ (1-2-3) phase. An examination of the evolution of microstructures during the transformation that occurs when x is varied from 0 to 1 by increasing the partial pressure of oxygen in a material with excess copper present indicates that the transformation is heterogeneously nucleated at existing grain boundaries and surfaces within the material as can be seen in the simulation snapshot shown in Figure 6. Due to the elastic compensation available at surfaces and grain boundaries, the resulting extra-layer defects are effectively pinned and hence are not observed to move far into the grain interior. This results in an inhomogeneous distribution of these beneficial stacking faults. When a transformation is induced by *increasing* x, an additional factor limiting the extent of double CuO layer formation is the ability of the system to deliver copper to grain boundaries and free surfaces.

Figure 3: Y or BaO stacking fault (indicated by arrow). This defect most likely corresponds to the $Y_2Ba_2Cu_4O_9$ structure of YBaCuO.

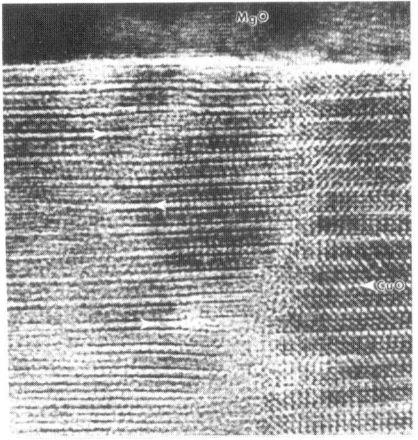

Figure 4: "Mismatch" boundary in YBaCuO film, generated by small step in the interface surface. Non-stoichiometry can be seen where atomic planes begin to split or bifurcate (arrows).

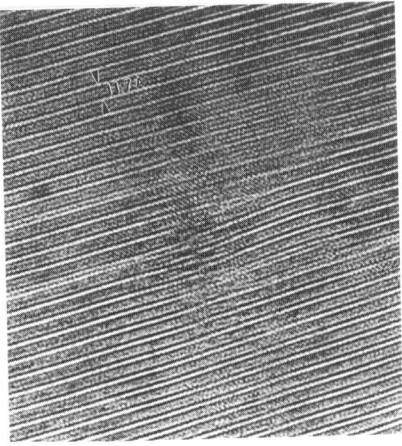

Figure 5: Complex, helical defect in YBaCuO film, possibly interpreted as a non-conservative screw dislocation produced as a result of a spiral growth mechanism. The presence of some double CuO layers can also be seen in this region

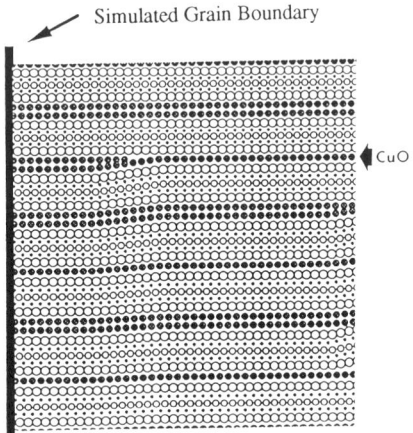

Figure 6: Kinetic "snapshot" obtained by Monte Carlo simulation of CuO (black circles) intercalation from simulated grain boundaries. Here the overall composition is close to $Y_2Ba_4Cu_7O_{15}$ and planar defects similar to those in Figures 1 and 2 can be seen.

A better distribution of defects is produced in the simulation by inducing the 1-2-4 to 1-2-3 and 2-4-7 to 1-2-3 transformations, where x is decreasing rather increasing. A study of the kinetic evolution of microstructures during these transformations reveals that CuO diffuses outward to grain boundaries and surfaces. However, due to microstructural and elastic inhomogeneities present in the film and induced by the motion of CuO planes out of the grain interior, many remnant double CuO layers are retained producing a much more homogeneous distribution of defects throughout the film than was created by the reverse transformation. Thus it is expected that films prepared with a nominal cation stoichiometry equal to 2-4-7 and then ex-situ annealed under conditions where the 1-2-3 phase is stable should produce films with superior critical current densities. In addition, films with a moderate distribution of grain boundaries and planar defects are expected to produce a more homogeneous distribution of the CuO layer defects than single crystal films as boundaries and surfaces are required throughout the film for outward diffusion of copper and oxygen.

5. DISCUSSION

The heavy-cation stacking fault seen in Figure 2 consists of either an extra Y plane or an extra BaO plane in the $YBa_2Cu_3O_7$ structure. The presence in YBaCuO films of a structure corresponding to the addition of an yttrium plane has in fact been reported by Ramesh *et al.*[13] This group used detailed simulation of high-resolution TEM images to distinguish the addition of a Y plane from that of a BaO plane, and concludes that the defect stoichiometry is $Y_2Ba_2Cu_4O_9$ (a "2-2-4" structure). Therefore, it is likely that the defect observed in the present study also corresponds to $Y_2Ba_2Cu_4O_9$. It should be noted, however, that the work by Ramesh *et al.* also suggests that the addition of a BaO plane to the $YBa_2Cu_3O_7$ structure also occurs, though less frequently than the 2-2-4 defect. The presence of CuO stacking faults is thought to enhance critical current density in YBaCuO, and this idea is further supported by the close match between the typical fault size and the superconducting coherence lengths in the 1-2-3 phase of YBaCuO (both within the a-b plane and perpendicular to it)[14], indicating that the faults should act as effective flux pinning sites.[14,15] Because of its geometrical similarity, the heavy cation stacking defect observed here may be another type of potential flux pinning site (though the electrical properties of the 2-2-4 material are unknown at present).

A recently published study[16] using scanning tunneling microscopy (STM) shows that YBaCuO films can grow by means of an island mechanism, with each island consisting of a series of expanding spiral ledges surrounding a screw dislocation core. In light of this classic growth mechanism, first proposed by Frank[17], the interpretation of the image in Figure 5 as a type of screw dislocation is particularly attractive. The dimensions of the dislocation core are comparable to the YBaCuO coherence length in the a-b plane, and thus these defects are also expected to contribute to flux pinning (especially since the dislocations extend through the entire thickness of the film). It should be noted that, based on a single image, it is impossible to conclusively determine the nature of a defect as complex as that shown. Nevertheless, the appearance of this image allows us to form a hypothesis which suggests a direction for future study investigating the actual structure of this helical defect and its possible role in the film growth mechanism.

It is of course expected that the geometry of the surface will influence the growth of a film, and from the high resolution TEM images, it is clear that this is indeed the case. What is interesting to note in the present context, however, is that the defects generated by irregularities in the interface surface are not always only structural in nature. As seen in this study, there are many defects present which also serve to incorporate deviations in the stoichiometry of a YBaCuO film. The substrate surface is also important in determining how a film grows, since surface steps and other features are preferred nucleation sites for film formation and may in fact serve as the source of screw dislocations produced during growth

6. SUMMARY AND CONCLUSIONS

The non-stoichiometric defects observed in YBaCuO films can be classified into four types: CuO stacking faults, Y (or BaO) stacking faults, complex, helical defects, and non-stoichiometric "mismatch" boundaries Of these, the first three types apparently have the potential to act as flux pinning sites in YBaCuO while mismatch boundaries, grain boundaries and other types of defects are detrimental to superconductivity. Thus, control of defect type and density in a film is essential for desired electrical properties. Achieving this control depends critically on both maintaining stoichiometry during deposition and on preparation of the substrate surface.

The combination of high resolution TEM imaging and computer simulations leads to an interplay of results which has proved extremely useful for the study of defects in YBaCuO thin films. TEM provides an experimental basis for the development of microscopic models of growth kinetics. Computer simulation of these models then provides the time resolution and control over "fabrication" conditions necessary to investigate the effects of transformation kinetics on defect formation. Simulations of transformations leading to the development of flux pinning CuO extrinsic stacking faults suggest specific ex-situ annealing conditions for producing films with improved superconducting properties. More detailed characterization of the possible screw dislocation observed here is needed in order to gain more information on growth mechanisms of YBaCuO films.

ACKNOWLEDGEMENTS

The authors would like to extend their thanks to S. Quong and E.C. Nelson for helpful assistance, and to Conductus, Inc. for supplying the materials used. This work is supported by a University of Houston subcontract under DARPA Grant No. MDA972-88-J-1002, using facilities at Lawrence Berkeley Laboratory funded by the Director, Office of Energy Research, Office of Basic Energy Sciences, Materials Sciences Division of the U.S. Department of Energy under Contract Number DE-AC03-76SF00098. One of us (L. T. W.) is supported by an Internal Research Grant from the Division of Sponsored Research at Florida Atlantic University.

REFERENCES

1   J. Evetts, Physics World **3**, 24 (1990).
2   M. Fendorf, E. Kvam, and R. Gronsky, in *High Temperature Superconductors: Fundamental Properties and Novel Materials Processing*, edited by J. Narayan, C.W. Chu, L.F. Schneemeyer, and D.K. Christen (Materials Research Society, Pittsburgh, 1990), MRS Symp. Proc. **169**, 789.
3   S. Jin, T.H. Tiefel, S. Nakahara, J.E. Graebner, H.M. O'Brien, R.A. Fastnacht, and G.W. Kammlot , Appl. Phys. Lett. **56**, 1287 (1990).
4   K. Char, N. Newman, S.M. Garrison, R.W. Barton, R.C. Taber, S.S. Laderman, and R.D. Jacowitz, Appl. Phys. Lett. **57**, 409 (1990).
5   R. Kilaas, in: *Proceedings of the 45th Annual Meeting of the Electron Microscopy Society of America*, edited by G. W. Bailey (San Francisco Press, Inc., San Francisco, 1987).
6   H.W. Zandbergen, R. Gronsky, and G. Thomas, Phys. Stat. Sol. A **105**, 207 (1988).
7   K. Char, R.W. Barton, A.F. Marshall, A. Kapitulnik, and S.S. Laderman, Physica C **152**, 475 (1988).
8   J. Karpinski, E. Kaldis, E. Jilek, S. Rusiecki, and B. Bucher, Nature **336**, 660 (1988).
9   D.E. Morris, J.H. Nickel, J.Y.T. Wei, N.G. Asmar, J.S. Scott, U.M. Scheven, C.T. Hultgren, A.G. Markelz, J.E. Post, P.J. Heaney, D.R. Veblen, and R.M. Hazen, Phys. Rev. B **39**, 7347 (1989).
10   M.E. Tidjani, R. Gronsky, J.J. Kingston, F.C. Wellstood, and J. Clarke, Appl. Phys. Lett. **58**, 765 (1991).
11   M. Fendorf, C.P. Burmester, L.T. Wille, and R. Gronsky, J. Less-Common Metals **164-165**, 84 (1990).
12   M. Fendorf, C.P. Burmester, L. T. Wille, and R. Gronsky, Appl. Phys. Lett. **57**, 2481 (1990).
13   R. Ramesh, D.M. Hwang, T. Venkatesan, T.S. Ravi, L. Nazar, A. Inam, X.D. Wu, B. Dutta, G. Thomas, A.F. Marshall, and T.H. Geballe, Science **247**, 57 (1990).
14   J.C. Phillips, *Physics of High-T_C Superconductors*, (Academic Press, San Diego, 1989).
15   A.C. Rose-Innes and E.H Rhoderick, *Introduction to Superconductivity*, (Pergamon Press, Oxford, 1978).
16   M. Hawley, I.D. Raistrick, J.G. Beery, and R.J. Houlton, Science **251**, 1587 (1991).
17   F.C. Frank, Disc. Farad. Soc. **5**, 48 (1949).

High T$_c$ Superconductor Thin Films
L. Correra (Editor)
© 1992 Elsevier Science Publishers B.V. All rights reserved.

X-ray photoemission studies on c-axis oriented high-T$_C$ superconducting YBa$_2$Cu$_3$O$_{7-x}$ thin films

W.A.M. Aarnink, J. Gao, H. Rogalla and A. van Silfhout,

University of Twente, P.O. Box 217, 7500 AE Enschede, the Netherlands.

Abstract

Angle resolved X-ray-photoemission spectroscopy (ARXPS) was performed on c-axis oriented high-T_C superconducting YBa$_2$Cu$_3$O$_{7-x}$ thin films. The layered structure of the YBa$_2$Cu$_3$O$_{7-x}$ films was used to model the quantitative analysis of the ARXPS experiments. Our XPS results may be compared to spectra taken on YBa$_2$Cu$_3$O$_{7-x}$ single crystal surfaces. On the spectra features are superposed that are assigned to a thin non-superconducting surface layer. For the first time, relative ARXPS measurements show that the interface between the superconducting YBa$_2$Cu$_3$O$_{7-x}$ film and the surface layer is formed by the Y layer. The surface layer consists mainly of BaCO$_3$ and C. A small volume fraction ($\approx 20\%$) contains BaCuO$_2$. Its thickness typically equals 1 nm. Good agreement between theory and experiment has been found in this report.

1. INTRODUCTION

Angle Resolved X-ray Photoemission Spectroscopy (ARXPS) is a widely used non-destructive tool to study the composition and thickness of the outermost top layer of solid materials (d $\approx$ 6 nm) [1-2]. For the high-T$_C$ superconducting Cu-O based ceramics, ARXPS was used to study the spatial distribution and chemical environment of elements in the surface layer of these materials and many groups have reported measurements [3-13].

However, mostly qualitative results were presented. Here we develop a model for the quantitative analysis of ARXPS spectra that can be applied to c-axis oriented high-T$_C$ superconducting YBa$_2$Cu$_3$O$_{7-x}$ thin films, covered with a non-superconducting surface layer. The layered structure of the YBa$_2$Cu$_3$O$_{7-x}$ films is used to model the quantitative analysis of the ARXPS experiments. The results enable us to determine the layer that forms the interface between the YBa$_2$Cu$_3$O$_{7-x}$ thin film and the non-superconducting surface layer. The chemical composition of the surface layer has been determined. A reliable estimate of the surface layer thickness has been calculated.

2. QUANTITATIVE ANALYSIS OF X-RAY PHOTOEMISSION SPECTRA

For the photoelectron energy distribution $dI_{i,k,p,q}(\varphi,E)$ from a core level k in atom q of element i in primitive unit cell p at depth z_p in the superconducting material (see figs. 1 and 2) we may write [1-8,14,25]:

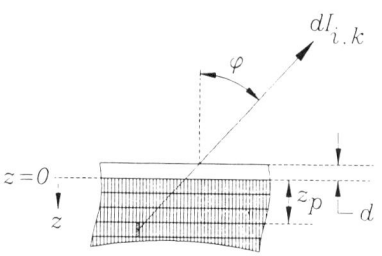
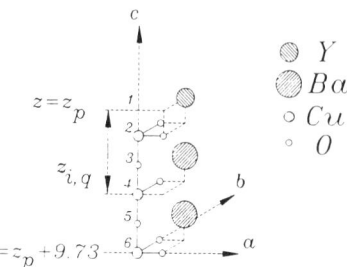

Fig. 1: X-ray photoemission spectroscopy. The YBa$_2$Cu$_3$O$_{7-x}$ thin film has a layered structure parallel to the surface.

Fig. 2: Primitive unit cell of YBa$_2$Cu$_3$O$_{7-x}$. Layer 1 is the Y layer, layer 2 the CuO$_2$ layer etc.

$$dI_{i,k,p,q}(\varphi,E) = \eta \; T_{tot} \; F \; \Delta\Omega \; \gamma_{i,k} \; \exp\left[\frac{-(z_p + z_{i,q})}{\lambda_i \; \cos\varphi}\right] \exp\left(\frac{-d}{\lambda_i \; \cos\varphi}\right) dE. \tag{1}$$

The photoelectron take-off angle is denoted by φ. The electron detector has an efficiency $\eta(E_d)$ with E_d being the kinetic energy of the photoelectrons after being transmitted through the lense and just before entering the electron detector. $T_{tot}(E)$ is the total transmission of the analyser, depending on the kinetic energy E of the photoelectrons. The X-ray flux inside the sample is given by $F(x,y,\varphi)$, with the x- and y-direction parallel to the sample surface. F is assumed not to depend on the depth z in the region where the photoelectrons can escape from the material. The photoelectron energy analyser acceptance angle is denoted by $\Delta\Omega$. The probability of an electron to absorb an X-ray dose and being ejected from orbital k in element i with an kinetic energy (E,dE), is given by $\gamma_{i,k}(\vartheta_k,E)$, the angular differential photoionization cross-section [15,16]. ϑ_k is the angle between X-ray photon and photoelectron path. The inelastic mean free path (IMFP) of the photoelectrons inside the material is given by $\lambda_i(E)$ [17]. The superconducting material may be covered with a non-superconducting surface layer with thickness d, as in the case of c-axis oriented high-T_C superconducting YBa$_2$Cu$_3$O$_{7-x}$ thin films. This surface layer differs in chemical composition or structure. A term describing the transmission of the photoelectrons through this layer is included.

To obtain the total photoelectron energy distribution $dI_{i,k}(\varphi,E)$ from a core level k in an atom of element i, first eq. (1) is summed over one primitive unit cell and, subsequently, summed over all primitive unit cells in the lattice. The X-ray flux F and transmission T_{tot} do not depend on depth z in the region where the photoelectrons escape. T_{tot} may be assumed to depend only on kinetic energy E of the photoelectrons inside the analysis area A_a and to equal 0 outside A_a. Since F may vary over the analysis area, we defined a mean X-ray flux $F(\varphi)$. After summation, the photoemission intensity $I_{i,k}(\varphi)$ may be written as:

$$I_{i,k}(\varphi) = \eta \; T_{tot}(E) \; \overline{F(\varphi)} \; A_a(\varphi) \; \Delta\Omega \; \cdot$$

$$\sigma_{i,k} \frac{1}{4\pi} \left[1 - \frac{1}{4} \beta_{i,k}(3\cos^2\vartheta_k - 1)\right] \exp\left(\frac{-d}{\lambda_i \cos\varphi}\right) S_i(\varphi) \; g_i(\varphi) \tag{2}$$

with $\sigma_{i,k}$ being the photoionization cross-section [15] and $\beta_{i,k}$ the asymmetry factor [16]. A structural factor $S_i(\varphi)$ and geometrical factor $g_i(\varphi)$ were defined as:

$$S_i(\varphi) = \sum_{\substack{unit \\ cell}} \exp\left[\frac{-z_{i,k}}{\lambda_i \cos\varphi}\right], \qquad g_i(\varphi) = \sum_{z_i} \exp\left[\frac{-z_p}{\lambda_i \cos\varphi}\right] = \left[1 - \exp\left(\frac{-c}{\lambda_i \cos\varphi}\right)\right]^{-1}$$

with c being the lattice constant in the c-direction. Eq. (2) gives us the possibility to compare the measured photoelectron intensity ratios $R_{i,j}(\varphi)$ of two elements in the superconducting $YBa_2Cu_3O_{7-x}$ material with the ratio of the structural factors $S_i(\varphi)$ and $S_j(\varphi)$, obtained from theory:

$$R_{i,k}(\varphi) \frac{\sigma_{j,k}\left[1 - \frac{1}{4}\beta_{j,k}(3\cos^2\vartheta_k - 1)\right] T_{tot}(E_j) \exp\left(\frac{-d}{\lambda_j \cos\varphi}\right) g_j(\varphi)}{\sigma_{i,l}\left[1 - \frac{1}{4}\beta_{i,k}(3\cos^2\vartheta_k - 1)\right] T_{tot}(E_i) \exp\left(\frac{-d}{\lambda_i \cos\varphi}\right) g_i(\varphi)} = \frac{S_i(\varphi)}{S_j(\varphi)} \qquad (3)$$

Since the structural factors strongly depend on the choice of the primitive unit cell, we can determine the primitive unit cell that describes relative ARXPS measurements optimal. This yields the plane at which the c-axis oriented high-T_C superconducting $YBa_2Cu_3O_{7-x}$ thin film starts below a non-supercomducting surface layer. A good estimate for the thickness d of the surface layer can be found.

3. EXPERIMENTAL

The high-T_C superconducting $YBa_2Cu_3O_{7-x}$ thin films have been deposited by a modified off-axis rf-magnetron sputtering technique. Experimental details can be found elsewhere [18,19]. As substrates Yttria Stabilized ZrO_2 (YSZ) (100) single crystals were used. After deposition of the $YBa_2Cu_3O_{7-x}$ thin films, the samples were transported through ambient environment to the XPS instrument, a Kratos XSAM 800 spectrometer [21]. Except for the photoelectron take-off angle, all the measurements were done with the same instrumental settings.

4. RESULTS

By means of the modified rf-magnetron sputtering technique, high-T_C superconducting $YBa_2Cu_3O_{7-x}$ thin films with a transition temperature $T_{C,zero}$ of about 90 K were obtained routinely for film thicknesses of 8-300 nm. The critical current density j_c at 77 K of these films is found to be higher than 1×10^6 A/cm^2. With X-ray diffraction analysis, besides the substrate reflections only the (00l) reflections could be observed [18-20]. In this report XPS spectra are given that were acquired on a typical high-T_C superconducting $YBa_2Cu_3O_{7-x}$ thin film on a YSZ (100) single crystal. The spectra were measured on the sample as received. The film had a thickness of 10 nm, its $T_{C,zero}$ was 88 K. In studying films with comparable thicknesses on $SrTiO_3$ and MgO (100) single crystals, no essential differences in the spectra shown here were observed.

For the elements Y, Ba, Cu, O and C, ARXPS core level spectra were recorded at photoelectron take-off angles of 0,10,..,70°. In fig. 1a-d the simulations of the spectra of the Y 3d, Ba 4d, Ba $3d_{5/2}$ and O 1s core level spectra are presented. Before simulating, a Shirley type background was subtracted [22]. The spectrum of Y 3d, measured at 0° and 60° take-off angle (see fig. 1a), can be simulated with two gaussians representing the Y $3d_{5/2}$ and Y $3d_{3/2}$ orbitals of Y in the superconducting $YBa_2Cu_3O_{7-x}$ thin film respectively. The Ba 4d spectrum, shown in fig. 1b, has been simulated with two sets of two gaussians where the two gaussians represent the Ba $4d_{5/2}$ and Ba $4d_{3/2}$ orbitals respectively. In fig. 1c the Ba 3d 5/2 core level spectrum has been simulated

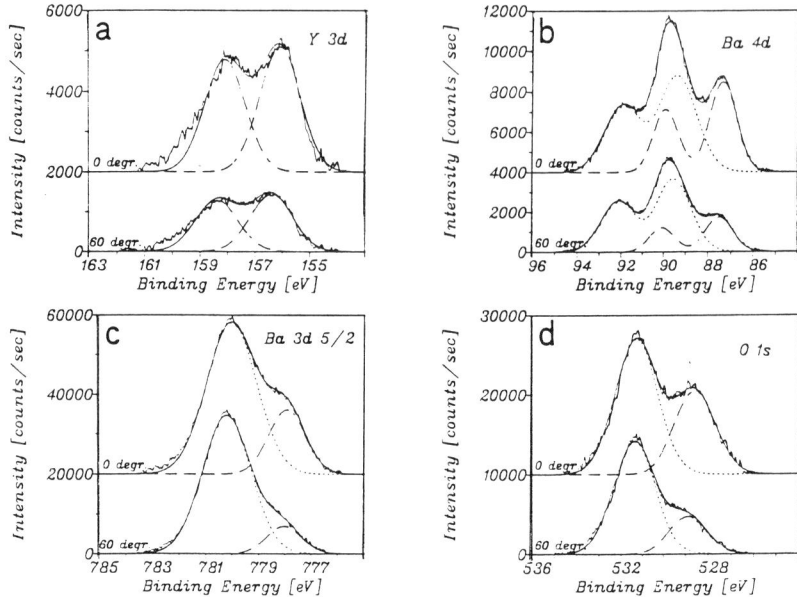

Fig. 3: Simulations of a) the Y 3d, b) the Ba 4d, c) the Ba 3d$_{5/2}$ and d) the O 1s core level spectra. For easy comparison the spectra recorded at 0° photoelectron take-off angle have been given an offset.

with two gaussians. The Ba 3d$_{3/2}$ level has also been simulated. Because the results are similar except for the BEs, they are not shown here. For O 1s, the simulations can be found in fig. 1d. In the Ba 4d, Ba 3d and O 1s core level spectra, the low BE gaussians represent the element in the YBa$_2$Cu$_3$O$_{7-x}$, the high BE gaussian the element in the surface layer, containing BaCuO$_2$, BaCO$_3$ and C. For Cu no simulations were performed. The intensity I_{sat} of the satellite near 943 eV is a measure for the Cu 2p$_{3/2}$ intensity I_{Cu} $_{2p3/2}$ of Cu in the YBa$_2$Cu$_3$O$_{7-x}$, since the valence of Cu is 2+ in this material. For the ratio $I_{sat}/I_{Cu \ 2p3/2}$, in the literature a value of 0.55 can be found for YBa$_2$Cu$_3$O$_{7-x}$ [12,13]. In the surface layer, the Cu valence is +1, as could be deduced from the photoelectron dependence of the Cu 2p core level spectra and comparison with other results [23].

The geometrical factors $g_i(\varphi)$ calculated at 0° and 60° take-off angle, can be found in table I. For the lattice constant c in the c-direction 1.168 nm was chosen [24]. The structural factor $S_i(\varphi)$ strongly depends on the choice of the primitive unit cell and, therefore, on the choice of the plane at which the high-T_C superconducting YBa$_2$Cu$_3$O$_{7-\delta}$ thin film electronically ends. In fig. 2, the Y-plane is chosen. For 0° and 60° photoelectron take-off angle, $S_i(\varphi)$ values have been calculated and the results are presented in table I.

In the last four columns of table I, the experimental values $X_{i,O1s}^{exp}(\varphi)$ as defined by eq. (3), and the ratios $S_i(\varphi)/S_{O1s}(\varphi)$ can be compared. As indicated by the subscript O1s, these values are given relative to those of the O 1s core level. From table I a remarkable good agreement between the experimental values and those predicted by theory can be observed, especially for the outlined quantities. The deviations between theory and experiment for the Ba 3d and Cu 2p levels are ascribed to effects of surface roughness. To determine the structural factors $S_i(\varphi)$ that describe these measurements optimal, we calculated the error function χ^2 as a function of the layer ending the high-T_C superconducting YBa$_2$Cu$_3$O$_{7-x}$ thin film using the outlined quantities in table I.

Table I: Relative ARXPS measurements. The Y layer forms the interface between the superconducting $YBa_2Cu_3O_{7-\delta}$ thin film and the non-superconducting surface layer.

XPS peak	σ $[\sigma_{C1s}]$	β	λ [nm]	E_{kin} [eV]	T_{tot}	$g_i(0)$	$S_i(0)$	$g_i(60)$	$S_i(60)$	$X_{i,O1s}^{exp}$	S_i/S_{O1s}	$X_{i,O1s}^{exp}$	S_i/S_{O1s}
						$\varphi = 0°$		$\varphi = 60°$		$\varphi = 0°$		$\varphi = 60°$	
Y3d	6.24	1.18	1.76	1087	0.66	2.062	1.000	1.361	1.000	0.247	0.214	0.253	0.306
Ba4d	5.41	1.33	1.76	1166	0.62	2.062	1.444	1.361	1.055	0.306	0.310	0.249	0.323
Ba3d5/2	24.75	1.10	0.82	476	1.52	1.317	1.009	1.062	0.537	0.205	0.216	0.283	0.164
Ba3d3/2	17.04	1.10	0.82	460	1.57	1.317	1.009	1.062	0.537	0.216	0.216	0.294	0.164
Cu2p3/2	15.87	1.30	0.66	321	2.26	1.205	1.386	1.030	0.777	0.281	0.297	0.062	0.238
O1s	2.85	2.00	1.35	725	1.00	1.727	4.664	1.215	3.270	1.000	1.000	1.000	1.000

The error function χ^2 is defined as:

$$\chi^2 = \sum (X_{i,O1s}^{exp} - \frac{S_i}{S_{O1s}})^2,$$

where $X_{i,O1s}^{exp}$ is defined in eq. (3). As can be seen in fig. 1 and 2, 6 different primitive unit cells can be chosen, this corresponds with 6 different layers ending the $YBa_2Cu_3O_{7-x}$ thin film. The error function χ^2 equaled 0.01, 0.06, 0.1, 0.04, 0.04 and 0.03 for layer 1, 2,.. and 6, respectively. The Y-layer is indicated as layer 1, the Ba-O layer as layer 2 etc., see figure 2. For the Y layer χ^2 is minimal, so this layer forms the interface between the superconducting $YBa_2Cu_3O_{7-x}$ thin film and the non-superconducting surface. It also was found that χ^2 is minimal for a thickness of the surface layer of $d \approx 1.0$ nm.

5. DISCUSSION AND CONCLUSIONS

Comparison of our spectra with those reported in the literature [3,5,6,9-11] reveals that our $YBa_2Cu_3O_{7-x}$ thin films are of high quality. Our XPS results can be compared with spectra taken on $YBa_2Cu_3O_{7-x}$ single crystal surfaces. On the spectra features are superposed that are assigned to elements in a thin (thickness $\approx$ 1.0 nm) non-superconducting surface layer, mainly consisting of $BaCO_3$ and C. The surface layer also contains some $BaCuO_2$.

It is important to note that in our model we descibed the $YBa_2Cu_3O_{7-x}$ film as a single crystal. If we do not so and assume that in the superconducting $YBa_2Cu_3O_{7-x}$ thin film the elements are homogeneously distributed and that the film is covered with a non-superconducting surface layer, we find: $n_Y : n_{Ba} : n_{Cu} : n_O = 1.4 : 1.8 : 2.8 : 7$, for the Y concentration an error of 40%. In our model the strongest deviation between theory and experiment arises for the Y 3d core level, with an relative error of 15%, as can be seen in table I for $\varphi = 0°$. This large improvement in agreement of theoretical and experimental values can be easily understood with our finding that the material has a layered structure with different composition. In the high-T_C superconducting $YBa_2Cu_3O_{7-x}$ film the first Y layer partly shields the photoelectrons of the elements Ba, Cu and O below it. The results show that in the qualitative analysis of XPS spectra, the effect of a layered structure of the investigated material may not be neglected.

A good agreement between theory and experiment was found in this investigation. Including the effect of the layered structure of *c*-axis oriented $YBa_2Cu_3O_{7-x}$ thin films in a model for quantitative analysis of photoemission spectra, relative ARXPS measurements provide structural and chemical information about the top layer of an

$YBa_2Cu_3O_{7-x}$ thin film. For the first time, relative ARXPS measurements show that electronically the interface between the superconducting $YBa_2Cu_3O_{7-x}$ film and the surface layer is formed by the Y layer. The thin surface layer with a thickness of ≈ 1 nm, consists mainly of $BaCO_3$ and C. A small volume fraction ($\approx 20\%$) contains $BaCuO_2$.

The model we developed here may be considered as a large improvement in the understanding of ARXPS experiments on c-axis oriented high T_C superconducting $YBa_2Cu_3O_{7-x}$ thin films.

ACKNOWLEDGEMENTS

This work is part of the research program of the "Stichting voor Fundamenteel Onderzoek der Materie (FOM)", which is financially supported by the "Nederlandse organisatie voor Wetenschappelijk Onderzoek (NWO)", and was made possible by support of the "Centrum voor Materialen Onderzoek (CMO)", Enschede. The authors would like to thank J. Halbritter for stimulating discussions.

REFERENCES

[1] Photoemission in Solids I, edited by M. Cardona and L. Ley (Springer-Verlag, New York, 1978).
[2] Practical Surface Analysis by Auger and X-ray Photoelectron Spectroscopy, edited by D. Briggs and M.P. Seah (Wiley, Chichester, 1983).
[3] P.A.P. Lindberg, Z.-X. Chen, W.E. Spicer, I. Lindau, Surf. Sci. Rep. 11 (1990) 1.
[4] J. Halbritter, P. Walk. H.-J. Mathes, B. Häuser, H. Rogalla, Z. Phys. B 73 (1988) 277.
[5] X.D. Wu, A. Inam, M.S. Hedge, T. Venkatesan, C.C. Chang, E.W. Chase B. Wilkens, J.M. Tarascon, Phys. Rev. B 38 (1988) 9307.
[6] C.C. Chang, M.S. Hedge, X.D. Wu, B. Dutta, A. Inam, T. Venkatesan, B.J. Wilkens, J.B. Wachtman Jr, J. Appl. Phys. 67 (1990) 7483.
[7] P.A.P. Lindberg, I. Lindau, W.E. Spicer, Phys. Rev B 40 (1989) 6822.
[8] W.A.M. Aarnink, J. Gao, H. Rogalla, A. van Silfhout, J. Less Common Metals 164 & 165 (1990) 321.
[9] H.M. Meyer, D.M. Hill, T.J. Wagener, Y. Gao, J.H. Weaver, D.W. Capone, K.C. Goretta, Phys. Rev. B 38 (1988) 6500.
[10] S.L.T. Andersson, J.C. Otamiri, Appl. Surf. Sci. 45 (1990) 1.
[11] D.E. Fowler, C.R. Brundle, J. Lerczak, F. Holtzberg, J. Elec. Spec. 52 (1990) 323.
[12] D. van der Marel, J. van Elp, G.A. Sawatzky, D. Heitmann, Phys. Rev. B 37 (1988) 5136.
[13] J. Ghijsen, L.H. Tjeng, J. van Elp, H. Eskes, J. Westerink, G.A. Sawatzky, M.T. Czyzyk, Phys. Rev. B 38 (1988) 11322.
[14] W.A.M. Aarnink, A. Weishaupt, A. van Silfhout, Appl. Surf. Sci. 45 (1990) 37.
[15] J.H. Scofield, J. Electron Spectrosc. 8 (1976) 129.
[16] R.F. Reilmann, A. Msezane, S.T. Manson, J. Elec. Spec. 8 (1976) 389.
[17] M.P. Seah, W. Dench, Surf. Interf. Anal. 1 (1979) 2.
[18] J. Gao, B. Häuser, H. Rogalla, J. Appl. Phys. 67 (1990) 2512.
[19] J. Gao, W.A.M. Aarnink, G.J. Gerritsma, H. Rogalla, Appl. Surf. Sci. 46 (1990) 74.
[20] J. Gao, W.A.M. Aarnink, G.J. Gerritsma, and H. Rogalla, Physica C 171 (1990) 126.
[21] XSAM800 Operators Handbook, Kratos Analytical Instruments.
[22] D.A. Shirley, Phys. Rev. B 5 (1972) 4709.
[23] P. Steiner, V. Kinsinger, I. Sander, B. Siegwart, S. Hüfner, C. Politis, R. Hoppe, H.P. Müller, Z. Phys. B 67 (1987) 497.
[24] J.D. Jorgensen, B.W. Veal, A.P. Paulikas, L.J. Nowicki, G.W. Crabtree, H. Claus, W.K. Kwok, Phys. Rev. B 41 (1990) 1863.
[25] W.A.M. Aarnink, J. Gao, H. Rogalla, A. van Silfhout, submitted to Phys. Rev. B.

High T$_c$ Superconductor Thin Films
L. Correra (Editor)
783

Structural and electrical characterization of Y Ba$_2$Cu$_3$O$_{7-x}$ thin films grown on different substrates

M. Bianconi[a], L. Correra[a], A. Lamagna[b], S. Nicoletti[a], R. Nipoti[a],
M. Sánchez Balmaseda[c] and M. Servidori[a]

[a]CNR-Istituto LAMEL, via Castagnoli 1, I-40126 Bologna, Italy

[b]FRURE Dept., C.N.E.A., 1429 Buenos Aires, Argentina

[c]Dept. Optica, Facultad de Ciencias Fisicas, Universidad Complutense,
28040 Madrid, Spain

Abstract

X – ray diffraction analyses using symmetrical and asymmetrical reflections were performed on Y Ba$_2$Cu$_3$O$_{7-x}$ films grown *in-situ* by pulsed laser deposition. Monocrystalline
(100) oriented SrTiO$_3$, MgO and YSZ were used as substrates. The measurements allowed
the calculation of the film lattice parameters and the determination of the film – substrate
epitaxial relationships. The in – plane and out – plane film textures were studied as well.
The critical current density between 17 K and 87 K was measured in narrow strips patterned by a direct – writing laser system. A comparative discussion of the results obtained
for the transport properties and the structural characteristics was carried out for films
grown on MgO.

1. Introduction

The preparation and characterization of high temperature superconducting oxides are
of great interest in view of their application to devices. To this end, high quality films are
required to grow on a suitable substrate with a high degree of reliability. Thin films of
Y Ba$_2$Cu$_3$O$_{7-x}$ (in the following referred to as YBCO) have been deposited on different
substrates by a variety of techniques. As the transport properties are strictly related to
the degree of structural perfection of the material, substrates with a good lattice match
with the superconductor are generally used. SrTiO$_3$, which has a perovskite structure
with a mismatch of about 2% between YBCO and substrate a – axes, has been demonstrated to be suitable to obtain high-critical currents, but it has poor dielectric properties.
Other crystals, such as rare-earth orthogallates or aluminates and ZrO$_2$–yttrium stabilized (YSZ), are also currently investigated. Thin films with excellent properties have also
been grown on MgO, despite the more than 9% lattice mismatch. For all the mentioned
monocrystalline substrates it has been established that c – axis perpendicular films are
grown when a relatively high deposition temperature is used (700 ÷ 800° C). The film tex-

ture is a very important feature to understand the transport properties of the material, that are likely to be affected by defects and grain misorientation.

In this study we report the results obtained by X-ray diffraction measurements of YBCO thin films deposited by laser ablation. Based on the analysis of the relative orientation of the film and substrate cells, the critical current density values obtained are qualitatively discussed in relation to the actual structure of the superconducting film.

2. Experimental

Thin (~ 1500 Å) films of YBCO were deposited by XeCl laser ablation of a massive, sintered pellet (95% density). Single crystals of MgO, SrTiO$_3$ and YSZ, (100) oriented, were used as substrates. All samples were grown at a dynamical oxygen pressure of 25 Pa and at a substrate temperature of 750°C. The laser fluence was about 2 J/cm^2. A detailed description of the experimental set-up and the processing conditions has been published elsewhere[1].

The crystal structure of the films and of the substrates was studied by X-ray diffraction. The orientation relationships between the YBCO film and the substrates were determined by using symmetrical and asymmetrical reflections. The former (Bragg-Brentano geometry) allowed us to determine which YBCO (hkl) planes are parallel to the (001) substrate surface, the interplanar distance of these planes and the angular spread of the <uvw> distribution, while the latter (Field-Merchant geometry) is suitable to determine the in-plane orientations[2].

The critical current J$_c$ at zero magnetic field was measured between 17 K and 87 K. For this purpose narrow strips were patterned using a direct-writing laser system. The strips were 40 μm wide and 300 μm long. Four gold contacts in line geometry were deposited by sputtering and subsequently annealed at 450°C in flowing oxygen for 20 min. The J$_c$ measurements were stopped when the voltage drop across the strip was > 1 μV. This is equivalent to keep a critical current criterion of 3 μV/cm[3].

3. Results and Discussion

The analysis of the X-ray symmetrical reflections showed that all the YBCO films grew with the (001) planes parallel to the substrate surface, independently of the substrate. The c-axis length was determined from the θ position of the 00l film reflection peaks in the $\omega - \theta$ scan diffraction spectra. The spread of the c-axis length was deduced from the FWHM$_{\omega-\theta}$ of the 00l peaks. The c-axis lengths are in the $11.67 \div 11.71$ Å range, in agreement with the values reported for the orthorhombic Y Ba$_2$Cu$_3$O$_7$ phase. The FWHM$_{\omega-\theta}$ values ranged between 0.08° and 0.29°, indicating a high degree of homogeneity in the c-axis lengths. The FWHM$_\omega$ of the 00l rocking curves gives the c-axis angular spread, that is, the out-plane texture of the YBCO subgrains. This parameter was found to range between 0.13° and 0.68°. The higher values were measured for films grown on MgO substrates as discussed thereafter.

The relative orientation of the c-axis of YBCO and [001] substrate direction has been determined by comparing the angular positions of 00l rocking curve pairs taken at ϕ positions differing by 180° from one another. Misalignments between these two axes were found to be of $0.2° \div 0.8°$. We measured that the [001] direction of the substrates was often slightly misoriented in relation with the surface. Since this angular misalignment is of the same order of that between the film and substrate c–axes, we concluded that YBCO always grows perpendicular to the substrate surface within $0.2°$. This behaviour was observed even in the case of a MgO substrate cut $4°$ off–axis. Similar results were obtained for YBCO films deposited by sputtering on vicinally polished MgO substrates[4].

The relative orientation of the a– and b– crystallographic axes of YBCO and of the substrate was measured using the asymmetrical X-ray reflections. For films grown on YSZ, the 224 reflection from the substrate and the 117 one from YBCO were used, while for SrTiO$_3$ and MgO the reflections were the 013 and the 224 for the substrate respectively, and the 038 for the film. It is worth mentioning that a perfect orthorhombic crystal shows a twofold symmetry for both the 038 and the 308 reflections, rotated with respect to each other by $90°$ about the c–axis. The expected $\Delta\omega$ angular separation between the two asymmetrical reflections is $0.27°$. Since, in our system, the divergence of the incident beam in the scattering plane was $0.12°$ and the angular acceptance of the detector was $0.04°$, at a given θ position, both reflections could be resolved.

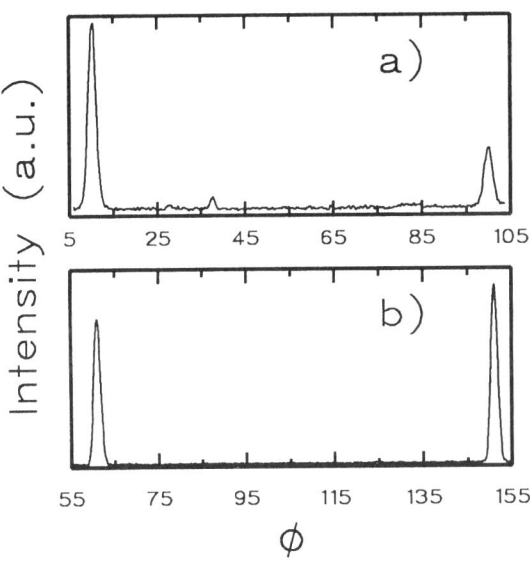

Figure 1. ϕ–scans of the YBCO 038 reflection for films deposited on MgO (a) and SrTiO$_3$ (b).

Fig. 1(a,b) shows the diffractograms of the YBCO 038 reflection when the sample is rotated about the axis normal to its surface (ϕ scan) for a film deposited on MgO (Fig. 1.a) and SrTiO$_3$ (Fig. 1.b). The presence of two well defined peaks about $90°$ apart in the ϕ scan reveals a twofold in–plane orientation. In particular, for the film grown on MgO the in–plane a– and b–axes of YBCO grew parallel to the same [110] direction of the substrate, whereas for the $SrTiO_3$ substrate the $a-$ and $b-$axes of YBCO grew parallel to the same [100] direction. The fraction of the two grain populations can be estimated by the relative intensities of the two peaks in the ϕ scan. For the angular position of the peaks, a tilt between the two orientations was found. This angle was $(0.31 \pm 0.01)°$ for the film grown on MgO and $(0.20 \pm 0.01)°$ for the one on $SrTiO_3$. This twofold in–plane orientation can also be interpreted as the occurrence of twins, being (1-10) the composition plane. To further investigate the epitaxy, the in–plane relative orientation of the YBCO and substrate cells was studied comparing the ϕ scan of the 038 and 224 asymmetrical reflections in the case of YBCO on MgO. The superposition of the two ϕ scans evidenced a peak shift of $(1.06 \pm 0.01)°$. This angle is the tilt of the YBCO a–axis with respect to the [110] direction of MgO. Using the the 308 reflection, a tilt of $(1.37 \pm 0.01)°$ for the YBCO $b-$axis and the [110] direction of the substrates was found. The same analysis was performed for the YBCO on $SrTiO_3$ samples using the 038 and 013 reflections. In this case it was found that the YBCO $a-$ and $b-$axes were tilted of $(0.64 \pm 0.01)°$ and $(0.44 \pm 0.01)°$, respectively, with respect to the $SrTiO_3$ [100] axis. Finally, distortion of the YBCO cell was never observed within the experimental error of $0.01°$.

Two aspects can be underlined considering the above results. First, the misalignment between the YBCO cell and the substrate cell is a common feature for all the samples considered in this work. This could likely be attributed to the fact that YBCO grows always oriented with the $c-$axis perpendicular to the substrate surface, which very often is not parallel to the (001) plane. Second, a tilt between the $a-$ and $b-$axes of the two different populations of grains aligned along the same direction of the substrate was always measured. This feature is under investigation, particularly in relation with a twinned grain growth.

The results of the $\phi-$scan analysis for YBCO deposited on different substrates showed that all films exhibited a grain structure with the $a-$ and $b-$axes locked into several preferred orientations. It was found that the <100> directions of the YBCO films grew completely aligned with the <110> of YSZ and with the <100> of $SrTiO_3$, as expected when examining the lattice match between film and substrate. For films deposited on MgO substrates the case is more complex. Some authors showed that YBCO grains grow on MgO along different in–plane preferential orientations[2, 5]. In particular, Moeckly et al. [6] obtained different film textures as an effect of different surface preparation. As for our samples, we observed so far two different texture patterns. The first shows grain axes mainly oriented along the <100> and <110> directions of MgO and others oriented along intermediate angles. The second shows most of the grain axes oriented along the <110> directions of MgO (see Fig. 1.a). This last result, already obtained by Ravi et al.[5], may be explained taking into account the lattice mismatch of 2.3% between [100] YBCO and [110] MgO where a long range accomodation occurs by the formation of periodic misfit dislocations.

For films grown on MgO and YSZ the $a-$ and $b-$axes of the unit cell were determined to be in agreement with those of the orthorhombic $YBa_2Cu_3O_7$ phase, whereas $a = (3.84 \pm 0.01) Å$, $b = (3.86 \pm 0.01) Å$ were found for a 1000 Å film grown on $SrTiO_3$. In this case, the a and b lengths are closer to that of the tetragonal phase. This result could be correlated with the misfit strain at the film–substrate interface in the different cases of lattice match considered.

The influence of the film structure on the transport properties is currently investigated for different substrates. As for MgO, it has been reported that the main limitation to the superconducting properties of the YBCO is the coexistence of multiple in–plane preferential orientations, which increases the number of grain boundaries[6]. A further limitation is given by the concurring in– and out– plane mosaic spread.

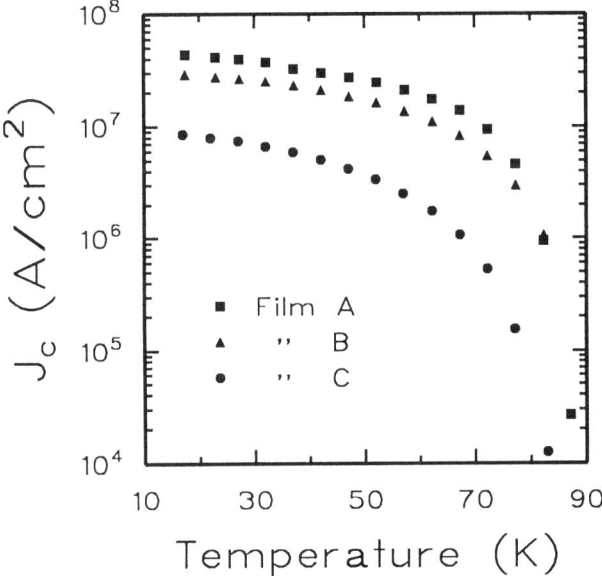

Figure 2. Critical current density *vs* temperature for three different films grown on MgO.

Fig. 2 reports the J_c values as a function of temperature in the $17 \div 87\,K$ range for three different YBCO films on MgO. The c–axis lengths are the same within the experimental errors evidencing the same oxygen stoichiometry. Excellent transport properties are exhibited by films A and B, whereas a relatively poorer behaviour is displayed by film C. The J_c values reflect the structural properties of the considered samples. In fact, film A and film B are both highly ordered and display a well–defined structure, as shown for film A in Fig. 1*a*. They only differ in the c–axis mosaic spread, as the FWHM_ω are $(0.26 \pm 0.01)°$ and $(0.52 \pm 0.01)°$, respectively. Film C was grown on a $4°$ off–axis substrate and displays a multiple–grain in–plane orientation. Moreover, a $\text{FWHM}_\omega = (0.68 \pm 0.01)°$ was measured. This disorder degree results in lower J_c values.

Similar J_c's were obtained for films grown on $SrTiO_3$, YSZ and $LaAlO_3$. A complete structural characterization for these films is in progress.

This work was financially supported by the *Progetto Finalizzato: Tecnologie Supercon-duttive e Criogeniche* of the Consiglio Nazionale delle Ricerche (CNR).

Dr. A. Lamagna undertook this work with the support of the *ICTP Programme for Training and Research in Italian Laboratories*, Trieste, Italy. Dr. M. Sánchez Balmaseda undertook this work with the support of *Programa Nacional de Formación de Personal Investigador*, Ministerio de Educación y Ciencia, Spain.

References

[1] M. Bianconi, G. Bobbio, L. Correra, A. Lamagna, S. Nicoletti, M. Sánchez Balmaseda and V. Soncini, Materials Science and Engineering, B1 (in press).

[2] J.P. Zheng, S.Y. Dong and H.S. Kwok, Appl. Phys. Lett. 58 (1991) 540.

[3] K. Watanabe, N. Kobayashi, H. Yamane, H. Kurosawa, T. Hirai, H. Kawabe and Y. Muto, Jpn. J. Appl. Phys. 28 (1989) L1417

[4] S. K. Streiffer, B. M. Lairson and J. C. Bravman, Appl. Phys. Lett. 57 (1990) 2501.

[5] T. S. Ravi, D. M. Hwang, R. Ramesh, Siu Wai Chan, L. Nazar, C. Y. Chen, A. Inam, and T. Venkatesan, Phys. Rev. B 42 (1990) 10141.

[6] B. H. Moeckly, S. E. Russek, D. K. Lathrop, R. A. Buhrman, Jian Li, and J. W. Mayer, Appl. Phys. Lett. 57 (1990) 1687.

High T$_c$ Superconductor Thin Films
L. Correra (Editor)

Atomic Resolution and Nanostructure of $Y_1Ba_2Cu_3O_{7-\delta}$ Laser-Ablated Epitaxial Thin Films on Various Substrates Studied by Scanning Tunneling Microscopy

H.P.Lang[a], T.Frey[a,b], R.Sum[a] and H.-J. Güntherodt[a]

[a] Institut für Physik, Universität Basel, Klingelbergstrasse 82, 4056 Basel, Switzerland

[b] IBM T. J. Watson Research Center, Yorktown Heights, New York 10598

Abstract

We applied Scanning Tunneling Microscopy (STM) to investigate the surface structure of high quality c-axis oriented laser-ablated thin films of the high T$_c$ superconductors $Y_1Ba_2Cu_3O_{7-\delta}$ and $RE_1Ba_2Cu_3O_{7-\delta}(RE = Eu, Er, Ho)$. To the best of our knowledge we show the first atomic resolution STM images in the a-b plane of the $Y_1Ba_2Cu_3O_{7-\delta}$ compound. The approximately square lattice indicates an interatomic distance of 0.38 nm. This lattice will be discussed in relation to the unit cell structure. The main features of the surface structure on a nanometers scale are steps of one unit-cell height proofing the two-dimensional growth-process of the epitaxial film. However, on a micrometer scale the dominating growth process is island growth. Islands with two appearances are observed: stacked circular or rather rectangular layer hills with or without lattice defect structures as screw dislocations or loops.

1. INTRODUCTION

Since the discovery of high T$_c$ superconductors (HTCSC) by Bednorz and Müller [1] in a La-Ba-Cu-O compound (1986), several other superconducting, layered ceramics with T$_c$ up to 125 K in a Tl-Ba-Ca-Cu-O compound were synthesized. Sample quality was improved by choosing synthesis roots different from sintering powders, e.g. melt-textured growth, growth of single crystals, deposition of thin epitaxial films by magnetron sputtering and laser-ablation [2].

The technique of scanning tunneling microscopy (STM) developed by Binnig and Rohrer [3] is well-suited to study clean surfaces in the nanometer range. Impressive results were achieved in the field of semiconductor surfaces and layered materials as graphite and graphite intercalation compounds, which can be easily cleaved to produce fresh surfaces for STM investigations. This method was soon applied to high T$_c$ superconductors. The most interesting results were achieved with cleaved Bi-Sr-Ca-Cu-O and Tl-Ba-Ca-Cu-O single crystals. Clear atomic resolution by STM was first reported in

1988 [4] respectively in 1989 [5].

The most investigated high T_c compound as far as its superconducting properties are concerned, is $Y_1Ba_2Cu_3O_{7-\delta}$. However, $Y_1Ba_2Cu_3O_{7-\delta}$ has a quite reactive surface due to its barium content. This results in the formation of barium-hydroxides on the surface, if exposed to humidity.

Established methods to study the atomic arrangement of $Y_1Ba_2Cu_3O_{7-\delta}$ were up to now high resolution electron microscopy, x-ray, electron, and neutron diffraction experiments. There is only one report on corrugations in the atomic scale on single crystals of $Y_1Ba_2Cu_3O_{7-\delta}$ [6] by STM.

Recently, two studies [7], [8] have investigated the nano- and microstructure of magnetron-sputtered epitaxial thin films of $Y_1Ba_2Cu_3O_{7-\delta}$ by STM. In this study, we report on atomic resolution of freshly prepared $Y_1Ba_2Cu_3O_{7-\delta}$ epitaxial laser-ablated thin films (see also [9]). Furthermore we discuss the nanometer morphology of laser-ablated epitaxial thin films by means of STM.

2. EXPERIMENTAL

Films investigated in this paper were prepared on different substrates such as MgO (100), $SrTiO_3$ (100) and $LaAlO_3$(100). The substrates were optically polished and ultrasonic-cleaned in acetone. Thereafter they were attached to a heating block which is placed in a vacuum chamber. The heating block can be heated up to 900 °C. Epitaxial film growth by laser-ablation at 200 mTorr oxygen partial pressure is achieved at a deposition rate of 0.1-0.2 nm/s, a typical substrate temperature between 740 and 760 °C, followed by subsequent slow cooling to 450 °C   in 200 Torr oxygen partial pressure. Film thickness is about 200 nm. We used an excimer laser with KrF filling (Lambda Physik LPX 300i, $\lambda = 248$ nm, 1.2 J/shot) run at 5-10 Hz.

The STM used in this study is a commercially available system [10] that was operated in air. The images were recorded using mechanically sharpened $Pt_{80}Ir_{20}$ tips. Bias voltage between tip and sample was about 0.8 V and tunneling current was typically about 300 pA. The STM was operated in the constant current mode mapping topography. If not indicated otherwise, the images represent non-filtered data.

3. RESULTS AND DISCUSSION

Films of $RE_1Ba_2Cu_3O_{7-\delta}$, RE= Er, Eu, Ho and Y, prepared under the conditions described above, have excellent superconducting and structural properties. Inductively measured critical transition temperature T_c above 89 K with a 10 % - 90 % transition width of less than 1 K, critical current densities J_c of about $5 \cdot 10^6 A/cm^2$ at 77 K and up to $3 - 4 \cdot 10^7 A/cm^2$ at 4.2 K [11]. X-ray diffraction shows that these films are c-axis oriented.

Scanning electron micrographs of the films taken with a SEM operated at 20 keV excitation energy show at all attainable magnifications a flat surface with very few crystallites (see Figures 1a, 1b). Simultaneous chemical analysis by energy dispersive scattering of X-rays (EDS) gives the expected 1-2-3 stoichiometry.

 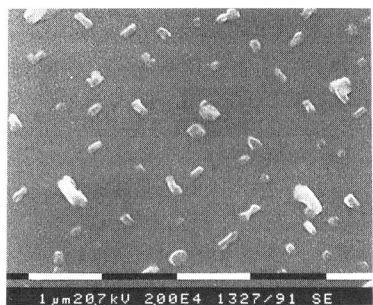

Figure 1: (a) SEM micrograph of an epitaxial $Y_1Ba_2Cu_3O_{7-\delta}$ film. The scale bar is 0.1 mm. (b) Same film at higher magnification. The scale bar is 1 μm.

In order to investigate the atomic structure of these films, we need larger magnification. This is attained by Scanning Tunneling Microscopy.

The topography of the films on a submicron scale mapped by STM is not as flat as expected from electron micrographs. Clear features of the surface are steps of unit-cell height (1.18 nm) in direct agreement with the x-ray diffraction data. The dominating growth process during laser-ablation is not two-dimensional layer by layer growth as in single crystals, but island growth. These islands have the appearance of circular or rectangular shaped stacked disks. Frequently, a screw dislocation is observed on top of these islands. The occurrence of such lattice defects seems to be closely related to the deposition parameters, especially to the deposition temperature on the substrate. The submicron structure of magnetron sputtered thin, epitaxial films of $Y_1Ba_2Cu_3O_{7-\delta}$ has been investigated by STM recently by two groups ([7], [8]) independently. Similarities between laser-ablated and magnetron-sputtered thin films are discussed elsewhere [12].

The scanning tunneling micrograph shown in figure 2a gives a illuminated threedimensional bird view of a 708 nm x 708 nm large area of a c-axis oriented laser-ablated $Y_1Ba_2Cu_3O_{7-\delta}$ film on MgO (100) deposited at 750 °C. The island-grown stacked layer structures (in the following referred to as *Towers of Hanoi*) of a diameter of 100-200 nm can be clearly seen. The steps in the image are of one unit cell height. Figure 2b shows an 400 nm x 400 nm large area of a $Y_1Ba_2Cu_3O_{7-\delta}$ film on $LaAlO_3(100)$ deposited at 770 °C. The STM bird view shows interesting features: epitaxial layers with unit cell steps, in the foreground a 6 unit cell deep hole (diameter 80 nm) and near the right border a presumably non c-axis oriented crystallite grown out of the epitaxial layers. Recently, Venkatesan et al. [13] have studied the growth of these crystallites by SEM.

Figure 3a is a STM topview (450 nm x 450 nm) showing two screw dislocations with opposite sign on a $Ho_1Ba_2Cu_3O_{7-\delta}$ film on MgO(100) deposited at 740 °C. Island growth with a screw dislocation on many of the stacked layer hills is another observed basic epitaxial growth mode (referred to as *Screw Dislocation Growth*) of laser-ablated thin films besides *Tower of Hanoi* growth.

 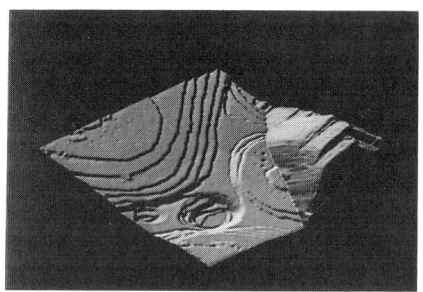

Figure 2: (a) STM bird view showing *Tower Of Hanoi* structures (708 nm x 708 nm).
(b) STM bird view showing epitaxial growth and an emanating crystallite (400 nm x 400 nm)

Figure 3a also illustrates both appearances of screw dislocations, namely circular and rather rectangular. Figure 3b is a illuminated bird view of a 294 nm x 294 nm large area of the same $Ho_1Ba_2Cu_3O_{7-\delta}$ film on MgO(100) as in Figure 3a. In the center of the image, a loop can be observed besides unit cell steps. A loop is an introduced extra lattice plane consisting of two screw dislocation of opposite sign. The loop had broadened out producing extra unit cell disks during growth.

This completes the survey of the growth phenomena of laser-ablated $RE_1Ba_2Cu_3O_{7-\delta}$ films in the submicron range. The next section deals with the atomic structure of the films observed by STM.

Figure 4a shows the approximately square atomic arrangement of the (00ℓ) face of a $Y_1Ba_2Cu_3O_{7-\delta}$ film on $SrTiO_3 < 100 >$ deposited at 720 °C. The interatomic distance of about 0.38 nm corresponds to the a- or b-axis length.

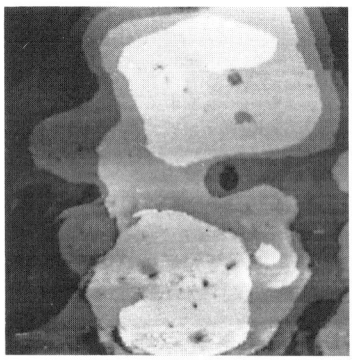 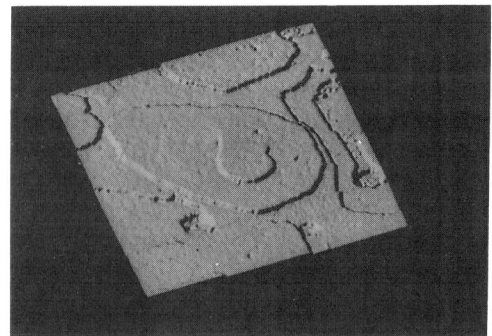

Figure 3: (a) STM topview showing two screw dislocations of opposite sign (450 nm x 450 nm), (b) STM bird view showing epitaxial growth and loop (294 nm x 294 nm)

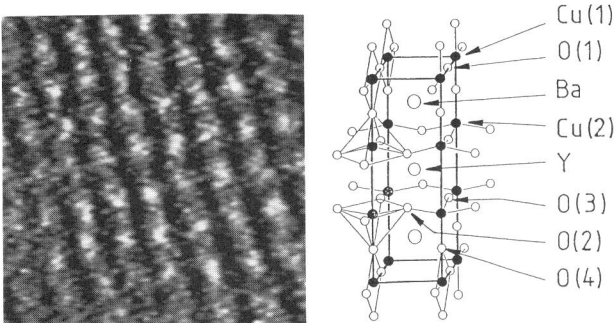

Figure 4: (a) STM topview (4 nm x 4 nm) showing atomic resolution on $Y_1Ba_2Cu_3O_{7-\delta}$. The image is slightly processed. (b) Crystal structure of $Y_1Ba_2Cu_3O_7$.

Possible distances of 0.38 nm in the $Y_1Ba_2Cu_3O_7$ crystal structure include Y-Y, Ba-Ba, Cu(1)-Cu(1) respectively O(1)-O(1) in the planes, Cu(2)-Cu(2) and O(4)-O(4) in the chains. For the STM, O(2) and O(3) are equivalent atoms (see Figure 4b). Therefore another distance would be expected for these species of atoms. The twofold symmetric position of the Ba atom in the unit cell makes it little probable that Ba is the top layer of this unit-cell, unless there exists a so called *dead endlayer* consisting of parts of the Cu-O chains and the barium layer as is was suggested by a group performing angle resolved photoemission experiments on that kind of thin film HTCSC samples [14]. Corresponding STM experiments on $RE_1Ba_2Cu_3O_{7-\delta}$ resulted in the same atomic arrangement as in $Y_1Ba_2Cu_3O_{7-\delta}$, with slightly different corrugations. We believe that STM cannot solve the problem what is the toplayer of $RE_1Ba_2Cu_3O_{7-\delta}$. Suitable experiments might be more accurate angle resolved photoemission or inelastic ion scattering.

4. CONCLUSION

Scanning tunneling microscopy was found to be an appropriate technique to image the nanostructure and atomic arrangement of laser-ablated thin films of $RE_1Ba_2Cu_3O_{7-\delta}$, RE= Eu, Er, Ho and Y. Although the X-ray spectrum of these films is similar to that of a single crystal, films show a large variety of structures in the submicron range inaccessible to scanning electron microscopy. Features of this nanostructure are at least two types of basic growth modes, namely *Tower of Hanoi* stacked layer island growth and *screw dislocation* island growth. Less frequently, loop structures are observed resulting from a system of two screw dislocations having opposite sign. Non c-axis oriented crystallites emanating from the epitaxial c-axis oriented film can be observed as well in STM on micron scale as in SEM.

On a nanometer scale, an approximatively square lattice with an interatomic spacing of 0.38 nm is observed. It was not possible to determine the species of these atoms only by the results of STM. Further work on these HTCSC laser-ablated films has to be done to understand the growth process more accurately.

Acknowledgement

We would like to thank S. Siegmann for providing the suszeptibility and critical current data. J.P. Ramseyer has prepared the targets and polished the substrates what is acknowledged. Scanning electron microscopy was performed at the Labor für Rasterelektronenmikroskopie of the University of Basel, and we are grateful to Prof. R. Guggenheim and his SEM team, especially E. Zuberbühler who has operated the SEM. We would like to thank D. Brodbeck and R. Hofer for their help with image processing. Valuable discussions and proof reading are acknowledged to Th. Jung and H. Haefke. Furthermore, we are greatful to the Kommission zur Förderung der Wissenschaftlichen Forschung and to The Swiss National Science Foundation for their financial support.

References

[1] J.G. Bednorz and K.A. Müller, Z. Phys. B **64**, 189-193 (1986)

[2] D. Dijkamp, T. Venkatesan, X.D. Wu, S.A Shaheen, N. Jisrawi, Y.H. Min-Lee, W.L. McLean, and M. Croft, Appl. Phys. Lett. **51**, 619-621, (1987)

[3] G. Binnig and H. Rohrer, Helv. Phys. Acta 55, 726 (1982)

[4] M.D. Kirk, J. Nogami, A.A. Baski, D.B. Mitzi, A. Kapitulnik, T.H. Geballe, and C.F. Quate, Science Vol. 242, 1673 (1988)

[5] X.L. Wu, C.M. Lieber, D.S. Ginley, and R.J. Baughmann, Appl. Phys. Lett. **55**, 2129 (1989)

[6] L.E.C. van Leemput, P.J.M. van Bentum, L.W.M. Schreurs, and H. van Kempen, Physica C **152**, 99-104, 1988

[7] C. Gerber, D. Anselmetti, J.G. Bednorz, J. Mannhart, and D. G. Schlom, Nature, Vol. 350, 28 March 1991, 279-280

[8] M. Hawley, I.D. Raistrick, J.G. Beery, and R.J. Houlton, Science, Vol. 251, 29 March 1991, 1587-1589

[9] H.P. Lang, T. Frey, H.-J. Güntherodt: accepted for publication in Europhys Letters.

[10] Nanoscope-II, Digital Instruments, Santa Barbara (CA)

[11] T. Frey, S. Siegmann, H.-J. Güntherodt, L. Drabeck, G. Grüner, U. Ebels and C. Quitmann, in *Science and Technology of thin Film Superconductors 2*, 197-203, Plenum Publishing Corporation (1991)

[12] D. G. Schlom, D. Anselmetti, J. G. Bednorz, R. Broom, A. Catana, T. Frey, Ch. Gerber, H.J. Güntherodt, H.P. Lang, J. Mannhart, and K.A. Müller, to be published

[13] R. Ramesh, T.S. Ravi. D.M. Hwang, C.C. Chang and A. Inam, T. Venkatesan, X.D. Wu, R.E. Münchausen, S. Foltyn and N.S. Nogar, Physica C **173**, 163-172 (1991)

[14] G. Frank, Ch. Ziegler, and W. Göpel, Phys. Rev. B **43**, 2828 (1991)

High T$_c$ Superconductor Thin Films
L. Correra (Editor)
© 1992 Elsevier Science Publishers B.V. All rights reserved.

795

ACOUSTIC MICROSCOPY OF HTSC MICROCRYSTALLITES AND FILMS

Roman Gr.Maev, Leonid A.Chernosatonskii

Center of Acoustic Microscopy, Institute of Chemical Physics
Academy of Sciences, Kosygina str., 4, Moscow, 117334, USSR

Abstract
     The new method of examination of microstructure - acoustic
microscopy - was used for studying HTSC-materials (ceramics,
monocrystals and thin films). The obtained high-resolution
images (with resolution up to ˉ0.4um) visualize the specimen
topography, variations of the local acoustic properties, in
particular, surface and subsurface defects - cracks, twins,
phase and structural discontinuities, peelings in films. For
bulk specimens (monocrystals and ceramics) we obtained a
quantitative characteristic - a local (on a section of
10x10um²) Rayleigh wave velocity, which makes it possible, in
particular, to characterize individual crystallites in
ceramics.

1.INTRODUCTION

     Nowadays the acoustic properties of high-temperature
superconductors are measured by integral methods and serve as a
characteristic of the whole volume of the sample.
     We demonstrate the use of acoustic microscopy in
investigation of both local acoustic properties of individual
crystallites and various defects in HTSC-materials. Important
is that the acoustic microscopy provides for: (a) viewing the
defects in the depth of a non-transparent sample, unlike the
optical methods; (b) determining the Rayleigh wave velocity on
a small (≲10um) section of surface (for example, on an
individual crystallite); (c) presence the defects in the
presurface layer by attenuation of such waves.
     In our work we used a commercial acoustic microscope ELSAM
(Ernst Leitz Scanning Acoustic Microscope).
     In a reflecting scanning acoustic microscope the microwave
acoustic signal (1-2GHz) is converted into a plane acoustic
wave, which is focused by the lens into a spot on the surface,
or under the surface of the subject under investigation,
reflected from it, received by the lens and converted into an
output electric signal of the microscope. The dimensions of the
focal region (acoustic wave length in an immersion liquid -
about 1μm) determine the resolving power of the acoustic

microscope. The lens is scanned mechanically in the plane
parallel to the subject, and image is formed on the display
synchronously with its motion. The brightness of each point is
proportional to signal amplitude V(Z) at the respective
position of the lens in the scanning plane, Z - distance
between the lens focus and the subject surface. The amplitude
of the acoustic wave reflected from the subject and,
consequently, V(Z) depends both on distance Z, and local
acoustic properties of the subject:

$$V(Z) = \int_0^{\Theta_m} U(\Theta)R(\Theta)e^{2ikz\cos\Theta}\sin\Theta d\Theta$$

where U(Θ) = lens aperture function, k = wave vector in
immersion liquid, Θ_m = lens aperture, R(Θ) - factor of
reflection from subject, taking the latter to be isotropic in
the surface plane . The more the difference between the
acoustic properties in the given point and the properties of
the immersion liquid, the higher reflection factor R(Θ), the
brighter the respective point on the subject image; V(Z) is
maximum, as a rule when Z=0.

2.EXPERIMENTAL RESULTS

Ceramics We have investigated 123-ceramics - $ABa_2Cu_3O_x$
(A=Y, Yo, Ho), BiSrCaCuO, both hot-pressed up to 70 kbar and
ordinary ones. Water heated up to 333K or methanol at a room
temperature was used as an immersion liquid.

Regions of crystallites are seen as bright sections, dark
sections - are pores.
The focus being brought onto the bottom of the pores by
the maximum of the acoustic wave reflected from them, makes it
possible to determine their depth accurate to 0.05um. Usually
it lies within 1 to 3um.
Using the ELIAS image processing system of the ELSAM
microscope, we can carried out an automatic count of the
quantity of pores and crystallites, plotted histograms of their
distribution over the area. For example, on the image of high-
dense hot-pressed 123-ceramics pores occupy 5.5% of the area,
the average area of the crystallites was approximately equal to
190um² .
The value of Rayleigh velocity V_R was determined at a
frequency of 1.7GHz according to V(z)-dependence at variation
of Z within 0 to 20μm, consequently, the maximum distance
covered by the Rayleigh wave made up d=Zxtgθ~10μm ., therefore
measured V_R characterizes the surface section with a diameter
of about 10μm. Of course, V_R is an average Rayleigh velocity in
all directions in the plane of the sample. Measurements of V_R
were taken at sufficiently large crystallites, so that their
boundaries do not influence the propagation of the Rayleigh

wave. On an area of about 2mm² there were selected 10-15 crystallites, on which V(z)-curves were taken. The obtained curves were compared in a computer, and the majority of them (70-80% for some samples) coincided.

It indicates an absence of occasional factors influencing the value of the microscope output signal: boundaries of crystallites, subsurface defects, small thickness of crystallites that could cause a reflection of the acoustic wave from the crystallite rear surface, which, in turn, would inevitably cause an accidental distortion of the V(z)-curve due to random variations of the crystallite thickness. Thus, on different crystallites of the same sample velocity V_R was found to be the same. Since it is know that in the perovskite structures under investigation there is a considerable difference between the sound velocity along axis c and the sound velocities in planes (a,b): $V_{[100]}$ $V_{[010]} < V_{[001]}$, and the crystallites in ceramics are plates with dimensions in directions [100] and [010] considerably exceeding their thickness in direction [001], one can suppose that axis c of the crystallites under examination is perpendicular to the ceramics surface.

For ceramics we obtained the following values of Rayleigh velocity: 3.01km/s for BiCaSrCuO, 2.80km/s for YBaCuO, 2.86km/s for HoBaCuO and YbBaCuO accurate to 0.05km/s, and made up 0.8-0.9 of the C_t values in plane [001] for the respective materials.

Absorption value α_R of the Rayleigh wave is determined by the slope to the side maxima of V(z)-curve. This makes it possible to determine value of α_R for every crystallite. So, with z<0, an absence of the side maxima caused by the Rayleigh wave testifies to its considerable attenuation on the surface of the crystallite under investigation. For example, the absence of the maxima on the V(z)-curve, with z<-15μm and working frequency 1.5GHz, indicates a complete absorption of $\alpha_R > 0.1k_R$, taking the signal-to-noise ratio for the ELSAM to be equal to 20 dB². Such an attenuation may be caused by a great number of microcraks, even if the value of their exposure (for example, 0.01μm) is far beyond the range of the microscope resolving power (0.4μm). This is associated with the fact that water does not penetrate into microcracks, and the hypersound of the used gigahertz frequencies practically fully gets attenuated in the air at the wave length. Such an estimation of α_R characterizes the integral quantity of attenuation of the Rayleigh wave on a section with a diameter of 10μm, and is rather sensitive to the magnitude of density of crack distribution on the crystallite.

Such high magnitudes of absorption were observed on microcrystallites of many ceramics, especially of high-density ones. The characteristic magnitude of crack exposure, as it is seen from the electron-microscopic photograph of the same sample, but already pickled, not to exceed 1000A, their presence causes an abnormally high attenuation of the Rayleigh wave. It should be pointed out that the accuracy of determining

of magnitude was about 50%, since the magnitude of the side
maxima determining angle Θ greatly depends on the instrument
radio engineering parameters, in particular, on the magnitude
of delay of the beginning of reception of the reflected pulse
with respect to the transmitted pulse (the so-called "gate").

Not the total value of attenuation of the Rayleigh wave is
a matter of practical interest, for low-defective and not too
porous samples it is mainly determined by the rate of its
outflow, but that part α_s of the whole absorption α_R that is
conditioned by the acoustic attenuation and structural
scattering. Attenuation $\alpha = \alpha_R - \alpha_s$ caused by the wave outflow
proper can be estimated, using the magnitudes for C_l, C_t -
given above, and taking water as an immersion liquid.

Using for α expression $\alpha \sim (\rho_1 C_1 / \rho C_t)(k_R / 2\pi)$ where ρ -
dense of the sample, we obtain $\alpha_R \sim 0.01 K_R$. Thus, the presence
of great number of microcracks results in an essential increase
of attenuation.

Observation of separate sections of crystallites reveals
sections that are notable for brightness, which may correspond
to a changed phase or structural composition. Different ceramic
phases feature different acoustic properties and, consequently,
different factors of reflection from them. Let us estimate the
magnitude of this effect, using for reflection factor R a
simplest expression for reflection of a normal incident
longitudinal wave from a liquid half-space. Then, for instance,
for phases $YBa_2Cu_3O_6$ and $YBa_2Cu_3O_7$. whose longitudinal
velocities are equal to 4.9km/s and 4.4km/s, respectively, we
obtain the difference in the reflection factors equal only to
2%.

We also observed a twin structure on some crystallites,
which, generally speaking, should not be observed with the aid
of a spherical lens featuring an axial symmetry. Apparently,
such a picture is associated with formation of strained regions
near the twinning boundaries. This is confirmed by the fact
that the twin structure is invisible with Z=0, and it becomes
visible only when Z<0, when the lens starts receiving the
outflowing Rayleigh wave.

Films   We have also investigated HTSC-films. The immersion
liquid was methanol, to which the samples were inert. Current
of different intensity was passed through the films being
studied. No structural changes have been detected in the films,
through which the subcritical current was passed, whereas in
those films through which current close to the critical one was
passed structural heterogeneities were detected; a
characteristic "noise" appeared on the V(z)-dependences, i.e.
there appeared random variations of the signal, associated with
the scattering and re-reflections of the acoustic wave on the
film heterogeneities, whose characteristic dimensions were 0.3-
0.5 μm.

Such destruction of the film may be the consequence of
large elastic stresses caused by the gradients of magnetic
forces upon passage of large electric current. This is

conformed indirectly by the appearance of strong acoustic pulses under the effect of current pulse.

We have also investigated HTSC-films after they were exposed to impulse Ar laser radiation when the film became non-superconductive. Such degradation of the film was reversible; after annealing the film had the same properties as before. On the fig.1 (1mm*0.8mm area) the right part of the film was exposed to the laser radiation. On the fig.2 one can see a part of exposed film with a higher magnification (60um*40um).

Films obtained by successive thermal deposition of Y, Ba and Cu on a silicon substrate at 720K in the oxygen atmosphere were investigated. The obtained multilayer film was then annealed in the oxygen atmosphere at 1120K during 4 minutes to form a superconducting phase. Investigation of the obtained films has shown that they are multi-phase ones. Along with the YBaCuO phase there are sections of Y_2BaCuO_5 - "green" phase (dark spots) and inclusions of CuO, Y_2O_3 and $BaCuO_2$ (dark points). As to the kind of V(z)-dependences, phases 123 and 211 are identified unambiguously. It is also possible to see light spots with interference rings -peelings caused by an insufficient adhesion and by the difference of the thermal coefficients of expansion at a thermal cycling.
Crystals Beside ceramic samples, we investigated also separate crystals YBaCuO and BiSrCaCuO. We revealed on them both surface (Fig.3) and subsurface (Fig.4) defects of structure. Surface of BiCaSrCuO monocrystal, 60x40 um (Fig.3). Surface cracks and inclusions are visible. It was used so-called quasi-three-dimensional image formation, at which oscilloscope beam while scanning on the scree besides variation in brightness, deflects from the scanning line proportionally with the magnitude of the output signal of the SAM. Same as in Fig.4 Subsurface (-6.6μm)of the BiCaSrCuO monocrystal, 60x40 μm. Quasi-three-dimensional mode. Subsurface defects are visible. It is delamination in right low corner. Twinning with a period of about 1um was observed on $YBa_2Cu_3O_7$ samples. The values of V_R measured on monocrystals were close to V_R measured on non-strained crystallites of the respective ceramics.

Thus, it was shown that the acoustic microscopy features rather wide opportunities to investigate and control the qualities of HT-superconductors, provides for revealing the specific characteristics of the microstructure, determining the local acoustic properties and various defects in crystallites, crystals and films.

3.ADDITIONAL CRYOGENIC EQUIPMENT

For examination of specimens in an acoustic microscope within a temperature range from 85 to 240K we developed a cryogenic thermostated chamber providing an accuracy of 0.1K in the specimen area.

The chamber operates on the principle of a circulating cryostat with a two-step temperature control. The working space of the chamber is filled with liquid propane, which is used as an immersion. The microscope acoustic lens is inserted into the chamber installed on the microscopic stage after the optical adjusting the latter.

4.CONCLUSIONS

Acoustic microscopy has been shown to offer ample opportunity for investigating and monitoring the quality of HTSC materials. It reveals specific features of the microstructure, determines the local acoustic properties, and allows detection of various defects in crystallites, crystals and films.

The authors are much grateful to M.Bukhny for obtaining the acoustic images, for Ya.M.Soifer and A.N.Khodan for samples preparation and to G .S.Abilov for assistance in preparing the cryogenic chamber.

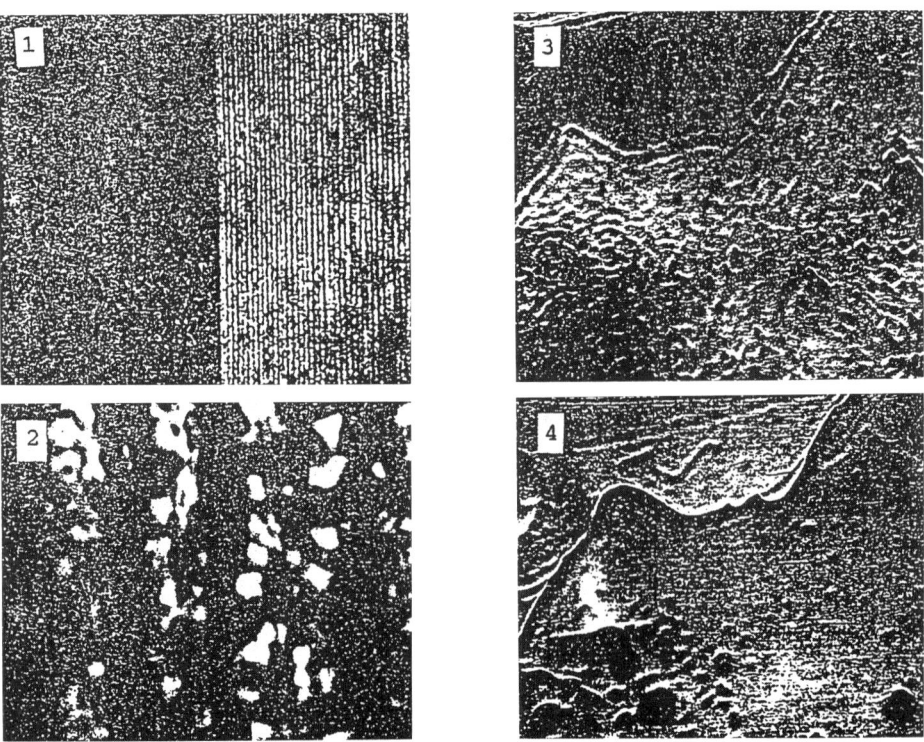

High T$_c$ Superconductor Thin Films
L. Correra (Editor)
801

A SPECTROSCOPIC STUDY OF FLUORINE DIFFUSION IN Y-Ba-Cu MULTILAYERS

P. DANESI, A. MEHDAOUI, J.C. PERUCHETTI, D. BOLMONT AND B. LOEGEL

Laboratoire de Physique et de Spectroscopie Electronique, URA CNRS 1435,
Faculté des Sciences et Technique, 4, rue des frères Lumière, 68093 Mulhouse Cedex.

Abstract

We have focused our attention on fluorine diffusion in Y / Ba / Cu multilayers deposited by Molecular Beam Epitaxy technique on Silicon or Yttrium Stabilized Zirconium (YSZ) and studied by Auger Electron Spectroscopy (AES). Our multilayers obtained by thermal sequential evaporation of Yttrium, Copper and BaF$_2$ are the precursors of High Temperature Superconductor (HTSC) thin films. Different time dependences for AES spectra are observed both with and without electron bombardment . AES depth profiling shows the selective diffusivity of fluorine through the layer of Yttrium.

INTRODUCTION

Spectroscopic studies have shown aging effects on fluorinated silicon surfaces [1] and on various high Tc superconductors (HTSC) like YBaCuO [2,3], LaSrCuO [3] and BiSrCaCuO [4] compounds, leading to time and temperature dependant AES, EELS, UPS and / or XPS spectra. This is especially the case for the F$_{KLL}$, Cu$_{LLM}$ and O$_{KLL}$ Auger spectra.

We have focused our attention on fluorine diffusion in Y / Ba / Cu multilayers deposited on Silicon or YSZ and studied by AES. These preliminary measurement where initiated in order to obtain a better insight in the passivation effects against contaminations when introducing fluorine [5]. Our multilayers obtained by thermal sequential evaporation of Yttrium, Copper and BaF$_2$ are the precursors of High Temperature Superconductor (HTSC) thin films, where it is well known that fluorine plays an essential role in the final stability of the films as Y-Ba-Cu-O thin films made using BaF$_2$ instead of metallic Baryum have a reduced sensitivity to fabrication and environmental conditions [6]. Fluorination has more generally a strong influence on electrical and magnetic properties, both for the thin films and the bulk samples [7, 8].

EXPERIMENTAL PROCEDURE

Molecular Beam Epitaxy (MBE) deposition system consists in an Ultra High Vaccum (UHV) preparation chamber linked with a characterization chamber.

The former contains three SEJ MECA 2000 cells, loaded respectively with Yttrium, Copper and BaF_2. The cells are driven by MECA 2000, ACT 3 type power supplies with either current or temperature feedback controlling system. Temperature regulation is achieved by a numerical P.I.D regulator coupled with K and W_5 thermocouples. Substrate support can translate and rotate samples of size 12x12x0,5 mm^3. Samples can be heated up to about 700°C (at the thin films) by resistive heating of a silicium platelet. Evaporation control rates are made by two sensor crystals.

The analysis chamber contains an AUGER type EA 150 RIBER analyser coupled with an channeltron, allowing recording of the spectra in the integrated mode. An electron gun type LEG 61 and an ion gun type CI 10 RIBER permits AES depth profiling. For a pression of 10^{-5} torr, 500 eV electron acceleration energy and 300 V sample polarisation, cleaning rate is about 100 Å / h .

Pumping system consists of a primary pump type DUO 016 B BALZERS, a turbomolecular type TPU 240 BALZERS and an ionic pump type PID 400 MECA 2000. An external N_2 trap and a titanium sublimation pump yield a base pressure of 3.10^{-10} torr.

RESULTS

We have first realized different AES spectra with and without continuous electron bombardment in order to know its influence when observing fluorine's peak, which has to be taken into account for AES depth profile.

A film of 200 Å of BaF_2 was deposited over Y on a Silicon substrate at room temperature (R.T). AES analysis was made at different times without stopping the electron gun. A spectrum which reflects time dependence of O_2, Ba and F peaks can be visualized in figure 1 . Electron gun current is 100 mA with 3,5 kV source acceleration. A vaccum of 3.10^{-9} torr and 3.10^{-8} torr during the deposition for respectively BaF_2(or Cu) and Y layers is reached.

The same experience with similar deposition conditions were realized, but electronic bombardment was reduced to only the time needed to take AES spectra.

A strong decrease of the F_{KLL} peak is observed in both cases (figure 1 and 2), but EDX analysis shows persistence of fluorine even after vanishing of the F_{KLL} peak, indicating diffusion in the bulk. The nearly logarithmic decay of intensity of the F_{KLL} peak with time is moreover greatly enhanced by electron bombardment, reflecting the influence of electron beam irradiation on fluorine peak intensities, as was already pointed out by Pantano and Madey[9].

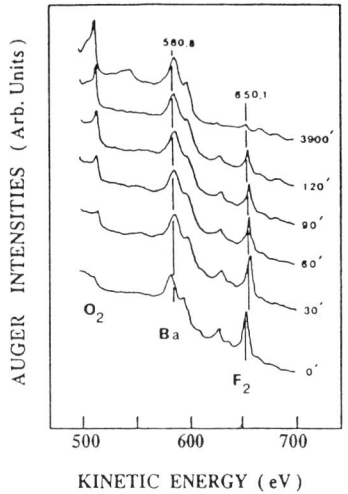

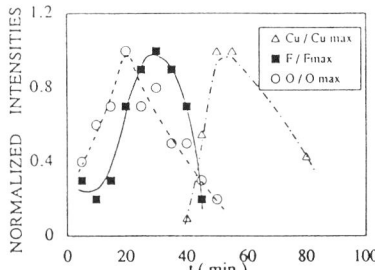

Figure 1: Time dependence of O, Ba, F AES spectra for Si / Y (30Å) / BaF$_2$ (200Å) multilayer with electron bombardment

Figure 2: Time dependence of the normalized intensities I(F) / I(Ba) both with and without electron bombardment respectively for Si / Y (30Å) / BaF2 (200Å) and YSZ / Y (30Å) / BaF$_2$ (320A) multilayers.

In order to obtain some information in relevance to the mechanisms involved in the time and irradiation dependences, diffusion of fluorine was studied on multilayer deposited in two different sequences on substrates like Si or YSZ , maintained at room temperature . Figure 3 gives an example of depth profiling for a sequence Cu(100 Å) / BaF$_2$(100 Å) / Y(50 Å) and figure 4 for a

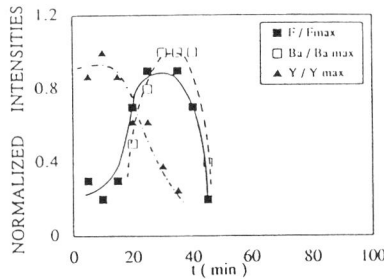

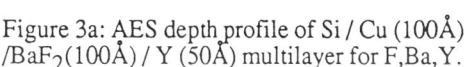

Figure 3a: AES depth profile of Si / Cu (100Å) /BaF$_2$(100Å) / Y (50Å) multilayer for F,Ba,Y.

Figure 3b : AES depth profile of Si/ Cu(100Å) / BaF$_2$(100Å) /Y (50Å) multilayer for F, O, Cu .

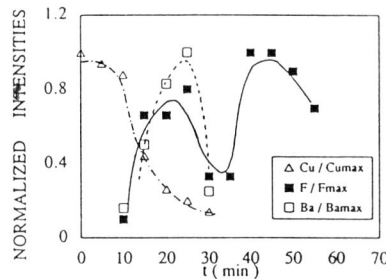

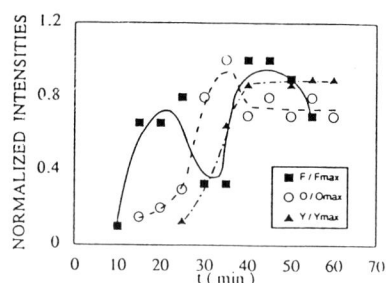

Figure 4a: AES depth profile of YSZ / Y
(100Å)/BaF$_2$(100Å) / Cu (50Å) multilayer
for F, Cu, Ba.

Figure 4b: AES depth profile of YSZ / Y
BaF$_2$(100Å) / Cu (50Å) multilayer for F,
O, Y.

sequence Y(100 Å) / BaF$_2$(100 Å) / Cu(50 Å) (Relative intensities for F$_{KLL}$,
Cu$_{LLM}$, O$_{KLL}$, Y$_{MNN}$, Ba$_{MNN}$ lines)

From these results one can point out the following observations:
- Fluorine diffusion in an Yttrium layer is greater than in Copper layers .
- Interdiffusion is important between Y and Ba layers, but not
  between Ba and Cu layers.
- A barrier of oxygen is present at the interface Y-Ba .
- The oxydation of Yttrium , which is significant at the Y-Ba interface, doesn't
prevent fluorine from diffusing.

CONCLUSION

Different , time dependences for AES spectra are observed both with and
without electron bombardment and such effects can lead to some systematic
discrepancy between measured elemental (relative) concentrations and actual
concentrations as deduced from depth profiling. As a result, current intensity
and exposition time of electronic bombardment must be limited to avoid
modifications of sample's properties .
By achieving Cu / BaF2 / Y and Y / BaF2 / Cu multilayers on Silicon and
YSZ substrates, AES depth profiling shows that the detailed evolution at the
surface of the multilayers and its dependence on sequential orders reflects the
selective diffusivity of fluorine through Yttrium's layer.
These observations should be related with (i) the increased stability of
YBCO thin films when BaF2 is used instead of metallic Baryum [6] (ii) the
great sensitivity of the quality of thin films to the order in which sequential
evaporation is realized [10].

REFERENCES

[1] G.Scott, K. Ninomiya, C.R Helms and I.Lindau, Surf. Sci., 221, 253 (1989)

[2] T.Hashemi, Z.T.Al-Dhhan, F.Golestani-Fard, C.A. Hogarth, J.Mat.Science, 23, 4513 (1988)

[3] M.G.Ramsey and F.P.Netzer, Mat. Sci. and Eng. B2, 269 (1989)

[4] S.Kishida, H. Tokutaka, S.Nakanishi, K.Nishimori, N.Ishihara and H.Fijimoto, Jpn.J.Appl. Phys., 28, L406 (1989)

[5] C. Perrin et M. Sergent, Nova sciences 6, (1990)

[6] P.M. Mankiewich, J.H Scofield, W.J. Skocpol, R.E Howard, A.H Dayem and E. Good, Appl. Phys. Lett. 51 (21) , (1987)

[7] S. Kao, S.Lee and K.Y.S Ng*, Solid State Comm. ,vol.72 n°5, 479-474 (1989)

[8] A. Tressaud, B. Chevalier, B. Lepine, K. Amine, L. Lozano, E. Marquestaut et J. Etourneau, Eur. J. Solid State Inorg. Chem 27, 309-325 (1990)

[9] C.G.Pantano and Theodore E. Madey, Appl. Surf. Sci. 7, 115-141 (1981)

[10] Nir Hess, Leandro R. Tessler, Uri Dai and Guy Deutscher , Appl. Phys. Lett., 53 (8), (1988).

High T$_c$ Superconductor Thin Films
L. Correra (Editor)
807

In situ characterization of YBaCuO thin films during rapid thermal annealing

F. Hosseini Teherani, E. Caristan, F. Carrié, T. Pech, J. Baixeras, A. Kreisler

Laboratoire de Génie Electrique des Universités Paris 6 et Paris 11
Unité de Recherche Associée 127 du CNRS, Ecole Supérieure d'Electricité,
Plateau de Moulon, 91192 GIF SUR YVETTE CEDEX- FRANCE

Abstract
Rapid Thermal Annealing (RTA) is rather unusual in the field of high Tc superconductor processing, in spite of beeing a classical procedure for semiconductor technologies. Nevertheless, when looking at the few results reported on YBaCuO thin films annealed by RTA, this thermal treatment appears as an attractive way, either to increase the efficiency of a diffusion barrier, when substrates such as silicon are used, or as a rapid step to improve YBaCuO grain growth after deposition. The first characterization of YBaCuO thin films during RTA is reported here, based on four probe resistivity measurements. The influence of two parameters are studied: i) the dwell temperature of the RTA cycle, ii) the gas nature during the process.

1. INTRODUCTION

Rapid Thermal Annealing (RTA) has been recently shown to be a potentially useful technique for the processing of thin dielectric layers[1], annealing of contacts and ion implantation damage [2], growth of thin chemical-vapour-deposited layers[3], and the modification of the electronic properties of surface layers[4]. Therefore RTA appears to be a very suitable process for the semiconductor device manufacturing[5].

Moreover, we have shown in previous papers[6] that the extreme rapidity of this process allows the significant advantage of minimizing the substrate-superconducting film interdiffusion. We have thus obtained superconducting YBa$_2$Cu$_3$O$_\delta$ (YBCO) thin films (of thickness 0.5 to 1 µm) on SiO$_2$/Si$_3$N$_4$/Si with a zero resistance at Tc$_{(R=0)}$=65 K. Furthermore, the RTA procedure enhances grain growth along the YBCO c-axis[7,8] (direction for which the superconducting properties of this coumpound are improved, due to its 2D structure). This second point is another important advantage, because the RTA of YBCO thin films on polycrystalline substrates produces highly c-axis oriented films with Tc$_{(R=0)}$≈85 K[7]. Another consequence of the particular growth mechanism during the RTA process, is the relatively high critical current density of spun wires (Jc=1100 A/cm^2 at 76 K) [9].

Up to date, many studies have been performed to characterize YBCO thin film crystallization during classical long thermal annealing[10], and more recently, during *in situ* deposition of YBCO[11]. These characterizations principally consist in RHEED, X-ray diffraction during the annealing and measurement of the film resistivity during elaboration. However, when considering the use of these last two characterizations during RTA, delicate assembling of apparatus must be set up, required by the rapidity of the transformations involved, and (in most cases) by the heat radiation.

In this work we focus on characterizing the RTA process of YBCO thin films performing *in situ* resistivity measurements. Amorphous films were obtained by rf diode sputtering at low temperature on polycristalline YSZ substrates[11] and the superconducting phase is obtained after RTA. We have studied the influence of two parameters during the RTA cycle: i) the dwell temperature, ii) the nature of the gas.

2. EXPERIMENTAL

The elaboration procedure of the YBCO amorphous films on polycrystalline YSZ slabs ($1.4 \times 3 \times 23$ mm^3) has been detailed elsewhere[12]. The films studied here were deposited at 375°C during the same run, under an argon atmosphere; their thickness was about 9000Å, and the composition for Y/Ba/Cu was 1/2.2/3.2.

The RTA furnace (ADAX, model R1000) made use of 12 high power incandescent lamps, allowing the sample (which was placed on a silicon substrate wafer inside a quartz enclosure) to reach 1000°C within 3 seconds. The RTA cycle studied here consisted of: i) 10s of heating up to T_{max}, ii) 60s dwell at Tmax, iii) cooling down within 3 minutes. We have carried out RTA cycles with different values of Tmax (940, 930, 900, 800, 700, 600, 500 and 400°C). The annealing cycles were performed under Ar or O$_2$ flowing gas at atmospheric pressure.

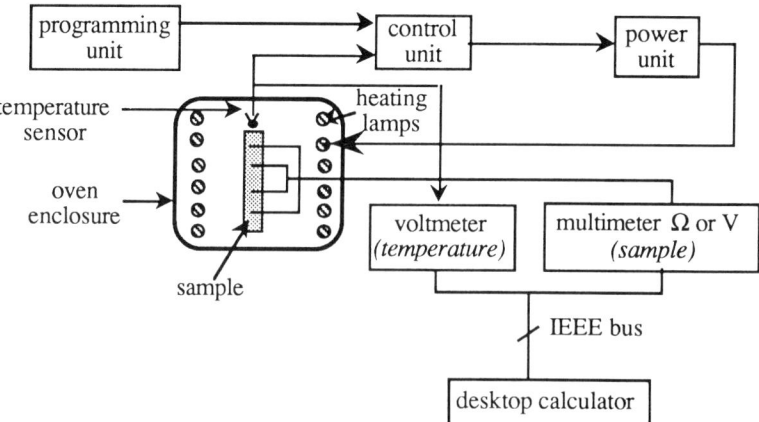

Figure 1. Schematic representation of the experimental system for the resistivity measurement during the RTA process of YBCO films.

Considering the heating process, the electrical contacts in view of *in situ* four-probe resistivity measurements had to be made as small as possible, so that no shadowing affected the radiation absorption by the material. Furthermore, the size of the potential probes had to be similar, so that the cooling of the material through the probes was symmetrical. By this way, we could neglect the thermocouple effects during the measurements. The contacts were realized by using (50 µm diam.) gold wires as probes directly thermal-bonded onto the amorphous YBCO film. This process for making contacts was successful only on polycrystalline substrates. Figure 1 shows the schematic representation of the system for the *in situ* four-probe resistivity measurements. Each measure have been done with 0.18s interval.

3. RESULTS AND DISCUSSION

First, we have performed several RTA cycles which only differed by the T_{max} value. During the heating and the dwell, samples were kept under flowing Ar. O_2 was introduced immediately after the plateau for the cooling. Figures 2 and 3 show respectively the temperature and time dependences of resistance during the annealing procedure for several values of T_{max}. On the figure 3 we have also reported a typical variation of the temperature with the time during the cycle.

For $T_{max} \leq 600°C$, the resistance variations are similar, revealing no notable crystallization. In fact, the films surfaces were shiny after RTA and no grains could be observed by optical microscopy.

For $600 < T_{max} < 930°C$, the resistance drops significantly when O_2 is introduced; this behaviour is attributed to the tetragonal-orthorhombic transition (T-O) i.e. semiconductor-metallic transition of the YBCO grains. These latter, appear during the dwell under Ar atmosphere, and grow in the tetragonal structure as the oxygen content is low under these conditions.

The room temperature resistance of the samples after the RTA process is reported on figure 4. It appears clearly that crystallization during this RTA cycle is significant only for $T_{max} \geq 700°C$. However, from optical microscopy observations, no crystalline structure was visible; moreover their low temperature resistance behaviour was semiconductor like. This can be explained by the presence of amorphous phase(s) embedding the crystalline phase.

When T_{max} was close to 930°C or higher, the resistance drops drastically during the dwell at T_{max}. The interesting point, in contrast to $T_{max} < 900°C$, is the increase of the resistance when oxygen is introduced. Moreover, after annealing, the films consisted of large grains such as needles. The low temperature resistance behaviour was consistant with a metallic-type conductivity. This unusual resistance behaviour during annealing can be principally attributed to a liquid-solid transformation. However, this point remains an open question, still several phenomena simultaneously interact, such as growth, T-O transition...

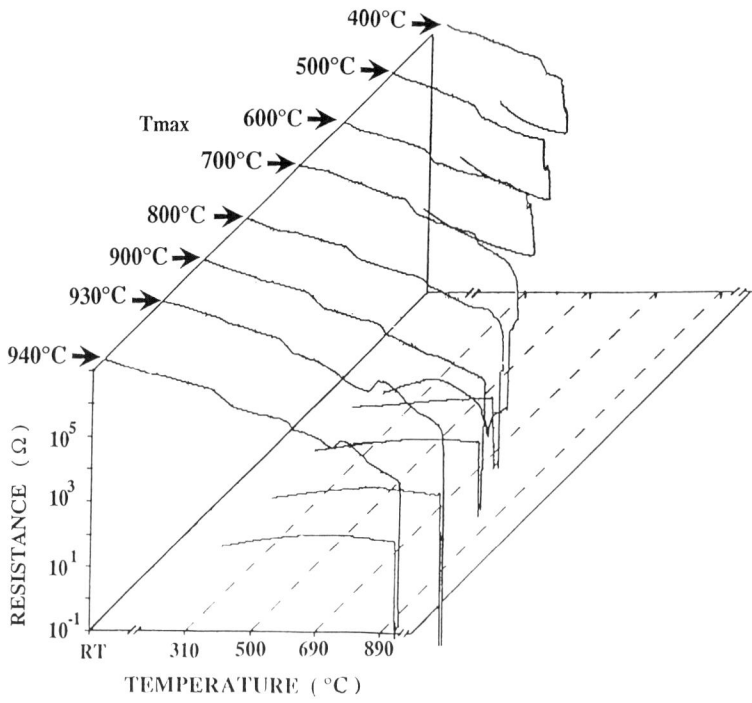

Figure 2 (up): Sample resistance versus annealing temperature for various plateau temperature.
Figure 3 (bottom): Sample resistance versus annealing time for various dwell temperature. The annealing cycle is shown at the bottom.

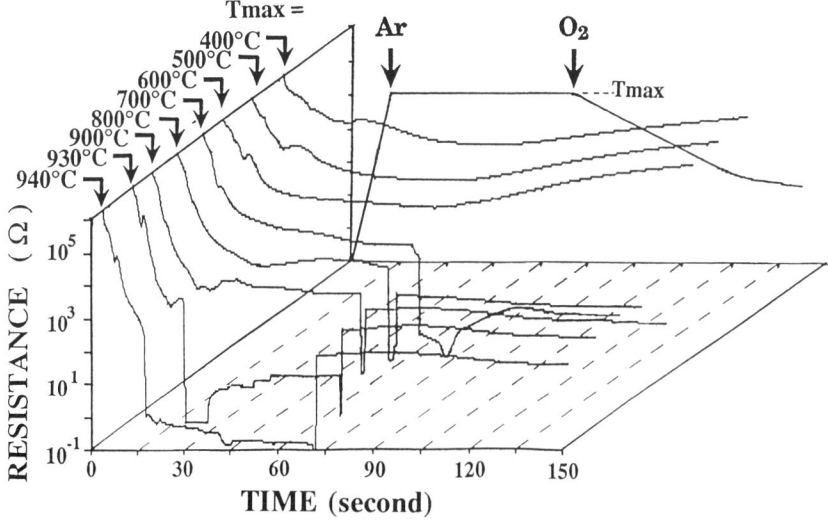

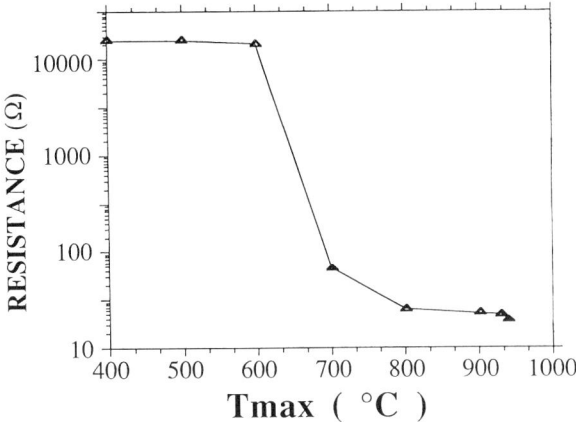

Figure 4: Room temperature resistance after annealing versus plateau temperature

The second parameter studied here is the influence of the gas nature during the annealing process. Figure 5 presents the variation of the resistance during the time associated to the RTA cycle, and also the variation of the temperature during the same time. Two films have been annealed at the same T_{max} (930°C): for sample (a), Ar was present until the end of the temperature plateau, whereas the whole cycle was performed under O_2 for sample (b).

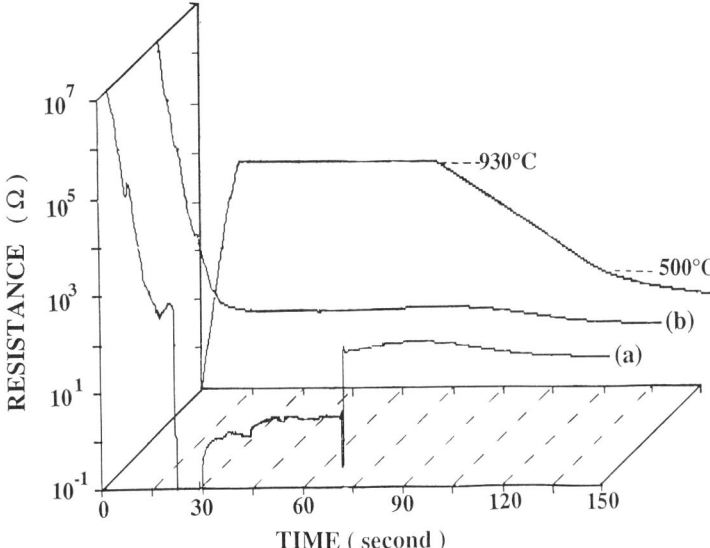

Figure 5: Sample resistance versus annealing time: (a) heating up and dwell under Ar, cooling down under oxygen, (b) whole cycle under oxygen.

Films obtained by this latter treatment consisted in small grains with no preferential orientation, whereas those annealed under Ar were textured along the YBCO c-axis, with large grains. The difference in resistance behaviour confirms that lowering the O_2 content during annealing, decrease the temperature of the phase transition. In quest of comparing the relative influences of the various phenomena which interact during annealing, X-ray diffraction and thermoelectric power measurement are now in progress.

4. CONCLUSION

We have presented the *in situ* characterization of YBCO thin films during the RTA process. The results show that the most important factor of an RTA cycle is presumably the value of Tmax during the dwell.

Although it has been reported[13] that the crystallization of the YBCO start at $\approx$550°C, and reaches a maximum at 750°C, both of these temperatures and also the T-O temperature depend on the heating rate[14] and on the O_2 content in the annealing atmosphere[15]. According to this, our results show that both T_{max} and the nature of the gas control the growth mechanism during the annealing process.

REFERENCES

(1) M.O. Thompson, S.T. Picraux, J.S. Williams, Mater. Res. Soc. Symp. Proc. (1987) 74

(2) M.J. Hart, A.G.R. Evans, Semicond. Sci. Technol., 3, (1988) 421

(3) S.K. Lee, Y.H. Ku, D.L. Kwong, Appl.Phys.Lett. 54, (1989) 1775

(4) K. Masri, J.P. Boryeaux, S.N.K. Umar, L. Mayet, G. Chaussenay, A. Laugier, Appl. Surf. Sci., 36, (1989) 554

(5) G.A. Paz de Araujo, Y.P. Huang, R. Gallegos, J. Electrochemical Society, Inc. (1988) 2035

(6) J. Baixeras, F. Hosseini Teherani, A. Kreisler, Proc. Int. Conf., SPIE, Vol 1362, "Physical Concepts of Materials for Novel Optoelectronic Device Applications II" edited by M. Razeghi (1990), 117

(7) F. Hosseini Teherani, J. Baixeras, A. Kreisler, Proc. Int. Conf., SPIE, Vol 1362, "Physical Concepts of Materials for Novel Optoelectronic Device Applications II" edited by M. Razeghi (1990) 921

(8) Y.L. Chen, J.V. Mantese, A.H. Hamdi, A.L. Micheli, J. Mater.Res. Vol. 4, (1989) 1065

(9) D.S. Gingeley, E.L. Venturini, J.F. Kwak, M.A. Mitchell, B. Morosin, R.J. Baughmen, J.W. Halloran, M.J. Neal, J. Appl. Phys. 67 (10), (1990), 6382

(10) D. Dubreuil, G. Garry, Y. Lemaître, L. Rogier, D. Dieumegard; J. Less-Common Metals, 151 (1989) 303

(11) K. Yamamoto, B.M. Lairson, C.B. Eom, R.H. Hammond, J.C. Bravman, T.H. Geballe, Appl.Phys.lett. 57 (18), (1990) 1936

(12) J. Baixeras, F. Carrié, F. Hosseini Teherani, A. Kreisler, J. Less-Common Metals, 164&165, (1990) 366

(13) A. Perrin, Le Vide-Les Couches Minces, Supl.n°241, 63 (1988)

(14) K.N. Tu, N.C. Yeh, S.I. Park, C.C. Tsuei, Phys. Rev. B, Vol. 39, N°1, (1989),304

(15) E.D. Specht, C.J. Sparks, A.G. Dhere, J. Brynestad, O.B. Cavin, D.M. Kroeger, H.A. Oye, Phys. Rev. B, Vol.37, N°13, (1988), 7426

High T$_c$ Superconductor Thin Films
L. Correra (Editor)
813

Changes in the Transport and Structural Properties of ReBaCuO HTSC Thin Films Induced by Ion Irradiation

T. Kroener, J. Geerk, G. Linker, O. Meyer, B. Strehlau, R. L. Wang [*]

Kernforschungszentrum Karlsruhe, Institut für Nukleare Festkörperphysik,
P. O. Box 3640, W-7500 Karlsruhe, Germany

Abstract

Thin single crystalline REBaCuO (RE = Y, Gd, Eu) HTSC films (a-axis and c-axis grown) were irradiated with 300 keV p and He$^+$ ions at RT. The irradiation lead to T$_c$-degradation depending on initial sample quality and irradiation conditions. A linear increase of the a- and c-axis lattice parameters with fluence was observed by x-ray diffraction measurements, revealing an orthorhombic - tetragonal phase transition reaching the tetragonal values at a fluence of 6×10^{16} p/cm^2. The irradiated samples are strained. The strain shows a linear increase with fluence with an intermediate release at a dose coinciding with the onset of the amorphization process.

1. Introduction

The superconducting state of the high-T$_c$ oxide superconductors was shown to be very sensitive to ion-irradiation (e. g. 1-4); this effect can be used to change the transport and structural properties in a controllable way. A good knowledge of such changes is needed for applications, e. g. for patterning of bridges or use in a radiation environment. The degradation of the superconducting properties like T$_c$ and the structural changes at comparatively low fluences are thought to be connected with the oxygen order in the perovskite - like structures. The changes depend on many characteristics of the unirradiated sample, e. g., growth quality and direction, transport properties, but also on the ions used. Especially the a-axis films now available allow an extension of defect studies in the superconducting oxides, because investigations of the structural changes in the CuO - chains became possible.

Our irradiation experiments are now focused on the structural phase transformation accompanying the metal-semiconductor transition. In this contribution we first report about the low dose irradiation experiments and then discuss the amorphization process occuring at higher doses.

[*] on leave from Chinese Academy of Science, Beijing, P. R. China

2. Experimental details

Thin single crystalline REBaCuO (RE = Y, Gd, Eu) films (c-axis or a-axis grown) were produced by dc sputtering in an inverted cylindrical magnetron (5) on $SrTiO_3$ - substrates. Details of the preparation processes for the different growth directions are described elsewhere (5, 6). Irradiations were performed with 300 keV p and He^+ ions at RT in a vacuum better than 10^{-6} mbar at low beam current to avoid sample heating and recovery of the films. The thicknesses of the films were determined by Rutherford Backscattering Spectrometry (RBS). Irradiations were performed such that the particles penetrated the films and came to rest in the substrates. X-ray diffraction in the Bragg-Brentano geometry and ω-scans were performed to investigate the structural changes using a Cu rotating anode generator with $CuK\alpha_1$ radiation ($\lambda = 1.5405$ Å). Line positions, their intensities and widths were analyzed for determining the lattice parameters, strain, and amorphized fractions of the films.

3. Results and discussion

In fig. 1 the transition temperatures, T_c^{mid}, and the transition widths, ΔT_c, after proton irradiations for films grown in different directions and irradiated under two different angles of incidence are summarized. T_c^{mid} was determined by the temperature at which the resistance is half the value of the onset of superconductivity and ΔT_c by the temperatures at which 10 % and 90 % of the value at the onset was observed.

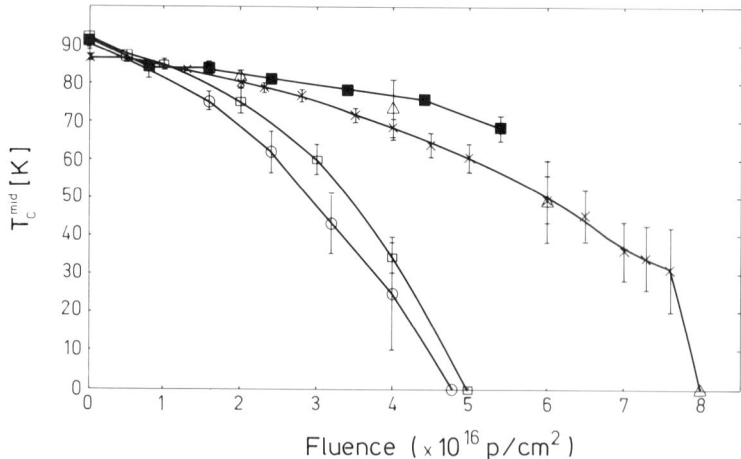

Fig. 1   Transition temperatures, T_c^{mid}, and transition widths (error bars), ΔT_c, for different proton irradiation experiments:
Irradiations under 0^0: $\Delta, \times,$ ■ YBaCuO-films (c-axis grown)
Irradiations under 15^0: □ GdBaCuO-film (c-axis grown)
                              o EuBaCuO-film (a-axis grown)

The results obtained can be seperated into two groups, depending on the angle of incidence of the irradiating ions, which was either 0 or 15 degrees with respect to the substrate normal. The c-axis grown YBaCuO-films irradiated under zero degree are the most unaffected samples due to channeling. The c-axis grown GdBaCuO-film and the a-axis grown EuBaCuO - film were irradiated under 15 degrees. In comparison to the irradiations performed under zero degree, T_c^{mid} is much stronger affected and ΔT_c increased. The a-axis grown film shows broader transition widths after irradiation than the c-axis films. This may be caused by the influence of the b-c twinning present in the a-axis films (6).

The lattice parameters a and c vs. fluence for the GdBaCuO sample mentioned above are displayed in fig. 2. The values increase nearly with the same slope, revealing a change at a fluence of 40×10^{15} p/cm^2. For the EuBaCuO film the values also increase linearly with a change in the slope at a fluence of 24×10^{15} p/cm^2. This fluence corresponds to an oxygen deficiency δ of 0.5 (7), as determined from the lattice parameters. The values of the tetragonal 123-phase (7) are reached at a fluence of 6×10^{16} p/cm^2. It is interesting to note that the mosaic spread deduced from ω-scans through the (005)-plane and the (200)-plane, respectively, shows an increase with fluence and then a decrease reaching the values of the unirradiated samples at the fluence of 6×10^{16} p/cm^2, when the orthorhombic-tetragonal phase transition is completed. This may be caused by a rearrangement of oxygen atoms and vacancies at this fluence, leading to a good alignment of the lattice planes.

Considering the elastic strain (increase in percent of the lattice parameters) in the samples, we found that for all fluences and both growth directions ([100], [001]) the samples are more strained in a-direction. If we calculate the stress ($\sigma_x = C_{xx} e_x$) using the elastic constants from ref. 8 ($C_{11} = 373$ GPa, $C_{33} = 185$ GPa), we obtain double stress values in a - direction than in c-direction.

Several effects occuring during irradiation of the a-axis grown EuBaCuO film can be seen in fig. 3, where the Theta-scan range from 50 to 54 degrees is shown for several fluences.

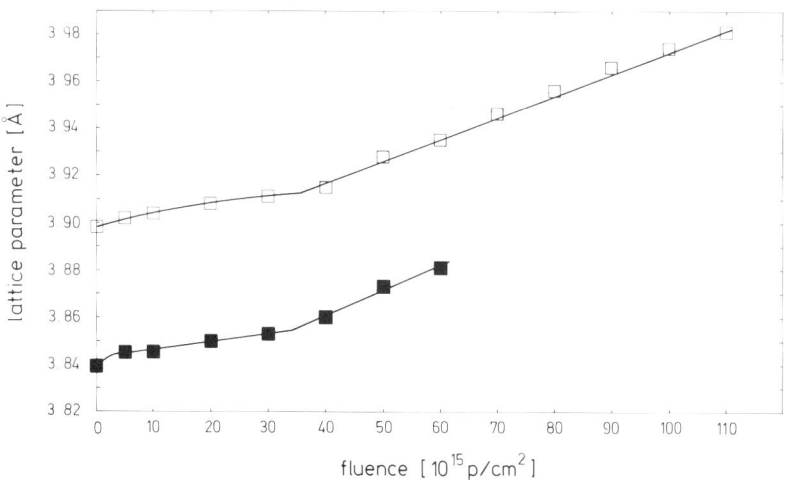

Fig. 2   Lattice parameters a (■) and c/3 (□) vs. fluence for the c-axis grown GdBaCuO-film.

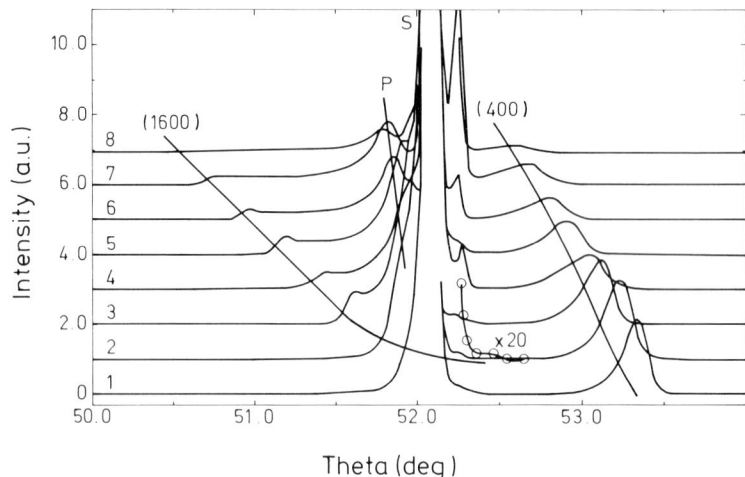

Fig. 3   Section of x-ray spectrum showing the (400)-peak of the $EuBa_2Cu_3O_7$ -
            phase, the (1600)-peak of the superstructure, the substrate peak S, and
            the perovskite peak P for different doses:
            1: unirradiated   2: 8×10^{15} p/cm²   3: 16×10^{15} p/cm²
            4: 24×10^{15} p/cm²   5: 32×10^{15} p/cm²   6: 40×10^{15} p/cm²
            7: 48×10^{15} p/cm²   8: 56×10^{15} p/cm²

First, the (400)-peak on the right side of the figure shifts to lower angles,
resulting in an increase of the lattice parameter a as already discussed before.
This sample contained small amounts of c-axis grown grains, as concluded from
the presence of weak (00l)-peaks in the x-ray diagram. The intensities of the (h00)-
peaks decreased much faster than those of the (00l)-peaks; for example, if one
compares the integrated intensities after an irradiation with 4×10^{16} p/cm² to
those of the unirradiated sample, the ratio for the (200)-peak becomes 0.36, but for
the (005)-peak it was only 0.85. This effect may be explained by a change in the
structure factor caused by Ba-relaxations towards the Y-atom or the CuO-chains
initiated by the O-redistributions. The amount of the Ba - relaxations is
comparable to the change of the Ba-positions during the orthogonal-tetragonal
phase transition (9).

Second, starting at a fluence of 8×10^{15} p/cm² we observed a peak of the
$Re_1Ba_2Cu_3O_{6.875}$ superstructure detected also in bulk samples (e. g. 10, 11). The
oxygen deficiency $\delta = 0.125$ agrees well with the value $\delta = 0.13$ calculated from
the increased a-axis lattice parameter (7) at this fluence. In this superstructure
with a $2\sqrt{2} a_c \times 2\sqrt{2} a_c \times 3 a_c$ ($a_c \approx 3.8$ Å) unit cell every fourth O(4) atom from
every Cu(1) O(4) - chain is removed, i. e. diffused out of the sample or moved
(periodically) to an initially vacant site (e. g. O(5) - site), leading to a periodicity of
$4 \times a$ along the a-axis. The peak also shifted resulting in periodicities ranging from
$4.04 \times a$ at a fluence of 8×10^{15} p/cm² to $4.11 \times a$ at a fluence of 48×10^{15} p/cm².
The deviation from the nominal value $4 \times a$ may also be caused by cation
displacements or by extra rows of ordered and disordered oxygen vacancies, which
would also explain the weakness of these peaks. As Alaro-Franco (10, 11)
observed, this vacancy ordering is not correlated along the c-direction, which
explains that only in the a-axis grown EuBaCuO-film this superstructure could be
observed.

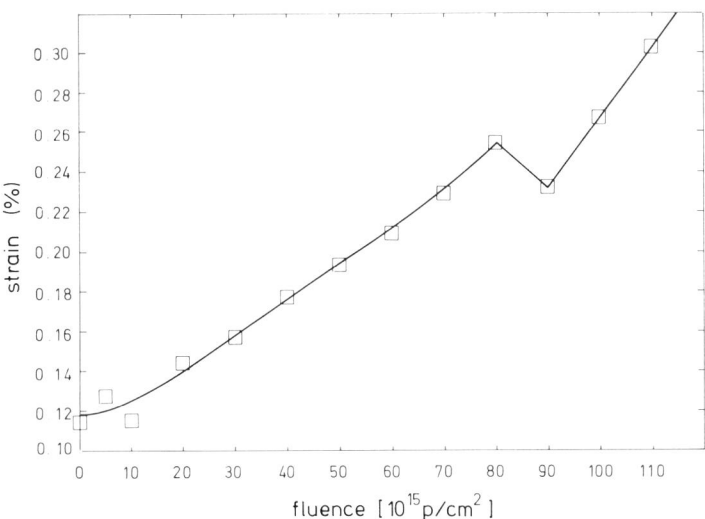

Fig. 4   Strain vs. fluence for the GdBaCuO-film showing a release at 9×10^{16} p/cm^2 coinciding with the beginning of the amorphization process.

Third, a splitting of the SrTiO$_3$-substrate peak induced by irradiation can be seen. We controlled this effect by irradiating a pure substrate with protons at different fluences. The substrate transformed partly into another perovskite starting at a fluence of 24×10^{15} p/cm^2.

Strain analyses were performed based on a method described by Klug and Alexander (12). This method relies on line broadening as a measure of strain. The results for the c-axis grown GdBaCuO sample are shown in fig. 4, where the calculated strain vs. fluence is shown after several doses. Already the unirradi-

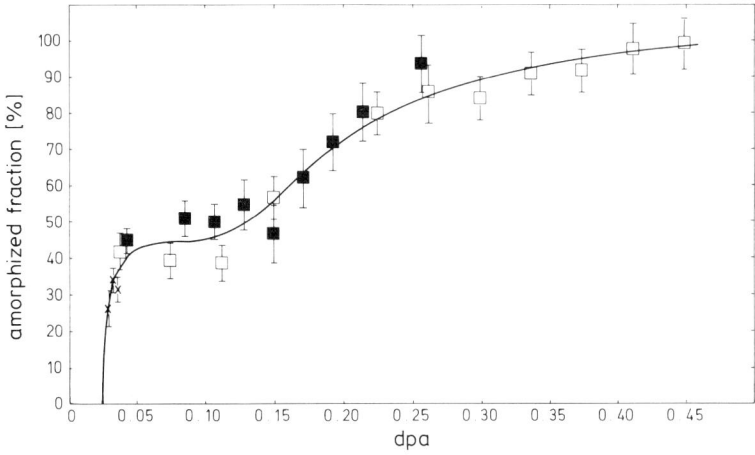

Fig.5   Amorphized fraction vs. dpa for different irradiation experiments:
□, ■ 300 keV He$^+$ → YBaCuO
×   300 ke V p → GdBaCuO

ated films on the SrTiO$_3$-substrate are strained. This may be caused by the imperfect lattice matching between the ReBaCuO-films and the substrates.

The strain shows an almost linear increase with an intermediate release at a fluence of 9×10^{16} p/cm^2. It is interesting to note that this fluence coincides with the beginning of the amorphization process which we discuss below. The a-axis films show nearly the same behaviour, i. e. they are already strained before irradition and the strain increases with increasing fluences.

As an example for the amorphization process in the superconducting oxides, we show in fig. 5 an amorphization curve (amorphized fraction vs. displacement per atom (dpa)) deduced from x-ray intensity measurements of ReBaCuO samples after p and He$^+$ ion irradiations. The transformation sets in rather abruptly at dpa $= 0.025$ (corresponding to 9×10^{16} p/cm^2), where the strain released. This indicates an accumulation type of amorphization (13), implying the existence of long range forces driving the transformation. For higher dpa´s we detected a plateau at an amorphized fraction of about 40 % which is not yet understood. The amorphization process is completed for dpa $= 0.35$.

In conclusion, the degradation of the superconducting properties induced by ion irradiation is influenced by the channeling effect. X-ray measurements revealed a linear increase of the a- and c-axis lattice parameters with a change in the slope where the values corresponding to $\delta = 0.5$ were reached. From this increase we deduced a highly anisotropic stress in the samples. The formation of the Re$_1$Ba$_2$Cu$_3$O$_{6.875}$ superstructure in an a-axis film after proton irradiation was observed. So we conclude that ion irradiation induces a reordering of oxygen atoms and vacancies. The strain in the films shows a linear increase with doses, revealing a small decrease coinciding with the onset of the amorphization process.

References:

(1)   O. Meyer et al., Mat. Res. Soc. Proc. Vol. 157 (1990), p. 493
(2)   G. Linker et al., Nucl. Instr. Meth., 1991 accepted
(3)   S. Vadlamannati, P. England, N. G. Stoffel, A. Findikoglu, Q. Li, T. Venkatesan, W. L. Mc Lean, Phys. Rev. B 43 (1991) 5290
(4)   B. Hensel, B. Roas, S. Henke, R. Hopfengärtner, M. Lippert, J. P. Ströbel, M. Vildic, G. Saemann-Ischenko, S. Klaumünzer, Phys. Rev. B 42 (1990) 4318
(5)   J. Geerk, G. Linker, O. Meyer, Mat. Sci. Rep. (1989), 193
(6)   G. Linker et al., this proceedings
(7)   M. Buchgeister, W. Hiller, S. M. Hosseini, K. Kopitzki, D. Wagener, Proceedings of the Int. Conference on Transport Properties of Superconductors, World Scientific Publishing, Singapore, 1990
(8)   R. C. Baetzold, Phys. Rev. B 38 (1988) 11304
(9)   J. D. Jorgensen, B. W. Veal, A. P. Paulikas, L. J. Nowicki, G. W. Crabtree, H. Claus, W. K. Kwok, Phys. Rev. B 41 (1990) 1863
(10)   M. A. Alario-Franco, C. Chaillout, J. J. Caponi, J. Chenavas, M. Marezio, Physica C 156 (1988) 455
(11)   M. A. Alario-Franco, C. Chaillout, J. J. Caponi, J. Chenavas, Mat. Res. Bull. Vol. 22 (1987) 1685
(12)   H. P. Klug, L. E. Alexander, X-Ray Diffraction Procedures for Polycrystalline and Amorphous Materials, chapter 9, Wiley, New York (1974)
(13)   G. Linker, Mat. Sci. Eng. 69 (1985) 105

High T$_c$ Superconductor Thin Films
L. Correra (Editor)
819

In-situ spectroscopic analysis for pulsed laser deposition of superconductive Y-Ba-Cu-O thin films

G. Masciarelli, F. Fuso, A. Iembo, M. Allegrini and E. Arimondo

Dipartimento di Fisica, Università di Pisa, Piazza Torricelli 2, I-56100 Pisa, Italy

Abstract

The deposition process of superconductive thin films by laser evaporation of YBaCuO stoichiometric pellets is investigated in-situ through the spectral and temporal analysis of radiative emission from the plasma plume. Oxygen flux and target-substrate position are confirmed to be critical parameters for the occurrence of superconductive behavior.

1. INTRODUCTION

Deposition of high-T$_C$ superconductive materials by laser evaporation of a target of superconductive material is a process technologically very important. Following the early observations [1], this process has allowed the production of thin films with large critical current in conditions of well controlled reproducibility [2]. While the applied side of thin film production is now well established, knowledge of physical phenomena involved in the different steps of the process itself, i.e. evaporation, material transport with exchange of temperature, substrate deposition and annealing, is incomplete [3,4]. This situation is typical of processes technologically new with important applications and stimulates further investigations of the basic physics involved in the process.

It has been well established that the final result of the laser deposition process depends strongly on the composition of the target, on the parameters of laser irradiation and on the parameters governing the propagation of the created plasma from the target toward the substrate of deposition. Establishment of the conditions required for the film deposition has been determined by the electrical and structural characteristics of the final superconductive film. On the contrary, investigation of the physical processes taking place in the evaporation and deposition can be obtained only through the application of techniques belonging to atomic and molecular physics. Analysis based on detection of atomic and molecular fluorescence have been already performed to derive different information about the deposition process [5,6]; we have extended this technique and in this work we report for the first time combined observations of time and space resolved fluorescence.

2. EXPERIMENT

Our investigation, starting from the realization of superconductive thin films, is aimed to a diagnostics in situ of evaporation and deposition processes in order to determine the essential parameters and to improve the quality of deposited products. By monitoring the plasma created in the laser evaporation, we are able to follow space-temporal evolution of the produced species and to analyze their collisional reaction mechanisms.

The status of the present research has involved construction of the deposition chamber and the set-up of the diagnostics in situ. The apparatus used for deposition of superconductive thin films and for detection of the fluorescence emitted by photo-ablated species is schematically described in Fig. 1. Radiation from an excimer XeCl laser at 308 nm wavelength is focused on a YBaCuO superconductive target at $1.5 \div 2$ J/cm^2 intensity . As important requirements for realization of good quality superconductive thin films, target rotation during laser irradiation, heating to 750 ^{o}C of the MgO substrate, heated oxygen flux on the substrate and thermal cycles of the film at higher oxygen pressure in the post deposition stage are needed.

To detect the fluorescence emitted by atomic and molecular species produced in the photo-evaporation, an optical system provides an image of the plasma region, and by scanning an optical fibre, different area of the plasma

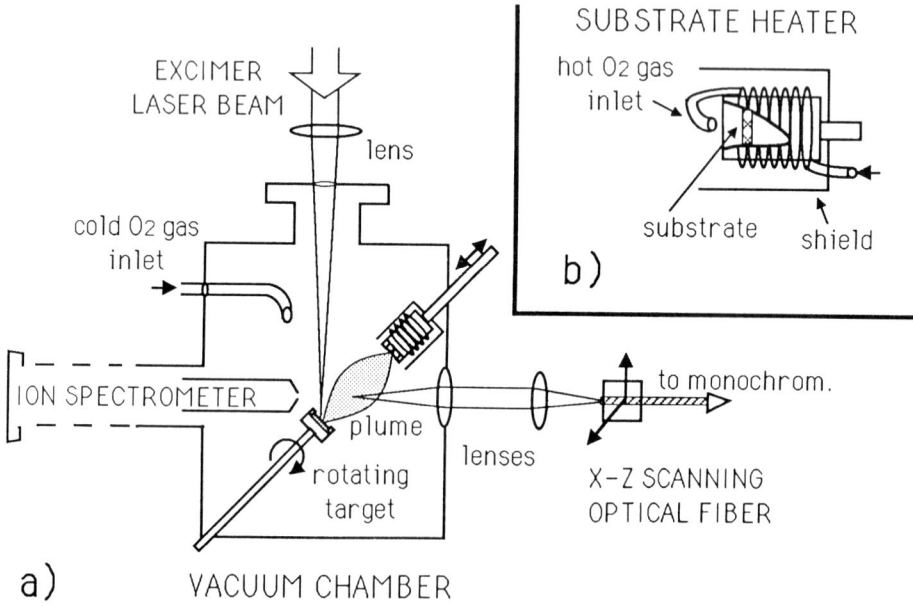

Figure 1. Experimental set-up (a) and details of substrate heater (b).

are focused onto the entrance slit of a monochromator. Applying a boxcar detection, at a given point of observation the time evolution of the emitted fluorescence is monitored. Installation of a mass spectrometer in the deposition chamber to detect ions produced in the evaporation step, is in progress, thanks to a collaboration with the Department of Physics of the University of Naples.

3. RESULTS

The deposition chamber has been used to prepare good quality superconductive thin films. Superconductive behavior is checked by measuring the resistance with a standard four-probe technique (nominal I_{short}=100µA), while cooling down the films using a liquid nitrogen cryogenic system (limit temperature ≈80K) and monitoring the temperature by a Pt calibrated resistance. As shown in Fig. 2, which reports the different results

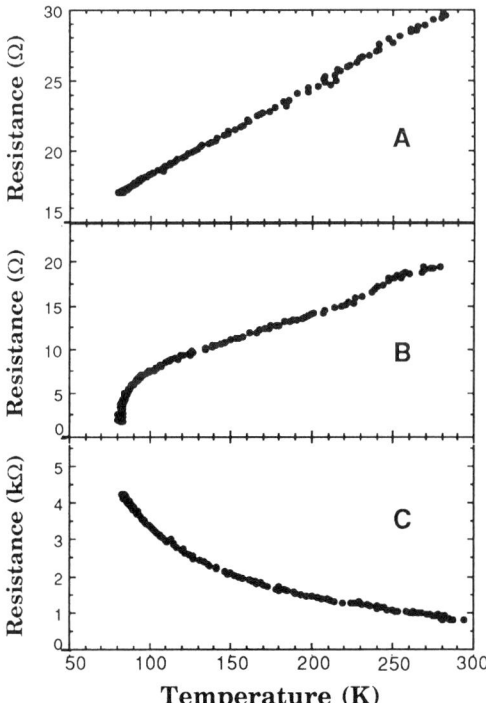

Figure 2. Four probe resistance vs. temperature measurements for three thin films obtained in the following conditions: A. cold oxygen in the chamber, substrate distance 38mm; B. hot oxygen flux on the substrate, substrate distance 38mm; C. hot oxygen flux on the substrate, substrate distance 32mm.

obtained for three deposition samples, we have observed a strong dependence of the superconductive behavior on experimental parameters like oxygen flux in the deposition chamber and target-substrate distance.

In order to correlate the role of those parameters with plume evolution, we have detected fluorescence emitted from Y, Ba, Cu and their oxides ablated from the target under different conditions of vacuum in the chamber and for different distances from the target. Then we have compared these spectra to those produced by laser evaporation of the single species to identify atomic species from their fluorescence lines. As an example, Fig. 3 reports spectra obtained for two different positions of the optical fiber. In the total light collected from regions far away from the target surface, the background emission becomes weaker, while the intensity ratio of atomic and molecular lines remains nearly the same. Electronic temperatures of the different species in the plasma were obtained from observed spectra in order to test Local Thermodynamic Equilibrium. Fig. 4 shows the results for Ba^+ ions fluorescence emission: experimental data are scattered around best fit straight line, indicating that probably the conditions assuring validity to LTE approximation are not completely fulfilled.

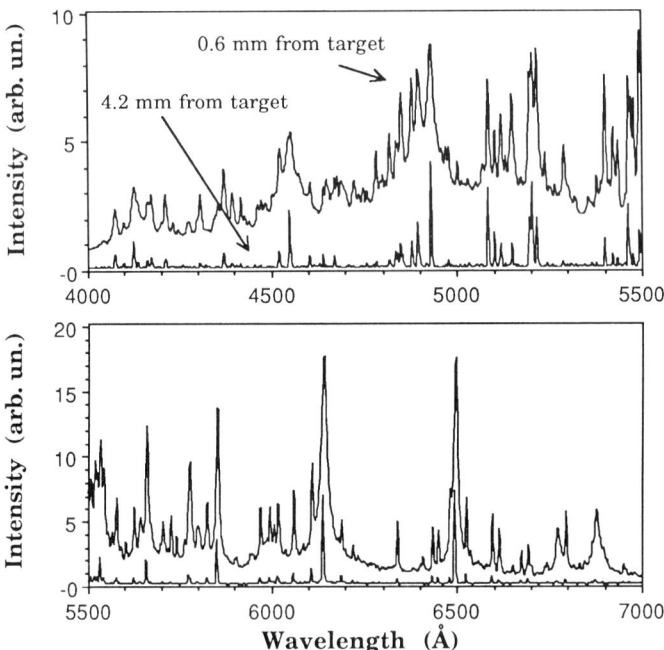

Figure 3. Fluorescence spectra in the range 4000÷7000 Å obtained for two optical fiber positions (oxygen pressure in the chamber 0.2mbar).

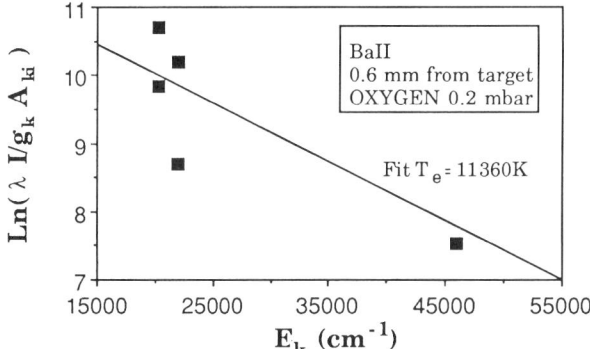

Figure 4. Results obtained applying LTE approximation to experimental data corresponding to BaII fluorescence lines: x-axis reports the energy of the considered k upper level, y-axis the $\mathrm{Ln}(\lambda\ I/g_k\ A_{ki})$ value , where λ is the transition wavelength, I the measured fluorescence intensity (in arbitrary units) , g_k the degeneracy of k level and A_{ki} the k->i transition spontaneous emission rate. The straight line results from a best fit of the data and corresponds to an electronic temperature $\approx 10^4 K$.

Optical time of flight (OTF) measurements give information about plume evolution. As an example, Fig. 5 reports time resolved fluorescence signal for Y^+ ions emitting 5663Å light at a distance 6mm from target. A best fit procedure was applied to experimental data according to the velocity distribution function $f(v) = v^3\ \exp(-m(v-v_s)^2/2kT_t)$, where stream velocity v_s

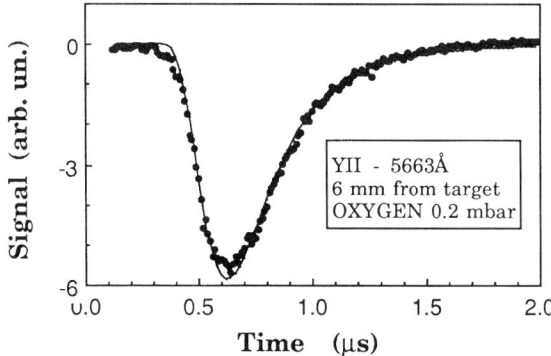

Figure 5. Time resolved fluorescence signal (dots) along with fitting curve (solid line) for YII 5663Å fluorescence line.

and translational temperature T_t are fit parameters. From OTF measurements on several atoms and ions at different distances from target, it results that both the translational temperature and the stream velocity evolve, with increasing target-substrate distances, toward a value common for the different species (see Fig. 6). This seems to be an important condition because depositions made with target-substrate distances outside the range where this effect holds did not show the superconductive transition. We have also experimental evidence that the behavior is enhanced by the oxygen presence.

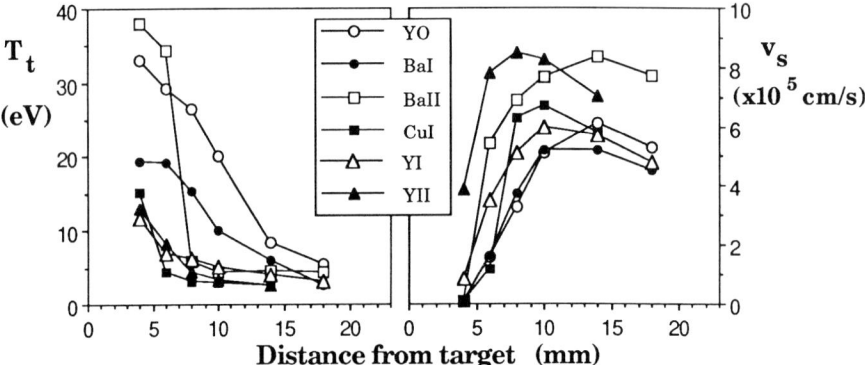

Figure 6. Results of best fit procedure for T_t and v_s (oxygen pressure in the chamber 0.2mbar), corresponding to different species, as a function of distance from target.

5. ACKNOWLEDGMENTS

We wish to thank L. Correra and his colleagues at LAMEL for their encouragement and invaluable advises, A. Diligenti and his collaborators for several discussions and generous disposal of their facilities, R. Cagnolati and P.A. Rolla for the loan of their resistance measurement apparatus. This work is supported by CNR through Comitato delle Scienze Fisiche e Comitato Ricerche Tecnologiche ed Innovazione.

6. REFERENCES

1. D. Dijkkamp et al., Appl. Phys. Lett. 51 (1987) 619
2. X.D. Wu et al., Appl. Phys. Lett. 57 (1990) 523
3. P.E. Dyer, A. Issa and P.H. Key, Appl. Surf. Sci. 46 (1990) 89
4. R.K. Singh and J. Narayan, Phys. Rev. 41B (1990) 8843
5. J.P. Zheng et al., Appl. Phys. Lett. 54 (1989) 954
6. A. Hoffman et al., Supercond. Sci. Technol. 3 (1990) 395

High T$_c$ Superconductor Thin Films
L. Correra (Editor)

MICROSTRUCTURAL FEATURES OF Y-Ba-Cu-O THIN FILMS GROWN ON SINGLE CRYSTAL MgO

M. V. Sidorov and S. R. Oktyabrsky

P. N. Lebedev Physical Institute, Department of Solid State Physics,
53 Leninsky prospect, 117924 Moscow, USSR

Abstract

The microstructure of superconducting Y-Ba-Cu-O thin films grown *in-situ* by magnetron sputtering on mechanically polished (001)MgO has been studied by transmission electron microscopy. The main microstructural feature of the films investigated is a number of grain boundaries formed by 90° misoriented epitaxial domains. The films also contain mirror twins, planar stacking defects and c$_0$ lattice parameter variations. Domains with *a*- and *c*-normal orientations have rather different microstructure owing to differences in growth mechanisms.

1. INTRODUCTION

A single crystalline MgO is a one of the most popular substrate materials for YBa2Cu3O7-x (YBCO) films deposition. In spite of the fact that (001)MgO is not closely matched with YBCO it is possible to obtain a high quality thin films with excellent superconductive properties which are comparable with the films grown on (001) SrTiO3 (T$_c$ up to 89K, J$_c$ at 77K up to 10^7A/cm^2) [1-9]. At the same time MgO is much more unexpensive and suitable for real application than SrTiO3. However, due to large lattice mismatch (8-9%) the microstructure (and therefore properties) of YBCO-on-MgO films is greatly influenced by growth conditions (growth temperature and rate, supersaturation, stoichiometry, oxygen pressure, substrate quality, deposition technique etc.).

In-situ grown films is known as not structurally ideal; they can contain twin and grain boundaries, stacking faults, c$_0$ variations, dislocations, local compositional deviations, extraneous phases and point defects.

In this study the typical microstructural defects of YBCO superconducting thin films grown on MgO are investigated by convenient (TEM) and high resolution (HREM) transmission electron microscopy. A particular attention is paid in microstructure of the epitaxial domains with *c*- and *a*- normal orientations which are typical to the films grown on MgO.

2. EXPERIMENTAL

The films were prepared by *in-situ* dc magnetron sputtering in the oxygen-argon mixture (at a pressure 0.2-1.0 Torr) from YBCO ceramic pellets on the mechanically polished (001)MgO substrates at relatively low substrate-holder temperature (T$_s$=650-700°C). Normally, the deposited films have a thickness of about 400 nm and exhibit a zero-resistance at 86-87K and J$_c$>10^5A/cm (77K). A detailed preparation techniques are described elsewhere [1].

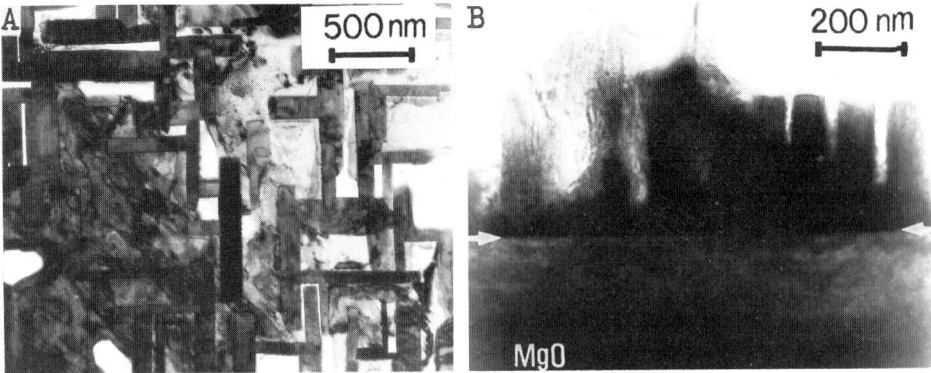

Figure 1. Typical plan-view (a) and cross-sectional (b) TEM images. Note "basket weave" texture (a) as well as "comb-like" shape of the real surface of the film (b) due to a,b-oriented domains existence. YBCO/MgO interface marked by arrows (b).

Plane and cross-sectional specimens for TEM were prepared by the standard method without using a water with ion-milling at the final stage of preparation. A PHILIPS CM30 electron microscope operated at 300 kV was used in this study.

3. RESULTS AND DISCUSSION

3.1. Epitaxy

According to our numerous TEM- observations, YBCO thin films grown in our preparational conditions commonly display a mixed a,b,c- orientation. The films appear to consist of three main types of epitaxial single crystalline domains (a-, b- and c-oriented domains) with their c-axis parallel to one of three equivalent orthogonal <001> directions of MgO respectively and thus $90°$ misoriented from each other. As one can see in TEM plan-view images (fig. 1a) the films domain structure occurs as bar-shaped YBCO-grains of approximately ~800×100 nm size (depending on growth condition) with c-axis parallel to the substrate surface (a- and b-domains) immersed in YBCO matrix with c-axis perpendicular to the substrate (c-domains). Electron microscopy cannot allow to distinguish [001] from [010] directions of YBCO due to small difference in lattice parameters and further we shall not draw a line between a- and b-domains (hereafter refer to as a,b-domains).

On the contrary to [2] very seldom grains with a- and b-axis parallel to <110> directions of MgO exist in our films. Therefore only so-called cube-on-cube epitaxial relationships has been observed without significant in-plane rotation. These relationships are:
1. $(001)_{ybco}//(001)_{mgo}$; (100) or $(010)_{ybco}//(010)_{mgo}$
2. (100) or $(010)_{ybco}//(001)_{mgo}$; $(001)_{ybco}//(100)$ or $(010)_{mgo}$

In a cross-sectional TEM-images the a,b-domains appear as a huge pillars spreading from the bottom to the top of the film. The c-domains look like a flat slabs which have a smaller thicknesses than a,b-domains have (fig. 1b). The real surface of the films thus is not smooth resulting from the different thicknesses of domains with different orientations.

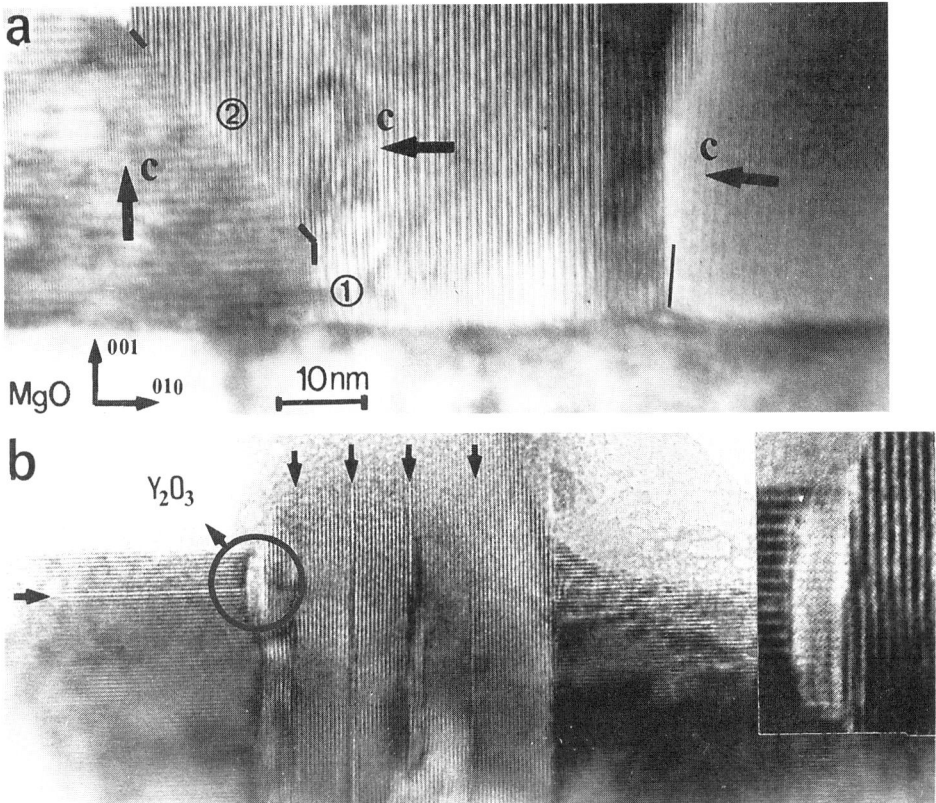

Figure 2. Cross-sectional HREM images.
(a) Lower part of the film. Three domains are visible but one of them has c-axis slightly deviated from normal orientation. 1)- and 2)-type grain boundaries are visible.
(b) Upper part of the film. Two grain boundaries of 3)-type are visible. Note the existence of the small inclusion of Y2O3 at very film surface (inset).

This films morphology results from growth mechanism peculiarities such as growth rate anisotropy of the YBCO grains (usually growth rate along c-direction is about 5-10 times slower than growth rate along perpendicular directions depending on Ts during deposition) and large lattice mismatch between YBCO and MgO (8-9%). Growth mechanism of such films is described in details elsewhere [10].

It is important to note that a,b- and c-domains have some differences in their growth mechanism. Growth of the a,b-domains occurs everywhere along the growth interface while growth of the c-domains is limited by availability of the "right" cationic species [11]. Moreover, oxygen diffusion along c-axis is known to be ~10^6 times slower than in a,b-planes. This phenomena leads to some differences in microstructure of a,b- and

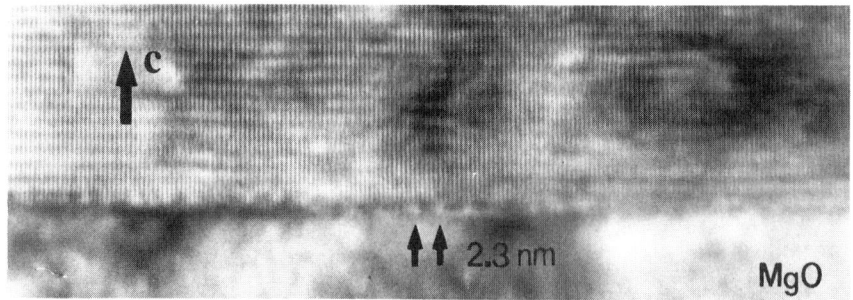

Figure 3 Cross-sectional HREM image of the film/substrate interface. Note periodical contrast variation due to misfit dislocations existence.

c-domains. Furthermore, as we have shown earlier by X-ray microanalysis [12] the a,b-domains usually slightly Cu-rich.

3.2. Typical Defects
The typical microstructural defects in the films investigated are the grain boundaries formed by 90° misoriented domains, misfit dislocations, planar stacking defects and c₀-parameters variations.

In addition, some of the well-known (110) mirror-twins of 30-50 nm width exist in c-domains. Such twins are formed to accommodate the strain energy of the tetragonal-to-orthorhombic transition during cooldown procedure.

3.2.1. Grain boundaries
The main microstructural feature of the films investigated is a large number of grain boundaries formed by 90° misoriented domains. Cross-sectional observations allow us to distinguish the three types of such boundaries which planes are : 1)- perpendicular (fig.2a), 2)- inclined by ~45°-75° to the substrate surface (fig.2a), and 3)- perpendicular but far away from substrate (fig.2b).
At initial stage of the growth process the three-dimensional nuclei of both a,b- and c- alternative orientations are to be formed with quite equal probability which then grow anisotropically (c-nuclei grow rapidly along substrate surface while a,b-nuclei grow rapidly in perpendicular direction). When two orthogonal domains meet and consequently coalescence they form the 1)-type boundary. The height of this boundary increases with increasing of the distance between neighboring nuclei e.g. with increasing of Ts. The 2)-type boundaries results from further growth of the two domains with quite equal rates. The inclination angle of such boundaries have a tendency to increase with increasing of Ts i.e. c-domains grow more rapidly than a,b-ones in (001) direction. The 3)-type boundaries are formed owing to nucleation of the c-oriented domains on the (001) planes of the already formed a,b-domains.
As proved out by both cross-sectional and plan-view observations all of three types of 90° boundaries are microscopically clean and contain neither any extraneous phases nor misfit dislocations (fig.2,4). These boundaries

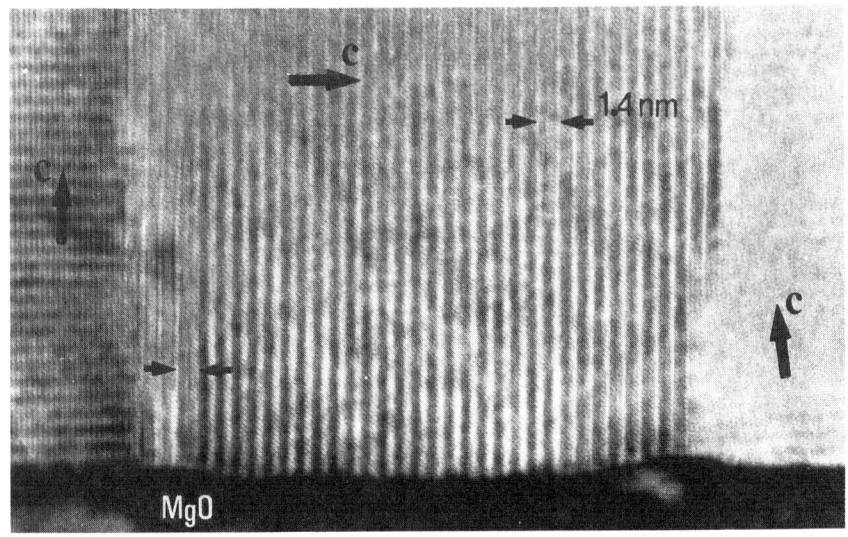

Figure 4 Cross sectional HREM image. Three domains are visible. Note the planar stacking defects with $c_0 \sim 1.4$nm existence.

are known as crystallographically special, low-energy boundaries [13]. Thus, if some impurities exist in the growing film they will not capturing by 90° boundaries but diffuse out from film volume commonly to the film surface (fig.2b). Such boundaries do not affect significantly superconductive transport properties [13]. To our opinion, these coherent boundaries are similar to [100] 90° rotation twins readily observed in dense YBCO material [14] but on the contrary have an epitaxial nature.

3.2.2. Film/substrate interface

The interface between YBCO and MgO is also microscopically clean without any intermediate layer. However a periodic array of misfit dislocations of 6-7 a_0-parameters of YBCO apart resulting from stress caused by ~8% lattice mismatch relaxation exists at the film/substrate interface (fig.3). As far as our films grown on mechanically polished MgO their structure appear to be influenced by the roughness of the substrate surface with the feature heights of ~2-3 nm. Such surface roughness may lead to deviation of a, b, c-axis of YBCO from normal orientation and thus may produce low-symmetry boundaries (fig.2a,4). These boundaries are known as rather a shortcoming for superconductive transport properties [15].

3.2.3. Planar defects

Two types of a intragranular planar stacking defects are frequently available in our films. The first is insertion of an extra CuO layer thus forming $(CuO)_2$ double layer between two BaO layers ("124" structure) (fig.2b). The density of such defects in a, b-domains is rather higher than in c-domains. On the other hand c-domains contain a number of short in-plane *fragments* of "124" structure resulting in quite poor average crystallinity of these domains. The second type of the planar defect is a fault with c_0 of

~1.45 nm which appears only in a,b-domains (fig.4). Y.Matsui et. al. [16] have considered that formation of these defects results from introducing an additional metal (Cu and/or Y) and oxygen layers near the Y layers that increases the separation of two CuO layers.

In addition other polytypoidic variations [17] are also observed but not often.

3.2.4. c_0-Parameter variations

Another microstructural feature of the films investigated is the c_0 variations. The average c_0-parameter varied from film to film between 1.16 and 1.19 nm depending on growth condition (mainly on oxygen pressure); this range extends the possible variation in bulk materials (1.168-1.182 nm). Such extended range is explained by chain oxygen content variations and/or by Cu/Y cation disorder [7]. Due to fast oxygen diffusion from the growing surface into a,b-domains they commonly have an average c_0 smaller than c-domains. At the same time a,b-domains display the local c_0 variations.

4. CONCLUSION

In general, we found that YBCO films *in-situ* grown on MgO substrates usually display the mixed a,b,c-orientations resulting from cube on cube epitaxy. The $90°$ misoriented epitaxial grains produce a number of coherent boundaries. We are also found that orthogonal grains have some microstructural differences which may lead to differences in their superconducting properties. This fact probably explains the lower T_c of the YBCO/MgO in comparison to YBCO/SrTiO3.

5. REFERENCES

1 A.I.Golovashkin, E.V.Ekimov, S.I.Krasnosvobodsev et.al. Physica C, 162-164 (1989) 715.
2 R.Ramesh, D.Hwang, T.S.Ravi et.al. Appl.Phys.Lett., 56 (1990) 2243.
3 L.A.Tietz, C.B.Carter, D.K.Lathrop et.al. J.Matter.Res., 4 (1989) 1072.
4 J.Chang, M.Nakajima, K.Yamamoto, A.Sayama, Appl.Phys.Lett., 54 (1989) 2349.
5 S.T.Lee, S.Chen, L.S.Hung, G.Braunsen, Appl.Phys.Lett., 55 (1989) 286.
6 C.B.Eom, J.Z.Sun, B.M.Lairson et.al. Physica C, 171 (1990) 354.
7 V.Matijasevic, P.Rosenthal, K.Shinohara et.al. subm. to J. of Mater. Res. (1990).
8 Q.Li, O.Meyer, X.X.Xi, J.Geerk, G.Linker, Appl.Phys.Lett., 55 (1989) 310.
9 O.Eibl, H.E.Hoenig, J.-M.Triscone et.al. Physica C, 172 (1990) 373.
10 M.V.Sidorov, S.R.Oktyabrsky, subm. to physica status solidi (1991).
11 R.Ramesh, C.C.Cang, T.S.Ravi et.al. Appl.Phys.Lett., 57 (1990) 1064.
12 M.V.Sidorov, S.R.Oktyabrsky, in: Proc. of C-MRS International'90,ed. by H.D.Li, Elsevier Science Publ. (1990)
13 S.-W.Chan, D.M.Hwang, R.Ramesh et.al. AIP Conf. Proc. No.200 (1989) 172.
14 H.W.Zandbergen,G.Thomas, Phys.Stat.Sol.(a), 107 (1988) 875.
15 D.Dimos, P.Chaudari, J.Mannhart, Phys.Rev.B 41 (1990) 4038.
16 Y.Matsui, E.Takayama-Muromachi, A.Ono et.al. J. of Appl. Phys., 26 (1987) L777.
17 R.Ramesh, D.M.Hwang, T.S.Ravi et.al. Physica C 171 (1990) 14.

Part V
Applications

High T$_c$ Superconductor Thin Films
L. Correra (Editor)
© 1992 Elsevier Science Publishers B.V. All rights reserved.

Basic elements of high T$_C$ superconductor wiring

B. Roas, G. Friedl*), B. Holzapfel+), H. Busch*), A. Müller, and L. Schultz

  Siemens Research Laboratories, D-8520 Erlangen, Germany
*)also at Phys. Inst., University of Augsburg
+)also at Phys. Inst., University of Erlangen

Abstract

   High T$_C$ superconductor films and multilayers were prepared by laser deposition using a SIEMENS XP 2020 excimer laser or by CVD. Epitaxial YBaCuO films show high dc critical current densities as $6 \cdot 10^6$ A/cm^2 at 77 K or $6 \cdot 10^7$ A/cm^2 at 4.2 K and excellent microwave properties. Using identical laser deposition parameters as for YBaCuO, SrTiO$_3$ and LaAlO$_3$ interlayers grow heteroepitaxially onto the YBaCuO films and are, therefore, used as insulating layers in multilayer structures. The YBaCuO and insulator layers can be patterned by means of a new lift-off technique, using a high-temperature-resistant CaO process mask. YBaCuO striplines crossing substrate steps have been prepared to study the current carrying capability of interconnections between conducting planes in an YBaCuO – insulator multilayer system. These striplines can either be formed as weak links or as high current leads depending on the geometrical parameters of the step.

1. INTRODUCTION

   For applications of HTSC thin films in future electron devices it is essential to have films with high current carrying capability and a good surface morphology on large substrates by economic preparation techniques. Since, in most cases, superconducting leads at different levels are needed, also multilayered films with insulating interlayers [1-4] and transitions from one level to the other across step edges [5] as well as insulating crossings are required [6] – without a tremendous reduction of the critical current density. In this contribution, we report on our experiments on growth, patterning, properties and surface morphology of YBa$_2$Cu$_3$O$_x$ films pared by laser deposition or MOCVD. A new process for patterning of $YBa_2Cu_3O_x$ - insulator multilayers is described, using a high-temperature-resistant CaO process mask, which can be evaporated directly onto previously patterned $YBa_2Cu_3O_x$ layers and removed after film deposition [7]. A crossover contact of two 3 μm wide and 30 μm long $YBa_2Cu_3O_x$ striplines has been fabricated by laser deposition, using this patterning technique. Also results for multilayers with insulating $SrTiO_3$ or $LaAlO_3$ interlayers and for current leads across step edges are presented.

2. $YBa_2Cu_3O_x$ FILMS BY LASER DEPOSITION

$YBa_2Cu_3O_x$ thin films have been prepared by in-situ laser de-position using a Siemens XP2020 excimer laser with 308 nm wavelength and a pulse energy of 2 J. Epitaxial growth and c-axis orientation were achieved by heating the <100> $SrTiO_3$ or $LaAlO_3$ substrates to about 800°C and applying a 0.4 mbar oxygen pressure during deposition. These films show reproducibly T_c values between 88 K and 91 K, critical current densities of 2 - $6 \cdot 10^6$ A/cm^2 at 77 K and 3 - $5 \cdot 10^7$ A/cm^2 at 4.2 K [8]. Although the film surfaces are relatively rough (Fig. 1), 77 K surface resistances < 10 mΩ at 87 GHz and 0.1 mΩ at 10 GHz can be observed [9]. With regard to application of HTSC materials in microelectronics (superconducting chip interconnections), $YBa_2Cu_3O_x$-striplines with widths < 2 μm are required. Laser-deposited $YBa_2Cu_3O_x$ films were patterned to 1.5 μm wide and 30 μm long striplines (Fig. 2) by wet-etching with EDTA (Ethylenediaminetetraacetic acid). The critical current density of these striplines is $2 \cdot 10^6$ A/cm^2 at 77 K.

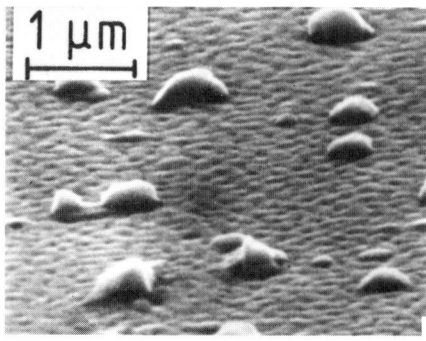
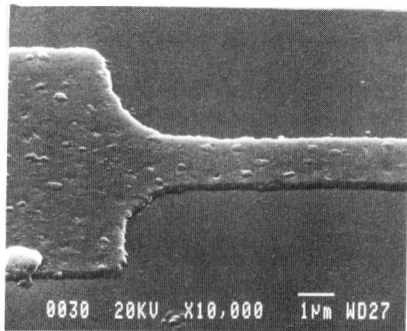

Figure 1. SEM micrograph of a laser-deposited film surface (tilted by 15°).

Figure 2. Etch pattern of a 1.5 μm wide stripline in a laser-deposited film.

3. YBa₂Cu₃Oₓ FILMS BY CVD

For technical application we use chemical vapor deposition of $YBa_2Cu_3O_x$ on an area up to 20 cm². The superconducting data are well reproduced with T_c up to 93 K (ac-inductively measured) and with critical current densities above 10^6 A/cm² at 77 K in zero magnetic field. Our CVD process works with 2.2.6.6-tetramethyl-3.5-heptanedionates as precursor materials [10]. Carrier gas is Ar at a low total flow rate up to 100 sccm. The operation time of the evaporator units ranges from 10 to 30 h without interruption. The chemical yield of the process is close to 100% for Ba and Cu and about 75% for Y. The deposition upon the whole area of 20 cm² shows low scatter as is demonstrated with samples arranged continuously across the susceptor diameter (Fig. 3). The film thickness can be described by a standard deviation of 5% from the mean value (about 250 nm). Inspite of somewhat lowered transition temperatures at the very edge of the susceptor, the mean value for the midpoint of the transition temperature is still 89.6 K. The critical current density in zero magnetic field at 77 K is correlated with the film thickness, the data range from $6 \cdot 10^6$ to $1.6 \cdot 10^6$ A/cm² [10] corresponding to film thicknesses of 100 and 730 nm, respectively. At a film thickness of 120 nm the critical current density at 7 T and 77 K (configuration $\underline{B}\|\underline{c}$, "worst case") is still about 10^5 A/cm². On highly c-axis oriented films, CuO particles are observed with diameters up to 1 μm with a density of about 10^7/cm² besides small (ab)-oriented YBaCuO crystals (Fig. 4)

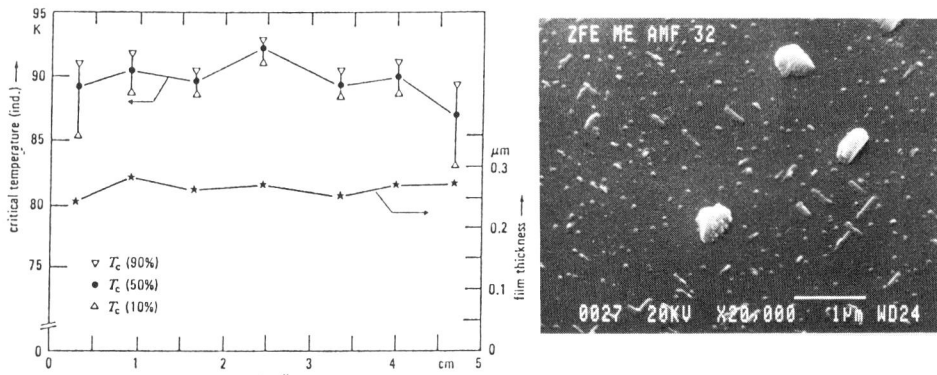

Figure 3. $YBa_2Cu_3O_x$ by CVD on large area; inhomogeneity of critical temperature and thickness referring to film location on the susceptor diameter.

Figure 4. SEM micrograph of a typical film surface for a CVD film.

4. PREPARATION OF YBaCuO - INSULATOR MULTILAYERS

$YBa_2Cu_3O_x/SrTiO_3$ and $YBa_2Cu_3O_x/LaAlO_3$ multilayers were prepared in-situ on (100) $SrTiO_3$, MgO and $LaAlO_3$ substrates by
pulsed laser deposition using a multi-target holder. Identical
deposition conditions were used for the deposition of the
$YBa_2Cu_3O_x$ as well as the $SrTiO_3$ and $LaAlO_3$ layers. Both symmetric and non-symmetric multilayer structures with modulation
wavelengths between 100 and 600 Å and an overall thickness of
about 2500 Å were prepared. The multilayers were analyzed using
X-ray diffraction and helium ion beam channeling. X-ray diffraction patterns revealed purely c-axis oriented $YBa_2Cu_3O_x$
film growth. For the $SrTiO_3$ and $LaAlO_3$ layers only (h00) peaks
could be observed. The existence of satellite lines up to third
order clearly indicates the high quality of the superlattice
(Fig. 5). 2.7-MeV He^{++} ion channeling with a minimum yield of
about 13 % shows the good crystallinity of the multilayers. Inductive T_C measurements as function of the modulation wavelength and the $YBa_2Cu_3O_x$ layer thickness of YBaCuO/SrTiO multilayers showed only a sligthly reduced T_C of 88 K down to a modulation wavelength of 150 Å with a $YBa_2Cu_3O_x$ layer thickness
of 100 Å. Below this $YBa_2Cu_3O_x$-single-layer thickness, T_C is
decreasing and the transition width is increasing, probably
caused by strain effects. A (50Å YBaCuO/50Å $SrTiO_3$) × 15.5
multilayer is still superconducting, but with a transition
temperature as low as 33 K (Fig. 6).

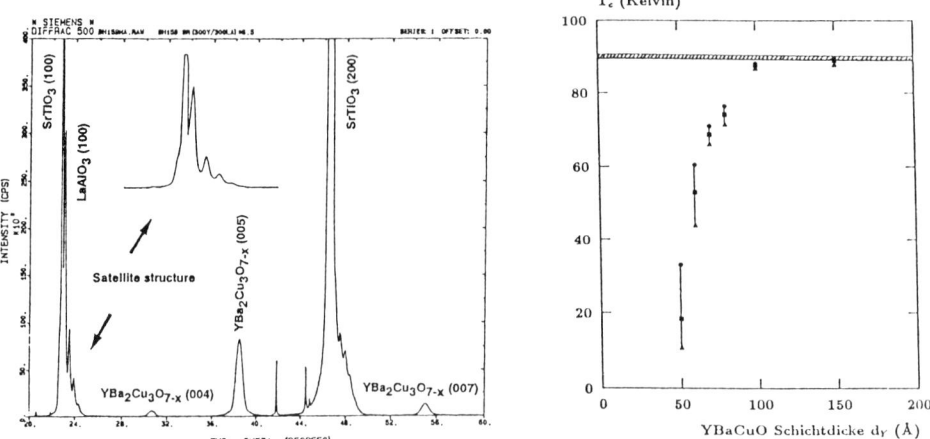

Figure 5.   X-ray diffraction pattern of an $YBaCuO/LaAlO_3$
multilayer on a $SrTiO_3$ substrate, showing satellite lines up to
third order.

Figure 6. T_C of   (d_y YBaCuO/50Å $SrTiO_3$) × 15.5 multilayers as
a function of the YBaCuO layer thickness d_y.

5. PATTERNING OF YBaCuO - INSULATOR MULTILAYERS

Selective patterning of individual layers of heteroepitaxial superconductor-insulator multilayer systems is a prerequisite for application of high-T_c-superconductors in microelectronics. For that purpose a lift-off process has been developed, using a heat-resistant and water-soluble CaO mask for in-situ deposition of perovskite films at high substrate temperatures. With this technique it is possible to deposit successively epitaxial $YBa_2Cu_3O_x$ or insulator layers onto a device, without destruction of the previously evaporated multilayer structure. The patterning process is described by five steps, shown in Fig. 7:

(i)   A $SrTiO_3$ or $LaAlO_3$ substrate, which may be coated with an $YBa_2Cu_3O_x$ film, is patterned with photoresist (thickness ≈ 2.5 μm) by conventional photolithography (Fig. 7a).

(ii) Subsequently a 1 μm thick, amorphous CaO layer is evaporated onto the substrate at room temperature, for example by pulsed laser deposition from a pure calcium target in 0.1 mbar oxygen atmosphere. The CaO layer is terminated by a 500 Å thick zirconium barrier (Fig. 7b).

(iii) After that the sample is put into an acetone ultrasonic bath to remove the photoresist and the CaO/Zr above (standard photoresist lift-off process). As CaO is not soluble in acetone, a CaO negative mask remains on the substrate (Fig. 7c).

(iv) The next step is the deposition of $YBa_2Cu_3O_x$, $SrTiO_3$ or $LaAlO_3$ layers through the mask. At $\approx 800^\circ$C substrate temperature and ≈ 0.4 mbar oxygen pressure they are growing epitaxially onto the uncovered parts of the substrate (Fig. 7d). Note that there is no chemical reaction of CaO and $YBa_2Cu_3O_x$ also at high substrate temperatures, necessary for the epitaxial growth of perovskite materials. A possible interdiffusion of the deposited materials through the relatively rough and porous CaO mask is inhibited by the dense zirconium barrier on top of the CaO.

(v)   As CaO is water-soluble, the mask with the layers on top can be subsequently removed in an ultrasonic bath in distilled water (Fig. 7e). The CaO lift-off takes about 15 minutes. A slightly alkaline solution will be formed by the reaction $CaO + H_2O \rightarrow Ca(OH)_2$. Epitaxially grown $YBa_2Cu_3O_x$ or insulator layers free of grain boundaries are not destroyed in that solution. Sometimes some small water-insoluble particles remain on the substrate, which can easily be wiped off with a soft tissue. Finally the sample is cleaned again in acetone. The patterning procedure can be repeated beginning with step (i) shown in Fig. 7a to grow the next epitaxial layer onto the device.

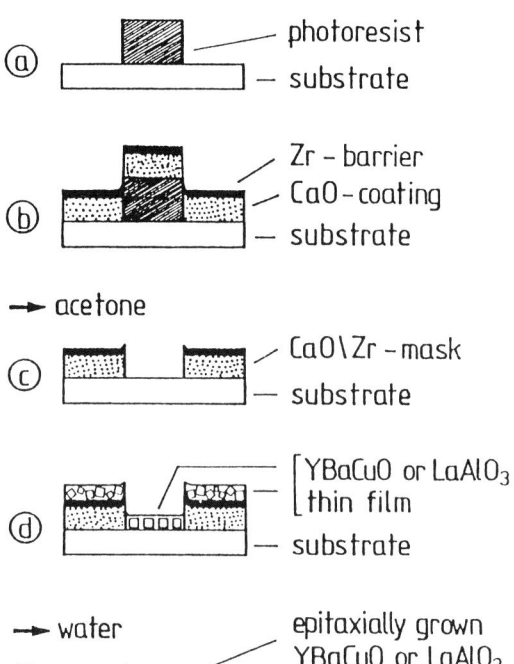

(a) — photoresist / substrate

(b) — Zr – barrier / CaO-coating / substrate

→ acetone

(c) — CaO\Zr – mask / substrate

(d) — [YBaCuO or LaAlO₃ thin film] / substrate

→ water   epitaxially grown YBaCuO or LaAlO₃ / substrate

Figure 7. Process of CaO lift-off technique for patterning of $YBa_2Cu_3O_x$-insulator multilayers.

To demonstrate the efficiency of this new technique, a cross-over contact of 3 μm wide $YBa_2Cu_3O_x$ striplines has been prepared. A CaO mask for deposition of an upper stripline was evaporated onto a previously patterned lower stripline on a <100> $LaAlO_3$ substrate. A SEM micrograph (Scanning Electron Microscope) of the mask and the YBaCuO stripline beneath it is presented in Fig. 8, after lift-off of the photoresist pattern. The edges of the mask are well defined. Fig. 9 shows a SEM micrograph of the complete contact after lift-off of the process mask in water and final cleaning. The transport properties of the upper and the lower stripline were investigated by $\rho(T)$ and $j_c(T)$ measurements. The contact (current from A to C in Fig. 9) was characterized by measuring the voltage between A and C as well as between B and D. In Fig. 10 resistivity versus temperature of the upper stripline (full triangles), the lower stripline (full circles) and the total path between A and C is given (open squares), measured with a current of 2 μA. Superconductivity can be observed at temperatures < 88.8 K. The resistivities $\rho(91\ K)$ are < 80 $\mu\Omega$cm. The contact resistance (voltage between B and D) was 630 mΩ at 100 K, measured with a current of 1 mA. Assuming a current carrying contact area of 3μm·3μm, this corresponds to a resistivity of $5\cdot10^{-8}$ Ωcm². At

87 K the contact is assumed to be superconducting, as the resistivity decreased to values $< 10^{-11}$ Ωcm^2 (lower limit of voltage measurement). The reduced T_c may be caused by the relatively high measuring current through the contact. The critical current density j_c as a function of temperature of the upper stripline (full triangles), the lower stripline (full circles) and the total path from A to C (open squares) is shown in Fig. 11. The film thickness of the striplines was 250 nm and 230 nm, respectively. The lower stripline, previously covered with CaO, heated to 800°C in 0.4 mbar oxygen for 15 min and finally cleaned in a water ultrasonic bath, shows still a j_c(77 K) of $4 \cdot 10^6$ A/cm^2. Its superconducting properties were not affected by interdiffusion or chemical reaction with the mask. The j_c of the path from the upper to the lower stripline is identical to that of the upper stripline, leading to the assumption that the current carrying capability of the path from A to C is not limited by the contact but by the upper stripline. Therefore, the critical current through the contact is estimated to be larger than 14 mA.

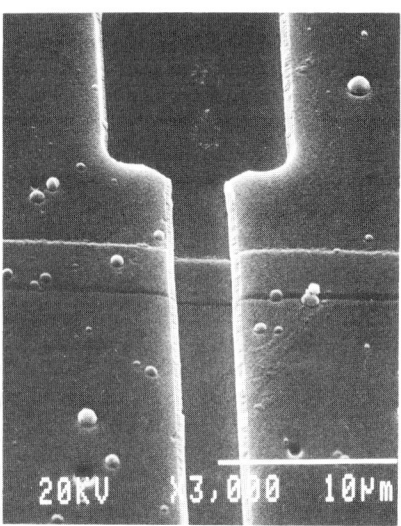 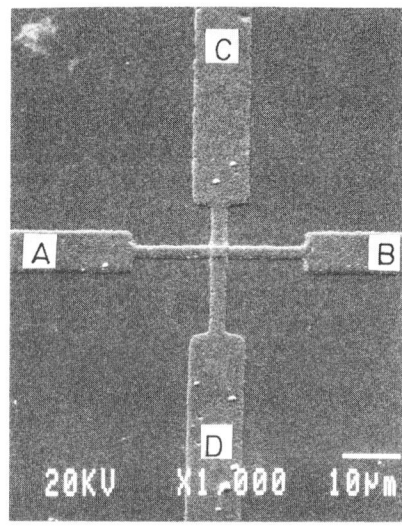

Figure 8. SEM micrograph of a CaO mask for fabrication of a 3 μm wide and 30 μm long upper stripline of an YBa$_2$Cu$_3$O$_x$ crossover contact. The lower stripline had been prepared in a previous process and can be seen beneath the CaO mask.

Figure 9. Overlap area of two 3 μm wide and 30 μm long YBa$_2$Cu$_3$O$_x$ striplines crossing each other. On a 200 nm scale the edges of the striplines are not well defined.

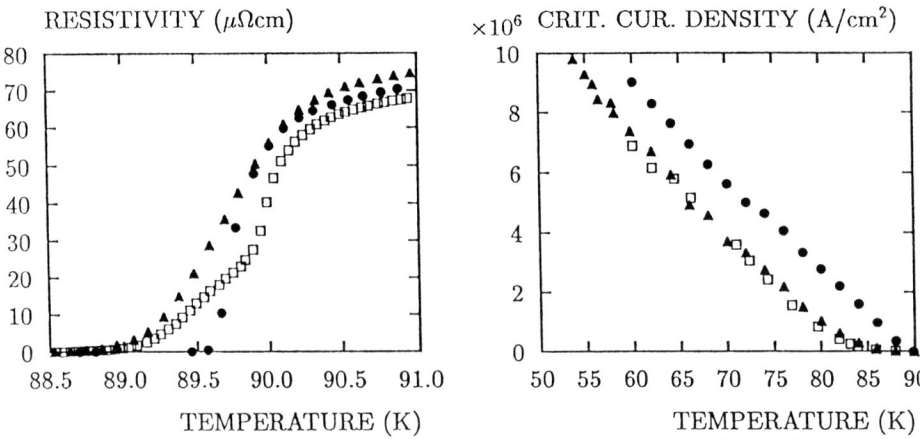

Figure 10. Resistivity versus temperature of the upper stripline (full triangles), the lower stripline (full circles) and a path from the upper to the lower stripline including the contact (open squares)

Figure 11. Critical current density versus temperature of the upper stripline (full triangles), the lower stripline (full circles) and a path from the upper to the lower stripline including the contact (open squares).

7. CURRENT LEADS OVER STEP EDGES

HTSC transmission lines are characterized by a very fast and dispersionless pulse transmission, giving rise to application of HTSC in microelectronics as interconnects in integrated circuits. In order to fabricate basic elements of an all epitaxial HTSC-wiring (crossings of striplines separated by insulator material, via holes, etc.) a high current transition over steps in a $YBa_2Cu_3O_x$ - insulator multilayer system is required. To simulate the current carrying capability between conducting planes of such a system the transport properties of $YBa_2Cu_3O_x$ striplines at artificial edges in (100) $SrTiO_3$ substrates were investigated. The step edges were fabricated by Ar^+ ion milling. Using a modified thermal treatment of the resist mask, different step heights between 80 nm and 800 nm and step edge angles with 2.5°, 10°, 25°, 40° and 65° could be prepared. The angles were determined by SEM (Fig. 12). Epitaxial c-axis oriented $YBa_2Cu_3O_x$ films were deposited in-situ onto these step edges by laser deposition. For measuring the critical current density three 50 μm wide striplines were patterned by wet etching, one across the step edge, one in the unetched, and one in the etched part of the substrate (Fig. 13). In the etched and unetched area T_c, j_c and $\rho(100$ K) were essentially

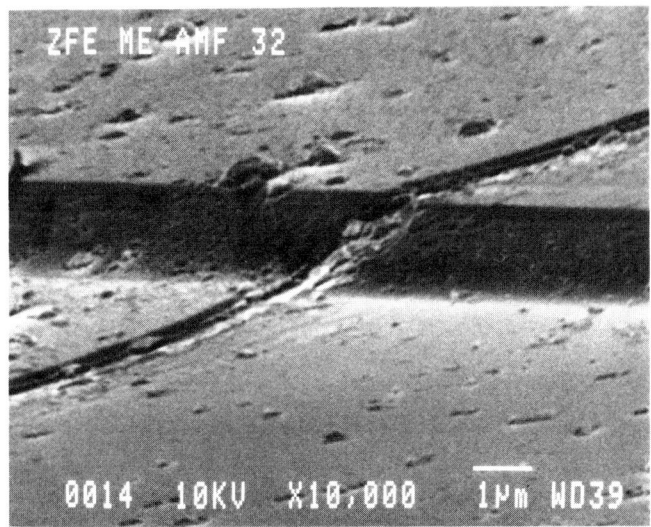

Figure 12. SEM micrograph of a step-edge stripline. Only the peripheral section of the 50 μm wide path is shown.

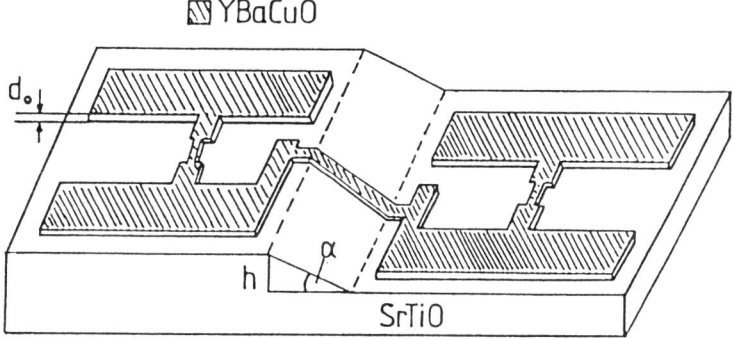

Figure 13. Schematic representation of the test arrangement for measuring the current carrying capability at step edges.

identical, indicating that the superconducting properties are not effected by the previous ion milling of the substrate. The transport properties of the step edge stripline were compared to the values of the striplines in the planar parts of the substrate. Up to an angle of 10° no degradation of the super-conducting properties could be observed, as can be seen in Fig. 14. Up to an angle of 30°, the current density is only slightly reduced (i.e. in the range $> 10^6$ A/cm^2), indicating that at flat steps a high current transition between two conducting planes is possible. For steeper steps ($a > 45°$) the

growth direction of the film across the step edge changes [11].
The formation of grain boundaries causes a reduction of the
critical current density by two orders of magnitude, thus
creating a weak link. Therefore, thin films over step edges can
be used either as high current interconnects in integrated circuits or as weak links, for example in SQUIDs, simply by adjusting the angle.

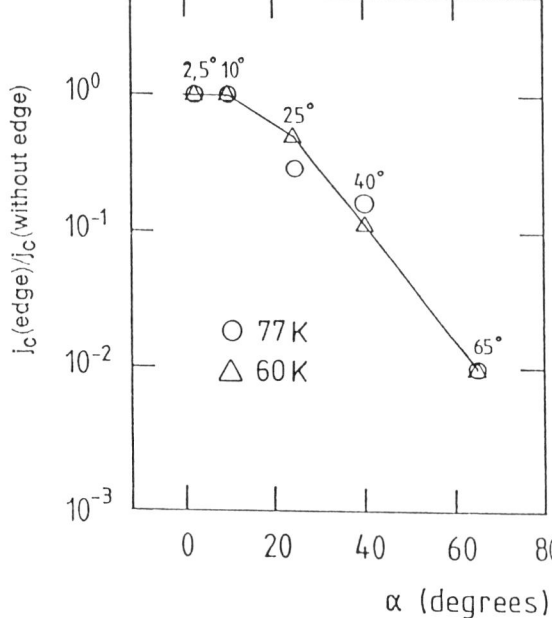

Figure 14. Standardized
critical current density
across the step edge vs.
the angle of the slope.

8. CONCLUSIONS

High quality HTSC films can be prepared both, by laser
deposition and CVD. Whereas laser deposition is a very flexible
technique, allowing many experiments within a short time (multilayer growth), the carefully controlled CVD process has a
high potential to prepare films with a good surface morphology
on large-size substrates. By means of CaO process masks it is
possible to deposit successively epitaxial $YBa_2Cu_3O_x$ or insulator layers onto a device, without destruction of the previously evaporated multilayer structure. A crossover contact
with a cross-section of $3\mu m \cdot 3\mu m$ was fabricated, carrying a
critical current > 14 mA. High critical currents were observerd
in striplines at substrate steps for small step edge angles.

Acknowledgement

The authors are grateful to L. Bär, G. Daalmans, A. Fink, H.E. Hoenig, D. Illner, W. Jutzi, M. Kühnl, G. Küsebauch, M. Römheld, G. Saemann-Ischenko, K. Samwer, R. Seeböck, K.J. Schmatjko, P. Schmitt, H. Seidel and B. Stritzker for assistance and stimulating discussions. This work was supported by the German Ministry for Research and Technology (TK 03327).

9. REFERENCES

[1] J.M. Triscone, ϕ. Fischer, O. Brunner, L. Antognazza, A.D. Kent, and M.G. Karkut, Phys. Rev. Lett. <u>64</u>, 804 (1990).

[2] X.D. Wu, X.X. Xi, Q. Li, A. Inam, B. Dutta, L. DiDomenico, C. Weiss, J.A. Martinez, B.J. Wilkens, S.A. Schwarz, J.B. Barner, C.C. Chang, L. Nazar, and T. Venkatesan, Appl. Phys. Lett. <u>56</u>, 400 (1990).

[3] L. Schultz, B. Roas, P. Schmitt, and G. Endres, Processing of Films for High T_C Superconducting Electronics, SPIE Vol. 1187, 204 (1990).

[4] A. Gupta, R. Gross, E. Olsson, A. Segmüller, G. Koren, and C.C. Tsuei, Phys. Rev. Lett. <u>64</u>, 3191 (1990).

[5] G. Friedl, B. Roas, M. Römheld, L. Schultz, and W. Jutzi, submitted to Applied Physics Letters (1991).

[6] F.C. Wellstood, J.J. Kingston, and J. Clark, Appl. Phys. Lett. <u>56</u>, 2336 (1990).

[7] B. Roas, submitted to Appl. Phys. Lett. (1991).

[8] B. Roas, L. Schultz and G. Saemann-Ischenko, Phys. Rev. Lett., <u>64</u>, 479 (1990).

[9] N. Klein, H. Chaloupka, G. Müller, S. Orbach, H. Piel, B. Roas, L. Schultz, U. Klein and M. Peiniger, J. Appl. Phys. <u>67</u>, 6940 (1990).

[10] H. Busch, A. Fink and A. Müller, submitted to Appl. Phys. Lett. (1991).

[11] C.L. Jia, B. Kabius, K. Urban, K. Herrmann, G.J. Cui, J. Schubert, W. Zander, A.I. Braginski, and C. Heiden, to be published Physica C (1991).

High T$_c$ Superconductor Thin Films
L. Correra (Editor)
845

HTSC multilayers for tunnel and proximity effect junctions

I. Iguchi, T. Kusumori and Z. Wen

Institute of Materials Science, University of Tsukuba, Tsukuba, Ibaraki 305, Japan

Abstract

The quasiparticle tunneling and Josephson characteristics of reproducible high Tc YBCO/I/Pb(or Au) tunnel junctions (I: artificial barrier of a few nm thick or native barrier) are presented. Both the c-axis and a,b-axes oriented YBCO films are grown for this study. In addition, the proximity-effect tunneling structure YBCO/Al/AlOx/Pb is also investigated. The observed results are consistently interpreted by a proposed tunneling model based on gap anisotropy and surface degradation effects.

1. INTRODUCTION

The investigation of quasiparticle tunneling and Josephson effects of high Tc tunnel junctions is quite important for the clarification of superconductivity mechanism and the application to superconducting electronics. The measurements on high Tc tunnel junctions yielded a variety of results due to the involvement of the oxygen deficiency and the extremely short coherence length at the early stage of this research [1]. The fabrication of tunnel junctions has been, however, remarkably improved recently and the high reproducibility of junctions has been reported by several groups [2-6].

In this paper, the gap structures of YBCO/I/Pb tunnel junctions with an in situ artificial barrier or a native barrier are presented, which are discussed in connection with the other published data. To interpret the observed data consistently, a phenomenological tunneling model taking the gap anisotropy, the extremely short coherence length and the surface degradation of

YBCO material into account is proposed. The results on Josephson
effect observed for the YBCO/I/Pb junctions with both the c-axis
and a,b-axes oriented YBCO films are also presented. The possible
occurrence of proximity effect of YBCO material is also
investigated using the YBCO/Al/AlO$_x$/Pb multilayers.

2. EXPERIMENTAL

The fabrication of tunnel junctions was performed using a
vacuum chamber installed with two electron beam guns of 10kV with
6 crucibles and two thermal heaters. An in situ metal mask
changing system was also equipped with this chamber. The gap
between a metal mask and a substrate was a few tenth of mm. The
reactive coevaporation of YBCO films was done by two electron
guns for Y and Ba and a thermal crucible heater for Cu. The
typical evaporation rate was 0.25nm/sec. We used single crystals
of MgO, SrTiO$_3$, LaGaO$_3$, NdGaO$_3$ and LaSrGaO$_4$ as substrate
materials. LaSrGaO$_4$ is a new substrate with the tetragonal
crystal structure. The lattice parameters for a-axis and c-axis
are 0.3843nm and 1.2861nm respectively [7]. With the a-axis
oriented substrate, the a,b-axes oriented YBCO films were
obtained. Tc of YBCO films was 80- 88K. It is important to use
the superconducting YBCO film above 80K in order to avoid the
possible formation of high resistance layer on the film surface
which prevents from obtaining the relevant tunnel data for high
Tc material. The artificial barrier was formed by depositing
either MgO or Y$_2$O$_3$ or AlO$_x$ with a few nm thick.

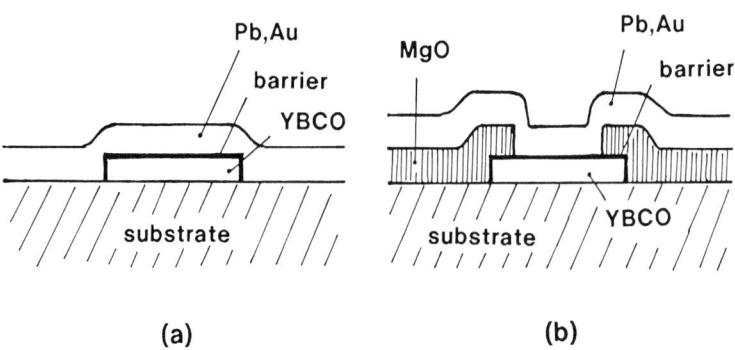

(a)   (b)

Figure 1. Cross-sectional view of a fabricated tunnel junction.

The line width of the junction was 0.2mm-0.4mm. Figure 1 shows the cross sectional view of two types of fabricated junctions. Figure 1(a) depicts a simple cross-type sandwiched junction structure, while Figure 1(b) does a window-type junction structure with the thick insulating layer of MgO to prevent the possible film edge effects.

3. TUNNEL CHARACTERISTICS

The tunnel resistance of YBCO/MgO/Pb tunnel junctions exhibited an exponential dependence on MgO barrier thickness and the estimated barrier height was 1.6eV in good agreement with the potential barrier height of MgO, whose details were reported in the previous paper [6]. The results indicate that the junction current is indeed due to the quasiparticle tunneling process.

Figure 2 shows a typical example of the observed tunnel characteristic (dV/dI vs V curve). The thickness of Y_2O_3 insulating layer was about 3nm. The YBCO film was strongly c-axis oriented perpendicular to the substrate surface. The observed features contain a broad gap structure, a Pb/I/N (N: normal metal) like tunneling characteristic and a zero bias anomaly, which are not expected for the conventional BCS characteristics. Above Tc of Pb, a sharp Pb/I/N structure disappeared and the zero bias aomaly was clearly seen as the conductance peak around zero bias. This zero bias became small as the junction resistance was increased. Typically, for the junctions with their tunnel resistance above 50Ω, it became almost diminished away. The result in Fig. 2 was quali-

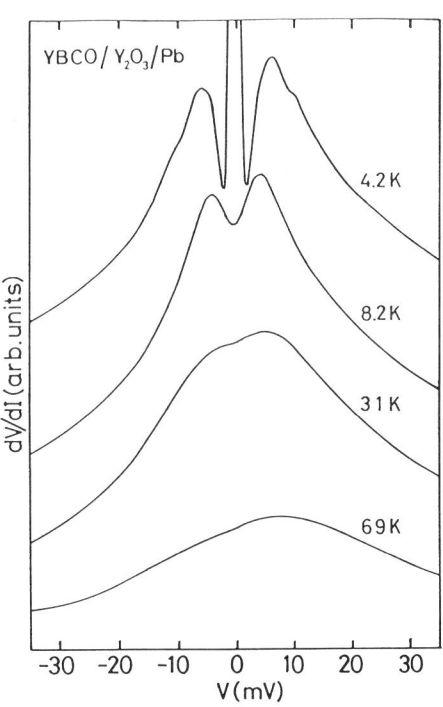

Figure 2. dV/dI vs V character-istics of a YBCO/Y_2O_3/Pb tunnel junction.

tatively similar to those for YBCO/I/Pb junctions with the different artificial barriers and also with the a,b axes oriented YBCO films [6]. The dV/dI vs V curves of YBCO/I/Au junctions were also qualitatively similar to Fig. 2 except the absence of Pb/I/N like tunneling behavior.

The Pb/I/N like tunneling curve was found to be almost ideal, suggesting that the fabricated tunnel junctions were almost free from pinholes. This happened probably because YBCO material and Pb were quite reactive at the interface and some kind of insulating barriers were formed between them. Hence the pinholes in an ultrathin MgO film did not cause troubles, even if they existed. The use of Au as a counterelectrode yielded a relatively weak dependence on MgO barrier thickness as compared with that of Pb, which reflects the effect of pinholes in MgO barrier since YBCO and Au are considered to be nonreactive and an appreciable leak current is expected through them.

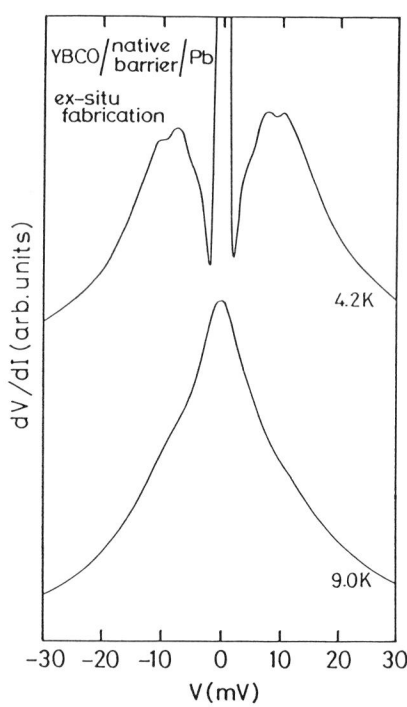

Next, the fabrication of the junctions with an ex-situ native barrier was performed by deliberately exposing a YBCO film to the ambient atmosphere for 20 - 40 min after its deposition and evaporating a counterelectrode Pb. The native barrier was formed probably by the oxidation and degradation effects of the YBCO surface by the external atmosphere and the reaction between YBCO and Pb. Hence the barrier material is unknown (perhaps PbO and additional degraded layer). The junction fabricated by in situ oxygen annealing of a YBCO film without breaking vacuum system and direct deposition of Pb did not yield a relevant tunneling behavior probably due to the low

Figure 3. dV/dI vs V curves of a YBCO/native barrier/Pb tunnel junction.

barrier height of an insulating layer.

Figure 3 shows the observed tunnel characteristics for a YBCO/native barrier/Pb junction fabricated on a $LaSrGaO_4(100)$ substrate. The X-ray diffraction method showed that the YBCO film was dominantly along the a-axis oriented perpendicular to the substrate surface. It is remarkable to point out that the zero bias anomaly disappeared for this junction and the resistance peaking behavior (conductance dip) was observed, although a very weak zero bias anomaly was occasionally observed. The result is quite similar to those of previously published reports [2,3,8] and is typical to YBCO junctions with a native barrier. The observed V-shaped characteristic might be related to the surface degradation effect in the external atmosphere.

To interpret the observed various results, a phenomenological tunneling model based on the gap anisotropy and the extremely short coherence length with taking the surface degradation effect into account is proposed [9,10]. Figure 4 shows the examples of the actual situation in the tunneling process for the junctions with artifical and native barriers. Although the YBCO film has good crystal orientation along the c-axis or a,b-axes by the X-

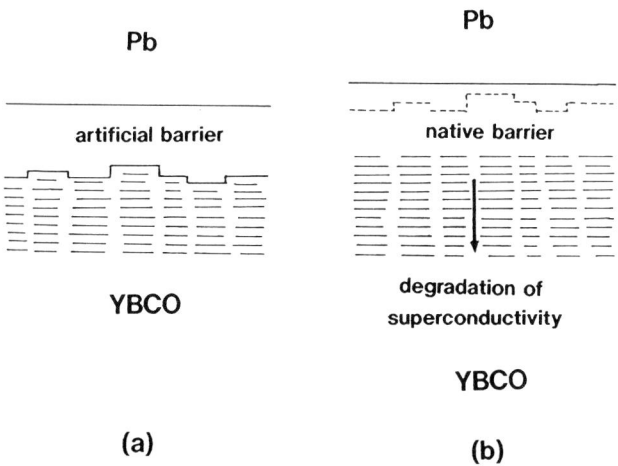

Figure 4. Possible cross-sectional view of a real tunnel junction.

ray analysis, its surface will not be completely smooth in an
atomic scale, which may be conjectured from the fact that the
observed tunnel characteristics were qualitatively similar
between the junctions with the c-axis and a,b-axes oriented YBCO
films. It is also conceivable that there will exist the
microstructures with crystal orientation different from the
preferred orientation. In this situation , it is possible to
consider the occurrence of tunneling for both the c-axis and ab-
plane directions. Moreover, there may exist normal regions on the
surface of a YBCO film, especially in the c-axis oriented
regions.

    In the actual tunneling process, it is reasonable to assume a
semiconductor tunneling model as shown in Fig. 5. Because of the

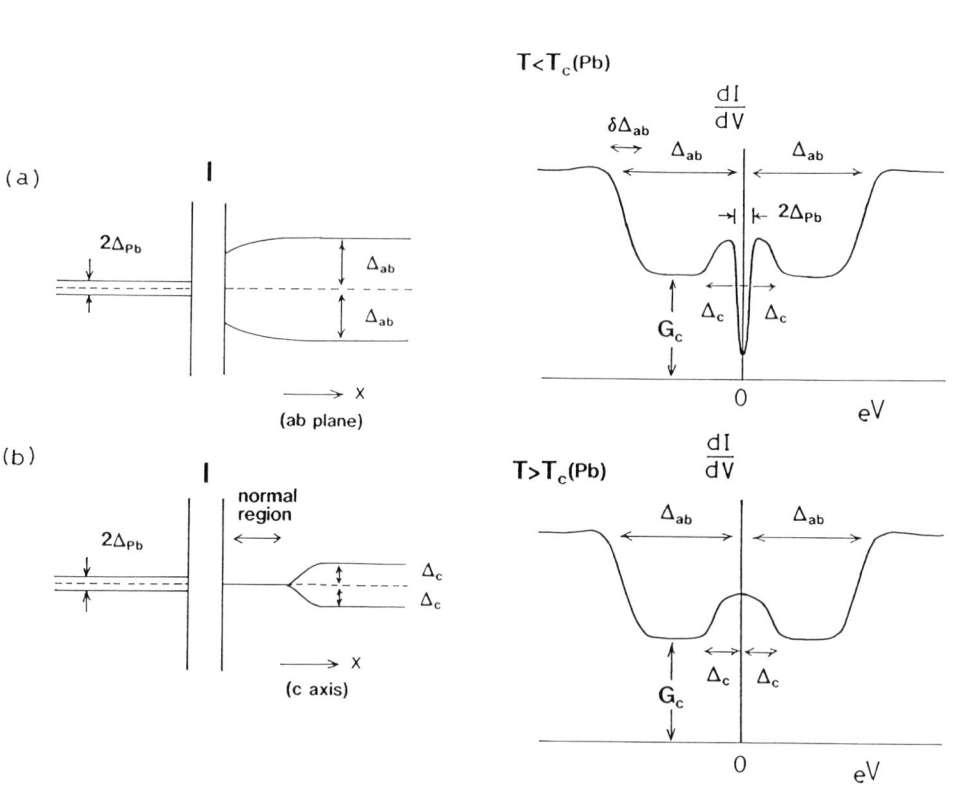

Figure 5. A tunneling scheme
for a YBCO/in situ barrier/Pb
junction.

Figure 6. An example of
expected tunnel conductance
for the scheme of Fig. 5.

gap anisotropy, two processes (a) and (b) are involved. The superconductivity in the ab plane is considered to be strong, which will survive up to the top of YBCO surface, although it is appreciably weakened in the close vicinity of the surface. The process (a) produces a conventional quasiparticle tunneling. On the other hand, the superconductivity along the c-axis will be destroyed to yield normal regions on the surface. The gap parameter at the appreciable depth from the film surface will be, however, well defined. Then the tunneling process consists of two different series processes, ie. the quasiparticles are first injected from the Pb side into the normal regions of a YBCO film, then experience the Andreev reflection by the pair potential Δ_c [11]. The extent of these two contributions depends on the detailed crystal structure on the surface.

Figure 6 depicts an example of the expected tunnel conductance corresponding to Fig. 5, which very much resembles the observed characteristics of Fig. 2. The finite conductance within the gap region appears due to the process (b) in Fig. 5, whose dominant contribution comes from the Pb/I/normal tunneling conductance. The gap sum and difference structures are not visible due to the broadness gap structure of Δ_{ab} $(\Delta_{ab} > \Delta_{Pb})$.

In case of YBCO/native barrier/Pb junctions, a long range variation of the order parameter in the ab plane from the surface toward the inside of a YBCO film is expected due to the effect of water, impurity gases etc. which were contained in the external atmosphere. The gap parameter along the c-axis will be strongly suppressed, being completely normal or barely surviving. The expected tunnel conductance in this case will become a V-type characteristic against the junction bias voltage, which is quite consistent with the results of Fig. 3.

4. JOSEPHSON EFFECT

The Josephson tunneling phenomena were observed for the YBCO/I/Pb junctions whose tunnel resistance less than 0.5Ω. They were observable for the junctions with both the c-axis and a axis oriented YBCO films. Figure 7 shows the I-V characteristics of the observed Josephson junctions with both the c-axis and a-axis oriented YBCO films. For the c-axis oriented film in whose case Josephson current will flow along the c-axis direction, the I-V curve looked like a weak-link type with small output voltage, which eventually be connected with the I-V curve of Pb/I/N gap structure. On the other hand, it exhibited strongly nonlinear

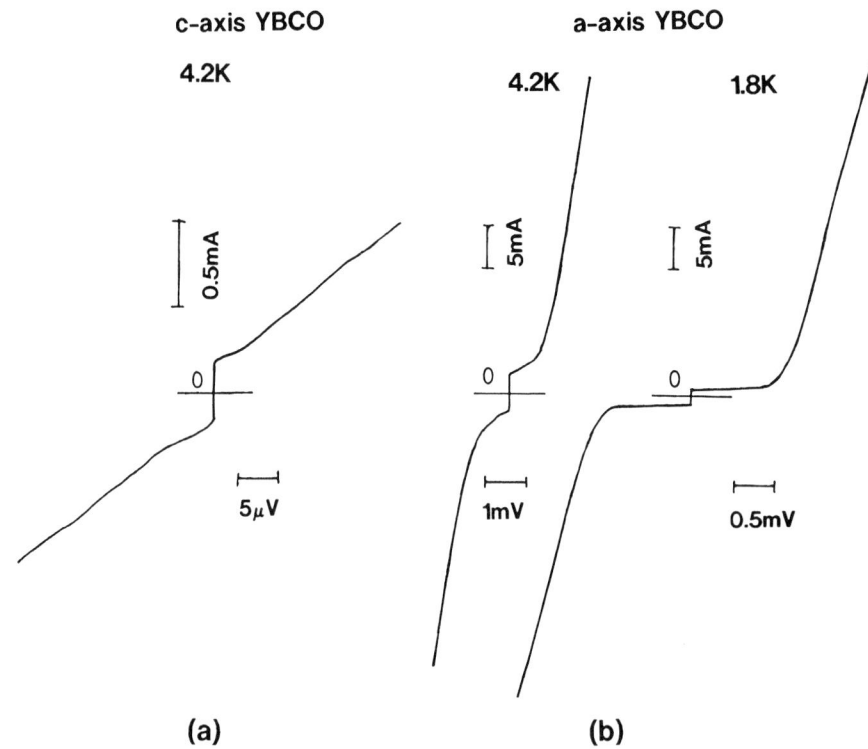

(a)   **(b)**

Figure 7. Josepshon I-V curves for YBCO/native barrier/Pb fabricated on the c-axis (a) and a-axis oriented YBCO films.

behavior for the use of a-axis oriented film. The nearly sudden appearance of voltage of about 1mV above the critical current of was recognized. Even a hysteretic behavior was sometimes observable. It seems that the Pb gap characteristic served as a load line switching for Josephson tunneling.

The magnitude of Josephson current was typically 0.05-0.2mA for the c-axis film and 0.5 - 3mA for the a-axis oriented film. The I_cR product was 0.01-0.02mV for the c-axis oriented film while it was 0.1-0.2mV for the a-axis oriented film. When the Josephson current was large, the voltage output became also large and sharp. This means that the load line switching behavior is governed by the vertically rising characteristic of Pb/I/N curve. According to the model given in Fig. 5, Josephson current will not flow along the c-axis due to the surface degradation. Therefore the relatively small Josephson current flow along the c-axis oriented film would arise from the imperfect surface morphology.

Figure 8 depicts the I-V curves of YBCO/native barrier/Pb Josephson tunnel junctions with the c-axis oriented YBCO film under the microwave irradiation of 9GHz at different power strength. The native barrier was formed by an in situ growth technique in this case. The induced Shapiro steps with the voltage interval of about 18μV are clearly visible whose step height changed with the microwave power variation. The current step height as a function of microwave voltage amplitude exhibited Bessel function like behavior which is consistent with the conventional theory, reflecting that the fabricated junctions really behaved tunnel like [6,12].

The Josephson maximum current exhibited a Fraunhofer-like diffraction pattern against external magnetic field. The observed diffraction pattern was sometimes very periodic, sometimes not. The asymmetric pattern against the direction of applied magnetic field was also often observed. Figure 9 shows an example of the observed characteristic. We believe that a variety of observed diffraction patterns were neither due to the crystal orientation of YBCO films nor due to the intrinsic property of YBCO material. It might be due to the imperfect surface morphology.

A puzzling nature of Josephson maximum current was found in its temperature dependence. Figure 10 shows such an example. With decreasing bath temperature below Tc of Pb, Josephson maximum current first increased but decreased again forming a broad maximum whose behavior is quite different from that of conventional Josephson junctions. It was found that the diffraction pattern itself shrinked as a whole [13].

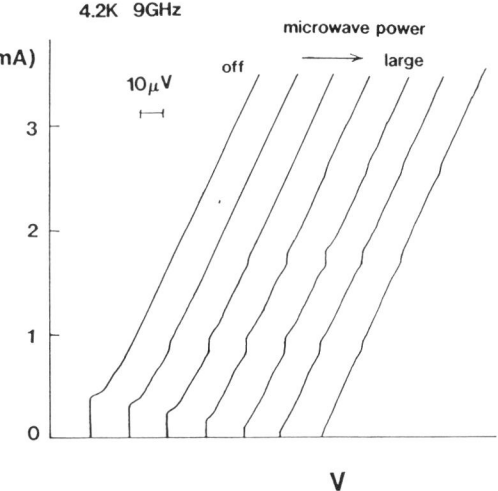

Figure 8. Josephson I-V curves under microwave irradiation of 9GHz.

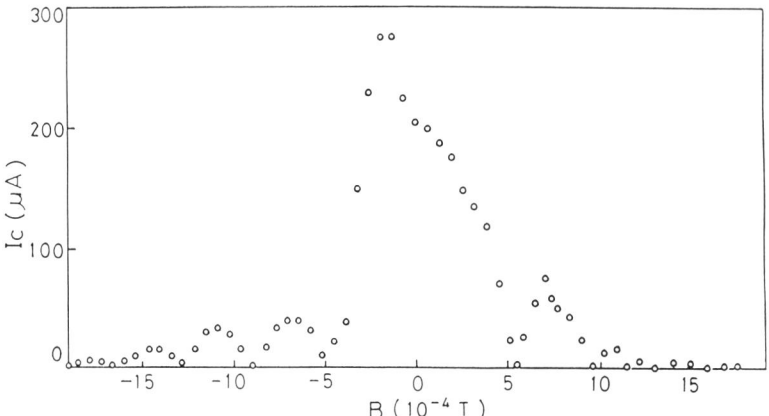

Figure 9. An example of Josephson maximum current as a function of the applied magnetic field.

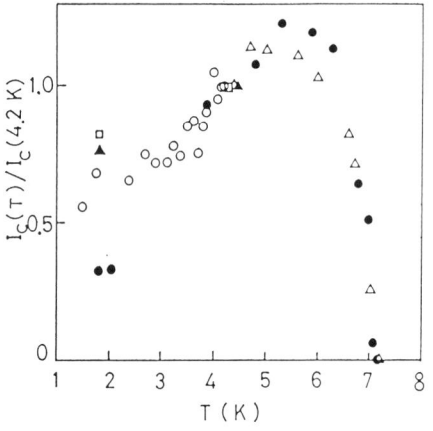

Figure 10. Temperature dependence of Josephson maximum current.

5. PROXIMITY EFFECT MULTILAYERS

Finally we mention about the results on the possible proximity effect of YBCO material by fabricating a SN/I/S' junction geometry. With this configuration, the exudation of the YBCO gap paramater is studied. The result on a YBCO/Al/AlOx/Pb multilayer tunnel junction (dV/dI vs V) is shown in Fig. 11. The samples were fabricated by an all in situ process, ie. directly depositing an Al film on a YBCO film, then oxidizing the surface of Al film by rf glow discharge technique (rf power 100W) in the oxygen pressure of 2×10^{-4} Torr for 10-30min, finally depositing a

Pb film [14]. The junction
resistance was typically
around 1kΩ.

The dV/dI vs V chara-
cteristics significantly
changed as the thickness
of aluminum was increased.
In the figure, the curve
corresponding to the
absence of aluminum layer
was that of a YBCO/MgO/Pb
junction. For the junctions
whose Al thickness was
less than 10nm, quali-
tatively similar charac-
teristics were obtained
although the Pb gap
structure around the zero
bias was apparently
broadened. For d_{Al}=15nm,
the Pb gap structure
completely disappeared to
yield one broad gap structure.

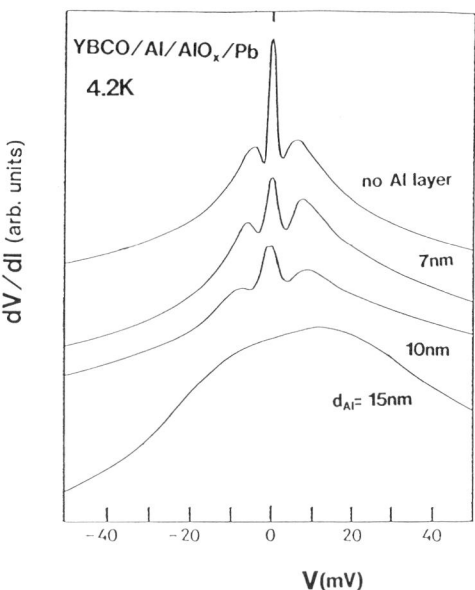

Figure 11. dV/dI vs V curves
of proximity-effect tunnel
junctions with the Al layer
with different thickness.

We understand the observed situation in the following way. On
the surface of a YBCO film, as stated above, the observable order
parameter distribution is inhomogeneous due to the possible mixed
tunneling of c-axis and ab plane. The gap parameter induced in
the Al film, however, tends to smear out these inhomogeneities as
the Al thickness is increased. The broadening and disappearance
of Pb gap structure are interpreted by this idea since , when the
tunneling into normal regions disappears, the junction I-V
characteristic should change to a conventional S/I/S' type curve
in which the Pb/I/N structure is not expected. In the transient
region of this phenomenon, the broadening of Pb gap structure is
of course expected due to the contribution of small YBCO order
parameter.

6. CONCLUSIONS

We have reported the tunneling gap and Josephson effect of
high Tc YBCO/I/Pb(or Au) tunnel junctions with artifical and
native barriers. The observed results were apparently quite
different from those expected from the BCS theory. Considering
the gap anisotropy, the short coherence length and the surface
degradation effect characteristic to YBCO material into account,
however, they may be well interpreted in terms of a proposed
tunneling model. Moreover, the result on the proximity-effect
tunnel junction was also consistent with this model.

7. REFERENCES

1 J.R. Kirtley, Intern. Journ. of Mod. Phys., B4 (1990) 201.
2 J. Geerk, X.X. Xi and G. Linker, Z. Phys., B73 (1988) 329.
3 M. Gurvitch, J.M. Valles, Jr., A.M. Cucolo, R.C. Dynes, J.P.
   Garno, L.F. Shneemeyer and J.V. Waszcak, Phys. Rev. Lett., 63
   (1989) 1008.
4 M. Furuyama, I. Iguchi, K. Shirai, T. Kusumori, H. Ohtake,
   S. Tomura and M. Nasu, Jpn. J. Appl. Phys., 29 (1990) L459.
5 I. Iguchi, M. Furuyama, T. Kusumori, K. Shirai, S. Tomura, M.
   Nasu and H. Ohtake, Jpn. J. Appl. Phys., 26 (1990) L614.
6 I. Iguchi and Z. Wen, IEEE Trans. on MAG, 27 (1991) 3102.
7 Z. Wen, I. Iguchi and K. Nakamura, J. Appl. Phys. 69 (1991)
   1 May.
8 J. Kwo, T.A. Fulton, M. Hong and P.L. Gammel, Appl. Phys.
   Lett., 56 (1990) 788.
9 I. Iguchi and Z. Wen, Physica C to be published.
10 I. Iguchi, Z. Wen and M. Nasu, Ext. Abstr. on 3rd Workshop on
   High-Temperature Superconducting Electron Devices, 1991, p.166.
11 G.E. Blonder, M. Tinkham and T.M. Klapwijk, Phys. Rev., B25
   (1982) 4515.
12 I. Iguchi, Z. Wen and M. Furuyama, Physica B, 165&166
   (1990) 73.
13 Z. Wen and I. Iguchi, Jpn. J. Appl. Phys. 30 (1991) L188.
14 T. Kusumori, Z. Wen, H. Ohtake, M. Kamiya and I. Iguchi, Proc.
   3rd Int. Symp. on Superconductivity (ISS'90), 1990, to be
   published.

High T_c Superconductor Thin Films
L. Correra (Editor)
© 1992 Elsevier Science Publishers B.V. All rights reserved.

CW and pulsed IR detection of aerosol deposited YBCO thin films

Q. Tang, A. Driessen, L.H. Hilderink and Th.J.A. Popma

MESA-Research Institute, Department of Applied Physics,
University of Twente, P.O.Box 217, 7500 AE Enschede, The Netherlands

Abstract

A CW and pulsed near-infrared (IR) detector made of aerosol-deposited YBCO thin films has been developed. The main characteristic of these films is the fine-grain structure, which will increase the detector sensitivity due to the presence of a high density of boundary weak links in a bridge or meander. Under certain conditions, a responsivity ($\mathcal{R}$) and detectivity (D^*) of about 10 V/W and 10^8 cm$\sqrt{\text{Hz}}$/W respectively are yielded. The response pulsewidth is less than 100 ns at 77 K.

1. INTRODUCTION

The mechanism of the optical response of high T_c superconducting (HTS) films is still a controversial issue whether it is a purely bolometric [1] or at least partially nonequilibrium [2,3] (breaking of the Cooper pairs and generation of quasiparticles) effect. However from a practical point of view, there is no doubt that a broadband, reasonably fast and sensitive optical detector can be constructed using HTS films.

A potentially commercial HTS optical detector should meet the following requirements: (1) broadband, from the visible to the far-IR; (2) response time in the order of nanoseconds; (3) $\mathcal{R} > 10^2$ V/W; (4) operating temperature at 77K; (5) low intrinsic detector noise. A prototype detector which can fulfil all these requirements has not yet been developed. However, several reports have shown that some parameters of the HTS detectors have been close to or even better than the expected values. For example, Frenkel *et al.* [2] reported a nonbolometric optical response as fast as 1 ns to a Nd:YAG (λ=1.06μm) laser radiation in nongranular YBCO films at 40 K. Zheng *et al.* [3] claimed the response time of the laser-deposited YBCO film is less than 10 ns to a CO_2 laser (λ=10.6μm) at 10 K. Nevertheless, somewhat discouragingly, the detectors with high speed always showed a very low responsivity. On the other hand, Richards *et al.* [4] demonstrated that HTS transition edge bolometers potentially had much better sensitivity at λ>20μm than any competing sensor at or above 77K. Noise equivalent powers in the range of $(1-20)\times10^{-12}$W Hz$^{-1/2}$ should be achievable.

In the following, we shall show that a reasonably fast as well as sensitive optical detector can be fabricated using fine-grain HTS films. The principle of the improved sensitivity can be described as follows: (1)

Films having fine grains (d≅200 nm) can form a high density of boundary weak links in a structured bridge or meander. The boundary weak links may contribute coherently to the detected signal, but incoherently to the noise, so that the signal-to-noise ratio will be high [5]. (2) It should be noted that, being similar to BPBO films, the reflection coefficient of the polycrystalline materials (including fine-grain films) is smaller than that of single crystals [5,6]. Low reflectivity and high absorption will increase the detection efficiency.

In our previous work we examined the optical response of a fine-grain detector to chopped visible light [7]. In this paper, we shall report on the response of bridges or meanders to CW and pulsed IR radiation (λ from 0.9 to 1.3μm). The detector parameters will be presented.

2. EXPERIMENTAL PROCEDURES

A fine-grain $YBa_2Cu_3O_{7-x}$ thin film with 250-nm thickness on a $SrTiO_3$ (100) substrate was prepared by metalorganic aerosol deposition. After deposition and post-annealing, the film with a c-axis orientation showed a good quality, typically, $T_{c,0} \cong$ 90K and $J_c > 10^4$ A/cm² at 77K. The grain size was about 200 nm. The details of the film preparation are published elsewhere [8,9].

The devices, bridge and meander, were fabricated by photolithographic technology and Ar ion-beam milling. Afterwards oxygen plasma etching was performed to remove the photoresist. Except for the development of photo-resist in the developer (Shipley Europe type 351), all processes were kept dry. This avoided the extra contamination of the YBCO film by chemical solutions. Using this process different size devices could be made as shown in Figs.1(a), (b) and (c). After structuring, $T_{c,0}$ showed a slight decrease of a few degrees, which was dependent on the film quality and the size of the device. More details about The morphology of the fine-grain films are given in Refs. [7,8].

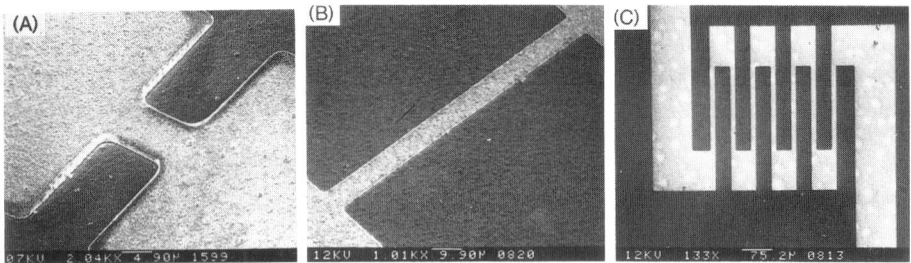

Fig. 1. Fine-grain YBCO devices on $SrTiO_3$ substrates. (a) a 10×10 and (b) a 10×100 μm² bridge, (c) 15×1400 μm² meander.

A gold layer with a thickness of 200 nm was evaporated onto the YBCO film to make electrical contacts for the electro-optical measurements. The experimental setup for the detection of a pulsed laser is shown in Fig.2; and the one for the detection of a CW laser is given in Ref.7. Two kinds of pulsed IR sources are used in the present work. One is a GaAs laser diode (λ=0.904μm) with 100-ns pulsewidth at a repetition rate that can be varied from 10 kHz to 30 kHz; the other is a mode-locked Nd:YLF laser (λ=1.31μm) with 75ps pulsewidth at a repetition rate of 80 MHz. Both lasers can be taken as CW radiation and a average power (=peak power $\times$ pulse duty cycle) is measured by a powermeter (Scientech type 737). The output energy of the GaAs laser diode is about 60 nJ per pulse.

A 100 MHz oscilloscope was used to display the pulse response to the GaAs laser. The fall time was measured by a fast oscilloscope (Tektronix CSA803) equipped with an input sampling head (Tektronix SD-26), having a rise time of about 25 ps.

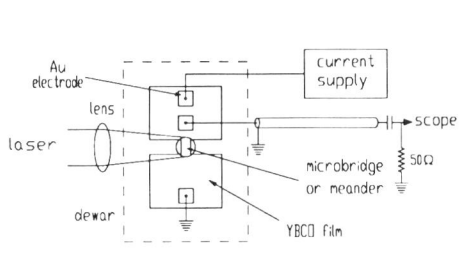

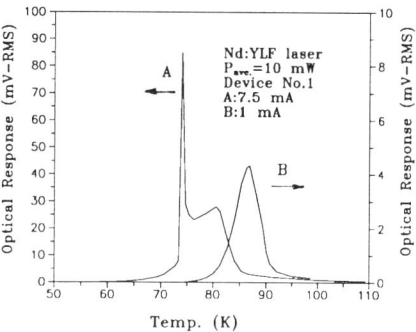

Fig.2. Experimental setup for the detection of a pulsed laser.

Fig.3. CW optical response versus temperature. 270Hz chopping.

3. RESULTS

3.1. CW IR detection

Fig.3 shows the optical response (device 1, see Table I) as a function of temperature at two bias currents (7.5mA and 1mA) to a Nd:YLF laser. At low bias current a single peak can be found in the transition temperature region. However, at a high bias current, besides this peak, a sharp response peak also appears, which is due to an abrupt voltage step in the current-voltage (I-V) curve. This result has been described in Ref.7, in which the sharp peak is referred to as a junction-related response.

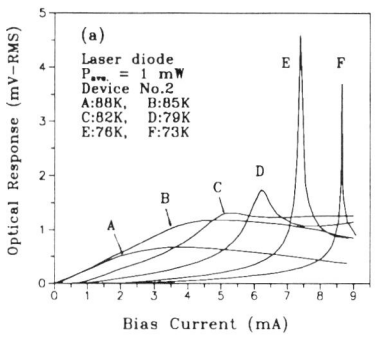

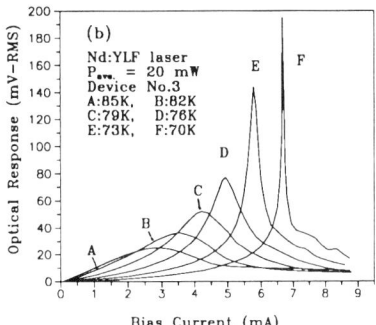

Fig.4. CW optical response versus bias current at different temperatures. (a) device 2 and (b) device 3.

Figs.4(a) and (b) show the optical responses as a function of bias current at different temperatures. We choose a temperature from the beginning of the transition to one low enough to observe the sharp response peak. Fig.4(a) shows the response of device 2 to the GaAs diode. The average incident power is approximately 1mW. Again it is shown that in a low bias current there is no sharp peak, while in the high bias current, a sharp peak is present. We repeat this measurement on device 3, using a Nd:YLY laser with an average power of about 20 mW. The results indicate that they are qualitatively the same. Due to the presence of the sharp optical response peak, a high $\mathcal{R}$ can be obtained in the vicinity of the peak. Normally, the $\mathcal{R}$ at the peak is one to two orders of magnitude higher than that at the position away from the peak region.

3.2. Pulsed IR detection

Fig.5 shows the temporal optical responses to a 100-ns pulsed GaAs diode at 5 mA current at different temperatures. In order to see clearly how the trace and the fall time change with the operating temperature, we remove the terminating 50Ω resistance in Fig.2 and directly connect the output signal to an oscilloscope. In the set of traces in Fig.5, it is found that their rise time is almost the same in all the response regions. The magnitude of the voltage output is proportional to the value of the CW response in Fig.4(a). The fall time at a high temperature (88K) is much longer than that of the real optical pulse, being about 1-2 μs. With the decrease of the operating temperature, a significant decrease of the fall time can be seen. From 82 K to 73 K the fall time continously reduces, meanwhile the magnitude of the response reduces as well. At a further reduced temperature no detectable response appears. We can therefore conclude that the waveform of response traces significantly changes with temperature, and that the fall time at high temperature is much longer than that at low temperature.

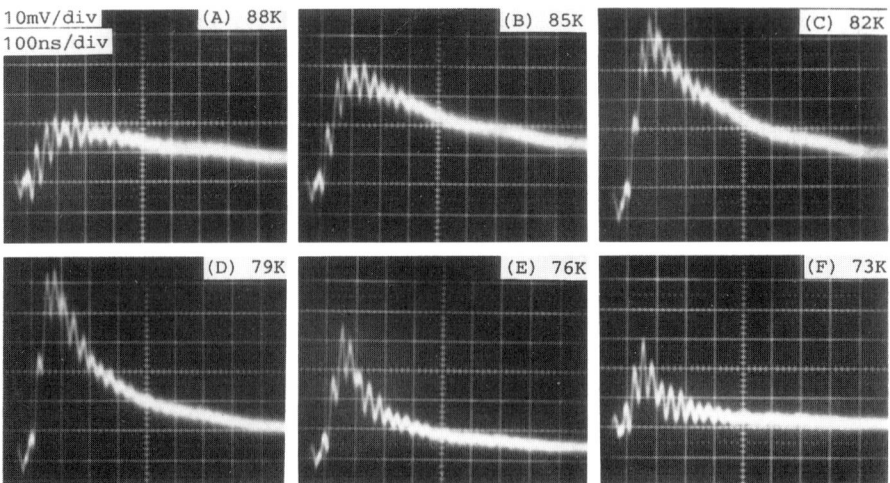

Fig.5. Oscilloscope traces of the voltage output of device 2 at different temperatures. X: 100ns/div; Y: 10mV/div.

The pulse performance of the HTS detector at 77K and 5 mA is given in Fig.6(b) when terminated with 50Ω. In order to compare its trace with that of a semiconducting photodetector, a reverse output trace of a InGaAs PIN photodiode (Antel AR-D25) at 295K is shown in Fig.6(a). This is an ultra high-speed photodetector (rise time <35 ps, spectral response from 0.85 to 1.7μm). Fig.6(b) indicates that the HTS detector is able to display a 100-ns pulse. The fall time is about 80 ns under optimal conditions, which corresponds to a bandwidth of 12 MHz. This value is of the same order of magnitude as commercial photovoltaic IR detectors, such as InAs, InSb and PbTe, with a bandwidth of 40 MHz [10].

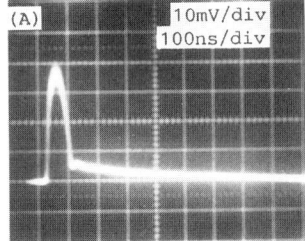
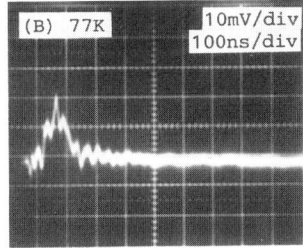

Fig.6. Comparison of the trace of (a) an InGaAs PIN photodiode at 295K (in reverse) with that of (b) a HTS detector (device 2) at 77K and 5.0 mA when terminated with 50Ω. X:100ns/div; Y:10mV/div.

4. DISCUSSION

We report the experimentally determined performance of a fine-grain HTS detector. The D^* can be estimated according to the following expression:

$$D^* = (A \ bw)^{1/2} \mathcal{R} / V_n, \tag{1}$$

where A is the area of the detector, bw is the bandwidth which is obtained by the inverse of the detector response time, $\mathcal{R}$ is the responsivity in units of volts per watt and V_n is the noise voltage. In our case, $V_n \approx 1\mu V$, and the rest of parameters can be found in Table I. Under certain conditions, $\mathcal{R}$ and D^* of 10 V/W and 10^8 cm$\sqrt{\text{Hz}}$/W respectively are yielded.

Table I Typical parameters of fine-grain HTS detectors.

Device No.	Active area (μm^2)	Resist. (at 95K) (kΩ)	Operating conditions Temp(K)	Bias(mA)	Light λ (μm)	$\mathcal{R}$ (V/W)	D^* cm$\sqrt{\text{Hz}}$/W	Fall time (μs)
1	10×10	0.5	88	2.0	0.9	0.5	0.4×10⁶	1-2
2	10×10	0.6	77	7.0	0.9	6.0	2.1×10⁷	0.08
3	10×100	1.0	70	6.7	1.3	15	0.9×10⁸	0.3
4	15×1400 (meander)	11	77	5.0	0.9	30	1.9×10⁸	2-5

With regard to fine-grain HTS films a peculiarity is that, at lower temperatures and larger currents, a voltage step in the I-V curve, a resistive step in the resistance-temperature curve and a sharp peak in the optical response are observed [7,8,11]. The nature of these reproducible steps is still not fully understood, but we believe that the properties are closely related to the presence of boundary-type weak links within a bridge. A possible interpretation will be given elsewhere [11]. Although there is a significant decrease of the fall time with decreasing temperature in Fig.5, it cannot be uniquely used as a criterion of nonequilibrim effect. The changes of the heat capacity and conductivity with temperature, for both the YBCO film and the substrate, also affect the thermal diffusion time. In fact, as a high bias current flows through a bridge, which is being exposed to a laser radiation, the bolometric component is inevitably involved in the measurement. It could be possible that the response of our detectors is mainly dominated by the heating of the radiation and is influenced by boundary weak links.

The intrinsic noise of a fine-grain film is a crucial problem for its application. We do not stress on it in the present study. A preliminary result shows that a device prepared by metalorganic aerosol deposition has a low noise, which is comparable to that of the polycrystalline devices. The extrinsic rf interference in Fig.5 can be eliminated by well designed shielding and grounding.

Further work on the improvement of detector sensitivity and response speed will be carried out. First, with a thinner fine-grain film ($\leq$ 100 nm) it may be possible to display a fast optical pulse, such as our picosecond mode-locked Nd:YLF laser pulse. Second, a meander structure shows a high sensitivity at the expense of the response speed. With proper optimization of the detector size and focussing conditions, the realization of a more sensitive and fast HTS detector is expected. Third, the performance of the fine-grain detector under far-IR radiation ($\lambda > 20\mu$m) will be investigated systematically.

5. REFERENCES

1   W.S.Brocklesby, Don Monroe, A.F.J.Levi, M.Hong, S.H.Liou, J.Kwo, C.E. Rice, P.M.Mankiewich and R.E.Howard, Appl.Phys.Lett., **54** (1989) 1175.
2   A.Frenkel, M.A.Saifi, T.Venkatesan, P.England, X.D.Wu and A.Inam, J.Appl.Phys., **67** (1990) 3054.
3   J.P.Zheng, Q.Y.Ying and H.S.Kwok, Physica C **168** (1990) 322.
4   P.L.Richards, J.Clarke, R.Leoni, Ph.Lerch, S.Verghese, M.R.Beasley, T.H.Geballe, R.H.Hammond, P.Rosenthal and S.R.Spielman, Appl.Phys.Lett., **54** (1989) 283.
5   Y.Enomoto and T.Murakami, J.Appl.Phys., **59** (1986) 3807.
6   S.Uchida, Physica **148B** (1987) 185.
7   Q.Tang, A.Driessen, P.Hoekstra, L.Hilderink, A.Van Silfhout and Th.J.A.Popma, J.Less-Common Metals, **164-165** (1990) 1587.
8   Q.Tang, H.Albers, A.Driessen, L.T.Hilderink, P.V.Lambeck and Th.J.A.Popma, J.Aerosol Sci., **21**, Suppl.1 (1990) S737.
9   A.Driessen, Q.Tang, L.Hilderink and Th.J.A.Popma, In:Proc.MRS **169** (1990) 601.
10 For example, Laser Monitoring Systems, product sheets.
11 A.Driessen, Q.Tang, P.Hoekstra and Th.J.A.Popma, submitted to J.Low Temp.

High T$_c$ Superconductor Thin Films
L. Correra (Editor)
1992 Elsevier Science Publishers B.V.

YBCO based multilayers for optoelectronic devices

J.C. Villegier[a], H. Moriceau[a], H. Boucher[a], L. Di Cioccio[a], A. Ghis[a], A. Jaeger[a], M. Levis[a], F. Pourtier[a], M. Schwerdtfeger[a], M. Vabre[a], C. Villard[a], R. Chicault[b]

[a]LETI-DTA-DOPT - 85X - 38041 Grenoble Cédex, France

[b]Laboratoire de Spectrométrie Physique, USMG, St Martin d'Hères, France

Abstract

YBCO based multilayers have been deposited independently by three techniques : laser ablation, inverted cylindrical target sputtering (ICM) and on-axis planar D.C. Magnetron Sputtering. The last technique is used to cover uniformly R-plane sapphire and LaAlO3 2 inch wafers with YBCO or multilayers to achieve optoelectronic devices such as infrared detectors. Very thin (~ 3 nm) YSZ and MgO dielectric films have been studied as tunnel barriers for making such high Tc tunnel junctions.

1. INTRODUCTION

In order to develop practical devices using high Tc superconductors for optoelectronic and signal processing components, several conditions have to be fulfilled. The possibility to make reliable integrated multilayer based circuits composed of active devices and broadband superconductive lines on conventional substrates such as silicon or sapphire of at least 2 inches diameter. It has been shown recently that good quality YBaCuO films with low surface resistance can be done on R-plane sapphire using a YSZ or SrTiO3 buffer layer [1][2][3]. In this study we compare three deposition techniques in order to deposit good quality YBCO films on large areas (up to 30 cm2) and show that it is possible to achieve uniform SIN (YBCO/YSZ/Ag), and SIS (YBCO/MgO/YBCO) trilayers. Such trilayers have been observed by HRTEM cross sections and patterned using a whole wafer "SNOP" process and ion milling.

2. COMPARISON OF YBa2Cu3O7-x LAYERS DONE BY LASER ABLATION AND SPUTTERING

Thin superconductive (30 nm to 300 nm), YBa2Cu3O7-x films have been deposited from a single sintered YBa2Cu3O7-x target on heated substrates by the "direct" process , meaning that the oxidation stage of the films occured during the cooling phase after the deposition. Three deposition techniques have been tested in three different vacuum systems :
- a laser ablation system using an KrF excimer laser (λ = 248 nm),
- an inverted Cylindrical Magnetron (ICM) sputtering system,
- a planar magnetron sputtering equipment.

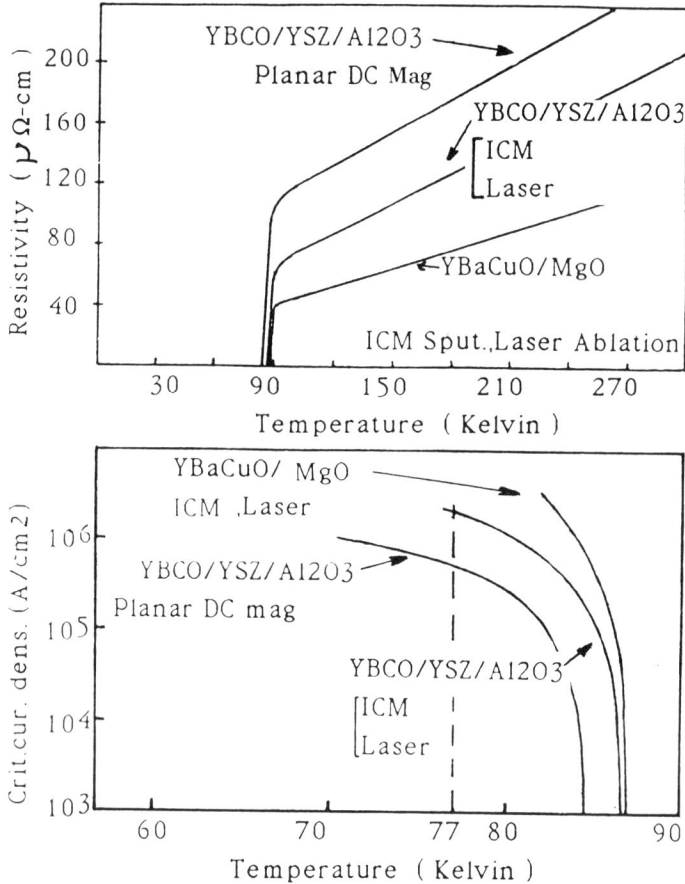

Figure 1. Typical performances of YBCO films done by laser ablation, ICM and DC Magnetron on MgO and YSZ/R-plane sapphire.

Each vacuum system contains other targets in order to deposit buffer layers or tunnel barriers in situ. For comparison we studied YBCO films on MgO (100) and on R-plane sapphire with a YSZ buffer layer deposited at the same substrate temperature than for the YBCO deposition. In such a case, a thin buffer layer is necessary both to avoid the diffusion from the substrate and to fit the physical parameters of the film to those of the sapphire. The YSZ buffer layer are grown at ~ 750°C either by RF sputtering or by laser ablation depending on the system with a thickness ranging from 30 to 100 nm. The YSZ layer is (002) oriented on the sapphire (Rocking curve FWHM is ~ 1.5°) with in-plane orientations as observed by pole figures.

Table 1
Deposition conditions of YBCO/YSZ on R-plane sapphire

DEPOSITION BY LASER ABLATION	DEPOSITION BY SPUTTERING
YTTRIUM STABILIZED ZIRCONIA (YSZ) film	**YTTRIUM STABILIZED ZIRCONIA (YSZ) FILM**
	R.F. Magnetron planar cathode
. Substrate : Al2O3 (R plane)	. Substrate : Al2O3 (R plane)
. Thickness : 150 nm	. Thickness : ~ 100 nm
. Excimer laser (λ=248 nm ; 3.2 J/cm2 ; 3 Hz)	. Deposition temperature : 750°C
. Deposition temperature : 700°C	. Atmosphere : Ar+O2 ; 12 mTorr
. Atmosphere : N2O ; 0.2 Torr	
YBaCuO SUPERCONDUCTIVE FILM	**YBaCuO SUPERCONDUCTIVE FILM**
	D.C. MAGNETRON PLANAR CATHODE
. Substrate : Al2O3 + YSZ buffer layer	. Substrate : Al2O3 + YSZ buffer layer
. Excimer laser (λ=248 nm ; 2.8 J/cm2 ; 2 Hz)	. Deposition temperature : 750°C
. Deposition temperature : 730°C	. Atmosphere : O2;Ar ; 12 mTorr
. Atmosphere : N2O ; 0.7 Torr	. Thermal annealing : O2 ; 1 atm
. Thermal annealing : O2 ; 300 Torr	. ρ100 K = 150 $\mu\Omega$.cm
. ρ100 K = 75 $\mu\Omega$ .cm	. Thickness : 300 nm
. Thickness : 170 nm	
	INVERTED CYLINDRICAL MAGNETRON
	. Substrate : Al2O3 + YSZ buffer layer
	. Deposition temperature : 750°C
	. Atmosphere : O2+Ar ; 0.6 Torr
	. Thermal annealing : O2 ; 1 atm
	. ρ100 K = 100 $\mu\Omega$.cm
	. Thickness : 120 nm

Table 2
Comparison of YBCO/YSZ/Sapphire ($1\bar{1}02$) c-axis films.
(Film thickness : YBCO ~ 200 nm. YSZ ~ 100 nm.

		Laser Ablation	Sputtering (Planar)	Sputtering (ICM)
Rocking curves	YSZ	< 2°	< 1.5°	-
	YBCO	< 2°	< 1.5°	< 2°
YBCO deposition rate		~6nm/mn	~6nm/mn	~10nm/mn
$R_{\square\,300K}$ (±20%) uniformity area		$\emptyset$ 10 mm	$\emptyset$ 55 mm	$\emptyset$ 25 mm
ρ_{100K}		~80 $\mu\Omega$cm	~120 $\mu\Omega$cm	~80 $\mu\Omega$cm
Tco		~ 87 K	~ 84 K	~ 86 K
Jc (77 K)		>10^6Acm^{-2}	>10^5Acm^{-2}	>10^6Acm^{-2}

A good c axis and in plane orientation of the YBCO layer on top of YSZ buffer layer is found both in the laser deposited films and in the sputtered films. Typical deposition conditions for the YSZ and YBCO layers are reported in table 1. YBCO films deposited on MgO(100) substrate close to ~ 750°C by ICM and planar magnetron sputtering and close to 720°C by laser ablation also present a very good in-plane epitaxial behavior when observed by pole figures and by plane view TEM, and show low dechanneling: Xmin < 10% at 1.5 MeV. The comparison of the YBCO/YSZ films deposited by the three techniques is listed in table 2. It appears that the films deposited by a large planar magnetron target (on axis) present some degradation in comparaison to ICM and laser ablated films, but they are uniform on a large diameter (ϕ 55 mm) with a good texture and rather good superconductive properties (figure 1). Such YBCO films deposited on two inches diameter sapphire substrates have been patterned by positive photoresist and ion-milled with Argon (~ 500 eV) in order to make striplines and infrared detectors as shown on figure 2.

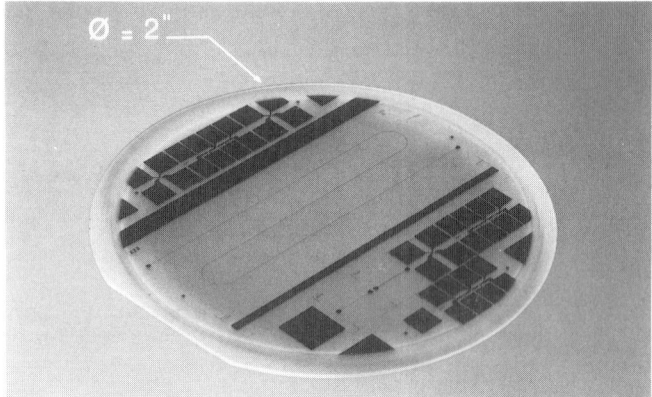

Figure 2. YBCO devices patterned on a 2" R-plane sapphire wafer covered by a YSZ buffer layer.

3. TRILAYERS "SNOP" PROCESS FOR YBCO-YSZ-Ag (SIN) AND YBCO-MgO-YBCO (SIS) TUNNEL JUNCTIONS AND ARRAYS

YBCO-YSZ barrier-Ag (SIN) and YBCO-MgO-YBCO (SIS) junctions and arrays have been fabricated using the "SNOP" process. Such a trilayer process has been developped first for NbN-oxide-NbN and Nb-AlOx-Nb junctions [4] and extended later for submicrometer areas NbN-MgO-NbN junctions [5]. The main advantages of "SNOP" (Selective Niobium Overlap Process) (figure 3) on other "trilayer" processes are :
- it needs only two stencil masks,
- simple "cross type" junctions are accurately defined in area far below 1 μm2,
- there is no parasitic capacity because there is no overlapping extra-insulating layer deposition.

The trilayers are patterned using conventional photolithography (Shippley AZ 1400-31) and Ar ion milling. Then a YSZ self aligned dielectric layer 50 nm thick is RF sputtered at room temperature on a rotating substrate before base electrode resist stripping in acetone (YSZ Lift off). The interconnecting Ag layer is then DC sputtered at room temperature, insuring a low resistance contact. The interconnecting and counter-electrodes layers are patterned using argon ion milling. The etching process is stopped on the surface of YBCO base electrode. Such a process shown in figure 4 precisely define junctions electrodes down to micrometer sizes and allows to make arrays of 100 junctions in series of 10 x 10 μm2 areas.

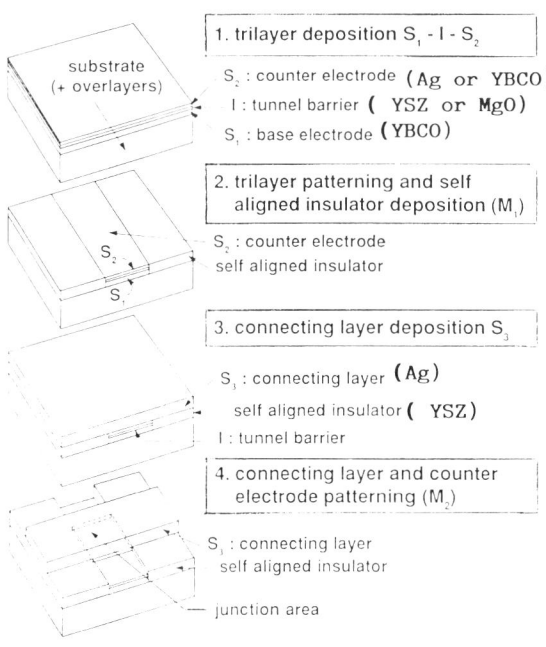

1. trilayer deposition S_1 - I - S_2

substrate (+ overlayers)

S_2 : counter electrode (Ag or YBCO)
I : tunnel barrier (YSZ or MgO)
S_1 : base electrode (YBCO)

2. trilayer patterning and self aligned insulator deposition (M_1)

S_2 : counter electrode
self aligned insulator

3. connecting layer deposition S_3

S_3 : connecting layer (Ag)
self aligned insulator (YSZ)
I : tunnel barrier

4. connecting layer and counter electrode patterning (M_2)

S_3 : connecting layer
self aligned insulator

junction area

Figure 3. SNOP process used for HTS junctions associated with argon ion milling for etching.

YBCO-YSZ-Ag on R-plane sapphire :
The deposition sequence of the YBCO-YSZ-Ag trilayer is :
a) YSZ (60 nm) + $YBa_2Cu_3O_{7-x}$ (200 nm) sequential deposition at ˜ 700°C followed by a slow-cooling phase down to room temperature in high oxygen pressure.
b) Light oxygen RF-plasma is followed by the YSZ barrier (˜ 3 nm thick) RF-sputter deposition at room temperature on a rotating (10 t/min) substrates holder.
c) Ag (˜ 50 nm thick) counter-electrode deposition at room temperature by DC-Magnetron.

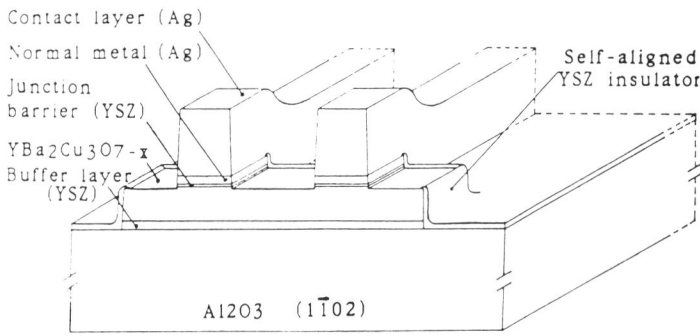

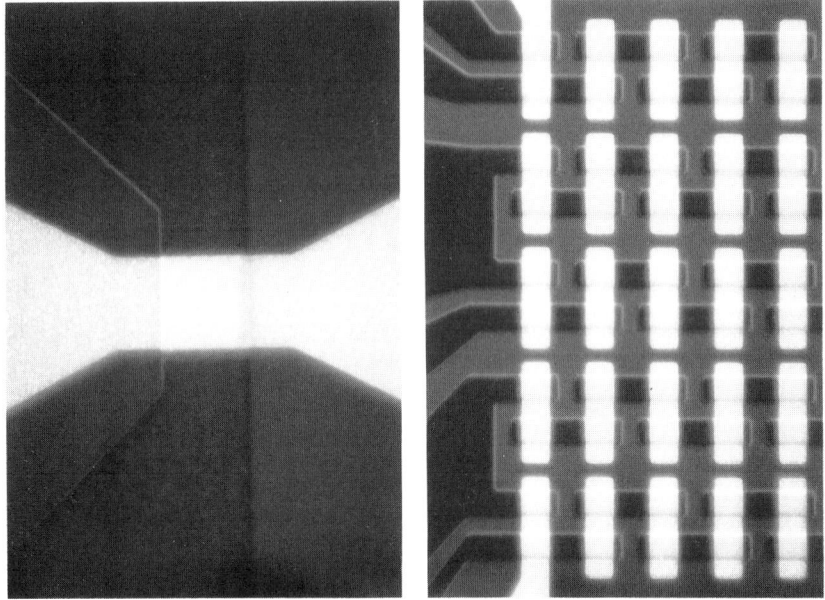

Figure 4. Cross-section of YBCO-HTC "snop" junctions multilayer. Plane view of single $20 \times 20 \mu m^2$ junction and arrays of $10 \times 10 \mu m^2$ SIN diodes.

YBCO-MgO-YBCO on MgO substrate :

In order to study the growth of HTSC (YBCO-MgO-YBCO) SIS structure using a MgO ultrathin film (˜ 3 nm) as a tunnel barrier we choose an alternate approach to the heteroepitaxial MgO growth leading to a shorted and discontinuous barrier [6][7] : a thin (˜ 75 nm) YBCO base electrode was deposited by DC Magnetron Sputtering using a hollow target (HITEC inc) at about 750°C in a mixture of O_2/Ar (0.2/0.4 mbar) followed by cool down in ˜ 300 mbar O_2 to 50°C where a thin (3 nm) amorphous and continuous MgO layer is RF sputtered. Then the films are heated again to ˜ 650°C where the YBCO (100 nm) counterelectrode is deposited with the same conditions as for the base electrode. After cool down in oxygen a thicker (10 nm) amorphous MgO layer is deposited as a passivation layer before patterning and junctions processing described elsewhere [4]. Such a YBCO-MgO-YBCO (+ αMgO) structure has been studied by cross section TEM (figures 5,6,7).

TEM Results :

Cross sections specimens for TEM observation were prepared by mechanical grinding up to 45 mm and then ion milling to electron transparency (4 kV Ar+). They were cooled to nitrogen temperature during ion milling and the ion beam was blocked parallel to the epoxy to minimize differential milling. TEM was operated in a Akashi 200 kV microscope with a resolution of 1.8 nm.

The YBaCuO base layer is oriented with the c axis perpendicular to the interface, but some zones (less than 20%) are a-axis oriented. The film has a regular thickness of 74 nm suggesting that a and c-axis oriented zones grew at the same growth rate at √750°C. The grain size of the mosaic microstructure is more than 250 nm. Very few antiphase domain boundaries were observed. On the other hand, structural edge dislocations[8] are present in the "123" structure.

The MgO intermediate layer is made of 6 nm polycristalline or/and amorphous islands. This non continuous layer is still parallel to the substrate interface and the MgO grew with no regards to the orientation of the YBaCuO base layer. So the uniform 3 nm amorphous MgO deposited layer recristallized when the whole structure was heated up to ∼650°C for the last YBaCuO layer.

The upper YBaCuO layer is of lower cristallinity than the previous one. The mosaic granularity is more important. Antiphase domains boundaries are numerous corresponding to changes in structural phase[9]. The MgO layer doesn't change the orientation from one YBaCuO layer to the other one ; but a-axis zones have increased and end on the surface of the layer by quite square hillocks of 30 nm height. Near the layer surface there is a lack of cristallinity between a and c axis zones. This can be explained by the fact that the growth temperature choosen for this second YBaCuO layer favours a-axis zones. So these ones have a higher growth rate.

4. INFRARED AND MICROWAVE CHARACTERIZATIONS OF YBCO FILMS AND JUNCTIONS

Photodetection :

The discovery of high Tc supraconductors has sparked interest for infrared photodetectors. For this investigation in a first approach, we have measured the variation of the current-voltage characteristic of patterned YBaCuO films under optical irradiation. The light source is a HeNe laser at 1.15 μm chopped by an accousto-optical modulator. The focalised beam has a diameter of 30 μm delivering a power of 2.5 mW. Two samples have been investigated by DC-magnetron sputtering technique :

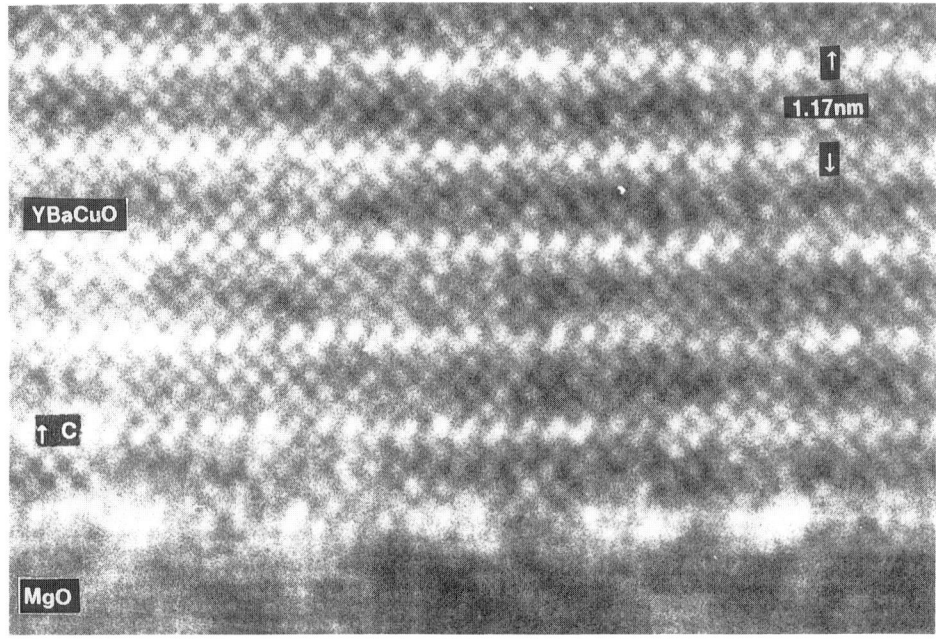

Figure 5. View of the YBaCuO/MgO substrate interface.

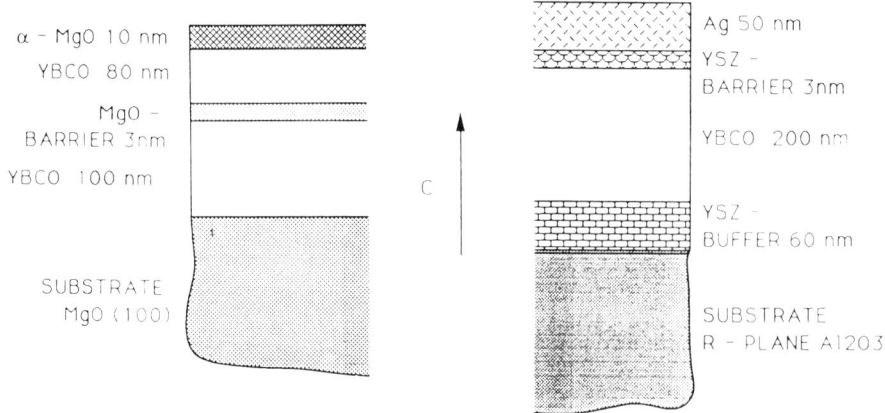

Figure 6. SIS and SIN trilayers cross-sections.

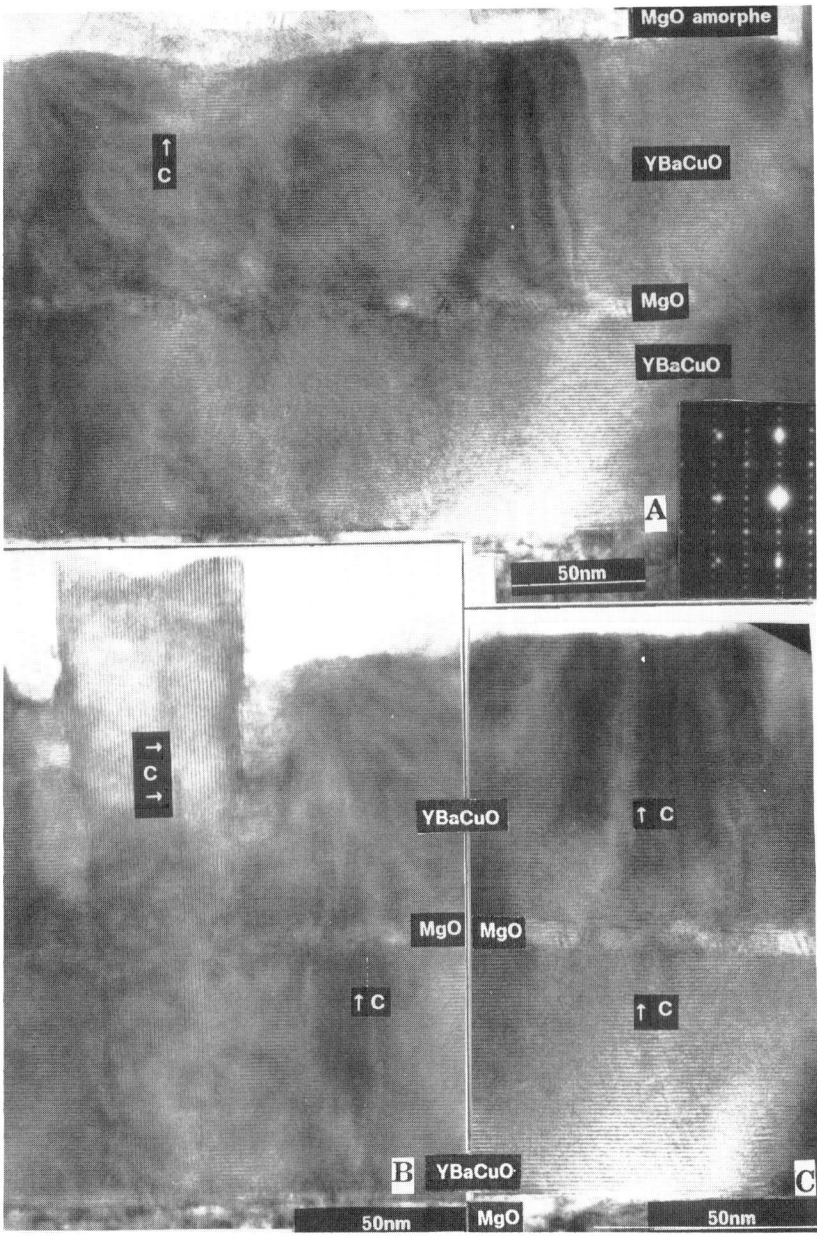

Figure 7. The MgO/YBaCuO/MgO/YBaCuO/MgO structure:the diffraction pattern shows the presence of the two orientations c// and c⊥ the interface. b) Enlarged view of a a-axis oriented zone. This zone grows from the MgO substrate. c) The MgO barrier is non uniform. The MgO Islands are polycristalline or amorphous.

- the granular sample "gran" consists of YBaCuO (300 nm) / YSZ (70 nm) layers grown on an Al2O3 (330 μm) substrate. A 100 μm long and 30 μm wide superconducting line has been etched by ion milling.

- A c-axis textured sample "text" consisting of a 340 nm thick YBaCuO layer grown on a 250 μm thick MgO substrate, which has been chemically etch before YBCO deposition. A 40 μm long and 10 μm wide line structure has been patterned.

The critical current density is 5.3 10^6 A/cm^2 at T = 20 K. The measured voltage variations ΔV for several currents due to the optical irradiation are shown in figure 8.

Since the maxima of the photoresponse and of the derivated resistivity occur at the same temperature, we consider the behaviour of the sample "gran" as purely bolometric [10][11].

The sample "text" exhibits a difference of 16K between the maximum of the photoresponse at $\sim$60K and the maximum of the differential resistivity at 76K. This temperature shift, opposed to a purely bolometric explanation, is neither caused by the delivered electrical power nor by a simple heating effect due to the optical irradiation. This and the fact that there is no frequency dependence of the photoresponse till 300 kHz suggest that we have likely measured the behaviour of weak-links photodetector under irradiation. The bandwith of the photoresponse is 1 MHz. We have found a linear dependence between the optical power and the response measured with different bias currents and at different temperatures. The corresponding detectivity is $\sim$ 2.6 10^7 cm (Hz)$^{1/2}$/W. We found by TEM a rather regular array of c oriented grains (180 nm grain size) with about 10° in plane grain misorientation probably induced by the chemical etching of MgO surface (figure 9) and some c axis spread (rocking curve 2°) indicating the presence of Josephson type grain boundaries. Such a high bandwidth non bolometric photoresponse of the "text" YBCO on MgO sample can be associated to the rather regular array of high Tc (10^5 to 10^6 A cm^{-2}) Josephson coupled grains of the film and might be interpreted as a photofluxonic detection described by Kadin et al [12].

Microwave characterizations :

40 mm long straight YBCO microstrip lines (300 nm thick) have been patterned on a two inches R-plane sapphire (250 μ m thick) covered on the backside by 1.2 μm thick silver ground plane. Such superconductive lines show a lower attenuation ($<$10 dB/m) and rather constant (0.107 mm/ps) phase velocity between 30K and 70K in the 0.5 GHz-10 GHz microwave band in comparaison of Ag lines[13]. A 200 nm YBCO film deposited on MgO(100) tested with a sapphire cavity resonator have a low surface resistance at 20.2 GHz as shown on figure 10 [14].

Tunnel junction characteristics :

YBCO-YSZ-Ag (SIN) tunnel junctions have been characterized in the 4.2K to 300K temperature range. The tunnel resistance RN of a single diode is $\sim$ 5 10^{-5} Ω.cm^2. The junctions are stable upon cycling with a resistance ratio lower than two between room temperature and 100K insuring a high enough barrier height. It is found from comparison between various diode sizes that there is a low leakage level due to the process and uniform I-V curves. The I-V curves are non linear between 4.2K to 80K as shown in figure 11 and present typical features at $\sim$ 25 meV. However, a high level of conducting state at Fermi level on the YBCO is probably due to the degradation of the YBCO, at the YSZ interface. The YBCO-MgO-YBCO (SIS) junctions are less reproducible probably due to a axis outgrowths and show a partially shorted MgO barrier due to recristallization.

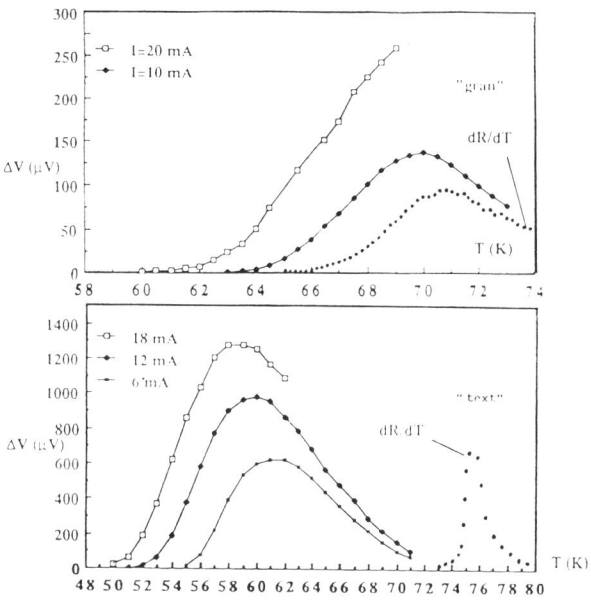

Figure 8. Photoresponse of YBCO samples "gran" and "text" as function of bias current and temperature. The chopping frequency is 1 kHz; dR/dT versus T is inserted (arbitrav unit)

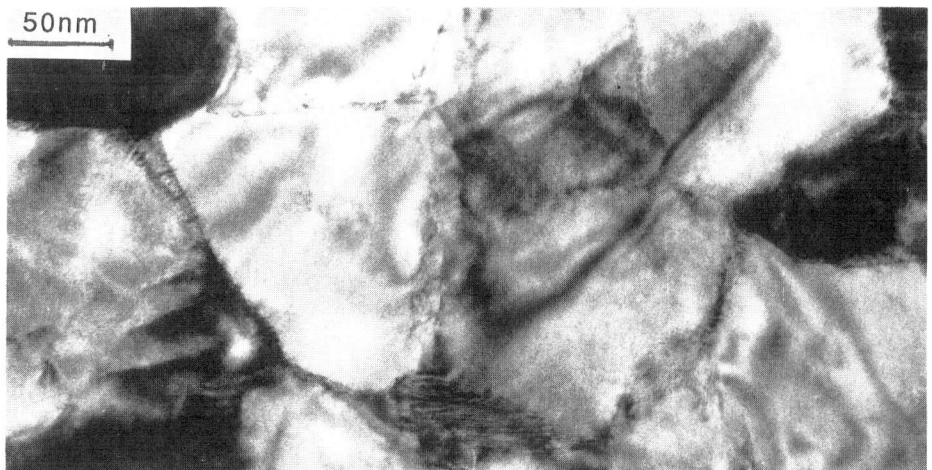

Figure 9. Plane view TEM of $\vec{c}$ "text" YBCO layer deposited on etched MgO(100).

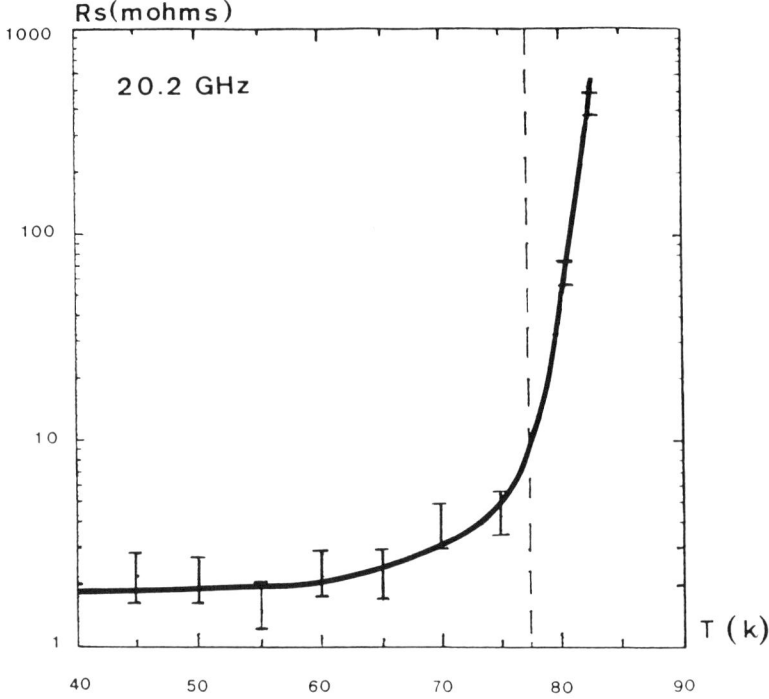

Figure 10 : Rs versus temperature of a YBCO film
deposited on MgO (see ref. 14).

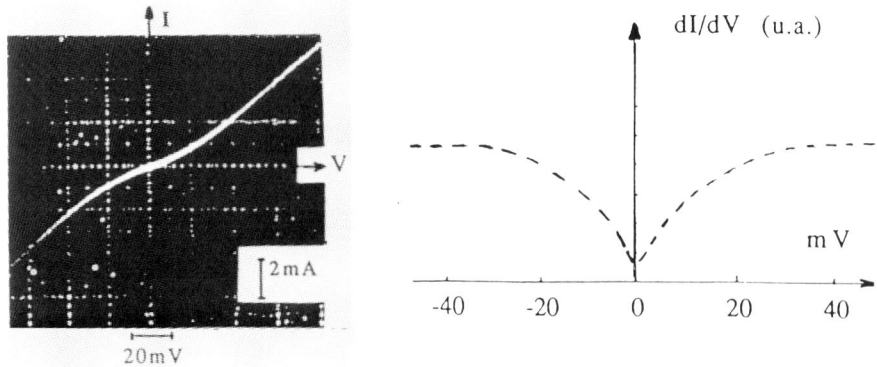

Figure 11 - I-V curve and derivative of YBCO-YSZ-Ag (SIN) junction at 4.2 K

CONCLUSION

YBCO based multilayers have been successfully deposited on various substrates (R-plane-sapphire, Si, MgO, LaAlO3, YSZ...) by three techniques (laser ablation, ICM and planar DC Magnetron Sputtering) with good in-plane heteroepitaxial and c-axis orientation behaviors. Such YBCO layers have low Rs and high Jc and has been selectively etched by argon ion milling in order to achieve sensitive and fast infrared detectors. It has been shown that a two inches R-plane sapphire covered by a thin (60 nm) RF-sputtered YSZ buffer layer is a suitable wafer for making uniform YBCO, as well as SIN and SIS trilayers where a~3 nm thick α-YSZ and α-MgO R.F sputtered film is use for the tunnel barrier. HRTEM cross sections reveal that the amorphous MgO barrier recrystallized at ~500°C before YBCO counterelectrode deposition but did not interrupt the YBCO epitaxy. There are different ways undertaken in order to improve the uniform coverage of a very thin tunnel barrier without breaking the epitaxial growth of the SIS HTS junctions.

Acknowledgements :

This work is partly supported by a DRET (Direction et Recherche des Etudes Techniques) research contract. We would like to thank G. Rolland, P. Martin for their assistance in characterization and J. Mounier for her technical support.

REFERENCES

1   D. Fork, K. Char, F. Bridges, S. Tahara, B. Lairson, J. Boyce, G. Gonnel, T. Geballe Physica C 162, (1989) 121

2   K. Char, N. Newman, S.M. Garrison, R.W. Barton, R.C. Taber, S.S. Laderman, R.D. Jacowitz, Appl. Phys. Lett. (57), (1990) 409

3   J.C. Villegier, H. Moriceau, H. Boucher, R. Chicault, L. Di Cioccio, A. Jäger, M. Schwerdtfeger, M. Vabre, C. Villard, IEEE Transactions on Magnetics, Vol 27 (1991) 1552

4   J.C. Villegier, L. Vieux-Rochaz, M. Goniche, P. Renard, M. Vabre, IEEE Transactions on Magnetics, MAG-21, (1985) 498

5   J.C. Villegier, M. Radparvar, L.S. Yu, S.M. Faris, IEEE Transactions on Magnetics, MAG-25, (1989) 1227

6   K. Sakuta, M. Iyori, T. Kobayashi, M. Matsui, N. Nakajima, IEEE Transactions on Magnetics, Vol 27, n°2, (1991) 1361

7   S. Tanaka, H. Nakanishi, T. Matsuura, K. Higaki, H. Itozaki, S. Yazu, IEEE Transactions on Magnetics, Vol 27, n°2, (1991) 1607

8   R. Ramesh, D.M. Hwang, J.B. Barner, L. Nazar, T.S. Ravi, A. Inam, B. Dutta, X.D. Wu, T. Venkatesan, J. Mater. Res., Vol 5, n°4, Ap. 1990

9   L.A. Tietz, C.B. Carter, D.K. Lathrop, S.E. Russek, R.A. Buhrmar, J.R. Michael, J. Mater. Res., Vol 4, n°5, Sept. Oct. 1989

10   D.P. Osterman, R. Drake, R. Pratt, E.K. Track, M. Radparvar, S.M. Faris, "Optical response of YBCO thin films and weak-links", IEEE Transactions on Magnetics, MAG 25, (1989) 1323

11   M.G. Forrester, M. Gottlieb, J.R. Gavaler, A.I. Braginski, IEEE Transaction on Magnetics MAG 25, (1989) 1327

12   A.M. Kadin, M. Leung, A.D. Smith, J.M. Murduck, "Nonbolometric infrared detection in thin superconducting films via photoproduction of fluxon pairs"

13   B. Cabon, B. Fléchet, J. Chilo, J.C. Villegier, "Caractéristiques de lignes de transmission en YBaCuO et argent", Proceeding JNM-91, (1991) 227

14   O. Llopis, J. Graffeuil, "Etude d'une cavité saphir-superconductrice", Proceedings JNM 91 - Grenoble (1991) 427

AUTHOR INDEX